Stability of Materials

NATO ASI Series

Advanced Science Institutes Series

A series presenting the results of activities sponsored by the NATO Science Committee, which aims at the dissemination of advanced scientific and technological knowledge, with a view to strengthening links between scientific communities.

The series is published by an international board of publishers in conjunction with the NATO Scientific Affairs Division

A Life Sciences **B Physics**	Plenum Publishing Corporation New York and London
C Mathematical and Physical Sciences **D Behavioral and Social Sciences** **E Applied Sciences**	Kluwer Academic Publishers Dordrecht, Boston, and London
F Computer and Systems Sciences **G Ecological Sciences** **H Cell Biology** **I Global Environmental Change**	Springer-Verlag Berlin, Heidelberg, New York, London, Paris, Tokyo, Hong Kong, and Barcelona

PARTNERSHIP SUB-SERIES

1. Disarmament Technologies	Kluwer Academic Publishers
2. Environment	Springer-Verlag
3. High Technology	Kluwer Academic Publishers
4. Science and Technology Policy	Kluwer Academic Publishers
5. Computer Networking	Kluwer Academic Publishers

The Partnership Sub-Series incorporates activities undertaken in collaboration with NATO's Cooperation Partners, the countries of the CIS and Central and Eastern Europe, in Priority Areas of concern to those countries.

Recent Volumes in this Series:

Volume 352 — History of Original Ideas and Basic Discoveries in Particle Physics
edited by Harvey B. Newman and Thomas Ypsilantis

Volume 353 — Hadron Spectroscopy and the Confinement Problem
edited by D. V. Bugg

Volume 354 — Physics and Chemistry of Low-Dimensional Inorganic Conductors
edited by Claire Schlenker, Jean Dumas, Martha Greenblatt, and Sander van Smaalen

Volume 355 — Stability of Materials
edited by A. Gonis, P. E. A. Turchi, and Josef Kudrnovský

Series B: Physics

Stability of Materials

Edited by

A. Gonis and
P. E. A. Turchi

Lawrence Livermore National Laboratory
Livermore, California

and

Josef Kudrnovský

Czech Academy of Sciences
Praha, Czech Republic

Plenum Press
New York and London
Published in cooperation with NATO Scientific Affairs Division

Proceedings of a NATO Advanced Study Institute on
Stability of Materials,
held June 25 – July 7, 1994,
in Corfu, Greece

NATO-PCO-DATA BASE

The electronic index to the NATO ASI Series provides full bibliographical references (with keywords and/or abstracts) to about 50,000 contributions from international scientists published in all sections of the NATO ASI Series. Access to the NATO-PCO-DATA BASE is possible in two ways:

—via online FILE 128 (NATO-PCO-DATA BASE) hosted by ESRIN, Via Galileo Galilei, I-00044 Frascati, Italy

—via CD-ROM "NATO Science and Technology Disk" with user-friendly retrieval software in English, French, and German (©WTV GmbH and DATAWARE Technologies, Inc. 1989). The CD-ROM contains the AGARD Aerospace Database.

The CD-ROM can be ordered through any member of the Board of Publishers or through NATO-PCO, Overijse, Belgium.

Library of Congress Cataloging-in-Publication Data

Stability of materials / edited by A. Gonis, P.E.A. Turchi, and Josef Kudrnovský.
p. cm. -- (NATO ASI series. Series B, Physics ; v. 355)
"Published in cooperation with NATO Scientific Affairs Division."
"Proceedings of a NATO Advanced Study Institute on Stability of Materials, held June 25-July 7, 1994, in Corfu, Greece"--T.p. verso.
Includes bibliographical references and index.
ISBN 0-306-45311-8
1. Electronic structure--Congresses. 2. Alloys--Mathematical models--Congresses. 3. Materials--Mathematical models--Congresses. I. Gonis, Antonios, 1945- . II. Turchi, Patrice E. A. III. Kudrnovský, Josef. IV. North Atlantic Treaty Organization. Scientific Affairs Division. V. NATO Advanced Study Institute on Stability of Materials (1994 : Kerkyra, Greece) VI. Series.
QC176.8.E35S73 1996
620.1'1299--dc20 96-18486
CIP

ISBN 0-306-45311-8

A Division of Plenum Publishing Corporation
233 Spring Street, New York, N. Y. 10013

10 9 8 7 6 5 4 3 2 1

Printed in the United States of America

This project was sponsored by

The North Atlantic Treaty Organization (NATO)
through the Scientific Affairs Division

and was co-sponsored by

The U. S. Department of Energy through
Lawrence Livermore National Laboratory

and by

The National Science Foundation through
The University of Kentucky Center for Computational Sciences

PREFACE

Engineering materials with desirable physical and technological properties requires understanding and predictive capability of materials behavior under varying external conditions, such as temperature and pressure. This immediately brings one face to face with the fundamental difficulty of establishing a connection between materials behavior at a microscopic level, where understanding is to be sought, and macroscopic behavior which needs to be predicted. Bridging the corresponding gap in length scales that separates the ends of this spectrum has been a goal intensely pursued by theoretical physicists, experimentalists, and metallurgists alike.

Traditionally, the search for methods to bridge the length scale gap and to gain the needed predictive capability of materials properties has been conducted largely on a trial and error basis, guided by the skill of the metallurgist, large volumes of experimental data, and often ad hoc semiphenomenological models. This situation has persisted almost to this day, and it is only recently that significant changes have begun to take place. These changes have been brought about by a number of developments, some of long standing, others of more recent vintage.

First, the introduction of quantum mechanics brought about the realization that most, if not all, properties of materials can be traced to the behavior of electrons, the "glue" that holds atoms together to form a solid. Second, methods for calculating the electronic structure of matter have been continually sharpened and refined to the point of determining the energetics of solid materials to within a fraction of a millirydberg. Third, in parallel with these developments, extremely accurate models, based on first-principles electronic structure calculations have evolved which allow the determination of thermodynamic and mechanical properties of materials, such as alloy phase stability, elastic constants, and bonding characteristics. These models make use of energetic parameters which can be directly obtained from the results of electronic structure calculations. Such calculations have been carried out by a number of means and various approximations, chosen to suit particular materials or specific properties. One focus of this effort is to understand the mechanisms by which materials, *e.g.*, intermetallic alloys and compounds, can be made to have technologically desirable mechanical properties such as ductility, and the deformation characteristics that bring about this behavior.

This brings us to the point of very recent developments, on both the theoretical/computational and experimental fronts, and the main reason for convening a NATO Advanced Study Institute (ASI) in the field of Stability of Materials. Namely, computational methods have been refined and also developed so as to provide a unique and powerful analytical and computational tool for the study of electronic structure. Also, experimental techniques have been developed which allow increasingly deeper probes into the microstructure of matter. Finally, computer simulation methods such as Monte Carlo simulations and molecular dynamics are often implemented in terms of parameters extracted out of first-principles calculations. Specifically, these new developments are characterized by the following features.

1. Removal of the shape approximation (often assumed to be spherical) of the charge distribution and the potential in the Wigner-Seitz cell of a material. A proper treatment of the nonspherical cells in a material, the charge distribution and the potential, are necessary for the accurate study of bonding mechanisms and phase stability.

2. Exact treatment of the Schrödinger and Poisson equations in materials with three-dimensional periodicity as well as with reduced symmetry, such as surfaces, grain

boundaries, and dislocations. This effectively removes the use of artificial boundary conditions, such as super cells or slabs in the study of such systems (which are used in many non-Green-function-based techniques).

3. A unified treatment of ordered and substitutionally disordered materials within the same formal and computational method. This makes possible the direct comparison of results obtained for various systems which can only be carried out in a highly questionable manner if different techniques are used for different systems.

4. Large-scale simulations of defect microstructures and their impact on mechanical properties. In many such applications, the parameters used, *e.g.*, interatomic potentials, have been extracted from the results of electronic structure calculations, effectively providing at least a small bridge across part of the length-scale spectrum.

5. On the experimental side, new techniques based on synchrotron radiation, photoemission, positron annihilation, and electron microscopy are used to provide a deeper insight into materials properties than has been possible in the recent past.

6. Well-known experimental procedures, such as molecular beam epitaxy and chemical vapor deposition, as well as newer methods, such as ball milling and mechanical alloying, have been developed to the point of being able to produce selected low-symmetry structures, surfaces and interfaces, with novel and desirable characteristics.

7. Finally, developments in computational technology make possible the performance of calculations that would not be carried out even as recently as five years ago.

Based on these observations, the topics of the ASI were divided into three main categories: (1) Numerical Simulations and Phenomenology; (2) Electronic Structure Theories; (3) Materials Properties and Characterization.

Every attempt was made to assure continuity of presentations between theory and experiment, with ample opportunity for questions and subsequent discussions. In addition, we hope to consolidate a serious dialogue between groups of experimental scientists and metallurgists on one hand and theoretical physicists on the other that will hopefully continue long into the future.

As directors of this ASI and as editors of the proceedings, we would like to take this opportunity to thank all those whose efforts contributed to the success of this endeavor. We greatly appreciate the work of the invited speakers to produce lecture materials that were of tutorial character, and of being present during the ASI to answer questions and hold discussions with the participants to the ASI. We also thank the participants for their attention and attendance. Of equal importance to the success of the ASI was the competent handling of the conference onsite by Mania Bessieri and her staff of the Congress Center "Organization Idea" in Athens. They arranged every detail, from meeting arriving participants at the airport, duplicating and distributing materials during the conference, programming social events for accompanying guests, and assuring a smooth departure of the participants at the conclusion of the ASI.

In addition to the main sponsorship provided by the NATO Scientific Affairs Division, the directors wish to express their gratitude for the co-sponsorship of the meeting by the US Department of Energy through the Department of Chemistry and Materials Science of the Lawrence Livermore National Laboratory, and the US National Science Foundation through the Center for Computational Sciences of the University of Kentucky. These institutions provided financial, and in the case of the Lawrence Livermore National Laboratory, technical support that were essential to the success of this ASI. Finally, the staff at Plenum did an excellent work in bringing forth this proceedings.

A. Gonis
P. E. A. Turchi
J. Kudrnovský

November 1995

CONTENTS

NUMERICAL SIMULATIONS AND PHENOMENOLOGY

Invited papers

Monte Carlo Simulations of Surfaces and Interfaces in Materials..........3
K. Binder

Continuum Diffuse-Interface Model for Modeling Microstructural Stability..........37
L. Q. Chen, D. N. Fan, Y. Z. Wang, and A. G. Khachaturyan

Atomistic Studies of the Structure of Grain Boundaries and Dislocations..........53
V. Vitek

Dislocation Patterns: Experiment, Theory and Simulation..........99
Ladislas P. Kubin

Computer Simulation of Fracture, Dislocations and Martensitic Transformations in Solids..137
Philip C. Clapp

Contributed papers

Continuous Phase Transitions at Surfaces of CuAu Alloy Models-A Monte Carlo Study of Surface Induced Order and Disorder..........165
W. Schweika, D. P. Landau, and K. Binder

Surface Ordering and Surface Segregation in Binary Alloys..........173
F. Schmid

Relaxed Monte Carlo Simulations on Au-Ni Alloy..........179
R. Tétot and A. Finel

Continuous Monte Carlo Simulation of Surfaces and of Interfaces of Mismatched Crystals..........185
Jeff Baker and Per-Anker Lindgård

Kinetics at Early Stages of Phase Separation and Ordering in Alloys..........191
S. V. Beiden, V. Yu Dobretsov, G. Martin, F. Soisson, and V. G. Vaks

The Gaussian Cluster Variation Method and its Application to the Thermodynamics of Transition Metals..........197
A. Finel and R. Tétot

Kinetic Path for Disorder-$L1_2$ Transition Studied by the PPM..........205
Tetsuo Mohri, Yoichi Ichikawa, and Tomoo Suzuki

Investigation of H-H (D-D) and H (D)-N Interactions in Nb and Ta by Means of Computer Simulation of Diffusion....211
M. S. Blanter and A. V. Vasiljev

A Molecular Dynamics Simulation of Vibrational Properties and Diffusion of Copper Adatoms on a Copper (001) Surface....217
N. I. Papanicolaou, G. C. Kallinteris, and G. A. Evangelakis

The Order-Disorder Transition at a Σ=17 Tilt Boundary in Cu_3Au....223
H. M. Polatoglou

Monte Carlo Simulation of MBE Growth of GaAs Analysis of RHEED....229
P. J. van Hall, H. Kökten, M. R. Leys, and M. Bosch

ELECTRONIC STRUCTURE THEORIES

Invited papers

Self-Consistent Green's Function Method for Random Surfaces and Interfaces....237
J. Kudrnovský, I. Turek, V. Drchal, and M. Šob

Electronic Structure and Physical Properties....265
Ingrid Mertig

Classical and First Principles Molecular Dynamics Simulations in Material Science: Application to Structural and Dynamical Properties of Free and Supported Clusters....295
Carlo Massobrio and Philippe Blandin

Contributed papers

New Tight-Binding Methodology for Calculating Total Energy of Solids....325
D. A. Papaconstantopoulos and M. J. Mehl

Contributions to the Total Energy of Random Alloys....333
J. S. Faulkner, Yang Wang, and G. M. Stocks

Competing Mechanisms for Ordering Tendencies in BCC $CuAuZn_2$ and FCC AuFe Alloys....339
D. D. Johnson, J. D. Althoff, J. B. Staunton, M. F. Ling, and F. J. Pinski

A Novel Full Potential Contracted Plane Wave (FCPW-) Method for Electronic Structure Calculations on Complex Materials....347
L. Fritsche, J. Koller, and J. Noffke

Effective Ising Hamiltonian for Surfaces of Metallic Alloys....355
Václav Drchal, Josef Kudrnovský, and Ilja Turek

The Influence of Quasi-Particle Lifetimes and Electronic Topological Transitions on the Deviations from Vegard's Rule in Ag-Pd Alloys....361
E. Bruno, B. Ginatempo, and E. S. Giuliano

Electronic Structure and Stability of Al-Ge Alloys Under Pressure....367
P. E. A. Turchi, Prabhakar P. Singh, and G. M. Stocks

Prediction of Unsuspected Ordering Tendencies in Pd-Pt and Rh-Pt Alloys....375
Barry M. Klein, Z. W. Lu, and Alex Zunger

Energetic Effects in the Au-Ni System....381
T. Deutsch and A. Pasturel

Electronic Topological Transitions and Phase Stability in Ag-Pd and Cu-Pt Alloys....387
E. Bruno, B. Ginatempo, and E. S. Giuliano

Energetics of the Light Actinides in a Full Charge Density Scheme....393
L. Vitos, J. Kollár, and H. L. Skriver

Impurity Effects on Bonding Charge in Ni_3Al....401
Sheng N. Sun, Nicholas Kioussis, Say-Peng Lim, A. Gonis, and W. Gourdin

Electronic Structure of Stacking Faults in $Pd_{1-x}Cu_x$, $Pd_{1-x}Ag_x$ and $Ag_{1-x}Au_x$ Alloys....407
J. M. MacLaren

Effects of Boron and Sulfur on the Ideal Yield Stress of Ni_3Al - A First-Principles Approach....413
Sheng N. Sun, Nicholas Kioussis, Mikael Ciftan, and A. Gonis

Ab-Initio Calculation of the Lattice Relaxation in Dilute Alloys....419
N. Papanikolaou, N. Stefanou, R. Zeller, and P. H. Dederichs

Prediction of Lattice Vibrations in Metastable and Unstable Transition Metal Systems....425
Göran Grimvall

Itinerant Magnetism of (001) Surfaces of Random Nickel-Copper and Iron-Vanadium Alloys....431
I. Turek, J. Kudrnovský, M. Šob, and V. Drchal

Calculation of Spontaneous Resistance Anisotropy of Disordered Ferromagnetic Alloys....437
John Banhart

Ab-Initio Based Atomistic Potentials and Application to Metallic Surface and Interface Structures....443
Bernard R. Cooper, J. Mei, and S. P. Lim

On the Role of Non-Pair Potential Terms in Semiempirical Quantum-Mechanical Simulations....449
M. Šob and V. Vitek

A Comparative Ab Initio Study of Small Si and C Clusters....455
Aristides D. Zdetsis

Brillouin Zone Concept and Crystal Symmetry of Intermetallic High Pressure Phases....465
V. F. Degtyareva

MATERIALS PROPERTIES AND CHARACTERIZATION

Invited papers

TEM Characterization of Structural Defects....473
Gustaaf Van Tendeloo

Dislocations and the Plasticity of Crystals....509
Patrick Veyssière

Mechanical Properties of TiAl and TiAl-Base Alloys....547
Kyosuke Kishida, Shigenobu Yokoshima, David R. Johnson, Haruyuki Inui, and Masaharu Yamaguchi

Structure-Property Relationship of Metal-Ceramic Interfaces....581
J. Th. M. De Hosson, W. P. Vellinga, X. B. Zhou, and V. Vitek

Stability and Instability of Crystalline and Amorphous Phases at High Pressure....615
E. G. Ponyatovsky

Contributed papers

Distributions of Dislocation Dissociation Widths....645
Václav Paidar

An Atomistic Study of the Image Force on a Dislocation in a Bimaterial....651
P. Beauchamp, J. Lépinoux, I. Lardon, and H. Kentabli

Slip Induced Configurational Energy Change in Binary Alloys....657
B. Schönfeld and P. Schwander

Cyclic Creep Deformation Characteristics of Single Crystal of Nickel Base Superalloy CMSX 3....663
J. Zrník, Z. G. Wang, Y. Yu, J. A. Wang, M. Žitňanský, and P. Horňak

An Electrochemical Model for Environmental Cracking of Metals....669
Raicho Raicheff

Instability Phenomena in Al-Zn Alloys....675
Ryszard Ciach

The Effect of Alloying Additions on the Interfacial Interactions at the Fe-Al Interface During Coating....681
Amudulla O. Mekhrabov, M. Vedat Akdeniz, and Ilhami Aktürk

Radiation-Induced Disordering and Dissolution of LRO Precipitates in Disordered Matrix....687
H. Wollenberger, S. Matsumura, S. Müller, and C. Abromeit

Non-equilibrium Solidification Microstructures in Alloys....693
E. Ramous, A. Zambon, B. Badan, and M. Magrini

Variations of Long-Range Order in Ni_3Al+B after Deformation by Cold-Rolling...699
R. Kozubski, M. Migschitz, and W. Pfeiler

Surface Disorder and Surface Segregation in Cu_3Au: An X-Ray Scattering Study....705
Helmut Dosch and Harald Reichert

Short-Range Order Kinetics in α-AuFe after Deformation and Recrystallization....713
M. Migschitz and W. Pfeiler

Stabilisation and Irreversibility of Martensite in Copper Base Shape Memory Alloys....719
J. Dutkiewicz

Crystal Structure and Physical Properties of High Pressure Hydrides........................725
V. E. Antonov

Invited Lecturers..733

Participants..735

Authors..739

Index...741

Oral presentations were given by the authors whose names are underlined.

NUMERICAL SIMULATIONS AND PHENOMENOLOGY

MONTE CARLO SIMULATIONS OF SURFACES AND INTERFACES IN MATERIALS

Kurt Binder

Institut für Physik
Johannes Gutenberg-Universität Mainz
Staudinger Weg 7
D - 55099 Mainz, Germany

INTRODUCTION

Many applications of materials are controlled by their surface and interface properties. In particular, metallic alloys (but also mixed dielectric materials and amorphous polymer blends) are not homogeneously mixed on a microscopic length scale, although they are macroscopically homogeneous. Depending on the preparation of the sample, there exists a heterophase microstructure, with typical domain sizes, e.g. in the 1 to 10^2 μm range, separated by interfaces between them. The physical properties of such intrinsic interfaces (grain boundaries between small crystallites, antiphase domain boundaries in ordered alloys, Bloch walls in magnetic materials, etc.) are not only an important controlling factor for many properties, but also a challenging fundamental problem of condensed matter science. The same fact is true for external surfaces of materials, e.g. a surface of a metallic alloy against vacuum or air, or a surface of a fluid mixture against the wall of a container, etc.: there may exist either a microscopic layer (thickness about 1 nm) where composition and structure differ significantly from the bulk (e.g., surface enrichment or surface induced disordering in alloys), or even mesocopic wetting layers (thickness 10-10^2 nm) may form.

These phenomena form a very wide area of research, and the present lectures hence can focus only on a very small section. We shall neither be concerned with the chemistry of interfaces and surfaces, nor with their electronic structure and crystallographic structure (for reviews dealing with such topics, see e.g. some recent books[1-5]). Rather we deal only with the properties of simple lattice models for interfaces and surfaces, from the point of view of statistical mechanics and the related phase transition phenomena: critical phenomena at surfaces[6-8]; roughening of interfaces[8-10]; wetting[8,11-14] and related phenomena such as surface-induced ordering or disordering[15]; of course, many of these phenomena require understanding of the structure of interfaces between phases coexisting in thermal equilibrium, and the related interfacial tension[8,14,16,17]. Since we restrict interest to lattice models from the beginning, we shall not discuss such interesting phenomena as surface reconstruction[18], surface melting[19], etc. Moreover, we shall focus on static phenomena throughout, and thus discuss neither ordering and growth at surfaces[20], nor the dynamics of interfaces during phase separation in the bulk[21-24] or near surfaces[25-27] nor dynamics of wetting[28-30], etc. Even within this narrowed scope the field is still huge, and hence the present lecture notes cannot provide a complete review, but shall attempt to explain the basic concepts only and illustrate their study by Monte Carlo simulation with examples taken from the author's research group.

In the next section, the basic Ising-type model will be formulated[6,8] and the corresponding Ginzburg-Landau type theory will be sketched. Fluctuation effects beyond this mean field theory will be discussed without detailed derivations. In the third section, Monte Carlo methods will be described for obtaining the interfacial tension between coexisting phases (and interfacial profiles, etc.),[31-40] discussing phenomena such as capillary waves[16,17] in simple Ising models,[34-36,39] models of body-centered cubic alloys of the Fe-Al type,[37,38] and symmetric polymer blends[40]. In the fourth section, we turn to simulations studying the unbinding of interfaces from free surfaces, again drawing a connection from wetting in simple Ising models[41-45] to surface-induced disordering in models for CuAu-alloys[46,47] to wetting in polymer mixtures[48,49] and surface-induced order of blockcopolymer melts[50-53]. Finally in the fifth section some conclusions are summarized.

THEORETICAL MODELS AND CONCEPTS

The Basic Ising-Type Lattice Model

Let us restrict attention for the moment to a model of a binary (AB) solid alloy in contact with a wall, assuming pairwise interactions φ_{AA}, φ_{AB}, φ_{BB} between atoms at sites $\vec{r}_i, \vec{r}_j$. The only degrees of freedom considered are local concentration variables c_i , $c_i = 1$ if site i is occupied by an A atom, and $c_i = 0$ if it is occupied by a B-atom. By $v_A(\vec{r}_i)$ [$v_B(\vec{r}_i)$] we denote forces exerted on A(B) atoms at site $\vec{r}_i$ due to the hard wall at $z = 0$ (cf. Fig.1). Due to the breaking of the translational symmetry of the surface, one must expect that in general the pairwise interactions φ_{AA}, φ_{AB}, φ_{BB} not only depend on the relative distance $\vec{r}_i - \vec{r}_j$, but also on $\vec{r}_i, \vec{r}_j$ separately (e.g., different interactions $\varphi^{s}_{AA}, \varphi^{s}_{AB}, \varphi^{s}_{BB}$ may occur {Fig. 1} if both sites i,j are in the layer right adjacent to the surface[6]. The Hamiltonian is

$$\mathcal{H} = \sum_{i \neq j} \Big[c_i c_j \varphi_{AA}(\vec{r}_i, \vec{r}_j) + c_i(1-c_j)\varphi_{AB}(\vec{r}_i, \vec{r}_j) + (1-c_i)c_j\varphi_{AB}(\vec{r}_i, \vec{r}_j) +$$

$$(1-c_i)(1-c_j)\varphi_{BB}(\vec{r}_i, \vec{r}_j)\Big] + \sum_i \big[v_A(\vec{r}_i)c_i + v_B(\vec{r}_i)(1-c_i) \big] \qquad (1)$$

Here the sum ($i \neq j$) runs over all pairs once, and lattice sites exist in the positive half-space $z > 0$ only.

At this point, we are not interested in the "correct" description of atomistic detail, but rather Eq. (1) (and Fig. 1) only serve as a generic model. In particularly, from this model we shall derive a continuum description which holds for a much larger class of systems[6,25]. Thus we simplify the problem by restricting the range of all interactions to nearest neighbors, and take all interactions φ_{AA}, φ_{AB}, φ_{BB} to be independent of their sites i,j except if both sites are in the surface layer $n = 1$. Finally, we also assume $v_A(\vec{r}_i)$, $v_B(\vec{r}_i)$ to be nonzero only if i is in the layer $n = 1$. (This latter assumption implies that we restrict attention to wetting with short range forces. It is well known that this situation differs in important qualitative aspects from wetting with long range forces,[12] e.g. van der Waals forces that decay proportional to z^{-3} with the distance z from the surface).

It is convenient to translate the model into the Ising spin representation[6] via the transformation $c_i = (1-S_i)/2$ {note $S_i = \pm 1$}, which yields

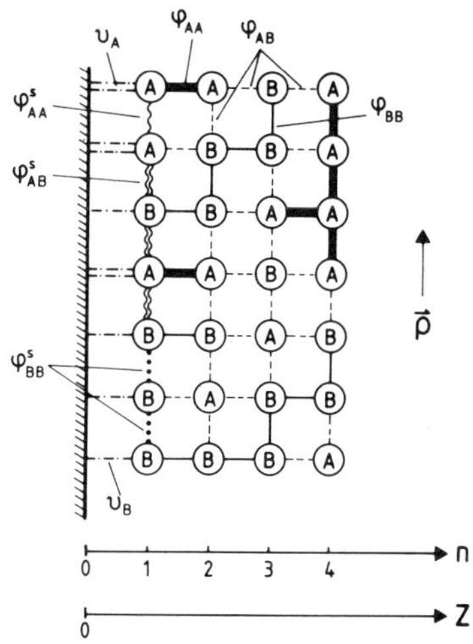

Figure 1. Schematic picture of the surface of a binary (AB) alloy at z = 0 (the shading indicates that this may represent an inert hard wall). Different atoms (circles) and nearest-neighbor interactions between them (and the wall) are indicated by different types of lines. For a discrete description, lattice planes parallel to the surface are labelled by a positive integer n, while in the continuum description coordinates parallel (ρ) and perpendicular (z) to the surface are used, as indicated. From Puri and Binder[25].

$$\mathcal{H} - \sum_i c_i \mu_A - \sum_i (1-c_i)\mu_B = -\sum_{\langle i,j \rangle} J_{ij} S_i S_j - H \sum_i S_i$$

$$-H_1 \sum_{i \in 1\text{st layer}} S_i + \mathcal{H}_0 \quad . \tag{2}$$

Here μ_A, μ_B are the chemical potentials of both species, and $\mathcal{H}_0$ is a constant which only affects the energy scale. The pairwise "exchange" energy J_{ij} is

$$J_{ij} = J = \frac{1}{2}\varphi_{AB} - \frac{1}{4}(\varphi_{AA} + \varphi_{BB}) \quad , \tag{3}$$

when at least one of the sites i,j is not in the surface layer n = 1. If both sites are in the layer n – 1, we have

$$J_{ij} = J_S = \frac{1}{2}\varphi^S_{AB} - \frac{1}{4}\left(\varphi^S_{AA} + \varphi^S_{BB}\right) \quad . \tag{4}$$

The bulk "magnetic field" H (= thermodynamic variable conjugate to the "pseudospin" magnetization $m_i = \langle S_i \rangle$) is

$$H = \frac{1}{2}(\mu_B - \mu_A) + \frac{1}{2}\sum_{j(\neq i)}(\varphi_{AA} - \varphi_{BB}) \quad . \tag{5}$$

In the first layer, however, there acts a field $H + H_1$, with the "surface field" being

$$H_1 = \frac{1}{2}(v_B - v_A) + \frac{1}{2} \sum_{\substack{j(\neq i) \\ j \in 1st\ layer}} \left[\left(\varphi^s_{AA} - \varphi_{AA}\right) - \left(\varphi^s_{BB} - \varphi_{BB}\right)\right] - \frac{1}{2} \sum_{\substack{j(\neq i) \\ j \in 0th\ layer}} (\varphi_{AA} - \varphi_{BB}) \quad (6)$$

We emphasize that a nonzero surface field H_1 arises even for the case where $v_B = v_A = 0$ (a natural assumption for a surface against air or vacuum, of course) and interactions unchanged near the surface, i.e. $\varphi^s_{AA} = \varphi_{AA}$ and $\varphi^s_{BB} = \varphi_{BB}$, as long as $\varphi_{AA} - \varphi_{BB} \neq 0$. This is a result of the "missing neighbors" of sites in the first layer. This term would only be absent for the "most symmetric" case $\varphi_{AA} = \varphi_{BB}$, which clearly has little physical relevance for actual mixtures. While the "bulk field" H, Eq. (5), is easily eliminated from the problem by fixing the average concentration,

$$\bar{c} = (1/N) \sum_i \langle c_i \rangle = (1/2N) \sum_i (1 - m_i) \quad , \quad (7)$$

which is also the concentration $c_b = (1 - m_b)/2$ in the bulk of our semiinfinite system, the additional surface field H_1 remains as a parameter in the problem. This term is responsible for both surface enrichment and wetting phenomena in mixtures[3,6,8,11-14,53,54].

Derivation of a Mean Field Theory

Eq. (2) is the starting point for the computer simulations described in the later sections. It also is the starting point for the basic theoretical concepts on the subject. The first step is the construction of a continuum theory, starting from a layerwise molecular field approximation (note that the result is much more general than this derivation)[6,55-57]. Denoting in a lattice model such as that of Fig. 1 the coordination number of atoms in a plane parallel to the wall as $q_{\parallel}$, and the coordination number of atoms in adjacent planes as $q_{\perp}$, the average magnetization of pseudospins in the n'th layer is, for $n \geq 2$

$$m_n = \tanh\left(\frac{H}{k_B T} + \frac{q_{\parallel} J}{k_B T} m_n + \frac{q_{\perp} J}{k_B T} m_{n+1} + \frac{q_{\perp} J}{k_B T} m_{n-1}\right) \quad , \quad (8)$$

while in the first layer we have

$$m_1 = \tanh\left(\frac{H + H_1}{k_B T} + \frac{q_{\parallel} J_S}{k_B T} m_1 + \frac{q_{\perp} J}{k_B T} m_2\right) \quad . \quad (9)$$

Assuming that the temperatures of interest are close to T_{cb} and the fields H, H_1 small enough such that $|m_n| \ll 1$, we can expand artanh $m_n \approx m_n + \frac{1}{3} m_n^3$ and also replace differences by differentials, denoting $m_n = m(z)$

$$m_{n\pm1} = m(z) \pm \frac{\partial m(z)}{\partial z} + \frac{1}{2} \frac{\partial^2 m(z)}{\partial z^2} \quad , \quad (10)$$

using the lattice spacing of Fig. 1 as our unit of length. As discussed in Ref. 57, Eq. (10) is accurate for large correlation length ξ_b, since it amounts to steps such as $\exp(1/\xi_b) \approx 1 + 1/\xi_b$ in the linearized limit.

Then Eqs. (8), (9) get replaced by a second order nonlinear differential equation,

$$\left[1-\frac{(q_{II}+2q_1)J}{k_BT}\right]m(z)+\frac{1}{3}m^3(z)-\frac{q_1J}{k_BT}\frac{\partial^2 m(z)}{\partial z^2}=\frac{H}{k_BT} \tag{11}$$

supplemented by a boundary condition at the surface, $z = 0$,

$$\left[1-\frac{q_{II}J_S+q_1J}{k_BT}\right]m(0)+\frac{1}{3}m^3(0)-\frac{q_1J}{k_BT}\frac{\partial m(z)}{\partial z}\Big|_{z=0}=\frac{H+H_1}{k_BT} \tag{12}$$

Due to the nonlinear character of Eqs. (8), (9), the results obtained from the discrete set of difference equations are not equivalent to the results following from the differential equation: Eqs. (8,9) yield an infinite sequence of layering transitions,[58] while Eqs. (11,12) yield a single wetting transition[56]. But a treatment of the problem beyond mean field implies,[43,58] that the infinite sequence of layering transitions can occur only for $T < T_R$, T_R being the roughening transition temperature of an interface between coexisting bulk phases [9,10]. For $T > T_R$, the interface is a smooth object, delocalized due to capillary waves,[6,8-17,59,60] and the description in terms of Eqs. (11), (12) is more appropriate than the set Eqs. (8), (9).

It is convenient to rescale Eqs. (11), (12) in terms of m_b and ξ_b, noting $T_{cb} = (q_{II} + 2q_1)\,J/k_B$ and

$$m_b = \sqrt{3(1-T/T_{cb})}\,,\ \xi_b = (1-T/T_{cb})^{-1/2}\sqrt{q_1/[2(q_{II}+2q_1)]} \quad , \tag{13}$$

$$h \equiv \frac{H/k_BT_{cb}}{\sqrt{3}}(1-T/T_{cb})^{-3/2}\,,\ h_1 \equiv \frac{H_1/k_BT_{cb}}{\sqrt{3}}(1-T/T_{cb})^{-3/2} \quad , \tag{14}$$

$$g \equiv 1+2\xi_b^2[q_{II}J_S/(q_1J)-1-q_{II}/q_1] \quad , \tag{15}$$

putting

$$\mu(\zeta) = m(z)/m_b, \qquad \zeta = z/(2\xi_b) \quad . \tag{16}$$

This yields a rescaled bulk equation free of any parameters,

$$\frac{1}{2}\frac{\partial^2\mu(\zeta)}{\partial\zeta^2}+\mu(\zeta)-\mu^3(\zeta)+h=0 \tag{17}$$

the only parameters appear in the boundary condition,

$$\xi_b\frac{\partial\mu(\zeta)}{\partial\zeta}\Big|_{\zeta=0}+g\mu(0)-\mu^3(0)+h+h_1=0 \quad . \tag{18}$$

As is well known,[6-8] Eqs. (17), (18) can also be thought of as the result of minimizing an effective free energy functional, $\delta\,\Delta\mathcal{F}\{\mu(\zeta)\}/k_BT_c/\delta\mu(\zeta) = 0$, with

$$\frac{\Delta\mathcal{F}}{k_BT_c}=\int_0^\infty d\zeta\left\{\frac{1}{2}\left[\frac{\partial\mu(\zeta)}{\partial\zeta}\right]^2-\mu^2(\zeta)+\frac{1}{2}\mu^4(\zeta)-h\mu(\zeta)-\frac{1}{\xi_b}\left[h_1\mu(\zeta)+\frac{1}{2}g\mu^2(\zeta)\right]\delta(\zeta)\right\} \tag{19}$$

Near T_{cb} the correlation length $\xi_b >> 1$, and therefore $|g| >> 1$ {Eq. (15)}, if we are not very close to the surface-bulk multicritical point,[6,7,55] which in mean field theory occurs for

$$J_{sc}/J = 1 + q_1 / q_{11} \quad . \tag{20}$$

In the absence of fields h, h_1, we note $g > 0$ for $J_s > J_{sc}$: then Eq. (18) shows that the surface layer is strongly ordered (since the surface then orders already before the bulk). Conversely, g is very strongly negative for $J_s < J_{sc}$, the situation that shall be considered in the following. Then the term $\mu^3(0)$ in Eq. (18) can be neglected.

Mean Field Theory of Interfaces

Before we analyse the behavior near a free surface, it is instructive to consider the related problem of describing an interface between two bulk phases of opposite sign of the order parameter (they can coexist for h = 0). Then Eq. (17) is replaced by

$$\frac{1}{2}\frac{\partial^2 \mu(\gamma)}{\partial \gamma^2} + \mu(\gamma) - \mu^3(\gamma) = 0 \quad \text{, interface,} \tag{21}$$

and the boundary conditions are

$$\mu(\gamma \rightarrow \pm\infty) \rightarrow \pm 1 \,, \quad \lim_{\gamma\rightarrow\pm\infty} (\partial\mu/\partial_\gamma) = 0 \tag{22}$$

This problem is formally analogous to a problem in classical mechanics, namely the motion of a point particle of unit mass in a potential $U(x) = x^2 - \frac{1}{2}x^4, \ddot{x} = -dU/dx$, if we identify x with μ and time t with γ. The energy of this problem then is chosen as E = 1/2, so the particle starts at $t = -\infty$ at the left potential well with zero velocity and comes again to rest for $t \rightarrow \infty$ at the right potential well. The velocity $\dot{x}$ (corresponding to the slope $d\mu/d_\gamma$ of the interfacial profile) is maximal for x = 0 (μ = 0, respectively). Since conservation of energy implies $E = U + \dot{x}^2/2$ = const, multiplication of Newton's law by $\dot{x}$ and integration over time from $t = -\infty$ to t yields $\dot{x}^2/2 = -(x^2 - 1) + \frac{1}{2}(x^4 - 1) = \frac{1}{2}(1 - x^2)^2$. Analogously, we find the scaled order parameter profile

$$[1 - \mu^2(\gamma)]^2 = (d\mu/d_\gamma)^2 \quad , \quad \mu(\gamma) = \tanh\gamma \,, \quad m(z) = m_b \tanh\left(\frac{z}{2\xi_b}\right) \tag{23}$$

The interfacial free energy, defined as the excess contribution of a system containing one interface relative to that of a homogeneous system {where $m(z) = m_b$ everywhere} can be shown to be (normalized per unit of the interfacial area A)

$$f_{int} \equiv \frac{\Delta F_{int}}{k_B TA} = \frac{1}{3} m_b^4 \xi_b \int_{-\infty}^{+\infty} d_\gamma \left(d\mu / d_\gamma\right)^2 = \frac{2}{9} m_b^4 \xi_b \tag{24}$$

If we define a critical exponent μ and amplitude $\hat{f}_{int}$ of this interfacial tension via $f_{int} = \hat{f}_{int}\,(1 - T/T_{cb})^{\mu}$, this mean field theory yields $\mu = 3/2$, and the "intrinsic width" of the interfacial profile clearly is the correlation length ξ_b.

Mean Field Theory of Wetting

Now for the problem of a free surface of a system which is at phase coexistence in the bulk (h = 0) the problem is still described by the first equation of Eq. (23), since Eq. (17) and Eq. (21) are identical. The order parameter profile of the free surface problem still is of the form $\mu(\zeta) = \pm \tanh(\zeta + \zeta_0)$, for a state in the incompletely wet regime, where ζ_0 (and the sign) have to be chosen such that the boundary conditions at $\zeta = 0$ and $\zeta \rightarrow \infty$ are fulfilled. For the wet case the profile rather has the form $\mu(\zeta) = \pm \coth(\zeta + \zeta_0')$, cf. Fig. 2. Note that the spin up - spin down symmetry of the bulk part of our Ising model {Eq. (2)} implies that both "wetting" and the opposite phenomenon, "drying", can occur: by "wetting" we denote here a macroscopically thick film of upspins {fluid atoms, for the lattice gas; B-atoms, for the A-rich AB mixture; etc.} intruding at the wall; "drying" means a film of downspins {gas layer at the surface of the "lattice fluid"; A-atoms, for the B-rich AB mixture; etc.}. The constants ζ_0, ζ_0' are then found from requesting that $\partial\mu(\zeta)/\partial\zeta = - |\mu^2(\zeta) - 1|$ must also satisfy

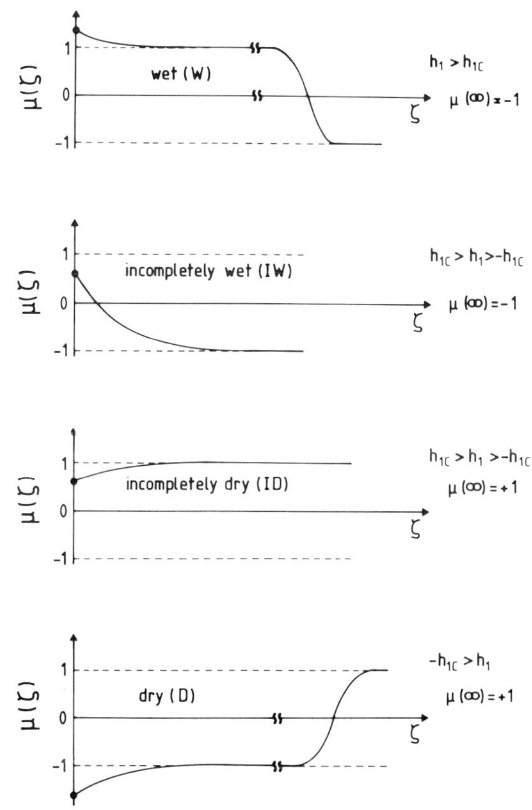

Figure 2. Order parameter profiles as a function of the distance from the wall, for different states of the surface and/or the bulk: wet (W), incompletely wet (IW), incompletely dry (ID), and dry (D). From Puri and Binder[56].

the boundary condition, Eq. (18). From a graphical analysis of these equations one derives the wetting phase diagram, Fig. 3, which displays both second order and first-order wetting transitions, separated by a wetting tricritical point[56].

In a real system, the parameters h_1, g that characterize the wall effects (Fig. 1) are not control variables but rather fixed. It then is of interest what the consequences of the "surface phase diagram" (Fig. 3) are for the phase diagram in the space of physically accessible control variables, such as temperature T and density ρ (for a liquid gas-system), or (equiva-

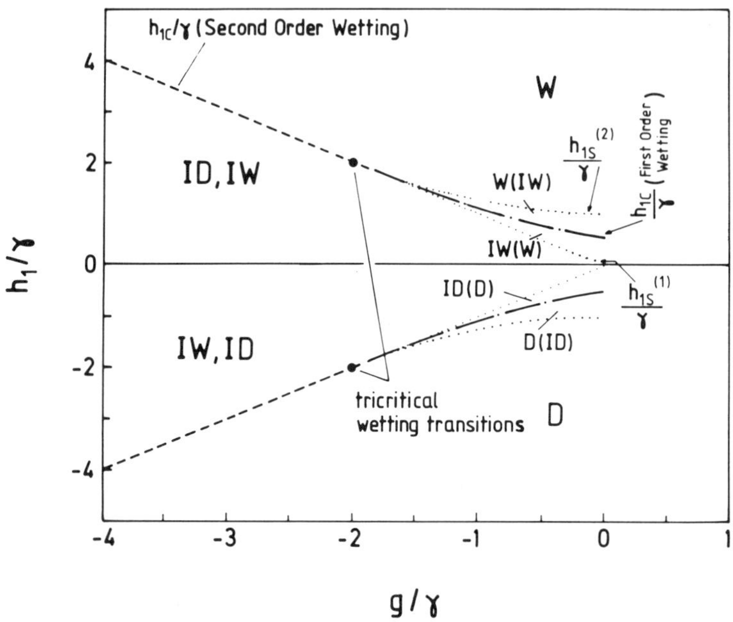

Figure 3. Phase diagram of the Ginzburg-Landau model for wetting and drying, in (g/ξ , h_1/ξ) space. Regions are labelled according to the stable static solution for those parameter values (metastable states, if any, are listed in brackets): wet (W), dry (D), incompletely wet (IW) and incompletely dry (ID). Note that the phase diagram is symmetric around the abscissa, if the meaning of W and D is interchanged. Second order phase boundaries are shown as dashed lines and given by $h_{1c} = \pm g$ (for $g/\xi < -2$). First order phase boundaries are shown as dash-dotted curves; surface spinodals are shown as dotted curves, $h_{1S}^{(1)} = -g$ and $h_{1S}^{(2)}/\xi = 1 + (g/2\xi)^2$, $g > -2\xi$ (upper half plane). From Puri and Binder[56].

lently), concentration c_b, for a binary mixture. This is shown in Fig. 4, assuming the "language" appropriate for a gas-liquid system - though this is readily translated to other systems. In Fig. 4, we have assumed that the parameters of Fig. 3 are such that we have a first order wetting transition. Note that Fig. 3 has used h = 0, i.e. we follow a path exactly along the coexistence curve in Fig. 4. In the one phase region (h > 0) then another singularity occurs, which shows up as a jump in the surface excess order parameter m_S (in the example of Fig. 4, m_S is denoted as surface excess density ρ_S).

General Surface Thermodynamics

For the discrete lattice model of Fig. 1, the surface excess order parameter can be obtained from the layer order parameters m_n as

$$m_S = \sum_{n=1}^{\infty}(m_n - m_b) \tag{25}$$

Alternatively, one can derive m_S as a derivative of the surface excess free energy with respect to the bulk field, analogously to the order parameter in the bulk {both $\mathcal{F}_S$ and $\mathcal{F}_b$ are normalized per lattice site},

$$m_S = -\frac{\partial \mathcal{F}_S(T,H,H_1)}{\partial H}\bigg|_{T,H_1}, \quad m_b = -\frac{\partial \mathcal{F}_b(T,H)}{\partial H}\bigg|_T \tag{26}$$

The relation between Eqs. (25), (26) is readily understood by considering a thin film of thickness D, since the total free energy of the film can be split in bulk and surface terms

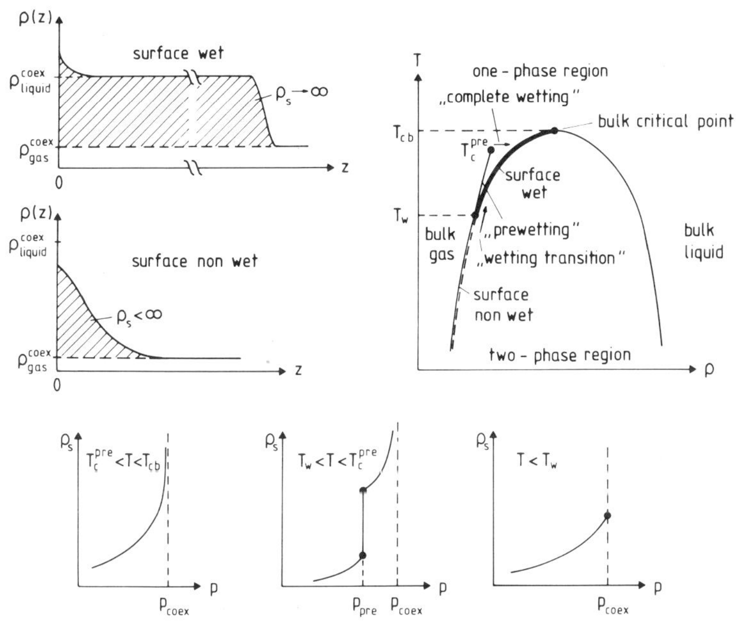

Figure 4. Schematic phase diagram of a semi-infinite fluid in contact with a wall (right part) and corresponding profiles of the local density ρ(z) as functions of the distance z from the wall at z = 0 (left part). In the temperature (T)-density (ρ) plane, the coexistence curve separates saturated gas with liquid in the bulk. The enclosed two-phase region ends at the bulk critical point (T_{cb}). The wetting transition is a singularity of the surface excess free energy associated with the wall and occurs at T_w for a path along the coexistence curve. The corresponding density profile may change from a nonwet state of the surface (surface excess density ρ_S, the integral of the shaded region, being finite) to a wet state in either a continuous fashion ("critical wetting") or via a discontinuous jump at T_w. In the latter case, as assumed here, there occurs a further discontinuity at the "prewetting transition" in the one-phase region. The schematic "adsorption isotherms" in the lower part of the figure show "complete wetting" of the wall (when the pressure p approaches its value p_{coex} at the gas-liquid coexistence curve) and a further discontinuity at the prewetting transition for $T_w < T < T_c^{pre}$. In contrast, for $T < T_w$ (surface nonwet), ρ_S reaches a finite limit for $p \to p_{coex}$.

(A is the area of a free surface of this film), for $D \to \infty$,

$$F_{film}(T,H,H_1,D)/(AD) = \mathcal{F}_b(T,H) + \frac{2}{D}\mathcal{F}_S(T,H,H_1) \tag{27}$$

$$m_{film} = \sum_{n=1}^{D} m_n = -\frac{\partial F_{film}(T,H,H_1,D)}{\partial H}\Big|_{T,H_1,D} /(AD) \quad . \tag{28}$$

The factor 2 in Eq. (27) simply comes from the fact that we assume two equivalent free surfaces in this geometry. From Eq. (27), (28) we then recognize that m_S can be written as $m_S = \lim\limits_{D\to\infty} \sum\limits_{n=1}^{D/2} (m_n - m_b)$, i.e. Eq. (25).

The local order parameter at the surface then results as a derivative with respect to the local field H_1 conjugate to it,

$$m_1 = -\partial\mathcal{F}_S(T,H,H_1)/\partial H_1 \Big|_{T,H} \quad . \tag{29}$$

It then also is of interest to consider "susceptibilities", defined in analogy to the bulk response function $\chi_b = (\partial m_b/\partial H)_T = -\partial^2\mathcal{F}_b(T,H)/\partial H^2\Big|_T$, but since there are two fields three susceptibilities χ_S, χ_1, χ_{11} result,

$$\chi_S = \left(\frac{\partial m_S}{\partial H}\right)_{T,H_1} = -\left(\frac{\partial^2 \mathcal{F}_S}{\partial H^2}\right)_{T,H_1} , \quad \chi_1 = \left(\frac{\partial m_1}{\partial H}\right)_{T,H_1} = -\left(\frac{\partial^2 \mathcal{F}_S}{\partial H \partial H_1}\right) = \left(\frac{\partial m_S}{\partial H_1}\right)_{T,H} , \tag{30}$$

$$\chi_{11} = \left(\frac{\partial m_1}{\partial H_1}\right)_{T,H} = -\left(\frac{\partial^2 \mathcal{F}_S}{\partial H_1^2}\right)_{T,H} \quad . \tag{31}$$

The surface excess susceptibility can also be written analogously to Eq. (25), in terms of susceptibilities $\chi_n = (\partial m_n/\partial H)_{T,H_1}$ related to the n'th layer,

$$\chi_S = \sum_{n=1}^{\infty} (\chi_n - \chi_b) \tag{32}$$

At a second order phase transition, singularities of such second order derivatives of the free energy occur: just as the bulk susceptibility χ_b diverges at the critical point T_{cb} of the bulk with an exponent γ_b,

$$\chi_b \Big|_{H=0} = \hat{\Gamma}_\pm |1 - T/T_{cb}|^{-\gamma_b} \quad , \tag{33}$$

where $\hat{\Gamma}_\pm$ is a "critical amplitude" and the $\pm$ signs refer to above (+) and below (-) T_{cb}, one defines critical exponents associated with the singularities of the various surface quantities at a critical wetting transition temperature T_w,[6-8,12]

$$\chi_S \propto \left|1 - T/T_w\right|^{-\gamma_S} \quad , \quad \chi_1 \propto \left|1 - T/T_w\right|^{-\gamma_1} \quad , \quad \chi_{11} \propto \left|1 - T/T_w\right|^{-\gamma_{11}} \quad . \tag{34}$$

From the above mean field theory of wetting one can derive with a little algebra that[12]

$$\gamma_S = 1,\ \gamma_1 = 1,\ \gamma_{11} = 0 \text{ (jump singularity)} . \tag{35}$$

Eq. (34) refers to a case similar to Fig. 4 (but for a second rather than a first order wetting transition), where one approaches the transition at $H = 0$, H_1 and J_S (or h_1 and g, respectively) are assumed to be held fixed. In the general case, where a variation of these surface parameters is considered, T_w is a function of these variables, e.g. in the (T,H_1) plane there is a line of critical wetting, $T_w = T_w(H_1)$, or alternatively one can consider the inverse function, $H_{1c} = H_{1c}(T)$. One then can also approach this wetting transition line at fixed T as a function of H_1, of course, e.g. $\chi_1 \propto \left|H_1 - H_{1c}\right|^{-\gamma_1}$, and the associated exponents are the same as in Eqs. (34), (35).

The Sharp Kink Approximation for Interface Unbinding Transitions

Very near to a critical wetting transition, as well as in a situation close to complete wetting at $T > T_w$ (Fig. 4), the thickness ℓ of the wetting layer is much larger than the intrinsic thickness ξ_b of the interface. On a coarse-grained length scale, the interfacial profile can thus be replaced by a step function {the scaled order parameter in Fig. 2 top then is put $\mu(\gamma) = 1$ for $\gamma < \ell$ and $\mu(\gamma) = -1$ for $\gamma > \ell$ }[12]. The surface free energy f_S can then be derived as[61] (τ denotes the distance from the line $H_{1c}(T)$ in the T,H plane, and ℓ is measured in units of ξ_b)

$$f_S(\tau,h,\ell) = h\ell - A\tau\exp(-\ell) + B\exp(-2\ell) \quad , \tag{36}$$

where A, B are constants (that can be related to the constants of the above Ginzburg-Landau model, if desired). Now the minimum of $\mathcal{F}_S$ with respect to ℓ has to be sought, i.e.

$$\frac{\partial f_S(\tau,h,\ell)}{\partial \ell} = 0 \Rightarrow h = 2B\exp(-2\ell) - A\tau\exp(-\ell) \tag{37}$$

Thus one finds that for $h = 0$ and $\tau > 0$ (i.e. on the nonwet side of the transition) the thickness ℓ diverges logarithmically,

$$\ell = -\ln(A\tau/2B) \quad , \tag{38}$$

while for $h = 0$, $\tau < 0$ the minimum occurs for $\ell \to \infty$ (unbound interface). Complete wetting then is described by

$$\ell = -\ln h + \text{const} \tag{39}$$

both for $\tau = 0$ (at critical wetting) and for $\tau < 0$ ($T > T_w$ in Fig. 4). Either from the Ginzburg-Landau treatment or from Eqs. (36), (37) one can show that the singular part of $\mathcal{F}_S$ satisfies a scaling law

$$f_S^{\text{sing}}(\tau,h) = \tau^2\,\Omega_1(h\tau^{-2}) + h\ln\left[h\Omega_2(h\tau^{-2})\right] \quad , \tag{40}$$

where $\Omega_1(X)$, $\Omega_2(X)$ are scaling functions whose explicit form is not needed here. Remembering that h is the rescaled field H {Eq. (14)}, and that a variation of H_1 means a variation of τ, the quantities m_S, m_1, χ_S, χ_1, χ_{11} {Eqs. (26)-(31)} are readily obtained from the present treatment. In particular, the singular part of the surface excess is (remembering the rescaling)

$$m_S \approx 2\ell\, m_b \approx m_b \ln[h\, \Omega_2(h\tau^{-2})] \quad . \tag{41}$$

If we hence define a critical exponent β_S as $m_S \propto \tau^{\beta_S}$, one concludes $\beta_S = 0$ due to the logarithmic singularity of Eqs. (38) or (41), respectively.

Fluctuations of the Interface

The above treatment clearly is oversimplified as it treats the interface by a single coordinate ℓ only, its average distance form the wall. Clearly, it is necessary to allow local fluctuations of the height $\ell(x,y)$ of the interface from its average position (Fig. 5).

A basic concept is then the "interfacial stiffness" and the description in terms of the "capillary wave Hamiltonian"[8,14,16,17,62]. To introduce these terms, we consider the one-dimensional interface $z = \ell(x)$ of a two-dimensional system for simplicity. Noting that in lattice systems the interfacial energy E_{int} will depend on the angle θ between the tangent to the interface and the x-axis, we write $[\theta = \text{artan}\,(d\ell/dx)]$

$$E_{int}/k_BT = \int d\lambda\, f_{int}(\theta) = \int dx\, f_{int}(\theta)\sqrt{1+(d\ell/dx)^2} \quad , \tag{42}$$

using the fact that the line element $d\lambda$ along the interface satisfies $(d\lambda)^2 = (d\ell)^2 + (dx)^2$. Of course, in this coarse-grained description of the interface both overhangs and hubbles are deliberately ignored, cf. Fig. 5, and we even assume that the interface is rather flat, such that $|d\ell/dx| << 1$.

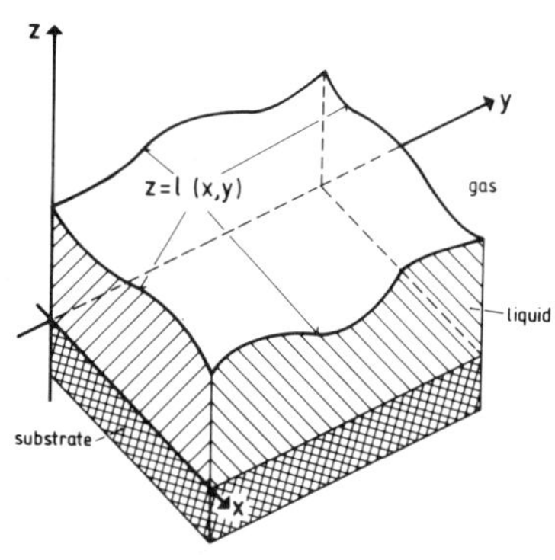

Figure 5. Coarse-grained description of a liquid-gas interface, where the "intrinsic" profile and local structure of the interface is disregarded, and one rather treats the interface as an "elastic membrane" at position $z = \ell(x,y)$ above a (substrate) surface at $z = 0$. Note that the description for a binary mixture AB of two coexisting phases would simply amount to reinterpret "liquid" as A-rich and gas as B-rich phases, for instance.

We then expand $(1 + (d\ell/dx)^2)^{1/2} \approx 1 + (d\ell/dx)^2/2$, $l_{int}(\theta) \approx f_{int}(0) + f'_{int}(d\ell/dx) + f''_{int}(d\ell/dx)^2/2$. The linear term in $d\ell/dx$ only yields boundary terms to the integral in Eq. (42) and can thus be omitted. One obtains

$$E_{int}/k_BT = f_{int}(0) \int dx + \frac{\kappa}{2} \int dx (d\ell/dx)^2 \quad , \qquad \kappa = f_{int}(0) + f''_{int} \quad . \tag{43}$$

While $k_BT f_{int}(0)$ tends to a finite constant for $T \to 0$ {e.g., in the nearest-neighbor Ising square lattice[63] $k_BT f_{int}(0) = 2J - k_BT \ln\{[1 + \exp(-2J/k_BT)] / [1 - \exp(-2J/k_BT)]\} \to 2J$ for $T \to 0$}, the interfacial stiffness diverges, $k_BT\,\kappa \to \infty$ $(d = 2)$, cf. Fig. 6.

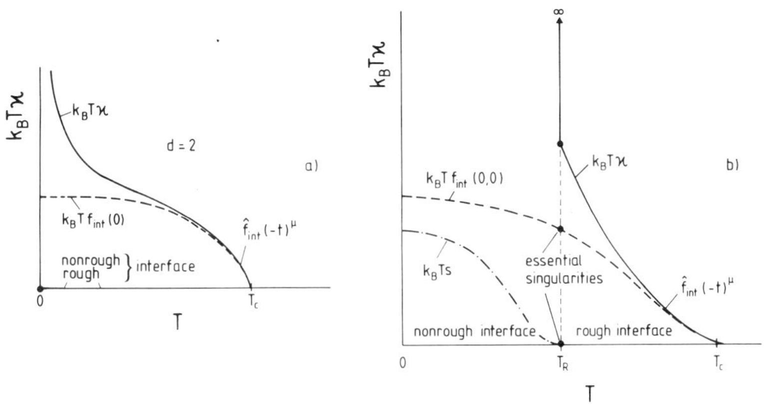

Figure 6. Schematic temperature variation of interfacial stiffness $k_BT\,\kappa$ and interfacial free energy, for an interface oriented perpendicularly to a lattice direction of a square (a) or simple cubic (b) lattice. While for $d = 2$ the interface is rough for all nonzero temperatures, in $d = 3$ it is rough only for temperatures T exceeding the roughening transition temperature T_R. For $T < T_R$ there exists then a nonzero free energy k_BTs of interfacial steps, which vanishes at T_R with an essential singularity. While κ is infinite throughout the nonrough phase, $k_BT\,\kappa$ reaches a universal value as $T \to T_R^+$. Note that κ and f_{int} to leading order in their critical behavior become identical near the bulk critical temperature T_c.

In d = 3 dimensions one finds analogously the "capillary wave Hamiltonian" $\mathcal{H}_{cw}$,

$$\frac{E_{int}}{k_BT} = f_{int}(0,0) \int dxdy + \frac{\mathcal{H}_{cw}}{k_BT} \, , \quad \frac{\mathcal{H}_{cw}}{k_BT} = \frac{\kappa}{2} \int dxdy \left[\left(\frac{d\ell}{dx} \right)^2 + \left(\frac{d\ell}{dy} \right)^2 \right] \quad , \tag{44}$$

if T exceeds the roughening transition temperature T_R (Fig. 6).

Eqs. (43), (44) are harmonic in terms of the Fourier components $\ell_{\vec{q}}$ of the local height variables $\ell(x)$ or $\ell(x,y)$,

$$\mathcal{H}_{cw}/k_BT = (\kappa/2)\,(2\pi)^{-(d-1)} \int d^{d-2}q\, q^2 |\ell_{\vec{q}}|^2 \quad . \tag{45}$$

Since equipartition implies

$$(k_BT\,\kappa/2)\,(2\pi)^{-d-1}\, q^2 \left\langle |\ell_{\vec{q}}|^2 \right\rangle_T = k_BT/2 \quad , \tag{46}$$

it is straightforward to obtain the local meansquare displacement of the interface,

$$\left\langle \ell^2(\vec{x}) \right\rangle - \left\langle \ell(\vec{x}) \right\rangle^2 = (2\pi)^{-(d-1)} \int d^{d-2}q \left\langle \left| \ell_{\vec{q}} \right|^2 \right\rangle \propto \kappa^{-1} \int q^{d-4} dq \quad . \tag{47}$$

One sees that for $d \leq 3$ this integral diverges for large wavelengths. Thus the interfacial width $W(L_{||})$ depends on the length scale $L_{||}$ along the interface,

$$W^2(L_{||}) \equiv \left\langle \left(\ell - \langle \ell \rangle\right)^2 \right\rangle = \kappa^{-1} \int_{2\pi/L_{||}}^{2\pi/\xi_b} dq/q = \kappa^{-1} \ln\left(L_{||}/\xi_b\right), \quad d = 3, \tag{48}$$

while in d = 2 a power law divergence results, $W^2(L_{||}) \propto \kappa^{-1} L_{||}$. This latter result can simply be interpreted by the "random walk picture" of the fluctuating one-dimensional interface[64].

The Interfacial Roughening Transition

While interfaces in fluids are always rough, interfaces in crystals may be atomistically smooth and localized in lattice planes. As the temperature is raised, thermal exitations do create local deviations of the interface from its average height, but the correlation function of these fluctuations decays exponentially to zero. It then may happen, as anticipated in Fig. 6, that one approaches a transition temperature T_R where this correlation length $\xi_{||}(T)$ of the height-height correlations diverges. An equivalent[9,10] description considers the vanishing of the step free energy $k_BT\, s(T)$. One considers the interfacial tension of interfaces which are tilted through an angle θ relative to a low-index lattice plane. For small θ in the non-rough phase the angular-dependent interfacial tension becomes (a = lattice spacing, k_BTs = step free energy)

$$f_{int}(\theta) = f_{int}(0) + (s/a)\left|\theta\right| + \tfrac{1}{2} f^{||}\theta^2 + \ldots \tag{49}$$

While in the previous section (describing the rough interface) we have assumed that $f_{int}(\theta)$ is analytic at $\theta = 0$, this is not so when Eq. (49) holds. Of course, any inclined interface can be described as a staircase of regularly spaced steps (neighboring steps are a distance D apart such that $|\theta| = a/D$ for small $|\theta|$. The (free) energy cost of these steps does not depend on the sign of θ, of course.

While the picture of the tilted interface developed so far is pretty obvious for $T \to 0$, at $T > 0$ thermal fluctuations need consideration. Those on the scale ξ_b can be eliminated by coarse graining. However, when viewed from above the steps will not be just parallel straight lines, they can have kink-type shifts either to the right or to the left and will therefore behave random-walk like. Even though "kinks" cost energy, entropy causes large excursions of the steps (the problem is somewhat analogous to fluctuations of one-dimensional interfaces, as considered above), and thus the step wandering due to these "soft mode"-like fluctuations creates long range step-step interactions.

Here we shall not elaborate on the theory of the roughening transition[9,10] beyond these qualitative comments, but quote only some of the key results. It turns out that the roughening transition is of Kosterlitz-Thouless[65] character, governed by essential singularities:

$$\xi_{||}^{-1}(T) \propto s(T) \propto \exp\left[-c/(T_R-T)^{1/2}\right], \quad T \to T_R \quad , \tag{50}$$

where c is a constant. The interfacial width is finite in the nonrough phase,

$$W^2(L_{\parallel}) \propto (T_R - T)^{-1/2} \quad , T \to T_R^- \quad , \tag{51}$$

while for $T > T_R$ a behavior of the type of Eq. (48) applies but κ^{-1} tends to a universal value as $T \to T_R^+$,

$$W^2(L_{\parallel}) \propto [\pi^{-2} + c^1 (T - T_R)^{1/2}] \ln (L_{\parallel}/\xi_b) \quad , \tag{52}$$

c^1 being another constant.

At this point we recall that the description of wetting phenomena given above heavily relies on the interface being a smooth, delocalized object, i.e., one implicitly assumes that the wetting transition temperature T_w clearly exceeds T_R. However, our basic Ising model {Eq. (2)} allows also for the inverse case,[8,41,43,58] and then the smooth increase of m_S as $H \to 0$ in the complete wetting regime {Eq. (39)} is replaced by an infinite sequence of layering transitions (Fig. 7). Note that the discrete version of mean field theory {Eqs. (8), (9)} would yield layering always, but with $\lim_{n\to\infty} T_c(n) = T_{cb}$, since in molecular field theory the interface is always nonrough, capillary wave-type fluctuations being not included. Thus, to some extent the surface phase diagrams shown in Fig. 7 are speculative. However, arguments have been given[66] to predict

$$T_R - T_c(n) \propto (\ln(n))^{-2} , \quad n \to \infty \ . \tag{53}$$

Fluctuation Effects on the Critical Wetting Transition

It turns out that the wandering of the interface due to capillary waves significantly

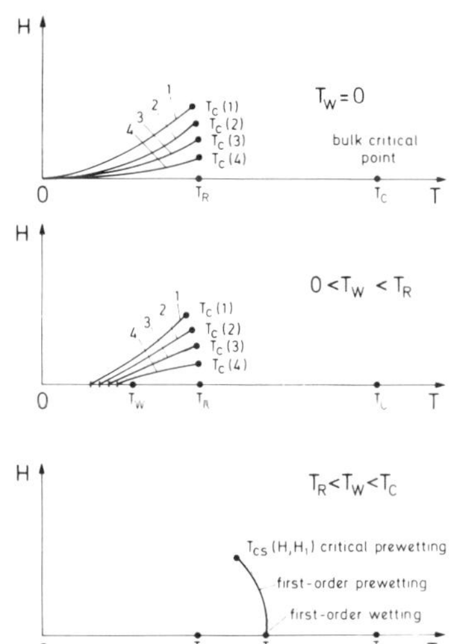

Figure 7. Schematic phase diagram of a semi-infinite Ising model {Eq. (2)} as a function of bulk field and temperature T. Three possible "scenarios" are shown [which of them is realized depends on the ratios between the surface and bulk interactions, H_1/J and J_S/J, see the section on the simulation results]. The surface is wet for $H \to 0$ if $T > T_w$. Assuming J_S/J is large enough, one has a first order wetting transition, if $T_w > T_R$, and a prewettting line ending in a prewetting critical point $T_{cs}(H,H_1)$. If $T_w < T_R$, one has an infinite sequence of first order layering transitions (labeled by the number n = 1,2,3,4... of the layer in the figure). These layering transitions end in layering critical points $T_c(n)$, with $\lim_{n\to\infty} T_c(n) = T_R$. After Pandit et al.[58].

changes the simple mean field picture, where we have minimized $\mathcal{F}_S(\tau,h,\ell)$ with respect to the mean interface position ℓ {Eqs. (36)-(41)}. Rather one has to use $\mathcal{F}_S(\tau,h,\ell) = V(\ell)$ as a potential in the capillary wave Hamiltonian, with $\vec{x}$ a(d-1) - dimensional coordinate at the surface

$$\mathcal{F}_S\{\ell(\vec{x})\} = \int d^{d-1}x\left\{\frac{\kappa}{2}(\nabla\ell)^2 + V[\ell(\vec{x})]\right\} \tag{54}$$

Unfortunately, the validity of Eq. (54) is somewhat questionable: it may be necessary to treat the interfacial stiffness κ not as a constant but as a function of $\ell(\vec{x})$ itself[61]; it also may be necessary to consider a second collective coordinate $\ell_1(\vec{x})$ which describes where the profile $\mu(z)$ near the wall becomes flat (cf. Fig. 2): although $<\ell_1>$ is of order ξ_b, it has been suggested that the coupling between $\ell(\vec{x})$ and $\ell_1(\vec{x})$ changes the theory of complete wetting significantly[67].

Disregarding these caveats here, we note that the treatment of Eq. (54) with renormalization group methods[68-73] predicted that some critical exponents of second order wetting transitions may deviate markedly from their mean field counterparts, and are nonuniversal, depending on a parameter ω

$$\omega = 1/[4\pi\kappa\xi_b^2] \quad . \tag{55}$$

In particular, the exponent $\nu_{||}$ describing the correlations of fluctuations of $\ell(\vec{x})$ deviates from its mean field value[12] $\nu_{||} = 1$ according to[68]

$$\nu_{||} = (1-\omega)^{-1},\ 0 < \omega < 1/2;\quad \nu_{||} = \left(\sqrt{2}-\sqrt{\omega}\right)^{-2},\ 1/2 < \omega < 2;$$
$$\nu_{||} = \infty\ ,\ \omega > 2 \quad . \tag{56}$$

As a consequence the singular part $f_s^{sing}(\tau,h)$ of the surface excess free energy no longer is given by Eq. (40) but rather by

$$f_s^{sing}(\tau,h) = \tau^{2\nu_{||}}\Omega_1\left(h\tau^{-2\nu_{||}}\right) + h\ln\left[h\Omega_2\left(h\tau^{-2\nu_{||}}\right)\right] \tag{57}$$

While the logarithmic singularities of m_S {which can still be identified with the mean distance $<\ell>$ of the interface from the wall, cf. also Eqs. (38),(39),(41)} are preserved, the surface layer susceptibilities χ_1, χ_{11} now get a different scaling behavior

$$\chi_1 \propto -\partial^2 f_S^{sing} / \partial h\partial\tau = \tau^{-1}\tilde{\chi}_1\left(h\tau^{-2\nu_{||}}\right) \quad , \tag{58}$$

$$\chi_{11} \propto -\partial^2 f_S^{sing} / \partial\tau^2 = \tau^{2\nu_{||}-2}\tilde{\chi}_{11}\left(h\tau^{-2\nu_{||}}\right) \quad , \tag{59}$$

where $\tilde{\chi}_1$, $\tilde{\chi}_{11}$ are scaling functions. χ_1 for $h = 0$ still has a Curie-Weiss type singularity for $\tau \to 0$, while χ_{11} has a weaker singularity than in mean field theory {Eq. (35)}. However, a more interesting behavior occurs for χ_1 at $\tau = 0$: to cancel the τ^{-1} factor, $\tilde{\chi}_1$ $(X >> 1)$ must behave as $\tilde{\chi}_1\ (X) \propto X^{-1/2\nu_{||}}$, and hence

$$\chi_1 \propto h^{-1/2\nu_{\parallel}} \quad , \tau = 0 \quad . \tag{60}$$

For a test of Eqs. (56)-(60) a knowledge of ω {Eq. (55)} obviously is required - but the estimation of this parameter also is a matter of debate[74]. Even for the nearest neighbor Ising lattice the interfacial stiffness is not known very accurately, and it is under discussion whether the length scale ξ_b in Eq. (55) is the true correlation length in a lattice direction.

Surface-Induced Ordering (SIO) and Disordering (SID)

We now consider the surface of a system that undergoes a first order (i.e., discontinuous) order-disorder transition at a temperature T_{cb} in the bulk. Due to "missing neighbors" at a surface, we expect that the order parameter at temperatures $T < T_{cb}$ is slightly reduced in comparison with its bulk value (Fig. 8). If this situation persists up to T_{cb} (Fig. 8a), the

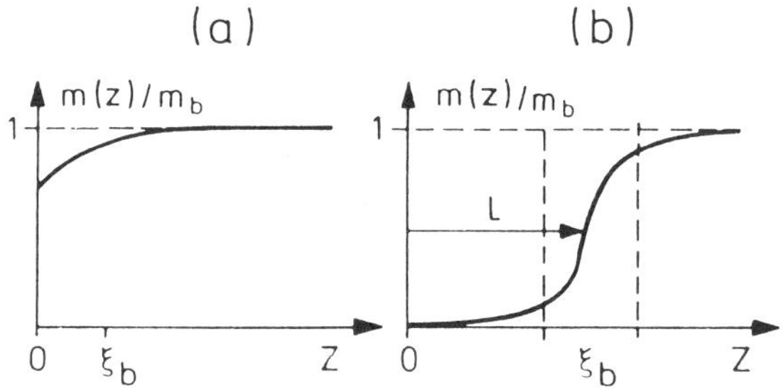

Figure 8. Order parameter profiles $m(z)/m_b$ associated with surface-induced disorder (SID). If case (a) persists up to T_{cb}, the surface stays ordered up to T_{cb}, while case (b) shows surface-induced disordering: a layer of thickness L has disordered already at $T < T_{cb}$, and as $T \to T_{cb}^-$ the (delocalized) interface at mean position z = L from the surface advances into the bulk, $L\ (T \to T_{cb}^-) \to \infty$. The surface order parameter $m_1 = m\ (z = 0)$ then vanishes continuously, $m_1 \propto (1 - T/T_{cb})^{\beta_1^{SID}}$. From Dosch at al.[75].

surface order parameter $m_1 = m\ (z = 0)$ vanishes discontinuously as the bulk order parameter m_b does - a situation that is not of particular interest. However, it may happen[15,76] that the surface region disorders somewhat already at $F < T_{cb}$, and this disordered layer grows as $T \to T_{cb}^-$. This leads to a continuous vanishing of the surface layer order parameter $m_1 \propto (1 - T/T_{cb})^{\beta_1^{SID}}$, although the bulk order parameter $m_b = m\ (z \to \infty)$ vanishes discontinuously at T_{cb}. This can be considered as a wetting phenomenon: the disordered phase wets the interface between the vacuum and the ordered phase as $T \to T_{cb}^-$, where the disordered phase becomes thermodynamically stable. The approach $T \to T_{cb}^-$ is comparable to complete wetting, the interface between the disordered surface layer and the ordered bulk being completely unbound from the surface. One can treat this problem again in terms of an one-component Landau theory analogous to Eq. (19), but now one has to choose the bulk free energy density f_b (m) appropriate for a first order transition in the bulk:

$$\Delta \mathcal{F}\ \{m(z)\}/k_B T_{cb} = \int_0^\infty dz \left\{ f_b(m) + \frac{1}{2d} R^2 \left(\frac{d\phi}{dz} \right)^2 \right\} - \frac{H_1 m_1}{k_B T_{cb}} + \frac{1}{2} g m_1^2 \quad , \tag{61}$$

with

$$f_b(m) = -\frac{H}{k_B T_{cb}} + \frac{r}{2}m^2 - \frac{b}{n}m^n + \frac{c}{n^l}m^{n^l} \quad , \tag{62}$$

where the coefficients r,b,c > 0 and $(n,n^l) = (3,4)$ or $(4,6)$, respectively. This problem can be studied with an approach analogous to Eqs.(17),(18) and the subsequent discussion. For $H = H_1 = 0$ the behavior is controlled by a comparison of the parameters g and r (T_{cb}): if $g \geq \sqrt{r(T_{cb})}$ the disordered phase wets the interface,[76] while if $g < -\bar{x}\sqrt{r(T_{cb})}$, the surface coupling is so strong that the ordered phase wets the interface between the vacuum and the disordered phase, for $T \to T_{cb}^+$, and one has surface-induced order instead of disorder. Here the constant $\bar{x} = 2^{1/3}$-1 for $(n,n^l) = (3,4)$ and $\bar{x} = \sqrt{2} - 1$ for $(n,n^l) = (4,6)$. In the intermediate region $-\bar{x}\sqrt{r(T_{cb})} < g < \sqrt{r(T_{cb})}$ there is no wetting, but rather a behavior as in Fig. 8a). In all cases, the theory predicts a logarithmic growth of the thickness of the wetting layer L(T) in Fig. 8 b) for short range wall forces (as assumed in Eq. (61), L(T) $\propto |\ln|T - T_{cb}||$).

From Fig. 8 and Eqs. (62),(62) it is clear that SID belongs to the class of interface unbinding transitions. We apply again the sharp kink approximation, the interface position ℓ being the relevant degree of freedom.

For short range surface forces one then has an effective potential[15]

$$V_{eff}(\ell) = cf_{int} \exp(-2\ell/\xi_b) + T_{cb}(S_{dis} - S_{ord})\, t\ell , \tag{63}$$

where c is a constant, f_{int} the interfacial tension between the ordered (ord) and disordered (dis) phases, S_{ord} , S_{dis} are their bulk entropies, ξ_b is the correlation length is the disordered phase, $t = 1 - T/T_{cb}$. In Eq. (63), the last term simply results from expanding the free energy difference of the disordered surface layer relative to the bulk, $(F_{dis} - F_{ord})\,\ell$, linearly in temperature. Comparing Eqs. (63),(36) we recognize that SID is equivalent to complete wetting at the critical wetting temperature: i.e., we approach the transition not along a path varying τ at h = 0, but along a path varying h at $\tau = 0$ in Eqs. (40) or (57)-(60), respectively.

The temperature-like variable t of SID in Eq. (63) corresponds to the field-like variable h for wetting {Eq. (60)}. Since for wetting the correlation length scales as

$$\xi_{\parallel} = \tau^{-\nu_{\parallel}}\tilde{\xi}\left(h\tau^{-2\nu_{\parallel}}\right) = h^{-1/2}\tilde{\tilde{\xi}}\left(h^{-1/2\nu_{\parallel}}\tau\right)\underset{\tau=0}{\propto} h^{-1/2} \tag{64}$$

where $\tilde{\xi}$, $\tilde{\tilde{\xi}}$ are appropriate scaling functions, the equivalence h (wetting) ↔ t (SID) yields the critical exponent of the correlation length for SID,

$$\xi_{\parallel}^{SID} \propto t^{-1/2}, \nu_{\parallel}^{SID} = 1/2 \quad , \tag{65}$$

irrespective of whether fluctuation effects are considered or not. While for wetting the critical part of m_1 scales like χ_1 h (cf. Eq. (58)),

$$\Delta m_1 = h\tau^{-1}\tilde{m}_1\left(h\tau^{-2\nu_{\parallel}}\right) = h^{1-1/2\nu_{\parallel}}\tilde{\tilde{m}}_1\left(h^{-1/2\nu_{\parallel}}\tau\right)\underset{\tau=0}{\propto} h^{1-1/2\nu_{\parallel}} \quad , \tag{66}$$

$\tilde{m}_1$, $\tilde{\tilde{m}}_1$ again being scaling functions, one finds for SID

$$m_1^{SID} \propto t^{\beta_1^{SID}}, \quad \beta_1^{SID} = 1 - \frac{1}{2\nu_{II}} \left\{ = \frac{1+\omega}{2} \text{ for } 0 < \omega < \frac{1}{2}, \text{cf. Eq. (56)} \right\} \tag{67}$$

The nonuniversal exponent ν_{II} of wetting {Eq. (56)} describes the continuous vanishing of the local order parameter in SID, and not the universal exponent ν_{II}^{SID} ! This description assumes, of course, that the interface between ordered and disordered states at T_{cb} is rough, such that the continuum description {Eq. (64)} is appropriate. From this translation rule $h \leftrightarrow t$ one can derive numerous scaling relations[47] for the exponents of SID, noting $\gamma_S^{SID} = 1$,

$$\gamma_1^{SID} + \beta_1^{SID} = 1, \quad 2\gamma_1^{SID} = \gamma_{11}^{SID} + 1, \quad \gamma_{11}^{SID} + 2\beta_1^{SID} = 1 \tag{68}$$

In contrast, the behavior of SIO is very different: the transition occurring at $T_{CS} > T_{cb}$ is a purely two-dimensional ordering phenomenon, with exponents of the corresponding universality class of bulk d = 2 criticality. Only when one approaches T_{cb} , the surface excess order parameter shows the same logarithmic behavior

$$m_S \propto \langle \ell \rangle \propto -(\xi_b/2) \ln|t| \quad . \tag{69}$$

MONTE CARLO SIMULATIONS OF INTERFACES BETWEEN COEXISTING PHASES

Simple Ising Models and the Roughening Transition

As discussed above, a quantity of basic interest in a crystal is the interfacial tension $f_{int}(\theta)$ of an interface tilted by an angle θ relative to a lattice direction: for a nonrough interface, $f_{int}(\theta) = f_{int}(0) + s/a\,|\theta| + \ldots$ {Eq. (49)} defines the step-free energy s, while for a rough interface $f_{int}(\theta) = f_{int}(0) + f_{int}^{II} \frac{1}{2}\theta^2$ yields the interfacial stiffness parameter $\kappa = f_{int}(0) + f_{int}^{II}$ {Eq. (43)}.

From Monte Carlo simulations, $f(\theta)$ can be obtained by a rather unusual combination of boundary conditions (Fig. 9). For the LxLxL lattices with the tilted interface one imposes

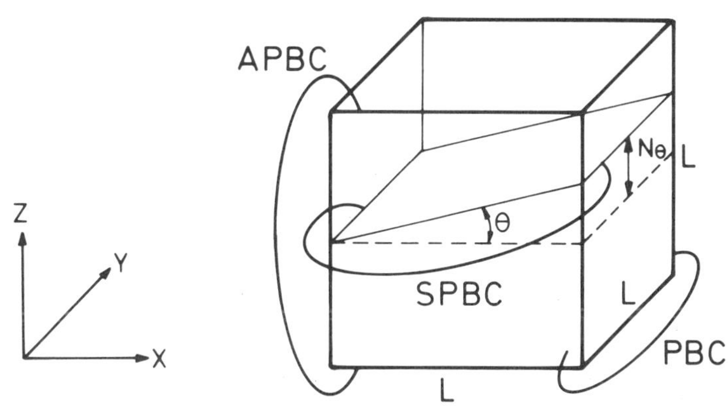

Figure 9. Boundary condition used to impose a tilted interface: antiperiodic in the z-direction, periodic in the y-direction, and screw periodic boundary condition in the x-direction. From Mon at al.[35].

an antiperiodic boundary condition (APBC) in the z-direction, and this has (for a ferromagnetic nearest neighbor Ising model) the effect of producing an interface. The interface is not pinned to a particular equilibrium position but it can wander and is translationally invariant. While an ordinary periodic boundary condition (PBC) is used in the y direction, the interface is tilted by applying a screw boundary condition (SPBC) in the x direction, with a shift of N_θ lattice constants. As a result, one produces a tilted interface with a tilt angle $\theta = \tan^{-1}(N_\theta/L)$. In order to isolate the interfacial contribution, one simulates also an identical reference system with the antiperiodic boundary condition replaced by a periodic one. The excess interfacial energy then is

$$U_{int}(\theta,T,L) = U_{APBC}(\theta,T,L) - U_{PBC}(\theta,T,L) \tag{70}$$

The excess interfacial free energy $\tau(\theta,T^*,L)$ {measuring temperature in units of T_{cb}, $T^* \equiv T/T_{cb}$} then is found from standard thermodynamic integration,[35,77] using as a reference state $T^* = 0$ where $\tau(\theta,0,L) = 2J(\sin\theta + \cos\theta)$. Fig. 10 shows the corresponding results. The slope of $\tau(\theta,T^*,L)$ as $\theta \to 0$ yields the step-free energy. As expected, the exponential divergence of $\xi_{\|}$ at the roughening transition temperature {Eq. (50)} leads to pronounced finite size effects of the step-free energy $f_S(L,T^*)$,

$$f_S(L,T^*) = \xi_{\|}^{-1}\tilde{f}_S(L/\xi_{\|}) \xrightarrow[T=T_R]{} L^{-1} \tag{71}$$

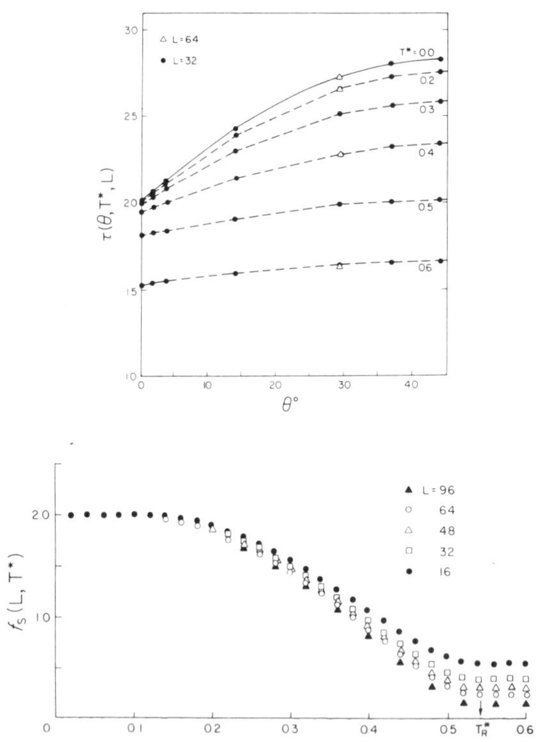

Figure 10. Angular dependence of the anisotropic interfacial tension of the nearest neighbor Ising model (upper part) and temperature dependence of the step-free energy (lower part). All free energies are given in units of J, and the lattice spacing is put to unity. Various lattice linear dimensions are shown as indicated. The location of the roughening transition is shown by an arrow. From Mon et al.[35].

A detailed analysis of the data of Fig. 10 in terms of Eq. (71) yielded good agreement, for a choice of $T_R^* = 0.54 \pm 0.02$ which is compatible with (but less accurate than) other recent estimates of the roughening transition temperature,[36,39] $J/k_BT_R = 0.4074 \pm 0.0003$ (note $J/k_BT_{cb} = 0.221660 \pm 0.000005$)[78]. These studies analyze the interfacial width, and assuming the validity of capillary wave theory, an estimate for the temperature dependence of the interfacial stiffness κ also was extracted[39].

Antiphase Domain Boundaries in Models for Fe-Al Alloys

The roughness of interfaces is not only important for simple models such as the nearest neighbor Ising model, but for complicated models of real materials as well. A study of antiphase domain boundaries of a model for FeAl alloys[37,38] (Fig. 11) demonstrated this. This model used interaction parameters deduced from experiment[79,80] on a $Fe_{0.8}Al_{0.2}$ alloy (V_1 = -12meV between nearest neighbors, $V_2/|V_1| = -0.167$ and $V_3/|V_1| = 0.208$ the interactions between nextnearest and third nearest neighbors, respectively), and $H/|V_1| = 4$ was chosen in Eq. (5), where the model[81] exhibits a second-order transition to the disordered A2-phase at $T_{cb}/|V_1| \approx 7.1$. Choosing a geometry $L_\perp x L_{||} x L_{||}$ with an even numer $L_{||}$ of lattice planes (containing the y,z directions), but an odd number $L_\perp$ in x-direction, and periodic

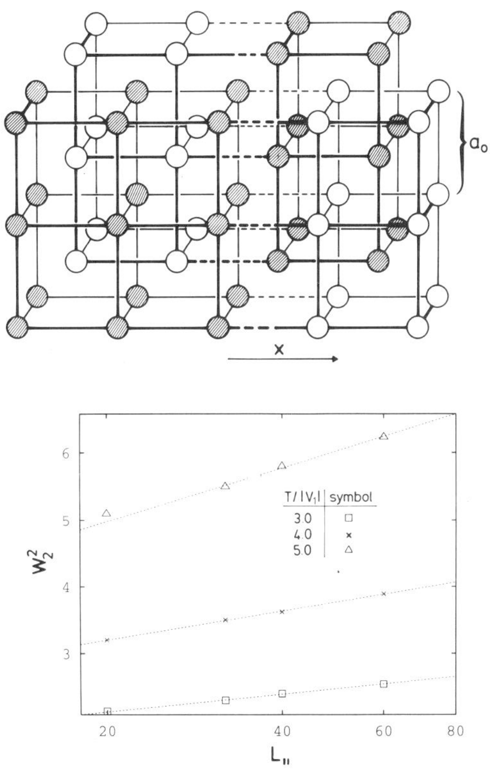

Figure 11. Schematic sketch of atomic arrangement in the ordered B2 structure on the body-centered cubic lattice across an antiphase domain boundary (upper part). A-atoms are shown by shaded circles, B-atoms by white circles Bonds across the wall are shown as broken straight lines. Lengths in this model are measured in units of $a_0/2$, a_0 being the lattice spacing. Lower part shows the mean square width W_2^2 (defined from the second moment of the order parameter profile) versus the logarithm of the linear dimension $L_{||}$ parallel to the interface. Three temperatures are shown as indicated. From Schmid and Binder[37].

boundary conditions throughout, one forces a wall (oriented perpendicular to the x-direction) into the system, see Fig. 11. This "misfit" has the consequence that in each configuration C of the system the order parameter profile $M_C(x)$ satisfies an antiperiodic boundary condition, $M_C(x \pm L_l) = -M_C(x)$, as imposed in the studies of the simple Ising model. Properties P not linearly related to the order parameter (e.g. energy, concentration, etc.) still satisfy a periodic boundary condition, $P_C(x \pm L_l) = P_C(x)$. With this boundary condition the domain wall in general is not localized at a particular position in the system (above the roughening temperature T_R it undergoes in fact a diffusive motion in x-direction during the simulation[37,38,82]). While this delocalization of the interface does not hamper the estimation of the interfacial free energy, for an investigation of the interfacial profile and its width the instantaneous interface position $\bar{\ell}(C)$ needs to be located in each configuration that is analyzed (the overbar indicates that an averaging over y,z directions is performed). This is done translating the system in x-direction until the total order parameter vanishes, i.e. $\sum_{x=-(L_l-1)/2}^{x=+(L_l-1)/2} M_C\left(x+\bar{\ell}(C)\right) = 0$ (we choose the origin of the x-coordinate in the middle of the

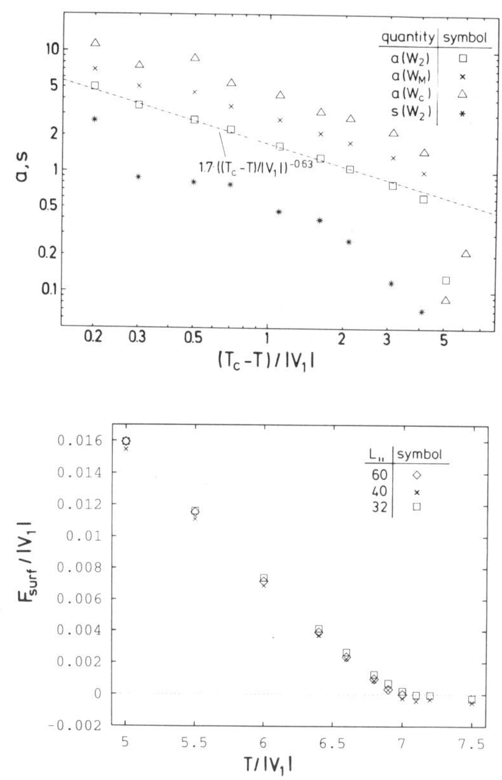

Figure 12. Log-log plot of the characteristic lengths a,s versus the distance from the bulk critical temperature (upper part) and the surface excess free energy $F_{surf}/|V_1|$ of the antiphase domain boundary (Fig. 11), normalized per lattice site, as a function of temperature near $T_{cb}/|V_1| = 7.1$ (lower part). The quantities $a(W_2)$, $s(W_2)$ are extracted from the second moment of the order parameter profile $\overline{M}(x)$, while $a(W_M)$ is extracted from $W_M \equiv \sum_x \left|\overline{M}(x)/M_{bulk} - 1\right|$, $a(W_C)$ from the relative excess of the concentration $W_C \equiv \sum_x \left[c(x) - c_{bulk}\right] / \left[c(0) - c_{bulk}\right]$. From Schmid and Binder[37].

simulation box). In this way interfacial order parameter profiles $M_C(x)$ taken in different configurations C can be averaged together, taking the position $\bar{\ell}$ (C) always as the origin of the x-axis, and similarly profiles of other quantities $P_C(x)$ are averaged, $\overline{P}(x) = \langle P_C(x + \bar{\ell}(C)) \rangle_C$.

This analysis shows[38,82] that the model exhibits an interfacial roughening transition at $T_R/|V_1| = 2.7 \pm 0.1$, i.e. far below the critical temperature ($T_R^* = T_R / T_{cb} \approx 0.3$ here). In the rough phase the expected logarithmic law {Eq. (48)} is indeed observed (Fig. 11). As a quantitative test, a fit of various widths W to the law $W^2 = a^2 \ln(L_{II}/s)$ was performed,[37] and the length a,s extracted (Fig. 12). While the length a is close to ξ_b {an analysis of the order parameter correlation function[37] yielded $\xi_b \approx 1.9 \left[(T_{cb} - T)/|V_1|\right]^{-\nu_b}$ with $\nu_b = 0.63$}, the length s seems to be significantly smaller than ξ_b , contrary to expectation {Eq. (48)}. The estimate for a disagrees with theoretical expectation either, since universality implies[74] for $\omega(T_{cb}) \approx 0.77$, and hence {Eq. (55)} $a^2 = \kappa^{-1}$ {Eq. (48)} should be $a^2 \approx 9.68\ \xi_b^2$ rather than $a^2 \approx 0.8\ \xi_b^2$. Also the critical amplitude of the interfacial free energy $\hat{f}_{int}$ (Fig. 12 yields $\hat{f}_{int}$ = $\rightarrow$ 0.417) is only in rough agreement with expectation ($f_{int} \approx \kappa$ near T_{cb} would then yield $\omega(T_{cb}) \approx 0.62$). No such problems were found for the simpler nearest neighbor Ising model,[39] where all these theoretical expectations are roughly satisfied. This problem may indicate that data of Fig. 11 are not in the asymptotic region ($L_{II} \rightarrow \infty$) where Eq. (48) applies and simulation of much larger systems would be required.

Interfaces in Polymer Mixtures

We now turn to interfaces in binary (AB) polymeric alloys. For simplicity, a lattice model of symmetrical mixtures (number of effective segments $N_A = N_B = N$) with rather short chain length N = 32 is considered[40]. Using the bond fluctuation model[83,84] where each effective monomer blocks 8 sites of an elementary cube of the simple cubic lattice for further occupation, bond lengths of b = 2, $\sqrt{5}, \sqrt{6}$, 3, and $\sqrt{10}$ lattice spacings can occur, it makes sense to choose an intermolecular interaction of square well type and range $\sqrt{6}$, $\varphi_{AA} = \varphi_{BB} = -\varphi_{AB} = -k_B T \varepsilon$ in Eq. (1). At a volume fraction of $\phi = 0.5$ occupied sites then each monomer has an effective coordination number $z_C \approx 2.65$ of monomers of other chains that it interacts with. The model then exhibits a critical point of unmixing at[85] ($\varphi_{AB}/k_B \equiv 1$) $T_{cb} = 69.3$. This large value is plausible since in polymer blends the configurational entropy of mixing is down by a factor of 1/N, in comparison with small-molecule mixtures: therefore polymer alloys typically are heterogeneous on mesoscopic scales, consisting of strongly segregated A-rich and B-rich regions separated by interfaces between them[86].

Obtaining the interfacial free energy from integrating energy differences (analogous to Eq. (70)) is cumbersome for complicated systems, and hence a different technique[32,33] is used,[40] based on a sampling of the order parameter distribution $P_L(\rho_A)$, where ρ_A is the relative density of A monomers {$\rho_A = \phi_A/(\phi_A+\phi_B)$ where ϕ_A, ϕ_B are volume fractions of sites taken by A or B monomers, respectively}. This quantity for $T < T_{cb}$ is very sharply peaked near the values $\rho_{A,coex}^{(1)} = \frac{1}{2}(1 - m_b)$ and $\rho_{A,coex}^{(2)} = \frac{1}{2}(1 + m_b)$, corresponding to the co-existing B-rich and A-rich phases. For ρ_A near ρ_A^{crit} = 1/2, on the other hand, the probability is minimal and this minimum $P_L^{min}(\rho_A)$ is reduced in comparison to the maxima $P_L^{max}(\rho_A)$, in a LxLxL system, see Fig. 13,

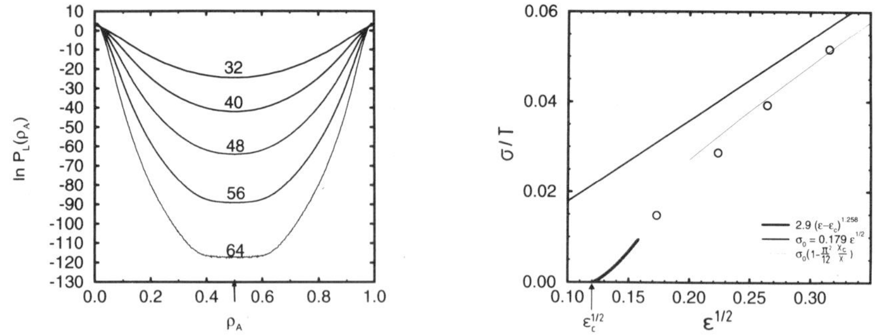

Figure 13. Plot of the logarithm of the probability of the relative density of A-monomers, ln $P_L(\rho_A)$, versus this density (ρ_A) at an interaction strength $\varepsilon = 0.03$ for various linear dimensions L of an LxLxL lattice, from L = 32 (left) to L = 64 (right); arrow shows ρ_A^{crit} (left part). Right part shows the normalized interfacial tension $f_{int} = \sigma/T$ vs. the square root of the normalized inverse temperature. The thick curve near the critical point (marked by an arrow) shows the function $f_{int} = 2.9\,(\varepsilon - \varepsilon_C)^{1.26}$ extracted from a finite size scaling analysis, while the circles result from directly using data such as of the left part, using Eq. (72). The thick straight line is Helfand's prediction,[87] $f_{int} = \rho a \sqrt{\chi / 6}$ where $\chi = 2z_C\,\varepsilon$, $a \approx 3.05$ is the effective length, $\rho = \phi/8 = 1/16$ the monomer density, and hence $f_{int} = 0.179\sqrt{\varepsilon}$. Thin curve is the result of Broseta et al.[88]. From Müller et al.[40].

$$P_L^{min}(\rho_A) / P_L^{max}(\rho_A) = \exp\left[-2L^2 f_{int}\right] \qquad (72)$$

Eq. (32) results from the fact[32] that for ρ_A near ρ_A^{crit} the dominating configurations are given by one A-rich domain separated by 2 LxL interfaces from the B-rich domain (remember the periodic boundary conditions). The resulting data for f_{int} (Fig. 13) can be used to test various theories. In the strong segration limit (SSL) of completely unmixed phases one expects[87] $f_{int} \propto \sqrt{\varepsilon}$, while at nonzero solubility of the minority species a correction factor $[1 - \pi^2 \chi_C/(12\chi)]$ with $\chi = 2z_C\varepsilon$ applies[88].

The behavior far from criticality ($\chi \gg \chi_C$) is controlled by the entropy of small random walk-like loops of the A-chains extending from the interface in the B-rich phase (and vice versa): each loop can contain $1/\chi$ monomers, and has a linear dimension $a/\sqrt{\chi}$. This argument hence yields an estimate for the intrinsic interfacial width. Using systems in a $L_{II}xL_yxL_{II}$ geometry with $L_y = 64$, $64 \leq L_{II} \leq 512$ the interfacial profile ρ_A (y) has been studied[40] with techniques as discussed above, see Fig. 14. While qualitatively the prediction $W \propto 1/\sqrt{\chi} \propto \varepsilon^{-1/2}$ is verified, the theoretical[87,88] prefactor is not verified. The reasons for this discrepancy are not entirely clear. The broadening due to capillary waves has also been studied, varying L_{II} , but found to be almost negligible, and again there is a serious disagreement with Eq. (48).

We emphasize that in complicated materials such as polymer alloys one can give reasonable order of magnitude estimates both of interfacial tension and interfacial width, but finer details are not understood. In addition, there are various effects[40] such as enrichment of chain ends in the interface, orientation of the coils parallel to the interface, a reduction of the monomer density in the center of the interface, etc. Even the effective coordination number gets drastically reduced (Fig. 15) due to deformations of the local coil structure:

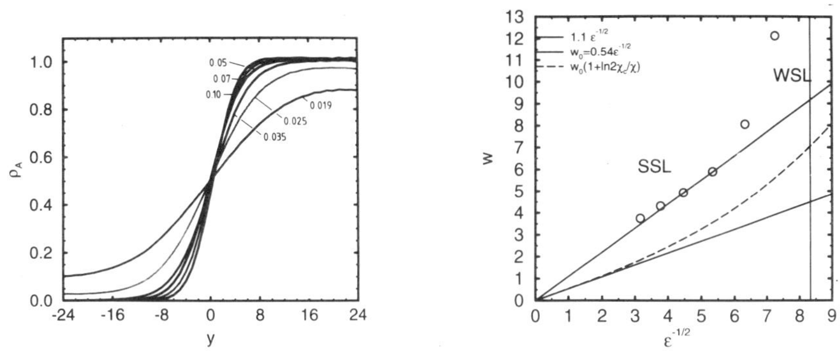

Figure 14. Concentration profile $\rho_A(y)$ plotted vs. y for the symmetrical polymer mixture with N = 32, for linear dimension $L_{||}$ = 512 parallel to the interface (left part). Six choices of ε are shown as indicated. Right part shows the interfacial width W{obtained from fitting a tanh $((y - y_0)/W)$ profile to $\rho_A(y)$} vs. $\varepsilon^{-1/2}$. The prediction $W = 0.54\ \varepsilon^{-1/2}$ resulting from Helfand[87] for the strong segregation limit is included, as well as the improvement due to Broseta et al.[88] (broken line). The critical coupling (ε_C) is indicated by a vertical straight line. The data for ε near ε_C are compatible with a law $W = 2\xi_b = 2\hat{\xi}\ (1 - \varepsilon_C/\varepsilon)^{-\nu}$, with ν = 0.63, $\hat{\xi} = 2.6$. From Müller et al.[40].

such effects cannot occur in the simple Ising model, but related effects could occur in real alloys when mechanical distortions in the interface are considered, and correpondingly unfavorable energies are reduced.

On the other hand, the behavior near the critical point is fully compatible with the Ising model: the data in Fig. 14 can be described as $w = \hat{w}\left(1 - \varepsilon_C / \varepsilon\right)^{-\nu_b}$ with $\nu_b = 0.63$, $\hat{w} = 5.2$ near T_{cb} , and assuming {eq. (23)} $w = 2\xi_b$, $f_{int} = \hat{f}_{int}\left(1 - \varepsilon_C / \varepsilon\right)^{2\nu_b}$ with $\hat{f}_{int} \approx 0.0139$ yields the universal invariant $\hat{f}_{int}\hat{\xi}^2 = 0.094$, compatible with the value 0.09 for the nearest neighbor Ising model[40].

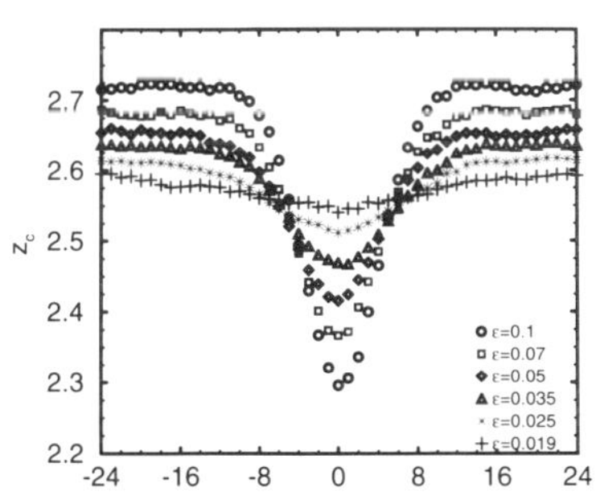

Figure 15. Effective coordination number z(y) in a symmetrical polymer mixture with chain-length N = 32 plotted as a function of distance y across the interfacial profile, for several choices of ε. From Müller et al.[40].

SIMULATIONS OF INTERFACE UNBINDING TRANSITIONS

Wetting in the Nearest Neighbor Ising Model

Monte Carlo simulations of the model Eq. (2) used LxLxD lattices, with two equivalent free LxL surfaces a distance D apart, and periodic boundary conditions in x and y directions parallel to these surfaces[41-43]. Figure 16 shows typical examples[42] of critical wetting at phase coexistence (H = 0) varying the surface magnetic field H_1. Near T_{cb} the distortion of the magnetization profile extends over many layers already in the nonwet phase. In the wet phase, the surface layer magnetization m_1 is even more negative than $-m_b$, and a well-developed domain of negative magnetization on top of a positive bulk occurs.

While the data are compatible with the expected behavior $\chi_1 \propto |H_1 - H_{1c}|^{-1}$, cf. Eqs. (34),(58), and also the logarithmic laws Eqs. (38),(39),(41) are verified (Fig. 17), the power laws involving the nontrivial exponent $\nu_{\parallel}$ {such as Eq. (60)} yielded problems: contrary to expectation ($\omega \approx 0.77$ at bulk criticality[74] and should be of order unity throughout) the exponent $\nu_{\parallel}$ seemed to be hardly distinct from its mean field value $\nu_{\parallel} = 1$, Eq. (56) (with $\nu_{\parallel} \approx 3.5$) could not be verified. The reasons for this failure are still unclear[61,67,72-74].

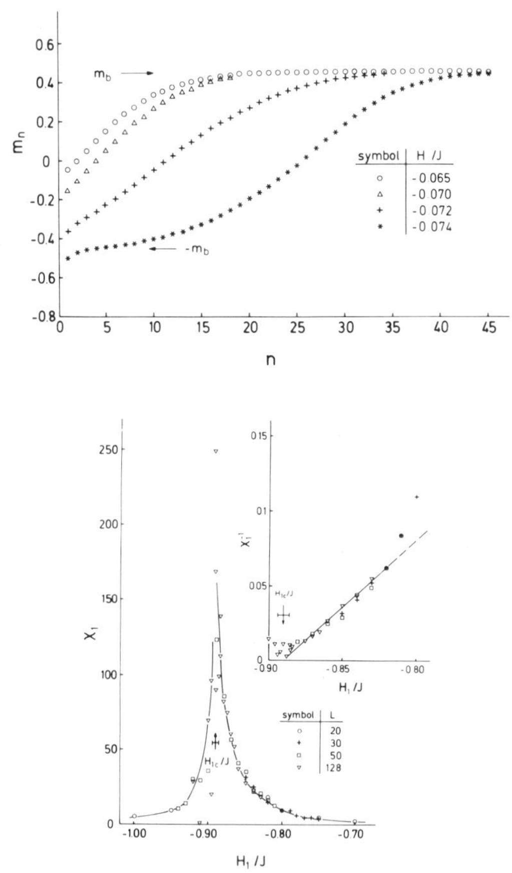

Figure 16. Magnetization profiles (layer magnetization m_n plotted vs. layer number n) for $J/k_BT = 0.226$, $J_S/J = 1.33$, a 128x128x160 system, and several values of the surface field H_1/J (upper part) lower part shows the surface layer susceptibility $\chi_1 J$ plotted vs. surface field H_1/J at $J_S/J = 1$, $J/k_BT = 0.35$, D = 40, and several choices of L as indicated. Insert shows $(\chi_1 J)^{-1}$ plotted vs. H_1/J on an expanded scale. The resulting estimate for the wetting transition H_{1c}/J is indicated by arrows. From Binder et al.[42].

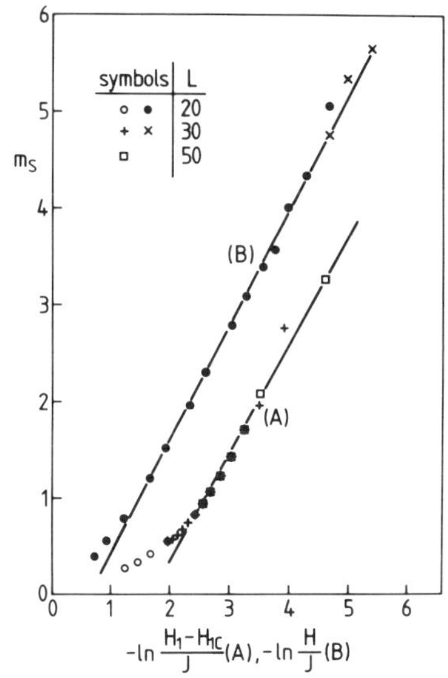

Figure 17. Surface excess magnetization m_S of the nearest neighbor Ising model plotted for $J/k_BT = 0.25$, $J_S/J = 1$, vs. $- \ln [(H_1 - H_{1c})/J]$ at $H = 0$ (A) and vs. $- \ln H/J$ at $H_1 = H_{1c} = - 0.55$ J (B). D = 40 was used throughout. From Binder and Landau[41].

Wetting Near the Roughening Transition Temperature T_R

"Surface transitions" at T_{CS} $(J_S/J) > T_{cb}$, where for zero field H = 0 the surface layer orders while the bulk stays disordered, and prewetting critical points T_C^{pre} $(J_S/J,H_1/J)$ are just special cases of two-dimensional Ising-type transitions forming a hypersurface in the space of variables $(T/J, H/J, J_S/J, H_1/J)$[89]. This fact implies that tricritical wetting {where T_C^{pre} $(J_S/J, H_1/J)$ coincides with the wetting transition for H = 0} for $T \to T_{cb}$ must coincide with the "special transition", where surface and bulk criticality coalesce[6]. This idea[89] could be verified by simulation,[42] see Fig. 18, and it was also shown that the associated critical (or tricritical) fields H_{1c} (H_{1tric}) vanish with power laws expected from scaling theory,[6,89] namely $H_{1c} \propto (1-T/T_{cb})^{\Delta 1}$ and $H_{1tric} \propto \left(1 - T/T_{cb}\right)^{\Delta_{1m}}$ where $\Delta_1 = 2\nu_b - \beta_1 \approx 0.48$ {$\beta_1 \approx 0.78$ describes the singularity of the surface layer magnetization at an "ordinary" second-order transition in the bulk, $m_1 \propto \left(1 - T/T_{cb}\right)^{\beta_1}$ }, $\Delta_{1m} = 2\nu_b - \beta_1^m \approx 1.04$ }. While the topology of the wetting phase diagram near bulk criticality is now clear,[42,89] it is not so clear how the smooth behavior of complete and critical wetting (Fig. 17, $m_S \propto - \ln H/J \to \infty$ for $H \to 0$) crosses over to the sequence of first-order layering transitions (Fig. 7) at $T< T_R$. Again simulations have contributed to clarify this issue partially[41,43] and have shown that for H = 0 one has several distinct first order layering transitions as function of H_1/J at low temperatures (Fig. 19). As T is near T_R, one encounters a layering critical point for the first layer, while higher layers still exhibit first order transitions. For $T > T_R$ a smooth behavior of m_S vs. H_1 is found, as expected for critical wetting. However, the behavior of layering for large n near T_R is still inaccessible by Monte Carlo.

Surface Induced Ordering and Disordering in Face-Centered Cubic (fcc) Alloys

Fcc Ising antiferromagnets with nearest neighbor interactions show orderings of A_3B, AB, AB_3 type as in the Au-Cu-system and have found longstanding attention[82]. Also the

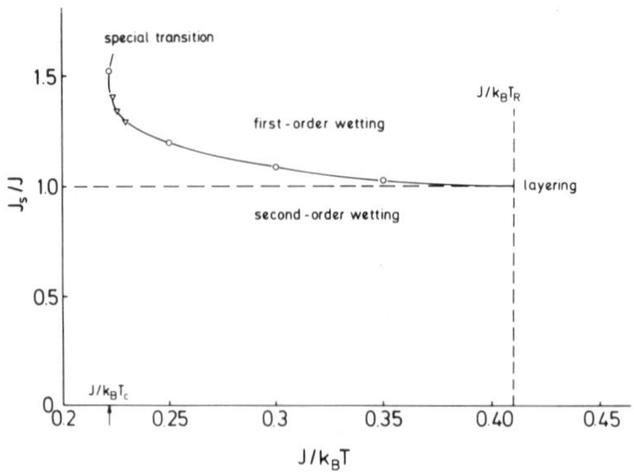

Figure 18. Surface phase diagram, in the plane of variables J_S/J and J/k_BT, exhibiting the line of tricritical wetting transitions (full curve). At bulk criticality (arrow) this line ends in the surface-bulk multicritical point ("special transition"). At lower temperatures it ends at the roughening temperature, where layering phenomena replace wetting. From Binder et al.[42].

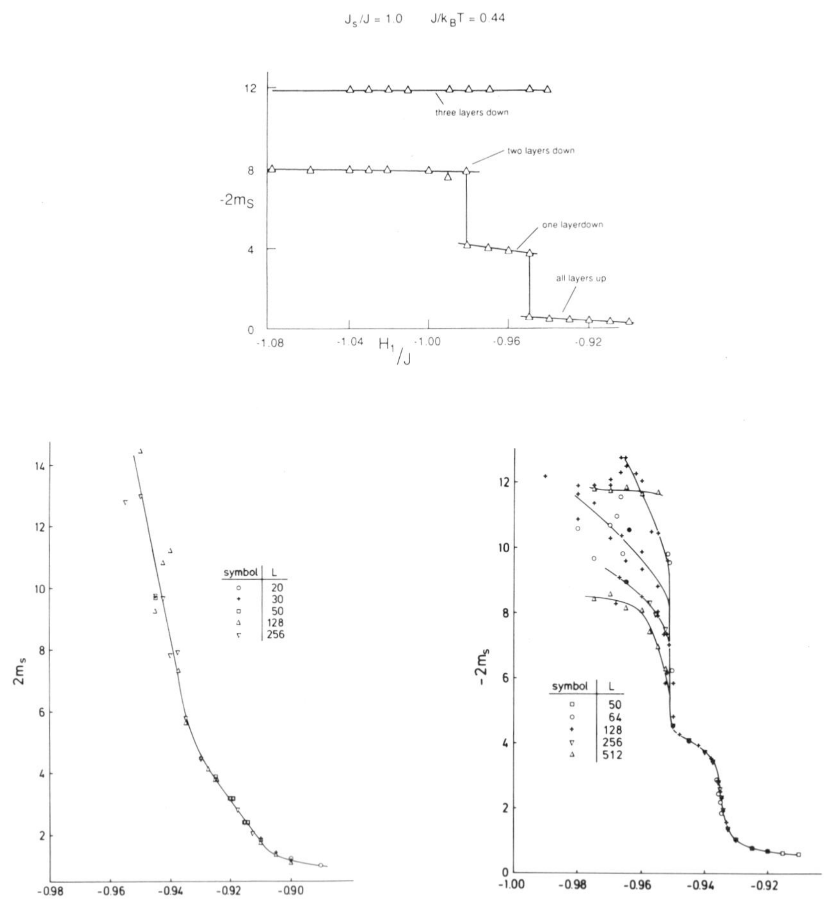

Figure 19. Surface excess magnetization $2m_S$ plotted vs. H_1/J at $H = 0$ for $D = 40$, $J_S/J = 1$ and three temperatures ($J/k_BT = 0.44$, upper part; $J/k_BT = 0.42$, lower part left; $J/k_BT = 0.40$, lower part right). Note[39] $J/k_BT_R = 0.407$. From Binder and Landau[43].

surface properties of this model are of great interest; even in the simplest case where H = 0 (both A,B species have then the same composition) and where $J_S = J$, $H_1 = 0$ (then no surface enrichment occurs!), one still has either SIO or SID: which of these phenomena occurs, depends on the orientation of the surface plane (Fig. 20)[47]. The reason for this surprising behavior is the "frustration" effect of the triangles with antiferromagnetic exchange J in the close-packed planes.

Therefore, T_{cb} for the (first-order) transition in this model is unusually low,[47] $k_B T_{cb}/|J| = 1.7380$. For a loose-packed 100 surface, oriented perpendicular to the stacking of A-rich and B-rich planes, there is no frustration and thus this surface orders at a temperature $T_{CS} > T_{cb}$. A close packed (111) surface, on the other hand, still suffers from

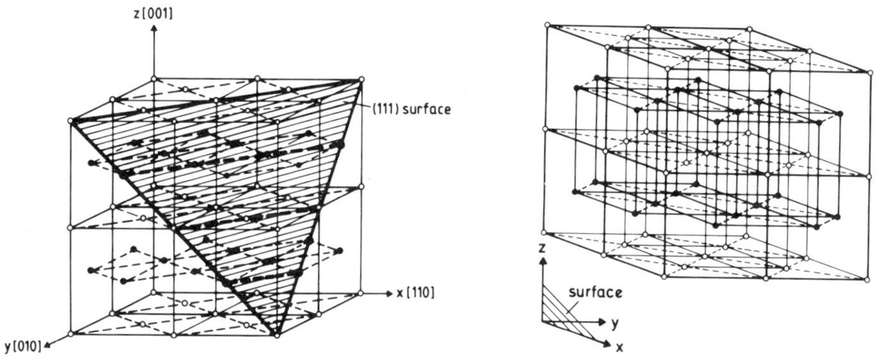

Figure 20. Sections of the ordered lattice of the fcc AB alloy in the $L1_0$ (CuAuI) structure. Frustrated nearest neighbor interactions are shown by dashed lines. Left part shows (100) surface oriented such that there is c(2x2) order in the surface layer. Right part shows a (111) surface (shaded). In this case a (2x1) order in the surface results, where one third of the bonds in the surface layer are still frustrated (thick broken lines). Unfrustrated nearest neighbor interactions are not drawn, in order to avoid overcrowding of these pictures. From Schweika et al.[47].

"frustration", and since the missing neighbors cause a further weakening of the ordering tendency, surface-induced disordering occurs[47].

Fig. 21 shows the order parameter profiles for both cases. The similarity with Fig. 16 is evident: in both cases an interface unbinds from the wall as the transition temperature (T_{cb} in the case of Fig. 21) is approached. While in Fig. 16 the interface separates phases with opposite orientation of the order parameter, in Fig. 21 we have a less symmetric situation, namely an interface between ordered and disordered phases. However, in the framework of a coarse-grained description of the interface (Fig. 5) this distinction is unimportant.

As for wetting, logarithmic growth of the surface excess order parameter occurs (Fig. 22), but one must approach T_{cb} rather closely ($|t| \leq 10^{-2}$). For SIO, the 2d-Ising character of the surface transition at T_{CS} has also been verified[47]. Most interesting, of course, is the continuous vanishing of the surface layer order parameter Ψ_1 {cf. Eq. (67)}, since the exponent β_1^{SID} depends on the non-universal parameter ω, Eq. (55). Indeed, one can see distinct deviations from the meanfield result ($\beta_1^{SID} = ½$), Fig. 23. However, an analysis of other quantities (namely the susceptibilities χ_s, χ_1, and χ_{11} considered above) does reveal difficulties with the scaling description summarized above; in particular, there seem to be[47] clear deviations from Eq. (68). The reasons for this discrepancy seem to be unclear.

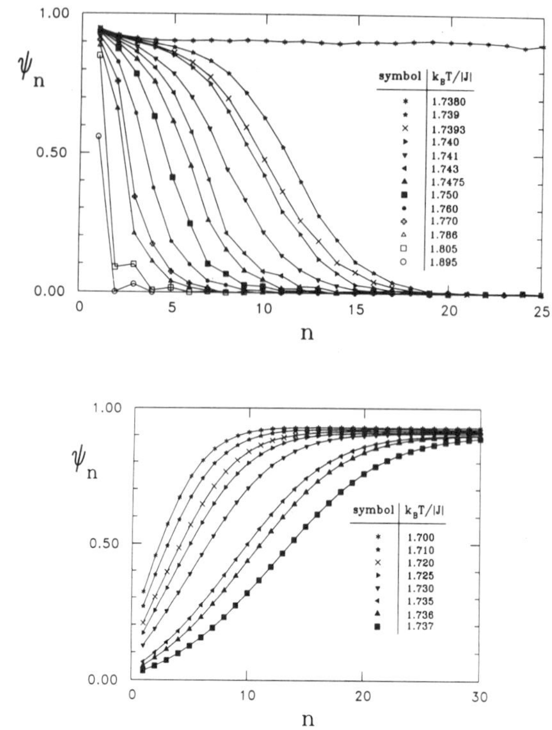

Figure 21. Order parameter profiles (order parameter in n'th layer Ψ_n plotted vs. layer number n) for the (100) surface (upper part) and the (111) surface (lower part). Several temperatures are shown as indicated. From Schweika et al.[47].

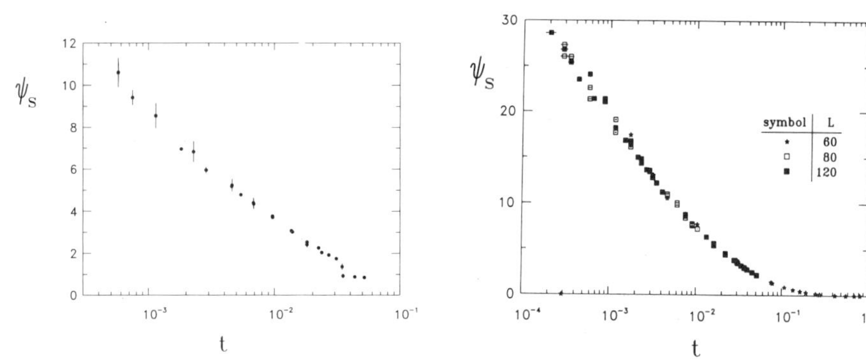

Figure 22. Semi-log plot of the surface excess order parameter ψ_S vs. the reduced temperature distance $t = |1-T/T_{cb}|$, for the (100) surface (left part) and the (111) surface (right part). From Schweika et al.[47].

Wall Effects in Soft Materials

Soft materials such as liquid crystalline systems, microemulsions, block copolymer melts, polymeric alloys, etc. exhibit a variety of surface effects that bear a close relationship to interface unbinding. One of the reasons why interfacial phenomena are particularly important in soft materials is the fact that typically the interfacial free energy f_{int} is very low. Polymer mixtures (Fig. 13) are a good example. One can show that within mean field theory[53]

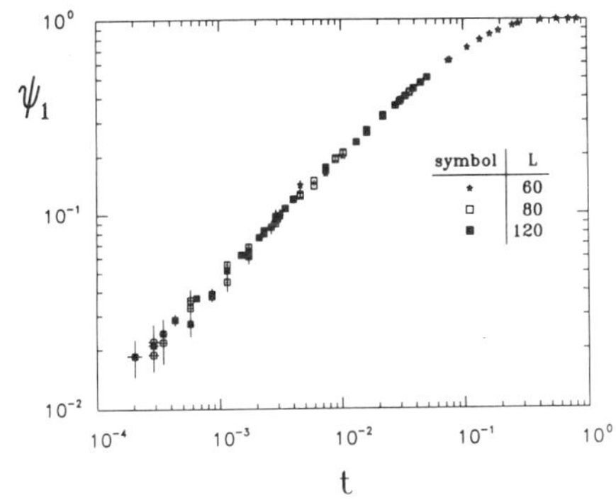

Figure 23. Log-log plot of the surface layer order parameter ψ_1 vs. t for the (111) surface. For $t \leq 5.10^{-2}$ the data are compatible with Eq. (67) with $\beta_1^{SID} \approx 0.64$. From Schweika et al.[47].

$$a^2 f_{int} = 9N^{-1/2}\, \tilde{f}(\chi/\chi_{crit}) \quad , \quad \tilde{f}(X>>1) \approx X^{1/2}\,[1-\pi/(12X)]/(9\sqrt{3}) \tag{73}$$

while near $\chi = \chi_{crit}$ one has $\tilde{f}(X) \propto (1-X)^{3/2}$. Thus near T_C the interfacial tension is down by a factor $\sqrt{N}$ - the same factor by which a polymer coil is larger than a small molecule. This argument is only slightly modified when one goes beyond mean field: according to the Ginzburg criterion[23,53,90,91] for $t = |1 - \chi_{crit}/\chi| \propto 1/N$ a crossover to the Ising-type critical behavior occurs ($\nu \approx 0.63$)

$$a^2 f_{int} \propto N^{-2(1-\nu)}\, t^{2\nu} \approx N^{-0.74}\, t^{1.26} \quad , \qquad Nt << 1 \quad . \tag{74}$$

Since the forces acting from walls at monomeric units are not necessarily that small, it follows that partially compatible polymer mixtures exposed to walls typically are under conditions where the wall is wet[48,53]. Although from the Ginzburg criterion one might conclude that the meanfield theory of wetting should be rather good, simulations[49] did reveal drastic deviations from the mean-field phase diagram[49]. Of course, as in the case of the interfacial profile[40] discussed above, there are special effects due to the intrinsic configurational entropy[53,90,91] of these large macromolecules: while in the bulk the orientation of a (soap-shaped) polymer coil is random, near surfaces and interfaces they are oriented[40,92] and chain ends get enriched,[40,93] etc. While in a small molecule mixture the concentration profile in the case of small surface enrichment has its maximal value right at the surface, in polymer mixtures the surface enrichment is nearly constant over a distance $\left\langle R_g^2 \right\rangle^{1/2} / 4 \approx a\sqrt{N}/\left(4\sqrt{6}\right)$ near the surface,[94] to avoid unfavorable elastic deformations of the polymer coils.

In the ordered mesophases of block copolymer melts[95] surface effects analogous to SIO occur[50-53]. E.g., if we have a lamellar mesophase where A-rich layers and B-rich layers alternate with a wavelength λ^*, typically the interfaces will arrange themselves parallel to external walls[50-53] (apart from cases of confinement between two parallel walls at a distance D comparable to λ^*, where frustration effects due to the incommensurability of D and λ^* may enforce a perpendicular orientation[51]). Then the preferential attraction of one species at the wall leads already above the transition to a surface induced weak lamellar-type ordering[50,53]

$$\psi(z) = \psi_1 \exp(-z/\xi)\cos(2\pi z/\lambda^* + \varphi)/\cos\varphi, \quad \tan\varphi = \lambda^*/(2\pi\xi) \quad , \tag{75}$$

where the correlation length ξ would diverge at an "ordering spinodal" χ_S as $\xi \propto a\sqrt{N}$ $(1 - \chi/\chi_S)^{-1/2}$. This divergence is cut off at the first order transition, at a correlation length of order $\xi_b \propto a\sqrt{N}$ $[N^{-1/3} + \text{const}\,(f - 1/2)^2]^{-1/2}$, where $f = N_A/(N_A + N_B)$ is the composition of the diblock copolymer. Then surface-induced ordering should have the character of a wetting phenomenon again, as for the (100) surface of the fcc antiferromagnet described above, with the thickness of the ordered region at the surface diverging as $\ell \propto \xi_b \ln|1-\chi_t/\chi|$ where χ_t is the value of the Flory-Huggins effective interaction parameter at the transition.

Again the large value of the prefactor ξ_b means a mesoscopic length scale over which the surface modifies the behavior of the system. On the other hand, these large length scales make a simulation approach very difficult[49,51,53].

CONCLUDING REMARKS

These lectures have been devoted to a discussion of free surfaces and interfaces in heterophase materials, considering metallic alloys and polymer mixtures as prototypical examples. We have attempted to draw a connection between real materials and the most simplified theoretical model descriptions, such as the nearest neighbor Ising model, Ginzburg-Landau theory of the "intrinsic" order parameter profile of an interface, mean field theory of wetting and surface-induced ordering or disordering, and fluctuation phenomena such as capillary waves and interfacial roughening. Basic ideas about the corresponding theoretical concepts were summarized, and it was pointed out that there are still important questions open, particularly concerning critical wetting, and the applicability of the "sharp-kink" approximation of interfaces to complete wetting.

The simulation studies described here only deal with drastically simplified lattice models, even in the case of the polymeric materials. Including capillary wave effects requires to include huge linear dimensions parallel to the interface. However, large length scales occur also in the intrinsic interfacial profile of Ising models near the critical point (Fig. 16) or of polymeric systems (Fig. 14). So system sizes with more than a million lattice sites have been used rather regularly - but even such systems are by far too small to clarify high order layering transitions near the roughening transition of the Ising model (Fig. 19). For the case of the polymer interface, a single short chain length $N = 32$ was used, and varying N over a wide range of larger values clearly would be desirable. However, already that study[40] has required a large computational effort on a massively parallel computer (CRAY-T3D). So it is not straightforward to push these efforts much further. Despite these limitations, the simulations have thrown considerable light on many of the theoretical issues, and clearly provide an important step towards the understanding of interfacial properties of materials.

ACKNOWLEDGEMENTS

Part of the research reviewed here was carried out at the Center for Simulation Physics, University of Georgia, in collaboration with D.P. Landau, K.K. Mon, and S. Wansleben, the other part in Mainz with I. Schmidt, F. Schmid, M. Müller, W. Oed, and S. Puri. SID and SIO in fcc alloys was studied in collaboration with W. Schweika (Jülich) and D.P. Landau. It is a great pleasure to thank all these coworkers for their fruitful efforts. This research is also partially supported by the Deutsche Forschungsgemeinschaft (DFG) under grant No. Bi 314/3, by the collaborative NATO grant No. CRG 921202, and by the Materialwissenschaftliches Forschungszentrum (MWFZ) at Mainz.

REFERENCES

1. D. Pettifor. "Cohesion and Structure of Surfaces," Elsevier, Amsterdam (1995).
2. B. Mutaftschiew , ed. "Interfacial Aspects of Phase Transformations," D. Reidel Publ. Comp., Dordrecht (1989).
3. W.C. Johnson and J.M. Blakely, eds. "Interfacial Segregation," American Society for Metals, Metals Park, OH (1979).
4. D. Wolf and S. Yip, eds. "Materials Interfaces: Atomic-Level Structure and Properties," Chapman & Hall, London (1991).
5. J.F. van der Veen and M.A. van Hove, eds. "The Structure of Surfaces II," Springer, Berlin (1988).
6. K. Binder, *in*: "Phase Transitions and Critical Phenomena, Vol. 8," C. Domb and J.L. Lebowitz, eds., p. 1, Academic Press, New York (1983).
7. H.W. Diehl, *in*: "Phase Transitions and Critical Phenomena, Vol. 10," C. Domb and J.L. Lebowitz, eds., p. 75, Academic Press, New York (1986).
8. K. Binder, *in*: Ref. 1, Chapter 3.
9. J.D. Weeks, *in*: "Ordering in Strongly Fluctuating Condensed Matter Systems," T. Riste, ed., p. 293, Plenum Press, New York (1980).
10. H. Van Beijeren and I. Nolden, *in*: "Structure and Dynamics of Surfaces II," W. Schommers and P. Blanckenhagen, eds., p. 259, Springer, Berlin (1987).
11. D.E. Sullivan and M.M. Telo da Gama, *in*: "Fluid Interfacial Phenomena," C.A. Croxton, ed., p. 45
12. S. Dietrich, *in*: "Phase Transitions and Critical Phenomena, Vol. 12," C. Domb and J.L. Lebowitz, eds., p. 1, Academic Press, New York (1988).
13. M. Schick, *in*: "Liquids and Interfaces," J. Charvolin, J.F. Joanny, and J. Zinn-Justin, eds., p. 415, North-Holland, Amsterdam (1990).
14. D. Henderson, ed. "Fundamentals of Inhomogeneous Fluids," M. Dekker, New York (1992).
15. R. Lipowsky, *J. Appl. Phys.* 55:2485 (1984); *Ferroelectrics* 73:69 (1987).
16. B. Widom, *in*: "Phase Transitions and Critical Phenomena, Vol. 2," C. Domb and M.S. Green, eds., p. 79, Academic Press, London (1972).
17. J.S. Rowlinson and B. Widom. "Molecular Theory of Capillarity," Clarendon, Oxford (1982).
18. For a review, see R.F. Willis, *in*: "Dynamical Phenomena at Surfaces, Interfaces, and Superlattices," F. Nizzoli, K.H. Rieder, and R.F. Willis, eds., p. 126, Elsevier, Amsterdam (1985).
19. For a review, see E. Tosatti, in Ref. 5, p. 535.
20. M.G. Lagally, ed. "Kinetics of Ordering and Growth at Surfaces," Plenum Press, New York (1990).
21. J.D. Gunton, M. San Miguel, and P.S. Sahni, *in*: "Phase Transitions and Critical Phenomena, Vol. 8," C. Domb and J.L. Lebowitz, eds., p. 267, Academic Press, New York (1983).
22. S. Komura and H. Furukawa, eds. "Dynamics of Ordering Processes in Condensed Matter," Plenum Press, New York (1988).
23. K. Binder, *in*: "Materials Science and Technology, Vol. 5: Phase Transformations in Materials," R.W. Cahn, P. Haasen, and E.J. Kramer, eds., p. 405, VCH Publ., Weinheim (1991).
24. G. Kostorz, *Physica Scripta* T49:636 (1993).
25. S. Puri and K. Binder, *Phys. Rev.* A46:4487 (1992); *Phys. Rev.* E49:5359 (1994).
26. H.L. Frisch, P. Nielaba, and K. Binder, *Phys. Rev.* E (1995, in press), and references therein.
27. R.A.L. Jones, L.J. Norton, E.J. Kramer, F.S. Bates, and P. Wiltzius, *Phys. Rev. Lett.* 66:1326 (1991).
28. R. Lipowsky, *J. Phys.* A18:L585 (1985); R. Lipowsky and D.A. Huse, *Phys. Rev. Lett.* 52:353 (1986).
29. K.K. Mon, K. Binder and D.P. Landau, *Phys. Rev.* B35:3683 (1987); I. Schmidt and K. Binder, *Z. Physik* B67:369 (1987); R. Bausch and R. Blossey, *Europhys. Lett.* 14:125 (1991).
30. U. Steiner, E. Eiser, J. Klein, A. Budkowski, and L.J. Fetters, *Science* 258:1126 (1992).
31. H.J. Leamy, G.H. Gilmer, K.A. Jackson, and P. Bennema, *Phys. Rev. Lett.* 30:601 (1973).
32. K. Binder, *Phys. Rev.* A25:1699 (1982).
33. A. Berg and T. Neuhaus, *Phys. Rev. Lett.* 68:9 (1992).
34. E. Bürkner and D. Stauffer, *Z. Physik* B55:241 (1983).
35. K.K. Mon, S. Wansleben, D.P. Landau, and K. Binder, *Phys. Rev. Lett.* 60:708 (1988); *Phys. Rev.* B39:7089 (1989).
36. K.K. Mon, D.P. Landau, and D. Stauffer, *Phys. Rev.* B42:545 (1990).
37. F. Schmid and K. Binder, *Phys. Rev.* B46:13553 (1992).
38. F. Schmid and K. Binder, *Phys. Rev.* B46:13565 (1992).
39. M. Hasenbusch and K. Pinn, *Physica* A192:342 (1993).
40. M. Müller, K. Binder and W. Oed, *Faraday Trans.* (1995, in press).
41. K. Binder and D.P. Landau, *J. Appl. Phys.* 57:3306 (1985); *Phys. Rev.* B37:1745 (1988); K. Binder, D.P. Landau, and D.M. Kroll, *Phys. Rev. Lett.* 56:2276 (1986).
42. K. Binder, D. P. Landau, and S. Wansleben, *Phys. Rev.* B40:6971 (1989).
43. K. Binder and D.P. Landau, *Phys. Rev.* B46:4844 (1992).

44. K. Binder, A.M. Ferrenberg, and D.P. Landau, *Ber. Bunsenges. Phys. Chem.* 98:340 (1994); K. Binder, D.P. Landau, and A.M. Ferrenberg, *Phys. Rev. Lett.* 74:298 (1995).
45. K. Binder, D.P. Landau, and A.M. Ferrenberg, *Phys. Rev.* E51:2823 (1995).
46. W. Schweika, K. Binder, and D.P. Landau, *Phys. Rev. Lett.* 65:3321 (1990).
47. W. Schweika, D.P. Landau, and K. Binder, preprint.
48. I. Schmidt and K. Binder, *J. Phys. (Paris)* 46:1631 (1985).
49. J.S. Wang and K. Binder, *J. Chem. Phys.* 94:8537 (1991).
50. G.H. Fredrickson, *Macromol.* 20:2535 (1987).
51. M. Kikuchi and K. Binder, *J. Chem. Phys.* 101:3367 (1994).
52. A. Menelle, T.P. Russell, S.H. Anastasiadis, S.K. Satija, and C.F. Majkrzak, *Phys. Rev. Lett.* 68:67 (1992).
53. K. Binder, *Acta Polymerica* 46:205 (1995).
54. J.M. Blakely, *in*: "Chemistry and Physics of Solid Surfaces Vol. 2," R. Vanselow, ed., p. 1, CRC Press, Boca Raton (1979).
55. K. Binder and P.C. Hohenberg, *Phys. Rev.* B6:3461 (1972); B9:2194 (1974).
56. I. Schmidt and K. Binder, *Z. Phys.* B67:369 (1987); S. Puri and K. Binder, *Z. Phys.* B86:263 (1992).
57. K. Binder and H.L. Frisch, *Z. Phys.* B84:403 (1991).
58. R. Pandit, M. Schick and M. Wortis, *Phys. Rev.* B26:5112 (1982); R. Pandit and M. Wortis, *Phys. Rev.* B25:3226 (1982).
59. F.P. Buff, R.A. Lovett, and F.H. Stillinger, *Phys. Rev. Lett.* 15:621 (1965).
60. J.D. Weeks, *J. Chem. Phys.* 67:3106 (1977); *Phys. Rev. Lett.* 52:2160 (1984).
61. A.J. Jin and M.E. Fisher, *Phys. Rev.* B47:7365 (1993); M.E. Fisher and J.A. Jin, *Phys. Rev. Lett.* 69:792 (1992).
62. V. Privman, *Int. J. Mod. Phys.* C3:857 (1992).
63. L. Onsager, *Phys. Rev.* 65:117 (1944).
64. M.E. Fisher, *J. Stat. Phys.* 34:667 (1984).
65. J.M. Kosterlitz and D.J. Thouless, *J. Phys.* C6:1181 (1973).
66. M.P. Nightingale, W.F. Saam, and M. Schick, *Phys. Rev.* B30:3830 (1984).
67. C.J. Boulter and A.O. Parry, *Phys. Rev. Lett.* 74:3404 (1995); *Physica*, in press.
68. E. Brézin, B.I. Halperin and S. Leibler, *Phys. Rev. Lett.* 50:1387 (1983).
69. R. Lipowsky, D.M. Kroll, and R.K.P. Zia, *Phys. Rev. Lett.* B27:4499 (1983).
70. D.S. Fisher and D.A. Huse, *Phys. Rev.* B32:247 (1985).
71. R. Lipowsky and M.E. Fisher, *Phys. Rev.* B36:2126 (1987); *Phys. Rev. Lett.* 57:2411 (1986).
72. E. Brézin and T. Halpin-Healey, *Phys. Rev. Lett.* 58:1220 (1987); *J. Phys. (Paris)* 48:757 (1987).
73. A.O. Parry and R. Evans, *Phys. Rev.* B39:12336 (1989).
74. M.E. Fisher and H. Wen, *Phys. Rev. Lett.* 68:3654 (1992); K. Binder, D.P. Landau, and D.M. Kroll, *Phys. Rev. Lett.* 68:3655 (1992).
75. H. Dosch, L. Mailänder, A. Lied, J. Peisl, F. Grey, R.L. Johnson, and S. Krummacker, *Phys. Rev. Lett.* 60:2382 (1988)
76. R. Lipowsky, *Phys. Rev. Lett.* 49:1575 (1982); *Z.Phys.* B51:165 (1983); R. Lipowsky and W. Speth, *Phys. Rev.* B28:3983 (1983).
77. K. Binder and D. Stauffer, *in*: "Applications of the Monte Carlo Method in Statistical Physics," K. Binder, ed., p. 1, Springer, Berlin (1984).
78. A.M. Ferrenberg and D.P. Landau, *Phys. Rev.* B44:5081 (1991).
79. V. Pierron-Bohnes, S. Lefebre, M. Bessiere, and A. Finel, *Acta Metall.* 38:249 (1990).
80. V. Pierron-Bohnes, M.C. Cadeville, A. Finel, and O. Schärpf, *J. Phys. (Paris)* I1:247 (1991).
81. F. Schmid and K. Binder, *J. Phys.: Condensed Matter* 4:3569 (1992).
82. K. Binder, *in*: "Statics and Dynamics of Alloy Phase Transformations," P.E.A. Turchi and A. Gonis, eds., p. 467, Plenum Press, New York (1994).
83. I. Carmesin and K. Kremer, *Macromol.* 21:2819 (1988).
84. H.-P. Deutsch and K. Binder, *J. Chem. Phys.* 94:2294 (1991).
85. H.-P. Deutsch and K. Binder, *Macromol.* 25:6214 (1992).
86. D.S. Walsh, J.S. Higgins, and A. Maconnachie, eds. "Polymer Blends and Mixtures," Martinus Nijhoff, Dordrecht (1985).
87. E. Helfand, *Accts. Chem. Res.* 8:295 (1975).
88. D. Broseta, G.H. Fredrickson, E. Helfand, and L. Leibler, *Macromol.* 23:132 (1990).
89. H. Nakanishi and M.E. Fisher, *Phys. Rev. Lett.* 49:1565 (1982).
90. P.G. De Gennes. "Scaling Concepts in Polymer Physics," Cornell Univ. Press, Ithaca (1979).
91. K. Binder, *Phys. Rev.* A29:341 (1984); *Adv. Polymer Sci.* 112:181 (1994).
92. J.S. Wang and K. Binder, *Macromol. Chem., Theory & Simul.* 1:49 (1992).
93. J.S. Wang and K. Binder, *J. Phys.* I1:1583 (1991).
94. Y. Rouault, B. Dünweg, J. Baschnagel, and K. Binder, preprint.
95. F.S. Bates and G.H. Fredrickson, *Ann. Rev. Phys. Chem.* 41:525 (1990).

CONTINUUM DIFFUSE-INTERFACE MODEL FOR MODELING MICROSTRUCTURAL STABILITY

L. Q. Chen,[1] D. N. Fan,[1] Y. Z. Wang,[2] and A. G. Khachaturyan[2]

[1]*Department of Materials Science and Engineering, The Pennsylvania State University, University Park, PA 16802*
[2]*Department of Materials Science and Engineering, Rutgers University, P. O. Box 909, Piscataway, NJ 08854*

1. INTRODUCTION

The majority of advanced materials in practical applications are multiphase and/or polycrystalline, and their physical properties depend not only on the atomic structure and properties of each constituent phase, but also crucially on the nanoscale or microscale microstructure. A microstructure is characterized by (1) the number of phases and their volume fractions, and (2) the size, shape, distribution and orientation of each phase. Generally, a microstructure is thermodynamically unstable, or at most, metastable, and at a given temperature, it will evolve as a function of time towards equilibrium. The driving force for the temporal evolution of a microstructure usually consists of one or more of the following: (1) reduction in the bulk chemical free energy; (2) decrease of the total interfacial energy between different phases or between different orientation domains or grains of the same phase; (3) homogeneous and heterogeneous relaxation of the elastic strain energy generated by the lattice mismatch between different phases or different orientation domains; and (4) external fields such as applied stress, electrical, temperature and magnetic fields. The phase changes and resulted microstructural development driven by the decrease of the bulk chemical free energy is usually referred to as phase transformations whereas the microstructural evolution due to reduction in the total excess free energies associated with interfaces is called coarsening (or grain growth in polycrystalline materials). In coherent systems, however, elastic strain energy may play an important, and very often, a controlling role in the microstructural evolution during both phase transformations and coarsening.

Stability of Materials
Edited by A. Gonis *et al.*, Plenum Press, New York, 1996

Traditional models of phase transformations and coarsening usually treat the interfaces between different phases or different orientation domains of the same phase (e.g., different grains in a polycrystal) as surfaces of zero thickness, i.e., sharp-interfaces. A set of differential equations are then solved in each phase with boundary conditions specified at the interfaces which are moving during a phase transformation or during coarsening. Therefore, traditional approaches present a moving- or free-boundary problem which is impossible to solve analytically and very difficult numerically, in particular for complicated microstructures. In order to overcome this difficulty, there has been increasing interest in the last few years in using the diffuse-interface field approach for modeling microstructural evolution. Typical Examples include the domain growth in a quenched system,[1-3] crystal growth during solidification,[4-7] morphological evolution in two-phase coherent systems,[8-11] and grain growth in single-phase and two-phase systems. Different from traditional continuum approaches, the field approach describes a microstructure using a set of spatially inhomogeneous field variables. The temporal evolution of these field variables, and thus the microstructural evolution, are described by the time-dependent Ginzburg-Landau equations. One of the main advantages of the field approach is that any arbitrary microstructure can be easily treated since the interfaces are not singular surfaces requiring imposition of special boundary conditions on them, but just a region where the fields have very high gradients. It is our belief that the diffuse-interface field model will find increasing applications in modeling temporal evolution of microstructures in both model systems and real materials of practical importance.

The main purpose of this report is, therefore, to present some recent applications of the diffuse-interface field model to microstructural evolution during phase transformation and coarsening. We will discuss some of the results primarily obtained from our own works: (1) microstructural evolution in coherent two-phase alloys with ordered intermetallic precipitates;[9] (2) twin formation during a cubic $\rightarrow$ tetragonal transformations;[10] (3) grain growth in a single-phase system.[11] Therefore, this is not intended to be a review for the entire subject of applications of the diffuse-interface field model. The main goal is to present the basic idea of how one can model a microstructure using the diffuse-interface field model, and hence the discussions of the examples will be as brief as possible. For details, readers are recommended to read the original papers. In the end, we will also point out some other potential important applications of the diffuse-interface field model.

2. THE DIFFUSE-INTERFACE FIELD MODEL

The basic idea behind the continuum field approach is to choose a small set of coarse-grained field variables whose dynamic evolutions are "slow" compared to the enormous number of microscopic degrees of freedom in a macroscopic system, and therefore can be handled by the current generation of supercomputers. For a two-component system, we define an array of field variables, $\eta_1, \eta_2, \ldots, \eta_p, c$, where p is the total number of nonconserved field variables, and c is composition. For multicomponent systems, additional compositional field variables have to be introduced.

In the diffuse-interface field model, the total free energy of a system, in which the field variables are spatially dependent, is given by [12]

$$F = \int \left[f_o\left(\eta_1(r), \eta_2(r), \ldots, \eta_p(r), c(r)\right) + \sum_{i=1}^{p} \frac{\kappa_i}{2} \left(\nabla \eta_i(r)\right)^2 + \frac{\kappa_c}{2} \left(\nabla c(r)\right)^2 \right] d^3 r \tag{1}$$

where f_o is the local free energy density which is a function of field variables η_i and c, κ_i and κ_c are the gradient energy coefficients. The origin of interfacial energy comes from the gradient energy terms, $\nabla\eta_i(r)$ and $\nabla c(r)$, which are nonzero only at and around the interfaces. For a given f_o, the boundary energy and thickness vary with κ_i and κ_c. The smaller the κ_i and κ_c, the thinner the boundary region, and smaller the boundary energy. The dependence of the free energy on the field variables should meet the permutation properties consistent with the parent (disordered) phase symmetry.

The relaxation kinetics of a set of initially nonequilibrium distributions of field variables towards equilibrium are described by the Langevin equations (the Ginzburg-Landau equation and the Cahn-Hilliard equation[13] plus noise terms),[1]

$$\frac{\partial c(\mathbf{r},t)}{\partial t} = M\nabla^2 \frac{\delta F}{\delta c(\mathbf{r},t)} + \xi(\mathbf{r},t) \tag{2}$$

$$\frac{\partial \eta_p(\mathbf{r},t)}{\partial t} = -L_{pq} \frac{\delta F}{\delta \eta_q(\mathbf{r},t)} + \zeta_p(\mathbf{r},t) \tag{3}$$

where $\xi(\mathbf{r},t)$ and $\zeta_p(\mathbf{r},t)$ are the Langevin noise terms which are taken to be Gaussian distributed and their correlation properties meet the fluctuation-dissipation theorem [14]:

$$\left\langle \xi(\mathbf{r},t)\xi(\mathbf{r}',t')\right\rangle = -2k_B TM\nabla^2\delta(\mathbf{r}-\mathbf{r}')\delta(t-t') \tag{4}$$

$$\left\langle \zeta_p(\mathbf{r},t)\zeta_q(\mathbf{r}',t')\right\rangle = 2k_B TL_{pq}\delta_{pq}\delta(\mathbf{r}-\mathbf{r}')\delta(t-t'), \tag{5}$$

where k_B is the Boltzmann constant and T is temperature. Physically, the noise terms are related to the microscopic degrees of freedom, e.g., the thermal vibrations (phonons), high-order correlations with short relaxation time, etc., which are assumed to be at equilibrium for a given temperature and a given set of field variables.

Since the field approach is phenomenological, it can potentially be applied to microstructural development produced by a wide variety of different material processes by defining different field variables. For example, using a single conserved field (the composition in a two-component system), the field equation describes the kinetics of isostructural spinodal decomposition.[2,13] With a single nonconserved field variable, the field equation describes the kinetics of ordering and subsequent antiphase domain coarsening for which the nonconserved field variable is the long-range order parameter,[15] or solidification in single-component systems for which the field variable is a parameter for distinguishing the structural difference between solid and liquid,[5,6] or the dynamics of 180° ferroelectric domain formation in ferroelectrics in which the field variable is the local electric polarization.[16] With one nonconserved field variable and the composition field, one can model the simultaneous ordering and decomposition process during precipitation of ordered intermetallics from a disordered matrix with the nonconserved field variable being the long-range order parameter for the ordered phase.[9,17] With two nonconserved field variables, the field equations may be employed to model the formation of 90° elastic domains in 2-dimensions[10] and the ordering and interface migration kinetics which produces two different orientation variants for adsorbed atoms on a surface, with the two field variables identifying the two different orientation domains. By defining three nonconserved order parameters for three possible orientation domains and/or a conserved order parameter for composition, the microstructural evolution during a cubic→tetragonal displacive and/or diffusional transformation can be simulated using three or four coupled field equations.[9,10] The set of equations can be employed to model the grain growth kinetics in a polycrystalline material with the non-conserved field variables representing the orientations of each grain, the number of variables being equal to the number of the grain orientations.[11]

The elastic energy generated by lattice mismatch between different phases or different orientation domains in coherent solids can be included in the total free energy functional for modeling the dynamics of strain-induced microstructural evolution as will be discussed in the specific examples presented below. One can, in principle, introduce the elastic displacement as another nonconserved field variable. However, with the assumption that the elastic constants are homogeneous, i.e., they are the same everywhere in the system, and with the fact the elastic equilibrium is established much faster than any diffusional processes such as ordering or phase separation, the equilibrium elastic displacements and thus the elastic energy of an inhomogeneous system with arbitrary distribution of nonconserved field variables and the composition can be analytically solved using the Green's function method in reciprocal space.[18] For the examples given below, the homogeneous moduli will be assumed. For problems which do not involve long-range elastic interactions, the set of kinetic equations are most conveniently discretized with respect to space and time and solved in real space. However, systems in which the elastic energies are involved, the set of equations are much more efficiently solved in reciprocal space through the Fast-Fourier Transformation (FFT) Technique.

3. EXAMPLES

3.1. Two-Phase Alloys With Intermetallic Precipitates[9]

Precipitation of an ordered intermetallic phase or phases from a disordered matrix is the basic process that underlies the processing of many advanced alloys such as high-temperature superalloys and ultralight aluminum alloys. Important examples include Al-Li alloy hardened by δ' ordered precipitates, Ni-Al alloy with γ' ordered precipitates, and Fe-Al alloy with B2 precipitates. The ordered precipitate and the disordered matrix phases always differ not only in compositions but also in structures. Therefore, in addition to the composition field variable, we need to define long-range order parameter fields to distinguish the structural differences between the ordered precipitate and disordered matrix as well as the different antiphase domains of the ordered phase.

As an example, a simple 2D prototype binary alloy will be considered. It is a homogeneous disordered cubic solid solution at high temperatures and a two-phase mixture of a solute-rich ordered phase and a solute-lean disordered phase at low temperatures. Both phases have cubic structures. For simplicity, we define only one nonconserved field variable, the long-order parameter field, implying only two kinds of antiphase domains (with $\eta > 0$ and $\eta < 0$, respectively) for the ordered phase. The local free energy density function, $f_o(\eta, c)$, was constructed in such a way to have the free energies of the ordered and disordered phases as shown in Fig. 1. The free energy of the disordered phase is obtained by simply set the value of the long-range order parameter equal to zero, i.e., $f_o(\eta = 0, c)$, and the free energy of the ordered phase is calculated by first calculating the equilibrium value of the long-range order parameter at a given composition and then substituting it back to the original free energy function, i.e., $f_o(\eta_e(c), c)$. Their common tangent determines their equilibrium compositions. The interfacial energy between the ordered particles and disordered matrix can be calculated similar to the original diffuse-interface theory of Cahn and Hilliard.[12]

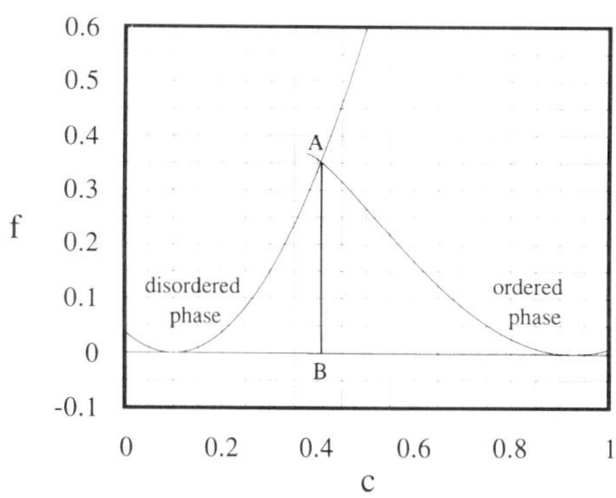

Figure 1. Free energy vs. composition curve obtained by projecting the specific free energy minima with respect to the long-range order parameter on the f-c plane. $\overline{AB}$ designates the typical driving force $|\Delta f|$.

Very often, the lattice parameters of the ordered precipitates and the disordered matrix are different. If a two-phase or multiphase mixture is coherent, which is almost always the case during the initial stages of precipitation process, this mismatch is accommodated by elastic lattice displacements. By assuming all the coexisting phases have the same elastic constants, the strain energy generated by an arbitrary concentration or structural heterogeneity can be analytically solved in Fourier space. The strain energy is a functional of the concentration and/or long-range order parameter fields. If the strain is predominantly caused by the concentration heterogeneity, which is the case for the Ni-based superalloys, and if the Vegard's law (the lattice parameter is a linear function of composition) is fulfilled, the morphology-dependent part of the strain energy is,

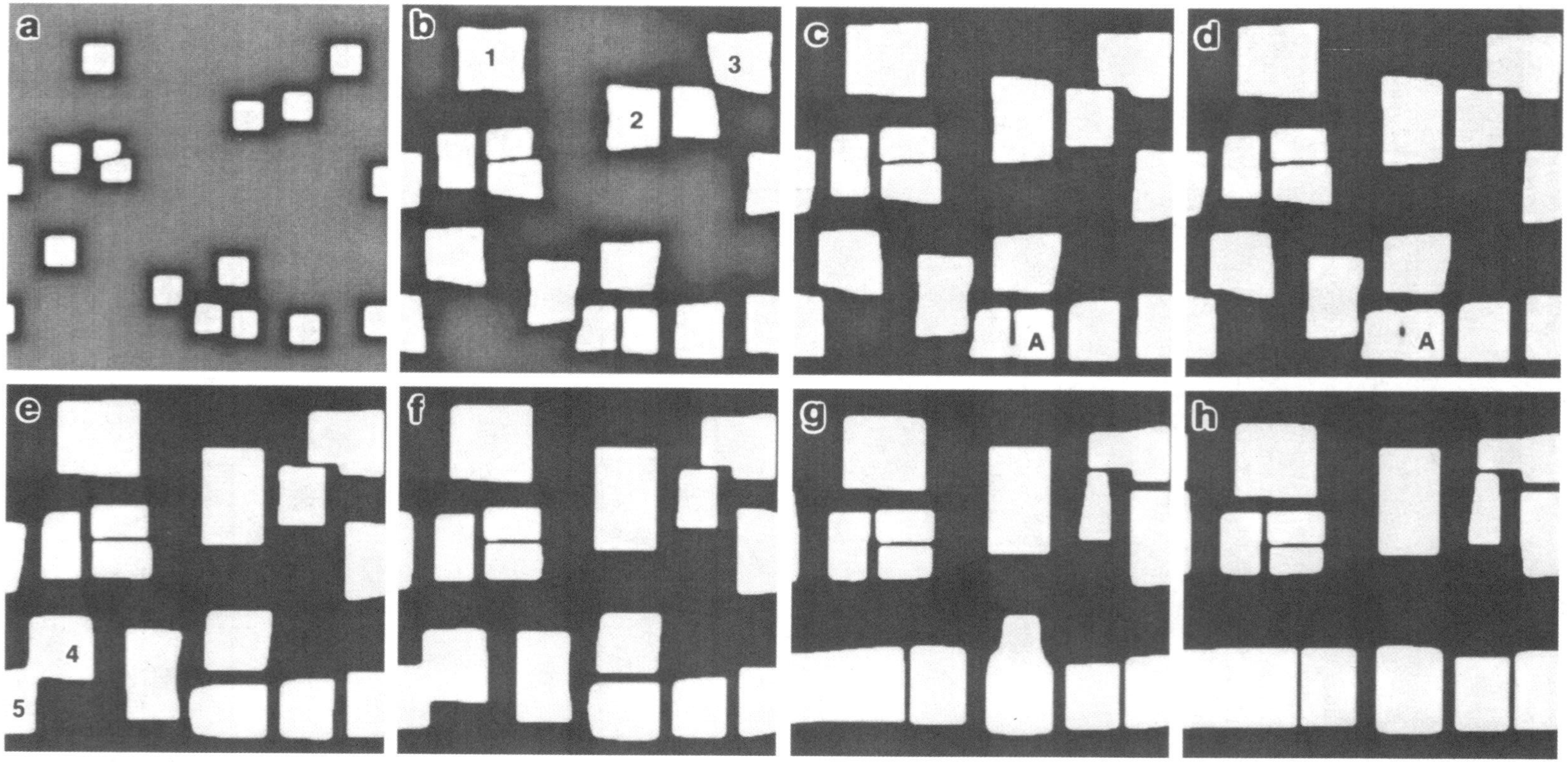

Figure 2. Shape evolution and spatial rearrangement of coherent ordered precipitates in a system with intermediate density of particles. See text for discussion. (a) $\tau = 50$; (b) $\tau = 200$; (c) $\tau = 400$; (d) $\tau = 450$; (e) $\tau = 1000$; (f) $\tau = 2000$; (g) $\tau = 5000$; and (h) $\tau = 10{,}000$.

$$E_{el} = \frac{1}{2}\int B(\mathbf{n})\left|\tilde{c}(\mathbf{k})\right|^2 \frac{d^3k}{(2\pi)^3} \tag{6}$$

where k is the wave vector, n $= \mathbf{k}/k$, $\tilde{c}(\mathbf{k})$ is the Fourier transform of the concentration field, c(**r**). The function B(**n**) is determined by the elastic moduli of the system and the stress-free strain,

$$B(\mathbf{n}) = -n_i\sigma_{ij}^0\Omega(\mathbf{n})_{jk}\sigma_{kl}^0 n_l + \max\left\{n_i\sigma_{ij}^0\Omega(\mathbf{n})_{jk}\sigma_{kl}^0 n_l\right\} \tag{7}$$

where $\Omega(\mathbf{n})_{ij}$ is the inverse to the tensor $\Omega(\mathbf{n})_{ij}^{-1} = c_{ijkl}n_k n_l$, σ_{ij}^0 is defined as $\sigma_{ij}^0 = c_{ijkl}\varepsilon_{kl}^0$, and c_{ijkl} and ε_{kl}^0 are the components of the elastic moduli tensor and the tensor of the stress-free concentration expansion coefficient.

For a 2-D model system with square elastic anisotropy, the function B(n) can be approximated as[19,20]

$$B(\mathbf{n}) \approx B_{el}\varepsilon_0^2 n_x^2 n_y^2 \tag{8}$$

where n_x and n_y are the components of n along the x and y axes in the reciprocal space and

$$B_{el} = -\frac{4(c_{11}+2c_{12})^2}{c_{11}(c_{11}+c_{12}+2c_{44})}\Delta \tag{9}$$

is a material constant in which $\Delta = c_{11} - c_{12} - 2c_{44}$ is a measure of elastic anisotropy. A positive value of B_{el} was used in our example, which implies a system with negative elastic anisotropy, $\Delta < 0$.

An example of microstructural development during precipitation and coarsening of ordered particles in an initially metastable disordered solution with the average composition c = 0.4 is shown in Fig. 2 in which the microstructures are visualized through the gray-levels representing different values of the local composition, c(r,t). The nucleation process of the ordered particles was promoted by thermal fluctuations introduced through the noise terms in the kinetic equations. The parameters were chosen in such as way to provide isotropic interfacial energy.

At the initial stage, when the size of precipitate particles is small and interfacial energy is dominant, the shape of particles appear to be circular. As the particles growth larger, the volume-dependent anisotropic elastic interactions becomes increasingly important and as a result, the particles transform to squares (see Fig. 2 (a)). Further growth of the precipitate particles results in the rather isolated ones developing into a concave "square"-like morphology with all four sides of negative curvatures (Fig. 2 (b)). However, for most of the particles, shapes are asymmetrical with only one or two concave sides, see, for example, the particles numbered 2 and 3 in Fig. 2 (b). These results seem to be in good agreement with the experimental observations[21,22] (see, e.g., Fig. 3 for comparison). The growth of particles at this stage is mainly driven by the bulk chemical free energy decrease.

After both the ordered precipitate and the disordered matrix reach their equilibrium compositions, further microstructural evolution is driven by a combination of elastic energy relaxation and interfacial energy reduction. During this coarsening stage, most of the concave interfaces disappear, indicating the concavity of the interfaces is a kinetic instability during growth driven by the bulk chemical energy decrease. However, some concave interfaces still remain, and they are due to the elastic interactions between closely spaced particles. During coarsening, coalescence was also observed (see the particle marked A in Fig. 2 (c) and (d)).

As we can see in the above example, the strain-induced coarsening process of coherent particles is very different from the normal coarsening process driven solely by reduction in interfacial energy. In the latter case, large particles grow at the expense of smaller ones while the shapes of the particles and their mutual arrangement are essentially time-independent. However, in strain-induced coarsening, change in particle shape and positions is a rule rather than an exception. For example, the two particles (4 and 5 in Fig. 2(e)) initially aligned along the [11] direction coalesce, forming a single particle with big steps at its interface. Then, the big steps gradually decompose into smaller ones at later stages, accompanied by a significant translational movement of the center of mass.

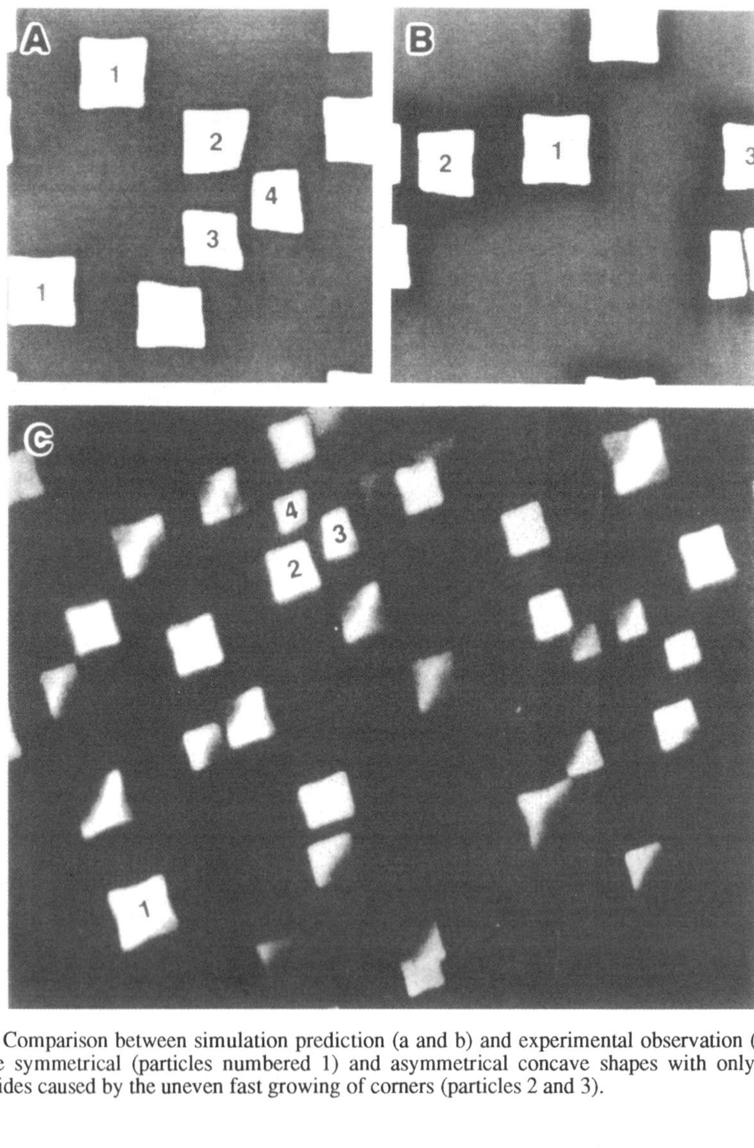

Figure 3. Comparison between simulation prediction (a and b) and experimental observation (c, from Ref. 22) of the symmetrical (particles numbered 1) and asymmetrical concave shapes with only one or two concave sides caused by the uneven fast growing of corners (particles 2 and 3).

3.2. Twin Formation

Many important ordering and displacive reactions in alloys generates several orientation domains which are related by the symmetry elements of the parent disordered phase. In a c→t transformation, the number of different orientation variants is three, with the tetragonal axes along crystallographic directions, [100], [010] and [001], of the parent cubic phase. Typical examples include the ordering transformation from a fcc disordered phase to an ordered tetragonal phase ($L1_0$) and the cubic → tetragonal displacive transformation in the ZrO_2-Y_2O_3 system. The crystal lattice rearrangement caused by the transformation produces a crystal lattice misfit at the boundaries between phases and between orientation domains. If the transformation is coherent, i.e., all crystalline planes run continuously from one phase (or orientation domain) to another, the misfit is accommodated by the elastic strain providing the lattice continuity across the interfaces. Overlap and interference of the directionally anisotropic strain fields generated by different transformed volumes result in the so-called strain-induced interaction.[23] Depending on their mutual position and orientation, this interaction is attractive if the overlapping strain fields cancel each other, or is repulsive if the field values are of the same sign and thus superimpose. The existence of several orientation domains of the tetragonal phase provides a configurational freedom to reduce the total elastic energy. In the case of a two-phase mixture of cubic and tetragonal phases, twin-related 'sandwiches' of different orientation domains of the tetragonal phase can be organized into plates in such a way that they eliminate the volume-dependent part of the elastic energy and the resulting elastic energy is only proportional to the interfacial area. If the transformation product is a tetragonal single phase, the final equilibrium state depends on the stress conditions. For example, a single crystal cubic phase with free surfaces transforms to a tetragonal single phase, the equilibrium state is expected be single domain tetragonal phase. However, if a cubic phase can not change its volume during a transformation, e.g., a cubic phase grain constrained in a polycrystalline assembly, it can be shown that the equilibrium state under this clamped condition is the twin structure comprised of two different orientation variants of the tetragonal phase.

As an example, we consider the case of *displacive* cubic → tetragonal transformation in the Y_2O_3-ZrO_2 system. In particular, we investigated the temporal microstructural evolution during a cubic → tetragonal transformation under different thermodynamic conditions, stress free or clamped.[10]

In order to model the microstructural evolution, and the twin formation in particular, we need to define three nonconserved field variables plus the composition, with each nonconserved order parameter representing a given orientation domain of the tetragonal phase. The local free energy density is constructed in such a way that it qualitatively describes the phase diagram of the ZrO_2-Y_2O_3 system. In our example, the free energies for the cubic and tetragonal phases obtained from a particular free energy density function are shown in Fig. 4. To provide stability of the tetragonal phase, the free energy $f(\eta_1, \eta_2, \eta_3, c)$ has triple-degenerated minima at $\{\eta_o, 0, 0\}$. The free energy of the cubic-phase was obtained by letting the nonconserved order parameters equal to zero, $f_o(\eta_1 = 0, \eta_2 = 0, \eta_3 = 0, c)$. The free energy of the tetragonal phase was calculated by, first, minimizing the local free energy density function with respect to a η_i, e.g., η_1, at a given c, and under the condition that the values of the rest of the order parameters are zero, and then expressing the order parameter as a function of composition c, $\eta_e(c)$, and hence express the free energy with c only, i.e., $f_o(\eta_1 = \eta_e(c), \eta_2 = 0, \eta_3 = 0, c)$.

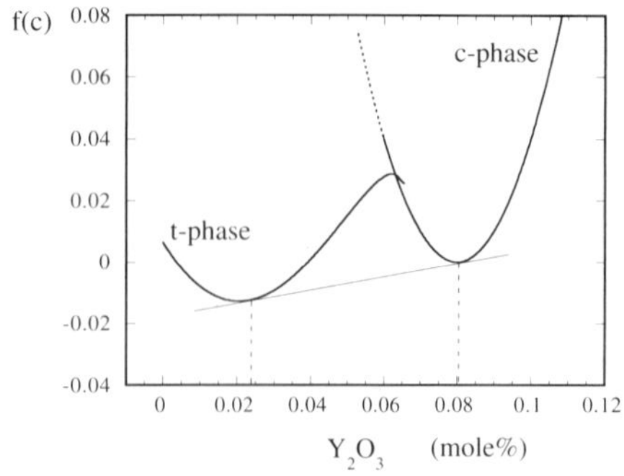

Figure 4. Specific free-energy vs. composition curves for the cubic and tetragonal phases.

During a cubic $\rightarrow$ tetragonal phase transformation, the local stress-free transformation strain $\varepsilon_{ij}^{0}(\mathrm{r})$ are related to the continuous nonconserved order parameter fields through

$$\varepsilon_{ij}^{o}(\mathrm{r}) = \sum_{p=1}^{q} \varepsilon_{ij}^{o}(p)\eta_p^2(\mathrm{r}) \tag{10}$$

where $\eta_p(\mathrm{r})$ is local order parameter field for a particular orientation domain p, and $\varepsilon_{ij}^{0}(p)$ is given by

$$\varepsilon_{ij}^{0}(1) = \begin{pmatrix} \varepsilon_{33}^{0} & 0 & 0 \\ 0 & \varepsilon_{11}^{0} & 0 \\ 0 & 0 & \varepsilon_{11}^{0} \end{pmatrix}, \ \varepsilon_{ij}^{0}(2) = \begin{pmatrix} \varepsilon_{11}^{0} & 0 & 0 \\ 0 & \varepsilon_{33}^{0} & 0 \\ 0 & 0 & \varepsilon_{11}^{0} \end{pmatrix}, \ \varepsilon_{ij}^{0}(3) = \begin{pmatrix} \varepsilon_{11}^{0} & 0 & 0 \\ 0 & \varepsilon_{11}^{0} & 0 \\ 0 & 0 & \varepsilon_{33}^{0} \end{pmatrix} \tag{11}$$

where $\varepsilon_{11}^{0} = (a_t - a_c)/a_c$, $\varepsilon_{33}^{o} = (c_t - a_c)/a_c$. a_c is the lattice parameter of cubic phase, and a_t and c_t are the lattice parameters of the tetragonal phase.

By assuming that the cubic and tetragonal phases have the same elastic moduli, a similar elastic energy to (6), which is a functional of the field variables, can be written down.

Because of the extensive computation involved in a three-dimensional (3D) kinetic study , we will show a two-dimensional (2D) simulation. In this case, only two equations are needed and $\varepsilon_{ij}^{0}(\mathrm{r})$ is a plain strain. In the computer simulation, a square cell with 256 $\times$ 256 grids is employed with periodic boundary conditions along both x and y directions.

The overall average composition, c, chosen to be 0.06 (mole fraction). The initial condition is a homogenous cubic phase in which all values for the order parameter fields are zero. The cubic phase is then quenched to a temperature below the absolute instability line of the cubic phase with respect to the transformation to the tetragonal phase, usually referred to as, T_o line. In order to initiate the kinetics, however, random perturbations to the order parameters fields are introduced at the beginning of our simulation. Since our main interest is the diffusionless transformation from a single c phase to a single tetragonal phase, we only solved the kinetic equations for the order parameter fields while keeping the composition homogeneous.

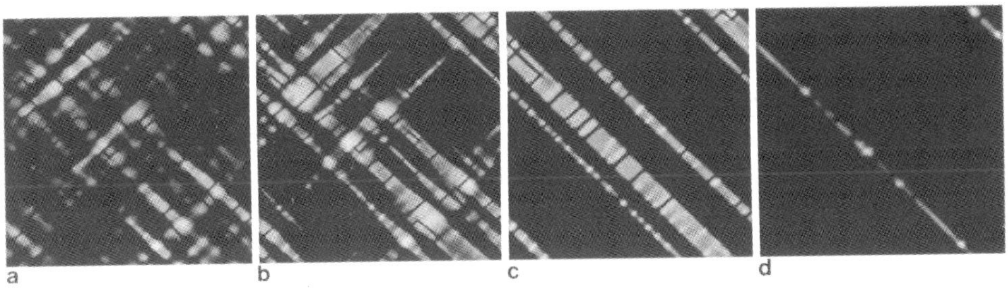

Figure 5. Temporal microstructure evolution during a cubic→tetragonal transformation under the stress-free condition.

3.3.1 Stress-free Condition. The elastic strain energy of the unconstrained system contains both the homogenous and heterogeneous strain relaxation. The morphological evolution from the initial cubic phase to the tetragonal phase and subsequent tetragonal domain coarsening is shown in Fig. 5. In the figure, the "gray levels" represent different magnitudes of the square of one of the two order parameter fields with the highest value represented by a completely white color and the lowest by a completely dark color. This representation corresponds to the dark field transmission electron micrographs using a reflection from one of the orientation variants.

From Fig. 5(a) and (b) it can be seen that the initial stage of transformation consists of growth of both order parameter fields at different locations as well as the increase in the domain size, which is primarily driven by the bulk chemical free energy decrease. After the cubic→tetragonal transformation, further microstructural evolution is driven by the antiphase domain boundary energy, the twin boundary energy as well as the elastic energy arising from the accommodation of different orientation domains. It is shown that the two orientation domains preferentially align along {1 1} planes, forming so-called the "tweed" structure. This tweed structure coarsens and develop into the twin (Fig. 5(c)) as time increases, with the amount of one orientation variant significantly more than the other. This twin structure is unstable and finally one orientation domain disappears. As the result, a single domain with anti-phase domain boundaries is formed (Fig. 5(d)). Eventually all the antiphase domain boundaries will vanish.

3.2.2. Clamped Condition. Under the clamped condition, the homogeneous strain relaxation is prohibited. The corresponding morphology evolution is shown in Fig. 6. On comparison of Fig. 6 with Fig. 5, the structure similarity is obvious. It turns out that the twin structure is stable if the homogeneous relaxation is not allowed. The final microstructure contains both orientation domains comprising the twins with the twin boundary being the (1,-1) plane. Many anti-phase domain boundaries still exist within the

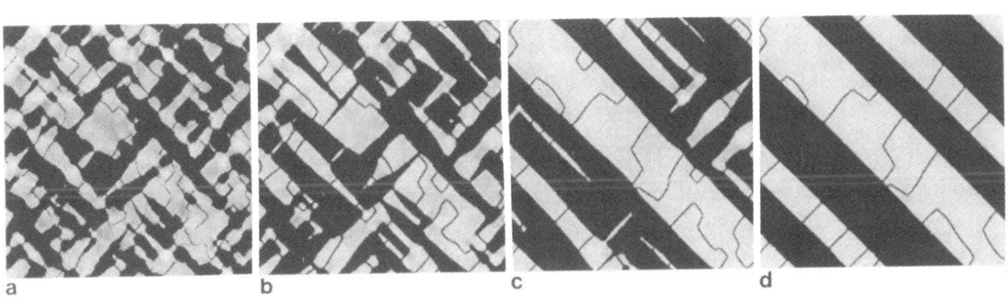

Figure 6. Temporal microstructure evolution during a cubic→tetragonal transformation under the clamped condition.

domains in Fig. 6 (d). The microstructure shown in Fig. 6(d) seems to agree well with the twin microstructure observed in the Y_2O_3-ZrO_2 system.[24]

In summary, in the stress-free state, a cubic phase transforms to a single-domain tetragonal phase through intermediate tweed and twin structures whereas in the clamped condition, the twin structure is thermodynamically stable.

3.3. Grain Growth in Pure Single-Phase Systems

Grain growth, due to its importance in controlling the physical properties of a wide variety of materials, has been extensively investigated.[24,25] Due to the difficulty of directly incorporating topological features into analytical theories of grain growth, there has been a variety of computer simulation models proposed. These include boundary dynamics models,[27-29] vertex models,[30,31] Potts models,[32-34] Voronoi tellselation,[35,36] and models based on mean-field theories.[37-40] A common feature of all these models is that all of them describe grain boundaries as sharp interfaces with zero thickness. Very recently, we showed that the microstructural evolution during grain growth can also be modeled by the diffuse-interface field model simply by introducing a large set of nonconserved order parameter fields, with each order parameter field representing grains of a given crystallographic orientation.[11]

The main requirement for f_0 in modeling grain growth in a pure single-phase is that it has p degenerate minima with equal depth, f_{min}, located at $(\eta_1, \eta_2, ..., \eta_p)$ = (1,0,...,0), (0,1,...,0), ..., (0,0,...,1) in p-dimensions space.

For example, the following simple free energy density functional satisfies this requirement if $\gamma > \beta/2$,

$$f(\eta_1, \eta_2, ..., \eta_p) = \sum_{i=1}^{p}\left(-\frac{\alpha}{2}\eta_i^2 + \frac{\beta}{4}\eta_i^4\right) + \gamma\sum_{i=1}^{p}\sum_{j>i}^{p}\eta_i^2\eta_j^2 \qquad (12)$$

where η_i are order parameters, and α, β, and γ are positive constants.

An example of microstructural evolution using a 512×512 cell with 36 nonconserved field variables is shown in Fig. 7. Periodic boundary conditions were applied. The initial condition is specified by assigning small random values to all field variables at every grid point, e.g., -0.001 < η_i (for i) < 0.001, simulating a liquid. The microstructure is represented by defining the following field variable

$$\varphi(r) = \sum_{i=1}^{p}\eta_i^2(r)$$

which are displayed by gray-levels with low and high values represented by dark and bright colors, respectively. Since the values within the domains are close to 1.0, while those at the boundaries are significantly less (~ 0.6), the bright regions are domains and the dark lines are grain boundaries. Since the initial values for η_i are essentially zero, the very early stage of the simulation corresponds to crystallization, i.e., the growth of η_i values at different locations driven by the bulk free energy change. A well defined grain structure is formed after a short time, about 200 time steps, at which essentially all the bulk driving force has been consumed. Further microstructural evolution is driven by the excess free energies associated with the grain boundaries, resulting in growth of overall microstructure scales or grain size.

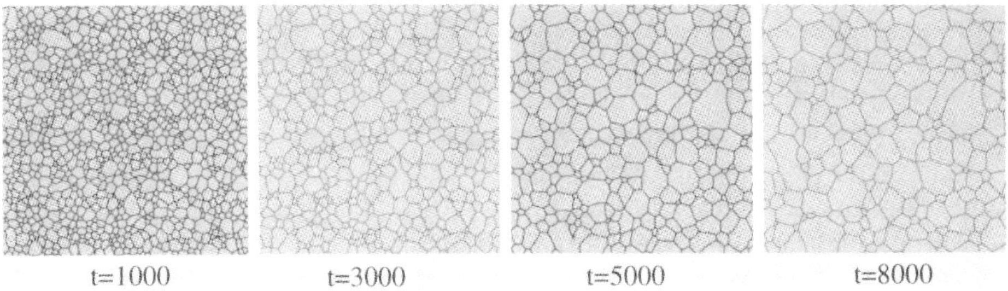

Figure 7. The microstructural evolution in the 512×512 system with 36 non-conserved field variables.

With the temporal microstructural evolution, all the statistical information about the microstructure such as the average grain size, size distribution, average number of sides, side distribution, local topological changes may be obtained. For example, the average grain radius as a function of time for the 512×512 system is shown in Fig. 8. The data were then fitted to the equation $\bar{R}_t^m - \bar{R}_0^m = kt$ by a multi-parameter nonlinear least-square fitting routine to extract growth exponent m and coefficient k. The data from the initial stage, about 500 to 1000 time steps, were removed before the fitting. The growth exponent, m, is shown to be almost exactly 2.0, in agreement with most of the previous mean-field theories and more recent large-scale Monte-Carlo Potts model simulations.[41] The corresponding grain size distribution at a certain time step is shown in Fig. 9 in which the probability of a grain having grain area $\bar{A}$ was plotted against $\log(\bar{A}/\bar{A}_o)$.

In addition to obtain the microstructural evolution and the statistical information about the microstructure, essentially all the classical theories and laws proposed for 2-D grain growth can be critically tested using the results obtained from the numerical simulation.[42]

4. Other Potential Applications

In addition to the examples discussed, there are many important microstructural problems that can be potentially studied using the diffuse-interface field model. These include: (1) solute segregation and second-phase precipitation during grain growth; (2)

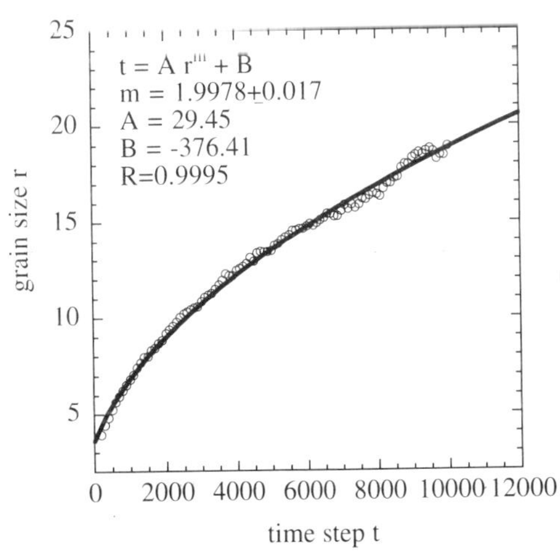

Figure 8. Time-dependence of average grain size.

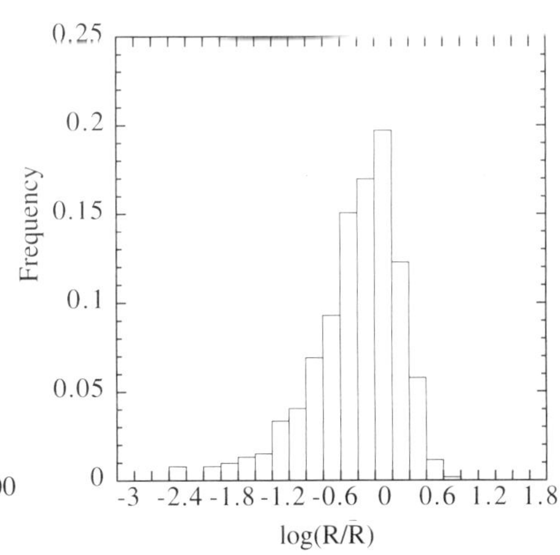

Figure 9. Grain size distribution.

sintering; (3) coupled-grain growth and Ostwald ripening; (4) coarsening of a solid-liquid mixture at high volume fractions of solids; and (5) thin film growth morphologies and microstructures. (1), (2), (4) and (5) can be modeled by coupling a large set of nonconserved parameter fields with the composition field where (3) can be described by two separate sets of nonconserved order parameter fields coupled with a composition field.

By these examples, we hope that we have demonstrated that the diffuse-interface field is a very powerful tool for microstructural modeling and we hope there will be more applications beyond the examples that we mentioned here.

5. Acknowledgments

The work is supported by National Science Foundation under grants No. DMR-9311898 (LQC and DNF) and NSF-DMR-9123167 (Y. W. and A. K.). The simulation results were obtained from the Pittsburgh Supercomputing Center.

References

1. J. D. Gunton, M. S. Miguel and P. S. Sahni, The dynamics of first-order phase transitions, in C. Domb and J. L. Lebowitz(eds), Phase Transitions and Critical Phenomena, Vol. 8., Academic Press, New York, pp. 267-466 (1983).
2. T. M. Rogers, K. R. Elder and R. C. Desai, Numerical study of the late stages of spinodal decomposition, Phys. Rev. B 37:9638 (1988).
3. Y. Oono and S. Puri, Computationally efficient modeling of ordering of quenched phases, Phys. Rev. Lett. 58:836 (1987).
4. A. A. Wheeler, W. J. Boettinger and G. B. McFadden, Phys. Rev. A45:7424 (1992).
5. G. Caginalp and E. Socolovsky Phase field computations of single-needle crystals, crystal growth, and motion by mean curvature, SIAM J. Sci. Comput. 15:106 (1994).
6. R. Kobayashi, A numerical approach to three-dimensional dendritic solidification, *Experim. Math.* 3:59 (1994).
7. J. A. Warren and W. J. Boettinger, Prediction of dendritic growth and microsegregation patterns in a binary alloy using the phase-field method, Acta metall. mater. 43:689 (1995) .
8. A. Onuki, Ginzburg-Landau approach to elastic effects in the phase separation of solids, J. Phy. Soc. Jpn. 58:3065(1989); Long-range interactions through elastic fields in phase-separating solids, J. Phy. Soc. Jpn., 58:3069 (1989); H. Nishimori and A. Onuki, Pattern formation in phase-separating alloys with cubic symmetry, Phys. Rev. B 42:980 (1990).
9. Y. Wang, Theoretical characterization and modeling of microstructure development during coherent phase transformations in metals and ceramics, Ph.D. dissertation, Rutgers, The State University of New Jersey, NJ, USA (1995); Y. Wang, L. Q. Chen, and A. G. Khachaturyan, Computer simulation of microstructure evolution in coherent solids, in Solid→ Solid phase Transformations, W. C.Johnson, J. M. Howe, D. E. Laughlin and W. A. Soffa (eds), The Minerals, Metals & Materials Society, pp245-265 (1994); Y. Wang, H. Y. Wang, L. Q. Chen and A. G. Khachaturyan, Shape evolution of a coherent tetragonal precipitate in partially stabilized cubic ZrO_2: a computer simulation, *J. Am. Ceram. Soc.* 76:3029 (1993); (1995) Microstructural development of coherent tetragonal precipitates in Mg-partially stabilized ZrO_2: a

computer simulation, J. Am. Ceram. Soc. 78, 657-661. Y. Wang and A. G. Khachaturyan, Effect of antiphase domains on shape and spatial arrangement of coherent ordered intermetallics, *Scipta Metall. et Mater.*, 31:1425 (1995); Theoretical characterization of the kinetics of martensitic transformation, to be submitted to Acta metall er mater (1995).
10. D. N. Fan and L. Q. Chen, Computer simulation of twin formation during the displacive c→t' phase transformation in the Zirconia-Yttria system, *J. Am. Ceram. Soc.* 78:769 (1995); Possibility of spinodal decomposition in ZrO_2-Y_2O_3 alloys: a theoretical investigation, ibid 78:1680 (1995).
11. L. Q. Chen and W. Yang, Computer simulation of the domain dynamics of a quenched system with a large number of nonconserved order parameters: the grain-growth kinetics, Phys. Rev. B50:15752 (1994); L. Q. Chen,, A novel computer simulation technique for modeling grain growth, Scripta metall. mater, 32:115 (1995); D. N. Fan, and L. Q. Chen, Computer simulation of grain growth using the diffuse-interface field model, submitted to Acta metall et mater (1995).
12. J. W. Cahn and J. E. Hilliard Free energy of a nonuniform system, J. Chem. Phys. 28:258 (1958).
13. J. W. Cahn, On spinodal decomposition, Acta metall. 9:795 (1961); On spinodal decomposition in cubic crystals, ibid, 10:179 (1962).
14. E. M. Lifshitz, and Pitaevskii, L. P. Statistical Physics, 3rd edition, Part 1, Landau and Lishitz Course of Theoretical Physics, Vol. 5, Pergamon Press (1980).
15. S. M. Allen and J. W. Cahn A microscopic theory for antiphase boundary motion and its application to antiphase domain coarsening, Acta metall. 27:1085 (1979).
16. W. Yang and L. Q. Chen, Computer simulation of the dynamics of 180° ferroelectric domain formation, to be published in the J. Am. Ceram. Soc. (1995).
17. T. Eguchi, K. Oki and S. Matsumura, Kinetics of ordering with phase separation, in MRS Symp. Proc., 21:589 (1984); K. Shiiyama, H. Kanemoto, H. NInomiya and T. Eguchi, Computer simulation of dynamics of the pattern formations in antiphase ordered structure and phase separation induced by ordering in binary alloys, in Proceedings of the International Workshop on Computational Materials Science (IWCMS), 103 (1990).
18. A. G. Khachaturyan, Some questions concerning the theory of phase transformations in solids, Sov. Phys. Solid State 8:2163 (1967); A. G. Khachaturyan and G. A. Shatalov, Elastic-interaction potential of defects in a crystal, Sov. Phys. Solid State 11:118 (1969).
19. Y. Wang, L. Q. Chen and A. G. Khachaturyan, Shape evolution of a precipitate during strain-induced coarsening: a computer simulation, *Scripta metall. mater.* 25:1387 (1991); Strain-induced modulated structures in two-phase cubic alloys, *ibid,* 25:1969 (1991); Kinetics of strain-induced morphological transformation in cubic alloys with a miscibility gap, *Acta metall. mater.* 41:279 (1993); Particle translational motion and reverse coarsening phenomena in multiparticle systems induced by a long-range elastic interaction, *Phys. Rev. B* 46:11194 (1992).
20. L. Q. Chen, Y. Wang and A. G. Khachaturyan, Transformation-induced elastic strain effect on the precipitation kinetics of ordered intermetallics, *Phil. Mag. Ltr.* 64:241 (1992).
21. Y. S. Yoo, D. Y. Yoon and M. F. Henry, The effect of elastic misfit strain on the morphological evolution of g'-precipitates in a model Ni-base superalloy, in press; Y. S. Yoo, The dendritic growth of g'-precipitates and grain boundary serration in Ni-based superalloys, Ph.D. dissertation, Korea Advanced Institute of Science and Technology, Taejon, Korea (1993).

22. A. Maheshwari and A. J. Ardell, Anomalous Coarsening of Small Volume Fractions of Ni_3Al Precipitates, An Explanation of Inhomogeneous Dispersions Observed at Small Undercoolings, *Scripta metall. mater.* 26:347 (1992); Anomalous coarsening behavior of small volume fractions of Ni_3Al precipitates in binary Ni-Al alloys, *Acta metall. mater.* 40:2661 (1992).
23. A. G. Khachaturyan, Theory of Structural Transformations in Solids, Wiley, New York, 1983.
24. M. Doi and T. Miyazaki, On the spinodal decomposition in Zirconia-Yttria (ZrO_2-Y_2O_3) alloys, Philos Mag. B68:305 (1993).
25. H. V. Atkinson, Theory of normal grain growth in pure single phase systems, *Acta metall.* 36:469 (1988).
26. D. Weaire and J. A. Glazier, Modeling grain growth and soap froth coarsening: past, present and future, *Materials Science Forum*, p27, vol. 94-96 (1992) Pt. 1.
27. D. Weaire and F. Bolton, Rigidity loss transition in a disordered 2D froth, *Phys. Rev. Lett.*. 65:3449 (1990).
28. D. Weaire and J. N.. Rivier, Soap, cells and statistics - random patterns in two dimensions, *Contemp. Phys.* 25:59 (1984).
29. D. Weaire and H. Lei, A note on the statistics of the mature two-dimensional soap froth, *Phil. Mag. Lett* 62:427 (1990).
30. R. L. Fullman, in *Metal interfaces*, p179, American Society for Metals, Cleveland, 1952.
31. K. Kawasaki, T. Nagai, and K. Nakashima, Vertex Model of Cellular Pattern Growth in Two and Three Dimensions, *Phil. Mag. B* 60:399 (1989).
32. M. P. Anderson, D. J. Srolovitz, G. S. Grest and P. S. Sahni, Computer simulation of grain growth - I. kinetics, *Acta metall.* 32:783 (1984).
33. D. J. Srolovitz, M. P. Anderson, P. S. Sahni and G. S. Grest, Computer simulation of grain growth - II. grain size distribution, topology, and local dynamics, *Acta metall.* 32:793 (1984).
34. M. P. Anderson and G. S. Grest, Computer simulation of normal grain growth in three dimensions, *Phil. Mag. B* 59:293 (1989).
35. S. Kumar, S. K. Kurtz, J. R. Banavar, and M. G. Sharma, Properties of a Three-Dimensional Poisson-Voronoi Tesselation, A Monte Carlo Study, *J. Statis. Phys.* 67:523 (1992).
36. S. K. Kurtz and F. M. A. Carpay, Microstructural and normal grain growth in metals and ceramics. *J. Appl. Phys.* 51:5125 (1980).
37. V. E. Fradkov, L. S. Shvindlerman and D. G. Udler, Computer simulation of grain growth in two dimensions, *Scripta. Met.* 19:1291 (1985).
38. C. W. J. Beenakker, Numerical simulation of a coarsening two-dimensional network, *Phys. Rev. A* 37:1697 (1988).
39. M. Marder, Soap-bubble growth, *Phys. Rev. A* 36:438 (1987).
40. C. V. Thompson, H. J. Frost, and F. Spaepen, The relative rates of secondary and normal grain growth, Acta Metall 35:887 (1987).
41. G S. Grest, M. P. Anderson and D. J. Srolovitz, Domain-growth kinetics for the Q-state Potts model in two and three dimensions, Phys. Rev. B 38:4752 (1988).
42. D. N. Fan and L. Q. Chen, to be submitted to Acta metall et mater.

ATOMISTIC STUDIES OF THE STRUCTURE OF GRAIN BOUNDARIES AND DISLOCATIONS

V. Vitek

Department of Materials Science and Engineering,
University of Pennsylvania, 3231 Walnut Street,
Philadelphia, PA 19104, U.S.A.

INTRODUCTION

Grain boundaries and dislocations are the most common extended defects in crystalline materials. The principal difference between them is that grain boundaries are planar imperfections which are not accompanied by long-range strain and stress fields, while dislocations are line defects that produce long-range elastic strains and stresses in the material. They both strongly affect, and to a great extent control, the physical and mechanical properties of materials. This is the reason why both dislocations and grain boundaries, and also more general types of interfaces, have been studied very extensively throughout the development of materials science. A common feature of both grain boundaries and dislocations is that their basic characteristics are of crystallographic, geometrical nature. Indeed, in earlier structural studies these characteristics were the main topic of experimental and theoretical investigations (Bollman, 1970; Hirth and Lothe, 1982; Sutton, 1984; Sutton and Balluffi, 1995). In the case of dislocations they were closely linked with studies of their elastic fields and associated long-range interactions (Nabarro, 1967; Hirth and Lothe, 1982). Since crystallographic attributes of dislocations and grain boundaries have to be defined prior to any study of their structures and properties we summarize these characteristics briefly in the following section.

No long-range elastic fields are associated with grain boundaries and, therefore, their atomic structure has always been considered as the principal characteristic which determines their properties. In grain boundaries, and interfaces in general, the atomic structure differs significantly from that of the bulk only in a narrow region, usually spanning not more than several lattice spacings, and all the significant interfacial phenomena occur in this region[1]. In contrast, earlier developments of the dislocation theory, and related theoretical analyses of material properties controlled by dislocations, concentrated on the effects of long range elastic fields. In particular, many mechanical properties

[1] Exceptions are some ceramic materials in which a relatively thick glassy phase occurs at grain boundaries (Rühle and Petzow,1981).

Stability of Materials
Edited by A. Gonis *et al.*, Plenum Press, New York, 1996

were analyzed in terms of long-range dislocation-dislocation, dislocation-point defects and dislocation-particles interactions (Friedel, 1964; Nabarro, 1967; Hirth and Lothe, 1982). It was common wisdom until late sixties that dislocation cores were of secondary importance in plastic deformation of metals and could only play a more significant role in covalently bonded solids. This attitude changed radically in the seventies and eighties when dislocation core phenomena were recognized to play at least as important role as long-range interactions in the deformation behavior of most materials. They are particularly important in ordered compounds and non-cubic crystalline structures. The most pronounced signatures of core effects are unusual changes in deformation modes and slip geometry, strong orientation and temperature dependencies of the yield stress, anomalous temperature dependencies of the yield and flow stresses etc. (Vitek, 1985; Duesbery and Richardson, 1991; Vitek, 1992).

Atomistic studies of grain boundaries and dislocations started in late sixties but such calculations became a common tool for investigation of structures of interfaces and dislocation cores only in the eighties (see proceedings of recent conferences Yoo et al., 1988; Aucouturier, 1990; Romig et al., 1991; Komninou and Rocher, 1993). The reason why such studies became very popular in the last ten years is not only the immense increase in the computing power which took place in recent years. Equally important, if not even more significant, has been the development of descriptions of atomic interactions which are both physically justified and usable in large scale simulations (Tersoff et al., 1989; Vitek and Srolovitz, 1989; Mark et al., 1992). An appropriate description of atomic interactions is, of course, the necessary precursor of any meaningful atomistic calculations and we summarize here the most commonly used schemes in metallic materials and describe briefly approaches employed in ionic solids and semiconductors. We then proceed to describe the methods of atomistic modeling of interfaces and dislocations and present several typical examples where atomistic studies revealed structural features that would not be discovered by other means and which at the same time led to explanation of observed macroscopic phenomena.

Finally, another very important development that provided significant impetus for atomistic modeling has been the advancement of experimental techniques, such as the high resolution electron microscopy (HREM), that achieve atomic resolutions. These experimental studies are not only capable to verify or refute the results of atomistic calculations but their interpretation requires a synergistic link with atomistic modeling of observed structures, in particular in the case of lattice defects (Sinclair et al., 1990; Clark et al., 1992). Several examples of such links are presented in this paper.

CRYSTALLOGRAPHY AND GEOMETRICAL CONCEPTS

Dislocations

The principal crystallographic characteristics of a dislocation are its Burgers vector and the direction of the dislocation line. In general, crystal dislocations occur in the form of loops but all the atomistic studies carried out so far have been made for straight dislocations and, therefore, in this paper we do not consider curved dislocations. While Burgers vectors of crystal dislocations are always lattice vectors and, therefore, uniquely defined, the orientation of the dislocation line may be arbitrary. Unless they are parallel to each other, the Burgers vector and the direction of the line determine the slip plane of the dislocation. In crystalline materials the possible slip planes are frequently limited to certain crystallographic planes which then limits possible directions of the dislocation line. The very important type of dislocations is the screw dislocation whose Burgers vector and dislocation line are parallel to each other. Screw dislocations do not possess any well defined slip plane and can, therefore, glide, at least, in principle, in any plane containing their Burgers vector. This distinct feature is the main reason why they play a unique, and often controlling role in the plastic behavior of many materials[2]. For this reason many atomistic studies have been devoted to investigation of screw dislocations.

Elastic strains and stresses associated with a straight dislocation are uniquely determined by its Burgers vector and the direction of its line. In crystalline solids these fields are generally anisotropic and when used as boundary conditions for atomistic studies of dislocations (see the following section), it is important to take the elastic anisotropy into account. These fields can be determined analytically for a given dislocation geometry and elastic moduli. For example, when the axis x_3 of the coordinate system is parallel to the dislocation line, the vector of elastic displacements, **u**, associated with the dislocation, has components

$$u_k = \frac{1}{2\pi}\sum_{n=1}^{3}\left\{G_{k,n}\arctan\left(\frac{x_1 + p_1^n x_2}{p_2^n x_2}\right) + F_{k,n}\ln\left[\left(x_1 + p_1^n x_2\right)^2 + \left(p_2^n x_2\right)^2\right]\right\} \quad (1)$$

where the coefficients $G_{k,n}$, $F_{k,n}$, p_1^n and p_2^n are determined in the framework of the Stroh's sextic formalism (Stroh, 1962); for more details see (Hirth and Lothe, 1982).

Grain Boundaries

A grain boundary is geometrically fully characterized by the following quantities: the relative misorientation of adjoining grains, orientation of the rotation axis, orientation of the boundary plane, and in the case of lattices with basis, position of the boundary plane; this represents six independent parameters.

Formally, any interface may be described as a network of dislocations (Christian, 1975). If the two adjoining crystals are obtained from a reference lattice by transformations described by matrices $\mathbf{R_1}$ and $\mathbf{R_2}$, and **P** is an arbitrary vector in the interface, then the total Burgers vector of dislocations crossing this vector is (Christian, 1975; Christian and Crocker, 1980)

$$\mathbf{B} = \left(\mathbf{R_1}^{-1} - \mathbf{R_2}^{-1}\right)\cdot\mathbf{P} \quad (2)$$

If the two transformations are equal and opposite rotations by an angle $\theta/2$ about a common direction **e**, the interface is a grain boundary separating grains misoriented by the angle θ. Equation (2) leads to the Frank formula (Frank, 1950)

$$\mathbf{B} = 2\sin(\theta/2)(\mathbf{P}\otimes\mathbf{e}) \quad (3)$$

---

[2] The most well known case is bcc metals (Christian, 1983; Duesbery, 1989) but other examples where screw dislocations are controlling are the prismatic slip in hexagonal metals (Vitek, 1985), anomalous yield behavior in $L1_2$ compounds such as Ni_3Al (Pope and Ezz, 1984) and deformation properties of other intermetallic compounds (Horton et al. 1995).

determining the dislocation content in this boundary. Since neither the choice of the reference lattice nor the choice of the transformation matrix are unique the dislocation content representing a given interface is not unique[3]. Hence, the dislocation model is a geometrical concept and cannot be directly used to analyze the structure and energy of interfaces. However, certain choices of dislocation contents may have a special physical significance. For example, in some cases the short range elastic field associated with the interface may be identified with the elastic field of a specific dislocation network that satisfies equation (2). In this case the part of the interfacial energy stored in the elastic field can be evaluated using such model. The most common example is the Read-Shockley model of low angle grain boundaries which are regarded as networks of lattice dislocations (Read and Shockley, 1950).

In general, grain boundaries may not possess any periodicities. However, boundaries with periodic structures have frequently been observed and, at least in cubic materials, any grain boundary may be approximated as precisely as desired by a boundary with the periodic structure. This then also allows application of periodic boundary conditions in atomistic studies of grain boundaries. Such periodic boundaries can be fully characterized and geometrically constructed using the concept of the coincidence site lattice (CSL), first introduced by Brandon et al (1964).

In order to construct a CSL we consider two interpenetrating ideal lattices which are misoriented with respect to each other and filling all space. In the cubic case, for certain relative orientations a three-dimensional lattice of coinciding lattice points is formed. This lattice is the CSL and it is characterized by the reciprocal density of the coincidence sites, Σ, which is the ratio of the crystal lattice sites to the coincidence sites found in a unit volume. While most misorientations do not lead to coincidence there is an infinite number of coincidence misorientations and their relation to all misorientations is the same as the relation of integers to all rational numbers (Warrington and Grimmer, 1974). Hence, any misorientation can be represented as closely as required by a coincidence one. For coincidence misorientations, if the boundary plane is parallel to a rational plane of the CSL, the boundary will be two-dimensionally periodic with the repeat cell coinciding with that of this CSL plane. The concept of coincidence was further generalized to include not only lattice points but interior cell points (Bollman, 1970). These coincidences define the so called 0-lattice but discussion of this analysis is beyond the scope of this paper.

An important concept coupled with the CSL is the notion of DSC (displacement shift complete) dislocations (Bollman, 1970). The Burgers vector of these dislocations is any vector joining the lattice points of either of the interpenetrating crystals defining the CSL. Displacement by such vector of one of the lattices recreates the pattern of the interpenetrating lattices though the origin of the CSL may be shifted. These dislocations are not, in general, lattice dislocations but for a given Σ the corresponding DSC dislocations may be present in grain boundaries since the structure far from them will not be altered by their presence; however, DSC dislocations may introduce steps into the boundaries (King and Smith, 1980). A boundary with the misorientation slightly deviating away from a coincidence Σ, can be regarded as the boundary with the misorientation

[3] For example, the transformation $\mathbf{R}$ can be combined with any homogeneous lattice invariant shear, $\mathbf{U}$, which does not alter the crystal lattice. The matrix $\mathbf{U} \cdot \mathbf{R}$ leads to the same transformation as the matrix $\mathbf{R}$ but when used in (2) a different value of $\mathbf{B}$ is obtained.

corresponding to this value of Σ containing a network of relevant DSC dislocations which accommodate the additional misorientation, similarly as do the lattice dislocations in the case of low angle boundaries. Their content in a boundary satisfies equation (3) where the angle θ is now the additional misorientation away from the coincidence Σ.

More detailed descriptions and analyses of the geometrical concepts described briefly in this section can be found in (Sutton, 1984; Sutton and Balluffi, 1995).

METHODS OF COMPUTER MODELING

Computer simulation of the atomic structure of a lattice defect always involves minimization of either the energy or free energy of the system of N particles with respect to their positions and, possibly, some internal coordinates (e. g. relative positions of atoms within molecules). Hence, the most important precursor of such calculations is an analysis determining the total energy of the system as a function of positions of the particles. This is discussed in the following section and at this stage we assume that such function is known.

Initially, a block of atoms is built in the computer and the defect studied is introduced into the center of the block geometrically. For example, in the case of a dislocation all the atoms of the block are initially displaced in accordance with the displacement field given by equation (1), assuming that the dislocation line passes through the center of the block. In the case of a grain boundary two appropriately misoriented blocks of ideal lattice are joined along the chosen plane to form a grain boundary in the center of the block. The block is then divided into two regions as shown schematically in Fig. 1 for the case of a dislocation and a grain boundary, respectively.

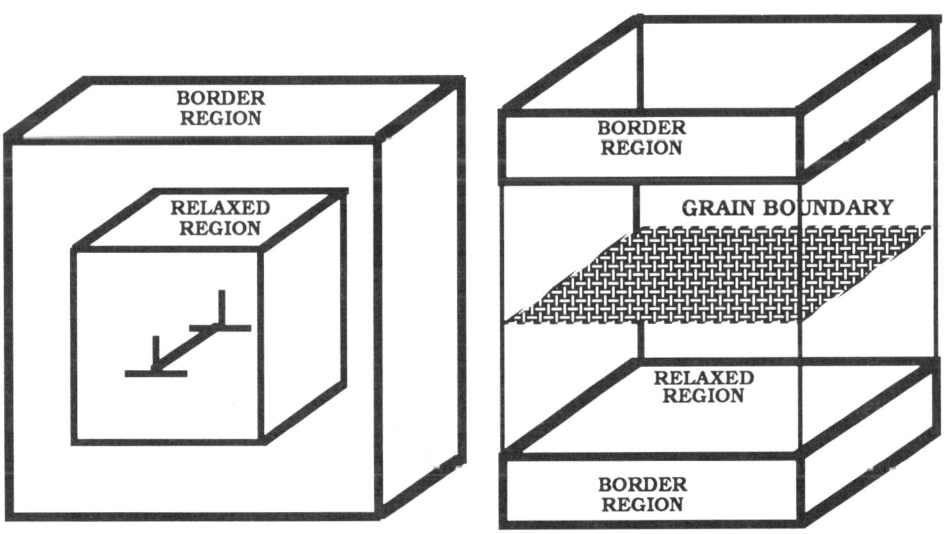

Fig. 1. Schematic picture of the blocks used when modeling a dislocation and a grain boundary, respectively.

Atoms in the relaxed region, which contains the *core of the defect*, are fully relaxed with respect to their positions using one of the methods described below. However, the border region may be treated in several different ways:

(i) No atoms are present in this region and thus the relaxed block is a cluster with free surfaces. In this case calculation of the free surface is carried out at the same time and the block must be large enough so that the surface does not affect the defect studied.

(ii) Atoms in this region are at fixed positions. In the case of a dislocation these positions are determined by its long-range elastic field and in the case of a grain boundary they are the ideal lattice positions corresponding to the adjoining grains. However, relative displacements of the two grains, both perpendicular and parallel to the boundary, must be allowed in order to attain full equilibrium in which there are no tensile stresses perpendicular and shear stresses parallel to the boundary. This allows for the expansion and/or contraction at the boundary and for relative rigid body displacements of the adjoining grains parallel to the boundary.

(iii) Positions of the atoms in this region are adjusted to follow the relaxation in the core of the defect but not using the atomistic relaxation algorithm. An example is the adjustment of the long range elastic field of the dislocation studied in accordance with a possible splitting of this dislocation into partials which may occur in the relaxed region. In general, continuum mechanics methods are used to adjust the positions of atoms in the border region in order to assure the continuity of displacements and tractions at the transition between the relaxed and border regions. A recent development which has been applied in atomistic studies of cracks, employs an overlap of the relaxed and border regions and the finite element method is used to evaluate the displacement field in the border region (Kohlhoff et al., 1991).

In studies of both dislocations and grain boundaries some periodic boundary conditions are usually employed. In the case of a straight dislocation with its line parallel to a crystallographic direction, the periodicity along this direction is applied. However, no periodicities are applied in the directions perpendicular to the dislocation line. In the case of a grain boundary, periodicity is imposed in two directions parallel to the boundary plane. The repeat cell is either the primitive cell determined by the crystal periodicity, for example the repeat cell of the CSL in the boundary plane, or a multiple of the primitive cell. In the latter case possible reconstructions of the boundary can be investigated. No periodicity is applied perpendicular to the boundary.

In most structural studies of dislocations and grain boundaries the molecular statics was used as the relaxation technique. The simplest version of this method is the steepest descent where the displacement of an atom is proportional to the force acting on it. However, mathematically more advanced versions, the conjugate gradient (Adby and Dempster, 1974) and the variable metric technique (Fletcher and Powell, 1963) have also been used frequently. It should be emphasized that in the case of grain boundaries the relaxation with respect to both the relative rigid body displacements and individual atom positions must be carried out. This can be conveniently performed simultaneously, as described by Sutton and Vitek (1983a).

Classical molecular dynamics and Monte Carlo techniques have also been used in structural studies. Since no algorithm can guarantee that global minimum is found, it is likely that relaxations become trapped in a variety of local minima in complex systems. The molecular dynamics performed at a relatively high temperature, but below melting, is then most likely to attain the global minimum. However, the main use of these techniques is in studies where thermally induced phenomena such as diffusion (e. g. Ma et al., 1993), melting (e. g. Nguyen et al., 1992; Besold and Mouritsen, 1994) and phase transformations in the boundary region (e. g. Pontikis, 1988) are investigated. However, it has been proposed recently (LeSar et al., 1989; Sutton, 1989b, 1992), that the entropic effects can be included explicitly via a quasi-harmonic approximation and static relaxations can then be carried out by minimizing the corresponding free energy. A more detailed assessment of this approach has recently been made by Hairie et al. (1994).

The Monte Carlo (MC) technique is particularly important in studies of segregation to grain boundaries which occurs commonly in alloys and is one of the most important interfacial phenomena. An efficient technique developed by Foiles (1985) (see also Foiles and Daw, 1987) utilizes a modified grand canonical ensemble in which the total number of atoms (N), temperature (T), volume and the differences of the chemical potentials between the species, corresponding to a given bulk composition, are fixed. The last condition assures that while the ratios of the species in the block can be altered, the bulk composition far away from the boundary plane remains fixed. The dependence of the chemical potentials on composition (at a given temperature) needs to be determined prior to the grain boundary calculations. It can be found from a series of MC simulations made for the block without any grain boundary in which, for chosen values of the differences of chemical potentials, corresponding compositions are obtained as a result.

Considering a binary alloy composed of elements A and B we define $\Delta\mu=\mu_A-\mu_B$, where μ_A and μ_B are the chemical potentials of the species A and B, respectively. Within each step of the MC calculation an atom is picked randomly and subjected to two possible configurational changes. The first corresponds to a small random displacement from its current position and the second to the change into the alternate species. The change of the total internal energy, ΔE, associated with each MC step is then evaluated and the decision whether to retain this configurational change or not depends on the quantity $\eta = \Delta E \pm \Delta\mu$, where the plus sign applies when species A is replaced by the species B and the minus sign applies in the opposite case. If η is smaller than zero the configurational change is always retained but if it is larger than zero the configurational change is retained with the probability $\exp[-\eta/k_BT]$, where k_B is the Boltzmann constant and T the temperature.

INTERATOMIC FORCES IN ATOMISTIC STUDIES

The description of atomic interactions needed in atomistic modeling requires, in general, knowledge of the total energy of the system studied as a function of the positions of the atoms. In principle, this can be provided by ab initio electronic structure calculations based on the local density functional theory which are at present the state of the art in structural studies involving total energies (Pettifor and Cottrell, 1992). While such calculations are now used routinely in studies of the stability of crystal structures, their application to extended defects is still very limited. The main reason is that they are ordinarily restricted to a much smaller number of atoms (usually not more than fifty) than that needed for a realistic representation of an extended defect such as a dislocation or the interface[4]. Furthermore, these calculations are very computer intensive and studies of dislocations and interfaces frequently require extensive relaxations and investigations of many different configurations so that full ab initio studies are not feasible.

This is one of the reasons why most atomistic studies of dislocations and grain boundaries have been made using relatively simple descriptions of atomic interactions. For metallic materials, pair-potentials were most common in the early developments (Gehlen et al., 1972; Lee, 1981). However, in recent years the embedded atom method (EAM) (Daw and Baskes, 1984; Daw et al., 1993) and many body potentials of the Finnis-Sinclair type (Finnis and Sinclair, 1984) have been most popular. Both these schemes are empirical central-force descriptions of the total energy. For ionic solids pair-potentials and the shell model have been used most frequently (Catlow et al., 1982; Lewis and Catlow, 1988). Potentials including non-central interactions have recently been developed on the basis of the tight-binding method (Tersoff, 1988a, b; Carlsson, 1990, 1991a; Pettifor, 1991; Aoki et al., 1993; Aoki and Pettifor, 1994; Xu and Adams, 1994). These are applicable to both metals with some directional bonding (transition metals) and covalently bonded solids but they have not yet been fully explored in atomistic studies. Similarly, the full tight-binding method has only been used to a limited extent.

These treatments of atomic interactions may appear as inadequate from the point of view of pure physics but it must be recognized that the purpose of the majority of atomistic studies of extended lattice defects is to elucidate their atomic structure and atomic level properties in materials with given: (a) crystal structure, (b) elastic properties and, possibly, phonon spectra, (c) values of certain material parameters such as vacancy formation energies, stacking fault and anti-phase boundary energies and (d) in alloys, alloying and ordering energies. This is in contrast with ab initio studies the objective of which is to determine all these properties from first principles.

General Framework

We deal with a system composed of N particles the positions of which are characterized by vectors $\mathbf{r}_i$ (i=1,..,N). The total energy, E_{tot}, of this system is a function of positions $\mathbf{r}_i$, and may also be an explicit function of the volume of the system, V. Without any significant loss of generality, we can assume that this function depends only on the relative positions of the atoms forming the system, i. e. on the vectors $\mathbf{r}_{ij} = \mathbf{r}_i - \mathbf{r}_j$; this does not imply that E_{tot} is only a function of the separation of the atoms i and j, $r_{ij} = |\mathbf{r}_i - \mathbf{r}_j|$. While there are N(N-1)/2 such vectors there are, of course, only N independent vectors.

In equilibrium E_{tot} must satisfy the following requirements which in empirical schemes are satisfied when fitting the functional form of E_{tot}:

(i) The force acting on any particle is zero, i. e. $\partial E_{tot}/\partial r_i^\alpha = 0$, where r_i^α is the α component of the vector $\mathbf{r}_i$ (α = 1,2,3).

(ii) The structure must be stable with respect to the application of a small homogeneous strain tensor, $\varepsilon_{\alpha\beta}$. When expanding the total energy into a Taylor series with respect to $\varepsilon_{\alpha\beta}$ the linear term represents the tensor of internal stresses

[4] Algorithms utilizing parallel computer architectures are being developed which allow use of blocks containing up to one thousand atoms (Payne et al., 1992).

$$\sigma_{\alpha\beta} = \frac{1}{V}\sum_{\substack{i,k \\ k\neq i}} \frac{\partial E_{tot}}{\partial r_{ik}^{\alpha}} r_{ik}^{\beta} + \frac{\partial E_{tot}}{\partial V}\delta_{\alpha\beta} \tag{4}$$

and in equilibrium $\sigma_{\alpha\beta} = 0$. The hydrostatic component of this tensor, which may be the only non-zero component in the case of the cubic symmetry, represents the pressure. The quadratic term in the Taylor series determines the tensor of elastic moduli $C_{\alpha\beta\gamma\delta}$. In the case of centrosymmetric structures this is[5]

$$\begin{aligned} C_{\alpha\beta\gamma\delta} = \frac{1}{4V}\sum_{\substack{i,j,k,l \\ k\neq i, l\neq j}} &\left(\frac{\partial^2 E_{tot}}{\partial r_{ik}^{\alpha}\partial r_{jl}^{\gamma}} r_{ik}^{\beta} r_{jl}^{\delta} + \frac{\partial^2 E_{tot}}{\partial r_{ik}^{\beta}\partial r_{jl}^{\gamma}} r_{ik}^{\alpha} r_{jl}^{\delta} + \frac{\partial^2 E_{tot}}{\partial r_{ik}^{\alpha}\partial r_{jl}^{\delta}} r_{ik}^{\beta} r_{jl}^{\gamma} + \frac{\partial^2 E_{tot}}{\partial r_{ik}^{\beta}\partial r_{jl}^{\delta}} r_{ik}^{\alpha} r_{jl}^{\gamma} \right) \\ &+ \frac{\partial E_{tot}}{\partial V}\left\{ \tfrac{1}{4}[\delta_{\alpha\beta}\delta_{\gamma\delta}(1-\delta_{\alpha\gamma}) + \delta_{\beta\alpha}\delta_{\gamma\delta}(1-\delta_{\beta\gamma}) + \delta_{\alpha\beta}\delta_{\delta\gamma}(1-\delta_{\alpha\delta}) + \delta_{\beta\alpha}\delta_{\delta\gamma}(1-\delta_{\beta\delta})] \right. \\ &\left. - \tfrac{1}{2}(\delta_{\alpha\gamma}\delta_{\beta\delta} + \delta_{\beta\gamma}\delta_{\alpha\delta})(1-\delta_{\alpha\beta})] \right\} + V\frac{\partial^2 E_{tot}}{\partial V^2}\delta_{\alpha\beta}\delta_{\gamma\delta} \end{aligned} \tag{5}$$

For non-centrosymmetric structures the auxiliary condition of zero forces on each atom of the strained medium leads to additional relaxations of atomic positions after application of the strain, and to an additional relaxation term in elastic moduli; however, the internal stresses are not affected by this relaxation (Born and Huang, 1954; Martin, 1975a, b). In equilibrium the tensor $C_{\alpha\beta\gamma\delta}$ must be positively definite.
(iii) The condition (ii) assures stability of the structure with respect to small homogeneous strains. A stricter condition is the requirement that all the phonon frequencies are real but it is usually sufficient to fulfill the condition (ii).
(iv) When studying dislocations and grain boundaries large distortions of the local environment away from equilibrium are present. Hence, it is important that the ideal lattice structure is stable with respect to large distortions. While this can never be fully examined, the most important tests are stability with respect to large shears parallel to some low index crystal planes and stability with respect to some large homogeneous strains. An example is tetragonal straining by gradually increasing the tetragonal c/a ratio that is equal to one for the fcc lattice and $1/\sqrt{2}$ for the bcc lattice (see, for example, Šob et al., 1992).
(v) Finally, the stability of the appropriate crystal structure relative to alternate structures with different symmetry or chemical order needs to be assured. In the case of alloys the basic features of the phase diagram should be reproduced.

If these conditions are satisfied unphysical structural or ordering instabilities are not likely to occur in the cores of lattice defects. Nevertheless, the validity of the structural features of lattice defects found in calculations using approximate empirical schemes for E_{tot} is best guaranteed if they can be related to fitted material properties and are not sensitively dependent on the details of the fittings and functional forms employed (Vitek and De Hosson, 1986).

---

[5] The second and third terms arrise because the volume change associated with the strain $\varepsilon_{\alpha\beta}$ is $\Delta V = V\left(\sum_{\alpha,\beta=1}^{3} \varepsilon_{\alpha\beta}\delta_{\alpha\beta} - \frac{1}{2}\sum_{\alpha,\beta,\gamma,\delta=1}^{3} \varepsilon_{\alpha\beta}\varepsilon_{\gamma\delta}\delta_{\alpha\gamma}\delta_{\beta\delta}(1-\delta_{\alpha\beta}) + \frac{1}{2}\sum_{\alpha,\beta,\gamma,\delta=1}^{3} \varepsilon_{\alpha\beta}\varepsilon_{\gamma\delta}\delta_{\alpha\beta}\delta_{\gamma\delta}(1-\delta_{\alpha\gamma}) \right)$.

Pair Potentials, EAM and Many-body Central Force Potentials for Metallic Systems

In all these schemes

$$E_{tot} = \frac{1}{2} \sum_{i \neq j=1}^{N} V_{ij}(r_{ij}) + U(\rho) \tag{6}$$

V_{ij} is the pair potential describing the part of the interaction between the atoms i and j that is strongly repulsive at short separations of these atoms. U(ρ) is the density dependent cohesive contribution to the energy. When only pair potentials describe the dependence of the energy on atom positions, ρ is the average density of the material. Such pair potentials belong to the class of *constant volume* potentials (Carlsson, 1990) and may reasonably describe the energy changes associated with the variation of atomic configurations at constant volume. For this reason most of the calculations using such potentials have been made at constant volume. Such potentials were derived from first principles for s- and p-bonded metals (e. g. alkali metals and aluminum) in the framework of the weak pseudopotentials (Harrison, 1980). They are relatively long-ranged and for large values of r exhibit so called Friedel oscillations of the form $\cos(2k_F r)/r^3$, where k_F is the Fermi vector. In this case ρ is the electron density but it is directly related to the material density. However, short-ranged empirical potentials of this type have frequently been used (Torrens, 1972; Gehlen et al., 1972; Lee, 1981).

The dependence of U on the average density of the material is a serious limitation since significant local changes of the density may occur in the cores of extended defects and this cannot be taken into account in pair potentials schemes. Furthermore, for non-cubic structures the conditions of the mechanical stability of the lattice cannot be usually attained in the pair potential framework. These problems have been alleviated to a significant extent in the EAM and many-body potentials. In the framework of these schemes which are to a great extent equivalent (Johnson, 1988),

$$U(\rho) = -\sum_{i=1}^{N} F_i(\rho_i) \tag{7}$$

where F_i is the many body and/or embedding part of the energy associated with the atom i. It is a function of a quantity ρ_i that is either equal or related to the electron density at the site i. The form of the function F_i is chosen on the basis of more fundamental physical considerations. In the EAM F_i is determined using arguments based on the local density functional theory (Daw and Baskes, 1984; Daw et al., 1993). In the case of Finnis-Sinclair potentials, which are based on a second-moment approximation to the density of states in the tight-binding theory incorporating charge conservation, $F_i = \sqrt{\ }$ (Finnis and Sinclair, 1984; Ackland et al., 1988). In both schemes

$$\rho_i = \sum_{j} \Phi_j(R_{ij}) \tag{8}$$

where Φ_j are short-ranged pair-potentials, generally dependent on species i and j. The physical interpretations of these potentials somewhat differ in the two schemes but both V and Φ, and in EAM also F, are fitted empirically to reproduce correct equilibrium lattice structure, cohesive energy, elastic moduli and other experimentally determined equilibrium properties and/or quantities obtain from ab initio calculations (Daw and

Baskes, 1984; Ackland and Thetford, 1987; Ackland et al., 1987; Baskes, 1992; Daw et al., 1993); in alloys alloying energies and other alloy properties are also fitted (Ackland and Vitek, 1990; Yan et al., 1993).

Ionic Solids

In ionic solids the cohesion arises from Coulomb interactions between anions and cations forming the solid and, therefore, the main part of the total energy originates in the Coulombic electrostatic energy. However, at small separations a strong repulsion takes place even between ions with opposite charges owing to the Pauli repulsion of the electrons. This repulsion is usually described empirically by a Born- Mayer potential (Born and Mayer, 1932). The total energy of an ionic solid is then given as

$$E_{tot} = \frac{1}{2}\sum_{\substack{i,j \\ i \neq j}} \left[\frac{Z_i Z_j}{r_{ij}} + A_{ij} \exp\left(-\frac{r}{\rho_{ij}} \right) \right] \tag{9}$$

where Z_i is the charge associated with the ion i and A_{ij} and ρ_{ij} are parameters of the Born-Mayer potential describing the interaction between ions i and j. In some studies additional short range term, $-C_{ij}/r_{ij}^6$, corresponding to a weak Van der Waals attraction, has been added to (9).

When the energy of an ionic system is given by equation (9) we are dealing with the so called rigid ion system in which no internal polarization of the ions is considered. The latter needs to be introduced when the model is to reproduce not only cohesive but also dielectric properties of ionic solids which is essential, for example, when defects possessing charges are studied. This is a common phenomenon in ionic solids and both dislocations and interfaces can be charged (Whitworth, 1975; Hirth and Lothe, 1982).

The polarization of individual ions is usually treated in the framework of the shell model first introduced by Dick and Overhauser (1958). In this model the charge of each ion is divided into two components: A charge of the core, Z^c, which has the same mass as the ion, and the charge, Z^s, of a massless particle called the 'shell'. The core and the shell are coupled by a harmonic spring with the force constant κ so that the force acting between them depends only on the magnitude of their relative displacement. Hence, in the shell model the following interactions need to be considered: (i) Shell-shell, shell-core and core-core Coulomb interactions. (ii) Short range shell - shell interactions described by the Born- Mayer, and possibly the Van der Waals potentials as in equation (9). No short range core - core and core - shell interactions are assumed. (iii) Core - shell interactions inside a particular ion modeled by a spring with the force constant κ.

Within the shell model the polarizability of an ion[6] is $\alpha = Z_s^2/\kappa$ and the polarizabilities of the ions forming the solid are related to the low and high frequency dielectric constants and the frequency of transverse optical phonons of the material (Brown, 1967). These relations are used to determine for a given ion the shell charge Z^s and the spring constant κ (see, for example, Dick and Overhauser, 1958; Sangster et al., 1978; Sangster and Stoneham, 1981). The parameters of the potentials describing short range interactions are determined similarly as in the case of potentials for metals by fitting the equilibrium lattice parameter, elastic

[6] If $\mathbf{p}$ is the dipole moment associated with an ion then the polarizability of this ion, α, is defined by the relation $\mathbf{p} = \alpha\mathbf{E}_{loc}$ where $\mathbf{E}_{loc}$ is the local electric field at the ion position.

moduli and other equilibrium properties and/or quantities obtained from ab initio calculations.

An important part of any atomistic study of ionic solids is the treatment of the Coulomb interactions. The direct summation converges very slowly and Ewald type methods (see, for example, Kittel, 1986), which transform the sum into rapidly convergent series, have to be employed. This summation method was originally developed for three-dimensionally periodic structures but in studies of interfaces and dislocations we are dealing with two and one-dimensional periodicities, respectively. For two-dimensional periodicities Ewald type summation of the Coulomb energies has been developed by Parry (1975) and Watson et al. (1981). In this case Ewald summations are first done plane by plane in the planes parallel to the interface; interactions between the planes are then decreasing exponentially with their separation. A similar approach has been developed for the case of one-dimensional periodicity by Puls and Norgett (1976) where Ewald summations are carried out row by row in the atomic rows parallel to the periodic direction.

Both the rigid ion model and the shell model are purely pair-potential models for the total energy. Such model implies the Cauchy relations between elastic moduli ($C_{12} = C_{44}$ for cubic symmetry) and these are not satisfied by most ionic solids. At the same time the model is not capable to reproduce correctly the optical phonons. A modification of the shell model, called breathing shell model, has been developed in which non-spherical deformation of the shell, i. e. an orientation dependence of the link between the shell and the core, is incorporated (Sangster, 1973; Catlow et al., 1976).

Covalent and Mixed Metallic-Covalent Bonding

The main feature of the descriptions of interatomic forces which consider the covalent character of bonding is that the total energy depends not only on the separations of the atoms but also on the angles between pairs of bonds. This is obviously essential in pure covalently bonded solids, such as silicon, but may also be important in transition metals and alloys containing transition metals (Pettifor et al., 1995). The approximate approach which can be employed ubiquitously to treat this problem is the tight binding method with the two-center orthogonal approximation (Harrison, 1980; Pettifor, 1992). In this case the Hamiltonian for a system composed of N atoms is written as

$$H_{TB} = \sum_{i,j=1}^{N} \sum_{\Gamma,\Lambda} h_{ij}^{\Gamma\Lambda} |i,\Gamma\rangle\langle j,\Lambda| \tag{10}$$

where $|i,\Gamma\rangle$ represents an atomic-like orbital of symmetry Γ at the site i and $h_{ij}^{\Gamma\Lambda}$ is the hopping integral of the Hamiltonian which couples the Γ-orbital at the site i with the Λ-orbital at the site j. The summations extend over all atomic sites and orbitals, respectively. In tight-binding schemes the atomic potentials are assumed fixed and thus an implicit assumption is made that the hopping integrals are transferable, with an appropriate scaling, from the ideal crystal to the defect environment. They are the Slater-Koster matrix elements of the Hamiltonian (Slater and Koster, 1954) corresponding to the orbitals employed. In transition metals with partially filled d-band the attractive part of the total energy originates principally from the formation of this band (Friedel, 1969; Pettifor, 1983). Hence, d-orbitals are used when studying transition metals while p-orbitals are utilized for covalently bonded semiconductors such as Si and Ge (Harrison, 1980).

Within the tight-binding framework the total energy of the system can be divided into two parts: a repulsive pair-potential part and an attractive bond energy part (Finnis et al., 1988; Sutton et al., 1988; Pettifor, 1990) defined as

$$U_{bond} = 2\sum_{i=1}^{N}\int_{-\infty}^{E_F}(E - E_i)\,\rho_i(E)\,dE \tag{11}$$

where $\rho_i(E)$ is the local density of electronic states (LDOS) associated with the atom i and E_i is the effective atomic energy level at site i. For a chosen form of the hopping integrals the bond energy of the system can be found after solving the one-electron eigenvalue problem for the Hamiltonian (10)[7]. This is the standard tight-binding calculation which has to be carried out employing the **k**-space representation and thus the system studied must be three-dimensionally periodic. This is particularly inconvenient when studying extended defects where the three-dimensional periodicity has to be imposed artificially. For this reason several approximate schemes, based on the tight-binding approach, have been developed in which the total energy is evaluated in the real space without resorting to **k**-space calculations (Carlsson, 1990, 1991a, b; Moriarty, 1990a; Pettifor, 1990, 1991; Moriarty and Phillips, 1991; Alinaghian et al., 1993; Aoki et al., 1993; Aoki and Pettifor, 1994; Xu and Adams, 1994).

The bond-order potentials advanced by Pettifor and his co-workers (Pettifor, 1990, 1991; Alinaghian et al., 1993; Aoki et al., 1993; Aoki and Pettifor, 1994) are one such scheme. In this approach U_{bond}, and the force on atom k arising due to U_{bond}, can be written as

$$U_{bond} = \sum_{\substack{i,j=1 \\ i\neq j}}^{N}\sum_{\Gamma,\Lambda} h_{ij}^{\Gamma\Lambda}\,\Theta_{i\Gamma,j\Lambda} \text{ and } F_k^{\alpha} = -\sum_{\substack{i,j=1 \\ i\neq j}}^{N}\sum_{\Gamma,\Lambda}\frac{\partial h_{ij}^{\Gamma\Lambda}}{\partial r_k^{\alpha}}\Theta_{i\Gamma,j\Lambda} \tag{12}$$

provided the orbital charge neutrality is satisfied. $\Theta_{i\Gamma,j\Lambda}$ is the bond order which is equal to the difference between the numbers of electrons in the bonding and anti-bonding states corresponding to the orbitals $|i,\Gamma\rangle$ and $|j,\Lambda\rangle$. It is, of course, dependent in a complex way on the hopping integrals $h_{ij}^{\Gamma\Lambda}$. When these are known, the main problem is to evaluate the bond order for a given configuration of atoms. In the framework of bond order potentials this is done numerically using the recursion scheme since $\Theta_{i\Gamma,j\Lambda}$ can be expressed by means of the Green's function of the Schrödinger equation corresponding to the Hamiltonian (10) (Alinaghian et al., 1993, 1994; Aoki et al., 1993; Aoki and Pettifor, 1994).

The bond order potentials can be used when studying d-bonded transition metals (Aoki and Pettifor, 1994) as well as sp-bonded semiconductors (Alinaghian et al., 1993, 1994). For silicon empirical potentials of the bond-order type have been proposed earlier by Tersoff (1988a, b) who suggested suitable empirical formulas for both $h_{ij}^{\Gamma\Lambda}$ and $\Theta_{i\Gamma,j\Lambda}$ that are functions of the separation of atoms as well as angles between pairs of bonds. These functions, together with a repulsive pair potential, were constructed so as to reproduce the dependence of the cohesive energy of Si on the coordination found in ab initio pseudopotential calculations (Yin and Cohen, 1982, 1984). These

[7] The pair potential, decribing a part of the total energy, has usually been fitted so as to reproduce the correct equilibrium density and the Bulk modulus.

potentials have superseded several angularly dependent potentials for Si proposed earlier (Keating, 1966; Stillinger and Weber, 1985).

A somewhat different approach to a real space description of the energy of transition metals that includes the dependence on angles between pairs of bonds, has been proposed by Carlsson (1991a, b). It is based on the fact that the LDOS entering equation (11) is fully determined by the complete set of its scalar moments referred to its center of gravity, E_i, (Akhiezer, 1965) defined as

$$\mu_n^{(i)} = \int_{-\infty}^{+\infty} (E - E_i)^n \rho_i(E)\, dE = \sum_{j_1, j_2, \ldots, j_{n-1}} \sum_{\Gamma, \Gamma_1, \Gamma_2, \ldots, \Gamma_{n-1}} h_{ij_1}^{\Gamma\Gamma_1} h_{j_1 j_2}^{\Gamma_1\Gamma_2} h_{j_2 j_3}^{\Gamma_2\Gamma_3} \ldots h_{j_{n-2} j_{n-1}}^{\Gamma_{n-2}\Gamma_{n-1}} h_{j_{n-1} i}^{\Gamma_{n-1}\Gamma} \tag{13}$$

for the tight-binding Hamiltonian (10); the summations extend over all the atoms j_k and atomic orbitals Γ_k but excluding diagonal elements of the type $h_{ll}^{\Gamma\Lambda}$. If only a finite number of moments are taken into consideration it represents a certain approximation to the shape of the LDOS and the bond energy can be evaluated in this approximation (Cyrot-Lackmann, 1967; Ducastelle and Cyrot-Lackmann, 1970).

The simplest is the second moment approximation in which the LDOS has one central peak and only its width affects the energy. This approximation forms a physical basis for the central force Finnis-Sinclair type potentials (Ackland et al., 1988) discussed above. The fourth moment, or more precisely, the dimensionless quantity $\gamma_4 = \mu_0(\mu_4\mu_2 - \mu_2^3 - \mu_3^2)/\mu_2^3$, characterizes different degrees of covalency of the DOS (Gaspar and Lambin, 1985). Hence, Carlsson (1991a, b) proposed to write the total energy as

$$E_{total} = E_{rep} + E_{bond}^{(2)} + E_{bond}^{(4)} \tag{14}$$

where

$$E_{bond}^{(4)} = B \sum_{i=1}^{N} \left(\frac{\mu_4^{(i)} - \mu_4^{(i)}(2)}{\left(\mu_2^{(i)}\right)^{3/2}} \right) \text{ and } E_{bond}^{(2)} = -\mathrm{Trace}\left[\left(\hat{\mu}_2^{(i)}\right)^{1/2}\right],$$

$\hat{\mu}_2^{\Gamma\Lambda(i)} = \sum_{j\neq i} \sum_{\Pi} h_{ij}^{\Gamma\Pi} h_{ji}^{\Pi\Lambda}$ and E_{rep} is the repulsive part described by a pair potential. $\mu_4^{(i)}(2)$ is the part of the fourth moment corresponding to the double hops between two atoms and is taken into account in the second moment contribution, $\mu_2^{(i)}$ is the scalar second moment and B is an adjustable parameter; the third moment was set equal to zero since only symmetrical LDOS, corresponding to the half-filled d-band, was considered. Unlike in the usual second moment approximation, the second moment is not taken as a scalar but a second moment matrix associated with atomic site i, $\hat{\mu}_2^{(i)}$, is used. Thus the angular dependence is included in both fourth and second moment terms. In this scheme the total energy, and thus the potentials, contain four adjustable parameters which were determined via a least-squares fit to a database consisting of the cohesive energy, equilibrium lattice constant and the three independent elastic moduli (Carlsson, 1991b).

DISLOCATIONS

General concepts

The Volterra dislocation is defined in the continuum theory via the cut, displace and glue process which does not allow for any other than linear elastic relaxations of the medium and stresses and strains are singular at the dislocation line. Therefore, the displacement is a step function in the plane of the cut when crossing the dislocation line (Nabarro, 1967; Hirth and Lothe, 1982). However, in reality inelastic relaxations will occur so that the displacement will vary continuously and the core will spread over a finite region. In the coordinate system with the x_3 axis parallel to the dislocation line, such core spreading can be described by a continuous distribution of dislocations with the density

$$\rho_{\alpha\beta}(x_1, x_2) = \frac{\partial u_\alpha}{\partial x_\beta} \quad (\alpha, \beta = 1,2,3) \tag{15.1}$$

where **u** is the displacement vector and

$$\sum_\beta \int_{-\infty}^{+\infty} \rho_{\alpha\beta}(x_1, x_2) dx_\beta = b_\alpha \tag{15.2}$$

where **b** is the Burgers vector of the dislocation. The reason for the core spreading is, of course, the decrease of the dislocation energy when it is redistributed into dislocations with smaller Burgers vectors. The dissociation into two or more well defined partial dislocations separated by a metastable stacking fault is, therefore, a special case of core spreading.

In crystals the core spreading is usually confined to certain crystallographic planes containing the dislocation line. If the core spreads into one such plane the core is *planar* while if it spreads into several non-parallel planes of the zone of the dislocation line it is a *non-planar* core. In general, the dislocation is glissile in the former case and moves easily in the plane of the core spreading, while in the latter case it is sessile. Examples of both glissile and sessile cores will be shown in the following. Since dislocation splitting is a special case of core spreading, an example of the planar core is dissociation of $1/2\langle 110\rangle$ dislocations in fcc materials into Shockley partials. The splitting and partial dislocation cores are confined into one $\{111\}$ plane and this is the reason why $\{111\}$ planes are the dominant slip planes in fcc metals. The first known example of the non-planar core is probably the Lommer-Cottrell lock corresponding to a dislocation with the Burgers vector $1/2\langle 011\rangle$ which splits into two Shockley partials on two different $\{111\}$ planes and a $1/6\langle 011\rangle$ stair-rod dislocation lying at the intersection of these planes (Nabarro, 1967; Hirth and Lothe, 1982). This dislocation is sessile and acts as a strong obstacle for other dislocations moving in the $\{111\}$ planes.

The first models of dislocation cores developed by Frenkel and Kontorova (1938), Peierls (1940) and Nabarro (1947) all make an implicit assumption that the core spreads into one plane which contains both the dislocation line and the Burgers vector. In the Peierls-Nabarro model, when the dislocation is parallel to x_3 and assuming that the core is spreading into the plane x_1, x_3, the core structure is determined by the integral equations

$$\sum_{\alpha=1,3} \int_{-\infty}^{+\infty} \frac{A^{\alpha\gamma}}{x_1 - x_1'} \rho_{\alpha 1}(x_1') dx_1' = F_\gamma(\mathbf{u}) \; (\gamma = 1,3) \tag{16}$$

where **F** is the periodic restoring force (per unit area) arising from the interaction of atoms on either side of the plane of core spreading when the part of the crystal above this plane is displaced relative to that below the plane by the vector **u**. $A^{\alpha\gamma}$ is the coefficient, depending on the elastic moduli, in the formula for the stress component, $\sigma_{\alpha\gamma}$, arising due to the dislocation parallel to the axis x_3 with the unit Burgers vector in the direction x_α. Hence, Equation (16) is a condition for the balance between the stresses induced by the continuous distribution of dislocations in the plane of the core spread and the restoring force **F**. In the original Peierls-Nabarro model **F** was taken as a sinusoidal function of the displacement in the direction of the total Burgers vector. However, in general this force is related to the energetics of shear faults in the plane of the core spreading.

Such faults are generalized stacking faults and can be defined as follows: We cut the crystal along the given crystal plane and displace the upper part with respect to the lower part by a vector **u**, parallel to the plane of the cut (see Fig. 2). The energy of such a fault, $\gamma(\mathbf{u})$, can be evaluated using an appropriate description of atomic interactions. In this calculation relaxation perpendicular to the fault must be allowed but no relaxations parallel to the fault are permitted. Repeating this procedure for various vectors **u** within the repeat cell of the given crystal plane, we obtain an energy-displacement surface which is commonly called the *γ-surface*. This concept was first introduced when searching for possible stacking faults in bcc metals (Vitek, 1968) since local minima on this surface determine possible metastable stacking faults on the crystal plane considered.

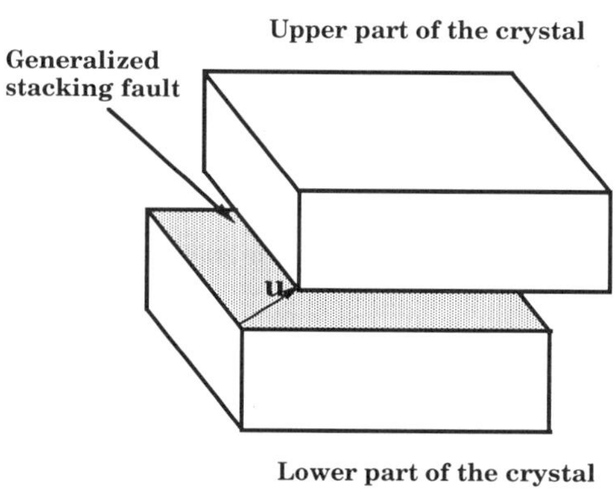

Fig. 2. Definition of the generalized stacking fault.

While all possible metastable stacking faults on a given crystal plane in a given material can only be found if the corresponding γ-surface is calculated, certain faults can be anticipated using the crystal symmetry. The symmetry arguments are based on the Neumann's principle (Neumann, 1885) which states that 'Any kind of symmetry which is possessed by the crystal structure of the material is also possessed by any physical property of this material'. For the generalized stacking fault this implies that if for a displacement vector, **u**, there is a mirror plane in the ideal lattice perpendicular to the plane of the fault, then the γ-surface will have a vanishing first derivative along the normal of this mirror plane. The γ-surface then possesses extrema (minima, maxima or inflexions) for such displacements for which there are at least two non-parallel mirror planes in the ideal lattice. Whether any of these extrema correspond to minima, and thus metastable faults, can often be decided by considering the first nearest neighbor interactions. Hence, by analyzing the symmetry properties of a γ-surface we can assess in advance the existence of possible metastable stacking faults. Such stacking faults are then common to all materials crystallizing in a given structure. However, for a particular material other minima than those associated with symmetry dictated extrema may exist. These cannot be anticipated on crystallographic grounds and their existence depends on the details of atomic interactions.

It follows from the definition of the γ-surface that $\mathbf{F(u)} = -\mathrm{grad}(\gamma(\mathbf{u}))$ and thus knowing the γ-surface the structure of the core spread in the plane of the cut can be found in the framework of the Peierls-Nabarro model using equation (16). The atomic structure enters into this model only indirectly via the restoring force **F** and the planarity of the core is a severe restriction of the model[8]. The main accomplishment of the model is the removal of the unphysical singularity of the Volterra dislocation in a physically consistent manner. This is the main reason for the recent revival of the model in the context of studies of emission of dislocations from cracks (Rice, 1992; Rice and Beltz, 1994). However, in studies of the effect of the dislocation cores upon plastic behavior this model is not adequate. First, the lattice friction stress (Peierls stress) evaluated in the framework of this model is (Nabarro, 1947; Christian and Vitek, 1970; Hirth and Lothe, 1982)

$$\tau_P = \frac{2G}{\alpha}\exp(-4\pi\zeta / b) \qquad (17)$$

where G is the shear modulus, α is a factor of the order of one and ζ is the width of the core. This stress is very small even when the core is very narrow; for example, if $\zeta = b$ $\tau_P \approx 10^{-5}G$. This is a general characteristics of the planar cores and, for example, in fcc metals where the cores are planar, spread into {111} planes, the Peierls stress is practically negligible.

The core effects are usually important only if the Peierls stress is high and represents a significant contribution to the yield stress. In metallic materials this usually means that the core is not planar and must undergo significant transformations before the dislocation can glide (Vitek, 1985, 1992; Duesbery, 1989; Duesbery and Richardson, 1991). In covalent solids the breaking and/or readjustment of the bonds in the core region may be responsible for a high lattice friction stress (Friedel, 1964; Hirth and Lothe, 1982) and in ionic solids the core charges strongly affect the dislocation mobility (Whitworth, 1975). These phenomena can only be investigated by direct atomistic modeling of dislocations. Notwithstanding, the γ-surfaces, which are relatively easily calculable properties of undislocated lattices, can provide a dependable guide to the nature of the dislocation core structure in all these cases (Duesbery, 1989; Duesbery and Richardson, 1991; Vitek et al., 1991c, 1995a; Vitek, 1992; Kaxiras and Duesbery, 1993; Joos et al., 1994) and, therefore, calculations of γ-surfaces should precede any detailed dislocation core studies.

Examples of atomistic studies

Body centered cubic metals. Some of the first atomistic studies of dislocations were done for bcc metals since their plastic behavior is strikingly different than that of fcc metals (Kubin, 1982; Christian, 1983; Duesbery, 1989). In the latter case the slip systems are always ⟨110⟩{111}, the critical resolved shear stress (crss) is small and essentially independent of orientation and temperature (of order $10^{-5}G$ at 4.2K) and

[8] With a general type of the restoring force only numerical solutions can be obtained but some universal conclusions about the shape of $\rho_{\alpha 1}$ determined by equation (16) can be drawn from knowledge of **F(u)** since from the mathematical point of view the relation (16) is a Hilbert transform between ρ and **F** (Kroupa and Lejček, 1972; Lejček, 1972; Vitek et al., 1972).

obeys the Schmid law[9]. In the former case the slip directions are the close-packed directions $\langle 111 \rangle$ but the slip surfaces, not always planar, vary with the orientation and sense of the applied stress. This clearly violates the Schmid law and suggests an intrinsic anisotropic resistance to the slip in these materials. The crss is strongly dependent on temperature, rising to values of order 10^{-2}G at 4.2K, and depends on the orientation of the applied stress. At this point it is noteworthy to mention that the main deformation features are common to alkali and transition metals (Duesbery, 1989) and have been found even in more exotic materials such as the molecular crystal hexamine (Dipersio and Escaig, 1977) and He_{III} (Sakai et al., 1979). This suggests that the main deformation phenomena are intrinsic to dislocations in the bcc lattice but not sensitively dependent on the details of bonding.

All distinctive aspects of the deformation behavior of bcc metals can be explained by the non-planar spreading of the cores of $1/2\langle 111 \rangle$ screw dislocations into several planes of the $\langle 111 \rangle$ zone, first proposed by Hirsch (1960). Already the early atomistic calculations, carried out using pair potentials, demonstrated this fundamental feature of the screw dislocation cores in bcc metals (Vitek et al., 1970; Basinski et al., 1971; Minami et al., 1972, 1974; Vitek, 1975). Furthermore, they revealed a complex response of these dislocations to applied stresses (Duesbery et al., 1973; Vitek, 1976; Duesbery and Basinski, 1993) and the intrinsic asymmetry of the cores connected with the asymmetry of shear in $\{112\}$ planes. This asymmetry is related to the well-known twinning-antitwinning asymmetry reflected in the tension-compression asymmetry of the yield stress.

Nevertheless, these model calculations have not been able to capture all the features of the dislocation cores, in particular those which are specific to certain metals rather than those common to all of them. The existence of such features is clearly reflected in the fact that the deformation behavior of all transition bcc metals is not identical. Large quantitative differences arise in the magnitudes of asymmetries of slip geometry and flow stress and qualitative differences occur in the tension-compression asymmetry which cannot be fully related to the twinning-antitwinning asymmetry of the core. Recently, the screw dislocation core structures were calculated using the Finnis-Sinclair type potentials for niobium and molybdenum, constructed by Ackland and Thetford (1987). The calculated core structures are shown in Figs, 3 and 4 using the method of differential displacements[10]. To facilitate understanding of these pictures the orientations of all the $\{110\}$ and $\{112\}$ planes belonging to the $[111]$ zone are shown in Fig. 5. Figs. 3a and 4a depict the screw components (parallel to $[111]$) while Figs. 3b and 4b depict the edge components (perpendicular to $[111]$). In the former case the length of the arrows is normalized by $|a/6[111]|$ and when the magnitude of the displacement is equal to this value the length of the corresponding arrow is equal to the separation of the neighboring atoms; in the latter case the vectors have been magnified by a factor of ten to make them clearly visible. While the main feature of the cores, the non-planar spreading, is the same

[9] It is not always recognized that the Schmid law contains two distinct assertions. The first, more widely known, is that plastic flow begins when the resolved shear stress on a possible slip system reaches a constant, critical value. The second part of the Schmid law states that this critical stress is not influenced by any other component of the applied stress tensor.

[10] The atomic arrangement is shown in the projection perpendicular to the direction of the dislocation line ($[111]$) where circles represent atoms within one period. The chosen component of the relative displacement of the neighboring atoms produced by the dislocation is depicted as an arrow between them. The length of the arrows is proportional to the magnitude of the displacements.

in both cases, it is clear that the calculated core structures shown in Figs. 3 and 4, respectively, are rather different.

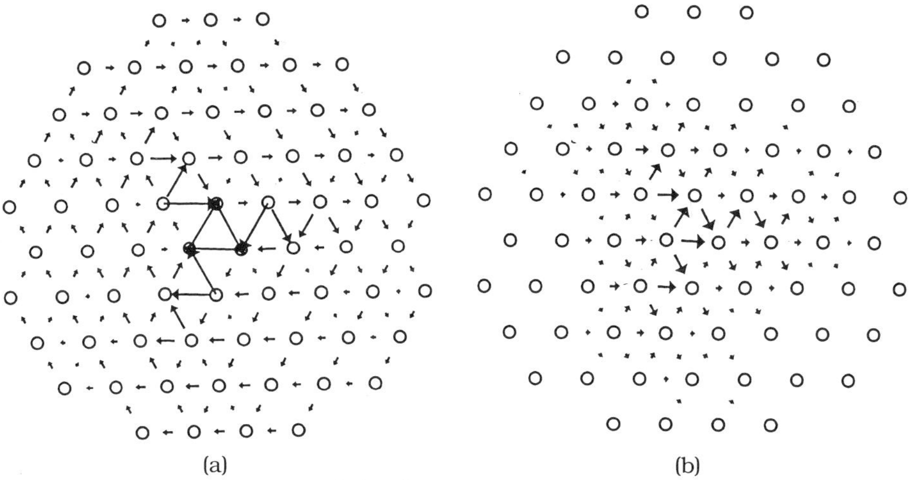

Fig. 3. Core structure of the 1/2[111] screw dislocation in molybdenum. (a) Screw component. (b) Edge component.

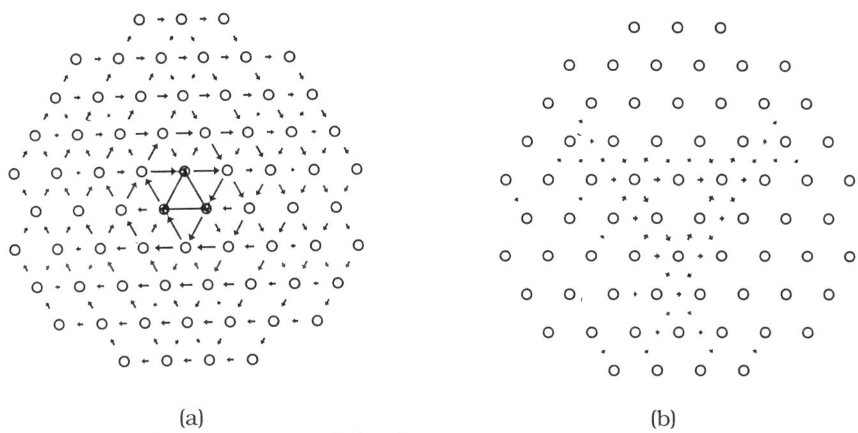

Fig. 4. Core structure of the 1/2[111] screw dislocation in niobium. (a) Screw component. (b) Edge component.

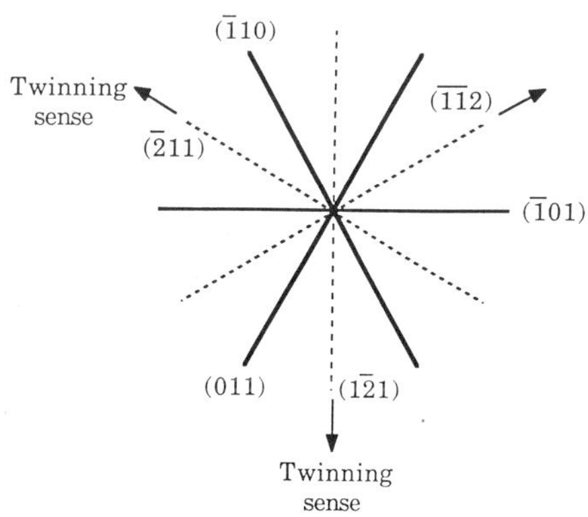

Fig. 5. Orientation of the {110} and {112} planes belonging to the [111] zone.

In the case of molybdenum the core is relatively widely spread into three {110} planes of the [111] zone with some further spreading into {112} planes which induces in these planes shears in the twinning sense. This is manifested in Fig. 3a by the fact that the arrows on those sides of the {110} planes where the corresponding {112} planes are sheared in the twinning sense, are larger than on the other sides. Furthermore, as seen from Fig. 3b, the displacements inside the core also possess significant edge components (the sum of which is, of course, zero). Since the core structure shown in Fig. 3 is not invariant with respect to the $[10\bar{1}]$ diad, a symmetry operation of the bcc lattice, another energetically equivalent configuration, related by this symmetry operation, must exist (for more details see Vitek, 1975). Hence, the dislocation core may exist in two symmetry related variants and, therefore, this core configuration is called a degenerate structure.

In contrast, in niobium the core structure is invariant with respect to all the symmetry operations of the lattice and thus this core structure is called non-degenerate. This core spreads symmetrically into the three {110} planes of the [111] zone but is much narrower than in the case of molybdenum. As seen from Fig. 4b, there are no significant edge components of the core displacements. Furthermore, twinning-antitwinning asymmetry is not visible in the core spreading but studies of the effect of external stresses on this core revealed this asymmetry in the orientation dependence of the crss (Vitek et al., 1995a).

Both the degenerate (Vitek, 1975; Duesbery, 1989) and non-degenerate (Rabier and Grilhe, 1973; Minami et al., 1974; Takeuchi and Kuramoto, 1975; Takeuchi, 1979; Van Heugten et al., 1979) core configurations were found in previous pair potentials studies. Calculations employing the scheme similar to the Finnis-Sinclair potentials found only non-degenerate cores (Masuda et al., 1981; Sato and Masuda, 1981). The results of the calculations presented here demonstrate that both the degenerate and non-degenerate states are viable forms for the bcc screw dislocation core structure and suggest specifically that in molybdenum the screw dislocations possess the former and in niobium the latter form. While the twinning-antitwinning asymmetry appears in both cases the edge component of the core displacements is present only in the degenerate case. Owing to this component the non-glide shear stresses (perpendicular to the Burgers vector) may have a strong influence on the dislocation core behavior and induce significant non-Schmid behavior (Vitek, 1992; Vitek et al., 1995a, b). The results presented here suggest that this effect will be pronounced in molybdenum while it will be weak or absent in niobium. The experimentally observed orientation dependence of the crss is, indeed, substantially less for niobium than for molybdenum (Bowen et al., 1967; Vesely, 1968; Duesbery and Foxall, 1969). However, in these calculations any angular character of bonding has been neglected while both niobium and molybdenum are transition metals with unfilled d-bands. Hence, only a study employing a scheme which includes the covalent character of d-bonding, such as the bond-order potentials discussed above, can unequivocally decide whether the suggested differences between niobium and molybdenum have been correctly predicted when using the Finnis-Sinclair potentials. Such calculations would at the same time provide a very valuable assessment of the limitations of the simpler schemes such as EAM and Finnis-Sinclair potentials.

Intermetallic compounds. Intermetallic compounds are very promising materials for high temperature structural applications owing to their high strength-to-weight ratio, high melting temperatures and very good oxidation resistance. However, a common serious limitation is that their room temperature ductility and toughness are usually very low. Since this and other features of their plastic behavior appear to be related to intrinsic properties of dislocations, atomic structure of dislocations in these alloys has been studied extensively in recent years (Yoo et al., 1993). The dislocation calculations were all made using either pair potentials or many-body central force potentials and EAM. However, ab initio calculations have been used extensively to calculate elastic moduli and fault energies and investigate relative stabilities of different crystal structures (Carlsson and Meschter, 1989; Fu and Yoo, 1990, 1992a, b; Hong and Freeman, 1991, 1992; Freeman et al., 1992; Woodward et al., 1993). These parameters have then been used in the construction of empirical potentials.

The compounds in which the atomic structure of dislocations was studied in most detail are the fcc based alloys crystallizing in the $L1_2$ (Cu_3Au) structure. The reason is that a number of these compounds, in particular nickel based alloys, such as Ni_3Al which is an important component of superalloys, display an anomalous yield behavior, i. e. an increase of the yield stress with increasing temperature (for reviews see Pope and Ezz, 1984; Dimiduk, 1991). This behavior is related to special features of the cores of $\langle 110 \rangle$ screw dislocations which glide in the anomalous regime on $\{111\}$ planes. In this case the $[\bar{1}01]$ superdislocation gliding on the (111) plane is dissociated into two superpartials with the $1/2[\bar{1}01]$ Burgers vectors separated by the antiphase boundary (APB). The core structure of these superpartials in screw orientation has been the object of a number of atomistic studies. Already the early calculations which were performed using pair potentials (Paidar et al., 1982; Yamaguchi et al., 1982) showed that there are two distinct metastable core configurations. These are presented in Figs. 6a and b employing the display method analogous to that used as in the case of screw dislocations in bcc metals; in this scheme rows of arrows of constant length mark planar faults such as stacking faults and APBs. The core shown in Fig. 6a is planar, spread in the plane of the APB, and in this form the dislocation is glissile. In contrast, the core shown in Fig. 6b is non-planar, extending principally onto the $(11\bar{1})$ plane and the dislocation is, therefore, sessile. The structure of the core on either the (111) or $(11\bar{1})$ plane resembles splitting into $1/6\langle 112 \rangle$ Shockley partials with the complex stacking fault (CSF) between them. Very similar results were obtained using many-body Finnis-Sinclair type potentials (Vitek et al., 1991b, c, d) and the EAM (Yoo et al., 1989; Pasianot et al., 1990, 1991; Parthasarathy et al., 1991).

Understanding that the $1/2[\bar{1}01]$ superpartials may possess both sessile and glissile configurations is the corner stone of our comprehension of the anomalous yield behavior in compounds like Ni_3Al. Assuming that the dislocations are generated in the glissile form in which they can glide away from the source, they will be immobilized by transforming from the glissile into sessile form since the sessile configuration is energetically favored. These transformations may be thermally assisted and thus immobilization will be increasing with increasing temperature, leading to the anomalous increase of the yield stress. This analysis of the anomalous yielding was developed in a number of studies (Takeuchi and Kuramoto, 1973; Paidar et al., 1984; Khantha et al., 1992a, b) which explain not only the anomalous temperature dependence of the yield stress but also its orientation dependencies and many other features of the anomalous regime (Vitek et al., 1995b).

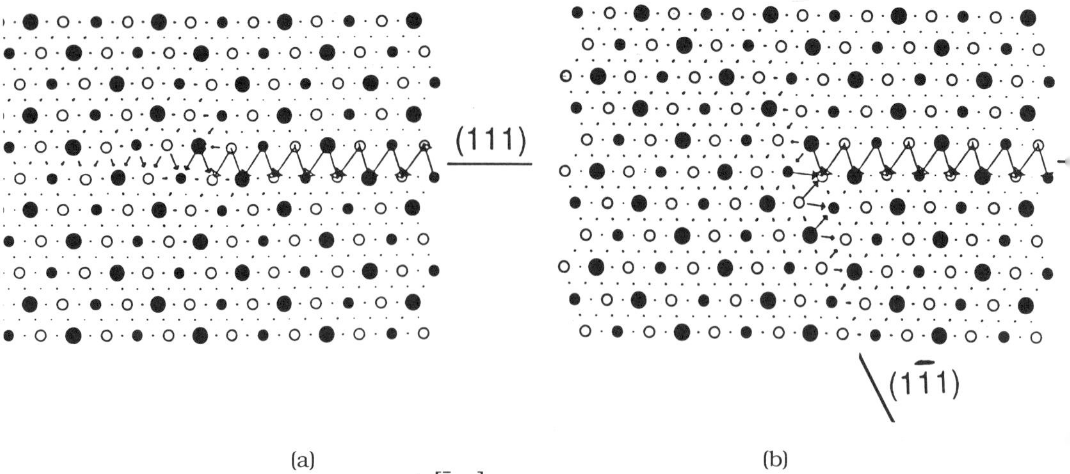

Fig. 6. Core structure of the $1/2[\bar{1}01]$ screw superpartial calculated using many-body potentials of the Finnis-Sinclair type for Ni_3Al. (a) Glissile core. (b) Sessile core.

Atomistic studies of γ-surfaces and dislocations in $L1_2$ compounds also revealed that another class of these alloys may exist in which a normal rather than an anomalous temperature dependence should be observed. First, the analysis of the γ-surfaces shows that stability of the APB and CSF on $\{111\}$ planes is not guaranteed by symmetry and thus there may be compounds where such faults are unstable (Yamaguchi et al., 1981; Vitek et al., 1991b, c, d). In this case dislocations will split on $\{111\}$ planes into $1/3\langle 112\rangle$ superpartials separated by the superlattice intrinsic stacking fault, or into $1/2\langle 110\rangle$ superpartials on $\{001\}$ planes, separated by the APB. The cores of all these superpartials were found to be non-planar, spread into two intersecting $\{111\}$ planes (Vitek et al., 1991b, c, d). Such dislocations are always sessile and the yield stress can be expected to be high at low temperatures and decrease with increasing temperature (Tichy et al., 1986a, b). Such behavior has, indeed, been found in Pt_3Al (Wee et al., 1980, 1984; Heredia et al., 1989) and in Al_3Ti transformed into the $L1_2$ structure (Wu et al., 1990, 1994).

Another compound which is an important candidate for high-temperature structural applications is NiAl which crystallizes in bcc based B2 structure (Darolia et al., 1992). The dominant observed slip system in this alloy is $\langle 001\rangle\{100\}$ but slip systems $\langle 001\rangle\{110\}$ and $\langle 110\rangle$ $\{110\}$ have also been observed. The latter system is particularly active for $\langle 001\rangle$ orientations of the tensile/compression axis since in this case the slip in the $\langle 001\rangle$ direction cannot be activated. However, the low temperature ductility of this alloy is very limited and this hinders its utilization (Forbes et al., 1993; Baker, 1995). Again, this seems to be related to an intrinsically low mobility of dislocations in this compound and, therefore, a number of atomistic studies of the core structure of dislocations in NiAl have been carried out in recent years.

All the recent calculations were performed using EAM potentials constructed for NiAl (Farkas and Vailhe, 1993; Mills et al., 1993; Parthasarathy et al., 1993; Xie et al., 1993; Pasianot et al., 1994). Screw dislocation with the $\langle 001\rangle$ Burgers vector was found to spread into $\{110\}$ planes. This spreading may be planar (Parthasarathy et al., 1993) or non-planar as observed in earlier pair-potential calculations of Yamaguchi and Umakoshi (1975). However, as pointed out by Farkas et al. (1991) the $\langle 001\rangle$ screw dislocations may be elastically unstable in NiAl and exhibit, therefore, a serrated morphology so that screw orientations rarely exist.

For the edge dislocation with the $\langle 001 \rangle$ Burgers vector lying in the $\{110\}$ plane the core was found to be planar while the same edge dislocation with the $\{100\}$ slip plane possesses a non-planar core spread into two $\{110\}$ planes. This implies that the motion of $\langle 001 \rangle$ dislocations will be easier on $\{110\}$ than on $\{100\}$ planes. This was substantiated by atomistic calculations studying the motion of these dislocations under the effect of applied stresses (Parthasarathy et al., 1993) and it agrees with experimental observations.

The cores of both screw and edge dislocations with the $\langle 110 \rangle$ Burgers vector were found to be non-planar with the Peierls stress substantially higher than that of the $\langle 001 \rangle$ dislocations (Parthasarathy et al., 1993). This suggests that if the slip is carried out by $\langle 110 \rangle$ dislocations the ductility will be much lower than when $\langle 001 \rangle$ dislocations operate. This again agrees with observations; in fact, the orientation for which $\langle 001 \rangle$ slip does not operate is commonly called hard. Furthermore, Mills and co-workers (Mills et al., 1993, 1995; Mills and Miracle, 1993) found in their EAM calculations, and confirmed by the electron microscopic observations, that the more stable core of the $\langle 110 \rangle$ edge dislocation corresponds to the splitting into two non-coplanar $\langle 010 \rangle$ dislocations. This dissociation renders the $\langle 110 \rangle$ edge dislocation even less mobile.

These examples demonstrate that calculations employing many-body central force potentials reveal a number of features of the dislocation cores both in Ni_3Al and NiAl that explain many observed characteristics of the plastic deformation of these compounds. This suggests that in these alloys bonding has only very weak covalent character in spite of the fact that nickel is a transition element with unfilled d-band, or, at least, that the covalent character, if present, does not affect significantly the dislocation properties. Nevertheless, this can only be inferred with full certainty if test calculations employing a scheme which includes the covalent character of d-bonding are carried out.

As a final example of intermetallics we discuss the Ti-Al alloys based on the TiAl compound crystallizing in the tetragonal $L1_0$ structure[11]. TiAl-based alloys display different structures when stoichiometric and when Al rich (McCullough et al., 1988). Nearly stoichiometric alloys exhibit two-phase microstructure composed of the $L1_0$ TiAl and a small volume fraction of Ti_3Al (hexagonal DO_{19} structure). These materials possess a lamellar structure and in the single crystalline form they are called the "polysynthetically twinned (PST) crystals". Aluminum rich alloys have single-phase $L1_0$ structure. Deformation properties of these two structures are substantially different (Yamaguchi and Inui, 1992; Yamaguchi et al., 1995).

Owing to the tetragonality of the $L1_0$ structure, both ordinary dislocations and super-dislocations may by present on $\{111\}$ type planes. For example, for the (111) plane, slip along $[10\bar{1}]$ or $[01\bar{1}]$ requires super-dislocations but only ordinary dislocations are needed for slip along $[1\bar{1}0]$. The single-phase TiAl deforms at room temperature by the glide of superdislocations but at high temperatures the slip is dominated by ordinary dislocations while superdislocations are not observed. However, twinning becomes active at temperatures around 800°C; this is an unusual phenomenon since twinning is commonly a low temperature deformation mode (Yamaguchi and Inui, 1992; Yamaguchi et al., 1995).

[11] This AB type structure is similar to fcc with the corner atoms of the cubic cell occupied by atoms A and four of the atoms in the faces occupied by atoms B. The tetragagonal axis c runs along the direction connecting the two faces the centres of which are occupied by atoms A. Cubic notation can be used for this structure.

This material also exhibits a pronounced yield stress anomaly i. e. increase of the yield stress with increasing temperature (Kawabata et al., 1985).

In contrast, in PST crystals twinning on $\{111\}$ planes along the $\langle 112]$ direction is the main deformation mode at room temperature. Slip by ordinary dislocations on $\{111\}$ planes is the complementary deformation mode while superdislocations are usually not observed at this temperature. At high temperatures twinning is still active, but its propensity decreases, and slip is accommodated by both ordinary dislocations and superdislocations (Inui et al., 1992a, b).

In the following we show results of a recent atomistic study of the cores of three types of screw dislocations, ordinary $1/2[1\bar{1}0]$ and $1/2[11\bar{2}]$ dislocations and $[10\bar{1}]$ superdislocation. In these calculations the atomic interactions were described by Finnis-Sinclair type empirical central-force potentials constructed to reproduce a number of equilibrium properties of the TiAl $L1_0$ crystals (Girshick and Vitek, 1995).

Two alternate core structures of the ordinary $1/2[1\bar{1}0]$ dislocation, shown in Figs. 7a and b, respectively, were obtained using different starting configurations. The core shown in Fig. 7a is planar, spread into the (111) plane, while the core shown in Fig. 7b is non-planar, spread concomitantly into the (111) and $(11\bar{1})$ planes and thus sessile. The sessile core is energetically favored since when a shear stress parallel to the $[1\bar{1}0]$ direction was applied in the (111) plane the planar core transformed into the non-planar one. However, in a similar study of $1/2[1\bar{1}0]$ dislocation Simmons et al. (1993) found only the planar core configuration.

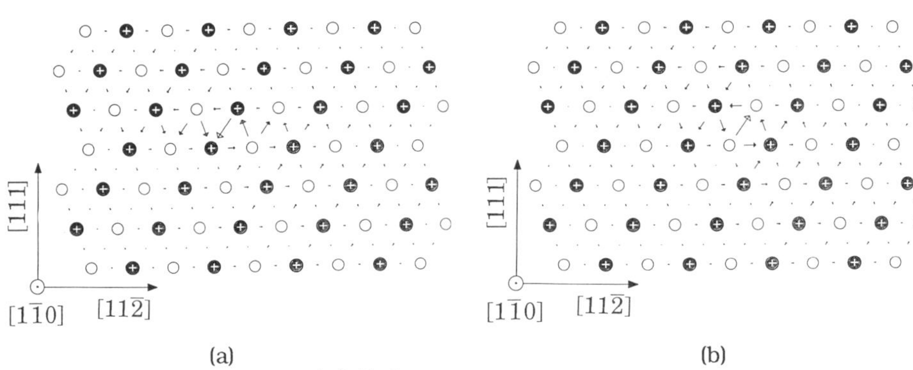

Fig. 7. Core structure of the $1/2[1\bar{1}0]$ screw dislocation in TiAl. (a) Planar core spread in the (111) plane; (b) Non-planar core spread in both (111) and $(11\bar{1})$ planes. The display method is the same as in previous examples.

For $[10\bar{1}]$ superdislocations two possible configurations of the core have also been found. They corresponding to two alternate dissociations of the superdislocation. The first is planar and corresponds to the following splitting $[10\bar{1}] = 1/6[11\bar{2}] + \text{SISF} + 1/6[2\bar{1}\bar{1}] + \text{APB} + 1/2[10\bar{1}]$ in the (111) plane. The alternative configuration is non-planar splitting according to the reaction $[10\bar{1}] = 1/6[11\bar{2}] + 1/3[20\bar{1}] + 1/6[1\bar{1}\bar{2}]$ into the (111) and $(\bar{1}1\bar{1})$ planes forming ribbons of the SISF in these planes. The partial dislocation with the Burgers vector $1/3[20\bar{1}]$, which lies at the intersection of the two $\{111\}$ planes, is entirely immobile. The latter configuration is favored when the energy of the SISF is low but the planar structure is metastable and does not spontaneously transform into the sessile one.

The core structure of the $1/2[11\bar{2}]$ dislocation is shown in Fig. 8. This core is spread into two adjacent $(\bar{1}11)$ planes and the superlattice extrinsic stacking fault (SESF) is formed within the core. Such faults have, indeed, been observed by electron microscopy (Hug et al., 1986). This fault represents a microtwin, two atomic layers wide, and it may serve as a nucleus for twinning. Application of the corresponding external shear stress, indeed, led at high enough stresses to the growth of the twin in the [111] direction. A similar core configuration was found in studies of dislocations in the tetragonal DO_{22} structure which deforms almost exclusively by twinning (Khantha et al., 1991a, 1994).

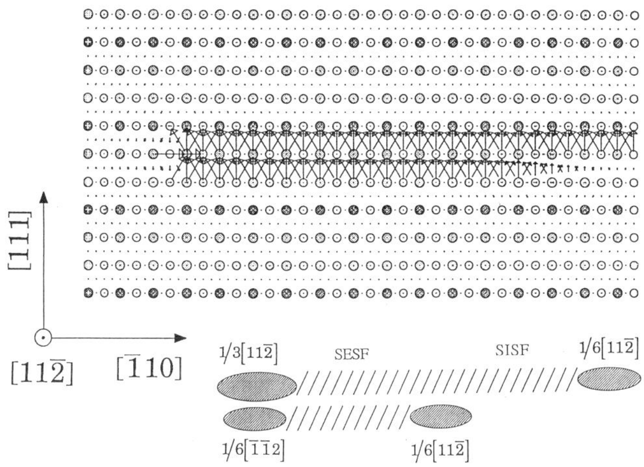

Fig. 8. Core structure of the $1/2[11\bar{2}]$ screw dislocation in TiAl.

Results of these calculations may provide a basis for our understanding of the plastic behavior of both single phase (Al rich) and PST TiAl assuming that the energies of the SISF and coherent twin boundary increase rapidly with increasing aluminum content. First, based on the results of the simulations, it is reasonable to assume that ordinary $1/2[1\,\bar{1}\,0]$ dislocations always have non-planar core and require substantial thermal activation to attain a significant mobility. In the single phase alloys the SISF energy is high and thus the difference between planar and non-planar core configurations of the superdislocations will not be large. Hence, even if the sessile form is energetically preferred, the superdislocations may possess metastable planar cores when gliding at room temperature. However, as the temperature increases superdislocations will be transforming into energetically more favorable sessile configurations involving SISF on both (111) and $(\bar{1}\,1\,\bar{1})$ planes. This is consistent with electron microscopic observations of non-planar dissociation of superdislocations in the single phase TiAl, observed by Hemker et al. (1993), though the details of the splitting involved may be different. Ordinary dislocations will then dominate the slip but since glide will still be more difficult at high temperatures than at low temperatures, twinning may occur as an alternative yielding mechanism. This also explains the anomalous increase of the yield stress with increasing temperature.

In PST crystals, which are nearly stoichiometric the SISF energy will be low and the sessile form of the superdislocations, corresponding to splitting into two non-parallel {111} planes with the SISF on these planes, will be strongly preferred. At the same time, since the SISF energy is low, the twin boundary energy is also low and wide splitting of the $1/2[11\bar{2}]$ dislocations on (111) planes can be expected. These serve as nuclei for twinning which then becomes the dominant deformation mode as observed. At high temperatures both the ordinary and superdislocations with non-planar cores become more mobile owing to thermal activations.

This example demonstrates how calculations employing relatively simple description of atomic interactions may provide valuable insights into the dislocation core structures on the basis of which the deformation behavior can be rationalized. Notwithstanding, the potentials employed in this study should only be regarded as model potentials for a generic $L1_0$ structure rather than as potentials for a specific material. The atomistic calculations of this type reveal a variety of possible core structures in the $L1_0$ lattice and their dependence on parameters such as the energy of the SISF. However, they are not capable to decide unequivocally which of these core structures applies in specific cases. In particular, the above discussion suggests that the most important factor controlling the types of dislocation cores, and thus the slip behavior in TiAl, is the dependence of the SISF energy upon the aluminum content. This dependence cannot be investigated in the framework of empirical potentials and more fundamental quantum mechanics based studies of alloying effects on this energy are needed. Furthermore, in alloys based on transition metals directional bonding may be significant. In the case of TiAl this is accentuated by the fact that the Cauchy pressure, C_{13}-C_{44}, is negative and this cannot be reproduced in the framework of the central force potentials (Pettifor et al., 1995). At the same time central force character of the potentials is also the main reason why the energy of the SISF is very low in this scheme. Hence, a significant further advancement of dislocation studies in such materials requires introduction of non-central forces. Some empirical attempts to include directionality of bonding have been made (Panova and Farkas, 1995) but only employment of rigorous quantum mechanics based potentials can reveal the importance of non-central forces unambiguously.

Semiconductors. Core structures of dislocations in silicon and germanium, which both crystallize in the diamond structure, have been studied since late fifties because dislocations may have a pronounced effect on electronic properties of semiconductors (for reviews see Amelinckx, 1979; Alexander, 1986; Duesbery and Richardson, 1991). The early models did not employ direct minimization of the energy but considered bond reconstructions for which no dangling bonds are present in the core (Hornstra, 1958; Hirsch, 1979; Jones, 1979). When minimizing the energy it is essential to include the covalent character of bonding and empirical potentials which incorporate directional bonding (Keating, 1966; Stillinger and Weber, 1985; Tersoff, 1988a, b) have been used in a number of dislocation studies (Jones, 1979; Chelikowsky and Spence, 1984; Nandekar and Narayan, 1990; Duesbery et al., 1991). However, recently the structure of the 90° partial dislocation[12] was for the first time studied using an ab initio method (Bigger et al., 1992). This calculation, based on the local density functional theory, utilized the pseudopotential method (King-Smith et al., 1991) and allowed for full atomic relaxation. In this calculation, which used 64 atoms per repeat cell, correct periodic boundary conditions are an essential ingredient.

The question asked in the ab initio study of Bigger et al. (1992) was which of the two reconstructions shown in Fig. 9 occurs in the cores of 90° partials. The asymmetric reconstruction (Fig. 9a) in which every atom is four fold coordinated but the mirror plane along the dislocation is broken,

[12] $1/2\langle 110\rangle$ dislocations in the diamond lattice dissociate on $\{111\}$ planes into $1/6\langle 112\rangle$ Shockley partials as in the fcc lattice. The 90° partials lie on $\{111\}$ planes in $\langle 110\rangle$ directions.

was proposed by Hirsch (1979) and Jones (1979). The symmetrically reconstructed structure (Fig. 9b) was found in calculations of Duesbery et al. (1991) who employed the Stillinger-Weber (1985) potential and the original potential of Tersoff (1988a). In ab initio calculations the asymmetric reconstruction was found to be favored. In this case there are no band gap states and sp hybrids pair up and form directed bonds to each other as envisaged by Hirsch (1979) and Jones (1979). Test calculations were then made by Bigger et al. (1992) using the tight-binding method with Chadi's Hamiltonian (Chadi, 1984) and the second Tersoff potential (Tersoff, 1988b). Both these calculations reproduce the ab initio structure reasonably well and, therefore, we can conclude that both Chadi's Hamiltonian and the second Tersoff potential are the most reliable empirical descriptions of atomic interactions in silicon.

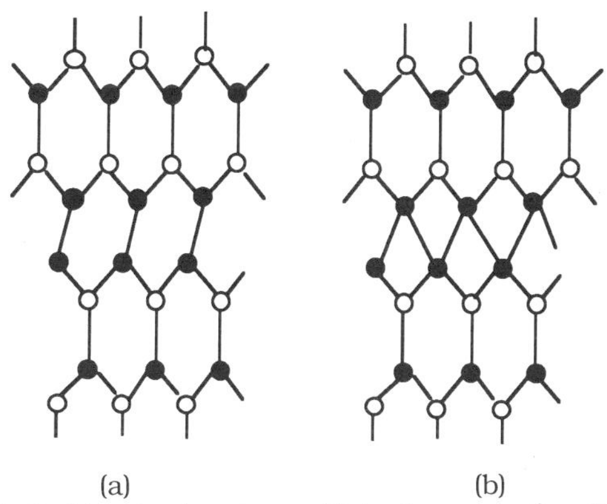

Fig. 9. Atomic structure within the core of a 90° partial in silicon. (a) Asymmetric reconstruction, (b) Symmetric reconstruction.

Recently Joos et al. (1994) used the Peierls-Nabarro type model to investigate the core structure of dislocations in silicon. The restoring force entering equation (16) was determined on the basis of the γ-surface calculated in ab initio studies of Kaxiras and Duesbery (1993). The calculations have been made for both 'shuffle' and 'glide' sets of total dislocations, which are distinguished by where the cut has been made in the sequence AaBbCc describing the stacking of $\{111\}$ planes in Si (Hirth and Lothe, 1982). Shuffle dislocations were found to be easier to form. However, dislocations in Si dissociate into Shockley partials and in this model, unlike in the full scale atomistic studies, the details of the cores of the partials cannot be investigated, in particular if bond reconstructions take place. Nevertheless, the γ-surfaces which are calculable relatively easily and provide a guide to possible alternative core structures.

Ionic crystals. Atomistic studies of dislocations in ionic crystals have been relatively limited and their main goals were estimates of the Peierls stress (Huntington et al., 1955; Granzer et al., 1968; Puls and Norgett, 1976; Woo and Puls, 1977b) and investigation of interactions of dislocations with point defects (Puls, 1980; Rabier and Puls, 1989). These calculations were carried out using either rigid ion or shell model with potentials described by equation (9). An important part of these calculations is evaluation of the Coulomb interactions using the Ewald type summations row by row in the atomic rows parallel to the dislocation (Puls and Norgett, 1976).

The most extensive were studies of dislocations in MgO (Puls and Norgett, 1976; Woo and Puls, 1977b) in which no major differences were found when using the rigid ion or shell model, respectively. Similarly, introduction of the breathing shell model did not alter the results significantly (Woo and Puls, 1977a). The dislocation cores were found to be very narrow and the Peierls stress is relatively high, factor of 3-6 higher than experimentally determined values. However, an overestimate of the Peierls stress seems to be a general feature of atomistic studies (Vitek,

1985; Duesbery, 1989). The goal of the studies of interactions between dislocations and anion and cation vacancies (Puls, 1980; Rabier and Puls, 1989) was to evaluate the corresponding binding energies. These were frequently found to concur with those estimated using the continuum elasticity models for separations of the dislocation and vacancies larger than a couple of lattice spacings (Puls, 1980; Rabier and Puls, 1989).

GRAIN BOUNDARIES

General structural studies

The earliest analyses of grain boundaries and other interfaces were principally geometrical and led to the formulation of the 'crystallography' of interfaces (Bollman, 1970; Christian, 1975, 1994; Christian and Crocker, 1980; Sutton, 1984). In recent years these analyses have been further advanced by putting them on group theoretical foundations (Pond and Bollmann, 1979; Gratias and Portier, 1982; Kalonji and Cahn, 1982; Pond and Vlachavas, 1983). However, the properties of interfaces are governed by their atomic structure and, therefore, atomistic modeling has been central for the development of our present understanding of interfacial behavior (Balluffi, 1982; Paidar et al., 1992; Finnis and Rühle, 1993; Sutton and Balluffi, 1995). The main goals of earlier calculations, usually made using pair potentials, was to analyze possible relations between the geometry of grain boundaries and their atomic structure and energy. These calculations were universal, possibly specific for a given crystal structure but not for a given material. Their results are, of course, most valuable when they are not dependent on the details of the description of atomic interactions used (Vitek and De Hosson, 1986).

While it was shown by Sutton and Balluffi (1987) that no general relationship between grain boundary geometry and energetics can be established, these studies revealed a variety of relations which apply at least to some classes of boundaries and allow certain generalizations of the results of atomistic studies. Examples are the structural unit model[13] (Sutton and Vitek, 1983a, b, c; Wang et al., 1984; Schwartz et al., 1985), structural multiplicity and related possibility of structural transformations in the boundaries (Vitek et al., 1985; Guillope, 1986; Oh and Vitek, 1986; Rottman, 1989) and the relationship between boundary expansion and

[13] The structural unit model, according to which the structure of a boundary can be described as an array of structural units of other boundaries sharing the same rotation axis, is the model which can predict the structure and some properties, given certain minimum information. When considering boundaries corresponding to different misorientations but sharing the same mean boundary plane, the structure of longer-period boundaries decomposes into an array of structural units of shorter period boundaries. The shorter-period boundaries are called delimiting boundaries because they delimit the range of misorientations in which this particular structural unit decomposition occurs. In this misorientation range all boundaries are composed of units of the delimiting boundaries and, conversely, the requirement that this decomposition must be valid for any boundary in this misorientation range imposes selection rules on the choice of delimiting boundaries (Sutton and Vitek, 1983a). Hence, knowledge of the structure of delimiting boundaries, which are usually short period, is sufficient for understanding structure of very long period boundaries sharing the same rotation axis and mean boundary plane. In fact it was shown by Sutton (1988) that this concept can be extended even to irrational boundaries. This model can predict not only the structure of boundaries within certain ranges of misorientations but also properties such as energy (Wang and Vitek, 1986) and self-diffusion (Balluffi, 1982). However, as shown by Sutton (1989a), the predictive capability of this model is very limited for mixed tilt and twist boundaries and even for pure tilt or twist boundaries it decreases with increasing indices of the rotation axis.

energy (Wolf, 1989a, b, c, 1990a, b, c; Sutton, 1991). A critical overview of these aspects of interfacial structures can be found in the recent book by Sutton and Balluffi (1995) and they will not be discussed further in this paper.

Contemporary trends in atomistic modeling

Recent atomistic studies of grain boundaries concentrate on the relation between the local interfacial structure and interfacial properties. These studies frequently employ descriptions of atomic interactions tailored to specific materials. Examples of properties studied are the local elastic response of grain boundaries (Adams et al., 1989; Wolf and Lutsko, 1989; Alber et al., 1992; Bassani et al., 1992), propagation of acoustic waves along interfaces (Alber et al., 1994; Vitek et al., 1994) and chemistry of grain boundaries, in particular segregation which habitually controls properties of grain boundaries in alloys (e. g. Srolovitz et al., 1992; Vitek et al., 1992; Menyhard et al., 1994; Seidman et al., 1994; Rittner et al., 1995).

Calculations linking modeling with the high resolution electron microscopy (HREM) are one important class of studies carried out for specific materials. These studies allow a direct comparison of atomistic modeling with experiments and, at the same time, are crucial for meaningful interpretation of HREM observations (Pennison, 1988; Self and O'Keefe, 1988; Sinclair et al., 1990). Closely connected are investigations of interfacial phase transformations and formation of two-dimensional boundary phases. Such phases, specific to interfaces, do not exist in the bulk and may occur in both pure elements and alloys. In the latter case they are usually associated with segregation that has been the object of extensive research in its own right. In the following section we present several examples of studies of this type. However, we first discuss briefly the use of different descriptions of atomic interactions in studies of grain boundaries.

Ab initio electronic structure calculations allowing for atomic relaxations have only been used in studies of very short period twist boundaries in germanium (Tarnow et al., 1989, 1990) and gold (Needels et al., 1992). These calculations have been made using the pseudopotential method with a plane-wave basis set (King-Smith et al., 1991). The relaxed configuration was found to differ only slightly from that constructed geometrically, and in the case of gold, it is very similar to the structure found using the EAM. One reason is the small period of the boundaries studied that exerts severe constrains on the structure. Several ab initio calculations of grain boundary energies which do not allow for atomic relaxations have also been made.

A number of calculations of grain boundaries in semiconductors have been made using the parametrized tight-binding method (Harrison, 1980). The first such study was performed by Thomson and Chadi (1984) for the $(221)/[1\bar{1}0]$ symmetrical tilt boundary. All atoms in the boundary region are four-fold coordinated and this structure was reproduced using the Stillinger-Weber potential (Phillpot and Wold, 1988) which strongly favors this coordination. $(111)/[1\bar{1}0]$, $(112)/[1\bar{1}0]$ and $(310)/[001]$ symmetrical tilt boundaries in silicon were studied by Paxton and Sutton (1989) and Kohyama et al. (1988). The same Hamiltonian was used in these two studies but different computational methods were employed and different repulsive terms and self-consistency conditions applied. Nevertheless, the two calculations are in an excellent agreement. The most interesting result was the finding that the reconstructed structure of the (112) boundary, possessing double the normal crystal periodicity along the $[11\bar{1}]$ and $[1\bar{1}0]$ directions, is favored. This agrees with the HREM observations of Bourret and Bacmann (1986). The main reason for this reconstruction is most likely the best manner in which tetrahedral bonding is preserved in the boundary.

Grain boundaries in alkali helides (Wolf, 1984b), NiO (Duffy and Tasker, 1983a, b; Duffy, 1986) and MgO (Wolf, 1980, 1984a) were studied using the shell and rigid ion models. Many of the structural features were found to be very similar to those found for boundaries in metals. This is not surprising since the total energy is expressed in terms of pair potentials just like in many studies for metals. Recently the rigid ion model was used in a more complex case of grain boundaries in α-Al_2O_3 and an excellent agreement between calculated and HREM observed structures was attained (Kenway, 1994). An important part of all these calculations is evaluation of Coulomb interactions using the Ewald type summations (Parry, 1975; Watson et al., 1981).

An interesting problem arose in studies of (001) twist boundaries in MgO where calculations suggested that these boundaries were unstable unless either a strong Van der Waals attraction is present or the boundaries are stabilized by the presence of impurities (Wolf, 1984a). On the other hand the (001) boundaries in MgO, manufactured by Sun and Balluffi (1982), displayed a strong cohesion. This paradox was resolved by Tasker and Duffy (1983) who found that the weak bonding is due to close pairing of like-charges across the interface but a much more stable structure is obtained when one anion and one cation per unit cell of the boundary are removed. This is an important general finding which suggests that removal and/or insertion of atoms into the boundary region should be considered when seeking the most stable structures.

In most studies dealing with metallic systems the EAM or Finnis-Sinclair potentials have been employed. In general, no very significant differences between structures calculated using these methods and those found when employing pair potentials were detected (Wolf et al., 1989). This is again not surprising because many of the basic structural features are of crystallographic nature and thus insensitive to the details of atomic interactions. However, the EAM and Finnis-Sinclair type potentials allowed a many studies of those structural features that are specific to given materials. Moreover, they are much more easily adaptable for studies of alloys and calculations can be carried out at constant pressure without any need for a special treatment of the $U(\rho)$ term in equation (6).

The non-central many body potentials for transition elements have not yet been used systematically in studies of grain boundaries. Moriarty's (1990b) non-central potentials for niobium were used to calculate the structure of the $\Sigma 5(310)$ and a good agreement with the HREM observations was attained, in particular for relative displacements of the grains (Campbell et al., 1993). However, reasons for differences encountered when using the EAM are not fully understood. Recently, a more detailed study of several symmetrical tilt boundaries in tungsten and molybdenum have been carried out (Marinopoulos et al., 1995) comparing Carlsson's non-central potentials (Carlsson, 1991a) described by equation (14) and Finnis-Sinclair type potentials for the same elements (Ackland and Thetford, 1987). Significant differences have been found both in the local atomic relaxations and in the rigid-body displacements of adjoining grains when using the two sets of potentials in spite of the fact that the same geometrical constraints, such as misorientation of the grains, orientation of the boundary plane, interplanar spacing and boundary periodicity, were imposed in both cases. In general, the central-force potentials tend to stabilize structures with first and second nearest neighbor interatomic separations and coordinations as close as possible to those in the bulk. In contrast, non-central potentials prefer cluster geometries with bond angles close to those found in the bulk even when neighbor separations are shorter than in the bulk.

Examples of recent studies

Grain boundary phases in copper. Atomistic studies employing EAM and many-body Finnis-Sinclair type potentials for copper revealed that the atomic structure of certain $\Sigma=3$ boundaries[14] consisted of thin layers of material that possessed other than the fcc crystal structure. At the same time it is well known that Cu is stable in the fcc structure up to its melting point. These findings were subsequently confirmed by HREM (Ernst et al., 1992; Wolf et al., 1992a, b; Schmidt et al., 1995).

The first such case was found when studying a series of $\Sigma=3$ tilt boundaries with the $[1\bar{1}0]$ rotation and varying inclinations of the boundary plane. If we measure the inclination angle Φ_{110} with respect to a common $\{111\}$ plane of the two half crystals than for $\Phi_{110}=0^\circ$ we obtain the usual $(111)/(11\bar{1})$ coherent twin and for $\Phi_{110}=90^\circ$ another symmetrical tilt boundary $(112)/(11\bar{2})$. First both calculations and experimental measurements of the dependence of the boundary energy on the angle Φ_{110} show a deep minimum at 0° and a less pronounced minimum at 82° (Wolf et al., 1992a, b). The latter is surprising because the $(112)/(11\bar{2})$ boundary has all characteristics normally associated with a low energy structure and yet possesses a higher energy. Secondly, the structure of the $(112)/(11\bar{2})$ boundary calculated in (Wolf et al., 1992b), shown in Fig. 10a, displays a broad transition region between the upper and lower grain. This is in contrast with the structure of the same boundary calculated using a potential for aluminum (Dagens et al., 1975) that is shown in Fig. 10b; this structure is practically the same as that found by Pond and Vitek (1977).

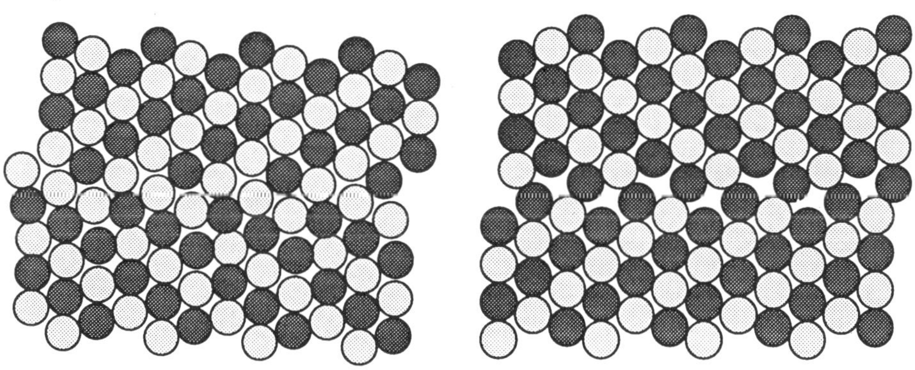

(a) (b)

Fig. 10. Calculated structure of the $(112)/(11\bar{2})$ boundary. (a) Copper; (b) Aluminum. Differently shaded circles depict atoms differently positioned along the $[1\bar{1}0]$ direction.

In the $(112)/(11\bar{2})$ boundary the $\{111\}$ type planes which are continuous through the boundary are the $(11\bar{1})$ planes in the upper grain and these become (111) planes in the lower grain. In both cases shown in Fig. 10 there is a relative rigid body translation of the upper grain with respect to the lower grain so that these $\{111\}$ planes are not contiguous

[14] We recall that bicrystals with this misorientation of the grains correspond to coherent twins when the boundary is parallel to $\{111\}$ planes in both grains.

but a step occurs in the boundary region. Considering the stacking of the $(11\bar{1})$ planes above the boundary as ...ABCABCABC ..., the stacking of the (111) planes below the boundary is ...ACBACBACB..., corresponding to the $\Sigma=3$ twin misorientation between the upper and lower grains. The transition between the two stackings is very sharp in the case of aluminum. However, in the case of copper the transition region is broad and stacking of $\{111\}$ planes in this zone is ...ABCBCACABC... which corresponds to creation of an intrinsic stacking fault on every third $\{111\}$ plane. This stacking corresponds to the rhombohedral 9R (α-Sm) structure (Wolf et al., 1992a, b) which has been observed as a bulk phase in CuZn and CuAl alloys as a product of martensitic transformations (Kajiwara, 1971; Cornelis and Wayman, 1974). The 9R structure was also found in other $\Sigma=3$ tilt boundaries with inclinations of the boundary plane close to 90°, in particular in the boundary inclined by 82°. The existence of this structure in the boundary region was confirmed experimentally by HREM (Wolf et al., 1992b); the same structure was also found and observed in silver (Ernst et al., 1992).

The difference between the structures of the $(112)/(11\bar{2})$ boundary in copper (and also silver) and aluminum is explained by the very different energies of intrinsic stacking faults in these materials. In aluminum this energy is high and thus formation of 9R structure is not favored and a sharp transition between the upper and lower grains is preferred. On the other hand, in copper and silver the stacking fault energy is low and a gradual transition involving the 9R structure is energetically favorable. The existence of the 9R structure also explains the energy minimum at 82° inclination of the boundary plane. While in the $(112)/(11\bar{2})$ boundary the $\{111\}$ planes of the 9R phase are geometrically constrained to lie closer together than in the bulk, this compression is relieved by tilting the boundary by about 8°. This makes the boundary with the boundary plane inclined by 82° energetically more favored (Wolf et al., 1992b). Recently a combined atomistic and HREM study of near $\Sigma=3$ boundaries in silver further confirmed the presence of the 9R structure in these boundaries and analyzed the dependence of this structure on both grain boundary inclination and grain misorientation (Hofmann and Finnis, 1994; Hofmann and Ernst, 1995). Grain misorientation induces further stacking faults in the transition region, so that a faulted 9R structure (a stacking polytype) is produced.

However, the 9R structure is not the only phase which can be part of the structure of $\Sigma=3$ boundaries in copper. It was discovered in recent atomistic studies that a particular $\Sigma=3$ tilt boundary with a $\langle 112 \rangle$ rotation axis stabilizes copper in the bcc structure (Schmidt et al., 1995). Copper has been observed before in the bcc structure either as an epitaxial layer on a substrate of bcc iron (Celinski et al., 1991) or as small precipitates in a bcc iron matrix (Othen et al., 1991) but this study demonstrates that the bcc structure also occurs under purely internal constraints in grain boundaries. The grain boundary in question is an asymmetrical tilt boundary with inclination of the boundary plane 84° with respect to the usual $(111)/(11\bar{1})$ coherent twin. Thermal grooving experiments (Laub et al., 1994) have indicated that the boundary energy of $\Sigma 3$ $\langle 112 \rangle$ tilt boundaries has a minimum at this inclination of the boundary plane. This was the motivation for the detailed atomistic simulations which were carried out using Finnis-Sinclair type potentials for copper (Ackland et al., 1987; Ackland and Vitek, 1990). These calculations showed that a layer of bcc structure forms in this boundary and $\{110\}_{bcc}$ planes connect the $\{111\}_{fcc}$ planes of each grain. The structure of the phase boundaries between bcc and fcc is a near coincidence type, involving a compromise in lattice strains. It is best suited to this particular boundary inclination and accounts for its lower energy with respect to neighboring inclinations but an essential precursor is that the energy difference between bcc and fcc Cu is particularly small (0.023eV/atom). This was shown by recent ab initio calculations (Paxton et al., 1990; Kraft et al., 1993) and it is reproduced by the potential used in this study. The structure has been confirmed experimentally by HREM on specially prepared bicrystals (Schmidt et al., 1995).

Grain boundary phases in copper-bismuth alloys. Copper-bismuth system is a very suitable model material for investigation of the segregation and embrittlement phenomena. A remarkable phenomenon observed in this system is the segregation induced faceting (Ference and Balluffi, 1988; Menyhard et al., 1989; Blum et al., 1990) which was shown to be a reversible transformation driven by addition or removal of bismuth (Ference and Balluffi, 1988). When the misorientation of the grains corresponds to $\Sigma = 3$ (rotation axis $[1\bar{1}0]$) the facets which contain a high concentration of bismuth are $(111)/(11\bar{1})$ boundaries. In pure copper these are the coherent twin boundaries and no segregation takes place to them if they are present in the sample prior to segregation (Luzzi, 1991). Furthermore, in studies of faceting of $\Sigma = 3$ boundaries in pure copper, made by Wolf et al. (1992a, b), no unusual preference for $(111)/(11\bar{1})$ facets was found. The $(111)/(11\bar{1})$ facets containing large amount of bismuth must, therefore, be formed by the transformation of general curved boundaries which already contain segregated bismuth. This was, indeed, confirmed by a recent conventional TEM study (Urdaneta et al., 1992). Hence, the segregation induced faceting in the Cu-Bi system appears to be associated with the formation of a new two-dimensional phase. This was the motivation for a combined theoretical and HREM study which, indeed, revealed the existence of such phase and determined in detail its structure (Luzzi, 1991; Yan et al., 1993).

The atomistic calculations were carried out using Finnis-Sinclair type central force many-body potentials. In this scheme the Cu-Bi interaction was fitted so as to reproduce the lattice parameter and the bulk modulus of the theoretical Cu_3Bi compound in the $L1_2$ structure and the enthalpy of mixing for the liquid copper-bismuth solution at 1200K. The parameters of the $L1_2$ Cu_3Bi were determined using ab initio full potential LMTO method (Yan et al., 1993).

The structure of the $(111)/(11\bar{1})$ facets containing Bi, which was first deduced on the basis of the HREM observations and then found to be the stable structure by molecular statics calculations, can be described as follows. In the framework of the 'ABC' notation for the stacking of $\{111\}$ atomic planes in fcc crystals, the $\Sigma = 3$ $(111)/(11\bar{1})$ twin in pure copper can be represented as

A-B-C-A-B-C-B-A-C-B-A-C

where the vertical lines mark the position of the boundary. The structure with Bi could be represented in the same way but in the marked layer C one third of the copper atoms are replaced by hexagonally arranged Bi atoms. However, the Bi atoms are centered outside this atomic plane so that a physically more appropriate representation of the model structure is

A-B-C-A-B-C'-C(Bi)-B-A-C-B-A

where C' denotes a plane of copper atoms in which one third of the atoms were replaced by hexagonally arranged vacancies, and C(Bi) the plane of bismuth atoms positioned above these vacancies. Thus the structure can be interpreted to a good approximation as splitting of one of the {111} planes, C, into two, C' and C(Bi) planes, containing Cu and Bi, respectively. While the bismuth atoms (i. e. the whole layer C(Bi)) are contracted towards the plane of copper containing vacancies (C'), there is a substantial overall expansion between the copper layer C' and the copper layer above the bismuth layer.

In order to make a detailed comparison of the calculated and observed structures, a series of images, corresponding to the same defocus conditions and thickness as in the HREM study, were calculated for the relaxed structure using the multislice formalism (Self and O'Keefe, 1988). There is an almost perfect match between the calculated and observed image and even quantitative features such as total expansion, i.e. the relative displacement of the two grains in the direction perpendicular to the boundary, are within the experimental limits of the accuracy (Yan et al., 1993).

This study shows that a two-dimensional ordered phase may form at the $(111)/(11\bar{1})$ twin boundaries in Cu-Bi provided a sufficient amount of Bi is available at this boundary. This is enabled by segregation to a general $\Sigma = 3$ boundary which then undergoes faceting. In order to assess whether this structure is not related to some possible three-dimensional phases which could then grow into the bulk, ab initio calculations of possible related structures have been made using the full-potential LMTO method (Siegl et al., 1995). These calculations unequivocally demonstrate that the structure found is purely a grain boundary phase.

It should be noted that these results provide a firm evidence that empirical central force many-body potentials are capable to describe with a sufficient accuracy atomic interactions even in relatively complex systems such as Cu-Bi provided an appropriate and sufficient input is used in their construction. For this purpose the use of ab initio electronic structure calculations to provide 'empirical' data was of principal significance since this allows sampling of crystal configurations which are not attainable experimentally. In this case the most important is the correct estimate of the 'size' of the Bi atom when surrounded by copper atoms.

Grain boundary structure in NiAl. An example of a synergistic approach using HREM and atomistic calculations is a recent study of the structure of the $\Sigma = 5$ (310)/[001] grain boundary in NiAl (B2 structure) (Fonda et al., 1995). The calculations were carried out using Finnis-Sinclair type potentials constructed for Ni-Al system (Vitek et al., 1991a) and modified so as to reproduce that Al enrichment in NiAl is associated with the formation of constitutional vacancies on Ni sites (Bradley and Taylor, 1937). The experimental HREM image of the boundary studied is shown in Fig. 11a. The two grains were imaged along their common [001] direction which, for NiAl, is along columns of pure Ni and pure Al. For the given thickness and defocus used the atomic columns exhibit white contrast, with the Ni columns brighter than the Al columns. An important feature of this image is an asymmetry produced by a rigid-body translation of the upper grain with respect to the lower grain parallel to the boundary and perpendicular to the [001] rotation axis.

The calculated structure of a fully stoichimetric boundary was found to be symmetrical with no relative displacement of the grains parallel to the boundary but with a relatively large expansion at the boundary. This is clearly in conflict with the experimental results. However, a structure which leads to an excellent agreement with observations was obtained by introducing Ni antisite defects into two adjacent Al columns in the boundary region. These columns then become alternating sequences of Ni and Al. This Ni-rich boundary is shown in Fig. 11b where the large circles represent aluminum atoms and small circles Ni atoms. Different shading depicts positions of atoms in two adjacent (002) planes. The atoms marked as squares corresponds to columns in which Ni and Al alternate. The group of atoms marked by the solid lines corresponds to the similarly marked group in the experimental image. This structure possesses a rigid-body displacement parallel to the boundary plane and expansion which both agree almost exactly with the observed ones. The match of individual atomic positions is clearly visible by comparison of the atom positions within the marked regions in Figs. 11a and b. This was fully corroborated by simulation of the image using the multislice technique (Fonda et al., 1995).

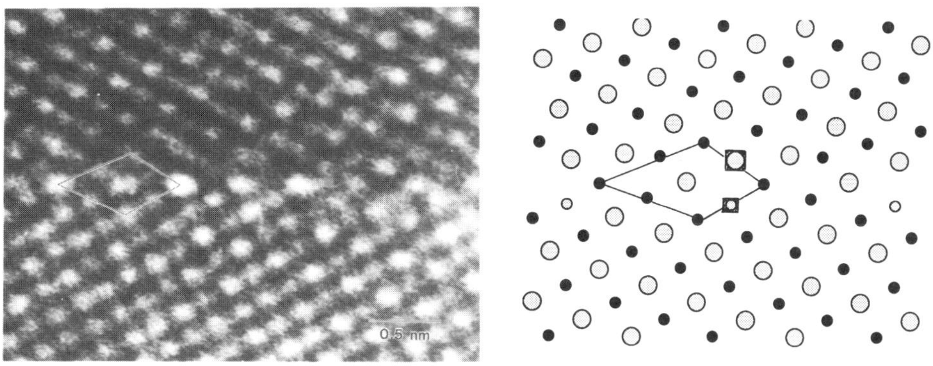

(a) (b)

Fig. 11. Structure of the $\Sigma = 5$ (310)/[001] grain boundary in NiAl. (a) Observed HREM image, (b) Calculated structure.

CONCLUDING REMARKS

Dislocations and grain boundaries, as well as other types of interfaces, are extended crystal defects and in the past a large effort has been expended on their crystallographic characterization. However, properties and behavior of interfaces, and to a great extent dislocations, cannot be determined solely on the basis of these characteristics since it is the atomic structure in their cores which is the controlling factor. Investigation of this structure requires atomistic modeling which has become very popular in recent years. Understanding of the crystallographic characteristics of extended defects is, of course, a necessary forerunner for atomistic studies just as the geometrical analysis of ideal crystals is for physical investigations of crystalline materials. Another precursor of atomistic studies is a description of atomic interactions, or in general, a method for evaluation of the total energy of the system of many particles as a function of their positions.

In principle, the present state of the art ab initio electronic structure calculations based on the local density functional theory may provide such description of the total energy for any system. However, in practice these calculations are still limited to relatively small systems and cannot be routinely employed to study atomic structure and atomic level behavior of extended defects. This is the reason why the bulk of atomistic studies of dislocations and interfaces have been made using much simpler descriptions of atomic interactions. The simpler schemes for evaluation of the total energy are usually empirical but based on some fundamental characteristics of bonding in given materials.

An important recent evolution in the development of empirical schemes describing atomic interactions is the close link with ab initio calculations. First, results of these calculations provide data bases which are much more comprehensive than what is available experimentally and these can be utilized when constructing empirical schemes. In particular, various non-equilibrium and metastable structures can be studied which are entirely inaccessible to experiments. Secondly, empirical treatments of the total energy can be directly tested against ab initio calculations for simpler situations. In the case of interfaces and dislocations the most important are tests of stability with respect to large straining and with respect to alternate structures. Such tests and fittings give us much more confidence when transferring the empirical schemes to more complex situations than when only a few equilibrium properties that can be measured experimentally have been fitted.

In parallel, empirical schemes are becoming more and more based on proper quantum mechanical treatments of atomic interactions. For example, pair potentials, though often successful, were not appropriately justified for metals other than those with purely sp-bonding. The EAM and Finnis-Sinclair potentials have a better foundation in the density functional theory, though they neglect any possible covalent character of bonding. The latter is addressed in the approaches which are grounded in the tight-binding treatment, such as the bond-order potentials. Hence, atomistic studies are employing more and more frequently schemes that are more reliable. However, when employing a more complex scheme, there are several reasons why the calculations should be compared with simpler approaches available. First, many properties of lattice defects are insensitive to the details of atomic interactions and these should be unambiguously identified. Secondly, it is very important to understand at what point the use of more sophisticated analyses of atomic interactions leads to uncovering structural features which could not be found when using simpler schemes. Finally, the simple schemes are computationally much less demanding and they should be used when they are sufficient.

In connection with the advancement of the descriptions of atomic interactions the atomistic studies of interfaces and dislocations can be more reliably made for specific materials. This is certainly one of the future trends of such studies when compared with the past when most reliable studies were always those least sensitive to the details of atomic interactions. At the same time, owing to rapidly increasing computational power, the modeling can be made for much larger systems that mimic more closely real situations. In parallel, experimental techniques with atomic level resolution are becoming a commonplace and computer modeling of structures studied is then an integral part of a combined experimental and theoretical effort to understand atomic level structures and properties of extended defects in materials.

Acknowledgment

The author would like to thank Dr. M. W. Finnis for most valuable comments and making available the structure shown in Fig. 10a. The micrograph in Fig. 11a was supplied by Dr. D. E. Luzzi. This research was supported by the NSF Grant No. DMR92-19089 (dislocation studies) and by the NSF MRSEC Program DMR92-20668 (interfacial studies). The financial support of the NEDO (Japan) Grant is acknowledged.

REFERENCES

Ackland, G.J., Finnis, M.W., and Vitek, V., 1988, *J.Phys. F: Metal Phys.* 18:L153.

Ackland, G.J., and Thetford, R., 1987, *Philos. Mag. A* 56:15.

Ackland, G.J., Tichy, G., Vitek, V., and Finnis, M.W., 1987, *Philos. Mag. A* 56:735.

Ackland, G.J., and Vitek, V., 1990, *Phys. Rev. B* 41:10324.

Adams, J.B., Wolfer, W.G., and Foiles, S.M., 1989, *Phys. Rev. B* 40:9479.

Adby, P.R., and Dempster, M.A.H., 1974,""Introduction to Optimization Methods", Chapman & Hall, London.

Akhiezer, N.I., 1965,""The Classical Moment Problem", Hafner Publ. Co., New York.

Alber, E.S., Bassani, J.L., Vitek, V., and Wang, G.J., 1994, *Modelling and Simulation in Mat. Sci. Eng.* 2:455.

Alber, I., Bassani, J.L., Khantha, M., Vitek, V., and Wang, G.J., 1992, *Phil. Trans. Roy. Soc. London A* 339:555.

Alexander, H., 1986, in "Dislocations in Solids", F.R.N. Nabarro, ed., Vol. 7, North-Holland, Amsterdam, p. 114.

Alinaghian, P., Gumbsch, P., Skinner, A.J., and Pettifor, D.G., 1993, *J. Phys.: Condens. Matter* 5:5795.

Alinaghian, P., Nishitani, S.R., and Pettifor, D.G., 1994, *Philos. Mag. B* 69:889.

Amelinckx, S., 1979, in "Dislocations in Solids", F.R.N. Nabarro, ed., Vol. 2, North-Holland, Amsterdam, p. 294.

Aoki, M., Gumbsch, P., and Pettifor, D.G., 1993, in "Interatomic Potentials and Structural Stability", K. Terakura and H. Akai, eds., Springer, Berlin, p. 23.

Aoki, M., and Pettifor, D.G., 1994, *Mat. Sci. Eng. A* 176:19.

Aucouturier, M., editor, 1990, "Intergranular and Interphase Boundaries in Materials", *J. Phys. France* 51:C1.

Baker, I., 1995, *Mat. Sci. Eng. A* 192/193:1.

Balluffi, R.W., 1982, *Metall. Trans. A* 13:2069.

Basinski, Z.S., Duesberry, M.S., and Taylor, R., 1971, *Can. J. Phys.* 49:2160.

Baskes, M.I., 1992, *Phys. Rev. B* 46:2727.

Bassani, J.L., Vitek, V., and Alber, I., 1992, *Acta Metall. Mater.* 40:S307.

Besold, G., and Mouritsen, O.G., 1994, *Phys. Rev. B* 50:6573.

Bigger, J.R.K., McInnes, D.A., Sutton, A.P., Payne, M.C., Stich, I., Kingsmith, R.D., Bird, D.M., and Clarke, L.J., 1992, *Phys. Rev. Lett.* 69:2224.

Blum, B., Menyhard, B., Luzzi, D.E., and McMahon, C.J., 1990, *Scripta. Metall. Mater.* 24:2169.

Bollman, W., 1970,""Crystal Defects and Crystalline Interfaces", Springer, Berlin.

Born, M., and Huang, K., 1954,""Dynamical Theory of Crystal Lattices", Clarendon Press, Oxford.

Born, M., and Mayer, J.E., 1932, *Z. Phys.* 75:1.

Bourret, A., and Bacmann, J.J., 1986, in "JIMIS-4", Y. Ishida et al., eds., Grain Boundary Structure and Related Phenomena, The Japan Institute of Metals, Minakami, Japan, p. 125.

Bowen, D.K., Christian, J.W., and Taylor, G., 1967, *Can. J. Phys.* 45:903.

Bradley, A.J., and Taylor, A., 1937, *Proc. Roy. Soc. London A* 159:56.

Brandon, D.G., Ralph, B., Ranganathan, S., and Wald, M.S., 1964, *Acta Metall.* 12:813.

Brown, F.C., 1967,""Physics of Solids", Benjamin, New York.

Campbell, G.H., Foiles, S.M., Gumbsch, P., Rühle, M., and King, W.E., 1993, *Phys. Rev. Lett.* 70:449.

Carlsson, A.E., 1990, in "Solid State Physics", H. Ehrenreich and D. Turnbull, eds., Vol. 43, Academic Press, New York, p. 1; 1991a, *Phys. Rev. B* 44:6590; 1991b, in "Defects in Materials", P.D. Bristowe, J.E. Epperson, J.E. Griffith and Z. Liliental-Weber, eds., Vol. 209, Materials Research Society, Pittsburgh, p. 165.

Carlsson, A.E., and Meschter, P.J., 1989, *J. Mater. Res.* 4:1060.

Catlow, C.R.A., Dixon, M., and Mackrodt, W.C., 1982, in "Computer Simulation of Solids", C.R.A. Catlow and W.C. Mackrodt, eds., Lecture Notes in Physics, Vol. 166, Springer, Berlin, p. 130.

Catlow, C.R.A., Faux, I.D., and Norgett, M.J., 1976, *J. Phys. C : Solid State Phys.* 9:419.

Celinski, Z., Heinrich, B., Cochran, J.F., Myrtle, K., and Arrott, A. S., 1991, in "Science and Technology of Nanostructured Magnetic Materials", G. C. Hadjipanayis and G. A. Prinz, eds., Plenum Press, New York, p. 77.

Chadi, D.J., 1984, *Phys. Rev. B* 29:785.

Chelikowsky, J.R., and Spence, J.C.H., 1984, *Phys. Rev. B* 30:694.

Christian, J.W., 1975,""The Theory of Transformations in Metals and Alloys", Pergamon Press, Oxford; 1983, *Metall. Trans. A* 14:1237; 1994, *Metallurgical and Materials Transactions A* 25:1821.

Christian, J.W., and Crocker, A.G., 1980, in "Dislocations in Solids", F.R.N. Nabarro, ed., Vol. 3, North-Holland, Amsterdam, p. 165.

Christian, J.W., and Vitek, V., 1970, *Rep. Prog. Phys.* 33:307.

Clark, W.A.T., Dahmen, U., and Briant, C.L., editors, 1992, "Structure and Properties of Interfaces in Materials", Mat. Res. Soc. Symp. Vol. 238, Materials Research Society, Pittsburgh.

Cornelis, I., and Wayman, C.M., 1974, *Acta Metall.* 22:291.

Cyrot-Lackmann, F., 1967, *Adv. Phys.* 16:393.

Dagens, L., Rasolt, L.M., and Taylor, R., 1975, *Phys. Rev. B* 11: 2726.

Darolia, R., Lahrman, D.F., Field, R.D., Dobbs, J.R., Chang, K.M., Goldman, E.H., and Konitzer, D.G., 1992, in "Ordered Inmtermetallics – Physical Metallurgy and Mechanical Behaviour", C.T. Liu, R.W. Cahn and G. Sauthoff, eds., Kluwer Academic Publishers, Dordrecht, p. 679.

Daw, M.S., and Baskes, M.I., 1984, *Phys. Rev. B* 29:6443.

Daw, M.S., Foiles, S.M., and Baskes, M.I., 1993, *Mater Sci Rep* 9:251.

Dick, B.G., and Overhauser, A.W., 1958, *Phys. Rev.* 112:90.

Dimiduk, D.M., 1991, *Journal de Physique III* 1:1025.

Dipersio, J., and Escaig, B., 1977, *Phys. Stat. Sol. (a)* 40:393.

Ducastelle, F., and Cyrot-Lackmann, F., 1970, *J. Phys. Chem. Sol.* 31:1295.

Duesbery, M.S., 1989, in "Dislocations in Solids", F.R.N. Nabarro, ed., Vol. 8, North Holland, Amsterdam, p. 67.

Duesbery, M.S., and Basinski, Z.S., 1993, *Acta Metall. Mater.* 41:643.

Duesbery, M.S., and Foxall, R.A., 1969, *Philos. Mag. A* 20:719.

Duesbery, M.S., Joos, B., and Michel, D.J., 1991, *Phys. Rev. B* 43:5143.

Duesbery, M.S., and Richardson, G.Y., 1991, *CRC Critical Reviews in Solid State and Materials Science* 17:1.

Duesbery, M.S., Vitek, V., and Bowen, D.K., 1973, *Proc. Roy. Soc. London A* 332:85.

Duffy, D.M., 1986, *J. Phys. C : Solid State Phys.* 19:4393.

Duffy, D.M., and Tasker, P.W., 1983a, *Philos. Mag. A* 47:817; 1983b, *Philos. Mag. A* 48:155.

Ernst, F., Finnis, M.W., Hofmann, D., Muschik, T., Schonberger, U., Wolf, U., and Methfessel, M., 1992, *Phys. Rev. Lett.* 69:620.
Farkas, D., Pasianot, R., Savino, E.J., and Miracle, D.B., 1991, in "High-Temperature Ordered Intermetallic Alloys IV", L.A. Johnson, D.P. Pope and J.O. Stiegler, eds., Mat. Res. Soc. Symp. Vol. 213, Materials Research Society, Pittsburgh, p. 223.
Farkas, D., and Vailhe, C., 1993, *J. Mater. Res.* 8:3050.
Ference, T.G., and Balluffi, R.W., 1988, *Scripta Metall.* 22:1929.
Finnis, M.W., Paxton, A.T., Pettifor, D.G., Sutton, A.P., and Ohta, Y., 1988, *Philos. Mag. A* 58:143.
Finnis, M.W., and Rühle, M., 1993, in "Materials Science and Technology", V. Gerold, ed., Vol. 1, VCH, Weinhein, p. 533.
Finnis, M.W., and Sinclair, J.E., 1984, *Philos. Mag. A* 50:45.
Fletcher, R., and Powell, M.J.D., 1963, *Computer J.* 6:16.
Foiles, S.M., 1985, *Phys. Rev. B* 32:7685.
Foiles, S.M., and Daw, M.S., 1987, *J. Mater. Res.* 2:5.
Fonda, R.W., Yan, M., and Luzzi, D.E., 1995, *Philos. Mag. Lett.* 71:221.
Forbes, K.R., Glatzel, U., Darolia, A., and Nix, W.D., 1993, in "High-Temperature Ordered Intermetallic Alloys V", I. Baker, R. Darolia, J.D. Whittenberger and M.H. Yoo, eds., Mat. Res. Soc. Symp. Vol. 288, Materials Research Society, Pittsburgh, p. 45.
Frank, F.C., 1950, in "Symposium on the Plastic Deformation of Crystalline Solids", Office of Naval Research, Pittsburgh, p. 150.
Freeman, A.J., Xu, J.-H., Hong, T., and Lin, W., 1992, in "Ordered Inmtermetallics - Physical Metallurgy and Mechanical Behaviour", C.T. Liu, R.W. Cahn and G. Sauthoff, eds., Kluwer Academic Publishers, Dordrecht, The Netherlands, p. 1.
Frenkel, J., and Kontorova, T., 1938, *Phys. Z. Sowj. Un.* 13:1.
Friedel, J., 1964,""Dislocations", Pergamon Press, Oxford.
Friedel, J., 1969, in "The Physics of Metals", J.M. Ziman, ed., Cambridge University Press, Cambridge, p. 340.
Fu, C.L., and Yoo, M.H., 1990, *Philos. Mag. Lett.* 62:159.; 1992a, *Mater Chem Phys* 32:25.; 1992b, in "Ordered Inmtermetallics - Physical Metallurgy and Mechanical Behaviour", C.T. Liu, R.W. Cahn and G. Sauthoff, eds., Kluwer Academic Publishers, Dordrecht, The Netherlands, p. 155.
Gaspar, J.P., and Lambin, P., 1985, in "The Recursion Method and its Applications", D.G. Pettifor and D.L. Weaire, eds., Springer Series in Solid State Sciences, Springer, New York, p. 72.
Gehlen, P., Beeler, J., and Jaffe, R., 1972, "Interatomic Potentials and Simulation of Lattice Defects", Plenum Press, New York.
Girshick, A., and Vitek, V., 1995, in "High-Temperature Ordered Intermetallic Alloys VI", J. Horton, I. Baker, S. Hanada, R.D. Noebe and D. Schwartz, eds., Mat. Res. Soc. Symp. Vol. 364, Materials Research Society, Pittsburgh, p. 145.
Granzer, F., Wagner, G., and Eisenblatter, J., 1968, *Phys. stat. sol.* 30:587.
Gratias, D., and Portier, R., 1982, *J. Phys. France* 43:C6-25.
Guillope, M., 1986, *J. Phys. France* 47:1347.
Hairie, A., Hairie, F., Lebouvier, B., Nourt, G., Paumier, E., Ralantoson, N., and Sutton, A.P., 1994, *Interface Sci.* 2:17.
Harrison, W.A., 1980,""Electronic Structure and Properties of Solids", Freeman, San Francisco.
Hemker, K.J., Viguier, B., and Mills, M.J., 1993, *Mat. Sci. Eng. A* 164:391.
Heredia, F.E., Tichy, G., Pope, D.P., and Vitek, V., 1989, *Acta Metall.* 37:2755.

Hirsch, P.B., 1960, in "Proc. 5th Int. Conf. Crystallography", Cambridge University, oral communication, p. 139; 1979, *J. Phys. France* 40:C6.

Hirth, J.P., and Lothe, J., 1982,""Theory of Dislocations", Wiley-Interscience, New York.

Hofmann, D., and Finnis, M.W., 1994, *Acta Metall. Mater.* 42:3555.

Hofmann, F., and Ernst, F., 1995, *Interface Sci.* 2:201.

Hong, T., and Freeman, A.J., 1991, *Phys. Rev. B* 43:6446; 1992, *J. Mater. Res.* 7:68.

Hornstra, J., 1958, *J. Phys. Chem. Sol.* 5:129.

Horton, J., Hanada, S., Baker, I., Noebe, R. D., and Schwartz, D., editors, 1995, "High-Temperature Ordered Intermetallic Alloys VI", Mat. Res. Soc. Symp. Vol. 364, Materials Research Society, Pittsburgh.

Hug, G., Loiseau, A., and Lasalmonie, A., 1986, *Philos. Mag. A* 54:47.

Huntington, H.B., Dickey, J.E., and Thomson, R., 1955, *Phys. Rev.* 100:1117.

Inui, H., Nakamura, A., Oh, M.H., and Yamaguchi, M., 1992a, *Philos. Mag. A* 66:557.

Inui, H., Oh, M.H., Nakamura, A., and Yamaguchi, M., 1992b, *Acta Metall. Mater.* 40:3095.

Johnson, R.A., 1988, *Phys. Rev. B* 37:3924.

Jones, R., 1979, *J. Phys. France* 40:C6.

Joos, B., Ren, Q., and Duesbery, M.S., 1994, *Phys. Rev. B* 50:5890.

Kajiwara, Z., 1971, *J. Phys. Soc. Japan* 30:1757.

Kalonji, G., and Cahn, J.W., 1982, *J. Phys. France* 43:C6-33.

Kawabata, T., Kanai, T., and Izumi, O., 1985, *Acta Metall.* 33:1355.

Kaxiras, E., and Duesbery, M.S., 1993, *Phys. Rev. Lett.* 70:3752.

Keating, P.N., 1966, *Phys. Rev.* 145:637.

Kenway, P. R., 1994, *J. Am. Ceram. Soc.* **77**, 349.

Khantha, M., Cserti, J., and Vitek, V., 1992a, *Scripta. Metall. Mater.* 27:487; 1992b, *Scripta. Metall. Mater.* 27: 481.

Khantha, M., Vitek, V., and Pope, D.P., 1991a, in "High-Temperature Ordered Intermetallic Alloys IV", L.A. Johnson, D.P. Pope and J.O. Stiegler, eds., Mat. Res. Soc. Symp. Vol. 213, Materials Research Society, Pittsburgh, p. 229; 1994, *Modelling and Simulation in Mat. Sci. Eng.* 2:587.

Khantha, M., Vitek, V., and Pope, D.P., 1991b, in "High-Temperature Ordered Intermetallic Alloys IV", L.A. Johnson, D.P. Pope and J.O. Stiegler, eds., Mat. Res. Soc. Symp. Vol. 213, Materials Research Society, Pittsburgh, p. 229.

King, A.H., and Smith, D.A., 1980, *Acta Crystallogr.* 36:335.

King-Smith, R.D., Payne, M.C., and Lin, J.-S., 1991, *Phys. Rev. B* 44:13063.

Kittel, C., 1986,""Introduction to Solid State Physics", John Wiley, New York.

Kohlhoff, S., Gumbsch, P., and Fischmeister, H.F., 1991, *Philos. Mag. A* 64:851.

Kohyama, M., Yamamoto, R., Watanabe, Y., Ebata, Y., and Kinishita, M., 1988, *J. Phys. C : Solid State Phys.* 21:3205.

Komninou, Ph., and Rocher, A., editors, 1993, "Intergranular and Interphase Boundaries in Materials: iib92", *Materials Science Forum*, Trans Tech Publications, Aedermannsdorf.

Kraft, T., Marcus, P.M., Methfessel, M., and Scheffler, M., 1993, *Phys. Rev. B* 48:5886.

Kroupa, F., and Lejček, L., 1972, *Czech. J. Phys. B* 22:813.

Kubin, L.P., 1982, *Rev. Deform. Behav. Materials* 4:181.

Laub, W., Oswald, A., Muschik, T., Gust, W., and Fournelle, R.A., 1994, in "Solid-Solid Phase Transformations", W.C. Johnson, J.M. Howe, D.E. Laughlin and W.A. Soffa, eds., TMS, Warrendale, Pennsylvania, p. 1115.

Lee, J. K., 1981, "Interatomic Potentials and Crystalline Defects", TMS, Warrendale, Pennsylvania.

Lejček, L., 1972, *Czech. J. Phys. B* 22:802.

LeSar, R., Najafabadi, R., and Srolovitz, D.J., 1989, *Phys. Rev. Lett.* 63:624.

Lewis, G.V., and Catlow, C.R.A., 1988, *J. Phys. C : Solid State Phys.* 18:1149.

Luzzi, D.E., 1991, *Ultramicroscopy* 37:180.

Ma, Q., Liu, C.L., Adams, J.B., and Balluffi, R.W., 1993, *Acta Metall. Mater.* 41:143.

Marinopoulos, A., Vitek, V., and Carlsson, A.E., 1995, *Philos. Mag. A*, to be published.

Mark, J. E., Glicksman, M. E., and Marsh, S. P., editors, 1992, "Computational Methods in Materials Science", Mat. Res. Soc. Symp. Vol. 278, Materials Research Society, Pittsburgh.

Martin, J.W., 1975a, *J. Phys. C : Solid State Phys.* 8:2837; 1975b, *J. Phys. C : Solid State Phys.* 8:2858.

Masuda, K., Kobayashi, K., Sato, A., and Mori, T., 1981, *Philos. Mag. B* 43:19.

McCullough, C., Valencia, J. J., Mateous, H., Levi, C. G., Mehrabian, R., and Rhyne, K. A., 1988, *Scripta Metall.* 22:1131.

Menyhard, M., Blum, B., and McMahon, C.J., 1989, *Acta Metall.* 37:549.

Menyhard, M., Yan, M., and Vitek, V., 1994, *Acta Metall. Mater.* 42:2783.

Mills, M.J., Daw, M.S., Angelo, J.E., and Miracle, D.B., 1995, in "High-Temperature Ordered Intermetallic Alloys VI", J. Horton, I. Baker, S. Hanada, R.D. Noebe and D. Schwartz, eds., Mat. Res. Soc. Symp. Vol. 364, Materials Research Society, Pittsburgh, p. 401.

Mills, M.J., Daw, M.S., Foiles, S.M., and Miracle, D.B., 1993, in "High-Temperature Ordered Intermetallic Alloys V", I. Baker, R. Darolia, J.D. Whittenberger and M.H. Yoo, eds., Mat. Res. Soc. Symp. Vol. 288, Materials Research Society, Pittsburgh, p. 257.

Mills, M.J., and Miracle, D.B., 1993, *Acta Metall. Mater.* 41:85.

Minami, F., Kuramoto, E., and Takeuchi, S., 1972, *Phys. Stat. Sol. (a)* 12: 581; 1974, *Phys. Stat. Sol. (a)* 22:81.

Moriarty, J.A., 1990a, *Phys. Rev. B* 42:1609; 1990b, in "Many Atom Interactions in Solids", R.M. Nieminen, M.J. Puska and M.J. Manninen, eds., Springer, Berlin, p. 158.

Moriarty, J.A., and Phillips, R., 1991, *Phys. Rev. Lett.* 66:3036.

Nabarro, F.R.N., 1947, *Proc. Phys. Soc. London B* 59:256; 1967,""Theory of Crystal Dislocations", Clarendon Press, Oxford.

Nandekar, A.S., and Narayan, J., 1990, *Philos. Mag. A* 61:873.

Needels, M., Rappe, A.M., Bristowe, P.D., and Joannopoulos, J.D., 1992, *Phys. Rev. B* 46:9768.

Neumann, F., 1885,""Vorlesungen über die Theorie der Elasticität", Leipzig.

Nguyen, T., Ho, P.S., Kwok, T., Nitta, C., and Yip, S., 1992, *Phys. Rev. B* 46:6050.

Oh, Y., and Vitek, V., 1986, *Acta Metall.* 34:1941.

Othen, P.J., Jenkins, M.L., Smith, G.D.W., and Phythian, W.J., 1991, *Philos. Mag. Lett.* 64:383.

Paidar, V., Gemperlová, J. , Lejček, P., and Vitek, V., 1992, *Mat. Sci. Eng. A* 154:113.

Paidar, V., Pope, D.P., and Vitek, V., 1984, *Acta Metall.* 32:435.
Paidar, V., Yamaguchi, M., Pope, D.P., and Vitek, V., 1982, *Philos. Mag. A* 45: 883.
Panova, J., and Farkas, D., 1995, in "High-Temperature Ordered Intermetallic Alloys VI", J. Horton, I. Baker, S. Hanada, R.D. Noebe and D. Schwartz, eds., Mat. Res. Soc. Symp. Vol. 364, Materials Research Society, Pittsburgh, p. 151.
Parry, D.E., 1975, *Surf. Sci.* 49:433.
Parthasarathy, T.A., Dimiduk, D.M., Woodward, C., and Diller, D., 1991, in "High-Temperature Ordered Intermetallic Alloys IV", L.A. Johnson, D.P. Pope and J.O. Stiegler, eds., Mat. Res. Soc. Symp. Vol. 213, Materials Research Society, Pittsburgh, p. 337.
Parthasarathy, T.A., Rao, S.I., and Dimiduk, D.M., 1993, *Philos. Mag. A* 67:643.
Pasianot, R., Farkas, D., and Savino, E.J., 1990, *Scripta. Metall. Mater.* 24:1669; 1991, *Journal de Physique III* 1:997.
Pasianot, R., Xie, Z., Farkas, D., and Savino, E.J., 1994, *Modelling and Simulation in Mat. Sci. Eng.* 2:383.
Paxton, A.T., Methfessel, M., and Potaglou, H.M., 1990, *Phys. Rev. B* 41:8127.
Paxton, A.T., and Sutton, A.P., 1989, *Acta Metall.* 37:1693.
Payne, M.C., Clarke, L.J., and Stich, I., 1992, *Phil Trans Roy Soc London A* 341:211.
Peierls, R., 1940, *Proc. Phys. Soc. London B* 52:34.
Pennison, J.-M., 1988, *J. Phys. France* 49:C5087.
Pettifor, D.G., 1983, in "Physical Metallurgy", R.W. Cahn and P. Haasen, eds., Elsevier, Amsterdam, p. 147; 1990, in "Many Atom Interactions in Solids", R.M. Nieminen, J. Puska and M. Manninen, eds., Springer, Berlin, p. 64; 1991, *Phys. Rev. Lett.* 63:2480; 1992, in "Electron Theory in Alloy Design", A.H. Cottrell and D.G. Pettifor, eds., The Institute of Metarials, London, p. 81.
Pettifor, D.G., Aoki, M., Gumbsch, P., Horsfield, A.P., Nguyen Manh, D., and Vitek, V., 1995, *Mat. Sci. Eng. A* 192/193:24.
Pettifor, D. G., and Cottrell, A. H., editors, 1992,""Electron Theory in Alloy Design", The Institute of Materials, London.
Phillpot, S.R., and Wold, D., 1988, in "Interfacial Structure, Properties and Design", M.H. Yoo, W.A.T. Clark and C.L. Briant, eds., Mat. Res. Soc. Symp. Vol. 122, Materials Research Society, Pittsburgh, p. 103.
Pond, R.C., and Bollmann, W., 1979, *Phil. Trans. Roy. Soc. London A* 292:449.
Pond, R.C., and Vlachavas, D.S., 1983, *Proc. Roy. Soc. London A* 385:95.
Pond, R.C., and Vitek, V., 1977, *Proc. Roy. Soc. London A* 357:453.
Pontikis, V., 1988, *J. Phys. France* 49:C5.
Pope, D.P., and Ezz, S.S., 1984, *Int. Met. Rev.* 25:233.
Puls, M.P., 1980, *Philos. Mag. A* 41:353.
Puls, M.P., and Norgett, M.J., 1976, *J. Appl. Phys.* 47:466.
Rabier, J., and Grilhe, J., 1973, *J. Phys. Chem. Sol.* 34: 1031.
Rabier, J., and Puls, M.P., 1989, *Philos. Mag. A* 59:533.
Read, W.T., and Shockley, W., 1950, *Phys. Rev.* 78:275.
Rice, J.R., 1992, *J. Mech. Phys. Sol.* 40:239.
Rice, J.R., and Beltz, G.E., 1994, *J. Mech. Phys. Sol.* 42:333.
Rittner, J.D., Udler, D., Seidman, D.N., and Oh, Y., 1995, *Phys. Rev. Lett.* 74:1115.
Romig, A. D., Fowler , D. E., and Bristowe, P. D., editors, 1991, "Structure/Property Relationships for Metal/Metal Interfaces", Mat. Res. Soc. Symp. Vol. 229, Materials Research Society, Pittsburgh.

Rottman, C., 1989, *Scripta Metall.* 23:1037.
Rühle, M., and Petzow, G., 1981, in "Surfaces and Interfaces in Ceramic and Ceramic-Metal Systems", J. Pask and A. Evans, eds., Plenum Press, New York, p. 167.
Sakai, A., Nishioka, Y., and Suzuki, H., 1979, *J. Phys. Soc. Japan* 46:881.
Sangster, M.J.L., 1973, *J. Phys. Chem. Sol.* 34:355.
Sangster, M.J.L., Schröder, U., and Arwood, A.M., 1978, *J. Phys. C : Solid State Phys.* 11:1523.
Sangster, M.J.L., and Stoneham, 1981, *Phil. Magazine* 43:597.
Sato, A., and Masuda, K., 1981, *Philos. Mag. B* 43:1.
Schmidt, C., Ernst, F., Finnis, M.W., and Vitek, V., 1995, *Phys. Rev. Lett.*, to be published.
Schwartz, D., Vitek, V., and Sutton, A.P., 1985, *Philos. Mag. A* 51: 499.
Seidman, D.N., Krakauer, B.W., and Udler, D., 1994, *J. Phys. Chem. Sol.* 55:1035.
Self, P.G., and O'Keefe, M.A., 1988, in "High Resolution Transmision Electron Microscopy and Associated Techniques", P.R. Buseck, J.M. Cowley and L. Eyring, eds., Oxford University Press, Oxford, p. 244.
Siegl, R., Vitek, V., Luzzi, D.E., and Yan, M., 1995, in "Structure and Phase Stability", G. M. Stocks, C. T. Liu and P. E. A. Turchi, eds., to be published, TMS, Warrendale, Pennsylvania.
Simmons, J.P., Rao, S.I., and Dimiduk, D.M., 1993, in "High-Temperature Ordered Intermetallic Alloys V", I. Baker, R. Darolia, J.D. Whittenberger and M.H. Yoo, eds., Mat. Res. Soc. Symp. Vol. 288, Materials Research Society, Pittsburgh, p. 335.
Sinclair, R., Dahmen, U., and Smith, D. J., editors, 1990, "High Resolution Electron Microscopy of Defects in Materials", Mat. Res. Soc. Symp. Vol. 183, Materials Research Society, Pittsburgh.
Slater, J.C., and Koster, G.F., 1954, *Physical Review* 94:1498.
Šob, M., Vitek, V., and Oh, Y., 1992, in "Computational Methods in Materials Science", J.E. Mark, M.E. Glicksman and S.P. Marsh, eds., Mat. Res. Soc. Symp. Vol. 278, Materials Research Society, Pittsburgh, p. 205.
Srolovitz, D.J., Wang, H.Y., and Najafabadi, 1992, in "Intergranular and Interphase Boundaries in Materials: iib92", P. Komninou and A. Rocher, eds., Trans. Tech Publications, Aedermannsdorf, p. 23.
Stillinger, F., and Weber, T., 1985, *Phys. Rev. B* 31:5262.
Stroh, A.N., 1962, *J. Math. Phys.* 7:625.
Sun, C.P., and Balluffi, R.W., 1982, *Philos. Mag. A* 46:49.
Sutton, A.P., 1984, *Int. Met. Rev.* 29:377; 1988, *Acta Metall.* 36:1291; 1989a, *Philos. Mag. Lett.* 59:53; 1989b, *Philos. Mag. A* 60:147; 1991, *Philos. Mag. A* 63:793; 1992, *Phil Trans Roy Soc London A* 341:233.
Sutton, A.P., and Balluffi, R.W., 1987, *Acta Metall.* 35:2177; 1995,""Interfaces in Crystalline Materials", Oxford University Press, Oxford.
Sutton, A.P., Finnis, M.W., Pettifor, D.G., and Ohta, Y., 1988, *J. Phys. C : Solid State Phys.* 21:35.
Sutton, A.P., and Vitek, V., 1983a, *Phil. Trans. Roy. Soc. London A* 309:1; 1983b, *ibid* 309:37; 1983c, *ibid* 309:55.
Takeuchi, S., 1979, *Philos. Mag. A* 39:661.
Takeuchi, S., and Kuramoto, E., 1973, *Acta Metall.* 21:415; 1975, *J. Phys. Soc. Japan* 38:480.
Tarnow, E., Bristowe, P.D., Joannopoulos, J.D., and Payne, M.C., 1989, *J. Phys. C : Solid State Phys.* 1: 327.
Tarnow, E., Dallot, P., Bristowe, P.D., Joannopoulos, J.D., Francis, G.P., and Payne, M.C., 1990, *Phys. Rev. B* 42:3644.

Tasker, P.W., and Duffy, D.M., 1983, *Philos. Mag. A* 47:L45.

Tersoff, J., 1988a, *Phys. Rev. B* 37:6991; 1988b, *Phys. Rev. B* 38:9902.

Tersoff, J., Vanderbilt, D., and Vitek, V., editors, 1989, "Atomic Scale Calculations in Materials Science", Mat. Res. Soc. Symp. Vol. 141, Materials Research Society, Pittsburgh.

Thomson, R.E., and Chadi, D.J., 1984, *Phys. Rev. B* 29:889.

Tichy, G., Vitek, V., and Pope, D.P., 1986a, *Philos. Mag. A* 53:467; 1986b, *Philos. Mag. A* 53:485.

Torrens, I.M., 1972,""Interatomic Potentials", Academic Press, New York.

Urdaneta, E.C., Luzzi, D.E., and McMahon, C.J., 1992, in "Structure and Properties of Interfaces in Materials", W.A.T. Clark, U. Dahmen and C.L. Briant, eds., Mat. Res. Soc. Symp. Vol. 238, Materials Research Society, Pittsburgh, p. 201.

Van Heugten, W.F.W.M., Caspers, L.M., and De Hosson, J.T.M., 1979, *Phys. Stat. Sol. (b)* 92:199.

Vesely, D., 1968, *Phys. Stat. Sol. (a)* 29:675.

Vitek, V., 1968, *Philos. Mag. A* 18:773; 1975, *Crystal Lattice Defects* 5:1; 1976, *Proc. Roy. Soc. London A* 352:109; 1985, in "Dislocations and Properties of Real Materials", M. Lorretto, ed., The Institute of Metals, London, p. 30; 1992, *Prog. Mater. Sci.* 36:1.

Vitek, V., Ackland, G.J., and Cserti, J., 1991b, in "Alloy Phase Stability and Design", G.M. Stocks, D.P. Pope and A.G. Giamei, eds., MRS Symposia Proceedings, Vol. 186, Materials Research Society, Pittsburgh, p. 237.

Vitek, V., Ackland, G.J., and Cserti, J., 1991a, in "Alloy Phase Stability and Design", G.M. Stocks, D.P. Pope and A.G. Giamei, eds., Mat. Res. Soc. Symp. Vol. 186, Materials Research Society, Pittsburgh, p. 237.

Vitek, V., Yan, M., and Ackland, G. J., 1992, in "Ordered Inmtermetallics – Physical Metallurgy and Mechanical Behaviour", C. T. Liu, R. W. Cahn and G. Sauthoff, eds., Kluwer Academic Publishers, Dordrecht, The Netherlands, p. 335.

Vitek, V., Cserti, J., and Duesbery, M.S., 1995a,""Professor Sir Peter Hirsch Festschrift", C. J. Humpreys, ed., to be published.

Vitek, V., and De Hosson, J.T.M., 1986, in "Computer-Based Microscopic Description of the Structure and Properties of Materials", J. Broughton, W. Krakow and S.T. Pantelides, eds., Mat. Res. Soc. Symp. Vol. 63, Materials Research Society, Pittsburgh, p. 125.

Vitek, V., Khantha, M., Cserti, J., and Sodani, Y., 1991c, in "Int. Symp. on Intermetallic Compounds", O. Izumi, ed., The Japan Institute of Metals, Sendai, p. 3.

Vitek, V., Lejček, L., and Bowen, D.K., 1972, in "Interatomic Potentials and Simulation of Lattice Defects", P.C. Gehlen, J.R. Beeler and R.I. Jaffee, eds., Plenum Press, New York, p. 493.

Vitek, V., Minonishi, Y., and Wang, G.J., 1985, J. Phys. France 46:C4-171.

Vitek, V., Perrin, R.C., and Bowen, D.K., 1970, *Philos. Mag. A* 21:1049.

Vitek, V., Pope, D.P., and Bassani, J.S., 1995b, in "Dislocations in Solids", F.R.N. Nabarro, ed., to be published, North Holland, Amsterdam.

Vitek, V., Sodani, Y., and Cserti, J., 1991d, in "High-Temperature Ordered Intermetallic Alloys IV", L.A. Johnson, D.P. Pope and J.O. Stiegler, eds., Mat. Res. Soc. Symp. Vol. 213, Materials Research Society, Pittsburgh, p. 195.

Vitek, V., and Srolovitz, D. J, 1989, editors, "Atomistic Simulation of Materials: Beyond Pair Potentials", Plenum Press, New York.

Vitek, V., Wang, G.J., Alber, E.S., and Bassani, J.L., 1994, *J. Phys. Chem. Sol.* 55:1147.

Wang, G.-J., and Vitek, V., 1986, *Acta Metall.* 34:951.

Wang, G.J., Sutton, A.P., and Vitek, V., 1984, *Acta Metall.* 32:1093.
Warrington, D.H., and Grimmer, H., 1974, *Philos. Mag.* 30:461.
Watson, R.E., Davenport, J.W., Perlman, M.L., and Sham, T.K., 1981, *Phys. Rev. B* 24:1791.
Wee, D.M., Noguchi, O., Oya, Y., and Suzuki, T., 1980, *Trans. Japan Inst. Met.* 21:237.
Wee, D.M., Pope, D.P., and Vitek, V., 1984, *Acta Metall.* 32:829.
Whitworth, R.W., 1975, *Adv. Phys.* 24:203.
Wolf, D., 1980, J. Phys. France 41:C6-142; 1982, J. Phys. France 43:C6-45; 1984a, J Amer Ceram Soc 67:1; 1984b, Philos. Mag. A 49:823; 1989a, Scripta Metall. 23:1913; 1989b, Acta Metall. 37:1983; 1989c, Acta Metall. 37:2823; 1990a, J. Mater. Res. 5:1708; 1990b, Acta Metall. Mater. 38:781; 1990c, Acta Metall. Mater. 38:791.
Wolf, D., and Lutsko, J.F., 1989, J. Mater. Res. 4:1427.
Wolf, D., Lutsko, J.F., and Kluge, M., 1989, in "Atomistic Simulations of Materials: Beyond Pair Potentials", V. Vitek and D. Srolovitz, eds., Plenum Press, New York, p. 245.
Wolf, U., Ernst, F., Muschik, T., Finnis, M.W., and Fischmeister, H.F., 1992a, in "Structure and Properties of Interfaces in Materials", W.A.T. Clark, U. Dahmen and C.L. Briant, eds., Mat. Res. Soc. Symp. Vol. 238, Materials Research Society, Pittsburgh, p. 177; 1992b, *Philos. Mag. A* 66:991.
Woo, C.H., and Puls, M.P., 1977a, *Philos. Mag.* 35:727; 1977b, *Philos. Mag.* 35:1641.
Woodward, C., MacLaren, J.M., and Dimiduk, D.M., 1993, in "High-Temperature Ordered Intermetallic Alloys V", I. Baker, R. Darolia, J.D. Whittenberger and M.H. Yoo, eds., Mat. Res. Soc. Symp. Vol. 288, Materials Research Society, Pittsburgh, p. 171.
Wu, Z.L., Pope, D.P., and Vitek, V., 1990, *Scripta Metall.* 24:2191; 1994, *Philos. Mag. A* 70:159.
Xie, Z.Y., Vailhe, C., and Farkas, D., 1993, *Mat. Sci. Eng. A* 170:59.
Xu, W., and Adams, J.B., 1994, *Surf. Sci.* 301:371.
Yamaguchi, M., and Inui, H., 1992, in "Ordered Intermetallics – Physical Metallurgy and Mechanical Behaviour", C.T. Liu, R.W. Cahn and G. Sauthoff, eds., Kluwer Academic Publishers, Dodrecht, The Netherlands, p. 217.
Yamaguchi, M., Inui, H., Kishida, K., Matsumoro, M., and Shirai, Y., 1995, in "High-Temperature Ordered Intermetallic Alloys VI", J. Horton, I. Baker, S. Hanada, R.D. Noebe and D. Schwartz, eds., Mat. Res. Soc. Symp. Vol. 364, Materials Research Society, Pittsburgh, p. 3.
Yamaguchi, M., Paidar, V., Pope, D.P., and Vitek, V., 1982, *Philos. Mag. A* 45:867.
Yamaguchi, M., and Umakoshi, Y., 1975, *Scripta Metall.* 9:637.
Yamaguchi, M., Vitek, V., and Pope, D.P., 1981, *Philos. Mag. A* 43:1027.
Yan, M., Šob, M., Luzzi, D.E., Vitek, V., Ackland, G.J., Methfessel, M., and Rodriguez, C.O., 1993, *Phys. Rev. B* 47:5571.
Yin, M.T., and Cohen, M.L., 1982, *Phys. Rev. B* 26:5668; 1984, *Phys. Rev. B* 29:6996.
Yoo, M. H., Clark, W. A. T., and Briant, C. L., editors, 1988, "Interfacial Structure, Properties and Design", Mat. Res. Soc. Symp. Vol. 122, Materials Research Society, Pittsburgh.
Yoo, M.H., Daw, M.S., and Baskes, M.I., 1989, in "Atomistic Simulations of Materials: Beyond Pair Potentials", V. Vitek and D. Srolovitz, eds., Plenum Press, New York, p. 401.
Yoo, M.H., Sass, S.L., Fu, C.L., Mills, M.J., Dimiduk, D.M., and George, E.P., 1993, *Acta Metall. Mater.* 41:987.

DISLOCATION PATTERNS:

EXPERIMENT, THEORY AND SIMULATION

Ladislas P. Kubin

LEM, CNRS-ONERA (OM)
29, Av. de la Division Leclerc, BP 72
92322 Châtillon Cedex, France

I. INTRODUCTION: DISLOCATIONS DYNAMICS AND INTERACTIONS

Dislocations are the elementary carriers of plastic flow. It follows that understanding the mechanical properties of crystalline solids involves the knowledge of the basic properties of dislocations, of their dynamics and of their interactions with various types of obstacles. Among the latter, the most troublesome are dislocation themselves, as the mutual interactions of dislocations are both short and long ranged. When dislocation-dislocation interactions are predominant, the flow properties of a material depend on individual as well as collective dislocation properties in a complex manner which is still not well understood. The most striking feature is, then, the spontaneous formation of organized patterns consisting of alternating dislocation-rich and dislocation poor regions (e.g., cells, subgrains, veins or walls, channels...). These microstructures exhibit more or less well-defined periodicities, usually in the micrometer range. Such collective phenomena are relatively unimportant at small strains, since the total dislocation density is small. They may, however, play a prominent role on flow properties as soon as the total dislocation density stored in the deforming crystal reaches a critical value. It follows that investigations on dislocation patterning are important in two respects, -i) as an example of self-organization in a dynamical system and -ii) as a necessary intermediate step in the view of developing more physical approaches of plasticity, based on dislocation theory. For an extensive review of these questions, see Kubin (1993).

Schematically, the theory of dislocations yields access to two types of properties. Dislocation core properties and their consequences should ideally be treated at atomistic scale. Within the present context, however, such important mechanisms as cross-slip or dislocation intersection are still beyond the possibilities of molecular dynamics simulations. By considering dislocations no longer as atomic configurations but as elastic singularities embedded in a continuum, the elastic theory allowed to compute the stress and stress fields of dislocations outside the core region and all the related mechanisms or properties such as long range interactions, line tension effects and the formation of stable dipolar or multipolar configurations. By necessity, the elastic theory was also used to model cross-slip (Escaig,

Stability of Materials
Edited by A. Gonis *et al.*, Plenum Press, New York, 1996

1968) or the formation of junctions at dislocations intersections (Schoeck and Frydman, 1972). These elaborations serve as input for the mesoscopic approaches dealing with dislocation ensembles. Therefore, the duality between local or core mechanisms and elastic interactions is necessarily reflected in the models for dislocation patterning. The fact that dislocation stress fields are Coulombian poses serious problems since the motion of a dislocation affects an extended neighbouring volume and is affected by any dislocation movement within this same volume. Whether or not self-screening phenomena are sufficiently efficient to allow for a simplification of this many-body problem is a question still under discussion.

In what follows, a few types of patterns are discussed and, for each of them, the elementary dislocation mechanisms involved are indicated. Early models attempting to describe pattern formation are then examined with emphasis on some serious oversimplifications. Such models are, therefore, unable to answer questions such as how and in which conditions organized structures emerge during plastic flow and which are the properties which drive the corresponding intrinsic length scales. More recent dynamical models are, then, examined with emphasis on some of their weak points.

The development of simulations of dislocations microstructures is reviewed in historical perspective, starting with simulations of elementary mechanisms and properties of a dislocation gliding in its slip plane. In more recent studies, 2-D simulations where the dislocation lines are seen edge on yielded a few interesting results regarding essentially the self-organization properties of assemblies of dislocations in conditions of easy glide. Finally a brief account is given of the most recent 3-D simulation of dislocations and plastic flow.

II. DISLOCATION MICROSTRUCTURES AND PATTERNING

In a deforming crystal, there are two types of dislocation populations. The mobile dislocations produce the strain rate and the immobile or weakly mobile ones, also called stored dislocations, are responsible for strain hardening as they constitute an obstacle for the moving ones. Slip patterns, i.e. the patterns produced at the surface of crystals by the emergence of the mobile dislocations will not be discussed here (cf. Kubin, 1993, for review and reference). The dislocation patterns observed by transmission electron microscopy are formed by the (essentially) immobile density. In this last case, one can distinguish between patterns formed in single slip and involving only one active Burgers vector (dipolar or multipolar patterns, persistent slip bands ..) from those formed in multislip (cellular patterns). The basis of this distinction is that dislocation intersections play an important role in the last case, while the trapping of dislocations in dipolar configurations is essentially dominated by elastic interactions. In what follows, these patterns are discussed with particular reference to the mechanical behaviour of f.c.c. single crystals around room temperature and below. This involves the well-know hardening stages in uniaxial straining (cf. Rollett and Kocks, 1994) as well as the cyclic stress-strain curves (cf. Suresh, 1993) in which the maximum stress at saturation is plotted as a function of the imposed plastic strain amplitude per half-cycle (cf. fig. 1).

2.1. Dipolar Patterning

2.1.1 Stability of Multipolar Structures, Critical Annihilation Distance. Dislocations moving in parallel slip planes can mutually trap themselves by their elastic interactions. This leads to the formation of dipolar and multipolar configurations. Under a resolved applied stress τ_a, an edge dipole is stable provided that its height is smaller than a critical value $y_c = \chi\mu b/8\pi\tau_a$, where μ is the shear modulus, b is the Burgers vector and χ depends on the dislocation's character. In a similar manner, any multipolar configuration has

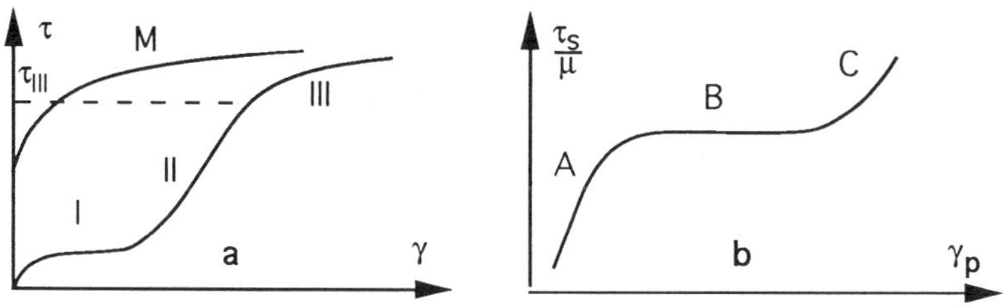

Figure 1. a) - The three-stage stress-strain curve typical of f.c.c. single crystals. b) - The typical three-stage cyclic stress-strain curve of fatigued copper single crystals at room temperature (τ_S is the saturation stress and γ_p the plastic strain amplitude per half-cycle). In both cases the crystals are initially oriented for easy glide. The curve M corresponds to crystals initially oriented for multiple slip. Stages I, A and B correspond to dipolar patterns, stages III, C and curve M to cellular patterns.

a stability limit under an applied stress and it is usually assumed that its density permanently tends to the critical value deduced from the stability limit. There is also a lower bound to the stability of dipolar configurations, which depends on the character of the dislocations involved. Screw dislocation dipoles may annihilate by cross-slip. This defines a critical annihilation distance, y_S, which of the order of 40-50 nm in copper at room temperature (Essmann and Mughrabi, 1979), and much larger for aluminium where cross-slip is comparatively easier. For edge dislocations, and assuming that the temperature is not too high so that no thermal vacancies are available to induce annihilation by climb, the destruction of a dipole is more difficult. The only possible annihilation mechanism is the mechanical breakdown of the configuration under the effect of its self-stresses when the latter are of the order of the theoretical strength (typically $\mu/10$). Essmann and Mughrabi (1979) measured values of $y_e \approx 1.5\text{-}1.6$ nm in very dense arrays of edge dislocations formed in fatigued copper at room temperature. Because of the strong difference between the two critical annihilation distances for screw and edge dislocations, it is usually possible to accumulate much larger densities of edges in deformed crystals. These critical annihilation distances serve, in addition, as a base for the discretization methods used in numerical simulations (cf. Sections 5.2.1. and 6.1.2.). Little is known about this rather important length parameter which certainly deserves more investigation, particularly at atomistic scale.

2.1.2. Dipolar and Multipolar Microstructures. During stage I of the deformation curve of pure f.c.c. metals (cf. fig. 1-a), multipolar clusters essentially containing edge dislocations are formed with a spacing of a few micrometers. This stage ends after moderate strains, before a well-defined pattern can be characterized. There are, however, two situations where it is possible to accumulate significant multipolar densities. One is concerned with the extended stage I of dilute alloys, where more strain is available before the onset of secondary slip due to the occurrence of strain softening phenomena. The other case is that of the fatigue of single crystals oriented for easy glide, which leads to the accumulation of very large dislocation densities in the primary system (Mughrabi, 1979). In the first stage of this curve (A on fig. 1-b) dense multipolar veins with a volume fraction of about 0.5 are formed. With increasing plastic strain amplitude (stage B on fig. 1-b) the vein structure is progressively destabilized and the specimen gets filled with an increasing number of Persistent Slip Bands (PSBs) which concentrate the slip activity. As illustrated by fig. 2-a, the structure of PSBs exhibit a striking regularity, with channel widths of the order of 1 μm in copper at room temperature. Such configurations are able to remain plastically active for a large number of cycles, so that they correspond to a dynamic steady state. This is one of the reasons why PSBs attracted lot of theoretical attention, the other one being that the stress field of multipolar configurations decays rather fast with increasing distance.

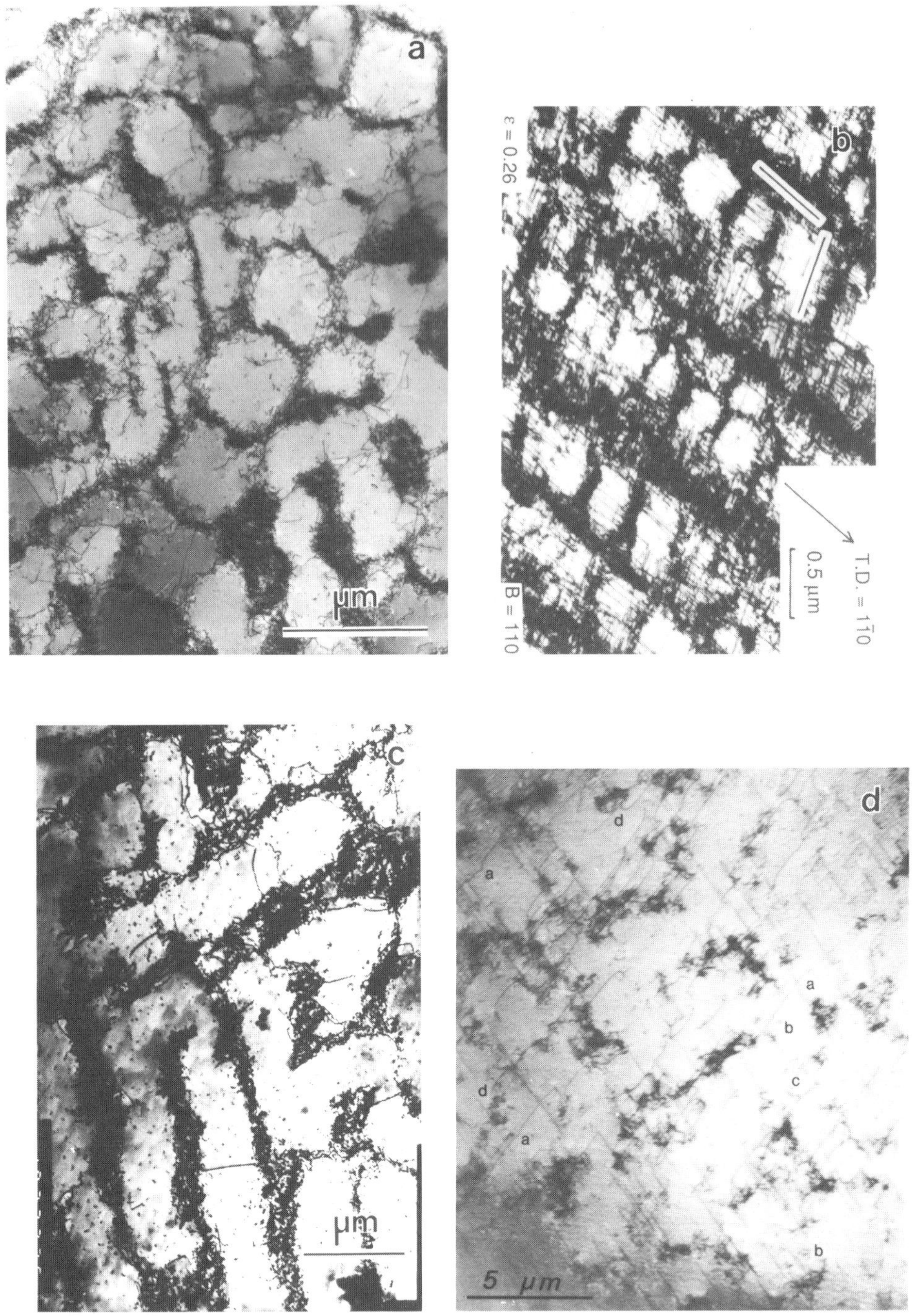

Figure 2. Types of dislocation patterns. a - A PSB (from bottom left to top right) in a copper crystal fatigued at room temperature (Courtesy J. Lépinoux). b - Cell structure in Fe-3%Si deformed at room temperature (After E. Rauch (1992), Courtesy J. Mech. Behav. of Mats.). c - Labyrinth structure in copper fatigued at room temperature (Courtesy J. Lépinoux). d - Uniform dislocation structure (under stress) in α-Fe strained at 180 K.

2.2. Cell Patterning

2.2.1. Stage II and the Forest Mechanism. Stage I of the deformation curve ends with the beginning of secondary slip activity (cf. fig. 1-a). During most of stage II, the secondary dislocations, although in density almost comparable to that of the primary ones, do not move over large distances. They contribute substantially to the total strain only towards the end of this stage. Secondary dislocations are, however, responsible for the strong, approximately linear, strain hardening coefficient (about $\mu/300$) which persists over a quite large domain of temperatures. The microstructures which develop in this domain are rather complicated and are dominated both by dipolar interactions and dislocation cutting (see the review of Basinski and Basinski, 1979, for detail). This is one reason why patterns developed in stage II have not attracted a lot of attention, the other one being that the strain hardening properties are satisfactorily represented by simple phenomenological laws (Kocks, 1976).

As soon as two or more Burgers vector interact (i. e. starting from stage II and beyond), within a good approximation, a simple relation holds between the resolved applied stress τ_a and a density which can be either the total dislocation density, or the forest density :

$$\tau = \alpha\mu b\sqrt{\rho_f} \tag{1}$$

α is a constant whose value is 0.3-0.4 in the f.c.c. structure [8]. This relation can be understood via the consideration of simple elementary dislocation mechanisms. If the forest density is ρ_f, the average spacing of forest obstacles in a glide plane is about $\rho_f^{-1/2}$. Through line tension effects, the cutting of such obstacles requires a stress inversely proportional to the obstacle spacing, which leads to eq. 1. An additional contribution to the glide resistance (about 20% of the total value, cf. Gil Sevillano, 1993) stems from the long range elastic stresses which also scale as the inverse of the distance between dislocations. The strong increase in flow stress recorded during stage II appears to be directly related to the increased density of the forest obstacles (forest hardening) and to the storage or immobilization of an increasing density of dislocations (cf. Kocks' model in Section 3.4). A few important features are, however, not well understood: - The strain hardening coefficient is approximately the same for all f.c.c. crystals despite large differences in stacking fault energy and critical stresses for obstacle cutting. - Eq. (1) seems to hold whenever obstacle cutting takes place and irrespective of the spatial distribution of the dislocations, at least within the precision of experimental measurements (see Neuhaus and Schwink, 1992, for a discussion).

2.2.2. Cell structures. Three-dimensional cell structures (cf. fig. 2-b) are most commonly observed in multislip conditions. They occur in polycrystals, where several slip systems are required in principle to maintain the compatibility of deformation between the grains, during stage C of the cyclic stress strain curve, during the deformation of single crystals oriented for multiple slip (curve M of fig. 1-a) and from the onset of stage III on the deformation curve of single crystals, i.e. when the stress reaches a critical value τ_{III} (cf. fig. 1-a). This last case is the most informative. Beyond τ_{III}, the strain hardening coefficient is reduced and the stress strain curve assumes a pseudo-parabolic shape. In phenomenological approaches, this critical stress marks the onset of recovery processes, leading to the annihilation of some forest obstacles and to a reduction of the rate of increase of the forest density.

Although there has been some debate about the nature of the recovery process involved during stage III (not to mention further stages which are not discussed here, see e.g. Rollett and Kocks, 1994; Gil Sevillano, 1993), there is little doubt that in fcc metals and at room temperature the effective mechanism involves a partial annihilation of the stored density by cross-slip. Indeed, experimental observations indicate that the critical stress τ_{III} is thermally activated, with activation parameters coherent with the predictions of cross-slip models (Bonneville et al., 1988). In copper at room temperature, the value of τ_{III} is about 30 MPa. In aluminium and at room temperature, there is no stage II. Stage III and the formation of a cell structure occur as soon as stage I ends, i.e. at a rather low stress. This is in full agreement with the fact that cross-slip is easy in this material as the dislocations are not dissociated.

A low temperature cell wall such as those shown on fig. 2-b contains a more or less complex entanglement of dislocations of different Burgers vectors. As soon as it is formed and although it is not totally inpenetrable, it constitutes an obstacle to dislocation motion. For each active slip system, dislocations of a given sign tend to accumulate on one side of the wall and dislocations of the other sign, which move in the opposite direction, on the other side. This gives rise to internal stresses of dipolar character. As the dislocation-poor regions enclosed by the walls do not necessarily experience the same strain, there is a statistical accumulation of differential glide which progressively leads to the building up of misorientations across the cell walls. Therefore, the dislocation density in a wall may be thought as containing two populations: the "geometrically necessary" dislocations which accommodate the misorientation and the redundant accumulated dislocations which induce a long range dipolar stress field. While cross-slip is efficient at moderate temperatures to eliminate the redundant screw dislocations, climb phenomena which are favoured by low strain rates and higher temperatures typical of creep experiments lead to the elimination of edge dislocations. Therefore, and quite generally, with increasing temperature the cell walls tend to become less fuzzy, while the geometrically necessary dislocations tend to form perfect subgrains devoid of long range stresses. Therefore, subgrain formation can be seen as occurring in situations where the critical time associated with recovery processes is small compared to the dynamic time interval associated with the arrival of dislocations at the walls.

The sequence of events leading to the formation of a cell structure are not really understood. Some are discussed below, in particular in relation with Kocks' kinematical model (cf Section 3.4). On the other hand, it is clear from the brief outline given above that several mechanisms concur to the formation of such three-dimensional structures. First, several slip systems must be active which means that the forest mechanism (i.e., both forest cutting and long range elastic interactions) is involved. A second necessary ingredient is provided by the observation that cell structures are formed simultaneously with the onset of recovery mechanisms, in particular cross-slip (Hirsch, 1975). This fact has been recognized very early, but the classical argument following which cross-slip provides (planar) dislocation glide with the necessary degree of freedom needed to form a three-dimensional microstructure is not totally convincing. Indeed, a confusion is often made between occasional cross-slip leading to the mutual annihilation of two mobile screw segments and recovery by cross-slip which involves the large scale annihilation of immobile screws in the dislocation rich areas. Several other questions have been the object of debates: - What is the exact role of cross-slip ? - What is the internal stress field associated with a cell structure (cf. e.g. Mughrabi, 1987) ? - What are the parameters or mechanisms governing the cell size (cf. Section 2.2.5.) ?

2.2.3. Wall and Channel Structures. With two simultaneously active Burgers vectors, other types of periodic or pseudoperiodic wall patterns can be formed. If cross-slip is not too active, such structures contain channels parallel to the direction common to the two slip systems where no forest obstacle is found. Thus, a blend of dipolar and forest mechanisms will be effective in such types of patterns. In some cases, the possible coexistence of two types of walls leads to the formation of a labyrinth structure (cf. fig. 2-c).Wall and channel patterns are formed during the deformation of suitably oriented single crystals and during the fatigue of single or polycrystals.

2.2.4 Effective vs. Internal Stresses. According to Seeger (1955), a linear superimposition law applies to all the stresses acting on a dislocation segment:

$$\tau_a = \tau_\mu + \tau^* \qquad (2)$$

τ_a is the resolved applied stress and τ^* is the thermally activated component of the flow stress which governs dislocation motion through short-ranged obstacles like Peierls barriers or small impurity clusters. τ_μ is the athermal component of the flow stress and it is made up of several contributions. One stems from the long range elastic interactions of the considered segment with all other dislocation lines present in the microstructure. Global stress equilibrium imposes that the spatial average of this internal stress is zero, but its average on dislocations is always resistive, as dislocations tend to arrange themselves in such a way as to produce more or less self-screening configurations. The intersections of dislocations of different Burgers vector gliding in different slip planes may lead to the formation of sessile reaction products which, as discussed above, govern strain hardening in multislip. Such reactions usually involve large energies (typically a few eV per atomic length of dislocation line) so that the breaking through such obstacles is almost athermal. Therefore, the main two types of dislocation-dislocation interactions contribute to the athermal stress. A last component is the resistive stress deriving from the line energy, which opposes the increase in length of the dislocation segments.

As plastic flow proceeds and the density of dislocations present in the crystal grows, the dislocation-dislocation interactions (both long and short ranged) induce a transition from a rather randomly distributed microstructure towards an organized one. Opposed to this tendency is, in some crystals, the interaction with a large and constant density of strong, localized obstacles. With increasing strain, this last component of the flow stress is rather constant while the former one increases. A transition to cell patterning may, therefore, occur for a certain critical strain, stress or dislocation density. This behaviour is illustrated by fig. 3 which shows the critical condition for the formation of a cell structure in (bcc) α-iron as a function of temperature. In this material, like in other bcc transition metals, the yield stress is governed by a strong lattice friction on screw dislocations which decays with increasing temperature. Fig. 2-d shows the uniform structure formed in α-iron after straining at low temperature (180 K). At largers strains, the small clusters grow and eventually form a cell structure.

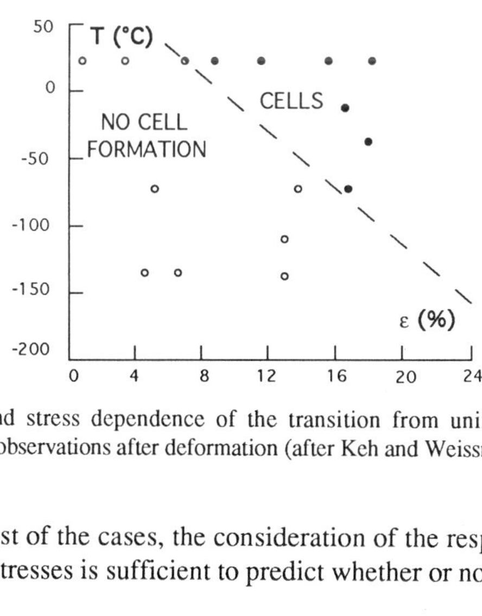

Figure 3. Temperature and stress dependence of the transition from uniform to cellular dislocation structures in bcc iron. TEM observations after deformation (after Keh and Weissmann, 1963).

In summary, in most of the cases, the consideration of the respective magnitude of the effective and athermal stresses is sufficient to predict whether or not dislocation patterns are formed in multislip. As indicated above (cf. Section 2.2.2.) the relaxation of internal stresses via recovery mechanisms at high temperatures or high stresses or low strain rates, i.e., in conditions where the effective stress is small, leads to a gradual transition between cells and subgrains.

2.2.5. Cell Dimensions: the "Similitude Principle". Up to now we have mentioned several types of dislocation patterns: veins, PSBs, walls and channels, cells, which may be characterized by essentially two length scales: the characteristic width of the dislocation-rich areas and the average width or diameter, d, of the dislocation-poor regions. This last quantity has attracted lot of attention and measurements have yielded a very simple relation between d and a stress σ, usually the applied stress:

$$\sigma = K\mu b / d \tag{3}$$

This relation is approximately valid for many materials but not all of them. In particular in some alloys with low stacking fault energy like certain stainless steels, glide remains planar all through the deformation curve. The constant K is probably not a “universal constant", although its value is often of the order of 20. Fig. 4 is extracted from the impressive compilation of data of Raj and Pharr (1986) and shows that a reasonable fit of eq. (3) can be obtained in such common materials as Al or α-Fe. The shrinking of dislocation patterns with increasing stress is known under the name of the "similitude principle", although it constitutes nothing else than an experimental fact. The origin of this very simple relation and the calculation of the value of K has been the object of many models which are discussed below, but none appears to be really convincing.

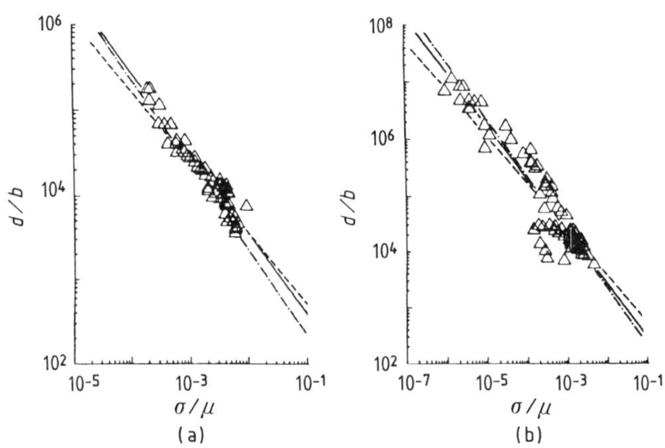

Figure 4. Plots of the reduced cell or subgrain size, d/b, vs. the reduced stress σ/μ in a - α-Fe and b - Al. After Raj and Pharr (1986).

3. CLASSICAL MODELS

3.1. General Assessment

By "classical models" is meant models elaborated within the conventional frame of dislocation theory and which essentially attempt to explain the charactristic features of cell structures. Implicitly or explicitly, most of the earlier models are based on dimensionality arguments and make more or less hidden assumptions about mechanical or thermal equilibrium. In addition, they assume most of the time that dislocation configurations are self-screening ones, so that the resistive component stemming from long range stresses can be treated as a local quantity (cf. e.g. Kocks et al., 1975). Long range stresses depend, indeed, on the correlation function of dislocations of opposite sign which defines the screening properties of a given microstructure. In other words, the long range stress fields of a multipole, of a perfect tilt wall, of a random array are substantially different from each other and a consistent model of patterning should not make any hypothesis about the type of organized structures that is bound to emerge. Further, the similarity between the scaling laws associated with forest interactions and eq. 3 has suggested that the scaling laws governing both individual and collective dislocation behaviour are of same origin. This is equivalent to assuming that dislocations structures are in mechanical equilibrium under stress, i.e., that the length scale of the microstructure smoothly adjusts to a change in stress through the individual displacement of each dislocation. As remarked by Seeger (1988), this cannot be true because if this was the case, TEM examinations would never lead to the observation of any srtucture in thin foils extracted from deformed and subsequently unloaded specimens. It follows that dislocation configurations are trapped in rather deep metastable positions. Their evolution during a change in applied stress necessarily involves steps of destruction and reconstruction of the dense configurations. Finally, thermodynamical arguments have also been invoked to explain the formation of dislocation patterns but it is now understood, from numerous studies performed in the last fifteen years on dynamical systems, that plasticity is a dissipative phenomenon among many others, where self-organization results from collective dynamic effects involving complex interactions.

3.2. Energetical Aproaches: LEDS.

In this controversial model, a principle is assumed according to which a given configuration of density ρ is arranged in such a way as to minimize the elastic energy per unit length of dislocation (Kuhlmann-Wilsdorf, 1989, 1992), hence the denomination of LEDS (Low Energy Dislocation Structures). Early versions of this purely static model have involved justifications in terms of thermal equilibrium and hypotheses of mechanical equilibrium and perfect self-screening. Although there is no broad physical justification to the LEDS principle, one may find a few situations where it applies. This requires two conditions: first, the temperature must be high enough to allow dislocations to move along the direction of the net applied force (even when the latter is not in the glide plane), in order to effectively reduce the local gradients in elastic energy. Second, a constraint must exist which locally fixes the density of dislocations in order to avoid total annihilation. A typical example is that of a perfect subgrain accommodating a misorientation between two grains. These geometrically necessary dislocations are precisely those which conform to the LEDS principle.

3.3. Energetical Aproaches: Holt's Model

Holt's model (1970) is based on an analogy between dislocation patterning and the spinodal decomposition which explains in energetical terms the spontaneous emergence of periodic modulations of composition in dilute alloys. The system considered by Holt is bi-dimensional and consists of a constant density of infinite, parallel, screw dislocations, in the absence of applied stress and of local mechanisms like annihilation or multiplication. The thermodynamic system considered is a closed, conservative one, so that it is justified to look for a configuration which minimizes an energy functional. The initial configuration consists of a random distribution of dislocations but a very strong hypothesis is made: it is assumed that the correlation function of dislocations of opposite sign decreases with increasing distance and exhibits a cut-off or self-screening distance R_c which scales with $\rho^{-1/2}$. In such conditions, the response to a periodic fluctuation in dislocation density yields a preferential wavelength λ_m proportional to R_c, i.e. to $\rho^{-1/2}$. This result has been considered for a long time as providing an explanation for the formation of cell structures, as $\rho^{-1/2}$ is proportional to the inverse of the applied stress (cf. eq. 1), so that $\lambda_m \propto (1/\sigma)$. This explanation may, however, seem strange for several reasons. One is that there is in fact no applied stress involved in the model, an other is that eq. 1 derives from the forest mechanism which is also absent from the model. It is now understood, and checked by two-dimensional simulations (cf. Section 5.2.2.), that the patterns thus obtained are just a direct consequence of the assumed self-screening. The system effectively minimizes its total interaction energy by forming non-interacting walls with a spacing of the order of R_c.

3.4. Kocks' Model

Kocks' kinematic or statistical model (1966, 1985) constitutes the first attempt to go beyond purely static or energetical approaches. It describes a sequence of events leading to the formation of a cell structure. The starting point was provided by two-dimensional simulations of the motion of dislocations through arrays of point obstacles in constant density ρ, representing the forest dislocations (cf Section 4.1.2.). During its motion, a mobile dislocation necessarily encounters fluctuations in the glide resistance opposed by the obstacles. In particular, impenetrable clusters may be met which can be by-passed only by the Orowan mechanism. The concave, planar, Orowan loops left by this process contribute to the stored density and behave as obstacles for other dislocations gliding in the same slip plane. This process leads to the formation of dislocation tangles around the hard spots and to subsequent dislocation storage and strain hardening. The model focusses on the building up of internal stresses in these dislocation-rich areas and their relaxation by recovery and by activation of secondary slip systems. When the tangles extend to the point that they become three-dimensionally connected, a cell structure is formed. In Kocks' simulations, the mean distance between the hard spots, d, and the obstacle spacing $\rho^{-1/2}$ are empirically found to follow the relationship $d \approx 10\,\rho^{-1/2}$. As the cell structure is formed around these impenetrable clusters, d also represents the cell size, so that with the help of eq. (1), the relation (3) between the cell size and the stress is recovered with $K \approx 10\,\alpha$.

The interest of this model is that it provides a natural explanation for the first stages of the cell formation. Further, the similitude principle is derived from a combination of two different phenomena, the forest mechanism on the one hand, the statistics of the obstacle distribution on the other hand. Finally, the role played by the primary stored density on strain hardening is emphasized. This contribution stems from the stored loops left by dislocations of same Burgers vector as the mobile ones, which adds up to the hardening effect of the forest. A discussion of this model has been given by Hirsch (1975) who pointed out at several difficulties. This model is essentially planar and privileges one single system. In conditions of multiple slip, not only the dislocation densities must initially be comparable in all the slip systems, but the initial distribution of the hard spots may be three-dimensional which could destroy the correlation found between d and ρ. Furthermore, cross-slip plays no role in this model, in contrast with experience and numerical simulations (cf. Section 6.2.).

3.5. Long Range Stresses: the Composite Model

Mughrabi's composite model (1987) is not actually a model of dislocation patterning. However, it emphasizes a possible origin for the long range internal stresses associated with dislocation patterns. Further, it points out at a very basic difficulty associated with the dynamic models which we describe in the next part. Within a continuum description of the microstructure, the latter can be seen as a biphased material containing a hard elastic phase (the dislocation-rich regions) and a soft plastic one (the channels or cell interiors). The requirement of strain continuity at the interfaces necessarily involves additional compatibility stresses. This model has the merit of explaining the origin of the internal stresses which accumulate during plastic flow, even in conditions of strong recovery and when the walls are perfectly self-screening structures.The question of the origin of such long range stresses has been the object of many discussions in the past. Within the present context, the point of importance is that when one attempts to build up a continuous model of the dislocation microstructure under stress, compatibility conditions must necessarily be taken into account. Finally, one can note that within a discrete description of the microstructure, like the one used in numerical simulations, compatibility conditions are naturally accounted for.

4. DYNAMIC MODELS

4.1. Dynamic Approach by Reaction-Transport-Diffusion

There have been several attempts to establish a physically founded approach of the question of dislocation patterning (see Goriatchev, this book). Recently, Hähner (1995) has proposed a stochastic theory of dislocations. In it, the stress or strain rate fluctuations are not in thermal equilibrium, as in classical equilibrium statistical mechanics, but derive in a "geometrically necessary" manner from the long range elastic interactions. At present, the existing models for pattern formation involve non linear evolutionary laws for the dislocation populations and spatial couplings. Both the non linearities and the spatial couplings strongly influence the obtained solutions and the main difficulty stems from the treatment of the non local internal stress. The models described below essentially deal with dipolar patterning, in contrast with the models discussed in the previous section. The reason is that multislip conditions involve a substantially larger number of mechanisms and of dislocation populations (one, at least, per active slip plane).

4.1.1. Reaction-transport-diffusion. Assuming that the total density is split into several populations, for instance mobile and immobile dislocations or mobile dislocations and dipoles, a classical continuity or dynamic balance equation can be written for the population i:

$$\partial\rho_i / \partial t + \mathrm{div} J_i = g_i(\rho_i, \rho_j, \tau) \tag{4}$$

$J_i = \rho_i v_i$ is transport term, i.e., the flux associated with the density ρ_i moving with the velocity v_i. The function g_i contains a set of reaction terms describing the short range mechanisms associated with this population and its couplings via local interactions with other populations j. In some of these reaction terms the stress τ, or any other relevant quantity, may play the role of a control parameter describing the influence of the externally applied conditions. If the divergence term is omitted, familiar evolutionary laws for the evolution of average dislocation densities are recovered in principle. The omission of the reaction terms leads to conservation equations such as those used in the continuum theory of dislocations or in Holt's model (1970), where local interactions were not taken into account. Therefore, the duality between local and long range interactions is also found in this reaction-transport scheme.

Such sets of partial differential and non linear equations have been widely used to describe the coupled behaviour of chemical species. In such cases, the divergence term can be transformed with the help of Fick's law into a Laplacian of the density/concentration. Then, a set of reaction-diffusion equations is obtained, which exhibits various types of solutions including spontaneous forms of spatial organization which are called Turing structures (cf. Misbah, 1988; Walgraef, 1988, for an account of such solutions). In the case of dislocations, it is not quite easy to establish the diffusive character. Let's adopt the simplifying assumption that the dislocation velocity is proportional to the effective stress. With the help of eq. (2), and neglecting the line tension, we have

$$v_i = M(\tau_a - \tau_\mu) \tag{5}$$

where M is a mobility (for the sake of simplicity, all equations are written in scalar form). As done by Walgraef and Aifantis (1985), it is convenient to express the total interaction energy $E_i(r)$ in the form:

$$E_i(r) = \int_V \rho(r')f(r')I(|r-r'|)d^3r' \tag{6}$$

where I is the pair interaction and f(r') a distribution function which takes into account the relative distributions of dislocations of both signs. The internal stress is deduced by $\tau_\mu \propto -\nabla E_i(r)$, but it is clear that its knowledge requires that of the solution which is searched for. Further, the internal stresses are non local, which introduces an additional complication. Alternatively, a Taylor expansion of the distribution function can be attempted yielding at first order $\tau_\mu = -\beta\nabla\rho_i(r)$ and a diffusion-like behaviour when inserting this result into eqs. (5) and (4):

$$\partial\rho_i / \partial t - D_i\nabla^2\rho_i = g_i(\rho_i, \rho_j, \tau) \tag{7}$$

The pseudo-diffusion coefficient, $D_i = \beta M\rho_i$, depends on both the mobility tensor and the dislocation densities. To date, only dipolar patterning has been examined by such methods. Indeed, there is a conceptual difficulty in attributing a diffusion coefficient to immobile dislocation walls which may have, in addition, long range stress fields.

4.1.2. The reaction-diffusion model of Walgraef and Aifantis. In this model (Walgraef and Aifantis, 1985; Aifantis, 1986, 1987, 1988, 1990; Walgraef, 1990; Walgraef et al. 1987; Schiller and Walgraef, 1988) two populations are considered, mobile dislocations (m) and a population of almost immobile dipoles (i), which are submitted to cyclic stress conditions. In its unidimensional form the set of equations is written:

$$\partial\rho_i / \partial t - D_i\partial^2\rho_i / \partial x^2 = g(\rho_i) - p\rho_i + \beta\rho_m\rho_i^2 \tag{8}$$

$$\partial\rho_m / \partial t - D_m\partial^2\rho_m / \partial x^2 = p\rho_i - \beta\rho_m\rho_i^2 \tag{9}$$

Here, the pseudo-diffusion coefficient D_m for the mobile dislocations is derived from the coupled evolution of the mobile dislocations of both signs. The pseudo-coefficient D_i for the "almost immobile" dislocations stems from a Taylor expansion of the internal stress as described above. The reaction term $g(\rho_i)$ describes the stress-assisted generation and annihilation of the slow dislocation species while the parameter p governs the freeing rate of the immobile dislocations and appears as a production term in the equation for the mobile density. The last term at the right-hand side is supposed to represent the rate of locking of immobile dislocations by slow dipoles. This set of equations is a simple transposition of a classical reaction-diffusion form (Walgraef, 1988), which is known to yield patterns (Turing instabilities) when the ratio D_m/D_i is large enough, as is the case here. As such, it has been criticized on the grounds that it contains some unrealistic assumptions in terms of dislocation processes (Salazar et al. 1995). The derivation of a diffusion coefficient for the dipolar structure, which in practice does not drift in the crystal, is perhaps questionable. The Taylor expansion of the dislocation distribution function is equivalent to an hypothesis of effective screening and needs to be justified. Further, the meaning of some reaction terms is a bit uncertain. The cubic non linearity plays an essential role in the solutions obtained for chemical reactions, but if the species i is dipolar, an hypothesis which is used to justify the convergence of the Taylor expansion, its interaction with mobile dislocations should only lead to a quadratic term at the right-hand side of eqs. 8 and 9.

The solutions reproduce well the patterns obtained in the fatigue of copper or nickel crystals at room temperature. With increasing maximum stress of the fatigue cycles, the predicted microstructures bifurcate from a homogeneous solution to a polygonal rod-like structure, akin to the matrix structure and, further, to a layered structure similar to that of PSBs. As the formation of PSBs in fatigue is not an universal feature, even in fcc metals and alloys, it is not clearly understood why this reaction-diffusion frame is so well adapted to the description of this important, but particular, case.

4.2. Kratochvil's Model

This elaboration does not attempt to follow closely a reaction-diffusion scheme. It consists of a dislocation dynamics part, emphasizing the behaviour of the dipolar density, complemented by a continuum approach of the stress and strain fields (Kratochvil, 1994). The microstructural part essentially incorporates a dynamic law for the evolution of the dipolar density which contains both a diffusion coefficient and a non local hardening term. In unidimensional form, this evolutionary law is written (Kratochvil, 1988a; Kratochvil and Libovicky, 1986; Franek, Kalus and Kratochvil, 1991):

$$\partial\rho / \partial t + D\partial^2\rho / \partial x^2 \int_{-\infty}^{+\infty} M(x - x')\rho(x')dx' = a - b\rho^n \qquad (10)$$

D is a (positive) diffusion coefficient for dipoles. In contrast with the model of Walgraef and Aifantis, this coefficient takes into account that dipoles do not move under uniform stress fields but under stress gradients. The latter are expressed by the non local term at the left-hand side of eq. 10. More specifically, the motion of dipoles has been assumed to occur through a drift mechanism towards stress concentrators under long range stresses stemming from the dipole population (Kratochvil and Libovicky, 1986) or via the encounter with mobile dislocations which "sweep" the dipoles towards regions of high dipolar density (Kratochvil and Saxlova, 1992a-b). M is a positive, symmetrical "influence" function which incorporates an unknown length scale defining its width. The reaction terms at the right-hand side take into account the generation and destruction of dipoles via the constants a, b and n. Although much remains to be justified in this phenomenological form of the sweeping model, eq. 10 predicts that the dipole loop distribution is unstable. Several patterning geometries emerge as a consequence: A pattern of veins consisting of dislocation patches approximately perpendicular to the primary slip direction, resembling the structures formed in Fe-Si at the end of stage I in unidirectional straining and the vein structure during stage A of cyclic deformation. In multiple slip various dipolar structures are obtained, ladder, labyrinth or cellular, provided that the structural aspects of the model predominate. One may note that in this model, the local interactions of dislocations of different Burgers vectors are not taken into account. No length scale can be predicted, as the patterns scale with the unknown width of the influence function. A self-consistent calculation of the influence function would, however, solve this difficulty. There is an other difficulty, as it is not certain that a gliding dislocation can sweep a dipolar loop of same Burgers vector without destroying it.

The second part of the model contains the conditions of stress equilibrium and Hooke's law (Kratochvil, 1989; Kratochvil and Orlova, 1990; Kratochvil, 1994). Different versions are used for small strain and large strains, which incorporate in the last case non linearities stemming from a continuum mechanical approach. It is argued that when this last aspect predominates over the structural one, geometrical considerations govern the length scale of the patterns, e.g. the width of the PSB chanels or the cell size in multislip (Kratochvil, 1994). For instance, the formation of subgrains at high temperatures stems, in the continuum approach from the large difference in glide and climb velocities, inducing an instability of the internal bending type (Kratochvil and Orlova, 1990). PSBs, on the countrary would result from an instability of the low strain hardening type, like shear bands (Kratochvil, 1994). Although some aspects remain to be justified, several dipolar patterning geometries emerge which are consistent with experiment.

The two original points of this model are the consideration of a non local hardening and of compatibility conditions. The points which may lead to discussion are concerned with the possible dual origin of the predicted wavelengths, microstructural or geometrical, the absence of numerical prediction about the numerical values of these wavelengths, the nature of the mechanism underlying the evolutionary equation which describes dipolar patterning and the absence of forest and cross-slip mechanisms in multislip conditions.

The two models discussed above illustrate well the present limitations of a too broad theoretical approach to dislocation patterning. Beyond the phenomenological character, and the need for more transparency in terms of dislocation mechanisms, several interesting questions can be raised: What are the basic dislocation mechanisms which must absolutely be incorporated, even in simplified form, and what are those which can safely be ignored ? Will a global model for dislocation patterning be too complicated to be tractable and should the emphasis bear instead on the simpler patterns occurring at small strains ? How far is it necessary to go when incorporating properties stemming from the continuum mechanical approach ?

4.3. The Model of Essmann and Differt

This model (Essmann and Differt, 1988; Differt and Essmann, 1993) is for the moment two-dimensional. Less ambitious than the two preceeding ones, it accounts more carefully for the basic properties of dipoles and their interactions with mobile dislocations. The coupled evolution of populations of individual dislocations and of dipoles are examined in order to understand the properties of the wall component of PSBs. The development is based, perhaps more specifically than in the two previous models, on the properties of annihilation and stability under stress of dipoles (which are described through critical distances) and on their interaction with mobile dislocations. Numerical values deduced from experimental studies on PSBs are used to calibrate the parameter values. One important point in the derivation consists of accounting for the fact that mobile edge dislocations can cross the walls and interact with the dipoles. This property characterizes PSB walls with respect to the veins formed during earlier stages of the fatigue life. The set of rate equations is written:

$$\partial\rho_1 / \partial t + \mathrm{div}(\rho_1 v_1) = -4\rho_1\rho_2 v y_d \tag{11}$$

$$\partial\rho_2 / \partial t + \mathrm{div}(\rho_2 v_2) = -4\rho_1\rho_2 v y_d \tag{12}$$

$$\partial\rho_d / \partial t = 8\rho_1\rho_2 v(y_d - y_e) - 2y_e\rho_d v(\rho_1 + \rho_2) \tag{13}$$

ρ_1 and ρ_2 are populations of mobile dislocations of opposite sign and ρ_d is the density of dipoles, the latter being assumed to be immobile. y_d is the critical distance for the annihilation of a dipole under the effect of an applied stress (cf. Section 2.1.1.) and depends on the stress level in the walls. Thus, the terms at the right-hand side of eqs. 11 and 12 represent the losses in mobile density due to mutual annihilation or dipole formation. They appear as a production term in eq. 13. The last term at the right-hand side of eq. 13 accounts for the destruction of dipoles through the encounter with mobile dislocations. The rate of creation of mobile dislocations does not appear here, as this process takes place in the channels of PSBs. It is accounted for via the boundary conditions.

As such, this model has a simple steady state solution representing a dipole-rich wall. Another solution is a travelling one, occurring when the dislocation fluxes into a wall are not equilibrated on its two sides. This accounts for the fact that the wall spacings become more and more regular as their formation proceeds. The main advantage of this model is that it is not based on broad statements but rather on a detailed model of cyclic glide in PSBs. The main difference with respect to the two models dicussed above stems from the fact that no use is made of a diffusion coefficient associated with the dipolar density. The walls move only via the destruction of the dipoles on one side and their creation on the other one. The dipoles are assumed to be immobile and long range stresses are not deemed essential in this treatment. One may notice a similarity between the situation treated in this model and the two-dimensional simulations of dipolar walls (cf. Section 5.2.2.). The latter could be useful in checking whether or not dipole drift can be neglected, which would bring up a considerable simplification into the models. Finally, the length scales associated with dislocation patterning and a consistent treatment of the multiplication rates of the mobile densities cannot be reached without introducing screw dislocations in this model. Such modelling is in progress (Essmann and Differt, 1995) and here also, it would be interesting to parallel it with numerical simulations.

4.4. Summary

It is now understood that dislocation patterning is a dynamic phenomenon due to the non linear character of dislocation dynamics and interactions. The main two types of patterns, dipolar and cellular ones, are associated with characteristic types of local and/or long ranged interactions. The fact that a well-defined type of dislocation microstructure corresponds to each deformation stage simply indicates that both are governed by the same dislocation processes. Although some characteristic features of dislocation patterns can be understood, the most striking one, viz. the occurrence of a scaling law between the size of the dislocation-poor regions and the stress has not been explained as yet.

Whereas it is now clear that theoretical efforts should aim at investigating the coupled dynamics of dislocation populations, the situation in this domain is still far from being satisfactory. The recent models differ in the nature of the elementary mechanisms they take into account and in the way they treat long range stresses and self-screening effects. As will be shown below, some of the questions raised by experiment as well as theory can be approached with the help of mesoscopic simulations.

5. 2-D SIMULATIONS

2-D simulations deal with very idealized and simplified situations. They have, however, been and can still be very useful in several respects. As illustrated by fig. 5, there are two different manners to describe dislocation behaviour in two dimensions, "in plane" simulations where the simulation plane is a slip plane and "edge on" simulations where the slip planes are perpendicular to the slip plane. The first case is convenient for the examination of local interactions of dislocations with various obstacles, because line tension effects can be taken into account. The second case is more adapted to the examination of simple collective effects involving elastic interactions. In this section, essentially mesoscopic simulations are described, i.e., dislocations are considered as line defects moving in an elastic continuum. Indeed, atomistic simulations cannot deal with systems of size larger than a few nanometers and time scales larger than a nanosecond.

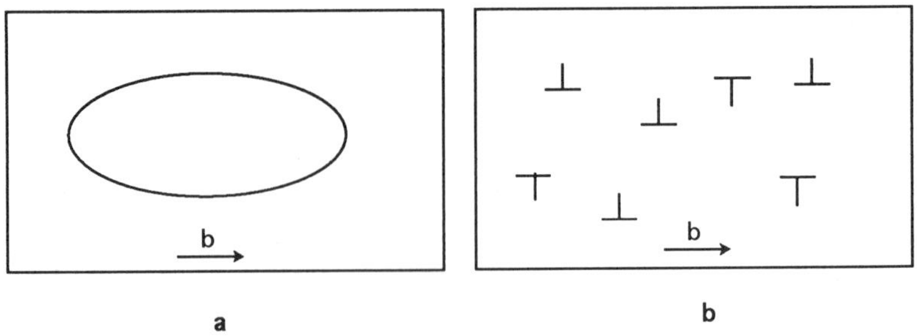

Figure 5. The two different types of 2-D simulations, a - in plane, b - edge on.

5.1 "In plane" 2-D simulations

Such simulations have evolved in three steps. During a first period, at the end of the sixties, the focus was on describing correctly the shape of a bent dislocation segment in order to check the line tension approximation. The interaction of a moving dislocation with a random distribution of point obstacles was then investigated and refinements were introduced in order to treat more accurately the interaction with forest dislocations. Contemporary studies investigate more complex interactions or mechanisms with the help of very sophisticated simulation codes.

5.1.1. Line tension and curvature. Computing the exact shape of a dislocation segment or loop bent under stress is a question which is now fully solved by elasticity theory, both in the isotropic and anisotropic cases (Kirchner, 1984). The only restriction to this statement is concerned with the inner cut-off radius for elastic interactions, or core radius. Its value is more or less arbitrarily fixed to b in most cases. The same holds for the outer cut-off radius (or screening distance) whose value depends on the arrangement of the surrounding microstructure. Various levels of approximation are available depending on the type of application considered (cf. Kocks et al., 1975, for a review). The simplest one consists of attributing to the dislocation segments a local line tension. The elastic interactions of each portion of the considered line with the other segments are then neglected. When they are not, the backstress (or self-stress) becomes a non local property and its computation requires an integration over the whole dislocation configuration (Brown, 1964). The first in plane simulations were, therefore, devoted to the computation of exact curved shapes, of the critical (Frank-Read) shapes and to an assessment of the line tension approximation. The methods used consist basically of finding the equilibrium position of a finite number of points on the dislocation lines by letting them to relax until they experience a zero total stress (Foreman, 1967; Bacon, 1967; Bacon et al. 1973; see also Shtolberg, 1971). These results confirm rather well the theoretical prediction for the critical stress τ_c of a segment of length L pinned at its ends:

$$\tau_c = A\frac{\mu b}{2\pi L}[\log(\frac{L}{r_o}) + B] \qquad (14)$$

A is a constant which depends on the character of the line. A ≈ 1 for edge and 1.5 for screw when the Poisson's coefficient is $\nu = 1/3$ (Foreman, 1967), in good agreement with the line tension model which predicts $A = 1/(1-\nu)$ and 1 for screw and edge, respectively. L and r_o are the outer and inner cut-off radii for elastic interactions with L being the length of the bowing out segment. B is a constant which depends on long range interactions. When this term is neglected, eq. 14 takes the form predicted by the line tension model.

5.1.2. Interaction with point obstacles. Historically, the next step consisted of introducing point-like obstacles (i.e., obstacles whose range of interaction with the dislocation lines are small compared to their average distance). Fig. 6 schematically depicts the critical configuration for breaking through the obstacle. The strength of the latter can, thus, be characterized either by the critical bowed-out radius R or the critical bowing angle φ_c. If L is the average length of the dislocation segments, we have $L = 2R\cos(\varphi_c / 2)$. In the line tension approximation, the curvature radius and the critical applied stress are related by $R = T/\tau_c b$, where T is the line tension. It follows that the critical stress is of the form:

$$\tau_c \approx (2T / bL)\cos(\varphi_c / 2) \tag{15}$$

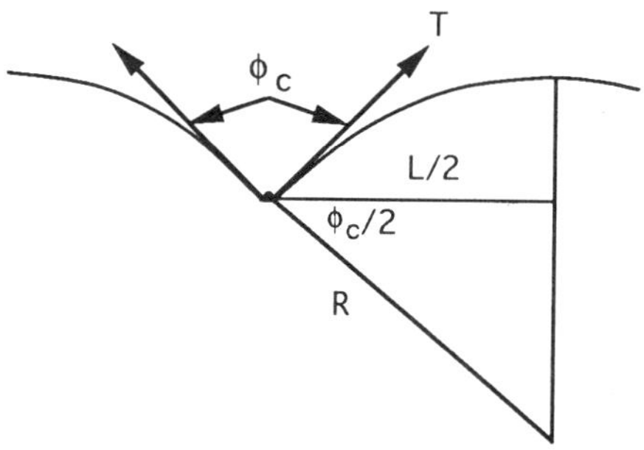

Figure 6. Critical bowed-out segment shape for the cutting of point obstacles.

The maximum obstacle strength is reached for impenetrable obstacles associated with a breaking angle $\varphi_c = 0$. The flow stress corresponding to a given areal density of obstacles N is obtained by finding a relation between L and N. This is not a straightforward exercise, as this relation is not the same for weak ($\varphi_c \geq 100°$) or strong obstacles and depends on the statistics of the obstacle distribution. The comparison of analytical and simulated results, the latter being usually obtained with infinitely strong obstacles, has been the object of numerous studies (cf. Foreman and Makin, 1966, and the review by Gil Sevillano, 1993).

If the point defects are assimilated to forest dislocations piercing the slip plane in total (volumic) density ρ, it is possible to investigate their interaction with mobile segments and the resulting critical stress. As one half of the trees (the attractive ones) lead to the formation of sessile locks, the density of forest obstacles is $\rho_f \approx \rho/2$ (a bit less in reality, since forest dislocations gliding in slip systems coplanar with the simulation plane have to be omitted). The average distance between strong obstacles is then $L \approx (\rho/2)^{-1/2}$ and eq. 15 is rewritten:

$$\tau_c \approx [\frac{T\sqrt{2}}{b}\cos(\varphi_c / 2)]\sqrt{\rho_f}\,, \tag{16}$$

i.e., the well-known relation between the flow stress and the square root of the dislocation density is recovered (cf. eq. 1 and Section 2.2.1.). Kocks areal glide model (1966, 1985, see also Section 3.4) investigates the flow stress and strain hardening properties deriving from a 2-D simulation where the point obstacles have maximum strength (cf. Section 6.2.). Figure 7-a shows the Orowan loops left by a mobile dislocation around impenetrable zones where the glide resistance is high. Further simulations of areal glide with different distributions of the obstacles and with several successive mobile dislocations simulating the activity of a source have been performed by Gil Sevillano et al. (Hernandez Olivares and Gil Sevillano, 1987; see Gil Sevillano, 1993 for a review). In general, estimates of the athermal strain hardening rate stemming from dislocation storage yield a value comparable to, but systematically smaller than, the strain hardening coefficient during stage II of the deformation curve of fcc crystals.

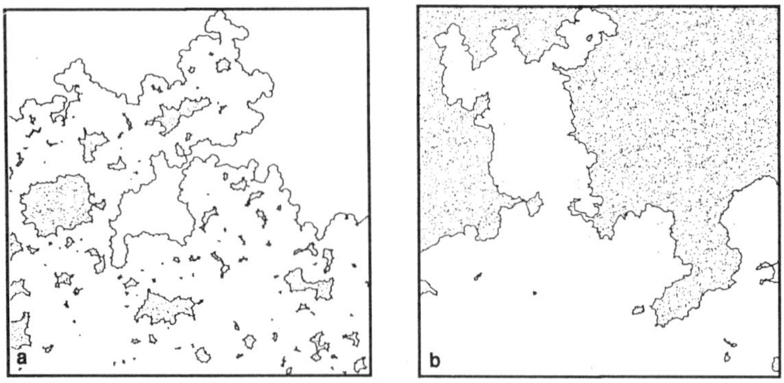

Figure 7. Critical shapes of dislocation lines moving through a random distribution of point obstacles (after Gil Sevillano et al., 1991) a - $\phi_c = 0$, notice the islands left by the gliding line. b - $\phi_c = 50°$.

There are, indeed, many simplifications in such simulations of forest hardening. The strength of the obstacles is often assumed to be maximum, while the stress required to break a sessile junction is certainly smaller than the Orowan stress. Indeed, with very strong obstacles the value found for the constant α in eq. 1 is of the order of 0.8 i.e., more than twice the experimental value for fcc crystals (cf Section 2.2.1.). As shown in fig. 7-b, below a certain threshold which is estimated to $\varphi_c \approx 50°$ (Gil Sevillano et al., 1991), no loops or debris are formed and dislocation storage no longer occurs. Therefore, the strength of the junctions must be defined accurately before a check of Kock's model can be attempted. In these early 2-D simulations, the forest obstacles have no long range elastic stress fields and the various segments pertaining to the mobile dislocation line do not mutually interact. The trees are immobile so that no relaxation of the local configurations into more stable ones is possible and no multiplication of the forest is allowed to occur during the successive passing of the mobile segments. As mentioned in Section 3.4, the approximation of purely planar glide is perhaps questionable, but relaxing this assumption means in practice going to three dimensions.

From a quantitative point of view, an interesting outcome of such 2-D simulations can be an estimate of the constant α involved in the relation between the flow stress and the square root of the dislocation density (cf eqs. 1 and 16). Several groups from the former Soviet Union have attempted to relax some of the assumptions mentioned above by introducing flexibility of the trees (Loginov, 1991), the possibility of forming reaction junctions (Loginov and Predvoditelev, 1982; Predvoditelev and Loginov, 1985), by taking into account the long range stress fields of the trees and introducing a self-interacting mobile dislocation (Predvoditelev and Nichugovskii, 1972; Tyupkina, 1992). The main result are that eq. 1 is still valid in such conditions. The value of α is somehow modified, but not drastically.

2-D simulations of dislocation processes. As emphasized by Duesbery and Sadananda (1991; see also Duesbery Louat and Sadananda, 1992), 2-D simulations performed within a linear elastic framework can solve mesoscale problems for which no atomistic treatment exists and whose analytical computation by continuum elasticity is untractable. Typical examples of such situations, which are qualitatively understood but cannot be modelled without strong approximations, are the intersection of attractive non coplanar dislocations (Schoeck and Frydman, 1972) and the motion of dissociated dislocations and in particular their cross-slip probability in fcc crystals (Escaig, 1968). For instance, Duesbery, Louat and Sadananda (1992) have numerically simulated cross-slip in the fcc structure within the self-stress model and compared their results with the available experimental data and analytic models. Their computation yields the saddle point configuration for the activation of cross-slip and the corresponding activation volume and enthalpy. It is shown that the Friedel-Escaig model (Escaig, 1968) provides the most physical picture of the process. However, it underestimates the activation energy, probably as a result of the use of the line tension approximation. A good agreement is found with experimental data, except for the orientation dependence of the cross-slip stress. This points out towards a deficiency in the above mentioned model, although the latter cannot be attributed to a secific approximation.

The present state of the art in the domain of the simulation of specific mechanisms is represented by the sophisticated simulation of Rönnpagel et al. (1994) where the mobile dislocations are described within the self-stress model but are also in thermal equilibrium with their surrounding and may be subject to any type of obstacle stress in their glide plane. This simulation yields data which are directly comparable to experiment. This simulation has been applied to the overcoming of a single obstacle by thermal activation and to the athermal motion of dislocation pairs through a random array of ordered precipitates. Fig. 8 shows an example of the striking similarity between the simulated and experimental results in a nickel-based superalloy.

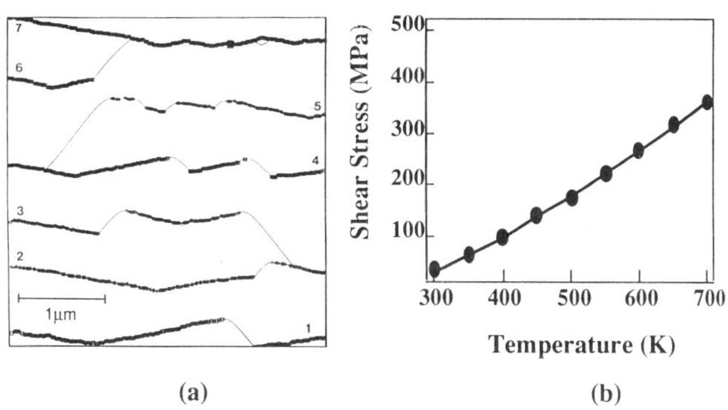

Figure 8. Motion of paired dislocations through a random distribution of ordered precipitates (mean diameter 7.5 nm, concentration 0.089). a - Simulated configuration (the precipitates are not drawn), b - *In situ* observation of Nembach and coworkers in a high voltage electron microscope (After Pretorius and Rönnpagel, 1994; Courtesy the Japan Institute of Metals).

The dynamical simulation of Mills et al. (Mills and Chrzan, 1992; see also Mills, Chrzan and Miracle, 1994) is adapted to the treatment of superdislocation loops in $L1_2$ ordered alloys. These alloys are known to exhibit an anomalous increase of the flow stress with temperature, whose origin is attributed to the cross-slip of <110> superdislocations from their {111} glide planes to {100} planes where glide is difficult. There is, however, no general agreement at present on the details of this mechanism nor on any of the proposed models for the resulting anomalous properties. In this simulation, the central point is the thermally activated formation of pinning points by cross-slip of the screw segments and their further destruction. The objective is to draw from the observed configurations some consequences regarding the anomalous mechanical properties. As in the previous example, dislocation dynamics are included, but through a phenomenological stress vs. velocity law for the superdislocations. As shown in Fig. 9, the anomalous stress vs. temperature dependence stemming from the increased probability of activation of cross-slip pinning with temperature is recovered. Configuration 7 of fig. 9-a is totally immobilized under the applied stress. This type of exhaustion mechanism may explain the extremely high strain hardening rates (typically $\mu/20$-$\mu/50$) recorded in these compounds.

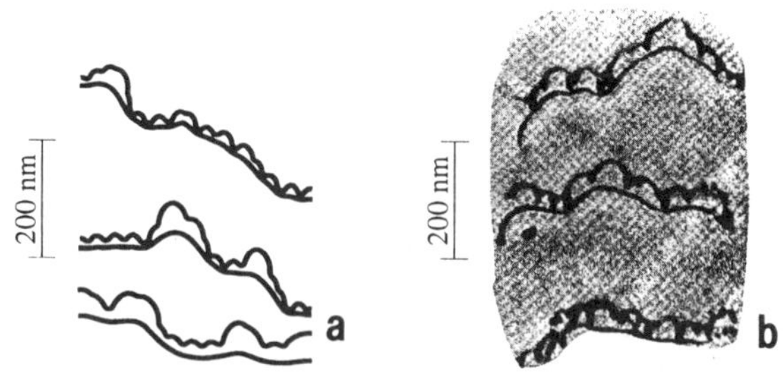

Figure 9. a - Sequential configuration showing the propagation of a <110> superdislocation in a {111} plane of Ni_3Al. The Burgers vector is horizontal and the square symbols denote pinning sites. The motion of near-edge dislocations takes place by the lateral motion of superkinks. b - Anomalous temperature dependence of the critical shear stress required for the motion of a near-edge dislocation at a constant velocity (After Mills, Chrzan and Miracle, 1994; Courtesy the Japan Institute of Metals).

In conclusion for this part, mesoscopic in plane 2-D simulations are quite useful for the exact numerical treatment of various dislocation mechanisms and interactions. A potentially interesting application is concerned with the mechanisms of dislocation motion in materials where dislocation core properties govern the flow stress (as in the $L1_2$ alloys). These simulations may also serve to yield numerical estimates for the critical stresses or activation energies associated with various mechanisms (like cross-slip or forest interactions). Such results are particularly useful for calibrating the input values for 3-D mesoscopic simulations (cf. Section 6).

5.2 "Edge on" 2-D simulations

In 2-D edge on simulations, the accent is on dynamics and elastic interactions rather than on an exact description of local mechanisms. The objects treated are punctual but interact through the elastic stress fields of infinite, parallel and straight dislocations, usually of edge character. *Ad hoc* rules are introduced to describe local properties which cannot be treated in a consistent manner. All such simulations incorporate dynamics, meaning that at every step a stress balance is computed for each segment (cf. Section 2.2.4.). The resulting effective strees is related to the velocity either through a viscous glide coefficient, representing the phonon and electron drag, or through a phenomenologic power law representing interactions with obstacles not materialized in the microstructure. Similar laws can be introduced to account for climb or cross-slip. It may seem surprising for the reader accustomed to atomistic simulations that the equations of motion involve no inertia term. Actually, the effective mass of a dislocation segment is that of a row of atoms of same length (Friedel, 1967), so that the steady-state velocity under stress is reached after a time interval of about 10^{-10}-10^{-11} s, smaller than the time scales involved in mesoscopic simulations (10^{-9} s or more).

Short range interactions like annihilation or the trapping in dipolar position may necessitate specific rules which depend on the detail of the technique used. The same holds for multiplication which takes place usually by the introduction of pairs of dislocations of opposite sign at certain places and under certain conditions. For more detail, in particular on the computational aspects, the reader is referred to the references given below.

As a consequence of the long range of elastic interactions, the number of pair interactions to be computed is $O(N^2)$ where N is the number of objects in the simulated area (N can be as large as a few thousands). Various algorithms deriving from the study of the many-body problem are discussed by Wang and Le Sar (1995, see also Ghoniem and

Amodeo, 1990a) and various simplifications are possible. With the help of such methods, the computation of pair interactions reduces to O(NlogN) or O(N) with a reasonable accuracy. For instance, Wang and Le Sar (1995) have used a fast-multipole method, leading to O(N) calculations for examining the relaxation of about 17×10^4 edge dislocations. This question is particularly critical in 3-D.

5.2.1 Methods. The first simulations were initiated a few years ago by Lépinoux et al. (Lépinoux and Kubin, 1987; Lépinoux, 1988; Kubin and Lépinoux, 1988) and Amodeo and Ghoniem (Amodeo, 1988; Ghoniem and Amodeo, 1988; Amodeo and Ghoniem, 1990a, b; Ghoniem and Amodeo, 1990). These simulations were based on methods deriving from cellular automata techniques and molecular dynamics, respectively. However, as emphasized by Ghoniem and Amodeo (1990a), the simulation techniques developed so far have little in common with existing methods, so that they represent a new methodology which may be called Dislocation Dynamics (DD).

Classical cellular automata are based on a set of five rules (Wolfram, 1984), where advantage is taken of the discretization of space and time to provide simplicity and of the use of local dynamic rules which provide flexibility:

- Space is discretized and the simulated map is made up of cells or sites.
- Each site has a finite number of states.
- The state of each site depends on its previous state and on that of its neighbourhood.
- All the sites evolve according to the same set of evolutionary rules (which can be deterministic or stochastic).
- The simulated configurations evolve by discrete time steps. Often, but not necessarily, all the sites are simultaneously updated, leading to successive "generations" of the simulated configurations. The most well-known and recent application, which led to a revival of this method is to be found in the domain of hydrodynamics and the study of lattice gases (D'Humières and Lallemand, 1986). A detailed account of this technique and of its applications to materials science can be found in the recent reviews by Mareshal (1995) and Lépinoux (1995).

The adaptation to dislocations was performed by Lépinoux et al (Lépinoux, 1987; Lépinoux and Kubin, 1987). The method was further improved by Hesselbarth and Seck (1992) and a 2-D version of the 3-D simulation, which also makes use of discretization methods, was developed by Devincre et al. (Devincre 1993; Devincre and Pontikis, 1993). The simulated area is divided into cells, each of them having three possible states: empty, or containing a dislocation of Burgers vector +b or -b. The elastic stress fields are those of parallel, straight and infinite dislocations. The interactions involve an extended neighbourhood, i.e. either the whole simulated area or a smaller area included within a cut-off distance. The main originality, and difference, with respect to conventional cellular automata is that the dimension y of the cell size is specified, so that a scaling length is introduced. The argument is as follows: if every cell contains one dislocation, the corresponding density must be the maximum dislocation density admissible in a crystal. As shown in Section 2.1.1., y is, then, the critical annihilation distance for the dislocation character considered, either screw and edge. Two dislocations of opposite sign entering the same cell are annihilated and this mimicks both direct annihilation and the mechanism of dipole annihilation.

In their discussion on the Dislocation Dynamics methodology, Ghoniem and Amodeo (who do not make use of discretization of space) stress the main two differences with respect to particle methods in statistical dynamics (e.g., Molecular Dynamics or Monte-Carlo). A DD system contains a comparatively smaller density of defects (a few thousands compared to a few hundred thousands or more). It follows that statistical thermodynamical principles cannot be applied, all the more as their applicability is not established in the case of dislocations. In addition, energy conservation is difficult to implement because of many non conservative reactions and the considerable energy dissipation which takes place during plastic flow. Long range (1/r) interactions are of course not common in MD, while the presence of short-range interactions (annihilation and multiplication) destroy the principle of particle conservation. The small number of deterministic events leads in DD to a complete representation of the evolution of the system and no statistical averaging is required. As described by Ghoniem and Amodeo (Amodeo, 1988; Ghoniem and Amodeo, 1988; Amodeo and Ghoniem, 1990b; Ghoniem and Amodeo, 1990) the DD method incorporates the treatment of dislocation dynamics on the base of the calculation of an effective stress. Various short range interactions like mutual annhilation of dislocations of unlike sign, multiplication and the formation of dislocation junctions (this last mechanisms is implemented when double coplanar slip is introduced in the simulation) are implemented. Most 2-D simulations published to date (Fang et al., 1992; Groma and Pawley, 1993a, b ; Gulluoglu et al. 1989, 1990; Gulluoglu and Hartley, 1992, 1993; Fournet 1994, Fournet and Salazar, 1995) have much in common regarding the elastic interactions and the dynamics. They, however, differ on the set of local rules employed, which usually define the shortest time interval and, hence, the elementary time step. Depending on the level of sophistication, more or less care is taken to ensure that the dynamics of events are correctly implemented (definition of the elementary time step, of the stress velocity law ...). For instance the way multiplication is mimicked is certainly a weak point in all 2-D simulations, as it strongly influences the obtained microstructures.

Boundary conditions are often of the usual type (e.g., periodic conditions or mapping on a sphere). However, it can be interesting to treat the simulated area or volume as a small crystal rather than a portion of an infinite body. The requirement is, then, that the external lines or surfaces must be traction free in order to ensure mechanical equilibrium. Van der Giessen and Needleman (1994, 1995) have recently proposed a full treatment of this question in which the stress fields of dislocations in an infinite medium is superimposed to a complementary solution for a body of finite size. The latter is computed from a finite element solution of a linear elastic boundary value problem. This treatment is also applicable to internal boundaries, like second phases of different elastic properties.

5.2.2 Results. We focus here on a few fundamental problems like self-screening and the sponatneous occurrence of a cut-off radius for elastic stresses, the formation of periodic, dipolar walls and other organized structures and a critical check of the predictions of Holt's (1970) model (cf. Section 3.3). Schematically two types of experiments were performed. In what follows, we define as relaxation experiments simulations performed in the absence of an applied stress where the initially random configurations evolve under of the effect of the long range interaction stresses. In such cases local mechanisms and interactions (such as multiplication, annihilation and cross-slip) are usually switched off. In simulations of plastic flow, an applied stress is added and truly dynamic structures are examined. It is interesting to note that the results obtained with different techniques and different sets of local rules converge to a few clear-cut and reliable conclusions.

Relaxation experiments. Gulluoglu et al. (1989, 1990) and Fournet and Salazar (1994, 1995) have examined the influence of an imposed cut-off radius for the long range elastic stresses on relaxed configurations of edge dislocations. In both cases, the simulated areas are rather small (1 μm^2) and the densities rather large, corresponding to dislocation densities in the range 4×10^{14} - 4×10^{15} m^{-2}. When a cut-off value is imposed, the final configuration obtained when all dislocations are immobilized by their mutual interactions consists of periodic arrays of dipolar walls perpendicular to the slip plane, as illustrated by fig. 10. By contrast, when all interactions are effectively taken into account, the final structure does not present any obvious sign of self-organization (cf. fig. 10-b). These relaxation experiments exactly reproduce the problem investigated by Holt (cf. Section 3.3), with the only difference that long range dislocation motion is permitted. The introduction of an artificial wavelength drives the result and the observed periodicity is exactly the cut-off distance.

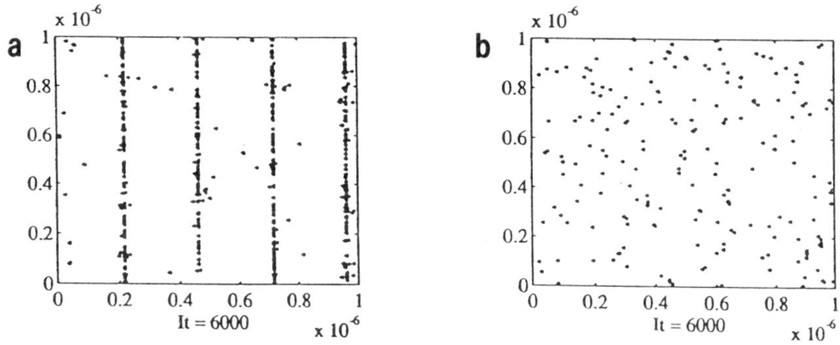

Figure 10. a- An artificial wall structure of edge dislocations obtained by truncating the long range elastic stresses. b - Same conditions but with no cut-off. No periodic structure is formed (After Fournet, 1994, by courtesy).

With smaller dislocation densities (of the order of 10^{12} m^{-2}) and larger simulated areas, this artefact is less spectacular, as the smaller density is not sufficient to allow for the formation of regular walls. In the relaxation experiments performed by Lépinoux (Kubin and Lépinoux, 1988), the simulated area is 31x13 μm^2 and an imposed cut-off of 6 $\rho_0^{-1/2}$ = 6 μm is imposed. The final microstructure is composed of an apparently random distribution of dipolar clusters. The average elastic interaction energy per dislocation is computed on the final configuration by adding up, for each dislocation, the contributions of concentric shells of increasing radius R. The plot of fig. 11 shows that beyond a value of about 4 μm, the average elastic energy no longer changes, so that the relaxed structure exhibits a spontaneous cut-off smaller than the imposed one. The same type of microstructure was obtained by Devincre (1993), this time with no imposed cut-off, and the computation of the spatial correlation function associated with the final microstructure confirmed that the preferential orientations are those of the elementary dipoles (at 45° or 90° of the glide direction).

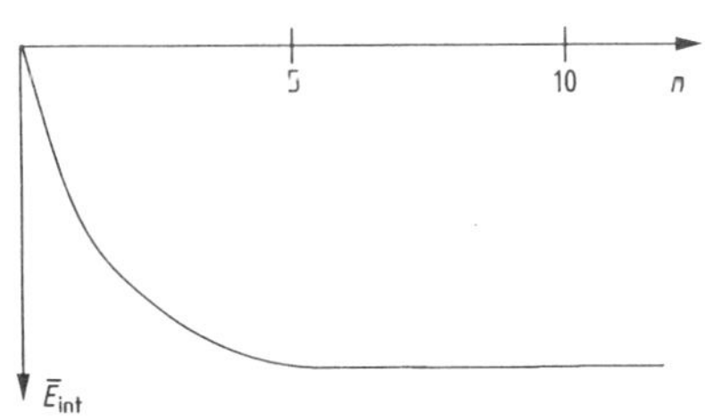

Figure 11. Average elastic interaction energy of dislocations in a relaxed configuration vs. the reduced distance n = r $\rho^{-1/2}$ (Courtesy J. Lépinoux).

From such sets of experiments the following conclusions are drawn: no periodic microstructure appears with one set of dislocations and in the absence of applied stress, contrarily to what was predicted by Holt's (1970) model. Further, caution must be taken when truncating long range elastic stresses. Artificial patterns are obtained, in theory as well as in simulation, when the imposed cut-off is smaller than the one which spontaneously appears in the final microstructure. It follows that the assumption of perfect self-screening and of the internal stress being a local quantity is absolutely not tenable. On the other hand, spontaneous but imperfect screening occurs, as expected.

2-D microstructures in monotonic loading. Because of the highly simplified treatment of local interactions and of dynamics, not too much significance is perhaps to be attributed to the evolution of the dislocation densities or to the simulated stress-strain behaviour. In addition the coupling between the screw and edge portions of an expanding loop (the motion of edges increases the screw density and vice-versa) cannot be reproduced. For this reason the questions of dislocation multiplication, of slip propagation along a direction normal to the slip plane, or of the instability of multipolar edge walls yield interesting but qualitative results (see Kubin, 1993 for a review). The most two sophisticated simulations from the point of view of the treatment of local rules are those of Ghoniem and Amodeo (1990) and van der Giessen and Needleman (1995 a, b). In single slip and with edge dislocations, a weak organization was observed and the development of densities of mobile and immobile dislocations could be examined. As already mentioned, the treatment of van der Giessen and Needleman allows to treat in a rigorous manner the problem of an inclusion and may be extended to the treatment of composite materials. In this respect the comparison attempted beween a Finite Element simulation of the problem and the corresponding mesoscale simulation looks quite promising for bridging the continuum and discrete approaches of plasticity.

Dislocation walls under cyclic stress. Under an applied stress, the most interesting patterns are obtained with edge dislocations. This is because the latter have an anisotropic stress field which allows for the formation of a variety of stable multipolar configurations. Dislocation walls can be obtained under monotonic stress or strain control (Hesselbarth and Steck, 1992). However, as seen in Section 2.1.2, fatigue tests provide the best testing conditions for the observation of dipolar patterning. The reason is that the microstructures formed in fatigue must be stable with respect to the two possible directions of the applied stress and correspond to configurations of lower energy than those formed in monotonic straining. In addition, conditions can be used in the simulations such that the mean distance travelled by dislocations during one half-cycle is small compared to the maximum glide distance, so that boundary effects can be minimized. Structures consisting of well-defined walls normal to the glide direction were obtained by Hesselbarth and Steck (1992) in the following conditions: a simulated area of $10 \times 4\ \mu m^2$, a cut-off distance of 5 μm in the glide direction and of $1 \mu m$ in the perpendicular direction, an (equivalent) initial density of $1.5 \times 10^{14}\ m^{-2}$ and an imposed plastic strain of 2×10^{-4} per half-cycle. Direct annihilation was permitted and pairs of fresh dislocations of opposite sign were introduced to compensate for the losses by direct annihilation. The microstructure formed consists of equispaced monopolar walls of alternating sign. In this case again, the wavelength was substantially smaller than the imposed cut-off and really corresponded to an emerging structure.

By suppressing direct annihilation and multiplication, i.e., by imposing a constant dislocation density, it is possible to further analyze these wall structures. In such conditions, Devincre (1992) and Fournet (1994) obtained the formation of dipolar walls after a few fatigue cycles in simulated areas of $20 \times 20\ \mu m^2$ with equivalent densities of the order of 10^{12} m^{-2}. Both authors have noticed that the elastic energies of such configurations are indiscernable from those of the (disordered) relaxed configurations obtained with the same dislocation density. This confirms that energetic criteria are of little use for characterizing dislocation patterns. Combining these results with those obtained from relaxation experiments, the main conclusion is that patterning in 2-D is, indeed, a dynamic phenomenon which only occurs in the presence of an applied stress.

The dependence of the wall spacing on the dislocation density and the applied shear stress τ_a has been examined and modelled by Devincre (1993). The numerical data drawn from the simulations can be summarized by the following expression:

$$\lambda = \kappa \tau_a / \rho \tag{17}$$

where λ is the wall spacing, ρ is the (constant) density and κ is a constant. This result can be understood as follows. During the first steps, isolated dipoles are formed together with small dipolar walls which are mobile and grow by accretion of the dipoles until they form extended dipolar walls (cf. fig. 12). The stability of these configurations implies that the dipole width is smaller than a critical value y_d (cf. Section 2.1.2.). Assuming that the width of the dipoles inside the subgrains is in equilibrium under the applied stress, a classical calculation yields:

$$y_d = \frac{\mu b}{16\pi(1-\upsilon)} \frac{1}{\tau_a} \tag{18}$$

where ν is the Poisson's coefficient. The dislocation density in the walls is $\rho_w \approx y_d^{-2}$ and for a periodic arrays of walls of wavelength λ, the average density per wall is $\rho \approx 1/\lambda y_d$, i.e.,:

$$\lambda = \frac{16\pi(1-\upsilon)}{\mu b} \frac{\tau_a}{\rho} \tag{19}$$

This expression is formally identical to eq. 17 and yields a correct estimate for the value of the constant κ. This model thus provides an explanation for the formation of periodic dipolar wall patterns: the walls have long range stresses ($1/r^2$) and consecutive walls have different signs and repel each other. More stable structures, involving more efficient self-screening like quadrupoles, cannot be formed because this would imply the mutual crossing of walls of opposite sign. The hypothesis of marginal stability of the structure under stress which was widely used in the past (cf. e.g., Neumann, 1986) may also prove very useful in this context. As a result, we notice that the wall periodicity is modelled without any reference to the long range stresses, which bears some analogy with the model of Essmann and Differt (cf. Section 4.3.). The relation between the wavelength, the stress and the dislocation density implied by eqs. 17 or 19 differs from what can be deduced from eqs. 1 and 3. This is obviously due to the absence of a multiplication mechanism and of forest interactions. Even if this 2-D situation is not too realistic, it yields an interesting case where patterning can be understood in simple terms.

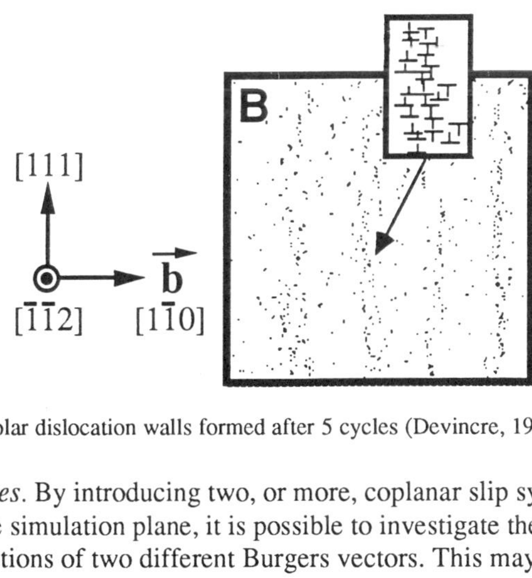

Figure 12. Dipolar dislocation walls formed after 5 cycles (Devincre, 1993, by courtesy).

2-D cell structures. By introducing two, or more, coplanar slip systems with their glide directions lying in the simulation plane, it is possible to investigate the formation of patterns formed by the interactions of two different Burgers vectors. This may yield a first approach to the formation of cell structures. The elastic interaction stresses are, then, those of infinite, straight dislocations, so far with edge character. As indicated in Section 2.2.1., to obtain patterning it is necessary to introduce short range dislocation-dislocation interactions, i.e., a conditional locking process at some dislocation intersections. In addition, a supplementary degree of freedom by climb has to be introduced (Ghoniem and Amodeo, 1990). Fournet and Salazar (1995), Fang and Dahl (1993), Groma and Pawley (1993) and Gulluoglu and Hartley (1993-b) have also attempted to simulate multislip structures under stress, while Gulluoglu and Hartley (1993-a) focused on relaxed multislip structures simulating annealing experiments. It seems dificult to find common points between these various investigations, probably beause the short range interactions and mechanisms are treated in widely different manners. The latter may sometimes involve oversimplifications, particularly multiplication and the treatment of the intersection mechanisms.

Ghoniem and Amodeo (1990) have simulated the case of creep deformation at various temperatures (the latter essentially governs the ratio of climb to glide velocities) with two orthogonal slip systems and a condition for the formation of reaction junctions at attractive intersections. Configurations akin to cellular structures are formed only in conditions of low stresses and with simulated temperatures in the range 400-600°C. The cell walls consist of immobilized dislocations, dislocation dipoles and junction dislocations. At higher temperatures, multipolar clusters are formed, while annihilation of the mobile dislocations by climb becomes efficient. At low temperatures a too large stress prevents the immobilization of the dislocations and the formation of a pattern. Fournet and Salazar (1995) examined the same problem but without introducing a condition for dislocation locking at the intersections, so that only immobilization and dipole formation were allowed to occur. One interesting point is that stable pseudo-cells are formed during relaxation experiments which, then, mimick the annealing of a dislocation structure leading to the formation of a Frank net. Thus, with two coplanar slip systems a pattern can be obtained in 2-D and in the absence of applied stress, while we have seen that this is not possible with only one slip system. Such a result had been previously obtained by Gulluouglu and Hartley (1993-a) and was confirmed recently in 3-D (Fivel and Canova, 1995, unpublished result). This emphasizes the predominant influence of short range interactions and mechanisms on cell formation.

For reasons that are not clear to date, it seems more difficult to form stable cells in 2-D under a constant applied stress, as was already noticed by Ghoniem and Amodeo (1990). This question of cell formation in 2-D certainly deserves more examination and a systematic study, essentially because such simulations are considerably lighter than the 3-D ones. A first point consists of understanding why the expected patterns do not form so easily, which points out to either deficiencies in the treatment of local mechanisms or to a lack of control of the dynamics since the creep experiments lead to either a complete immobilization of the mobile density or to a complete destruction of the structure. A possible origin for these difficulties stems, at least in some numerical experiments, from the assumed form of the stress vs. climb velocity law. If controlled conditions can be defined for the formation of stable bidimensional cell structures, it would be desirable to measure the relation between the total density and the applied stress (cf. eq. 1), in order to check whether the forest mechanisms operates correctly. Further, a check of the cell size vs. the reciprocal of the applied stress could shed some light into the origin of the relation of similitude (eq. 3). Such tests were already attempted by Gulluoglu and Hartley (1993-b) and Fournet and Salazar (1995), yielding encouraging results.

In summary, edge on simulations have yielded valuable results on the formation of patterns in highly idealized conditions and may still prove useful for dealing with fundamental problems. With the present computational power presently available, these 2-D simulations no longer pose a technical challenge. Coherent results have been obtained as far as dipolar patterning is concerned. As yet, this is not so for cellular patterning, probably because of some roughness in the dynamics and in the treatment of the local rules.

6. 3-D SIMULATION

6.1. 3-D Simulation of Dislocations and Plasticity

6.1.1. Introduction. This simulation combines the possibilities of in plane and edge on simulations in that sense that it accounts for both elastic and local interactions. The ouput consists of microstructures and mechanical properties of small model crystals of typical size (15-20 μm)^3. Fig. 13b presents a view of the simulated crystal, here with a fcc structure, in which a density of dislocations has been produced by a few initial dislocation sources (cf. fig. 13a). The dislocation content of such simulated microstructures can be investigated and various types of information, either local (e.g. internal stresses) or global (dislocation densities, stress-strain curves) can be extracted in parallel.

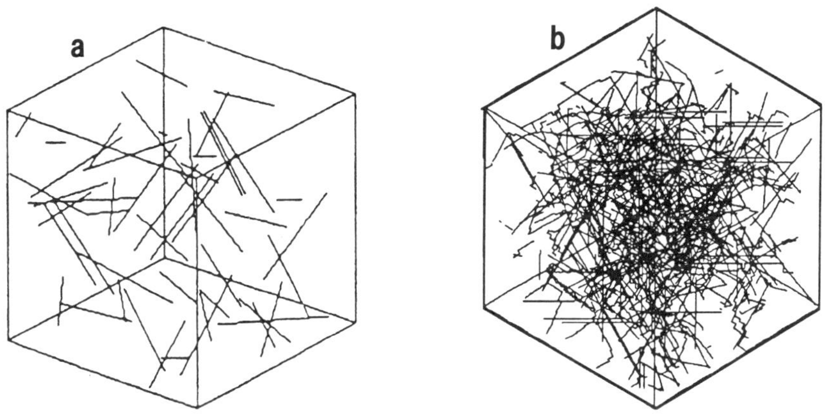

Figure 13. a - Initial configuration consisting of a random distribution of sources in a simulated fcc crystal of size (15 μm)^3. b - A dislocation configuration produced after a strain of 0.1%.

The basic principles of the simulation have already been described in some detail (Kubin et al., 1992; Devincre, 1993, 1995) and will be recalled only briefly. In what follows, we describe two types of applications: the first one is concerned with the question of strain hardening in fcc crystals in multiple slip and the second one with strain localizations arising from glide softening phenomena. Other numerical experiments, in particular on the brittle to ductile transition in (bcc) molybdenum, are discussed by Devincre and Roberts (1995).

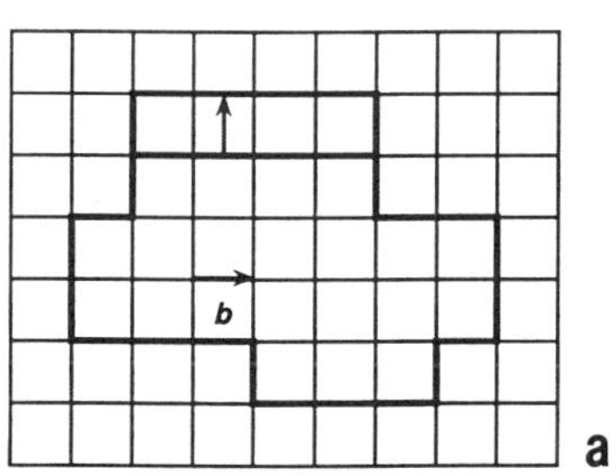

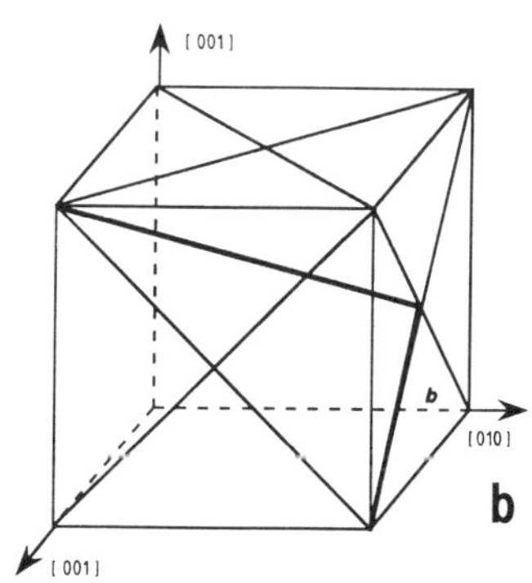

Figure 14. a - Decomposition of a dislocation loop into a succession of screw and edge segments. b - The fcc lattice underlying the 3-D simulation. The bold lines represent sub-segments of screw and edge character glissile in a (111) plane.

6.1.2. Discretisation, Elastic Interactions. Dislocation segments or loops are decomposed into a succession of screw and edge sub-segments (cf. fig. 14a). These segments move on an underlying 3-D lattice (cf. fig. 14b) whose parameter is defined from the critical annihilation distance of an edge dipole. For a (copper) fcc lattice, the lattice parameter is 5.54 nm, and dislocations of Burgers vector 1/2 <110> glide on the {111} planes of this lattice. The optimum conditions for discretization have been examined carefully in order to ensure that the long range stress field of a discrete loop reproduces within a reasonable accuracy that of a continuous loop (Devincre and Condat, 1992). For N interacting segments, the number of pair interactions to be computed is proportional to N^2, and no cut-off is introduced for these long range elastic stresses. As the value of N increases continuously with increasing strain and dislocation density, the calculation of the pair interactions has been a major limitation to the maximum strain achieved by the simulation; beyond a few tens of thousands segments, the computation of steps require a prohibitive time. This difficulty is partially dealt with by adopting a technique deriving from the study of the many-body problem: the simulated cube is decomposed in a number of boxes (typically 6^3 or 9^3) and for each segment the updating of the internal stress is performed at every step for the interactions with segments contained in the same box and the neighbouring ones, and with a larger time period for the more distant ones. The number of interactions to be computed is then reduced to $\propto$ NlogN. In their principle, the stress balance, the local estimates of the effective stress and of the resulting velocity are performed much in the same manner as in 2-D in plane simulations. With respect to the latter, however, additional mechanisms are introduced, some of which are described in what follows.

The peculiarities of mesoscopic simulations can now be illustrated by the example of the line tension. When a segment moves, the length of its two connected neighbours is altered (cf. fig. 14-a). The line tension is estimated within the framework of linear, isotropic elasticity, as the derivative of the corresponding line energy with respect to the line length. A small non local contribution which stems from the interactions of other segments in the same bowed-out configuration is automatically taken into account via the calculation of the pair interactions. For a segment of given character and length and considering only the local line tension, the latter reduces to $T = C\ Gb^2$, where the constant C depends on the outer and inner cut-off radii for long range elastic interactions. As already mentioned, dislocation theory does not provide a numerical estimate for this constant, particularly for dislocations immersed into a microstructure whose elastic screening properties are unknown. For this reason, the literature quotes values of C ranging between 0.5 and 1 (Friedel, 1967; Bacon, 1967; Foreman, 1967). This numerical uncertainty was removed by adopting values determined experimentally by Mughrabi (1973) in the case of dislocation loops of micrometer size in copper crystals: $T_s \approx 2.5\ 10^{-9}$ N for the screw parts and $T_e \approx 10^{-9}$ N for the edge edge parts. A a result, we have $C \approx 0.715$ and 0.535 for the screw and edge parts of the loops, respectively, and $C \approx 0.62$ for the (geometric) average value.

6.1.3. Local rules: Stability of Junction Segments. While all properties deriving from the dislocation elastic stress fields can be accounted for in a rather rigorous manner, those deriving from the core properties of dislocations require a particular treatment. The main two such mechanisms are cross-slip, through which dissociation properties and the influence of temperature are introduced in the simulation ((Kubin et al., 1992; Devincre, 1993, 1995) and the formation and destruction of sessile segments at the intersection of non coplanar dislocations. The stability of such segments is an important dislocation parameter; it governs the strain hardening properties of fcc crystals, as further discussed below. The breaking down of a junction requires that the local effective stress produces a work equal to the elastic energy gained when the junction was formed. In the approximation of the isotropic forest, the stability of all junctions is assumed to be the same and their destruction requires the same threshold stress. This mean stress can be estimated through experimental data, so that no arbitrary assumption deriving from elasticity theory is used. The procedure used derives from eq. 16. The knowledge of the constant α (cf. Section 2.2.1. and eq. 1) and of the constant C involved in the line tension (cf. Section 5.1.3.) allows to estimate the value of the critical angle to $\phi_C \approx 60$-70°. Hence, forest obstacles are weaker than has been assumed in many in plane simulations. To determine the corresponding critical stress for the breaking of a junction, τ_j, several methods are available. The most reliable consists of using the 3-D simulation to reproduce simulations of areal glide like those of fig. 7 but with more realistic obstacles (Devincre and Kubin, 1994). In such simulations the cross-slip probability is set to zero and the forest dislocations are immobilized, which allows for a comparison with Kocks' model. The critical stress for long range dislocation motion is then defined as the stress at which one segment emitted by the source reaches the upper part of the simulated area.

Fig. 15-a shows a plot of the critical stress vs. τ_j for a given configuration of density 10^{11} m^{-2}. The critical stress increases from the stress needed to activate the source up to a plateau corresponding to the Orowan stress for obstacle overcoming. We note that the strength of a forest obstacle is a fraction $\cos\phi_C \approx 0.4$ of the Orowan stress $\tau_o = 2T/bl$. This defines the value of τ_j. Finally, fig. 15-b shows a plot of the critical stress vs. the square root of the number of dislocation obstacles for various configurations. Although the slope of this graph is adjusted through the fitting of the value of τ_j, the occurrence of a straight line, which is the expected result (cf. eq. 3), naturally comes out from the simulation.

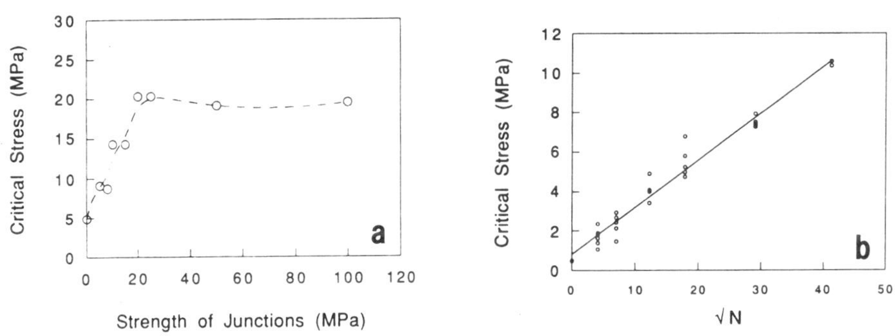

Figure 15. a - The critical stress for long range dislocation motion through a given configuration vs. the strength of junctions τ_j. b - The critical stress for long range dislocation motion vs. the square root of the number of intersecting segments N. The value of junction strength is fixed.

6.2. Strain Hardening and Cell Formation in fcc Crystals

The geometric and energetic factors governing dislocations intersections being established, an attempt can be performed to understand the early stages of dislocation cell formation in [100] crystals in relation with strain hardening properties. A preliminary attempt to form cell structures in simulated aluminium crystals has been performed (Kubin, 1993), but in the absence of a precise calibration for the line tension and the forest strength its conclusions are only qualitative. Quite generally, two types of explanations are invoked to explain the hardening of fcc crystals in multiple slip. They are illustrated by the classical in plane 2-D simulations of a dislocation moving through a density of point obstacles of variable strength (cf. fig. 7-a, b). In fig. 7-a, the obstacle strength is maximum ($\phi_C = 0$) while in fig. 7-b, the value is $\phi_C = 50°$. The main difference between these two configurations resides in the occurrence of dislocation debris and loops left in the slip planes by moving dislocations around impenetrable fluctuations of the forest density. Such debris are present in fig. 7-a, but almost absent from fig. 7-b. Kocks' model for areal glide (cf. Section 3.4.) describes strain hardening as being due to the storage of such debris in the active slip planes. This is at variance with the classical forest models (Friedel, 1967) which assume that strain hardening arises not from the storage of dislocations in a slip plane but from the increased number of forest obstacles, as forest dislocations move and multiply.

The simulation of such processes yields interesting and somehow unexpected results (Devincre, 1993; Devincre and Kubin, 1994). Dislocation loops left by the moving dislocations are, indeed, formed. However, they systematically collapse within a few steps under the effect of their self-stresses. The same behaviour is observed during 3-D simulations of multiple slip performed in the same conditions, i.e. in the absence of cross-slip. One would, then, expect a strain hardening coefficient of about $\mu/300$, typical of forest interactions in the absence of cross-slip/recovery effects. In contrast, and as shown on fig. 16, the strain hardening coefficient gradually drops to zero, while the dislocation density reaches a saturation value. On the other hand, introducing a non zero cross-slip probability has the effect of yielding a non zero strain hardening coefficient of the right order of magnitude, while the dislocation density increases without saturating (cf. fig. 16). This can be explained as follows. In the absence of cross-slip, the collapse of the dislocation debris results in the absence of dislocation storage. This is why no strain hardening occurs once the stress has reached such a level that junctions can be formed and destroyed at a constant rate by a constant density of moving dislocations. In such conditions, the formation of a cell structure cannot occur. Indeed, the stored dislocations viz. loops and debris are thought to constitute the nucleus of the dislocation tangles from which the cell walls progressively emerge. Cross-slip is seen to stabilize Kocks' loops by bringing them into non planar, stable, configurations which can no longer collapse. Then, a stored density builds up which results in strain hardening in the Kocks' sense. Strain hardening also occurs through the forest model, since the density of mobile obstacles must continuously increase to compensate for the losses due to storage. Further, double cross-slip events lead to increased multiplication and further hardening. All these processes are seen to occur during the early stages of plastic flow as simulated in 3-D.

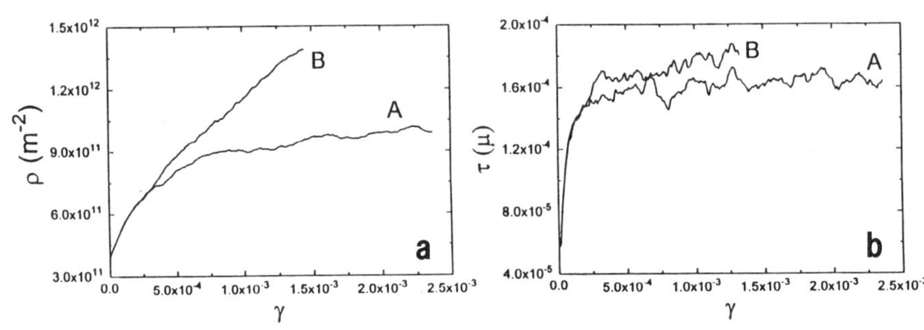

Figure 16. [100] copper crystal deformed with an imposed strain rate of 20 s^{-1}. A: without cross-slip, B: with cross-slip. a - Total dislocation density vs. shear strain, b - resolved stress-strain curves.

This explanation may look paradoxical as cross-slip is usually thought to induce softening (i.e., recovery) rather than hardening. However, it must be kept in mind that recovery mechanisms operate only once a cell structure has been established (i.e., in a region of high dislocation density), while the stabilization of dislocation tangles is essential in the early stages of deformation. The well established experimental fact that cell structures are only formed in the presence of cross-slip is, then, interpreted as stemming from the need to stabilize Kock's loops.

6.3. Strain Softening and Strain Localization in Alloys

Strain softening phenomena and the associated strain localizations have recently drawn a lot of attention and we give here an example of an investigation related to slip patterning. The questions investigated are related to the occurrence of static or propagating strain localizations and to the elaboration of constitutive forms taking into account spatial couplings and which may have propagating solutions. The coupling terms sought for are likely to be gradient terms, but their physical origin, either geometrical or microstructural, is still a matter of debate (cf. Kubin, 1995, for a review).

3-D simulations of strain softening in copper alloys, in single and duplex slip, have been performed and earlier results are reported by Bréchet et al. (1993) and Canova et al. (1993). Since what is searched for is essentially the spatial distribution in space of the active slip zones and its evolution with time, the strain softening properties can be described in a rather simple manner. An alloy friction force is introduced in the simulation of pure copper to account for the hardening effect of solute atoms in a copper alloy. Further, a softening rule is introduced in each slip plane, which describes a progressive reduction of the alloy friction with increasing shear in the considered slip plane. The shape of this softening curve can in principle be computed, in particular for the case of the shearing of coherent precipitates (cf. Canova et al. 1993). However, for a generic description of these effects the knowledge of this functional is not necessary. Rather, a simple exponential decay is assumed and the flow stress τ_n in the slip plane of index n is assumed to decrease with the local shear strain γ_n as:

$$\tau_n = \tau_0 + \Delta\tau_0 \mathrm{Exp}[-\gamma_n/\gamma_0] \tag{20}$$

τ_0 is the initial flow stress and the softening process is characterized by two parameters. One, $\Delta\tau_0$, is the amount of alloy friction which can be removed by the moving dislocations, the other is the characteristic strain γ_0 for this destruction to occur. To fix the ideas, $\Delta\tau_0$ is typically about a few tens of MPa for the shearing and dissolution of small coherent precipitates, while $\gamma_0 \approx d/b$ where d is the precipitate diameter. For the destruction of short-range order, it is known that the shearing by one or a few dislocations is sufficient to soften the alloy. The case of dislocation avalanches, either through dislocation unpinning (as in the classical Lüders band mechanism) or due to multiplication (as in covalent semiconductors) can also be elaborated to yield a local softening function. Therefore, mesoscopic simulations of strain softening have been carried out, using the two parameters $\Delta\tau_0$ and γ_0 to represent the set of possible experimental situations. Simulations of easy glide were performed with an initial density of sources 10^{11} m^{-2} and with applied strain rates in the range of 20 s^{-1}.

Several situations are met according to the values of $\Delta\tau_0$ and γ_0 and these results can be put in the form of an instability diagram which is reproduced on fig. 17. Globally, the tendency to strain localization can be rationalised through the value of the dimensionless coefficient $s = \Delta\tau_0/h\gamma_0$, where h is the strain hardening coefficient (here $h = \theta_I$, the hardening cofficient during the stage I of deformation in fcc crystals). For small values of s ($s < 1$), i.e., for small amounts of softening or large characteristic strains, the deformation is homogeneous. During straining, strain softening is always counteracted by strain hardening, so that the net result is always a stable deformation with a positive hardening slope.

For larger values of s, deformation is heterogeneous (cf. fig. 18). A few source segments start bowing out at small stresses but at the yield stress all the applied strain rate is concentrated in one set of active slip planes and the strain localizes in one narrow band.

As discussed by Canova et al. (1993), two situations can then be met. With intermediate values of s, one observes successive strain localisations at random places in the simulated specimen. Plastic strain is permanently transferred from glide bands rendered inactive by strain hardening to new active sites. For larger values of s, i.e. in the upper right portion of the instability diagram (cf. fig. 17), band propagation takes place with a constant velocity. It

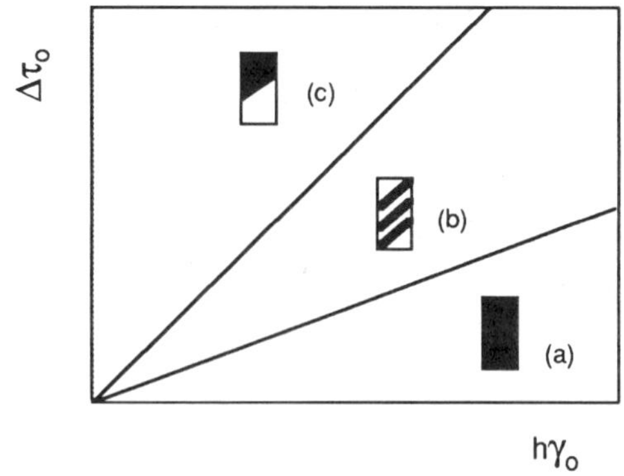

Figure 17. Instability diagram showing the various types of behaviour obtained in the simulation of strain softening. a - Uniform deformation, b - successive strain localizations, spatially distributed at random, c - propagation of a Lüders-like band with a constant velocity. The boundaries are those yielded by a preliminary model of these effects based on a spatial coupling via long range elastic interactions.

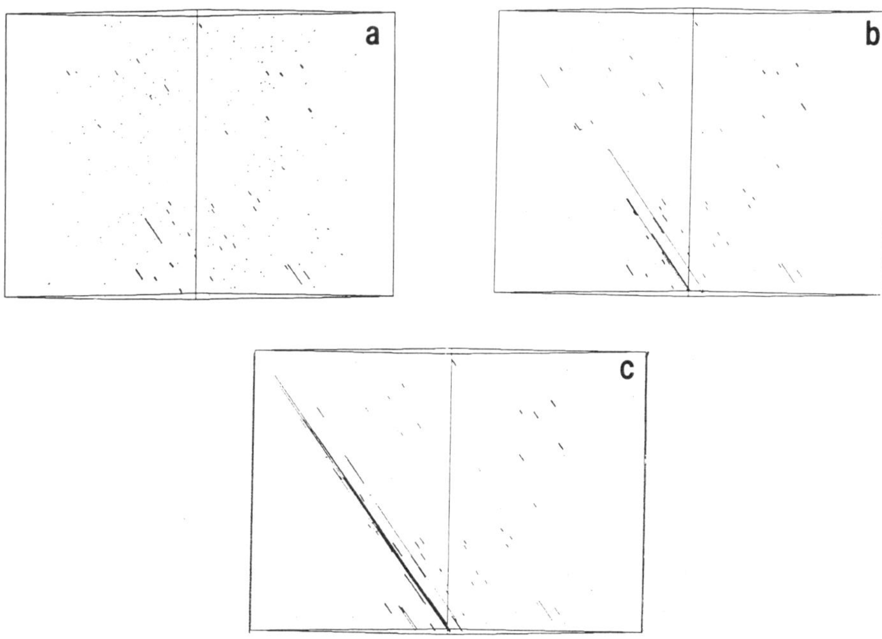

Figure 18. [110] view of a crystal of copper alloy of size (15 μm)^3 deformed in constant strain rate along a [100] axis. The initial conditions consist of a random distribution of screw source segments, perpendicular to the view plane. The edge component of the active sources are in the view plane. Development of strain localization strain softening. a - Below the yield stress, b - at the upper yield point, c - during the stress plateau ($\varepsilon \approx 4 \times 10^{-3}$).

can be shown that the propagation mechanism remains unaltered when cross-slip is switched off. This means that, within the simulation, there is no other possible mechanism than long range elastic stresses to induce slip propagation.

This was confirmed by a detailed examination of the simulated dynamic configurations at the band front and behind. Dislocation sources are activated by the expansion of sources of same sign at the front. Figure 19 shows snapshots of dynamic sequences at the rear of the band, where elastic interactions between dislocations of same and unlike sign play a major role. Actually, the interactions of dislocations of opposite sign are responsible for hardening, while softening plays the same role all throughout the band, inducing the coupling of dislocations of same sign in addition to source activation. The spatial coupling is, therefore, identified as stemming from the interactions of dislocations of same sign. This allows to reconstruct a constitutive form which incorporates time and space as evolutionary variables. The model elaborated to account for the observed behaviour is not presented here. It recovers in simple terms the instability diagram of fig. 17 as well as the band velocities and seems in good agreement with the available experimental results.

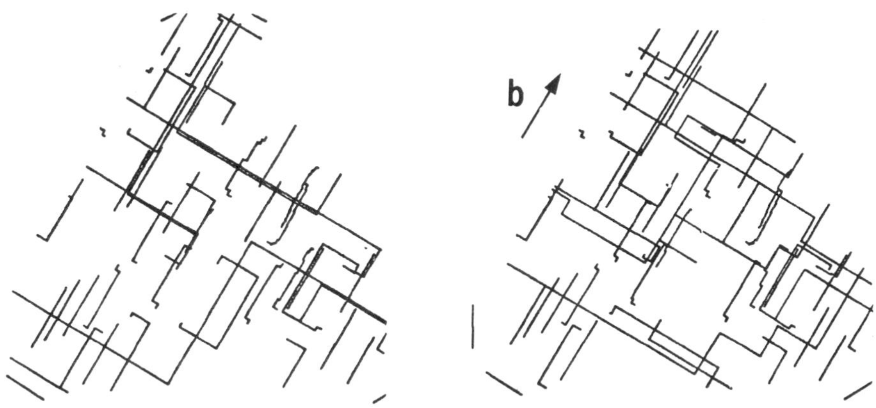

Figure 19. Dislocation interactions at the rear of the slip band (sections of thickness 1 μm). The evolution of mobile dislocations in the same area shows that edge segments of same sign and of and unlike sign (dipoles) are strongly coupled. The former interaction leads to a spatial coupling and the latter to hardening.

CONCLUSION

The two examples of application briefly presented here show the potential of such studies for bridging the gap between the microscopic and macroscopic aspects of plasticity. In the case of strain hardening in fcc crystals in multiple slip, a new factor emerges, the beneficial influence of cross-slip on the stabilization of the stored density and its apparent relation to the formation of cell structures. The simulations of strain softening, on the other hand, represent the first situation which could be investigated in closed form, yielding appropriate answers for the construction of a model for strain localizations and band propagation.

This summary of the recent progress in the development of the 2- and 3-D simulations of dislocation dynamics and interactions is necessarily incomplete. The treatment of local interactions has proved to be the major difficulty of such mesoscopic elaborations. Mechanisms involving dislocation core properties cannot be rigorously taken into account. They even involve phenomena which are not well described by dislocation theory in its present state. As illustrated by the example of forest interactions, the use of well-defined experimental data allows to set coherent input values through simple micro-macro connections. It must be emphasized that this procedure does not drive the output of the simulations, as the main properties looked for (strain hardening, evolution of dislocation densitites, internal stresses, patterning) evolve by themselves as a result of collective behaviour.

More generally, the two approaches of dynamic patterning, by modelling and by simulations, that have been developed in the past ten years are still in their infancy. They have started yielding both generic types of differential equations leading to patterning and simple simulated patterns. The general trend is to pursue these two approaches in parallel and in connection with experimental reality. The latter is conveniently represented by mesoscopic quantities which integrate atomistic properties like critical annihilation distances, cross-slip properties, dislocation junction strength ... Therefore, in addition to the continuous effort to understand dislocation microstructures in dynamical terms and their consequences on mechanical properties, there is also a strong motivation to close the narrowing gap between atomistic and mesoscopic studies.

ACKNOWLEDGEMENTS

The author wishes to gratefully thank B. Devincre, E. Rauch, R. Fournet, J. Lépinoux, M. Mills and D. Rönnpagel for authorizing partial reproduction of their works. The first parts of the present review reproduce with some modifications a paper presented at the NATO ASI on "Numerical Simulations in Materials Science" (F-Oléron, June 1995), to be published by Kluwer Academic Publishers (N-Dordrecht).

REFERENCES

E.C. Aifantis, 1986, On the dynamical origin of dislocation patterns, *Mat. Sci. Eng.* 81:563.

E.C. Aifantis, 1987, On the heterogeneity of plastic deformation, *in:* Constitutive relations and their physical bases (Proc. 8th Risø Symposium), p. 205, S.I. Anderson et al., eds., Risø Natl. Lab., Dk-Roskilde.

E.C. Aifantis, 1988, On the gradient approach to deformation patterning and fracture, *Solid State Phenomena* 3&4:355.

E.C. Aifantis, 1990, Nonlinearity and selforganization in plasticity and fracture, *in:* Patterns, Defects and Materials Instabilities, p. 221, D. Walgraef and N.M. Ghoniem, eds., Kluwer Acad. Publ., N-Dordrecht.

R.J. Amodeo, 1988, Doctoral dissertation, University of California, Los Angeles.

R.J. Amodeo and N.M. Ghoniem, 1990a, Dislocation dynamics, I. A proposed methodology for deformation mechanisms, *Phys. Rev. B* 41:6958.

R.J. Amodeo and N.M. Ghoniem, 1990b, Dislocation dynamics, II, *Phys. Rev. B* 41:6968.

D.J. Bacon, 1967, A method for describing a flexible dislocation, *Phys. Stat. Sol.* 23:527 .

D.J. Bacon, U.F. Kocks and R.O. Scattergood, 1973, The effect of dislocation self-interaction on the Orowan stress, *Phil. Mag.* 28:1241.

S.J. Basinski and Z.S. Basinski, 1979, Plastic deformation and work hardening, *in:* Dislocations in Solids, Vol. 4, p. 263, F.R.N. Nabarro, ed., North-Holland, Amsterdam.

J. Bonneville, B. Escaig and J.L. Martin, 1988, A study of cross-slip activation parameters in pure copper, *Acta metall.* 36: 1989.

L.M. Brown, 1964, *Phil. Mag.* 10:441.

Y. Bréchet, G. R. Canova and L.P. Kubin, 1993, Static versus propagative plastic strain localizations, *Scripta metall. mater* 29:1165.

G. Canova, L.P. Kubin and Y. Bréchet, 1993, Glide softening in alloys: a simulation, in *Large Plastic Deformations*, p. 27, C. Teodosiu et al. (Eds.), A.A. Balkema, Rotterdam.

B. Devincre, 1993, Doctoral Thesis N° 2838 (in French), University of Paris-Sud (Orsay).

B. Devincre and M. Condat, 1992, Model validation of a 3D simulation of dislocation dynamics: discretization and line tension effects, *Acta metall. mater.* 40:2629.

B. Devincre and L.P. Kubin, 1994, Simulations of forest interactions and strain hardening in FCC crystals, *Modelling Simul. Mater. Sci. Eng.* 2:559.

B. Devincre and V. Pontikis, 1993, Computer modelling of dynamically-induced dislocation patterning, *in* Mater. Res. Symp. Proc., Materials Research Society, Pittsburgh.

B. Devincre and S. Roberts, 1995, 3-D simulation of dislocation crack interactions at the mesoscopic scale, in Computer Simulations in Materials Science, L.P. Kubin et al. (eds.), Kluwer Acad. Publ., N-Dordrecht, *in press*.

K. Differt and U. Essmann, 1993, Dynamical model of the wall structure of persistent slip bands of fatigued metals. I: dynamical model for edge dislocation walls, *Mat. Sci. Eng.* A164:295.
M.S. Duesbery and K. Sadananda, 1991, The numerical simulation of continuum dislocations, *Phil. Mag. A* 63:535.
M.S. Duesbery, N.P. Louat and K. Sanandana, 1992, The interaction of dislocations with coherent inclusions I. Perfect edge and screw dislocations, *Phil. Mag. A* 65:311 .
M.S. Duesbery, N.P. Louat and K. Sanandana, 1992, The mechanics and energetics of cross-slip, *Acta metall. mater.* 40:149.
B. Escaig, 1968, Sur le glissement dévié des dislocations dans la structure cubique à faces centrées, *J. de Physique* 29:225.
U. Essmann and K. Differt, 1988, The nature of the wall structure in persistent slip bands of fatigued metals, *Scripta metall.* 22:1337.
U. Essmann and K. Differt, 1995, Dynamical model of the wall structure in persistent slip bands of fatigued metals II. The wall spacing and the temperature dependence of the yield stress in saturation, *Mat. Sci. Eng.* submitted.
U. Essmann and H. Mughrabi, 1979, Annihilation of dislocations during tensile and cyclic deformation and limits of dislocation densities, *Phil. Mag. A* 40:731.
X.F. Fang, J. Hagedorn and W. Dahl, 1991, Computer simulation of dynamic dislocation organization and the strain hardening in metals, *in* Strength of Metals and Alloys (Proc. ICSMA 9), D.G. Brandon, R. Chaim and A. Rosen, eds., Freund Publ. House, London.
X.F. Fang and W. Dahl, 1993, Investigation of the formation of dislocation cell structure and the strain hardening of metals by computer simulation, *Mat. Sci. Eng.* A164:300.
A.J.E. Foreman, 1967, The bowing of a dislocation segment, *Phil. Mag.* 15:1011.
A.J.E. Foreman and M.J. Makin, 1966, Dislocation movement through random arrays of obstacles, *Phil. Mag.* 14:911.
R. Fournet, 1994, Doctoral thesis (in French), Université de Bourgogne.
R. Fournet and J.M. Salazar, 1995, Computer simulation on dislocation patterning, *Solid State Phenomena* 42&43:205.
J. Friedel, 1967,"Dislocations", Pergamon Press, Oxford.
A. Franek, R. Kalus and J. Kratochvil, 1991, Model of early stage of dislocation structure formation in cyclically deformed metal crystals, *Phil. Mag.* 64:497.
N.M. Ghoniem and R.J. Amodeo, 1988, Computer simulation of dislocation pattern formation, *Solid State Phenomena* 3&4:377.
N.M. Ghoniem and R.J. Amodeo, 1990, Numerical Simulation of dislocation patterns during plastic deformation, *in P*atterns, Defects and Materials Instabilities, p. 303, D. Walgraef and N.M. Ghoniem, eds., Kluwer, NL-Dordrecht.
E. van der Giessen and A. Needleman, 1994, On the solution of two-dimensional plasticity problems with discrete dislocations, *in* Computational Material Modelling, AD-Vol. 42/PVP-Vol. 294, p. 53, ASME.
E. van der Giessen and A. Needleman, 1995, Discrete dislocation plasticity: a simple planar model, *Modelling Simul. Mater. Sci. Eng.* in press.
J. Gil Sevillano, 1993, Flow stress and work hardening, *in:* Treatise in Materials Science and Technology, Vol. 6, p. 19, H. Mughrabi, ed., VCH, D-Weinberg.
J. Gil Sevillano, E. Bouchaud and L.P. Kubin, 1991, *Scripta metall. mater.* 25:355.
I. Groma and G.S. Pawley, 1993a, Role of the secondary slip system in a computer simulation model of the plastic behaviour of single crystals, *Mat. Sci. Eng. A* 164:306.
I. Groma and G.S. Pawley, 1993b, Computer simulation of plastic behaviour of single crystals, *Phil. Mag. A*, 67:1459.
A.N. Gulluoglu, D.J. Srolovitz, R. LeSar and P.S. Lomdahl, 1989, Dislocation distributions in two dimensions, *Scripta met.* 23:1347.
A.N. Gulluoglu, D.J. Srolovitz, R. LeSar and P.S. Lomdahl, 1990, Dynamical simulation of dislocation microstructure, *in* Simulation and theory of evolving microstructures, M.P. Anderson, A.D. Rollet, eds., TMS, Warrendale (PA).
A.N. Gulluoglu and C.S. Hartley, 1992, Simulation of dislocation microstructures in two dimensions: I. Relaxed structures, *Modelling Simul. Mater. Sci. Eng.* 1:1.
A.N. Gulluoglu and C.S. Hartley, 1993, Simulation of dislocation microstructures in two dimensions: II. Dynamic and relaxed structures, *Modelling Simul. Mater. Sci. Eng.* 1:383.
P. Hähner, 1995, Habilitation dissertation (in English), University of Lille I.
F. Hernandez Olivares and J. Gil Sevillano, 1987, A quantitative assessment of forest-hardening in F.C.C. metals, *Acta metall.* 35:631.

H.W. Hesselbarth and E. Steck, 1992, A simulation of dislocation patterning derived from cellular automata, *Solid State Phenomena* 23&24:445.
P.B. Hirsch, 1975, "The Physics of Materials", Vol. 2, p. 189, Cambridge University Press.
D.L. Holt, 1970, Dislocation cell formation in metals, *J. Appl. Phys.* **41**: 3197.
D. D'Humières and P. Lallemand, 1986, Lattice gas automata for fluid mechanics, *Physica* 140 A:326.
H.O.K. Kirchner, 1984, The concept of line tension: theory and experiment, *in:* " Dislocations 1984", p. 53., P. Veyssière, L. Kubin and J. Castaing, eds., Editions du CNRS, Paris.
U.F. Kocks, 1966, A statistical theory of flow stress and work-hardening, *Phil. Mag.* 13:541.
U.F. Kocks, 1976, Laws for work-hardening and low temperature creep, *J. Eng. Mat. and Technology* 98:76.
U.F. Kocks, 1985, Dislocation interactions, flow stress and work-hardening, *in:* "Dislocations and properties of real materials", p. 125, The Institute of Metals, London.
U.F. Kocks, A.S. Argon and M.F. Ashby, 1975, "Thermodynamics and Kinetics of Slip", Progress in Materials Science, Vol. 19, B. Chalmers, J.W. Christian and T.B. Massalski, eds., Pergamon Press, Oxford.
J. Kratochvil, 1988a, Dislocation pattern formation in metals, *Rev. Phys. Appl.* 23:419.
J. Kratochvil, 1988b, Dislocation structure instability and fatigue, *in:* Basic Mechanisms of Fatigue in Metals, p. 15, P. Lukas and J. Pollack, eds. Academia, Prague.
J. Kratochvil, 1989, Stability approach to problem of work hardening in metals, *Rev. Def. Behav. of Metals* 2:353.
J. Kratochvil, 1994, Continuum mechanics approach to collective behaviour of dislocations, *Solid State Phenomena* 35-36:71.
J. Kratochvil and S. Libovicky, 1986, Dipole drift mechanism of early stages of dislocation pattern formation in deformed metal single crystals, *Scripta metall.* 20:1625.
J. Kratochvil and A. Orlova, 1990, Instability origin of dislocation structure, *Phil. Mag.* 61:281.
J. Kratochvil and M. Saxlova, 1992a, Sweeping mechanism of dislocation pattern formation, *Scripta metall. mater.* 26:113.
J. Kratochvil and M. Saxlova, 1992b, A model of formation of dipolar dislocation strcutures, *Solid State Phenomena* 23&24:369.
L.P. Kubin, 1995, Strain and strain rate softening instabilities: length scales and spatial couplings, *in* : Plasticity of Metals and Alloys (ISPMA 6), p. 219, P. Lukac ed., Trans Tech Publications, CH-Aedermannsdorf.
L.P. Kubin, 1993, Dislocation patterning during multiple slip of fcc crystals: a simulation approach, *Phys. Status Solidi* (a) 135: 433.
L.P. Kubin, 1993, Dislocation patterning, *in:* Treatise in Materials Science and Technology, Vol. 6, p. 138, H. Mughrabi, ed., VCH, D-Weinberg.
L.P. Kubin and J. Lépinoux, 1988, The dynamic organization of dislocation structures, *in* Strength of Metals and Alloys (Proc. ICSMA 8), Vol. 2, p. 35, P.O. Kettunen et al, eds., Pergamon Press, Oxford.
L.P. Kubin, G. Canova, M. Condat, B. Devincre, V. Pontikis and Y. Bréchet, 1992, Dislocation structures and plastic flow: a 3-D simulation, *Solid State Phenomena* 23&24, 455 (1992).
D. Kuhlmann-Wisldorf, 1989, Theory of plastic deformation: - properties of low energy dislocation structures, *Mat. Sci. Eng. A* 113: 1.
D. Kuhlmann-Wilsdorf, 1992, Fundamentals of cell and subgrain structures in historical perspective, *Scripta metall. mater.* **27**: 951.
J. Lépinoux, 1987, Doctoral Thesis N° 456 (in French), University of Poitiers.
J. Lépinoux and L.P. Kubin, 1987, The dynamic organization of dislocations: a simulation, *Scripta met.* 21:833.
J. Lépinoux, 1988, Simulation of the dynamic organization of dislocation microstructures, *Solid State Phenomena*, 3&4:389.
J. Lépinoux, 1995, Application of cellular automata in materials science, in "Computer Simulations in Materials Science", L.P. Kubin et al. (eds.), Kluwer Acad. Publ., N-Dordrecht, *in press*.
B.M. Loginov, 1991, The role of forest dislocation flexibility properties in the process of crystal work hardening, *Phys. Stat. Sol. (a)* 125:481.
M. Mareshal, 1995, Cellular automata: a review, 1995, in "Computer Simulations in Materials Science", L.P. Kubin et al. (eds.), Kluwer Acad. Publ., N-Dordrecht, *in press*.
M.J. Mills and D.C. Chrzan, 1992, Dynamical simulation of dislocation motion in $L1_2$ alloys, *Acta metall. mater.* 40:3051.
M.J. Mills, D.C. Chrzan and D.B. Miracle, 1994, Influence of dislocation fine structure on the strength and flow behavior of ordered intermetallic compounds, *in:* Strength of Materials (Proc. ICSMA 10), p 41, H. Oikawa et al., eds., The Japan Institute of Metals.

C. Misbah, 1988, Dynamics of nonequilibrium systems in the weakly nonlinear regime, *Solid State Phenomena* 3&4:29.

H. Mughrabi, 1973, in *Proc. ICSMA 3*, Vol. 1, The Institute of Metals, Cambridge, p. 407.

H. Mughrabi, 1979, Microscopic mechanisms of metal fatigue, *in* Strength of Metals and Alloys (Proc. ICSMA 5), Vol. 3, p. 1615, P. Haasen, V. Gerold and G. Kostorz, eds., Pergamon Press, Oxford.

H. Mughrabi, 1987, A two-parameter description of heterogeneous dislocation distribution in deformed metal crystals, *Mat. Sci. Eng.* 85: 15.

R. Neuhaus and Ch. Schwink, 1992, On the flow stress of [100] and [111]-oriented Cu-Mn single crystals, *Phil. Mag. A* 65: 1463.

P. Neumann, 1986, Low energy dislocation configurations: a possible key to the understanding of fatigue, *Mat. Sci. Eng.* 81:465.

A.A. Predvoditelev and B.M. Loginov, 1972, Laws of passage of slipping dislocations through flexible and reacting dislocation ensembles, *Sov. Phys. Crystallogr.* 30:433 .

A.A. Predvoditelev and G.I. Nichugovskii, 1972, A model for dislocation motion through a dislocation forest, *Sov. Phys. Crystallogr.* 17:132.

Th. Pretorius and D. Rönnpagel, 1994, Dislocation motion in Ni-basis superalloys, a quantitative comparison between simulation calculations, TEM observations and bulk measurements, *in:* Strength of Materials (Proc. ICSMA 10), p 689, H. Oikawa et al., eds., The Japan Institute of Metals.

S.V. Raj and G.M. Pharr, 1986, A compilation and analysis of data for the stress dependence of the subgrain size, *Mat. Sci. Eng.* 81:217.

A.D. Rollett and U.F. Kocks, 1994, A review of the stages of work hardening, *Solid State Phenomena* 35-36, 1.

D. Rönnpagel and V. Mohles, 1994, Simulation calculations of solid solution hardening, *in:* Strength of Materials (Proc. ICSMA 10), p 137, H. Oikawa et al., eds., The Japan Institute of Metals.

J.M. Salazar, R. Fournet and N. Banai, 1995, Dislocation patterns from reaction-diffusion models, *Acta metall. mater.* 43:1127.

C. Schiller and D. Walgraef, 1988, Numerical simulation of persistent slip band formation, *Acta met.* 36:563.

A. Seeger, 1955, *Phil. Mag.* 46:1194.

A. Seeger, 1988, Thermodynamics of open systems, self-organization and crystal plasticity, *in* Strength of Metals and Alloys (Proc. ICSMA 8), p. 463, P.O. Kettunen et al, eds., Pergamon Press, Oxford.

G. Schoeck and R. Frydman, 1972, The contribution of dislocation forest to the flow stress, *Phys. Stat. Sol. (b)* 53:661.

A.A. Shtolberg, 1971, A method for computation of equilibrium dislocation configurations, *Phys. Stat. Sol. (b)* 43:523.

S. Suresh, 1993, Cyclic deformation and fatigue, *in:* Treatise in Materials Science and Technology, Vol. 6, p. 509, H. Mughrabi, ed., VCH, D-Weinberg.

O.G. Tyupkina, 1992, Dislocation ensemble movement through random arrays of obstacles, *Phil. Mag.* 65:111.

D. Walgraef, 1986, Reaction diffusion equations: an application to the formation of dislocation patterns, *in:* Mechanical properties and behaviour of solids: plastic instabilities, p. 354, V. Balakrishnan and C.E. Bottani, eds., World Scientific, Singapore.

D. Walgraef, 1988, Instabilities and patterns in reaction-diffusion dynamics, *Solid State Phenomena*, 3&4:77.

D. Walgraef, 1990, Kinetic models for defect populations in driven materials, *in:* Patterns, Defects and Materials Instabilities, p. 73, D. Walgraef and N.M. Ghoniem, eds., Kluwer Acad. Publ., N-Dordrecht.

D. Walgraef and E.C. Aifantis, 1985, Dislocation patterning in fatigued metals as a result of dynamical instabilities, *J. Appl. Phys.*, 58:688.

D. Walgraef, C. Schiller and E.C. Aifantis, 1987, Reaction-diffusion approach to dislocation patterns, *in:* Patterns, Defects and Microstructures in Nonequilibrium Systems, p. 257, D. Walgref, ed., Martinus Nijhoff Publ., NL-Dordrecht.

H.Y. Wang and R. LeSar, 1995, O(N) algorithm for dislocation dynamics, *Phil. Mag.* 71:149.

S. Wolfram, 1984, Computation theory of cellular automata, *Commun. Math. Phys.* 96:15.

COMPUTER SIMULATION OF FRACTURE, DISLOCATIONS AND MARTENSITIC TRANSFORMATIONS IN SOLIDS

Philip C. Clapp

Center for Materials Simulation
Institute of Materials Science
University of Connecticut
Storrs, CT 06269-3136, USA

1. Introduction

In the past decade or so computer simulation has become a principal method for investigating and predicting a wide variety of material properties. In fact it is now routinely used to interpret experiments, or to test theoretical models, or to make new links between theory and experiment. This "third branch of investigation" will continue to expand in usefulness, as the power of computers continues to increase at an astounding rate. It has been noted that from 1960 to 1990, computers have doubled their power per dollar about every 18 months, but in the last several years there has been an abrupt increase by roughly a factor of 2000! This is because computers with parallel architecture of immense computational capability and reasonably user friendly interfaces are now becoming available. As just one example of this remarkable advance, simulations of arrays of 100 million atoms are now being carried out by several groups around the world, whereas three years ago 100 thousand atom arrays were at the extreme upper limit of possibility.

The primary techniques of computer simulation applied to solids are: (a)Molecular Dynamics, (b) Monte-Carlo and (c) a variety of discrete element methods. Molecular Dynamics usually consists of calculating the motions of all the atoms of an array, which interact via some semi-empirical force law, and obey the laws of Newtonian classical mechanics. The size of the arrays can be as much as 100 million atoms but are more typically less than 100 thousand. The simulations are in "real time" and the longest runs usually correspond to not much more than a nanosecond. This fact naturally restricts the application of Molecular Dynamics simulations to very rapid kinetics events, such as fracture or displacive transformations. In order to study phenomena involving substantially longer times and/or larger samples (e.g. long range diffusion, precipitation, etc.) Monte-Carlo or discrete element methods need to be employed. These latter methods however lack a true time measure, and of course cannot accurately represent atomic scale phenomena, but there are now new developments in which these various methods can be used in combination to span larger space and time scales.

In the past decade the ability to accurately simulate specific metals and alloy systems has been very significantly improved by the advent of the Embedded Atom Method[1] (or the very similar Finnis-Sinclair method[2]). These are semi-empirical approaches, incorporating some of the fundamental theoretical concepts of first principles quantum mechanical calculations. They can be used to construct interatomic potentials (including volume dependent many body interactions), based upon experimental (and other) data of a given binary alloy system. Then by using a supercomputer equipped with a Molecular Dynamics program to simulate the motion of an atomic array interacting according to the fitted EAM potentials, many different details of atomistic behavior can be observed and predicted. Considerable success has been achieved in predicting (via simulation) such properties as bulk and surface phonons, surface structure, impurity segregation, phase stability, liquid structure, grain boundary structure and thermal expansion[3,4,5].

The Monte-Carlo method, as introduced by Metropolis et al.[6] , estimates the equilibrium properties of a system by averaging over a finite sequence of quasi-randomly generated microscopic states where the "transition probability" for the trajectory through phase space is guided by the value of a Boltzmann factor determined from an appropriate Hamiltonian. Good overviews of the method with illustrative examples are contained in the excellent text on computer simulation of Allen and Tildesley[7] , the detailed monograph by Binder and Heerman[8] or the wide ranging volume edited by Binder[9].

In essence, the method proceeds by choosing some starting configuration of interacting objects (atoms, spins, electrons, etc.), calculating the energy of this configuration via the chosen Hamiltonian, and then generating a new trial configuration (with its associated change in energy, ΔE) by a quasi-random shift of the initial configuration. Depending on the configurational variable of interest this would involve displacing atoms, interchanging atomic species, flipping spins, etc. A "simulation temperature" Υ is then chosen and the analog of a Boltzmann factor, $\exp(-\Delta E/k\Upsilon)$ is calculated. If this factor is found to be greater than a number, δ, chosen at random from the interval (0,1), then the trial configuration is accepted; otherwise the computer returns to the initial state and trys again. The measurement of equilibrium averages over the configurations generated in this fashion begins when it is believed that the residual influence of the starting configuration has become negligible, and it is stopped when it appears that no further accuracy in the measured quantities is worth the additional computer costs. In one modified strategy called "simulated annealing" the value of Υ is slowly reduced towards zero, which it is hoped will yield the ground state. This may or may not lead to the lowest energy configuration depending on the multi-valleyed topology of the energy surface in phase space and also a degree of luck.

Chief amongst the discrete element techniques (which simply use a grid on which to perform numerical calculations of some property of a continuum fluid or solid) is the "finite element method" which imposes specific topological connections and continuity conditions on the grid to obtain well behaved approximations to the solutions of the differential equations that are being modelled with the grid, such as those governing heat or fluid flow, stress distribution, etc. Many standard references and commercially available software packages exist for implementing this approach today that are too numerous to mention here.

There are, however, new combinations developing between different methods of investigation. For example it is now possible to get the data necessary to construct the semi-empirical potentials used in Molecular Dynamics from either experimental results or first principles quantum mechanical calculations, or some combination of the two. This now makes it possible to simulate many interesting alloy systems for which there is not yet adequate experimental data to construct a potential, if the appropriate quantum mechanical calculations have been done. Another important combination is the use of Molecular Dynamics simulations to obtain the parameters for a Ginzburg-Landau type of strain free energy functional, which in turn can be used in Monte-Carlo/ discrete element simulations of macroscopic sized samples. In this way a hierarchy of simulation methods is developing rapidly to cover the full range of time and size scales of potential interest. To illustrate some of these techniques, a variety of studies (mostly performed at our Center for Materials Simulation) in the fields of dislocation dynamics, fracture and displacive phase transformations will now be described.

2. Fracture Simulations

We have simulated the fracture behavior of the metals Al and Nb and the intermetallic alloys Nb_3Al, RuAl and NiAl under mode I loading (applied stress normal to the crack plane). Among the properties studied are the value of stress (and the critical stress intensity factor) required to initiate fracture in a material for a given size of initial crack, the identities of the activated slip systems, and the value of several material parameters (the surface energy and "unstable stacking" energy) needed to predict the critical stress intensity by theoretical means. While information about these properties can be obtained from experiment, Molecular Dynamics simulation offers the ability to precisely control the material and extract information about atomic motions. Furthermore, some quantities such as the unstable stacking energy are not experimentally measurable, and can only be determined by computation.

2.1 Conditions of the simulations

The simulations were performed on arrays of approximately 8,000 particles with several different orientations. In the array an initial crack was created and a tensile load normal to the crack applied. In the Z direction, taken parallel to the crack edge, periodic boundary conditions were used, but otherwise the outer surfaces of the array were free. The dimensions of the arrays were on the order of 70 Å by 70 Å by 20 Å, where the shorter direction was the one with the repeating boundary.

The interatomic potentials (which describe the forces between atoms) are from Voter and Chen[10] for the Al-Al and Ni-Al interactions, and from Rifkin et al.[11] for the Ru-Ru, Nb-Nb, Ru-Al and Nb-Al interactions. All potentials were computed using the Embedded Atom Method (EAM) of Daw and Baskes. These potentials were empirically fitted to the cohesive energies, vacancy energies, lattice constants and elastic constants for the simulated systems. Atomic motion was obtained from the calculated forces via numerical integration (Gear's algorithm[12]), which was performed in discrete time steps of 10^{-15} seconds.

A crack was formed by displacing the atoms of the otherwise perfect array according to the continuum mechanics equations for anisotropic elasticity due to Sih et al.[13]. The initial displacements are linearly proportional to the stress intensity factor $K_I \equiv \sigma \sqrt{\pi a}$, :where σ is the tensile applied stress (very far from the crack) and $\boldsymbol{a}$ is the length of the crack.

After the atoms had been positioned to form a crack of given length (about 25Å) under a given value of σ, the lattice was allowed to relax for 2,000 steps. During these 2,000 steps, the atoms moved according to the Molecular Dynamics algorithm (that is, according to the classical equations of motion) for some chosen temperature. To prevent the crack from closing, the outermost two layers of atoms on the upper and lower traction surfaces (normal to Y) and on the free surface ahead of the crack (normal to X) were constrained to remain in their initial positions (given by equation 1). Since these immobile atoms neither release nor absorb thermal energy, they do not affect the temperature distribution of the interior, which will be essentially uniform after equilibration. These constrained atoms are referred to as the *static layers* (see Fig. 1).

During the relaxation period the crack either closed, stabilized or propagated. When a value of the applied tensile load σ had been found at which the crack remained stable, the stress intensity factor was slowly increased by displacing the atoms of the "static layers" by 1% of their original displacements (given by equation 1) every 20 steps. With this scheme, the stress intensity factor was doubled in 2,000 computational steps. Under the increasing load, the crack could propagate (brittle behavior), be blunted by dislocation emission (ductile behavior) or show both phenomena (see Fig. 2-4). The value of the stress intensity factor when a crack propagated was taken to be the critical value. Depending upon the orientation of the crack plane and of the tensile load, different behaviors were observed as will be described below.

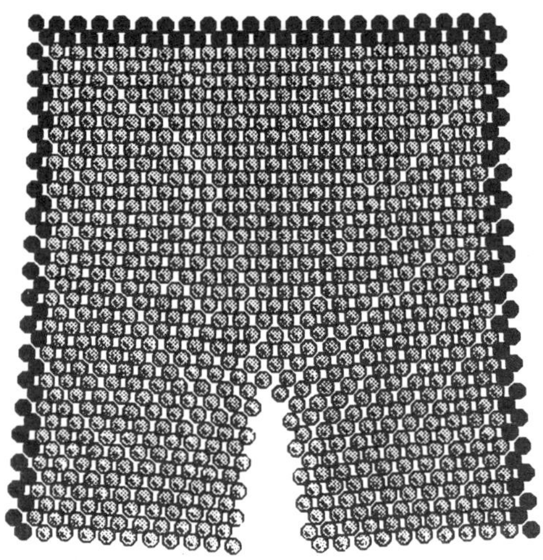

Fig. 1. Crack-embedded array. Dark atoms are at the interface with the continuum and are frozen during the simulations.

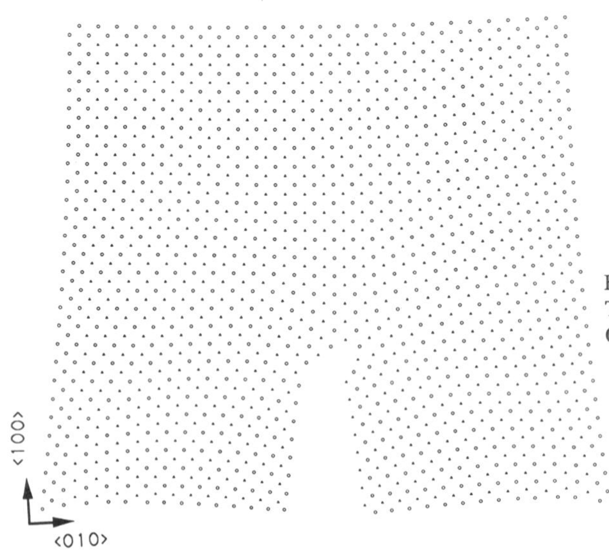

Fig. 2. RuAl in [100](010) orientation. Triangles are Al, circles Ru. Crack tip veers onto the (110) plane:

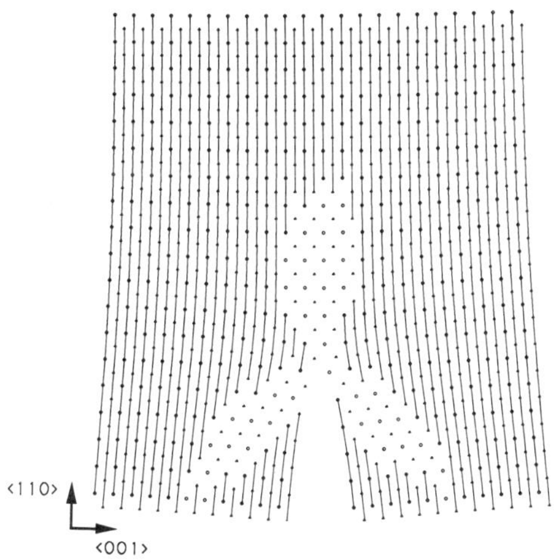

Fig. 3. Strained region around crack tip in RuAl. *Bonds* are drawn connecting atoms separated by $\mathbf{a}_0 \pm 3\%$.

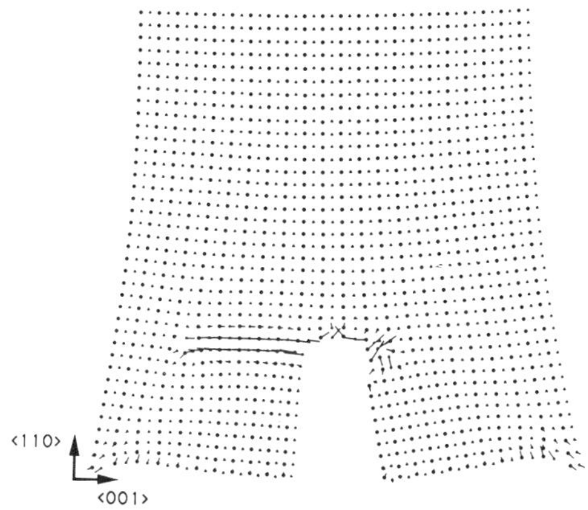

Fig. 4. RuAl in [110](001) orientation. Emission of dislocation at right angles to crack plane has blunted the crack tip. Tails represent displacement of atoms from initial positions.

The question of the degree to which transient stress waves are present in these simulations, given the fairly rapid loading conditions, bears examination. At 300K and above it was found that stress waves damp out in a few hundred computational steps (i. e. about five atomic vibrational periods), but at 10K, the damping time was more like a thousand steps. As a result, for simulations at or above 300K, the stress at the crack tip contained a relatively small stress ripple superimposed on a much larger quasistatic component. At 1K or 10K, the stress wave ripple represented a much larger fraction of the total, which requires that interpretation of the low temperature simulations under dynamic loading conditions be taken with caution.

It should also be mentioned that since the array in the Z direction was only about 20Å before periodic boundary conditions were applied, double kink nucleation will be very difficult because the minimum distance between a metastable kink-antikink pair in these metals can be roughly estimated to be of this same magnitude. In order to observe dislocation nucleation via a kink nucleation process, it would probably be necessary to increase the array thickness by a factor of four or more (which is planned in the near future). However, since straight line dislocation motion is so much faster than the stepwise advance of a dislocation mediated by kink motion, we expect that once the stress at the crack tip has reached the level capable of nucleating straight line dislocations, these will dominate the stress relief processes in the near crack tip region.

Furthermore, the stress intensity required to nucleate a kink-antikink pair, may not be very much lower that what we have found necessary for the nucleation of straight dislocations (since it will presumably require slip of at least one-half **b** over a distance along the crack front of about five unit cells, though this has not yet been tested).

2.2 Techniques for Analyzing the Simulations

The raw output from the computer simulations was a listing of energies and atomic positions at discrete times. These atomic positions were passed through visualization programs to produce several types of visual output. The basic representation was a plot of atom positions. Because of the large number of atoms in the simulation a subset of atoms was drawn, typically a "slice" through the atom array oriented normal to the direction of the repeating boundary. Figure 2 is an example of this kind of plot. Enhancements to this basic image include *bonds* and *tails*.

"Bonds" are lines drawn between every pair of atoms whose radial separation is within a given range. Such lines highlight several types of defects in a lattice. Cracks are visible as a region void of bonds. Stacking faults are seen as a misalignment of bonds across the fault plane. Bonds can also be used to highlight strained regions. For example, in figure 3, bonds are drawn connecting all atoms separated by a RuAl B2 unit cell edge ($\mathbf{a_0}$ = 4.285Å) with a tolerance of ±3%. The strained region where the unit cell dimensions are elongated beyond this tolerance show no bonds, and appears as a three lobed region centered on the crack tip.

"Tails" are drawn to enhance the perception of displacements. A "tail" is a line drawn from an atom to its past (or future) position. A refinement of this technique is made by subtracting from the simple displacement the *average background displacement*, that is the average displacement of the atom's neighbors. An atom's neighbors are all atoms (including the atom itself) within a given radius (typically the second nearest neighbor shell). When this average background displacement is removed what remains is the component of the atom's motion which differs from its neighbors. The benefit of this procedure is to filter out those displacements and strains that are locally uniform. What remain are inhomogeneous displacements which typically result from the passage of a dislocation core. In figure 4 we can see the result of this technique. The slip plane is clearly marked by the several rows of displacement vectors. The signs of the displacements reverse as the slip plane is crossed, which indicates the relative motion of the half lattices.

A third analysis technique used was computer animations (or movies). Atom positions (with or without bonds) were displayed on a computer monitor. By updating the display with new atom positions from later times a simple animation was produced. The animations had minimal depth cueing (shown only by those atoms which overlapped on the display). However this simple animation was an effective means of visualizing the dynamics of the phenomena simulated.

2.3 Results of the simulations

The Molecular Dynamics simulations were run typically for 10,000 or 20,000 time steps. Some cases showed only crack propagation and therefore the material being simulated was considered purely brittle. Others showed only dislocation emission and crack tip blunting and were considered purely ductile. Still others showed a simultaneous combination of both crack propagation and dislocation emission. In all instances of dislocation emission, the relevant slip system was determined. For those cases which showed crack propagation a value of the critical stress intensity factor was measured .

2.3.1 RuAl:

In the notation below, orientations are given as: [direction of crack propagation] (crack plane). Four orientations were studied:[110](1-10), [1-10](111), [100](010) and [110](001). The first two exhibited purely brittle behavior under mode I loading, the crack simply propagating on the original crack plane at some value of the stress intensity factor. Orientation [100](010) showed a combination of events with the crack attempting to move onto the lowest surface energy {110} planes from the original crack plane (Fig. 2). Almost simultaneously {100}<010> dislocations were emitted if the temperature was above 100 K. At higher temperature (600K) the crack started to propagate on (010) for one or two bond lengths, but shifted to {110} before emitting {100}<010> dislocations.

For the fourth orientation [110](001), the crack did not propagate and dislocation emission (on {110} planes with <001> type Burgers vector) blunted the crack tip (Fig. 4). However, just before dislocation emission, a broken bond appears on each side of the crack very near the tip as if the crack was attempting to propagate at 90 degree angles to the normal crack direction. This may be due to local surface reconstruction (which has been observed in other studies[14]), or to the fact that the {110} planes have lower surface energies than the {100} planes. This bond rearrangement, however, apparently plays a significant role in changing the stress concentration pattern near the crack tip.

The values of stress intensity factor $\mathbf{K}_{Ic}$ at which the crack started to propagate in RuAl at different temperatures are summarized in Table 1 . $\mathbf{K}^{G}_{Ic}$ is the theoretical critical stress intensity factor according to the Griffith criterion. It is calculated assuming that the energy required to propagate the crack is only used to create the two new free surfaces of the crack, and ignores the additional energy required for dislocation emission or blunting of the crack tip.

For the [100](010) orientation, it was difficult to determine the critical stress intensity factor. At lower temperatures especially, the array was very brittle and it was almost impossible to increase the stress by displacing the atoms of the "static" layers without having them tear away from the rest of the array. At higher temperatures many events (such as dislocation emission and deformation of the crack tip) were taking place almost simultaneously, also making it difficult to determine the critical value of the applied stress intensity factor.

For the other orientations listed, these difficulties were much reduced. It was found that the crack either propagated at stress intensity factors somewhat higher than $\mathbf{K}^{G}_{Ic}$ or, in the case of the [110](001) orientation, did not propagate at all due to crack tip blunting as a result of dislocation emission. If one discounts the very low temperature results because of possible effects of stress waves, it appears that the value of $\mathbf{K}_{Ic}$ is essentially temperature independent up to the highest temperature tested, 1000K, predicting that a ductile-brittle transition temperature would have to be even higher.

2.3.2 Pure Al:

For pure Al, only two orientations were studied: [100](010) and [110](1-10). From the fracture point of view, they exhibited the same behavior: the crack never propagated and was always blunted by copious dislocation emission. Two types of emitted dislocations were observed with either an *a/2* <110> or *a/6*<211> Burgers vector. As the cracks never propagated in the simulations, no values of the critical stress intensity factor $\mathbf{K}_{Ic}$ could be determined. To the best of our knowledge, no experimental values of $\mathbf{K}_{Ic}$ exist for pure Al, for much the same physical reason.

2.3.3 Pure Nb:

Three orientations were studied: [100](010), [110](1-10) and [110](1-11). For the first orientation, the stress built up until it was relieved by {110}<111> type dislocations emitted at 45 degrees from the crack plane. After the first dislocation emission, this cycle of stress build up and dislocation emission repeated indefinitely. Consequently, no crack propagation was observed in this orientation. For the two remaining crack orientations, the same slip system ({110}<111>) was activated. However, in these cases the slip system made an angle of 90 degrees with the crack plane. Therefore the production of dislocations was apparently not as effective in relieving the stress intensity building up in front of the crack, and crack propagation occurred.

2.3.4 Nb_3Al:

For all orientations studied: [100](010), [110](1-10) and [110](1-11), the crack was observed to propagate without any sign of dislocation emission. The values of the stress intensity factors at which the crack started to propagate are summarized in Table 2. Although there appears to be a slight trend towards easier crack initiation with increasing temperature, the 10K results undoubtedly have a larger error because of stress wave transients, and probably should be discounted. It can be seen that as the temperature rose it became easier to propagate the crack. Thus there was no indication of a ductile-brittle transition within the temperature range simulated. Given the fact that this alloy is known experimentally to be notoriously brittle, perhaps these simulation results are not surprising.

2.3.5 NiAl:

This system was the first one chronologically in which fracture simulations were performed at our Center[15] . A more primitive method of applying tractions was employed, wherein no use was made of continuum displacement calculations, but atomic tensile forces were simply applied to the outer surfaces and normal to the crack plane. As a result, it is not possible to easily analyse the NiAl simulations from the point of view of continuum mechanics and stress intensity factors, but it did provide information on operable slip systems, which we include here.

Three orientations were studied: [100](010), [110](1-10) and [11-1](112). The [100](010) orientation showed ductile behavior from 100K to 1000K in that copious dislocation generation and crack blunting was seen. The last two orientations exhibited a brittle to ductile transition somewhere between 600K and 1000K as evidenced by significant dislocation emission and crack blunting at the higher temperature. Intermediate temperatures were not tested. It is worth noting that experimental studies indicate that the ductile-brittle temperature in NiAl appears to be about 700K.

The slip system observed in the [100](010) orientation was {110}<100>, the same observed experimentally. The dislocations observed in the [11-1](112) orientation were type {1-10}a/2 <111>. Interestingly, Loretto et al.[16] observed dislocations of this same type under high stress levels. Thus the high stresses used in our simulations may account for the occurrence of these dislocations.

2.3.5 Comparison with Fracture Theory

A theory on the competition between brittle and ductile behavior of a crack has been recently published by Rice[17]. This theory assumes a periodic relationship between the applied shear stress and the atomic displacement. Because of this relation, when a small load is applied a small amount of slip can take place at the crack tip which can eventually lead to a fully formed dislocation. The crack will be blunted by dislocation emission if the resistance to emission is overcome before the resistance to cleavage is. The resistance to dislocation nucleation is taken to be proportional to a parameter which Rice calls the *unstable stacking energy,* while the resistance to cleavage is the *surface energy*. The *unstable stacking energy* is the energy barrier encountered in a rigid block-like sliding of one half of a lattice relative to the other half along a slip plane, with no change in interplanar spacing. This concept was originated by Vitek[18], who referred to it as a "generalised stacking fault energy". If relaxation of the distance between the two blocks is allowed, and a saddlepoint energy is found, this produces a second number: the *relaxed unstable stacking energy.* Both quantities are relatively easy to obtain with a computer, and are quite important theoretically, but there is so far no known experimental technique for measuring them. It is not yet clear in the theoretical models whether the relaxed or unrelaxed energy is the relevant activation energy for dislocation nucleation. This is perhaps a question of how rapidly the stress increases at the crack tip, and if there is sufficient time to allow relaxation in the material before bond fracture intervenes. Consequently, both types of energy were obtained from simulations, and both were used in the theory to provide a prediction that would bracket the two dynamic extremes for all the metals discussed here. They are listed in Table 3.

In all cases covered here, no prediction depended on which type of stacking energy was inserted in the formulae, but it is quite possible that other systems may depend sensitively on which stacking energy is used. Surface energies were calculated from the change in energy of an array when two parallel relaxed free surfaces are eliminated by imposing periodic boundary conditions, and are listed in Table 4.

Using the surface energy and stacking energies as input in Rice's theory, predictions on the competition between crack propagation and dislocation nucleation can be made for each material and for each orientation[19,20]. Al and Nb were predicted to be ductile, and NiAl, RuAl and Nb_3Al to be brittle for all the orientations studied. As was shown earlier, these predictions were correct for pure Al, for pure Nb and for Nb_3Al. For RuAl and NiAl, it was shown that some orientations behaved in a ductile way. The source of this discrepancy between prediction and simulation is as yet unknown.

2.3.6 Summary of Fracture Simulations

We have seen that the EAM potentials which were fitted only to static lattice properties (cohesive energies, vacancy energies, lattice constant and elastic constants) successfully predict dynamic properties such as ductile and brittle behavior. As an example, the Al and Nb potentials correctly exhibited ductile behavior for the pure metals, with Al more ductile than Nb, and these same potentials also yield the expected brittle behavior for the Nb_3Al alloy. NiAl exhibited a brittle to ductile transition in the simulations, somewhere between 600K and 1000K. Such a transition has been observed experimentally between 600K and 700K[21]. Furthermore all three intermetallics showed less ductility than the metals, as would be expected.

The extra slip systems observed in our simulations of Al and RuAl may be explained by the results for NiAl. In this case two slip systems were observed during simulations, the same two having been observed experimentally. While one is the normally observed slip system, the second has been observed only under high stress. This suggests that the anomalous Al and RuAl slip systems that appear may be due to the very high rates of stress loading in the simulations, which are comparable to shock loading conditions.

The comparison of the simulation results with Rice's theory is very good for Al, Nb and Nb_3Al but shows discrepancies for NiAl and RuAl which will require future work to explain.

Table 1. Variation of the critical stress intensity factor with temperature for RuAl

Crack orientation	K_{Ic}^{G} (Mpa√m)	K_{Ic} (Mpa√m) from simulations at different temperatures				
		1K	100K	300K	600K	1000K
[100](010)	**1.26**	2.2-2.4			<2.1	<2.1
[110](1-10)	**1.10**	1.3	1.7	1.7	1.7	1.7
[1-10](111)	**1.26**	2.0	1.6	1.6	1.6	1.6
[110](001)	**1.31**	**NO CRACK PROPAGATION**				

Table 2. Variation of the critical stress intensity factor with temperature for Nb_3Al

Crack orientation	K_{Ic}^{G} (Mpa√m)	K_{Ic} (Mpa√m) from simulations at different temperatures				
		10K	300K	600K	900K	1200K
[100](010)	**0.810**	1.1	0.89	0.85	0.79	0.77
[110](1-10)	**0.745**	1.08	1.01	.99	0.99	0.99
[1-10](111)	**0.84**	1.18	1.05	1.05	1.0	0.92

Table 3. Unstable stacking energies

Material	Slip orientation	Unrelaxed unstable stacking energy (mJ/m^2)	Relaxed unstable stacking energy (mJ/m^2)
	{100}<010>	4654	349
	{110}<1-10>	579	278
Al	{110}<1-11>	16823	3901
	{111<1-10>	453	195
	{100}<010>	1800	1780
	{110}<1-10	1120	1120
Nb	{110}<011>	1120	1120
	{110<1-11>	620	620
	{100}<010>	>36920	2960
	{110}<1-10>	>26520	2920
Nb_3Al	{110}<001>	>48720	3000
	{111}<1-10>	>40000	3060
	{110}<1-11>	17760	2320
	{100}<010>	3070	2498
	{100}<011>	7224	3981
RuAl	{110}<001>	2268	1882
	{110}<1-10>	5014	2190
	{110}<1-11>	1202	1130
	{100}<010>	5420	1200
NiAl	{110}<001>	3920	1020
	{110}<1-10>	4940	1260
	{110}<1-11>	1220	720

Table 4. Surface energies

Material	Surface orientation	Surface energy (mJ/m^2)
Nb	{100}	2220
	{110}	1890
	{100}	865
	{110}	828
Al	{111}	1002
	{211}	981
	{100}	950
	{110}	1600
NiAl	{111}	2000
	{211}	1900
	{100}	2420
Nb_3Al	{110}	1890
	{111}	2040
	{100}	3203
	{110}	2446
	{111}	3029
RuAl	{210}	2998
	{211}	2873
	{521}	3103

3. Dislocation Dynamics

The theory of the dynamic response of a dislocation in a material to stress and temperature underlies much of the current thinking about deformation and fracture mechanisms of solids. The presently accepted "classical" theory presumes that dislocations carry mass and reach terminal velocities that are proportional to the driving stress. Experimental data bearing on this matter measure only coarse averages over a large population of dislocations in varying stress states. Computer simulations are virtually the only method of obtaining detailed data on the dynamic characteristics of single dislocations as a function of stress and temperature.

The minimum stress to initiate and maintain the motion of dislocations (the Peierls stress), and the mobility of dislocations (as measured by their steady state velocity under stress) are quantities that frequently enter equations predicting different properties of materials, yet they are numbers that are extremely difficult to measure directly by experimental means. In addition, theoretical estimates of these quantities heretofore have depended on the inherent approximations of linear continuum elasticity theory which are known to seriously fail in describing the details of the dislocation core, and consequently err in predicting properties dependent on the core structure, such as the Peierls stress. As a result, only a very few crude estimates are available in the experimental or theoretical literature for these numbers at this writing [22] .

Furthermore, it is now generally agreed[23,24] that the actual atomistic structure of dislocations may produce very significant effects in the plasticity of crystalline materials, with the possible exception of the most ductile fcc metals where it is believed that the Peierls stress is so low that thermal fluctuations effectively wash out any significance of the core structure[25] . The symptoms which generally signal the importance of the core structure in material plasticity are the appearance of at least one of the following: (a) a strong temperature dependence of the flow stress, (b) a crystallographic orientation dependence of the critical resolved shear stress (crss), (c) different, or multiple, slip systems operating at different temperatures.

Another issue of considerable current interest and debate revolves around the question of whether the core configuration of a static dislocation is a good representation of the same dislocation in motion. A related question is whether the core configuration at absolute zero (0 K) remains stable at finite temperatures, or whether important configurational transformations might occur as the temperature is increased and be a significant contributing factor to the well known ductile-brittle transition that occurs in a wide variety of materials. Virtually all of the current predictions regarding the relative mobility of different dislocations at finite temperatures are based on static 0 K core configuration calculations and so may require substantial revision.

The "classical" theory of dislocation dynamics[26] based on continuum mechanics concepts asserts that the steady state velocity (V_S) of a dislocation under an applied resolved shear stress (σ) will be:

$$V_S = \mathbf{b}\,\sigma/B \qquad (1)$$

where **b** is the Burger's vector of the dislocation, and B is a material dependent damping coefficient. The magnitude of B is postulated to be governed by a combination of phonon damping, electron damping and "flutter radiation" contributions, with the phonon damping thought to be the dominant mechanism at temperatures above about 10% of the Debye temperature. Efforts have been made to calculate the interaction of the moving non-linear displacement field of a dislocation core with the linear elastic vibrational waves (phonons) in order to estimate the phonon contribution to B. This is an extremely difficult calculation, requiring many approximations which leave the validity of the final result in some question. The general trend of such calculations however is to indicate that B should increase approximately linearly with absolute temperature.

An additional prediction of the classical theory is that dislocations possess inertia (arising from the strain energy field carried along by the dislocation), so that when the external stress is first applied there will initially be an acceleration stage which proceeds until the terminal velocity given in Eq. (1) is reached. Thus the velocity-time characteristics are projected to be of the form:

$$V(t) = V_S [1 - \exp (-t/T)] \qquad (2)$$

where t is the time, and T is a relaxation time proportional to the "mass" of the dislocation.

Although some attempts to experimentally determine the dynamics of individual dislocations have been made, this is in principle a very difficult measurement. Almost any sample investigated will have a rather large number of different kinds of dislocations, which will interact in varying ways with an unknown distribution of crystalline defects and pinning points. One of the most frequently cited measurements is that of Johnston and Gilman[27] in LiF, where they used the technique of monitoring the change of the surface etch pit patterns to deduce how far the dislocations had moved when under stress. These measurements showed quite a strong dependence of velocity on stress. However, subsequent measurements by Baker[28] on the same material using an ultrasonic internal friction technique found a very different velocity versus stress relationship, which he ascribed to the average velocity of dislocations as they bowed out between fixed pinning points on successive stress cycles. He assumed that at the low applied stress levels used in the ultrasonic measurements, no dislocations broke free from their anchors, and that the population of dislocations contributing were the number that could be seen from etch pit determinations. The inferred velocities could change by orders of magnitude if either of these assumptions in his analysis were wrong. However, the fact that dislocations were obviously moving at stresses an order of magnitude or more below those used in the Johnston-Gilman study does strongly suggest that the latter were not measuring the instantaneous dislocation velocities between pinning points, but primarily some average velocity of the pinning points themselves. One must conclude that even in this much cited system, the instantaneous velocity of dislocations moving at stresses in excess of the Peierls stress is not known, even to within several orders of magnitude, nor can the velocity versus stress characteristic be given any measure of validity.

More direct visual measurements of dislocation dynamics have been made by Louchet in Si[29] and Ge[30] by in situ straining in a high-voltage electron microscope. In the case of Si, he was able to obtain the velocity versus stress (and temperature) characteristic, and his Arrhenius plot of the dislocation velocity versus inverse temperature gave an activation energy of about 1.9 eV, which he asserted could be associated with the activation energy for double kink nucleation (although such double kinks were not resolvable in the microscope images). The model for dislocation motion at these stress levels was then proposed in terms of a kink nucleation limited process requiring thermal activation of a double kink, followed by rapid spreading of same, thus producing an advance of the dislocation by 1 **b**. Each advance of the dislocation would require repeating this basic nucleation process. The velocities measured were on the order of 10^{-6} cm/s and clearly in the sub-Peierls stress regime, given the importance of thermal activation and the very low resulting velocities. Thus, one still does not have any sort of reliable measurement of the dynamic characteristics of individual dislocations in the super-Peierls stress regime, a regime which is bound to be of great importance in the understanding of dislocation behavior in metals (where Peierls stresses are thought to be very low) or near crack tips in any material (where stresses can become very high).

The Peierls stress itself has only been very crudely inferred from scattered measurements with rough estimates[31] being in the range of about 1% of the shear elastic constant for covalently bonded crystals to 0.01% of the shear constant for close packed metals, with bcc metals and ordered alloys lying somewhere in between. There appears to be no direct experimental method of obtaining this quantity with any degree of accuracy, despite its considerable importance as a fundamental parameter in theories of dislocation motion. In addition, there is virtually no experimental information bearing on the question of whether the Peierls stress increases or decreases if the lattice is put under compression, nor can the continuum mechanics theories address this problem since the answer depends very sensitively on the anharmonic nature of the atomic interactions.

3.1 Computer Simulations of Dislocations in Motion

A number of computer simulations have been carried out over the past twenty five years or so to investigate different aspects of dislocation dynamics. Reviews of the field have appeared periodically[32,33,34] and two excellent reviews including this topic have appeared lately[35,36]. Until fairly recently simulations were carried out on small arrays of a few hundred atoms, were often two dimensional, and employed simple pair potentials. Even so, some very significant advances have been achieved with these limited means. For instance, the essential correctness of Hirsch's hypothesis[37] that the 1/2<111> screw dislocation core in bcc metals would not be confined to the slip plane, but extend into several {110} planes, thereby causing it to be sessile was confirmed (at least at 0 K) by a series of static relaxation simulations[38,39,40] . This result also appeared to be relatively insensitive to the specific details of the interatomic potential used[41], which added additional confidence to the demonstration, and provided a reasonable basis for understanding the large Peierls stresses, and complex orientation dependences of the critical resolved shear stress (crss) in bcc metals. Although one may conjecture that this complex core structure may also provide the key to understanding the strong temperature dependence of the flow stress observed in bcc metals, simulations at temperatures above absolute zero largely remain to be done.

One of the more extensive "modern" studies of dislocation dynamics using EAM potentials was performed via Molecular Dynamics simulations by Daw, Baskes and co-workers[42,43] on <1-10>(111) edge dislocations in pure fcc Ni and in H and He doped Ni. They used arrays of about 700 atoms in a slab approximately 75x25x4Å with periodic boundary conditions in all directions, except normal to the (111) glide plane. The edge dislocation was found to split into Shockley partials separated by about 12-15 Å, and when a sufficient external shear stress was applied to the top and bottom (111) planes, the partials would move at an ever increasing velocity until some limit was reached depending upon the stress level and temperature. They found that by using four adjustable parameters these dynamic characteristics could be well described by the classic equations of dislocation motion (including phonon damping and relativistic effects), i. e.:

$$\frac{d}{dt}\frac{mv}{\sqrt{1-(v/c)^2}} = (\sigma - \sigma_0)b - \frac{3\alpha kT}{10b^2}\frac{(v/c)}{\sqrt{1-(v/c)^2}} \tag{3a}$$

or from Eq. (3a), when the left hand side is zero, (and with the variable replacement: $\omega \equiv \mathbf{3\alpha kT/10b^3}$) one has for the steady state velocity, $\mathbf{V_S}$:

$$V_S = \frac{c(\sigma - \sigma_0)}{\sqrt{\omega^2 + (\sigma - \sigma_0)^2}} \tag{3b}$$

where m is the dislocation mass per unit length, c is a limiting velocity, σ is the applied stress, σ_0 is the friction stress (or Peierls stress), b is the magnitude of the Burger's vector of the dislocation, and α is a dimensionless constant. b and σ were known and the other four parameters were fit to the simulation results. They reported the values: c = 2000 m/s, m = 0.2 atoms/b, α = 0.98, and σ_0 = 4 MPa, which they argued were sensible numbers although only c (the transverse sound velocity) is known with any accuracy experimentally. That number is about 3340m/s in a macroscopic sample, but would be reduced to about 1840 m/s in the very small sample size used in the simulations, thus achieving reasonable agreement. We were thus encouraged to follow this same approach, but to use larger arrays in order to minimize the possible size and boundary effects.
Our first results have been dramatically different from those to be expected from classical dislocation theory and suggest an entirely different mechanism for dislocation motion, at least in some stress and temperature regimes, as we shall detail in the next section.

3.2 Non-classical Dislocation Dynamics ?

A limited set of Molecular Dynamics simulations of edge dislocations in $L1_2$ Ni_3Al using Voter-Chen EAM potentials[44] has been carried out. The edge dislocation $\mathbf{b}$ = $\mathbf{a}_0$<100> moving on a {001} glide plane was studied between the temperatures of 10 K and 1000 K with levels of applied shear stress ranging from 0.01 to 0.08 in units of C_{44} (where C_{44} = 131 GPa). The $\mathbf{a}_0$<100> type of edge dislocation, although not seen experimentally in Ni_3Al, in the simulations formed a stable dislocation core only slightly spread out along the glide plane with an easily recognizable structure. It therefore seemed to represent a good example of the Peierls model of a dislocation (in that the core was relatively compact and essentially confined to the glide plane), and so was used as a "starter case" to simplify the analysis and to test pattern recognition routines. The equilibrated structure projected along the dislocation line is shown in Fig. 5, and consisted of 6,720 atoms (24x14x5 unit cells). The array extended 5 unit cell distances (5$\mathbf{a}_0$, where $\mathbf{a}_0$ = 3.57 Å) along the dislocation line before periodic boundary conditions were applied in that direction. Free surface conditions were used in the two other orthonormal directions, thereby permitting a natural bending of the crystal. We felt this was a better choice than imposing periodic boundary conditions on the end surfaces, which would have the effect of warping the array back to an approximately straight configuration and lead to very large artificial tensile and compressive stresses in the vicinity of the dislocation core.

After the array was fully equilibrated (about 5000 time steps, each being $4.0x10^{-16}$s, and the total being equivalent to about 100 atomic vibrational periods) shear stresses were applied instantaneously on the four outer free surfaces by the application of tangential atomic forces to the two outer atomic layers in such a manner that there was zero net torque on the array. The applied stress was held constant for the duration of the run, and a "temperature clamp" which rescaled the velocities of all atoms at each step by 1/33 of the value necessary to reach the target temperature exactly was used to keep the temperature approximately constant, without introducing noticeable transients in the vibration spectrum. Also, it was found to be important to allow the array to adjust to its

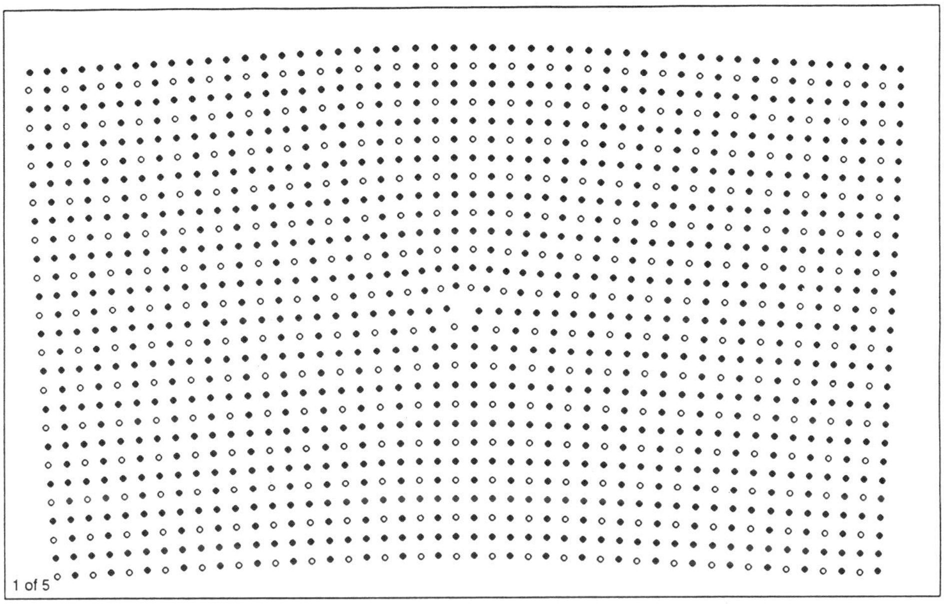

Figure. 5 - Relaxed stress free configuration of Ni_3Al edge dislocation with $\mathbf{b} = \mathbf{a_0[100]}$ on **(001)** glide plane after 12,000 time steps (300 atomic vibration periods) of thermal equilibration at T = 10 K. Solid circles = Ni atoms; open circles = Al atoms. Without stress the core remains undissociated.

correct equilibrium lattice parameter at each temperature before applying the stress. It was found that a stress of about 1,310 MPa (or 1% of C_{44}) was necessary to get the dislocation to move at all, so this would appear to be about the magnitude of the Peierls stress for a straight edge dislocation of this type in Ni_3Al.

Immediately after the application of the stress no motion occurred for about 600 time steps, which corresponds almost exactly with the time necessary for a shear wave initiated at the surface to reach the position of the dislocation core. A plot of the displacement of both the leading and trailing partial as a function of time thereafter (see Fig. 6) shows essentially no sign of an acceleration stage, indicating that the dislocation is behaving as a massless object. At the lower stress levels, the partial dislocations

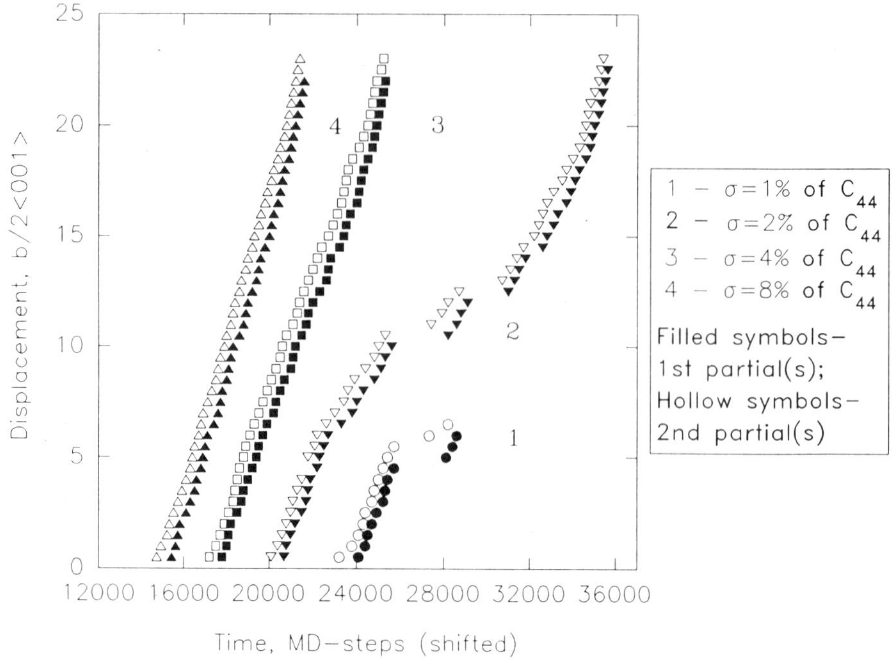

Figure 6 - **[100](001)** edge dislocation mobilities in Ni_3Al at T = 10 K versus applied shear stress.

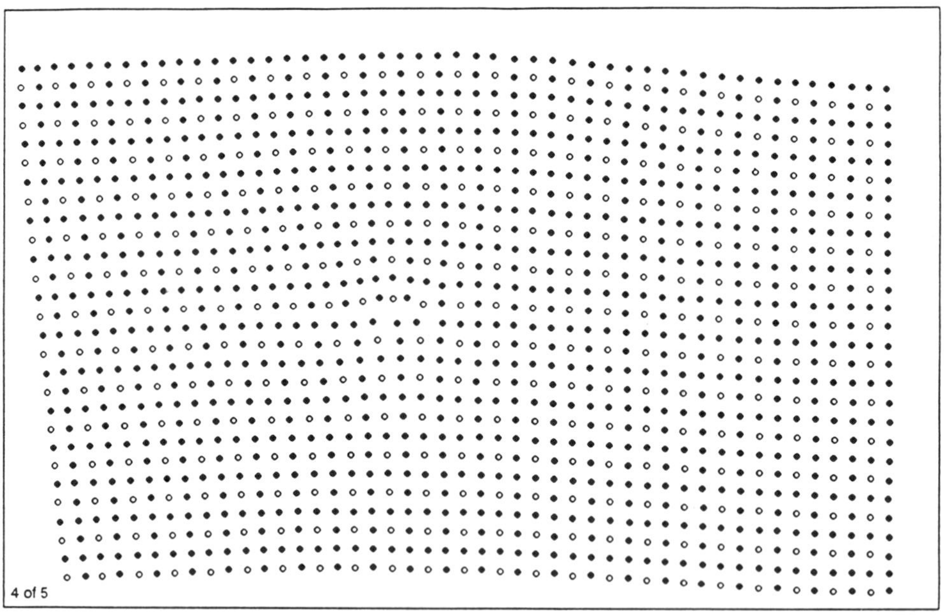

Figure 7 - **[100](001)** glissile edge dislocation in Ni_3Al under stress showing dissociation of the core structure into partials. T = 1000 K and σ = 1% of C_{44}.

would occasionally recombine into a compact core which is apparently more sessile and causes a temporary interruption of forward motion, as seen in Fig. 2. After a certain time the core would redissociate and the partials would move on as before. This "interruption time" was subtracted out in the calculation of the dislocation velocities. The dislocation was only observed to separate into partials while being driven under stress, but this separation is never more than about 1 $\mathbf{a}_0$ (see Fig. 7).

The dislocation dynamics data for temperatures at 300 K and above was difficult to analyse because in the higher stress ranges multiple dislocation generation often occurred, emanating from the original core and /or glide plane. An example of this phenomenon is shown in Fig. 8. An even more dramatic manifestation of dislocation multiplication was the appearance (at temperatures of 600 K and above) of an amorphisation transition, as shown in Fig. 9. We do not offer here any theoretical explanation for either of these phenomena, other than to point out that very high strain rates, easily in the realm of shock loading, are being used in the simulations. However, these complications did not arise at 10 K or 100 K, and the data from these temperatures is displayed in Fig. 10.

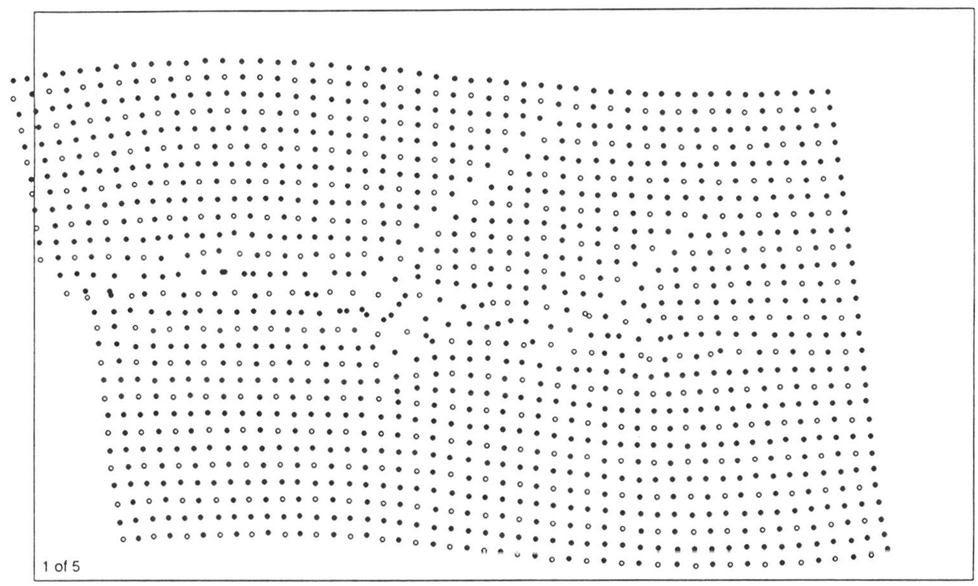

Figure 8 - Multiple nucleation of dislocations from the original **(001)** glide plane onto **{110}** glide planes. MD time step = 9,900; T = 300 K; σ = 8% of C_{44}.

Several surprising features appear in these low temperature simulations of individual straight dislocations in Ni_3Al. First, the velocity at 10 K shows only a very weak dependence on stress. A previous study on edge dislocation motion in B2 NiAl , which however had the complicating interference of a martensitic transformation, also found quite similar phenomena[45]. Second, for a given stress level (in the higher stress range) the velocity at 100 K is greater than at 10 K, contrary to expectations from classical damping concepts. Third, there is a lower velocity regime which apparently is forbidden, and attempts to fit the data to the classical Eq. 3b (the solid lines in Fig. 10), lead to absurd parameter values, such as negative numbers for the Peierls stress.

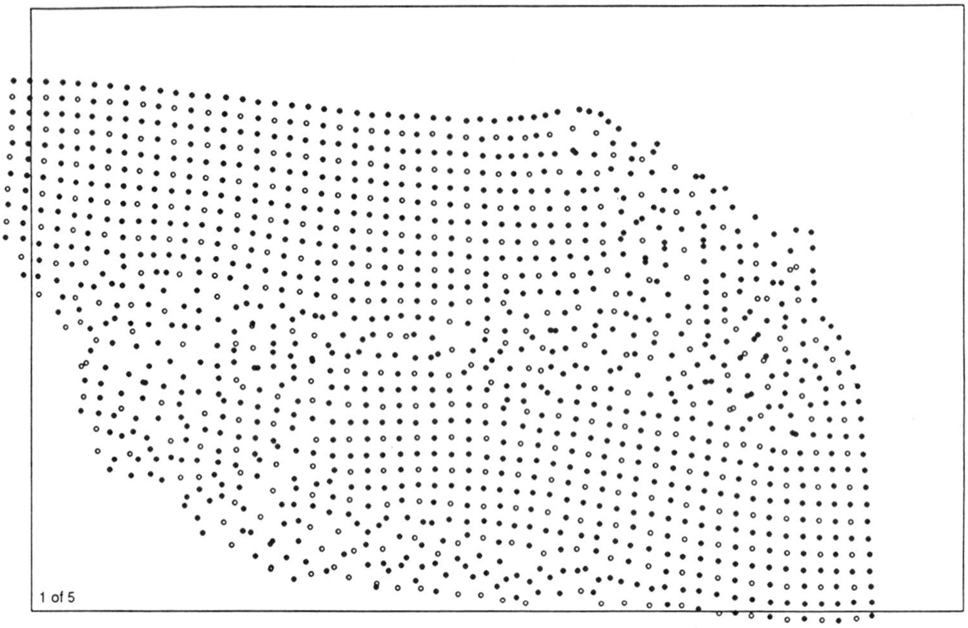

Figure 9 - Partial amorphisation of the array as a result of copious dislocation generation. MD time step = 13,500; T = 600 K; σ = 8% of C_{44}.

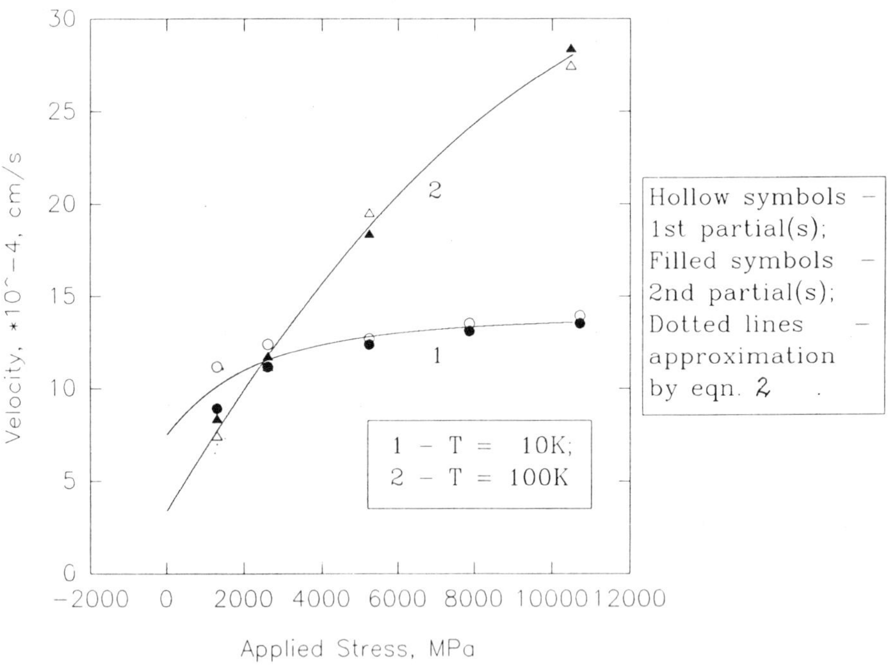

Figure 10 - Velocities of the partials of a Ni_3Al **b=a[100]** edge dislocation gliding on a **(001)** plane versus stress at T = 10 K and 100 K. The partials, with **b = (a/2)[110]**, remain within a unit cell distance of each other during the simulation.

3.3 Summary of Dislocation Dynamics Simulations

Neither the stress nor temperature dependence of dislocation velocities predicted by the standard continuum mechanics theories of dislocation dynamics are observed in Molecular Dynamics simulations of edge dislocation motion in Ni_3Al. Furthermore, in the simulations, the dislocations reach their terminal velocities virtually instantaneously, violating the classical concept that dislocations should act as if they have a mass.

Although Daw & Baskes[46] found reasonably classical dislocation dynamics in their Ni simulations, whereas we have found decidedly non-classical dynamics in our studies of Ni_3Al (and NiAl), there are potentially important differences between the two cases which may be relevant to an understanding of this disparity:

a) The Peierls stress in the intermetallics (Ni_3Al & NiAl) is at least a factor of a thousand higher than in Ni, as a percentage of the shear moduli.

b) The edge dislocation cores in the intermetallics remained very compact, whereas the edge dislocation in Ni dissociated into well-separated partials.

All of these phenomena together suggest that the motion of straight edge dislocations with compact cores may be more satisfactorily modelled with soliton concepts[47,48,49]. At the very least, these initial results are quite provocative and we believe are worth exploring much more extensively, as we propose to do in the near future.

4. Martensitic Transformations

Another topic which is almost ideally suited for study by means of Molecular Dynamics simulations is that of displacive transformations because they occur at speeds approaching the velocity of sound. The heterogeneous nucleation process for these transformations also presumably occurs very rapidly, and therefore should fall within the timeframe constraints of this simulation technique. This question was explored for the case of NiAl alloys[50,51] in the ordered B2 phase near equiatomic stoichiometry, using the Voter-Chen EAM potentials which lead to a slightly lower energy for an ordered $L1_0$ fct phase at absolute zero than the B2 metastable phase (Fig. 11). The $L1_0$ phase is found experimentally as a martensitic phase, but only for Ni rich compositions of greater than 60 atom % Ni. Although the Voter-Chen potentials do correctly show that the $L1_0$ becomes less stable relative to the B2 as the Ni content decreases, the composition at which the two are equal in energy at 0K is not quite accurate. Nevertheless it provides a very interesting case to explore in simulations, and to account for this composition inaccuracy one may regard the simulations at 50 at% Ni as corresponding to the experimental system at 60 at% Ni.

In the simulations aimed at understanding the martensitic nucleation process, arrays of 8,000 to 10,000 atoms were used (or about 30x30x5 in terms of unit cell dimensions). Periodic boundary conditions were used in the short direction but free surfaces were established normal to the long directions. It was discovered that in simulations at relatively low temperatures (10-100K) after a short amount of time a corner of the array where two free surfaces intersected would begin to nucleate the transformation by a coherent shear process. Although the thermal motion of the atoms was always started with a different set of random numbers for each simulation run, it was found that, surprisingly, the same corner always started the nucleation process, although at varying times. This was ultimately traced to the fact that the B2 array had been cut such that the outermost atomic layer on two of the free surfaces was entirely Al atoms, and the other two were purely Ni atoms. The corner which always initiated the transformation was that where the two Ni surfaces met (Fig. 12).

This suggested the idea of repeating the simulations but stripping off the two outer Al layers so that all four free surfaces, and therefore all four corners would be pure Ni. The result then was that all four corners initiated the transformation, but at slightly different times and with different geometrical variants of the martensite phase. These different variants would grow coherently toward the center of the array until they encountered another growing martensite region, and then after some time one of the regions would manage to push the other three out of the crystal, thus forming finally a single crystal of martensite (Fig. 13).

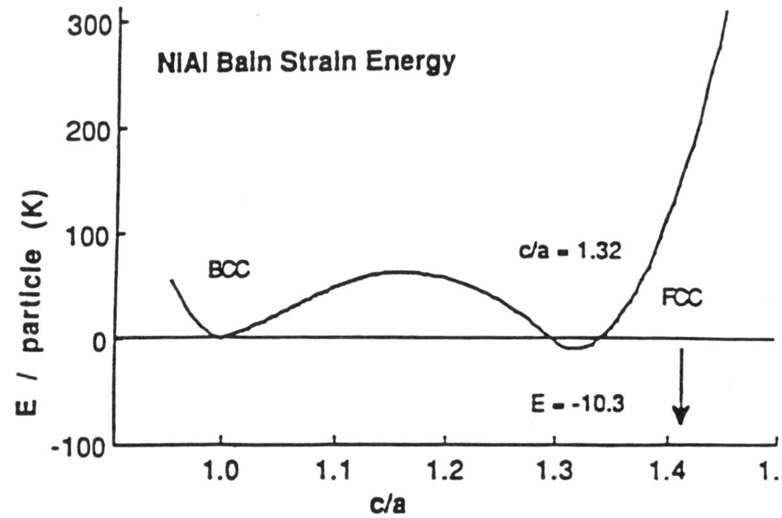

Figure 11. Energy as a function of tetragonal strain at constant volume at 0 K. Minimum energy occurs for the tetragonal $L1_0$ phase at c/a ~1.32 and is 10.3 K/atom (20 cal/mole) below the B2 minimum. An ordered FCC phase would correspond to c/a = 1.414 and is not even metastable in this system.

On the other hand, if all four free surfaces were set up to be pure Al atoms, then one could let the simulations run for many thousands of steps (equivalent to many hundreds of atomic vibrations) and not see any martensitic nucleation whatsoever. An important clue to this difference in behavior was found in comparing the average mean square displacement of each species at a given temperature in the Molecular Dynamics simulations. Contrary to the expectation one would have based on their relative masses, it was discovered that Ni atoms had an approximately 25% larger mean square displacement than Al atoms (Fig. 14). This simulation result agrees almost exactly with the experimental X-ray results on the same alloy [52]. This larger amplitude of vibration characteristic of the Ni atoms can perhaps be understood from the relative size of the charge clouds of the two species, the Al being the larger such that Al-Al pairs are essentially in contact in the B2 structure even though they are only second neighbors. The

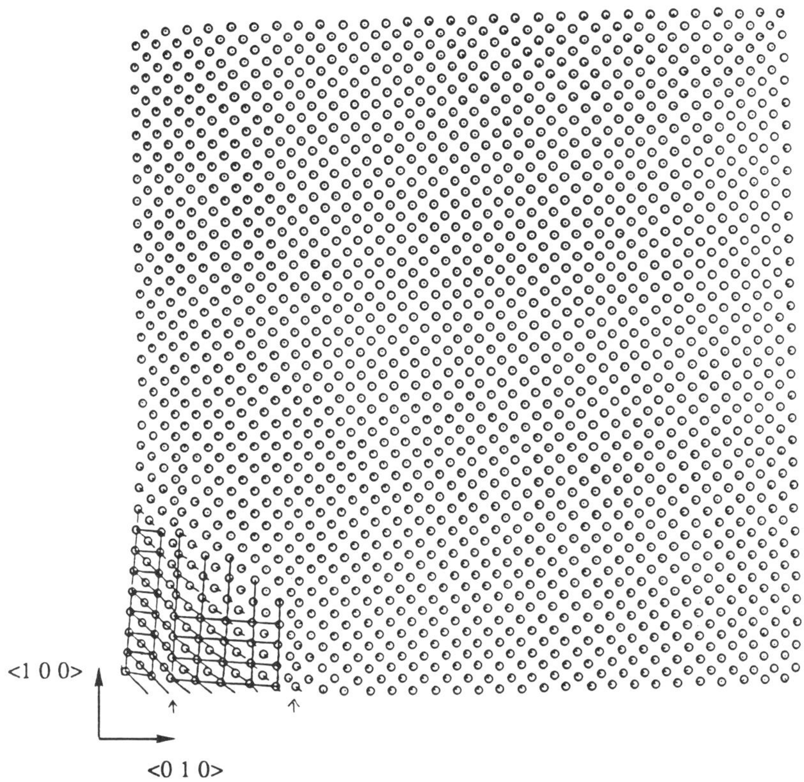

Figure 12. Atomic displacement plot at step 2000 (5.0 picosecs.) of two adjacent (0 0 1) layers (from a total of ten) at T=100K. Free surfaces with Ni outermost are at left and bottom and those with Al outermost are at right and top. Time=5 picoseconds. "Tails" on atoms show movement from start. Surface upheaval is evident. The unit cell of the new phase is outlined and the long c axis has chosen the <0 1 0> orientation. Arrows indicate location of lattice invariant shears (periodic slip).

smaller Ni atoms therefore find themselves in a loose cage formed by their Al nearest neighbors.

This loosely bound character of the Ni atoms becomes even more emphasized at a free surface, or especially at a free corner, where very large amplitude ("soft") local vibrational modes can be observed in the simulations (Fig. 15). There are also large amplitude local modes at an Al-Al corner, but they are on the order of only half the amplitude of those at a Ni-Ni corner and apparently are sub-critical for the purposes of nucleating the martensitic transformation. The fairly rapid decay of the large amplitude corner vibrations as one moves in toward the interior of the crystal can also be seen in the simulations, underlining the fact that these are truly <u>localized</u> soft modes.

In watching the atomistic dynamics of the nucleation process in the simulations it is possible to see that two independent localized soft modes interact at the corner to accomplish the nucleation process (Fig. 15). These are a <1-10>(110) transverse shear

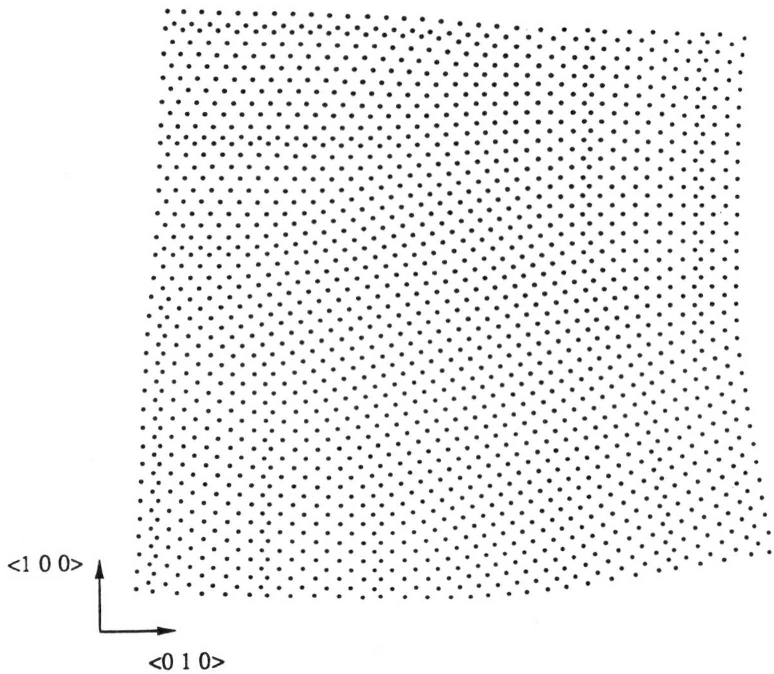

Figure 13. Atomic position plot at step 3000 in the simulation of the array with all four free surfaces occupied by Ni atoms. This view of two (0 0 1) planes of the original B2 array is now mostly transformed to four different variants of the tetragonal $L1_0$ structure. T=100K, Time=7.5 picosecs.

mode and a <110> longitudinal mode. On one half cycle of the longitudinal mode the local (110) interplanar spacing increases and this reduces the shear resistance between these planes, thus making the <1-10>(110) shear motion easier energetically and therefore larger amplitude. This phase correlation can be seen to increase occasionally until after only a few atomic vibration periods the process goes critical and initiates a shear that is large enough to start the transformation. This shear then propagates inward from the corner maintaining a coherent interface with the parent B2 phase, until it encounters some obstacle such as another martensite variant or the other side of the crystal. The shape change and surface upheaval are increasingly visible as the transformation proceeds.

The interaction between different martensite variants when they encounter each other in the process of growth is also interesting to observe (Fig. 13). In the NiAl simulations it is apparent that the interfaces formed between variants is also completely coherent but spreads over a distance of two to three lattice parameter distances (about 10 Å) normal to the interface, in much the same manner that the austenite-martensite coherent interface is seen to be diffuse. In most cases whichever variant had achieved the largest volume fraction prior to impingement is the one which continues to grow at the others expense until the array has become a single crystal of that variant.

4.1 Monte-Carlo Simulations of Martensitic Transformations

The strain free energy of a material is often written [53,54] in the Ginzburg-Landau form:

$$f(\eta) = A\eta^2 + B\eta^3 + C\eta^4 + ... + K1\,(\nabla\eta)^2 + K2\,(\nabla^2\eta)^2 + ... \tag{4}$$

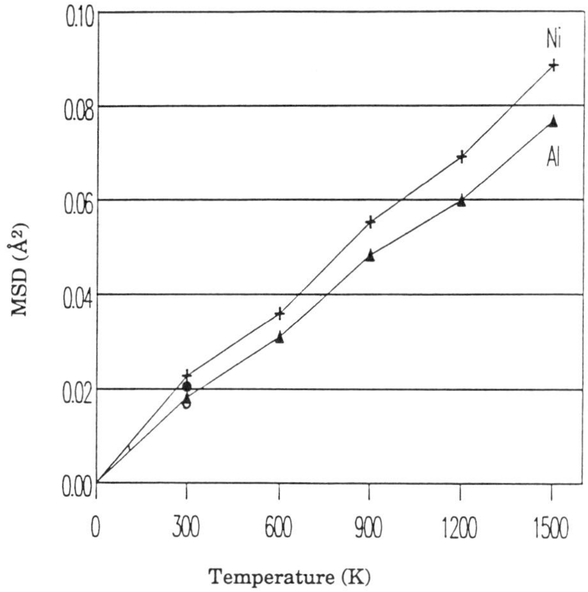

Figure 14. Mean square displacement of each atomic species in 50-50 NiAl as a function of temperature measured in large arrays with no free surfaces. Note the close agreement with the experimental data of Georgopoulus and Cohen[19] at 300K (solid circle is for Ni and hollow circle for Al).

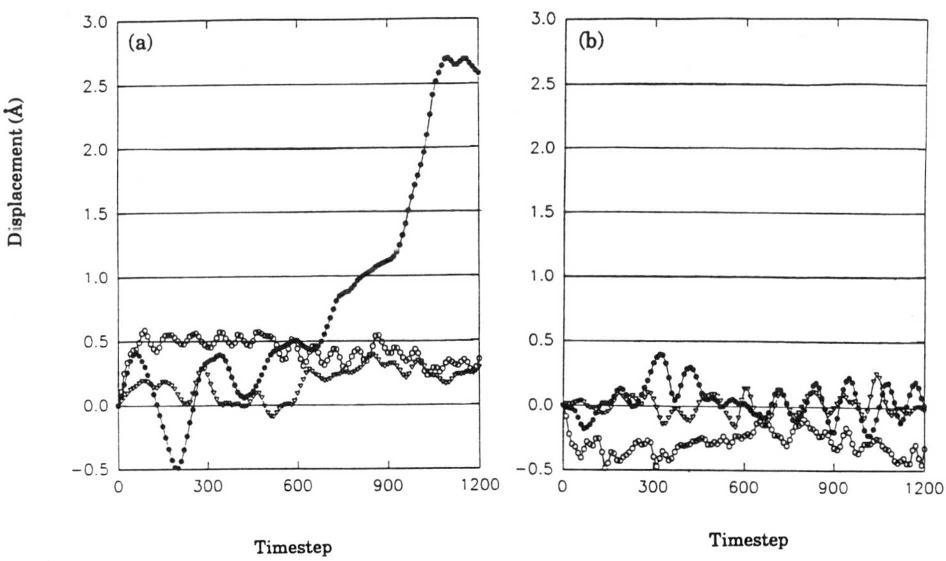

Figure 15. Displacement vs. time for particle 1 (the extreme corner atom) from the start of the simulation. T=100K, 1 timestep=2.5 femtosecs. Solid circles are for (1 1 0) <1-10> shear motion which goes critical at about timestep 600~700. Hollow circles are for (1 1 0) <1 1 0> longitudinal motion. Triangles are for (1 1 0) <0 0 1> transverse motion.
(a) in the transforming region at Ni-Ni corner (lower left).
(b) in the non-transforming region at Al-Al corner (upper right).

where A, B, C, K1, K2 are material dependent constants, η is a Lagrangian strain variable, and $\nabla\eta$ is the spatial strain gradient, etc. In most cases where this formalism has been applied, the values of the constants have simply been guessed at because there is usually insufficient experimental data for any given material to determine them. As a result any conclusions drawn from calculations based on this approach are of necessity extremely qualitative. However, what has not generally been realized before is that Molecular Dynamics (or rather "molecular statics") can be used in combination with an appropriate EAM potential to directly determine all of the above constants via computer simulation. The technique is to set up a suitably large array, equilibrate it at a low temperature (say 1 deg K) via Molecular Dynamics using periodic boundary conditions, and then apply a variety of homogeneous strains to the array, noting the resulting change in energy of the array. A least squares fit can then be made to the parameters: A, B, C, ... in Eq. (4). The strain gradient coefficients, K1, K2, etc. can then be obtained in a similar fashion by imposing a variety of sinusoidal strain fields (whose period must of course be compatible with the periodic boundary conditions), and again fitting these energy variations to the formula above. In this way we have obtained all of the relevant material parameters for the Ginzburg-Landau description of stoichiometric B2 NiAl, which includes the energy characteristics of the L1o martensite phase as well[55,56,57].

Having this in hand, one may then set up a discrete element grid representing points in the solid of interest (NiAl in this case) of whatever size one wishes to simulate, with the energy of any configuration of this grid being determined by the Ginzburg-Landau strain free energy functional. One may then use this grid in a Monte-Carlo simulation to minimize the energy under various conditions of stress and temperature. In this way we have studied the stress induced martensitic (SIM) transformation in the vicinity of a crack tip under an applied external stress in order to examine the SIM contribution to toughening. We find that the pattern of martensitic transformation near the crack tip is quite complicated and generates a non-intuitive pattern of backstresses which needs to be analysed further before any conclusions can be drawn as to whether the transformation toughens or embrittles the material[58].

5. Conclusions

Computer simulations, whether of the atomistic Molecular Dynamics type or of the larger scale Monte-Carlo variety, are rapidly becoming of greater and greater use in understanding the properties of solids under a wide range of temperature and stress conditions. We have illustrated some of the uses in this report where new discoveries and insights have emerged from this approach. With the rapidly escalating power of computers, it is to be expected that this avenue of investigation will become a broad boulevard displaying many remarkable sights.

6. Acknowledgements

Various portions of this work have been supported by the National Research Foundation, the Department of Energy, the Office of Naval Research and the University of Connecticut Research Foundation, all of whose generosity is gratefully appreciated.

7. References

1 M. S. Daw and M. I. Baskes, Phys. Rev. **B 29**, 6443 (1984)
2 M. W. Finnis and J. E. Sinclair, Phil. Mag. **A50**, 45 (1984)

3 M. S. Daw and M. I. Baskes, in "Proceedings of 4th International Conference on Hydrogen Effects on Material Behavior, Jackson Lake Lodge, WY, Sept. 1990

4 M. I. Baskes, M. S. Daw and S. M. Foiles, Mat. Res. Soc. Symp. Proc. **141**, 31-41 (1989)

5 M. S. Daw, M. I. Baskes, C. L. Bisson and W. G. Wolfer in "Modeling Environmental Effects on Crack Growth Processes", R. H. Jones and W. W. Gerberich, eds. The Metallurgical Society, Inc. Warrendale, PA (1986) pp. 99- 124

6 N. Metropolis, A. W. Rosenbluth, M. N. Rosenbluth, A. H. Teller and E. Teller, J. Chem. Phys. **30**, 1087 (1953)

7 M. P. Allen and D. J. Tildesley, "Computer Simulation of Liquids", Clarendon Press, Oxford (1987)

8 K. Binder and D. W. Heermann, "Monte Carlo Simulation in Statistical Physics", Springer-Verlag, Berlin (1988)

9 "Monte Carlo Methods in Statistical Physics", 2nd edition, K. Binder, ed., Springer-Verlag, Berlin (1986)

10 A.F. Voter and S.P. Chen, *Mat. Res. Soc. Symp. Proc.* **82**, 175 (1987).

11 J.A. Rifkin, C.S. Becquart, D. Kim and P.C. Clapp, *Mat. Res. Soc. Symp. Proc.* **278**, 173.(1992)

12 C. W. Gear, *Numerical Initial Value Problems in Ordinary Differential Equations,* Prentice-Hall, New Jersey, 1971.

13 G.C. Sih and H. Liebowitz, in H. Liebowitz (ed.) *Fracture: An Advanced Treatise, Vol.2*, Academic Press, New York, 1968.

14 C. S. Becquart, P. C. Clapp & J. A. Rifkin - , J. Mater. Res. 9, 548-552 (1994)

15 S. Charpenay, Master Thesis, University of Connecticut, (1989).

16 M.H. Loretto and R.J. Wasilewski, *Phil. Mag.*, **23**, 1311. (1971)

17 J.R. Rice, J. Mech. *Phys. Solids*, **40**, 239. (1991)

18 V. Vitek, *Crystal Lattice Defects* **5**, (1974) 1. ; *Phil. Mag.* **18,** 773. (1968)

19 C. S. Becquart, Ph. D. Thesis, University of Connecticut, (1993).

20 D. Kim, Ph. D. Thesis, University of Connecticut, (1993).

21 R. Darolia, D. F. Lahrman, R. D. Field, A. J. Freeman, *Mat. Res. Soc. Symp. Proc.* **133**, 113.(1989)

22 J. P. Hirth and J. Lothe - "Theory of Dislocations", 2nd edn., Wiley Interscience, NY, (1982) pp. 182-217, especially.

23 M. S. Duesbery and G. Y. Richardson, Critical Reviews in Solid State and Materials Sciences **17**, 1-46 (1991)

24 V. Vitek, Prog. in Mat. Sci. **36**, 1-27 (1992)

25 J. Friedel, Phil. Mag., **A45** 271 (1982)

26 Hirth & Lothe, op. cit. (ref. 22)

27 W. G. Johnston and J. J. Gilman, J. Appl. Phys., **30**, 129 (1959)

28 G. S. Baker, J. Appl. Phys., **33**, 1730 (1962)

29 F. Louchet, Phil. Mag. A, **43**, 1289-1297 (1981)

30 F. Louchet, D. Cochet Muchy, Y. Brechet and J. Pelissier, Phil. Mag. A, **57**, 327-335 (1988)

31 Hirth & Lothe,p. 241, op. cit. (ref. 22)

32 A. Englert and H. Tompa, J. Phys. Chem. Solids **21**, 306 (1961)

33 See articles in "Interatomic Potentials and Simulation of Defects", P. C. Gehlen, J. R. Beeler, Jr. and R. I. Jaffee, eds. Plenum Press, N.Y. 1972 , especially those of P. C. Gehlen pp. 475-492, V. Vitek, L. Lejcek and D. K. Bowen pp. 493-508 and Z. S. Basinski, M. S. Duesbery and R. Taylor pp. 525-552

34 D. M. Esterling, in "Computer Simulation in Materials Science", R. J. Arsenault, J. R. Beeler, Jr. and D. M. Esterling, eds. ASM International, Metals Park, OH (1986) pp. 149-163

35 Duesbery & Richardson, op. cit. (ref. 23)
36 Vitek, op. cit. (ref. 24)
37 P. B. Hirsch, *5th Int. Conf. on Crystallography*, p. 139 Cambridge (oral communication) (1960)
38 V. Vitek, R. C. Perrin and D. K. Bowen, Phil. Mag. **21**, 1049 (1970)
39 Z. S. Basinski, M. S. Duesbery and R. Taylor, Can. J. Phys. **49**, 2160 (1971)
40 M. S. Duesbery, V. Vitek and D. K. Bowen, Proc. Roy. Soc. (Lond.) **A332**, 85 (1973)
41 V. Vitek, Phil. Mag. **A58**, 193 (1988)
42 Daw & Baskes, op. cit. (ref. 3)
43 Daw et al., op. cit. (ref. 5)
44 A. F. Voter and S. P. Chen, "Accurate Interatomic Potentials for Ni, Al and Ni3Al", MRS Symp. Proc. **82**, 175-180 (1987)
45 A. Moncevicz, P. C. Clapp and J. A. Rifkin, Mat. Res. Soc. Symp. Proc. **209**, 213-218 (1991)
46 Daw & Baskes, op. cit. (ref. 3)
47 J. Frenkel and T. Kontorova, J. Phys., Moscow **1**, 137 (1939)
48 A. Seeger and P. Schiller, in "Physical Acoustics", W. P. Mason, ed., Academic Press, NY (1966) vol. IIIA, pp. 361-505
49 G. L. Lamb, Jr., "Elements of Soliton Theory", J. Wiley and Sons, NY (1980)
50 P. C. Clapp, Y. Shao and J. A. Rifkin MRS Symposia Proceedings, 246, (1992)
51 Y. Shao, Ph. D. thesis, University of Connecticut (1992)
52 P. Georgopoulus and J. B. Cohen, Scripta Met. 11, 147 (1977)
53 G. Guenin and P. C. Clapp, Proc. of the International Conference on Martensitic Transformations (ICOMAT), Japan Institute of Metals, pp. 171-179 (1986)
54 P. C. Clapp, "Phase Transformations '87", Inst. of Metals, London, pp. 391-393 (1988)
55 Y. Zhao, Ph. D. thesis, University of Connecticut (1993)
56 P. C. Clapp, C. S. Becquart, Y. Shao, Y. Zhao and J. A. Rifkin - Modelling Simul. Mater. Sci. Eng. 2, 551-558 (1994)
57 P. C. Clapp, Y. Zhao & J. A. Rifkin - MRS Symposia Proceedings: 350, 299-311 (1994)
58 R. Guenin, P. C. Clapp, Y. Zhao and J. A. Rifkin, Mat. Sci. & Eng. B. (in press).

CONTINUOUS PHASE TRANSITIONS AT SURFACES OF CuAu ALLOY MODELS - A MONTE CARLO STUDY OF SURFACE INDUCED ORDER AND DISORDER

W. Schweika,[1] D.P. Landau,[2] and K. Binder[3]

[1]Institut für Festkörperforschung
Forschungszentrum Jülich, D-52425 Jülich, Germany
[2]Center of Simulational Physics
The University of Georgia, Athens, Georgia 30602, U. S. A.
[3]Institut für Physik
Johannes Gutenberg Universität Mainz, D-55099 Mainz, Germany

INTRODUCTION

The influence of surfaces on phase transitions has found significant attention in recent years, and a number of excellent reviews exists.[1, 2, 3] A variety of complex phenomena occur which are also related to the physics of adsorption and wetting. The scenario of wetting requires three distinct phases, for instance the vacuum, the bulk phase and a third phase intervening in between at equilibrium. In case of surface induced disorder (SID), a film of disordered layers at the surface "wets" the bulk phase as the temperature approaches the bulk transition temperature $T_{c,b}$. The transition at the surface may be continuous (standard critical wetting phenomena), and, as theoretically investigated in detail by Lipowsky[3], it may remain to be continuous even in the case of a *discontinuous bulk transition.* It may appear natural that missing bonds at a surface induce more disorder near the surface, as it is also observed for the equivalent phenomena of surface induced melting, e. g. for metals like lead and aluminum.[4] In the present study of a CuAu alloy model (using only nearest neighbor interactions) SID is for instance found for a (111) surface, but it will certainly occur at some other facets as well. This particular case serves to test the actual theories of wetting. However, SID is not necessarily what occurs and instead surface induced order (SIO) can be observed in our model for instance at a (100) surface.

There has been only one example of surface induced ordering reported so far, at least if one is confined to only temperature driven transitions without external fields. This is an alloy model with CuAu-type of order as previously studied by us using

Monte Carlo(MC) simulations.[5] The phenomenon of SIO is reconsidered here in more detail and for the nearest neighbor interaction model. The frustration effects which occur for the ordered structure on the fcc lattice are the origin of the first order bulk phase transition. However, a (2×2) superstructure (which is a part of the CuAu order) at a (100)-surface feels less frustrated and is stable above the $T_{c,b}$. This holds generally for interactions which lead to a small antiphase boundary energy in [100] direction. Although both cases SID and SIO were considered to be similar it is obvious that the surface critical behaviour for SIO is independent of the wetting at the lower temperature $T_{c,b}$. Therefore, the critical exponents and scaling relations have to be transferred with caution to this case.

In real alloys, it is likely that there are more surface effects than only missing bonds. For instance, there maybe effective surface fields leading to the enrichment of one component in the top surface layer. The origin of such fields is due to another phase, where the vacuum appears only as a special choice of the more general solid on solid models. Furthermore, one could expect that the effective interactions are modified close to the surface. Both aspects have been taken into regard by Kroll and Gompper in theoretical investigations[6] of surface induced disordering at (100) surfaces of Cu_3Au alloys. In the present model and for a (100) surface the effects of surface fields are more drastic than for the Cu_3Au model. Here, the surface field acts as a conjugate field to an order parameter which is usually suppressed. A likely preference of say Au at the surface favors a layering AuCuAuCu... perpendicular to the (100) surface and, if sufficiently strong, this can be expected to destabilize the other two cubic ordering variants which prefer the (2×2) ordered surface.

In comparison to Cu_3Au (and other alloys of this type of order), where wetting phenomena at surfaces as well as at anti-phase boundaries have been the subject of a number of experimental investigations,[7] it is noteworthy that the situation in alloys of CuAu type of order is even more interesting since it may show the full variety of the above mentioned phenomena of SID and SIO.

A more detailed description of our study will be presented elsewhere.[8]

MODEL AND MONTE CARLO TECHNIQUE

We study a model of a stoichiometric fcc alloy, which orders in the $L1_0$ structure (CuAu) at low temperatures, using the Ising Hamiltonian with only nearest neighbor interactions:

$$\mathcal{H} = -J \sum_{\langle i,j \rangle} S_i S_j, \qquad S_i = \pm 1 .$$

The interactions $J < 0$ are not further modified at the surface in our model.

In the particular case of the CuAu type of order, one needs in principle a three dimensional order parameter refering to the three possible orientations of ordered domains, which are alternate layering of pure "Cu" and pure "Au" planes in one of the three $\langle 100 \rangle$ directions. Using the four simple cubic sublattice "magnetizations" for the fcc lattice we have

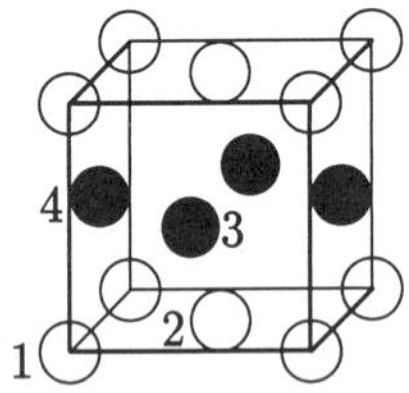

CuAu structure

$$\vec{\psi} = \begin{pmatrix} \psi_x \\ \psi_y \\ \psi_z \end{pmatrix} = \frac{1}{4} \begin{pmatrix} m_1 - m_2 - m_3 + m_4 \\ m_1 - m_2 + m_3 - m_4 \\ m_1 + m_2 - m_3 - m_4 \end{pmatrix} .$$

However, a scalar order parameter ψ based on the occupations of only two sublattices appears to be a proper choice if one variant of $\vec{\psi}$ dominates. In such cases, and these will be the typical ones here, ψ_n denotes the order in a single layer n, while $\psi_{x,n}$ *etc.* always refers to a bilayer.

Further useful quantities to describe SID and SIO are the excess quantities induced by the presence of a single surface. One can define an excess order parameter $\psi_s = \sum_n\{\psi_n - \psi_b\}$ and similar parameters for the excess energy, susceptibility, and specific heat: E_s, χ_s, and C_s respectively. The (bi-)layer susceptibilities χ_n and $\chi_{n,n}$ are obtained from the appropriate fluctuation relations[9]

In the simulations the systems have two surfaces, a geometry with layers of $L \times L$ sites with periodic (cyclic) boundary conditions and a thickness D, which needs to be sufficient to insure that the two surfaces behave essentially independent of each other. We used a Metropolis, single spin-flip method with preferential layer sampling which was determined by the nature of the order parameter profile. We implemented an efficient, vectorized single spin flip algorithm on a CRAY-YMP computer. Typically the MC data were obtained for parameters L and D varying from 40 to 200, and an average number of MC steps per site varying from 10^4 to 10^5. Of course, the data close to the transitions required most of the computing time. The bulk transition temperature was determined with very high precision, $T_{c,b} = 1.738005(25)\ |J|/k_B$, using large systems, full periodic boundary conditions, and standard thermodynamic integration methods.

SURFACE INDUCED ORDER AT THE (100) SURFACE

What happens near the (100) surface if the temperature is slightly *above* the first order bulk transition temperature $T_{c,b}$? The bulk is still disordered, however, order has developed near the (100) surface as can be seen in Fig. 1

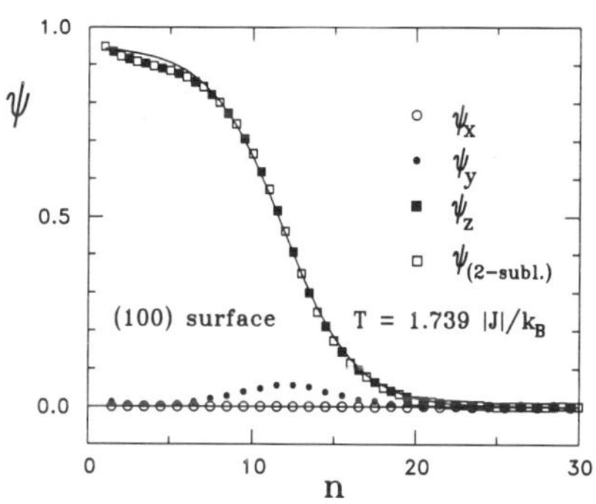

Figure 1. Order parameter profile for layers n at the (100) surface slightly above $T_{c,b}$

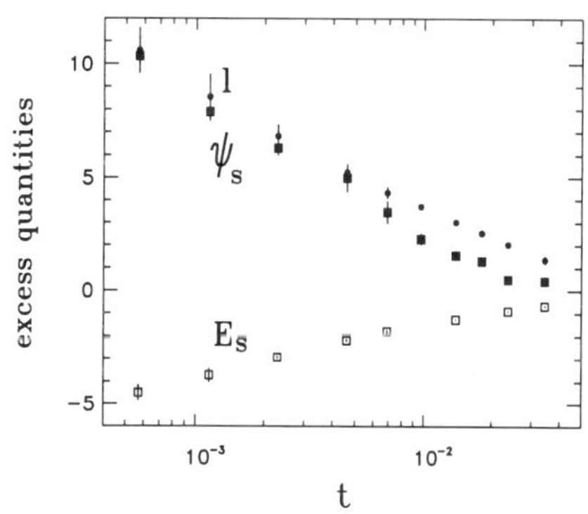

Figure 2. Logarithmic divergence of the excess quantities l, ψ_s, and E_s versus $t = 1 - T_{c,b}/T$.

A (2×2) superstructure characterized by ψ_n can be found in the first and a few subsequent layers. In terms of the the full three component order parameter, one of the components, either $\psi_{y,n}$ or $\psi_{z,n}$, dominates, while $\psi_{x,n}$ is negligble. The order parameter profiles can be nicely approximated by a *tanh* function apart from slight deviations near the surface. As T approaches $T_{c,b}$, the interface separating the ordered surface from the disordered bulk moves into the bulk. The position of the interface l, as well as the excess quantities ψ_s and E_s show a weak and theoretically expected divergence $\propto ln|T - T_{c,b}|$, see Fig. 2. With an increase of T, we found that the interface approaches the surface while its gradient increases. Finally, the order in the first layer vanishes continuously at a temperature $T_{c,s} = 1.8964\ |J|/k_B$. More precisely, we found, by finite size analysis of the order parameter and its 4th order cumulant, that the surface transition belongs to the universality class of the 2-dim Ising model. The transition of the second layer at $T_{c,n=2} = 1.791\ |J|/k_B$ is also continuous and the critical exponents (β and ν) are also still compatible with those of the 2-dim Ising model.

However, as is already apparent from the order parameter profile in Fig. 1 there is no further sequence of continuous layer by layer transitions. A careful analysis shows, that the onset of ordering takes places in the 3rd layer (and in the following odd layers) simultaneously with the 1st layer, however, with exponentially decreasing amplitudes. The same is found for the even layers upon ordering of the 2nd layer. This peculiar behavior is due to the competition in the even layers between the ordering variants ψ_y and ψ_z.

Fig. 3 gives a survey of the different ordering situations we studied. In particular, it can be seen how the continuous transitions in the different layers approach the discontinuous transition of the bulk.

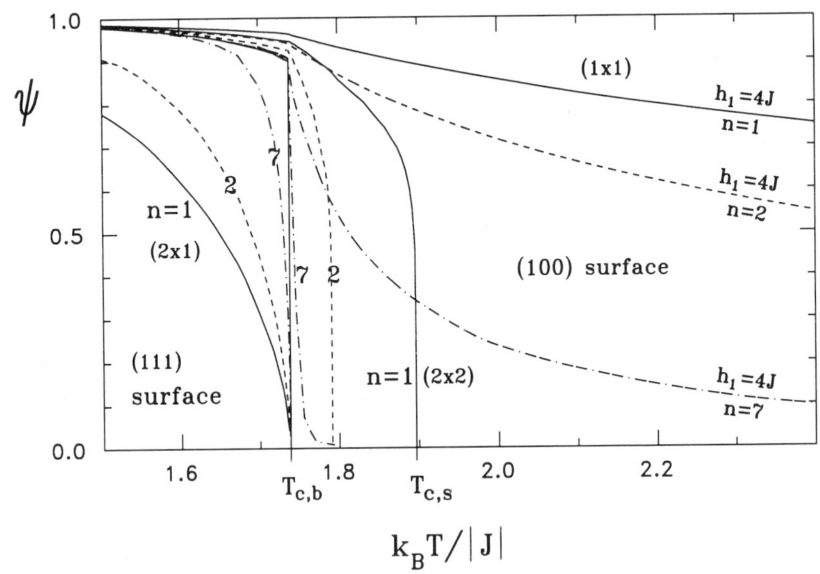

Figure 3. Order parameter variation with temperature at surfaces of a CuAu alloy model.Three distinct cases are shown:
(i) SID at a (111)-surface, $\psi = \psi^{(2x1)}$. There is a continuous transition of the surface layer, as well as of the subsequent layers at the discontinuous bulk transition $T_{c,b}$.
(ii) SIO at a (100)-surface, $\psi = \psi^{(2x2)}$, (without external fields). The continuous transition of the surface layer is at $T_{c,s}$, i. e. above $T_{c,b}$. The first and second layer order continuously $\propto t^{\beta}$, with $\beta = 1/8$ the value for the 2 dim Ising universality class.
(iii) SIO at a (100)-surface and a surface field $h_1 = 4J$ (which favors eithers "Cu" or "Au" in the surface layer), $\psi = \psi^{(1x1)}$.

THE EFFECT OF SURFACE FIELDS

A more realistic situation can be considered if one takes into account a (uniform) surface field (– therefore, one needs to add a term $-h_1 \sum_i s_i$ to the Hamiltonian, where the sum is only over the 1st layer–). At first, this will cause a tendency to enrich the surface with one of the components. If the surface field can act as a conjugate field to one of the possible order parameters, it may also cause further surface induced order. This situation is discussed in Fig. 4. Indeed, at a (100) surface of our CuAu model a sufficiently strong surface field favors a layering "AuCuAuCu.." perpendicular to this surface. This is described by the variant $\psi_{\mathrm{x},n}$.

While a comparison of the ground state energies yields a critical field $h^0_{c,1} = 2|J|$ for the stability of the variant $\psi_{\mathrm{x},n}$, around $T_{c,s}$, by MC we found a lower value of $h_{c,1} \approx 1.15|J|$. However, SIO (wetting) by $\psi_{\mathrm{x},n}$ seems to occur only for $h_1 \geq 3|J|$. Otherwise, only a finite thin film of order appears at the surface above $T_{c,b}$ (prewetting). Note also the temperature dependence of $\psi_{\mathrm{x},n}$ (equals (1×1)) for $h_1 = 4|J|$ in Fig. 3. This case, $h_1 = 4|J|$, also refers to a situation where the surface is completely covered by one component, say Au, which induces an effective field of $4|J|$ on the next layer. Therefore, the result contains the bordering case of an infinite surface field; one only needs to shift the order profile by one layer. For $h_1 > 0$ the conjugate order parameter $\psi_{\mathrm{x},1}$ appears continuously, of course (the response to h_1 at all finite temperatures is nonzero) there is no phase transition.

More generally, this type of SIO induced by surface fields can be expected for a number of other situations like at (111) surfaces of a CuPt alloy to give only one example. One may note that adsorbates give rise to effective surface fields, and that a covered, and not a free, surface (solid on solid model) is the more general case. For small h_1 we expect the transition to the (2×2) surface phase to persist, at $T_{c,s}(h_1)$, up to a critical field $h_{c,1}$ for which $T_{c,s}(h_{c,1}) = T_{c,b}$.

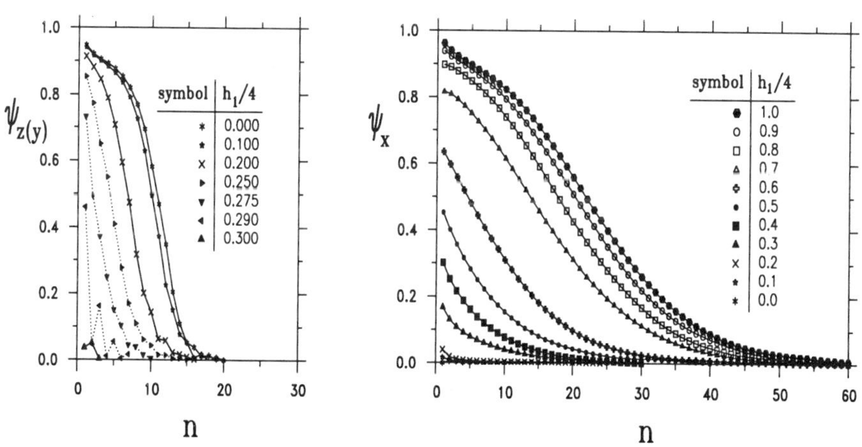

Figure 4. Dependence of the order parameter profiles, $\psi_{z(y)}$ (left) and ψ_x (right), versus layers n on a surface field h_1 slightly above $T_{c,b}$, $T = 1.74\ |J|/k_B$.

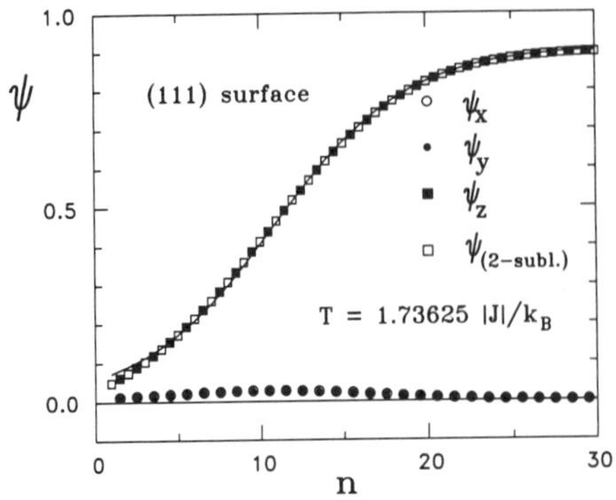

Figure 5. Order parameter profile for layers n at the (111) surface slightly below $T_{c,b}$

At the (111) surface (in absence of any surface field) one can observe the phenomena of surface induced disorder (SID) which arises from the incomplete bonding. A typical order parameter profile just below $T_{c,b}$ is shown in Fig. 5.All three ordering variants would be energetically equivalent by symmetry at the (111) surface. In other words, the behavior is equivalent for all (111) surfaces. As T approaches $T_{c,b}$, the order of the surface layer vanishes continuously and the interface penetrates into the bulk as theoretically predicted.[3]

In Fig. 6 one can see a power law behavior not only for ψ_1 but also for subsequent layers all with the same exponent β. The value $\beta = 0.64$ differs from the mean-field prediction $\beta^{MF} = \frac{1}{2}$, which implies that fluctuation corrections proposed for critical wetting are indeed relevant here. It also implies that the interface becomes rough as it becomes unbound from the surface as $T \to T_{c,b}$. This can be verified by the behavior of the interfacial width $\xi_\perp$: we find that $\xi_\perp^2 \propto ln\ |T_{c,b} - T|$ and also that $\xi_\perp^2 \propto ln\ L$ at $T_{c,b}$.[8]

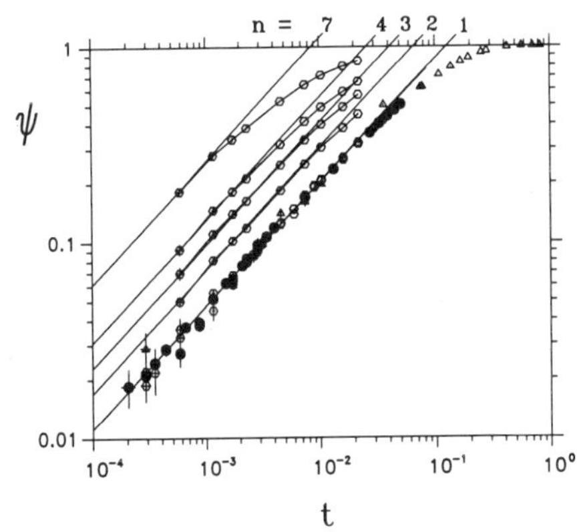

Figure 6. Log-log plot of the layer order parameters versus $t = 1 - T/T_{c,b}$ at a (111) surface. $\beta_1 = \beta_n = 0.64$

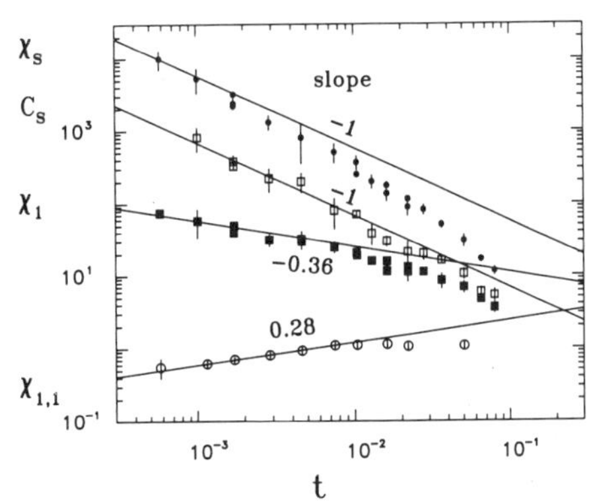

Figure 7. Log-log plot of the excess specific heat C_s and the susceptibilities versus $t = 1 - T/T_{c,b}$. Straight lines indicate the theoretically expected exponents, $\gamma_s = 1$, $\alpha_s = 1$, $\gamma_1 = 0.36$, $\gamma_{1,1} = -0.28$.

Although one can define quite a number of critical exponents to describe the various singularities at this transition for SID, there is only a single independent exponent[2], and scaling laws can be derived to determine other (non-universal) exponents from the value of β. It is certainly a good challenge for our simulations to check the actual theories of SID and critical wetting. Of course, it is not only difficult but also expensive to study the critical asymptotic behavior of the susceptibilities and the specific heat as shown in Fig. 7. The MC results confirm the essential theoretical predictions as indicated by the straight lines in the Figure.

SUMMARY

An alloy model with CuAu-type of order and free surfaces has been studied by Monte Carlo simulations which examine possible and typical surface critical phenomena in the case of a discontinuous order-disorder bulk transition. Surface induced disorder (SID) is found at a (111) surface, where the continuous surface transition coincides with the first order bulk transition. Our results for the critical exponents are in agreement with actual theories of wetting. Surface induced order (SIO) occurs at a (100) surface, while the continuous surface transition occurs above the bulk transition temperature because of reduced frustration effects. The critical exponents of the 2 dimensional Ising universality class are consistent with the surface transition. Furthermore, we examined the influence of effective surface fields at a (100) surface, which may cause pre-wetting and, for sufficiently strong fields, also SIO by another ordering variant which is usually suppressed.

REFERENCES

[1] For reviews of surface critical phenomena see
K. Binder, in *Phase Transitions and Critical Phenomena, Vol. 8* (C. Domb and J.L. Lebowitz, eds.) Ch. 1 (Academic Press, New York 1983); H.W. Diehl, in *Phase Transitions and Critical Phenomena, Vol. 10* (C. Domb and J.L. Lebowitz, eds.) p. 75 (Academic Press, New York 1986); D.E. Sullivan and M.M. Telo da Gama, in *Fluid Interfacial Phenomena* (C.A. Croxton, ed.) p. 45 (Wiley, New York 1986); P.G. de Gennes, Rev. Mod. Phys. **57**, 825 (1985); M.E. Fisher, J. Stat. Phys. **34**, 667 (1984); J. Chem. Soc. Faraday Trans. **282**, 1569 (1986); H. Dosch, *Critical Phenomena at Surfaces and Interfaces (Evanescent X-ray and Neutron Scattering)*, Springer Tracts in Modern Physics Vol. 126 (Springer, Berlin, 1992); K. Binder, *present volume*, and Ref. 2 and 3.

[2] For reviews of wetting phenomena see S. Dietrich, in *Phase Transitions and Critical Phenomena, Vol. 12* (C. Domb and J.L. Lebowitz, eds.) Ch. 1 (Academic Press, New York 1988)

[3] R. Lipowsky, J. Appl. Phys. **55**, 2485 (1984); Ferroelectrics **73**, 69 (1987).

[4] B. Pluis, A.W. Denier van der Gon, J.F. van der Veen, and A.J. Riemersma, Surf. Sci. **239**, 205 (1990); B. Pluis, D. Frenkel, and J.F. van der Veen, Surf. Sci. **239**, 282 (1990).

[5] W. Schweika, K. Binder, and D.P.Landau, Phys. Rev. Lett. **65**, 3321 (1990).

[6] D.M. Kroll and G. Gompper, Phys. Rev. B**36**, 7078 (1987); G. Gompper and D.M. Kroll, Phys. Rev. B**38**, 459 (1988).

[7] V.S. Sundaram, B. Farrell, R.S. Alben, and W.D. Robertson, Phys. Rev. Lett.**31**, 1136 (1973); V.S. Sundaram, R.S. Alben, and W.D. Robertson, Surf. Sci. **46**, 653 (1974); E.G. McRae and R.A. Malic, Surf. Sci. **148**, 551 (1984); S.F. Alvarado, M. Campagna, A. Fattah and W. Uelhoff, Z. Phys. **B66**, 103 (1987); H. Dosch, L. Mailänder, A. Lied, J. Peisl, F. Grey, R.L. Johnson, and S. Krumnacher, Phys. Rev. Lett. **60**, 2382 (1988); H. Dosch, L. Mailänder, H. Reichert, J. Peisl, R.L. Johnson, Phys. Rev. B**43**, 13172 (1991) Ch. Ricolleau, A. Loiseau, F. Ducastelle and R. Caudron, Phys. Rev. Lett.**68**, 3591 (1992).

[8] W. Schweika, D.P.Landau, and K. Binder, to be submitted to Phys. Rev. B.

[9] K. Binder and D.P. Landau, Phys. Rev. Lett. **52**, 318 (1984); D.P. Landau and K. Binder, Phys. Rev. **B41**, 4768 (1990).

SURFACE ORDERING AND SURFACE SEGREGATION IN BINARY ALLOYS

F. SCHMID

Institut für Physik
Johannes Gutenberg Universität Mainz
D-55099 Mainz
Germany

Many technologically relevant properties of metallic alloys are determined by the structure of their surfaces, especially in the field of catalysis and corrosion. One important aspect of a surface or grain boundaries is, that the stoichiometry of the alloy close to the surface normally differs from the bulk stoichiometry. Due to different interaction energies and different atom sizes of the components, one of them will get enriched at the surface, a phenomenon called surface segregation[1].

The presence of a surface may produce many other effects, particularly in the vicinity of a bulk phase transition. In alloys which display an order-disorder phase transformation, the surface often has a lower, sometimes a higher degree of ordering than the bulk. The statistical mechanics of phase transitions in semi-infinite systems have received enormous attention[2], and binary alloys have provided excellent model systems for the experimental test of theoretical predictions[3, 4]. However, those studies mostly focused on the ordering behavior of the systems, and relatively little work has been concerned with the interplay of surface order and surface segregation. Since order parameter and concentration are generally coupled to each other, surface ordering effects may have a pronounced effect on the concentration profile, especially close to a phase transition. For example, the order parameter profile decays towards its bulk value with a decay length proportional to the bulk correlation length, which diverges close to a critical point. On approaching the critical temperature in the ordered phase, the decay length of the segregation profile therefore diverges too, and surface enrichment may extend over many layers[8]. In the disordered phase, on the other hand, surface segregation may induce surface ordering[3] or partial surface ordering[5]. This induced order parameter profile will in return influence the segregation. In this paper, some of the possible consequences of the interplay between surface segregation and surface ordering in binary alloys with order-disorder phase transitions will be discussed.

From a very idealizing point of view, one can map such an alloy on an Ising-type lattice spin model, where the Ising spin ± 1 corresponds to a lattice site being occupied by atom A or B, respectively[6]. With the simplest assumption that it is sufficient to

Stability of Materials
Edited by A. Gonis *et al.*, Plenum Press, New York, 1996

consider pair energies $W_l^{\alpha\beta}$ between lth nearest neighbor atoms ($\alpha, \beta = A$ or B), one gets a bulk Hamiltonian

$$\mathcal{H}_0 = -\frac{1}{2}\sum_{i,j} V_{ij} S_i S_j - H \sum_i S_i, \qquad (1)$$

where the effective interactions V_{ij} are given by $V_{ij} = -\frac{1}{4}(W_l^{AB} + W_l^{BA} - W_l^{AA} - W_l^{BB})$ for lth nearest neighbor sites i and j. The "field" H is related to the chemical potential difference $(\mu^A - \mu^B)$ by $H = \frac{1}{2}(\mu^A - \mu^B) + \frac{1}{4}\sum_l z_l (W_l^{AA} - W_l^{BB})$, where z_l is the number of neighbors in the lth coordination shell. The presence of a surface has the effect of reducing the coordination number z_l, thus introducing surface fields $H^s(n)$, which act on atoms in the nth layer below the surface. The Hamiltonian then gets an additional surface term $\mathcal{H} = \mathcal{H}_0 + \mathcal{H}_s$ with

$$\mathcal{H}_s = -\sum_n H^s(n) \sum_{n\text{th layer}} S_i, \qquad H^s(n) = \frac{1}{4}\sum_l (z_l(n) - z_l^{\text{bulk}})(W_l^{AA} - W_l^{BB}). \quad (2)$$

This already explains the effect mentioned in the beginning, that different interaction energies between alloy components cause surface segregation (bond-breaking). The interaction energies may also be altered at the surface, but we shall disregard this possibility here.

The Monte Carlo results discussed in this paper are based on a bcc Ising model of the type (1), which has been shown to successfully reproduce the topology of the bcc-phases of Iron-Aluminium[7]. Ising spins $S_i = -1$ correspond to Aluminium atoms, $S_i = +1$ to Iron atoms, and the latter are equipped with additional Heisenberg spins $\vec{\sigma}_i$ of length 1, which account for the magnetic properties of Iron. Hence the full bulk Hamiltonian is given by $\mathcal{H}_0 = \mathcal{H} + \mathcal{H}_{\text{magn.}}$ with the magnetic contribution

$$\mathcal{H}_{\text{magn.}} = -\frac{1}{2}\sum_{i,j} J_{ij} \frac{(1+S_i)}{2}\frac{(1+S_j)}{2} \vec{\sigma}_i \vec{\sigma}_j. \qquad (3)$$

The configurational interactions V_{ij} were taken from experimental data ($V_1 = -12$ meV, $V_2/|V_1| = -0.167$, $V_3/|V_1| = 0.208$, where V_l is the interaction between l-nearest neighbors), and the magnetic interaction $J_1/|V_1| = 1.65$ was chosen such that magnetic transition temperatures match configurational transition temperatures.

The phase diagram of this model in the $(H - T)$ plane is shown in Figure 1. In this work, we shall deal only with B2-type ordering. The order parameter can be defined in terms of average sublattice concentrations $M = \frac{1}{2}(\langle S\rangle_a - \langle S\rangle_b)$, and the stoichiometry is essentially given by the average spin $P = \langle S\rangle \equiv (2c_{Fe} - 1)$ (see Fig. 1).

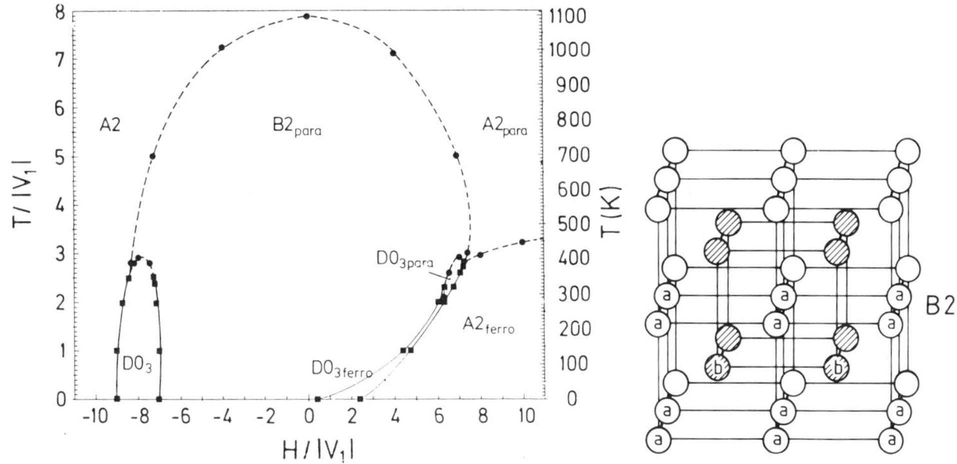

Figure 1. Bulk phase diagram of our model in the grand canonical ensemble (from Ref. [7]). A2 denotes the configurationally disordered structure of the bcc alloy, B2 and DO_3 different ordered phases. A2- and DO_3 phases can be both ferro- or paramagnetic.
Right part shows the bcc lattice with B2 structure and assignment of sublattices a and b.

When studying surfaces, one is interested in determining layer dependent order parameter or segregation profiles. This may require to combine adjacent layers. In the case of (100)-surfaces and B2-type ordering, all lattice sites in one layer belong to the same sublattice – the definition of a layer dependent order parameter involves at least two subsequent layers: $M_{2n} = \frac{1}{2}(\langle S \rangle_{2n} - \langle S \rangle_{2n+1})$.

The surface field $H^s(n)$ couples to the concentrations of the components in the nth layer. If subsequent layers are not equivalent with respect to the bulk order, this may induce an effective coupling to the order parameter at the surface: $H^s(n)$ may cause an *ordering* surface field. To illustrate this, let us consider a (100)-surface with a surface field $H^s(0) \equiv H_1$ acting on the top layer. For convenience, we define a layer dependent stoichiometry parameter $P_{2n} = \frac{1}{2}(\langle S \rangle_{2n} + \langle S \rangle_{2n+1})$. The surface contribution to the Hamiltonian $\mathcal{H}^s \propto -H_1 \langle S \rangle_0$ decomposes into two terms $\mathcal{H}^s \propto -\frac{1}{2} H_1 M_0 - \frac{1}{2} H_1 P_0$, i.e. one gets an effective staggered ordering field $h^s(0) = \frac{1}{2} H_1$ at the surface. Figure 2 illustrates the effect of this field: Above the transition temperature, the surface still orders, whereas the bulk is disordered[8].

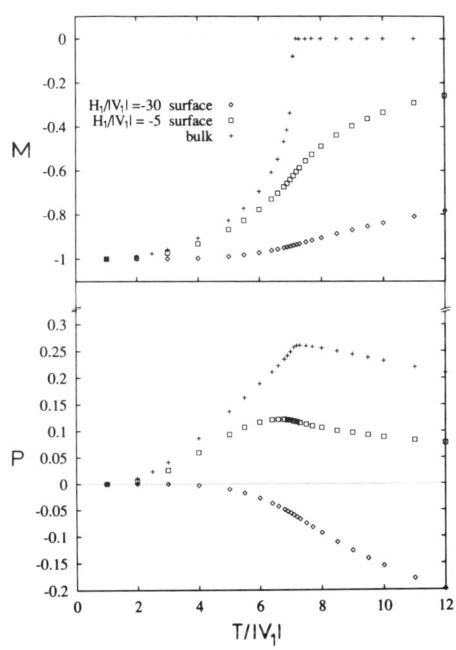

Figure 2. Order parameter M and stoichiometry parameter P in the top layer and in the bulk, versus temperature, for different surface potentials H_1, at $H/|V_1| = 4$.

H. Reichert et al have recently measured an oscillating segregation profile at the (100)-surface of Cu_3Au, which can be interpreted in similar terms. At the temperature 663 K, the fcc alloy undergoes a first order transition from a disordered phase into the ordered structure shown in Figure 3. The order parameter $\vec{\eta}$ for this transition has three components (where sublattices a,b,c and d of the fcc lattice are defined as indicated in Fig. 3)

$$\vec{\eta} = \begin{pmatrix} \eta_1 \\ \eta_2 \\ \eta_3 \end{pmatrix} = \frac{1}{4} \begin{pmatrix} \langle S\rangle_a + \langle S\rangle_b - \langle S\rangle_c - \langle S\rangle_d \\ \langle S\rangle_a - \langle S\rangle_b + \langle S\rangle_c - \langle S\rangle_d \\ \langle S\rangle_a - \langle S\rangle_b - \langle S\rangle_c + \langle S\rangle_d \end{pmatrix}.$$

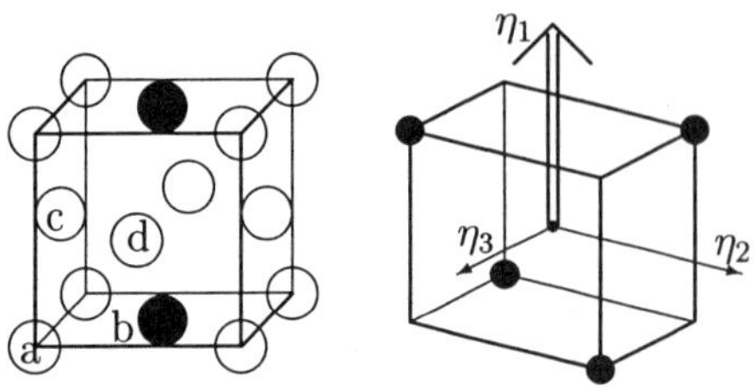

Figure 3. Left: fcc-lattice with ordered Cu_3Au-structure and assignment of sublattices a,b,c,d.
Right: Location of the different Cu_3Au-type ordered phases in order parameter space (circles). Thick arrow indicates direction of the surface field at a (100)-surface.

The stoichiometry parameter is given by $P = \frac{1}{4}(\langle S\rangle_a + \langle S\rangle_b + \langle S\rangle_c + \langle S\rangle_d)$. Cu_3Au-type ordered phases are described by vectors (-1,-1,-1), $(-1,1,1)$, $(1,-1,1)$ or $(1,1,-1)$ in order parameter space. A surface field acting on a single (100)-layer couples to P and η_1, i.e. it excites a combination of the ordered states $(1,1,-1)$ and $(1,-1,1)$. As a result, the segregation profile oscillates in the (100)-direction ([5], see also [9]). The surface field does not necessarily induce full Cu_3Au order, since this requires a further symmetry breaking in the (η_2, η_3)-plane.

Next, let us consider an "ideal" system with components of the same size, same interaction energies etc. Although no direct surface fields $H^s(n)$ act on such a system, there still exist mechanisms which may induce surface segregation. In order to discuss those, we return to the (100)-surface of a bcc-alloy close to a transition between disordered and B2-ordered phase, and first consider the symmetric case $P_{\text{bulk}} = 0$ $(H = 0)$.

Due to the lack of direct neighbors in the surface layer, the ordering tendency there is smaller than in the bulk, and the order parameter decreases at the surface. This characterizes the surface as one with "ordinary behavior", where the surface is less ordered than the bulk, and the transition temperatures for surface and bulk are the same. If the top layer (sublattice a) has an excess of component A, the next layer (sublattice b) will have a higher excess of B, the third layer still more of A and so on, until the values of the bulk sublattice concentrations are reached. As a result, the integral concentration of A in the surface region will be lower than that of B, and one has an effective enrichment of component B. Again, this kind of mechanism only works if layers are not equivalent with respect to the ordering. A non monotonic order parameter profile will then induce a segregation profile, in our case $P_n \propto -\frac{d}{dn}M_n$[8]. The effect is illustrated in Figure 4. Below the transition temperature, the excess order parameter at the surface $M_{\text{exc}} = \sum(M_{2n} - M_{\text{bulk}})$ causes an integral segregation $P_{\text{exc}} = \sum(P_{2n} - P_{\text{bulk}})$. Above the transition temperature, the order parameter is zero, i.e. constant, and no segregation is induced.

From the theory of critical phenomena, one knows that the order parameter profile decays exponentially towards its bulk value with a decay length λ, which diverges according to a power law $\lambda \propto |T - T_c|^{-\nu}$, $\nu = 0.63$, as the temperature approaches the (critical) transition temperature T_c. This has consequences for the induced segregation profile: Being proportional to the derivative of the order parameter profile, it also decays exponentially with the same decay length, and the amplitude ratio

$(P_n - P_{\text{bulk}})(M_n - M_{\text{bulk}}) \propto \lambda^{-1} \propto |T - T_c|^{\nu}$ follows a power law with the same critical exponent. The simulation data confirm this prediction (Figure 5).

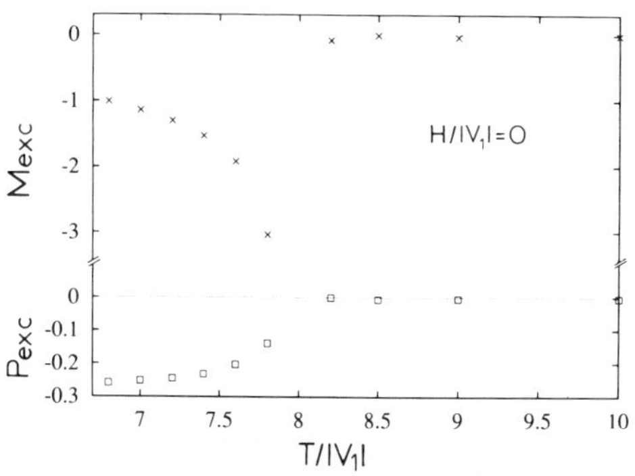

Figure 4. Excess order parameter and excess concentration parameter P at the (100)-surface, vs. temperature, at $H/|V_1| = 0$.

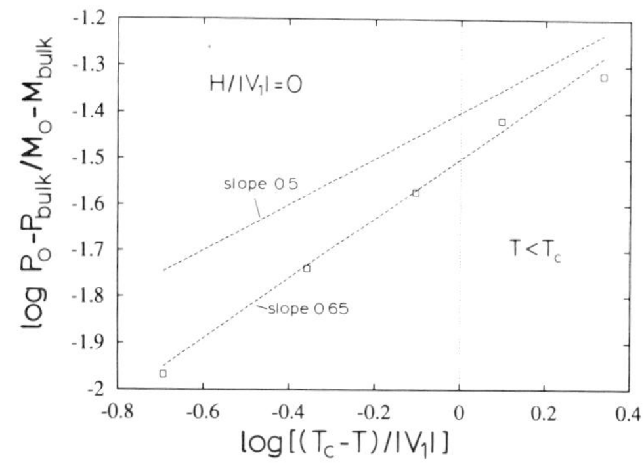

Figure 5.
Scaling of $(P_0 - P_{\text{bulk}})/(M_0 - M_{\text{bulk}})$ below T_c at $H/|V_1| = 0$ [8].

Finally, we discuss the asymmetric case $H \neq 0$, where the bulk does not have the ideal stoichiometry of the ordered phase. On the average, atoms of the majority component have more energetically costly neighbors of their own type, than atoms of the minority component. Therefore, it is most favorable for them to stay right at the surface, where the number of neighbors is reduced: Non-stoichiometry of the alloy induces a effective surface field $H_1 \propto P$. At surfaces, where layers are not equal with respect to bulk ordering, this field may again couple to the order parameter profile, and act as an ordering surface field[8].

Varying H has therefore the same effect on the surface behavior, than introducing and varying a (relatively weak) surface field H_1. We consider the case $H, P > 0$, i.e. the induced surface field favors component A (spin $S = 1$): One has an excess of A in the top layer, and the order parameter at the surface is positive. Let us discuss the question, what surface segregation results from there.

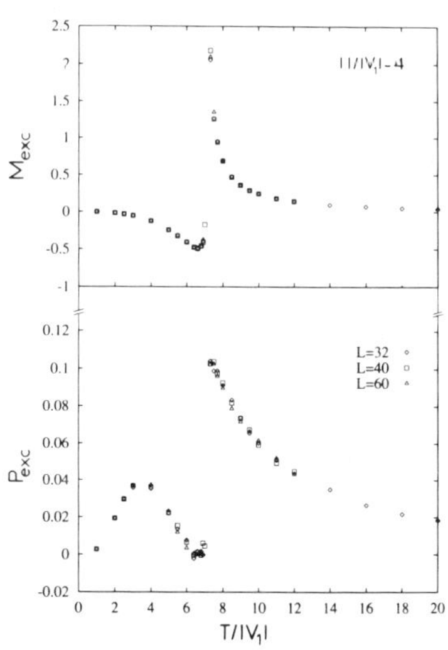

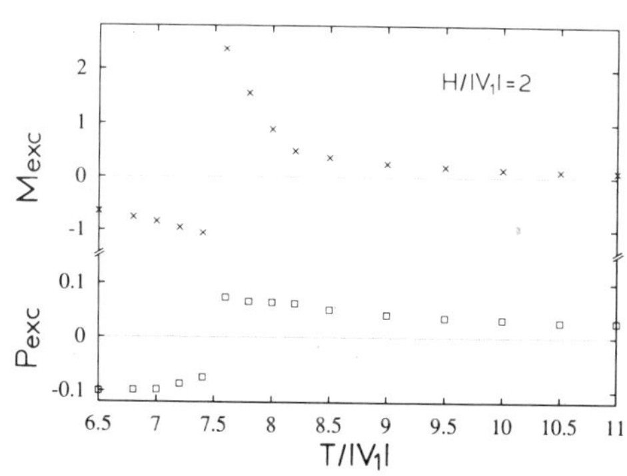

Figure 6. Excess order parameter and excess concentration parameter P at the (100)-surface, vs. temperature, at $H/|V_1| = 4$ (left) and $H/|V_1| = 2$ (right).

Above the ordering temperature, the answer is straightforward. The order parameter decays towards zero in the bulk, hence $-\frac{d}{dn}M_n > 0$: From the coupling of order parameter and segregation, one expects a positive segregation profile. The surface field itself favors positive segregation as well. Thus the resulting segregation is positive, component A gets enriched.

Below T_c, the situation is less clear. Here the order parameter is smaller at the surface than in the bulk, i.e. $-\frac{d}{dn}M_n < 0$. We noted earlier for the symmetric system, that one finds a total enrichment of B in the surface region, if the top layer contains mainly A-atoms. In the symmetric situation, A and B are interchangeable, and one will also encounter the inverse situation. However, a small surface field breaks the symmetry: The system will therefore respond to a surface field favoring A with surface enrichment of the *other* component, B. On the other hand, one still has the direct coupling of the surface field to the concentration profile, which favors A. Both effects counteract each other. At small surface fields, the former wins and B segregates, at high surface fields, the latter wins and A segregates. The simulation data show nicely, how the system switches from one behavior to the other between $H/|V_1| = 2$ and 4 (Figure 6).

In sum, we have seen that the interplay of surface segregation and surface ordering gives rise to a phenomenology in the surface behavior of alloys, which is considerably richer than in simpler systems. This study has been far from exhausting the subject. Other regimes of surface ordering behavior should be considered ("extraordinary" and "surface" transition). Interesting effects can also be expected in the context of dynamics at surfaces, kinetics of segregation (see [10]) and of ordering, or in the context of wetting.

It is a pleasure to thank K. Binder for suggesting the topic, and to acknowledge stimulating discussions with him, B. Dünweg and H. Reichert. The Monte Carlo Simulations have involved the use of extensive computer time at the Siemens-Fujitsu VP100 at the RHRK Kaiserslautern. This work was supported in part by the Deutsche Forschungsgemeinschaft, SFB 262.

References

[1] for reviews see J. Du Plessis, Solid State Phenomena, Sci. Tech. 11 (1990); P.A. Dowben, A. Miller (eds.), *Surface Segregation and related Phenomena*, CRC Press, Boca Raton (1990).

[2] for reviews see K. Binder in *Phase Transitions and Critical Phenomena*, Vol. 8, p. 1 (1983), C. Domb and J.L. Lebowitz eds., Academic Press, London; S. Dietrich, ibid, Vol. 12, p.1. (1988).

[3] L. Mailänder, H. Dosch, J. Peisl, R.L. Johnson, Phys. Rev. Lett. **64**, 2527 (1990).

[4] S.F. Alvarado, M. Campagna, F. Ciccacci, M. Hopster, J. Appl. Phys. **53**, 7920; B.H. Dauth, S.F. Alvarado, M. Campagna, Phys. Rev. Lett. **58**, 2118 (1987); C. Rau, M. Robert, Phys. Rev. Lett. **58**, 2714 (1987).

[5] H. Reichert, P.J. Eng, H. Dosch, I.K. Robinson, Phys. Rev. Lett. **74**, 2006 (1995).

[6] D. De Fontaine in *Solid State Physics* **34**, p 73 H. Ehrenreich, F. Seitz and D. Turnbell eds., Academic Press, New York 1979; K. Binder in *Festkörperprobleme (Advances in Solid State Physics)* **26**, p 133, P. Grosse ed., Vieweg, Braunschweig 1986.

[7] F. Schmid, K. Binder, J. Phys.: Condens. Matter **4**, 3569 (1992).

[8] F. Schmid, Z. Phys. **B 91**, 77 (1993).

[9] K.R. Mecke, S. Dietrich, Phys. Rev. B **52**, 2107 (1995).

[10] Cs. Cserhati, H. Bakker, D.L. Bakker, Surf. Sci. **290**, 345 (1993).

RELAXED MONTE CARLO SIMULATIONS ON AU-NI ALLOY

R. Tétot,[1] and A. Finel[2]

[1]URA 446 CNRS, Bâtiment 415, Université de Paris-Sud, Centre d'Orsay 91405 Orsay Cedex, France
[2]Office National d'Etudes et de Recherches Aérospatiales (ONERA) BP 72, 92322 Châtillon Cedex, France

INTRODUCTION

This work deals with statistical thermodynamics of binary systems including, in addition to the classical Ising model, the lattice relaxation (atomic and volume). More precisely, it is a contribution to the study of the consequences of the atomic size effect on thermodynamic properties of binary alloys. We have taken as an example the Au-Ni system in which the size difference between the two constituents is very large (≅ 15%). This system presents a miscibility gap below 1100K (Hultgren et al., 1963; Massalsky, 1986) which is strongly asymmetric, its maximum temperature corresponding to 70.6% Ni. The reason why this miscibility gap occurs has stimulated many thermodynamic and atomistic investigations (see for example Hansen, 1958; Hultgren et al., 1963, and references therein). It is well known now that the lattice-distorsion energy rather than the energetic favoring of bond between like atoms is responsible for the phase segregation (Wu et al., 1982, 1983, 1984). Recently, Lu and Zunger (1994) have calculated the two contributions of the formation energy of the alloy, a pure chemical one (reflecting charge transfert and atomic relaxation) and a volume deformation energy reflecting the size effect. Their results show that the chemical term is negative and then corresponds to an ordering tendency, whereas the second one, always positive but here very large is responsible for the phase separation. These features make Au-Ni a very convenient system for our purpose.

We have performed Monte Carlo simulations using a many body tight-binding potential. The code has been checked by comparing zero temperature quantities already computed by an other technic by Emery et al. (1992). The original contribution of this work lies in the calculation of the phase diagram in the solid state region which compares quite well with the experimental one.

POTENTIAL MODEL

It is well known that the cohesive properties of transition metals can be correctly described with the second-moment approximation (SMA) of the tight-binding scheme (TB) (Ducastelle, 1970). According to this scheme, the total cohesive energy of the system is:

$$E_c = \sum_n \left\{ A \sum_m \exp\left[-p\left(\frac{r_{nm}}{r_0} - 1\right)\right] - \left[\sum_m \xi^2 \exp\left[-2q\left(\frac{r_{nm}}{r_0} - 1\right)\right]\right]^{1/2} \right\}$$

where r_{nm} is the distance between atoms n and m, and r_0 is the first-neighbour distance. The first term is a repulsive interaction of the Born-Mayer type and the second one is the band energy contribution. The quantity ξ is an effective hopping integral and the parameter q describes its distance dependence. In the present work, the summations over the index *m* are limited to the first neighbours of an atom n. For a given metal, the parameters A, ξ, p and q are fitted to reproduce some experimental quantities *at zero temperature* (cohesive energy, lattice parameter, bulk modulus and elastic constants). We use here the parameters A, ξ, p and q determined by Rosato et al. (1989) for a lot of transition metals.

Monte Carlo results concerning the mean square displacement of pure Cu and Ni as a function of temperature are reported in a paper by Finel and Tétot in this volume. The agreement with experimental data is noteworthy. Note that the Gaussian CVM described in the same paper yields comparable results with a remarkable gain in computer time.

In order to deal with alloys, it is necessary to evaluate heteroatomic potential. When the system presents an ordered compound (or more), like Au-Cu for example, it is possible to fit the heteroatomic potential parameters to the properties of this compound (Cleri and Rosato, 1993). In the Au-Ni system, no such compound exists. In this case, a usual procedure consists in averaging the parameters obtained for pure metals. Although this procedure is not rigorous, it can be acceptable when the energy levels of the two components are nearly the same (which is achieved when they lies in two columns next to each other in the periodic table, as Ni and Au). We apply this method in this paper, using an arithmetic average of r_0, p and q and a geometric average of A and ξ (see Table 1).

Table 1. Parameters A, ξ, p and q of the TB-SMA potentials for Au, Ni and Au-Ni alloy. We give also the experimental values of the cohesive energy E_c and of the lattice parameter a (after Kittel, 1972).

	A (eV)	ξ (eV)	p	q	E_c (eV)	a (Å)
Au	0.2161	1.8398	10.15	4.13	- 3.78	4.08
Ni	0.1368	1.7558	10.00	2.70	- 4.44	3.52
Au-Ni	0.1720	1.7970	10.07	3.415		

MONTE CARLO PROCEDURE

We have used the Metropolis algorithm (Metropolis et al., 1953) in the grand canonical ensemble. A Monte Carlo step (MCS) consists in trying to exchange the nature (Ising step) and the position of each atom and to move the total volume once. The calculations are made at zero pressure. The simulation box contains $N=4 \times L^3$ atoms, with

L=8 to 12 (N=2048 to 6912). Helicoidal periodic boundary conditions have been applied. The number of MCS per site (MCS/s) is typically of the order of 2000, including 500 MCS/s used for putting the system at equilibrium. Note that in our calculations, the kinetic entropy is not taken into account. One can thought that this term may distord a little the obtained results. A study of this subject is planned.

RESULTS

We have checked the Monte Carlo program and the potential model by computing the nearest neighbour distances and the parameter in the $Au_{1-x}Ni_x$ solid solution at zero temperature. These values are obtained by extrapolating at T=0K the variations of these quantities with temperature. The results, plotted in Figs.1-2, coincide very well with experimental data as well as with the calculations by Eymery et al. (1992) which have performed energy minimizations directly at 0K. Note that these authors have not used exactly the same parameters than those used in this work and have extended the range of interactions between the second and third neighbours.

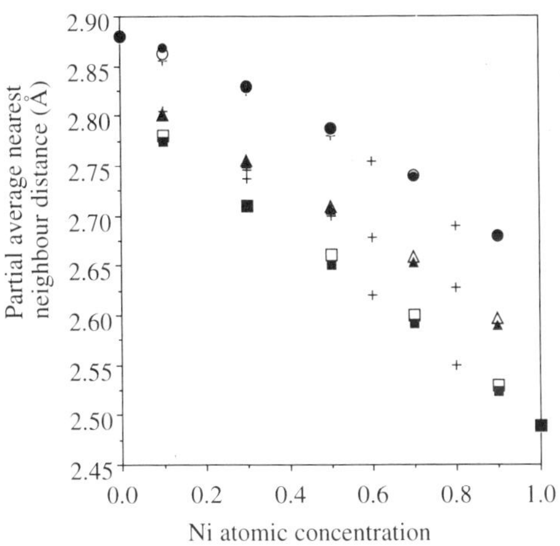

Figure 1. Mean Au-Au (circles), Au-Ni (triangles) and Ni-Ni (squares) nearest neighbour distance in the $Au_{1-x}Ni_x$ solid solutions *versus* Ni concentration at OK. Crosses represent experimental data (Renaud et al. 1987), open symbols the results of Emery et al. (1992) and close symbols our results.

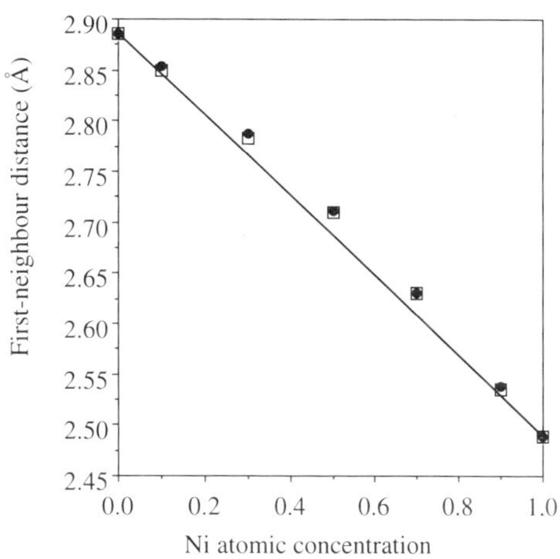

Figure 2. Deviation from Vegard's law of the $Au_{1-x}Ni_x$ solid solutions *versus* Ni concentration at 0K. Open squares represent the experimental results of Renaud et al.(1987) and close circles our results. The results of Emery et al.(1992) coincide with ours and are not represented here.

We have then proceeded to calculate some quantities at finite temperature. The formation energy of $Au_{1-x}Ni_x$ solid solution *versus* Ni concentration at 1150K is reported in Fig.3. The energy curve lies below the experimental curve from Hulgren et al. (1963). This is probably due to the short range of our potentials, limited to first neighbours. As a matter of fact, calculations of the formation energy by Eymery et al. (1992) at zero temperature, with longer ranged potentials lead to a good agreement. These authors have also calculated that the unrelaxed alloys have an energy about twice as big as the energy of relaxed alloys.

The taking into account of more distant neighbours tends to rigidify the lattice, which should lead to higher energies.

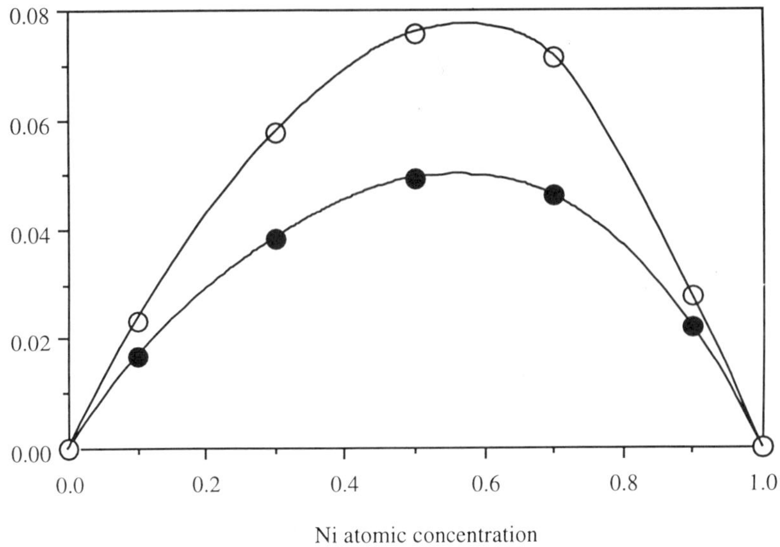

Figure 3. Formation energy of the $Au_{1-x}Ni_x$ solid solutions *versus* Ni concentration at 1150K. Open circles represent the experimental data from Hultgren et al. (1963) and close circles our results.

The miscibility gap have been determined by computing concentration-chemical potential (which is in fact the difference μ_{Au} - μ_{Ni}) curves at constant temperature. The demixing corresponds to a discontinuity in the concentration. In order to show the hysterisis due to the first order character of the transition, two isotherms are calculated, one at increasing Ni concentration and the other at decreasing Ni concentration. In Fig. 4, an example of such curves at 800K is plotted.

We can see also in this figure that the solubility of Ni in Au is larger than the solubility of Au in Ni, which is in agreement with the asymmetry of the miscibility gap. For each concentration an upper and a lower limit are obtained. The values reported in the Fig.5 are the mean values of these two limits. The uncertainties attached to these values, not reported in the Fig. 5 for sake of clarity, are of the order of ±2% in Ni concentration. When the temperature increases, the discontinuity becomes less and less sharp and, consequently, it is very difficult to obtain accurate values above 900K. It is the reason why we have drawn a broken line in this region. Above 950K, we obtain a continuous solid solution.

The calculated phase diagram (Fig.5) reproduces well the essential features of the experimental data, particularly the strong asymmetry of the miscibility gap. In a preliminary work (Tétot and Finel, 1995), we have performed Monte Carlo simulations using quasi harmonic pair potentials. This model yields a phase diagram which is also asymmetric but in the opposite direction. The many body character of the potentials is therefore essential to describe the Au-Ni system.

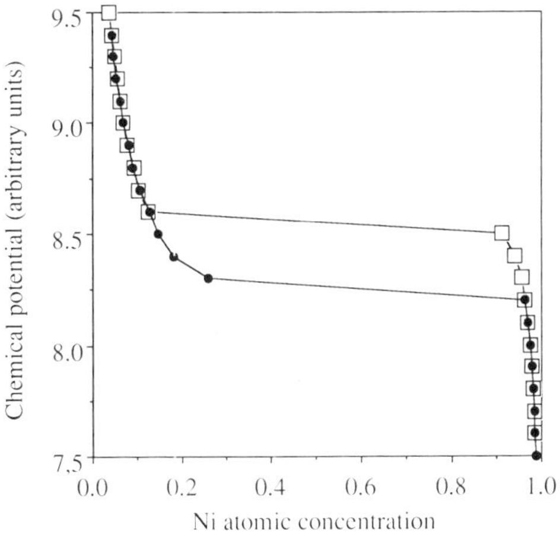

Figure 4. Chemical potential *versus* Ni concentration at 800K. Close circles and open squares represent respectively the results obtained at increasing Ni concentration and decreasing Ni concentration. The calculation have been performed on 2048 atoms with 2000 MCS/s.

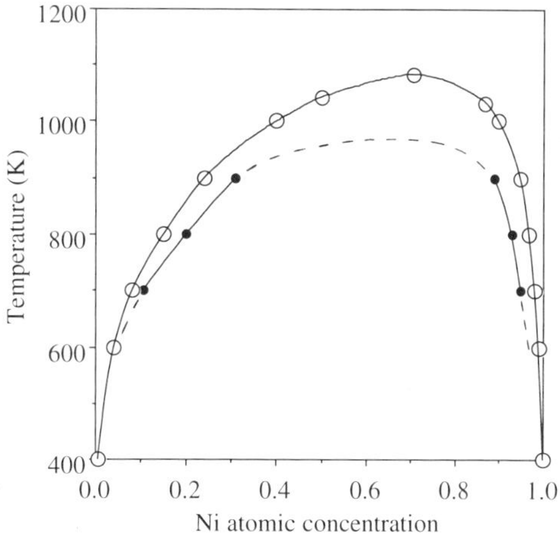

Figure 5. Miscibility curve in the Au-Ni system. Open circles represent experimental data of Massalsky (1986) and close circles our results.

CONCLUSION

We have performed Monte Carlo simulations on the Au-Ni system using TB-SMA interatomic potentials. Our aim was to study the effect of a strong size difference between unlike atoms on thermodynamic properties of alloys. Our model has allowed us to account for the essential features of the phase diagram in the solid state region. The use of many body potentials is necessary, pair potentials yielding erroneous results. However, the model can probably be improved by extending the range of the potentials. Morever, a more physical treatment of the heteroatomic potential is needed, for example by fitting the parameters to the total energies of ordered compounds (even unstable) calculated by *ab-initio* methods (see the paper by Deutsch and Pasturel in this volume). A more complete study, in connection with diffuse neutron scattering experiments, is in progress.

On the other hand, the Monte Carlo simulations presented here are computer time consuming. So, we are obliged to limit the size of the simulation box to relatively small numbers of atoms (about 4000), which limits the accuracy of the results. In order to resolve this problem, a Gaussian CVM program adapted to alloys is being developped (see Finel and Tétot in this volume).

REFERENCES

Cleri, F. , and Rosato, V., 1993, Lattice dynamics of ordered and disordered Cu_3Au with a tight-binding potential model, *Philos. Mag.* 67:369.

Ducastelle, F., 1970, Modules élastiques des métaux de transition, *J. Phys.* (Paris) 31:1055.

Eymery, J., Lançon, F., and Billard, L. , 1992, Au-Ni solid solutions studied by numerical relaxation, *J. Phys.* (Paris) 3:787.

Hansen, M., 1958, "Constitution of Binary Alloys," McGraw-Hill, New York.

Hultgren, R., Orr, R.L. , Andersen, P.D. , and Kelley, K.K., 1963, "Selected values of the Thermodynamic Properties of Binary Alloys," Wiley, New York.

Kittel, C., 1972, "Introduction à la Physique de l'Etat Solide," Dunod Université, France.

Lu, Z.W., and Zunger, A., 1994, Unequal wave vectors in short- versus long-range ordering in intermetallic compounds, *Phys. Rev.* B 50:6626.

Massalsky, T.B., 1986, "Binary Alloys Phase diagrams," A.S.M. Metal Park, Ohio.

Metropolis, N.A., Rosenbluth, A.W., Rosenbluth, M.N., Teller, A.H., and Teller E.J., 1953, Equation of states calculations by fast computing machines, *J. Chem. Phys.* 21:1087.

Renaud, G., Motta, N., Belakhovsky, M., Lançon, F., and Billard, L., 1987, Studies of lattice distorsions in AuNi solid solutions by EXAFS and computer simulations, *Solid State Commun.* 63:659.

Rosato, V., Guillope, M., and Legrand, B., 1989, Thermodynamical and structural properties of f.c.c. transition metals using a simple tight-binding model, *Philos. Mag.* A 59:321.

Tétot, R., and Finel, A. , 1995, unpublished results.

Wu, T.B. , Cohen, J.B., and Yelon, W., 1982, Phonon dispersion and the thermodynamic properties of a Au-Ni alloy, *Acta Metall.* 30:2065.

Wu, T.B. , and Cohen, J.B., 1983, Clustering in a Au-Ni alloy above the miscibility gap, *Acta Metall.* 31:1929.

Wu, T.B. , and Cohen, J.B., 1984, The chemical and elastic contributions to the free energy of a Au-Ni alloy above the miscibility gap, *Acta Metall.* 32:861.

CONTINUOUS MONTE CARLO SIMULATION OF SURFACES AND OF INTERFACES OF MISMATCHED CRYSTALS

Jeff Baker[1,2] and Per-Anker Lindgård[2]

[1]Department of Physics,
Florida State University, Tallahassee, Florida 32306-3016, USA
[2]Department of Condensed Matter Physics,
Risø National Laboratory, DK-4000 Roskilde, Denmark.

INTRODUCTION

The relaxation phenomena encountered when an otherwise perfect crystal is terminated either by vacuum - *i.e.* at a surface, or by another crystal *i.e.* at an interface, is the topic of this study. The interface could be a grain boundary between two differently oriented crystals of the same material or between two different materials with different structures and/or lattice constants. In both cases there is an often incommensurate and large mismatch between the optimal atomic positions at the interface. The energy cost of this can be minimized by relaxing the atoms from the perfect positions in a region near the interface and sometimes in addition by a rotation of the adjacent crystals by a small epitaxial angle away from high symmetry relative orientation.

Epitaxis is of course Greek, $\epsilon\pi\iota\tau\alpha\xi\iota\xi$, simply meaning 'upon-arrangement'. As such, eptiaxial studies could cover simulation of interfaces and film growth as well as relaxation and reconstruction phenomena. In recent papers by Vives and Lindgård [1, 2] a continuous Monte Carlo method was developed and used to accurately describe the behavior of the ideal test example of rare gas mono layers on graphite. It is ideal from the point of view that the interatomic forces which enter in the simulations are well known and a large amount of experimental data exist for direct comparison [3]. The phenomena observed, in particular the so called defect stabilized phases which also might be occuring in binary alloys, were discussed in the last of the present series of meetings at Rhodos [4].

Here we shall discuss an extension of the method to deal with several layers of adatoms as well as of the behavior of free surfaces in the dimension perpendicular to the substrate, which is important since the relaxation may occur in complicated three dimensional patterns. We have chosen to simulate relatively large systems, in order not to predispose a superstructure by the computational setup.

The purpose is not to calculate a potential from first principles, or to study certain potentials; rather it is to choose a simple system where the forces are well known and where experimental data exists for comparison. One such system is the alkali halides. They all form the simple NaCl structure (except for the few that have the CsCl structure) but with substantially different lattice constants. Compared to the binary alloys mainly discussed at this meeting, the ionic crystals pose an extra all important question of how to treat the unscreened Coulombic forces. For the alkali halides it is on the other hand a good approximation to assume no fractional charge transfer, which again simplifies the problem. We do not expect that the details in the forces are very important for the generic aspects of the interface relaxation, so the results may be instructive also for the behavior of at least simple binary alloys without strong angular forces. In particular it would be instructive to clarify similarities and differences between ionic and metallic interfaces. Of course it should be very interesting subsequently to study the interfaces precisely between metals and ionic or covalent crystals, which a problem of current technical interest. This requires a further development of the relevant interaction potentials. A general discussion of the state of the art of various potentials is given by Vitek in these proceedings [5]

On the experimental side of these systems, KBr epitaxially grown on NaCl has been intensively studied by He-scattering [6] and a X-ray synchrotron study has been initiated. The lattice mismatch between KBr (a=6.592 Å) and NaCl (a=5.627 Å) is about 15% (very large). It is incommensurate with the nearest rational repeat distance behaving like 6:7. In spite of this large mismatch it is experimentally clear that KBr grows very well on NaCl although the first layer appears to be disordered, as judged from He-scattering. However, growth of perfect KBr surfaces is achieved after two or more layers, and remarkably the crystal is perfectly oriented according to the underlying NaCl crystal *i.e.* with no epitaxial rotation.

A problem with the reciprocal space techniques like He- or X-ray scattering is that a real space understanding of the phenomena is missing. Here is an important mission for computer simulations which can provide simultaneous information both in form of real space snapshots (and time averaged structures) and the corresponding scattering function in reciprocal space. As will be demonstrated, the interface studied is actually rumpled in two directions in a kind of waffle pattern. Thus, if this feature is generic and such distorted interfaces exist in materials which can be studied by a real space method like transmission electron microscopy (TEM), TEM would be likely to exhibit an apparent disordered interface in most directions of a projection and not the correct waffled pattern. Only diffraction techniques with wave vectors in the interface plane would clearly resolve the structure. We shall demonstrate this with the present model calculation.

INTERACTION POTENTIAL AND MONTE CARLO SIMULATION

At the interface of greatly mismatched layers, atomic displacements are likely to be outside the range of any harmonic potential approximations, such as are familiar in phonon lattice dynamical calculations. We therefore follow Gowda[7], where the potential between ions i and j with charge p_i and p_j at a distance r is very well approximated by the following phenomenological form

$$V_{ij}(r) = p_i p_j e^2 \left\{ \frac{1}{r} + \frac{(\alpha_i + \alpha_j)}{2r^4} + \frac{2\alpha_i\alpha_j}{r^7} \right\} - \frac{C_{ij}}{r^6} + A_{ij} exp(-\frac{r}{\rho_{ij}}). \quad (1)$$

In the study of KBr on NaCl we need the interaction potential between all four kinds of ions. The polarizabilities α_i, the Van der Waals coefficients C_{ij}, and the Born-Mayer repulsive parameters A_{ij} and ρ_{ij}, are from Celli[8] with small modifications in some of the C_{ij} in order to reproduce energy minima at the correct lattice spacings as consistently as possible. The table below lists the potential parameters used and the resulting lattice spacings.

Potential Parameters for Alkali Halide Molecules

	Parameters [a]			Expt. Values		Calculated	
ij	C_{ij}(eV-Å^6)[b]	A_{ij}(eV)[c]	ρ_{ij}(Å)[d]	R_{eq}(Å)[e]	$-U_{eq}$(eV)[f]	R_{eq}(Å)	$-U_{eq}$(eV)
KK	50.32	3360	0.278	-	-	-	-
NaNa	6.07	1622	0.240	-	-	-	-
KNa	17.34	2335	0.258	-	-	-	-
KBr	122.62	2878	0.365	3.296	7.068	3.295	9.970
KCl	89.70	2840	0.347	3.147	7.507	3.010	10.678
NaBr	41.11	2000	0.331	2.987	7.741	3.124	10.260
NaCl	30.16	1973	0.316	2.813	8.145	2.813	11.102
BrBr	2234*	2465	0.532	-	-	-	-
ClCl	530*	2400	0.461	-	-	-	-
BrCl	126*	2432	0.494	-	-	-	-

[a] Except where noted, from Ref. [8]
[b] From [8], Table 2, column 1.
[c] From [8], Table 4, column 2g.
[d] From [8], Table 3, column 2.
[e] Ref.[9]
[f] From Ref. [7]
*Modified to achieve correct crystalline lattice spacings for NaCl and KBr

A longstanding problem in the calculation of the structural properties of the alkali halides has been how to treat the slowly converging Coulomb interactions. Several methods have been devised which work well for perfect crystals [10], in particular the Ewald method which is a partial real and reciprocal space method[11], but also various fractional charge box methods[12]. We have devised a new simple method, which works both for perfect and distorted crystals. The idea is simply to group the ionic crystal as an assembly of formal dipoles. The distance between the members of a dipole can vary freely. An ions Coulomb energy is found by summing its interaction with neutral dipoles only. The method converges surprisingly rapidly to the exact Madelung result for the ideal NaCl structure [13]. This scheme accomplishes the charge neutrality necessary for convergence, without excess computational complexity.

The Monte Carlo calculation is a further development of the continuous method used for studying adsorbed rare gases on graphite [1, 2]. The atoms are formally recorded in a mesh, but are free to do any excursion in a small sphere from a given position. The energy of an atom is tested before and after a move, which is then accepted or rejected by the standard Metropolis rule [14], which introduces the temperature by the Boltzmann factor. The continuous method has the advantage over lattice gas models in treating the strains set up by the lattice mismatch exactly. The advantage of Monte Carlo methods, and in particular the lattice gas models, is that they are very efficient on a computer, thus allowing principally interesting *statistical* phenomena to be studied.

Because KBr and NaCl nearest neighbor distances fit nearly in a ratio of 6:7, simulations are performed with KBr films of $6n \times 6n$ on top of NaCl substrates that are $7n \times 7n$ atoms, where n is an even integer - because of the difference in the ionic charges. Excess strain caused by the inexactness of a fit to the 6:7 ratio is first minimized by using large n, and then altogether eliminated by periodic boundary conditions in the

Figure 1. Real-space view of a 24×24 atom KBr film on a 28×28 atom NaCl substrate (not shown). The larger, darker atoms are the Br^- ions. The film and the top two substrate layers were equilibrated and averaged as described in the text to give this picture. The vertical scale is exaggerated by a factor of 10.

planes parallel to the film which are slightly different for the film and the substrate. In practice, the final structure and equilibrium energy are independent of n for tried values $n = 2, 4, 6, 8, 10, 12$.

THE RUMPLING INTERFACE BETWEEN STRONGLY MISMATCHED CRYSTALS AND THE BEHAVIOR OF AN IONIC SURFACE

The real space picture of the structure that results from the relaxation of two monolayers of KBr on NaCl(001) at 200K is shown on Fig. 1, viewed down the $\langle 110 \rangle$ direction. The scale perpendicular (z) to the surface is magnified by a factor of ten. The starting configuration was a perfect 24×24 atom film on a perfect 28×28 atom substrate. A relatively small system is shown for ease of viewing in real space; however, systems up to 84×84 are used, partially to get high resolution in reciprocal space. The film and the two topmost layers of substrate are allowed to move. The configuration is equilibrated for 500 Monte Carlo cycles, then the structure is averaged over 90 snapshots taken during 4500 additional Monte Carlo cycles. For clarity, only the atoms of the film are shown, but the substrate is similarly distorted, which is expected since the forces are similar. First consider the interface layer (lowest). Along the $\langle 100 \rangle$ direction (along an edge in the picture), there is a rumpling with a periodicity of 6 KBr units corresponding to 12 atomic spacings. This gives a surface unit supercell of about 28×28 Å^2 with a width of 39 Å along $\langle 100 \rangle$. It is difficult to see the waffle pattern except edge-on, but the structure looks about the same when viewed along the $\langle 1\bar{1}0 \rangle$ direction. In spite of the time averaging done, there seems to be a considerable amount of small scale disorder. The rumpling is clearly transmitted, and in phase, to the neighboring layer. The amplitude of the rumple dies out after a few layers if more layers are put on. We remark that the interface layer of the substrate (not shown) is distorted similarly, in anti- phase with the film and with a period of 7 NaCl units. Naturally, the crystals repel each other in the regions where the the misfit would place like charged atoms closest to each other.

Turning to the surface layer (topmost), we notice a general tendency for the negative Br ions (big dark spheres) bulge upwards by ≈ 0.1Å. This feature remains even if there are several intervening layers - or when we consider a relaxed surface of a pure crystal. This has also been found in the calculations of surface relaxation for pure crystals by de Wette *et al*[15], where the Coulomb problem was solved exactly by means of an Ewald technique. This lends confidence to the reliability of our dipole method.

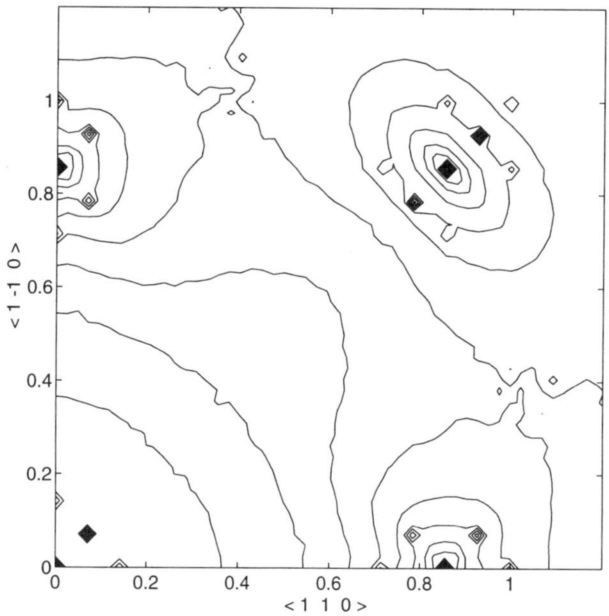

Figure 2. Logarithmic contour plot of the square of the 2-d Fourier transform of a structure similar to that of fig. 1. The axes are scaled so that the NaCl substrate Bragg peaks would fall at integer surface reciprocal lattice points (h, k), whereas the ones for KBr occur at $\sim 0.86(h, k)$. Notice the diagonal is in the $\langle 100 \rangle$ direction.

The general aspects of the distortion are more easily seen in the reciprocal space view of Fig. 2. This is a logarithmic-scale contour plot of the square of the 2-d Fourier transform of a configuration similar to that in Fig. 1, except that the system size is 42×42 atoms of NaCl ($n = 6$). Only the two film layers are included in the transform. The axes are scaled so the NaCl Bragg peaks would appear at integer surface reciprocal lattice points (h, k). KBr Bragg peaks occur at ~ 0.86 of these positions. There is clearly a large diffuse contribution indicating the above mentioned disorder, with an anisotropy indicating a weak tendency towards epitaxial rotation; see Refs. [1,2]. However, most interesting is the appearance of a multitude of satellite peaks. These peaks, which emerge after the equilibration and configurational averaging of the system, are offset from the the specular reflection and the Bragg peaks by a star with a characteristic vector $\delta\mathbf{q}$, which is 1/14 of the NaCl $(1, 1)$ surface reciprocal lattice vector (along the $\langle 100 \rangle$ direction). The peaks are arising from real-space periodicities corresponding to the superstructure seen in the real space picture.

He-scattering experiments were done in the $\langle 100 \rangle$ direction and peaks were observed which are consistent with our calculation. However, we predict that a much larger richness of features might be observable in other directions of $\mathbf{Q}$ with He- or X-ray scattering - in particular if we also include the scattering from the distorted substrate.

CONCLUDING REMARKS

Real space pictures and scattering functions are found by a continuous Monte Carlo simulation method. We have investigated the behavior of the interface between strongly mismatched ionic crystals. In less mismatched crystals one may have walls were the distortions are concentrated; in our case everything is such walls. These form a regular waffled pattern, with a periodicity which is twice that given by the difference in the lattice spacings. We think that these features are generic and also to be expected in other systems, such as binary alloys, although in those cases the effect may be smaller due to possible charge transfer. We have also investigated the surface and find that the large and negative ions have a tendency to relax outwards, which again seems to be a generic feature.

REFERENCES

[1] E. Vives and P. -A. Lindgård, *Surface Science Lett.* 284: L449 (1993).

[2] E. Vives and P. -A. Lindgård, *Phys. Rev. B* 47: 7431 (1993).

[3] K. J. Strandburg, *Rev. Mod. Physics* 60: 161 (1988).

[4] "Statics and Dynamics of Alloy Phase Transformations", eds. P. E. A. Turchi and A. Gonis, Plenum Press, New York (1994) p.683.

[5] V. Vitek, *These Proceedings.*

[6] J. Duan, G. G. Bishop, E. S. Gillman, G. Chern, S. A. Safron, and J. G. Skofronick, *Surface Science* 272: 220 (1992).

[7] B. T. Gowda and S. W. Benson, *J. Phys. Chem.* 86: 847 (1982).

[8] V. Celli and G Urzúa, *J. Phys.: Condens. Matter* 5: B91 (1993).

[9] N. W. Ashcroft and N. D. Mermin, "Solid State Physics", Saunders Company (1976).

[10] Kittel, C., "Introduction to Solid State Physics, Secon Edition", J. Wiley & Sons, Inc. London (1960).

[11] Ewald, P. P., *Ann. Physik* 64: 253 (1921).

[12] Evjen, H. M., *Phys. Rev.* 39: 675 (1932).

[13] The method converts the convergence problem from a $\frac{1}{r}$ case to a $\frac{1}{r^3}$ case. This converges certainly much faster, but still has a non-vanishing contribution from the dipoles on the overall sample surface. This makes the result sample-shape dependent in a well known way, effectively acting as external electric dipole field which may influence the general displacements of the positive and negative ions. We neglect this problem here, since it is not expected to influence the rumpeling behavior as such. The method could be improved *ad libitum* by instead considering higher order multipoles in the same way.

[14] N. Metropolis, A. Rosenbluth, M. Rosenbluth, A. Teller, E. Teller, *J. Chem. Phys.* 21: 1087 (1953).

[15] F.W. de Wette, W. Kress, and U. Schröder, *Phys. Rev. B* 32: 4143 (1985).

KINETICS AT EARLY STAGES OF PHASE SEPARATION AND ORDERING IN ALLOYS

S.V. Beiden[1], V.Yu. Dobretsov[1] G. Martin[2], F. Soisson[2] and V.G. Vaks[1]

[1]RRC "Kurchatov Institute", Moscow 123182, Russia
[2]SRMP, CEA-Saclay, 91191 Gif-sur-Yvette CEDEX, France

INTRODUCTION

Problems of structural evolution of nonequilibrium alloy states attract much attention. Theoretically, they are studied using either approximate kinetic equations[1−6] (AKE) for the local concentration or mean occupation number $c_i = c(\mathbf{r}_i, t)$ at lattice site i (say, for A-atoms in the binary A-B alloy), or direct simulation, e.g. Monte-Carlo methods[7,8]. AKE seem to be more transparent and universal than simulation methods, and many new qualitative results in this field[1−4] were obtained using AKE. However, the currently used phenomenological forms of AKE[3,4] are based on Onsager-type linear relations between time derivatives dc_i/dt and thermodynamic driving forces $\partial F/\partial c_j$ (where F is the free energy of an inhomogeneous alloy), which hold only for near-equilibrium states. Thus extrapolation of these relations to states far from equilibrium, such as those obtained by deep quenching, transient states at spinodal decomposition (SD), etc, can hardly be justified, while kinetic phenomena just in these states attract most interest[3,4,9].

Recently some of us[10,11] suggested to consistently derive AKE using the fundamental master equation for probability P to find the occupation number set $\{n_i\} = \alpha$ (where $n_i = n_i^{\mathrm{A}} = 1$ if atom A is at site i and $n_i = 0$ otherwise):

$$dP(\alpha)/dt = \sum_{\beta}[W(\alpha,\beta)P(\beta) - W(\beta,\alpha)P(\alpha)] \equiv \hat{S}P \tag{1}$$

while $W(\alpha, \beta)$ is the $\beta \to \alpha$ transition probability. Below we shortly describe this approach and its applications to some problems of kinetics of phase transformations.

Stability of Materials
Edited by A. Gonis *et al.*, Plenum Press, New York, 1996

EXACT AND APPROXIMATE KINETIC EQUATIONS

Accepting for probabilities $W(\alpha, \beta)$ in (1) the "thermally activated direct exchange model"[5], we can express the transfer matrix $\hat{S}$ via the microscopic inter-site jump probabilities $w_{kl} = \omega_{kl} \exp[-\beta(E_{kl}^s - \hat{E}_{kl}^{in})]$ where ω_{kl} is the attempt frequency, $\beta = 1/T$ is the reciprocal temperature, E_{kl}^s is the saddle point energy, while the initial (before jump) energy $\hat{E}_{kl}^{in}$ for the alloy A-B with the hamiltonian $H\{n_i^{\rm A}, n_j^{\rm B}\}$ can be written as the variational derivative: $\hat{E}_{kl}^{in} = n_k^{\rm B} \delta H/\delta n_k^{\rm B} + n_l^{\rm A} \delta H/\delta n_l^{\rm A} - n_k^{\rm B} n_l^{\rm A} \delta^2 H/\delta n_k^{\rm B} \delta n_l^{\rm A}$.

Multiplying eq. (1) by operators n_i, $n_i n_j$, etc, and summing over all configurations $\{n_k\}$ we obtain an exact system of equations[6] for mean occupations $c_i(t) = < n_i >$ and correlators $g_{i\ldots j}(t) = < n_i \ldots n_j >$. To approximately solve this system we can employ the regular approximate methods of statistical physics, such as the mean-field approximation (MFA), cluster methods, etc. In particular, using MFA (which is even quantitatively justified if the interaction range includes a sufficiently large number of sites, see[6] and refs. therein), we obtain the mean-field kinetic equation (MFKE)[10,11]

$$dc_i/dt = 2\sum_j \gamma_{ij}(c_i c_i' c_j c_j')^{1/2} \exp[(uc)_i + (uc)_j] \sinh(\partial f/\partial c_j - \partial f/\partial c_i) \quad (2)$$

(similar approach has been discussed by Gouyet[12]). Here $\gamma_{ij} = \omega_{ij} \exp(-\beta E_{ij}^s)$, $c_i' = 1 - c_i$, $(uc)_i = \frac{1}{2}\beta \sum_l U_{il} c_l$ where $U_{il} = v_{il}^{\rm AA} - v_{il}^{\rm BB}$ is the "asymmetric" interaction potential (that doesn't affect equilibrium properties but is important for kinetics[10]), and $f = \frac{1}{2}\beta F_{\rm MFA}$ where $F_{\rm MFA}\{c_l\}$ is the MFA free energy of an inhomogeneous alloy[10]. Comparing eq. (2) with the arbitrary extrapolation of linear Onsager equations to nonlinear region used by Khachaturyan et al.[3,4] we see that this extrapolation corresponds to (i) linearizing MFKE in $\partial f/\partial c_i$ (while this quantity is, generally, not small in nonlinear cases), and (ii) neglecting the c_l-dependence of the generalized kinetic coefficients - prefactors of sinh in eq. (2). The latter neglection can lead to nonuniform in time and space significant distortions of temporal evolution of nonequilibrium alloys, particularly at early stages of phase transformations when inhomogeneities in c_i are large.

APPLICATION OF MFKE TO STUDIES OF SPINODAL DECOMPOSITION AND ANTIPHASE BOUNDARY MOTION

Nonlinear effects in alloy morphology at first stages of SD. Linearizing MFKE in small deviation $x_i = c_i - c$ from the initial constant value c (after a quench of a homogeneous alloy below the spinodal curve $T_s(c)$), we obtain the linear equation of Cahn [1]. The linear Cahn's theory describes development at early stages of SD of approximately periodic structures with a characteristic wavelength $\lambda \sim \lambda_{\rm C}(c, T)$. However, the experimentally observed microstructures in most cases rather correspond to non-linear stages, and have usually not periodic, but a "tongue-like" character, see e.g. Figs. 5(a)-5(c) in[13] or Fig. 1, middle column in[14].

To study this and other problems of SD, we made two-dimensional (2D) simulations based on MFKE, for broad intervals of reduced time $t' = t/\tau$ where $\tau \sim 1/<\gamma>$ is the effective time for an intersite atomic jump. All characteristics of temporal evolution depend strongly on parameters c, $T' = T/T_c$ and U_{ij}/T_c where T_c is the critical temperature. For example, for the parameter values $c = 0.35$, $T' = 0.4$ and $U = 0$ (used in simulations illustrated by Figs. 1 and 2) the linear stage discussed by Cahn corresponds to $t' \lesssim 5 - 7$. At larger t' nonlinear interactions between Cahn's concentration waves become significant, which is illustrated in Fig. 1. The comparison of Figs. 1 and 2,a shows that the formation of tongue-like precipitates (being considered as a "label" of SD[13,14]) is just the result of coalescence of the concentration waves. At larger t' effects of strong interaction between the forming precipitates ("droplets") discussed below become important.

"Bridge" mechanism for droplet coalescence. Until recently only two coarsening mechanisms were discussed for droplets during phase separation: (i) Lifshits-Slyozov[15] evaporation-condensation mechanism for well-separated droplets, and (ii) Binder-Stauffer[7] mechanism for droplet coalescence caused by their thermal diffusion, without any interaction. Recently Tanaka[9] observed peculiar effects of inter-droplet interaction under their coalescence during SD in two-dimensional fluid mixture, which he attributed to features of droplet diffusion in liquids.

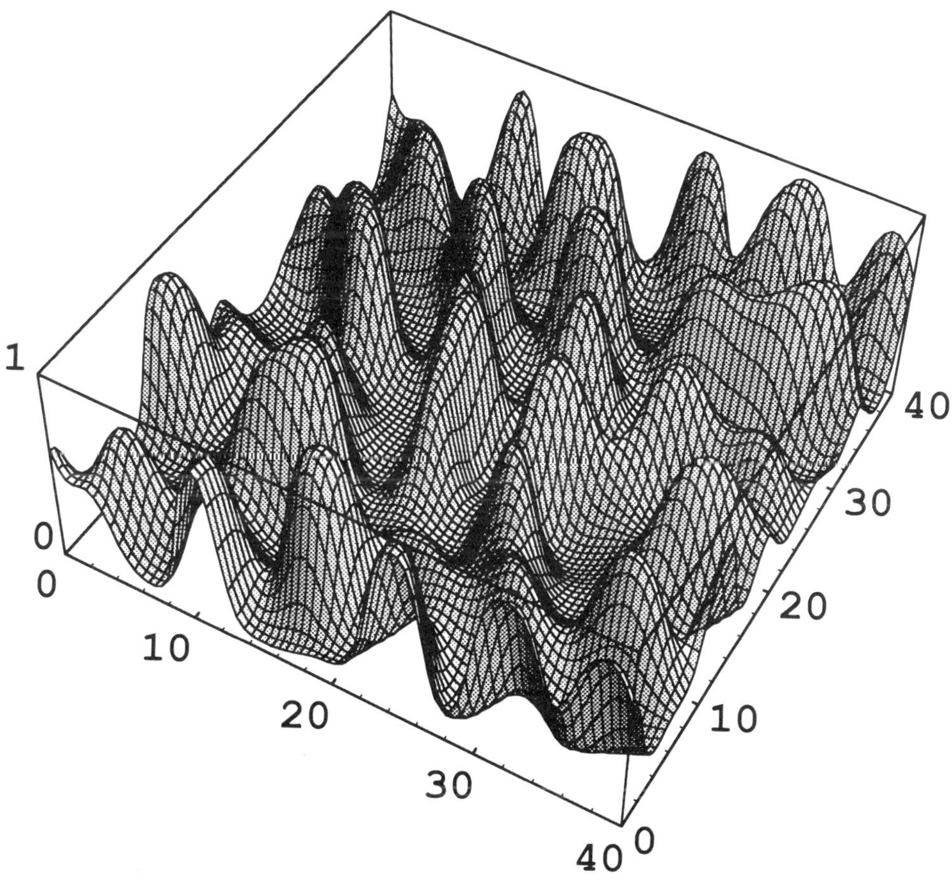

Fig. 1. Profile of concentration $c(\mathbf{r})$ at spinodal decomposition in the 2D model described in text for t'=10. Distances at horizontal axis are given in the interaction radius units[10].

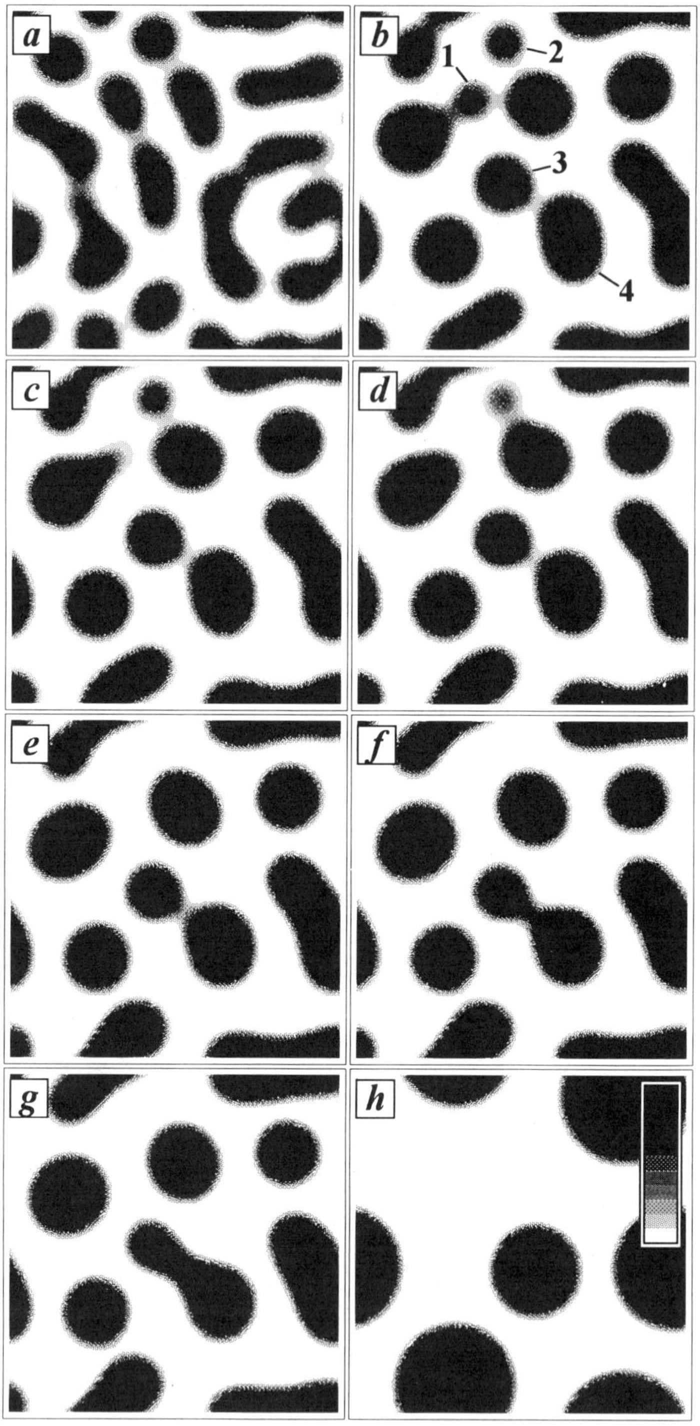

Fig. 2. Temporal evolution of concentration $c(\mathbf{r})$ for the 2D model described in text for following t': (a) 20, (b) 120, (c) 130, (d) 140, (e) 160, (f) 180, (g) 200, and (h) 5000. Insert in Fig. 2,h shows relation between darkness and c values varying between 0 and 1 from bottom to top.

In our simulations we observe similar effects for the solid alloy model, where we see no diffusion of droplets as a whole, but rather a strong dynamic coupling of diffusion fluxes that arises around droplets under certain conditions. This new mechanism of coalescence (to be called the "bridge" mechanism) dominates the first stage of coarsening of droplet pattern at intermidiate c values (e.g. at $c = 0.35$, $T' = 0.4$, $U_{ij} = 0$, for $t' = 20 - 500$), before the final, Lifshits-Slyozov stage (illustrated in Fig. 2,h) begins.

Figs. 2, $b - g$ illustrate two main versions of the bridge mechanism: (i) that for droplets 1 and 2 when a smaller droplet is consumed by its bigger neighbor(s) that pulls it over the bridge formed by diffusion fluxes, and (ii) that for droplets 3 and 4 of similar size when the "bridge" can occur for considerable time (e.g. $\Delta t' \sim 100$ for droplets 3 and 4) before its density begins to sharply rise and droplets coalesce. These effects seem to be rather similar to those shown by Tanaka[9] in his Figs. 1 and 4.

Kinetics of antiphase boundary (APB) motion. MFKE has also been applied[16] to study kinetics of APBs after a quench of a disordered alloy below the A2→B2 ordering spinodal. It was found that the strong segregation of the alloy majority component on APBs in nonstoichiometric alloys results in peculiar changes of the APB structure under its motion (being not considered in previous treatments[2,13]), and in very sharp variations of its mobility (typically, by several orders of magnitude) with the time, concentration or temperature.

This work was supported by the International Science Foundation under Grant N MQA000, and by the Russian Fund of Fundamental Research under Grant N 95-02-06426.

REFERENCES

1. J.W. Cahn, Acta Metall. **9**, 795 (1961).
2. S.M. Allen and J.W. Cahn, Acta Metall. **27**, 1085 (1979).
3. S. Semenovskaya and A.G. Khachaturyan, Phys. Rev. Lett. **67**, 2223 (1991).
4. L.-Q. Chen and A.G. Khachaturyan, Phys. Rev. B **44**, 4681 (1991).
5. G. Martin, Phys. Rev. B **41**, 2279 (1990).
6. V.G. Vaks and S.V. Beiden, ZhETF **105**, 1017 (1994).
7. K. Binder and D. Stauffer, Adv. Phys. **25**, 343 (1976).
8. J.D. Gunton, M. San Miguel and P. Sahni, in *Phase Transitions and Critical Phenomena*, ed. by C. Domb and J.H. Lebowitz (Academic, London, 1983), Vol. 8, p. 269.
9. H. Tanaka, Phys. Rev. Lett. **72**, 1702 (1994).
10. V.G. Vaks, S.V. Beiden and V.Yu. Dobretsov, Pis'ma v ZhETF **61**, 65 (1995).
11. G. Martin, Phys. Rev. B **50**, 12362 (1994).
12. J.-F. Gouyet, Europhys. Lett. **21**, 335 (1993); Phys. Rev. E, **51**, 1711 (1995).
13. S.M. Allen and J.W. Cahn, Acta Metall. **24**, 425 (1976).
14. K. Oki, H. Sagane and T. Eguchi, J. de Phys. **38**, C7-414 (1977).
15. I.M. Lifshits and V.V. Slyozov, J. Phys. Chem Solids **19**, 35 (1961).
16. V.Yu. Dobretsov, G. Martin, F. Soisson and V.G. Vaks, Europhys. Lett. (in press).

THE GAUSSIAN CLUSTER VARIATION METHOD AND ITS APPLICATION TO THE THERMODYNAMICS OF TRANSITION METALS

A. Finel,[1] and R. Tétot[2]

[1]Office National d'Etudes et de Recherches Aérospatiales (ONERA)
BP 72, 92322 Châtillon Cedex, France
[2]URA 446 CNRS, Bâtiment 415, Université de Paris-Sud, Centre d'Orsay
91405 Orsay Cedex, France

INTRODUCTION

The Cluster Variation Method (CVM) is an analytical technique that directly gives the free energy of a system. Proposed by Kikuchi in 1951, the CVM was originally designed for the statistical mechanics of the Ising model on a fixed lattice. Consequently, most of its subsequent formulations and applications concerned systems with discrete degrees of freedom.

In view of it success, some attempts have recently been made to extend the scope of the CVM to systems with continuous degrees of freedom. This is the case of the Gaussian CVM (Finel, 1994), which has been devised for the statistical mechanics of lattice site distortions, such as those due to thermal vibrations, thermal dilatation, mixture of atoms of different sizes ...

The aim of this paper is to present the first application of the Gaussian CVM. It will be focused on the thermodynamical properties of some transition metals. In particular, we compare the Gaussian CVM results to those obtained by Monte Carlo simulations, and show that the Gaussian CVM is very precise and powerful (the cpu time is several order of magnitude smaller than the one needed for numerical simulations). The extension of the Gaussian CVM to alloys is currently under way.

Another extension of the CVM to lattice site displacements has recently been proposed by Kikuchi and illustrated on a two-dimensional pure system (Kikuchi and Beldjenna, 1992). This extension is different in spirit from the Gaussian CVM and is based on a discretization of the atomic vibrations. This "lattice-gas" approach of the atomic displacements probably leads to cumdersome computations for three-dimensional systems. Another extension of the CVM to the statistical mechanics of fluids has recently been proposed by Schlijper (1994).

Stability of Materials
Edited by A. Gonis *et al.*, Plenum Press, New York, 1996

The paper is organized as follows. First, we recall the basics of the Gaussian CVM and we discuss some technical details of its implementation. Next, we present the potential used here for the transition metals. The finite temperature results of the gaussian CVM are then discussed, along with some comparison with Monte Carlo simulations. Finally, a brief conclusion is presented.

THE GAUSSIAN CLUSTER VARIATION METHOD

The CVM entropy

There are many reviews on the foundations of the CVM (see for example de Fontaine 1989, Ducastelle 1991, Inden and Pitsch 1991, Finel 1994). Hence, we just recall here the strictly necessary materials in a very condensed way. Basically, the CVM consists in approximating the entropy of an infinite system by a sum of entropies of finite clusters. More precisely, we select a maximum cluster Ω, referred to as the *basic* cluster, and we replace the exact entropy by a CVM entropy defined by :

$$S_{CVM} = \sum_{\alpha}' a_\alpha S_\alpha \tag{1}$$

where S_α is the entropy of cluster α and where the prime in the sum means that we keep only the clusters α included in at least one basic cluster Ω. Consider, as example, the triangular lattice and suppose that we select the first neighbor triangle as the basic cluster. Then, the sum in (1) is restricted to all the points, first neighbors pairs and triangles. The coefficient a_α that enter (1) are defined recursively by the relations:

$$\sum_{\alpha \supset \beta}' a_\alpha = 1 \ , \quad \forall \ \beta \tag{2}$$

where the sum runs other all the cluster α (included of course in at least one basic cluster) that contain a fixed cluster β.

This CVM formulation is common to any problem of statistical mechanics where the degrees of freedom can be referenced, in some way, to the sites of an underlying lattice. This is of course the case for distributions of atomic species over a fixed lattice. It is also the case for the thermal vibrations of atoms around their mean positions, for the static displacements of these mean positions due to thermal expansion or lattice distorsions ...

The Gaussian approximation

We decompose the actual position of a site n into its mean position $\mathbf{R}_n$ and a displacement $\mathbf{q}_n$ from this mean position. In the CVM spirit, we have to deal with density probability functions $\rho(\mathbf{q}_n,\mathbf{q}_m,...)$ of having the sites n,m ... at positions $\mathbf{q}_n,\mathbf{q}_m$ ... away from their mean positions. More precisely, the positional entropy of a cluster α is given by:

$$S_\alpha = - \int dQ_\alpha \, \rho_\alpha(Q_\alpha) \, \mathrm{Log} \, \rho_\alpha(Q_\alpha)$$

where $Q_\alpha = (\mathbf{q}_n, n \in \alpha)$ is a vector that collects the $n_\alpha = 3|\alpha|$ cartesian coordinates of the $|\alpha|$ sites of cluster α, and where $\rho_\alpha(Q_\alpha)dQ_\alpha$ is the probability for cluster α of having its sites in the volume dQ_α around the point Q_α.

The Gaussian CVM consists in approximating the density probability function ρ_Ω of a maximum cluster by a multivariate Gaussian distribution:

$$\rho_\Omega (Q_\Omega) = (2\pi)^{-n\Omega/2} (\det V_\Omega)^{-1/2} \exp(-\frac{1}{2} Q^t_\Omega . V^{-1}_\Omega . Q^t_\Omega) \qquad (3)$$

where V_Ω is the symmetric covariance matrix of the coordinates and ($\det V_\Omega$) its determinant. If $q_{m\mu}$ and q_{nv} are two components of the coordinate vector Q_Ω, their averages is equal to zero, by definition, and their covariance is given by the appropriate entry of the matrix V_Ω :

$$< q_{m\mu}\, q_{nv} > = \int dQ_\alpha\, q_{m\mu}\, q_{nv}\, \rho_\Omega (Q_\Omega) = (V_\Omega)_{m\mu\ nv}$$

In particular, the mean square displacements are given by the diagonal of V_Ω.

Now, a staightforward calculation shows that eq.(3) leads to a very simple form for the entropy of a subcluster α of Ω (Finel, 1994):

$$S_\alpha /k = \frac{1}{2} \text{Log} (\det V_\alpha) + \frac{n_\alpha}{2} (1 + \text{Log}\, 2\pi) \qquad (4)$$

where V_α is the covariance matrix of the sites of cluster α. In other words, V_α is obtained from the basic matrix V_Ω by keeping only the n_α lines and columns that correspond to the sites of α

An important point here is that the entries of the basic covariance matrix are the only free parameters of the theory. Besides, they are independant and there are only a few of them, even for relatively large basic clusters. Consider for example an homogeneous pure system (this will be the case below). All the basic clusters are equivalent (except in particular cases) and there is only one basic covariance matrix V_Ω. Then, the number of independant covariances is $3|\Omega|*(3|\Omega|+1)/2$. In the fcc lattice, this leads to 78 variational parameters if we select the first-neighbor tetrahedron as basic cluster. If we take into account the full space group symmetry, we are left with only six independant covariances.

Next, we consider the potential energy. As the natural parameters of the Gaussian approach are the covariances, the internal energy E should be written as an explicit function of them. This is done by expanding E to some powers of the displacements $\mathbf{q}_n$ around the (anticipated) mean positions $\mathbf{R}_n$:

$$E(\{\mathbf{R}_n + \mathbf{q}_n\}) = E_o (\{\mathbf{R}_n\}) + \frac{1}{2} \sum_{(n\mu),(m,v)} D^{\mu v}_{nm}(\{\mathbf{R}_n\})\, q_{n\mu} q_{mv} + \ldots. \qquad (5)$$

For an homogeneous system, there is no first order term, because the first derivatives of E vanishe identically. The quasi-harmonic (QH) approximation consists in retaining only the second order term in (5). In that case, the enthalpy U=<E>, defined as the thermodynamical average of E over the displacements $\mathbf{q}_n$, is a simple linear function of the covariances. Higher order terms in (5) can also be retained and the enthalpy will then depend also on the next moments of the displacements distribution function. But, within the gaussian approximation, these higher order moments are trivial functions of the second order ones (as example, $<q_{n\mu}^4> = 3*<q_{n\mu}^2>^2$) and, consequently, U is still a simple explicit function of the covariances.

Finally, the free energy functional F=U-TS must be minimized with respect to the covariances. This is easily done, because the first derivatives of F are easy to compute. Indeed, if $<q_i q_j>$ is a covariance between two cartesian coordinates that belong to a cluster α, the derivative of the entropy S_α with respect to $<q_i q_j>$ is given by:

$$\frac{\partial S_\alpha}{\partial <q_i q_j>} = \frac{1}{2} [(V_\alpha^{-1})^t]_{ij}$$

where $(V_\alpha^{-1})^t$ is the transpose of the inverse matrix of the covariance matrix V_α.

THE TIGHT-BINDING MODEL AND THE SECOND MOMENT APPROXIMATION

There are some basic objections against the use of pair potentials for metallic systems. It is now widely recognized that the cohesive effect of the conducting electrons requires the use of many-body potentials. Different approach have been proposed to correctly take into account the essential band character of the metallic bond. A relatively simple scheme, without resorting to the complex treatment of first-principles calculations, is the tight-binding (TB) method within the second moment approximation (SMA) (Ducastelle, 1970). According to the TB-SMA scheme, the contribution to the band energy of an atom n is given by:

$$E_n^b = -\xi \left[\sum_m \exp -2q\left(\frac{r_{nm}}{r_0} - 1\right) \right]^{1/2} \tag{6}$$

where r_{nm} is the distance between atoms n and m, and r_0 is the first-neighbor distance. The quantity ξ is an effective hopping integral and the parameter q describes its distance dependence. In order to ensure the stability of the system, we add, as usual, a repulsive ion-ion interaction of the Born-Mayer type:

$$E_n^r = A \sum_m \exp -p\left(\frac{r_{nm}}{r_0} - 1\right) \tag{7}$$

The total cohesive energy E_c of the system is then written as:

$$E_c = \sum_n \left(E_n^b + E_n^r\right) \tag{8}$$

The parameters A, ξ, p and q are then fitted to reproduce some experimental quantities *at zero temperature* (cohesive energy E_c, lattice parameter *a*, bulk modulus and elastic constants). In the present work, the summations over the index *m* in eqs. (6) and (7) are either limited to the Z first neighbors only, and in that case we use the parameters A, ξ, p and q determined by Rosato et al. (1989), or extended up to the fifth neighbors, and in that case we use the parameters of Cleri and Rosato (1993).

Table 1. Parameters A, ξ, p and q of the TB-SMA potentials for Cu, Ni and Ag. The range is limited to the first neighbors in case (1) (Rosato et al., 1989) and to the fifth neighbors in case (2) (Cleri and Rosato, 1993). We give also the experimental values of the cohesive energy E_c and of the lattice parameter (after Kittel, 1972).

	A (eV)	ξ (eV)	p	q	E_c (eV)	a (Å)
Cu (1)	0.0993	1.3543	10.08	2.56	- 3.50	3.61
Ni (1)	0.1368	1.7558	10.00	2.70	- 4.44	3.52
Ag (1)	0.1231	1.2811	10.12	3.37	- 2.96	4.09
Ag (2)	0.1028	1.1780	10.928	3.139	- 2.96	4.09

RESULTS

As explained in Sec.2, we use a quasi-harmonic expansion of the internal energy. We recall that the use of the Gaussian CVM is not restricted to this second order expansion.

The only requirement is that the internal energy must be written as an explicit function of the covariances between the coordinates of the sites that belong to the same basic cluster. The QH approximation leads to an internal energy that depends linearly on the covariances; any expansion of higher order will introduce extra terms which depend on some power of these covariances, but this will not change, in any way, the general implementation of the Gaussian CVM.

Within the QH (or any higher order) approximation of the internal energy, the atoms are supposed to vibrate around the sites of a mean lattice, whose parameter a is yet unknown. For a given temperature, the CVM free energy is then minimized with respect to the covariances, while a is kept fixed. As a result, we get a free energy minimum that depends on a. As example, we show in Fig.1 the free energy of Cu as a function of the lattice parameter, for T=800 K. We would like to mention that the computation of this curve requires only a few seconds on a very usual workstation. The true equilibrium lattice parameter, for the chosen temperature, is the one that minimizes this free energy curve.

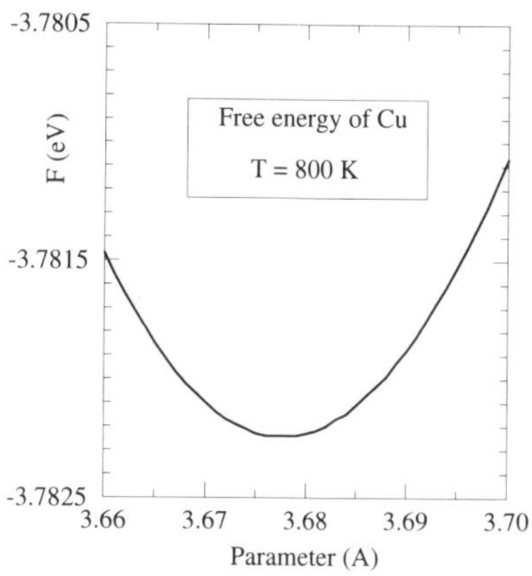

Figure 1. Free energy of Cu as a function of the lattice parameter, for T=800K, within the Gaussian CVM (tetrahedron approximation).

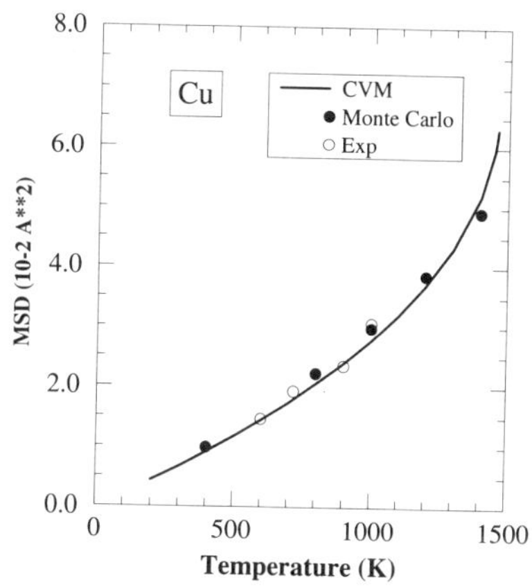

Figure 2. Mean square displacement (MSD) of Cu as a function of temperature. Comparison between the Gaussian CVM results (full line), the Monte Carlo simulations (full circles) and the experimental data (empty squares, Martin and O'Connor, 1977). The potential is limited here to the first neighbors.

We can then proceed to a complete study as a function of temperature and calculate various quantities (specific heat, thermal expansion coefficient ...). We can also get an estimate for the melting temperature T_m. More precisely, an upper bound T_m^{up} of T_m is given by the temperature at which the second derivative of the free energy with respect to the lattice parameter vanishes. Our result concerning the specific heat (at zero pressure) and the upper bound T_m^{up} are presented in Table 2. The results for Cu and Ni with the first neighbor potential compare very favorably with the experimental ones. But this is not the case for Ag, where a good agreement is obtained only with the fifth neighbor potential (we comment on that point latter on).

Another very stringent test for the Gaussian CVM is its ability to reproduce mean-square displacements (MSD). But first, before comparing our theoretical results to the experimental ones, we would like to test the precision of the Gaussian CVM against Monte Carlo simulations.

Table 2. Specific heat C_p (T=800K) and upper bound T_m^{up} of the melting temperature. (1): potential limited to the first neighbors, (2): potential extended up to the fifth neighbors (see text and Table 1). The theoretical values T_m^{up} are compared to the experimental melting temperature. Experimental data in parentheses are taken from the compilation by Barin (1993).

	Cu (1)	Ni (1)	Ag (1)	Ag (2)
C_p (kJ mol^{-1} K^{-1})	29.4 (28.3)	28.1 (31.1)	38.1 (28.3)	32 (28.3)
T_m^{up} (K)	1460 (1356)	1760 (1728)	865 (1234)	1130 (1234)

We present in Fig.2 the MSD for Cu, as a function of temperature, according to the Gaussian CVM (within the tetrahedron approximation and using the QH approximation for the cohesive energy) and according to Monte Carlo simulations (with of course the full cohesive energy). The Monte Carlo simulations have been performed on a lattice of $4*14^3$ sites, using a Metropolis algorithm and, after a preliminary equilibration, the time averages of various quantities are calculated over 4000 steps [for more details, see Tetot and Finel, *ibid*]. We see in Fig.2 that the agreement between the Gaussian CVM and the Monte Carlo simulations is excellent, over the whole range of temperature.

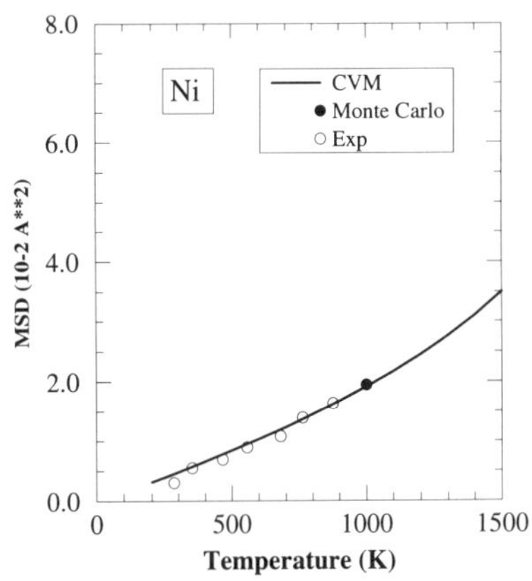

Figure 3. Mean square displacement (MSD) of Ni as a function of temperature. Comparison between the Gaussian CVM results (full line) and the experimental data (empty squares; after Simerska, 1962). A Monte Carlo point is also reported (full circle). The TB-SMA potential is limited here to the first neighbors.

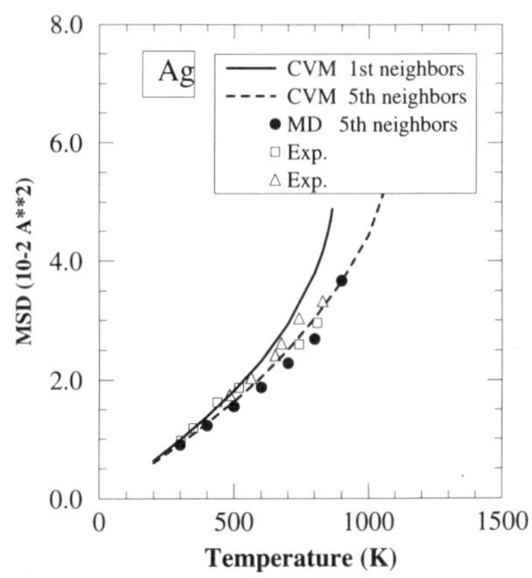

Figure 4. Mean square displacement (MSD) of Ag as a function of temperature. Comparison between the 1st neighbor potential within the Gaussian CVM (full line), the 5th neighbor potential (CVM results: dotted line, Molecular Dynamic results: full circles) and the experimental data (empty squares: Simerska 1961, empty diamond: Haworth 1960)

Our calculted MSD's for Cu and Ni are compared to the experimental ones in Fig.2 and 3. The agreement is very good. This means that the anharmonicity in the thermodynamic behaviour of Cu and Ni are correctly reproduced by the TB potentials used here, i.e. limited to the first neighbors in eqs. (6) and (7).

This is not the case for Ag, as shown in Fig.4: the MSD's calulated with the short range potential are too high. Hence, this first neighbor potential contains an excess of anharmonicity and should lead to anticipated metlting. This is the reason why the upper bound of the melting temperature, calculated with the same potential, is well below the melting temperature (see Table 1). Of course, anharmonic effects can be reduced by a potential of longer range. This is shown in Fig.4, where we display the MSD's calculated with the fifth neighbor potentials proposed by Cleri and Rosato (1993) (see also Table 1). The agreement with the experimental data is now very good. Concurrently, the calculated specific heat is close to the experimental one and the upper bound T_m^{up} comparable to the melting temperature (see Table 1). We compare also in Fig.4 our CVM results with the ones obtained by Molecular Dynamic simulations (Cleri and Rosato, 1995), using the same fifth neighbor potential. We note that the agreement between the two different techniques is again very good.

CONCLUSION

We have used the Gaussian CVM to calculate the thermodynamical behaviour of some transition metals. Our aim was to show the the Gaussian CVM is a very precise and powerful technic. We have shown that its results are very close to those obtained by "exact" methods (Monte Carlo and Molecular Dynamics simulations), with the advantage of being much less time consuming. We have also explained that the use of the Gaussian CVM is not restricted to the QH approximation. Its extension to alloys is currently under way.

REFERENCES

Barin, I., 1993, "Thermochemical Data of pure substances"

Cleri, F., and Rosato, V., 1993, Phys. Rev. B, **48**:22

de Fontaine, D., 1989, in "Alloy Phase Stability", eds. G.M. Stocks and A. Gonis (Kluwer Academic Publishers, Dordrecht) 177

Ducastelle, F., 1970, J. Phys., Paris, **31**:1055

Ducastelle, F., 1991,"Order and Phase Stability in Alloys", in "Cohesion and Structure", Vol.3, F.R. de Boer and D.G. Pettifor, eds (North-Holland, Amsterdam).

Finel, A., 1994, in "Statics and Dynamics of Alloy Phase Transformations", eds. P.E.A. Turchi and A. Gonis (Plenum Press, New York) 495

Finel, A., 1994, Prog. Theor. Phys., **115**:59

Haworth, C.W., 1960, Philos. Mag., **5**:1229

Inden, G., and Pitsch, W., 1991, in "Materials Science and Technology", eds. R.W. Cahn, P. Haasen, E.J. Kramer (VCH, Weinheim) **5**:497.

Kikuchi, R., 1951, Phys. Rev. , **81**:988.

Kikuchi, R., and Beldjenna, A., 1992, Physica A, **182**:617.`

Kittel, C., 1972, in "Introduction à la Physique de l'Etat Solide", eds Dunod Université.

Martin, C.J., and O'Connor, D.A., 1977, J. Phys. C, **10**:3521

Rosato, V., Guillope, M., and Legrand, B., 1989, Philos. Mag. A, **59**:321.

Schlijper, 1994, Prog. Theor. Phys., **15**:69

Simerska, M., 1961, Acta Crystallogr., **14**:1259

Simerska, M., 1962, Czech. J. Phys. B, **12**:858

Tétot, R., and Finel, A., *ibid.*

KINETIC PATH FOR DISORDER-$L1_2$ TRANSITION STUDIED BY THE PPM

Tetsuo Mohri, Yoichi Ichikawa and Tomoo Suzuki

Division of Materials Science and Engineering,
Graduate School of Engineering, Hokkaido University
Kita-13 Nishi-8, Kita-ku, Sapporo 060, Japan

INTRODUCTION

The authors' group has been studying ordering kinetics by the Path Probability Method (hereafter PPM) [1] of statistical mechanics. In those works [2, 3, 4, 5, 6], the main focus has been placed on the $disorder - L1_0$ transition. Various intriguing phenomena experimentally observed, however, are centred around $L1_2$ ordered phase as is indicated in the recent report [7] of phase separation with ordering phenomenon observed in Al-Li system. A clue to understand such a phenomenon is believed to be interplays between configurational kinetics and thermodynamic instabilities located in the $disorder + L1_2$ two-phase field. It is indispensable to employ reliable thermodynamic and kinetic models in order to clarify essential features behind the phenomenon.

The PPM is recognized as the natural extension of the Cluster Variation Method (hereafter CVM) [8] to the time domain, and, therefore, the PPM inherits various fruitful features from the CVM. Among them is the fact that not only the long range order but also the short range order kinetics are explicitly investigated. Also, noteworthy is that the derived quantities in the long time limit correctly converge to the equilibrium ones independently obtained by the CVM. Hence, the combination of the CVM and PPM is an invaluable tool to explore configurational thermodynamics and kinetics starting from non-equilibrium towards final equilibrium states.

The investigation of the kinetic behavior in the two-phase field by the PPM [5], however, is not trivial. This is because the synthesis of two kinetic behavior derived separately for each phase into a single framework leaves uncertainties. Hence, in the present study, our main focus is placed on the single $L1_2$ ordering reaction of a spin system, which is believed to be a precursor to the further elaborate study of the transition kinetics in the two-phase field of an alloy system. In this brief article, the abstract of the relevant items in our recent comprehensive article [6] is reported.

PHASE EQUILIBRIA AND ORDERING INSTABILITY

In the CVM, the atomic correlation of a given system is approximated by a selected set of finite clusters, which is most efficiently described by correlation functions [9, 10] forming an independent set of configurational variables. The maximum cluster explicitly considered in the free energy functional F specifies the level of the approximation. The

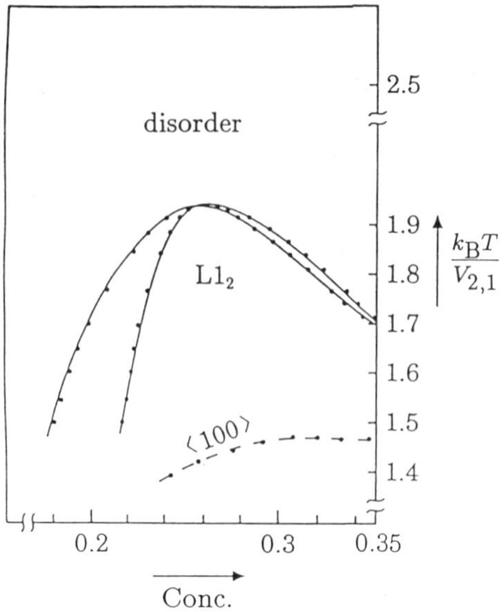

Figure 1. Disorder-$L1_2$ phase diagram in the vicinity of 1:3 stoichiometry. The broken line indicates the $\langle 100 \rangle$ Spinodal ordering locus.

common practice for an fcc-based system is the tetrahedron approximation in which nearest neighbor tetrahedron cluster is taken as the maximum cluster. Hence, within the tetrahedron approximation, the free energy of disordered and $L1_2$ ordered phases are symbolically expressed as

$$F_{dis} = F(T, \xi_1, \xi_2, \xi_3, \xi_4), \tag{1}$$

and

$$F_{L1_2} = F(T, \xi_1^\alpha, \xi_1^\beta, \xi_2^{\alpha\beta}, \xi_2^{\beta\beta}, \xi_3^{\alpha\beta\beta}, \xi_3^{\beta\beta\beta}, \xi_4^{\alpha\alpha\beta\beta}), \tag{2}$$

where T is the temperature, ξ_i the correlation function for an nearest neighbor i-point cluster and α and β are the sublattices introduced due to the breakdown of the symmetry associated with the ordering transition. Minimization of the free energies followed by the common tangent construction (or equivalent methods) determines the phase equilibria as is shown in Fig. 1.

The second order derivative of the free energy with respect to the deviation of the correlation functions from their equilibrium values, $\Delta\xi_i$, provides information of stability of a disordered phase.

$$F''_{dis} = \frac{\partial^2 F_{dis}}{\partial \Delta\xi_i \cdot \partial \Delta\xi_j}. \tag{3}$$

The vanishing condition of the above equation is regarded as the stability criterion of a homogeneous solid solution against an excitation and an amplification of a concentration wave which is most efficiently described by the Fourier transformation of $\Delta\xi_i$. The broken line in Fig. 1 is the stability locus, which is termed *Spinodal Ordering* (hereafter *SP*) [11], for $\langle 100 \rangle$ concentration wave. It is expected that the ordering reaction takes place by *Nucleation – Growth* mechanism (hereafter *NG*) above the locus, whereas the $\langle 100 \rangle$ concentration wave destabilizes the homogenous solid solution and leads to spontaneous ordering reaction below the *SP* locus. We place the main focus of the present study on the kinetic consideration of the two modes of ordering reactions, *NG* and *SP*, in the $L1_2$ phase field.

THEORETICAL BACKGROUND OF THE PPM

In the PPM, the state variables (cluster probabilities or correlation functions of the CVM) and their time derivatives are not explicitly dealt, which is in marked difference with other kinetic theories. The central quantities of the PPM are the *path variables* which describe transition among state variables. The essential feature of the PPM can be well grasped by comparing with the CVM in the following way.

The counter part of the grand potential of the CVM is the *Path Probability Function* (hereafter *PPF*) $P(t; t + \Delta t)$ which is an explicit function of time and is defined as the product of three factors P_1, P_2 and P_3. The logarithmic expression of each term relevant to $L1_2$ ordered phase is as follows.

$$\begin{aligned}\ln P_1 &= \frac{1}{4}\left(X^{\alpha}_{1,2} + X^{\alpha}_{2,1} + 3X^{\beta}_{1,2} + 3X^{\beta}_{2,1}\right)\ln(\theta\Delta t) \\ &+\frac{1}{4}\left(X^{\alpha}_{1,1} + X^{\alpha}_{2,2} + X^{\beta}_{1,1} + X^{\beta}_{2,2}\right)\ln(1-\theta\Delta t)\ ,\end{aligned} \tag{4}$$

$$\ln P_2 = -\frac{\Delta E}{2k_{\mathrm{B}}T}\ , \tag{5}$$

and

$$\begin{aligned}\ln P_3 &= -\frac{5}{4}\left[\sum_{i,j}\left\{L\left(X^{\alpha}_{i,j}\right) + 3L\left(X^{\beta}_{i,j}\right)\right\} - \sum_{i}\left\{L\left(x^{\alpha}_i(t)\right) + 3L\left(x^{\beta}_i(t)\right)\right\}\right] \\ &+3\left[\sum_{ij,kl}\left\{L\left(Y^{\alpha\beta}_{ij,kl}\right) + L\left(Y^{\beta\beta}_{ij,kl}\right)\right\} - \sum_{ij}\left\{L\left(y^{\alpha\beta}_{ij}(t)\right) + L\left(y^{\beta\beta}_{ij}(t)\right)\right\}\right] \\ &-2\left[\sum_{ijkl,mnop} L\left(W_{ijkl,mnop}\right) - \sum_{ijkl} L\left(w_{ijkl}(t)\right)\right]\ ,\end{aligned} \tag{6}$$

where α and β denote sublattices, $L(z)$ represents $z\ln z - z$, θ is the spin flip probability per unit time, $X^{\delta}_{i,j}$, $Y^{\delta\delta'}_{ij,kl}$ and $W_{ijkl,mnop}$ are the path variables for the flipping from one spin configuration to another designated by the subscript(s) (up spin or down spin) before (at time t) and after (at time $t + \Delta t$) the comma sign, on a point, pair and tetrahedron clusters over the specified sublattice(s), respectively, $x^{\delta}_i(t)$, $y^{\delta\delta'(t)}_{ij}$ and $w_{ijkl}(t)$ are state variables (cluster probabilities) describing the probabilities of finding a spin configuration designated by subscripts on a point, pair and tetrahedron clusters, respectively, over the specified sublattice(s) and ΔE is the change of the internal energy before and after the flipping events. The subscripts $i, j, k, \ldots\ldots, o, p$ take either 1 or -1 depending upon up or down spins, respectively. Within the nearest neighbor pair interaction model, the internal energy $E(t)$ of the $L1_2$ ordered phase can be readily derived in the following way,

$$E(t) = \frac{1}{4}\omega\sum_{i,j} e_{ij}\left(y^{\alpha\beta}_{ij}(t) + y^{\beta\beta}_{ij}(t)\right) + \frac{1}{4}\sum_{i}\left(\mu^{\alpha}_i x^{\alpha}_i(t) + \mu^{\beta}_i x^{\beta}_i(t)\right), \tag{7}$$

where ω is the coordination number, e_{ij} the nearest neighbor pair interaction energy between i and j, and μ^{δ}_i is the chemical potential of i-species. It is noted that the effective nearest neighbor pair interaction energy $v_{2,1}$ is related to the pair interaction energies e_{ij} through

$$v_{2,1} = \frac{1}{2}\{(e_{ii} + e_{jj}) - 2e_{ij}\} \tag{8}$$

by which the temperature axis of Fig. 1 is normalized. The first factor P_1 describes the statistical average of non-correlated spin flip events over entire lattice points, and the second factor P_2 is the conventional thermal activation factor. Hence, the product of P_1 and P_2 corresponds to the Boltzmann factor in the free energy and gives the probability that one of the paths specified by a set of path variables occurs. The third factor P_3 characterizes the PPM. One may see the similarity with the configurational entropy term of the CVM, which gives the multiplicity, i.e. the number of equivalent states. In a similar sense, P_3 can be viewed as the number of equivalent paths, i.e. the degrees of freedom of the microscopic evolution from one state to another.

The *path variables* of the PPM correspond to the cluster probabilities of the CVM by which the grand potential is minimized to obtain the most probable state. Likewise, the *PPF* is maximized with respect to the *path variables* for each time step, which yields the optimized set of *path variables*. Since a set of *path variables*, $\Xi_{\psi,\phi}(t;t+\Delta t)$, relates cluster probabilities $\chi_\psi(t)$ at time t and $\chi_\phi(t+\Delta t)$ at time $t+\Delta t$ through

$$\chi_\phi(t+\Delta t) = \chi_\psi(t) + \sum \Xi_{\psi,\phi}(t:t+\Delta t), \tag{9}$$

where ψ and ϕ represent clusters, one can pursue the time evolution of cluster probabilities χ (equivalently correlation functions) with a given initial set of cluster probabilities, $\chi(t=0)$.

RESULTS AND DISCUSSIONS

As was mentioned in the previous section, the criterion of SP is given as the vanishing condition of the second order derivative of the free energy, eq.(3). The second order derivative, on the other hand, describes the curvature of the free energy hypersurface in the configurational space. In the following, we first examine the free energy contour surface to clarify the thermodynamic implication of the NG and SP.

Shown in Figs. 2a and 2b are the free energy contour surfaces at 1:3 stoichiometric composition in the vicinity of the disordered phase (vanishing long range order parameter) for T=1.60 above the *Spinodal Ordering* temperature, T_s=1.4 at this stoichiometry (see Fig. 1), and T =1.20 below T_s, respectively. Note that the long range order parameter, in this study, is defined as $\xi_1^\alpha - \xi_1^\beta$, the difference of the point correlation functions of the two sublattices. The entire surface is also demonstrated above in each figure. These are derived by minimizing the free energy, eq.(2), under a given set of temperature T, long range order parameter, and the tetrahedron correlation function $\xi_4^{\alpha\alpha\beta\beta}$ (a short range order parameter) with respect to the rest of the correlation functions. One clearly recognizes that a saddle point exists for T=1.60 whereas the free energy monotonously behaves for T=1.20. It is such a nature of the appearance and disappearance of the saddle point configuration that characterizes the NG as an activated process and the SP as a spontaneous process.

Although the above discussions provide clear thermodynamic implications for the two modes of the ordering reaction, kinetic behavior during the transition process are not thoroughly grasped. In fact, the thermodynamic argument exclusively claims the steepest descent kinetic path, which is not the case for, in particular, $far-from-equilibrium$ transition as is discussed below.

It is recalled that the spin system generally does not conserve a composition. Hence, it is recommended to work with the grand canonical ensemble with a fixed chemical potential. In the present study, however, a slightly different attempt is made. The initial equilibrium state of the disordered phase corresponding to the 1:3 stoichiometric composition is realized at T=3.0 by assigning the chemical potential of μ=-8.4725. Then, right

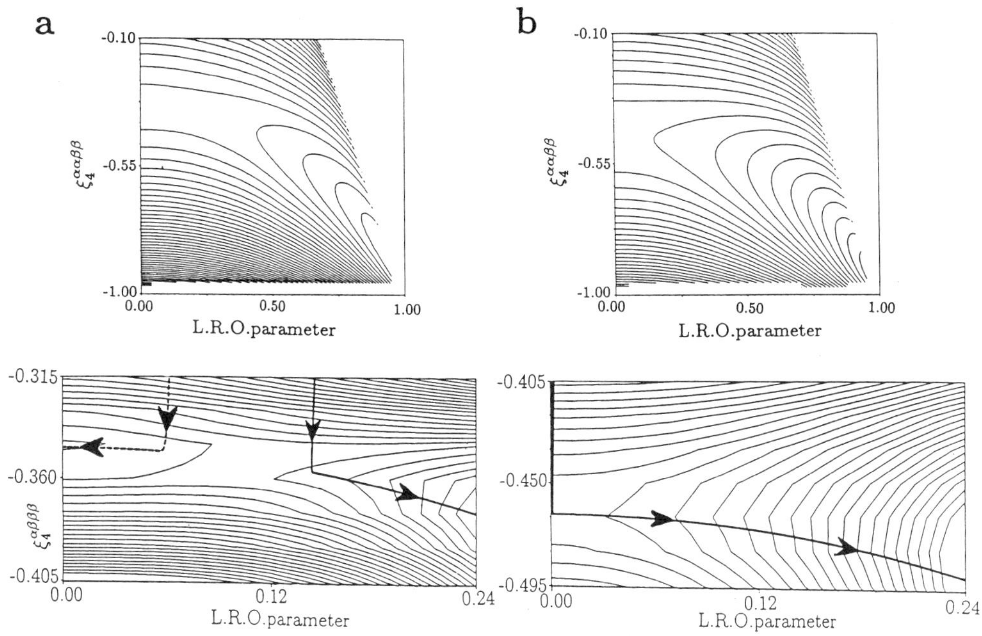

Figure 2. Free energy contour surface in the vicinity of disordered state (null value of the Long Range Order parameter; $\xi_1^\alpha - \xi_1^\beta = 0$) and kinetic path for T=1.60 (Fig. 2a) and 1.2 (Fig. 2b). The solid line in each figure indicates the kinetic path evolving towards the $L1_2$ ordered phase and the dotted line in the Fig. 2a is the one devolving towards disordered phase. The free energy contour surface for the entire configurational space is shown above for each temperature.

after quenching to a aging temperature, the chemical potential is switched to the one which yields the $L1_2$ ordered phase of 1:3 stoichiometry in the equilibrium state at that temperature. Those chemical potentials found by the CVM are μ=-8.4090 for T=1.60 and μ=-8.3263 for T=1.20, respectively.

Among various results obtained, demonstrated in Figs. 3a and 3b are the time evolution of the concentration of A species (*up* or *down* spin) on α and β sublattices at T=1.60 and 1.20, respectively. Note that the time axis is normalized by the spin flip probability. The final steady state value of the concentration is confirmed to be the equilibrium one independently obtained by the CVM for each case. The initial separation of the concentration shown in Fig. 3a is due to the perturbation of the chemical potential imposed to trigger the transition. It is realized that with sufficient amount of perturbation the ordering transition is processed (solid line) while the system devolves towards the initial disordered phase for the small perturbation (dotted line). These results clearly indicates the existence of the critical amount of fluctuation necessary to trigger the transition at T=1.6 above the *Spinodal Ordering* temperature, T_s, while the spontaneous ordering reaction triggered by an infinitesimal fluctuation is suggested in Fig. 3b at T=1.20 below the T_s. These are the kinetic evidence of the NG and SP clarified by the present PPM study.

The corresponding paths are indicated in Figs. 2a and 2b. It is understood, in Fig. 2a, that the imposed chemical potential perturbs the long range order parameter and that the critical fluctuation corresponds to the one necessary to overcome the saddle point configuration. Moreover, in both Figs. 2a and 2b, the path evolving towards the ordered phase which is drawn by a solid line deviates from the steepest descent direction. This can

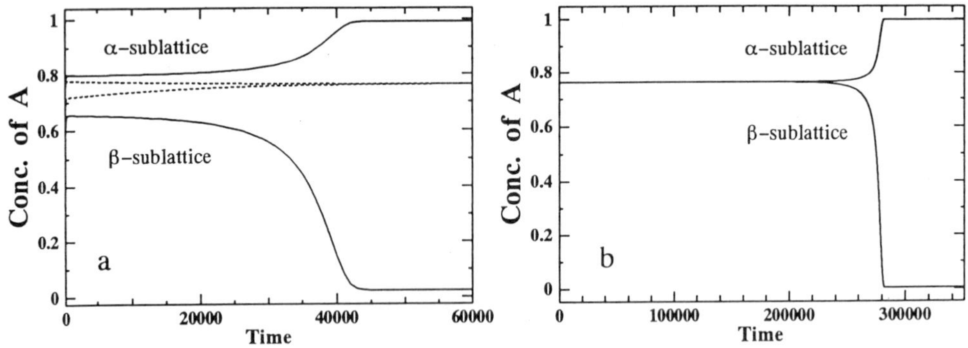

Figure 3. Time evolution of the concentration of A atom (*up* spin) on $\alpha(x_A^\alpha)$ and $\beta(x_A^\beta)$ sublattices at 25% when the system is quenched from T=3.0 to T=1.6 (Fig. 3a) and to T=1.2 (Fig. 3b) with the spin flip probability θ=0.001. The solid and broken lines in Fig. 3a indicate the ones with and without sufficient initial fluctuation, respectively [6].

be attributed to the kinetic effects due to the spin flipping events.

In this brief article, we report preliminary results of the PPM calculations for *disorder*–$L1_2$ ordering transition. More detailed study including the fluctuation analysis [5] and two-phase decomposition process of *disorder* + $L1_2$ will be reported in the future publication.

REFERENCES

1. R. Kikuchi, Prog. Theoret. Phys. Kyoto Suppl. **35**,1 (1966).
2. T. Mohri, Statistics and Dynamics of Alloy Phase Transformations, ed. by P. E. A .Turchi and A. Gonis, Plenum, New York,668 (1994).
3. T. Mohri and T. Ikegami, TMS EMPMD Monograph Series **3**, Diffusion in Ordered Alloys and Intermetallic Compounds ed. by B. Fultz et al. TMS Publication, Warrendale, PA,90 (1993).
4. T. Mohri, Acta metall. **38**,2460 (1990).
5. Tetsuo Mohri, Solid-Solid Phase Transformations ed. by W. C Johnson, J. M. Howe, D. E. Laughlin and W. A. Soffa, The Minerals, Metals and Materials Society,74 (1994), and references therein.
6. T. Mohri, Y. Ichikawa, T. Nakahara and T. Suzuki, Proc. of the International Workshop on the Theory and Applications of the Cluster Variation and Path Probability Methods, to be published.
7. M.-S. Yu and H. Chen, Kinetics of Ordering Transformations in Metals ed. by H. Chen and V. Vasudevan, TMS Publication, Warrendale, PA, 336 (1992).
8. R. Kikuchi, Phys. Rev. **81**,998 (1951).
9. J. M. Sanchez and D. de Fontaine, Phys. Rev. **B17**,2926 (1978).
10. J. M. Sanchez, F. Ducastelle and D. Gratias, Physica **128A**,334 (1984).
11. D. de Fontaine, Acta metall. **23**,553 (1975).

INVESTIGATION OF H-H(D-D) AND H(D)-N INTERACTIONS IN Nb AND Ta BY MEANS OF COMPUTER SIMULATION OF DIFFUSION

M. S. Blanter and A. V. Vasiljev

Moscow State Academy of Instrumental Engineering and Information Science,Stromynka 20,Moscow,107846,Russia

INTRODUCTION

The interaction of solute atoms in metals has been the subject of numerous experimental studies because such information is indispensable for understanding many basic physical processes such as short-range order, segregation, ordering, diffusion, etc. The interaction energies are necessary for calculations of phase equilibria and phase diagrams,and mechanical and physical properties of solid solutions.

The theory of solute interactions is yet imperfectly developed and it is so far impossible to obtain these energies for many alloys. For interstitial solid solutions in b.c.c. metals,the elastic interaction is essential [1].The theory of strain-induced (elastic) interaction in the framework of the model of a discrete crystal lattice has been developed [1] and the energies of pair interactions of H, D, O, N, C in b.c.c. metals (V, Nb, Ta, Cr, Mo, W and α-Fe) evaluated [2-6]. The interaction is strong and long-range. However, calculations of atomic structure and internal friction showed [1,7,8] that it is necessary to supplement the long-range elastic interaction by a short-range repulsion of interstitials. In the case of H-H interaction, the existence of this repulsion is evident from analysis of hydride structures [9].

A possible source of the repulsion is a screened Coulomb interaction of charged interstitials [3]. Rough estimations of energies of this interaction were carried out for H-H pairs in Nb [3] but for further progress it is necessary to find methods for investigation of the interaction. Previously we investigated the distance of H-H(D-D) and H(D)-O(N) repulsive interaction in Nb and Ta by means of calculation of Snoek-type hydrogen internal friction peak [7].It was shown that H-H(D-D) repulsion extends to at least the third coordination shell.In the case of H(D)-O(N) interaction the repulsion extends up to the third shell.

In the case of internal friction measurement "diffusion under stress" has two peculiarities:H or D atoms diffuse at a short distance of the order of a lattice spacing and the internal friction maxima appear at low temperatures at which H atoms diffuse probably by tunnelling [10].However there is extensive experimental data on long-range diffusion of H and D in Nb and Ta observed by Gorsky effect [11,12].According to the data the diffusion rate depends on concentration of H or D and nitrogen impurities.Taking into account that the concentration dependence of diffusion coefficients is a result of interstitial interaction in this paper the tracer diffusion coefficients of H(D) atoms and the activation energy of diffusion were calculated taking into account energies of long-range strain-induced interaction and varying the radius of the repulsion. The purpose was:a) to determine the radius of the repulsion; b) to verify the applicability of the applied model of solute interaction; c)to refine the diffusion mechanism.

INTERACTION ENERGIES

As mentioned above, the model of H-H(D-D) and H(D)-N interactions is a long-range elastic interaction supplemented by a repulsive interaction in the nearest coordination shells due to probably a screened Coulomb interaction [7]. The energies of strain-induced pair interaction of interstitials calculated within the model of a discrete crystal lattice were taken from previous papers [3-5] and corrected in [7] by taking into account new more precise values of the coefficients of expansion of the crystal lattice per unit H,D or N concentration.The description of the calculation method can be found in Khachaturyan's book [1].

The full list of these energies was displayed in Tables 1 and 2 of Ref. [7]. Positive values of the interaction energy $W(\mathbf{r})$ mean repulsion ($\mathbf{r}$ is a vector separated two interstitials). This interaction is strong, long-range and anisotropic. The elastic interaction is strongly attractive at the nearest distances up to the third coordination shell.In the case of H(D)-N interaction there are two types of tetrahedral intersticies in the third shell near octahedral interstices with different interaction energies (3-a and 3-b).Their disposition is shown in Fig.1 of Ref. [7].

The repulsion was taken into account by suppression of elastic attraction in the nearest coordination shells by putting the positive values $W(\mathbf{r})$ (=1 eV) or by lowering of the attractive elastic energies in the shells blocked partially. We used different radii of the repulsion and calculated energies of H-H(D-D) and H(D)-N elastic interaction for calculation of activation energies of H(D) diffusion. The radii of the repulsion were obtained by comparing cases of coincidence of the calculated activation energies with the experimental ones. When the repulsion radii were obtained, energies of elastic interaction outside these radii could be used for calculation of structure and properties of solid solutions.

CALCULATION METHOD

The long-range tracer diffusion coefficients D of H or D atoms have been determined by Monte-Carlo simulation on the base of random walk theory [13]. It was suggested that the interatomic interaction affects diffusion by changing the arrangement (a short-range order) and the energy of H or D atoms in tetrahedral interstices and therefore the activation energy of diffusion. H or D atoms were randomly placed into tetrahedral interstices of a $12 \cdot 12 \cdot 12\ a_0^3$ (a_0 is the lattice parameter of the host lattice) b. c. c. crystal with periodic boundary conditions. In some cases fixed nitrogen atoms were also randomly placed in octahedral interstices. A mobile H or D interstitial and one of its four neighbouring tetrahedral interstices were randomly chosen at a constant temperature. When the chosen interstice was empty, the activation energy H_p of the corresponding jump of p-th interstitial was calculated: $H_p = (H_D - \Delta E_p)$ where H_D is equal to the activation energy of diffusion of an isolated interstitial atom(at very low concentration and therefore without interaction with other interstitials).ΔE_p is an energy of interaction of the p-th atom located in tetrahedral interstice with other interstitials, fixed and mobile. For the atomic configuration

$$\Delta E_p = \sum_j W(\mathbf{r_p} - \mathbf{r_j}) C(\mathbf{r_j}) + \sum_m W(\mathbf{r_p} - \mathbf{r_m}) \tag{1}$$

where $\mathbf{r_p}$ and $\mathbf{r_j}$ are positions of tetrahedral interstices, and $\mathbf{r_m}$ are positions of fixed nitrogen atoms. $W(\mathbf{r_p} - \mathbf{r_j})$ are pairwise (D-D) or (H-H) interaction energies, $W(\mathbf{r_p} - \mathbf{r_m})$ are pairwise H(D)-N interaction energies. $C(\mathbf{r_j})$ are the occupation numbers for tetrahedral intersticies. $C(\mathbf{r_j}) = 1$ if the interstice is occupied, $C(\mathbf{r_j}) = 0$ otherwise. We neglected the change of energy of the saddle point of the diffusion barrier.

The crystal lattice is considered to act as a heat bath for the migrating H and D atoms.Then the energy of an interstitial fluctuates around its average thermal energy $k_B T$ (k_B is the Boltzman constant) and the probability $f(E)$ of having energy E is given by $f(E) = exp(-E/k_B T)$. Thus by choosing a random number X with $0 < X < 1$ we can provide an energy for the mobile interstitial via the following equation: $X = exp(-E/k_B T)$. If energy $(E + k_B T)$ is larger than the energy barrier H_p a jump is possible.The jump will take place only if the chosen tetrahedral interstice is free.In this way site saturation,blocking of sites and correlation effects in diffusion are taken into account.

For calculation of diffusion coefficients one traces the displacement of each p-th H(D) atom from their initial position $\mathbf{R_{0p}}$ to the final position $\mathbf{R_p}$.If an interstitial was jumping across the borders of the model crystal,periodic boundary conditions were applied to bring it back to the crystal.Thus diffusion takes place in an infinite lattice with a repetition cell.

An attempt of an interstitial to jump is called a Monte Carlo step independent of whether it has been successful or not and the growing number of Monte Carlo steps is a measure of the elapsed time t.The repeating loops were interrupted when a definite number of jumps (usually $1000N$,where N is a number of H(D) atoms in the model crystal) had occurred.A diffusion coefficient $D_p^{(1)}$ was calculated for each p-th mobile interstitial according to random walk theory

$$D_p^{(1)} = \gamma(\mathbf{R_p} - \mathbf{R_{0p}})^2/t \tag{2}$$

The factor γ has been neglected in the following (=1) because units of length and time are arbitrary in Monte Carlo calculations. By neglecting the dimensionality the prefactor for the diffusion coefficient D_0 is unity.

The tracer diffusion coefficient from this first run ($j = 1$) was calculated as an average over all H(D) atoms

$$D^{(j=1)} = \frac{1}{N}\sum_{p=1}^{N} D_p^{(j=1)} \tag{3}$$

The whole procedure was repeated with the H(D) atoms starting from the final position of the previous run j until the average over all runs k remained constant.

$$D = \sum_{j=1}^{k} D^{(j)} t^{(j)}/t \tag{4}$$

The jump rate V as the ratio of the total number of jumps and the total number of attempts was also calculated.From temperature dependence of D and V: $D = D_0 exp(-\Delta H_{TD}/k_B T)$ and $V = V_0 exp(-\Delta H_V/k_B T)$ the activation energy of tracer diffusion $\Delta H_{TD}^{(cal)}$ and jump rate $\Delta H_V^{(cal)}$ were determined. The values $\Delta H_{TD}^{(cal)}$ are equal to $\Delta H_V^{(cal)}$ within the accuracy of the computer simulation (± 0.004 eV) but values $\Delta H_V^{(cal)}$ were determined more precisely.Both of these activation energies were used for comparison with experimental values of the activation energy of H or D tracer diffusion $\Delta H_{TD}^{(exp)}$.

According to [13] we used the next values H_D: 0.106; 0.127; 0.136 and 0.153 eV for H and D in Nb and H and D in Ta respectively.As known there are two different temperature ranges of hydrogen diffusion:at the temperatures less than 250 K we have probably the tunneling mechanism with a low activation energy while at temperatures higher than 250 K there is a high activation energy.We have to note that above mentioned values of activation energies of hydrogen diffusion in Nb and Ta do match the high temperature range while our simulation of hydrogen internal friction [7] was performed for the low temperature range.

RESULTS AND DISCUSSION

We know the experimental data on the tracer diffusion of H and D in Nb and Ta over wide concentration range [11] but we have used the alloys with minimum concentration less than 10 % (Table 1).For such low concentration it is possible to neglect the concentration dependence of the H_D value.

In all cases a consideration of interstitial interaction results in the increasing of the activation energy of diffusion $\Delta H_{TD}^{(cal)}$ due to attractive interaction and the decreasing of a tracer diffusion coefficient.The effect is stronger as the attractive elastic energy is bigger.While the repulsion is not taken into account the temperature dependence of a diffusion coefficient diverges from the exponential one.In the case of the repulsion in the first coordination shell in the alloy Nb-4.8 % (D/Nb)(Fig.1/1) the dependence is exponential but the calculated $\Delta H_{TD}^{(cal)}$ value (=0.8 eV) is much larger than the experimental $\Delta H_{TD}^{(exp)}$ value (=0.140 $\pm$ 0.007 eV).With an

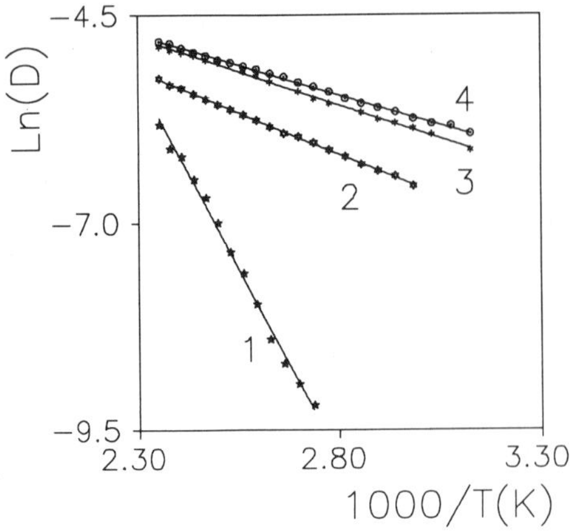

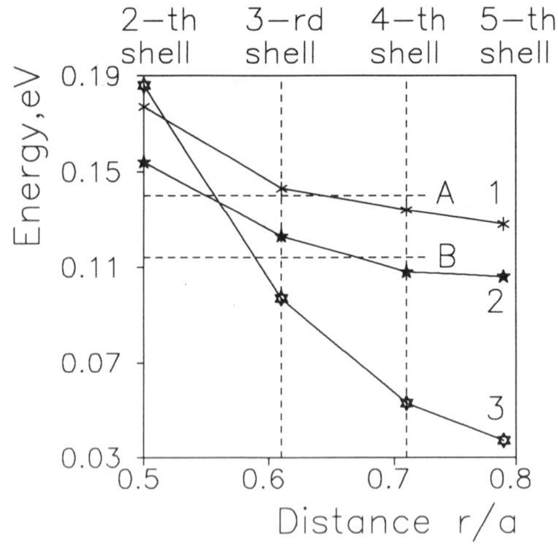

Figure 1.(left)Temperature dependence of diffusion coefficients D of deuterium in the alloy Nb-4.8%(D/Nb) calculated with different distance of repulsion:1-the first shell; 2-two first shells;3-three first shells;4-without D-D interaction.

Figure 2.(right)Dependence of the calculated activation energy $\Delta H_{TD}^{(cal)}$ of H(D) diffusion from repulsion distance in Nb:1-4.8%(D/Nb);2-7.0%(H/Nb).3-the H-H(D-D) strain-induced interaction energy $[-W(\mathbf{r})]$;A and B-the experimental activation energy $\Delta H_{TD}^{(exp)}$ of D and H respectively.

increase of the repulsion distance, $\Delta H_{TD}^{(cal)}$ decreases (Fig.2) and is in coincidence with $\Delta H_{TD}^{(exp)}$ in the case of blocking of the three first shells (Table 1.).By doing so the diffusion coefficients of D atoms in Nb are lower than ones in Nb with very low D concentration (Fig.1/3,4).

Such dependence of $\Delta H_{TD}^{(cal)}$ on the repulsion distance is observed for other alloys (Table 1.) but there is an isotopic effect. That means that the repulsion distance is more for H-H interaction than for D-D interaction.In the case of D the distance is not more than the third shell whereas for H the repulsion extends to the fourth shell.The isotopic effect for hydrogen in Nb and Ta -based solid solutions is well known and was discussed(see for example [9]).However it was first observed for the case of the H-H(D-D) repulsion in the present investigation.Such effect is a demonstration that dynamic processes influences on H-H interaction.

One can see in Table 1 (the alloys with 5.6 and 9.6 % (H/Ta)) that over the investigated area of concentration the H concentration does not have a considerable influence on the repulsion distance.

The results observed are in a good agreement with our previous results observed by computer simulation of internal friction [7] which showed that the repulsion expands not less than up to the third shell.In this paper we have realized a more concrete result:the third (D) or fourth (H) shell.

The strain-induced interaction energies of H-H pairs are equal to D-D ones in Nb and are close in Ta [7].But the larger distance of repulsion for H-H pairs results in the less residual elastic

Table 1. Results of investigation of H-H and D-D interaction.

Alloy	Temperature range, K	$\Delta H_{TD}^{(exp)}$, eV [11]	The last blocked shell	Full or partial blocking	Energy of max. elastic attraction after blocking,eV
Nb-7.0%(H/Nb)	375–475	0.114	4	Partial	-0.037(5th shell
Nb-4.8%(D/Nb)	325–425	0.140	3	Full	-0.053(4th shell)
Ta-5.6%(H/Ta)	320–425	0.138	4	Partial	-0.036(5th shell)
Ta-9.6%(H/Ta)	300–400	0.146	4	Partial	-0.036(5th shell)
Ta-3.8%(D/Ta)	320–420	0.169	3	Partial	-0.043(3th shell

attraction energies (which are the strain-induced energies of atoms located out of the repulsion range ,Table 1) for the H-H interaction than D-D one.That means that the temperature of long-range ordering of hydrides are bound to be lower than ones of deuterides.Not much difference was observed for NbH(D), TaH(D), Ta$H_{0.8}(D_{0.8})$, Ta$H_{0.5}(D_{0.5})$ [9,14].Previously this difference was explained by an isotopic change of vibration energy at long-range ordering [14].But this effect can explain only a small part of the experimentally observed differences of the ordering temperatures.The isotopic effect of the static interaction observed in the present paper can give the remaining contribution to this difference.A posible explanation of the isotopic effect consists in a decrease of the electron screening of the Coulomb interaction due to the higher frequency of the hydrogen vibration.

Experimental data for the diffusion of H(D) in Nb and Ta with N impurities were displayed in Ref.[13] only in graphical form.Therefore the $\Delta H_{TD}^{(exp)}$ values in Table 2 were observed by our treatment of these data.

Table 2. Results of investigation of H-N and D-N interaction.

Alloy (at.%)	Temperature range, K	$\Delta H_{TD}^{(exp)}$, eV [12]	The last blocked shell	Full or partial blocking	Energy of max. elastic attraction after blocking,eV
Nb-0.6H-0.44N	250–330	0.113	3b	Partial	-0.06(3b-th shell)
Nb-0.4H-0.8N	250–330	0.119	3b	Partial	-0.06(3b-th shell)
Nb-0.44D-0.26N	210–400	0.139	3b	Partial	-0.10(3b-th shell)
Nb-0.44D-0.44N	210–400	0.154	3b	Partial	-0.10(3b-th shell)
Ta-0.3D-0.9N	210–300	0.160	3b	Partial	-0.02(3b-th shell)
Ta-0.38D-0.41N	210–300	0.160	3b	Partial	-0.05(3b-th shell)
Ta-0.8D-0.41N	210–300	0.160	3b	Partial	-0.05(3b-th shell)
Ta-1D-0.41N	210–300	0.160	3b	Partial	-0.05(3b-th shell)

For simulation of diffusion in Nb(Ta)-H(D)-N alloys it is necessary to use the H-H(D-D) and H(D)-N pairwise interaction energies.We used the H-H(D-D) energies with the repulsion obtained in present paper for the binary alloys Nb(Ta)-H(D) and described above.In the case of alloys with nitrogen only the distance of the H(D)-N repulsion was changed.

Computer simulation showed that the mobile H(D) atoms stay near immobile nitrogen atoms (the short-range order parameters were varied between -10 and -100).The activation energy of H(D) diffusion are very sensitive to the interatomic interaction.Without H(D)-N repulsion the temperature dependence of the H(D) diffusion coefficient diverges from the exponential one.It is necessary to block at least two nearest coordination shells to have a normal exponential dependence.In all investigated alloys $\Delta H_{TD}^{(cal)}$ are coincident with $\Delta H_{TD}^{(exp)}$ for the full repulsion in the first,second and third-a shells and the partial one in the third-b shell (Table 2).

The above-mentioned isotopic effect of interstitial interaction in the case of H(D)-N interaction appears as the different degree of decreasing of the energy of elastic attraction in the third-b shell:by two times for H and only by 1.12 times for D in Nb.

The observed distance of the repulsion (the third-b shell) is close to one observed by computer simulation of internal friction [11] in spite of the difference in temperature range:"high" temperatures without tunneling for diffusion simulation and "low" temperatures with tunneling of H for internal friction simulation.

This computer simulation of long-range diffusion and the previous simulation of short-range diffusion (internal friction) of H and D atoms [7] demonstrate that the diffusion can be described in the wide temperature range on the basis of the model of the long-range strain-induced (elastic) interaction supplemented by a repulsive interaction in the nearest shells (up to the third or fourth shells).

CONCLUSIONS

1.The diffusion coefficients of H and D in b.c.c. metals are very sensitive to interstitial interaction energies.

2.The model of long-range strain-induced (elastic) interaction of interstitials , supplemented by a repulsive interaction in the nearest coordination shells,are useful for a description of the solid solutions.

3.The repulsive interaction of interstitial atoms in b. c. c. metals extends up to the third or fourth coordination shell and an isotopic effect appears.

REFERENCES

1.A. G. Khachaturyan, *Theory of structural transformations in solids*, Willey, New York (1983).

2.M. S. Blanter and A. G. Khachaturyan,Metall.Trans.**9A**: 753,(1978).

3.V. G. Vaks, N. E. Zein, V. Zinenko and A. G. Orlov , Zh. Eksp. Teor. Fiz. **87**: 2030 (1984) [English translation: Sov. Phys. JETP].

4.A. I. Shirley, C. K. Hall and N. J. Prince, Acta. Metall. **31**: 985 (1983).

5.H. Horner and H. Wagner, J. Phys. C **7**: 3305 (1974).

6.M. S. Blanter, Fiz. Met. Metalloved. **60**: 50 (1985). [English translation: Phys. Met. Metallogr.]

7.M. S. Blanter,Phys.Rev.B,**50**:3606 (1994).

8.M. S. Blanter and M. Ya. Fradkov,Acta Met.et Mat.**40**:220 (1992).

9.V. A. Somenkov and S. S. Schilschtein, Progress in Material Science, **24**: No. 3/4, 267 (1979).

10.J. Volkl and G. Alefeld, in *Hydrogen in Metals*, **1**, Springer-Verlag, New York (1978).

11.H. C. Bauer,J. Volkl,J. Tretkowski and G. Alefeld,Z.Physik B,**B29**:17 (1978).

12.Zh. Qi,J. Volkl and H. Wipf,Scr.Met.**16**:859 (1982).

13.R.Kirchheim,Acta Met.,**35**:271 (1987).

14.E.Entin,V.A.Somenkov and S.Sh.Shilshtein,Fix.Tverd.Tela **16**: 2412 (1974) [English translation:Sov.Solid State Phys.].

A MOLECULAR DYNAMICS SIMULATION OF VIBRATIONAL PROPERTIES AND DIFFUSION OF COPPER ADATOMS ON A COPPER (001) SURFACE

N.I. Papanicolaou, G.C. Kallinteris, and G.A. Evangelakis

Department of Physics, University of Ioannina, P.O. Box 1186, GR-45110 Ioannina, Greece

INTRODUCTION

In recent years, there has been considerable interest in the dynamical behaviour and diffusion of adatoms on metal surfaces. The presence of adatoms on surfaces modifies drastically most of the surface dynamical properties.[1,2] In addition, the diffusion of adatoms plays an important role for the understanding of homoepitaxial growth, heterogeneous catalysis, etc.[3,4]

In this communication we focus our attention to the Cu(001) surface containing Cu adatoms, for which we studied the vibrational properties and the diffusion mechanisms. The main reasons that motivated this work can be summarized as follows.

One of the major tools for the study of the vibrational properties of a surface, is the phonon density of states (DOS) and the phonon spectral density (SD). While there has been a lot of work concerning the clean surface of Cu(001),[5,6,7] studies referring to the surface with adatoms are very scarce.[8,2] On the other hand, concerning the self-diffusion on a Cu(001) surface, the experimental methods such as helium-atom beam scattering [9,10] and low-energy ion scattering[11] do not distinguish which diffusion mechanism is dominant (hopping or exchange). Theoretical calculations by means of Embedded-Atom-Method (EAM) and closely related schemes give contradictory results. In ref. 12 the hopping mechanism is found dominant. However, when the kinetic-exchange-correlation energy is included, the exchange mechanism is found to be favorable.[13,14] Recently, ab initio calculations[3] based on the Car-Parrinello molecular dynamics method, found the hopping mechanism to be favorable over the exchange.

We report here results of the phonon SD of Cu adatoms on a Cu(001) surface, obtained by molecular dynamics (MD) simulations and n-body potential[15,16] adapted to copper. We found that the presence of adatoms on the surface is manifested mainly by new dispersionless vibrational modes. In addition, the mean-square-displacements (normal to the surface) of the adatoms are equal to those of the surface atoms up to 700 K, while for higher temperatures they become much larger. Furthermore, we have calculated the concentration of the spontaneously created adatom-advacancy pairs (defects), which exhibit Arrhenious behaviour and we deduced the formation energy of this process; this value is found to be in very good agreement with static calculations.

Finally we present simulations of the self-diffusion of adatoms on a Cu(001) surface. Detailed trajectory analysis revealed a lot of diffusion mechanisms, the more frequent of which were counted and further analysed. It comes out that the hopping mechanism is energetically favored over the exchange and that the more complicated exchange type mechanisms (more than one surface atom involved) require almost the same migration energy that the simple one. These complicated exchange mechanisms have to be taken into account in the calculation of the total diffusion coefficient, when the high temperature region is concerned.

COMPUTATIONAL DETAILS

The dynamical behaviour of Cu(001) surface has been studied using standard MD in the microcanonical ensemble and a system made of 4500 particles arranged on a fcc lattice. Periodic boundary conditions are imposed in the three space directions parallel to the $[1\bar{1}0]$, [110] and [001] crystallographic axes. By fixing the dimensions of the computational box at a value twice larger than the thickness along the z-direction, a pseudoinfinite slab is produced delimited by two free surfaces parallel to the (001) planes. The model is made up of 20 atomic layers parallel to the free surfaces, containing respectively 225 atoms. The atomic interactions derive from a n-body potential, [15,16] which reproduces successfully bulk and surface properties of copper.[16,17] The parameters of the potential have been fitted to the cohesive energy, equilibrium equation and the elastic constants of copper, including interactions up to 5th neighbor. The newtonians equations of motion are integrated by means of the Verlet algorithm and the chosen time step, 10^{-14} s, ensures the conservation of the total energy within $\delta E/E \sim 10^{-5}$. To increase statistical accuracy, we used 10 independent trajectories of 10 ps, produced from different initial conditions.

The vibrational properties of the surface with adatoms were studied for a coverage of 4% on each face of the slab, which excludes possible interactions between the adsorbates. The phonon SD for a given polarization and a specific vector of surface Brillouin zone was calculated by Fourier transforming the velocity dependent autocorrelation function.[18] The temperature of the system was 300 K.

The mean-square-displacements of surface atoms and adatoms normal to (001) plane have been determined on a layer by layer basis at various temperatures from equilibrium averages of the atomic density profiles along the z-direction.

In order to study the diffusion of a single Cu adatom, we used a system made of 2560 particles (40 atomic layers parallel to the free surfaces, each containing 64 atoms). One adatom was placed on each face of the slab and the time step was $2\text{x}10^{-15}$ s. Simulations were performed in the range 700-1100 K. For the low temperatures, the simulation length was about 1 ns, in order to ensure a satisfactory number of diffusion events. The jump frequency of each mechanism was computed from the number of diffusion events of the adatoms, while the diffusion coefficient associated with each process was deduced by using the standard expressions of the transition rate theory.

RESULTS AND DISCUSSION

Vibrational Properties

In figure 1a we present the phonon SD of the clean surface of Cu(001) along the [100] direction at the point $\bar{M}$ for the longitudinal polarization at 300 K. We observe a pronounced peak at 5.05 THz, corresponding to the L_1 mode, in excellent agreement with ab initio calculations[5] and MD-EAM simulations.[6]

In figure 1b we show the phonon SD of the Cu adatoms deduced under the same conditions as in fig. 1a. The structure of this spectrum is completely different from that appearing in fig. 1a. A main peak emerges at about 2 THz, suggesting weaker interactions from those existing between surface atoms. It should be noted here that this peak shows a dispersionless behaviour.

Figure 1c shows a zoom of fig. 1a around 2 THz for the clean surface phonon SD (solid line), along with for the surface containing 4% adatoms (dashed line). We note that the whole shape of the SD for the surface with adatoms has roughly the same structure as that for the clean surface for all frequencies, except for a new peak appearing at about 2 THz, due to the presence of adatoms. A similar additional peak appears at the same frequency at the point $\bar{X}$ along the [110] direction for the longitudinal polarization. Thus, the presence of adatoms is manifested by the appearance of new vibrational modes in the phonon SD's.

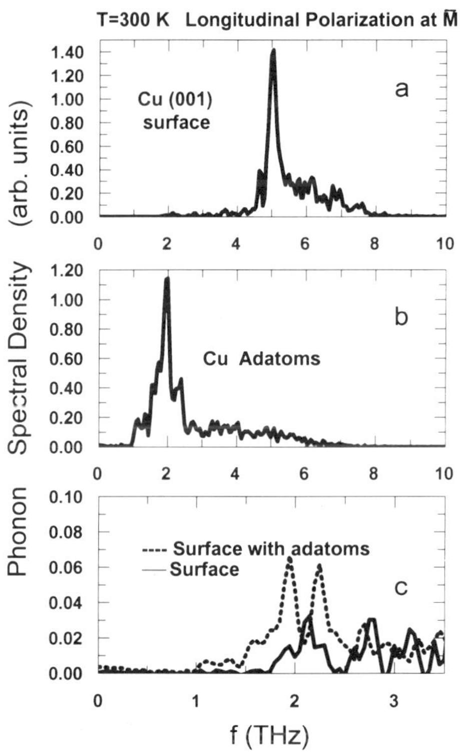

Figure 1. Spectral density of phonon states at 300 K at $\bar{M}$. a) clean surface of Cu (001). b) 4% Cu adatoms on Cu (001) face. c) Zoom around 2 THz.; solid line: Cu (001) clean surface, dashed line: Cu (001) surface with 4% Cu adatoms.

In figure 2 we present the computed mean-square-displacements normal to the (001) surface for the surface atoms and adatoms at various temperatures. In the same figure we show also for comparison the experimental values for the surface atoms, derived from thermal helium beam scattering measurements.[19] For temperatures below 700 K the vibrational amplitudes of the adatoms are comparable to those of the surface, while for higher temperatures the adatoms present much larger displacements, consistent with the appearence of various diffusion mechanisms.

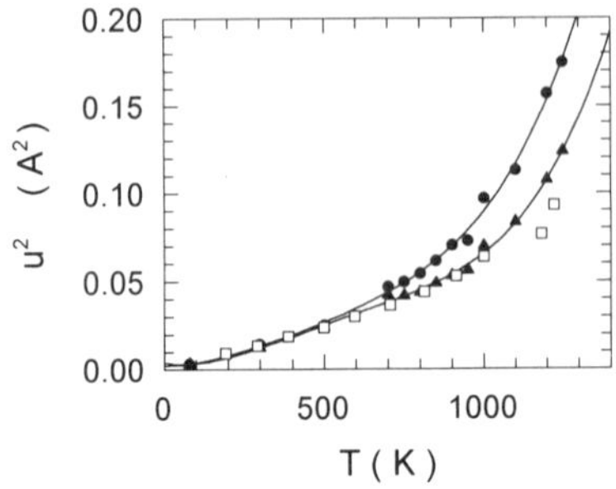

Figure 2. Mean-square-displacements normal to the Cu (001) surface; filled triangles : surface atoms, open squares : experimental data[19]. filled circles : adatoms.

Furthermore, we have computed the concentration of the thermally created defects. It is known that at high temperatures in the step-free MD model there is a spontaneous creation of adatom-advacancy pairs on the surface. In figure 3 we show the concentration of adatoms on Cu(001) as a function of the temperature. From an Arrhenius least squares fit to the simulated data we obtain the formation energy of defect pairs to be E=1.28 eV, in good agreement with our static evaluation for a dissociated adatom-advacancy pair (E'=1.16 eV). The later has been computed from static energy minimization and was found to be the sum of the adatom formation energy (0.62 eV) and advacancy formation (0.54 eV) both created from a kink.

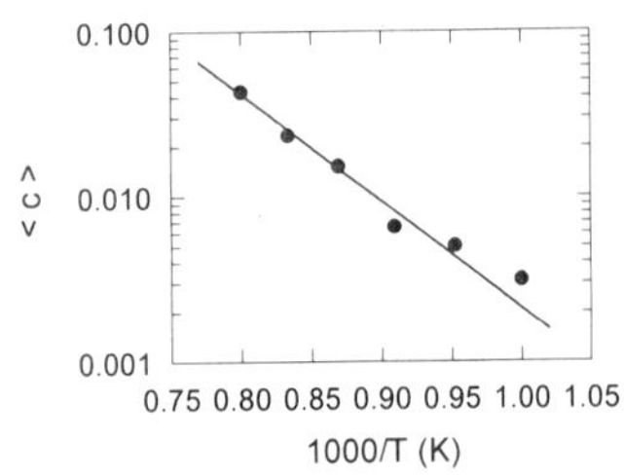

Figure 3. Concentration of thermally created adatoms on Cu (001) as a function of temperature.

Self-diffusion of Cu adatoms

We have also studied the self-diffusion mechanisms of an adatom on Cu(001) surface. In figure 4 we present the more frequent diffusion process, the hopping or bridge mechanism, according to which an adatom jumps over the saddle point of the potential barrier between two adjacent sites and migrates to the next four-fold site (T=750K). We have also detected the double hopping and the diagonal hopping mechanism, acting along the diagonal of two binding sites. The later is shown in figure 5 at T=950 K.

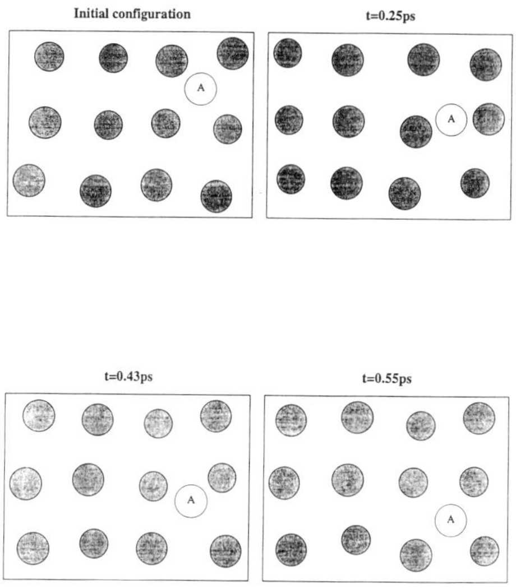

Figure 4. Sequence of snapshots showing the the simple hopping mechanism at 750 K; the adatom is labeled by A.

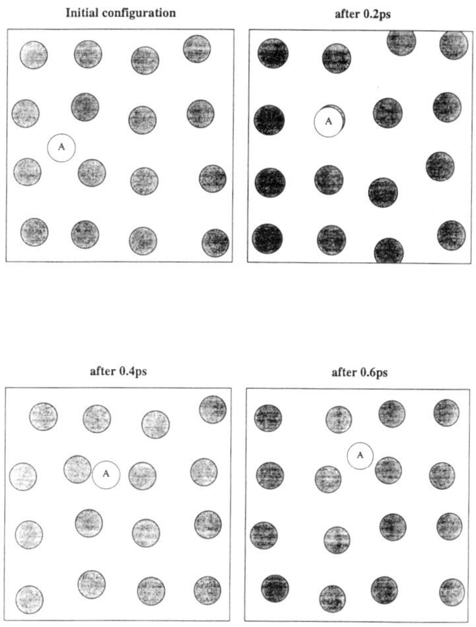

Figure 5. Diagonal hopping mechanism at 950 K; the moving adatom is labeled by A.

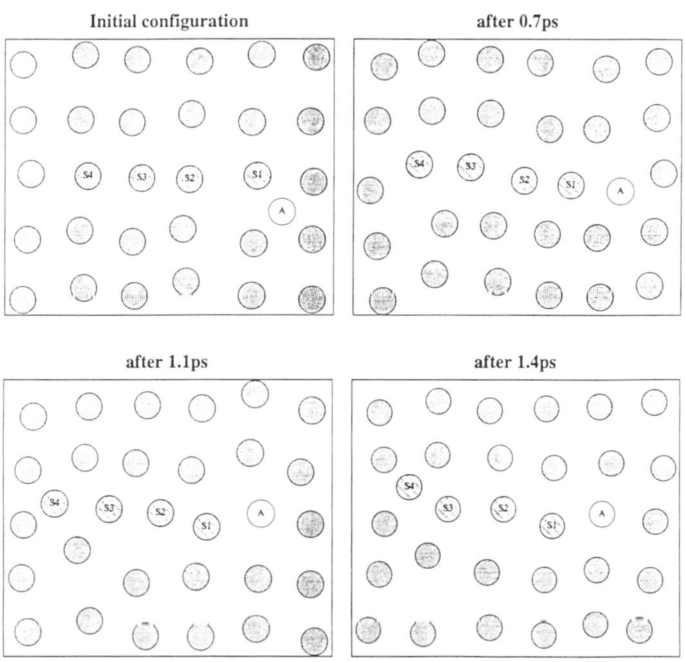

Figure 6. Complicated exchange mechanism at 950 K. The adatom (labeled by A) moves to the S1 surface atom position, the S1 towards S2, which in turn towards to S3 and finally the S4 surface atom goes to an adatom position.

In addition to the above processes, we have also observed various exchange type diffusion mechanisms, involving one or more surface atoms. The exchange mechanism has found experimentally to be the dominant mechanism for the self-diffusion on Ir(001)[20] and Pt (001)[21] surfaces. According to the above process, an adatom moves to the position of an atom in the surface, which, in turn, becomes adatom on the next-nearest binding site. In figure 6 we present a complicated exchange mechanism at 950 K involving four surface atoms. From the above example it is obvious that the nearest-neighbor rows are quite mobile.[4] These complicated events are of course less frequent than the simple exchange mechanism.

From the Arrhenius behaviour of the various mechanisms we obtained the activation energies associated with each of them. Thus, it is found that the simple hopping mechanism is favorable over the other processes, requiring a migration energy of 0.43 eV. This value is in good agreement with experimental data (0.48 eV,[9] 0.28 eV,[10] 0.36 eV[11]) and in qualitative agreement with ab initio calculations.[3] One of the most important results is that all the exchange type mechanisms have almost the same migration energy : simple exchange - 0.7 eV, complicated exchange (fig. 6) - 0.75 eV.

Finally, it comes out that above 950 K the effective contribution of the multiple exchange mechanisms to the total diffusivity is of the same order of magnitude compared to the simple hopping or simple exchange mechanisms, because they are acting at longer distances (or the pre-exponential factors are more important). Therefore, they have to be taken into account in the estimation of the total diffusion coefficient.

Acknowledgments

We would like to thank Dr V. Pontikis for fruitful discussions.

REFERENCES

1. H. Häkkinen and M. Manninen, Phys. Rev. B 46: 1725 (1992).
2. F. Fabre, G. Armand, J.R. Manson, and J. Lapujoulade, Vacuum 41: 424 (1990).
3. C. Lee, G.T. Barkema, M. Breeman, A. Pasquarello, and R. Car, Surf. Sci. Lett. 306: 575 (1994).
4. J.E. Black and Zeng-Ju Tian, Phys. Rev. Lett. 71: 2445 (1993).
5. Y. Chen, S.Y. Tong, J.-S. Kim, L.L. Kesmodel, T. Rodach, K.P. Bohnen, and K.M. Ho, Phys. Rev.B 44: 11394 (1991).
6. L. Yang, T.S. Rahman, and M.S. Daw, Phys.Rev.B 44: 13725 (1991).
7. D.D. Koleske and S.J Sibener, Surf. Sci. 290: 179 (1993).
8. J. Ellis and J.P. Toennies, Phys. Rev. Lett. 70: 2118 (1993).
9. J.J. de Miguel, A. Sanchez, A. Cebollada, J.M. Gallego, J. Ferron, and S. Ferrer, Surf. Sci. 189/190: 1062 (1987).
10. H.J. Ernst, F. Fabre, and J. Lapujoulade, Phys. Rev. B 46: 1929 (1992).
11. M. Breeman and D.O. Boerma, Surf. Sci. 269/270: 224 (1992).
12. C.L. Liu, J.M. Cohen, J.B. Adams, and A.F. Voter, Surf. Sci. 253: 334 (1991).
13. L. Hansen, P. Stoltze, K.W. Jacobsen, and J.K. Norskov, Phys. Rev.B 44: 6523 (1991).
14. L.S. Perkins and A.E. DePristo, Surf. Sci. 294: 67 (1993).
15. M. W. Finnis and J.E. Sinclair, Philos. Mag. A 50: 45 (1984).
16. V. Rosato, M. Guillope, and B. Legrand, Philos. Mag. A 59: 321 (1989).
17. B. Loisel, J. Lapujoulade, and V. Pontikis, Surf. Sci. 256: 242 (1991).
18. N.I. Papanicolaou, I.E. Lagaris, and G.A. Evangelakis, to appear in Surf. Sci. Lett. (1995).
19. G. Armand, D. Gorse, J. Lapujoulade, and J.R. Manson, Europhys. Lett. 3: 1113 (1987).
20. C. Chen and T.T. Tsong, Phys. Rev. Lett. 64: 3147 (1990).
21. G. Kellogg and P.J. Feibelman, Phys. Rev. Lett. 64: 3143 (1990).

THE ORDER–DISORDER TRANSITION AT A Σ=17 TILT BOUNDARY IN Cu_3Au

H.M. Polatoglou

Physics Department
Aristotle University of Thessaloniki
GR-54006 Thessaloniki, Greece

INTRODUCTION

Internal boundaries influence the properties of materials and have therefore attracted due attention[1–4]. Of particular importance are the internal boundaries in load-bearing intermetallics as they effect their mechanical properties[5]. Many interesting phenomena can occur at the boundaries, like segregation[6], wetting[7,8], and for the softly-ordered intermetallics, wetting by the disordered phase[9–11]. A problem that hidders the study of the boundaries is that they are not easily accessible to the experimental probes, or different probes are needed to access their different aspects, like the structure of the boundary, the presence of segregated atoms and their chemical indentity, and their relation to the material properties[2,12]. On the other hand theoretical methods have been progressed to a point where predictions of material properties of bulk crystals and planar defects like boundaries and surfaces are possible[1–3]. It is certain that, using more powerful computers and order-N methods, it will be routine to study new materials. At present, the study of the thermodynamics of alloys and intermetallics and their defects necessitates, simpler methods. A simple method which was successful to address the effects of segregation at boundaries is the Monte-Carlo method utilizing a N-body semiempirical potential[6].

The aim of the present work is to apply the constant temperature, presure and chemical potential difference Monte–Carlo method to the problem of a tilt boundary in Cu_3Au. The method was previously applied to the case of [001] twist boundaries[13,14] in the same material, the [001] surface[15], and $\Sigma = 3$ {111} twin[16].

METHOD

The method used was described previously[13,14]. Here we will give a brief description of the unit cell and the boundary conditions. The boundary we chose is the $\Sigma = 17$ [010] (401) tilt boundary. A coincidence lattice can be defined for such a boundary, which has a two dimensional periodicity along the (401) planes. In the presence of the boundary the periodicity perpendicular to the (401) planes is broken and we must find a way to terminate the crystal. We choose to terminate it with [401] surfaces. The size of the

Stability of Materials
Edited by A. Gonis *et al.*, Plenum Press, New York, 1996

simulation cell is $\sqrt{17}$ X 4 X $6\sqrt{17}$ in units of the lattice constant of Cu_3Au. The (401) planes of the crystallites that form the boundary are parallel to the x-y plane of the simulation cell, and as a consequence the z-axis of the simulation cell is perpendicular to the boundary plane.

Special consideration should be given to the quantities we measure during the simulation. To make contact to the [001] twist boundaries calculations we measure plane-averaged quantities for planes parallel to the boundary, such as stoichiometry, structure factor and strains. Going to the present case it can be noted that the density of the atoms on the (401) planes is smaller that that for the (001) planes and therefore the statistics for the plane-averaged quantities are expected to be worse. From the present calculations we have found that only the structure factor needs special attention.

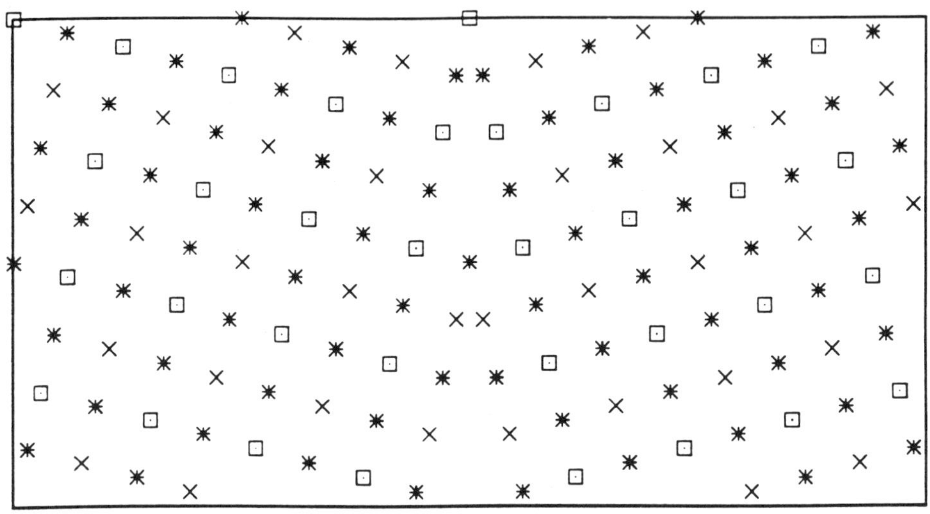

Figure 1: A part of the simulation cell which includes one period of the boundary along the x-axis. The open squares and the crosses represent the Au and Cu atoms, respectively, with y = 0. Similarly, the open triangles and the stars for atoms with y = 1/2.

As in the case of the [001] twist boundaries we choose to evaluate the structure factor for the wavevector $\mathbf{G} = (2\pi/a)(2,1,1)$, which monitors the ordering on the (001) planes. Since the (001) planes of the two crystallites are inclined relative to the z axis, we make an appropriate rotation of the reciprocal vector at which we evaluate the structure factor. In addition we sum not only the atoms that belong to a given (401) plane but all the atoms which belong to a slab containing four (401) planes. Thus we effectively sample two successive (001) atomic planes. For the perfect lattice the plane-averaged structure factor on the (001) planes is zero for the pure Cu planes and 1 for the mixed Cu-Au planes. As we sum over two successive (001) planes, i.e. a pure Cu and a mixed Cu-Au, the plane-averaged structure factor for the ordered phase has the value of 1/2. In addition, to avoid complex numbers we use the square of the norm of the structure factor as the order parameter. From all the above we expect the order parameter to take the following values: 1/4 for the ordered phase and 0 for the disordered one. The above definition for the order parameter gives meaningless values for slabs which contain (401) planes from the two crystallites. For this reason we do not present in the following the order parameter for such slabs.

In figure 1 we present the initial unrelaxed structure of the $\Sigma = 17$ [010] (401) tilt boundary at T = 0K. It can be observed that the (401) planes contain either Cu or Cu and Au in equal proportion. As a consequence the stoichiometry profile perpendicular to the (401) planes takes the values 1 and 1/2 respectively. For the strains we define a positive value if the average z-component of the atoms belonging to a given (401) plane

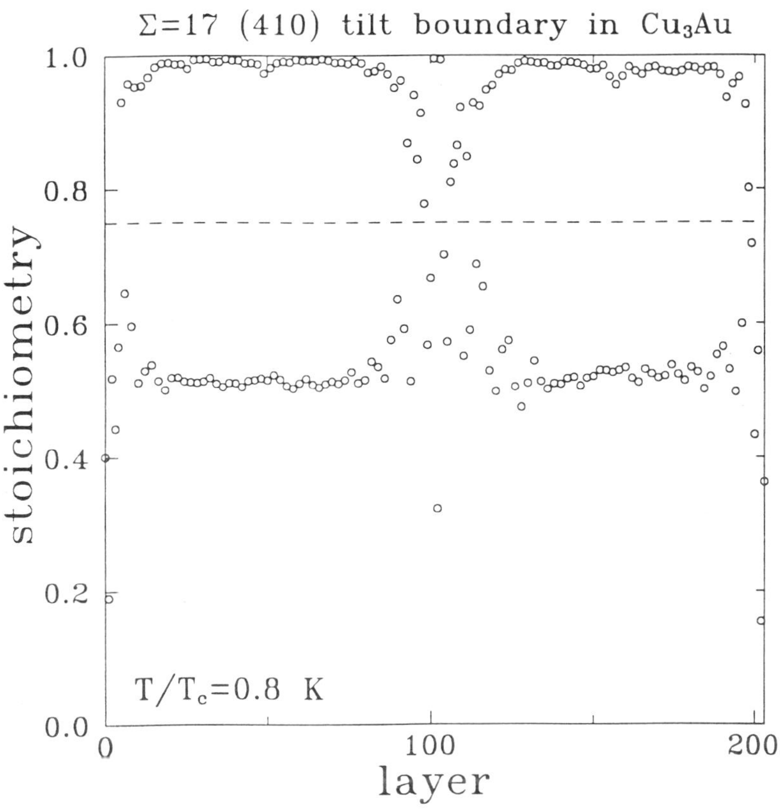

Figure 2: Plane-averaged stoichiometry profile of the $\Sigma = 17$ [010] (401) boundary in Cu_3Au for $T/T_c = 0.8$.

is larger than the position of the plane in the initial structure and negative the other way around.

RESULTS AND DISCUSSION

In figure 2 we present the results for the stoichiometry profile at $T/T_c = 0.8$. We notice that away from the surface and the boundary the plane-averaged stoichiometry takes alternative the values 1/2 and 1. This pattern breaks close to the surfaces and the boundary. Approaching the surface from the bulk we observe that the values tend to the average value, indicated by the dashed horizontal line, and very close to the surface

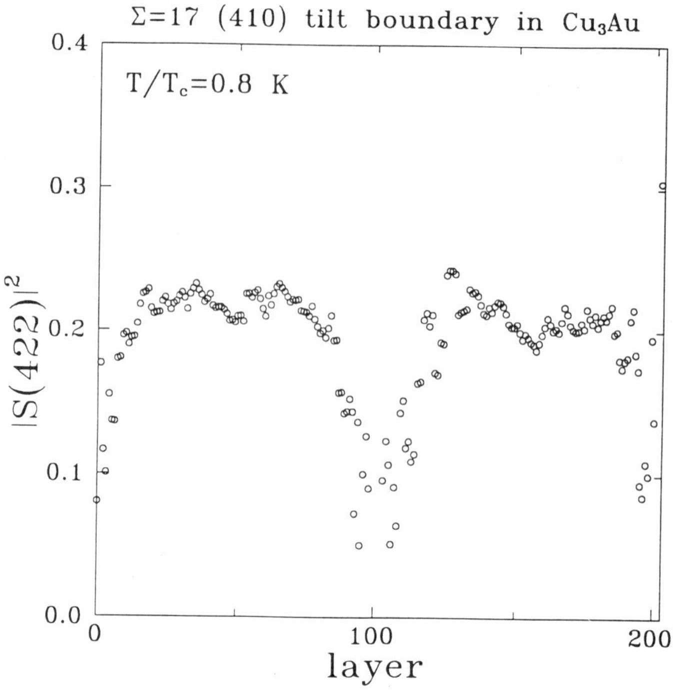

Figure 3: Order parameter of the $\Sigma = 17$ [010] (401) tilt boundary in Cu_3Au for $T/T_c = 0.8$.

the values of the stoichiometry are below 1/2, having their lowest value at the surface. Clearly, Au atoms segregate to the surface. Close to the boundary, the plane-averaged stoichiometry displays similar behavior, although it shows a lot of variation very close to the interface, tending similarly to the average. At the interface it exhibits two values equal to one and one at about 0.3. One can not tell directly from the figure whether Au or Cu atoms segregate to the boundary. To address the question we average the stoichiometry of pairs of successive planes. Away from the boundary these averaged stoichiometries are equal to the average value, i.e. 3/4, while at the surfaces they have values below the average and at the boundary they take values below and above, without noticeable preference to values below or above the average. Quantitatively we find that on the average all the planes in the neighborhood of the boundary have a stoichiometry equal to 3/4. Therefore we come to the conclusion that Au atoms segregate to the surface but not to the $\Sigma = 17$ [010] (401) tilt boundary.

The order parameter is depicted in figure 3. As we have mentioned above, away from the boundary the order parameter takes values close to 1/4, i.e. the values for the ordered phase. Large variations can be observed at the boundary and the surface. Some values are missing, according to the discussions given above. Since the order parameter close to the surface and the boundary shows values below 1/4 with the lowest values at the defects we can conclude that the material around the defects has lower ordering than at the rest of the material. Similar behavior was observed for the [001] twist boundaries[13,14], and [001] surface[15]. These disordered region was found to get wider as the temperature approached the transition temperature, indicating wetting[13,14]. Calculations are in progress to examine whether this case holds also for the present tilt boundary.

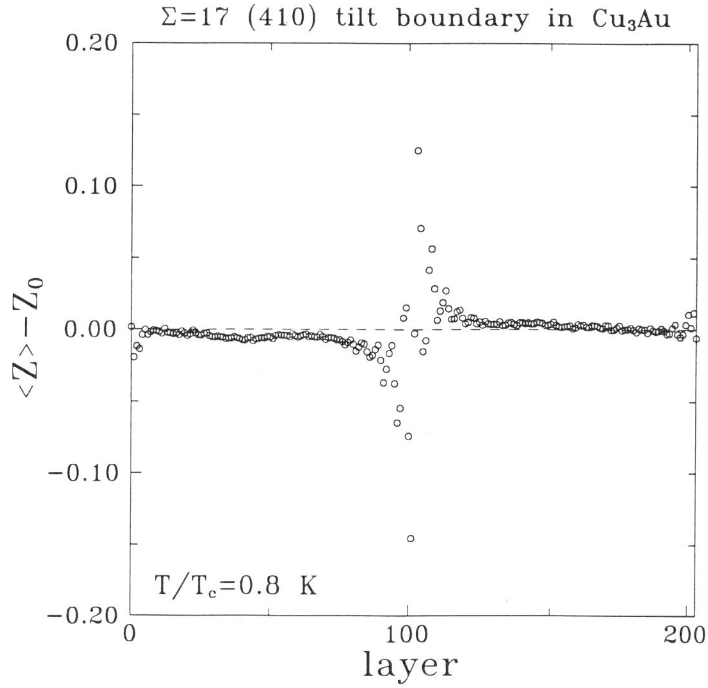

Figure 4: Plane-averaged strains of the $\Sigma = 17$ [010] (401) tilt boundary in Cu_3Au at $T/T_c = 0.8$.

From figure 4, showing the average strain of the (401) planes, we observe that on the average the first surface plane does not relax, while the other planes show an outward relaxation. The relaxations are small, i.e. some percent of the lattice parameter. On the other hand the boundary shows relaxations as large as 15 % of the lattice parameter. Curiously enough one plane at the boundary shows almost zero relaxation, a fact which correlates with what we found for the (401) surface. In addition the relaxation is oscillatory and has the same behavior on both sides of the boundary. The extend of the strain is small, about two lattice parameters on each side of the boundary. This magnitude is similar to the one observed for the [001] twist boundaries in Cu_3Au[13,14].

In summary, we have studied the Σ 17 tilt boundary using atomistic simulations. We found that the boundary affects the ordering at its neighborhood indicating the possible occurrence of wetting phenomena in the proximity of T_c. No segregation of Cu or Au atoms to the tilt boundary could be found, in contrast to the case of the surface, where segregation of Au atoms to the surface is observed.

References

[1] M.C. Payne, P.D. Bristowe, and J.D. Joannopoulos, Phys. Rev. Lett. **58**, 1348 (1987).

[2] D.E. Luzzi, Min Yan, M. Sob, and V. Vitek, Phys. Rev. Lett, **67**, 1894 (1991).

[3] F. Ernst, M.W. Finnis, D. Hofmann, T. Maschik, U. Schonberger, U. Wolf, and M. Methfessel, Phys. Rev. Lett., **69**, 620 (1992).

[4] J.L. Putaux, J. Thibault, A. Jacques, and A. George, Materials Science Forum, **126-128**, 73 (1993).

[5] R.W. Cahn MRS Bulletin, May 18 (1991).

[6] H.Y. Wang, R. Najafabadi, D.J. Srolovitz, and R. LeSar, Interface Science, **1**, 31 (1993).

[7] G. Ciccotti, M. Guillope, and V. Pontikis, Phys. Rev. B**27**, 5576 (1983).

[8] S.R. Phillpot, S. Yip, and D. Wolf, Computers in Physics, Nov./Dez., 20 (1989).

[9] H.M. Polatoglou, Computational Materials Science **3**, 109 (1994).

[10] J.G. Antonopoulos, F.W. Schapink, and F.D. Tichelaar, Phil. Mag. Lett. **61**, 195 (1990).

[11] L. Potez, and A. Loiseau, Interface Science **2**, 91 (1994).

[12] B.W. Krakauer and D.N. Seidman, Phys. Rev. B**48**, 6724 (1993).

[13] H.M. Polatoglou, J. Phys. Condens. Matter **6**, 5621 (1994).

[14] H.M. Polatoglou, Interface Science **2**, 67 (1994).

[15] H.M. Polatoglou, Proceedings of the 6th Joint EPS-APS International Conference on Physics Computing, 605 (1994).

[16] El M. Azzaoui, M. Hou, and V. Pontikis, Interface Science **2**, 95 (1994).

MONTE CARLO SIMULATION OF MBE GROWTH OF GaAs ANALYSIS OF RHEED

P.J.van Hall[1], H.Kökten[2], M.R.Leys[1] and M.Bosch[1]

[1] Physics Dpt. Eindhoven University of Technology
Eindhoven, Netherlands
[2] Physics Dpt. Middle East Technical University
Ankara, Turkey

INTRODUCTION

Modern optoelectronic devices such as quantum well lasers contain thin layers (20-100 Å) of different semiconductor materials. Mostly one uses III-V compound semiconductors for example GaAs, AlGaAs, InP, InGaAs etc. These structures are grown using MOCVD, MBE or CBE growth technologies[1]. All these techniques have in common that one operates under a large excess of group V elements, which serves to stabilize the surface.

It is clear, that the performance of these devices will largely depend on the quality of the interfaces between the layers. Often one tries to improve the interface by a growth interrupt when switching from one material to the other. This is mostly accomplished by turning off the flow of the group III elements while maintaining the group V fluxes. If it is neccessary to change also the group V element the switching procedure can be quite complicated[2].

A method to monitor in situ the structure of the surface is RHEED[3], where one looks at the (specular) reflection of a high energy electron beam incident at grazing angle. The recorded intensity is a good measure for the smoothness, since a rough surface will provide low reflectivity.

In our institute (E.U.T) both a MBE machine and a CBE machine are operational. We therefore considered it worthwile to start a theoretical investigation of the interface properties and the growth processes. A suitable method for this is the Monte-Carlo simulation[4]. Quite substantial work has already been done in this field, especially by Vvedensky and Clarke[5,6], who applied a solid on solid model. As a measure of the interface quality they used the step density. In our work we explicitly account for the III-V zincblende structure and we really calculate the electron reflection.

In this field of layered III-V semiconductors the GaAs/AlGaAs system has been used as a model system to test experimental and theoretical concepts and techniques. Therefore this first report of our work is on the epitaxial growth of GaAs under MBE conditions. Recently Hopkins et al.[7], also at our institute, have published extensive data and a thorough analysis of the damping and recovery of the RHEED intensity. In this paper we concentrate on the analysis and interpretation of their data. We had to restrict ourselves to the data below 610 ^{0}C, because above this temperature one enters the regime of the 3x1 reconstruction with a Ga terminated surface (Ga atoms on As sites), a possibility at present not allowed in our program.

CALCULATIONS

For our simulations we used a 256x256 matrix with periodic boundary conditions in the (001) plane of a zincblende structure. In this plane we have alternating group III (Ga) and group V (As) layers. As said before MBE and CBE conditions imply a large excess (up to 20 times) of group V elements. In these circumstances the growth is determined by the kinetics of the group III elements i.c. Ga. For these atoms we included the following mechanisms: deposition, desorption, migration and - at the edge of an island - the choice between desorption and falling to a lower layer. In principle the time scale in this type of simulations is relative. Only the ratio of probabilities is important. A crucial parameter turned out to be the migration to deposition ratio denoted below by M. In order to match as closely as possible the actual situation we fixed our timescale to a growth rate of about 1 ML/sec.

The As-rich conditions were accounted for in the following way. Whenever two Ga atoms, whether by migration or by deposition, arrived at two neighbouring positions in the [110] direction thus creating an As site, we immediately placed an As atom on top of this pair making them immobile. This As film on top of the Ga's, however, is in a dynamical equilibrium with the As vapour. This gives a certain probability that an As atom is absent, thus making the underlying Ga atoms mobile again. This effect is of course especially important at the edge of an island. Normally one specifies the degree of coverage θ; we shall also use the degree of absence A ($=1-\theta$).

The RHEED signal was calculated by investigating the specular reflections of electrons impinging on the surface under 8 deg. It was averaged over the <110>, <100> and <010> directions. The large dimensions of our matrix permitted us to do this with reasonable precision.

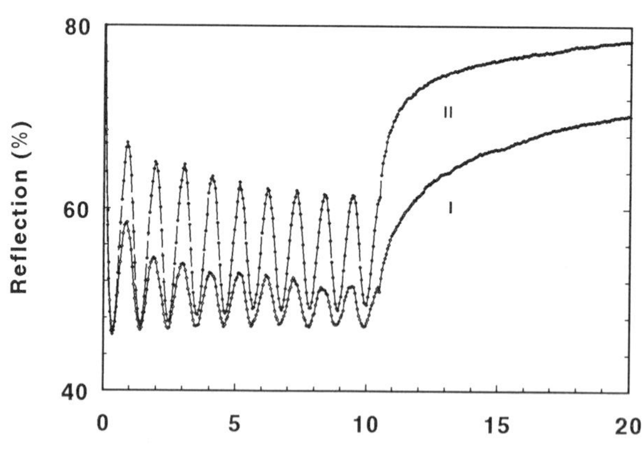

Figure 1. Two examples of calculated RHEED signals. Curve I has been calculated with A=0.10 and M=50, while for curve II we have A=0.10 and M=250

The aim of this paper is to investigate the combined effect of the parameters A and M. As under normal MBE conditions we kept all the desorption channels for the Ga atoms closed. An example of our results is given in fig.1, where we show the RHEED oscillations calculated for two different sets of parameters. One can easily discern the decrease of the amplitude with time. After 10.5 seconds the Ga flux was stopped, after which the RHEED signal recovered with a fast and a slow component. This damping and recovery will be the subject of the next two sections.

DAMPING

We started our analysis with the calculations of the damping, defined as the ratio between the amplitudes of the first and the tenth oscillation ($D = I_1 / I_{10}$). We obtained the value of D in two different ways: either by taking the direct ratio or by deriving it from an exponential least squares fit to all the amplitudes. Generally the two values agreed quite well. In the case of large damping viz. ill defined oscillations, these values differed. Under these circumstances also the results of separate runs with the same parameters but with other random sequences could show considerable differences. From this we could make an estimate of the error in D.

In fig. 2 we show the dependence of D on the migration/flux ratio M for various values of the incompleteness A ($=1-\theta$) of the As layer. When the As layer is filled maximally (A=0) the damping is large (4 to 5) and independent of M (we consider the minimum around M=150 as purely accidental). It follows from the calculations, that nearly all the Ga atom have an As on top making them immobile. This results in a rough surface with a corresponding high damping. It should be emphasized, however, that the RHEED amplitudes are not equal. There is an increase by a factor of 3 when M changes from 25 to 250 (A=0). Obviously the damping is a measure of the *rate* of the roughening not of the roughness itself.

Above A=0.05 the damping varies slowly with M and does not show a significant dependence on A. It is possible therefore to take average values for A>0.05. It seems that once the Ga atoms possess a certain amount of mobility - a combined effect of M and A - the damping is rather insensitive to the actual values of the parameters. We have investigated this threshold effect in more detail by calculating the

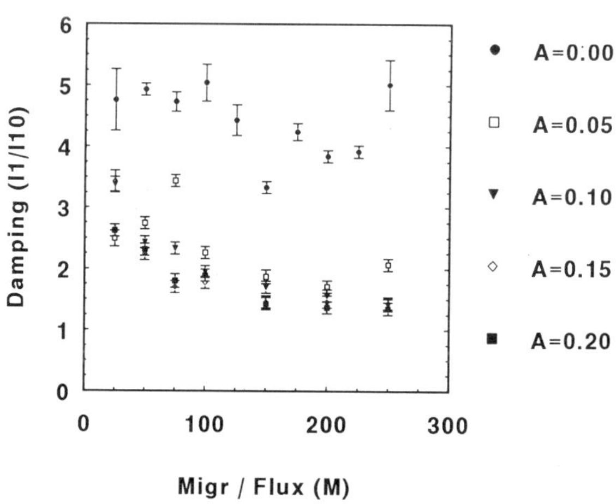

Figure 2. The dependence of the damping of the RHEED oscillations on M for various values of A

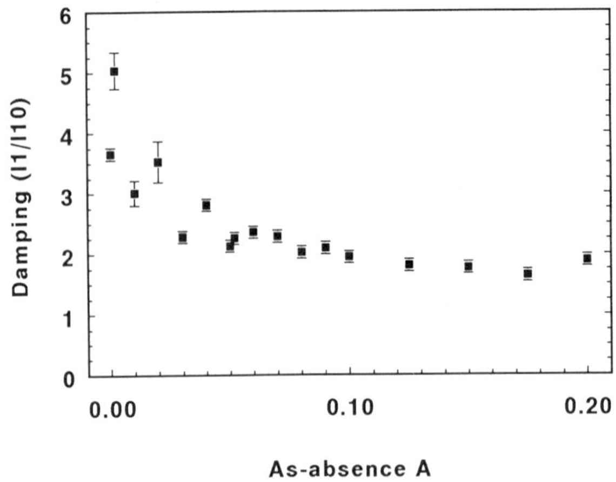

Figure 3. The dependence of the damping on A for a fixed value of M (M=100)

damping as a function of A, while keeping M fixed at the value of 100. The results displayed in fig. 3 indeed clearly show a sharp initial decrease in the damping followed by a nearly constant D for A > 0.05, thus confirming our ideas.

These results have been used to interpret the results of Hopkins et al.[7]. These authors found the damping changing from 2.75 at a substrate temperature of 550 ^{0}C to 1.75 for T= 610 ^{0}C. Using average values derived from the data in fig.2 we arrived at a value of M=35 for T=550 ^{0}C, M=85 for T=580 ^{0}C and M=125 for T=610 ^{0}C.

RECOVERY

The next item we studied was the recovery of the RHEED intensity during a growth interrupt. We did this by stopping the Ga flux at the maximum of the 10th oscillation (viz. at 10.5 sec.) and following the RHEED signal up to 20 sec. As is already clear from fig.1 the recovery shows a fast and a slow component in nice agreement with the experimental findings[7]. First of all it should be emphasized that the temporal removal of the As atoms on top of the Ga atoms is essential to obtain any recovery. If we fix A to zero, we obviously freeze the situation.

We now shall try to identify the two underlying mechanisms giving a fast and a slow component. There are two possibilities. At the growth interrupt the occupation of the 10th Ga layer is about 80%, while the filling of the 11th layer is around 7%. So one possibility is the falling of the Ga atoms from the 11th layer into the 10th layer.This we investigated by recording the change in the occupations. The other possibility is the rearrangement of the islands in the 10th layer - coalescence of small islands to larger ones. Such an effect can be made visible by counting the Ga atoms with four nearest neighbours viz. the Ga atoms inside an island. In fig.4 the change of both quantities with time is shown. The conclusion is evident: we can identify the fast component with the rearrangement of the islands and the slow one with the Ga atoms falling from the islands.

We have analysed the recovery data using a combination of exponential least squares fits followed by a steepest descent minimum search. An important quantity is the ratio of both contributions. A high fast/slow ratio is favorable, because a short growth

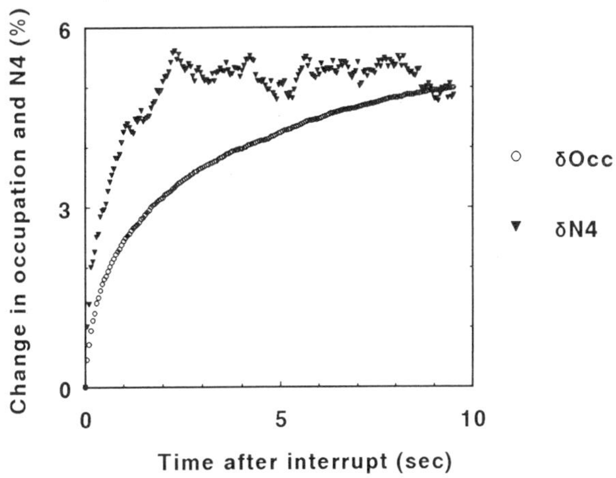

Figure 4. The dependence on the time after the growth interrupt of the change in the occupation of the 10th layer and of the fraction of Ga atoms in this layer having four nearest neighbours. The parameters were A=0.10 and M=50 (curve I in fig. 1)

interrupt already results in a large improvement. This ratio varied with M from about 0.25 at M=25 to about 1.0 for M=250. Again these values were quite independent of the amount of absence of As (the parameter A). When we compare these values with the results of Hopkins et al.[7], we see rather nice agreement. We now also understand why Hopkins et al. found that a growth interrupt before the maximum of the RHEED oscillation gives the best fast/slow ratio. In this case there are hardly any Ga atoms in layer N+1, so the slow component vanishes to a large extent. Moreover there are plenty of sites available in layer N making the process fast.

The time constant of the slow component had a value of about 5 seconds, showing an indication of a minimum around M=175. We can understand this in the following way. For low values of M the speed of recovery will be determined by the speed of migration. For higher values of M, however, the initial occupation of the underlying layer is larger, so the number of available places is less. This will retard the recovery. Hopkins et al. do not give in their paper the values of this time constant. We learned[8] however, that the values amounted to well above 5 seconds. So the predictions of our calculations seem somewhat too low.

The fast component had a time of about 0.5 seconds, a value rather independent of the choice of the parameters as long as the value of M was above 50. For M=25 the fast component is rather small and it became difficult to extract the values for the fast component. Nevertheless we have the impression, that again our values are too low and do not show the implicit temperature dependence via M as seen experimentally. Obviously we have to invoke a retarding mechanism for as well the fast and the slow component. It is well known that the As atoms form dimers as a more stable configuration[9]. Incorporating this mechanism will greatly influence the effect of our parameter A. Investigations also accounting for this mechanism are under way and will be published in the near future.

CONCLUSIONS

Summarising: we have obtained a semiquantitative description of experimental RHEED data. From this we conclude, that the most important physical ingredients have been included in our model. We were also able to identify the two mechanisms governing the stabilisation and smoothening of the surface as seen in the RHEED recovery.

REFERENCES

1 G.Bauer et al. (eds.)
Two dimensional systems: Physics and New devices Springer Berlin 1986.
2 J.Camassel et al. Journ. de Physique II 3 (1993) 99.
3 B.A.Joyce et al. Appl. Surf. Sci. 60/61 (1992) 200.
4 A.Madhukar and S.V.Ghaisas Crit.Rev. Solid. St. Mat. Sci. 14 (1988) 1.
5 S.Clarke and D.D.Vvedensky Journ. Appl. Phys. 63 (1988) 2272.
6 D.D.Vvedensky and S.Clarke Surf. Sci. 225 (1990) 373.
7 J.Hopkins et al. Appl. Surf. Sci. 84 (1995) 299.
8 J.Brübach private communication.
9 H.H.Farrel et al. Journ. Vac.Sci. Techn. B5 (1987) 1482.

ELECTRONIC STRUCTURE THEORIES

SELF-CONSISTENT GREEN'S FUNCTION METHOD FOR RANDOM SURFACES AND INTERFACES

J. Kudrnovský[1,2], I. Turek[3], V. Drchal[1,2], and M. Šob[3]

[1]Institute of Physics, Academy of Sciences of the Czech Republic, Na Slovance 2, CZ-180 40 Prague 8, Czech Republic
[2]Institute of Technical Electrochemistry, Technical University, Getreidemarkt 9, A-1060 Vienna, Austria
[3]Institute of Physics of Materials, Academy of Sciences of the Czech Republic, Žižkova 22, CZ-616 62 Brno, Czech Republic

INTRODUCTION

The last decade is characterized by a fast development of materials science with special emphasis on production of new materials, often with unique properties. We just mention thin magnetic films on non-magnetic substrates, metallic multilayers, or semiconductor superlattices. Surfaces and interfaces of solids are also of great interest from both theoretical and technological points of view, as many of material properties depend critically on the properties of surface and interface regions. Catalysis, chemisorption, crystal growth, grain boundaries, and segregation of atoms to the surface or interface are just a few such examples.

Knowledge of electronic structure of solids is of vital importance for understanding their macroscopic properties on the microscopic ground. At present, there exist a lot of efficient schemes to determine electronic characteristics of pure elements and ordered alloys, which are based on the density-functional formalism (see for example Refs. 1 and 2). The presence of structural and compositional inhomogeneities requires the use of techniques based on Green's functions which differ from those used in the conventional band structure calculations employing wave functions. The Green's functions correspond to quantum-mechanical probability, while the wave functions correspond to its amplitude. The observables depend linearly on Green's functions, but are bilinear forms in wave functions. Consequently, the Green's functions are the relevant physical quantities to be evaluated for materials without the translational symmetry as random alloys and their surfaces. Once the Green's function of the studied system is found, any physical quantity of interest can be determined. It should be noted, that the conventional band structure methods can still be used to study some specific problems for surfaces or random alloys by modelling them via suitably chosen slabs or supercells. By construction, the Green's function depends rather on the charge density than directly

Stability of Materials
Edited by A. Gonis *et al.*, Plenum Press, New York, 1996

on the wave function, and it is thus much less sensitive to the imposed boundary conditions. In addition, the Green's function language allows to understand the nature of the physical problems and approximations used in a systematic way.

In this paper we give a brief account of an efficient self-consistent Green's function method for the calculation of electronic properties of disordered alloys, their surfaces and interfaces. The main features of the presented approach can be summarized as follows: (i) the application of the all-electron tight-binding linear muffin-tin orbital (TB-LMTO) method[1,3] within the local density approximation (LDA) to describe the electronic structure from the first-principles; (ii) the description of the semi-infinite geometry of the system using the surface Green's function (SGF) formalism;[4,5] (iii) the use of the coherent potential approximation (CPA) approach[6–8] extended to inhomogeneous systems such as surfaces[9,10] or interfaces; (iv) the characterization of the vacuum region by empty spheres which represent the continuation of a semi-infinite solid lattice to infinity,[11] (v) the description of atomic potentials in the atomic sphere approximation (ASA), and (vi) the inclusion of both monopole and dipole terms of the charge density in the calculation of the Madelung potential[12] at the solid surface.

The advantage of the present approach is a unified treatment of both bulk alloys and their surfaces. This is achieved by using the same starting Hamiltonian in both cases, which in the orthogonal LMTO representation γ has the following form[1]:

$$H^{\gamma}_{RL,R'L'} = C_{RL}\delta_{RR'}\,\delta_{LL'} + \Delta^{1/2}_{RL}\,S^{\gamma}_{RL,R'L'}\,\Delta^{1/2}_{R'L'},\quad O^{\gamma}_{RL,R'L'} = \delta_{RR'}\,\delta_{LL'}\,. \tag{1}$$

In Eq. (1), H^{γ} and O^{γ} are, respectively, the Hamiltonian and overlap matrices. Further, R and R' are the site indices denoting positions of atoms or empty spheres, whereas L and L' label the orbitals located at the corresponding sites and are equal to usual angular momentum indices, $L = (\ell m)$. The geometry of the problem enters the Hamiltonian via the structure constant matrix S^{γ} expressed in terms of the canonical structure constant matrix S^0, which is known analytically[1,13]:

$$S^{\gamma}_{RL,R'L'} = [\, S^0\,(1-\gamma\,S^0)^{-1}\,]_{RL,R'L'}\,. \tag{2}$$

The elements $S^0_{RL,R'L'}$ depend only on $(R-R')/w$, where w is the average Wigner-Seitz radius of a solid. Here we consider the simplest case, namely an ideal periodic lattice randomly occupied by atoms A and B. We thus neglect the lattice relaxations in the bulk which are connected with different sizes of alloy constituents as well as possible inward or outward shifts of surface layers. We can then use the ideal bulk structure constants S^0 also for bulk alloys and surface or interface problems.

The properties of individual atoms occupying the ideal lattice sites are characterized by the potential parameters X_{RL} ($X = C, \Delta$, and γ), which are randomly X^A_{RL} and X^B_{RL} in a solid, and X^v_{RL} in empty spheres in the vacuum region. The potential parameters have a simple physical interpretation:[1,13] they describe the positions C_{RL}, the widths Δ_{RL}, and the distortions γ_{RL} of the 'pure' RL-bands. The potential parameters are obtained from the solutions of the radial scalar-relativistic Schrödinger equation for the LDA potential in a corresponding atomic sphere and can be combined into the potential function matrix

$$P^0_{RL,R'L'}(z) = P^0_{RL}(z)\,\delta_{RR'}\,\delta_{LL'}\,,\quad P^0_{RL}(z) = \frac{z - C_{RL}}{\Delta_{RL} + \gamma_{RL}(z - C_{RL})}, \tag{3}$$

which plays a central role in the LMTO formalism.[1,13] For $\ell \leq 2$, the basis set consists of s-, p-, and d-orbitals on each site, which allows to describe all transition metals. The relativistic effects and higher angular momenta ($\ell > 2$) can also be included. We note that the potential function matrix can be defined within the ASA more generally in terms of the logarithmic derivatives of the radial wave function in the atomic sphere evaluated at the sphere radius.[13–15] The parametrized form (3) is consistent with the Hamiltonian (1) used in this paper.

Note that the potential parameters, in general, depend also on the site index R due to the charge self-consistency. They are the same for each group $\{R_p\}$ of geometrically and electronically equivalent sites. The number of sites in each group is N_p, and the total number of sites in the system is $N = \sum_p N_p$. For example, $\{R_p\}$ can be one of inequivalent sublattices of a complex lattice, or a layer of atoms close to the surface with properties different from those in the bulk. On a simple lattice, e.g. fcc lattice, there is only one set of potential parameters. The concentrations c^Q of alloy constituents $Q = A, B$ could also depend on the site index $R \in \{R_p\}$, $c_R^Q \equiv c_p^Q$. The set of numbers c_R^Q thus defines the concentration profile in the solid. In a homogeneous bulk alloy all c_R^Q are independent of the site index R. On the other hand, in the case of the segregation of one of the alloy constituents at the solid surface the local concentrations c_R^Q may differ from the bulk values in several surface layers.

All layers parallel to the surface or the interface could have different local physical properties as a consequence of the the semi-infinite nature of the surface or interface problems. In order to overcome this difficulty, we assume that from a certain layer on, the electronic properties of all subsequent layers are identical to those of the corresponding infinite systems, namely either to a homogeneous alloy or to the vacuum. The solid is thus considered to be divided into three regions: (i) a homogeneous semi-infinite bulk alloy, (ii) a homogeneous vacuum region represented by empty spheres and characterized by flat potentials, and (iii) an intermediate region consisting of several layers where all inhomogeneities (chemical or electronic) are concentrated, and which contains also a few layers of empty spheres of the vacuum-solid interface. In case of solid-solid interface, we consider the intermediate region (iii) sandwiched by two (generally different) semi-infinite bulk systems.

In conclusion of this Section, we mention one important feature of the LMTO basis set, namely its transformation property. The starting Hamiltonian (1) was defined in the so-called orthogonal LMTO representation γ the name of which is self-explanatory: the orbitals in the basis set are mutually orthogonal (see Eq. (1)). This representation is convenient both for formulation of the problem and evaluation of the physical quantities of interest, as the overlap matrix O^γ is the unit matrix. The orthogonal LMTO basis set can be transformed exactly into many other basis sets the properties of which can be tailored to a specific problem. For example, it is convenient to perform the configurational averaging[7,8] or to introduce the surface[9–12] in the so-called tight-binding LMTO representation. The idea of the LMTO representations, introduced by Andersen *et al.*,[3,1] is a powerful tool for the applications of the LMTO formalism to random alloys and their surfaces. In the next Section we give a brief account of this idea in the language of the Green's functions suitable for further applications.

GREEN'S FUNCTIONS IN THE LMTO THEORY

The Green's function matrix corresponding to the Hamiltonian (1) is defined as

$$G_{RL,R'L'}(z) = [(z - H^\gamma)^{-1}]_{RL,R'L'} \,. \tag{4}$$

For systems with broken translational symmetry it is, however, advantageous to reformulate the theory by using different LMTO representations. A general LMTO representation α is specified by a set $\alpha_{RL} = \alpha_{R\ell}$ of the so-called screening constants,[3] which form a diagonal matrix α and define a new structure constant matrix $S^{\alpha}_{RL,R'L'}$ and corresponding potential functions $P^{\alpha}_{RL}(z)$ by means of the following transformations:

$$S^{\alpha}_{RL,R'L'} = [\, S^0\,(1 - \alpha S^0)^{-1}\,]_{RL,R'L'}\,, \tag{5}$$

and

$$P^{\alpha}_{RL}(z) = \frac{P^0_{RL}(z)}{1 - \alpha_{RL}\, P^0_{RL}(z)} = \frac{z - C_{RL}}{\Delta_{RL} + (\gamma_{RL} - \alpha_{RL})(z - C_{RL})}. \tag{6}$$

We note that for spherically symmetric potentials, the potential parameters C_{RL}, Δ_{RL}, γ_{RL} and the potential functions $P^{\alpha}_{RL}(z)$ can be represented by the site- and orbital-diagonal matrices C, Δ, γ, and $P^{\alpha}(z)$. In fact, they depend only on the orbital quantum number ℓ rather then on the whole angular momentum index L. The non-diagonal matrix S^{α} is defined in terms of the original, canonical structure constant matrix S^0.

The Green's function matrix $G(z)$ corresponding to the Hamiltonian (1) can be expressed using the LMTO representation α as

$$G_{RL,R'L'}(z) = \lambda^{\alpha}_{RL}(z)\,\delta_{RR'}\,\delta_{LL'} + \mu^{\alpha}_{RL}(z)\,g^{\alpha}_{RL,R'L'}(z)\,\mu^{\alpha}_{R'L'}(z), \tag{7}$$

where the auxiliary Green's function $g^{\alpha}(z)$ is defined by

$$g^{\alpha}_{RL,R'L'}(z) = [(P^{\alpha}(z) - S^{\alpha})^{-1}]_{RL,R'L'}\,. \tag{8}$$

The site-diagonal matrices $\lambda^{\alpha}(z)$ and $\mu^{\alpha}(z)$ are given by

$$\begin{aligned} \lambda^{\alpha}_{RL}(z) &= -\frac{1}{2}\frac{\ddot{P}^{\alpha}_{RL}(z)}{\dot{P}^{\alpha}_{RL}(z)} = \frac{\gamma_{RL} - \alpha_{RL}}{\Delta_{RL} + (\gamma_{RL} - \alpha_{RL})(z - C_{RL})}, \\ \mu^{\alpha}_{RL}(z) &= [\dot{P}^{\alpha}_{RL}(z)]^{1/2} = \frac{\Delta^{1/2}_{RL}}{\Delta_{RL} + (\gamma_{RL} - \alpha_{RL})(z - C_{RL})}. \end{aligned} \tag{9}$$

In Eqs. (9), the symbols $\dot{P}^{\alpha}(z)$ and $\ddot{P}^{\alpha}(z)$ denote, respectively, the first and second derivative with respect to the variable z. Let us remark that the expressions containing the potential function $P^{\alpha}(z)$, Eqs. (6) and (9), are generally valid within the ASA,[14,15] while the parametrized forms using potential parameters are consistent with the Hamiltonian H^{γ}, Eq. (1). The term $\lambda^{\alpha}(z)$ in Eq. (7) compensates for the poles of $P^{\alpha}(z)$, Eqs. (6), and guarantees a proper analytical behavior of the Green's function $G(z)$ for large z.

The advantage of the above Green's function formulation lies in the possibility to choose an LMTO representation α[15] which is the most suitable for a particular physical problem. The transformation between two representations involves a rescaling of the auxiliary Green's function $g^{\alpha}(z)$ and a trivial change of the potential functions. The only non-trivial change concerns the transformation of the structure constants. This, however, can be done once for a given problem. Clearly, the full Green's function $G(z)$, Eq. (7), is independent of the particular choice of the LMTO representation α.

The most important LMTO representations are: (i) the canonical representation with $\alpha \equiv 0$, (ii) the orthogonal representation with $\alpha \equiv \gamma$, and (iii) the tight-binding representation with $\alpha_{R\ell} \equiv \beta_{\ell}$ given by the site independent but ℓ-dependent particular values β_{ℓ},[1,3,15] which lead to the shortest spatial extent of the structure constants S^{β}. In the orthogonal representation ($\alpha{=}\gamma$) the potential functions $P^{\gamma}(z)$ are linear functions of energy z, the scaling factors $\lambda^{\gamma}(z)$ and $\mu^{\gamma}(z)$ become trivial, and the true resolvent $G(z)$ differs from the auxiliary one $g^{\gamma}(z)$ only by an energy independent rescaling.

However, the structure constant S^γ depends on the actual values of the potential parameters $\gamma_{R\ell}$. This leads, in random alloys, to the off-diagonal randomness in the alloy Hamiltonian (1) which cannot be treated within the CPA unless additional approximations are adopted. This undesirable feature can be avoided using either the canonical or tight-binding representations ($\alpha = 0$ or β). In this way a complete separation of the structural and atom-dependent parts can be achieved. The problem of chemical disorder can be then reduced to the configurational averaging of the auxiliary resolvent $g^\alpha(z)$ by means of the CPA since the off-diagonal structure matrix S^α is non-random and the random quantities $P^\alpha(z)$ are site-diagonal.

A detailed discussion of the CPA approach will be given in the next Section, but the convenience of the Green's function approach can be illustrated in the case of an isolated substitutional impurity.[14] For the sake of simplicity we neglect the lattice relaxations and the change of potential parameters in the host near the impurity. The auxiliary Green's function $g^\alpha(z)$ for this problem satisfies the Dyson equation with the perturbation confined to the impurity site, which can be solved easily by inverting the local Green's function matrix. On the other hand, the perturbation in the Hamiltonian (1) extends, in principle, over the whole crystal due to the extended nature of the structure constants S^γ.

The main advantage of the tight-binding representation over the canonical one for the case of random substitutional alloys is in the numerical feasibility of the evaluation of the Fourier transform of the structure constants. The elements $S^0_{RL,R'L'}$ of the canonical structure constant behave like $d^{-(\ell+\ell'+1)}$ (where $d = |R - R'|$) as contrasted with the exponential-like decay of the tight-binding structure constants in the real space.[1,3,15] Consequently, the lattice sums must be done using the Ewald technique[13] in the canonical representation. In the tight-binding representation, the lattice sums extend essentially up to the second nearest-neighbor shell for close-packed lattices and are easy to perform.

A profound difference between the canonical and tight-binding representations occurs in the surface-related problems.[5,9–12] A short-range character of the tight-binding structure constants allows to view the semi-infinite solid as a stack of the so-called principal layers with only nearest-neighbor interactions between them. As the principal layer contains only a few atomic layers, and as the electronic structure changes are usually confined to a few top surface layers, the problem can be described by finite matrices. On the contrary, a large spatial extent of the canonical structure constants requires the use of specific, numerically highly involved techniques.[16]

Summarizing, the notion of the physical Green's function $G(z)$ and the auxiliary Green's function $g^\alpha(z)$ expressed in different LMTO representations α leads to a unified framework for the treatment of isolated impurities, concentrated disordered alloys as well as their surfaces. In all these cases, the auxiliary resolvent $g^\alpha(z)$ is the quantity that can easily be evaluated and configurationally averaged. The scaling character of the connection between the auxiliary Green's function $g^\alpha(z)$ and the physical Green's function $G(z)$ is preserved also after the configurational averaging for the case of random alloys[7,8] and their surfaces.[9] This allows to calculate all physically relevant quantities such as the local densities of states or the Bloch spectral densities for bulk alloys or surfaces. The local atom- and layer-dependent Green's functions $G_{RL,RL'}(z)$ are also necessary for electronic charge and spin densities to achieve complete LDA self-consistency of the theory.

CONFIGURATIONAL AVERAGING

In this Section we describe the configurational averaging of the Green's function $G(z)$, Eq. (7), within the CPA and discuss its basic properties. The CPA is an effective medium theory and its rigorous derivation is based on the multiple-scattering formalism.[6] Here we present its simple, intuitive derivation. After the configurational averaging, the scattering properties of a site R can be characterized by an effective coherent potential function $\mathcal{P}_R^\beta(z)$ in a direct analogy with the potential functions $P_R^{\beta,Q}(z)$ characterizing the scattering properties of the constituent atoms $Q = A$ and B. All calculations will be done in the tight-binding LMTO representation, hence the superscript β. The coherent potential functions are the same in each group $\{R_p\}$ of geometrically and electronically equivalent sites, $\mathcal{P}_R^\beta(z) \equiv \mathcal{P}_p^\beta(z)$, $R \in \{R_p\}$, $p = 1, 2, .., M$. These different groups $\{R_p\}$ refer either to different sublattices of a complex bulk solid, or to different layers in the intermediate region in surface or interface problems. This means that the configurational averaging recovers the translational symmetry of the system. The corresponding configurationally averaged Green's function, or equivalently, the Green's function of effective medium, is

$$\bar{g}^\beta_{RL,R'L'}(z) \equiv \langle\, g^\beta(z)\,\rangle_{RL,R'L'} = [(\mathcal{P}^\beta(z) - S^\beta)^{-1}]_{RL,R'L'}\,. \tag{10}$$

The barred quantity and the symbol $\langle \ldots \rangle$ denote the configurational averaging. The unknown $\mathcal{P}_p^\beta(z)$ can be determined within the single-site approximation in the following manner. We embed an atom of the type Q ($Q = A, B$) on a given site $R \in \{R_p\}$ into the effective medium $\mathcal{P}_{R'}^\beta(z)$ of all other sites. The scattering from such a single impurity object is described by a single-site T-matrix $t_p^{\beta,Q}(z)$ defined as

$$\begin{aligned} t^{\beta,Q}_{p,LL'}(z) &= \{(P_p^{\beta,Q}(z) - \mathcal{P}_p^\beta(z))[1 + \Phi_p^\beta(z)(P_p^{\beta,Q}(z) - \mathcal{P}_p^\beta(z))]^{-1}\}_{LL'}\,, \\ \Phi^\beta_{p,LL'}(z) &\equiv \bar{g}^\beta_{RR}(z) = \frac{1}{N_p}\sum_k \{[(\mathcal{P}^\beta(z) - S^\beta(k))^{-1}]_{pp}\}_{LL'}\,. \end{aligned} \tag{11}$$

Here, $P_p^{\beta,Q}(z)$ denotes the potential function of a given atom, Eq. (6), which is generally different on inequivalent sites as a result of the charge self-consistency. It is important to note that a simple form of the scattering matrix, Eq. (11), is due to the single-site character of alloy scattering in the tight-binding LMTO representation. The quantity $\Phi_p^\beta(z)$ is the on-site element of the configurationally averaged auxiliary Green's function $\bar{g}^\beta(z)$, Eq. (10), corresponding to a given subset $\{R_p\}$ of lattice sites. In Eq. (11), N_p is the number of the lattice sites in the subset $\{R_p\}$, and the sum runs over the Brillouin zone of k-vectors corresponding to the recovered translational symmetry (two- or three-dimensional) of the system after averaging. We note that $\Phi_p^\beta(z)$ depends on all inequivalent $\mathcal{P}_q^\beta(z)$, ($q = 1, 2, ..., M$). The determination of $\Phi_p^\beta(z)$ is of central importance for the CPA. Its evaluation is rather straightforward for the bulk case, but nontrivial for surfaces and interfaces. This will be discussed in the next Section.

The configurational averaging within the single-site approximation is equivalent to the requirement of vanishing of the resulting scattering. This requirement leads to the following equation for each inequivalent subset $\{R_p\}$

$$\sum_{Q=A,B} c_p^Q\, t^{\beta,Q}_{p,LL'}(z) = 0, \quad (p = 1, 2, ..., M). \tag{12}$$

In Eq. (12), c_p^Q denote the concentrations of atoms Q on a subset of sites $\{R_p\}$ ($c_p^A + c_p^B = 1$ for each p). The conditions (12) are the CPA equations for a generally inhomogeneous medium. As $\Phi_p^\beta(z)$ depends on all inequivalent $\mathcal{P}_q^\beta(z)$, Eqs. (12) represent the set of M coupled CPA equations for unknown $\mathcal{P}_p^\beta(z)$. Once $\mathcal{P}^\beta(z)$ and $\bar{g}^\beta(z)$ are determined, we go over to the physical, orthogonal LMTO representation γ, to evaluate the quantities of interest. To this end, the configurationally averaged $\bar{G}(z)$ must be determined. We introduce the occupation index η_R^Q: $\eta_R^Q = 1$ if an atom of the type Q is at the site R, and $\eta_R^Q = 0$ otherwise. The configurationally averaged physical resolvent $\langle G(z) \rangle = \bar{G}(z)$ can be expressed as[8]

$$
\begin{aligned}
\bar{G}_{RL,R'L'}(z) &= \sum_Q \lambda_{RL}^{\beta,Q}(z) \langle \eta_R^Q \rangle \delta_{RR'} \delta_{LL'} + \sum_Q \sum_{Q'} \mu_{RL}^{\beta,Q}(z)\, \bar{g}_{RL,R'L'}^{\beta,QQ'}(z)\, \mu_{R'L'}^{\beta,Q'}(z), \\
\bar{g}_{RL,R'L'}^{\beta,QQ'}(z) &= \langle \eta_R^Q \, g_{RL,R'L'}^\beta(z)\, \eta_{R'}^{Q'} \rangle .
\end{aligned} \tag{13}
$$

The first term, proportional to η_R^Q, averages trivially as

$$
\sum_Q \lambda_{RL}^{\beta,Q}(z) \langle \eta_R^Q \rangle = \sum_Q c_p^Q \lambda_{pL}^{\beta,Q}(z) \equiv \langle \lambda_{pL}^\beta(z) \rangle . \tag{14}
$$

In Eq. (14), we assumed that $R \in \{R_p\}$. The evaluation of $\bar{g}^{\beta,QQ'}(z)$ is non-trivial, but it is facilitated by the site-diagonal character of quantities η_R^Q in this product. The details of derivation can be found in the Appendix of Ref. 8. Assuming that $R \in \{R_p\}$ and $R' \in \{R_q\}$, the final result with an explicit separation of site-diagonal and site off-diagonal terms can be summarized as

$$
\begin{aligned}
\bar{g}_{RL,R'L'}^{\beta,QQ'}(z) &= c_p^Q \,[\bar{g}_{RR}^\beta(z)\, f_p^{\beta,Q}(z)]_{LL'}\, \delta_{RR'}\, \delta_{QQ'} \\
&\quad + c_p^Q c_q^{Q'} [\tilde{f}_p^{\beta,Q}(z)\, \bar{g}_{RR'}^\beta(z)\, f_q^{\beta,Q'}(z)]_{LL'} (1 - \delta_{RR'}), \\
f_{q,LL'}^{\beta,Q}(z) &= [\{1 + (P_q^{\beta,Q}(z) - \mathcal{P}_q^\beta(z))\, \Phi_q^\beta(z)\}^{-1}]_{LL'} , \\
\tilde{f}_{p,LL'}^{\beta,Q}(z) &= [\{1 + \Phi_p^\beta(z)\, (P_p^{\beta,Q}(z) - \mathcal{P}_p^\beta(z))\}^{-1}]_{LL'} .
\end{aligned} \tag{15}
$$

In Eq. (15) and below, the symbol $\tilde{X}$ denotes the transposed matrix to the matrix X. The site-diagonal scaling factors $f_p^{\beta,Q}(z)$ also enter the expressions for the single-site T-matrix, Eq. (11), and the local conditionally averaged resolvent (see Eq. (17) below). We note that quantities $\bar{g}^{\beta,QQ'}(z)$ are all expressed solely via $\bar{g}^\beta(z)$, Eq. (10). Combining Eqs. (13 – 15), and assuming again that $R \in \{R_p\}$ and $R' \in \{R_q\}$ we can derive for the physical configurationally averaged resolvent $\bar{G}(z)$ the following expression:

$$
\begin{aligned}
\bar{G}_{RL,R'L'}(z) &= \{\Lambda_p^\beta(z)\, \delta_{RR'} + \tilde{M}_p^\beta(z)\, \bar{g}_{RR'}^\beta(z)\, M_q^\beta(z)\}_{LL'} , \\
\Lambda_{p,LL'}^\beta(z) &= \langle \lambda_{pL}^\beta(z) \rangle\, \delta_{LL'} \\
&\quad + \sum_{Q \neq Q'} c_p^Q c_p^{Q'} \left[\mu_p^{\beta,Q}(z)\, \tilde{f}_p^{\beta,Q}(z)\, \Phi_p^\beta(z)\, f_p^{\beta,Q'}(z)\, (\mu_p^{\beta,Q}(z) - \mu_p^{\beta,Q'}(z)) \right]_{LL'} , \\
M_{q,LL'}^\beta(z) &= \sum_Q c_q^Q\, f_{q,LL'}^{\beta,Q}(z)\, \mu_{qL'}^{\beta,Q}(z), \\
\tilde{M}_{p,LL'}^\beta(z) &= \sum_Q c_p^Q\, \mu_{pL}^{\beta,Q}(z)\, \tilde{f}_{p,LL'}^{\beta,Q}(z).
\end{aligned} \tag{16}
$$

Eqs. (13 – 16) are the main result of this Section. They give the expression for the full configurationally averaged Green's function of disordered bulk alloy and its surface, from which the physical quantities of interest can be obtained.

To get charge densities, only the local quantities related to the so-called conditionally averaged physical Green's function are necessary. For this reason we give below the expression for the atom- and orbital-resolved density-of-states matrix $D^{Q}_{p,LL'}(E)$ for the sites $R \in \{R_p\}$:

$$\begin{aligned} D^{Q}_{p,LL'}(E) &= -\frac{1}{\pi}\,\mathrm{Im}\,\mathcal{D}^{Q}_{p,LL'}(E+i0), \\ \mathcal{D}^{Q}_{p,LL'}(z) &= \lambda^{\beta,Q}_{pL}(z)\,\delta_{LL'} + [\dot{P}^{\beta,Q}_{pL}(z)]^{1/2}\,\Phi^{\beta,Q}_{p,LL'}(z)\,[\dot{P}^{\beta,Q}_{pL'}(z)]^{1/2}, \\ \Phi^{\beta,Q}_{p,LL'}(z) &= \left\{ \Phi^{\beta}_{p}(z)\,\left[1 + (P^{\beta,Q}_{p}(z) - \mathcal{P}^{\beta}_{p}(z))\,\Phi^{\beta}_{p}(z)\right]^{-1} \right\}_{LL'}. \end{aligned} \tag{17}$$

The first term in the expression for $\mathcal{D}^{Q}_{p}(z)$ cancels the unphysical poles of the potential function $P^{\beta,Q}_{p}(z)$. The quantities $\mathcal{D}^{Q}_{p}(z)$ and $\Phi^{\beta,Q}_{p}(z)$ are, respectively, the local conditionally averaged physical and auxiliary Green's functions.[6,8,9] The L-diagonal components of $D^{Q}_{p}(E)$ enter the expression for the spheridized charge density while the parts off-diagonal in L are needed for the full non-spheridized charge density.

Similarly, the spectral density $A_p(k,E)$, or the k-resolved density of states, corresponding to the subset $\{R_p\}$ of lattice points, is related to both site-diagonal and site off-diagonal elements of the configurationally averaged physical and auxiliary Green's functions, Eq. (16) and (10), respectively. The result is given by

$$\begin{aligned} A_p(k,E) &= -\frac{1}{\pi}\,\mathrm{Im}\sum_{L} \mathcal{A}_{p,L}(k,E+i0), \\ \mathcal{A}_{p,L}(k,z) &= \Lambda^{\beta}_{p,LL}(z) + [\tilde{M}^{\beta}_{p}(z)\,\bar{g}^{\beta}_{pp}(k,z)\,M^{\beta}_{p}(z)]_{LL}, \end{aligned} \tag{18}$$

where $\bar{g}^{\beta}_{pp}(k,z)$ is the Fourier transform of $\bar{g}^{\beta}_{RR'}(z)$ over the Brillouin zone corresponding to the translational symmetry of the averaged system. The notion of the spectral density substitutes in random alloys and the surface problems the notion of energy bands in the conventional band theory.

We conclude this Section by some remarks concerning the symmetry of the CPA-related quantities with respect to the angular momentum index L. Let us stress that for spherical potentials, the potential parameters and related quantities λ^{β}, μ^{β}, and potential functions $P^{\beta}(z)$ are diagonal with respect to the index L. On the other hand, all the quantities $\Lambda^{\beta}(z)$, $M^{\beta}(z)$, the coherent potential function $\mathcal{P}^{\beta}(z)$, and the Green's function $\Phi^{\beta}(z)$ have the point symmetry of the system. They are, therefore, generally non-diagonal matrices with respect to indices L, L'. Specifically, for cubic lattices and $\ell \leq 2$, they are still diagonal with respect to L. For cubic lattices and $\ell > 2$, as well as for complex lattices and surfaces where the symmetry of the system is lowered, they are not diagonal any more. Consequently, the calculations can be facilitated by using the group analysis.

SURFACE GREEN'S FUNCTIONS

In this Section we discuss the evaluation of $\bar{g}^{\beta}_{pp}(z)$, Eq. (10), for the case of surfaces of random alloys[9] or random overlayers on non-random substrates.[10,17] For the surface-related problems, the subset $\{R_p\}$ of structurally and electronically equivalent sites is simply a layer of atoms parallel to the solid surface. Let us start with the definition of the notion of the principal layer[4] (PL): a semi-infinite solid can be partitioned into PLs parallel to the surface such that only nearest-neighbor PLs are coupled by the structure constants. A principal layer can include one or more atomic layers depending on the solid face, the lattice type, and the spatial extent of the structure constants S^{α}. It defines the characteristic dimension D of the problem, D=n$(\ell_{max}+1)^2$, where n is the number of atoms in the two-dimensional primitive cell of the PL and ℓ_{max} is the maximum angular momentum. The advantage of the tight-binding representation, with the structure constants S^β having the steepest descent with respect to interatomic distances, is thus obvious. Even the simplest case, when the PL consists of one atomic layer with one atom per primitive cell, includes a number of important low-index surfaces, e.g. the fcc(001) and fcc(111) faces within the first nearest-neighbor terms in S^β, or bcc(110) face within the first and second nearest-neighbor terms in S^β. The generalization to more complex PLs can also be made[18] but in this paper we confine ourselves to the simplest case with one atom in the PL basis. We can further simplify the problem by neglecting possible outward or inward relaxations of surface layers, as well as lattice relaxations due to different sizes of alloy constituents. Accordingly, $S^\beta_{RL,R'L'}$ has a block-tridiagonal form with respect to layer indices p and q, $R \in \{R_p\}, R' \in \{R_q\}$. Employing further the translational symmetry parallel to the surface, one gets

$$S^\beta_{pq}(k_\|) = S^\beta_{00}(k_\|)\,\delta_{pq} + S^\beta_{01}(k_\|)\,\delta_{p+1,q} + S^\beta_{10}(k_\|)\,\delta_{p-1,q}\,, \tag{19}$$

where

$$S^\beta_{pq}(k_\|) = \sum_{R\in\{R_{pq}\}} \exp\,(ik_\| \cdot R)\, S^\beta(R). \tag{20}$$

Here, $k_\|$ is a vector from the surface Brillouin zone (SBZ), and the symbol R_{pq} denotes the set of vectors that connect one lattice site in the p-th layer with all lattice sites in the q-th layer. In Eqs. (19,20) we made use of the fact that the bulk structure constants S^β depend only on the difference vector $R = R_p - R_q$.

The coherent potential function $\mathcal{P}^\beta(z)$ is a site-diagonal matrix, which is layer-dependent due to the charge self-consistency and possible concentration inhomogeneities:

$$\mathcal{P}^\beta(z) = \begin{cases} \mathcal{P}^{\beta,v}(z) & \text{for layers in the vacuum region,} \\ \mathcal{P}^\beta_p(z) & \text{for layers in the intermediate region } (p = 1, 2, ..., M), \\ \mathcal{P}^{\beta,a}(z) & \text{for layers in the bulk alloy region.} \end{cases} \tag{21}$$

The superscripts v and a refer to the vacuum and bulk alloy regions, respectively, as discussed in the Introduction. The values $\mathcal{P}^\beta_p(z)$ are obtained from the solution of the set of coupled CPA equations (11, 12) for M layers in the intermediate region. On the other hand, the quantities $\mathcal{P}^{\beta,v}(z)$ and $\mathcal{P}^{\beta,a}(z)$ are found from separate calculations for corresponding infinite systems, namely for the homogeneous vacuum represented by empty spheres, or the homogeneous bulk alloy. In the latter case, the value $\mathcal{P}^{\beta,a}(z)$ is obtained from a charge self-consistent bulk LMTO-CPA calculation.

The configurationally averaged Green's function matrix $\bar{g}^\beta_{pq}(k_\|, z)$, which is the two-dimensional Fourier transform of $\bar{g}^\beta_{RR'}(z)$, Eq. (10), $R \in \{R_p\}$, $R' \in \{R_q\}$, is represented by an inverted infinite tridiagonal matrix (similarly as the structure constants $S^\beta_{pq}(k_\|)$) with respect to the PL indices p, q. The effect of the homogeneous semi-infinite systems (bulk alloy or vacuum) with the physical properties identical for each layer can be characterized by a single quantity, the surface Green's function (SGF).[4,5] By definition, the SGF is the top PL projection of the Green's function of the homogeneous semi-infinite bulk alloy or vacuum, and its knowledge is of central importance for the present formalism. The SGF can be determined directly in a real space by the technique developed in Refs. 5 and 19, which avoids the use of the bulk Green's function and related integration over the surface-normal component $k_\perp$ of the k-vector in the bulk Brillouin zone, common to other approaches.[11,12,20] The basic idea is nearly self-evident: by adding (removing) a PL of atoms to (from) the surface we recover the same semi-infinite solid. The above condition is easily expressed mathematically, and one obtains the following equations for the SGFs of the bulk alloy $\mathcal{G}^{\beta,a}(k_{\|}, z)$ and of the vacuum $\mathcal{G}^{\beta,v}(k_{\|}, z)$[5,10]:

$$
\begin{aligned}
\mathcal{G}^{\beta,a}(k_{\|}, z) &= [\mathcal{P}^{\beta,a}(z) - S^{\beta}_{00}(k_{\|}) - S^{\beta}_{01}(k_{\|})\, \mathcal{G}^{\beta,a}(k_{\|}, z)\, S^{\beta}_{10}(k_{\|})]^{-1}, \\
\mathcal{G}^{\beta,v}(k_{\|}, z) &= [\mathcal{P}^{\beta,v}(z) - S^{\beta}_{00}(k_{\|}) - S^{\beta}_{10}(k_{\|})\, \mathcal{G}^{\beta,v}(k_{\|}, z)\, S^{\beta}_{01}(k_{\|})]^{-1},
\end{aligned}
\tag{22}
$$

which have to be solved self-consistently for each $k_{\|}$ and energy z. These SGFs provide the necessary coupling of the intermediate region to the semi-infinite bulk alloy and vacuum. The equations for the SGFs (22) have a simple physical interpretation: the first two terms in the brackets of Eqs. (22) describe the inverse Green's function of an isolated layer of atoms, $[\mathcal{G}^{\beta,l}(k_{\|}, z)]^{-1} = \mathcal{P}^{\beta}(z) - S^{\beta}_{00}(k_{\|})$, which is coupled to the semi-infinite systems characterized by corresponding SGFs, $\mathcal{G}^{\beta,a}(k_{\|}, z)$ and $\mathcal{G}^{\beta,v}(k_{\|}, z)$. This coupling is mediated by the interlayer structure constants $S^{\beta}_{01}(k_{\|})$ and $S^{\beta}_{10}(k_{\|})$. The resulting semi-infinite system is the same because it consists of an infinite number of identical PLs.

For the (p, q) block $(1 \leq p, q \leq M)$ of the inverse configurationally averaged Green's function $\bar{g}^{\beta}(k_{\|}, z)$ one thus gets[10]

$$
\begin{aligned}
\{(\bar{g}^{\beta}(k_{\|}, z))^{-1}\}_{pq} &= \{\mathcal{P}^{\beta}_{p}(z) - S^{\beta}_{00}(k_{\|}) - \Gamma^{\beta}_{p}(k_{\|}, z)\}\, \delta_{pq} \\
&\quad - S^{\beta}_{01}(k_{\|})\, \delta_{p+1,q} - S^{\beta}_{10}(k_{\|})\, \delta_{p-1,q}\, ,
\end{aligned}
\tag{23}
$$

where

$$
\begin{aligned}
\Gamma^{\beta}_{1}(k_{\|}, z) &= S^{\beta}_{10}(k_{\|})\, \mathcal{G}^{\beta,v}(k_{\|}, z)\, S^{\beta}_{01}(k_{\|}), \\
\Gamma^{\beta}_{p}(k_{\|}, z) &= 0 \qquad \text{for } p = 2, 3, ..., M-1, \\
\Gamma^{\beta}_{M}(k_{\|}, z) &= S^{\beta}_{01}(k_{\|})\, \mathcal{G}^{\beta,a}(k_{\|}, z)\, S^{\beta}_{10}(k_{\|}).
\end{aligned}
\tag{24}
$$

The quantities Γ^{β}_{1} and Γ^{β}_{M} have thus the meaning of embedding potentials that couple the intermediate region to the vacuum region and to the bulk alloy, respectively. In other words, the concept of the SGF allows one to reduce the original problem of the infinite order in the PL indices to an effective problem of the finite order M in the PL indices.

The desired quantity $\Phi^{\beta}_{p}(z)$, Eq. (11), which enters both the CPA equations (12) and the expression for the density-of-states matrix, Eq. (17), needed for the charge self-consistency within the LDA, is obtained by integrating over the SBZ of the (p, p) block of $\bar{g}^{\beta}(k_{\|}, z)$, namely

$$
\Phi^{\beta}_{p}(z) = \frac{1}{N_{\|}} \sum_{k_{\|}} \bar{g}^{\beta}_{pp}(k_{\|}, z).
\tag{25}
$$

In this equation, $N_{\|}$ is the number of atoms in a given layer. Due to the block-tridiagonal structure of $\{\bar{g}^{\beta}(k_{\|}, z)\}^{-1}$, Eq. (23), the explicit evaluation of the block-diagonal parts of $\bar{g}^{\beta}(k_{\|}, z)$ is possible using the partitioning technique.[21] The result is[9,17]

$$
\bar{g}^{\beta}_{pp}(k_{\|}, z) = [\mathcal{P}^{\beta}_{p}(z) - S^{\beta}_{00}(k_{\|}) - \Pi^{\beta}_{p}(k_{\|}, z) - \tilde{\Pi}^{\beta}_{p}(k_{\|}, z)]^{-1},
\tag{26}
$$

where

$$
\begin{aligned}
\Pi^{\beta}_{p}(k_{\|}, z) &= S^{\beta}_{01}(k_{\|})\, [\mathcal{P}^{\beta}_{p+1}(z) - S^{\beta}_{00}(k_{\|}) - \Pi^{\beta}_{p+1}(k_{\|}, z)]^{-1}\, S^{\beta}_{10}(k_{\|}), \\
\tilde{\Pi}^{\beta}_{p}(k_{\|}, z) &= S^{\beta}_{10}(k_{\|})\, [\mathcal{P}^{\beta}_{p-1}(z) - S^{\beta}_{00}(k_{\|}) - \tilde{\Pi}^{\beta}_{p-1}(k_{\|}, z)]^{-1}\, S^{\beta}_{01}(k_{\|}).
\end{aligned}
\tag{27}
$$

The initial values for the above set of recursive equations are given by the embedding potentials of the intermediate region (compare Eqs. (24))

$$\begin{aligned} \Pi^{\beta}_{M}(k_{\parallel},z) &= \Gamma^{\beta}_{M}(k_{\parallel},z)\,, \\ \tilde{\Pi}^{\beta}_{1}(k_{\parallel},z) &= \Gamma^{\beta}_{1}(k_{\parallel},z)\,. \end{aligned} \tag{28}$$

We give now another example illustrating a deep connection between the Green's functions of the bulk and its surface. Namely, we determine the Green's function of the bulk layer $\mathcal{G}^{\beta,b}(k_{\parallel},z)$, i.e. of the layer situated infinitely deep below the solid surface, from the SGF. We couple an isolated layer of bulk alloy atoms via $S^{\beta}_{01}(k_{\parallel})$ and $S^{\beta}_{10}(k_{\parallel})$ to the ideal 'left' and 'right' semi-infinite alloy subspaces characterized by the corresponding SGFs. The result is, of course, the infinite bulk alloy Green's function. Formally, this construction is again a special case of Eqs. (26, 27) with $M=1$, $\tilde{\Pi}^{\beta}_{1}(k_{\parallel},z)= S^{\beta}_{10}(k_{\parallel})\,\tilde{\mathcal{G}}^{\beta,a}(k_{\parallel},z)\,S^{\beta}_{01}(k_{\parallel})$, and $\Pi^{\beta}_{1}(k_{\parallel},z)= S^{\beta}_{01}(k_{\parallel})\,\mathcal{G}^{\beta,a}(k_{\parallel},z)\,S^{\beta}_{10}(k_{\parallel})$. The $\tilde{\mathcal{G}}^{\beta,a}(k_{\parallel},z)$ denotes the SGF of the conjugated semi-infinite subspace of the random alloy (compare Eq. (22)). For the GF of the bulk alloy layer we thus obtain

$$\begin{aligned} \mathcal{G}^{\beta,b}(k_{\parallel},z) &= [\,\mathcal{P}^{\beta,a}(z) - S^{\beta}_{00}(k_{\parallel}) \\ &\quad - S^{\beta}_{01}(k_{\parallel})\,\mathcal{G}^{\beta,a}(k_{\parallel},z)\,S^{\beta}_{10}(k_{\parallel}) - S^{\beta}_{10}(k_{\parallel})\,\tilde{\mathcal{G}}^{\beta,a}(k_{\parallel},z)\,S^{\beta}_{01}(k_{\parallel})\,]^{-1}. \end{aligned} \tag{29}$$

In this way, the maximum internal consistency for the bulk and surface calculations can be obtained, which is of importance when charge self-consistent calculations are performed.

The formalism developed in this Section can also be applied to more complicated structures. Consider, as an example, a superlattice consisting of periodically repeated n layers of A-atoms and of m layers of B-atoms. Eqs. (23, 24) are still valid, but we have now $M = n + m$, and $\mathcal{P}^{\beta}_{p}(z) = P^{\beta,Q}_{p}(z)$ with $Q = A$ for $p = 1, 2, ..., n$; $Q = B$ for $p = n+1, n+2, ..., M$. Here, $P^{\beta,Q}_{p}(z)$ is the potential function of an atom of the type Q in a given layer p within the superlattice period of M layers. The embedding potentials are modified to

$$\begin{aligned} \Gamma^{\beta,A}_{M}(k_{\parallel},z) &= S^{\beta}_{01}(k_{\parallel})\,\mathcal{G}^{\beta,A}(k_{\parallel},z)\,S^{\beta}_{10}(k_{\parallel}), \\ \Gamma^{\beta,B}_{1}(k_{\parallel},z) &= S^{\beta}_{10}(k_{\parallel})\,\mathcal{G}^{\beta,B}(k_{\parallel},z)\,S^{\beta}_{01}(k_{\parallel}). \end{aligned} \tag{30}$$

Here, $\mathcal{G}^{\beta,A}(k_{\parallel},z)$ and $\mathcal{G}^{\beta,B}(k_{\parallel},z)$ are, respectively, the SGFs of the semi-infinite superlattice terminated by A and B layers. These SGFs can be, in turn, determined from the self-consistency conditions

$$\mathcal{G}^{\beta,A}(k_{\parallel},z) = g^{\beta,A}_{11}(k_{\parallel},z), \qquad \text{and} \qquad \mathcal{G}^{\beta,B}(k_{\parallel},z) = g^{\beta,B}_{MM}(k_{\parallel},z), \tag{31}$$

applied now to the system of Eqs. (23, 24) with Γ^{β}_{1} set to zero for the calculation of $\mathcal{G}^{\beta,A}$, and Γ^{β}_{M} set to zero for the calculation of $\mathcal{G}^{\beta,B}$. The quantities $g^{\beta,A}_{pp}(k_{\parallel},z)$, $p = 1, 2, .., n$ and $g^{\beta,B}_{pp}(k_{\parallel},z)$, $p = n+1, n+2, .., M$ are then found from a set of recursive equations (26 – 28). The physical interpretation of the first of Eqs. (31) is analogous to that of Eqs. (22): after addition of first n layers of A-atoms and then m layers of B-atoms to the semi-infinite superlattice terminated by A layers we recover the same semi-infinite superlattice (and similarly for the semi-infinite superlattice terminated by B layers after addition of first m layers of B-atoms and then of n layers of A-atoms). Once $\mathcal{G}^{\beta,Q}$, $Q = A, B$ are determined, the embedding potentials $\Gamma^{\beta,B}_{1}$ and $\Gamma^{\beta,A}_{M}$ are known and the layer-diagonal elements $g^{\beta}_{pp}(k_{\parallel},z)$ of the superlattice GF matrix (23), which are needed for both the LDA and the CPA part of the problem, can be found again using Eqs. (26 – 28).

CHARGE SELF-CONSISTENCY

The all-electron charge density within the individual atomic sphere centered at the site $R \in \{R_p\}$ and occupied by an atom of the type Q can be evaluated according to

$$\varrho_p^Q(r) = 2 \sum_{LL'} \int_{-\infty}^{E_F} \psi_{pL}^Q(r,E)\, D_{p,LL'}^Q(E)\, \psi_{pL'}^Q(r,E)\, dE \;+\; \varrho_p^{Q,core}(r)\,, \tag{32}$$

where the prefactor 2 accounts for the spin degeneracy. The density-of-states matrix $D_{p,LL'}^Q(E)$ is given by Eq. (17), while the wave functions $\psi_{pL}^Q(r,E) = R_{p\ell}^Q(|r|,E)Y_L(\hat{r})$ are found from the radial solutions of the Schrödinger equation $R_{p\ell}^Q(|r|,E)$ (normalized to unity within the corresponding atomic sphere) multiplied by the real spherical harmonics $Y_L(\hat{r})$. The second term in Eq. (32) is the spherically symmetric density due to the core electrons. The spherically averaged charge density can be obtained by a simpler formula

$$\varrho_p^Q(|r|) = \frac{2}{4\pi} \sum_{L} \int_{-\infty}^{E_F} D_{p,LL}^Q(E)\, [R_{p\ell}^Q(|r|,E)]^2\, dE \;+\; \varrho_p^{Q,core}(|r|), \tag{33}$$

containing the L-diagonal elements $D_{p,LL}^Q(E)$ of the local density-of-states matrix, Eq. (17).

The wave functions in Eqs. (32, 33) obey the Schrödinger equation with the effective one-electron potential given by

$$V_p^Q(|r|) = -\frac{2Z_p^Q}{|r|} + V_p^{Q,H}(\varrho_p^Q(|r|)) + V_{p,xc}^Q(\varrho_p^Q(|r|)) + V_p^{Mad}, \tag{34}$$

which depends both on the site $R \in \{R_p\}$ and on the type of atom Q. The first three terms in Eq. (34) have a standard form and meaning common to all LDA calculations using spherical charge density (see e.g. Refs. 1, 2, 13, and 15). The first term describes the Coulomb attraction to the nucleus with the charge Z_p^Q, the second term is the Hartree potential due to the spheridized electronic charge density $\varrho_p^Q(|r|)$, and the third term is the exchange-correlation contribution. The last term in Eq. (34), the Madelung term, describes the spherically averaged part of the electrostatic potential coming from the full, non-spheridized charge density $\varrho_p^Q(r)$ in the other atomic spheres. Quite generally, this term can be expressed as

$$V_p^{Mad} = \sum_L \sum_q M_{pq}^{sL}\, \bar{q}_q^L\,, \tag{35}$$

where

$$\bar{q}_q^L = \sum_Q c_q^Q \left\{ \sqrt{\frac{4\pi}{2\ell+1}} \int_0^{s^Q} |r|^\ell\, Y_L(\hat{r})\, \varrho_q^Q(r)\, dr - Z_p^Q\, \delta_{\ell,0} \right\} = \langle Q_q^L \rangle - \langle Z_q \rangle\, \delta_{\ell,0}\,. \tag{36}$$

We note that the Madelung term at a site $R \in \{R_p\}$ is a linear combination of the configurationally averaged multipole moments $\bar{q}_q^L$ at all nonequivalent sites $R' \in \{R_q\}$. The quantities M_{pq}^{sL} are called Madelung constants.

In a bulk system consisting of several sublattices with one or more sublattices occupied randomly, the averaged net charges $\bar{q}_p^s$ (Eq. (36), terms $L \equiv s$ or $\ell = m = 0$) will be nonzero thus describing a possible charge transfer among different sublattices with different electronic, magnetic, and chemical nature. Since the configurationally averaged environment of each site usually has a certain degree of symmetry (e.g. tetrahedral), it can be expected that the higher multipole moments can be neglected and that the monopole-monopole interaction will play a decisive role.

In bulk systems with all sites equivalent (e.g. in substitutionally disordered fcc or bcc alloys), the above arguments together with the condition of overall charge neutrality lead to the vanishing Madelung term, since the averaged charge transfer at each site is equal to zero. One can go a step further and include the effect of the local fluctuations of the electronic charge by changing the atomic sphere radii s^Q so that each atomic sphere is charge neutral while preserving the total volume of the random alloy.[8,15] In this way not only the average value of the Madelung term, Eq. (35), is zero but also the local fluctuations around this value are minimized. The absence of the Madelung contribution to the one-electron potentials, Eq. (34), is compensated by the different values s^Q up to which the radial Schrödinger equation is integrated thus leading to modified potential parameters.

In the surface-related problems, the Madelung term must describe not only the charge redistribution among different layers (due to the presence of the surface and possible concentration inhomogeneities), but also the strong deviations from the spherical symmetry of the charge density at the surface. It has been shown[10,12] that the inclusion of the term $\bar{q}_p^z$ (Eq. (36), $L \equiv z$ or $\ell = 1, m = 0$, i.e. the dipole moment, the quantization axis being perpendicular to the surface) in Eq. (35) is necessary to get results comparable to full-potential slab calculations. In addition to the Madelung term in Eq. (34), nonzero net charges and dipole moments in the surface layers lead to an electrostatic potential barrier across the solid-vacuum interface. This quantity (and the closely related work function) is also very sensitive to the proper treatment of the charge nonsphericity.[10,12] The technical points concerning the evaluation of V_p^{Mad} are discussed later.

At this stage it is worth mentioning some differences between the bulk and surface alloys. The most notable difference is the pinning of the Fermi level at the surface by the underlying bulk. Consequently, the overall charge neutrality cannot be achieved by shifting the Fermi level to accommodate valence electrons as in the bulk case but via a reconstruction of the electronic structure in the region of the solid-vacuum interface (the intermediate region). As a result, as already mentioned, the properties of a top few layers differ from each other as well as from those which are deeper in the solid (the infinite medium).

GENERALIZATIONS OF THE FORMALISM

In this Section we briefly describe two important generalizations of the present formalism, namely its extension to the spin-polarized case within the local spin-density approximation (LSDA)[13] and the fully-relativistic treatment, both for random bulk alloys and their surfaces. We also briefly discuss evalution of interatomic interactions in the effective Ising model and of interlayer interactions in layered systems.

Spin-Polarized Formalism

The study of bulk magnetic systems, but especially of their two-dimensional counterparts as surfaces, interfaces, multilayers, or ultrathin films offers a number of promising applications in the area of magnetic recording and new device applications. Investigations of bulk magnetic properties of materials with varying local environment as alloys of different compositions enriched significantly our understanding of the nature of magnetism. This is even more true for systems with reduced dimensionality.

The application of the formalism to the case of collinear spin-structures is rather straightforward.[22] All relevant quantitites such as the potential parameters $X_{RL}^{Q\sigma}$ ($X = C, \Delta$, and γ), the Green's functions $G_{RL,R'L'}^{\sigma}(z)$, $g_{RL,R'L'}^{\beta,\sigma}(z)$, and $\mathcal{G}^{\beta,a,\sigma}(k_{\parallel},z)$, Eqs. (7), (10), and (22), as well as the potential functions $P_{RL}^{\beta,Q\sigma}(z)$ and $\mathcal{P}_p^{\beta,\sigma}(z)$, Eqs. (6) and (21), respectively, are trivially spin-resolved ($\sigma = \uparrow, \downarrow$ for the spin-up and spin-down quantities). Note, however, that the tight-binding structure constants $S_{RL,R'L'}^{\beta}$ are spin-independent by definition. The CPA part of the problem, Eqs. (11), can be then solved for each spin orientation σ separately whereby the spin-up and spin-down charge densities (compare Eqs. (32, 33)) or, equivalently, the component total charge $\varrho_p^Q(|r|)$ and magnetization $m_p^Q(|r|)$ densities are needed in the exchange-correlation part of the LSDA potential. Thus, Eq. (34) is modified as follows:

$$V_p^{Q,\sigma}(|r|) = -\frac{2Z_p^Q}{|r|} + V_p^{Q,H}(\varrho_p^Q(|r|)) + V_{xc,p}^{Q,\sigma}(\varrho_p^Q(|r|), m_p^Q(|r|)) + V_p^{Mad}, \tag{37}$$

where

$$\varrho_p^Q(|r|) = \varrho_p^{Q,\uparrow}(|r|) + \varrho_p^{Q,\downarrow}(|r|), \quad m_p^Q(|r|) = \varrho_p^{Q,\uparrow}(|r|) - \varrho_p^{Q,\downarrow}(|r|). \tag{38}$$

Note that the electrostatic part of the LSDA potential, namely its Hartree part and the Madelung part, depend on the total charge densities.

Relativistic Formalism

Metallic alloys and their surfaces, particularly those containing heavy transition and noble metal elements, exhibit a number of properties interesting both from physical and technological points of view. The phenomena of ordering and disordering at CuAu surfaces, of the face-dependent segregation in NiPt alloys, or the use of platinum or other heavy transition metals and their alloys in heterogeneous catalysis are just a few examples. Self-consistent fully-relativistic studies of random bulk alloys and, in particular, of random alloy surfaces are still rare.

Formally, there are two ways for including relativististic effects into the formalism. One possibility is to start from the scalar-relativistic theory (e.g. as that developed in previous Sections) and to add perturbatively the spin-orbit coupling. The other possibility is to reformulate a recently developed fully-relativistic LMTO method (see Ref. 23) in the GF language and apply it to the case of random alloys and their surfaces.

A key point is the solution $\phi_\Lambda(E,r)$ of the single-site Dirac equation for a spherically symmetric potential $V_p^Q(|r|)$ confined to the atomic sphere[24] and normalized here to unity:

$$\phi_{p\Lambda}^Q(E,r) = i^{\,\ell} \begin{pmatrix} g_{p\kappa}^Q(E,|r|)\,\Omega_{\kappa\mu}(\hat{r}) \\ i\,f_{p\kappa}^Q(E,|r|)\,\Omega_{-\kappa\mu}(\hat{r}) \end{pmatrix}, \quad \Lambda = (\kappa\mu). \tag{39}$$

The radial functions $g_{p\kappa}^Q(E,|r|)$ and $f_{p\kappa}^Q(E,|r|)$ are regular solutions of the radial Dirac equation in an atomic sphere on a given subset $\{R_p\}$ of lattice sites and the $\Omega_{\kappa\mu}(\hat{\mathbf{r}})$

are spin-spherical harmonics.[24] The relativistic quantum numbers are $\kappa = -\ell - 1$ for $j = \ell + 1/2$, $\kappa = \ell$ for $j = \ell - 1/2$, and $-j \le \mu \le j$.

Once the radial functions $g^Q_{p\kappa}$ and $f^Q_{p\kappa}$ are calculated, the potential parameters $X^Q_{R\Lambda}$ ($X = C, \Delta$, and γ) and the potential fuctions $P^{\beta,Q}_{R\Lambda}(z)$ may be found.[23] The potential functions and the potential parameters are matrices diagonal in Λ:

$$\begin{aligned} P^{\alpha}_{R\Lambda,R'\Lambda'}(z) &= P^{\alpha}_{R\Lambda}(z)\,\delta_{RR'}\,\delta_{\Lambda\Lambda'} \equiv P^{\alpha}_{R\kappa}(z)\,\delta_{RR'}\,\delta_{\Lambda\Lambda'}\,, \\ P^{\alpha}_{R\kappa}(z) &= (z - C_{R\kappa})/\left[\Delta_{R\kappa} + (\gamma_{R\kappa} - \alpha_{R\kappa})(z - C_{R\kappa})\right], \end{aligned} \tag{40}$$

where $\alpha_{R\kappa} = \alpha_\kappa$. The relativistic (canonical) structure constants

$$\begin{aligned} S^0_{R\Lambda,R'\Lambda'} &= \sum_{s=\pm\frac{1}{2}} c(\ell j 1/2; \mu - s, s)\, S^0_{RL,R'L'}\, c(\ell' j' 1/2; \mu' - s, s)\,, \\ L &= (l, \mu - s) \quad \text{and} \quad L' = (l', \mu' - s)\,, \end{aligned} \tag{41}$$

are defined using their non-relativistic counterparts $S^0_{R'L',\,RL}$ (2) and Clebsch-Gordan coefficients. Thus, in the relativistic case, our starting Hamiltonian is still of the form of Eq. (1), but the potential parameters, the potential functions, the structure constants and, consequently, the Green's functions are now matrices with indices R and Λ. For example, for a basis set consisting of s-, p-, and d-orbitals on each site, the corresponding matrices are of order 18×18 rather than 9×9 as in the non-relativistic treatment. Keeping this in mind, all developments in previous Sections, in particular those concerning the CPA and the SGF, remain formally unchanged. As concerns the LDA part of the problem, the component all-electron charge density in a given atomic sphere, which is needed for the construction of effective one-electron potential, Eqs. (34 – 36), is given by

$$\varrho^Q_p(r) = \sum_{\Lambda\Lambda'} \int_{-\infty}^{E_F} (\phi^Q_{p\Lambda}(E,r))^\dagger\, D^Q_{p,\,\Lambda\Lambda'}(E)\, \phi^Q_{p\Lambda'}(E,r)\, dE \;+\; \varrho^{Q,core}_p(r)\,. \tag{42}$$

The local density-of-states matrix $D^Q_{p,\,\Lambda\Lambda'}(E)$ is a corresponding counterpart of Eq. (17). Its diagonal elements, $D^Q_{p,\,\Lambda\Lambda}(E)$, are the local densities of states, and the corresponding spherically averaged charge density is then simply given by

$$\varrho^Q_p(|r|) = \frac{1}{4\pi} \sum_{\Lambda} \int_{-\infty}^{E_F} D^Q_{p,\,\Lambda\Lambda}(E)\,[(f^Q_{p\kappa}(E,|r|))^2 + (g^Q_{p\kappa}(E,|r|))^2]\, dE \;+\; \varrho^{Q,core}_p(|r|)\,. \tag{43}$$

We note that the definition of the Madelung term, Eqs. (35, 36) remains unchanged. Otherwise, the fully relativistic theory for bulk alloys and their surfaces can be developed in an analogy to its scalar-relativistic counterpart. For further details we refer to Ref. 25.

Interatomic Interactions

The configurational dependence of the total energy of a semi-infinite disordered binary alloy or random overlayer on the non-random substrate can be expressed in the form of an effective Ising type Hamiltonian

$$H_I = E_0 + \sum_R \sum_Q D^Q_R\, \delta\eta^Q_R + \frac{1}{2} \sum_{R \ne R'} \sum_{Q,Q'} V^{QQ'}_{RR'}\, \delta\eta^Q_R\, \delta\eta^{Q'}_{R'} + \ldots\,. \tag{44}$$

The parameters entering (44) are the configurationally independent part of the total energy E_0, the on-site energies D^Q_R, the pair interatomic interactions $V^{QQ'}_{RR'}$, and generally, the interatomic interactions of higher orders. The pair interactions play a central role in the study of order-disorder phenomena while the on-site energies are important for the surface segregation problems.[26] A particular configuration of the alloy is determined by a set of occupation indices η_R^Q, where $\eta_R^Q = 1$ if the site R is occupied by an atom of the type Q, $\eta_R^Q = 0$ otherwise, and $\delta\eta_R^Q = \eta_R^Q - \langle\eta_R^Q\rangle$. Using the transformation to the lattice gas model,[27] ($\eta_R^A = \eta_R$, $\eta_R^B = 1 - \eta_R$), Eq. (44) can be recast into a simpler form

$$H_I = E_o + \sum_R D_R\, \delta\eta_R + \frac{1}{2} \sum_{R \neq R'} V_{RR'}\, \delta\eta_R\, \delta\eta_{R'} + \ldots\,, \tag{45}$$

where $\delta\eta_R = \eta_R - \langle\eta_R\rangle$ and

$$\begin{aligned} D_R &= D_R^A - D_R^B\,, \\ V_{RR'} &= V_{RR'}^{AA} + V_{RR'}^{BB} - V_{RR'}^{AB} - V_{RR'}^{BA}\,. \end{aligned} \tag{46}$$

The parameters D_R and $V_{RR'}$ can be determined by using various methods. Here we mention the Generalized Perturbation Method (GPM),[27] which is based on the force or structural energy theorem,[27] according to which it is sufficient to map only the band term of the total energy onto the Ising Hamiltonian. The band term depends on the alloy configuration in two ways: via the density of states, and via the Fermi energy E_F. The configurational dependence of the E_F can be eliminated by employing the grandcanonical rather than the canonical ensemble. The band energy term is replaced by the grandcanonical potential for non-interacting particles at $T = 0$,

$$\Omega = E_{band} - E_F N_{el} = -\int_{-\infty}^{E_F} N(E)\, dE\,. \tag{47}$$

Here E_F is the Fermi energy, N_{el} is the number of electrons in a system, and $N(E)$ is the integrated density of states. The Fermi energy is fixed to its configurationally averaged value. This average is a well defined quantity as follows from the self-averaging property of E_F, whose fluctuations are of order $O(N^{-\frac{1}{2}})$. Within the present formalism the expression for the integrated density of states $N(E)$ is[8,28]

$$N(E) = \frac{1}{\pi}\, \mathrm{Im}\, \mathrm{Tr} \left(\frac{1}{2} \ln[\dot{P}^\beta(z)] + \ln g^\beta(z) \right), \tag{48}$$

where $z = E + i0$ and Tr denotes the trace over the site and angular momentum indices. The first term compensates the extra singularities in $\ln g^\beta(z)$ which originate from the poles of $P^\beta(z)$ (compare also Eq. (7)). In the tight-binding representation β, as a rule, these singularities lie well outside the occupied part of the spectrum and need not be considered.[8,28]

Within the GPM, the Ising Hamiltonian parameters D_R^Q and $V_{RR'}^{QQ'}$ for a semi-infinite alloy are given by[28]

$$D_R^Q = -\frac{1}{\pi}\, \mathrm{Im} \int_{-\infty}^{E_F} \mathrm{tr} \ln \left[1 - \left(P_R^{\beta,Q}(z) - \mathcal{P}_R^\beta(z) \right) \bar{g}_{RR}^\beta(z) \right] dE\,, \tag{49}$$

and

$$V_{RR'}^{QQ} = -\frac{1}{\pi} \mathrm{Im} \int_{-\infty}^{E_F} \mathrm{tr}\ \ln \left[1 - t_R^{\beta,Q}(z)\, \bar{g}_{RR'}^\beta(z)\, t_{R'}^{\beta,Q'}(z)\, \bar{g}_{R'R}^\beta(z) \right] dE\,. \tag{50}$$

Here $z = E + i0$, and $t_R^{\beta,Q}$ is the single-site T-matrix for $Q = A, B$ atoms in the efective medium (compare Eq. (11)). The contributions from the other terms in the total energy (double counting, core and electrostatic terms) may become important for some alloy systems. Their role is elucidated in an improved theory based on fully selfconsistent calculations.[29]

Interlayer Interactions

The magnetic state of metallic magnetic multilayers is governed by the exchange interactions. Let us consider a simple system, composed of two ferromagnetic slabs (e.g. Co) separated by N atomic layers of a non-magnetic spacer (e.g. Cu), and sandwiched between two semi-infinite non-magnetic half-spaces made again from Cu.[30] We denote the spacer by index $\mathcal{C}$, the left part (i.e. left halfspace and left magnetic slab) of the system by $\mathcal{L}$, and the right part (i.e. right halfspace and right magnetic slab) by $\mathcal{R}$.

According to the force theorem, the energy difference between the states with parallel and anti-parallel orientation of magnetization is determined solely by the band energy terms and can be expressed via the grand canonical potentials of the antiferromagnetic (AF) and ferromagnetic (F) alignments, $E_x = \Omega_{AF} - \Omega_F$, defined as

$$\Omega_\lambda = -\frac{1}{\pi}\,\mathrm{Im}\int_{-\infty}^{\infty} f(E) \sum_\sigma \mathrm{Tr}\,\ln \mathcal{G}_\lambda^\sigma(z)\, dE\,. \tag{51}$$

Here $\lambda = F$ or AF denotes the type of alignment, $z = E + i0$, $f(E)$ is the Fermi-Dirac distribution, σ is the spin index, and $\mathcal{G}_\lambda^\sigma(z)$ is the Green's function of the system. Note that Eq. (51) is a particular form of the Lloyd's formula.[31]

The expression for $\mathrm{Tr}\,\ln \mathcal{G}_\lambda^\sigma(z)$ can be split into parts corresponding to the spacer, the magnetic subsystems, and to the single interfaces. The coupling of the magnetic interfaces can be then extracted as

$$\delta \mathrm{Tr}\,\ln \mathcal{G}_\lambda^\sigma(z) = -\,\mathrm{Tr}\,\ln\left[1 - \tau_{\lambda\mathcal{L}}^\sigma(z)\, \mathcal{G}_{\mathcal{LR}}(z)\, \tau_{\lambda\mathcal{R}}^\sigma(z)\, \mathcal{G}_{\mathcal{RL}}(z)\right]. \tag{52}$$

The quantities $\tau_{\lambda\mathcal{L}}^\sigma$, $\tau_{\lambda\mathcal{R}}^\sigma$ are the T-matrices at the two interfaces $\mathcal{L}/\mathcal{C}$ and $\mathcal{C}/\mathcal{R}$, respectively, and $\mathcal{G}_{\mathcal{LR}}$ ($\mathcal{G}_{\mathcal{RL}}$) are corresponding projections of the unperturbed Green's function. The result (52) is quite general and can be used to evaluate the energy differences in any system divided into unperturbed part and two localized perturbations. For example, if we consider the decoupled subsystems $\mathcal{L}, \mathcal{C}, \mathcal{R}$ as the unperturbed system, and their interactions at the interfaces $\mathcal{L}/\mathcal{C}$ and $\mathcal{C}/\mathcal{R}$, given by the interlayer structure constants S_{01} and S_{10} as perturbations, the Eq. (52) can be cast into the form

$$\delta\, \mathrm{Tr}\,\ln \mathcal{G}_\lambda^{\beta\sigma}(z) = -\frac{1}{N_\parallel}\sum_{k_\parallel} \mathrm{tr}\,\ln\left[1 - \tau_{\lambda 1}^{\beta\sigma}(k_\parallel, z)\, g_{1N}^\beta(k_\parallel, z)\, \tau_{\lambda N}^{\beta\sigma}(k_\parallel, z)\, g_{N1}^\beta(k_\parallel, z)\right]. \tag{53}$$

The effective T-matrices $\tau_{\lambda i}^{\beta\sigma}$ $(i = 1, N))$

$$\tau_{\lambda i}^{\beta\sigma}(k_\parallel, z) = \Gamma_{\lambda i}^{\beta\sigma}(k_\parallel, z)\,\left[1 - g_{ii}^\beta(k_\parallel, z)\, \Gamma_{\lambda i}^{\beta\sigma}(k_\parallel, z)\right]^{-1} \tag{54}$$

are defined in terms of the embedding potentials (24)

$$\begin{aligned} \Gamma_{\lambda 1}^{\beta\sigma}(k_\parallel, z) &= S_{10}(k_\parallel)\, \tilde{\mathcal{G}}_\lambda^{\beta\sigma}(k_\parallel, z)\, S_{01}(k_\parallel)\,, \\ \Gamma_{\lambda N}^{\beta\sigma}(k_\parallel, z) &= S_{01}(k_\parallel)\, \mathcal{G}_\lambda^{\beta\sigma}(k_\parallel, z)\, S_{10}(k_\parallel)\,, \end{aligned} \tag{55}$$

that describe scattering of spacer electrons at $\mathcal{L}/\mathcal{C}$ and $\mathcal{C}/\mathcal{R}$ interfaces. Here, $g_{11}^\beta(z)$, $g_{1N}^\beta(z)$, $g_{N1}^\beta(z)$, and $g_{NN}^\beta(z)$ are the corresponding blocks of the spacer slab Green's function, tr denotes the trace over the angular momentum index (L), and $S_{01}(k_\parallel)$ and $S_{10}(k_\parallel)$ are the interlayer structure constants coupling the neighboring PLs (compare

Eqs. (19, 20)). Finally, $\tilde{\mathcal{G}}_{\lambda}^{\beta\sigma}(k_{\parallel}, z)$ and $\mathcal{G}_{\lambda}^{\beta\sigma}(k_{\parallel}, z)$ are the SGFs of magnetic subsystems $\mathcal{L}$ and $\mathcal{R}$, respectively.

The coupling between the two magnetic subsystems is due to the layer off-diagonal projections $g_{1N}^{\beta}(k_{\parallel}, z)$ and $g_{N1}^{\beta}(k_{\parallel}, z)$ of the Green's function of the finite spacer. The oscillatory behavior of interlayer coupling follows from the oscillatory behavior of the quasi-one-dimensional spacer Green's functions $g_{1N}(k_{\parallel}, z)$ and $g_{N1}(k_{\parallel}, z)$ that connect the first and the last layer of the spacer, while the effective T-matrices determine the phases and amplitudes of the oscillations. This formulation can describe systems with arbitrary thickness of the magnetic slabs including two limiting cases, namely the model of two magnetic monolayers in an ideal bulk spacer, and a model of two semi-infinite magnetic subsystems separated by a finite non-magnetic spacer. Even more important is a possibility to employ Eq. (53) to describe the effect of randomness in the spacer, magnetic slabs, or at $\mathcal{L}/\mathcal{C}$ and $\mathcal{C}/\mathcal{R}$ interfaces between magnetic and non-magnetic layers within the CPA.

COMPUTATIONAL IMPLEMENTATION

Here we present some details concerning the numerical implementation of the formalism developed in previous Sections.

Solution of the CPA Equations

The CPA equations (11, 12) constitute a set of coupled nonlinear equations for the coherent potential functions $\mathcal{P}_p^{\beta}(z)$. The physical solution must preserve the analytical properties of the Green's function $\Phi^{\beta}(z)$. The form of CPA equations as expressed by (12) is not suitable for numerical implementation as it does not converge in the strong scattering case. Therefore, we rewrite the system of CPA equations to a numerically more stable form by introducing the notion of the coherent interactor $\Omega_p^{\beta}(z)$.[8] It is defined with the help of the configurationally averaged Green's function $\Phi_p^{\beta}(z)$, Eq. (11), as follows:

$$\Phi_{p,LL'}^{\beta}(z) = \{[\mathcal{P}_p^{\beta}(z) - \Omega_p^{\beta}(z)]^{-1}\}_{LL'}\,. \tag{56}$$

The physical meaning of $\Omega_p^{\beta}(z)$ is simple: it describes an energy-dependent coupling of an effective atom in a random alloy, as characterized by the coherent potential function $\mathcal{P}^{\beta}(z)$, to all other sites. Using the definition (56), the expression for the local conditionally averaged Green's function $\Phi_p^{\beta,Q}(z)$, Eq. (17), becomes surprisingly simple and physically transparent:

$$\Phi_{p,LL'}^{\beta,Q}(z) = \{[P_p^{\beta,Q}(z) - \Omega_p^{\beta}(z)]^{-1}\}_{LL'}\,. \tag{57}$$

It has formally the same structure as the averaged Green's function $\Phi_p^{\beta}(z)$, Eq. (56), but the coherent potential function $\mathcal{P}^{\beta}(z)$ is substituted by the potential function $P_p^{\beta,Q}(z)$ of a given atom $Q = A, B$. This result is just in the spirit of the single-site approximation. The CPA equations (12) can be then put into the form

$$\Phi_{p,LL'}^{\beta}(z) = \sum_{Q} c_p^Q\, \Phi_{p,LL'}^{\beta,Q}(z). \tag{58}$$

The set of equations (56 – 58) is suitable for numerical implementation. It is solved by simple iterations starting with $\Omega_p^{\beta}(z) = 0$ and the virtual-crystal approximation for the coherent potential function $\mathcal{P}_p^{\beta}(z)$ (i.e. Eq. (6) with concentration-weighted values of species potential parameters C_p^Q, Δ_p^Q, and γ_p^Q). Iterations proceed in the following steps: (i) the $\mathcal{P}_p^\beta(z)$ known from the previous iteration is used to find the $\Phi_p^\beta(z)$, Eq. (11); (ii) a new $\Omega_p^\beta(z)$ is found from Eq. (56) and using it new conditionally averaged Green's functions $\Phi_{p,LL'}^{\beta,Q}(z)$, Eq. (57), are calculated; (iii) an updated value of $\Phi_p^\beta(z)$ is calculated from the CPA condition, Eq. (58); and (iv) a new $\mathcal{P}_p^\beta(z)$ is found employing $\Phi_p^\beta(z)$, updated in the previous step, from Eq. (56). The LDA part follows the CPA iteration: the potential functions $P_p^{\beta,Q}(z)$ appearing in Eq. (57) are updated with the help of just calculated $\Phi_{p,LL'}^{\beta,Q}(z)$ (compare Eqs. (17)). The CPA and LDA iterations are thus performed in an alternating manner, one CPA iteration followed by one LDA iteration, until the full selfconsitency to required accuracy is achieved. The above system of CPA equations converges reliably for both bulk and surface cases, and independently for each energy. This 'parallel' feature can be used in future, together with an obvious parallelism inherent to the k-space Brillouin zone (BZ) integration, to speed up the calculations significantly.

Complex Contour Integration and Deconvolution

The LDA part of the problem requires an evaluation of integrals over the energy interval (E_b, E_F), where E_b holds for the bottom of the energy spectrum of the system and E_F is the Fermi level. We substitute this integral by a contour integral in the complex energy plane

$$\int_{E_b}^{E_F} f(E+i0)\,dE = \oint_C f(z)\,dz\,, \tag{59}$$

where $f(z)$ is some analytical function, e.g. the Green's function. The applied semi-circle contour C starts at some minimum energy E_{min} below the bottom of the valence spectrum E_b and ends at E_F. The integration is performed using the Gaussian quadrature method, which avoids the knowledge of the integrand at E_F. The k-space integration takes advantage of energy variables with finite imaginary parts, thereby smoothing out the integrand and speeding up the iteration processes involved in the evaluation of the CPA equations (56 – 58) and of the surface Green's functions (22). Typically 10 to 15 energy points are sufficient for an accurate energy integration over the energy interval of about 1 Ry.

In some situations, the quantities as density of states or the spectral density on the real axis are needed. Then the Green's function can be evaluated on a dense mesh of energy points $z_k = E_{min} + h(k-1) + i\Delta$, $f_k = f(z_k)$ slightly above the real axis, and then, using the analytical deconvolution method,[32] its values on the real axis can be found. There are many versions of that approach, but it seems that the simplest one (linear deconvolution) is the most efficient and reliable. Its idea is quite simple: the function $f(z)$ is approximately represented by its Taylor expansion in the neighbourhood of the point $\bar{z}$ up to linear term, i.e. $f(z) \approx f(\bar{z}) + f'(\bar{z})(z-\bar{z})$ with $\bar{z} = (z_k + z_{k+1})/2$, $f(\bar{z}) \approx (f_k + f_{k+1})/2$, and $f'(\bar{z}) \approx (f_{k+1} - f_k)/h$. Then the value of the function $f(z)$ in a new point $\bar{z} - i\delta$, $\delta \leq \Delta$, closer to the real axis is calculated. The procedure can be repeated several times.

Solution of the Equation for the SGF

A natural way of solving the nonlinear equation (22) for the SGF seems to be a simple iteration of Eq. (22), using the layer Green's function $\mathcal{G}^{\beta(0)}(k_\|, z) = [(\mathcal{P}^\beta(z) - S_{00}^\beta(k_\|)]^{-1}$, as a starting value. The next iteration adds to an initial layer another one, so that after n iterations we have a slab consisting of $(n+1)$ layers. At some $(k_\|, z)$

points hundreds of such iterations, particularly for the complex energies z close to the real axis, are needed to obtain the converged value. This problem may be overcome by the renormalization-decimation technique proposed in Ref. 33. The basic idea of this method, which we use to solve Eqs. (22), is the following. We replace the original set of PLs by an effective one with doubled lattice constant, where a pair of neighboring PLs is replaced by one effective layer in a new set. These new layers are coupled to each other by effective energy-dependent interlayer structure constants with weaker interactions than those in the original set. This replacement can be repeated iteratively until the effective interlayer interactions vanish. After the i-th iteration one thus represents a semi-infinite solid by a stack of 2^i original PLs. In other words, the expression for $\mathcal{G}^{\beta}(k_{\|}, z) \equiv \mathcal{G}^{\beta,a}(k_{\|}, z)$ after the i-th renormalization-decimation iteration is

$$\mathcal{G}^{\beta\,(i)}(k_{\|}, z) = (\mathcal{P}^{\beta}(z) - \tilde{S}_{00}^{\beta\,(i)}(k_{\|}, z))^{-1}, \tag{60}$$

where

$$\begin{aligned} \tilde{S}_{00}^{\beta\,(i)}(k_{\|}, z) &= \tilde{S}_{00}^{\beta\,(i-1)}(k_{\|}, z) + \tilde{S}_{01}^{\beta\,(i-1)}(k_{\|}, z) \left(\mathcal{P}^{\beta}(z) - \hat{S}_{00}^{\beta\,(i-1)}(k_{\|}, z)\right)^{-1} \tilde{S}_{10}^{\beta\,(i-1)}(k_{\|}, z), \\ \tilde{S}_{01}^{\beta\,(i)}(k_{\|}, z) &= \tilde{S}_{01}^{\beta\,(i-1)}(k_{\|}, z) \left(\mathcal{P}^{\beta}(z) - \hat{S}_{00}^{\beta\,(i-1)}(k_{\|}, z)\right)^{-1} \tilde{S}_{01}^{\beta\,(i-1)}(k_{\|}, z), \end{aligned} \tag{61}$$

and similar equation holds for $\tilde{S}_{10}^{\beta\,(i)}(k_{\|}, z)$. Finally,

$$\begin{aligned} \hat{S}_{00}^{\beta\,(i)}(k_{\|}, z) &= \hat{S}_{00}^{\beta\,(i-1)}(k_{\|}, z) + \tilde{S}_{01}^{\beta\,(i-1)}(k_{\|}, z) \left((\mathcal{P}^{\beta}(z) - \hat{S}_{00}^{\beta\,(i-1)}(k_{\|}, z)\right)^{-1} \tilde{S}_{10}^{\beta\,(i-1)}(k_{\|}, z) \\ &+ \tilde{S}_{10}^{\beta\,(i-1)}(k_{\|}, z) \left((\mathcal{P}^{\beta}(z) - \hat{S}_{00}^{\beta\,(i-1)}(k_{\|}, z)\right)^{-1} \tilde{S}_{01}^{\beta\,(i-1)}(k_{\|}, z). \end{aligned} \tag{62}$$

Here, $\tilde{S}_{00}^{\beta\,(i)}(k_{\|}, z)$ and $\tilde{S}_{01}^{\beta\,(i)}(k_{\|}, z)$, $\tilde{S}_{10}^{\beta\,(i)}(k_{\|}, z)$ are renormalized intralayer and interlayer couplings, respectively. Clearly, this method represents a significant improvement over the simple iteration scheme described at the beginning of this subsection. For further details we refer to Ref. 33.

There exists a simpler and sometimes more effective way of solution of Eqs. (22). One can still use a simple iteration procedure, but the starting value $\mathcal{G}^{\beta\,(0)}(k_{\|}, z)$ should be chosen as close to the exact solution as possible. We can use the known solution for z_0 in the vicinity of the energy point z, namely $\mathcal{G}^{\beta\,(0)}(k_{\|}, z) = \mathcal{G}^{\beta}(k_{\|}, z_0)$. For the initial energy on the complex contour, which is outside the band spectrum, the Green's function is a smooth function without singularities, and it is sufficient to use a simple iteration scheme as described at the beginning of this subsection. This method in its efficiency supersedes the renormalization-decimation technique, provided that the SGF need not to be evaluated too close to the real axis and the energy points z and z_0 are sufficiently close to each other in the complex energy plane. This latter condition can be relaxed if the energy points z and z_0 are sufficiently deep in the complex energy plane. These requirements are usually fulfilled in the case of complex contour integration. In other words, we pay for the simplicity of this method by loosing the 'parallel' property with respect to the energy which, on the other hand, is preserved by the renormalization-decimation technique.

Solution of the LDA Equations

The problem of the charge self-consistency is essentially the same for ordered and disordered systems, for bulk and surfaces, and for non-relativistic and relativistic cases. The basic quantity for the charge density calculations is the local density-of-states matrix $D^Q_{p,LL'}(E)$, Eq. (17). It is this quantity which contains the information on mutual positions of atoms in a solid and on their occupations. On the other hand, the radial wave functions $R^Q_{p\ell}(|r|, E)$ in the non-relativistic case and $g^Q_{p\kappa}(E, |r|)$ and $f^Q_{p\kappa}(E, |r|)$ in the relativistic approach for both the valence and core orbitals can be found for each atomic sphere separately. The evaluation of the spheridized electron charge densities $\varrho^Q_p(|r|)$ according to Eq. (33) therefore represents only a negligible numerical effort compared to the determination of the density-of-states matrix $D^Q_{p,LL'}(E)$. The charge densities are expressed in terms of the energy moments $m^{Q,k}_{pL}$ $(k = 0, 1, 2)$ of the local atom- and momentum- resolved density of states.[1,13] Their form suitable for the complex contour integration is[34]

$$m^{Q,k}_{p,L} = -\frac{1}{\pi} \mathrm{Im} \oint_{C+} (z - E^Q_{\nu p\ell})^k \, \mathcal{D}^Q_{p,LL}(z) \, dz. \tag{63}$$

Here, the symbol C^+ denotes a contour C that is closed in the upper part of the complex energy plane ($\mathrm{Im}\, z > 0$) and the local conditionally averaged physical Green's function $\mathcal{D}^Q_{p,LL}(z)$ is defined by Eq. (17). These moments appear as a result of approximating the energy dependence of the radial wave functions by three terms of its Taylor expansion around a chosen value $E^Q_{\nu p\ell}$. The non-spheridized charge density, Eq. (32), need not be evaluated throughout the whole atomic sphere, since the nonspherical parts of the density enter the one-electron potentials (34) only via the multipole moments, Eq. (36), which are integral characteristics. These integrals are expressed using the moments $m^{Q,kk'}_{p,LL'}$ $(k, k' = 0, 1, 2)$ of the local density-of-states matrix, Eq. (17), as

$$m^{Q,kk'}_{p,LL'} = -\frac{1}{\pi} \mathrm{Im} \oint_{C+} (z - E^Q_{\nu p\ell})^k \, \mathcal{D}^Q_{p,LL'}(z) \, (z - E^Q_{\nu p\ell'})^{k'} \, dz. \tag{64}$$

The resulting multipole moments are thus reduced to radial integrals up to the Wigner-Seitz radius s^Q and angular integrals over the unit sphere.

The charge density of core electrons $\varrho^{Q,core}_p(|r|)$ is recalculated at each LDA iteration instead of the usual frozen-core approximation.[12,13] This has only negligible effect on the resulting valence charge densities, but the corresponding changes in the core orbitals and their eigenvalues are of importance for the core-level shifts[10] or hyperfine magnetic fields.

To achieve the self-consistency within the ASA, the effective one-electron potentials (34) are needed. The Madelung constants, due to the long range nature of the electrostatic interactions, must be evaluated by means of the Ewald technique. For complex bulk lattices consisting of several sublattices this is a standard problem.[13] For surfaces, the two-dimensional Ewald techniques must be used.[12,16] Moreover, the problem of the charge neutrality of the finite number of self-consistently treated layers must be overcome. One possibility is to introduce additional energy shifts to potentials in the intermediate region which restores the perfect charge neutrality.[12] Another possibility is to redefine reference points for electrostatic potentials.[10] Instead of reference points at infinite distances from the surface one can choose reference points in the first bulk and vacuum layers which are not treated self-consistently. Details of derivation of Madelung constants can be found in the next subsection (see also Ref. 35). We note that in the relativistic case the quantities $D^Q_{p,LL'}(E)$ are substituted by their relativistic counterparts, $D^Q_{p,\Lambda\Lambda'}(E)$, but otherwise we proceed in the same way.

The calculation of the potential parameters X_{pL} or $X_{p\Lambda}$ $(X = C, \Delta, \gamma)$ from the wave functions and their radial derivatives at $|r| = s^Q$ is again a procedure well documented in the literature[13,23,25] and is omitted here. The input of these potential parameters into the Green's function problem closes up the loop and makes the theory fully self-consistent in the sense of CPA and LDA.

The Fermi level is a bulk property even for surface problems. Its evaluation requires the knowledge of the total bulk density of states, which is found by integrating over the subset of k-points corresponding to an infinite solid (the bulk Brillouin zone). Similarly, the layer-resolved densities of states needed for the charge self-consistent solution within the LDA are obtained by integrating over the subset of k-points corresponding to a layer (the surface Brillouin zone). The internal consistency between both calculations can be achieved by evaluating the bulk Green's function (and the Fermi level) by integrating the Green's function of a bulk layer, Eq. (29), over the surface Brillouin zone. The Green's function determined in this way then enters the self-consistent bulk TB-LMTO-CPA codes.

Madelung Constants for Surface Problems

For low-index surfaces of homogeneous bulk systems (pure metals or substitutional alloys), the Madelung contributions V_p^{Mad} to the one-electron potentials (34) in the individual atomic layers indexed by p can be constructed from all net charges $\bar{q}_q^s$ and dipole moments $\bar{q}_q^z$ corresponding to the electronic charge densities in the atomic spheres in the intermediate region[10,35]

$$V_p^{Mad} = \sum_q \left(M_{pq}^{ss}\, \bar{q}_q^s \,+\, M_{pq}^{sz}\, \bar{q}_q^z \right) . \tag{65}$$

The calculation of the surface Madelung constants M_{pq}^{ss} and M_{pq}^{sz} employs a special function $\varphi_2(r) = \varphi_2(r_{||}, r_{\perp})$ which is equal to the electrostatic potential at the point $r = (r_{||}, r_{\perp})$ generated by a two-dimensional (2D) lattice of unit point charges. Here $r_{\perp}$ is the coordinate along the normal to the plane of the point charges and $r_{||}$ is the 2D projection of the radius vector r onto this plane. The point charges are placed at the 2D lattice points $(r_{||}, r_{\perp}) = (T_{||}, 0)$ where $T_{||}$ runs over the 2D translation vectors. The resulting electrostatic potential for points outside the plane ($|r_{\perp}| > 0$) is equal to

$$\varphi_2(r) = -\frac{4\pi}{A}\,|r_{\perp}| + \frac{4\pi}{A} \sum_{G_{||}\neq 0} \frac{\cos(G_{||}\cdot r_{||})}{|G_{||}|}\, \exp(-|G_{||}|\,|r_{\perp}|)\ , \tag{66}$$

where $G_{||}$ denotes the 2D lattice points in the reciprocal space and A is the area of the 2D cell in the real space. For points near and inside the plane, a superposition of Gaussian charge distributions centered at the 2D lattice points and an $r_{\perp}$-dependent charge density with a Gaussian profile must be considered. The Fourier analysis then leads to the electrostatic potential

$$\begin{aligned}
\varphi_2(r) \;&=\; -\frac{4\pi}{A}\, r_{\perp} \operatorname{erf}\left(\frac{r_{\perp}}{2\sigma}\right) \;-\; \frac{8\sigma\sqrt{\pi}}{A}\, \exp\left(-\frac{r_{\perp}^2}{4\sigma^2}\right) \\
&+\; \sum_{T_{||}} \frac{2}{\sqrt{(r_{||}-T_{||})^2 + r_{\perp}^2}}\, \operatorname{erfc}\left(\frac{\sqrt{(r_{||}-T_{||})^2 + r_{\perp}^2}}{2\sigma}\right) \\
&+\; \frac{2\pi}{A} \sum_{G_{||}\neq 0} \frac{\cos(G_{||}\cdot r_{||})}{|G_{||}|} \left[\exp(|G_{||}|\, r_{\perp}) \operatorname{erfc}\left(\sigma\,|G_{||}| + \frac{r_{\perp}}{2\sigma}\right) \right. \\
&\qquad \left. +\; \exp(-|G_{||}|\, r_{\perp}) \operatorname{erfc}\left(\sigma\,|G_{||}| - \frac{r_{\perp}}{2\sigma}\right) \right] ,
\end{aligned} \tag{67}$$

where erfc x stands for the complementary error function

$$\operatorname{erfc} x \;=\; \frac{2}{\sqrt{\pi}} \int_x^\infty \exp(-t^2)\, dt \;, \tag{68}$$

$\operatorname{erf} x = 1 - \operatorname{erfc} x$, and $\sigma > 0$ is an Ewald's parameter proportional to the Gaussian width. The function $\varphi_2(r)$ exhibits the 2D translational periodicity and has singularities for $r = (T_\parallel, 0)$. The removal of the singularity at $r = 0$ leads to the value

$$\tilde{\varphi}_2(0) = -\frac{2}{\sigma\sqrt{\pi}} + \sum_{T_\parallel \neq 0} \frac{2}{|T_\parallel|} \operatorname{erfc}\left(\frac{|T_\parallel|}{2\sigma}\right) - \frac{8\sigma\sqrt{\pi}}{A} + \frac{4\pi}{A} \sum_{G_\parallel \neq 0} \frac{1}{|G_\parallel|} \operatorname{erfc}(\sigma\,|G_\parallel|) \;, \tag{69}$$

which enters the expressions for the Madelung constants M_{pq}^{ss} as well. The partial derivative

$$\varphi_2'(r) \;=\; \frac{\partial}{\partial r_\perp} \varphi_2(r) \;, \tag{70}$$

which is related to the electrostatic field due to unit dipoles placed at the 2D lattice points $(r_\parallel, r_\perp) = (T_\parallel, 0)$ and oriented along the normal to the plane $r_\perp = 0$, is necessary for the evaluation of the Madelung constants M_{pq}^{sz}.

In realistic calculations, the condition of charge neutrality at the surface cannot be exactly fulfilled with finite number of atomic layers in the intermediate region which leads to unbounded values for $r_\perp \to \pm\infty$ of the electrostatic field $\chi(r)$ generated by the net charges $\bar{q}_p^s$ and the dipole moments $\bar{q}_p^z$ in the intermediate region. To overcome this difficulty, one can choose[35] a point r^{bulk} in the bulk region as a reference point for the electrostatic field $\chi(r)$. The reference point r^{bulk} can be taken in the semi-infinite bulk at a lattice point neighbouring the intermediate region. This construction leads to the vanishing Madelung term at the reference bulk site and to the following surface Madelung constants:

$$\begin{aligned} M_{pq}^{ss} &= \varphi_2(R_p - R_q) \;-\; \varphi_2(r^{bulk} - R_q) \quad \text{for} \quad p \neq q \;, \\ M_{pp}^{ss} &= \tilde{\varphi}_2(0) \;-\; \varphi_2(r^{bulk} - R_p) \;, \\ M_{pq}^{sz} &= \varphi_2'(r^{bulk} - R_q) \;-\; \varphi_2'(R_p - R_q) \quad \text{for} \quad p \neq q \;, \\ M_{pp}^{sz} &= \varphi_2'(r^{bulk} - R_p) \;. \end{aligned} \tag{71}$$

Here R_p stands for a lattice site in the p-th atomic layer of the intermediate region.

Total Energy

Once the self-consistent charge density is determined by solving coupled systems of the CPA and LDA equations, one can calculate the total energy of the system composed of electrons and fixed nuclei. Below we derive such an expression for the case of random alloy and within the ASA. This is reasonable approximation for the evaluation of ground state properties such as the equilibrium volumes of random alloys and their surface energies. On the other hand, when the derivatives of the total energy with respect to positions of atoms (forces) are concerned the validity of the ASA is limited and a more accurate expression for the total energy which uses a true non-spheridized charge densities has to be used.

Within the density functional theory, the total energy of a system composed of electrons and fixed nuclei is given by

$$\mathcal{E}^{tot} = T[\varrho] + U[\varrho] + E_{xc}[\varrho] + \frac{1}{2} \sum_{R \neq R'} \frac{2 Z_R Z_{R'}}{|R - R'|} \;, \tag{72}$$

where $T[\varrho]$ is the kinetic energy of electrons, $U[\varrho]$ is the electron-electron and electron-nuclei interaction energy, $E_{xc}[\varrho]$ is the exchange and correlation energy, and the last term is the electrostatic energy of the nuclei. For one configuration of the system the total energy is given within the ASA as[36] (for notation see also Eqs. (33 – 36)

$$\begin{aligned}\mathcal{E}^{tot} = \sum_i^{occ} E_i \; &- \; \sum_R \int d^3r \int d^3r' \, \frac{\varrho_R(|r|)\, \varrho_R(|r'|)}{|r-r'|} \\ &+ \; \sum_R \int d^3r \, \varrho_R(r) \left[\epsilon_{xc}(\varrho_R(r)) - V_{xc}(\varrho_R(r))\right] \\ &+ \; \frac{1}{2}\left[\sum_{R\neq R'} \frac{2 Z_R Z_{R'}}{|R-R'|} - \sum_{R\neq R'} \sum_{LL'} Q_R^L \mathcal{M}_{RR'}^{LL'} Q_{R'}^{L'}\right] .\end{aligned} \tag{73}$$

The first term in (73) can be rewritten within the ASA as

$$\sum_i^{occ} E_i = \mathcal{E}^{core} + \mathcal{E}^{band} = \sum_R \sum_j^{occ} \epsilon_{Rj}^{core} + \sum_R \sum_L \int_{E_b}^{E_f} E \, D_{R,\,LL}(E) \, dE \,, \tag{74}$$

where j runs over the occupied core states with energies ϵ_{Rj}^{core}, $D_{R,\,LL}(E)$ is the valence density of states, and E_b is the bottom of the valence band. The second and third terms in (73) represent corrections for the electrostatic and exchange-correlation double-counting terms, $\epsilon_{xc}(\varrho_R(r))$ is the local density of the exchange-correlation energy, and $V_{xc}(\varrho_R(r))$ is the exchange-correlation potential. The fourth term corresponds to the electrostatic energy of the nuclei, and the fifth term describes the electrostatic energy of interacting electron densities in different Wigner-Seitz spheres expressed via their multipole expansion ($\mathcal{M}_{RR'}^{LL'}$ are the corresponding Madelung constants). The sum of the fourth and fifth term is usually called the Madelung energy, $\mathcal{E}^{Mad}$. The first three terms can be written as a sum of contributions from individual atoms, however, not the Madelung energy, which arises from electrostatic interactions of pairs of atoms. In the case of disordered systems, the total energy $\mathcal{E}^{tot}$, Eq. (73), has to be averaged over all configurations.[37]

Besides the total energy $\mathcal{E}^{tot}$ of the whole system, one can also define in general a total energy normalized per atom $E^{tot} = \mathcal{E}^{tot}/\,N$ for homogeneous system, or per one atom in a given subset $\{R_p\}$, $E^{tot} = \mathcal{E}^{tot}/\,N_0$ for inhomogeneous systems.

The configurationally averaged sum of the first three terms, $\langle \mathcal{E}^{ind} \rangle$, is given within the CPA as

$$\begin{aligned}\langle \mathcal{E}^{ind} \rangle \;&=\; N_0 \sum_p \langle E_p^{ind} \rangle \,, \quad \langle E_p^{ind} \rangle = \sum_Q c_p^Q \, E_p^{ind,\,Q} \,, \\ E_p^{ind,\,Q} \;&=\; \sum_j^{occ} \epsilon_{pj}^{core,\,Q} + \sum_L \int_{E_b}^{E_F} E \, D_{p,\,LL}^Q(E) \, dE \\ &+ \; \int_{s_p^Q} d^3r \, \varrho_p^Q(r) \left[\epsilon_{xc}(\varrho_p^Q(r)) - V_{xc}(\varrho_p^Q(r)) - \int_{s_p^Q} d^3r' \, \frac{\varrho_p^Q(r')}{|r-r'|}\right] ,\end{aligned} \tag{75}$$

where $R \in \{R_p\}$.

For homogeneous isotropic alloys the configurationally averaged total energy can easily be calculated, because the multipoles Q_R^L for $\ell > 0$ vanish on average, and the averaged Madelung energy $\langle \mathcal{E}^{Mad} \rangle$ is given as

$$\langle \mathcal{E}^{Mad} \rangle = \sum_{R\neq R'} \frac{\langle Z_R Z_{R'} \rangle - \langle Q_R^s Q_{R'}^s \rangle}{|R-R'|} \,, \tag{76}$$

since $\mathcal{M}^{ss}_{RR'} = 2|R - R'|^{-1}$. In this case $\langle \mathcal{E}^{Mad} \rangle = 0$, because the Z_R's and the Q^s_R's average independently, $\langle Z_R Z_{R'} \rangle = \langle Z_R \rangle \langle Z_{R'} \rangle$ and $\langle Q^s_R Q^s_{R'} \rangle = \langle Q^s_R \rangle \langle Q^s_{R'} \rangle$ for $R \neq R'$, and $\langle Q^s_R \rangle = \langle Z_R \rangle$ for all R.

For inhomogeneous systems the electronic multipoles and the nuclear charges at different sites are averaged independently, but their averages $\langle Z_R \rangle = \langle Z_p \rangle$ and $\langle Q^L_R \rangle = \langle Q^L_p \rangle$ depend on their position $R \in \{R_p\}$. In this case one finds that the Madelung energy can be expressed in terms of the layer dependent Madelung constants $M^{LL'}_{pq}$ with the monopole-monopole term $L = L' = s$ excluded from the (L, L') sum:

$$\langle \mathcal{E}^{Mad} \rangle = -\frac{1}{2} \sum_{pq} \left[\left(2\langle Z_p \rangle + \langle q^s_p \rangle \right) M^{ss}_{pq} \langle q^s_q \rangle + \sum_{L,L'}{}' \langle Q^L_p \rangle M^{LL'}_{pq} \langle Q^{L'}_q \rangle \right] . \tag{77}$$

The configurationally averaged total energy $\langle \mathcal{E}^{tot} \rangle$ for a disordered alloy with a surface is then given by the sum of $\langle \mathcal{E}^{ind} \rangle$ and of $\langle \mathcal{E}^{Mad} \rangle$. A similar expression for the crystal surface using the frozen-core approximation was given in Ref. 12. The generalization of above expressions to the relativistic case is straightforward.[25]

The surface energy, i.e. the energy needed to the formation of a surface, is defined then by $\Delta E^{surf} = E^{tot}(surf) - mE^{tot}(bulk)$, namely as the difference of the total energy $E^{tot}(surf)$ of a system with surface represented by m atomic layers in the intermediate region (excluding the layers of empty spheres), and the m-multiple of the total energy $E^{tot}(bulk)$ of the corresponding bulk system per one site.

APPLICATIONS

We omit completely the discussion of the results of numerical calculations, but for the reader's convenience we give below references to calculations done for particular physical problems and systems. The bulk transition metal alloys represent a necessary first step in the study of alloy surfaces and interfaces. Systems with the fcc,[8] bcc,[38] hcp,[39] as well as with complex lattice structures of the $D0_3$-[40] and $L1_2$-type[41] were studied. Random overlayers of transition metals on a non-random substrate were the subject of investigations of Refs. 10, 42 while the study of surface ordering was performed in Ref. 43. The properties of magnetic overlayers on non-magnetic substrates were investigated in Ref. 44 whereas the ordering induced by the presence of magnetic moments in the system was examined in Ref. 45. The effect of the surface roughness and interdiffussion on the overlayer magnetism was also studied.[46] Surfaces of random transition metal alloys were the subject of study in Ref. 35 and the phenomenon of the surface segregation was treated in Ref. 47. The extension of the present formalism to study the surface magnetism of random binary transition metal alloys can be found in another contribution in these Proceedings.[48] The studies of surface ordering and surface segregation are based on the knowledge of the effective Ising model for the corresponding inhomogeneous systems[27]. The parameters of the effective Ising Hamiltonian for surfaces of random alloys were determined within the GPM in Ref. 28. The contribution[29] in these Proceedings concerns a recent attempt to improve the GPM for such inhomogeneous systems, in particular as concerns the application to the surface segregation problem. The applications of the fully-relativistic formalism to the case of random bulk alloys and their surfaces can be found in Ref. 25 while the subject of Ref. 30 was the exchange oscillatory coupling in magnetic metallic multilayers. We refer the reader interested in specific applications to the above papers.

SUMMARY AND CONCLUSIONS

We have given a brief review of an efficient self-consistent Green's function method to calculate properties of disordered alloys and their surfaces. The approach is based on the local density approximation within the all-electron first-principles tight-binding linear muffin-tin orbital method. In this approach both the intralayer and interlayer interactions are short ranged, which greatly simplifies the evaluation of the surface Green's function needed for a proper description of the semi-infinite systems, such as surfaces and overlayers. A simple technique was described which allows to evaluate the surface Green's function directly in the real space without the knowledge of the bulk resolvent. The potentials of constituent atoms are treated as spherically symmetric ones within the atomic sphere approximation. In the case of surface-related problems, we include both the monopole and dipole terms in the multipole expansion of the charge density, which accounts for its nonspherical shape at the vacuum-solid or solid-solid interface and which yields the correct value of the electrostatic surface barrier, the work function, and of the surface energy. The method makes possible to study many systems, as e.g. simple metals, transition metals, or semiconductors and insulators on the same footing. In the latter case, the concept of empty interstitial spheres to describe electron states of open-structure solids is used together with the downfolding technique to take full advantage of the minimal basis set.[49] The effect of disorder in the bulk or at the surface is included via the coherent potential approximation capable of describing various situations occurring in experiment: ideal crystal and its surface completely disordered alloy as well as the case of the partial ordering on one sublattice of a complex lattice, various stages of adsorption ranging from a low coverage limit to the case of adsorbed multilayers, as well as such complex systems as the surface of disordered alloy with a possible segregation of one of alloy components at the surface. The theory can be improved in some aspects, of which we mention the effect of lattice relaxations around atoms of different sizes and the inclusion of inward or outward layer shifts at the surface, the extension of the formalism to the relativistic spin-polarized case, the effect of electron correlations beyond the local density approximation, or the study of more complex structures containing several atoms in the unit cell.

Acknowledgements

The financial support for this work was provided by the Grant Agency of the Czech Republic (Projects No. 202/93/0688 and 202/93/1174), the Grant Agency of the Academy of Scinces of the Czech Republic (Project No. 110437), the Austrian Science Foundation (Project No. P10231) and the Center for Computational Materials Science (GZ 45.384).

References

1. O. K. Andersen, O. Jepsen, and D. Glötzel, in *Highlights of Condensed-Matter Theory*, edited by F. Bassani, F. Fumi, and M. P. Tosi (North-Holland, New York, 1985), p. 59.
2. V. L. Moruzzi, J. F. Janak, and A. R. Williams, *Calculated Electronic Properties of Metals*, (Pergamon Press, New York, 1978).
3. O. K. Andersen and O. Jepsen, Phys. Rev. Lett. **53**, 2571 (1984).
4. F. Garcia-Moliner and V. R. Velasco, Prog. Surf. Sci. **21**, 93 (1986).

5. B. Wenzien, J. Kudrnovský, V. Drchal, and M. Šob, J. Phys. Condens. Matter **1**, 9893 (1989).
6. B. Velický, S. Kirkpatrick, and H. Ehrenreich, Phys. Rev. **175**, 747 (1968).
7. J. Kudrnovský, V. Drchal, and J. Mašek, Phys. Rev. B **35**, 2487 (1987).
8. J. Kudrnovský and V. Drchal, Phys. Rev. B **41**, 7515 (1990).
9. J. Kudrnovský, P. Weinberger, and V. Drchal, Phys. Rev. B **44**, 6410 (1991).
10. J. Kudrnovský, I. Turek, V. Drchal, P. Weinberger, N. E. Christensen, and S. K. Bose, Phys. Rev. B **46**, 4222 (1992).
11. W. Lambrecht and O. K. Andersen, Surf. Sci. **178**, 256 (1986).
12. H. L. Skriver and N. M. Rosengaard, Phys. Rev. B **43**, 9538 (1991).
13. H. L. Skriver, *The LMTO Method* (Springer, Berlin, 1984).
14. O. Gunnarsson, O. Jepsen, and O. K. Andersen, Phys. Rev. B **27**, 7144 (1983).
15. O. K. Andersen, O. Jepsen, and M. Šob, in *Electronic Band Structure and its Applications*, ed. M. Yussouff (Springer, Berlin, 1987), p. 1; an extended version was presented in *Lecture Notes of the Workshop on Interatomic Forces*, (Centre European de Calcul Atomique et Moleculaire (CECAM), Paris 1987), unpublished.
16. J. M. Mac Laren, S. Crampin, D. D. Vvedensky, and J. B. Pendry, Phys. Rev. B **40**, 12164 (1989).
17. J. Kudrnovský, B. Wenzien, V. Drchal, and P. Weinberger, Phys. Rev. B **44**, 4068 (1991).
18. M. Šob and J. Kudrnovský, in *Defects in Materials*, edited by P. D. Bristowe, J. E. Epperson, J. E. Griffith, and Z. Liliental-Weber, MRS Symposia Proceedings vol. **209** (Materials Research Society, Pittsburgh, 1991), p. 177.
19. J. Kudrnovský and V. Drchal, in *Studies in Surface Science and Catalysis* **36**, edited by J. Koukal (Elsevier, Amsterdam, 1988), p. 74.
20. J. Pollmann and S. T. Pantelides, Phys. Rev. B **18**, 5524 (1978).
21. P. O. Löwdin, J. Chem. Phys. **19**, 1936 (1951).
22. J. Kudrnovský, I. Turek, V. Drchal, and P. Weinberger, Prog. Surf. Sci. **46**, 159 (1994).
23. I. V. Solovyev, A. I. Liechtenstein, V. A. Gubanov, V. P. Antropov, and O. K. Andersen, Phys. Rev. B **43**, 14414 (1991).
24. P. Weinberger, *Electron Scattering Theory for Ordered and Disordered Matter* (Clarendon Press, Oxford, 1990).
25. V. Drchal, J. Kudrnovský, and P. Weinberger, Phys. Rev. B **50**, 7903 (1994).
26. G. Trèglia, B. Legrand, and F. Ducastelle, Europhys. Lett. **7**, 575 (1988).
27. F. Ducastelle, *Order and Phase Stability*, (North Holland, Amsterdam, 1991).
28. V. Drchal, J. Kudrnovský, L. Udvardi, P. Weinberger, and A. Pasturel, Phys. Rev. B **45**, 14328 (1992).
29. V. Drchal, J. Kudrnovský, and I. Turek, these *Proceedings*.
30. J. Kudrnovský, V. Drchal, I. Turek, and P. Weinberger, Phys. Rev. B **50**, 16105 (1994).
31. P. Lloyd and P. V. Smith, Adv. Phys. **21**, 69 (1972).
32. K. C. Hass, B. Velický, and H. Ehrenreich, Phys. Rev. B **29**, 3697 (1984).
33. M. P. López Sancho, J. M. López Sancho, and J. Rubio, J. Phys. F: Metal Phys. **15**, 851 (1985).
34. J. Kudrnovský, V. Drchal, S.K. Bose, I. Turek, P. Weinberger, and A. Pasturel, Comp. Mater. Sci. **2**, 379 (1994).
35. J. Kudrnovský, I. Turek, V. Drchal, P. Weinberger, S. K. Bose, and A. Pasturel, Phys. Rev. B **47**, 16525 (1993).
36. J. F. Janak, Phys. Rev. B **9**, 3985 (1974).
37. D. D. Johnson, D. M. Nicholson, F. J. Pinski, B. L. Gyorffy, and G. M. Stocks, Phys. Rev. B **41**, 9701 (1990).

38. J. Kudrnovský and V. Drchal, Z. Phys. B **73**, 489 (1989).
39. J. Kudrnovský, V. Drchal, M. Šob, and O. Jepsen, Phys. Rev. B **41**, 10459 (1990).
40. J. Kudrnovský, N. E. Christensen, and O. K. Andersen, Phys. Rev. B **43**, 5924 (1991).
41. J. Kudrnovský, S. K. Bose, and O. K. Andersen, Phys. Rev. B **43**, 4613 (1991).
42. M. V. Ganduglia-Pirovano, J. Kudrnovský, I. Turek, V. Drchal, and M. H. Cohen, Phys. Rev. B **48**, 1870 (1993).
43. J. Kudrnovský, S. K. Bose, and V. Drchal, Phys. Rev. Lett. **69**, 308 (1992).
44. I. Turek, J. Kudrnovský, V. Drchal, and P. Weinberger, Phys. Rev. B **49**, 3352 (1994).
45. J. Kudrnovský, I. Turek, A. Pasturel, R. Tetot, V. Drchal, and P. Weinberger, Phys. Rev. B **50**, 9603 (1994).
46. I. Turek, J. Kudrnovský, M. Šob, V. Drchal, and P. Weinberger, Phys. Rev. Lett. **74**, 2551 (1995).
47. A. Pasturel, V. Drchal, J. Kudrnovský, and P. Weinberger, Phys. Rev. B **48**, 2704 (1993).
48. I. Turek, J. Kudrnovský, M. Šob, and V. Drchal, these *Proceedings*.
49. J. Kudrnovský, V. Drchal, M. Šob, N. E. Christensen, and O. K. Andersen, Phys. Rev. B **40**, 10029 (1989).

ELECTRONIC STRUCTURE AND PHYSICAL PROPERTIES

Ingrid Mertig

Technische Universität Dresden
Institut für Theoretische Physik
D-01062 Dresden, Germany

INTRODUCTION

In solid state theory two different approaches exist. Starting from a model Hamiltonian (e.g. Hubbard model, Ising model, Heisenberg model, Anderson model, ...) one can elucidate a large number of electronic and magnetic properties qualitatively. On the other hand, during the last decades first principle methods have been developed which are able to describe both the qualitative and quantitative behaviour of the considered systems and to predict its properties. In first principle calculations the only input parameters are (besides some fundamental physical constants) the atomic numbers and masses of the constituent atoms. These parameters are inserted into equations derived from the Schrödinger equation, the fundamental equation of quantum theory. Using the techniques of band structure theory, the one-electron Schrödinger like Kohn-Sham equation is solved to calculate the electronic structure and other properties of the system, like total energy, cohesive energies, bulk modulii and atomic volumes. Although it is possible to calculate the equilibrium crystal structure from first principles this process is mostly shortened by taking the lattice spacings as input values from experiment. Most of the considerations are limited to translationally invariant systems (pure metals, intermetallic compounds). In crystalline materials the Bloch theorem holds due to the periodicity of the corresponding lattice and the electronic states can be described in terms of Bloch waves, band structure and Fermi surface. But in real systems different kinds of disorder exist. At $T > 0$ the ideal crystal is perturbed by phonons. At $T = 0$ imperfections of an otherwise perfect crystalline host, as impurities and interstitials have a great influence on its physical properties. Furthermore, lattice imperfections like dislocations, stacking faults or grain boundaries occur in real materials. There is a large number of strongly disordered systems like substitutional alloys or amorphous structures. Although a large number of different types of disorder exist two limiting cases are considered which were tackled very successfully by means of first principle methods. One case is the limit of weak disorder that is achieved in dilute alloys. The electronic properties of dilute alloys, i.e. of point defects in metals are successfully treated by

Stability of Materials
Edited by A. Gonis *et al.*, Plenum Press, New York, 1996

Green function methods. The limit of strong disorder is embodied by substitutional alloys and can be described in terms of the coherent potential approximation (CPA). Taking into account disorder a variety of new macroscopic properties of the system appears.

The aim of this paper is to show that first principle methods well established in the theory of electronic properties of pure crystals can also be used successfully for the description of disordered systems and can be extended straightforwardly to treat macroscopic properties as well. In this paper we concentrate on the first principle description of transport properties. For this purpose we need transport theory in addition to the electronic structure methods. The transport theory will be illustrated for dilute and substitutional alloys. For the case of dilute alloys the quasiclassical description using the linearized Boltzmann equation is applied. The appropriate transport theory for substitutional alloys starts from the Kubo formula.

The paper is organized as follows: Section 1 is a survey of the macroscopic transport coefficients whereas Section 2 presents a review of the theoretical methods. Section 3 is an application of the formalism which has been designed to elucidate the microscopic origin of the recently discovered phenomenon of giant magnetoresistance in magnetic multilayers [Baibich *et al.*, 1988; Binash *et al.*, 1989].

MACROSCOPIC TRANSPORT COEFFICIENTS

Assuming linear response of the considered system to the influence of external fields, we shall always find that the densities of the electrical current $\mathbf{j}$ and of the thermal current $\mathbf{q}$ are linear functions of the electric field $\mathbf{E}$ and of the temperature gradient ∇T

$$\mathbf{j} = \hat{L}_{EE}\mathbf{E} + \hat{L}_{ET}\nabla T, \tag{1}$$

$$\mathbf{q} = \hat{L}_{TE}\mathbf{E} + \hat{L}_{TT}\nabla T. \tag{2}$$

All the generalized transport coefficients $\hat{L}$ are tensors which can be reduced to scalars L in case of cubic systems. The generalized transport coefficients are however not observed directly. Depending on the arrangement of the experiment the following properties are measured [Ziman, 1962].

Electrical conductivity $\hat{\sigma}$ An electric field is applied and the specimen is kept at constant temperature. The electric current is measured. Under this conditions eq. 1 becomes

$$\mathbf{j} = \hat{L}_{EE}\mathbf{E}. \tag{3}$$

Comparison with Ohm's law

$$\mathbf{j} = \hat{\sigma}\mathbf{E} \tag{4}$$

leads to

$$\hat{\sigma} = \hat{L}_{EE}, \tag{5}$$

where $\hat{\sigma}$ is the electrical conductivity. The electrical resistivity is given by $\hat{\rho} = \hat{\sigma}^{-1}$.

Thermal conductivity $\hat{\kappa}$ The specimen is electrically insulated to prevent any electric current flowing through it. A thermal gradient is maintained, and the flux of heat measured. Under this conditions eq. 1 and 2 can be transformed into

$$\mathbf{E} = -\hat{L}_{EE}^{-1}\hat{L}_{ET}\nabla T, \quad (6)$$

$$\mathbf{q} = (\hat{L}_{TT} - \hat{L}_{TE}\hat{L}_{EE}^{-1}\hat{L}_{ET})\nabla T. \quad (7)$$

Comparison with the empirical law

$$\mathbf{q} = -\hat{\kappa}\nabla T \quad (8)$$

leads to the thermal conductivity

$$\hat{\kappa} = \hat{L}_{TE}\hat{L}_{EE}^{-1}\hat{L}_{ET} - \hat{L}_{TT}. \quad (9)$$

What happens is that, in order to stop the flow of electric current, carried by the flux of heat, an electric field has to build up along the specimen. This field slightly reduces the thermal current so that the thermal conductivity is not simply the coefficient $-\hat{L}_{TT}$.

Thermopower S If instead of the thermal current the electric field caused by a thermal gradient is measured for the same experimental conditions

$$\mathbf{E} = S\nabla T \quad (10)$$

we obtain the absolute thermopower of a metal

$$S = -\hat{L}_{EE}^{-1}\hat{L}_{ET}. \quad (11)$$

Peltier coefficient Π Using the conditions under which the electrical conductivity was measured there will also be a thermal current associated with the electric field and hence proportional to the electric field

$$\mathbf{q} = \Pi\mathbf{j}. \quad (12)$$

Comparison with eq. 1 and 2 leads to the relation

$$\Pi = \hat{L}_{TE}\hat{L}_{EE}^{-1}. \tag{13}$$

With this four measurements, we have determined the generalized transport coefficients in the linear relations between the fluxes and the fields.

Kelvin-Onsager relation The thermopower and the Peltier coefficient are related by

$$S = \Pi\ T. \tag{14}$$

That is, the coefficients $\hat{L}_{TE}$ and $\hat{L}_{ET}$ are not independent

$$\hat{L}_{ET} = -\hat{L}_{TE}/T. \tag{15}$$

Wiedemann-Franz law As long as the thermal conductivity κ can be approximated by

$$\kappa \approx -\hat{L}_{TT} \tag{16}$$

the thermal conductivity is proportional to the electrical conductivity

$$\kappa \approx L_o\sigma \tag{17}$$

where L_o is the Lorenz number.
Although the transport coefficients are caused by the external fields, the electronic structure is nearly unchanged (Landau levels are neglected for the consideration of conductivity) by external fields because the atomic electric and magnetic fields are larger by several orders of magnitude than the external fields. An estimation of the order of the atomic electric field leads to

$$E_o \approx \frac{e}{4\pi\varepsilon_o a_o^2} \sim 10^{11}\frac{V}{m} \tag{18}$$

with the electron charge e, the dielectric constant ε_o and the Bohr radius a_o. The atomic magnetic field is

$$B_o \approx \frac{E_o}{v_F} \sim \frac{\hbar}{ea_o} \sim 10^5\ Tesla, \tag{19}$$

where v_F is a typical Fermi velocity. The external electric field in metals can be estimated by

$$E \approx \frac{j}{\sigma} \sim 10^{-2} \frac{V}{m} \tag{20}$$

and the range of external magnetic fields amounts to

$$B = 10 \ - \ 100 \ Tesla. \tag{21}$$

MICROSCOPIC THEORY

Dilute Alloys

Electronic Structure of the ideal system In the framework of density functional theory the electronic structure of a translationally invariant crystal is determined by the one particle Schrödinger equation for a metal electron moving in the periodic crystal potential $\overset{\circ}{V}$

$$\left[-\frac{\hbar^2}{2m}\frac{\partial^2}{\partial \mathbf{r}^2} + \overset{\circ}{V}(\mathbf{r})\right] \overset{\circ}{\varphi}_k(\mathbf{r}) = E_k \overset{\circ}{\varphi}_k(\mathbf{r}) \ . \tag{22}$$

E_k is the dispersion relation or bandstructure of the system and $\overset{\circ}{\varphi}_k(\mathbf{r})$ the one particle wave function where k is a short hand notation for the momentum quantum number $\mathbf{k}$ and the band index ν. The Fermi surface is defined by $E_k = E_F$. The electronic states at the Fermi surface are important for the understanding of the transport properties. It is the "visiting card" of metals. The Green function of the translationally invariant system can be defined by

$$\left(-\frac{\hbar^2}{2m}\frac{\partial^2}{\partial \mathbf{r}^2} + \overset{\circ}{V}(\mathbf{r}) - E\right) \overset{\circ}{G}(\mathbf{r}, \mathbf{r}'; E) = -\delta(\mathbf{r} - \mathbf{r}'). \tag{23}$$

The Green function includes all information of the electronic states for a potential $\overset{\circ}{V}$ and the description by eq. 23 is equivalent to eq. 22.

Electronic Structure of the perturbed system The Green function is especially advantageous for the description of perturbed systems like point defects in an ideal host [Zeller *et al.*, 1980; Mertig *et al.*, 1987]. The impurity atom leads to a change of the potential

$$V(\mathbf{r}) = \overset{\circ}{V}(\mathbf{r}) + \Delta V(\mathbf{r}). \tag{24}$$

Translational invariance is broken and it becomes impossible to solve a Schrödinger equation (eq. 22). But we are able to calculate the Green function of the perturbed system by means of the Dyson equation

$$G(\mathbf{r},\mathbf{r}';E) = \mathring{G}(\mathbf{r},\mathbf{r}';E) + \int d^3r''\mathring{G}(\mathbf{r},\mathbf{r}'';E)\Delta V(\mathbf{r}'')G(\mathbf{r}'',\mathbf{r}';E). \tag{25}$$

The analogous equation to determine the perturbed wave function is the Lippmann-Schwinger equation

$$\varphi_k(\mathbf{r}) = \mathring{\varphi}_k(\mathbf{r}) + \int d^3r''\mathring{G}(\mathbf{r},\mathbf{r}'';E)\Delta V(\mathbf{r}'')\varphi_k(\mathbf{r}'). \tag{26}$$

Scattering process The scattering of an unperturbed Bloch wave at the potential perturbation ΔV can be described by the T-matrix (see Fig. 1)

$$T_{kk'} = \frac{1}{\mathcal{V}}\int d^3r\, \mathring{\varphi}_k(\mathbf{r})\Delta V(\mathbf{r})\varphi_{k'}(\mathbf{r}). \tag{27}$$

The microscopic transition probability for the elastic scattering of an electron from

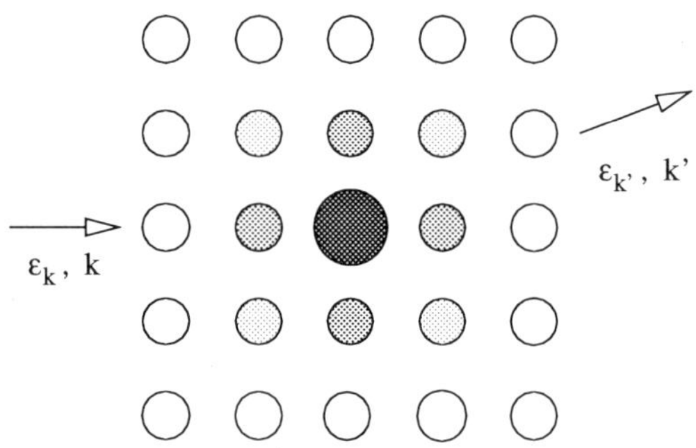

Figure 1. Scattering of a Blochwave at a cluster of perturbed potentials

a state k into a state k' is given by Fermi's golden rule

$$P_{kk'} = \frac{2\pi}{\hbar}xN\,|T_{kk'}|^2\delta(E_k - E_{k'}). \tag{28}$$

xN is the number of defects. The energy conservation is taken into account by $\delta(E_k - E_{k'})$. For the scattering process the principle of microscopic reversibility is valid

$$P_{kk'} = P_{k'k}. \tag{29}$$

By summation over all final states k', the inverse relaxation time is obtained

$$\tau_k^{-1} = \sum_{k'} P_{kk'}. \tag{30}$$

τ_k is the time the electron stays in a state k between two scattering processes.

Linearized Boltzmann Equation The transport properties will be described within the quasiclassical theory using the Boltzmann equation. We assume that a distribution function $f_k(\mathbf{r})$ exists, which measures the number of carriers in the state k in the neighbourhood of $\mathbf{r}$. The distribution function can change through diffusion of charge carriers due to the influence of external fields and due to scattering. These processes give rise to a net rate of change

$$\frac{\partial f_k}{\partial t}.$$

which must vanish in the steady state. The resulting equation is the Boltzmann equation

$$(\frac{\partial f}{\partial t}) + (\frac{\partial f}{\partial t})_{Diff} + (\frac{\partial f}{\partial t})_{Field} = (\frac{\partial f}{\partial t})_{Scatt}. \tag{31}$$

If we restrict our further considerations to the influence of a homogeneous, external electric field $\mathbf{E}$, eq. 31 reduces to

$$e\left(\frac{\partial f_k^o}{\partial E_k}\right) \mathbf{v}_k \mathbf{E} = \left(\frac{\partial f_k}{\partial t}\right)_{Scatt}. \tag{32}$$

$\mathbf{v}_k$ is the velocity of the electrons. f_k^o is the Fermi-Dirac distribution function

$$f_k^o = \frac{1}{e^{(E_k - E_F)/k_B T} + 1} \tag{33}$$

for $T = 0$. k_B is the Boltzmann's constant. The local change of charge carriers due to elastic scattering of independent particles is given by

$$\left(\frac{\partial f_k}{\partial t}\right)_{Scatt} = \sum_{k'} [f_{k'}(1 - f_k)P_{k'k} - (1 - f_{k'})f_k P_{kk'}]. \tag{34}$$

The first term in eq. 34 characterizes the scattering of electrons from occupied states k' into an unoccupied state k (scattering-in term) while the second term characterizes the inverse process, the scattering of an electron from an occupied state k into unoccupied states k' (scattering-out term). Assuming for small external fields $\mathbf{E}$ that the deviations of the electron distribution function g_k in the steady state are small relative to the equilibrium distribution function f_k^o, we can write

$$f_k = f_k^o + g_k, \tag{35}$$

where $g_k \ll f_k^o$. According to the principle of microscopic reversibility (eq. 29) and the ansatz for the local change of charge carriers (eq. 35), eq. 34 simplifies to

$$\left(\frac{\partial f_k}{\partial t}\right)_{Scatt} = \sum_{k'} P_{kk'}(g_{k'} - g_k). \tag{36}$$

Assuming a linear response for the deviations of the electron distribution function means $g_k \sim E$, hence we make the ansatz

$$g_k = -e\left(\frac{\partial f_k^o}{\partial E_k}\right)\mathbf{\Lambda}_k\mathbf{E}, \tag{37}$$

where $\mathbf{\Lambda}_k$ is the vector mean free path. With eq. 36 and eq. 37 the Boltzmann equation (eq. 32) can be transformed into an integral equation for $\mathbf{\Lambda_k}$

$$\mathbf{\Lambda}_k = \tau_k\left[\mathbf{v}_k + \sum_{k'} P_{kk'}\mathbf{\Lambda}_{k'}\right]. \tag{38}$$

τ_k is the relaxation time for electron impurity scattering introduced by eq. 30. The current density is given by

$$\mathbf{j} = 2\frac{e}{\mathcal{V}}\sum_k f_k\mathbf{v}_k. \tag{39}$$

$\mathcal{V}$ is the crystal volume. The factor 2 arises from the spin summation. Comparing eq. 35, 37 and 39 with Ohm's law (eq. 4), the conductivity tensor $\hat{\sigma}$ is derived by

$$\sigma_{\mu\nu} = \frac{2e^2}{\mathcal{V}}\sum_k \delta(E_k - E_F)\, v_{k_\mu}\Lambda_{k_\nu}. \tag{40}$$

μ and ν correspond to the Cartesian components. If cubic systems are considered $\sigma_{\mu\nu} = \sigma\delta_{\mu\nu}$.

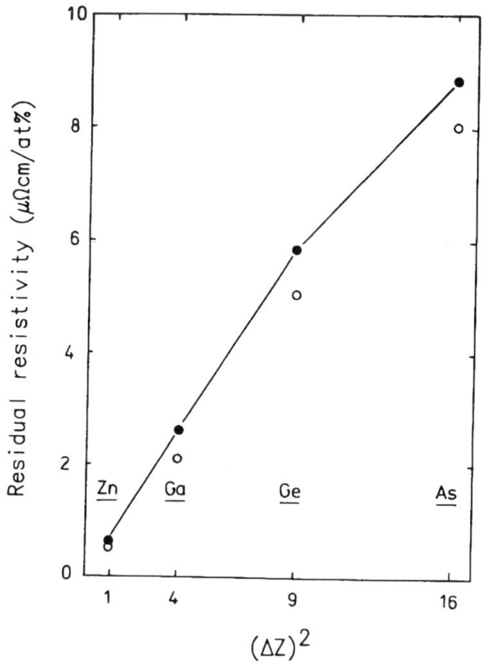

Figure 2. Residual resistivity of dilute $Ag(4sp)$ alloys [Vojta *et al.*, 1992] (closed circles - calculated values, open circles - experimental results).

An application of the above described procedure to calculate the residual resistivity of dilute Ag alloys with $4sp$ impurities is shown in Fig. 2. Considering that the calculations are performed without any free parameter the agreement between theory and experiment is almost perfect. Moreover one can see from Fig. 2 that the residual resistivity of sp impurities in noble metal hosts is proportional to the square of the valence difference $(\Delta Z)^2$ between host and impurity which is well-known as **Linde's rule** [Ziman, 1962; Vojta *et al.*, 1992].

The impurity contribution to the thermoelectric power S_{imp} can be calculated by means of Mott's formula [Mott and Jones, 1936; Ziman, 1962],

$$S_{imp} = -\frac{\pi^2 k_B^2 T}{3e} \left[\frac{\partial ln\sigma}{\partial E}\right]_{E=E_F}. \tag{41}$$

According to eq. 41 the knowledge of the energy dependence of the host and impurity properties is necessary to calculate the thermoelectric power S_{imp}. To determine the energy derivative, the integrations in eq. 40 have to be performed over the Fermi surface and also over surfaces of constant energy near E_F. The results of the thermopower of dilute Ag alloys are shown in Fig. 3 [Vojta *et al.*, 1992]. Also for the thermopower the agreement between experimental and calculated values is quite satisfactory.

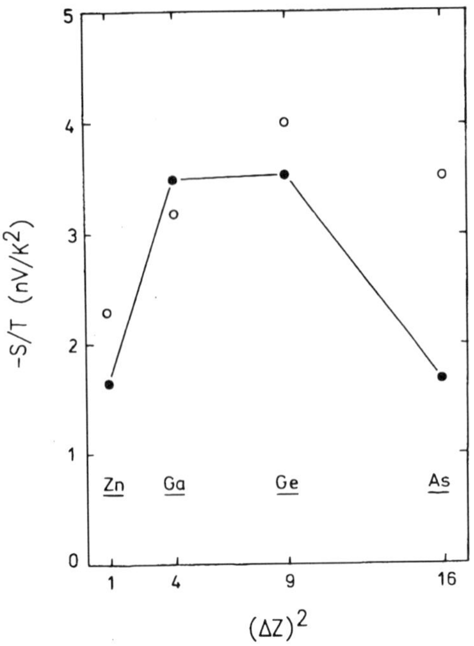

Figure 3. Thermopower of dilute $Ag(4sp)$ alloys [Vojta *et al.*, 1992] (closed circles - calculated values, open circles - experimental results).

Concentrated alloys

The presently available theory to treat the electronic properties of substitutionally disordered alloys A_xB_{1-x} (see Fig. 4) for arbitrary composition x is based on the coherent-potential approximation (CPA).
The alloy is assumed completely disordered, and the electrons are described by a one-

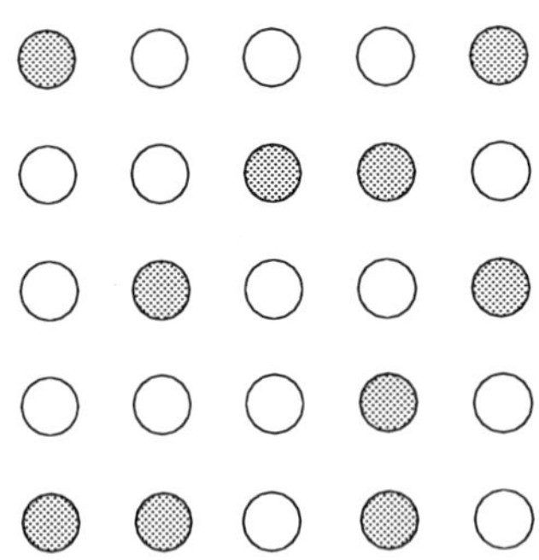

Figure 4. Substitutional disordered alloy

electron Hamiltonian whose random part is composed of atomic contributions of two types, A and B

$$H = H_o + \sum_n V_n. \tag{42}$$

The operator H_o is the periodic part of H. The random part is formed by single site contributions V_n which assume one of the two possible forms V_n^A, V_n^B in correspondence with the kind of atom occupying site n.

Kubo-Greenwood formula The basis for the calculation of the conductivity of a disordered alloy system is the Kubo formula [Kubo, 1957] in a form derived by Greenwood [Greenwood, 1958]. For zero temperature this so-called Kubo-Greenwood formula reads

$$\sigma_{\mu\nu} = \frac{\pi\hbar}{N\mathcal{V}_{at}} \langle \sum_{ii'} < i|j_\mu|i' >< i'|j_\nu|i > \delta(E - E_i)\delta(E - E_{i'}) \rangle \tag{43}$$

where

$$j_\mu = -i\hbar \frac{e}{m} \frac{\partial}{\partial \mathbf{r}_\mu} \tag{44}$$

denotes the current operator in the μ-th direction. $|i>$ and $|i'>$ are the eigenfunctions of the particular configuration of the random system. The average, indicated by $\langle ... \rangle$ has to be taken over all configurations of the system. $\mathcal{V}_{at}$ is the atomic volume and N is the number of atoms. Usually the electronic structure of the alloy is expressed in terms of one-particle Green functions rather than in terms of wave functions and energy eigenvalues. Using the side limits of the Green function

$$\begin{aligned} \sum_i |i >< i|\delta(E - E_i) &= -\frac{1}{\pi} \lim_{\delta \to 0} Im G(E - i\delta) \\ &= -\frac{1}{2\pi i} \lim_{\delta \to 0} (G(E + i\delta) - G(E - i\delta)) \\ &= -\frac{1}{2\pi i} \lim_{\delta \to 0} (G(z_1) - G(z_2)), \end{aligned} \tag{45}$$

where z_1 and z_2 are the complex energies $E + i\delta$ and $E - i\delta$, the conductivity (eq. 43) can be expressed as [Velicky, 1969; Butler, 1985]

$$\sigma_{\mu\nu} = \frac{1}{4} \lim_{\delta \to 0} [\tilde{\sigma}_{\mu\nu}(z_1, z_1) + \tilde{\sigma}_{\mu\nu}(z_2, z_2) - \tilde{\sigma}_{\mu\nu}(z_1, z_2) - \tilde{\sigma}_{\mu\nu}(z_2, z_1)] \tag{46}$$

with

$$\tilde{\sigma}_{\mu\nu}(z_1, z_2) = -\frac{\hbar}{\pi N \mathcal{V}_{at}} Tr\langle j_\mu G(z_1) j_\nu G(z_2) \rangle. \tag{47}$$

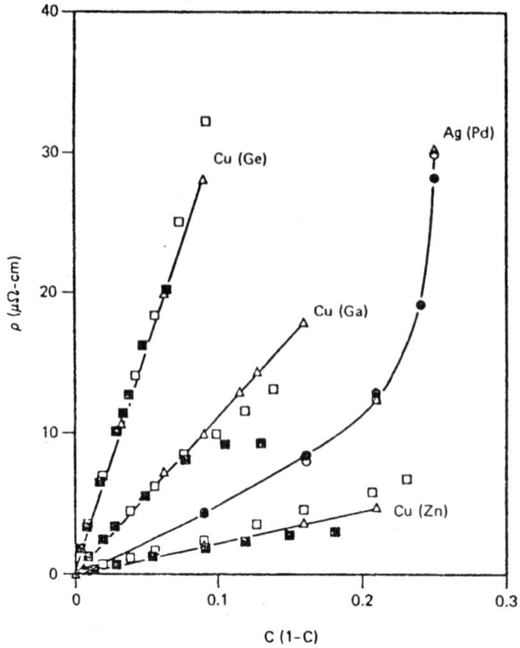

Figure 5. Calculated and experimental resistivities of $Cu(Zn)$, $Cu(Ga)$, $Cu(Ge)$ and $Ag(Pd)$ alloys. Triangles represent the calculated values. All the other symbols are experimental results. The straight lines are the Nordheim relation. [Swihart *et al.*, 1986]

CPA-approximation In order to evaluate the ensemble average of the product of Green functions in eq. 47, the CPA-approximation is used [Faulkner and Stocks, 1980]. In this formulation, the Green function does not occur explicitly. The information is carried by the scattering path operator τ^{CPA} of the CPA-medium which is closely related to the Green function $\langle G \rangle$ [Faulkner and Stocks, 1980]. The quantity $\tilde{\sigma}$ (eq. 47) can then be expressed as

$$\tilde{\sigma}_{\mu\nu}(z_1, z_2) = -\frac{4m^2}{\pi N \mathcal{V}_{at}\hbar^3} \sum_{ii'} c^i c^{i'} \tilde{J}^i_\mu(z_1, z_2)(1+\chi\omega)^{-1}\chi(z_1, z_2)\tilde{J}^{i'}_\nu(z_1, z_2) \tag{48}$$

$$+ \sum_i c^i \tilde{J}^i_\mu(z_1, z_2)\tau^{CPA}(z_1)\tilde{J}^i_\nu(z_1, z_2)\tau^{CPA}(z_2).$$

The matrix of the scattering path operator τ^{CPA} is related to the CPA-t-matrix and the structure constants by a Brillouin zone integral of the inverted KKR-matrix $\tau(\mathbf{k}, z)$

$$\tau^{CPA} = \frac{1}{\mathcal{V}_{BZ}} \int_{BZ} d^3k \, \tau(\mathbf{k}, z). \tag{49}$$

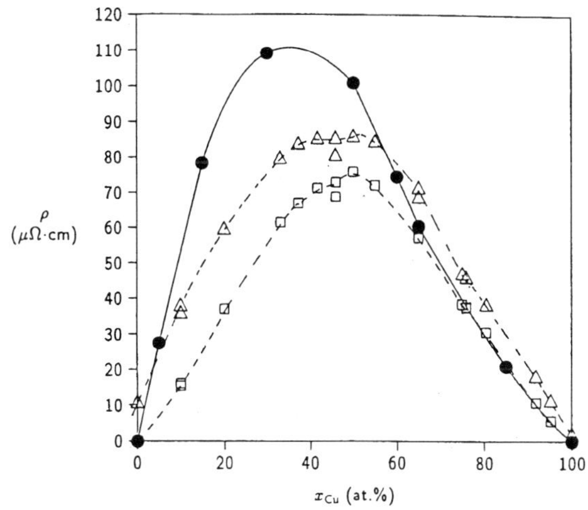

Figure 6. Residual resistivity of Cu_xPt_{1-x} alloys [Banhart *et al.*, 1994]. Full circles are calculated values. Triangles and squares are experimental values.

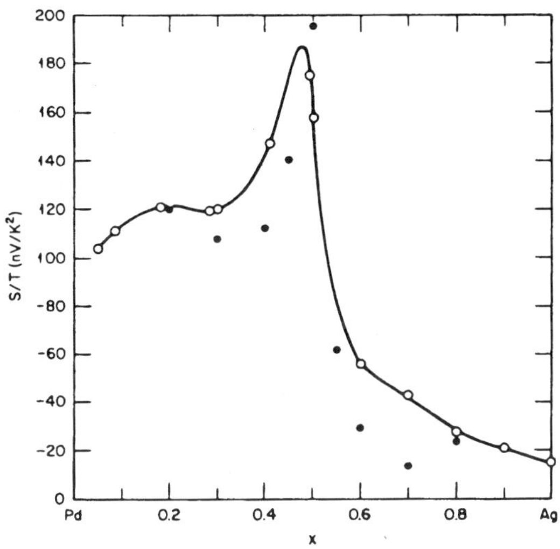

Figure 7. Calculated (solid circles) and experimental (open circles) values of the diffusion thermopower for Ag_xPd_{1-x} [Butler and Stocks, 1984].

χ is essentially a Brillouin zone integral of a pair of $\tau(\mathbf{k}, z)$

$$\chi(z_1, z_2) = \frac{1}{\mathcal{V}_{BZ}} \int_{BZ} d^3k \; \tau(\mathbf{k}, z_1)\tau(\mathbf{k}, z_2) - \tau^{CPA}(z_1)\tau^{CPA}(z_2). \tag{50}$$

The operator ω in eq. 49 contains the single-site t-matrices and the CPA scattering operator [Butler, 1985]. $\tilde{J}$ is related to the matrix elements of the current operator by $\tilde{J} = D^t J D$, where D is the CPA impurity operator and D^t the transposed operator.

The matrix elements of the current operator are of the form

$$J^i_{\mu,LL'}(z_1,z_2) = -\frac{ie\hbar}{m}\int_{V_{WS}} d^3r_n Z^i_L(\mathbf{r}_n,z_1)\frac{\partial}{\partial r_\mu}Z^i_{L'}(\mathbf{r}_n,z_2). \tag{51}$$

The integration is performed over the Wigner-Seitz cell. Z^i_L is the regular solution of the Schrödinger equation. L is a short hand notation of angular momentum quantum number l and magnetic quantum number m. The above equations represent the non-relativistic formalism. For a relativistic generalization see [Banhart *et al.*, 1994].
Once the CPA equations have been solved the Kubo-Greenwood conductivity (eq. 46 and 48) can be calculated. Since the evaluation of eq. 48 is extremely difficult the procedure is often simplified by neglecting vertex-corrections. That is, $(1+\chi\omega)^{-1}$ is set to unity. This approximation is equivalent to neglecting the scattering-in term in the Boltzmann equation (see eq. 34) and will be discussed in detail below (section: relaxation time approximation).
Numerical results for the residual resistivity of concentrated Cu and Ag alloys are compared to experimental values in Fig. 5. The agreement between theory and experiment is excellent. The straight lines in Fig. 5 represent the **Nordheim relation**, that is, the residual resistivity of concentrated alloys is proportional to $x(1-x)$. Furthermore, **Linde's rule** is fulfilled if we compare the results of $Cu(Zn)$, $Cu(Ga)$ and $Cu(Ge)$. Results for the residual resistivity of Cu_xPt_{1-x} alloys using the relativistic version of the CPA and the Kubo-Greenwood formula [Banhart *et al.*, 1994] are shown in Fig. 6. Fig. 7 presents the thermopower for Ag_xPd_{1-x} alloys [Butler and Stocks, 1984] in comparison with experimental results. From the examples (Fig. 5, 6 and 7) it becomes obvious that also in the case of the concentrated alloys a first principle description of the transport properties leads to satisfactory results.

Magnetic dilute alloys

Following Mott's idea [Mott, 1964] several transport properties of ferromagnetic alloys (see Fig. 8) can be explained by assuming conduction in parallel by electrons in the majority bands (spin-up electrons) and by electrons in the minority bands (spin-down electrons). The physical basis of this two current model is the dominance of

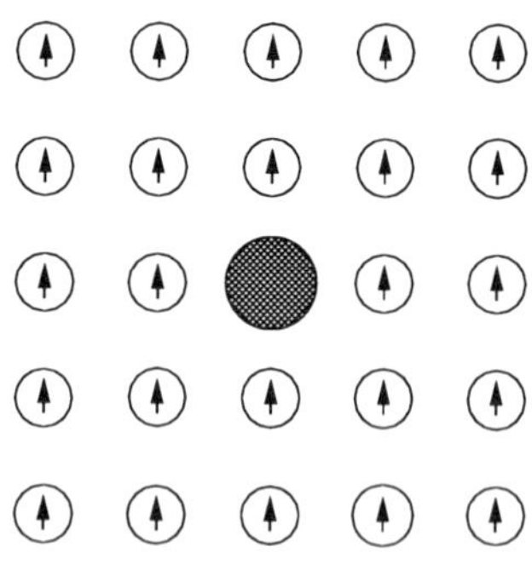

Figure 8. Ferromagnetic dilute alloy

spin conserving-potential-scattering and the weakness of spin-flip collisions in a fer-

romagnetic alloy at low temperatures. It seems that the model provides a good basis for the discussion of a wide range of alloy properties [Campbell and Fert, 1982]. Owing to the developments of density-functional theory and sophisticated numerical techniques we are now able to perform realistic $ab-initio$ calculations which puts us in position to check the reliability of the above-mentioned model studies. Extensive theoretical studies of the electronic structure of ferromagnetic dilute alloys have been performed [Blügel *et al.*, 1987; Stefanou *et al.*, 1987; Zeller, 1987; Drittler *et al.*, 1989; Stepanyuk *et al.*, 1994; Mertig *et al.*, 1993; Mertig *et al.*, 1994].
In ferromagnetic alloys we consider two subgroups of charge carriers, i.e. the spin-up and the spin-down electrons, characterized by the quantum numbers $k = (\mathbf{k}, \nu)$ and spin index σ. Both spin directions contribute separately to the resistivity. With respect to the dependence on the spin-quantum number, the scattering probability (eq. 28) consists of four parts

$$P_{kk'}^{\sigma\sigma'} = \begin{pmatrix} P_{kk'}^{\uparrow\uparrow} & P_{kk'}^{\uparrow\downarrow} \\ P_{kk'}^{\downarrow\uparrow} & P_{kk'}^{\downarrow\downarrow} \end{pmatrix} \tag{52}$$

two spin-conserving scattering processes $P_{kk'}^{\uparrow\uparrow}$ and $P_{kk'}^{\downarrow\downarrow}$ and two spin-flip contributions $P_{kk'}^{\uparrow\downarrow}$ and $P_{kk'}^{\downarrow\uparrow}$ (see Fig. 9). From this point of view the spin-flip processes lead to a momentum transfer between the currents.
The linearized Boltzmann (eq. 38) equation becomes a system of coupled integral

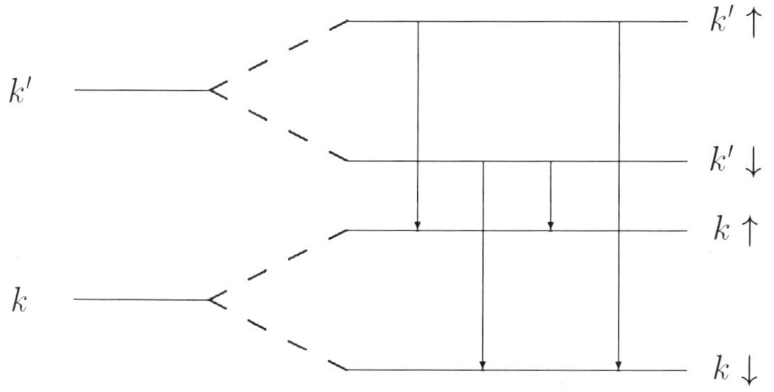

Figure 9. Spin-dependent scattering in ferromagnetic systems.

equations

$$\mathbf{\Lambda}_k^\sigma = \tau_k^\sigma [\mathbf{v}_k^\sigma + \sum_{\sigma'} \sum_{k'} P_{kk'}^{\sigma\sigma'} \mathbf{\Lambda}_{k'}^{\sigma'}] \tag{53}$$

for the vector mean free path $\mathbf{\Lambda}_k^\sigma$ of the majority and minority electrons. τ_k^σ is the spin-dependent life time.

If we assume that the spin-flip scattering contributions are small in comparison to the spin-conserving scattering contributions, which is well established by experiments [Fert, 1969; Bourquart *et al.*, 1968], eq. 53 decouples for both spin directions and can be solved by iteration [Mertig *et al.*, 1993]. Finally, we obtain two currents with the density

$$\mathbf{j}^\sigma = e^2 \sum_k \delta(E_k - E_F)\, \mathbf{v}_k^\sigma\, (\mathbf{\Lambda}_k^\sigma\, \mathbf{E}). \tag{54}$$

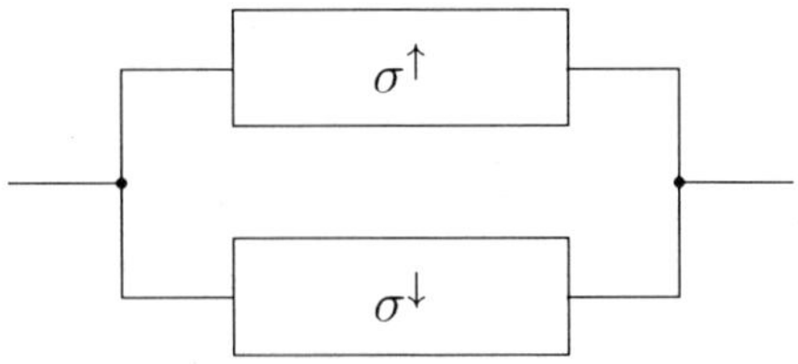

Figure 10. Total conductivity within the two current model

Comparing with Ohm's law (eq. 4) two conductivity contributions are derived

$$\sigma^{\sigma}_{\mu\nu} = e^2 \sum_k \delta(E_k - E_F)\, v^{\sigma}_{k\,\mu}\, \Lambda^{\sigma}_{k\,\nu}. \tag{55}$$

$\hat{\sigma}^{\uparrow}$ for the majority electrons and $\hat{\sigma}^{\downarrow}$ for the minority ones.

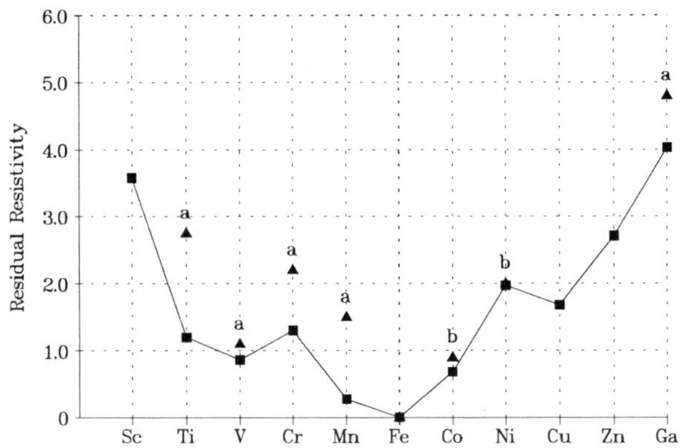

Figure 11. Residual resistivity of dilute $Fe(3d)$ alloys [Mertig *et al.*, 1994] (squares - calculated values, triangles - experimental results).

Due to the parallel addition of the two currents (Fig. 10), the total conductivity for cubic systems becomes

$$\sigma = \sigma^{\uparrow} + \sigma^{\downarrow} \qquad \text{with} \qquad \hat{\sigma}^{\sigma}_{\mu\nu} = \sigma^{\sigma}\,\delta_{\mu\nu}. \tag{56}$$

$\alpha = \sigma^{\uparrow}/\sigma^{\downarrow}$ is the spin anisotropy ratio of the conductivity and $\beta = \tau^{\uparrow}/\tau^{\downarrow}$ is the spin anisotropy of the relaxation times.

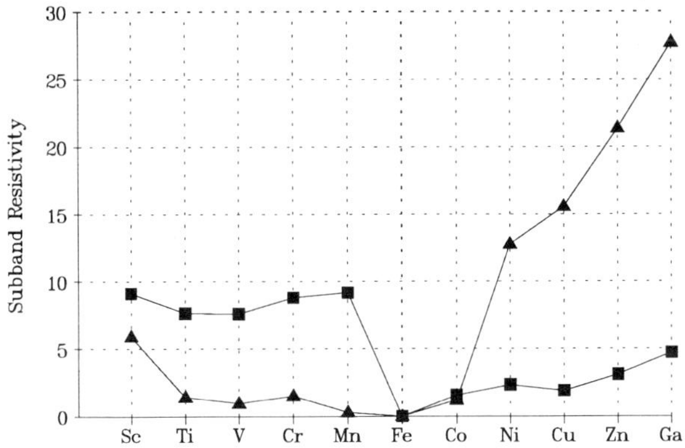

Figure 12. Subband resistivities of dilute $Fe(3d)$ alloys [Mertig *et al.*, 1994] (squares - majority electrons, triangles - minority electrons).

Table 1. Anisotropy ratios $\alpha = \sigma^{\uparrow}/\sigma^{\downarrow}$ of dilute Fe alloys (^{a}[Fert and Campbell, 1976], ^{b}[Dorleijn and Miedema, 1975; Dorleijn, 1976]).

Impurity	α_{theor}	α_{exp}
Sc	0.65	-
Ti	0.19	$0.25^a, 0.66^b$
V	0.13	$0.12^a, 0.13^b$
Cr	0.17	$0.17^a, 0.37^b$
Mn	0.03	$0.09^a, 0.17^b$
Co	0.77	$1.^a, 3.7^b$
Ni	5.48	$3.^a, 7.^b$
Cu	8.20	-
Zn	6.87	-
Ga	5.87	-

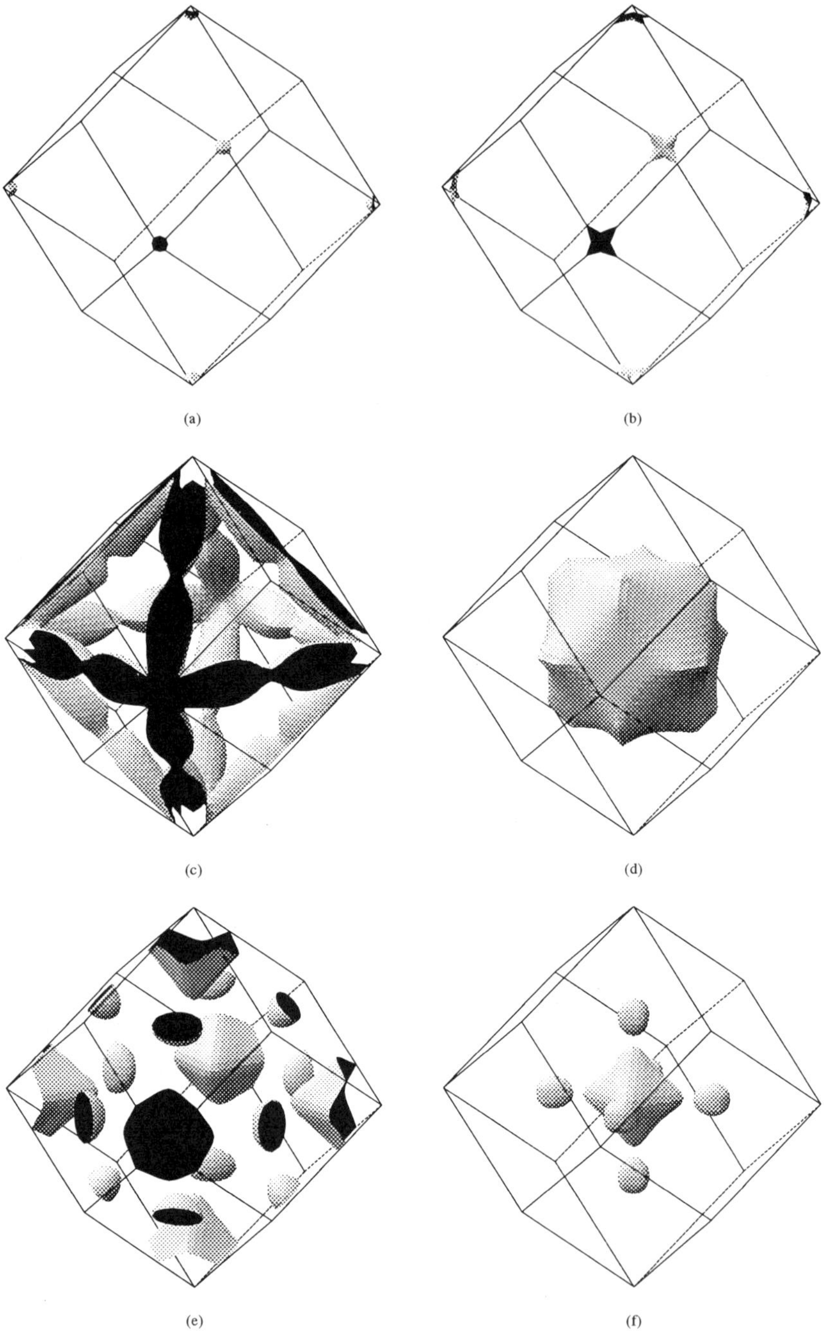

Figure 13. Fermi surface of Fe. Majority electron sheets (a, b, c and d) and minority electron sheets (e and f).

Results for the residual resistivity of dilute $Fe(3d)$ alloys in comparison with experimental results are shown in Fig. 11. A more detailed understanding follows from the subband resistivities $\rho^{\uparrow}$ and $\rho^{\downarrow}$ which are shown in Fig. 12. In dilute Fe alloys with a lack of electrons due to the impurities, the current is carried by the minority electrons (see Fig. 13 e, f). For impurities with $\Delta Z > 0$, the majority electrons dominate the transport properties (see Fig. 13 a, b, c, d).
This behaviour is well reflected in the anisotropy ratio α (see Tab. 1). The criterion is that the anisotropy ratio is either smaller or larger than 1. $\alpha < 1$ means strong scattering of the majority electrons, i.e., the current is carried by the minority electrons. For $\alpha > 1$ the situation is opposite.
The $3d$ defects in Fe lead to an anisotropy ratio smaller than one except for Ni and the $4sp$ defects Cu and Zn where α is extremely large. For Co impurities α is nearly one, so that both bands have a comparable resistivity.

Relaxation time approximation

Since the solution of the Boltzmann equation (eq. 38 and 53) is difficult, several approximations of the integral equation exist. The approximations are known as relaxation time approximations. From the structure of eq. 38, it becomes obvious that the vector mean free path $\mathbf{\Lambda_k}$ and the velocity $\mathbf{v_k}$ are related by a tensor, that is, the direction and the amount of the velocity of an electron in a state k are affected by scattering.

Table 2. Comparison of the exact solution of the Boltzmann equation (I - eq. 38 or 53), the state dependent relaxation time approximation (II - eq. 57) and the relaxation time approximation (III - eq. 58) of the spin-dependent residual resistivity ρ^{σ} in $\mu\Omega cm/at\%$.

System		ρ^{σ} in $\mu\Omega cm/at\%$		
			Näherung	
		I	II	III
Fe(Cr)	$\uparrow$	8.81	8.64	9.65
	$\downarrow$	1.52	1.49	1.88
Cu(Co)	$\uparrow$	3.52	4.24	4.25
	$\downarrow$	28.59	30.21	30.33

Neglecting the scattering-in terms in eq. 38 or 53 the Boltzmann equation reduces to

$$\mathbf{\Lambda}_k = \tau_k \mathbf{v}_k. \tag{57}$$

This approximation takes into account only the part of the scattering in direction of the electronic velocity. It is equivalent to neglecting vertex-corrections in the Kubo conductivity (eq. 48). Another common approximation is the restriction

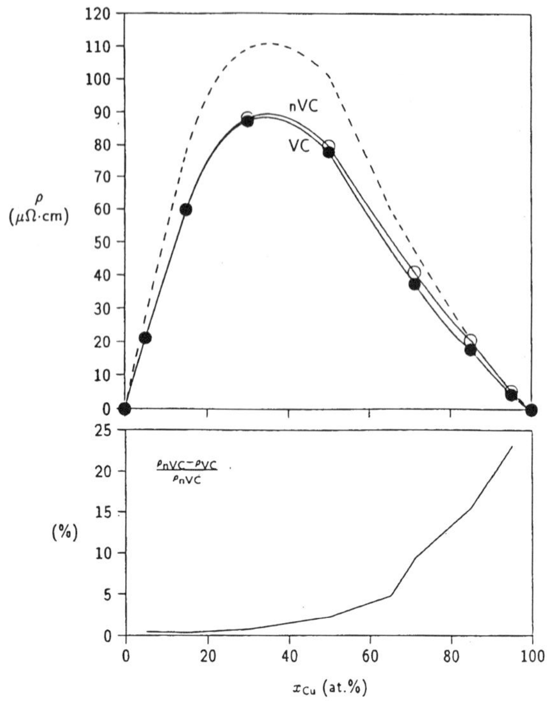

Figure 14. Comparison between the resistivity of Cu_xPt_{1-x} without (nVC) and including vertex-corrections (VC) [Banhart *et al.*, 1994] and results from [Banhart *et al.*, 1991] (dashed line). In the lower part the relative differences between the two approximations are shown.

to a k-independent relaxation time τ

$$\mathbf{\Lambda}_k = \tau \mathbf{v}_k \tag{58}$$

where τ is chosen to be a Fermi surface average

$$\tau = \frac{\sum_k \delta(E_k - E_F)\, \tau_k}{\sum_k \delta(E_k - E_F)}. \tag{59}$$

The underlying assumption of this approximation is that the electrons are scattered similarly in each state k.

The quality of the approximation depends strongly on the scattering anisotropy. If the inversion symmetry of the Bravais lattice is conserved in the presence of the impurity atom, that is, $P_{kk'} = P_{-k-k'}$, then

$$\mathbf{\Lambda}_{-\mathbf{k}'} = -\mathbf{\Lambda}_{\mathbf{k}'}. \tag{60}$$

Consequently,

$$\begin{aligned} \sum_{k'} P_{kk'}\mathbf{\Lambda}_{k'} &= \sum_{k'} \tfrac{1}{2}\left(P_{kk'}\mathbf{\Lambda}_{k'} + P_{k-k'}\mathbf{\Lambda}_{-k'}\right) \\ &= \sum_{k'} \tfrac{1}{2}\left(P_{kk'} - P_{k-k'}\right)\mathbf{\Lambda}_{k'} \end{aligned} \tag{61}$$

the scattering-in contribution is determined by the antisymmetric part of the scattering operator

$$P^a_{kk'} = \frac{1}{2}\left(P_{kk'} - P_{k-k'}\right). \tag{62}$$

The life time, however, is connected to the symmetric part

$$P^s_{kk'} = \frac{1}{2}\left(P_{kk'} + P_{k-k'}\right). \tag{63}$$

That means, the relaxation time approximation is sufficient in case of isotropic scattering. For strongly anisotropic scattering, like pronounced forward or back scattering, the relaxation time approximation fails. A comparison of the different approximations is given in Tab. 2 for dilute $Fe(Cr)$ and $Cu(Co)$ alloys.

GIANT MAGNETORESISTANCE

Since the discovery of giant magnetoresistance (GMR) in magnetic multilayer systems (like Fe/Cr or Co/Cu) [Baibich *et al.*, 1988; Binash *et al.*, 1989] several experimental and theoretical studies have been carried out to elucidate the microscopic origin of the phenomenon. Here, *ab initio* calculations of the electronic structure and of the electron scattering are of great interest. The GMR effect occurs when successive fer-

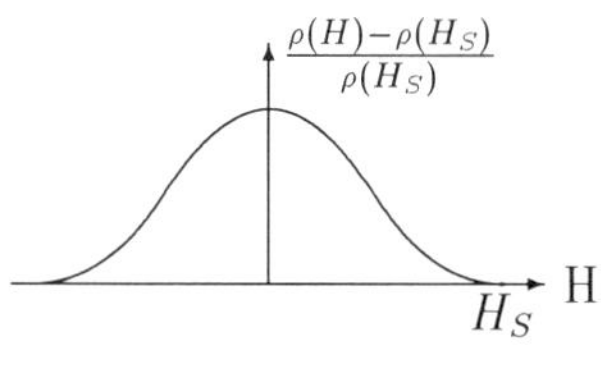

Figure 15. Giant magnetoresistance. The resistivity is maximum when the magnetic moments of successive ferromagnetic layers are antiparallel. It drops off as the applied field aligns the magnetic moments H_S, denoted by the right and left diagrams for positive and negative fields, respectively.

romagnetic layers exhibit anti-parallel magnetization. The application of an external

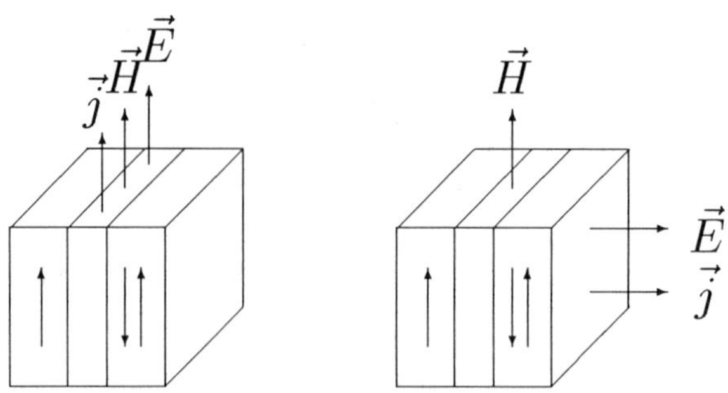

Figure 16. Geometrical arrangements of current density **j**, electrical field **E** and magnetic field **H** in CIP and CPP measurement of GMR.

magnetic field brings the magnetization of the ferromagnetic layers into alignment and causes a decrease of resistivity (Fig. 15) for both current-in-plane (CIP) and current-perpendicular-to-plane (CPP) geometries (see Fig. 16). Most theories [Camley and Barnaś, 1989; Levy *et al.*, 1990; Inoue *et al.*, 1991; Hood and Falicov, 1992; Valet and Fert, 1993; Levy, 1994] try to explain the GMR by spin-dependent scattering at interfaces or bulk defects but neglect the electronic structure of the multilayer system. Attempts to include the electronic structure have been made by several authors [Oguchi, 1993; Butler *et al.*, 1993; Schep *et al.*, 1995; Zahn *et al.*, 1995]. All calculations predict a strong influence of the electronic structure of the multilayer on the GMR. The GMR in magnetic multilayers is defined as

$$\mathrm{GMR} = \frac{\sigma^P}{\sigma^{AP}} - 1, \tag{64}$$

where σ^P and σ^{AP} are the conductivities of the multilayer for parallel and anti-parallel alignment of successive ferromagnetic layer magnetization, respectively. Experimentally, one can be sure parallel alignment is achieved, in view of the marked transition from strong to weak magnetic field dependence of the conductivity at the saturation field (see Fig. 15).

Our considerations focuse on (100) oriented Fe_mCr_n layer sequences (m monolayers Fe followed by n monolayers Cr), where the Fe layers are intrinsically ferromagnetic and the Cr layers are intrinsically antiferromagnetic. *Ab initio* electronic structure calculations have been performed using spin-density functional theory for initially parallel and antiparallel spin configurations of subsequent Fe layers [Zahn *et al.*, 1995].

Starting from the bandstructure E_k^σ of the situation, that we have just addressed, potential scattering causes transitions from a state k into a state k' of the multilayer system. The scattering can be caused by impurity atoms in the ferromagnetic or in the nonmagnetic spacer layer. The scattering is described by spin-dependent relaxation times (see eq. 30, 58) with a spin anisotropy $\beta = \tau^\uparrow/\tau^\downarrow$.

The conductivity can be calculated in the same way as for the dilute alloys (eq. 55) using the relaxation time approximation (eq. 57). Consequently, the conductivity tensor becomes

$$\hat{\sigma}^\sigma = e^2\tau^\sigma \sum_k \delta(E_k^\sigma - E_F)\, \mathbf{v}_k^\sigma\, \mathbf{v}_k^\sigma. \tag{65}$$

The integration is performed over the Fermi surface of the parallel configuration of the layered system. Summation of the two currents yields the total conductivity

$$\hat{\sigma}^P = \hat{\sigma}^{\uparrow} + \hat{\sigma}^{\downarrow}. \tag{66}$$

In the anti-parallel configuration the electronic states are spin degenerate, and the conductivity becomes

$$\hat{\sigma}^{AP} = 2\, e^2 \tau^{AP} \sum_k \delta(E_k^{AP} - E_F)\, \mathbf{v}_k^{AP}\, \mathbf{v}_k^{AP}. \tag{67}$$

Assuming that the layered system is grown in the z direction, CIP corresponds to the xx or yy component of the conductivity tensor and CPP to the zz component in eq. 65 and 67. τ^{AP}, the relaxation time in the anti-parallel configuration, can be obtained by summation of the scattering operators. For equal concentration of defects in adjacent ferromagnetic layers, the relaxation time is then given by

$$\frac{1}{\tau^{AP}} = \frac{1}{2}\left(\frac{1}{\tau^{\uparrow}} + \frac{1}{\tau^{\downarrow}}\right). \tag{68}$$

The simplest situation occurs, when all the scattering centers are localized in the anti-ferromagnetic Cr layers (for example Fe impurities). In this case the lifetimes τ^{σ} for both spin directions must be equal since the spin up wavefunction amplitudes of one half of the Cr atoms is equal to the spin down wavefunction amplitudes of the other half. This corresponds to $\beta = 1$. Eq. 68 now implies $\tau^{AP} = \tau^{\uparrow} = \tau^{\downarrow}$. Hence the lifetime completely disappears from the GMR expression

$$\mathrm{GMR} = \frac{\sum_\sigma \sum_k \delta(E_k^\sigma - E_F)\, v_{k\,\mu}^\sigma\, v_{k\,\mu}^\sigma}{2\, \sum_k \delta(E_k^{AP} - E_F) v_{k_\mu}^{AP}\, v_{k_\mu}^{AP}} - 1. \tag{69}$$

Here, the GMR is fully determined by the Fermi surface and Fermi velocities as functions of the magnetization configuration. Hence it is a pure bandstructure effect. The obtained GMR values are about 200% in CIP and 700% in CPP (dots in Fig. 17 and 18), which is in agreement with the experimentally obtained maximum CIP-GMR (220%) for ultrathin Fe layers [Schad *et al.*, 1994].
Simplifiying eq. 69 for further discussion of the origin of GMR leads to

$$GMR = \frac{\sum_\sigma N^\sigma(E_F) v_\mu^{\sigma\,2}}{N_{AP}(E_F)\ v_\mu^{AP\,2}} - 1 \tag{70}$$

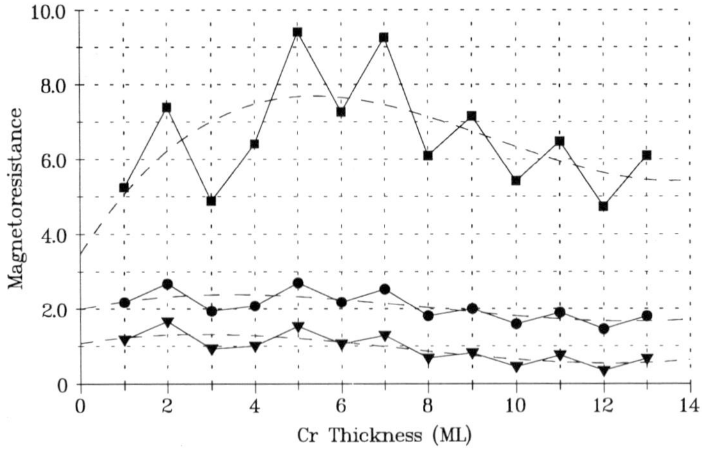

Figure 17. Calculated CIP-GMR for Fe_3Cr_n versus Cr layer thickness $n = 1, 13$. The results assuming Cr defects in the Fe layers are indicated by squares ($\beta = 0.11$). Dots mark the results for spin-independent relaxation times $\beta = 1$ and triangles for $\beta = \beta_{min}$.

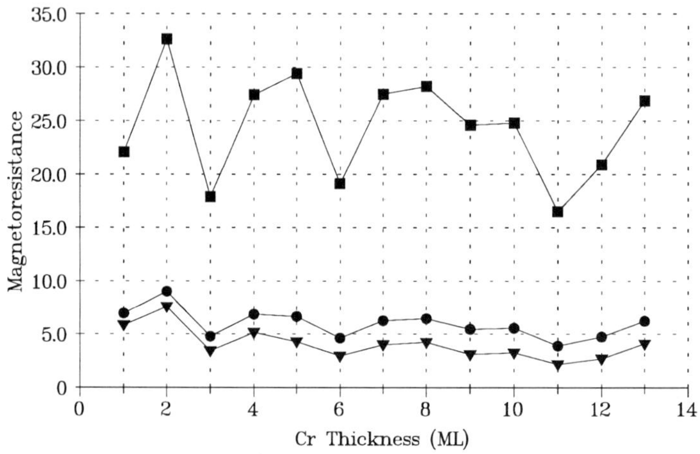

Figure 18. Calculated CPP-GMR for Fe_3Cr_n versus Cr layer thickness $n = 1, 13$. The results assuming Cr defects in the Fe layers are indicated by squares ($\beta = 0.11$). Dots mark the results for spin-independent relaxation times $\beta = 1$ and triangles for $\beta = \beta_{min}$.

with the density of states of the superlattice for parallel alignment

$$N_P(E_F) = \sum_{\sigma} N^{\sigma}(E_F) = \sum_{\sigma} \sum_{k} \delta(E_k^{\sigma} - E_F) \tag{71}$$

and for anti-parallel alignment

$$N_{AP}(E_F) = 2\sum_k \delta(E_k - E_F). \tag{72}$$

v_μ^σ and v_μ^{AP} are Fermi surface averages of the Cartesian components of the velocity for parallel

$$v_\mu^\sigma = \sqrt{\frac{\sum_k \delta(E_k^\sigma - E_F)\, v_{k_\mu}^{\sigma\;2}}{\sum_k \delta(E_k^\sigma - E_F)}} \tag{73}$$

and anti-parallel alignment

$$v_\mu^{AP} = \sqrt{\frac{\sum_k \delta(E_k^{AP} - E_F)\, v_{k_\mu}^{AP2}}{\sum_k \delta(E_k^{AP} - E_F)}}. \tag{74}$$

The densities of states in the parallel and anti-parallel configuration (Fig. 19) are of

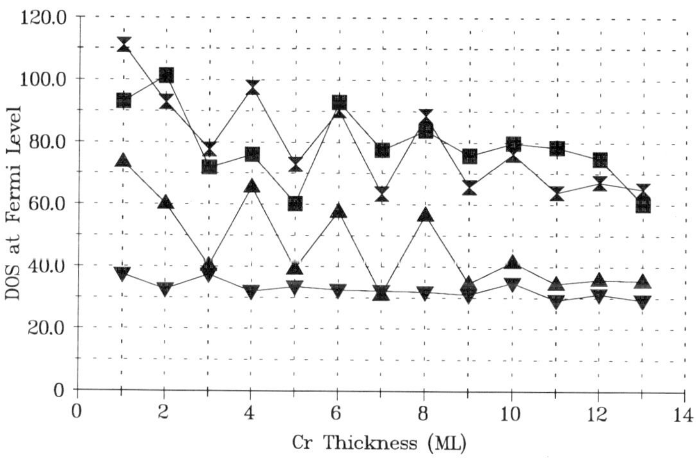

Figure 19. $N_P(E_F)$ (squares) and $N_{AP}(E_F)$ (double triangles) of Fe_3Cr_n superlattices versus Cr layer thickness in arbitrary units. Triangles mark spin-projected densities of states, $N^\uparrow(E_F)$ (uparrows) and $N^\downarrow(E_F)$ (downarrows) of $N_P(E_F)$ [Mertig *et al.*, 1995].

the same order and they oscillate in phase with the Cr layer thickness, that is, they do not account for the 200% CIP-GMR and 700% in CPP-GMR (Fig. 17 and 18). Obviously, GMR is originated by the differences in the Fermi velocities in the parallel and anti-parallel configurations (Fig. 20).

The spin dependence of the relaxation time ($\beta \neq 1$) due to scattering centers in the

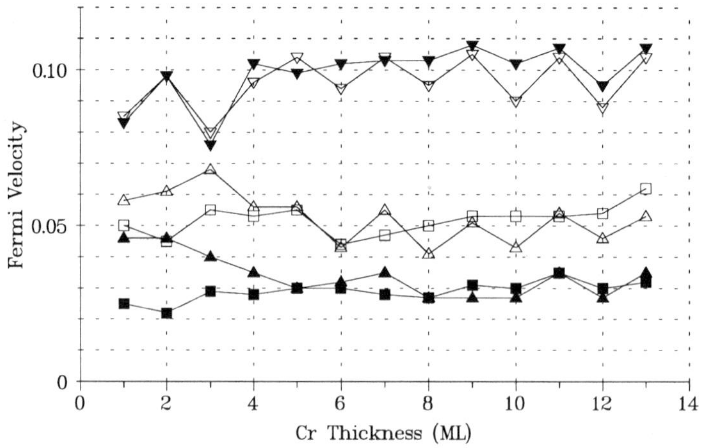

Figure 20. Fermi surface averages of the velocity components. Open symbols indicate v_x and full symbols v_z. Upwards triangles mark the velocity of majority electrons, downwards triangles the velocity of minority electrons and squares the velocity in the anti-parallel aligned configuration [Mertig *et al.*, 1995].

ferromagnetic layers (Cr impurities in Fe layers, for example) leads to modifications. GMR is now to be calculated via eq. 64, 65 and 67. Under the assumption that the extension of the impurity is small compared to the layer thickness, the scattering properties are described by the spin-dependent relaxation times calculated for the Cr impurities in Fe [Mertig *et al.*, 1993; Mertig *et al.*, 1994]. For this case $\beta = 0.11$, that is, the majority electrons are scattered strongly at a Cr defect and the minority electrons are just weakly scattered as they pass the defect.
The results assuming Cr defects located in the Fe layers lead to a GMR of about 600% in CIP (squares in Fig. 17) and 2500% in CPP (squares in Fig. 18) which is larger than the experimental results [Gijs and Okada, 1992; Gijs *et al.*, 1993; Pratt Jr. *et al.*, 1991; Gijs *et al.*,]. The anisotropy β obtained in [Mertig *et al.*, 1993; Mertig *et al.*, 1994] for a variety of defects scatter from 0.1 to 10 (see Tab. 1). Therefore, GMR is discussed as a function of β in relaxation time approximation. If only one type of scatterer is included, GMR takes on a minimum for

$$\beta_{min} = \frac{\tau^{\uparrow}}{\tau^{\downarrow}} = \sqrt{\frac{\sum_k \delta(E_k^{\downarrow} - E_F)\, v_{k\mu}^{\downarrow 2}}{\sum_k \delta(E_k^{\uparrow} - E_F)\, v_{k\mu}^{\uparrow 2}}}. \tag{75}$$

That is, GMR can be enhanced or reduced by spin-dependent impurity scattering. The results for $\beta = \beta_{min}$ are shown in Fig. 17 and 18 (triangles).
Finally, we discuss the combination of alternating Fe layers with Cr defects ($\beta = 0.11$) and with Cu defects ($\beta = 3.68$). For the sake of simplicity, equal concentration is assumed. The relaxation times are in parallel configuration

$$\frac{1}{\tau^{\sigma}} = \frac{1}{2}\left(\frac{1}{\tau_{Cr}^{\sigma}} + \frac{1}{\tau_{Cu}^{\sigma}}\right), \tag{76}$$

and in anti-parallel configuration

$$\frac{D_{\uparrow}^{+} + D_{\uparrow}^{-}}{\tau^{\uparrow}} = \left(\frac{D_{\uparrow}^{+}}{\tau_{Cr}^{+}} + \frac{D_{\uparrow}^{-}}{\tau_{Cu}^{-}} \right), \tag{77}$$

and

$$\frac{D_{\downarrow}^{+} + D_{\downarrow}^{-}}{\tau^{\downarrow}} = \left(\frac{D_{\downarrow}^{-}}{\tau_{Cr}^{-}} + \frac{D_{\downarrow}^{+}}{\tau_{Cu}^{+}} \right), \tag{78}$$

respectively where the superscripts $+, -$ correspond to majority and minority electrons. D_{σ}^{+} and D_{σ}^{-} is the local spin-dependent density of states in a ferromagnetic layer where spin-σ electrons are majority or minority electrons, respectively. In the considered configuration the factors have been set to unity. The conductivity in the anti-parallel state including both defects is given by

$$\hat{\sigma}^{AP} = e^2(\tau^{\uparrow} + \tau^{\downarrow}) \sum_k \delta(E_k^{AP} - E_F)\, \mathbf{v}_k^{AP}\, \mathbf{v}_k^{AP}. \tag{79}$$

The results are shown in Fig. 21. The main message from this calculation is that the

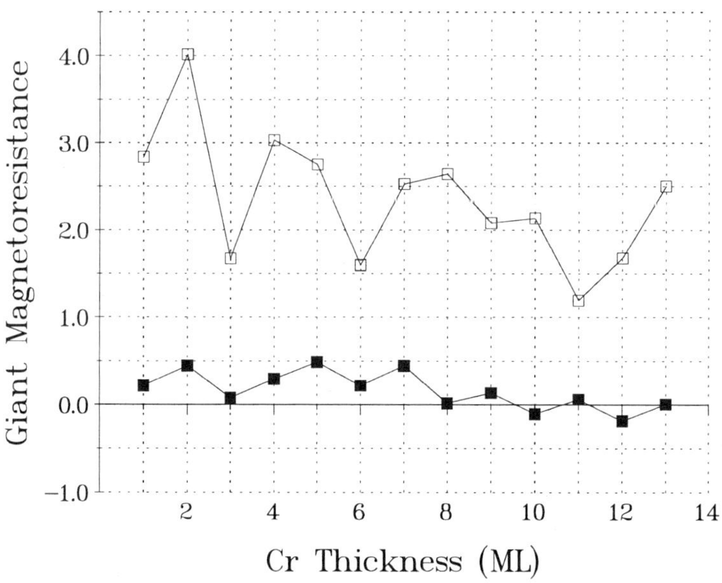

Figure 21. Calculated GMR for Fe_3Cr_n with Cr and Cu defects in adjacent Fe layers in CIP (closed squares) and CPP (open squares). For $n \geq 10$ the inverse GMR is obtained.

GMR can be reduced drastically or even change sign, if scatterers with different spin anisotropy are combined. (If they would be located in the same layer they would merely act as an effective scattering potential.) One should note that in Fig. 21 the inverse effect occurs in CIP but not in CPP. Experimentally, the inverse effect was obtained for $Fe/Cr/Fe/Cu/Fe/Cu$ multilayers [George *et al.*, 1994], as well as in $Fe_{1-x}V_x/Au/Co$ multilayers [Renard *et al.*, 1995], both in CIP geometry.
Generally, it should be noted that CPP-GMR for comparable scattering is always larger than CIP by a factor of about 4 in agreement with experimental results [Gijs *et al.*,]. This factor stems from the difference of the Fermi velocity components in plane and perpendicular to the plane, and it finds its natural explanation in the fact that the carriers with large momentum components in direction of the current are more influenced by the superstructure in CPP.
Moreover, GMR (Fig. 17, 18, 21) shows characteristic variations with layer thicknesses. On Fig. 21 these variations are reminiscent of experimentally found oscillations [Okuno and Inomata, 1994].
In conclusion, we have shown that GMR under the assumption of a coherent multilayer system and a spin-independent impurity scattering is determined by the changes of the electronic structure as a function of the magnetization direction. Spin-dependent impurity scattering can enhance or reduce the effect. The combination of impurities with different spin anisotropy can in particular cause the inverse GMR.
The aim of the last section was to show how first principle methods are able to elucidate the microscopic origin of a new phenomenon, especially if the phenomenon depends strongly on the composition of the system and cannot easily be described by simple models.

Acknowledgement

The author would like to thank P. H. Dederichs and H. Eschrig for valuable discussions and is grateful to J. Binder and P. Zahn for technical assistance.

References

M. N. Baibich, J. M. Broto, A. Fert, F. Nguyen van Dau, F. Petroff, P. Etienne, G. Creuzet, A. Friederich, and J. Chazelas, Phys. Rev. Lett. **61**, 2472 (1988).

J. Banhart, R. Bernstein, J. Voitländer, and P. Weinberger, Sol. Stat. Comm. **77**, 107 (1991).

J. Banhart, H. Ebert, J. Voitländer, and P. Weinberger, Phy. Rev. B **50**, 2104 (1994).

G. Binash, P. Grünberg, F. Saurenbach, and W. Zinn, Phy. Rev. B **39**, 4828 (1989).

S. Blügel, H. Akai, R. Zeller, and P. H. Dederichs, Phy. Rev. B **35**, 3271 (1987).

A. Bourquart, E. Daniel, and A. Fert, Phys. Lett. **26 A**, 260 (1968).

W. H. Butler, Phy. Rev. B **31**, 3260 (1985).

W. H. Butler, J. M. MacLaren, and X.-G. Zhang, Mat. Res. Soc. Proc. **313**, 59 (1993).

W. H. Butler and G. M. Stocks, Phy. Rev. B **29**, 4217 (1984).

R. E. Camley and J. Barnaś, Phys. Rev. Lett. **63**, 664 (1989).

I. A. Campbell and A. Fert, in *Ferromagnetic Materials* (E. P. Wohlfahrt, North–Holland, New–York, 1982), Chap. 9, p. 747.

J. F. Dorleijn, Philips Res. Pepts. **31**, 287 (1976).

J. F. Dorleijn and A. R. Miedema, J. Phys. F **5**, 487 (1975).

B. Drittler, N. Stefanou, S. Bl gel, R. Zeller, and P. H. Dederichs, Phy. Rev. B **40**, 8203 (1989).

J. S. Faulkner and G. M. Stocks, Phy. Rev. B **21**, 3222 (1980).

A. Fert, J. Phys. C **2**, 1784 (1969).

A. Fert and I. A. Campbell, J. Phys. F **6**, 849 (1976).

J. M. George, L. G. Pereira, A. Barthelemy, F. Petroff, L. Steren, J. L. Duvail, and A. Fert, Phys. Rev. Lett. **72**, 408 (1994).

M. A. M. Gijs, S. K. Lenczowski, and J. B. Giesbers, Phys. Rev. Lett. **70**, 3343 (1993).

M. A. M. Gijs, S. K. J. Lenczowski, J. B. Giesbers, R. J. M. van de Veerdonk, M. T. Johnson, R. M. Jungblut, A. Reinders, and R. M. J. van Gansewinkel (unpublished).

M. A. M. Gijs and M. Okada, Phy. Rev. B **46**, 2908 (1992).

D. A. Greenwood, Proc. Phys. Soc. London **71**, 585 (1958).

R. Q. Hood and L. M. Falicov, Phy. Rev. B **46**, 8287 (1992).

J. Inoue, A. Oguri, and S. Maekawa, J. Phys. Soc. Jpn. **60**, 376 (1991).

R. Kubo, J. Phys. Soc. Japan **12**, 570 (1957).

P. M. Levy, Sol. Stat. Phys. **47**, 367 (1994).

P. M. Levy, S. Zhang, and A. Fert, Phys. Rev. Lett. **65**, 1643 (1990).

I. Mertig, E. Mrosan, and P. Ziesche, in *Multiple Scattering Theory of Point Defects in Metals: Electronic Properties* (Teubner–Verlag, Leipzig, 1987).

I. Mertig, P. Zahn, M. Richter, H. Eschrig, R. Zeller, and P. H. Dederichs, J. Mag. Mag. Mat. (accepted) (1995).

I. Mertig, R. Zeller, and P. H. Dederichs, Phy. Rev. B **47**, 16178 (1993).

I. Mertig, R. Zeller, and P. H. Dederichs, in *Metallic Alloys: Experimental and Theoretical Perspectives*, (Kluwer Academic Publishers, Dodrecht, Boston, London, 1994), p. 423.

N. F. Mott, Adv. Phys. **13**, 325 (1964).

N. F. Mott and H. Jones, *The theory of properties of metals and alloys* (Oxford University Press, Oxford, 1936).

T. Oguchi, J. Mag. Mag. Mat. **126**, 519 (1993).

S. N. Okuno and K. Inomata, Phys. Rev. Lett. **72**, 1553 (1994).

W. P. Pratt Jr., S. F. Lee, J. M. Slaughter, R. Loloee, P. A. Schroeder, and J. Bass, Phys. Rev. Lett. **66**, 3060 (1991).

J.-P. Renard, P. Bruno, R. Megy, B. Bartenlian, P. Beauvillain, C. Chappert, C. Dupas, E. Kolb, M. Mulloy, P. Veillet, and E. Veu, preprint (1995).

R. Schad, C. D. Potter, P. Beliën, G. Verbanck, V. V. Moshchalkov, and Y. Bruynseraede, Appl. Phys. Lett. **64**, 3500 (1994).

K. M. Schep, P. J. Kelly, and G. E. W. Bauer, Phys. Rev. Lett. **74**, 586 (1995).

N. Stefanou, A. Oswald, R. Zeller, and P. H. Dederichs, Phy. Rev. B **35**, 6911 (1987).

V. S. Stepanyuk, R. Zeller, P. H. Dederichs, and I. Mertig, Phy. Rev. B **49**, 5157 (1994).

J. C. Swihart, W. H. Butler, G. M. Stocks, D. M. Nicholson, and R. C. Ward, Phys. Rev. Lett. **57**, 1181 (1986).

T. Valet and A. Fert, Phy. Rev. B **48**, 7099 (1993).

B. Velicky, Phy. Rev. B **184**, 614 (1969).

T. Vojta, I. Mertig, and R. Zeller, Phy. Rev. B **46**, 15751 (1992).

P. Zahn, I. Mertig, M. Richter, and H. Eschrig, Phys. Rev. Lett. (accepted) (1995).

R. Zeller, J. Phys. F **17**, 2123 (1987).

R. Zeller, R. Podloucky, and P. H. Dederichs, Z. Phys. B **38**, 165 (1980).

J. M. Ziman, *Electrons and Phonons* (Claredon Press, Oxford, 1962).

CLASSICAL AND FIRST PRINCIPLES MOLECULAR DYNAMICS SIMULATIONS IN MATERIAL SCIENCE: APPLICATION TO STRUCTURAL AND DYNAMICAL PROPERTIES OF FREE AND SUPPORTED CLUSTERS.

Carlo Massobrio and Philippe Blandin

Institut de Physique Expérimentale,
Ecole Polytechnique Fédérale de Lausanne,
PHB-Ecublens, CH-1015 Lausanne

INTRODUCTION

The increasing availability of powerful computers has an enormous impact on the solution of a large variety of problems in modern physics and chemistry. Molecular Dynamics (MD) is particularly attractive since it provides an atomic-scale description of the dynamics of complex systems. This is achieved by exploiting the equivalence between thermodynamic quantities and time averages of appropriate variables of coordinates and velocities. In the context of material science topics, Molecular Dynamics has now achieved a firmly established role of useful tool which complements experimental findings, predicts behaviors not accessible to experiments and elucidates mechanisms which can only be understood by analyzing the atomic movements. In principle, the level of detail and accuracy of MD is only limited by the reliability of the model employed. This is a crucial issue which necessitates some historical remarks. As it was nicely described in the introductory paper of one of the first international conferences devoted to the applications of molecular dynamics to condensed matter problems [1], MD simulations began as a method aimed at testing statistical mechanics theories. Simple model potentials were employed with the intent of investigating generic static and dynamic properties of monoatomic fluids. The border between statistical mechanics and material science, giving rise to simulations more oriented toward complex systems featuring both fundamental and technological interest, was crossed in the early seventies with simulations of ionic solids and liquids for which the coulombic interaction is largely predominant. The search of a realistic modeling able to elucidate the structural and thermal behavior of real materials received in the eighties a great impulse from the introduction of empirical N-body potentials [2]. These are constructed by relying heavily on two factors, namely the fitting of the potential parameters to experimental quantities and a judicious choice of the analytical form for the interatomic interactions. Over the years, these potentials have proved to be successful in reproducing some basic features of chemical bonding by requiring only a limited increase of computational cost with respect to pair potentials. A further element of progress was introduced through the developments of equations of motions formally equivalent to the sampling of phase space configurations in the constant temperature and/or constant pressure ensembles, as done in the Nosé-Hoover [3, 4] and the Parrinello-Rahman [5] methods. A real breakthrough in the significance and reliability of molecular dynamics approaches has been represented by the idea of Car and Parrinello [6], which proposed to combine molecular dynamics and density functional theory to derive interatomic forces within a generalized Lagrangian scheme. In this framework, electronic, structural and thermal properties of a given system are accurately treated on equal footing without the approximations inherent in the model potentials approach. These issues have been largely discussed in the literature over the past ten years and a consistent number of excellent review papers and textbooks are now available to learn about the development and methodology of the MD approach [7, 8, 9, 10, 11, 12, 13]. In particular, first principle molecular dynamics implementations are carefully discussed in Ref. [14, 15, 16].

This article focuses on three main issues of materials science-oriented simulations by MD. In the first part we deal with the basics integration algorithms by Verlet [17] and we show how to implement it in the context of the extended equations of motion associated with the use of fictitious "thermostats". This example is particularly instructive since the establishment of first principles MD, which is based on the introduction of a "fake" adiabatic dynamics for the motion of the electronic wave functions, requires an effective control of the temperature associated with different set of degrees of freedom. Moreover it provides an example of construction of an extended hamiltonian, which can be associated to a desired statistical ensemble.

The second part of the paper is devoted to the motivations and the strategy underlying the construction of an empirical, n-body interatomic potential within the embedded atom method (EAM) scheme. This model potentials allow large scale simulations for a class of materials, the transition and noble metals, which are still out of the reach of the more elaborate first-principles treatments,especially if one is interested in thermal properties requiring extended time intervals. We focus on the conceptual foundations of these schemes and we present an explicit discussion of the implementation of the embedded atom potentials to the case of Ag and Pt. An application to the study of the early stages of growth of Ag on Pt(111) is given in detail. Finally we conclude with a section devoted to first principles, Car-Parrinello simulations. After a brief review of the ideas and equations relative to this approach, we discuss results concerning the investigation of structural and electronic properties of small copper clusters. These small objects play a role of increasing importance in material science. Indeed, understanding nanoscale physics is a necessary step to control the formation of new materials that can eventually be formed by depositing small aggregates on well defined substrates at given coverages and temperatures.

CLASSICAL MOLECULAR DYNAMICS: SIMPLE CONCEPTS

Velocity Verlet and the Thermostat of Nosé-Hoover.

Classical molecular dynamics is based on interatomic potentials, which can be

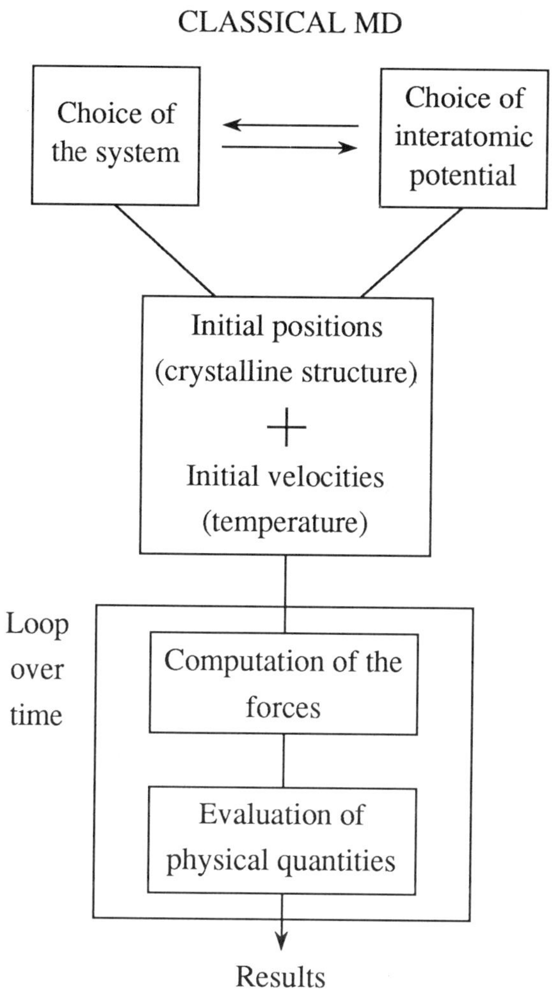

Figure 1: Classical molecular dynamics evolution.

either of the simple pair form, like the Lennard-Jones for the rare gases, or an N-body one like that used in the EAM description of metals. The typical scheme of a classical MD evolution is reproduced in Fig. (1). If we call V the system potential energy, that is dependent on the atomic positions $\mathbf{r}_i$, the classical equations of motion are:

$$\begin{cases} m\ddot{\mathbf{r}}_i = -\nabla_{\mathbf{r}_i} V \\ \mathbf{r}_i(t_0) = \mathbf{r}_{i0} \\ \dot{\mathbf{r}}_i(t_0) = \dot{\mathbf{r}}_{i0} \end{cases}$$

These equations may be solved using the velocity Verlet (vV) algorithm, which is closely related to the original scheme proposed by Verlet [17]. The vV scheme is particularly appealing due to its formal simplicity, its good stability and its reasonable requirements of memory storage. All these features make it particularly appealing for both classical and first principles simulations.

The first step of the algorithm consists in evaluating the positions at $t = t + \delta t$ and the velocities at the intermediate time $t = t + \frac{1}{2}\delta t$ by resorting to Taylor expansions

$$\mathbf{r}_i(t+\delta t) = \mathbf{r}_i(t) + \delta t\, \mathbf{v}_i(t) + \frac{\delta t^2}{2}\, \mathbf{a}_i(t)$$

$$\mathbf{v}_i(t+\frac{\delta t}{2}) = \mathbf{v}_i(t) + \frac{\delta t}{2}\, \mathbf{a}_i(t)$$

The force corresponding to this new configuration is then evaluated and the corresponding acceleration updated

$$\mathbf{a}_i(t+\delta t) = -\frac{\nabla_i U_{tot}(\mathbf{r})}{m}$$

The velocity at time $t + \delta t$ is finally calculated.

$$\mathbf{v}_i(t+\delta t) = \mathbf{v}_i(t+\frac{\delta t}{2}) + \frac{\delta t}{2}\, \mathbf{a}_i(t+\delta t)$$

If one is interested in investigating the behavior of a system at constant temperature, i.e. its time evolution within the canonical ensemble, the equations of motion have an additional contribution that accounts for the coupling of the system to a heat bath. Among the various methods used to control the temperature of the sample we describe here the thermostats of Nosé-Hoover [4]. The great advantages of this approach are its efficiency and the simplicity of the coupling mode between the system and the heat bath. This is determined by the evolution of one single parameter $\gamma(t)$.

This scheme is particularly useful in Car-Parrinello-like MD simulations, characterized by fictitious electronic degrees of freedom which are expected to evolve on the Born-Oppenheimer energy surface. Blöchl and Parrinello [18] have shown this can be achieved by imposing a Nosé-Hoover dynamics to both electrons and ions. Formally the equations of motion of the physical system are of the Langevin type [19] but without any stochastic term. Since the heat-bath have also be able to provide some additional kinetic energy to the system, the coupling coefficient $\gamma(t)$, that may be assimilated to a friction contribution in a Langevin equation, may become negative. The heat bath is characterized by $3(\mathcal{N}-1)$ degrees of freedom, where $\mathcal{N}$ is the number of particles of the system, and Q determines the rate of damping of the temperature fluctuations in the system. Instead of Q one may equally use a time of response parameter τ that is related to Q by: $Q = 3(\mathcal{N}-1)k_B T \tau^2$. The Nosé-Hoover Hamiltonian reads

$$H(\mathbf{p},\mathbf{q},\gamma) = U_{tot}(\mathbf{q}) + \sum \frac{\mathbf{p}^2}{2m} + 3(\mathcal{N}-1)k_B T \int_{t_0}^{t} \gamma(t')\, dt' + \frac{P_s{}^2}{2Q}$$

where $P_s = Q\gamma$ stands for the linear momentum of the heat bath and where T is the desired temperature. The equations of motion for the physical system are:

$$\begin{cases} \dot{\mathbf{q}} &= \dfrac{\mathbf{p}}{m} \\ \dot{\mathbf{p}} &= -\nabla_{\mathbf{q}} U_{tot}(\mathbf{q}) - \gamma\, \mathbf{p} \end{cases}$$

while the dynamics of the coupling coefficient is driven by:

$$\dot{\gamma} = \frac{1}{Q}\left[\sum \frac{\mathbf{p}^2}{m} - 3(\mathcal{N}-1)k_B T\right]$$

The latter equation induces an increase of the friction $\gamma(t)$ if the kinetic energy of the physical system is too large and its decrease in the opposite case resulting in negative values of $\gamma(t)$.

The former equations may be rewritten as

$$\dot{\mathbf{r}}_i(t) = \mathbf{v}_i(t)$$

$$\dot{\mathbf{v}}_i(t) = -\frac{\nabla_i U_{tot}(\mathbf{r})}{m} - \gamma(t)\mathbf{v}_i(t)$$

$$\dot{\gamma}(t) = \frac{1}{Q}\,[\sum_i m\mathbf{v}_i^2(t) - 3(\mathcal{N}-1)k_BT]$$

The Nosé-Hoover equation can then be solved as follows [19]:

$$\mathbf{r}_i(t+\delta t) = \mathbf{r}_i(t) + \delta t\,\mathbf{v}_i(t) + \frac{\delta t^2}{2}\,[\mathbf{a}_i(t) - \gamma(t)\,\mathbf{v}_i(t)]$$

$$\gamma(t+\frac{\delta t}{2}) = \gamma(t) + \frac{\delta t}{2}\,\dot{\gamma}(t)$$

$$\mathbf{v}_i(t+\frac{\delta t}{2}) = \mathbf{v}_i(t) + \frac{\delta t}{2}\,[\mathbf{a}_i(t) - \gamma(t)\,\mathbf{v}_i(t)]$$

$$\dot{\gamma}(t+\frac{\delta t}{2}) = \frac{1}{Q}\,[\sum_i m\mathbf{v}_i^2(t+\frac{\delta t}{2}) - 3(\mathcal{N}-1)k_BT]$$

$$\gamma(t+\delta t) = \gamma(t+\frac{\delta t}{2}) + \frac{\delta t}{2}\,\dot{\gamma}(t+\frac{\delta t}{2})$$

$$\mathbf{a}_i(t+\delta t) = -\frac{\nabla_i U_{tot}(\mathbf{r})}{m}$$

$$\mathbf{v}_i(t+\delta t) = \frac{2}{2+\delta t\,\gamma(t+\delta t)}[\mathbf{v}_i(t+\frac{\delta t}{2}) + \frac{\delta t}{2}\,\mathbf{a}_i(t+\delta t)]$$

$$\dot{\gamma}(t+\delta t) = \frac{1}{Q}\,[\sum_i m\mathbf{v}_i^2(t+\delta t) - 3(\mathcal{N}-1)k_BT]$$

We mention that another possible choice could be a purely predictor-corrector approach:

$$\mathbf{r}_i(t+\delta t) = \mathbf{r}_i(t) + \delta t\,\mathbf{v}_i(t) + \frac{\delta t^2}{2}\,\mathbf{a}_i(t)$$

$$\mathbf{v}_i(t+\delta t) = \mathbf{v}_i(t) + \delta t\,\mathbf{a}_i(t)$$

$$\gamma(t+\delta t) = \gamma(t) + \delta t\,\dot{\gamma}(t)$$

$$\mathbf{a}_i(t+\delta t) = -\frac{\nabla_i U_{tot}(\mathbf{r})}{m} - \gamma(t+\delta t)\,\mathbf{v}_i(t+\delta t)$$

$$\mathbf{v}_i(t+\delta t) = \mathbf{v}_i(t+\delta t) + \frac{\delta t}{2}[\mathbf{a}_i(t+\delta t) - \mathbf{a}_i(t)]$$

$$\dot{\gamma}(t+\delta t) = \frac{1}{Q}\sum_i m\mathbf{v}_i^2(t) - 3(\mathcal{N}-1)k_BT$$

$$\gamma(t+\delta t) = \gamma(t+\delta t) + \frac{\delta t}{2}\,[\dot{\gamma}(t+\delta t) - \dot{\gamma}(t)]$$

The conservation of the total energy of the extended system reads:

$$E_{tot} = U_{tot}(\mathbf{r}) + \frac{1}{2}\sum m\mathbf{v}^2 + 3(\mathcal{N}-1)k_BT\Gamma + \frac{Q\gamma^2}{2}$$

where the variable $\Gamma(t)$ is defined by the relation $\dot{\Gamma}(t) = \gamma(t)$. Finally the time dependence of $\Gamma(t)$ may be provided by:

$$\Gamma(t + \delta t) = \Gamma(t) + \delta t\, \gamma(t) + \frac{\delta t^2}{2}\, \dot{\gamma}(t)$$

REALISTIC POTENTIALS FOR MATERIALS: THE EAM SCHEME

The choice of the system and phenomena of interest determine the nature of the potential to be used in MD simulations, since a good compromise have to be found between the accuracy of the description and the time and size limitations typical of the MD approach. For example, physical properties of transition metals, such as diffusion of small clusters on metallic surface and the related characterization of the early stages of growth, cannot be addressed by simulations based on two-body interatomic potentials. The main deficiencies of two-body potentials (TBP) in reproducing the properties of metallic surfaces are the following: **a**) TBP predict a linear dependence of the cohesive energy per atom on the coordination number N, while, in the case of Ni, for example, a $\sqrt{N}$ dependence fits the experimental data better [20]. As a consequence, the vacancy formation energy E_v is badly reproduced: TBP predict energies of the order of the cohesive energy E_c while, in metals, they are typically less then half of it, **b**) to accommodate the abrupt symmetry breaking at the interface between the slab and the vacuum, many metallic systems display a contraction of the interlayer distance between the two outermost layers. TBP, however, erroneously predict an outward relaxation of the topmost layer, **c**) TBP do not reproduce the elastic constants of metals. In particular, for a cubic lattice, TBP lead to the Cauchy relation $C_{12} = C_{44}$ in disagreement with experiments . In principle, this problem can be overcome by adding a volume dependent contribution to the cohesive energy in order to describe the effects of conduction electrons. However this additional term is difficult to handle when dealing with surfaces or defects like vacancies and dislocations.

The simplest choice to circumvent the limitations of pair potentials, while keeping a computationally efficient scheme, is to replace at least part of the pair potential by a pair *functional* [20] that accounts for the N-body aspects of the binding, leading to total energy expressions of the type:

$$E_{tot} = \frac{1}{2} \sum_{\alpha,\beta\neq\alpha} V(R_{\alpha\beta}) + \sum_{\alpha} U_\alpha(\sum_{\beta\neq\alpha} f(R_{\alpha\beta}))$$

where $V(R)$ is a purely two-body contribution while U, that depends on the sum of pair functions f, accounts for the N-body aspects. In this article we illustrate this approach in the case of EAM interatomic potentials.

The EAM formalism can be understood within a modeling of metals containing a crystalline arrangement of ion cores surrounded by the gas of the conduction electrons. Within this picture, the ion cores interact repulsively with each other, as screened coulombian charges, while the crystal is stabilized by the attractive interaction between the ion cores and the electron gas. This applies best to simple metals and to transition metals with almost empty or almost completely filled d shells. These considerations suggest the following splitting of the system total energy:

$$E_{tot} = \frac{1}{2} \sum_{\alpha,\beta\neq\alpha} \phi_{\alpha\beta}(R_{\alpha\beta}) + \sum_{\alpha} F_\alpha(\rho_\alpha)$$

where $\phi_{\alpha\beta}$ is the pair potential that accounts for the ion-ion repulsion. The embedding function $F_\alpha(\rho_\alpha)$ depends only on the nature of element α and corresponds to the energy gain when embedding the ion core α into the electronic density due to the other atoms of the system.

The EAM is not the only computational methods based on a pair functional approach. Similar schemes have been developed over the last ten years, as, for example, the Effective Medium(EM) [21], the Glue Model of Ercolessi et al. [22] , the Finnis-Sinclair approach [23]. Again a closely related scheme was developed by V. Rosato, M. Guillope and B. Legrand, [24] based on a second moment expansion of the TB energy. Besides their computational efficiency, all these schemes provide a quite simple description of the interatomic interactions yielding a fairly good insight on the physical processes. For simple metals and transition metals with an almost empty or filled d band the trends and orders of magnitude are usually sufficiently well described although the energetics is not entirely reliable, as for example, in the case of surface energies and adatoms diffusion barriers [25, 26, 27]. For partially filled d band transition metals one has to take explicitly into account the additional attractive energy rising from the $d-d$ hybridization [21]. We recall that some authors adapted the EM and the EAM methods to account for covalent binding. Among the others, Nørskov included covalent effects in the EMT [28] to describe the hydrogen interaction with $3d$ metals. Baskes [29], and Carlsson et al. [30] applied a modified EAM scheme to treat semiconductors. In order to describe Si, Baskes [31] used a rather complicated embedding function legitimated by the good agreement found between calculated and experimental structural energies and bond lengths. Better elastic constants are obtained by Baskes [31] by adding an angular dependent term. For Si, Carlsson et al. [30] resorted to new energy terms in order to account for the non-spherical components of the local atomic environment and to describe the local ring topology. Their treatment is based on a moment expansion of the TB scheme [23].

Development of EAM Potentials for Ag and Pt

The pair potential that enters the EAM energy expression have to account for the core-core repulsion. Daw and Baskes [32] suggest a Coulomb-like form:

$$\phi_{\alpha\beta}(R_{\alpha\beta}) = \frac{Z_\alpha(R_{\alpha\beta})Z_\beta(R_{\alpha\beta})}{R_{\alpha\beta}}$$

where $Z_\alpha(R)$ is an effective charge. Since we are dealing with metals it looks reasonable to approximate the potential $\phi(R_{\alpha\beta})$ by a screened Coulomb interaction. The expression used by Foiles et al. [26] for the effective charge reads:

$$Z_\alpha(r) = \begin{cases} Z_0(1+\beta r^\nu)e^{-\alpha r} - \tilde{Z}_{Rcut} & \text{if } r \leq R_{cut} \\ 0 & \text{otherwise} \end{cases}$$

Z_0 is taken to be the number of the outershell electrons.
This expression for $\phi_{\alpha\beta}(r)$ follows the commonly assumed empirical combination rule for the description of the interaction between atoms of type A and B: $U_{AB} \approx \sqrt{U_{AA}U_{BB}}$. The second ingredient required to specify the EAM potential is the electron embedding density. The embedding electronic density at site α, $\rho_\alpha(\mathbf{r})$ is written as the superposition of the atomic-like densities from all the other atoms in the system. For simplicity, we will assume that each atom β contributes a spherical charge distribution:

$$\rho_\alpha = \sum_{\beta\neq\alpha} \rho_\beta^{at}(|\mathbf{R}_\alpha - \mathbf{R}_\beta|)$$

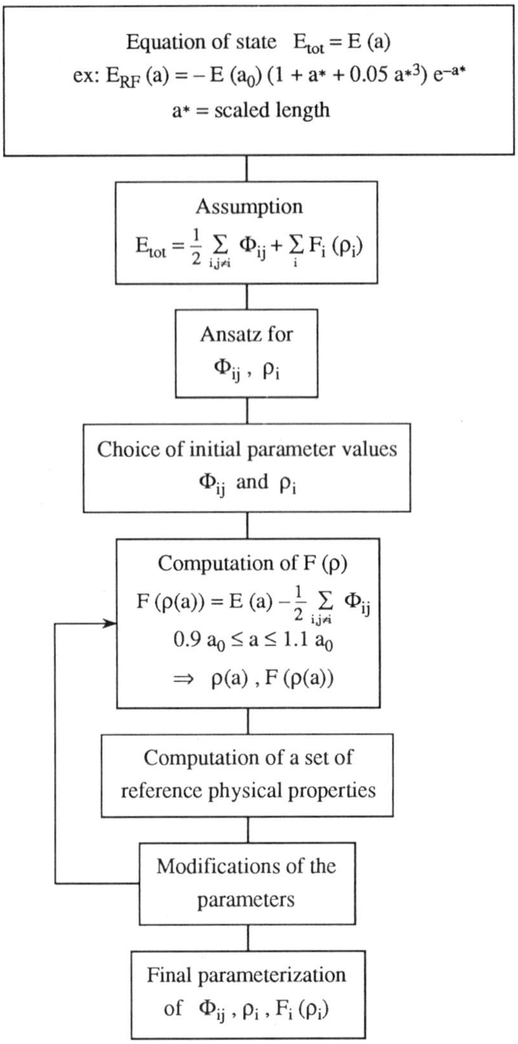

Figure 2: Implementation of an EAM interatomic potential.

Since we know that in transition metals the cohesion is determined mainly by the d electrons and, to a minor extent, by the s electrons, it is natural to write:

$$\rho_\beta^{at}(r) = n_s \rho_{\beta\ s}^{at}(r) + n_d \rho_{\beta\ d}^{at}(r)$$

where $n_s + n_d$ have to be equal to the number of valence electrons. In principle, n_s and n_d could be chosen on the basis of some simple prescription (like taking the ground state atomic configuration), or computed within an ab-initio scheme. In practice, n_s and n_d are usually considered as free parameters, to be fitted on some known property of the system. Since $\rho_{\beta\ d}^{at}(r)$ is short range, $\rho_{\beta\ s}^{at}(r)$ is the only relevant contribution to the embedding density ρ_α, and, for mono-atomic systems n_s is mainly a scale parameter for the embedding density. Note that changing n_s does not change the total potential, since the variation in n_s can be exactly compensated by a related transformation in the repulsive interaction. This statement is true for mono-atomic systems only. The third crucial ingredient of the scheme is an easy and flexible parameterization of the embedding energy. In the EAM implementation by Foiles et al. [26], used in our computations of the Ag/Pt(111) system, this is achieved by first specifying the zero temperature equation of state of the solid as a function of pressure (or, experimentally, volume). Rose and coworkers [33] showed that the cohesive energy E_{RF} of most metals can be expressed in terms of the lattice parameter a by a simple two parameter $(E(a_0), l)$ scaling of a universal function $\mathcal{E}^*(a^*)$:

$$E_{RF}(a) = E(a_0)\mathcal{E}^*(a^*)$$

where a^* is a scaled length. In what follows we give its expression in terms of the Wigner-Seitz radius r_{WS} (which is equal to r_{WSE} at equilibrium) and of the scaling length parameter l, $a^* = \frac{r_{WS} - r_{WSE}}{l}$. In our case the Wigner-Seitz radius and the lattice parameter are related by: $\frac{4}{3}\pi r_{WS}^3 = \frac{a^3}{N}$, N being the number of atoms per unit cell. l measures the range of the interactions in the crystal and it is therefore closely related to the screening length of the metal. It may be expressed in terms of the equilibrium cohesive energy $E(r_{WSE}) = E(a_0)$ and of the bulk modulus B: $l = \frac{E(r_{WSE})}{12\pi B r_{WSE}}$. Rose and coworkers approximated the universal function $\mathcal{E}^*$ by:

$$\mathcal{E}^*(a^*) \approx -(1 + a^* + 0.05a^{*3})\exp(-a^*)$$

Foiles et al [26] only kept the polynomial expansion up to the first order in a^*.
Once the form of the repulsive potential, of the embedding density, and of the equation of state are specified, the fit of the potential proceeds as follows. By starting from a trial set of parameters for the repulsive potential (α, β, ν) and for n_s, it is possible to compute the repulsive energy $E_{rep}(a)$ and the embedding density $\rho_\alpha(a)$ of a solid (assuming a fixed structure) for a series of different lattice parameters a. Then the embedding energy for each $\rho_\alpha(a)$ is computed as the difference of the equation of state provided by the Rose's et al's [33] and the repulsive energy: $F(\rho_\alpha(a)) = E_{RF}(a) - E_{rep}(a)$. At this point the EAM potential is fully specified and it is possible to compute elastic properties (like the bulk modulus and the lattice constants) corresponding to the input parameters. By comparing computed and measured elastic properties one can optimize the input parameters for the repulsive potential. This represents a non-linear fit, that may require a considerable effort and care to provide satisfactory results.

Finally, the alloy properties of the different elements are tuned by varying the n_s parameters. The construction of an EAM potential is schematized in figure (2). For the Ag and Pt of our computations, the electronic distribution was calculated using the Hartree-Fock orbitals tabulated by Clementi and Roetti [34] in the case of Ag and by McLean and McLean [35] in the case of Pt. The results by Foiles et al. [26] are reproduced exactly by applying a shift to the wave functions in order to force the electronic density and its first derivative to vanish smoothly at the cutoff distance R_{cut}.

The parameters that enter the effective charge and the electronic density of Ag and Pt are summarized in table (1). The resulting functions $Z_\alpha(R)$ and $F_\alpha(\rho)$ are smooth and $Z_\alpha(R)$ is monotonically decreasing. In agreement with Puska et al [36] and Jacobsen et al [21] findings, $F_\alpha(\rho)$ has a single minimum, is linear at high densities and has a positive curvature. The latter condition ensures that the bond strength decreases with increasing coordination.

Table 1: Foiles et al. parameterization of Ag and Pt (see text).

	Ag	Pt
Z_0	11.0	10.0
α	2.1395	1.2663
β	1.3529	0.1305
ν	2.0	1.0
n_s	1.6760	1.0571
$n_s + n_d$	11	10
R_{cut}	5.25Å	5.25Å

THE Ag/Pt(111) SYSTEM

Diffusion of Small Ag Clusters.

Diffusion properties of clusters of Ag on Pt(111) system have been investigated by performing both static and molecular dynamics computations [37]. We use the EAM potentials with the parameterization given in Ref. [26]

The key parameter to understand the growth mode of adsorbates on (111) fcc metallic surfaces is their diffusivity . As suggested by FIM studies [38] diffusivity is very enhanced in the case of metallic adatoms on the Pt(111).
The static diffusion barriers, for the Ag adatom, dimer and trimer on Pt(111) are reported in table (2). They have been obtained by considering two configurations, in the first a given Ag adatom is stable on its adsorption site, in the second it is constrained at the saddle point located in between two adsorption sites, the rest of the system being allowed to relax. The increase of the size of the aggregate results into a surprisingly low increase of the diffusion barrier in agreement with the experimental finding of Wang and Ehrlich [39]. These authors find even a smaller dependence on the size of the adcluster since the measured activation energy of Ir_2 and Ir_3 are respectively 1.6 and 2.3 times larger than in the case of the Ir monomer. Our diffusion barriers are typically within

Table 2: Diffusion barriers and activation energies.

	static [eV]	dynamical [eV]	experimental
Ag_1	0.05	0.06 ± 0.005	0.16 ± 0.01
Ag_2	0.10	0.09 ± 0.01	
Ag_3	0.14	0.13 ± 0.01	

the range of values obtained on the basis of N-body potentials, i.e. in between 0.05eV-0.15eV, (see the results on Cu/Cu(111) [40] and Ag/Ag(111) [41]) but much smaller than the experimental value measured via STM [42] and the value 0.20eV calculated by Feibelman within a fully ab-initio approach [27]. This is an indication of the fact that in order to get truly quantitative results the EAM potentials should to be re-optimized for the range of the electronic densities typical of surfaces, as explicitly observed in ref. [27, 43]. However, since most of the calculated barriers have been obtained statically, it is of interest to investigate the role played by dynamical effects. To address this issue, we computed the activation energy of diffusion by a dynamical approach. In this case the barrier is the slope of $\ln(D(T))$, where T is the temperature, with $D(T)$ evaluated as $\lim_{t\to\infty}\langle R^2(t)\rangle/4t$. The results are reported in the table (2). The values for the dimer and the trimer refer to the motion of their center of mass.

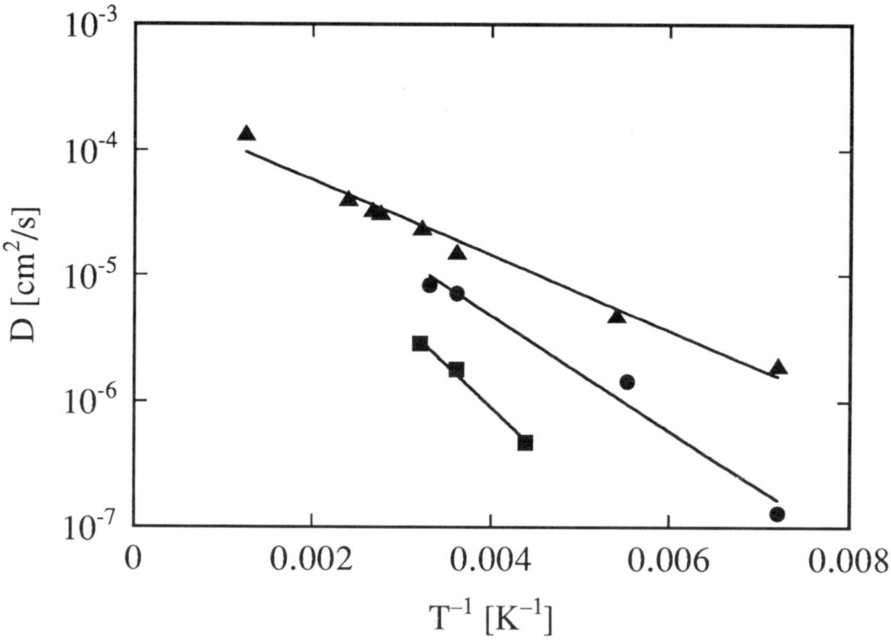

Figure 3: Diffusion coefficients of Ag adatom (triangles), dimer (circles) and trimer (square) on Pt(111), displayed in an Arrhenius plot as a function of temperature.

Figure (3) is the associated Arrhenius plot. The values of the diffusion coefficients we obtains for temperatures larger than 300K D are higher than 10^{-5}cm^2/s, value that corresponds to the diffusion in liquids. The activation energies extracted from the long time limit of $D(t)$ are in fairly good agreement with the static diffusion barrier. This confirms that the discrepancy we found between EAM and experiments cannot be attributed to neglected dynamical contribution, but it is due to an unsatisfactory behavior of the potential for situations involving undercoordinated atoms. A note of caution, however, is in order: our computations consider only ideal surfaces, without impurities and defects, that, in real experiments could affect the results. This point is analyzed in more detail below, where we simulate the diffusion process on Pt(111) in the presence of steps and a low coverage of hydrogen impurities.

Diffusion of Ag on Pt(111) in the Presence of Steps.

The investigation of the effect of steps on the diffusion of Ag on Pt(111) requires the use of a rather large system. Each of our (111) planes has 1150 Pt atoms. Two straight parallel step edges are created by removing 29 rows of Pt atoms in the uppermost plane. We resort to the frozen slab approximation, which consists in fixing the Pt positions after relaxation at T=0K, except for the Pt atoms belonging to the steps, which follow the same dynamics of the Ag monomer [44]. The Ag monomer is deposited at random on the lower terrace and equilibrated at T=300K. After a long wandering, the atom reaches a step edge and gets adsorbed (see Fig. (4)). A time interval as extended as 2.8 nsec is not sufficient to observe its desorption. Once the adatom is trapped, its motion becomes one-dimensional. Even though our simulated trajectory is not long enough to compute a fully reliable value of the one-dimensional diffusion coefficient D_1, we still can estimate its upper bound, which is $D_1 < 8\ 10^{-7}$ cm^2/sec. This value corresponds

Figure 4: Ag monomer in presence of a step.

to an activation energy of ~ 250meV. In order to investigate the effective potential felt by the adatom at the step we calculate the distribution of probability $\rho(z)$ of finding the Ag adatom at a distance z of the edge. Then, assuming that $\rho(z)$ may be written as

$$\rho(z) = \Theta \exp(-U_{eff}(z)/k_B T)$$

we extract the desired effective potential. In figure (5) we display the function $\rho(z)/\Theta$ and, in the inset, the associated effective potential. The latter is short range (< 3Å) and is characterized by a well depth of 250meV. We see that the effective potential has a clear repulsive part opposing the diffusion from both the upper and lower terrace.

Influence of Preadsorbed Hydrogen.

The main problem for the study of the influence of preadsorbed H on the evolution of a metallic surface is to find a potential that is both reliable and computationally efficient. The difficulty comes from the fact that H is neither a metal nor a noble gas but it is rather a molecular insulator, at least at moderate temperatures and pressures. However, an effective potential approach appears to give reasonable results in the case of atomic H adsorbed on a surface or dissolved in the bulk of metals [29, 45]. We cannot expect to obtain quantitatively reliable results from this application of the EAM well beyond its natural range of validity. For this reason we limit our analysis to a qualitative description of the system behavior. Furthermore, in this case the range of validity of MD is limited to the intermediate and high temperature regimes because of the complete neglect of quantum effects important for hydrogen.

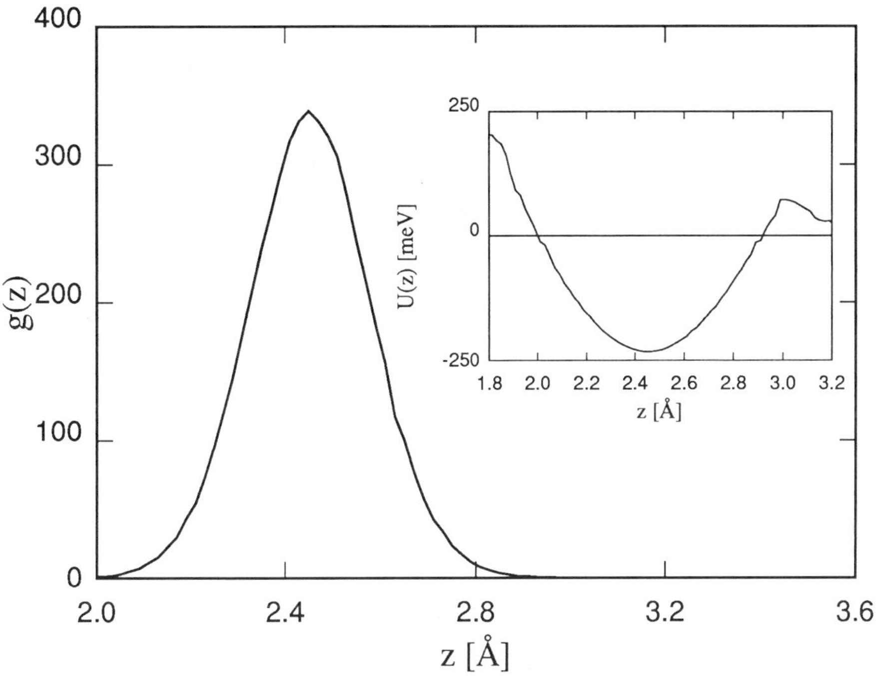

Figure 5: Density distribution (main plot) and effective potential (inset) for one Ag ad-atom adsorbed at a step at $T = 300$K. The distance z is measured from the step edge.

Experimentally a whole monolayer of H adsorbed on the Pt(111) surface is known to be stable in a (1×1) unreconstructed structure, it has an adsorption energy of 0.27eV per H atom [46] and does not dissolve in the bulk. At high T subsurface sites might be preferred [47]. Thermal desorption experiments show two clear peaks: one at 230K and another at 310K [48]. At both temperatures H desorbs by the formation of H_2. For the EAM parameterization of H we assume the following expressions for the distance dependence of the screened H core charge and for its electronic density:

$$\begin{cases} Z_H(r) = Z_0[\exp(-2r/a_0) - \exp(-2R_{cut}/a_0)] & \text{if } r < R_{cut} \\ Z_H(r) = 0 & \text{otherwise} \end{cases}$$

$$\begin{cases} \rho_H(r) = A[\exp(-2r/a_0) - \exp(-2R_{cut}/a_0)] & \text{if } r < R_{cut} \\ \rho_H(r) = 0 & \text{otherwise} \end{cases}$$

where a_0 is the Bohr radius (0.529 Å). Z_0 is taken equal to 1 while $R_{cut} = 4$Å. The embedding function $F(\rho)$ is optimized in order to reproduce the main properties of the free H_2 molecule, i.e., its equilibrium distance, binding energy and vibrational frequency. The constant A, that enters the ansatz of the electronic distribution, is determined in order to fit the heat of adsorption of one monolayer of H on the Pt(111) surface. We obtained $A = 0.7e/\text{Å}^3$. Our simulation cell is made of 9 moving layers of 120 atoms each. This corresponds to a surface of roughly 798Å^2. The Newton's equations of motion are integrated with a timestep of $2.8\ 10^{-16}s$. Our potential reproduces the adsorption energy as well as the stable (1×1) structure of the H overlayer. The H-Pt

distance is however overestimated by 10%, we obtain 1.89 Å instead of the experimental value 1.68 Å. The surface segregation is well reproduced, since the energy of an H atom in an interstitial position is higher by 0.5eV than on the surface. Upon heating the (1×1) monolayer of H up to 200K we observe the formation of dimers that, however, do not desorb during our short MD simulation. At 300K the whole H layer becomes unstable and massive desorption sets in. This is consistent with the experimental findings. A single atom on the surface is stable (i.e., it does not desorb) up to very high temperatures and is very mobile: we compute diffusion constants of the order of 10^{-3}cm^2/s ($D_{\rm H}(250K) = 1.4\ 10^{-3} cm^2/s$, $D_{\rm H}(300K) = 3.0\ 10^{-3} cm^2/s$, $D_{\rm H}(400K) = 6.0\ 10^{-3} cm^2/s$) that correspond to an activation energy of $\sim$ 0.12eV. This value is roughly one half of the experimental one.
The effect of H on the diffusion of a Ag monomer was studied by depositing a single H atom and a single Ag atom, well separated from each other, on the surface and by letting the system evolve. As a benchmark we simultaneously considered the evolution of the Ag alone starting from the same initial conditions. The results show that the H-Ag interaction reduces the diffusion of the Ag by approximately one order of magnitude: from $D_{\rm Ag}(300K) = 2.8\ 10^{-5} cm^2/s$ and $D_{\rm Ag}(400K) = 4.4\ 10^{-5} cm^2/s$ the diffusion constants drop to $D_{\rm Ag,\ H}(300K) = 0.23\ 10^{-5} cm^2/s$ and $D_{\rm Ag,\ H}(400K) = 0.9\ 10^{-5} cm^2/s$. Taking into account that the diffusion is a thermally activated process i.e. that $D(T) = D_0 \exp(-\Delta E/K_B T)$ and assuming that the preexponential factor D_0 is a constant, we observe an increase of $\sim 0.1eV$ of the adatom activation energy compared to the one of the Ag on the ideal and clean terrace. This may be rationalized by noticing that, when they meet, the Ag and the H form a dimer with a characteristic dissociation time of the order of 0.4ps. It is worth mentioning that taking into account the presence of a very low coverage of H makes our activation energies perfectly compatible with the experimental findings of Röder et al [42].

Ag/Pt(111) Heteroepitaxy.

In order to study the dynamics of growth of the Ag/Pt(111) for submonolayer coverages we resorted to a large simulation box by using the frozen slab approximation. This simplification is justified since the in-plane adsorbate-substrate interaction is quite weak, as confirmed by the high values of diffusion coefficients, and is consistent with the experimental evidence that Ag and Pt atoms mix only through the step edges and not at the center of the terraces. Furthermore these approximations results in an increase of the diffusion barrier to the value 0.1 eV, thereby improving the agreement with the experiments.

The initial configuration is a random distribution of Ag atoms on the available adsorption sites. The number of Ag atoms reproduces the desired coverage Θ as illustrated on the figure (6) (a) for the $\Theta = 0.25$ and $T = 1000K$ case. Our simulations are long enough to distinguish at least the nucleation and the early stages of the growth sequence. For the highest temperatures the occurrence of a dynamical equilibrium between a stationary population of islands and a vapor phase of small and highly mobile clusters is apparent. In order to monitor the clusters evolution on the substrate, we calculate the cluster size distribution P(N) and the probability p(N) that a Ag atom belongs to a cluster of size N, where p(N) is proportional to NP(N).

Nucleation and Early Stage Growth.

To evaluate the nucleation time τ_N of our system, we consider the number $\mathcal{N}$

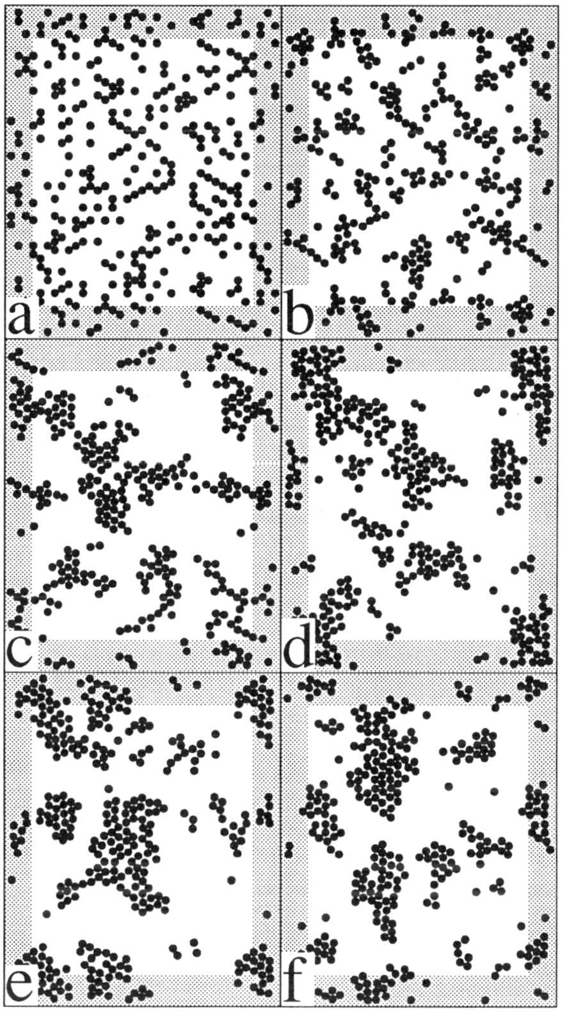

Figure 6: Snapshots of the adatom configurations, $\Theta = 0.25$, $T = 1000$K. (a) Starting (random) distribution; (b) $t = \tau_N = 5.6$ psec after deposition; (c) $t = 0.8$ nsec; (d) $t = 1.6$ nsec; (e) $t = 2.4$ nsec; (f) $t = 3.8$ nsec. The shaded area belongs to the periodic replicas of the central region.

of clusters whose size is larger than the critical size $\bar{N}$. A stable nucleus is defined according to its special stability towards dissociation and to its low diffusivity since it must grow by incorporating smaller and more mobile aggregates. In our case, we find that even at 1000K the trimers do not diffuse too much during our simulation times. Therefore we adopt the convention that $\bar{N} = 3$ for $\Theta = 0.1$ and $\bar{N} = 4$ for $\Theta = 0.25$.

We observe that initially $\mathcal{N}$ grows rapidly, until it reaches a maximum that corresponds to the end of the nucleation phase. Then, when the island growth sets in, $\mathcal{N}$ slowly decreases, inducing a reduction of the total number of aggregates on the surface. We defined the nucleation time τ as the time at which $\mathcal{N}$ reaches its maximum. The values obtained are reported in the table (3).

The snapshots displayed on fig. (6) show that the early stage growth does not end

Table 3: Nucleation and equilibration times.

	τ_N [psec]		τ_e [psec]	
T[K]	$\Theta = 0.10$	$\Theta = 0.25$	$\Theta = 0.10$	$\Theta = 0.25$
400	9.1	6.3	4550	3136
600	10.5	4.9	2240	1868
800	9.8	7.0	1383	1176
1000	10.5	5.6	1071	784

with the formation of a large and unique island but rather lead to the occurrence of a dynamical equilibrium between a population of moderately large aggregates and a fairly mobile gas phase of small clusters. We therefore estimate the end of the early stage growth τ_e as the time after which no systematic modifications of $P(N)$ can be detected. The values of τ_e as a function of Θ and T are reported in the table (3). They correspond to the end of the short-time aggregation process.

The dynamical equilibrium we observe is characterized by an important and steady population of monomers, indicating that their rate of condensation is in balance with their reevaporation. On a mesoscopic or a macroscopic time scale this equilibrium might still evolve because of long time relaxation processes, like Ostwald ripening and coalescence [49]. These long time behaviors, however, are out of the reach of the MD approach. Furthermore, on a macroscopic time scale, the influence of defects like vacancies, steps and dislocations cannot be neglected.

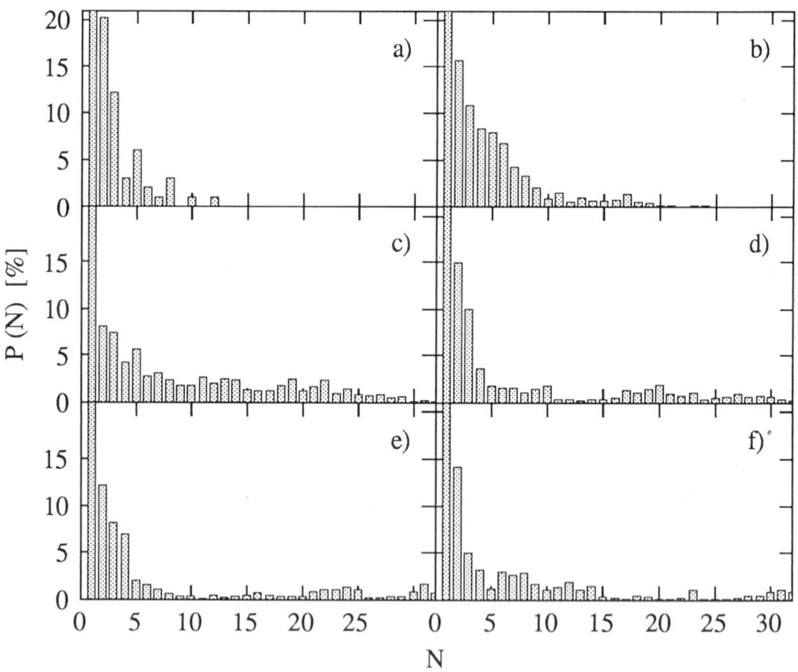

Figure 7: Cluster size distribution $P(N)$ as a function of time, $\Theta = 0.25$, $T = 1000$K. (a) Starting (random) distribution; (b) $t = \tau_N = 5.6$ psec after deposition; (c) $t = 0.4$ nsec; (d) $t = 1.2$ nsec; (e) $t = 2$ nsec; (f) $t = 3$ nsec.

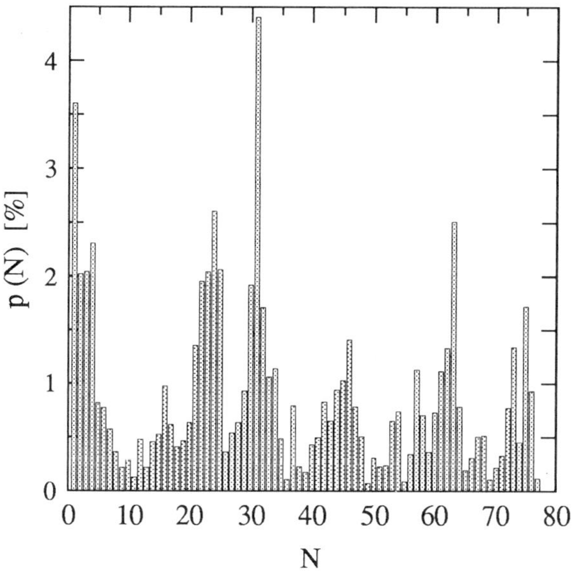

Figure 8: Adatom probability $p(N)$ to belong to a cluster of size N, $\Theta = 0.25$, $T = 1000\text{K}$, $t = 3$ nsec after deposition.

The panel (a) of figure (7) shows the size distribution function associated to our initial configuration for a coverage of Θ =0.25ML and a temperature T=1000K. Its approximately gaussian shape reflects the random character of our "deposition" process. The nucleation phase is illustrated in panel (b) (t=5.6ps): the number of clusters increases rapidly while the average size remains rather small ($\sim$ 4 atoms/cluster). The early stage growth, (panel (c)), displays a clear shift towards the larger aggregate sizes ($\sim$ 8 atoms/cluster). The remaining panels (d to f) correspond to the steady state we discussed, with some statistical fluctuations between them. At this point, as can be seen on the graph (8) of $p(N)$, the great majority of atoms belongs to large aggregates ($N \geq 30$). However, we see that the contribution due to small clusters is always present, highlighting the importance of the vapor phase.

Globally, this picture still holds for smaller coverages ($\Theta < 0.25$) and temperatures ($T < 1000K$). Smaller coverages just induce a shift of the size distribution towards smaller clusters while lower temperatures require a longer equilibration time, the nucleation being essentially unaffected. This results from the very different activation barriers of the processes involved in the nucleation (diffusion of adatoms) and early stage growth (diffusion of small clusters and dissociation events) stages. The former is indeed almost barrierless while the diffusion of small aggregates and their dissociation correspond to much larger activation energies. At each T, the nucleation of small clusters (up to 10 atoms) occurs within a time of the order of the picosecond while the end of the growth stage is much less clearly defined. Except in the 1000K and 800K cases, the values of τ_e in table (3) are mainly estimates: at 600K, P(N) still displays a slight drift and at 400K, once the nucleation is completed, the system is mainly frozen on the nsec scale accessible to MD.

The occurrence of this "freezing", once the nucleation sequence is at completion, can be monitored by computing an "average" diffusion constant accounting for the positions of all Ag atoms. The time evolution of the average diffusion at T=1000K and at T=600K is illustrated in the fig. (9). For both these temperatures the nucleation and equilibration phases corresponds to a significant reduction of the diffusion. But, while at 1000K the diffusion remains sizable during the equilibrium stage, it drops by two orders of magnitude at 600K. This residual diffusion is mainly related to the intra-cluster

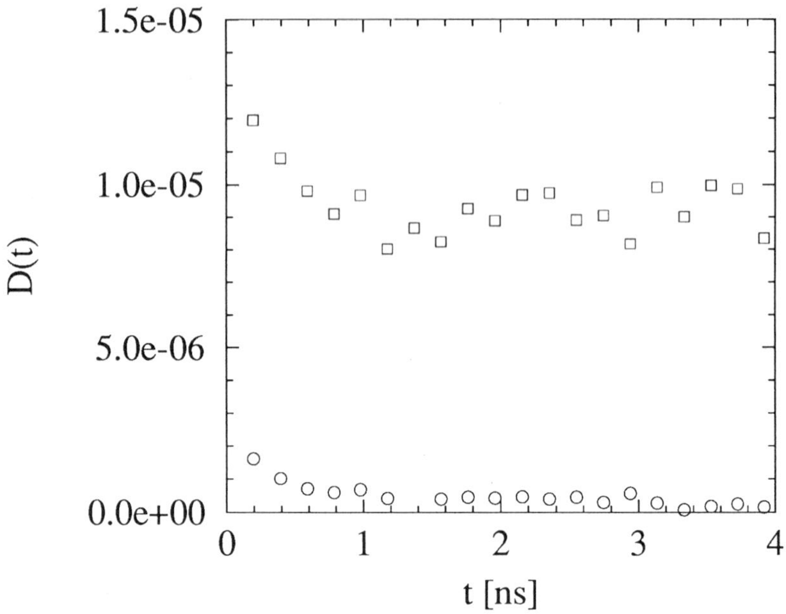

Figure 9: Time dependence of the diffusion coefficient D, $\Theta = 0.25$.

motion. For the nucleation and early stage growth phases available experimental information concerns the size of the stable nucleus at each temperature. This is extracted from flux dependent experiments [50]. These experimental results relative to the size of the stable Ag nuclei on Pt are summarized in the table (4).
We see that these results are compatible with our choice of the critical size.

Table 4: Stable nucleus as a function of temperature.

	stable nucleus
$35K < T < 45K$	Ag
$45K < T < 80K$	Ag_2
$T \geq 100K$	$\geq Ag_3$

Equilibrium Properties.

A dynamical equilibrium between the islands and a two-dimensional vapor phase sets in, for the large enough temperatures ($T \geq 600K$), once the early stage growth is at completion. At this point, the cluster size distribution is stationary and only displays some statistical fluctuations. At high temperatures, our simulations describe an equilibrium situation while at the lower ones, we might be trapped into metastable configurations. The fact that randomly distributed adatoms aggregate spontaneously indicates that an homogeneous distribution of adatoms is unstable at the (Θ, T) conditions of our study. In order to establish the coexistence curve between the condensed and the gas phase of our system we compute the density of the two phases as a function of Θ and T. The main problem is to have an operational definition of the area of a cluster. This task is complicated by the fact that the shape of the clusters is very irregular. We adopted the following convention. Due to its hexagonal symmetry, the Pt(111) surface can be easily partitioned into small triangles with their center of mass located on the adsorption sites. Each atom i entirely occupies the triangle at $\mathbf{R}_i$ as well as $\frac{1}{3}$ of all the triangles whose center of mass ξ_k is within a cutoff distance $\mathcal{R}$ of atom i ($|\mathbf{R}_i - \xi_k| < \mathcal{R}$). This distance corresponds to the exclusion radius of the radial distribution function (i.e. 2.25 Å). According to this convention, an atom at rest in its equilibrium position occupies 2 triangles while a larger area will be associated to an oscillating or diffusing adatom. The area occupied by the vapor is defined as the difference between the total cell area and the surface occupied by the condensed phase. For each coverage and temperature we report the densities of the gas and condensate phases, normalized to the density of a filled monolayer, in the figure (10). The liquid-vapor coexistence curve we obtain turns out to have an asymmetric bell shape , due to the asymmetrical behavior between adatoms and ad-vacancies described in Ref. [51].

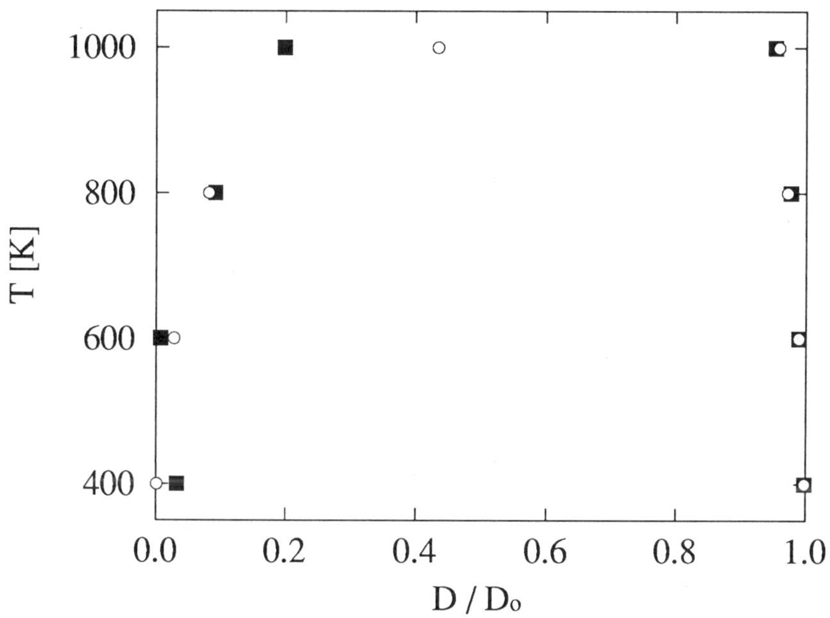

Figure 10: Vapor - condensate coexistence curve. Dots: $\Theta = 0.25$; squares: $\Theta = 0.10$. D_0 is the density of a filled monolayer.

Influence of Steps on Nucleation and Growth.

The main qualitative feature of experimental systems missing in the model described above is the interaction of the adatoms with the surface defects and especially with steps. Experiments clearly show that steps, always present even on the most compact and stable surfaces, act as sinks for the adsorbed species. After deposition, adatoms and clusters will migrate on the terraces, to reach eventually a step. As shown

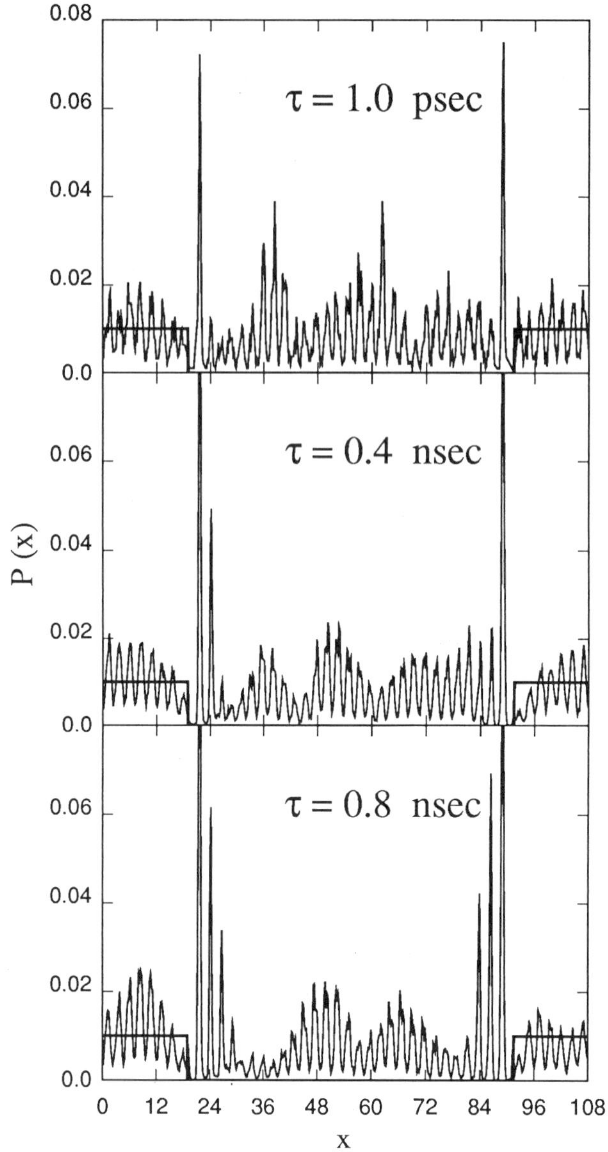

Figure 11: Adatom distribution probability $P(x)$ along the direction perpendicular to the steps, $\Theta = 0.25$, $T = 1000$K.

above, the increase of their coordination results into an attractive potential that, at moderate T, confines their movement to a one-dimensional motion. In the case of Ag/Pt(111), if the temperature exceeds 650K, some mixing occurs starting precisely at the step-edges [52]. To investigate the effect of steps we resorted to a large simulation cell, each (111) plane containing 1150 Pt atoms. Two different step edges were created on our surface by removing 29 rows of Pt atoms of the uppermost plane. These two steps differ by the relative position of their edge atoms with respect to the one of the second substrate layer. Despite this small difference, we prefer the simplicity of this configuration over more complex geometries involving only one kind of step (as in the case of triangular islands). The stepped Pt surface was relaxed and all Pt atoms were frozen, except those from the upper terrace.

Our simulation was performed at $\Theta = 0.25$ and $T = 1000K$. We monitored the effect of steps on the adsorbate distribution by computing the density profile $\rho_{Ag}(x,t)$, where x is the direction perpendicular to the steps (see figure 11).

The nucleation time $\tau_N =\sim 6$ psec is already sufficient to create a clear density peak in the first row of the lower terrace adjacent to the steps. After ~ 0.4 nsec the first row is almost filled by ad-atoms, and a peak starts to develop on the following row. This process of filling and the subsequent propagation of the perturbation to the next row (wetting) proceeds at a regular rate, with one added row every ~ 0.4 nsec. The process is limited by the adatom diffusion in the rectangular region (roughly three row wide) adjacent to the growing step, that appears depleted of adatoms.

In the presence of a constant flow of adatoms, the steps will move to close the lower terrace, until when a new layer will start to grow. In the condition of the present simulation at constant number of adatoms, the process terminates when all the Ag atoms are adsorbed at the steps, with only a residual distribution of ad-atoms rapidly diffusing in between. The upper terrace appears to be much less affected than the lower one due to the mobility of the step atoms, allowing both vertical and in-plane relaxation effects. Furthermore, the undercoordination of the edge atoms results, on this side of the step, into an effective potential that is slightly repulsive. Over times of the order of 3 nsec, the evolution of ad-atoms and clusters in the upper terrace is similar to that described in the previous sections. Only beyond this time the effect of the steps manifests itself also in the upper terrace configuration. At $\tau =\sim 4$ nsec the step repulsive potential (together with the small size of our system) confines the ad-atoms at the center of the upper terrace in a unique large islands surrounded by the vapor phase.

These observations set the limit for our description of the previous sections for the experimental systems. Up to ~ 1 nsec (i.e., comparable to the early stage growth time) the effect of the steps extends over few atomic rows in the lower terrace only. For terrace widths of the order of 1000 Å, therefore, our study of the ideal surface remains valid for most of the system during times longer than the early stage growth and extending well into the late stage aggregation.

Over times exceeding ~ 50 nsec the effect of the steps extends over most of the surface, and a realistic model has to take into account their presence to quantitatively describe the system evolution.

The times reported and discussed here all refer to the $\Theta = 0.25$, $T = 1000$K case, and we expect only a moderate dependence on Θ. For different temperatures, instead, these times must be rescaled, for instance by using the temperature dependence of τ_e as a guideline.

FIRST-PRINCIPLES MOLECULAR DYNAMICS

Molecular dynamics simulations based on empirical potentials suffer from one major, intrinsic limitation, which undermines their general validity and applicability in the context of realistic modeling of materials. Although in principle interatomic potentials fitted largely on experimental data may achieve a high degree of refinement, they cannot provide interatomic forces by accounting for an explicit knowledge of the electronic structure of the system. A major breakthrough in the direction of performing molecular dynamics based on a first principle determination of the forces was proposed in 1985 by Car and Parrinello (CP) [6], which showed how to extract in a manageable way ionic forces directly from the electronic structure, in a way which is totally independent on any empirical parameter. The method first described in Ref. [6] has rapidly become the ideal approach to tackle and solve a large number of issues related to the knowledge of the equilibrium configurations and the thermal properties of a wide class of materials. Moreover, it has the additional attractive feature of providing a detailed description of the electronic structure, which is valid within the limit of the local density approximation of density functional theory. In this article we limit ourselves to recall briefly the basics concepts underlying the CP scheme and we describe in more detail the application of the method to a typical problem of combined electronic and structural minimization, i.e. the investigation of the equilibrium geometries and electronic properties of small Cu clusters.

The MD-DF method is based on the evolution of a combined classical and quantum system which is reproduced by the evolution of a classical dynamical system defined by the Lagrangian:

$$\mathcal{L} = \sum_i \mu_i \int_\Omega |\dot{\psi}_i^* \dot{\psi}_i| d\mathbf{r} + \sum_I \frac{1}{2} M_I \dot{\mathbf{R}}_I^2 - E[\{\psi_i\}, \{\mathbf{R}_I\}] + \sum_{ij} \Lambda_{ij} \left(\int_\Omega \psi_i \psi_j^* d\mathbf{r} - \delta_{ij} \right) \quad (1)$$

and the corresponding equations of motion

$$\mu_i \ddot{\psi}_i(\mathbf{r}, t) = -\frac{\delta E}{\delta \psi^*(\mathbf{r}, t)} + \sum_k \Lambda_{ik} \psi_k(\mathbf{r}, t) \quad (2)$$

$$M_I \ddot{\mathbf{R}}_I = -\nabla_{\mathbf{R}_I} E \quad (3)$$

In (1) μ_i is the fictitious mass associated with the electronic degrees of freedom, M_I and R_I are the ionic masses and coordinates and $\mathcal{L}$ is the matrix of Lagrangian multipliers which ensure the orthonormality constraints on the one-particle orbitals $\psi_i(\mathbf{r}, t)$. In the norm-conserving pseudopotential description the electronic density is given by $n(\mathbf{r}, t) = \sum_i |\psi(\mathbf{r}, t)|^2$. The action of the classical potential appearing in (1) is taken by the density functional expression for the total energy

$$\begin{aligned} E[\{\psi_i\}, \{\mathbf{R}_I\}] &= \sum_i < \psi_i(\mathbf{r})| - \frac{\nabla^2}{2} |\psi_i(\mathbf{r}) > + \int n(\mathbf{r}) [V_{ext}(\mathbf{r}) + \frac{\Phi(\mathbf{r})}{2}] d\mathbf{r} \\ &+ E_{xc}[n(\mathbf{r})] + \frac{1}{2} \sum_{I \neq J} \frac{Z_I Z_J}{|\mathbf{R}_I - \mathbf{R}_J|} \end{aligned}$$

A fundamental point regarding the dynamical evolution expressed in (2, 3) concerns the accuracy by which the trajectories derived from the extended Lagrangian are close to those defined in the Born-Oppenheimer (BO) surface, as required by a derivation of interatomic forces from first principles. For insulators and semiconductors the BO surface is followed accurately, provided a very small μ_i is chosen with respect to the value of the physical mass M. In the case of metals, the absence of a gap causes instabilities in the Eqs.(2, 3), due to the transfer of energy flowing from the ions to the electrons. In most cases, this shortcoming can be avoided by using two different Nosés thermostats, so as to control separately the temperatures associated to the ionic and the fictitious electronic motion[18]. An attractive feature of the CP scheme is its close resemblance to a traditional classical molecular dynamics evolution, thereby allowing the use of all the techniques developed to mimic the effect of temperature, pressure and geometrical constraints on the system.

Application to the Study of Small Copper Clusters

Cluster physics plays an important role in materials science in many respects. The growing interest for nanotechnologies makes clusters the ideal prototype system to study the evolution of fundamental properties when the size of the system changes from the microscopic to the macroscopic dimensions. Over the last twenty years, clusters have acquired an identity on their own, and are now being investigated as the "smallest" form of materials which may bear some resemblance to their bulk counterpart, while still retaining many physical and chemical features pertaining to the isolated atom. Recent advances in cluster detection and analysis techniques, as for example photoelectron spectroscopy and mass selected soft landing on inert-gas matrices, are opening the way to direct comparisons of experimental results and theoretical predictions in the field of clusters [53]. From the theoretical point of view, this offers the opportunity to test complex theories on these small, more affordable systems. Moreover, understanding at an high level of accuracy the electronic and structural properties of clusters is necessary to build up consistently more simplified models of interaction of the atomic scale, which can be used to investigate material science problems involving a large number of atoms. First principles molecular dynamics is ideally suited to explore the complex potential energy surface of a cluster and to describe accurately the nature of the forces. In the following example, we focus on the determination of structural and electronic properties of small Cu clusters, which are particularly intriguing since they appear to share many similarities with alkali-metal clusters. This is due to the electronic structure of the Cu atom, consisting of one s electron outside a closed shell of d electrons. In order to understand the role played by the d electrons in determining the structural properties of Cu clusters and investigate analogies and differences with Na clusters, calculations requiring an explicit treatment of d valence electrons are crucial. Early results, available only for very small cluster sizes have suggested that the ground state geometries of small Cu_N [54] and Na_N [55] clusters are similar despite the existence of important hybridization effects between $3d$ and $4s$ valence electrons.

In the calculations presented here the Car-Parrinello method is used in in conjunction with Vanderbilt ultrasoft pseudo-potentials [56]. The Vanderbilt scheme allows to perform accurate computations for systems containing first-row elements and transition metals [57, 58]. We employ a plane-wave basis set with periodic boundary conditions with the clusters positioned in a cubic cell of side equal to 25.9 a.u.. The energy cut-offs are 15 Ry for the orbitals expanded in plane waves at the Γ point and 130 Ry for the augmented electron density. Exchange and correlation are treated using Perdew and Zunger's interpolation formulas [59]. In the following we describe the equilibrium geometries found for N = 6, 8, 10.

The initial geometries are selected from those indicated in the literature for Na_N [55] and Ag_N [60] cluster. We first relaxed the electronic degrees of freedom at fixed ionic positions. Ground state and local minima are then found by using a combination of steepest descent and CP molecular dynamics. In Fig. (12) we show the three lowest minima of Cu_6 which are the C_{5v} flat pentagonal pyramid (ground state), the C_{2v} structure composed of tetrahedral subunits (ΔE = 0.01 eV) and the planar D_{3h} structure (ΔE = 0.04 eV). We note that the rather compact structure of C_{2v} symmetry, which has the largest coordination number CN and the smallest average distance R from the center of the cluster, is extremely close to the ground state. CI calculations for Na_6 [61] give the C_{5v} and D_{3h} structures essentially degenerate, with the C_{2v} geometry is

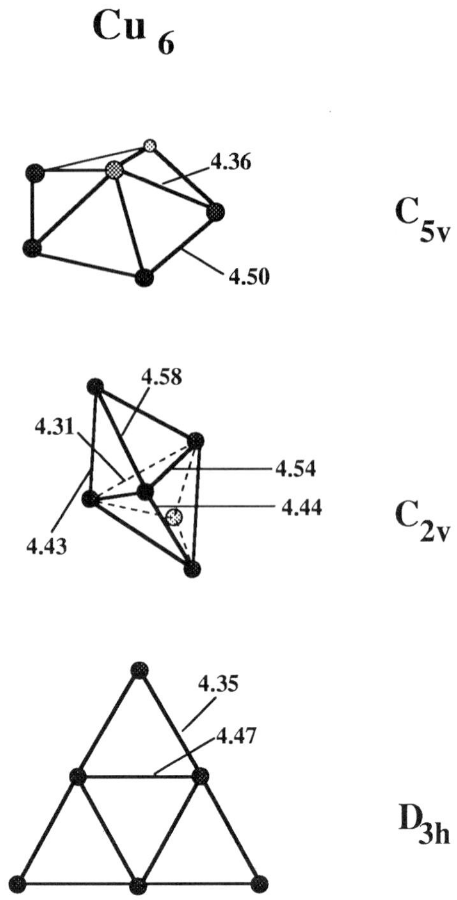

Figure 12: Ground state (C_{5v}) and equilibrium structures of Cu_6 (C_{2v} and D_{3h}). The distances are given in atomic units.

higher by at least 0.16 eV. A further study by LDA [55] reports a $\Delta E = 0.05$ eV between the C_{5v} (ground state) and D_{3h} structures, with an oblate elongated 3D structure found at 0.24 eV above the ground state.

The D_{2d} symmetry is the calculated ground state structure for Cu_8. The closest isomer, higher by 0.15 eV, has T_d symmetry, followed by the capped pentagonal bipyramid (PBP) of C_s symmetry with $\Delta E = 0.20$ eV, and by a structure with D_{4d} symmetry, with $\Delta E = 0.50$ eV (Fig. 13).

It is interesting to observe that the D_{4d} structure, which has the lowest coordination number, is found at higher energy than the D_{2d}, T_d and C_s structures, which all have a larger CN. For Na_8, CI calculations [62] yield the T_d structure as ground state, while according to the LDA results of Ref. [55] the ground state of Na_8 corresponds to the D_{2d} symmetry. Other isomers found in Ref. [55] are the D_{4d} ($\Delta E = 0.07$ eV) and the T_d ($\Delta E = 0.10$ eV) structures; the capped PBP turned out to be unstable and collapsed to the D_{2d} structure. The equilibrium geometries for Cu_{10} are essentially at the same energy and correspond to a bicapped antiprism (D_{4d} symmetry) and to a multicapped PBP (C_s symmetry), with the latter characterized by a higher coordination number (5.2 against 4.8) and a slightly larger average distance from the center of the

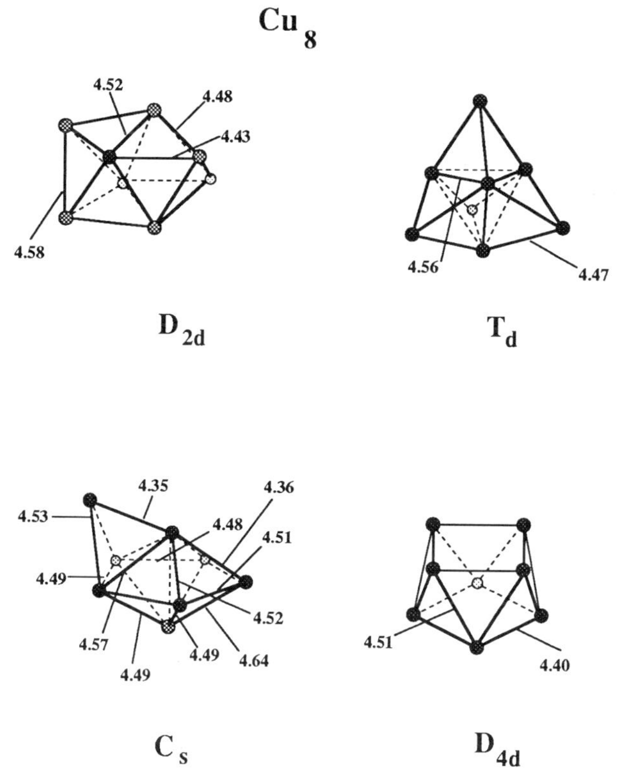

Figure 13: Ground state (D_{2d}) and equilibrium structures (T_d, C_s, D_{4d}) of Cu_8.

cluster (4.09 a.u. against 3.99 a.u). Analysis of the above structures show that bonding lengths within each Cu_N cluster show little variation, with $\sigma \leq 0.05 d_{av}$, where d_{av} is the average nearest neighbor distance and σ is the standard deviation of the nearest neighbor distances. By comparing Cu_N and Na_N clusters [63], systematic smaller values of σ/d_{av} are obtained for Cu_N. Indeed, from the results of Ref. [63] for Na_N clusters one obtains in this case $\sigma \leq 0.14 d_{av}$.

In Fig. (14) we display the electronic density of states (DOS) obtained from the single particle eigenvalues for Cu_6, Cu_8, and Cu_{10} in their ground state geometry. These results prove that a very strong $s-d$ hybridization exists in Cu_N clusters, and demonstrate the fundamental importance of an adequate treatment of d electrons. As shown in Fig. (14), the band edges of the DOS have mostly s character, while the d character is predominant in the middle portion of the band. The lowest occupied molecular orbitals are the most $s-$like, with less than a 20% (Cu_6), 15% (Cu_8) and 10% (Cu_{10}) $d-$character. Hybridization between s and d valence character can be found even in the HOMO states which are 45%, 50% and 25% $d-$like for Cu_6, Cu_8, and Cu_{10} respectively. Difference and analogies between Cu_N clusters and Na_N clusters can be better understood by analyzing the DOS in terms of the shell model. To this end, we decompose each eigenfunction in terms of $L = S, P,$ and D spherical harmonics centered at the center of the cluster, as detailed in Ref. [55]. Hereafter upper case S, P and D are used for the angular projections of the eigenstates with respect to the center of the clusters while lower case characters denote atomic angular components s and d. According to

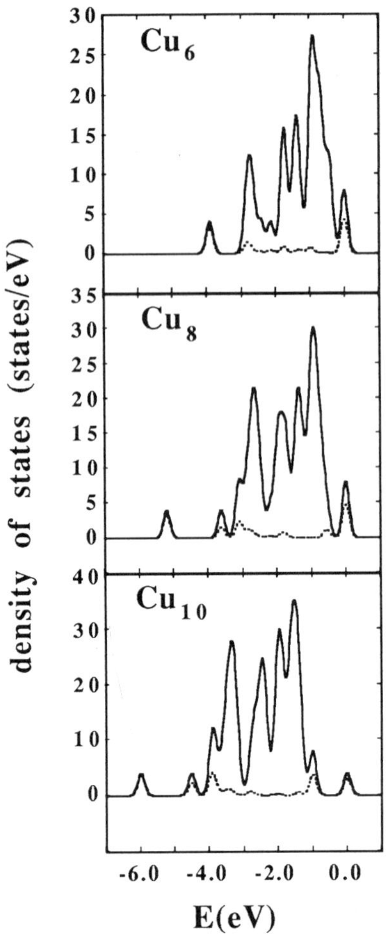

Figure 14: Calculated density of states for Cu_6, Cu_8, Cu_{10} (ground state structures), convoluted with Gaussians of width 0.1 eV. The dashed line gives the weight of the atomic s character. The Fermi level is at zero energy.

the shell model [53], one would expect to observe in each DOS a given number of peaks, well separated from the d band and corresponding to the $1S$ and the $1P$ levels (for Cu_6 and Cu_8) and to the $1S, 1P$ and $1D$ levels (for Cu_{10}). When the actual cluster geometry is accounted for the spherical symmetry breaks down and more peaks might appear. For example, in Na_N clusters, considering ground state structures, two peaks are observed in the DOS of Na_6, which correspond to a $1S$ and a degenerate $1P$ level. In Na_8 an additional $1P$ peak appears. Five shell model features are present in Na_{10} ($1S$, three $1P$ and $1D$). These general patterns are also recognizable in Cu_6, Cu_8 and Cu_{10} at the bottom and at the top of the energy band [64]. However, due to $s-d$ hybridization, the decomposed DOS are more complicated, since any state having a substantial s atomic character contributes to the shell model angular projections yielding additional shell model features. As expected, states in the middle of the band, mostly d–like, do not exhibit significant shell model character.

This investigation has proved that the ground state symmetries of Cu_N clusters are the same as those found for Na_N clusters by DF-LDA calculations. However, Cu_N clusters show a tendency to form more compact structures. From the point of view of the electronic structure, atomic $s-$like states are located mostly at the band edges and are strongly hybridized with $d-$states. Moreover, unlike in the case of Nan clusters for which the shell model applies to a large extent, at least for the structures determined at T=0K, Cu_N clusters can be described only qualitatively within a shell model picture.

References

[1] W.W. Wood. Early history of computer simulations in statistical mechanics. In G. Ciccotti and W. G. Hoover, editors, *Molecular dynamics Simulations of Statistical-Mechanics Systems.*, page 3. North Holland, Amsterdam, 1986.

[2] V. Vitek and D. J. Srolovitz, editors. *Atomistic Simulation of Materials, beyond pair potentials.* Plenum Press , New York and London, 1989.

[3] S. Nosé. A unified formulation of the constant temperature molecular dynamics methods. *J. Chem. Phys.*, 81:511, 1984.

[4] W.G. Hoover. Canonical dynamics: Equilibrium phase-space distributions. *Phys. Rev. B*, 31:1695, 1985.

[5] M. Parrinello and A. Rahman. Polymorphic transitions in single crystals: a new molecular dynamics method. *J. Appl. Phys.*, 52:7158, 1981.

[6] R. Car and M. Parrinello. Uniform approach for molecular dynamics and density-functional theory. *Phys. Rev. Lett.*, 55:2471, 1985.

[7] G. Ciccotti and W. G. Hoover, editors. *Molecular Dynamics Simulation of Statistical-Mechanical Systems.* North Holland, Amsterdam, 1986.

[8] J.P. Hansen and I.R. McDonald. Theory of simple liquids. *Academic Press, London,* 1986.

[9] M.P. Allen and D.J. Tildesley. Computer simulation of liquids. *Clarendon Press, Oxford*, 1986.

[10] G. Ciccotti, D. Frenkel, and I. R. McDonald. *Simulation of Liquids and Solids.* North Holland, Amsterdam, 1987.

[11] C.R.A. Catlow, S.C. Parker, and M.P. Allen, editors. *Computer Modeling of Fluids, Polymers and Solids.*, volume 293 of NATO ASI series C Kluwer. Dordrecht, 1990.

[12] M. Meyer and V. Pontikis, editors. *Computer Simulation in Material Science.*, volume 205 of NATO ASI series E Kluwer. Dordrecht, 1991.

[13] M.P. Allen and D.J. Tildesley, editors. *Computer Simulation in Chemical Physics.*, volume 397 of NATO ASI series C Kluwer. Dordrecht, 1993.

[14] G. Galli and A. Pasquarello. First principles molecular dynamics. In M. P. Allen and D. J. Tildesley, editors, *Computer Simulation in Chemical Physics*, volume 397 of NATO ASI series C Kluwer. Dordrecht, 1993.

[15] M.C. Payne, M.P. Teter, D.C. Allan, T.A. Arias, and J.D. Joannopoulos. Iterative minimization techniques for ab-initio total-energy calculations: molecular dynamics and conjugate gradients. *Rev. Mod. Phys.*, 64:1045, 1993.

[16] D.K. Remler and P.A. Madden. Molecular dynamics without effective potentials via the car-parrinello approach. *Mol. Phys.*, 70:921, 1992.

[17] L. Verlet. Computer "experiments" on classical fluids. i. thermodynamical properties of lennard-jones molecules. *Phys. Rev.*, 159:98, 1967.

[18] P.E. Blöchl and M. Parrinello. Adiabaticity in first-principles molecular dynamics. *Phys. Rev. B*, 45:9413, 1992.

[19] A.P. Sutton, J.B. Pethica, H. Rafii-Tabar, and J.A. Nieminen. *Electron Theory In Alloy Design*, chapter Mechanical properties of metals at the nanometre scale. London: The Institute of Materials, 1992.

[20] A.E. Carlsson. Beyond pair potentials in elemental transition metals and semiconductors. In Henry Ehrenreich and David Turnbull, editors, *Solid State Physics 43*, volume 43. 1990.

[21] K.W. Jacobsen, J.K. Nørskov, and M.J. Puska. Interatomic interactions in the effective-medium theory. *Phys. Rev. B*, 35:7423, 1987.

[22] F. Ercolessi, E. Tosatti, and M. Parrinello. Au(100) surface reconstruction. *Phys. Rev. Lett.*, 57:719, 1986.

[23] M.W. Finnis and J.E. Sinclair. A simple empirical n-body potential for transition metals. *Phil. Mag. A*, 50:45, 1984.

[24] V. Rosato, M. Guillope, and B. Legrand. Thermodynamical and structural properties of fcc transition metals using a simple tight-binding model. *Phil. Mag. A*, 59:321, 1989.

[25] N. Chetty, K. Stokbro, K.W. Jacobsen, and J.K. Nørskov. Ab-initio potentials for solids. *Phys. Rev. B*, 46:3798, 1992.

[26] S.M. Foiles, M.I. Baskes, and M.S. Daw. Embedded-atom-method functions for the fcc metals cu, ag, au, ni, pd, pt, and their alloys. *Phys. Rev. B*, 33:7983, 1986.

[27] P.J. Feibelman. Diffusion barrier for ag adatom on pt(111). *Surf. Sci.*, 313:L801, 1994.

[28] J.K. Nørskov. Covalent effects in the effective medium theory of chemical binding: Hydrogen heats of solution in the 3*d* metals. *Phys. Rev. B*, 26:2875, 1982.

[29] M.I. Baskes, S.M. Foiles, and C.F. Melius. Dynamical calculation of low energy hydrogen reemission off hydrogen covered surfaces. *Nucl. Mater.*, 145-147:339, 1987.

[30] A.E. Carlsson, P.A. Fedders, and Charles W. Myles. Generalized embedded-atom format for semiconductors. *Phys. Rev. B*, 41:1247, 1990.

[31] M.I. Baskes. Application of the embedded-atom method to covalent materials: A semi-empirical potential for silicon. *Phys. Rev. Lett.*, page 2666, 1987.

[32] Murray. S. Daw and M.I. Baskes. Embedded-atom method: Derivation and application to impurities, surfaces and other defects in metals. *Phys. Rev. B*, 29:6443, 1984.

[33] James H. Rose, John R. Smith, Francisco Guinea, and John Ferrante. Universal features of the equation of state of metals. *Phys. Rev. B*, 29:2963, 1984.

[34] Enrico Clementi and Carla Roetti. Roothaan-hartree-fock atomic wave functions. basis functions and their coefficients for ground and certain excited states of neutral and ionized atoms, $z \leq 54$. *Atomic Data and Nuclear Data Tables*, 14:177, 1974.

[35] A.D. McLean and R.S. McLean. Roothaan-hartree-fock atomic wave functions slater basis set expansions for $z = 55 - 92$. *Atomic Data and Nuclear Data Tables*, 26:197, 1981.

[36] M.J. Puska, R.M. Nieminen, and M. Manninen. Atoms embedded in an electron gas: Immersion energies. *Phys. Rev. B*, 24:3037, 1981.

[37] C. Massobrio and P. Blandin. Structure and dynamics of ag clusters on pt(111). *Phys. Rev. B*, 47:13687, 1993.

[38] D.W. Basset and P.R. Webber. Diffusion of single adatoms of platinum, iridium and gold on platinum surfaces. *Surf. Sci.*, 70:520, 1978.

[39] S.C. Wang and Gert Ehrlich. Structure, stability and surface diffusion of clusters: ir_x/ir(111). *Surf. Sci.*, 239:301, 1990.

[40] L. Hansen, P. Stoltze, K.W. Jacobsen, and J.K. Nørskov. Self-diffusion on copper surfaces. *Phys. Rev. B*, 44:6523, 1991.

[41] W.K. Rilling, C.M. Gilmore, T.D. Andreadis, and J.A. Sprague. An embedded-atom-method study of diffusion of an ag adatom on (111) ag. *Can. J. Phys.*, 68:1035, 1990.

[42] Holger Röder. Microscopic processes in heteroepitaxial growth: nucleation, growth and alloying of silver on the (111) surface of platinum. *Thèse 1288, Lausanne, EPFL*, 1994.

[43] Michael I. Haftel. Surface reconstruction of platinum and gold and the embedded-atom model. *Phys. Rev. B*, 48:2611, 1993.

[44] P. Blandin and P. Ballone. Diffusion of metal adatom on compact metal surfaces in the presence of defects and impurities. *Surf. Sci.*, to be published.

[45] J.K. Nørskov. Chemisorption on metal surfaces. *Rep. Prog. Phys.*, 53:1253, 1990.

[46] R.J. Madix, G. Ertl, and K. Christmann. Preexponential factors for hydrogen desorption from single crystal metal surfaces. *Chem. Phys. Lett.*, 62:38, 1979.

[47] W. Eberhardt, F. Greuter, and E. W. Plummer. Bonding of h to ni, pd and pt surfaces. *Phys. Rev. Lett.*, 46:1085, 1981.

[48] R.W. McCabe and L.D. Schmidt. Binding states of co and h_2 on clean and oxidized (111) pt. *Surf. Sci.*, 65:181, 1977.

[49] Martin Zinke-Allmang, Leonard C. Feldman, and Marcia H. Grabow. Clustering on surfaces. *Surf. Sci. Reports*, 16:377, 1992.

[50] Christoph Romainczyk. Struktur und kinetic von reinen und silberbedeckten platinoberflächen. *Thèse 1289, Lausanne, EPFL*, 1994.

[51] P. Blandin, C. Massobrio, and P. Ballone. Nucleation and growth of metallic submonolayers on compact metal surfaces. *Phys. Rev. B*, 49:16637, 1994.

[52] H. Röder, R. Schuster, H. Brune, and K. Kern. Monolayer-confined mixing at the $ag - pt(111)$ interface. *Phys. Rev. Lett.*, 71:2086, 1993.

[53] W.A. de Heer. The physics of simple metal clusters: experimental aspects and simple models. *Rev. of Modern Phys.*, 65:611, 1993.

[54] K.A. Jackson. First principles study of the structural and electronic properties of cu clusters. *Phys. Rev. B*, 47:9715, 1993.

[55] U. Rothlisberger and W. Andreoni. Structural and electronic properties of sodium microclusters (n=2-20) at low and high temperatures: New insight from ab-initio molecular dynamics studies. *J. Chem. Phys.*, 94:8129, 1991.

[56] D. Vanderbilt. Soft self-consistent pseudopotetials in a generalized eigenvalue formalism. *Phys. Rev. B*, 41:7892, 1990.

[57] A. Pasquarello, K. Laasonen, R. Car, C. Lee, and D. Vanderbilt. Ab-initio molecular dynamics for *d*-electron systems: liquid copper at 1500k. *Phys. Rev. Lett.*, 69:1982, 1992.

[58] K. Laasonen, A. Pasquarello, R. Car, C. Lee, and D. Vanderbilt. Car-parrinello molecular dynamics with vanderbilt ultrasoft pseudopotentials. *Phys. Rev. B*, 47:10142, 1993.

[59] J.P. Perdew and A. Zunger. Self-interaction correction to density-functional approximations for many-electron systems. *Phys. Rev. B*, 23:5048, 1981.

[60] V. Bonacic-Koutecky, L. Cespiva, P. Fantucci, and J. Koutecky. Effective core potential-configuration interaction study of electronic structure and geometry of small neutral and cationic ag_n clusters: Predictions and interpretation of measured properties. *J. Chem. Phys.*, 98:7981, 1993.

[61] V. Bonacic-Koutecky, J. Pittner, C. Scheuch, M.F. Guest, and J. Koutecky. Quantum molecular interpretation of the adsorption spectra of na_5, na_6, and na_7 clusters. *J. Chem. Phys.*, 96:7938, 1992.

[62] V. Bonacic-Koutecky, P. Fantucci, and J. Koutecky. Systematic ab-initio configuration-interaction study of alkali-metal clusters. ii. relation between electronic structure and geometry of small sodium clusters. *Phys. Rev. B*, 37:4369, 1988.

[63] I. Moullet, J.L. Martins, F. Reuse, and J. Buttet. Static electric polarizabilities of sodium clusters. *Phys. Rev. B*, 42:11598, 1990.

[64] C. Massobrio, A. Pasquarello, and R. Car. Structural and electronic properties of small copper clusters: a first principle study. *Chem. Phys. Lett.*, to be published.

NEW TIGHT-BINDING METHODOLOGY FOR CALCULATING TOTAL ENERGIES OF SOLIDS

D. A. Papaconstantopoulos and M. J. Mehl

Complex Systems Theory Branch
Naval Research Laboratory
Washington, D.C. 20375-5345

I. INTRODUCTION

Recently[1-3] we have introduced a tight-binding (TB) methodology for efficiently calculating the total energy of crystals. The method is based on a TB fit of both the energy bands and the total energies resulting from first-principles calculations. The value of the method rests in the fact that the TB parameters are determined from simple high symmetry crystal structures, and then used in an interpolative manner to compute the total energy of low symmetry structures such as those needed for the evaluation of elastic constants and phonon frequencies. This approach leads to computations that are orders of magnitude faster than those of first-principles density functional studies such as augmented plane wave (APW) calculations, with little loss of accuracy.

In Section 2 we discuss details of our method and in Section 3 we present results for the metal Nb as an example.

II. METHOD

One of us has published a book[4] providing TB parameters of most elements in the periodic table. These parameters result in a very good representation of the band structure and densities of states at a given crystal structure and lattice constant. However, the volume dependence of these parameters using scaling laws is not particularly accurate, especially if we wish to apply a substantial volume deviation. In addition, these parameters pose problems in obtaining total energy information. For these reasons the above parameters served only as a guide and starting values in the procedure of finding parameters for the new methodology we present here. We have adopted the two-center approximation and have written Slater-Koster (SK) parameters in the following form:

$$P_i(r) = (a_i + b_i r + c_i r^2) e^{-\lambda_i^2 r} f(r) \tag{1}$$

where P_i stands for the ten SK parameters ssσ, spσ, ppσ, ppπ, sdσ, pdσ, pdπ, ddσ, ddπ, and ddδ. The parameters a_i, b_i, c_i and λ_i are determined by a least square fit to APW calculations and f(r) is a cutoff function of the form

$$f(r) = \left\{ \exp\left[(r-r_o)/d\right] + 1 \right\}^{-1} , \tag{2}$$

Stability of Materials
Edited by A. Gonis *et al.*, Plenum Press, New York, 1996

where $r_o = 14.0$ a.u. and $d = 0.5$ a.u. This choice of f(r) allows the inclusion of approximately 5 shells of neighbors in a typical transition metal.

We also put a volume dependence on the on-site SK parameters in the following way.

Define a density $$\rho = \sum_{\text{neighbors}} e^{-\delta^2 r} \tag{3}$$

and write the on-site TB parameters h_ℓ as

$$h_\ell = \alpha_\ell + \beta_\ell \rho^{2/3} + \gamma_\ell \rho^{4/3} \tag{4}$$

where ℓ = s, p, d and δ, α_ℓ, β_ℓ and γ_ℓ are parameters to be determined by the fit.

The above formalism applies to either an orthogonal or a non-orthogonal Hamiltonian. The difference is that in the non-orthogonal case we have twice as many parameters from Eq. (1), the additional parameters describing the overlap matrix. In both cases we have a 9x9 matrix per atom.

In our method we simultaneously fit both the local density approximation (LDA) eigenvalues and the values of the total energy. In previous works,[5-7] with the exception of Ref. 8, only the total energy was fitted according to the following form

$$E[n(r)] = \sum_{\text{occ}} \varepsilon_n + F[n(r)] \tag{5}$$

where the first term is the eigenvalue sum and the second term is charge density dependent and includes the remaining contributions to the total energy expression as given by density functional theory. In other works[5-10] the term F[n(r)] has been approximated by a pair potential. However, we have instead shifted each eigenvalue ε_n by an amount

$$V_o = F[n(r)]/N_e \tag{6}$$

where N_e is the number of valence electrons. The result is that now,

$$E[n(r)] = \sum_{\text{occ}} \varepsilon_n' \tag{7}$$

where $$\varepsilon_n' = \varepsilon_n + V_o. \tag{8}$$

We have applied the above formalism according to the following procedure:

We perform first-principles APW calculations for the fcc and bcc structures for 4 or 5 lattice constants each. Then we choose the parameters in Eqs. 1-4 by a least squares method so that they accurately reproduce the shifted APW eigenvalues (ε_i') and total energies E for all the above structures and volumes. Typically our fit has an rms error of about 5 mRy for the occupied states and 0.5 mRy for the total energies. In the parametrization of transition metals we omitted the quadratic term in eq. (1). This choice results in a total of 40 parameters for an orthogonal Hamiltonian and 70 parameters for a non-orthogonal Hamiltonian. In the case of binary compounds (AB) the number of parameters increases by roughly a factor of three due to the inclusion of A-A, B-B and A-B interactions. The large number of "parameters" required in the present method should not be viewed in the usual parametrization sense in which one fits to experiment or applies adjustments to first principles calculations. What we have here is simply a change of basis set from APW to LCAO and therefore the SK parameters should be considered as intermediate quantities in this transformation.

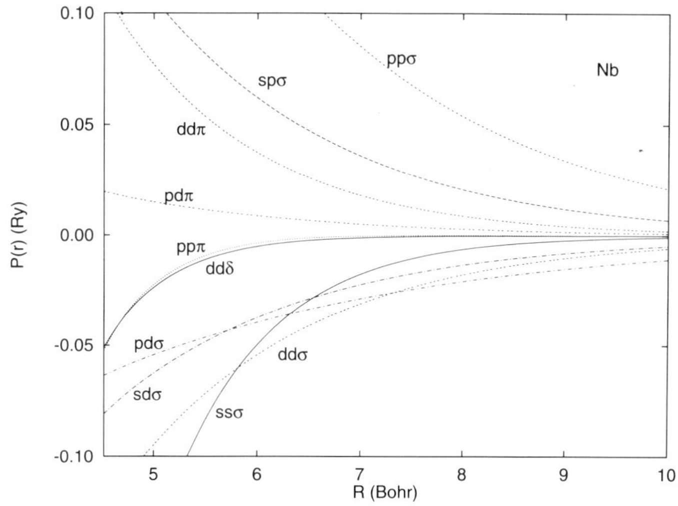

Figure 1. Two-center energy integrals as a function of distance.

III. RESULTS

Since many of our results are published elsewhere[1,2,3,11] we will illustrate our method by presenting our calculations for Nb as a typical example. Fig. 1 shows the variation of the SK parameters $P_i(r)$ with bond length. It should be noted that all $P_i(r)$ go to zero at large r and have smooth monotonic behavior without any sign changes. At very small r these parameters may become unphysically large but this happens at bond lengths that are well beyond the range needed to calculate the physical quantities discussed below.

In Fig. 2 we show a comparison of the Nb energy bands at the equilibrium lattice constant (a=3.30Å) between full potential LAPW results and the TB fit. It is clear that the agreement is very good and we note that we have the same level of agreement at another 5 lattice constants (from a=3.12Å to a=3.39Å) for the bcc lattice as well as 4 lattice constants of the fcc (from a=4.00Å to a=4.21Å). The very accurate fitting of the energy bands (as well as the total energy discussed below) is a unique feature of our method and it should be emphasized that methods such as the EAM[12] or those TB schemes based on the second moment approximation[6] are not capable of obtaining accurate energy bands. We argue that a method that fits the band structure in addition to the total energy provides a more solid foundation for further studies.

We turn now to our total energy results and remind the reader that our procedure requires fitting to the fcc and bcc first principles calculations. After we accomplish this fitting we calculate the total energy of most common phases as is shown in Fig. 3 for Nb. This figure shows that bcc is the ground state of Nb as, of course, is known experimentally. We have also checked the energetics of other crystal structures in addition to the bcc and fcc that we fit. It is satisfying to note that the hcp is found at higher energies, near the fcc. In addition, we have calculated the αMn and βMn structures to ensure that they are not lower in energy than the bcc. Incidentally, we have performed a similar calculation[2] for Mn and there indeed the αMn is confirmed as the ground state.

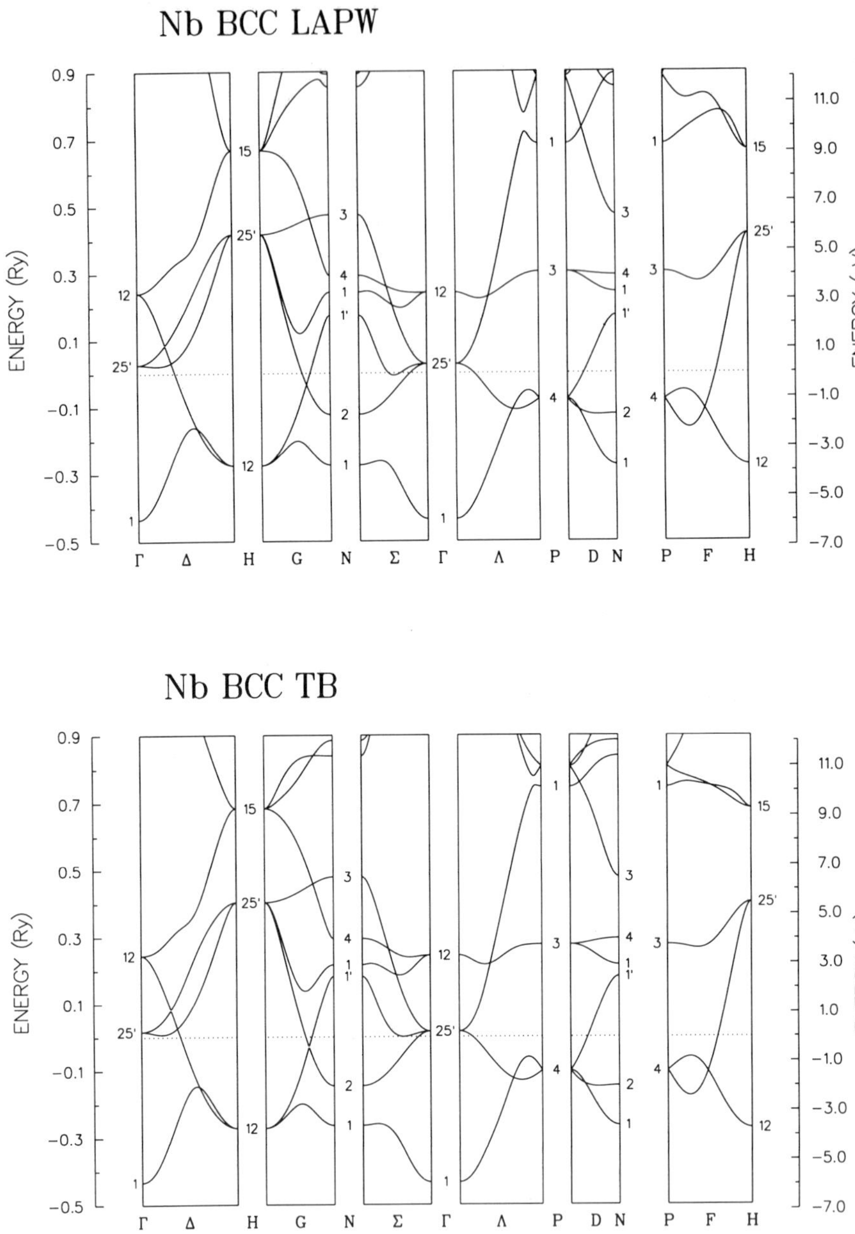

Figure 2. Band structure of bcc Nb at the LDA equilibrium value (a) full potential LAPW (b) tight-binding fit.

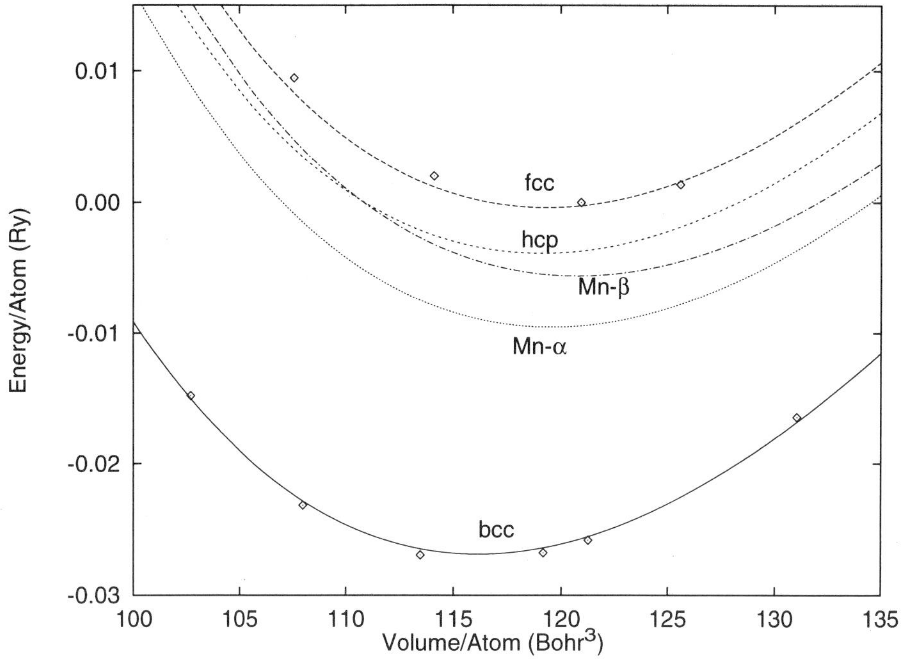

Figure 3. Equation of state for Nb. The "diamond" symbols indicate the LAPW values.

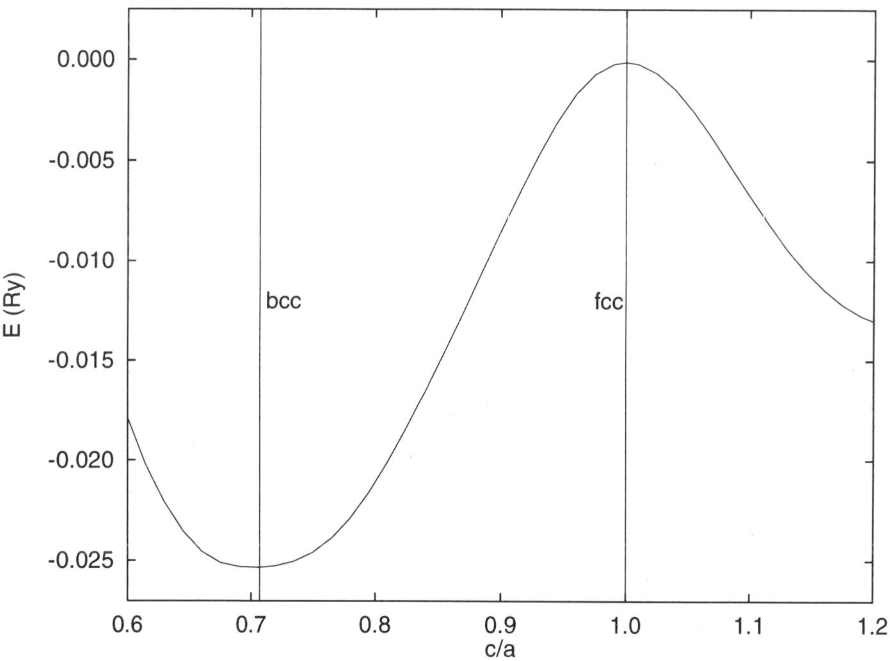

Figure 4. Bain path for Nb.

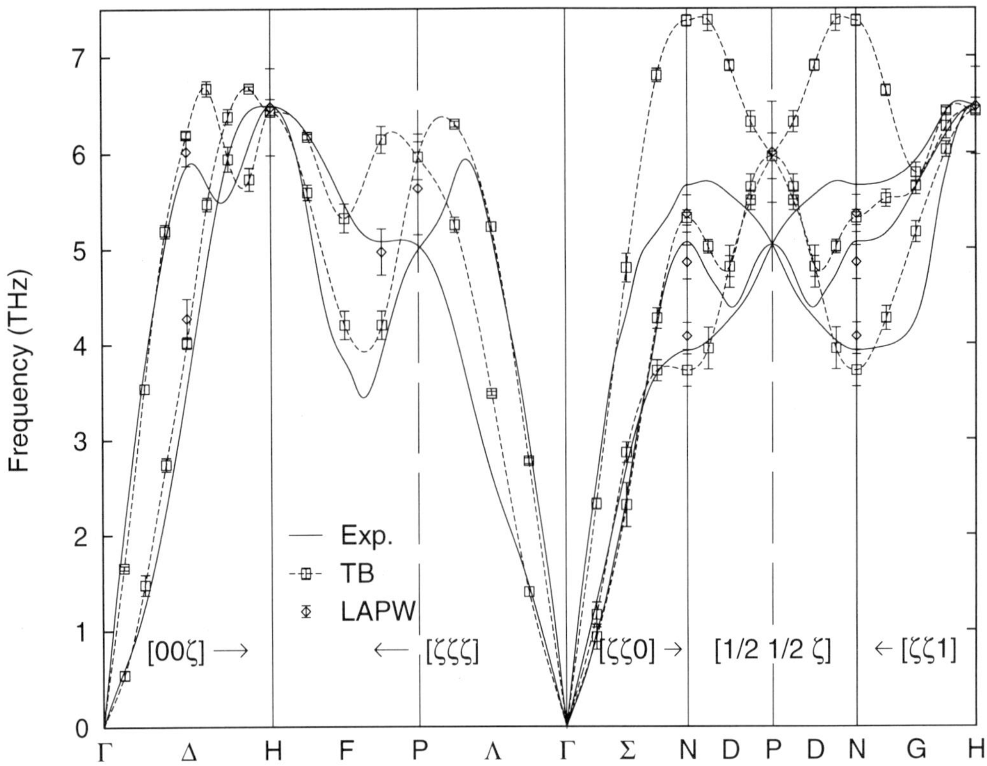

Figure 5. Phonon dispersion curves for Nb. Solid lines indicate experimental results and dashed lines the calculated TB results.

We have also examined the strain path from bcc to fcc, the so-called Bain path. As is shown in Fig. 4 the energy of the crystal Nb as a function of c/a is smooth and continuous with no surprising minima between the bcc and fcc structures.

Another application of our method is the calculation of the elastic constants C_{ij}. This is done by applying appropriate distortions to the crystal and recalculating the total energy.[13] The energy difference between the distorted and undistorted lattice yields C_{11}-C_{12} and C_{44}. Our results for Nb are shown in Table I, where the TB values are compared with those found from a direct LAPW calculation and with experiment.[14] We note that the TB C_{ij}'s are in just as good agreement with experiment as the LAPW results.

Table I. Elastic Constants of Nb in GPa.

	Method		
Elastic Constant	TB	LAPW	Exp.
C_{11}	204	230	236
C_{12}	137	122	139
C_{44}	34	25	29

It is also possible to use our methodology to perform frozen-phonon calculations by appropriate atom displacements. We have indeed calculated phonon dispersion curves for Nb as shown in Fig. 5. Both transverse and longitudinal modes are shown with the solid lines representing experimental results[15] and the dashed lines the TB calculations. The agreement between theory and experiment is satisfactory in most directions in the Brillouin zone. A serious disagreement occurs at the N point where the TB overestimates the longitudinal frequency. This discrepancy can be corrected if we include as input to our fit the energies and band structure found in the frozen-phonon calculation corresponding to the N symmetry points. In Fig. 5 we have indicated with a diamond symbol frequencies at high symmetry points obtained by direct LAPW calculations.

We calculated the vacancy formation energy of Nb by removing one Nb atom from a 108 atom supercell at the experimental lattice constant. This supercell is large enough to eliminate the interaction between vacancies in the periodic lattice. Atoms adjacent to the vacancy are allowed to relax, though this changes the total energy by only 0.02 eV in Nb. The vacancy formation energy is given by

$$E_{vacancy} = E(\text{supercell}) - (N-1)\, E(\text{bulk}) \qquad (9)$$

When E(supercell) is the total energy of the 107 atom + 1 vacancy relaxed supercell, N=108 is the number of sites in the supercell, and E(bulk) is the energy/per atom of bcc Nb at the experimental volume. We find a vacancy formation energy of 2.82 eV, close to the experimental range of 2.07-2.65 eV.[16]

The TB method is ideal for calculating surface energies. The speed of the method allows us to use a large number of atoms and to perform the large number of calculations necessary to determine the relaxation and reconstruction of the surface. As a preliminary step, we have calculated the formation energies of the unrelaxed (001), (011), and (111) surfaces of Nb within our TB model. These calculations were done by cutting the crystal twice along the appropriate direction, creating a 25 layer slab which was periodically repeated, with the distance between slabs so large that no interaction could take place between the slabs. The surface energy is then given by

$$E_{surface} = 1/2\, \left[E(\text{slab}) - N\, E(\text{bulk})\right] \qquad (10)$$

when E(slab) is the total energy of the slab, N=25 is the number of layers in the slab, and E(bulk) is the energy per atom of bulk bcc Nb. All calculations were performed at the experimental volume. In Table II we present the results and compare them to Modified Embedded Atom Method (MEAM)[17] calculations, as well as to experiment.[18] There is good agreement between calculations and experiment for all the three faces.

Table II. Surface Energies of Nb in ergs/cm^2

FACE	TB	MEAM	Exp
(001)	2370	2788	2300*
(011)	1540	1868	
(111)	2440	2018	

*averaged over all faces

ACKNOWLEDGEMENT

This work is partially supported by a grant from the U.S. Office of Naval Research. LAPW calculations were run on the CEWES Cray-YMP as part of the Department of Defense High Performance Computing Initiative.

REFERENCES

1. R.E. Cohen, M.J. Mehl and D.A. Papaconstantopoulos, Phys. Rev. B50, 14694 (1994).
2. M.J. Mehl and D.A. Papaconstantopoulos, Europhys. Lett. 31, 537 (1995).
3. M.J. Mehl and D.A. Papaconstantopoulos, Phys. Rev., to be published.
4. D.A. Papaconstantopoulos, "Handbook of the Band Structure of Elemental Solids," Plenum Press, New York, 1986.
5. L. Goodwin, A.J. Skinner and D.G. Pettifor, Europhys. Lett. 9, 701 (1989).
6. W. Zhong, Y.S. Li and D. Tomanek, Phys. Rev. B44, 13053 (1991).
7. C.H. Zu, C.Z. Wang, C.T. Chan and K.M. Ho, J. Phys. Condens. Matter 4, 6047 (1992).
8. M.M. Sigalas and D.A. Papaconstantopoulos, Phys. Rev. B49, 1574 (1994).
9. J.M. Mercer, Jr. and M.Y. Chou, Phys. Rev. B47, 9366 (1993); Phys. Rev. B49, 8506 (1994).
10. M. Menon, J. Connolly, N. Lathiotakis and A. Andriotis, Phys. Rev. B50, 8903 (1994); Europhys. Lett. 29, 135 (1995).
11. M.J. Mehl, D.A. Papaconstantopoulos and R.E. Cohen, Intern. J. Thermophysics 16, 67 (1995).
12. S.M. Foiles, M.I, Baskes and M.S. Daw, Phys. Rev. B33, 7983 (1986).
13. M.J. Mehl, B.M. Klein and D.A. Papaconstantopoulos, in "Intermetallic Compounds: Principles and Applications," eds. J.H. Westbrook and R.L. Fleischer, (John Wiley & Sons, Ltd., London, 1994).
14. G. Simmons and H. Wang, Single Crystal Elastic Constants and Calculated Aggregate Properties: A Handbook, 2nd Edition, (MIT Press, Cambridge, MA, 1971).
15. Y. Nakagawa and A.D.B. Woods, Phys. Rev. Lett. 11, 271 (1963).
16. H.-E. Schaefer, Phys. Stat. Sol. A 102, 47 (1987).
17. M.I. Baskes, Phys. Rev. B46, 2727 (1992).
18. W.R. Tyson and W.A. Miller, Surf. Sci. 62, 267 (1977).

CONTRIBUTIONS TO THE TOTAL ENERGY OF RANDOM ALLOYS

J. S. Faulkner[1], Yang Wang[2], and G. M. Stocks[2]

[1]Alloy Research Center and Department of Physics, Florida Atlantic University, Boca Raton, FL 33431.
[2]Metals and Ceramics Division, Oak Ridge National Laboratory, Oak Ridge, Tennessee 37830.

We have been calculating the electronic structure of solids with no long-range order using an order-N method.[1] The origin of the term order-N is that the computer time required to carry out conventional calculations on N atoms increases as N^3, but computational physicists have recently developed methods for which the increase in time is only linear in N. We call the order-N method that we use the locally self-consistent multiple scattering method (LSMS),[2] and it has been shown to be a completely converged method for calculations within the density functional theory local density approximation (DFT-LDA).[3]

We have been studying a model of a random substitutional copper-zinc alloy. The atoms are placed on the sites $\mathbf{a}_i$ of a body centered cubic Bravais lattice such that the probability of a Cu or Zn atom appearing on a given site is c or 1-c . We calculate the electronic structure, the total energy U, the net charge q^i on a given site, and the electrostatic potential V^i at that site. The only approximations that we make, beside the LDA, is that the crystal is divided into cells containing a very large but finite number of atoms and, at the present time, the muffin-tin approximation for the self-consistent potentials. The electrostatic contribution to the total energy per site may be written

$$U_C = \frac{1}{2N}\sum_{i=1}^{N} q^i V^i, \tag{1}$$

where N is the number of atoms in the cell. The cells are periodically reproduced to fill all space, so the electrostatic potentials are calculated for an infinite number of atoms from

$$V^i = \sum_j 2M_{ij}q^j, \tag{2}$$

where the formal expression for the M_{ij} is

$$M_{ij} = \frac{1}{N}\sum_p{}' \frac{1}{|\mathbf{a}_i-\mathbf{a}_j+\mathbf{R}_p|}\,. \tag{3}$$

The $\mathbf{R}_p$ are the lattice vectors for the cells, and $\mathbf{R}_p$=0 is included in the sum only when i≠j. The Ewald method is used to evaluate the M_{ij}, employing the standard trick of adding and subtracting the potential from a uniform charge of opposite sign. The factor 2 appears because of our use of atomic units in which the energy is measured in Rydbergs, distances in Bohr radii, and q^i is the number of electrons minus the atomic number of the atom on the site Z^i. The calculations that we show were done for cubic cells containing 432 atoms.

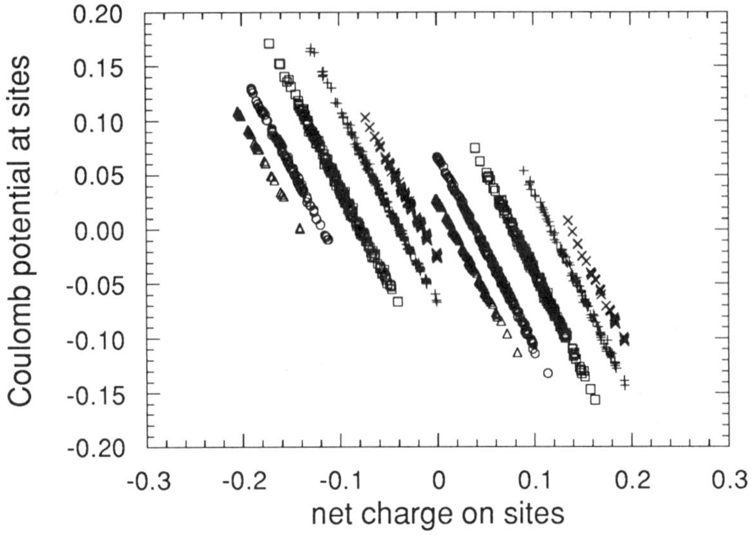

Figure 1. The Coulomb potential V^i versus the charge q^i on the 432 sites in the cell for a series of Cu-Zn alloy with the bcc crystal structure. The crosses are the data points for both the Cu and Zn sites in alloys that contain 10% Cu, the plus signs are for 25%, the squares are for 50%, the circles are for 75%, and the triangles are for 90% Cu.

The coulomb potentials V^i are plotted in Fig. 1 versus the charges q^i for a series of bcc Cu-Zn alloys with copper concentrations of 10%, 25%, 50%, 75%, and 90%. It can be seen that, for each concentration, the points fall on two lines, the one with positive values of q^i corresponding to sites in the alloy occupied by Cu atoms and the other by Zn atoms. This is in keeping with the sign convention that we have chosen, because the Cu atoms gain electrons and the Zn atoms lose them. The first point to be made is that the lines are truly straight. When a least squares fit is made to the data, no improvement in the fit is achieved by adding additional terms to allow for curvature. Also, the data points fall very close to the straight lines, the rms deviation for the examples in Fig. 1 being 0.002. These features are so striking that we believe we are looking at a rule, not just a trend.

The charge neutrality of the crystal leads to the conditions

$$\sum_i q^i = 0, \qquad \sum_i V^i = 0. \tag{4}$$

The average charge and Coulomb shift for a Cu site are defined by equations like

$$\langle q \rangle_{Cu} = \frac{1}{N_{Cu}} \sum_{i \subset Cu} q^i, \quad \langle V \rangle_{Cu} = \frac{1}{N_{Cu}} \sum_{i \subset Cu} V^i. \tag{5}$$

From charge neutrality it follows that

$$\begin{aligned} c\langle q \rangle_{Cu} + (1-c)\langle q \rangle_{Zn} &= 0, \\ c\langle V \rangle_{Cu} + (1-c)\langle V \rangle_{Zn} &= 0. \end{aligned} \tag{6}$$

By dividing the second of these equations by the first, it can be seen that another aspect of charge neutrality is

$$\langle V\rangle_{Cu} = -\alpha\langle q\rangle_{Cu}, \qquad \langle V\rangle_{Zn} = -\alpha\langle q\rangle_{Zn}, \tag{7}$$

and that the proportionality constant is the same for Cu and Zn.

Given the relationships in Eq. (7), the most convenient equations to describe the straight lines in Fig. 1 are

$$V^i = -\beta_{Cu}(q^i - \langle q\rangle_{Cu}) - \alpha\langle q\rangle_{Cu} \qquad i \subset Cu,$$
$$V^i = -\beta_{Zn}(q^i - \langle q\rangle_{Zn}) - \alpha\langle q\rangle_{Zn} \qquad i \subset Zn. \tag{8}$$

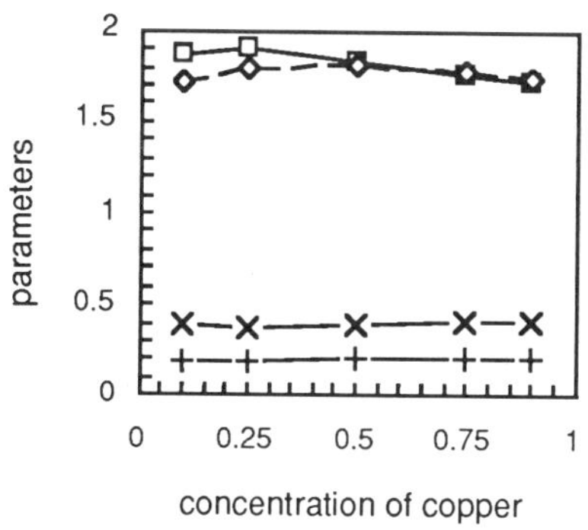

Figure 2 . The parameters that define the straight lines shown in Fig. 1 for five Cu-Zn alloys with the bcc crystal structure as a function of the concentration of Cu atoms in the alloys. The squates and diamonds are the slopes β_{Cu} and β_{Zn}. The proportionality constant α defined in Eq. (7) is shown by the crosses. The difference between the average Cu and Zn charges, Δ, is shown by the plus signs.

It can be seen from Fig. 1 that the slopes β_{Cu} and β_{Zn} for a given concentration appear to be approximately equal, but there is no condition that requires that as is the case for α. The concentration dependence of α, β_{Cu}, and β_{Zn} are shown in Fig. 2. This figure also shows that the difference between the average charges

$$\Delta = \langle q\rangle_{Cu} - \langle q\rangle_{Zn}, \tag{9}$$

depends very little on concentration in these alloys. This relation leads to an approximation for the average charges for any concentration

$$\langle q\rangle_{Cu} = (1-c)\Delta, \quad \langle q\rangle_{Zn} = -c\Delta. \tag{10}$$

Previous calculations on this alloy that we have published elsewhere[4] show that the only difference between the results obtained using cells containing 256 atoms and the ones for 432 atoms is that the points are packed in more densely. However, these same calculations show that the slopes β_{Cu} and β_{Zn} for an fcc crystal differ from the ones for the bcc crystal by over 20%.

Inserting the qV relations from Eq. (8) into Eq. (1) leads to an interesting equation for the Coulomb energy

$$U_C = \alpha\frac{1}{2}\langle q\rangle_{Cu}\langle q\rangle_{Zn} - \beta_{Cu}\frac{c}{2N_{Cu}}\sum_{i\subset Cu}^{N_{Cu}}(q^i - \langle q\rangle_{Cu})^2 - \beta_{Zn}\frac{1-c}{2N_{Zn}}\sum_{i\subset Zn}^{N_{Zn}}(q^i - \langle q\rangle_{Zn})^2. \tag{11}$$

Comparing this equation with the usual one

$$U_C = \sum_{i,j} q^i M_{ij} q^j, \tag{12}$$

allows us to comment on the theoretical implications of these results.

The version of the Coulomb energy in Eq. (12) leads to the standard interpretation that the energy arises from interactions of charges distributed over great distances. This is, of course, true, but at the same time charges that are produced by a self-consistent LDA calculation have the relationship to each other that allows U_C to be written by a sum of single-site terms. This may be looked upon as a kind of locality. It should be emphasized that these results are not just electrostatics. If one generates a set of charges q^i by some rule and then calculates the V^i from Eq. (2), there is not a linear relationship between the q^i and V^i. It is only when the LDA calculation is iterated to self consistency that the q^i,V^i points will converge to the straight lines shown in Fig. 1.

One way to look at all of these qV relations is within the theoretical framework that was developed by Harris [5] and Foulkes [6]. They pointed out that quite accurate total energies could be calculated without going to self-consistency by inserting charge densities from other calculations into a particular form of the LDA total energy equation. They proved algebraically that the correction to the total energy obtained with these charge density fragments is quadratic in the charges. Using the average charges as the fragments, there is an explicit quadratic correction to U_C in the second term in Eq. (11). Of course, this is a new use for the Harris-Foulkes algebra because their conclusion was that since the correction is quadratic, it can be ignored. From the calculations shown here, the quadratic term must be included, but there is no quartic term.

As is well known, the term local in the LDA refers to the approximation in which the exchange-correlation energy is replaced by the simplified form $E_{xc} = \int \rho \varepsilon_{xc}(\rho) dr$, where $\varepsilon_{xc}(\rho)$ is calculated for a uniform electron gas with the given density. It is not obvious that a DFT-LDA calculation must lead to a Coulomb contribution to the total energy U_C that is local in the sense that we use the term. All of the other contributions to the energy are manifestly local in that sense, so a conclusion that can be drawn from this observation is that the total energy has this property. To sum up, this locality property requires that V^i is a unique function of q^i, and the linearity of the qV relation is equivalent to the fact that there are no contributions to U_C beyond the quadratic.

Another aspect of the expression for the Coulomb energy U_C in Eq. (11) that has turned out to be of some interest is that it naturally breaks itself into two parts, $U_C = U_1 + U_2$, with

$$U_1 = \alpha \frac{1}{2} \langle q \rangle_{Cu} \langle q \rangle_{Zn} = - \alpha \frac{1}{2} \Delta^2 c(1-c), \tag{13}$$

and

$$U_2 = -\beta_{Cu} \frac{c}{2N_{Cu}} \sum_{i \subset Cu}^{N_{Cu}} (q^i - \langle q \rangle_{Cu})^2 - \beta_{Zn} \frac{1-c}{2N_{Zn}} \sum_{i \subset Zn}^{N_{Zn}} (q^i - \langle q \rangle_{Zn})^2. \tag{14}$$

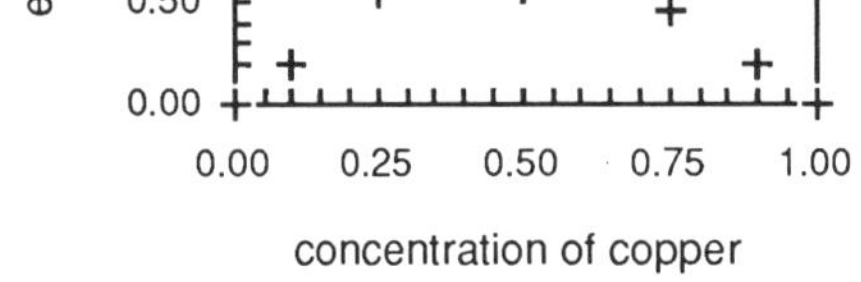

Figure 3 . Parts of the Coulomb energy U_1, solid rectangles, and U_2, open rectangles, versus the concentration of copper atoms in the alloy. The energy unit is milli-Rydbergs.

The second form in Eq. (13) is obtained assuming Δ is independent of c. These two parts of the Coulomb energy are plotted in Fig. 3. It can be seen that U_1 is approximately a parabolic function of c, as would be expected from Eq. (13). There is no reason for the contribution U_2 to be even approximately proportional to c(1-c), and it is not. However, both of these contributions must be zero for c equal to zero or one, and are expected to be smooth functions of c for intermediate values, so they will be approximately parabolic.

Only the average charges appear in the coherent potential approximation (CPA) for the electronic structure of alloys,[7] so it is logical that only U_1 would be included in U_C. This is done in some treatments.[8] Other CPA treatments also include U_2 in and approximate way.[9]

Another model for the Coulomb energy in alloys, proposed by Magri, Wei, and Zunger (MWZ),[10] asserts that the charge on a site is proportional to the number of atoms in the first nearest-neighbor shell of the opposite species. This model is not consistent with the correlation that we demonstrate in this paper, as has been described elsewhere.[11] The MWZ model makes very specific predictions for the parts of the Coulomb energy, which are that

U_1 and U_2 are precisely proportional to c(1-c) and that the ratio U_2/U_1 is a constant equal to 0.38313 for all bcc lattices and for all concentrations. Neither of these predictions agree with the data in Fig. 3.

In conclusion, the computational results in this paper combined with others that we have published elsewhere show that there is a linear relationship between the net charges on the sites and the Coulomb potentials at those sites. The observation of this relationship provides new insights into the Coulomb energy U_C, and demonstates an interesting locality in the total energy of the system. This observation in itself does not simplify the problem of calculating the Coulomb energy of a system of atoms, since one has to solve a fairly complicated electronic structure problem using conventional methods in order to find the slopes and intercepts of the lines. As is the case with the fundamental theorems of DFT, it simply makes a statement about the shape of the resulting energy. However, it should be clear that this knowledge about the shape of the Coulomb energy and the relationship between the Coulomb shift and the charge suggests approximate methods for dealing with the electronic structure of extended systems of atoms. It will be particularly useful when a set of charges is known for atoms in some region of an extended system, but less than complete knowledge is available about the charges on the other atoms. We are developing some of these approximate methods at the present time.

This work was supported by the U. S. Department of Energy, Division of Materials Science, Office of Basic Energy Sciences under subcontract DE-ACO5-84OR21400 With Martin Marietta Energy Systems, Inc., and by the Federal High Performance Computing and Communications program, Applied Mathematical Sciences Program, Office of Energy Research. We made use of the Intel Paragon XP/S-150 massively parallel supercomputer at the Oak Ridge National Laboratory.

REFERENCES

1. "High Performance Computing Symposium; Grand Challenges in Computer Simulation," Proceedings of the 1994 Simulation Multiconference, edited by A. M. Tentner, The Society for Computer Simulation, San Diego, Ca. (1994).

2. G. M. Stocks, D. M. C. Nicholson, Yang Wang, W. A. Shelton, W. M. Temmermann, Z. Szotek, and B. Ginatempo in the proceedings of the NATO-ARW Metallic Alloys: Experimental and Theoretical Perrspectives, NATO ASI Series E: Applied Sciences, Vol. 256, edited by J. S. Faulkner and R. G. Jordan, Kluwer Academic Publishers, the Netherlands, 1994; Phys. Rev. Letters, submitted.

3. Hohenberg and W. Kohn, Phys. Rev. **136**, B864 (1964), W. Kohn and L. J. Sham, Phys. Rev. **140**, A1133 (1965).

4. J. S. Faulkner, Yang Wang, and G. M. Stocks, "Coulomb energies in disordered alloys and improvements on the coherent potential approximation," in *Alloy Modeling and Design*, edited by G. M. Stocks, C. T. Liu, and P. E. A. Turchi, The Minerals, Metals, and Materials Society, Warrendale, Pennsylvania, 1995.

5. J. Harris, Phys. Rev. B **31**, 1770 (1985).

6. W. M. C. Foulkes and R. Haydock, Phys. Rev. B **39**, 12520 (1989).

7. P. Soven, Phys. Rev. 156, 809 (1967); **178**, 1136 (1969).

8. D. D. Johnson and F. J. Pinski, Phys. Rev. B **48**, 11553 (1993).

9. I. A. Abrikosov, Yu. H. Vekilov, P. A. Korzhavyi, A. V. Ruban, and L. E. Shilkrot, Solid State Comm. **83**, 867 (1992); P. A. Korzhavyi, A. V. Ruban, I. A. Abrikosov, and H. L. Skriver, Phys. Rev. B **51**, 5773, (1995); also Abrikosov et al. in this volume.
10. R. Magri, S. H. Wei, and A. Zunger, Phys. Rev. B 42, 11388 (1990).
11. Reference 4.

COMPETING MECHANISMS FOR ORDERING TENDENCIES IN BCC $CuAuZn_2$ AND FCC <u>Au</u>Fe ALLOYS

D.D. Johnson[1], J.D. Althoff[1],
J.B. Staunton[2], M.F. Ling[2],
and F.J. Pinski[3]

[1] *Computational Materials Science Dept. MS 9161, Sandia National Laboratories, Livermore, California 94551*
[2] *Dept. of Physics, University of Warwick, Coventry CV7 4AL, UK*
[3] *Dept. of Physics, University of Cincinnati, Cincinnati, Ohio 45221*

INTRODUCTION

We have implemented a first-principles theory of short-range order in high-temperature, disordered, multicomponent alloys (and magnetic and paramagnetic binary alloys) which determines the atomic (and magnetic) pair-correlations from the underlying local-density-functional electronic structure of the chemically (homogeneously) random alloy. Features in diffuse scattering intensities may be directly related to their electronic origins, for microscopic, system-dependent understanding of the ordering tendencies, or atomic short–range order (ASRO). We discuss briefly the electronic origins, and the novel competitions they produce, for the ordering tendencies in two different alloy systems. First, we discuss the ASRO indicating a high–temperature B2–type state in BCC (Heusler) $CuAuZn_2$; the ordering tendency is determined by a competition between the different types of order associated with each of the three independent pair–correlations, with the dominant ordering originating from band filling. Second, the controversial $< 1, \frac{1}{2}, 0 >$ ASRO observed in FCC <u>Au</u>Fe alloys (prepared at high–temperature and rapidly quenched) arises from a competition between ordering (from hybridization) and clustering (from band–filling) tendencies within the same pair-correlation.

We investigate the stability of the disordered solid solution phase of metallic N-component alloys by calculating the atomic (and/or magnetic) pair–correlation function(s) using a (mean-field) thermodynamic, linear-response approach based on a first-principles description of the finite-temperature, electronic grand potential of the disordered alloy[1–5]. The infinite–temperature limit (i.e., T $\rightarrow \infty$) of the theory (*i.e.*, no ASRO) is the homogeneously random state and is our linear–response reference state; hence the use of Korringa-Kohn-Rostoker Coherent-Potential-Approximation[2,3,6] (KKR-CPA) electronic-structure formalism. Because of its electronic–structure basis, a strength of the approach is that peaks in the alloy's pair–correlation functions (or Warren-Cowley parameters) may be directly related to specific electronic features. Such information is particularly valuable when the ASRO is a precursor of the low–T ordered state, as is the case for $CuAuZn_2$, e.g., since it may reveal much about the electronic origin of the long-range ordered state. From a practical standpoint, knowledge of the high-T behavior is relevant to most materials processing. Furthermore, it is the case that pair–correlations in the perhaps more technologically interesting N-component alloys may be calculated with little more effort than for binaries. The various applications, details, and improvements to this first–principles, concentration–wave approach may be found in refs.[1–5,7,8].

CONCENTRATION FLUCTUATIONS AND ASRO IN MULTICOMPONENT ALLOYS

For any N-component alloy, i.e., $A_{c_A}B_{c_B}...N_{c_N}$, any chemical configuration (whether ordered or disordered) can be described by a set of site–occupation variables $\xi_{\alpha,i}$. Here $\xi_{\alpha,i}$ is 1(0) if an α-type atom does (does not) occupy the ith site in the lattice. The thermodynamic average of $\xi_{\alpha,i}$, $<\xi_{\alpha,i}>$, is then the average concentration (or probability) of an α atom at that site, $c_{\alpha,i}$. At very high temperatures the alloy is (assumed to be) homogeneously disordered and $c_{\alpha,i} = c_\alpha$ for all sites i. The KKR-CPA electronic structure and energetics are then an appropriate basis for describing the configurational average over the electronic degrees of freedom for this reference state[1,3–5]. Furthermore, the *inhomogeneous* version of the KKR-CPA theory provides a formal expression for the electronic grand potential for any atomic arrangement, denoted by $\Omega_e(T, \mu, \{\xi^i_\alpha\})$, which can be used to study the properties of the chemical fluctuations and hence the ASRO. By applying small, inhomogeneous 'external' chemical potentials $\{\delta\nu_{\alpha,i}\}$ to induce changes in the site-occupational probabilities, $\{\delta c_{\alpha,i}\}$, we consider the linear response of the homogeneously disordered alloy. The compositional correlations can then be investigated via the fluctuation-dissipation theorem which connects these responses to the Warren–Cowley parameters, $\alpha^{\alpha\beta}_{ij} = \delta c_{\alpha,i}/\delta\nu_{\beta,j} = \beta(\langle\xi_{\alpha,i}\xi_{\beta,j}\rangle - \langle\xi_{\alpha,i}\rangle\langle\xi_{\beta,j}\rangle)$, where $\beta = (k_BT)^{-1}$.

It is important to note that the site probabilities (occupation variables) satisfy a constraint $\sum_\alpha c^i_\alpha = 1$ ($\sum_\alpha \xi^i_\alpha = 1$) on every site i. It is obvious then that one of the concentration variables may be eliminated by the constraint so as to only work with the independent chemical degrees of freedom. Effectively, the N–dimensional concentration space then can be fully described by a (N-1)–dimensional reduced ("host") space.

The experimentally–measured diffuse intensities can be expressed as linear combinations of the Warren–Cowley parameters $\alpha(\vec{k})$ which are related to the correlation function via

$$\alpha_{\alpha\beta}(\vec{k}) = \frac{q_{\alpha\beta}(\vec{k})}{\bar{c}_\beta\,(\delta_{\beta\alpha} - \bar{c}_\alpha)} \quad . \tag{1}$$

Experimentally *any* $N(N-1)/2$ independent SRO parameters can be determined by making as many measurements, say by varying the energies in an anomalous x-ray experiments. Traditionally, the off–diagonal elements of the Warren–Cowley matrix have been reported. The central result of the present first-principles theory for the pair-correlation function is

$$\left[q^{-1}\left(\vec{k},T\right)\right]_{\alpha\beta} = \left[\delta_{\alpha\beta}\frac{1}{\bar{c}_\alpha} + \frac{1}{\bar{c}_N}\right] - \beta\left[S^{(2)}_{\alpha\beta}(\vec{k},T) - \Sigma_{\alpha\beta}(T)\right] \quad . \tag{2}$$

Here $S^{(2)}_{\alpha\beta}(\vec{k})$ is the Orstein-Zernike direct-correlation function evaluated about the *homogeneous* state. It is defined in real-space as $S^{\alpha\beta}_{ij} = -\left(\frac{\partial^2\langle\Omega_e\rangle_0}{\partial c^i_\alpha \partial c^j_\beta}\right)$, and it is the chemical stability matrix of the disordered alloy. The temperature-dependent $\Sigma_{\alpha\beta}$ is the so-called Onsager cavity field and, for this mean-field theory, maintains the diagonal part of the fluctuation-dissipation theorem; i.e., $q^{ii}_{\alpha\beta} = \bar{c}_\alpha\left(\delta_{\beta\alpha} - \bar{c}_\beta\right)$, or $\alpha_{\mu\nu}(\vec{R} = (0,0,0)) = 1$[4,11,12]; it is the most important for long-ranged interactions[4,11]. Eq. (2) is the N-component generalization of the first-principles, concentration-wave formalism developed initially by Györffy and Stocks[1] and further by Staunton, Johnson, and Pinski[4] who elucidated the details and importance of charge-transfer effects and Onsager corrections. Details on the multicomponent version have been discussed in two recent papers on the ASRO in the FCC Cu-Ni-Zn alloys[7,8].

Within linear-response theory, $[q^{-1}]_{\alpha\beta}$ is proportional to the Hessian of the free energy, and most of the analysis of concentration waves in N-component alloys is related (in a somewhat complicated way[9,10]) to the eigenvectors and eigenvalues of this matrix, as is the definition of the spinodal temperature[10]. Notice that $S^{(2)}_{\alpha\beta}$ is a $(N-1)\times(N-1)$ matrix and can be roughly viewed as a *effective chemical pair-interaction* in this "host" picture. Thus, our mean-field theory permits a straightforward interpretation of the correlation function in terms of pairwise interactions (in real space, if so desired) – analogous to that of Gorsky-Bragg-Williams[13,14] but it is more robust owing to the electronic basis and concentration dependence of the theory[4,5].

SELECTED RESULTS IN A FCC AuFe BINARY AND BCC CuAuZn$_2$ TERNARY

For the ternary alloys to be discussed here, we have used the band-energy-only approximation (i.e., only band-filling effects were kept, no electrostatic and exhange-correlation (magnetic) effects), whereas, for AuFe[19], all terms were kept[4]. This approximation typically captures the main physics but overestimates the spinodal temperature, T_{sp}. The first-principles, Disordered Local Moment (DLM) model[3] was used to model the paramagnetic state of AuFe. In the ASRO calculations, potentials were obtained self-consistently from the scalar-relativistic KKR-CPA, using local-density-functional theory. Also, we imposed the constraint of a fixed Bravais lattice.

BCC CuAuZn$_2$ Heusler Alloy

In Fig. 1, we show the calculated $S^{(2)}_{CuZn}$, $S^{(2)}_{CuAu}$, and $S^{(2)}_{AuZn}$ for BCC CuAuZn$_2$ along the main symmetry directions in the BCC reciprocal lattice. The minimum of $S^{(2)}_{\mu\nu}$ results in a maximum in the corresponding $\alpha_{\mu\nu}(\vec{k})$, see eq. (2). The absolute maximum (dictated by $S^{(2)}_{AuZn}$) clearly falls at the (100) H-points, indicating a partially-ordered B2 (CsCl-type) ordering tendency, as is indeed observed[16,17].

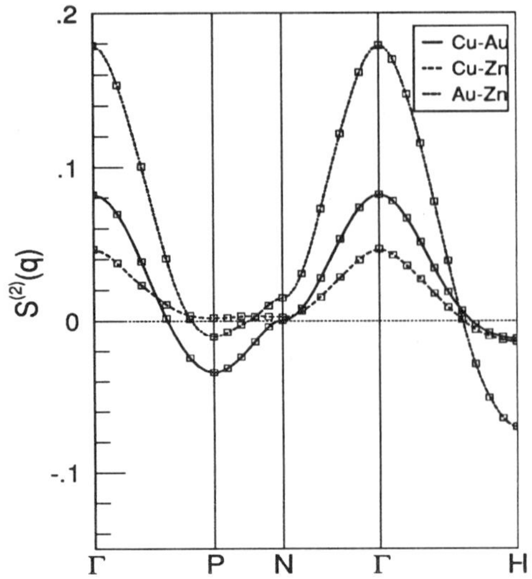

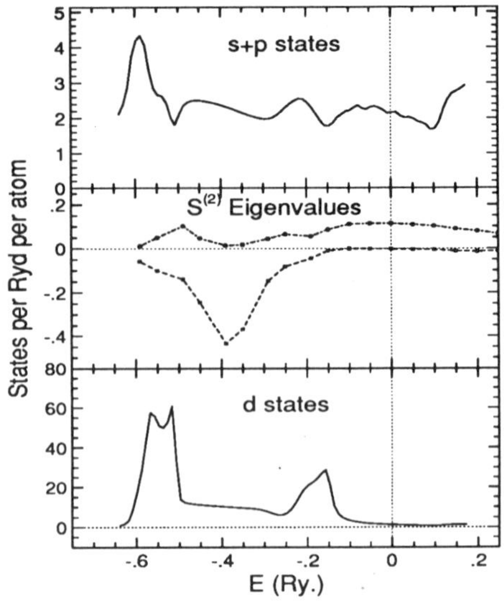

Figure 1. $CuAuZn_2$ $S^{(2)}$ pairs (in Ryd) along BCC Brillouin zone symmetry lines.

Figure 2. The two $< 1, 0, 0 >$ eigenvalues (in Ry) of $CuAuZn_2$ $S^{(2)}$ (middle) and L-decomposed DOS (top [bottom] s+p [d] states) versus energy (Ry). The vertical line is the Fermi level.

Interestingly, ternary $CuAuZn_2$ is a rather ideal case where its ordering properties are dictated completely by the behavior of the endpoint BCC binaries. The Cu-Zn pair clearly indicates a propensity for weak B2–type order, with no negative minimum (energetically favorable fluctuations) elsewhere, as observed in BCC CuZn[18]. At high temperatures, the Cu-Au pair shows a stronger propensity for B32–type order (i.e. $\vec{k} =< \frac{1}{2}, \frac{1}{2}, \frac{1}{2} >$, the P-point), as, in fact, we find for the binary CuAu alloy on a fixed BCC lattice. The Au-Zn pair, with its strong $< 1, 0, 0 >$ ordering tendency, also has a weak minimum at the P-point indicating a low-T propensity for a Heusler structure (formed by a $< 1, 0, 0 > + < \frac{1}{2}, \frac{1}{2}, \frac{1}{2} >$ ordering wave). With no positive contributions at the P-points (i.e. no energetically unfavorable dissent from the other pairs), $S^{(2)}$ supports a low-T Heusler phase, as is observed[18].

For $CuAuZn_2$, the electronic origins for this strong partially–ordered B2 (and subsequent Heusler) phase formation is rather straightforward. Fig. 2 shows the density of states (DOS) for the disordered alloy, along with the two (N-1, N=3, a ternary) $< 1, 0, 0 >$ eigenvalues of $S^{(2)}$ plotted versus energy. This was obtained by fictitiously moving the Fermi energy, in a rigid-band fashion, from the true Fermi level. The second eigenvalue is obviously not associated with $< 1, 0, 0 >$–type ordering because of its negative value at $< 1, 0, 0 >$; it, in fact, promotes clustering. This eigenvalue is nearly zero at the Fermi level, but is increases in significance when the energy is reduced to make d-band filling important. Roughly speaking (and a full analysis has yet to be completed), an almost half–filled s–p band is mostly responsible for maximum (positive) $< 1, 0, 0 >$ eigenvalue at the Fermi level. This eigenvalue governs the high–T behavior and, thus, the resulting B2–type ordering. The analysis is complicated because the eigenvalue in question is a function of the three independent pairs, each perhaps depending on different electronic effects. Because the eigenvalue of the $S^{(2)}$ dictates the temperature scale, the plot indicates that little increase in T_{sp} would be found if, through alloying, the Fermi level was altered. Of course, it is possible that alloying with some other element may significantly alter the electronic structure and allow for an increase in T_{sp} (a non–rigid–band effect). By understanding the role of the various electronic mechanisms responsible for the $\vec{k}$–dependent behavior of each of the independent pair–correlations, it should be possible to "design" a partially–ordered B2 state with a higher T_{sp} by, *e.g.*, emphasizing one pair at the H–point while de–emphasizing another at at the P–point, by, for instance, a further alloying addition. Such effects bear exploration for more technological important alloys, such as $NbAlTi_2$.

We note the above somewhat limited interpretation of ASRO in this Heusler alloy did not require the determination of the site–probability distributions, which is needed for a full description of the ASRO, or the partially–ordered state that is suggested by it. This complication requires a discussion beyond the scope of this paper, see ref.[10]. To determine uniquely the ASRO in N-component alloys more than just a wavevector is required. In contrast to a binary alloy, where the wavevector (due to the Ising symmetry: if not A, then B) determines uniquely the type of ordered state in stoichiometric alloys, concentration waves in in an ABC_2 alloy, for instance, may have numerous partially–ordered stoichiometric phases delineated only by the polarization: one can have, e.g., C ordering against half A and half B. These polarizations (the analog of normal modes in a phonon calculation) are related to the eigenvectors of $[q^{-1}]_{\alpha\beta}$ and, hence, are determined by the electronic interactions in the system[8,10]. It follows[10] that the data from diffuse-scattering experiments also provide this polarization information to determine the ASRO uniquely.

$< 1, \frac{1}{2}, 0 >$ ASRO IN $\underline{Au}$Fe ALLOYS

The DLM paramagnetic state corresponds to an alloy which is prepared and then rapidly quenched from high temperature, as done in the experiments which observe the $< 1, \frac{1}{2}, 0 >$ ASRO, see ref.[19]. With this in mind, we find[19] that the novel $< 1, \frac{1}{2}, 0 >$ ASRO in paramagnetic $\underline{Au}$Fe alloys has a similar origin to the ASRO found in FCC $Ni_{50}Pt_{50}$. Recall that contrary to standard tight-binding arguments, NiPt orders with $< 1,0,0 >$– (or $L1_0$–) type order even though the d–bands are very nearly filled[15]. However, as first noted for NiPt[15], there is an *electronic size* effect at play which results from a large d-bandwidth mismatch, and which competes with the band–filling effects. Owing to the disparity between the d-bandwidths of the elemental metals, low-lying, t_{2g}-type bonding d-states are created through alloy hybridization and then occupied at the expense of the antibonding states. These t_{2g} d-states, whose lobes point towards the FCC near-neighbors, promote $L1_0$–type order[15]. So, in NiPt, despite the alloy's d-states being nearly filled, this mechanism produces $L1_0$–type ASRO, a precursor to the observed low-T $L1_0$ phase.

Compared to NiPt, the ordering tendency in paramagnetic $\underline{Au}Fe$ alloys is somewhat complicated by the 'local exchange splitting' in the electronic structure associated with the Fe local moments, as shown in Fig. 3 by the double-peak in the Fe-associated DOS[19]. Fig. 3 also shows that the filled Au d-states hybridize strongly at low energies with about half of the local-exchange-split Fe states. This causes the occupation of the low-lying, t_{2g} states, as in NiPt, and promotes $< 1,0,0 >$–type order. To conserve electrons, the high–energy, exchange-split, Fe states must become partially filled. In filling those antibonding states near the Fermi level, a clustering effect is promoted. It is a competition between $< 1, 0, 0 >$ ordering and clustering that is responsible for the $< 1, \frac{1}{2}, 0 >$ special–point order in $\underline{Au}Fe$, including the ferromagnetic correlations found in the paramagnetic susceptibility[19]. This can be seen by studying the band–filling effects in $\alpha(\vec{k})$, as done for the ternary above. Raising the Fermi level by as much as 50 mRy does not shift the $< 1, \frac{1}{2}, 0 >$ peak, whereas lowering it by 50 mRy produces roughly equal intensity maxima at all $< 1, 0, 0 >$ and $< 1, \frac{1}{2}, 0 >$. With a reduction of 100 mRy, robust $< 1, 0, 0 >$ ordering is achieved because the Fermi level is now located at the trough of the Fe double-peak, see Fig. 3, and the antibonding states have been depopulated.

We find, due to the same mechanism described above, that ferromagnetic $\underline{Au}Fe$ would show only $< 1, 0, 0 >$ ASRO, suggesting that such long–range order may be created via magnetic annealing[19]. Finally, a similar explanation holds for the $< 1, \frac{1}{2}, 0 >$ magnetic SRO seen in $\underline{Cu}Mn$ alloys, see refs.[20,21].

SUMMARY AND FUTURE

We have briefly discussed the ASRO in $\underline{Au}Fe$ and $CuAuZn_2$. The general points to carry away are that (1) we have implemented a first-principles theory of ASRO in N-component alloys which allows determination of the electronic origins of said ASRO; (2) such calculations can provide much information on the high- and (sometimes) low-temperature alloys; and (3) this approach has identified the origin for the novel special–point ASRO in $\underline{Au}Fe$. Displacement effects, i.e., non–rigid lattice effects, as well as the other contributions beyond band–energy, are being incorporated into the multicomponent alloy calculations. Such improvements will allow us to investigate other alloys, where charge effects may play a role, to "design", for example, higher temperature intermetallics through alloying.

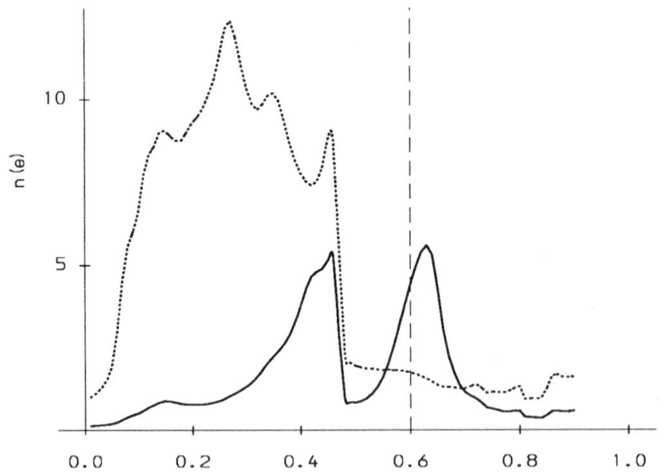

Figure 3. Compositionally- and orientationally–averaged DOS for DLM $Au_{75}Fe$ in states/Ryd-atom vs. Ryd. Contributions from Au (Fe) sites are given by dotted (full) curves.

ACKNOWLEDGEMENTS

This work is supported in part by the U.S. Department of Energy at Sandia National Laboratories (by a New Initiative with OBES, Division of Materials) under contract No. DE-AC04-94AL85000, and by the National Science Foundation under contract No. DMR 9217297, and, by the Engineering and Physical Sciences Research Council, U.K.

REFERENCES

[1] B.L. Györffy and G.M. Stocks, Phys. Rev. Lett. **50**, 374 (1983).
[2] B.L. Györffy *et al.*, in *Alloy Phase Stability*, edited by G. Stocks and A. Gonis (Kluwer Academic Publishers, Dordrecht, 1989), p. 421.
[3] J. Staunton *et al.*, in *Alloy Phase Stability*, edited by G. Stocks and A. Gonis (Kluwer Academic Publishers, Dordrecht, 1989), p. 469.
[4] J.B. Staunton, D.D. Johnson, and F.J. Pinski, Phys. Rev. B **50**, 1450 (1994).
[5] D.D. Johnson, J.B. Staunton, and F.J. Pinski, Phys. Rev. B **50**, 1473 (1994).
[6] D.D. Johnson *et al.*, Phys. Rev. B **41**, 9701 (1990).
[7] J.D. Althoff, D.D. Johnson, and F.J. Pinski, Phys. Rev. Lett. **74**, 138 (1995).
[8] J.D. Althoff, D.D. Johnson, J.B. Staunton and F.J. Pinski, to appear in Phys. Rev. B.
[9] D. de Fontaine, J. Phys. Chem. Solids, **34**, 1285 (1973).
[10] J.D. Althoff and D.D. Johnson, submitted to Phys. Rev. B.
[11] L. Onsager, J. Am. Chem. Soc. **58**, 1468 (1936).
[12] I. Masanskii, V. Tokar, and T. Grishchenko, Phys. Rev. B **44**, 4647 (1991).
[13] V. Gorsky, Z. Phys. **50**, 64 (1928).
[14] W. Bragg and E. Williams, Proc. R. Soc. London Ser. A **145**, 699 (1934); and A **152**, 231 (1935).
[15] F.J. Pinski *et al.*, Phys. Rev. Lett. **66**, 766 (1991).
[16] L. Muldawer, J. of Applied Physics **37**, 2062 (1966).
[17] Y. Murakami, N. Nakanishi, S. Kachi, Acta Metal. **19**, 93 (1970).
[18] *Binary Alloy Phase Diagrams*, 2nd ed., edited by B. Massalski, H. Okamoto, P. R. Subramanian, and L. Kacprzak (ASM International, New York, 1990).
[19] M.F. Ling, J.B. Staunton, D.D. Johnson, and F.J. Pinski, accepted Phys. Rev.: Rapid Comm.
[20] M.F. Ling, J.B. Staunton and D.D. Johnson, Europhys. Lett. **25**, 631 (1994).
[21] M.F. Ling, J.B. Staunton and D.D. Johnson, J. Phys.: Cond. Matt. **6**, 5981 and 6001 (1994).

A NOVEL FULL POTENTIAL CONTRACTED PLANE WAVE-(FCPW-) METHOD FOR ELECTRONIC STRUCTURE CALCULATIONS ON COMPLEX MATERIALS

L. Fritsche, J. Koller and J. Noffke

Institut für Theoretische Physik, Technische Universität Clausthal
Leibnizstraße 10, D-38678 Clausthal-Zellerfeld
Germany

1 INTRODUCTION

Modern electronic structure calculations on solids are with only a few exceptions based on the Kohn-Sham (KS-) version of density functional (DF-) theory. The key problem to be solved within this scheme consists in self-consistently solving the KS-one particle equations that have the form of an one-particle Schrödinger equation with an effective local energy-independent potential. None of the modern methods used in treating this effective one-particle problem really solves the KS-equation numerically, but rather replaces it by the equivalent variational problem for the expectation value of the associated one-particle Hamiltonian. In the process of meeting up with the increasing demands of accuracy, there are essentially three methods that have evolved as sufficiently dependable and technically practicable. All of them are based on this variational approach and do not require any shape approximation to the effective potential. The basis functions in terms of which the solutions $\psi_i(\underline{r})$ to the KS-equations are expanded are either linear combinations of muffin-tin orbitals (LMTO's) or similarly constructed augmented spherical waves (ASW's) or energy linearized augmented plane waves (LAPW's). The idea of the LMTO- and LAPW-method was first described in detail by Andersen [1]. The ASW-scheme whose development was not entirely independent, is due to Williams et al.[2]. In addition to these conceptually related schemes ab initio - pseudopotential methods play also an important role in practically treating the KS-problem. Their foundation requires, however, some additional assumptions which we omit to discuss here.

Practical applications of the LMTO-, LAPW- and ASW-method were first confined to muffin-tin potentials. The completion of full potential versions is relatively recent [3, 4, 5, 6, 7]. The numerical efficiency of the methods in terms of computing time depends strongly on the size of the Hamiltonian matrix $\underline{\underline{H}}$ which the variational problem reduces to and which has to be diagonalized to find the energy-eigenvalues ϵ_i and the associated one-particle states $\psi_i(\underline{r})$. For an elemental transition metal the LMTO- and the ASW-method lead to a matrix $\underline{\underline{H}}$ of size 9x9 in contrast to the LAPW-method where $\underline{\underline{H}}$ takes the size of about 60x60. Since one is not dealing with a sparse matrix in the latter case, the diagonalization time is by a factor of ~ 200 larger than in the LMTO-/ASW-case. On the other hand, the full potential FLAPW-method offers considerable advantages in calculating the additional matrix elements in $\underline{\underline{H}}$ that contain the portions of the full potential beyond the original muffin-tin reference.

The subject of the present article is to outline a modified FLAPW-method that utilizes symmetry adapted contracted plane waves (CPW's) as a set of basis functions rather than individual augmented plane waves. These CPW's are closely related to Andersen's LMTO's, and hence the resulting matrix $\underline{\underline{H}}$ has comparably small size. However, the full potential portions can be readily incorporated as in the FLAPW-case. Moreover, one can benefit considerably from another recent modification of the FLAPW-method[8, 9] which allows one to generate the Hartree potential at each iteration step by using the Fast Fourier Transform, thereby exploiting a highly efficient tool that is otherwise characteristic of the pseudopotential method. All in all, the new method seems to be well suited to treat considerably more complex systems than were so far accessible by using the standard FLAPW-method. We want to emphasize, that our method strongly resembles a modified version of the LMTO-method put forward by Weyrich [10]. The formal structure of the latter is however, considerably more involved.

In Section 2 we briefly describe the two steps that are required to construct the CPW's. Section 3 is concerned with the detailed form of the matrix $\underline{\underline{H}}$.

2 GENERATION OF CONTRACTED PLANE WAVES

For simplicity we shall confine ourselves to non-relativistic N-electron systems in which case the Kohn-Sham equations have the form (in Hartree units)

$$\left[-\frac{1}{2}\underline{\nabla}^2 + V_{eff}(\underline{r}, s)\right] \psi_{is}(\underline{r}) = \epsilon_{is}\psi_{is}(\underline{r}) \tag{2.1}$$

where the effective (in general spin-dependent) potential is rigorously local and consists of three portions

$$V_{eff}(\underline{r}, s) = V_{ext}(\underline{r}) + V_H(\underline{r}) + V_{xc}(\underline{r}, s) \tag{2.2}$$

with $V_{ext}(\underline{r})$ denoting the electrostatic potential set up by the atomic nuclei of the system under study, and $V_H(\underline{r})$, $V_{xc}(\underline{r}, s)$ stand for the Hartree potential and the exchange-correlation potential, respectively. The latter is a functional of the spin-dependent electronic densities defined by

$$\rho_s(\underline{r}) = \sum_i^{(N_s)} |\psi_{is}(\underline{r})|^2 \tag{2.3}$$

where the sum runs over the N_s lowest lying solutions to Eq.(2.1) for either spin orientation $s = \pm 1$. Hence the total charge density is given by

$$\rho(\underline{r}) = \sum_s \rho_s(\underline{r}) \tag{2.4}$$

which on real-space integration yields

$$N = \sum_s N_s \quad . \tag{2.5}$$

To keep the considerations as simple as possible we set $N_\uparrow = N_\downarrow$ thereby disregarding ferro- or antiferromagnetic order. Furthermore, for the purpose of the present paper the detailed functional dependence of $V_{xc}(\underline{r}, s)$ on $\rho_\uparrow(\underline{r})$ and $\rho_\downarrow(\underline{r})$ is irrelevant and may be assumed to have the form suggested by Gunnarsson and Lundqvist [11]. The spin index s is obviously dispensible then and will therefore be omitted from now on.

We now consider a periodic lattice whose unit cells contain N_{at} non-equivalent atoms. The solutions to Eq.(2.1) are characterized then by quantum numbers $\underline{k}$ and n that refer to the pertinent crystal momentum and the band index, respectively. For the purpose of generating basis functions in terms of which these solutions $\psi_n(\underline{k}, \underline{r})$ can be expanded, we temporarily modify the true lattice potential $V_{eff}(\underline{r})$ to attain a muffin-tin form by performing a spherical averaging within spheres centered at the atomic nuclei and by replacing the interstitial potential by its spatial average. The spheres may be viewed as being blown up from their associated centers at different rates that correspond to the different radii of the free atoms. As they grow, the ones that belong to the nearest neighbors touch each other first, thereby marking a certain point on the line interconnecting the pertinent nuclei. If there is more than one pair of nearest neighbors or neighbors of approximately the same distance, one would simultaneously have one or more such points on the respective connecting lines. In the case of an ideal fcc- or bcc-lattice there would be, respectively, 12 or 8 such those points. The ones that refer definitely to nearest neighbors define the radii of the largest non-overlapping spheres within which $V_{eff}(\underline{r})$ is spherically averaged. One can blow up the spheres further into the interstitial region until they touch each other for next nearest neighbors. Again, one obtains one or more points of contact along the lines connecting the pertinent nuclei. The following scheme implies that only those pairs of nearest and next nearest neighbors have to be accounted for whose associated points of contact on spheres around the same center have distances from the nuclei that differ no more than 20%.

Before we discuss the details of constructing contracted plane waves (CPW's) which are pivotal to the new method, we want to emphasize that the procedure is based on a first step where one generates an approximate set of basis functions by requiring boundary conditions at the points of contact defined above. The idea of this first step is closely akin to the early method of Wigner and Seitz [12]. The basis functions thus obtained are then shaped into CPW's in a second step. These are well defined in the sense that each CPW fullfills rigorously Bloch's theorem, is everywhere smooth and can be characterized (similar to LMTO's) by the angular momentum and magnetic quantum number l and m, respectively, which refer to the dominant partial wave of its expansion inside the atomic spheres. As will become apparent in the following, the first step leaves some latitude in the generation of the raw set of basis functions so that the resulting set of CPW's depends to some extent on details of the first step. However, this has - in principle - no effect on the final wave function in that differently generated CPW's combine differently to give the same $\psi_n(\underline{k}, \underline{r})$.

Inside the muffin-tin spheres (numbered by an index α that runs from 1 to N_{at}) the solution to Eq.(2.1) can be separated into a radial and angle-dependent part so that the associated expansion may be cast as

$$\hat{\chi}_\alpha(\underline{k}, \underline{r}) = \sum_{\nu=0}^{1} \sum_{L}^{(l_{max})} \hat{a}_{L\nu}^{(\alpha)}(\underline{k}) R_{L\nu}^{(\alpha)}(\epsilon_l, r_\alpha) Y_L(\hat{\underline{r}}_\alpha) \tag{2.6}$$

where $\underline{r}_\alpha = (r_\alpha, \hat{\underline{r}}_\alpha)$ is referenced to the center of the α-th sphere and $R_{L0}^{(\alpha)}(\epsilon_l, r)$ solves the radial Kohn-Sham equation in that sphere for an appropriately chosen $\epsilon = \epsilon_l$ somewhere close to the center of gravity of the bands that one wants to calculate. The function $R_{L1}^{(\alpha)}(\epsilon_l, r_\alpha)$ represents the energy derivative of $R_{L0}^{(\alpha)}(\epsilon, r_\alpha)$, renormalized to unity with respect to the sphere volume which is also the normalization volume of $R_{L0}^{(\alpha)}(\epsilon_l, r_\alpha)$. The two quantum numbers l, m have been abbreviated into L, and $Y_L(\hat{\underline{r}}_\alpha)$

denotes the associated spherical harmonics. At common points $(\underline{r}_i^{(\alpha')}, \underline{r}_i^{(\alpha)})$ of contact inside the lattice cell one has to require

$$\hat{\chi}_{\alpha'}(\underline{k}, \underline{r}_i^{(\alpha')}) - \hat{\chi}_{\alpha}(\underline{k}, \underline{r}_i^{(\alpha)}) = 0 \tag{2.7}$$

and there are similar equations for the radial and tangential derivative at these points. For spheres at the cell boundary having common points $(\underline{r}_\mu^{(\alpha')}, \underline{r}_\mu^{(\alpha)})$ of contact with spheres outside that correspond to equivalent spheres inside the cell one has instead of (2.7)

$$\hat{\chi}_{\alpha'}(\underline{k}, \underline{r}_\mu^{(\alpha')}) - \hat{\chi}_{\alpha}(\underline{k}, \underline{r}_\mu^{(\alpha)})\, e^{i\underline{k}\cdot\underline{R}_\mu} = 0 \tag{2.8}$$

where $\underline{R}_\mu = \underline{r}_\mu^{(\alpha')} - \underline{r}_\mu^{(\alpha)}$ denotes a lattice vector. Again, there are analogous equations to be fullfilled for the radial and tangential derivatives at these points. If the points of contact are not on a muffin-tin sphere (of radius $r_{0\alpha}$) but rather belong to a larger sphere, one has to generate $R_{L\nu}^{(\alpha)}(\epsilon_l, r)$ beyond $r_{0\alpha}$ by just continuing the integration of the radial Kohn-Sham equation into the interstitial region, so that Eq.(2.6) defines $\hat{\chi}_\alpha(\underline{k}, \underline{r})$ also for these points of contact.

If one breaks off expansion (2.6) at some l_{max} and sets $l_{max} = 2$ for spheres that contain a transition metal atom and $l_{max} = 1$ otherwise (provided that the cell does not contain rare earth atoms), the set of equations described by (2.7) and (2.8) cannot be satisfied because of the unsufficient number of coefficients $\hat{a}_{L\nu}^{(\alpha)}$. One therefore relaxes the conditions (2.7), (2.8) for the functions and spatial derivatives by requiring that the sum of all the square moduli of the differences on the left-hand side attains a minimum as a function of these coefficients. This procedure is closely related to that applied within the energy linearized rigorous cellular- (LRC-) method. For details we refer the reader to the pertinent articles [12,13,14]. The minimization of the mean square mismatch at the points of contact leads to an eigenvalue problem the solution of which yields as many solutions as there are atomic orbitals that characterize the chemical properties of the atoms in the cell. For the sake of clarity we concentrate in the following on the simplest case where the cell has the form of an fcc-Wigner-Seitz cell and contains one transition metal atom only. One obtains $2(l_{max}+1)^2$ functions $\hat{\chi}_{\hat{L}}(\underline{k}, \underline{r})$ in that case where $l_{max} = 2$ and $\hat{L}$ stands in place of (L, ν). These functions are distinctly different from each other in that each function possesses one expansion coefficient $a_{\hat{L}}(\underline{k})$ that is characteristically large for a particular $\hat{L}$ so that the expansion may be labeled accordingly. Since the functions fulfill Bloch's theorem only approximately in the vicinity of the points of contact, it takes a second step to remove this deficiency. To this end we use a plane wave expansion in the interstitial region

$$\chi_{\hat{L}}(\underline{k}, \underline{r}) = \sum_{\underline{G}}^{(G_{max})} B_{\underline{G}}^{(\hat{L})}(\underline{k})\, e^{i(\underline{k}+\underline{G})\cdot\underline{r}} \tag{2.9}$$

where $\underline{G}$ denotes vectors of the reciprocal lattice. We now determine the coefficients $B_{\underline{G}}^{(\hat{L})}$ by requiring that the expectation value of $\hat{\mathsf{H}}_{KS} - \epsilon_l$ in the interstitial region of volume Ω_{II}, i.e. the expression

$$\langle \chi_{\hat{L}} |\, \hat{\mathsf{H}}_{KS} - \epsilon_l | \chi_{\hat{L}} \rangle_{\Omega_{II}}$$

plus the mean square mismatch of $\chi_{\hat{L}}(\underline{k}, \underline{r})$ with respect to $\hat{\chi}_{\hat{L}}(\underline{k}, \underline{r})$ attains a minimum. Here $\hat{\mathsf{H}}_{KS}$ denotes the Hamiltonian associated with Eq.(2.1), but V_{eff} has been replaced by its muffin-tin approximation. The minimization leads to a set of inhomogeneous linear equations that can be solved for the coefficients $B_{\underline{G}}^{(\hat{L})}(\underline{k})$.

We are now in the position to modify the basis functions which so far consist of the preliminary partial wave expansion (2.6) and its Bloch-type continuation (2.9) by keeping the coefficients $B_{\underline{G}}^{(\hat{L})}(\underline{k})$ fixed and by smoothly continuing the plane waves backwards into the sphere. To this purpose one expands each plane wave in Eq.(2.9) in terms of spherical Bessel functions $j_l(|\underline{k}+\underline{G}|\, r)$ and spherical harmonics $Y_L(\widehat{\underline{k}+\underline{G}})$ and $Y_L(\hat{\underline{r}})$, respectively. On interchanging the $\underline{G}$- and L- dependent summations we can recast expression (2.9) as

$$\chi_{\hat{L}}(\underline{k},\underline{r}) = \sum_{L'} R_{L'}^{(\hat{L})}(\underline{k},r)\, Y_{L'}(\hat{\underline{r}}) \tag{2.10}$$

where the known functions $R_{L'}^{(\hat{L})}(\underline{k},r)$ consists of $B_{\underline{G}}^{(\hat{L})}$-weighted products $4\pi i^l j_l(|\underline{k}+\underline{G}|r) Y_L(\widehat{\underline{k}+\underline{G}})$ summed over all vectors $\underline{G}$. A smooth continuation of expression (2.9) into a new partial wave expansion of the type (2.6) inside the sphere requires

$$\sum_{\nu=0}^{1} a_{L'\nu}^{(\hat{L})}(\underline{k}) R_{l'\nu}^{\sigma}(\epsilon_l, r_0) = R_{L'}^{(\hat{L})\sigma}(\underline{k}, r_0) \tag{2.11}$$

where $\sigma = 0, 1$ refers,respectively, to the pertinent function itself and its radial derivative, and r_0 denotes the sphere radius. For each L' there are hence two equations that can easily be solved for the two coefficients $a_{L'0}^{(\hat{L})}$ and $a_{L'1}^{(\hat{L})}$ which then define a new partial wave expansion (2.6) for $\chi_{\hat{L}}(\underline{k},\underline{r})$ inside the sphere. Since we have now removed the original mismatch between the two expansions inside and outside the sphere we may omit marking $\chi_{\hat{L}}(\underline{k},\underline{r})$ and the coefficients in Eq.(2.6) by a hat which was only meant to indicate their preliminary character. In practice, these coefficients change relatively little on recalculating them from Eq.(2.11). The new basis functions $\chi_{\hat{L}}(\underline{k},\underline{r})$ which we shall henceforth refer to as CPW's, are smooth everywhere, they rigorously conform to Bloch's theorem, and they are built from linear combinations of radial functions $R_{l\nu}(\epsilon_l, r)$ in exact analogy to LMTO's. However, since $\hat{L}$ refers for a given L actually to two basis functions (associated with $\nu = 0, 1$) our basis set is by a factor of two larger which offers a gain in accuracy if one expands the sought-for function $\psi_n(\underline{k},\underline{r})$ in terms of the $\chi_{\hat{L}}(\underline{k},\underline{r})$.

3 DETERMINATION OF BAND STATES AND ENERGIES

If one expands the band states in terms of CPW's

$$\psi_n(\underline{k},\underline{r}) = \sum_{\hat{L}}^{(l_{max})} b_{n\hat{L}}(\underline{k}) \chi_{\hat{L}}(\underline{k},\underline{r})$$

and forms the expectation value of $\mathbf{H}_{KS}$ by using this expansion, the usual minimization of $\mathbf{H}_{KS}$ leads to an eigenvalue problem of the form

$$\underline{\underline{H}}_{KS}\underline{b}_n = \epsilon_n(\underline{k})\underline{\underline{S}}\,\underline{b}_n \tag{3.12}$$

where the coefficients $b_{n\hat{L}}(\underline{k})$ have been collected into a column matrix $\underline{b}_n$, and $\underline{\underline{S}}(\underline{k})$ denotes an overlap matrix the elements of which are defined

$$\left(\underline{\underline{S}}\right)_{\hat{L}'\hat{L}} = \int_{\Omega} \chi_{\hat{L}'}^{*}(\underline{k},\underline{r}) \chi_{\hat{L}}(\underline{k},\underline{r}) d^3r$$

with Ω standing for the volume of the lattice cell. This matrix has elements of large absolute value only along and close to the diagonal. The order of the matrices $\underline{\underline{S}}$ and $\underline{\underline{H}}_{KS}$ is $2(l_{max}+1)^2$, that is by a factor of two larger than in the LMTO- or ASW-method, but, on the other hand, by a factor of four smaller than in the FLAPW-method. The elements of $\underline{\underline{H}}_{KS}(\underline{k})$ have the form

$$(\underline{\underline{H}}_{KS})_{\hat{L}'\hat{L}} = \sum_{\underline{G}',\underline{G}} B_{\underline{G}'}^{(\hat{L}')*}(\underline{k}) B_{\underline{G}}^{(\hat{L})}(\underline{k}) H_{\underline{G}'\underline{G}}^{(II)}(\underline{k}) + \sum_{\hat{L}'''\hat{L}''} a_{\hat{L}'''}^{(\hat{L}')*}(\underline{k}) a_{\hat{L}''}^{(\hat{L})}(\underline{k}) H_{\hat{L}'''\hat{L}''}^{(I)}$$

where $H_{\underline{G}'\underline{G}}^{(II)}(\underline{k})$ relates to the interstitial region and is identical with the respective matrix elements of the FLAPW-method. The matrix elements in the second sum refer to the muffin-tin sphere and are defined

$$H_{\hat{L}'''\hat{L}''}^{(I)} = \int_{\Omega_I} R_{l'''\nu'''}(\epsilon_{l'''}, r) Y_{L'''}^{*}(\hat{\underline{r}}) \, \mathbb{H}_{KS} R_{l''\nu''}(\epsilon_{l''}, r) Y_{L''}(\hat{\underline{r}}) d^3r \quad .$$

The integral on the right-hand side can be similarly simplified as with the FLAPW-method. Note that $\mathbb{H}_{KS}$ contains the full potential $V_{eff}(\underline{r})$ without any shape approximation.

The eigenvalue problem (3.12) that has to be solved in order to obtain the band states and energies is practically identical to that of the LMTO- or ASW-method except that the full potential matrix elements can be determined as easily as in the FLAPW-method. Once the CPW-coefficients $B_{\underline{G}}^{(\hat{L})}(\underline{k})$ have been determined within the first iteration of a self-consistent calculation, they may be kept frozen for each individual $\underline{k}$. As the iterations are carried further, only the functions $R_{l\nu}(\epsilon_l, r)$ undergo changes which then lead to new coefficients $a_{\hat{L}}^{(\hat{L}')}(\underline{k})$. The latter can, however, easily be calculated from Eq.(2.11).

As a first test we have checked the new method for the case of fcc-copper against an equivalent FLAPW-calculation based on the same potential $V_{eff}(\underline{r})$ and standard specifications as regards G_{max} and l_{max}. The absolute value of the difference of corresponding band energies averaged over all occupied valence bands was less than 7 mRy. Further work in the direction of open and more complex structures is underway.

References

1. O. K. Andersen, Phys. Rev. B 12, 3060 (1975)
2. A. R. Williams, J. Kübler and C. D. Gelatt jr., Phys. Rev. B 19, 6094 (1979)
3. E. Wimmer, H. Krakauer, M. Weinert and A. J. Freeman, Phys. Rev. B 24, 3060 (1981)
4. H. J. F. Jansen and A. J. Freeman, Phys. Rev. B 30, 561 (1984)
5. P. Blaha, K. Schwarz and P. Sorantin, Computer Physics Communications 59, 399 (1990)
6. M. Methfessel, D. Hennig and M. Scheffler, Phys. Rev. B 46, 4816 (1992)
7. V. Eyert, Doctoral Thesis, TH Darmstadt 1991
8. C.-P. Beulshausen, Doctoral Thesis, TU Clausthal 1993
9. H. Grieb, Doctoral Thesis, TU Clausthal 1994

10. K. H. Weyrich, Phys. Rev. B 37, 10269 (1988)

11. O. Gunnarsson and B. J. Lundqvist, Phys. Rev. B 13, 4274 (1976)

12. E. Wigner and F. Seitz, Phys. Rev. 43, 804 (1933), Phys. Rev. 46, 509 (1934)

13. L. Fritsche, M. Rafat mehr, R. Glocker and J. Noffke, Z. Physik B 33, 1 (1979)

14. W. Wegener and L. Fritsche, phys. stat. sol. (b) 78, 585 (1976)

15. J. Noffke, W. Brunn and K. Hermann, Z. Physik B 29, 353 (1978)

EFFECTIVE ISING HAMILTONIAN FOR SURFACES OF METALLIC ALLOYS

Václav Drchal[1,2], Josef Kudrnovský[1,2] and Ilja Turek[3]
[1] Institute of Physics, Acad. Sci.,
Na Slovance 2, 180 40 Praha 8, Czech Republic
[2] Institute of Technical Electrochemistry,
Technical University Vienna,
Getreidemarkt 9/158, A-1060 Wien, Austria
[3] Institute of Physics of Materials, Acad. Sci.,
Žižkova 22, 616 62 Brno, Czech Republic

INTRODUCTION

Ab-initio studies of bulk and surface phase diagrams and surface segregation are based on the Ising-type Hamiltonian, whose parameters are given by the effective cluster interactions (ECI). In bulk alloys the ECIs are determined either by the Connolly-Williams inversion scheme[1] which extracts the ECIs from total energies of suitably chosen periodic structures, or by the generalized perturbation method (GPM)[2], based on the expansion of the band energy term of the total energy of the completely disordered state described within the coherent potential approximation (CPA). The Connolly-Williams method yields concentration independent ECIs, while those determined by the GPM depend on the alloy concentration. Both approaches were recently generalized to the case of alloy surfaces[3–7].

Within the GPM, in contrast to the Connolly-Williams scheme, the double-counting and electrostatic terms are neglected. First attempt to overcome this limitation in the bulk case already appeared[8]. In the case of macroscopically inhomogeneous systems, like disordered surfaces or interfaces, the CPA mean field of the homogeneous bulk alloy is used as a reference medium and, on the basis of the force theorem[9], only the band term is mapped onto the Ising model. This scheme can be improved by employing the inhomogeneous CPA medium from selfconsistent calculation for the surface of an disordered alloy. In such a case, the consistent treatment of the double-counting, core and electrostatic terms is necessary. It is the main purpose of the present work to develop the corresponding theory that would include these additional terms in a correct way. We neglect the local environment effects, i.e. we assume that the potential on each site is independent of the occupation of the neighboring sites. We thus employ the frozen-potential approximation. The calculations are done within the TB-LMTO-CPA method and the atomic sphere approximation (ASA) is used.

Stability of Materials
Edited by A. Gonis *et al.*, Plenum Press, New York, 1996

THEORY

The total energy of one particular configuration of a semi-infinite alloy within the local density approximation, is given by[10,11]

$$E_{tot} = E_{band} + E_{core} + E_{dc} + E_{Mad} . \tag{1}$$

The one-particle contributions are the band term

$$E_{band} = \int_{-\infty}^{E_F} E\, n(E)\, dE , \tag{2}$$

where $n(E)$ is the valence density of states, and the core term

$$E_{core} = \sum_{\mathbf{R}} \sum_{j} \varepsilon_{\mathbf{R}j}^{core} . \tag{3}$$

Here j runs over the core states with energies $\varepsilon_{\mathbf{R}j}^{core}$. The necessary corrections to the intraatomic electrostatic and exchange-correlation energies are given by the double-counting term

$$E_{dc} = \sum_{\mathbf{R}} \Phi[\rho_{\mathbf{R}}] , \; \Phi[\rho] = \int_{(s_{\mathbf{R}})} d^3r\, \rho(r) \left[\epsilon_{xc}(\rho(r)) - \mu_{xc}(\rho(r)) - \int_{(s_{\mathbf{R}})} d^3r' \frac{\rho(r')}{|\mathbf{r} - \mathbf{r}'|} \right] , \tag{4}$$

where $\rho_{\mathbf{R}}(r)$ is the spherically symmetric charge density in the Wigner-Seitz sphere of radius $s_{\mathbf{R}}$ centred at $\mathbf{R}$, $\epsilon_{xc}(\rho(r))$ is the local density of the exchange-correlation energy, and $\mu_{xc}(\rho(r))$ is the exchange-correlation potential. The electrostatic interaction energies of the nuclei and of electron densities in different Wigner-Seitz spheres (expressed via their multipole expansions) are included in the Madelung term

$$E_{Mad} = \sum_{\mathbf{R} \neq \mathbf{R}'} \frac{Z_{\mathbf{R}} Z_{\mathbf{R}'}}{|\mathbf{R} - \mathbf{R}'|} - \sum_{\mathbf{R} \neq \mathbf{R}'} \sum_{LL'} \left(Q_{\mathbf{R}}^{L} \right)^* \mathcal{M}_{\mathbf{R}\mathbf{R}'}^{LL'} Q_{\mathbf{R}'}^{L'} , \tag{5}$$

where $Z_{\mathbf{R}}$ is the nuclear charge at site $\mathbf{R}$, $L = (lm)$ is the angular momentum index, $Q_{\mathbf{R}}^{L}$ is the multipole moment of the non-spherical charge density in a sphere centered at $\mathbf{R}$, and $\mathcal{M}_{\mathbf{R}\mathbf{R}'}^{LL'}$ are the interaction constants. For example, the monopole-monopole interaction ($L = L' = s \equiv (l = 0, m = 0)$) is given by $\mathcal{M}_{\mathbf{R}\mathbf{R}'}^{ss} = |\mathbf{R} - \mathbf{R}'|^{-1}$.

The configurational dependence of the total energy of the semi-infinite disordered binary alloy can be expressed in the form of an effective Ising-type Hamiltonian

$$H_I = E_0 + \sum_{\mathbf{R}} D_{\mathbf{R}} (\eta_{\mathbf{R}} - \langle \eta_{\mathbf{R}} \rangle) + \frac{1}{2} \sum_{\mathbf{R} \neq \mathbf{R}'} V_{\mathbf{R}\mathbf{R}'} (\eta_{\mathbf{R}} - \langle \eta_{\mathbf{R}} \rangle)(\eta_{\mathbf{R}'} - \langle \eta_{\mathbf{R}'} \rangle) + \ldots . \tag{6}$$

The parameters entering (6) are the configurationally independent part of the total energy E_0, the on-site energies $D_{\mathbf{R}} = D_{\mathbf{R}}^{A} - D_{\mathbf{R}}^{B}$, the pair interatomic interactions $V_{\mathbf{R}\mathbf{R}'} = V_{\mathbf{R}\mathbf{R}'}^{AA} - V_{\mathbf{R}\mathbf{R}'}^{AB} - V_{\mathbf{R}\mathbf{R}'}^{BA} + V_{\mathbf{R}\mathbf{R}'}^{BB}$, and generally, the interatomic interactions of higher order. A particular configuration of the alloy is determined by a set of occupation indices $\eta_{\mathbf{R}}$, where $\eta_{\mathbf{R}} = 1$ if the site $\mathbf{R}$ is occupied by an atom of the type A, and $\eta_{\mathbf{R}} = 0$ otherwise. This form corresponds to the lattice gas model.

The parameters in (6) should be found by mapping the total energy (1) onto the Ising Hamiltonian (6). In other words, it should hold $H_I = E_{tot}$ for each configuration of the alloy. In general, it is a very difficult problem. Usually, it is sufficient to determine only the most important terms in the Ising Hamiltonian, say the on-site terms for several layers close to the surface, and the pair interactions within several coordination spheres. We will require that some properly chosen conditional averages of the total energy for a cluster of n atoms defined as

$$E_{tot}(\mathbf{R}_1 Q_1, ..., \mathbf{R}_n Q_n) = \frac{\langle \eta_{\mathbf{R}_1}^{Q_1} ... \eta_{\mathbf{R}_n}^{Q_n} E_{tot} \rangle}{\langle \eta_{\mathbf{R}_1}^{Q_1} ... \eta_{\mathbf{R}_n}^{Q_n} \rangle}, \tag{7}$$

where $\eta_{\mathbf{R}}^A = \eta_{\mathbf{R}}$ and $\eta_{\mathbf{R}}^B = 1 - \eta_{\mathbf{R}}$, $Q = A, B$, and similar averages $H_I(\mathbf{R}_1 Q_1, ..., \mathbf{R}_n Q_n)$ of the Ising Hamiltonian coincide.

In a completely disordered alloy, the occupation indices $\eta_{\mathbf{R}}$ average as independent variables: $\langle \eta_{\mathbf{R}} \rangle = c_{\mathbf{R}}$, $\langle \eta_{\mathbf{R}} \eta_{\mathbf{R}'} \rangle = c_{\mathbf{R}} c_{\mathbf{R}'}$ (for $\mathbf{R} \neq \mathbf{R}'$), etc., where $c_{\mathbf{R}}$ is a site-dependent concentration of A atoms. The result for on-site and pair interactions (higher order terms can be found in a similar manner) can be expressed in terms of differences of conditional averages

$$D_{\mathbf{R}} = E_{tot}(\mathbf{R}A) - E_{tot}(\mathbf{R}B), \tag{8}$$

$$V_{\mathbf{R}\mathbf{R}'} = E_{tot}(\mathbf{R}A, \mathbf{R}'A) + E_{tot}(\mathbf{R}B, \mathbf{R}'B) - E_{tot}(\mathbf{R}A, \mathbf{R}'B) - E_{tot}(\mathbf{R}B, \mathbf{R}'A). \tag{9}$$

The energy differences in (8), (9) can be found within the density functional theory with a good accuracy using the so-called force, or structural energy theorem[9]: the variation $\delta\rho$ of the ground state charge density ρ will change the ground state energy by an amount which is of second order in $\delta\rho$, $E_{tot}[\rho + \delta\rho] = E_{tot}[\rho] + O(\|\delta\rho\|^2)$.

Further analysis can be considerably facilitated by neglecting the local environment effects. which allows to replace the actual charge densities by the conditionally averaged charge densities $\bar{\rho}_{\mathbf{R}}^Q$. They are used to construct the potentials in each atomic sphere, the potential parameters and finally the one-electron Hamiltonian. Consequently, the charge densities, atomic potentials and potential parameters are identical for all equivalent atoms, while all the fluctuations induced by different environments of each atom are neglected.

The basic assumption of the single-site theory of electronic structure of disordered alloys is that the potential at any site $\mathbf{R}$ depends only on the occupation of this site by atom A or B, and is completely independent of the occupation of the other sites. This assumption can be fulfiled only, if the net charges $q_{\mathbf{R}} = Q_{\mathbf{R}}^0 - Z_{\mathbf{R}}$ and multipoles $Q_{\mathbf{R}}^L, L \neq s$ are independent of the occupation of the lattice sites by atoms A or B. Within the LMTO-ASA method, it is possible to change the atomic radii in such a way that the net charges $q_{\mathbf{R}}$ do not depend on the type of atom while preserving the total volume of the system[12,13].

In the bulk case, the atomic sphere radii s^Q are adjusted such that each atomic sphere is charge neutral. Then the Madelung potential and the Madelung contribution to the total energy (5) vanish which is compensated by a change of potential parameters induced by changes in sphere radii. Although this compensation is in principle exact, the approximations involved in the computational scheme, like the ASA, can cause certain error.

In the surface case, the situation is more complex due to the macroscopic charge redistribution within the region of the solid-vacuum interface. The net charges remain non-zero in the surface region (but identical for A and B atoms) and the semi-infinite alloy, from the electrostatic point of view, behaves as a semi-infinite ideal crystal.

The band term depends on the alloy configuration in two ways: via the density of states $n(E)$, and via the Fermi energy E_F. The configurational dependence of the E_F can be eliminated by employing the grandcanonical rather than the canonical ensemble. The band energy term is replaced by the grandcanonical potential

$$\Omega = E_{band} - E_F N_{el} = -\int_{-\infty}^{E_F} dE N(E) \tag{10}$$

for non-interacting particles at $T = 0$, where $N(E)$ is the integrated density of states and E_F is fixed and equal to its configurationally averaged value. This average is a well defined quantity as follows from the self-averaging property of E_F, the fluctuations of which are of order $O(N^{-\frac{1}{2}})$. The contributions from the band term to the Ising Hamiltonian are determined by using the standard GPM technique[6].

The core energy E_{core}, Eq. (3), and double counting terms E_{dc}, Eq. (4), have a single-site character. Consequently, if the local environment effects are neglected, they contribute only to the on-site terms, but not to the pair and higher order interatomic interactions. Inserting (3) into (8) we obtain

$$D_{\mathbf{R}}^{core} = \sum_j^{core} \left(\varepsilon_{\mathbf{R}j}^A - \varepsilon_{bulk,j}^A \right) - \sum_j^{core} \left(\varepsilon_{\mathbf{R}j}^B - \varepsilon_{bulk,j}^B \right) \tag{11}$$

The core state energies $\varepsilon_{bulk,j}^Q$ of bulk atoms can be either included into the configurationally independent part E_0 of the alloy internal energy, or into the $D_{\mathbf{R}}^{core}$. The latter choice, adopted here, has certain advantages, because the on-site terms defined in (11) express the net effect of surface inhomogeneity and are not obscured by large bulk contribution. The contribution of double counting terms to the on-site parameters is given by

$$D_{\mathbf{R}}^{dc} = \left(\Phi[\rho_{\mathbf{R}}^A] - \Phi[\rho_{bulk}^A] \right) - \left(\Phi[\rho_{\mathbf{R}}^B] - \Phi[\rho_{bulk}^B] \right) \tag{12}$$

where the functional Φ was defined in Eq. (4). Similarly as for the core contributions, the bulk atom terms are subtracted in (12).

The Madelung energy E_{Mad}, Eq. (5), contributes only to the on-site terms

$$D_{\mathbf{R}}^{Mad} = -2(Z^A - Z^B) \sum_{\mathbf{R}'}{}' \left\{ \frac{Q_{\mathbf{R}'}^s - Z_{\mathbf{R}'}}{|\mathbf{R} - \mathbf{R}'|} + \mathcal{M}_{\mathbf{R}\mathbf{R}'}^{sz} \langle Q_{\mathbf{R}'}^z \rangle \right\}, \tag{13}$$

while its contribution to the pair interactions is zero because $Z^A - Z^B = Q_{\mathbf{R}}^{sA} - Q_{\mathbf{R}}^{sB}$ and the dipole moments at A and B atoms are identical, $Q_{\mathbf{R}}^{zA} = Q_{\mathbf{R}}^{zB} = \langle Q_{\mathbf{R}}^z \rangle$.

RESULTS AND DISCUSSION

We have calculated the effective interatomic interactions for the (001) surface of the fcc $Cu_{50}Ni_{50}$ alloy using the all-electron fully relativistic version of the TB-LMTO-CPA method extended to semi-infinite systems[11].

First, we have applied the frozen potential approximation, i.e. we assumed a semi-infinite alloy with atomic potentials and the CPA medium taken from the homogeneous disordered bulk alloy[5] (model I). No redistribution of the electronic charge near the surface was allowed and the system was thus electrically neutral. The ECIs were determined by mapping only the band term onto the Ising Hamiltonian.

Second, we have performed a fully selfconsistent calculation of the surface electronic structure including the surface charge redistribution and determined ECIs according to the present theory (model II). The results are tabulated below.

Table 1. The on-site terms (in mRy) of the effective Ising Hamiltonian for (001) surface of the fcc $Cu_{50}Ni_{50}$ alloy. The contributions from the band energy, core electrons, double-counting terms, and Madelung energy are given. Their sum D^{tot} is compared with the result D^{nsc} of the approximate calculation (model I) based on the CPA reference medium corresponding to the homogeneous bulk alloy.

layer	D^{band}	D^{core}	D^{dc}	D^{Mad}	D^{tot}	D^{nsc}
1	54.88	88.80	79.74	-246.85	-23.43	-35.10
2	-12.74	-24.65	20.87	16.32	-0.20	-2.13
3	9.33	19.44	-23.57	-5.15	-0.06	-0.50
4	-3.44	-8.15	11.01	0.56	-0.02	-0.91

The non-selfconsistent theory (model I) yields already qualitatively correct values D^{nsc} for the on-site terms corresponding to the segregation of copper atoms at the surface in agreement with experimental data and with recent theoretical results[14]. This was confirmed[7] by a statistical Monte Carlo study of surface segregation in Cu-Ni alloy system based on the parameters calculated within the model I.

The selfconsistent theory (model II) leads to a reduction of the on-site terms in comparison with model I. In addition, when going from the surface into the bulk, they converge much faster to zero. This behavior can be expected as a result of efficient screening at high metallic electron densities. The screening mechanisms are not contained in model I, while the model II takes them into account by allowing for charge redistribution in surface region. In the selfconsistent theory, the on-site terms result from a delicate interplay between the band, core, double-counting and Madelung contributions. Some of these contributions are rather large, but they mostly cancel each other. It should be noted that within the selfconsistent theory, the on-site terms cannot be determined solely from the band energy as in model I. This conclusion is also confirmed by the results for fcc (001) surface of Cu-Ni alloys[14]. Such an attempt would lead to a serious error. For example, as follows from Table 1, $D_1^{band} = 54.88$ mRy, which corresponds to the segregation of Ni atoms.

The pair interactions converge quite rapidly to their bulk values. On the other hand, the difference between their surface and bulk values is quite pronounced and this could enhance the intra-layer segregation of Cu and Ni atoms in surface layers.

Table 2. The renormalized pair interactions (in mRy) in the effective Ising Hamiltonian for (001) surface of the fcc $Cu_{50}Ni_{50}$ alloy.

	layers									
neighbor	1-1	1-2	1-3	2-2	2-3	2-4	3-3	3-4	4-4	bulk
1	-3.60	-0.87	-	-0.50	-1.34	-	-1.31	-1.15	-1.09	-1.33
2	-0.86	-	0.95	1.25	-	0.97	0.99	-	1.14	1.14
3	-	-0.42	0.10	-	0.06	0.19	-	0.11	-	0.08
4	-0.26	-	-0.26	-0.20	-	-0.18	-0.07	-	-0.13	-0.14

CONCLUSIONS

In summary, we have developed a theory that allows to determine the effective cluster interactions for surfaces of disordered alloys. It is based on selfconsistent electronic structure and includes the charge redistribution at the surface. Its results can serve as an input in statistical mechanical studies of surface segregation, wetting and ordering. In contrast to a simplified theory based on the semi-infinite bulk effective medium (model I) which assumes fixed bulk concentrations of A and B atoms, it can yield effective cluster interactions for any concentration profile. This makes possible to determine the surface concentration profile from first principles in a self-consistent manner, including the construction of the surface Ising Hamiltonian with concentration dependent parameters for a given concentration profile and its statistical study providing the updated profile. In this way a complete and reliable solution of the surface segregation profile on an ab-initio level can be achieved.

Acknowledgements

Financial support for this work was provided by the Grant Agency of the Czech Republic (Project 202/93/0688), Grant Agency of the Academy of Sciences of the Czech Republic (Project 110 437) and the Austrian Science Foundation (P10231).

REFERENCES

1. J. W. Connolly and A. R. Williams, Phys. Rev. B **27**, 5169 (1983).
2. F. Ducastelle, *Order and Phase Stability* (North Holland, Amsterdam, 1991).
3. D. M. Wood and A. Zunger, Phys. Rev. B **40**, 4062 (1989); R. Osório, J. E. Bernard, S. Froyen, and A. Zunger, Phys. Rev. B **45**, 11173 (1992).
4. A. V. Ruban, I. A. Abrikosov, D. Ya. Kats, D. Gorelikov, K. W. Jacobsen and H. L. Skriver, Phys. Rev. B **49**, 11383 (1994).
5. B. Legrand, G. Tréglia, and F. Ducastelle, Phys. Rev. B **41**, 4422 (1990); G. Tréglia, B. Legrand, and F. Ducastelle, Europhys. Lett. **7**, 575 (1988).
6. V. Drchal, J. Kudrnovský, L. Udvardi, P. Weinberger, and A. Pasturel, Phys. Rev. B **45**, 14328 (1992).
7. A. Pasturel, V. Drchal, J. Kudrnovský, and P. Weinberger, Phys. Rev. B **48**, 2704 (1993).
8. A. Gonis, P. P. Singh, and P. E. Turchi, in *Metallic Alloys: Experimental and Theoretical Perspectives*, eds. J. S. Faulkner and R. G. Jordan, (Kluwer Academic, Dordrecht, 1994), p. 197.
9. A. R. Mackintosh and O. K. Andersen, Chapter 5.3 in *Electrons at the Fermi Surface*, edited by M. Springford (Cambridge University Press, Cambridge, England, 1980).
10. J. F. Janak, Phys. Rev. B **9**, 3985 (1974).
11. V. Drchal, J. Kudrnovský, and P. Weinberger, Phys. Rev. B **50**, 7903 (1994).
12. J. Kudrnovský and V. Drchal, Phys. Rev. B **41**, 7515 (1990).
13. P. P. Singh, A. Gonis, and P. E. A. Turchi, Phys. Rev. Lett. **71**, 1605 (1993).
14. T. Schulthess, R. Monnier and S. Crampin, Phys. Rev. B **50**, 18564 (1994).

THE INFLUENCE OF QUASI-PARTICLE LIFETIMES AND ELECTRONIC TOPOLOGICAL TRANSITIONS ON THE DEVIATIONS FROM VEGARD'S RULE IN AG-PD ALLOYS

E. Bruno, B. Ginatempo and E.S. Giuliano

Dipartimento di Fisica and Unità INFM
Università di Messina
Salita Sperone 31
98166 S. Agata di Messina (Italy)

INTRODUCTION

In this paper, we discuss about the variations of the equilibrium lattice parameter versus the alloy concentration in AgPd fcc solid solutions. We interpret the experimentally observed deviations from Vegard's rule in terms of the Electronic Topological Transition (ETT) related to the d band filling which occurs about the equiatomic concentration. Through simple Rigid Band Model (RBM) calculations, we show how the finiteness of quasi-particle lifetimes affects these deviations, namely reducing their amplitude. RBM calculations are also used to compare versus the experiment the predictions of ab initio Local Density Approximation, Korringa-Kohn-Rostoker Coherent Potential Approximation (LDA-KKR-CPA) calculations. On this basis we shall discuss the apparent underestimation of the relevant quasi-particle lifetime in the CPA theory.

The paper is organised as follows. In the next section we shall briefly recall the concept of ETT and outline our theory[1] about the influence of ETT's on lattice spacing variations. In the final section the role played in these phenomena by the finiteness of quasi-particle lifetimes shall be analysed and the results obtained shall be discussed.

ELECTRONIC TOPOLOGICAL TRANSITIONS AND DEVIATIONS FROM VEGARD'S LAW

According with the well-known Vegard's law, the volumes per atom in metallic random should vary linearly with the atomic concentrations. However, there is experimental evidence for a number of "exceptions" to this empirical rule. As we have already discussed in a previous paper[1], in certain cases, the observed deviations from Vegard's law can be related to the occurrence of ETT's, the so called Lifshitz's 2 1/2 "phase transitions"[2].

According with Lifshitz[2], an ETT occurs when, on varying the alloy concentration, the chemical potential crosses a Van Hove singularity. It is easily shown that in such case the Fermi Surface (FS) undergoes a change of its topology. Thus, accordingly with the ETT theory[1-5], the electronic density of states (DOS) may be written as the sum of a regular term, $n_0(E)$, related to metal bands not responsible of the ETT, and a special term, $n_1(E)$, i.e.,

$$n(E) = n_0(E) + n_1(E) \qquad (1).$$

Stability of Materials
Edited by A. Gonis *et al.*, Plenum Press, New York, 1996

The special term in Eq. (1) comes from states belonging to the band that originates the ETT and can be evaluated exactly in a neighbourhood of the ETT itself. In fact, neglecting the finiteness of the temperature and the quasi-particle lifetimes, we would have[2]

$$n_1(E) = \alpha |z|^{1/2} \Theta(z) \qquad (2).$$

In Eq. (2) the energies are measured from the Van Hove singularity occurring at $E=\varepsilon_c$,

$$z = s(E - \varepsilon_c) \qquad (3),$$

$\Theta(z)$ is the Heaviside step function and α is related to the mass tensor elements along the principal axes,

$$\alpha = q V |2 m_2 m_1 m_3|^{1/2} / (\pi^2 \hbar^3) \qquad (4).$$

In the most frequent cases, the stationary point in the band structure responsible of the singularity is a maximum (M), a minimum (m), a saddle point of index 1 (S1) or a saddle point of index 2 (S2). Accordingly, the quantities s and q in Eqs. (2) and (3) take the values given below,

$$s = \begin{cases} +1 & (m, S2) \\ -1 & (M, S1) \end{cases} \qquad q = \begin{cases} +1 & (m, M) \\ -1 & (S1, S2) \end{cases} \qquad (5),$$

and the corresponding changes in the FS shall be the filling of an hole pocket (m), the formation of an electron pocket (M), the disruption (S1) or the opening (S2) of a neck.

The presence of a special term in the DOS in a proximity of an ETT gives rise to a special term in the thermodynamic potentials as well and, therefore, affects equilibrium properties. In random alloys, the effects on the equilibrium volume per atom versus concentration curve, V(N), can be studied by minimisation of the free energy with respect to the volume at fixed atomic concentration. The result[1] is that the "compressibility" with respect to the electron per atom ratio (we shall call it briefly charge compressibility), can be written again as the sum of a regular and a special, ETT related, term,

$$dV/dN = dV_0/dN + dV_1/dN \qquad (6),$$

$$dV_1/dN \cong -(dV_0/dN)(d\mu/dN)\, n_1(z(N)) \qquad (7),$$

where N is the electron per atom ratio and $z(N)=s\,(\mu(N)-\varepsilon_c)$.

The above charge compressibility of Eq. (6) may be obtained in a RBM approximation[1], if the DOS is evaluated, for a single alloy concentration, at two different (close) values of the lattice parameter about the equilibrium. In Fig. 1 we plot the RBM charge compressibility for fcc random AgPd alloys, as we have obtained from fully relativistic muffin tin LDA-KKR calculations of the DOS in pure Pd. As we have shown[1], the variations versus the concentration of this quantity can be fully explained in terms of the ETT's occurring in the AgPd system as the electron per atom ratio is increased on going from pure Pd to pure Ag. However, the most evident feature in Fig. 1 is the very large variation occurring about the equiatomic concentration. This is due to the closure of a hole pocket at the point X in the Brillouin Zone (m). In fact, according with Eq. (7), the magnitude of dV_1/dN is proportional to $(d\mu/dN)\, n_1(z(N)) \approx n_1(z(N))/n_0(z(N))$. Now the relevant state is characterised by a large effective mass (cfr. Fig. 1 in Bruno et al.[6]), thus n_1 is large according with Eq. (4), while the normal DOS, n_0, that comes from a typical noble metal-like s-p band, is small.

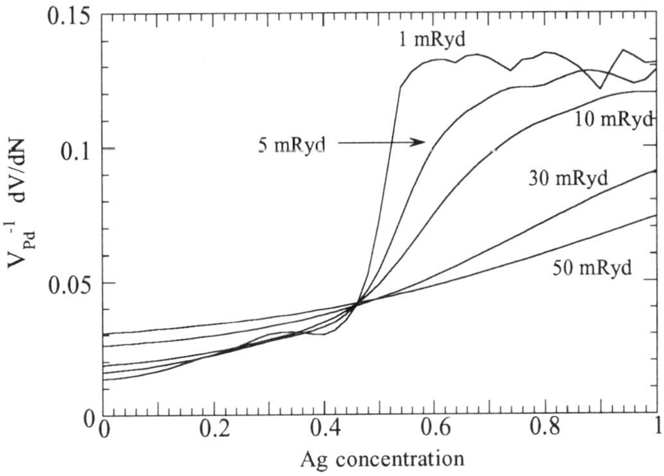

Figure 1. Rigid Band Model charge compressibility in AgPd alloys versus Ag atomic concentration at various quasi-particle lifetimes, τ. Labels on each curve indicate the values of $\eta = \hbar/\tau$.

THE EFFECTS OF FINITE QUASI-PARTICLE LIFETIMES

The presence of finite quasi-particle finite lifetimes, $\tau = \hbar/\eta$, and/or finite temperatures smears out the singularity in Eq. (2)[3,4]. This notwithstanding, as we have shown[1], the picture given by Eqs. (6) and (7) can still be used, provided that the special DOS of Eq. (2) is replaced by the following "effective DOS",

$$n_1(z,\eta,T) = \alpha\, \mathrm{Re}\{z + i\eta\} + O\left(\left(k_B T/\eta\right)^2\right) = \left(\alpha/\sqrt{2}\right)\left[\left(z^2+\eta^2\right)^{1/2} + z\right]^{1/2} + O\left(\left(k_B T/\eta\right)^2\right) \qquad (8).$$

The inclusion of finite lifetimes in the RBM calculations for the charge compressibility can be easily modelled, simply by adding an imaginary increment $\eta = \hbar/\tau$ in the calculation for the DOS. The ETT related contribution, proportional to the special DOS, are smeared out accordingly with Eq. (8). This behaviour is apparent in Fig. 1, where we have plotted the RBM calculation for several values of η. As expected, the almost sudden step visible at η=1 mRyd is smoothened out as η is increased.

Integrating the charge compressibility versus concentration curve, the deviations of the lattice parameter a from Vegard's rule,

$$\Delta a(N) = a(N) - \left(c_{Ag} a_{Ag} + c_{Pd} a_{Pd}\right) \qquad (9),$$

are easily obtained. These are plotted in Fig. 2, where the results for various η's are compared with the experiment[7] and our ab initio, fully relativistic, LDA-KKR-CPA, muffin tin, calculations[6]. Interestingly, all the curves shown, exhibit the same general trend. Namely, there is a marked minimum about the equiatomic concentration, which, therefore, could be related to the filling of the d hole pocket at X. This interpretation is supported also by thermoelectric power measurements[8] in the same system, which show a large ETT related[9,1] peak in the same range of concentrations. On the other hand, in Fig. 2, it is visible a disagreement between LDA-KKR-CPA calculations and the experiment, clearly significant beyond the errors in the determination of the lattice parameters. Namely, the calculations

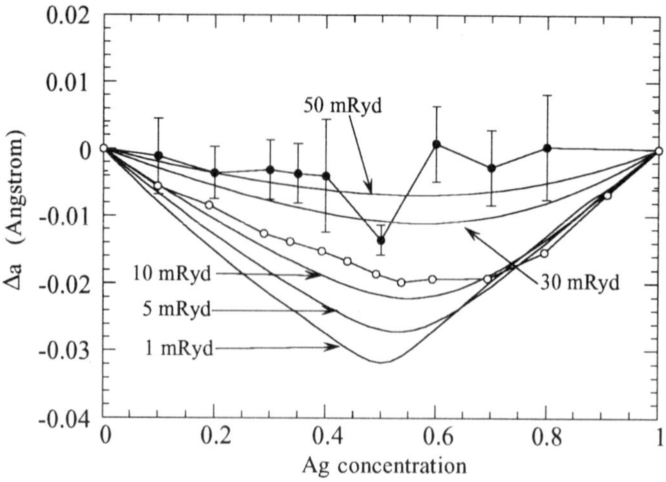

Figure 2. Deviations of the lattice parameter in AgPd versus Ag concentrations. Open symbols: experiment[7], closed circles with error bars: LDA-KKR-CPA calculations[1], full lines: RBM calculations for various quasi-particle lifetimes. Labels display the corresponding values of η.

appear to underestimate the overall size of the effect. Remarkably, by a look at the RBM curves, we see that as η is increased (or the quasi-particle lifetimes decreased), the RBM peak in Fig. 2 becomes less and less marked. Thus, at very large quasi-particle lifetimes (see, e.g. η=1 mRyd), the effect is overestimated, while at η=10 mRyd the RBM curve is very close to the experiment and at η=30÷50 mRyd it resembles the total energy calculation results. Since increasing η in the RBM is equivalent to increase some kind of effective anelasticity in the scattering, our simple RBM estimates quite naturally suggest to interpret the above discrepancy between the sophisticated LDA-KKR-CPA theory and the experiment as due to an overestimation of the amount of disorder (or an underestimation of the quasi-particle lifetimes) intrinsic to the definition of the CPA effective medium. This was somehow expected for. The CPA, in fact, deals with substitutional disorder in alloys in a mean field approximation, thus, as an intrinsic feature of the method, it misses to account for those fluctuations inducing local ordering.

We wish to elaborate further on the above suggestion. First, we observe that, since the inputs of our RBM machinery were the DOS's from ab initio LDA-KKR calculations, the only physical approximation which makes the difference with the full alloy calculations is that here the CPA is missing. At variance of the traditional RBM scheme in which lifetimes are assumed infinite, in the present RBM calculations we force the lifetimes to have the same, fixed, value for all the quasi-particle states, while, as in traditional RBM, the band shapes are not allowed to change with the concentration. This, of course, is not correct in principle and the CPA is actually an elegant, self-consistent, method for determining the lifetimes at each k point and the band shapes in a random alloy. Nevertheless, as we discussed when commenting Fig. 1, in the present case, most of the effect comes just from a single state, i.e. the Van Hove singularity at X. Thus, only the lifetime of this state is expected to be relevant. Moreover, the deformations of the band structure due to substitutional disorder should not be of qualitative importance since the same RBM provides fairly good estimates of the concentrations at which the ETT's occur in the system under consideration[6]. We wish to add that the above referred lattice parameters measurements have been done at room temperature, i.e. $k_BT \approx 2$ mRyd. Thus, the T-dependent corrections in Eq. (8) are expected to be small. Anyway, their only effect would be just an additional smearing of the features in the charge compressibility versus concentration curve. This could

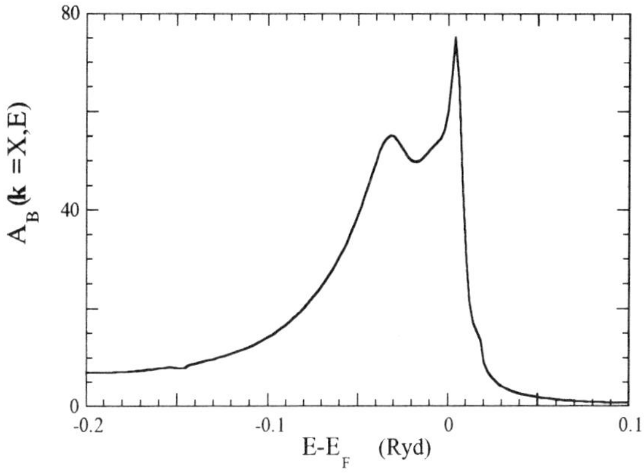

Figure 3. The BSF, $A_B(k,E)$, at the X point as a function of the energy from LDA-KKR-CPA calculations for $Ag_{0.50}Pd_{0.50}$ random alloy.

only worsen the discrepancy in the estimate of the relevant lifetimes between the CPA calculations, which have been performed at T=0, and the experiment.

On the basis of the above analysis, the RBM results quite naturally suggests some numbers for the lifetime of the relevant quasi-particle state in AgPd: namely, $\eta=30\div50$ mRyd for the LDA-KKR-CPA and $\eta=10$ mRyd for the experiment. In order to check these quantities, we have plotted in Fig. 3 the LDA-KKR-CPA Bloch Spectral Function (BSF) at the X point for AgPd about the relevant concentration. As it is known from the CPA theory[10], the half-width of the peaks in the BSF is just the imaginary part of the CPA "eigenvalues", that is to say, our η. Remarkably, the half-width of the broad peak visible in Fig. 3 compares favourably with the above value, $\eta=30\div50$ mRyd, at which RBM substantially reproduces the CPA curve. This would support the reliability of the above RBM estimates. However, we have to point out that our theoretical elaboration on the ETT's influence on the lattice parameters as well as RBM calculations consider only the band contribution to the total energies. The inclusion of other energetic contributions as, e.g. exchange and correlation, which are properly accounted for in the LDA-KKR-CPA calculations, although cannot suppress the ETT related term in the thermodynamic potential, nevertheless it is expected to renormalise the proportionality factor in Eq. (7). Anyway, the comparison with the BSF half-width suggests that, at least in the present case, the RBM estimate is sensible.

In summary, we have shown in this paper how, in presence of ETT's, the quasi-particle lifetimes can play a relevant role in determining the equilibrium volumes as a function of the atomic concentration in random alloys. We have also found that the CPA, although accurate[5] in the determination of the spectral properties in random solid solutions, nevertheless could lead to substantial underestimates of quasi-particle lifetimes. This feature, at least in the case studied, does not forbid the overall reproduction of the experimental trends, but has to be kept in mind when calculating quantities that depend appreciably on the lifetimes.

REFERENCES

1. E. Bruno, B. Ginatempo and E.S. Giuliano, Phys. Rev. B, in press (1995).
2. I.M. Lifshitz, *Zh. Eksp. Teor. Fiz.* **33**, 1569 (1960).
3. A.A. Varlamov, V.S. Egorov and A.V. Pantsulaya, *Adv. in Phys.* **38**, 469 (1989).
4. Ya. M. Blanter, M.I. Kaganov, A.V. Pantsulaya and A.A. Varlamov, *Phys. Rep.* **245**, 160 (1994).

5. E. Bruno, B. Ginatempo, E. S. Giuliano, A. V. Ruban and Yu. Kh. Vekilov, *Phys. Rep.* **249,** 353 (1994).
6. E. Bruno, B. Ginatempo and E.S. Giuliano, Phys. Rev. B, in press (1995).
7. W.B. Pearson, "A Handbook of Lattice Spacings and Structures of Metals and Alloys", Pergamon Press, Oxford (1958).
8. A.M. Guénault, Phil. Mag. **30**, 641 (1974).
9. W.H. Butler and G.M. Stocks, *Phys. Rev.* **B29**, 4217 (1984).
10. Gonis A., "Green Functions for Ordered and Disordered Systems", North-Holland (1992).

ELECTRONIC STRUCTURE AND STABILITY OF Al-Ge ALLOYS UNDER PRESSURE

P. E. A. Turchi[1], Prabhakar P. Singh[1,2], and G. M. Stocks[3]

[1]LLNL (L-268), P. O. Box 808, Livermore, CA 94551
[2]Present address: Dept. of Physics, ITT, Powai, Bombay 400-076, India
[3]ORNL, Metals and Ceramics Division, Oak Ridge TN 37831-6114

INTRODUCTION

The synthesis of new phases under pressure and the study of their structure-properties relationships yield valuable information on the general laws of phase formation in multicomponent alloys. They also provide an important insight in the conditions of occurrence of particular phases in systems which exhibit typical phase diagrams at ambient pressure [1]. Although pressure-temperature diagrams are available for most elements and some compounds [2], very little work has been done so far on the study of the possible competition between structural and chemical order effects in alloys as a function of pressure.

With this respect, Al-Ge alloys exhibit interesting stability properties [1,3-6]. This alloy system displays, at atmospheric pressure, a simple eutectic phase diagram with limited mutual solubility of the components in the solid state [7]. However, based on atomic size factors and electronegativity differences, favorable mutual solubility is expected. In fact, a substantial increase in solubility of Ge in Al has been observed by quenching from the melt under high pressure [5,6] as well as by rapid quenching from the liquid state [3,4,8]. An increase in the applied pressure extends the range of solubility of Ge in the fcc matrix of Al from 2.8 at. % Ge at atmospheric pressure to about 20 at. % Ge at 10 GPa [5,6]. Similarly, rapid quenching from the melt of Al Ge alloys ensures solubility of up to 13 at. % Ge in Al. For completeness, note that in these pressure experiments, a spontaneous solid state amorphization which may offer potential applications is observed upon annealing at high temperature [9].

The present study shows that, as far as the phase stability properties are concerned, the miscibility gap which characterizes the Al-Ge phase diagram at normal pressure, is not associated with a lack of chemical affinity of Ge for Al but, instead, finds its origin in the huge structural energy difference of Ge between its ground-state, which is the diamond structure, and the fcc structure. This energy difference decreases substantially by applying pressure, as expected since the coordination number usually increases with pressure. In addition to this "endpoint effect", the ordering tendencies in this alloy, as will be discussed in the following, increase with pressure, thus resulting in the possible stabilization of ordered fcc-based alloys, preferentially at low Ge concentration.

METHODOLOGY

With the recent development of efficient and accurate electronic structure methods, based on the local density approximation within the density functional theory, many

Stability of Materials
Edited by A. Gonis *et al.*, Plenum Press, New York, 1996

properties of metals and alloys can now be explained and computed with a reasonable degree of accuracy. In particular, it has become possible to investigate the thermodynamic properties of alloys from first-principles, *i.e.* from the sole knowledge of the underlying lattice and the atomic numbers of the alloy species [10].

To describe the electronic structure properties of fully ordered alloy configurations, we use the scalar relativistic Linear-Muffin-Tin-Orbital method in the Atomic Sphere Approximation (LMTO-ASA) [11]. In other words, all relavistic effects are accounted for, except the spin-orbit coupling. The so-called combined correction terms, correcting for the overlap of the atomic spheres, and for the higher partial waves inside the spheres l>2) are included in the calculations, and the exchange and correlation potential is parametrized as suggested by von Barth and Hedin [12]. In this approach, the total energy E_T^C of any alloy configuration C is directly calculated, thus resulting in a straightforward evaluation of the formation energy ΔE_{Form}^C of the binary alloy $A_{1-c}B_c$ according to

$$\Delta E_{Form}^C = E_T^C - (1 - c_C)E_T^A - c_C E_T^B \tag{1}$$

For the treatment of chemical order in alloys, the Generalized Perturbation Method (GPM) [13] is used. Within the GPM, the energy associated with a particular alloy configuration is given by the sum of two terms. First, a concentration-dependent, but configuration-independent energy, usually called the mixing energy, and second, a concentration and configuration-dependent energy, also called the ordering energy, ΔE_{Ord}^C. Hence, the formation energy associated with an alloy configuration C, which is based on a given crystalline structure, is written as

$$\Delta E_{Form}^C = \Delta E_{Mix}(c_C) + \Delta E_{Ord}^C \tag{2}$$

where $\Delta E_{Mix}(c_C)$ is the mixing energy, or the formation energy of the disordered alloy at the composition of the ordered configuration C, given by the difference between the total energy of the disordered state (here, defined within the coherent potential approximation, or CPA), and the sum of the concentration weighted averaged total energies of the pure species, E_T^i (i=A,B), *i.e.*,

$$\Delta E_{Mix}(c) = E_T^{CPA}(c) - (1 - c)E_T^A - cE_T^B \tag{3}$$

In the GPM framework, the ordering energy is expressed as an expansion in terms of pair and multi-site interactions. At zero temperature, and to the lowest order in perturbation, where only the pair interactions are retained in the expansion, we have

$$\Delta E_{Ord}^C \approx \sum_s q_s^C V_s \tag{4}$$

where the V_s are the so-called effective pair interactions (EPI's) which are, by definition, concentration-dependent, and given by the combination $V_s = V_s^{AA} + V_s^{BB} - 2V_s^{AB}$. According to this definition, a positive (negative) interaction V_s favors the formation of AB (AA or BB) pairs for the particular shell index s. The sum in Eq. (4) runs over the neighbor shell index s, and the q-coefficients are configuration specific and expressed in terms of the number of BB pairs, n_s^{BB}, and the total number of pairs, n_s, per site, associated with the sth-neighbor shell according to:

$$q_s^C = \frac{c}{2}(n_s^{BB} - cn_s) \tag{5}$$

In the following, we will refer to the implementation of the GPM in the charge self-consistent Korringa Kohn Rostoker-CPA (KKR-CPA) electronic description of the chemically random alloy based on multiple scattering theory [10,14]. The KKR-CPA itself [15] is described within the muffin-tin approximation, with an angular momentum expansion up to l=3, and a von Barth-Hedin description of the exchange and correlation potential [12]. The EPI's introduced in Eq. (4) are computed individually, in the GPM context, from the knowledge of the scattering properties of the disordered reference medium.

ORDER AND STABILITY AS FUNCTIONS OF PRESSURE

Electronic structure calculations were carried out for the fcc-based pure elements and the chemically random Al-Ge alloys, as functions of concentration and lattice parameter, on the basis of charge self-consistent KKR and KKR-CPA muffin-tin potentials. The equilibrium structural properties have been extracted from a least square fit to the Birch-Murnaghan's equation of state [16]. For the pure elements, the equilibrium properties are in good agreement with results from literature. Note that for this alloy system, the deviation from the Vegard's law is minimal, and this is reflected in a negligible charge transfer of about 10^{-2} (or less) electron from Ge to Al.

The densities of states are typical of almost free-electron like systems, mostly characterized by sp-derived bands, and the stability properties remain essentially controlled by the valence sp-bands, at least up to a pressure of about 50 GPa.

The EPI's were calculated in the GPM context, for the two stoichiometric compositions Al_3Ge and AlGe as functions of lattice parameter, and therefore pressure. Indeed, the lattice parameter-information can simply be converted to pressure P, since $P=-\partial E/\partial V$. In the following, the pressure studies will be limited to a maximum pressure of 50 GPa, to avoid having to consider, in the electronic treatment of the problem, core-core and core-polarization effects which, in principle, take place at very high pressure. For both alloy compositions, the EPI's converge rapidly with distance (beyond the 2nd neighbor shell, they are less than 0.2 mRy/atom), and the multi-site interactions, beyond the pairs, are negligible. The results are summarized in Figure 1 where the EPI's for the nearest neighbor shells are plotted as functions of pressure for the two alloy compositions. The most striking result is that, at both stoichiometries, the dominant EPI's which build up the ordering energy, see Eq. (4), increase dramatically with pressure. From an analysis of the angular momentum-decomposition of the EPI's, the ss, pp and the hybrid-sp components mainly contribute to the first and second total EPI's, for both stoichiometries, as expected. However, the pp and sp (ss) components favor ordering (clustering) in the first neighbor shell whereas the ss and pp (sp) components favor clustering (ordering) in the second neighbor shell.

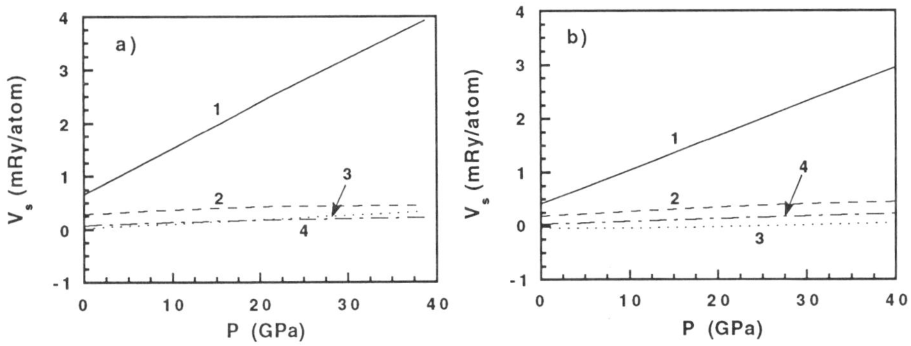

Figure 1. Effective Pair Interactions V_s (in mRy/atom) referred to by their shell index, as functions of pressure (in GPa) for Al_3Ge (a), and AlGe (b).

With these EPI's, one can deduce from the ground-state analysis of the Ising model the most probable ordered structures at zero temperature at any given pressure. For fcc-based systems, the most extensive analysis was performed with an Ising model which includes the first four pair interactions [17]. The final analysis leads to the following conclusions. For Al_3Ge the ground state is of the [1 1/2 0] family, namely $D0_{22}$, of Al_3Ti-type, up to about 40 GPa, with a possible change of order to the [100] family, *i.e.*, $L1_2$ of Cu_3Au-type at higher pressure. This is most easily analyzed by studying the variation of the (001)/a/2[110] APB energy with pressure. Up to the fourth neighbor EPI, this energy, denoted γ, is simply given by the ordering energy difference between $D0_{22}$ and $L1_2$, *i.e.*, $\gamma=-V_2/4+V_3-V_4$ (by definition, a positive γ favors the $L1_2$-type of ordering). As apparent in Figure 2a, the change of sign for γ indicates the possible existence of long period superstructures (LPS's) below 40 GPa, and of $L1_2$ at higher pressure.

For AlGe, phase "40" (see Ref.[17] for a description of this ordered phase) is the most stable superstructure in the domain of pressure considered here (*i.e.*, up to 50 GPa), and the competing ordered states of the [100] family ($L1_0$ of CuAu I-type) and of the 1/2[111] family ($L1_1$) are very unlikely around this stoichiometry (see Figure 2b).

When addressing the effect of a hydrostatic pressure on the stability properties of materials, the relevant quantity to deal with is the Gibbs free energy, given by $G=E_T+PV-TS$, where E_T, P, V and S refer to total internal energy, pressure, (atomic) volume, and entropy, respectively (note that, at T=0 K, this energy is simply equal to the enthalpy, $H=E_T+PV$).

For pure Ge, the energetics of its three modifications, namely, diamond, β-Sn and fcc, as derived from *ab initio* pseudo-potential calculations [18] were rescaled by using the KKR results for fcc-based Ge in order to define a common zero of energy for the pure elements and the alloy at various compositions. For pure Al, we only considered the energetics of its fcc-based structure in the range of pressure from 0 to 50 GPa, and ignored the possible structural transformation from fcc to hcp at about 20.5 GPa [19].

With these two known endpoints, we can calculate the Gibbs free energy as a function of lattice parameter, and therefore pressure, for both Al_3Ge and AlGe in any ordered alloy configuration, from Eq. (2), and also in the disordered state, from Eq. (3). The results are displayed in Figures 2a and b for both alloy compositions. The sum of the concentration weighted average Gibbs free energies of the pure fcc-based species at a given pressure, is taken as the reference energy so that

$$\Delta G^{\text{Alloy}}(P) = G_T^{\text{Alloy}}(P) - (1-c)G_T^{\text{Al,fcc}}(P) - cG_T^{\text{Ge,fcc}}(P) \tag{6}$$

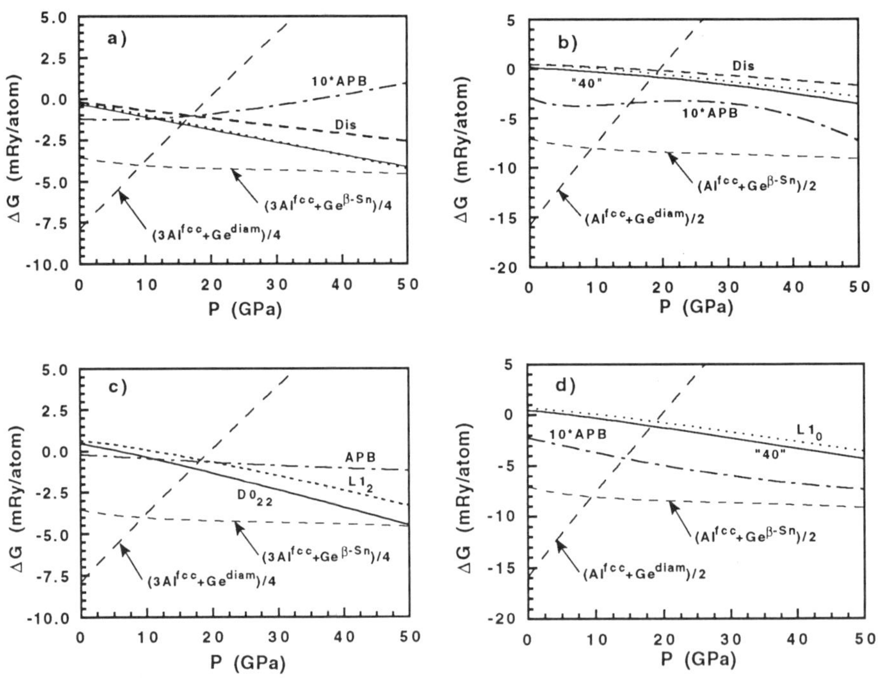

Figure 2. Gibbs free energies (in mRy/atom), at zero temperature, of fcc-based (a,c) Al_3Ge (solid line: $D0_{22}$; dotted line: $L1_2$; dashed line: disordered state; dashed-dotted line: APB energy multiplied by 10), and (b,d) AlGe (solid line: "phase 40"; dotted line: $L1_0$, dashed line: disordered state), as a function of pressure (in GPa). The calculations are performed within the KKR-CPA-GPM approach (a and b) and the LMTO-ASA approach (c and d). The sum of the concentration weighted average Gibbs free energies of the pure species (see text) is taken as the reference energy, at any given pressure.

It is clear from Figure 2 that, at zero pressure, the mixing energy (*i.e.*, the tendency -X(1/2 0 0)-M(1/2 1/2 0)-X(0 1/2 0)-Γ (in units of $2\pi/a$), unfolded in the irreducible wedge of the Brillouin zone of the corresponding disordered alloy: Γ-X (100)-W(1 1/2 0)-K(3/4 3/4 0)-X(0 1 0)-Γ. Note, in particular, the flat portion of Fermi surface perpendicular to the [110] direction, in reciprocal space. The value of M calculated for this portion is equal to about 1.20, at zero pressure, and slightly decreases with pressure to 0.92 at 30 GPa and 0.86 at 50 GPa. These values of M are compatible with the existence of SRO of $D0_{22}$-type. Since the Fermi level occurs in the midst of the Al-Ge hybridized bands of Al and Ge, mostly s and p in character, it is not surprising that we find the origin of the flat portion of Fermi surface in pure Al itself. And the addition of Ge to the Al matrix is responsible for flattening out this feature and moving it along the [110] direction so that a $D0_{22}$-LPS is stabilized with $M \approx 1$.

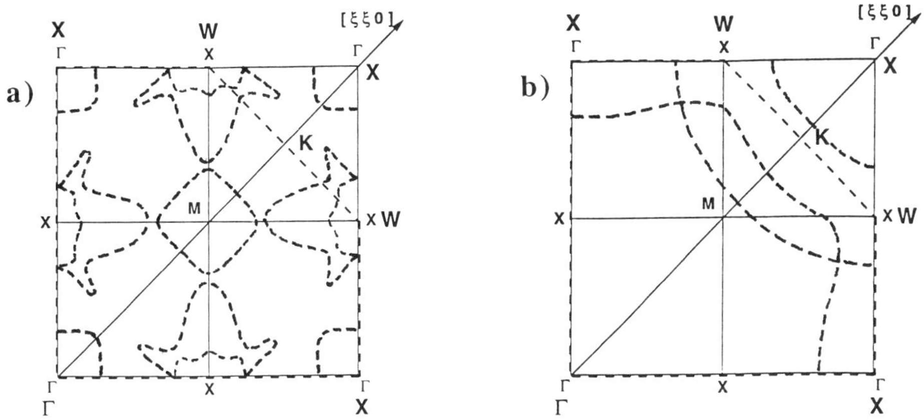

Figure 3. a) Projection of the Fermi surface in the (001) plane of a $L1_2$-Al_3Ge alloy, as obtained from LMTO-ASA, represented in the irreducible wedge of the simple cubic Brillouin zone (solid lines) unfolded in the irreducible wedge (dashed lines) of the fcc Brillouin zone. b) Maximum of the Bloch spectral function in the (001) plane, at the Fermi energy, of a disordered fcc-based Al_3Ge alloy, as obtained from KKR-CPA, represented in the irreducible wedge of the fcc Brillouin zone (dashed lines).

In the case of the chemically disordered alloy, what remains of the Fermi surface is described by the so-called Bloch spectral function $A(\mathbf{k},\varepsilon)$ which has sharp peaks in regions of wavevector **k** and energy ε where there are well defined bands. We show in Figure 3b the maximum of the Bloch spectral function at the Fermi energy in the (001) projection of the fcc Brillouin zone represented by dahed lines. Clearly, there are well defined peaks whose loci exhibit a relatively flat section normal to the Γ-K direction. This section is part of a flat sheet of Fermi surface. The spanning vector between two flat such portions of Fermi surface gives a M value which ranges from 1.0 at zero pressure to 0.8 at 50 GPa. From a close inspection of Figures 3a and 3b, we conclude that there is a one to one correspondence between the remaining features of the Fermi surface of the disordered alloy and the ones which were computed for the ordered alloy. And the extra features in the Fermi surface of the ordered alloy are the result of a small variation in the Fermi energy and of the folding process in an irreducible Brillouin zone which differs from the one of the disordered alloy.

CONCLUSION

Using first-principles electronic structure approaches, the effect of pressure on stability and order was examined in the case of Al-Ge substitutional alloys. Three main conclusions have been drawn. First, pressure prepares the semiconductor element, Ge, to lie in substitutional position in a fcc-Al matrix by adjusting the volume of Ge closer to the one of Al, therefore reducing the lattice mismatch which exists at ambient pressure. Second, pressure favors the alloy formation, *i.e.*, the mixing energy becomes more negative with increasing pressure. Third, pressure also enhances the ordering tendencies, leading, for toward alloy formation) and the ordering tendencies are rather weak, and the feature which dominates the existence of the miscibility gap is the huge structural energy difference between the diamond structure of Ge and its fcc-modification (about 33.8 mRy/atom). By increasing pressure, the structural energy difference between the β-Sn modification of Ge, which occurs above 9.6 GPa, and the fcc crystalline structure remains practically constant and equal to about 16 mRy/atom, up to at least 50 GPa. Thus, at 25 at. % Ge for example, the energy necessary to compensate for this structural energy difference is only 4 mRy/atom at 30 GPa (see Figure 2a).

In addition to this "endpoint effect", the ordering tendencies in the alloy increase substantially with pressure, resulting in the possible formation not only of a fcc-based solid solution at low Ge concentration, but also of the appearance of a definite short-range order (SRO). This SRO should be revealed by the existence of diffuse peaks in the diffraction pattern, located at the $2\pi/a(1\ 1/2\ 0)$ positions, in reciprocal space, which are associated with a SRO of the $D0_{22}$-type, up to 40 GPa, and at the $2\pi/a(100)$ positions which are the signature of the $L1_2$, at higher pressure.

Since the magnitude of the EPI's which drive the ordering tendencies is relatively weak, the order-disorder temperature at 25 at. % Ge is below room temperature. Indeed, a statistical treatment based on the application of the cluster variation method (CVM) in the tetrahedron-octahedron approximation [20], with an ordering energy described by the first and second EPI's computed at a pressure of 30 GPa, gives an order-disorder transition temperature of about 170 K. Therefore, even if the existence of a long-range order is unlikely, the SRO should be revealed quite easily, especially because of the first order nature of the order-disorder transformation. Also, note that, as is usually the case experimentally, the kinetics tends to favor more the formation of an ordered state, even in a metastable regime, than of phase separation. The ordering trends follow the same pattern at higher Ge concentration but, since the energy which is required to compensate for the structural energy difference of pure Ge becomes increasingly more important, the formation of a fcc-solid solution (or a superstructure) is less likely to occur. Part of these results were experimentally observed, but there is not yet any report on the observation of SRO of $D0_{22}$-type at low Ge concentration. Experimental studies to confirm such predictions are highly desirable.

Guided by the results obtained within the KKR-CPA-GPM approach, a series of possible ordered states was selected, namely $L1_2$, $D0_{22}$ and $D0_{23}$, and $L1_0$ and phase "40", at 25 and 50 at. % Ge. From the total energy calculations based on the LMTO-ASA, the equilibrium properties and the equation of states for each of these ordered states were obtained. In our calculations, the 3d, 4s and 4p of Ge, and the 3s, 3p and 3d of Al were treated in the band mode. The radii of all atomic spheres were taken to be the same. The equations of states V=f(P), where V is the atomic volume, for all ordered structures obtained from LMTO-ASA agree extremely well with the ones derived from KKR-CPA-GPM. Within the LMTO approach, the parameters which define the equation of states for each ordered configuration can be used to calculate the Gibbs free energy as a function of lattice parameter, and therefore pressure, at both atomic compositions of the Al-Ge system. The results are displayed in Figures 2c and d with reference, to the sum of the concentration weighted average Gibbs free energies of the pure species at any given pressure for both Al_3Ge and AlGe. In slight contrast with the KKR-CPA-GPM results, $D0_{22}$ and phase "40" are the only possible fcc-based ground-states at 25 and 50 at. % Ge, respectively, in the whole range of pressure considered here. Overall, both approaches clearly indicate that pressure enhances the stability of ordered phases in this alloy system.

ELECTRONIC ORIGIN OF THE ORDERING TENDENCIES

According to Sato and Toth, Krivoglaz and Moss [21], the origin of a LPS may be found in the existence of parallel flat sections of Fermi surface spanned by the vector $2\mathbf{k}_F$ which is related to the inverse of the period of modulation M of the LPS according to

$$M = [2\sqrt{2}(\sqrt{2} - 2k_F)]^{-1} \qquad (7)$$

where k_F is measured along the [110] direction and expressed in units of $2\pi/a$, a being the lattice parameter of the fcc structure.

Let us first examine the Fermi surface of a hypothetical ordered $L1_2$-Al_3Ge alloy, as obtained from LMTO-ASA. In Figure 3a, we show, at zero pressure, its projection in the (001) plane in the irreducible simple cubic Brillouin zone, *i.e.*, the wedge defined by Γ(0 0 0) example, to the possible existence of a metastable fcc-based order state of the Al_3Ti-type, or $D0_{22}$, around Al_3Ge. We suggest that the existence of flat sheets of Fermi surface accounts for the stabilization of $D0_{22}$, one of the simplest long period superstructure, at 25 at. % Ge, and this specific topology of Fermi surface originates from the one of pure Al. Similar results would have been obtained at higher Ge concentration; however, the energetics becomes less and less favorable to justify the possible stability (or even metastability) of an ordered fcc-based alloy, under pressure.

Acknowledgments
This work was performed under the auspices of the U.S. Department of Energy by the Lawrence Livermore National Laboratory under contract W-7405-ENG-48 (PT and PS), and under contract DEAC05-840R21400 at Oak Ridge National Laboratory with Martin Marietta Energy Systems Inc. (GMS).

REFERENCES

[1] V. F. Degtyareva and E. G. Ponyatowskii, Sov. Phys. Solid State **24** (9),1514 (1982).
[2] J. F. Cannon, J. Phys. Chem. Ref. Data **3**, 781 (1974); C. W. F. T. Pistorius, Prog. Solid State Chem **11**, 1 (1976); L. Merrill, J. Phys. Chem. Ref. Data **6**, 1205 (1977).
[3] H. Mii, M. Senoo, and I. Fujishiro, Japanese J. of Appl. Phys. **15**, 777 (1976).
[4] S. M. Banova, I. A. Korsunskaya, G. M. Kuznetsov, and V. A. Sergeyev, Phys. Met. Metall. **46**, 58 (1979).
[5] V. F. Degtyareva, G. V. Chipenko, E. G. Ponyatovskii, and V. I. Rashchupkin, Sov. Phys. Solid State **26** (4), 733 (1984); V. F. Degtyareva, G. V. Chipenko, I. T. Belash, O. I. Barkalov, and E. G. Ponyatovskii, Phys. Stat. Sol. (a) **89**, K127 (1985).
[6] O. I. Barkalov, I. T. Belash, V. F. Degtyareva, and E. G. Ponyatovskii, Sov. Phys. Solid State **29** (7), 1138 (1987).
[7] A. J. McAlister and J. L. Murray, Bull. of Alloy Phase Diagrams **5**, 341 (1984); H. Okamoto, J. of Phase Equilibria **14**, 118 (1993); J. L. Murray and A. J. McAlister, Bull. of Alloy Phase Diagrams **5**, 74 (1984).
[8] M. Itigari, B. C. Giessen, and N. J. Grant, Trans ASM **61**, 330 (1968).
[9] V. F. Degtyareva, I. T. Belash, and E. G. Ponyatovskii, Phys. Stat. Sol. (a) **124**, 465 (1991); see also E. G. Ponyatovskii in these Proceedings.
[10] P. E. A. Turchi, "*Electronic Theories of Alloy Phase Stability*", published in "*Intermetallic Compounds: Principles and Practice*", J. H. Westbrook and R. L. Fleischer eds. (John Wiley & Sons, Ltd., New York, 1995), vol. **1**, Chap. 2, pp. 21-54 ; and references cited therein.
[11] H. L. Skriver, in *The LMTO Method*, Springer series in Solid-State Sciences, vol. **41** (Springer, Heidelberg, 1983); O. K Andersen, O. Jepsen, and D. Glötzel, in *Highlights of Condensed Matter Theory*, F. Bassani, F. Fumi, and M. P. Tosi (eds.) (North-Holland, Amsterdam, 1985), p.59.
[12] U. von Barth and L. Hedin, J. Phys. C**5**, 1629 (1972).
[13] F. Ducastelle and F. Gautier, J. Phys. F**6**, 2039 (1976).
[14] P. E. A. Turchi, G. M. Stocks, W. H. Butler, D. M. Nicholson, and A. Gonis, Phys. Rev. B**37**, 5982 (1988).
[15] J. S. Faulkner, Progress in Materials Science **27**, 1 (1982); and references cited therein.
[16] J. Birch, J. Geophys. Res. **83**, 1257 (1978); F. D. Murnaghan, Proc. Natl. Acad. Sci. USA **30**, 244 (1944).
[17] J. Kanamori and Y. Kakehashi, J. Phys. (Paris) **38**, C7-274 (1977).
[18] M. T. Yin and M. L. Cohen, Phys. Rev. B**26**, 5668 (1982); K. J. Chang and M. L. Cohen, Phys. Rev. B**34**, 8581 (1986).
[19] N. N. Roy and E. G. Steward, Nature (London) **224**, 905 (1969).
[20] J. M. Sanchez and D. de Fontaine, Phys. Rev. B**21**, 216 (1980).
D. de Fontaine and J. Kulik, Acta Metall. **33**, 145 (1985).
[21] H. Sato and R. S. Toth, Phys. Rev. **124**, 1833 (1961); M. A. Krivoglaz, in *Theory of X-Ray and Thermal Neutron Scattering by Real Crystals*, (Plenum Press Inc., New York, 1969); S. C. Moss, Phys. Rev. Lett. **22**, 1108 (1969).

PREDICTION OF UNSUSPECTED ORDERING TENDENCIES IN Pd-Pt AND Rh-Pt ALLOYS

Barry M. Klein,[a] Z. W. Lu,[a] and Alex Zunger[b]

[a]Department of Physics, University of California, Davis, CA 95616
[b]National Renewable Energy Laboratory, Golden, CO 80401

INTRODUCTION

In 1959, Raub[1] suggested the existence of miscibility gaps in the Pd-Pt and Rh-Pt systems from the assumed correlation between the maximum miscibility gap temperature and the difference in melting points of the constituents, and from the experimentally observed miscibility gaps in "similar alloys" such as Pd-Rh, Ir-Pt, and Ir-Pd.[1] Attempts to observe these miscibility gaps have, however, consistently failed.[2–5] Phenomenological theories[6,7] are not conclusive on whether Pd-Pt and Rh-Pt should phase-separate or order: Miedema's model[6] predicts $\Delta H \sim \pm$ (1–3) kJ/mol for Pd-Pt and Rh-Pt, a value being close to the model's limit of accuracy. Thus, neither experiment[2–5] nor simple phenomenological theories[6,7] shed light on whether these systems order or not. Yet, all phase-diagram compilations including the latest assessments by Massalski *et al.*,[8] have adopted Raub's view, labeling Pd-Pt and Rh-Pt as phase-separating ("miscibility gap") systems.

In this paper we show, using a combination of *ab initio* calculations in conjunction with "cluster expansion" methods, that Pd-Pt and Rh-Pt are expected to exhibit ordering[9] rather than phase separation[1,8] at low temperatures. This conclusion is reinforced by experimental evidence[10–13] that seems to have been overlooked by the alloy phase diagram community. For Pd-Pt and Rh-Pt, the situation is similar to that in the Ag-Au system, where an ordering tendency has been recognized by ample experiments[14] and theoretical calculations,[15] although the ordering temperatures appear to be too low to grow ordered compounds. An experimental re-examination of the phase stability of Pd-Pt and Rh-Pt appears called for.

CALCULATIONAL METHODS

Cluster-Expanding Lattice Quantities

The excess energy $\Delta E(\sigma, V)$ of a configuration σ of an $A_{1-x}B_x$ alloy is given by:

Stability of Materials
Edited by A. Gonis *et al.*, Plenum Press, New York, 1996

$$\Delta E(\sigma, V) = E(\sigma, V) - [(1-x)E_A(V_A) + xE_B(V_B)] \ . \tag{1}$$

where V is the volume of the structure σ, and $E_A(V_A)$ and $E_B(V_B)$ are the energies of the pure solids A and B at their respective equilibrium volumes V_A and V_B. The formation energy is thus $\Delta H(\sigma) = \Delta E(\sigma, V_\sigma)$, where V_σ is the equilibrium volume of configuration σ. Whether a system will order or phase-separate is determined by the sign of the formation energy. ΔH can be accurately determined using the density functional formalism,[16] as implemented numerically by, for instance, the highly precise full-potential linearized augmented plane wave (LAPW) method.[17] However, for a binary system, quantum-mechanical calculations for 2^N (N is the number sites in the lattice) configurational energies $\Delta E(\sigma, V)$ is prohibitive. An approach to circumvent this problem is to perform such a "ground state search" by expanding Eq. (1) in a finite Ising-like "cluster expansion" (CE) fit to first-principles calculations of the form,[18]

$$E_{\mathrm{CE}}(\sigma, V) = \sum_f D_f J_f(V) \overline{\Pi}_f(\sigma) \ , \tag{2}$$

where $J_f(V)$ are volume-dependent interaction energies of some basic lattice "figures" f (e.g., nearest-neighbor pairs, next neighbor pairs, triangles, etc.). The "lattice averaged spin products" $\overline{\Pi}_f(\sigma)$ are the product over the figure f of the variables $\hat{S}_i$, averaged over all symmetry equivalent figures of the lattice. All of the terms on the right-hand side of Eq. (2) are trivially determined geometrical quantities, with the exception of $\{J_f\}$, the effective interaction energies, which are determined by fitting to $O(10)$ *ab initio* total energy calculations (details may be found in Refs. 15 and 19).

Having obtained a converged set of interactions, the $T = \infty$ mixing enthalpy of the random alloy can be readily evaluated: Denoting a configurational average for the random (R) state by angular brackets $\langle \ \rangle_R$ we obtain,

$$\Delta H_{\mathrm{mix}}(x) = \sum_f D_f J_f \langle \overline{\Pi}_f \rangle_R = \sum_f D_f J_f (2x-1)^{k_f} \ , \tag{3}$$

where $\langle \overline{\Pi} \rangle_R = (2x-1)^{k_f}$ and k_f is the number of vertices in figure f (e.g., $k_f = 2$ for pairs, 3 for triangles). The "ordering energy" $\delta E_{\mathrm{ord}}(\sigma)$, is defined as the difference between the formation enthalpy $\Delta H_{\mathrm{F}}(\sigma)$ of an ordered configuration σ and the mixing enthalpy $\Delta H_{\mathrm{mix}}(x)$ of a random alloy at the same x:

$$\delta E_{\mathrm{ord}}(\sigma) = \sum_f D_f J_f [\overline{\Pi}_f(\sigma) - (2x-1)^{k_f}] \ . \tag{4}$$

The order-disorder transition temperature can be calculated using the cluster-variational method (CVM)[20] or Monte-Carlo simulations.[21,22]

The Warren-Cowley SRO parameter[23] for the N-th atomic shell at distance R_{lmn} from the origin is

$$\alpha_{\mathrm{SRO}}(R_{lmn}) = \frac{\langle \overline{\Pi}_{0,N} \rangle - q^2}{1 - q^2} \ , \tag{5}$$

where $q = 2x - 1$ and the angular bracket denotes a configurational average. Note that $\alpha_{\mathrm{SRO}}(0) \equiv 1$ by definition. The Fourier transform of the real-space SRO is $\alpha_{\mathrm{SRO}}(\mathbf{k})$, which is proportional to the diffuse intensity due to SRO. Its value depends on the number, N_R, of real-space shells used in the transform:

$$\alpha_{\mathrm{SRO}}(\mathbf{k}, N_R) = \sum_N^{N_R} \alpha_{\mathrm{SRO}}(R_{lmn}) e^{i\mathbf{k}\cdot\mathbf{R}_{lmn}} \ . \tag{6}$$

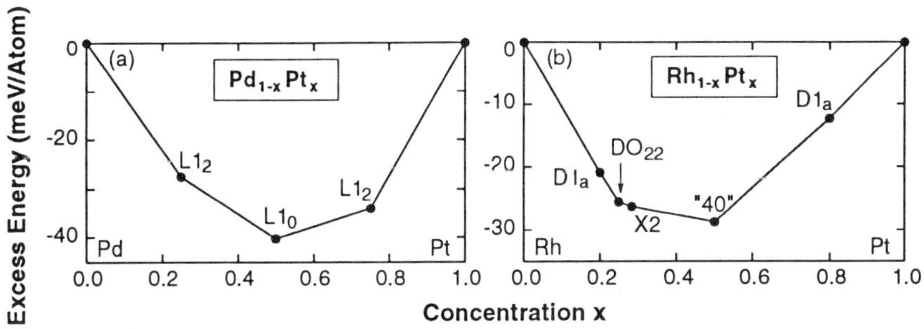

Fig. 1. Predicted $T = 0$ ground states (circles) as a function of composition for (a) Pd-Pt and (b) Rh-Pt systems.

Calculating the Cluster Expansion Interaction Energies

We have calculated the excess energy $\Delta E_{\mathrm{direct}}(\sigma, V)$ *vs* volume V for 12 ordered structures each in Pd-Pt and Rh-Pt using the full-potential LAPW method.[17] Here, we use the Wigner[24] form of exchange-correlation function. We avoid any shape approximation (e.g., muffin-tin) to the potential and charge density or any elimination of a subset of states (e.g., the pseudopotential approximation). The core states are treated fully-relativistically while the valence states are treated scalar relativistically.

The calculated formation enthalpies and fitted interaction energies, $\{J_f\}$, determined via Eq. (2), are given in the Tables of 1 and 2 of Ref. 15.

RESULTS AND DISCUSSIONS

Low-Temperature Long-Range Order

Pd-Pt: Pd-Pt and Rh-Pt all exhibit negative formation enthalpies ΔH_{F} for the 12 structures studied, thus unequivocally indicating ordering rather than phase-separating tendencies. Concomitantly, the dominant interaction energies, J_2, are positive, also indicating ordering tendencies. The predicted $T = 0$ ground states for $\mathrm{Pd}_{1-x}\mathrm{Pt}_x$ are shown in Fig. 1(a), where the vertices represent the $\mathrm{L1}_2$, $\mathrm{L1}_0$, and $\mathrm{L1}_2$ structures for $x = \frac{1}{4}$, $\frac{1}{2}$, and $\frac{3}{4}$, respectively. These ordered structures are characterized by a reciprocal-space wave-vector of $\mathbf{k} = \frac{2\pi}{a}\langle 001 \rangle$, where a is the lattice constant of the underlying fcc lattice. The peaks of the calculated high-temperature diffuse-scattering intensity due to SRO also occur at these wavevectors.

Rh-Pt: Figure 1(b) shows the predicted $T = 0$ ground states for $\mathrm{Rh}_{1-x}\mathrm{Pt}_x$. They are all superlattices along the [012] direction. These are denoted as $\mathrm{D1}_a$, DO_{22}, X2 ($\mathrm{Au_5Mn_2}$ type), "40", and $\mathrm{D1}_a$ for $x = \frac{1}{5}$, $\frac{1}{4}$, $\frac{2}{7}$, $\frac{1}{2}$, and $\frac{4}{5}$, respectively. Many of these predicted ground states ($\mathrm{D1}_a$ and $\mathrm{Au_5Mn_2}$, see Fig. 4 in Ref. 15) are absent from our input set (used to extract the interaction parameters), since such structures were not suspected to be stable. These ordered structures are characterized by a reciprocal-space wave-vector of $\mathbf{k} = \frac{2\pi}{a}\langle 1\frac{1}{2}0 \rangle$. Below we show that these indeed correspond to maxima in the calculated diffuse scattering intensity.

The order-disorder transition temperatures T_{c} for the ground state is found from Monte-Carlo simulated annealing procedures gives the $T_{\mathrm{c}}^{\mathrm{MC}} \sim 210$ K for $L1_0$ (PdPt) and "40" (RhPt). At such low transition temperatures, the atomic diffusion rates are

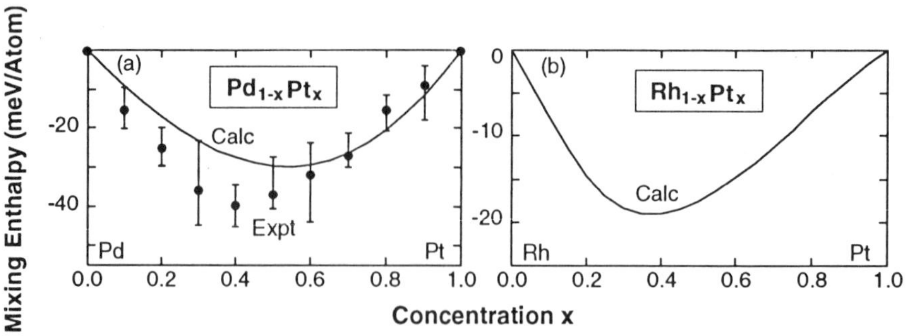

Fig. 2. Calculated and measured (where available) mixing energy for (a) $Pd_{1-x}Pt_x$ at $T = 300$ K and (b) $Rh_{1-x}Pt_x$ at $T = 300$ K. The measured data are cited from Ref. 12 for Pd-Pt.

clearly too slow to actually form these ordered ground states. Only solid solutions across the whole composition ranges can thus be observed. Of course, the observation of solid solutions above T_c is consistent with *both* phase-separation *or* ordering at lower temperature. Our theory suggests unequivocally that the equilibrium structures should be ordered. We will see that these should be detectable in the measured mixing enthalpies and SRO at higher temperatures.

Mixing energy of disordered alloys at finite temperatures

Figure 2 compares the calculated and available experimental mixing enthalpies[12] at finite temperatures. The calculated results were obtained using the cluster variation method (CVM). The negative signs of the mixing enthalpies for Pd-Pt and Rh-Pt indicate ordering tendencies of these alloys rather than phase-separation.

Pd-Pt: The calculated $\Delta H_{\rm mix}(x)$ for Pd-Pt agrees reasonably well with experiment.[12] The negative experimental mixing energy indicating an ordering tendency was, unfortunately, overlooked by the phase-diagram assessment community.

Rh-Pt: We are not aware of any measured $\Delta H_{\rm mix}$ data for solid solutions of Rh-Pt. However, the enthalpy of mixing of *liquid* $Rh_{1-x}Pt_x$ was recently estimated to be $\Delta H^{\rm L}_{\rm mix}(x) = -6x(1-x)$ kJ/mol, which was derived from the measured liquidus and solidus[13] together with a semi-empirical model.[25] Thus, $\Delta H^{\rm L}_{\rm mix} = -1.50$ kJ/mol$=$ -15.6 meV/atom at $x = \frac{1}{2}$, values that are consistent in sign and magnitude with our calculated mixing enthalpy (-11.8 meV/atom when $T \to \infty$ and -17.6 meV/atom when $T = 300$ K) of the solid solutions. Hence, accurate measurements on mixing energies of solid solution for Rh-Pt is called for to compare with our theoretical prediction in Fig. 2(b).

High-temperature short-range order

Pd-Pt Kidron[10] measured the SRO in a polycrystalline $Pd_{0.49}Pt_{0.51}$ sample annealed at 873 K. He found a *negative* $\alpha(R_{110}) = -0.10$ indicating an ordering tendency. Unfortunately, this work, as well as the subsequent measurements of $\Delta H_{\rm mix}(x)$[11,12] showing negative values, seems to have largely been overlooked by phase-diagram compilations that characterized this system as having a miscibility gap. Our calculations

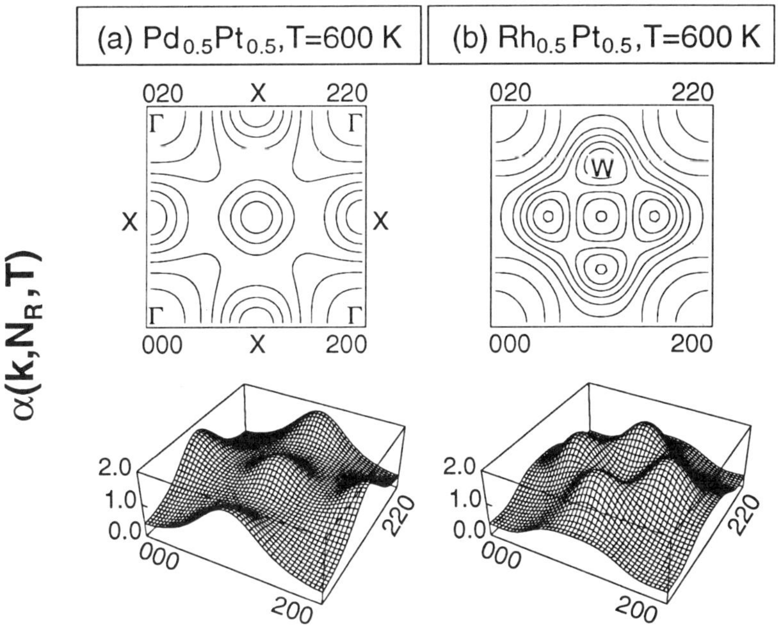

Fig. 3. Calculated diffuse scattering intensity due to short-range order $\alpha(\mathbf{k}, N_R)$ for (a) $Pd_{0.5}Pt_{0.5}$ (T = 600 K) and (b) $Rh_{0.5}Pt_{0.5}$ (T = 600 K). $\alpha(\mathbf{k}, N_R)$ were Fourier synthesized using $N_R = 13$ real-space $\alpha(R_{lmn})$ (including $\alpha(R_{000}) \equiv 1$). Note that Pd-Pt and Rh-Pt have peaks off the $\Gamma = \frac{2\pi}{a}\langle 000 \rangle$ point indicating ordering tendencies.

give $\alpha(R_{110}) = -0.040$ at $T = 873$ K for $Pd_{0.5}Pt_{0.5}$. Since the SRO parameters are very sensitive to the sample condition and heat treatment, careful SRO measurements on single crystal Pd-Pt and Rh-Pt samples are highly desirable. The predicted $\alpha(\mathbf{k})$ at T = 600 K for $Pd_{0.5}Pt_{0.5}$ is presented in Figure 3(a) and is awaiting experimental comparison. The maximum diffuse intensities occur at the $X = \frac{2\pi}{a}\langle 001 \rangle$ points for $Pd_{0.5}Pt_{0.5}$ corresponding to the wavevectors of the $T = 0$ predicted $L1_0$ ordered ground state.

Rh-Pt The predicted $\alpha(\mathbf{k})$ at T = 600 K for $Rh_{0.5}Pt_{0.5}$ is shown in Figure 3(b), which peaks at the $W = \frac{2\pi}{a}\langle 0\frac{1}{2}1 \rangle$ points for $Rh_{0.5}Pt_{0.5}$, corresponding to the wavevectors of the lower-temperature ordered ground states (structure "40"). Since the structure "40" rarely occurs in nature, it would be very interesting to see experimentally if the SRO indeed peaks at the $\frac{2\pi}{a}\langle 0\frac{1}{2}1 \rangle$ points for $Rh_{0.5}Pt_{0.5}$.

CONCLUSIONS

Using first-principles quantum-mechanical methods, which utilize *ab initio* LAPW total-energy calculations and a cluster-expansion approach, we have shown that Pd-Pt and Rh-Pt binary alloy systems will order, rather than phase-separate as assumed by many compilations of binary phase-diagram collections. However, the ordering temperatures for these two systems are rather low. In the case of Pd-Pt, our calculated mixing energy is in fair agreement with experiment. We further predict that $Pd_{1-x}Pt_x$ will order in the $L1_2$, $L1_0$, and $L1_2$ structures at $x = \frac{1}{4}$, $\frac{1}{2}$, and $\frac{3}{4}$, respectively, while ordered structures in $Rh_{1-x}Pt_x$ are all superlattices stacked along the [012] directions. Since the ordering temperatures are so low (below room temperature), these ordered phases may never be observed. However, peak positions in the measured SRO intensity would lend support to the existence of these ordered low-temperature LRO structures. Hence, accurate SRO measurements on single-crystal Pd-Pt and Rh-Pt as well as measured mixing energies for solid solution of Rh-Pt are highly desirable.

Acknowledgments

BMK and ZWL acknowledges support from University Research Funds of the University of California at Davis. AZ acknowledges support from the Office of Energy Research (OER) [Division of Materials Science of the Office of Basic Sciences (BES)], U. S. Department of Energy, under the contract No. DE-AC36-83CH10093.

REFERENCES

1. E. Raub, *J. Less Common Met.*, **1**, 3-18 (1959).
2. E. Raub and G. Falkenburg, *Z. Metallkd.* **55**, 392-397 (1964).
3. V. E. Antonov, T. E. Antonova, I. T. Belash, E. G. Ponyatovshii, and V. I. Rashupkin, *Phys. Stat. Sol.* (a) **78**, 137-146 (1983).
4. S. R. Bharadwaj, A. S. Kerkar, S. N. Tripathi, and S. R. Dharwadkar, *J. Less-Common Met.* **169** 167-172 (1991).
5. H. Noh, T. B. Flanagan, and Y. Sakamoto, *Scripta Metall. et. Mat.* **29**, 445-450 (1993).
6. *Cohesion in Metals: Transition Metal Alloys* F. R. de Boer *et al.* (Amsterdam, North Holland, 1988).
7. D. G. Pettifor, *New Scientist*, **110**(1510), 48-53 (1986).
8. *Binary Alloy Phase diagrams*, edited by T. B. Massalski *et al.* (ASM International, Materials Park, Ohio, 1990), see Vol. 3, p. 3033-3034 for Pd-Pt and p. 3020, and 3022 for Rh-Pt.
9. Z. W. Lu, S.-H. Wei, and A. Zunger, *Phys. Rev. Lett.*, **66**(13) 1753-1756 (1991).
10. A. Kidron, *Phys. Lett.*, **25**A, 112-113 (1967).
11. F. H. Hayes and O. Kubaschewskil, *Metal Sci. J.*, **5**, 37 (1971).
12. J. B. Darby and K. M. Myles, *Metall. Trans.* **3**, 653-657 (1972).
13. G. M. Kuznetsov, E. I. Rytvin, and I.V. Nikomova, *Izv. Akad. Nauk SSSR, Met.* **4**, 194-197 (1985) in Russian; TR: *Russ. Metall.* **4**, 189-192 (1984).
14. *Phase Diagrams of Binary Gold Alloys*, edited by H. Okamoto and T. B. Massalski, (ASM International, Materials Park, Ohio, 1987), p. 76.
15. Z. W. Lu, B. M. Klein, A. Zunger, *Journal of Phase Equilibria*, **16**, 36 (1995).
16. W. Kohn and L. J. Sham, *Phys. Rev.* **140**, A1133 (1965).
17. D. J. Singh, *Planewaves, Pseudopotentials, and the LAPW Method*, (Kluwer, Boston, 1994).
18. J. M. Sanchez, F. Ducastelle, and D. Gratias, *Physica* **A 128**, 334 (1984); J. M. Sanchez, *Phys. Rev. B* **48**, 14013 (1993).
19. A. Zunger, in *Statistics and Dynamics of Alloy Phase Transitions*, edited by P .E. A. Turchi and A. Gonis, NATO ASI Series (Kluwer, Dordrecht, 1994).
20. R. Kikuchi, *Phys. Rev.* **81**, 988 (1951).
21. K. Binder and D. W. Heermann, *Monte Carlo Simulations in Statistical Physics* (Springer-Verlag, Berlin, 1988).
22. Z. W. Lu, D. B. Laks, S.-H.Wei, and A. Zunger, *Phys. Rev. B* **50**, 6642 (1994).
23. J. M. Cowley, *J. Appl. Phys.* **21**, 24 (1950).
24. E. Wigner, *Phys. Rev.* **46**, 1002 (1934).
25. A. K. Niessen, F. R. de Boer, R. Boom, P. F. de Chatel, W. C. M. Mattens, and A. R. Miedema, CALPHAD **7**(1), 5 1-70 (1983).

ENERGETIC EFFECTS IN THE Au-Ni SYSTEM

T. Deutsch, [1] A. Pasturel [2]

[1]Département de Recherche Fondamentale sur la Matière Condensée,
CEA-CENG, SP2M/MP, 38054 Grenoble Cedex9, FRANCE.
[2]Laboratoire de Physique Numérique, Maison des Magistères, CNRS,
BP 166 38042 Grenoble Cedex9, FRANCE.

INTRODUCTION

The solid-state portion of the Au-Ni phase diagram is characterized by an extended asymmetrical miscibility gap at low temperatures. The critical point of the miscibility gap is at about 70.6 % Ni and 1083.3 K[1]. However despite its phase separation at low temperatures, NiAu exhibits $L1_0$-type short-range order (SRO) at high temperatures[2]. From STM (Scanning Tunneling Microscopy) studies in the Au/Ni(110) system[3], it has been shown that deposited Au atoms replace surface Ni atoms forming a Au Ni surface alloy. For Ni/Au(001) system, very recent experimental studies based on HREM (High Resolution Electronic Microscopy) and X-Ray diffractions[4, 5] have revealed an intermixing of Au and Ni atoms at the interface. Then the structural properties of the Au-Ni system are manifested through both a phase-separating LRO (long-range order) and an ordering-type SRO. The sign of the formation energy of compounds, $\Delta E(\sigma_x)$, indicates if the low-temperature LRO corresponds to compound formation (when $\Delta E(\sigma_x)<0$) or to phase separation (when $\Delta E(\sigma_x)>0$). The type (ordering or clustering) of high-temperature SRO is given by the sign of the ordering energy, ΔE_{ord}, defined as the difference between the formation energy $\Delta E(\sigma_x)$ of the ordered compound A/B in structure σ and the mixing energy, $\Delta E_{mix}(x)$ of a random alloy A_xB_{1-x} of composition x. If $\Delta E_{ord} <0$, the random alloy would develop SRO of the type underlying the structure σ. In this paper we decide to investigate the structural properties of some Au-Ni alloys with use of all-electron total-energy full-potential linear-muffin-tin-orbitals method. We present results for bulk alloys as well as for the Au(001)/Ni system. Both the elastic and chemical interactions are shown to be essential towards obtaining a correct description of the structural properties in the Au-Ni system. This scheme is also used to test the results provided by numerical simulations based on tight-binding second-moment approximation potentials.

COMPUTATIONAL MODELS

The two central energetic quantities used in our discussion are the formation energy $\Delta E(\sigma_x)$ and the mixing energy $\Delta E_{mix}(x)$ or their difference.Such quantities are calculated using two energetic models. The first one is based on first-principles local density calculations while the second one is developed within the second moment approximation of the tight-binding description of the metallic cohesion.

In the first approach, the formation energy $\Delta E(\sigma_x)$ of a configuration σ is calculated using the full-potential linear-muffin-tin-orbitals (FP-LMTO) as implemented by Methfessel[6]. The LMTO method employs a minimal basis set which consists of site-centered Hankel functions, augmented in a continuous and differentiable way by numerical solutions to the radial Schrödinger equation inside nonoverlapping muffin-tin spheres. Nonspherical and interstitial terms are calculated using the fitting procedures described in [6]. The local density potential of von Barth-Hedin is used. The basis set contains s, p, d orbitals from the LMTO envelopes with tail kinetic energies κ=-0.01Ry and -1.0 Ry respectively and s, p orbitals from the third LMTO envelope with κ=-2.3 Ry. The nonspherical terms of the potential and charge density are expanded in spherical harmonics up to l=6.To evaluate integrals over the Brillouin zone, we use a uniform mesh of sampling points which ensure that the total energy is converged to within 0.05 mRy per atom.

In the second approach , the total energy of the system of N atoms is expressed from two terms; the first one is a sum of pair-potential interactions while the second one gives the many-body attractive part of the cohesive energy[7] (TB). We choose a square root dependence of the latter term due to the approximation of the second moment of the density of states. For each type of homoatomic interactions, namely Au-Au and Ni-Ni, there are four parameters which are determined by a fitting procedure. We refer to [7] for more details. Essentially are fitted the experimental values of pure metals as the cohesion energies, the bulk modulus, the experimental volumes, and the shear elastic constants.To calculate heteroatomic interactions, i.e. Au-Ni , the same functional form is used. A classical procedure is to average the parameters of the homoatomic interactions to obtain the values of the corresponding unlike atom parameters. This gives a heteroatomic concentration-independent potential. This approximation has been largely used in the literature (see Ducastelle [8] for a review) but it is also shown that the second moment approximation is in principle not sufficient to account properly for chemical ordering effects. We shall turn back to this problem in the next section.

BULK ALLOYS

To calculate the ordering effects in the Au-Ni system, we use the "ε-G" representation[9]. The formation energy of a given configuration σ can be written as:

$$\Delta E\ (\sigma_x) = G(x) + \varepsilon(\sigma) \qquad (1)$$

where G(x) > 0 is essentially the elastic energy necessary to deform the constituents (1-x)A+xB from their equilibrium volumes V_A and V_B to the alloy volume V(x), and this term does not depend on the configuration σ. The second term describes the energy changes when A combines with B at the volume V(x) of the AB compound. It is labeled " chemical energy" and we use an Ising-like cluster expansion [10] to calculate it.

$$\Delta E(\sigma_x) = G(x) + {\sum_\alpha}' m_\alpha E_\alpha \overline{\Phi}_\alpha(\sigma) \qquad (2)$$

where E_α are the volume and composition-independent "spin-flip" interaction energies of basic lattice cluster α (e.g. nearest-neighbor pairs, next neighbor pairs, triangles, ..) and $\overline{\Phi}_\alpha(\sigma)$ the cluster functions, the overbar indicating that they are averaged over all symmetry-related clusters. The prime indicates that the sum is only over cluster distinct by symmetry, m_α is the multiplicity which is defined as the number of symmetry equivalent clusters.

The interaction energies are obtained by mapping eq.(2) into a set of directly calculated formation energies after the elastic energy G(x) has been substracted from them. We refer to [10] for the description of the procedure. Here we have used 10 configurations: fcc(A and B), $L1_2$ (A_3B and AB_3), $D0_{22}$ (A_3B and AB_3), $L1_0$, $L1_1$, "40" and Z2 structures. The clusters are i) E_0 and E_1 (empty and single site), the first four pair interactions between first through fourth neighbors, and iii) the first three- and four-body terms. The transferability of this set has been examined by using these cluster interactions to predict via eq.(2) the formation energy of a new set of structures σ' not included in the mapping procedure. Convergence is tested by comparing these predictions with the directly calculated $\Delta E(\sigma'_x)$ [11]. The energy of the perfectly random alloy is then defined as:

$$\Delta E_{mix} = G(x) + {\sum_\alpha}' m_\alpha E_\alpha (2x-1)^{n_\alpha} \qquad (3)$$

n_α is the number of vertices in cluster α.

The ordering energy at a given composition is given by:

$$E_{ord}(\sigma) = {\sum_\alpha}' m_\alpha E_\alpha [\overline{\Phi}_\alpha(\sigma) - (2x-1)^{n_\alpha}] \qquad (4)$$

The directly calculated formation energies $\Delta E(\sigma_X)$ are collected in Table 1. At this point all that can be said is that Au-Ni compounds have $\Delta E(\sigma_X) > 0$ indicating that the system is phase separating.

Now we can analyze the ordering energy obtaining from the cluster expansion. We can see that the sign of the ordering energy depends on the chosen configuration . At x=0.5 it is negative for the $L1_0$, 40 or Z2 structures but positive for the $L1_1$ structure. We find that the $L1_0$ structure displays the most negative ordering energy. This means that Au-Ni alloys exhibit a ordering-type SRO and that the fundamental reason of the phase segregation is due to the dominance of G(x). The values of α and β, which fit the elastic energy G(x) to the form $G(x)=4[\alpha+\beta(x-0.5)]x(1-x)$ are 582 and -211 meV/atom respectively. Note that all our results are in very nice agreement with the ones published in [12]. That gives a great confidence in the determination of such energetic data from first-principles calculations.

Table 1. Listing of the energetic values in the Au-Ni system (meV/atom)

	FPLMTO			TB		
	$\Delta E(\sigma_X)$	ΔE_{rel}	E_{ord}	$\Delta E(\sigma_X)$	ΔE_{rel}	E_{ord}
$L1_2$	78			54.1		
$D0_{22}$	68					
$L1_0$	79.4	-17	-57	57.9	-16	-24.8
$L1_1$	175.4					
"40"	89.9					
Z2	123.1	-177	-13.3	62.7	-65	-20
$D0_{22}$	81.5					
$L1_2$	80.7			58.0		
$\Delta E_{mix}(0.5)$	136.4	-54		82.7	-43.3	

In this table we have also reported values obtained by the use of the semi-empirical potentials based on the second moment approximation. The random configuration to simulate the solid solution consists of a large cell with N=4000 sites extended over the whole space by periodic boundary conditions. The minimization of the energy as a function of the atomic coordinates is achieved numerically using the conjugate gradient method. The first conclusion is that the formation energies of the different compounds as well as the formation energy of the random phase display values smaller than the FPLMTO ones. The second moment approximation is in principle not sufficient to account properly for alloying effects and this potential certainly not since no parameter has been used to obtain Au-Ni interactions. However this semi-empirical scheme captures the gross features of the first-principles based results since it gives the correct sign for the formation energies of the compound as well as for the ordering energy of the $L1_0$ structure. Let us mention that the relaxation effects are also correctly represented.

SURFACE ALLOYS

Several experimental features are indicative of an intermixing of Au and Ni atoms at the Au/Ni interfaces. During the growth of Ni film on Au (001) substrate, a transformation of the structure of the deposited Ni film has been observed by RHEED (Reflexion High Energy Electron Diffraction) [5]. This can be explained by a covering of the surface by a Au monolayer which drastically decreases the formation energies. As metallic Ni has a higher surface energy than does Au, it may be that Au atoms may segregate to the surface in order to lower the surface energy. To examine this possibility we have compared the total energies of two slabs; the first one consists of five layers of Au(001) with a Ni monolayer on each side of the slab while the second one has Ni as a subsurface layer. The two-dimensional lattice parameter and the Au-Au interlayer spacing are taken to be those of bulk Au. The Ni-Au interlayer spacing is determined from total-energy calculations for both slabs. Magnetic effects are not taken into account. In table 2 are collected the main results.

Table 2 : Energies in the Ni/Au(001) system (mRy/atom)

FPLMTO		Tight Binding (TB)	
Surface Energy:			
Au(001)	53.7	Au(001)	19.3
Ni(001)	75.7	Ni(001)	44.9
Formation energy:			
Ni/Au(001) (nrel.)	151.6		
Ni/Au(001) (rel.)	139.2	Ni/Au(001) (rel.)	87.3
Au/Ni/Au(001) (nrel.)	104.1		
Au/Ni/Au(001) (rel.)	83.2	Au/Ni/Au(001) (rel.)	56.7

For Au/Ni/Au(001) , the total energy is found to be lower than that of Ni/Au(001) by a relatively large amount (56 mRy/atom). Hence it is possible for deposited Ni atoms to diffuse into the Au substrate. However our results are not sufficient to develop a "ε-G" analysis similar to the one proposed for bulk alloys. At this point we can only discuss two results. First the chemical distribution in the top layer is driven by the lowest surface energy, namely Au one; the creation of a second Au/Ni interface does not seem to penalize this driving force. Then the relaxation of the stressed Ni layer is sensitive to the number of Au/Ni interfaces since the Au/Ni/Au(001) system displays a relaxation energy roughly two times as high as one of the Ni/Au(001) system. The tight-binding approach gives the same chemical trend even if it underestimates the FPLMTO values for the same reasons as explained above.

CONCLUSION

In summary, the "ε-G" method illustrated here for the Au-Ni system gives a way of analyzing trends in phase stability using first-principles calculations. A semiempirical approach based on the tight-binding approximation gives the same qualitative conclusions. This agreement is stimulating for employing this approach in more intricate problems like the growth of metallic layers on a metallic substrate.

REFERENCES

1. H. Okamoto and T.B. Massalski (ed.) Phase Diagrams of Binary Gold Alloys (ASM International, Materials Park, OH:1987).

2. J.E Woodilla and B.L. Averbach, Modulated Structures in Au-Ni alloys, *Acta Metall.* 16:255 (1968).

3. L. Pleth Nielsen, F. Besenbacher, I. Stensgaard and E. Laegsgaard, Dealloying Phase Separation during growth of Au on Ni(110), *Phys. Rev. Lett.* 74:1159 (1995).

4. B. Gilles, J. Eymery, A. Marty, J.C. Joud, and A. Chamberod, The growth of Ni overlayers on Au(100) buffers, Mat. Res. Soc. Symp. Proc. Vol 317, MRS, Pittsburg, Pennsylvania (1993).

5. T. Deutsch, P. Bayle, F. Lancon, and J. Thibault, Computer Simulation of Au(001)/Ni multilayers: comparison with experiments, *J. Phys. Cond. Matter* accepted for publication.

6. M. Methfessel, Elastic Constants and Phonon Frequencies of Si calculated by a fast full-potential linear-muffin-tin-orbital method, *Phys. Rev. B* 38:1537 (1988).

7. J. Eymery, F. Lancon, and L. Billard, Au-Ni solid solutions studied by numerical relaxation, *J. Phys. I France* 3:787 (1993).

8. F. Ducastelle, Order and Phase Stability in Alloys, ed. by F.R. de Boer and D.G. Pettifor, North Holland, Amsterdam(1991).

9. L.G. Ferreira, A.A. Mbaye, and A. Zunger, Chemical and elastic effects in isostructural phase diagrams: The ε-G approach, *Phys. Rev. B* 37:10547 (1988).

10. Z.W. Lu, S-H Wei, A. Zunger, S. Frota-Pessoa, and L.G. Ferreira, First-principles statistical mechanics of structural stability of intermetallic compounds, *Phys. Rev. B* 44:512 (1991).

11. A. Pasturel et al. to be published.

12. Z. W. Lu, and A. Zunger, Unequal wave vectors in short- versus long-range ordering in intermetallic compounds, *Phys. Rev. B* 50: 6626 (1994)

ELECTRONIC TOPOLOGICAL TRANSITIONS AND PHASE STABILITY IN AG-PD AND CU-PT ALLOYS

E. Bruno, B.Ginatempo and E.S. Giuliano

Dipartimento di Fisica and Unità INFM
Università di Messina
Salita Sperone 31
98166 S. Agata di Messina
ITALY

INTRODUCTION

Metallic random alloys can display Electronic Topological Transitions, i.e. changes of the Fermi Surface connectivity, on varying the atomic concentrations. We present a study of Ag-Pd and Cu-Pt Fermi Surfaces, based on the LDA-KKR-CPA theory, and argue that the different phase equilibria of such isoelectronic systems are related to the spectacular differences in their Fermi Surfaces.

Recent review papers[1,2] discussed on the Electronic Topological Transitions (ETT) and their effects on physical properties of metals and alloys. ETT's occur[3] as the electron chemical potential varies (by applying external pressure or just changing the e/a ratio by alloying) crossing a Van Hove singularity in the band structure, e.g. opening or closing electron or hole pockets, opening or disrupting necks in the Fermi Surface (FS). In two recent papers we have discussed about disorder and temperature effects on ETT's related phenomena[4] and calculated[5] by first principles the changes of topology of the FS of Ag_c-Pd_{1-c} alloys. We have found in this system five distinct ETT's, namely a neck disruption (c=0.06), a jungle-gym disruption at the L point (c=0.20), two hole pockets filling at the X point (c=0.35 and c=0.53) and a neck opening at the L point (c=0.70). In the same paper, we also pointed out the need of carrying out fully relativistic calculations. In fact, if the Spin Orbit interaction would be neglected, the degeneracy of the relevant bands along the ΓX line would force the two distinct hole pockets filling at the X point to occurr at the same concetration. In this paper we would like to explore the effects of ETT's on the phase stability of metallic random alloys, bearing in mind the cases of Ag-Pd and Cu-Pt systems.

FERMI SURFACE TOPOLOGY AND STABILITY OF ALLOYS

The experiment[6], as well as ab initio calculations[5,7] clearly indicate that the enthalpy of mixing in random Ag_c-Pd_{1-c} is negative in a wide range of concentration. Thus, at temperatures low enough, this system should be stable in an ordered phase. Unfortunately, the low temperature phase diagram about the relevant concentrations is not known yet. Lu et al[7] found the $L1_1$ phase more stable than the random one. However, on this basis, the existence of different ordered phases cannot be excluded. It is worth to recall here that Cu-Pt is the only metallic alloy known to be stable in the $L1_1$ phase.

Recently, Clark et al.[8] calculated the short range order Warren-Cowley parameter in $Cu_{0.50}$-$Pt_{0.50}$, within the Gyorffy and Stocks concentration functional approach[9] including

Onsager cavity field corrections[10], in order to account for the charge fluctuations. The Fourier transform of the concentration fluctuation susceptibility, $\chi_{cc}(\vec{q})$, can be written as:

$$\chi_{cc}(\vec{q}) \propto \int_{-\infty}^{\infty} d\varepsilon \int_{-\infty}^{\infty} d\varepsilon' \frac{f(\varepsilon,T) - f(\varepsilon',T)}{\varepsilon - \varepsilon'} \int_{BZ} d^3k \, A_B(\vec{k},\varepsilon) \, A_B(\vec{k}+\vec{q},\varepsilon') \quad (1),$$

where $f(\varepsilon,T)$ is the Fermi function and $A_B(\vec{k},\varepsilon)$ (see Section 2) is the Bloch Spectral Function (BSF), the density of states at the $\vec{k}$ point. In the limit of low temperatures, Eq. (1) might be approximated as:

$$\chi_{cc}(\vec{q}) \propto \int_{BZ} d^3k \, A_B(\vec{k},\varepsilon_F) \, A_B(\vec{k}+\vec{q},\varepsilon_F) \quad (2),$$

that shows that peaks in the susceptibility, as a function of $\vec{q}$, have to be expected when either (i) large, parallel and flat portions of the FS are connected by $\vec{q}$ or (ii) small portions of the FS are connected but both $A(\vec{k},\varepsilon_F)$ and $A(\vec{k}+\vec{q},\varepsilon_F)$ are sufficiently high, which is bound to occur when $\vec{q}$ links two Van Hove singularities. Usually (i) is referred to as the Fermi surface nesting mechanism and is related to the incommensurate concentration wave phenomena. The (ii) mechanism is the one found[8] for Cu-Pt, where, in particular, $\vec{q}$ is the LX vector. Thus, a concentration fluctuation along the (1,1,1) direction, i.e. a vector parallel to LX, destabilises the random phase. Such an ordering mechanism would lead to the $L1_1$ crystal structure, on lowering the temperature. We shall show, in Section 3, that this mechanism is related to the particular topology of the FS of $Cu_{0.50}$-$Pt_{0.50}$, which has both a neck at L and a pocket at X.

We like to notice, however, that Eq.(2) is only an approximation, and the complete calculation involves the double energy integrals indicated in Eq. (1). As a consequence, not only states at the Fermi level contribute to the k-space convolution, but also states in an energy interval around the Fermi level brings relevant contributions. It is easy to show that the width of this interval is at least $\Delta\varepsilon \approx k_B T + \eta$, where η is the inverse lifetime of the relevant quasi-particle states. In other words, the "peak to peak" correspondence type in Eq. (2) could be replaced by a "peak to tail" type contribution, which might produce a non zero susceptibility. In the last case, however, one should expect for a lower susceptibility and, therefore, for a weaker ordering propensity. It is interesting to note that the ordering temperature for Cu-Pt is 1085K[11].

As we are going to show in the following, in Ag-Pd random alloys, opposite to the case of Cu-Pt, there is a range of concentrations, in which the pocket around X is already filled while the neck at L is not yet opened. Thus, the FS topologies of Cu-Pt and Ag-Pd are very different. As a consequence, the relevant ETT's occur at different e/a values, in spite of the fact that these systems are isoelectronic. That has a consequence on the phase stability of Ag-Pd. In fact the above peak to peak mechanism cannot apply in this case and only the peak to tail mechanism might be active, suggesting for Ag-Pd the possibility of a much weaker $L1_1$ ordering tendency than for Cu-Pt. That implies either a lower ordering temperature or a different low temperature phase.

THE SEARCH FOR ETT IN RANDOM ALLOYS

Constant energy KKR is an efficient method for looking at FS's in metals[2]. By introducing the Bloch Spectral Function (BSF)[12] as

$$A_B(\varepsilon,\vec{k}) = -\frac{1}{\pi} \mathrm{Im}\, G(\varepsilon,\vec{k}) \quad (3),$$

where $G(\varepsilon,\vec{k})$ is the lattice Fourier transform of the Green's function, the FS is straightforwardly defined as the locus of the peaks (δ-function like for a pure system and real ε) of BSF at the Fermi level For a random alloy one has to perform the configurational averages of the Green's function in eq. (3). That gives:

$$\left\langle A_B\left(\varepsilon,\vec{k},\{\xi_\lambda\}\right)\right\rangle = -\frac{1}{\pi}\operatorname{Im}\left\langle G\left(\varepsilon,\vec{k},\{\xi_\lambda\}\right)\right\rangle \cong -\frac{1}{\pi}\operatorname{Im}G^{CPA}\left(\varepsilon,\vec{k}\right) \quad (4),$$

where by the second equality we approximated the configurationally averaged Green's function by means of the Coherent Potential Approximation. Following Bruno et al[2], we recognize that the meaning of eq. (4) is that each configuration produces a slightly different "band structure", whose average transforms the pure metal δ-functions in a superpositions of Lorentzian like peaks. The dispersion in the ε vs k plane of such peaks positions is the operative definition of the "alloy band structure". Therefore, the alloy FS can be defined, again, as the locus of the peaks of the BSF at the Fermi level. The last can be evaluated for a given direction ζ, and fitted by a sum of Lorentzians, in order to find the peaks positions and their halfwidths. The last has the meaning of the projection along the ζ direction of the inverse coherence lenght of an electron of that energy and momentum. Provided its value is small, on the scale of the Brillouin Zone size, the FS might be also detectable by, say, 2D-ACAR experiments. It can happen, however, that due to the finite inverse lifetime of the states, the "energy halfwidth" of the BSF peaks, when the bands get flat, the corresponding mean free path becomes small[2].

The last situation occurs when the Fermi level is close to an extremum of a band, i.e. a Van Hove singularity. In that case the fit procedure by Loretzians may not work accurately, and one might think that the BSF, as a tool for studying the possible ETT's, is not helpful. That notwithstanding, this procedure[2,5] is absolutely safe, as far as the ETT's search is concerned. In fact, if one follows the BSF's as a function of the concentration or the band filling one can see the BSF peak, say, disappearing as a clear sign of the fact that a band has been totally filled and therefore the FS is experiencing a change of shape.

Furthermore, in "split-band" or "common band" like alloys, as Ag-Pd or Cu-Pt, the disorder greatly affects the peaks halfwidths, and, as a consequence, an ETT does not take place suddenly, but rather "continuosly", i.e. in a broad range of concentration. That has the advantage that instead of carrying on a complete LDA-KKR-CPA calculation for many concentration values (including the estimation of the equilibrium lattice parameter) one can evaluate the constant energy BSF for several values of the energy and follow the peaks dispersion. This is a sort of rigid band approach which can give good estimates of the "critical concentrations" or "critical chemical potentials" values. A detailed discussion and pratical examples of this rigid band machinery have been given elsewhere[5].

ETT IN AG-PD AND CU-PT 50% ALLOYS

The main reason why FS's of Ag-Pd alloys and Cu-Pt alloys are greatly different originates from the different band structures of pure Pd and Pt. In Table 1 we show our results for the eigenvalues, relevant to our discussion, at L and X. and in Fig. 1 we plot the band structure for such systems in a neighborhood of the Fermi level. In order to find the peaks we moved the energy a little off the real axis by adding a constant imaginary part of 1mRyd to the energy.

Table 1. Energy Positions of L and X Van Hove singularities for Pd and Pt. The energy values are in Ryd and referred to the Fermi Level.

Metal	L point (neck)	X point (pocket)
Pd	0.0573	0.0359
Pt	0.0273	0.0765

By looking at the bands at the X points, one can see that by addition of noble metal impurities the Fermi level will move upwards and at two different values of concentration the hole pockets will get filled (two distinct ETT's). Analogously at the L point, the Fermi level eventually will pass through the saddle point.

It is at L that Pd and Pt have profound differences. By inspection of Fig. 1 and Table 1, the reader can notice that, for Pd, the top X eigenvalue (the closure of the second hole pocket) lies at an energy lower than the L saddle point (the neck opening), while it happens precisely the opposite for Pt. That means, in the language of the ETT's, that, on increasing the noble metal concentration, there will be a sensibly large gap where there is neither X pocket nor neck in Ag_c-Pd_{1-c}, while in Cu_c-Pt_{1-c}, there are both, as we are going to show in the following.

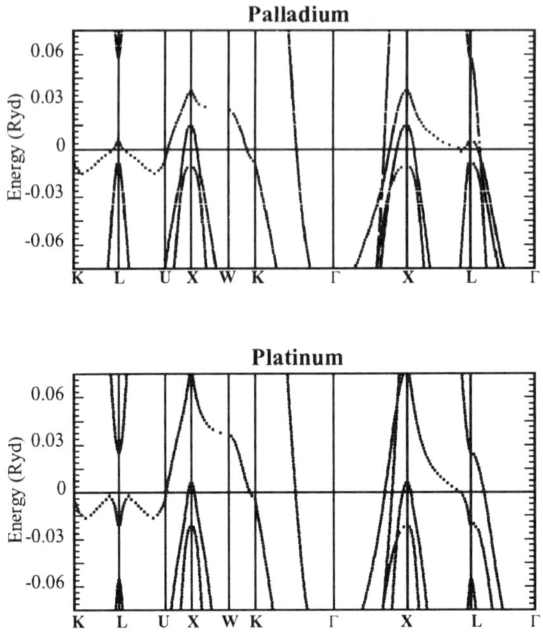

Figure 1. Band structure of Pd (a) and Pt (b), in a neighbourhood of the Fermi level. Energy units are Rydbergs.

In order to follow the topology changes of the FS for such system, we show in Fig. 2 the KKR-CPA BSF's along the FCC Brillouin Zone edges for some Ag concentration values. The energy values are the Fermi energies of the alloys of the indicated concentrations. As one can see, several changes do occur, on varying the concentration[5]. Here we want to remark that for c=0.50 no state is present, in practice, at L while at X one can see still broad states, indicating that the hole pocket is still open. A further increase of concentration, at c=0.60, causes such hole pocket to disappear, but one cannot find any more peaks on the Brillouin Zone boundaries, indicating that the FS is simply connected and all inside the Brillouin Zone. In other words Ag_c-Pd_{1-c} has a simple metal like FS, although not spherical, around c=0.60.

In order to illustrate better this point, we show in Fig. 3 the intersection of the Ag_c-Pd_{1-c} calculated FS with the plane ΓKLUX. The reader can see that at c=0.60 it happens precisely what we have said above, while at c=0.70 a tiny neck opens up and enlarges at c=0.80.

We display in Fig. 4 a calculation analogous to that of Fig. 2 for $Cu_{0.50}$-$Pt_{0.50}$, but this time the concentration is fixed and we let the energy to vary around the Fermi level. There the corresponding e/a values, as obtained by the integrated D.O.S,. are indicated.

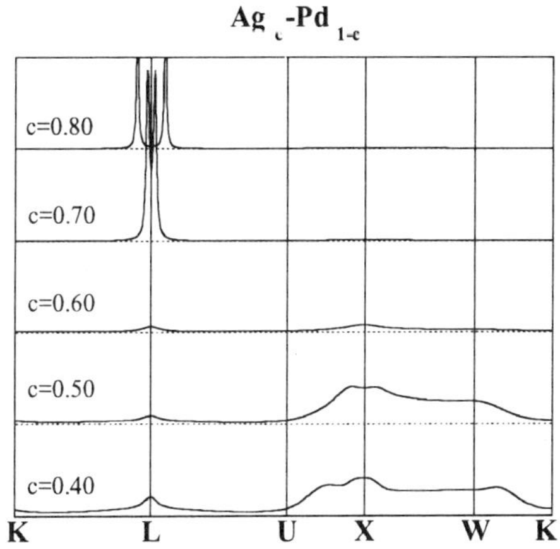

Figure 2. The Ag_c-Pd_{1-c} Bloch Spectral Functions at the Fermi energies along the Brillouin Zone edges. The labels indicate the concentrations, c.

The presence of both the neck and the X pocket, confirms the profound difference between the FS's of such systems. For example, one needs to increase the e/a ratio of at least 0.25 in order to close the X-pocket, and to reduce it at least of about 0.20 in order to disrupt the neck. Interestingly, the Cu-Pt $L1_1$ phase ranges[11] about in the concentration interval, 0.1÷0.7, i.e. about in the same range of e/a ratio where both the neck and the X pocket are present in the FS. It is worth to recall, however, that the calculations of Fig. 4 imply a sort of rigid band approximation. A CPA calculation is therefore required in order to confirm these numbers, although at c=0.50 the coesistence of the L and X structures is out of question.

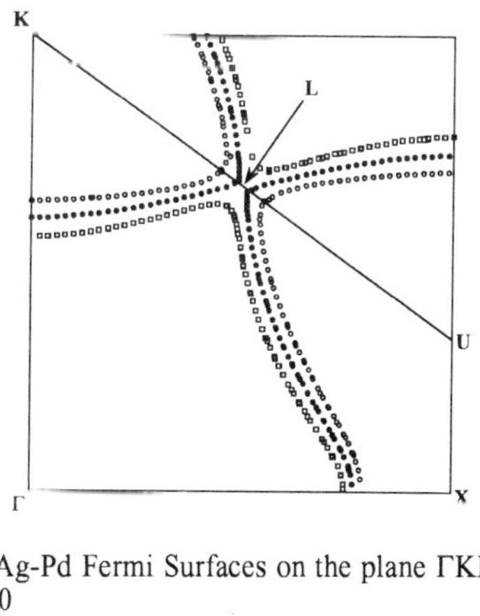

Figure 3. The neck opening in Ag-Pd Fermi Surfaces on the plane ΓKLUX. Open squares: c=0.60; full circles c=0.70; open circles c=0.80

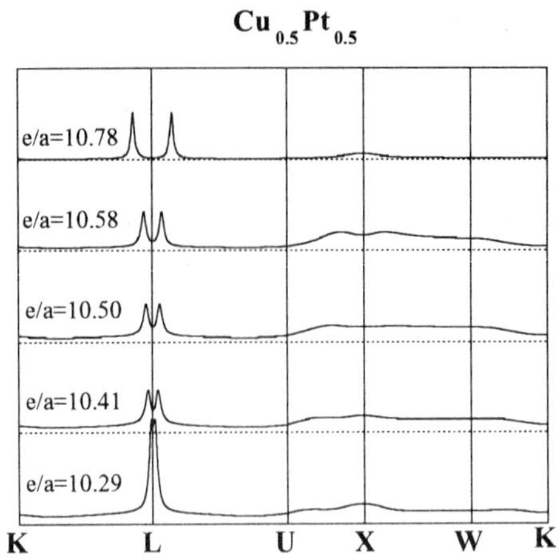

Figure 4. The Bloch Spectral Function along the Brillouin Zone edges for $Cu_{0.50}$-$Pt_{0.50}$ as a function of the band filling, as indicated.

As we have discussed above, the presence of high spectral density of states both at L and X for $Cu_{0.50}$-$Pt_{0.50}$ is the basis for claiming[8] a FS related ordering mechanism for such system. The results of Fig. 2 do not prevent the existence for a stable $L1_1$ phase for Ag-Pd, but, imply, at least, a lower ordering temperature. Whether the stable phase would be the $L1_1$, as suggested by Lu et al[7], or others is still an unanswered question.

We like to conclude this discussion remarking that a lot of physics is hidden in the Grand Potential excess produced[1,2,3,4] by an ETT's and the proposed examples might contribute to confirm the idea that also the alloy phase stability is related to such phenomena. In particular we like to draw some attention to the premartensitic phenomena in systems like Ni-Al[13] or In-Tl[14], where the FS topology is expected to play a fundamental role.

REFERENCES

1. Ya.M. Blanter, M.I.Kaganov, A.V. Pantsulaya and A.A. Varlamov, Phys. Rep., 245, 160 (1994).
2. E. Bruno, B. Ginatempo, E.S. Giuliano, A.V. Ruban and Yu.Kh. Vekilov, Phys. Rep., 249, 353(1994).
3. I.M. Lifshitz, Zh. Eksp. Teor. Fiz., 33, 1569 (1960).
4. E. Bruno, B. Ginatempo, E.S. Giuliano, Phys. Rev. B, to appear (1995).
5. E. Bruno, B. Ginatempo, E.S. Giuliano, Phys. Rev. B, to appear (1995).
6. R. Hultgren et al eds., "Selected Values of The Thermodynamics Properties of Random Binary Alloys", American Society for Metals, Metals Park, OHIO 1973
7. Z.W. Lu, S.-H. Wei and A. Zunger Phys. Rev. B 44, 10470 (1991).
8. J.F. Clark, F.J. Pinski, D.D. Johnson, P.A. Sterne, J.B. Staunton and B. Ginatempo, Phys. Rev. Lett., 74, 3225 (1995).
9. B.L. Gyorffy and G.M. Stocks, Phys. Rev. Lett., 50, 374 (1983).
10. See for example J.B. Staunton, Rep. Prog. Phys., 57, 1289 (1994) and references therein.
11. M. Hansen, "Constitution of Binary Alloys", McGraw-Hill, New York, USA (1958)
12. J.S. Faulkner and G.M. Stocks, 1980, Phys. Rev. B 21, 3222 (1980).
13. G.M. Stocks, W.A. Shelton, D.M. Nicholson, F.J. Pinski, B. Ginatempo, A. Barbieri, B.L. Gyorffy, D.D.Johnson, J.B. Staunton, P.E.A Turchi. and M. Sluiter, in "Ordered Intermetallics-Physical Metallurgy and Mechanical Behaviour", NATO-ASI Series E, Vol.213, Kluwer, Dordrecht, The Netherlands(1992), and references therein.
14. P.D. Norman, S.W. Wilkins, T.R. Finlayson, P. Goodman and A. Olsen, Scripta Met, 18, 575 (1984)

ENERGETICS OF THE LIGHT ACTINIDES IN A FULL CHARGE DENSITY SCHEME

L. Vitos[1], J. Kollár[1], and H. L. Skriver[2]

[1]Research Institute for Solid State Physics,
H-1525 Budapest, P.O.Box 49, Hungary
[2]Center for Atomic-scale Materials Physics and Physics Department,
Technical University of Denmark, DK-2800 Lyngby, Denmark

INTRODUCTION

During the last years many attempts have been made to explain the unusual equilibrium bulk properties of the light actinides using ab initio methods for the calculation of the total energy [1, 2, 3]. With the subsequent filling of the f-conduction band, for Np and especially for Pu, the observed equilibrium volumes are reasonably larger and the bulk moduli are smaller than the calculated ones; in contrast to the theoretical predictions the measured equilibrium volume for plutonium is larger than that of neptunium. According to the recent study by Wills and Eriksson [3] this discrepancy might be ascribed to the fact that in the calculations not the observed structures were used. To examine this possibility, in this paper we carried out a recently developed [4, 5] full charge density LMTO-ASA (linear muffin-tin orbital - atomic sphere approximation) calculation for these elements using the observed α-structures. We will show that our calculated values are fairly accurate for Th, Pa and U, but the anomalously large plutonium volume is not reproduced and our results show similar monotonic decrease in the atomic volumes as the earlier calculations. Therefore we conclude that this discrepancy reflects the limitations of the local density scheme, namely that the quasi-localized, relativistic f-conduction band is improperly treated in this approximation.

TOTAL ENERGY IN THE LMTO-ASA FULL CHARGE DENSITY SCHEME

Here we present a *full charge density energy functional* (FCD) based on a spherically symmetric atomic-sphere potential. Within the local density approximation we define

$$E_{FCD}[n(\mathbf{r})] = \sum_{\mathbf{R}} T^R_{ASA} + E[n(\mathbf{r})] \tag{1}$$

where

$$T^R_{ASA} = \int^{E_F} \epsilon N^R(\epsilon)d\epsilon - \frac{1}{\sqrt{4\pi}}\int_S n_0^{R,ASA}(r)V^R_{eff}(r)d^3r \tag{2}$$

Stability of Materials
Edited by A. Gonis *et al.*, Plenum Press, New York, 1996

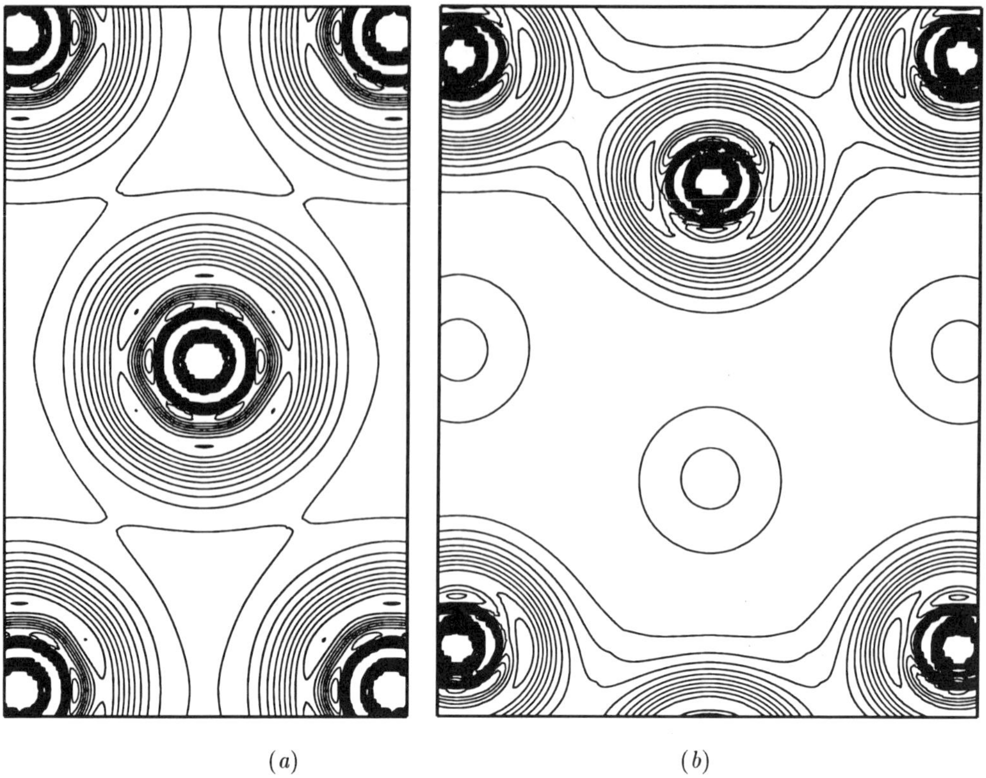

Figure 1: Charge density contours (in units *electron*/*a.u.*^{3}) for (*a*) *bct* Pa cut in the (110) plane and (*b*) for orthorhombic U cut in the (100) plane. The spacing between the lines is 0.015 for Pa and 0.02 for U.

is the kinetic energy per site evaluated in the ASA and

$$E[n(\mathbf{r})] = \int v_{ext}(\mathbf{r})n(\mathbf{r})d\mathbf{r} + \frac{1}{2}\int\int \frac{n(\mathbf{r})n(\mathbf{r}')}{|\mathbf{r}-\mathbf{r}'|}d\mathbf{r}d\mathbf{r}' + \int \epsilon_{xc}[n(\mathbf{r})]n(\mathbf{r})d\mathbf{r}. \tag{3}$$

In these equations E_F is the Fermi level, $N^R(\epsilon)$ the site-projected one-electron state density, S the atomic Wigner-Seitz radius, $n_0^{R,ASA}(r)$ the $l = 0$ component of the electronic charge density at the site $\mathbf{R}$, $V_{eff}^R(r)$ the effective one-electron potential corresponding to the ASA energy functional, $v_{ext}(\mathbf{r})$ the potential from the nuclei and ϵ_{xc} the exchange-correlation energy density. The total electronic charge density $n(\mathbf{r})$ reflects the proper symmetry of the lattice and is constructed from the output of an ASA calculation. In the self-consistent LMTO-ASA calculation we used a scalar-relativistic scheme. Since for the *fcc* structure we obtained the same trend in the atomic number dependence of the equilibrium volumes as that by Wills and Eriksson [3] in their fully relativistic calculation, we hope that the neglect of spin-orbit coupling does not influence remarkably our results. In our calculation we kept the core states frozen, while the *6p* semicore states (first panel) and the *5f,6d* and *7s* states (second panel) were considered as band states. In the first panel the *s*, *d* and *f*, while in the second the *p* states were downfolded. The number of *k*-points in the irreducible part of the Brillouin zone of the observed *fcc*, *bct* and α-structures were $675, 675, 392, 75$ and 48 for Th, Pa, U, Np and

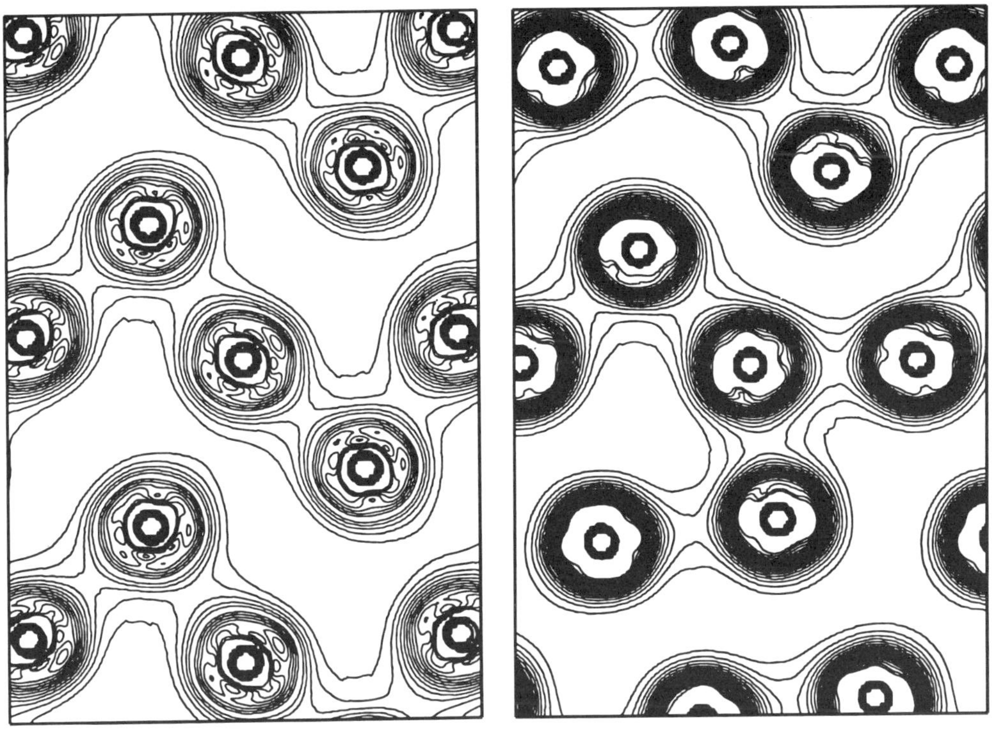

(a) (b)

Figure 2: Charge density contours (in units *electron/a.u.*^{3}) for (a) orthorhombic Np cut in the (001) plane and (b) for monoclinic Pu cut in the (001) plane. The spacing between the lines is 0.02.

Pu, respectively. The possibility of spin polarization was also examined (except Pu) and we have found no sign of local magnetic moments in these materials.

It is well known among the users of the LMTO-ASA method that using this approximation for the potential gives surprisingly good results for the charge density even for open structures like silicon [6, 7]. In this method, however, the outcoming charge density is given in a multi-centre form which is less suitable for further use in total energy calculation. In our method we rewrite the outcoming ASA charge density in a one-centre form

$$n^R(\mathbf{r}) = \sum_L n_L^R(r) Y_L(\hat{\mathbf{r}}), \tag{4}$$

where L is short-hand notation for (l, m), Y_L is a real harmonic and $\mathbf{r}$ is measured from the origin of the cell at $\mathbf{R}$. This expression is appropriate to create the charge density within one cell at $\mathbf{R}$. The details for the construction of the charge density is given in [8]. In Fig. 1 *a* and *b* we show the resulting charge density contours for Pa and α-U in the (110) and (100) plane using the observed body centered tetragonal and orthorhombic structures, respectively. We can see that the charge density contours are almost continuous going from one cell to the other and agrees reasonably well with the available full-potential LMTO results by Wills and Eriksson [3]. In the case of Pa the density contours approach to a hexagonal shape indicating that with the filling of the f-band the charge density is rearranged in the directions of the

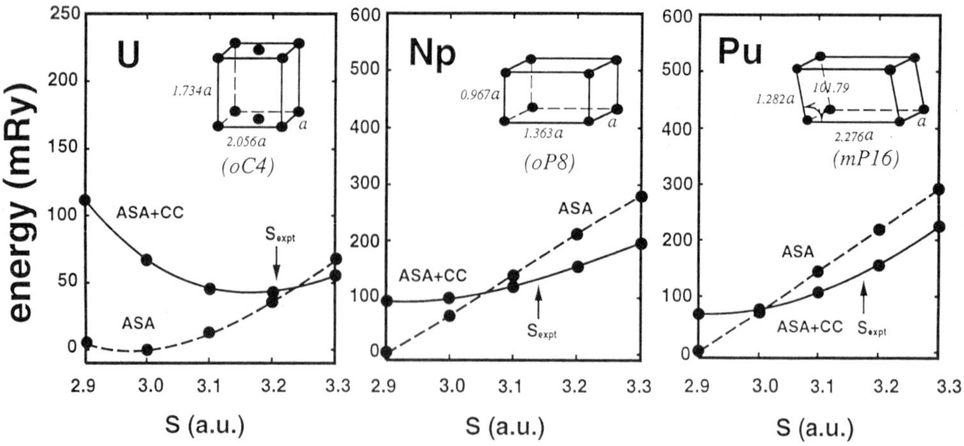

Figure 3: Total energy as a function of the atomic radius using the ASA with (ASA-CC) and without (ASA) the so called combined correction term for the observed α-structures of U, Np and Pu.

nearest neighbours which stabilizes the *bct* structure. Similar charge density maps are shown in Fig. 2 *a* and *b* for α-Np and α-Pu with orthorhombic ($oP8$) and monoclinic ($mP16$) structures, respectively. For the sake of easier comparison, approximately two unit cells are plotted for Np. Knowing the one-centre form of the charge density, the explicit charge density-dependent part of the total energy (3) can be calculated by dividing the solid into cells around each lattice site, where the electronic charge density is given in (4). Each cell is characterized by a set of multipole moments:

$$Q_L^R = \frac{(4\pi)^{1/2}}{2l+1} \int_{W-S\ cell\ at\ \mathbf{R}} \left(\frac{r}{S}\right)^l n^R(\mathbf{r}) Y_L(\hat{\mathbf{r}}) d\mathbf{r}. \tag{5}$$

To integrate within the W-S cell we use the so-called *shape function technique* (for the details see [4]. The total electrostatic energy of the system in (3) can be divided into contributions E^R belonging to the cell at $\mathbf{R}$ and the average electrostatic energy (per cell) can be written as $E^{Coulomb} = \frac{1}{N} \sum E^R$ (N denotes the number of cells). The cell-contribution E^R can be divided into *intra-cell* and *inter-cell* terms. For simplicity, we only give expressions for the contribution corresponding to the cell centered at the origin, i.e.

$$E^0 = E^0_{intra}[n^0] + E^0_{inter}[\{Q_L^R\}], \tag{6}$$

where we indicate that the inter-cell energy contribution depends on the multipole moments Q_L^R of all other cells as well. The intra-cell energy may be determined in the usual way, e.g. by solving the l-dependent Poisson equation, while the inter-cell interaction energy belonging to the cell at the origin may be written in the following form [9]:

$$E^0_{inter}[\{Q_L^R\}] = -\frac{1}{2S} \sum_L \sum_{\mathbf{R}} \frac{1}{2l+1} \left(\frac{b_R}{S}\right)^l Y_L(\hat{\mathbf{b}}_R)$$

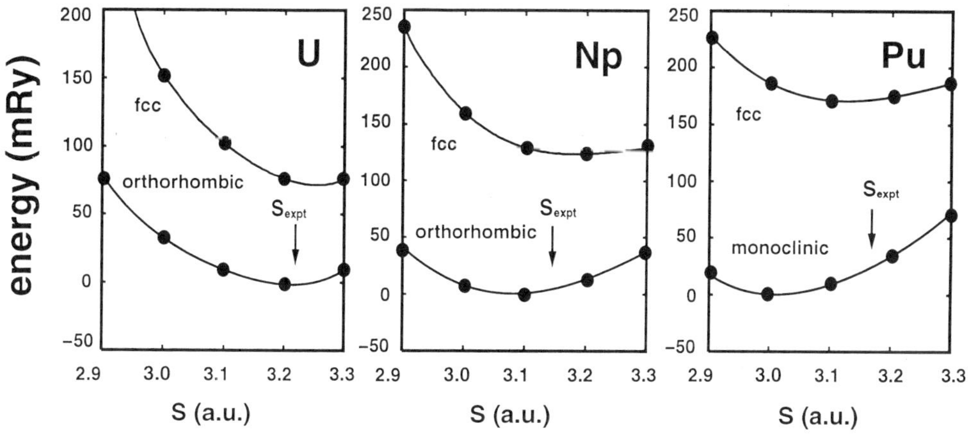

Figure 4: Total energy as a function of the atomic radius in our full charge density (FCD) scheme for fcc and the observed α-structures of U, Np and Pu.

$$\times \sum_{L',L''} Q^0_{L'} \frac{4\pi(2l''-1)!!}{(2l-1)!!(2l'-1)!!} C^L_{L',L''} \delta_{l'',l+l'} \sum_{L'''} S_{L'',L'''}(\mathbf{R}+\mathbf{b}_R) Q^R_{L'''} \tag{7}$$

where $S_{L,L'}(\mathbf{R})$ are the conventional LMTO structure constants and $C^L_{L',L''}$ denotes a real-harmonic Gaunt coefficient. In (7) $\mathbf{R}$ runs over the lattice vectors. For neighbouring cells with overlapping bounding spheres (with the radii ${S_C}^0$ and ${S_C}^R$) b_R has to be larger than zero to ensure the l-convergency of the summations; an optimal choice for b_R is described in [10]:

$$b_R + R = \left(1 + \frac{1}{2\alpha}\right)({S_C}^0 + {S_C}^R). \tag{8}$$

The parameter α was taken to equal 0.3 for Th, Pa and U and 0.6 for Np and Pu.

RESULTS

In Fig. 3. our ASA results are plotted for $\alpha - U$ (orthorhombic), $\alpha - Np$ (orthorhombic) and $\alpha - Pu$ (monoclinic) structures with (CC) and without the combined correction term. We can see that for $\alpha - Np$ and $\alpha - Pu$ the ASA is completely inadequate to describe the total energy curve; it does not show a minimum in the examined region of the Wigner-Seitz radius. The situation is somewhat better with the inclusion of the combined correction term. In these structures the W-S cells have non-vanishing, low order multipole moments which can not be described properly within the ASA. In Fig. 4. we compare the FCD total energy curves for *fcc* and the observed low temperature α-structures for U, Np and Pu. Our results for U are in good agreement with the full-potential LMTO calculation of Wills and Eriksson (the relative stability of the α-structure is about 70 *mRy/atom* in this case). For the relative stability of the $\alpha - Pu$-structure we obtained 160 *mRy/atom*, which is to be compared with the estimation by van Ek *et al.* [11], who obtained about 150 *mRy/atom* using the Ewald

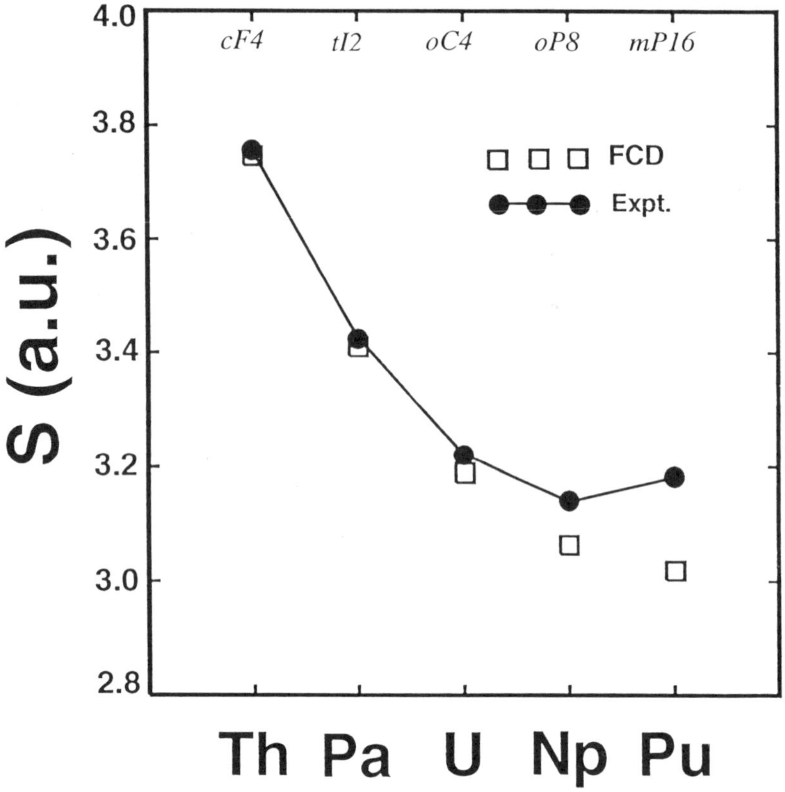

Figure 5: Calculated equilibrium volumes of the light actinides in our FCD scheme with the observed α-structures. The experimental values are also shown.

correction and the ASA total energy. In Fig. 5. we show our calculated W-S radii for the light actinides compared to the experimental values. One can see from Fig.5 that our FCD values agree fairly well with the measured ones from Th to U, but for Np and especially for Pu the equilibrium W-S radii are systematically underestimated in our local density calculation and we could not explain the observed trend that the W-S radius of $\alpha - Pu$ is larger than that of $\alpha - Np$. We mention that the inclusion of the gradient correction to the exchange correlation energy functional would increase the equillibrium W-S radius considerably, but the usage of this approximation is questionable for these elements. The most probable source of this discrepancy is the that the exchange correlation functional that we are using is unable to treat properly the almost localized f-electron states for these elements.

Finally in Fig. 6. we show the calculated Bain's-path for Th and Pa at normal pressure. In accordance with the observation the *fcc* structure is stable for Th (the energy difference is $5mRy/atom$ for $c/a = 0.81$), while for Pa the energy of the *bct* structure with $c/a \simeq 0.81$ lies $\simeq 5mRy/atom$ lower than that of the *fcc* structure. At extreme volume compression ($\simeq 50$ %) Th undergoes a $fcc \rightarrow bct$ phase transition [12] with a larger c/a ratio (1.58-1.65 depending on the pressure). This transition was studied theoretically by Eriksson *et al.* [13] and they found close agreement with the experiment. Our calculation supports their results; we have found that the energy of the *bct* structure with $c/a = 1.63$ lies lower than that of the *fcc* at $V/V_0 < 0.54$.

This work was supported in part by grant No ERB-CIPA-CT-92-2096 of the Commission of the European Communities and research project OTKA 2950 and 016740 of the Hungarian National Scientific Research Foundation.

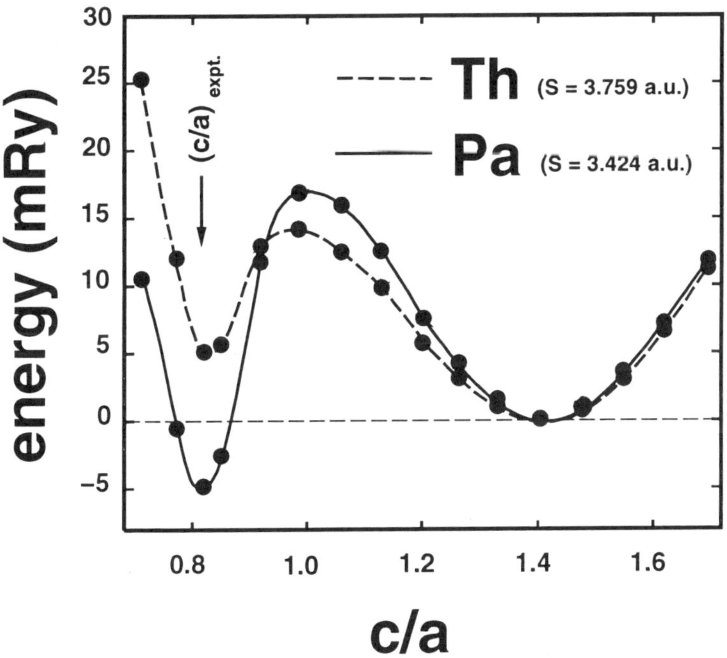

Figure 6: Calculated total energy for *bct* Th (dashed line) and Pa (solid line) as a function of c/a ratio (Bain's-path).

References

[1] H.L. Skriver, O.K. Andersen and B. Johansson, Phys. Rev. Lett. **41**, 42 (1978)

[2] M.S.S. Brooks, J. Phys. F: Met. Phys. **13**, 103 (1983)

[3] J.M. Wills and O. Eriksson, Phys. Rev. B **45**, 13879 (1992)

[4] L. Vitos, J. Kollar and H. L. Skriver, Phys. Rev. B, **49**, 16694 (1994)

[5] J. Kollár, H. L. Skriver and L. Vitos, Phys. Rev. B **49**, 11288 (1994)

[6] O.K. Andersen, Z. Pawlowska and O. Jepsen, Phys. Rev. B **34**, 5253, (1986)

[7] O.K. Andersen, A.V. Postnikov and S.Yu. Savrasov in: Applications of Multiple Scattering Theory in Materials Science, Eds. W.H. Butler, P.H. Dederichs, A. Gonis and R.L. Weaver, pp 37-70, (1992)

[8] L. Vitos, J. Kollar and H. L. Skriver, to be published

[9] A. Gonis, E. C. Sowa and P. A. Sterne, Phys. Rev. Lett., **66**, 2207 (1991)

[10] L. Vitos and J. Kollár, Phys. Rev. B, **51**, 4074, (1995)

[11] J. van Ek, P.A. Sterne and A. Gonis, Phys. Rev B, **48**, 16280, (1993)

[12] Y. K. Vohra and J. Akella, Phys. Rev. Lett., **67**, 3563, (1991)

[13] O. Eriksson, P. Söderlind and J. M. Wills, Phys. Rev. B, **45**, 12588, (1992)

IMPURITY EFFECTS ON BONDING CHARGE IN Ni_3Al

Sheng N. Sun, Nicholas Kioussis, and Say-Peng Lim

Department of Physics and Astronomy
California State University, Northridge, CA 91330

A. Gonis and W. Gourdin

Lawrence Livermore National Laboratory,
Department of Chemistry and Material Science
Livermore, CA 94550

INTRODUCTION

Microalloying studies have shown[1] that doping with certain impurities, which prefer to segregate towards grain boundaries, can significantly improve the ductility of polycrystalline Ni_3Al.[2] The addition of only 0.05 wt% (0.25 at%) of boron increases the room temperature elongation of polycrystalline Ni_3Al from a few percent to values of 45-50%. This effect, however, is only observed in specimens that contain excess Ni. Because boron is known to segregate preferentially to grain boundaries, one explanation for its dramatic effect on ductility is that it promotes better cohesion across the grain boundary planes.[3] More recent experiments have shown that boron also improves the ductility of single Ni_3Al crystals,[4] suggesting that a "bulk effect" should be considered in addition to the grain boundary strengthening effect of boron when explaining the improvement in ductility of polycrystalline Ni_3Al due to B additions.

While the *intrinsic* factors, such as poor grain boundary cohesion, are important, in many cases dominant, in limiting ductility, recent work by George *et al* has shown[5] that *extrinsic* factors, in particular the humidity, can be a major cause of low ductility in some systems. Their results demonstrate that the poor ductility commonly observed in air tests involves the dissociation of H_2O to generate atomic hydrogen, which diffuses into the region of the crack tip and promotes brittle crack propagation. George *et al* [5] have suggested that the principal role of boron in ductlizing Ni_3Al is to suppress the environmental embrittlement.

The purpose of the this work is to understand the electronic mechanism underlying the contrasting effects of the boron-induced strengthening and the hydrogen-induced embrittlement in Ni_3Al. To this end, we have carried out full potential linear-muffin-tin-orbital total energy calculations to investigate the effects of boron and hydrogen in Ni_3Al, and to study the changes of bonding associated with these impurities.

Stability of Materials
Edited by A. Gonis *et al.*, Plenum Press, New York, 1996

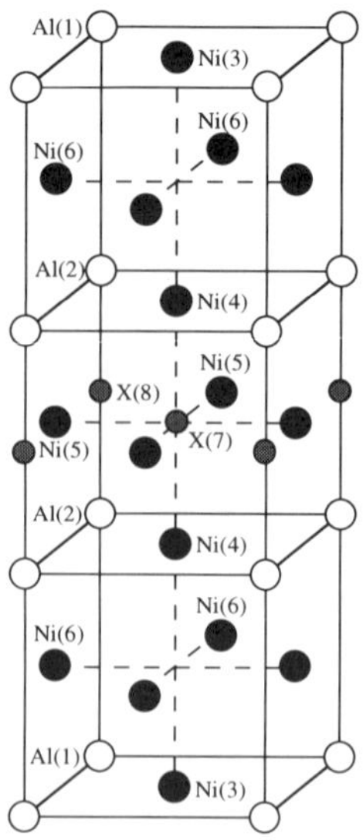

Figure 1. Supercell geometry used in the calculation. The filled, empty, and gray shaded circles represent Ni, Al, and X atom, respectively (X stands for the Boron or Hydrogen impurity). The two types of octahedral sites for the impurity, are denoted by X(7) and X(8), respectively.

COMPUTATIONAL METHOD

Our method of solving the electronic structure problem is based on the full-potential linear-muffin-tin-orbital (FLMTO) method, and has been described in detail elsewhere.[6] In treating intermetallic systems, the full potential approach is essential because of the large charge anisotropy of these systems. The supercell employed in our calculations is shown in Fig. 1. Inequivalent atoms in the cell are denoted by numerical labels (enclosed in parenthesis), depending on their point group symmetry. Based on X-ray diffraction intensity ratios, Masahashi *et al.*[7] conclude that boron occupies octahedral interstices in the Ni_3Al structure. Therefore, in order to study the effect of the local environment of the impurity on the electronic structure, the impurity (labeled by X) was placed at two different octahedral interstitial sites: 1) At the center of the octahedral cell [X(7)] and 2) at the center of the cube edge [X(8)]. We will refer to the X(7) and X(8) sites as Ni-rich and Ni-deficient octahedral-sites (O-sites), respectively. Such an arrangement corresponds to an impurity concentration of 7.7 at.% when one of the octahedral sites is occupied. In all calculations, atomic relaxation was ignored and the lattice constant was kept frozen at the value of 3.568$\mathring{A}$ for pure Ni_3Al.[8] The muffin-tin radii of 1.07$\mathring{A}$ were chosen to be equal for both Ni and Al atoms. The muffin-tin radii for the boron and hydrogen impurity were chosen also to be equal at 0.71$\mathring{A}$.

NUMERICAL RESULTS AND DISCUSSION

First we present the bonding charge density, $\Delta\rho(\mathbf{r}) = \rho_{solid}(\mathbf{r}) - \sum_\alpha \rho_\alpha(\mathbf{r}-\mathbf{r}_\alpha)$, where $\rho_{solid}(\mathbf{r})$ is the self-consistent charge density for the solid, $\rho_\alpha(\mathbf{r})$ is the atomic charge

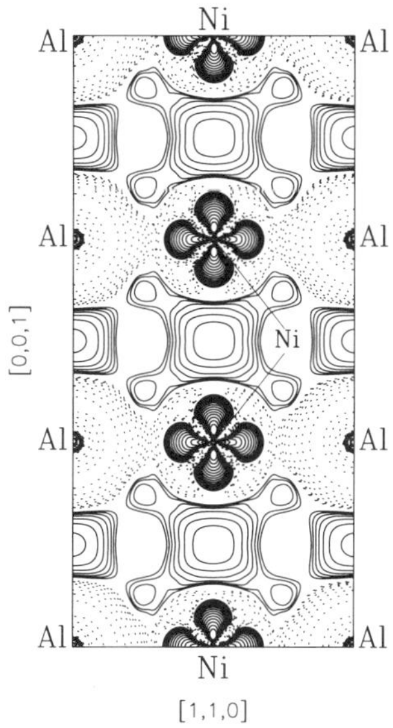

Figure 2. The charge density difference between Ni_3Al and the superposition of neutral Ni and Al atomic charge densities on the (110) planes. The solid (dotted) contours denote contours of increased (decreased) density as atoms are brought together to form the Ni_3Al crystal. Contours start from $\pm 8.0 \times 10^{-4}$ e/(a.u.)^3 and increase successively by a factor of $\sqrt{2}$.

density, and $\mathbf{r}_\alpha$ is the position of the atom, on the (110) plane in Fig. 2 in units of 10^{-3} e/(a.u.)^3. Here, solid and dotted curves represent contours of increased (accumulation) and decreased (depletion) electronic charge density. We find that the depletion of electron density at the aluminum sites is accompanied by significant *anisotropic* build-up of the directional d-bonding charge at the nickel sites. The bonding charge accumulation at the Ni site is along the nearest-neighbor (NN) Ni-Al and next-nearest-neighbor (NN) Ni-Ni directions. The bonding directionality is mainly caused by the polarization of p-electrons at the Al sites as a result of the p-d hybridization effect. One can also see a significant build-up of the interstitial bonding charge at the octahedral sites [sites between next-nearest-neighbor (NNN) Ni atoms and NNN Al atoms] between the (001) Ni-Al planes. These results are in qualitative agreement with those of recent warped-muffin-tin LMTO electronic structure calculations.[9] However, it is *the non-spherical corrections* to the spherical potential that yield the directionality of the d-bonding charge at the Ni sites. Thus, the bonding mechanism in Ni_3Al involves the combination of charge transfer and strong Al-p/Ni-d hybridization effects.

In order to study the effect of the local environment of the impurity on the electronic structure, we consider two different types of octahedral sites X(7) and X(8) in Fig. 1. To compare the energetics of the different impurity configurations we have calculated the impurity formation energy, $\Delta E_{imp} = E_{tot}(\mathrm{Ni_3AlX}) - E_{tot}(\mathrm{Ni_3Al}) - E_{tot}(\mathrm{X})$. Here, $E_{tot}(\mathrm{Ni_3AlX})$ is the total energy of the supercell with the impurity X placed at a certain octahedral site, $E_{tot}(\mathrm{Ni_3Al})$ is the total energy of the supercell without the impurity and $E_{tot}(\mathrm{X})$ is the total energy of an isolated impurity atom. We find that the impurity formation energies for a boron impurity placed at the Ni-rich [X(7)] and Ni-deficient [X(8)] octahedral sites in Ni_3Al, are -4.4 eV/(unit cell) and -0.08 eV/(unit cell),

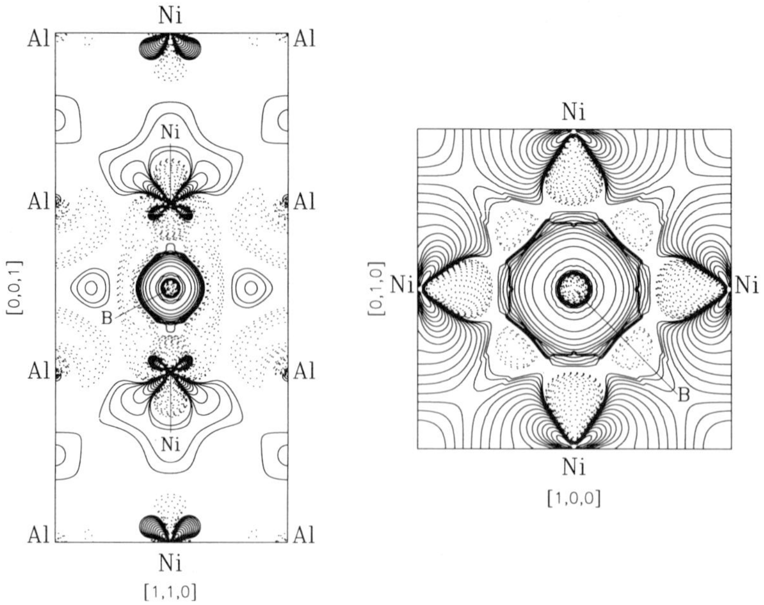

Figure 3. Boron-induced charge density of $Ni_3AlB_{\frac{1}{3}}$. Boron occupies the Ni-rich octahedral site [X(7)]. Solid (dotted) contours denote positive (negative) induced charge density. Contours start from $\pm 8.0 \times 10^{-4}$ e/(a.u.)^3 and increase successively by $\sqrt{2}$, (a) on the (110) plane and (b) on the (002) plane.

respectively. Thus, boron prefers to occupy octahedral sites which are surrounded by nearest-neighbor Ni atoms.

To understand the effect of absorbed boron on the bonding properties of Ni_3Al we next consider the redistribution of charge induced by the impurity atom when placed at the Ni-rich octahedral site [X(7)]. This can be best described by the *difference of bonding charge density* between Ni_3AlB and Ni_3Al, i.e., $\Delta\rho_{ind}(\mathbf{r}) = \Delta\rho_{solid}(\mathrm{Ni_3AlX}) - \Delta\rho_{solid}(\mathrm{Ni_3Al}) = \rho_{solid}(\mathrm{Ni_3AlX}) - \rho_{solid}(\mathrm{Ni_3Al}) - \rho_{atom}(\mathrm{X})$. We will refer to $\Delta\rho_{ind}(\mathbf{r})$ as the impurity-induced charge density. The impurity-induced charge density on the (110) and (002) planes is shown in Figs. 3(a) and 3(b), respectively. One can see that the enhancement of bonding charge around the Ni(4) site in Fig. 3(a) is quite asymmetric relative to the (001) plane through the atom. This is expected because the presence of boron breaks such a mirror symmetry. Though boron induces a decrease of the total MT valence charge in the Ni(3) site, the redistribution of the bonding charge about the Ni(3) site is significant. A comparison of Figs. 2 and 3(a) shows that boron induces a charge accumulation about Ni(3) in a direction where there is a depletion of the bonding charge $\Delta\rho(\mathbf{r})$ in the pure Ni_3Al system. Thus, the bonding-charge directionality of Ni(3) is *reduced* due to the presence of a small percentage of boron. More importantly, boron causes a significant build-up of interstitial charge near the (001) planes containing solely Ni atoms [Ni(6) atoms], which in turn enhances both the intraplanar bonding between the NN Ni(6) atoms and the interplanar bonding between the NNN Al(1) and Al(2) atoms, and the NNN Ni(3) and Ni(4) atoms. Note that boron also induces an additional charge depletion from all Al atoms. The boron-induced charge density on the (002) plane is shown in Fig. 3(b). Boron induces a substantial enhancement of interstitial bonding charge between the NN Ni(5) sites, increasing thus the intraplanar Ni-Ni bonding within the (002) planes. Finally, an interesting point

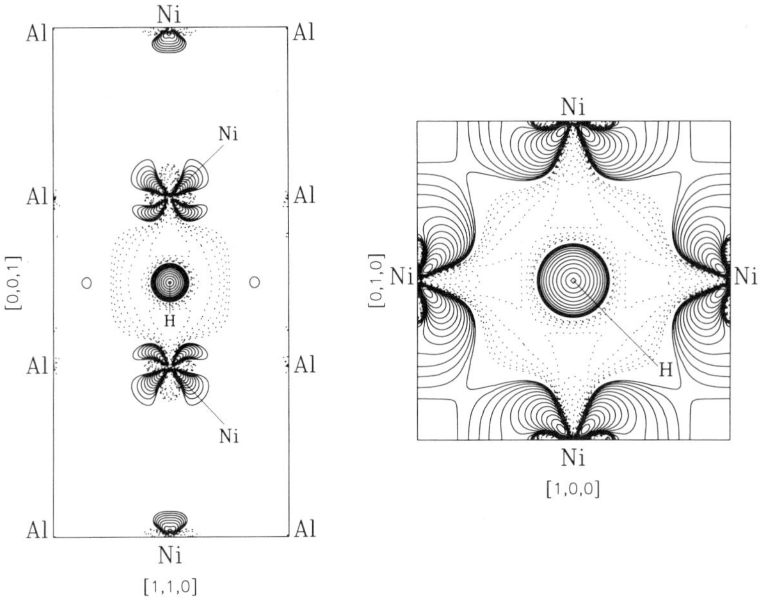

Figure 4. Hydrogen-induced charge density of $Ni_3AlH_{\frac{1}{3}}$. Hydrogen occupies the Ni-rich octahedral interstitial site [X(7)]. Solid (dotted) contours represent positive and negative induced charge density. $\pm 8.0 \times 10^{-4}$ e/(a.u.)^3 and increase successively by a factor of $\sqrt{2}$, (a) on the (110) plane (b) on the (002) plane.

to note is that electronic charge is depleted from the interstitial regions near the NiB plane in the [001] direction. This suggests a weaker local cohesion near these planes. This may be due to the fact that the local concentration of boron is high enough in our calculations (7.7 at.%) so as to weaken the local cohesion. This is consistent with the experimental results that Ni_3Al single crystals show a reduced ductility when doped with more than 0.8 at. % boron.[4]

We model the absorbed hydrogen in the supercell geometry of Fig. 1. The hydrogen formation energy is found to be -3.85 eV/(unit cell) and -2.60 eV/(unit cell) for placing the hydrogen impurity at the X(7)(Ni-rich) and X(8) (Ni-deficient) octahedral sites, respectively, indicating that hydrogen also prefers to occupy Ni-rich octahedral sites.

Figs. 4(a) and 4(b) show the impurity-induced charge on the (110) and (002) planes for $Ni_3AlH_{\frac{1}{3}}$. To compare with the boron-induced charge redistribution in Fig. 3(a) and Fig. 3(b), we use the same contour levels for both $Ni_3AlH_{\frac{1}{3}}$ and $Ni_3AlB_{\frac{1}{3}}$. It is clear that hydrogen induces a much weaker charge redistribution than boron. More importantly, the enhancement of interstitial bonding charge found near the pure Ni(6) plane in the [001] direction for the case of boron, is completely missing in the case of hydrogen. Thus, hydrogen does not provide the intraplanar and interplanar bonding enhancement found in the boron doped system. Furthermore, hydrogen affects the more distant nickel atoms [Ni(3)] quite differently. Hydrogen induces a significant build-up of charge along the [001] direction and a charge depletion within the (001) planes containing equal numbers of Ni(3) and Al atoms. This charge redistribution results in an *enhancement* of the *bonding-charge* directionality of Ni(3), which could be interpreted as the signature of a brittle system. Also note that hydrogen causes a larger redistribution of charge on the Ni(3) than on the Al(2) sites, a rather surprising result in view of the larger distance of the Ni(3) atoms from the hydrogen impurity. The charge redistribution, due to hydrogen, within the (002) plane (Fig. 4(b)) is similar to that induced by boron, which, however, is quantitatively weaker. Overall, while we cannot conclude directly that such a charge redistribution implies brittleness, a comparison of Figures 3(a), 3(b), 4(a), and 4(b) does suggest that hydrogen decreases local crystal cohesion. This is certainly consistent with its role as an embrittling agent.

In conclusion, we have studied the effect of boron and hydrogen on the charge density in Ni_3Al employing first-principles electronic structure calculations based on the FLMTO method. We find that the changes in the electronic structure induced by boron result from the hybridization of the *d* states of the nearest neighbor Ni atoms with adjacent B-*p* states. Thus, boron prefers to occupy Ni-rich octahedral interstices [X(7)]. Boron is found to greatly enhance the intraplanar metallic bonding between the Ni atoms, to enhance the interplanar bonding between the NiAl layers in the [001] direction, and to reduce the bonding-charge directionality near the Ni(3) atoms. Thus, we conclude that in such an environment boron acts to increase the cohesion of the crystal. In contrast, hydrogen is found to enhance the bonding-charge directionality near the Ni(3) atoms and provides virtually no interstitial charge enhancement. This suggests that hydrogen does not promote local cohesion. When both boron and hydrogen are present in the system,[10] the dominant changes in the electronic structure (DOS, induced charge densities) are induced by boron and hydrogen seems to have very little effect.

The research at California State University Northridge (CSUN) was supported through the U.S. Army Research Office under Grant No. DAAH04-93-G-0427, the Lawrence Livermore National Laboratory through Grant No. B157318, and the Office of Research and Sponsored Projects at CSUN. The research at Lawrence Livermore National Laboratory was supported through the U.S. Department of Energy under contract No. W-7405-ENG-48. We have benefited greatly from discussions with Ruqian Wu.

REFERENCES

1. C. T. Liu, in *Alloy Phase Stability*, edited by A. Gonis and G. M. Stocks (Kluwer Publications, 1989), 7.
2. K. Aoki and O. Izumi, J. Jap. Inst. Metals **43**, 1190 (1970).
3. C.T. Liu, C.L. White, C.C. Koch and E.H. Lee, in *High Temperature Materials Chemistry–II* (edited by L.A. Minir and D. Cubicciotti), Vol. **83**, p. 32, Electrochemical Society, Reminton, N.J. (1983); C.L. White, R.A. Padgett, C.T. Liu, and S. M. Yalisove, Scripta Metall. **18**, 1417 (1984); C.T. Liu and C.L. White, Scripta Metall. **35**, 643 (1987); G.S. Painter and F.W. Averil, Phys. Rev. Lett. **58**, 234 (1987); M.E. Eberhart and D. D. Vvedinsky, Phys. Rev. Lett. **58**, 61 (1987);
4. F.E. Heredia and D.P. Pope, Acta Metall. Mater. **39**, 2017 (1991).
5. E. P. George, C. T. Liu, and D. P. Rope, Scripta Metall. et Mater, **27**, 365 (1992).
6. David L. Price and Bernard R. Cooper, Phys. Rev. B **39**, 4945 (1989).
7. N. Masahashi, T. Takasugi, and O. Izumi, Atca Metall. **36** [7], 1815 (1988);
8. D. Hackenbracht and J. Kubler, J. Phys. F **13**, L179 (1983).
9. N. Kioussis, H Watanabe, R. G. Hemker, W. Gourdin, A. Gonis, and P. E. Johnson, Mat. Res. Soc. Symp. Proc. Vol. **319**, 363 (1994).
10. Sheng N. Sun and Nicholas Kioussis, unpublished.

ELECTRONIC STRUCTURE OF STACKING FAULTS IN $Pd_{1-x}Cu_x$, $Pd_{1-x}Ag_x$, AND $Ag_{1-x}Au_x$ ALLOYS

J.M. MACLAREN

Department of Physics, Tulane University, New Orleans, LA 70118.

INTRODUCTION

A face-centered cubic (FCC) crystal can be viewed as the ...ABCABCABC... stacking of (111) planes. Stacking faults are caused by the disruption of this perfect stacking sequence. The most common fault found in a face-centered cubic (FCC) crystal, the intrinsic stacking fault, results when one atomic plane (a C plane) is removed, *i.e.* ...ABCABABC.... When a $<110|$-type dislocation separates into two partial dislocations, the crystal contains an intrinsic stacking fault bounded by the two partial dislocation cores. The energy of the stacking fault determines the degree of separation and mobility of these partial dislocations. Dislocation mobility and core structure are some of the factors that influence ductility. Intrinsic stacking faults in FCC noble metals have been extensively studied experimentally,[1,2] and more recently theoretically.[3,4] $Pd_{1-x}Ag_x$ and $Ag_{1-x}Au_x$ alloys form a series of FCC solid solutions over the entire range of x, $Pd_{1-x}Cu_x$ also form a series of FCC solid solutions, though there is long range ordering at certain compositions. Thus, these three systems provide an excellent test of theoretical predictions of stacking fault energies in FCC alloys.

THEORY

The electronic structure for ordered and disordered stacking faults has been computed within the local spin-density approximation using the layer Korringa-Kohn-Rostoker method within the atomic sphere and coherent potential approximations (LKKR-CPA).[3,5] The LKKR method is a multiple scattering Green's function approach which can compute the electronic structure of an infinite solid without requiring three-dimensional translational symmetry. Rather only two-dimensional translational symmetry within each layer is assumed. The infinite solid is viewed as a sequence of layers of atoms. Computing times scale linearly in the number of layers in the interfacial region. The approach is thus ideally suited to a study of interfaces embedded in an otherwise perfect host crystal. For an ordered system, the site diagonal spin-restricted expansion of $\mathcal{G}$ is given in Hartree atomic units by

$$\mathcal{G}^{\alpha_i\alpha_i}(\vec{r},\vec{r},E) = -4i\sqrt{2E}\sum_{L,L'} Y_L^*(\vec{r})\,[Z_\ell^{\alpha_i}(r)\delta_{LL'}S_{\ell'}^{\alpha_i}(r) + Z_\ell^{\alpha_i}(r)\tau_{LL'}^{\alpha_i\alpha_i}Z_{\ell'}^{\alpha_i}(r)]\,Y_{L'}(\vec{r}). \tag{1}$$

$Y_L(\vec{r})$ is a complex spherical harmonic and τ the Scattering Path Operator (SPO). The composite angular momentum summation index L implies the double indices ℓ and m. The wavefunctions $Z_\ell^{\alpha_i}$ and $S_\ell^{\alpha_i}$ are the regular and irregular solutions to the scalar relativistic Schrödinger equation[6] for atom α is layer i. τ has both *intra* and *inter* layer contributions.

$$\tau_{LL'}^{\alpha_i\alpha_i} = \frac{1}{\Omega}\int_{\mathrm{BZ}} d\vec{k}\left[T_i^{-1}(\vec{k}) - \Delta_i(\vec{k})\right]^{-1}_{L\alpha_i L'\alpha_i}. \tag{2}$$

The $\vec{k}$-space integration in equation **(2)** is over the two dimensional Brillouin zone, area Ω, and is evaluated using the special point algorithm. $T_i(\vec{k})$ is the scattering path operator of the isolated layer i.

$$\left[T_i^{-1}\right]_{L\alpha_i L'\alpha_i'}(\vec{k}) = m_\ell^{\alpha_i}\delta_{LL'}\delta_{\alpha\alpha'} - G_{L\alpha_i L'\alpha_i'}(\vec{k}) \tag{3}$$

where $G_{L\alpha_i L'\alpha_i'}(\vec{k})$ is the two-dimensional Bloch free space Green's function and $m_\ell^{\alpha_i}$ is the inverse of the atomic $t-$matrix. $\Delta^i(\vec{k})$ includes all multiple scattering paths of the host crystal excluding layer i.

$$\Delta^i(\vec{k}) = (1 + R_i^{+-})(1 - R_i^{-+}R_i^{+-})^{-1}(1 + R_i^{-+}) - 1 \tag{4}$$

R_i^{+-} and R_i^{-+} are the reflection operators of the semi-infinite stacks of layers $i + 1, \ldots, \infty$ and $i - 1, \ldots, -\infty$ respectively. These are obtained in a plane wave basis using the layer doubling algorithm.

In a disordered alloy, the effective medium SPO is given by

$$\tau_c^{\alpha_i\alpha_i} = \sum_v c_v^{\alpha_i}\tau_v^{\alpha_i\alpha_i} \tag{5}$$

where the sum is taken to include all components on site α_i. The concentration of each species can be site dependent and is given by $c_v^{\alpha_i}$. The quantity $\tau_v^{\alpha_i\alpha_i}$ is the SPO for an atom of type v placed on site α_i in the effective medium.

$$\tau_v^{\alpha_i\alpha_i} = \left[(\tau_c^{\alpha_i\alpha_i})^{-1} + m^{\alpha_i} - m_c^{\alpha_i}\right]^{-1}, \tag{6}$$

$m_c^{\alpha_i}$ is the inverse of the single site coherent potential t-matrix. The SPO of the effective medium is obtained using the LKKR method and the inverse of the single site coherent potential t-matrices using equations **(2-4)**. The LKKR-CPA equations, equations **(2-6)**, are solved self-consistently at each energy.[3] The singe site Green's function for each component is computed from equation (1) with the appropriate SPO (equation **6**) and wavefunctions.

The valence electronic charge density is obtained from the imaginary part of $\mathcal{G}$. The integration over energy, from the bottom of the valence band up to the Fermi energy, is performed as a contour integral in the complex energy plane. New potentials are constructed from the charge density. The Ewald method, specialized to the case of a crystal with only 2-d translational symmetry, has been used to compute the long range part of the Coulomb potential. The one-electron potentials are then solved self-consistently. The total energy is obtained as a sum of kinetic, coulomb and exchange-correlation from the charge density. In all calculations reported here, equal radii were assumed for the different atoms, s, p and d partial waves, 16 energy points, 18 special k-points and 13 plane waves were used to evaluate $\mathcal{G}$. This basis set size is a compromise between numerical accuracy and reasonable computing times. The stacking fault energy was computed from the force theorem using self-consistent bulk calculations as a reference state.[4]

RESULTS FOR INTRINSIC STACKING FAULTS

Fig. 1a shows the theoretical and experimental fault energies for the $Pd_{1-x}Cu_x$ system. Experimental fault energies and error bars are taken from ref. [2]. Over the whole composition range from pure Pd to pure Cu, the agreement between theory and experiment can be seen to be excellent. Adding Cu to Pd causes a dramatic, and almost linear, drop in fault energy until the atomic percentage of Cu reaches about 60%. Beyond this composition the fault energy decreases much more slowly with increasing additions of Cu. In particular for alloys with compositions between $Pd_{.2}Cu_{.8}$ and pure Cu, the decrease in fault energy is also linear though with a smaller slope. The variation in fault energy divides naturally into these two linear regions. The differences between the slopes can be attributed to the significant differences in the electronic structure of the alloy in these composition ranges.

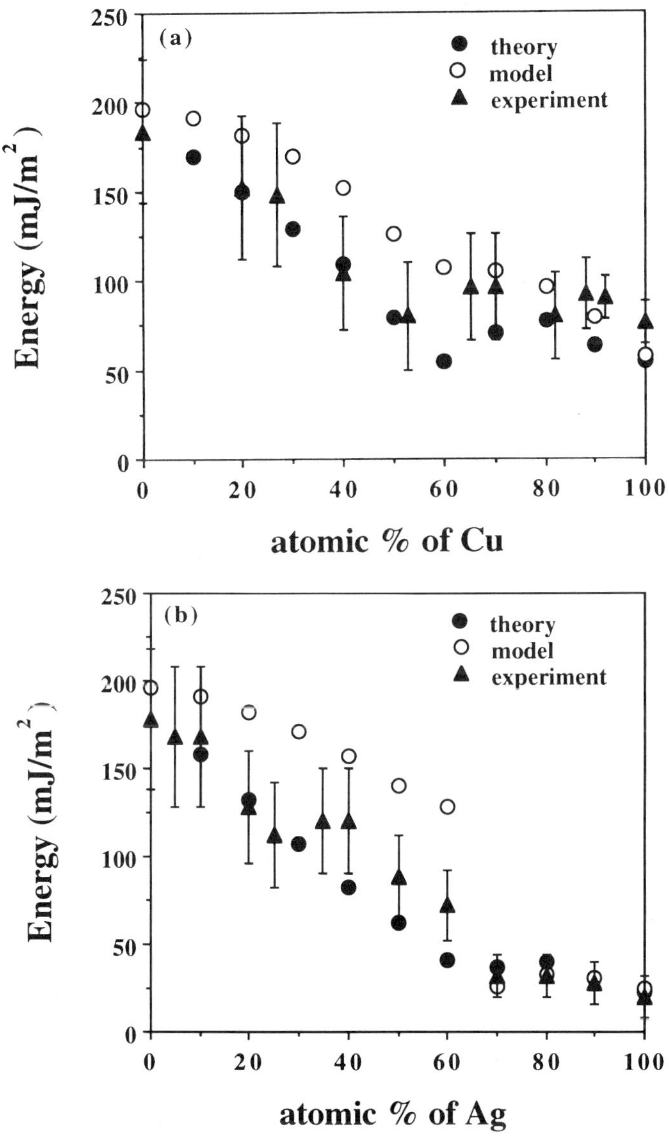

FIG.1. Theoretical and experimental stacking energies in (a) $Pd_{1-x}Cu_x$, and (b) $Pd_{1-x}Ag_x$ alloys.

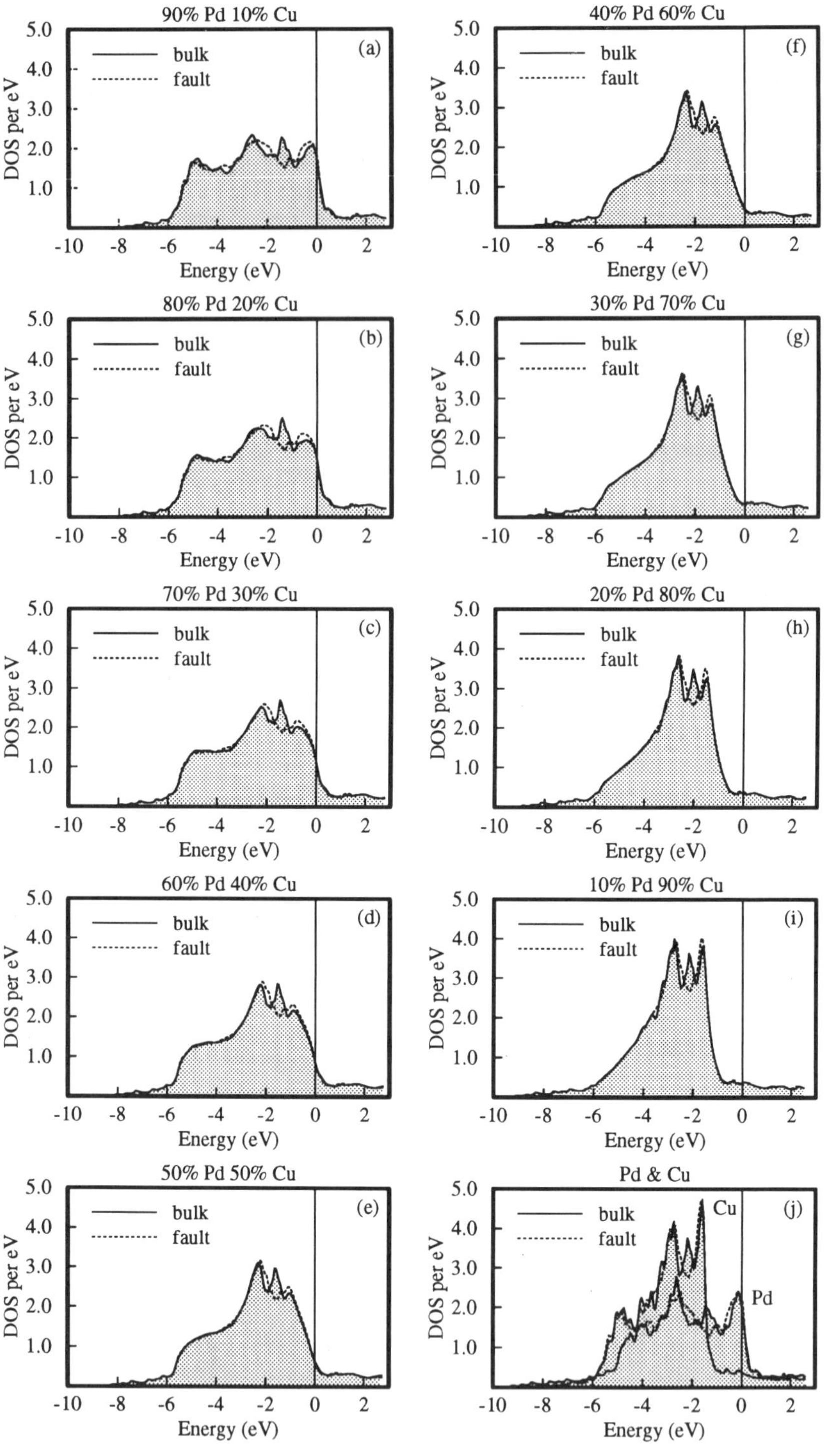

FIG.2. Densities of states for an atom in the bulk (shaded area) and an atom at the fault plane (dotted line) for (a) $Pd_{.9}Cu_{.1}$, (b) $Pd_{.8}Cu_{.2}$, (c) $Pd_{.7}Cu_{.3}$, (d) $Pd_{.6}Cu_{.4}$, (e) $Pd_{.5}Cu_{.5}$, (f) $Pd_{.4}Cu_{.6}$, (g) $Pd_{.3}Cu_{.7}$, (h) $Pd_{.2}Cu_{.9}$, (i) $Pd_{.1}Cu_{.9}$, and (j) Pd and Cu.

The region of rapid decrease in fault energy in $Pd_{1-x}Cu_x$ $(0 < x < 0.6)$ occurs when the Fermi energy passes through the top of the d-bands. When sufficient Cu has been added $(x > 0.6)$ the Fermi energy lies above the majority of the d-bands, and passes through a region of predominantly s-p electronic states. This is illustrated in Fig. 2 which shows the density of states for a bulk atom (shaded area) and an atom at the stacking fault (dotted line) for a range of alloy compositions.

To understand how a linear variation of fault energy can arise, we will consider the change in the band energy contribution to the fault energy upon changing the alloy composition. Changes in the Madelung energy will be neglected since the sphere charges are small and vary slowly with alloy composition. Defining $E_f(x)$ as the energy of the crystal $Pd_{1-x}Cu_x$ containing the stacking fault, $N_f(x)$ as the number of valence electrons, $\rho_f(x, E)$ as the corresponding density of states, and $\epsilon_f^F(x)$ as the local Fermi energy, then

$$E_f(x) = \int^{\epsilon_f^F(x)} dE\ E\rho_f(x, E) \quad \text{and} \quad N_f(x) = \int^{\epsilon_f^F(x)} dE\rho_f(x, E). \tag{7}$$

Similar expressions exist for the bulk alloy. The fault energy is defined as $E_{isf}(x) = E_f(x) - NE_b(x)$, where N is the number of layers included in the interface region. Thus

$$\begin{aligned} \frac{\partial E_{isf}(x)}{\partial x} &= \frac{\partial \epsilon_f^F(x)}{\partial x}\epsilon_f^F(x)\rho_f(x, \epsilon_f^F(x)) + \int^{\epsilon_f^F(x)} dE\ E\frac{\partial \rho_f(x, E)}{\partial x} \\ &- N\frac{\partial \epsilon_b^F(x)}{\partial x}\epsilon_b^F(x)\rho_b(x, \epsilon_b^F(x)) - N\int^{\epsilon_b^F(x)} dE\ E\frac{\partial \rho_b(x, E)}{\partial x} \end{aligned} \tag{8}$$

and

$$\frac{\partial \epsilon_f^F(x)}{\partial x}\frac{\rho_f(x, \epsilon_f^F(x))}{N} + \frac{1}{N}\int^{\epsilon_f^F(x)} dE\frac{\partial \rho_f(x, E)}{\partial x} = \frac{\partial \epsilon_b^F(x)}{\partial x}\rho_b(x, \epsilon_b^F(x)) + \int^{\epsilon_b^F(x)} dE\frac{\partial \rho_b(x, E)}{\partial x} = V \tag{9}$$

where V is the valence difference between the two alloy components. If the profile of the density of states does not change much with alloy concentration then equations (9) and (10) can be solved to give

$$\frac{\partial E_{isf}(x)}{\partial x} = NV(\epsilon_f^F(x) - \epsilon_b^F(x)). \tag{10}$$

The predictions of this limit of the model have been calculated starting both from pure Cu and from pure Pd, and are shown on Fig. 1a. At the Cu rich end, where this approximation is good, the model reproduces the observed behavior, while at the Pd rich end, the model reproduces the trends, but not the magnitude. Examination of the density of states of states profile, Fig. 2, at the Pd rich end shows that there is significant changing hybridization with composition between the d-states and thus the neglect of the $\frac{\partial \rho}{\partial x}$ is probably not justified.

Similar behavior is observed in the Pd/Ag system. Fig. 1b shows the fault energy variation with concentration. Again experiment and theory show good agreement, and the graph can be divided into two regimes where linear changes in fault energy are seen. As in the Pd/Cu system these correspond to concentration ranges where the d-bands are partially filled and regions where the d-band is filled and the Fermi energy passes through s-p electronic states.

The final alloy studied is the Au/Ag system. Fig. 3 shows the influence of alloy composition on fault energy. The fault energy varies smoothly between that found in pure Au and pure Ag. The algebraic model can also be solved in the limit of weakly interacting bands by assuming that the alloy density of states is approximately a compositional average of the constituent densities of states. The result of such an analysis is that the fault energy is just a compositional average of the pure element fault energies. The ab-initio results follow this trend satisfactorily. The one experimental data point for the 50/50 alloy also agrees with the calculations.

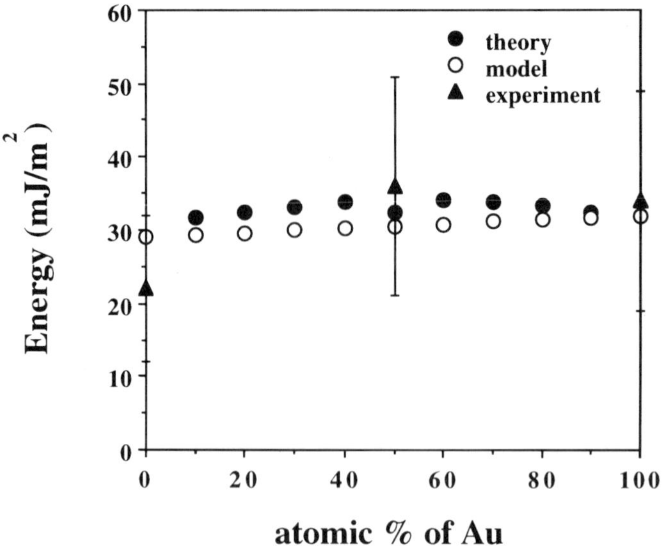

FIG.3. Theoretical and experimental stacking energies in $Ag_{1-x}Au_x$ alloys.

In conclusion, first principles electronic structure calculations using the LKKR-CPA method demonstrate the ability to compute the electronic structure and stacking fault energy in a wide variety of alloy systems. Important parameters in determining the variation of the fault energy with composition include valence difference, position of the Fermi energy, and changes in densities of states upon alloying. For dilute impurities in the noble metal hosts, band filling arguments alone suffice to explain the observed linear change in fault energy with dopant concentration, while for alloy compositions where the Fermi energy passes through the d-bands, and alloying strongly perturbs the density of states, these effects are seen to be as important. In alloys where the electronic structure of the constituents are similar, a simple linear variation between the two pure metal fault energies is expected and seen.

This work was supported in part by the Louisiana Quality Education Support Fund under grant number LEQSF (1993-95)-RD-B-15, the Department of Energy under contract number LEQSF (1993-95)-03.

REFERENCES

1. P.J.C. Gallagher, Metall. Trans. **1**, 2429 (1970).
2. I.R. Harris, I.L. Dillamore, R.E. Smallman, and B.E.P. Beeston, Phil. Mag. **14**, 325 (1966).
3. J.M. Maclaren, A. Gonis and G. Schadler, Phys. Rev. B**45**, 14392 (1992).
4. S. Crampin, K. Hampel, D.D. Vvdensky, and J.M. MacLaren, J. Mater. Res. **5**, 2107 (1990); S. Crampin, D.D. Vvedensky, and R. Monnier, Phil. Mag. A **67**, 1447 (1993).
5. J.M. MacLaren, S. Crampin, D.D. Vvedensky, R.C. Albers, and J.B. Pendry Comp. Phys. Comm. **60**, 365 (1990).
6. D.D. Koelling and B.N. Harmon, J. Phys. C **10**, 3107 (1977).

EFFECTS OF BORON AND SULFUR ON THE IDEAL YIELD STRESS OF Ni_3Al–A FIRST-PRINCIPLES APPROACH

Sheng N. Sun and Nicholas Kioussis

Department of Physics and Astronomy
California State University, Northridge, CA 91330

Mikael Ciftan

Physics Division, U.S. Army Research Office
Research Triangle Park, N.C. 27709-2211

A. Gonis

Lawrence Livermore National Laboratory
Department of Chemistry and Material Science
Livermore, CA 94550

INTRODUCTION

The past two decades have witnessed dramatic growth in both the experimental and theoretical activity in intermetallic compounds and ordered alloys, because of the potential use of these systems in a wide variety of technological applications.[1] Of central interest are their high melting temperature, low density, resistance to oxidation,[2] and the increase of yield stress it displays with increasing temperature,[3] in contrast to conventional compounds or disordered alloys. However, as with many other intermetallics, an inherent drawback to using polycrystalline ordered stoichiometric Ni_3Al alloys as a structural material is the tendency to undergo brittle intergranular fracture,[4] even though single crystals of Ni_3Al are highly ductile, having an elongation of 20%. Microalloying studies have shown[2] that doping with certain impurities, which prefer to segregate towards grain boundaries, can significantly improve the ductility of polycrystalline Ni_3Al.[3] At room temperature, the addition of only 0.05 wt% (0.25 at%) of boron increases elongation of polycrystalline Ni_3Al from a few percent to values of 45-50%. On the other hand, impurities such as sulfur or phosphorus act as strong embrittlers.[5]

Impurities which are known to be cohesive (embrittling) agents such as boron (sulfur) are believed to either increase (lower) the bond strength across the crack tip or to lower (increase) the barrier to dislocation nucleation and therefore crack blunting. Recent theoretical efforts have applied first-principles electronic structure calculations to study the role of impurities in the embrittlement of grain boundaries.[6] However, not much theoretical work has been done on correlating cleavage strength and fracture properties and the effect of impurities on these properties with electronic structure.

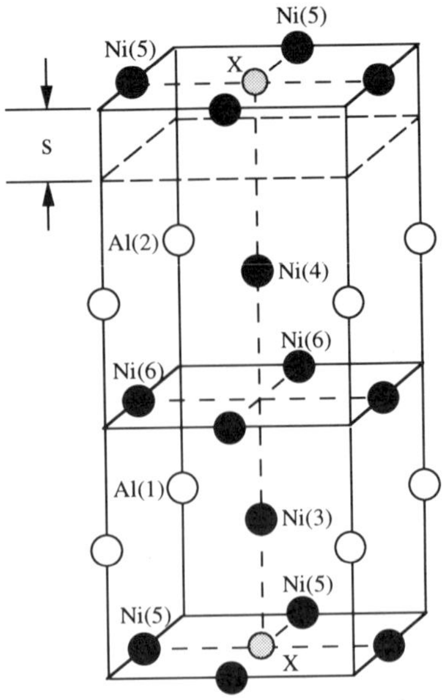

Figure 1. Supercell geometry used in the cleavage calculation. The filled, empty, and gray shaded circles represent sites for Ni, Al, and impurity (X stands for boron or sulfur) atoms, respectively. Atoms of different symmetry are distinguished by a type number.

The purpose of this work is to investigate on an atomistic scale the interactions underlying the contrasting effects of the boron-induced strengthening and the sulfur-induced embrittlement in Ni_3Al. We present results, from first-principles calculations, for the stress-strain relationship, cleavage energies, ideal yield stress and ideal strains with and without impurities, and discuss the electronic origins of the bonding behavior induced by boron and sulfur impurities across the cleavage plane. These calculation, although in and of themselves are not capable to address fully the complex phenomena associated with cleavage and fracture, provide insights into the bonding behavior and offer the potential for new directions in alloy composition for improvement of fracture-sensitive properties. A much more appropriate set of calculations should involve the geometry of a grain boundary and such calculations are currently being contemplated.

CALCULATION METHOD

Our method of calculation is the full-potential linear-muffin-tin-orbital (FLMTO) total-energy electronic structure method, and has been described elsewhere.[7] For intermetallic compounds, the bonding charge has both metallic and directional characteristics. Thus, a full potential treatment of the electronic structure is essential. The creation of a cleavage plane is achieved in this calculation by using a repeated slab construction. Each cell, shown in Fig. 1, contains one "slab" and each slab consists of four layers of (001) planes of Ni_3Al. The simulation of the separation of two neighboring planes (cleaved surfaces) is accomplished by varying the separation, s, between the slabs. Thus, for s=0 we simply have bulk Ni_3Al, and as s is increased the slab separation increases, until at large separations we have a repeated slab calculation of isolated surfaces. The impurity is placed at the octahedral interstitial site labeled by X, where the impurity has four Ni(5), one Ni(4), and one Ni(3) nearest neighbors. Such an arrangement corresponds to an impurity concentration of 11.1 at.%.[8]

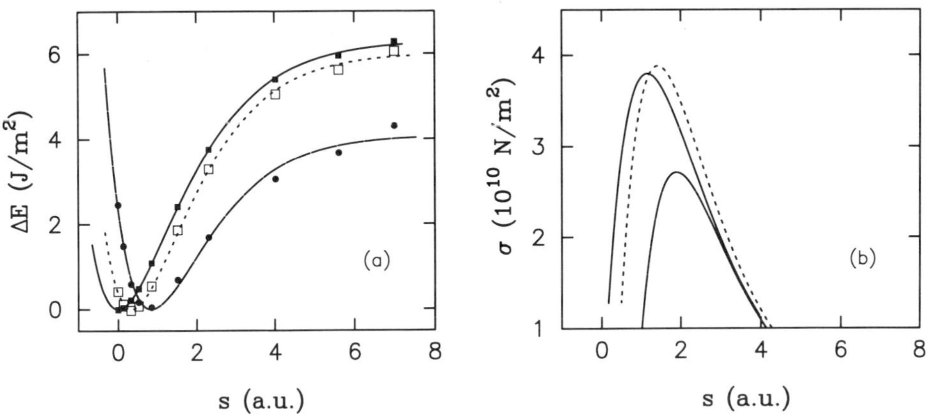

Figure 2. (a) The calculated total energy difference per unit area, ΔE, versus cleavage separation for Ni_3Al (solid squares), $Ni_3AlB_{\frac{1}{2}}$ (open squares), and $Ni_3AlS_{\frac{1}{2}}$ (solid circles). The solid, short- and long-dashed lines are the least-squares fits. (b) The cleavage stress versus cleavage separation for Ni_3Al (solid line), $Ni_3AlB_{\frac{1}{2}}$ (short-dashed line), and $Ni_3AlS_{\frac{1}{2}}$ (long-dashed line).

RESULTS AND DISCUSSION

In Fig. 2(a) we have plotted the difference in total energy per unit area, ΔE, of the four-layer unit cell versus the separation (s in Fig. 1) between neighboring slabs, for Ni_3Al (solid squares), for $Ni_3AlB_{\frac{1}{2}}$ (open squares), and for $Ni_3AlS_{\frac{1}{2}}$ (solid circles). Here, ΔE refers to the total energy of the slab at separation s minus the total energy at separation s_0, where s_0 represents the slab separation at the minimum of the total energy. Atomic relaxation between the (001) layers within the slab was ignored and the lattice constant was held fixed at the value of 3.568Å for pure Ni_3Al.[9] The difference in total energy ΔE converges to the ideal cleavage energy (Griffith energy) as the separation between neighboring slabs increases (the ideal cleavage energy is the total surface energy of the two cleaved surface planes). We find that sulfur reduces the ideal cleavage energy of the pure Ni_3Al system by 35%, while boron reduces it only by 5%. The nonzero value of s_0 for the $Ni_3AlB_{\frac{1}{2}}$ and $Ni_3AlS_{\frac{1}{2}}$ systems indicates a preferred outward (001) surface relaxation induced by the impurities. By fitting ΔE to a simple function (cubic polynomial times an exponential) by the nonlinear least-squares method, we have determined the variation of tensile stress with cleavage separation. The tensile stress, a quantity of central interest in fracture calculation, is given by $\sigma = \frac{1}{a^2}\frac{d\Delta E}{ds}$, where a is the lattice constant. Figure 2(b) shows the variation of tensile stress with separation between neighboring slabs for pure Ni_3Al (solid curve), for $Ni_3AlB_{\frac{1}{2}}$ (dotted curve), and for $Ni_3AlS_{\frac{1}{2}}$ (dashed curve). Calculated values for the ideal cleavage energy, the ideal yield stress, and the ideal yield strain for pure Ni_3Al, for $Ni_3AlB_{\frac{1}{2}}$, and for $Ni_3AlS_{\frac{1}{2}}$ are listed in Table 1, respectively. We find an ideal cleavage energy of 6.3 Jm^{-2} for pure Ni_3Al, which compares well with the value of 5.8 Jm^{-2} found by Fu and Yoo.[10] It is important to notice, that while boron increases the maximum cleavage stress (ideal yield stress) by about 2% compared to that of the pure Ni_3Al system, sulfur causes a significant reduction of σ_{max} by 29%. The rather small effect of boron on σ_{max} might be due to the fact that the local concentration of boron is high enough in our calculations so as to weaken the local cohesion across the cleavage

Table 1. Calculated values for the ideal cleavage energy, G_e, the maximum tensile stress σ_{max}, and the ideal strain ϵ corresponding to the maximum tensile stress, for fracture cleavage of the (001) planes of Ni_3Al, $Ni_3AlB_{\frac{1}{2}}$, and $Ni_3AlS_{\frac{1}{2}}$.

Compound	G_e (J m^{-2})	σ_{max} (10^{10} N m^{-2})	ϵ
Ni_3Al	6.31	3.80	35%
$Ni_3AlB_{\frac{1}{2}}$	6.02	3.89	30%
$Ni_3AlS_{\frac{1}{2}}$	4.08	2.71	25%

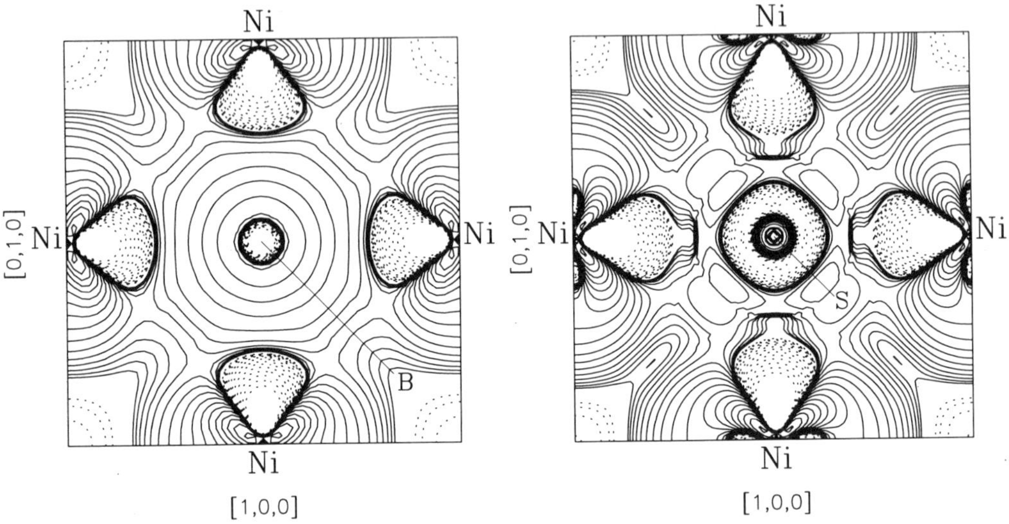

Figure 3. (a) Boron-induced charge density on the (001) plane for slab separation s_0. (b) Sulfur-induced charge density on the (001) plane for slab separation s_0. Solid (dotted) contours denote contours of increased (decreased) charge density. Contours start from $\pm 8.0 \times 10^{-4}$ e/(a.u.)^3 and increase successively by a factor $\sqrt{2}$.

plane. This is consistent with experimental results that Ni_3Al single crystals show a reduced cleavage strength when doped with more than 0.8 at.% boron.[11]

In order to understand the breaking and rearrangement of electronic bonds induced by the impurities within and across the cleavage plane, we have calculated the *impurity-induced* charge density, $\Delta\rho_{ind} = \rho_{solid}(Ni_3AlX_{\frac{1}{2}}) - \rho_{solid}(Ni_3Al) - \rho_{atom}(X_{\frac{1}{2}})$. (The first two terms are the self-consistent charge density of the slab with and without the impurity, and the last term is the superposition of atomic charge densities centered on impurity sites). In Figs. 3(a) and 3(b) we show the boron- and sulfur-induced charge density on the horizontal (001) plane (the plane which contains just the Ni(5) atoms and the impurity in Fig. 1) for slab separation s_0. The solid (dotted) contours represent accumulation (depletion) of electronic charge. We find that there is a significant build-up of charge at the region between boron and Ni(5), and a large depletion of charge at the region around the sulfur atom. This apparent reversal of charge transfer for the case of sulfur contradicts the simple estimate from electronegativity (2.58 for S and 1.81

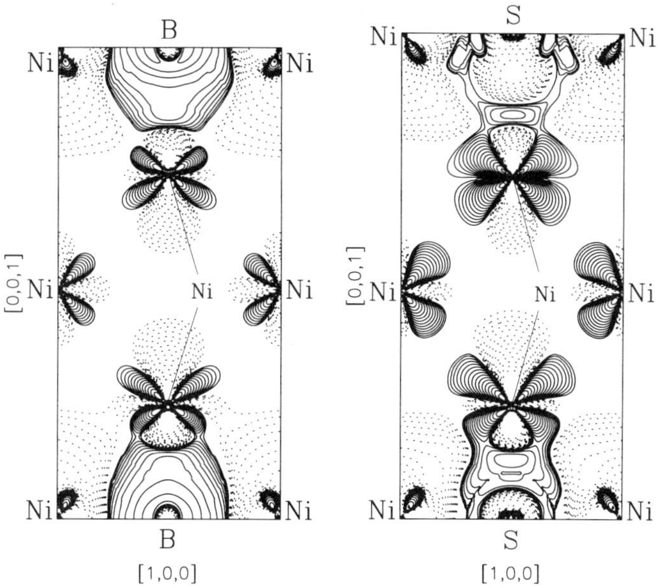

Figure 4. (a) Boron-induced charge density on the (200) vertical plane for slab separation s_0. (b) Sulfur-induced charge density on the (200) plane for slab separation s_0. Solid (dotted) contours denote contours of increased (decreased) charge density. Contours start from $\pm 8.0 \times 10^{-4}$ e/(a.u.)^3, and increase successively by a factor $\sqrt{2}$.

for Ni). On the other hand, both boron and sulfur impurities induce an accumulation of interstitial bonding charge between the nearest-neighbor Ni(5) atoms. In Figs. 4(a) and 4(b) we show the boron- and sulfur-induced charge density on the vertical (200) plane for the slab separation s_0. For $Ni_3AlB_{\frac{1}{2}}$, strong charge accumulation is seen in the region between B-Ni(4) across the cleavage plane, while there is large charge depletion in the region between S-Ni(4) for $Ni_3AlS_{\frac{1}{2}}$. This strong bonding charge that "hangs on" across the interface explains the fact that the ideal cleavage strength for $Ni_3AlB_{\frac{1}{2}}$ is larger than that of $Ni_3AlS_{\frac{1}{2}}$, and it is the microscopic basis for the resistance of the boron-doped system to cleavage fracture.

The different behavior of boron and sulfur can be understood from the spatial extension of the B-2p and S-3p wave functions in Ni_3Al and the different number of p electrons. Comparison of the boron 2p wave function in Ni_3Al with that in the free boron atom shows that the free 2p wave function has a greater spatial extent, i.e., B-2p wave function in the solid shrinks towards the nucleus, giving rise to a strong covalent-like $p-d$ hybridization. On the other hand, due to the existence of the node, the main peak of the sulfur 3p wave function lies mostly outside the sulfur muffin tin with long tails away from the S nucleus. As a result, the S-3p wave function is expected to be strongly affected by nearest-neighbor nickels (i.e., soft wave function in contrast to the boron 2p wave function, which lies mainly inside the muffin tin). Thus, the S-3p and Ni-d interaction is more embedded atom like via electrostatic potential instead of a covalent hybridization as in the case of boron. The S-3p electrons can directly feel the attraction from the nearest-neighbor nuclei, which pull the S-3p wave function outwards and so result in a reduction of the charge density in the region around the sulfur atom. However, the strong Coulomb and Pauli repulsions push the S-3p electrons away from the high density region around the nearest-neighbor Ni atoms. The net result of these two interactions is that the available Ni t_{2g} antibonding states are filled up and there is some charge accumulation in the region between S and Ni atoms, as can be seen in Figs. 3(b) and 4(b). These differences in electronic structure are responsible for the contrasting roles of boron and sulfur impurities in Ni_3Al.

To summarize, we have presented self-consistent total-energy FLMTO electronic structure calculations of the role of boron and sulfur on the ideal yield stree properties of Ni_3Al under tensile stress. For the impurity concentration used in the calculations, we find that boron has little effect on the fracture properties, whereas sulfur reduces greatly both the ideal cleavage energy and yield strength. The electronic bonding mechanism underlying the contrasting effects of the boron and sulfur impurities is a strong covalent-like hybridization for the boron impurity and a more embedded-like (electrostatic) interaction for the sulfur impurity due to the larger spatial extension of the S-3p wave function. These calculation are, of course, not able to address directly the physical mechanism that control fracture in real materials. More realistic calculations, involving the geometry of grain boundaries would be much more appropriate and we intend to proceed with such calculations in the near future.

We have benefited greatly from discussions with Dr. Ruqian Wu and Dr. Say Peng Lim. The research at California State University Northridge (CSUN) was supported through the U.S. Army Research Office under Grant No. DAAH04-93-G-0427 and the Office of Research and Sponsored Projects at CSUN.

REFERENCES

1. J.H. Westbrook, in *Ordered Alloys, Physical Metallurgy, and Structural Applications*, edited by B.H. Keaar, C.T. Sims, N.S. Stoloff, and J.H. Westbrook, Claitor's, Baton Rouge, Lao, 1970, p. 1.
2. C.T. Liu, in *Alloy Phase Stability*, edited by A. Gonis and G.M. Stocks (Kluwer Publications, 1989), 7.
3. K. Aoki and O. Izumi, J. Jap. Inst. Metals **43**, 1190 (1970).
4. K. Aoki and O. Izumi, Acta. Metall. **27**, 807 (1979).
5. C.L White and D.F. Stein, Metall. Trans. A **9A**, 13 (1978).
6. Ruqian Wu and A. J. Freeman, and G. B. Olson, Phys. Rev. B **50**, 75 (1994); R.P. Messmer and C.L. Briant, Acta Metall. **30**, 457 (1983); M.E. Eberhart and D.D. Vvedensky, Phys. Rev. Lett. **58**, 61 (1987).
7. David L. Price and Bernard R. Cooper, Phys. Rev. B **39**, 4945 (1989).
8. This concentration is certainly much larger than physically significant concentrations occuring in real material. Our aim is to gain initial insight through the present calculation and study lower, and more realistic concentrations, in future work.
9. D. Hackenbracht and J. Kubler, J. Phys. F **13**, L179 (1983).
10. C.L. Fu and M.H. Yoo, Mat. Chem. and Phys. Vol.**32**, 32 (1992).
11. F.E. Heredia and D.P. Pope, Acta Metall. Mater. **39**, 2017 (1991).

AB-INITIO CALCULATION OF THE LATTICE RELAXATION IN DILUTE ALLOYS

N. Papanikolaou[1], N. Stefanou[1], R. Zeller[2], P.H. Dederichs[2]

[1] University of Athens, Section of Solid State Physics, Panepistimioupolis GR-15784 Athens, Greece
[2] Institut für Festkörperforschung, Forschungszentrum Jülich D-52425 Jülich, Germany

INTRODUCTION

Electronic structure calculations for point defects in metals usually neglect the lattice distortion in the neighborhood of the defect. In fact, a reliable theoretical prediction of the perturbed lattice geometry around impurities is a rather complicated task because it requires accurate total-energy and/or force calculations. From the experimental point of view detailed information about the local geometry in the vicinity of a defect can be obtained by extended X-ray-absorption-fine-structure (EXAFS) measurements.[1] Inclusion of lattice distortion is important for a more realistic description of the electronic structure of point defects in bulk systems, while it becomes crucial in the case of defects on surfaces where the reduced coordination number induces a greater structural flexibility. Moreover, the study of the strain fields caused by the size mismatch between impurity and host atoms is interesting because they influence the interaction of solute atoms with other defects and the tendency towards long-range ordering or clustering.

Calculations of the lattice distortion around point defects in semiconductors using a pseudopotential approach have been reported by e.g. Scheffler et al..[2] Benedek et al.[3] carried out pseudopotential calculations of the lattice relaxation around a single vacancy in Al and Li and an Al impurity in Li, using a supercell approach.

Among the various first-principles computational formalisms, the Korringa-Kohn-Rostoker (KKR) Green function method offers an elegant and efficient framework for the treatment of the defect problem.[4,5] The extension of this method to include lattice relaxation effects was formally developed by Lodder and co-workers,[6] while Stefanou et al.[7] calculated the effect of lattice relaxation on the local moments of 3d impurities in Cu and on the electronic structure of a Pd impurity in Cu,[8] using the KKR formalism in the muffin-tin approximation. Gonis et al.[9] also developed a KKR scheme including lattice distortion, within the context of the coherent-potential approximation.

In this work, we extend the full potential version of the KKR Green function method in order to calculate lattice relaxation. The idea is to optimize the local geometry around a point defect using total energy and/or Hellmann-Feynman (H-F) forces, for various shifted atomic configurations. In the following sections we present a brief description of the theoretical method and report some preliminary results for the lattice relaxation around 3d substitutional impurities in Cu.

METHOD OF CALCULATION

Within the framework of the KKR multiple scattering theory the crystal is divided into non-overlapping, space-filling cells around each atom. Using a site-centered expansion, the crystal Green function can be written in the form[10]

$$G(\mathbf{r}+\mathbf{R}^n,\mathbf{r}'+\mathbf{R}^{n'};E)=\delta^{nn'}G^s(\mathbf{r}+\mathbf{R}^n,\mathbf{r}'+\mathbf{R}^n;E)+\sum_{L,L'}R_L^n(\mathbf{r};E)G_{LL'}^{nn'}(E)R_{L'}^{n'}(\mathbf{r}';E) \tag{1}$$

The first term on the right hand side represents the Green function for a single potential centered at the position $\mathbf{R}^n$ in an otherwise free space, while the second term gives the multiple scattering contribution. The vectors $\mathbf{r},\mathbf{r}'$ are restricted within the cells n, n′. The index L denotes the angular momentum quantum numbers ℓ,m and $R_L^n(\mathbf{r};E)$ are the regular wave functions for the potential in cell n. The matrix elements of the structural Green function $G_{LL'}^{nn'}$ of the defect system are connected with their counterparts of the ideal host $G_{LL'}^{0nn'}$ through the algebraic Dyson equation

$$G_{LL'}^{nn'}(E)=G_{LL'}^{0nn'}(E)+\sum_{n''}\sum_{L'',L'''}G_{LL''}^{0nn''}(E)\Delta t_{L''L'''}^{n''}(E)G_{L'''L'}^{n''n'}(E) \tag{2}$$

Here the summation extends over all cells and angular momenta where the difference $\Delta t_{L''L'''}^{n''}$ between the t-matrices of the system with and without defect is significant.

The charge density can be obtained from the Green function

$$n(\mathbf{r})=-\frac{1}{\pi}\mathrm{Im}\int_{-\infty}^{E_F}dEG(\mathbf{r},\mathbf{r};E) \tag{3}$$

For the calculation of the integral in (3) we take advantage of the analytical properties of the Green function and replace the integral on the real energy axis with a contour integral in the complex energy plane starting from the bottom of the valence band and ending at the Fermi energy. In this way a much smaller number of points are used for the integration, since the Green function is less structured as we move away from the real axis.[11]

In order to calculate the full-potential wave function we proceed as follows.[12] The wave function as well as the anisotropic potential within each cell are expanded in real spherical harmonics $Y_L(\mathbf{r})$

$$R_L^n(\mathbf{r};E)=\sum_{L'}R_{L'L}^n(r;E)Y_{L'}(\mathbf{r}),\qquad V^n(\mathbf{r})=\sum_L V_L^n(r)Y_L(\mathbf{r}) \tag{4}$$

The mixing of the L channels caused by the anisotropic potential leads to a system of coupled radial equations for the double-indexed radial wave functions $R_{L'L}^n(r;E)$. An efficient way to determine the full-potential wave functions is by first calculating the wave function $\overline{R}_\ell^n(r;E)$ for the spherical part of the potential $V_{\ell=0}^n(r;E)$ and then treating the non-spherical contribution as a perturbation

$$R_{L'L}^n(r;E)=\overline{R}_\ell^n(r;E)\delta_{LL'}+\int dr'r'^2g_{\ell'}(r,r';E)\sum_{L''}\Delta V_{L'L''}^n(r')R_{L''L}^n(r';E) \tag{5}$$

$g_\ell(r,r';E)$ is the single-cell radial Green function corresponding to the spherical part of the potential, and

$$\Delta V_{LL'}^n(r)=\sum_{L''\neq 0}C_{LL'L''}V_{L''}(r) \tag{6}$$

where $C_{LL'L''}$ are the Gaunt coefficients. Equation (5) can be solved iteratively. The second Born approximation is usually sufficient to obtain adequate convergence. The anisotropic part of the potential is important only in the outer region of the cells, whereas in the core region, where much more mesh points are needed, the potential is almost spherically symmetric. This implies that equation (5) needs only to be solved for rather few mesh points. Thus, using the second Born approximation and only a few mesh points to determine the anisotropic wave functions the computational time is considerably reduced.

The calculation of the total energy in a system with a point defect is a very delicate problem. Indeed, contrary to usual band structure calculations where charge neutrality is preserved by adjusting the Fermi energy E_F, in defect calculations the Fermi energy is fixed by the host. The usual deviation from charge neutrality is, typically, of the order of 0.01 electron if the perturbation is restricted only to the first neighbors, and drops to 0.0001 electrons if up to 5 shells around the impurity atom are taken into account. As a result of the long-range perturbation, the total energy converges slowly as a function of the size of the perturbed cluster. This problem can be overcome by using for the energy the following generalized functional, which is extremal for non-particle-conserving variations of the charge density[13]

$$\tilde{E}[n(\mathbf{r})] = E[n(\mathbf{r})] - E_F\left[\int n(\mathbf{r})d\mathbf{r} - N\right] \tag{7}$$

where $E[n(\mathbf{r})]$ is the usual density-functional total energy which is extremal only against density variations conserving the total number of particles N.

The total energy is separated into single particle and double-counting contributions. For the single-particle energy we calculate the integrated density of states by using the Lloyd formula generalised for full potentials.[12,13] The calculation of the double-counting terms which include the Coulomb and exchange-correlation energies, involves integrations over the Wigner-Seitz cell. This is done by introducing a shape function $\Theta(\mathbf{r})$ which equals to 1 inside a given cell and vanishes outside. The same shape function is used for the truncation of the anisotropic potentials outside a given Wigner-Seitz cell. Efficient algorithms for the calculation of the angular momentum components of this function for cubic cells as well as for an arbitrary polyhedron have been developed by Stefanou et al.[14]

In order to treat lattice relaxation, we need the expansion of the Green function around the new centers n, shifted by $\mathbf{s}^n$ with respect to the original ones. This can be obtained by the transformation[7]

$$\tilde{G}^{nn'}_{LL'}(E) = \sum_{L'',L'''} U_{LL''}(\mathbf{s}^n;E) G^{nn'}_{L''L'''} U_{L'L'''}(\mathbf{s}^{n'};E) \tag{8}$$

where the matrix U is given by

$$U_{L'L}(\mathbf{s};E) = 4\pi \sum_{L''} i^{\ell'+\ell''-\ell} C_{LL'L''} j_{\ell''}(s\sqrt{E}) Y_{L''}(\mathbf{s}) \tag{9}$$

$j_\ell(s\sqrt{E})$ are the spherical Bessel functions. From the above equation it follows directly that $\left[U^{-1}(\mathbf{s};E)\right]_{LL'} = U_{LL'}(-\mathbf{s};E) = U_{L'L}(\mathbf{s};E)$. The defect structural Green function can be obtained by solving the equation

$$\overline{G}^{nn'}_{LL'}(E) = \tilde{G}^{0\,nn'}_{LL'}(E) + \sum_{n''} \sum_{L'',L'''} \tilde{G}^{0\,nn''}_{LL''}(E) \left[t^{n''}_{L''L'''}(E) - \tilde{t}^{0\,n''}_{L''L'''}(E)\right] \overline{G}^{n''n'}_{L'''L'}(E) \tag{10}$$

where $\tilde{t}^0$ is the ideal host t-matrix expanded around the shifted centers

$$\tilde{t}^{0\,n}_{LL'}(E) = \sum_{L'',L'''} U_{LL''}(\mathbf{s}^n;E) t^{0\,n}_{L''L'''}(E) U_{L'L'''}(\mathbf{s}^n;E) \tag{11}$$

The transformations (8) and (11) do not involve major problems in the angular momentum convergence. It can be shown[7] that to a first order approximation the only non-vanishing off-diagonal elements of $U_{LL'}$ are those corresponding to $|\ell - \ell'| = 1$. Therefore, in the case of small displacements (typically a few percent) and for an angular momentum cutoff ℓ_{max}, an adequately accurate calculation of $\tilde{G}^0$ and $\tilde{t}^0$ up to $\ell_{max} - 1$ is ensured. Once the ideal host Green function and t-matrix are expanded around the new centers, local distortion of the lattice does not introduce any additional difficulty. Our previous results of total-energy and H-F force calculations using the full-potential KKR Green function method show that $\ell_{max} = 3$ is usually sufficient to obtain adequate convergence in the absence of lattice relaxation.[13,15] Therefore in the present calculations an angular momentum cutoff $\ell_{max} = 4$ in the expansion of the Green function is used. If we restrict ourselves to radial shifts only, the cubic symmetry is retained and equation (10) is solved utilizing a group theoretical decomposition of the matrices into irreducible submatrices, as in the case of no lattice relaxation.

The relaxed atomic positions can be calculated by minimizing the total energy. An alternative way is by calculating the configuration for which the forces on all atoms vanish. The H-F theorem provides an exact expression for the force on an atom, being given by the electric field on the nucleus due to all electrons and all the other nuclei. However in order to obtain reliable forces the potential must be calculated with high accuracy, which requires a full-potential treatment of both valence and core electrons. For this reason the application of the H-F theorem is limited by our spherical core treatment. This problem can be overcome by calculating the derivative of the total energy with respect to the atomic position if we assume that the Kohn-Sham equations are solved exactly only for the valence electrons. This finally leads to the following expression for the force on the atom at $\mathbf{R}^n$[15,16]

$$\mathbf{F}^n = Z^n \left.\frac{\partial V_{Coul}(\mathbf{r})}{\partial \mathbf{r}}\right|_{\mathbf{r}=\mathbf{R}^n} - \int d^3r\, n_c(\mathbf{r} - \mathbf{R}^n) \frac{\partial V_{eff}(\mathbf{r})}{\partial \mathbf{r}} \tag{12}$$

where Z^n is the nuclear charge, $n_c(\mathbf{r} - \mathbf{R}^n)$ is the core charge density in cell n, $V_{Coul}(\mathbf{r})$ is the Coulomb potential and $V_{eff}(\mathbf{r})$ is the Kohn-Sham effective potential. The first term describes the force of the valence electrons on the nucleus, while the second term gives the valence force on the core electrons, consisting of a Coulomb and an exchange-correlation part.

RESULTS AND DISCUSSION

We have used the full-potential KKR Green function method to determine the lattice distortion around 3d substitutional impurities in Cu. The calculations are carried out self-consistently within the framework of the density-functional theory. Exchange and correlation effects are included through the local-spin-density approximation, using the parameterization of Vosko, Wilk and Nusair.[17] First, the lattice constant of the Cu host is determined by minimization of the total energy and the ideal host Green function is obtained from the band structure. Since the relaxation in the case of the 3d impurities in Cu is small, the shape functions for the shifted configuration are obtained by considering the unshifted Wigner-Seitz cell of the ideal lattice but they are expanded into angular momenta around the new, shifted atomic positions. The total energy of a perturbed cluster of 5 shells (79 atoms in an fcc lattice) around the impurity together with the H-F force on each atom, are calculated for several radial shifts of the first nearest neighbors (n.n.).

The general trend that an outward relaxation enhances the impurity moment[7] was confirmed in the case of the magnetic impurities V, Cr, Mn, Fe. Typical results for the total energy and force are shown in figure 1 for a Ni impurity. The total energy exhibits a parabolic behavior with respect to the displacement of the first n.n., while the force on all atoms varies linearly. Our calculation predicts that the first n.n. around a Ni impurity in Cu are shifted inwards by 0.3% and the minimum energy configuration is in perfect agreement with that predicted from the zero H-F force condition.

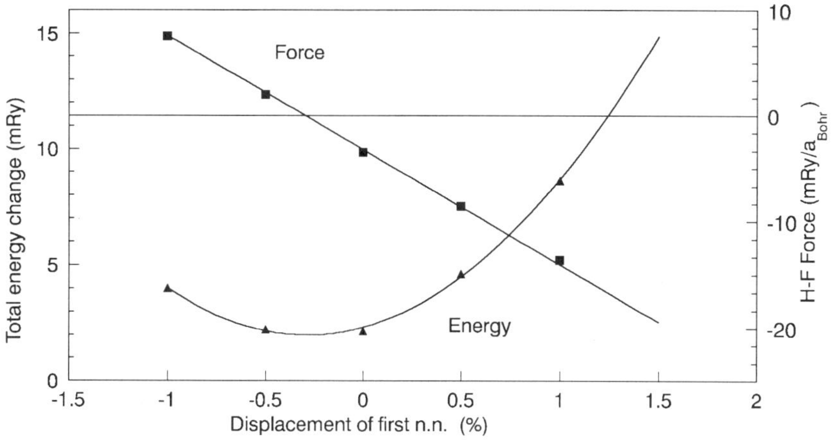

Figure 1. Change of the total energy and radial H-F force exerted on a first n.n. Cu atom, as a function of the displacement of the first shell calculated for a Ni impurity in Cu. A linear fit for the force and a parabolic one for the total energy change are shown.

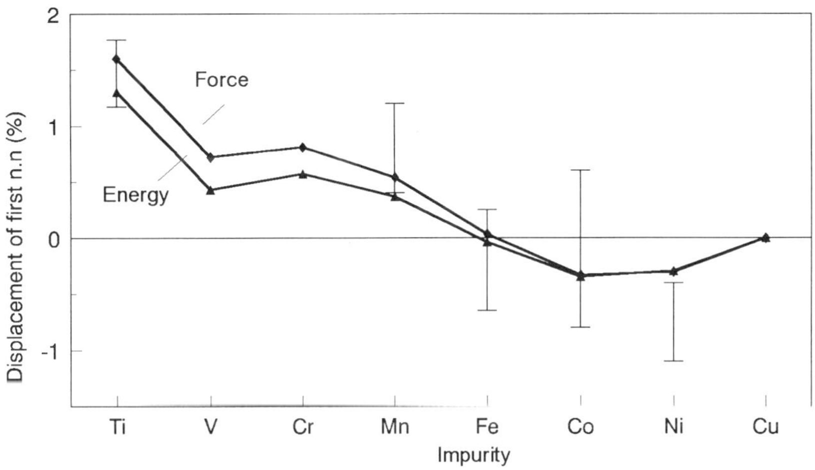

Figure 2. Displacement of the first n.n. around 3d impurities in Cu. The results obtained from both, total energy and H-F force, methods are shown. The error bars indicate the EXAFS experimental data.[1]

The calculated displacements of the first n.n. around the 3d impurities in Cu, together with the available experimental data are shown in figure 2. The agreement between theory and experiment is satisfactory. Albeit the displacement calculated from the total energy does not always agree with that predicted from the force. The discrepancy becomes more pronounced as we move to the beginning of the 3d series. For the total-energy calculation the generalized functional (equation (7)) was used. Since this is extremal against non-particle-conserving charge densities only second order errors of the charge density affect the calculated energies. On the other hand, the calculation of the force is very sensitive to any approximation for the charge density.[12] This fact could be the reason for the observed discrepancy.

In general, the calculation of the lattice distortion by considering only the displacement of the first n.n. leads to an underestimation of the outward shifts. An exact calculation of the lattice relaxation, would require to take into account the displacement of more atoms. The minimization of the total energy allowing more atoms around the impurity to relax is possible with our method, but at present it involves a large computational effort.

In summarising we have reported results for the lattice relaxation around 3d impurities in Cu, using the full-potential KKR Green function method. The atomic displacement was determined by total-energy and H-F force calculations. Our results are in good agreement with the available experimental data. Further work is being carried out in order to take into account the displacement of more atoms. This is especially important in the case of large lattice relaxation.

ACKNOWLEDGMENTS

This work was supported by a bilateral German-Greek cooperation grant. It has also benefited from collaborations within the HCM Network on "Ab-initio (from electronic structure) calculation of complex processes in materials" (Contract ERBCHRXCT 930369)

REFERENCES

1. U. Scheuer and B. Lengeler Phys. Rev. B **44**, 9883 (1991)
2. M. Scheffler, J.P. Vingeron and G.B. Bachelet, Phys. Rev. Lett. **49**, 1765 (1982); Phys. Rev. B **31**, 6541 (1985)
3. R. Benedek, L.H. Yang, C. Woodward and B.I.Min, Phys.Rev. B **45**, 2607 (1992)
4. P. Podloucky, R. Zeller and P.H. Dederichs Phys. Rev. B **22**, 5777 (1980)
5. P.J. Braspenning, R. Zeller, A. Lodder and P.H. Dederichs Phys. Rev. B **2**, 703 (1984)
6. A. Lodder, J. Phys. F **6** 1885 (1976); J. Molenaar and A.Lodder, J. Phys. F **13** (1983)
7. N. Stefanou, P.J. Braspenning, R. Zeller and P.H. Dederichs, Phys. Rev. B **36**, 6372 (1987)
8. N. Stefanou, R. Zeller and P.H. Dederichs, Solid State Commun. **62**, 735 (1987)
9. A. Gonis, A.J. Freeman and P. Weinberger, Phys. Rev. B **32**, 7713 (1985); ibid. **32**, 7720 (1985)
10. R. Zeller, J. Phys. C**20**, 2347 (1987)
11. R. Zeller, J. Deutz and P.H. Dederichs, Solid State Commun. **44**, 993 (1982)
12. P.H. Dederichs, B. Drittler and R. Zeller, Mat. Res. Soc. Symp. Proc. **253**, 185 (1992); B. Drittler, Ph.D. Thesis RWTH Aachen (1990) (unpublished)
13. B. Drittler, M. Weinert, R. Zeller and P.H. Dederichs, Phys. Rev. B**39**, 930 (1989)
14. N. Stefanou, H. Akai and R. Zeller, Computer Phys. Commun. **60**, 231 (1990); N. Stefanou and R. Zeller, J. Phys.: Condens. Matter **3**, 7599 (1991)
15. K. Abraham, Diploma Thesis RWTH Aachen (1990) (unpublished)
16. J. Harris, R.O. Jones and J.E. Muller, J. Chem. Phys. **75**, 3904 (1981)
17. S.H. Vosko, L.Wilk, and M. Nusair, J. Can. Phys. **58**, 1200 (1980)

PREDICTION OF LATTICE VIBRATIONS IN METASTABLE AND UNSTABLE TRANSITION METAL SYSTEMS

Göran Grimvall

Department of Physics
Royal Institute of Technology
S-100 44 Stockholm, Sweden

INTRODUCTION

Many materials of practical importance are not in their thermodynamic equilibrium, but contain metastable phases. Often thermodynamic data for such phases are difficult or virtually impossible to obtain experimentally, but the data may be required for instance in the modelling of materials processing. This paper presents some aspects of the prediction of the phonon contribution to the entropy, relying on extrapolations of known data for related systems. It is inherent in extrapolation methods that the desired quantity varies smoothly as a function of some other known quantity. Here we shall be particularly concerned with the failure of such a smooth behaviour.

THE ENTROPY DEBYE TEMPERATURE

The leading terms in a high-temperature expansion of the entropy S_{vib} of harmonic lattice vibrations are

$$S_{\mathrm{vib}} = 3Nk_{\mathrm{B}}\,[1 + \int_0^{\omega_{\max}} \ln(k_{\mathrm{B}}T/\hbar\,\omega)\,F(\omega)\,d\omega] = 3Nk_{\mathrm{B}}\,[4/3 + \ln(T/\theta_S)\,] \tag{1}$$

$F(\omega)$ is the phonon density of states. In the second equality, the logarithmically averaged phonon frequency ω has been represented by an entropy Debye temperature θ_S defined such that when it is used in a Debye model, the correct high temperature value of the entropy is reproduced. One can show that the description (1) of the entropy is adequate also when anharmonicity is included, within low-order perturbation theory.

The (squared) logarithmic average of the phonon frequencies has the important property that it factorises into the logarithmic average of the atomic masses, $M_{\log}$, and a factor k_S that has the dimension of a force constant and only depends on the electronic structure. Hence we can write

$$k_{\mathrm{B}}\theta_S = \hbar\,\sqrt{k_S/M_{\log}} \tag{2}$$

In our analysis we shall also use a quantity E_S with the dimension of energy. E_S can be obtained from information on the vibrational properties of a solid. It only depends on the electronic structure, and not on the atomic masses, and is defined

$$E_S \equiv k_S \Omega_a^{2/3} \tag{3}$$

where Ω_a is the average volume per atom in the solid.

The quantities θ_S and E_S have been discussed and utilised frequently; see for instance Refs. 1-4 and further references therein. The procedure to get E_S is first to calculate θ_S from the experimental S_{vib}, i.e., the entropy after subtraction of non-vibrational parts. Then k_S is obtained from eq. (2) and finally E_S from eq. (3).

EXTRAPOLATION OF E_S THROUGH ELECTRON BAND FILLING

In a rigid-band model, properties are related to the filling of the electron bands. The qualitative predictions of such a model usually hold also for real systems, when physical properties are considered as a function of the number of valence electrons per atom, n_e, and one compares chemically related solids with the same crystal structure. This is the case, e.g., for E_S of transition metal carbides[1,2], nitrides[1] and diborides[3]. Here we exemplify by 3*d*-transition metal carbides and oxides that have the NaCl lattice structure. Figure 1 shows E_S of ScC, TiC, VC and of TiO and VO. The figure also shows estimated E_S of CrC, MnC, FeC, and CrO in the NaCl structure. The qualitative variation of E_S with n_e can be understood from the electron density of states of these transition metal compounds in the NaCl structure[1,2]. Even complex carbides and nitrides, of the form M_mC_n and M_mN_n (M is a transition metal) show a regular behaviour in an E_S versus n_e plot, as discussed in detail elsewhere[2]. However, this type of interpolation or extrapolation to estimate E_S, and hence θ_S and the vibrational entropy, of course assumes that there is no abrupt change in the general character of the electronic band structure. The oxides MnO, CoO and NiO, which are also stable in NaCl-type or distorted NaCl-structures, are (antiferromagnetic) insulators and therefore have effective force constants k_S that do not extrapolate from those for the metallic oxides TiO, VO and CrO, cf. Fig. 1.

For MnO and CoO, S_{vib} was obtained after an approximate and graphical subtraction of the magnetic anomaly as measured in the heat capacity. NiO has such a high Néel temperature (640 K) that the vibrational E_S is well approximated by the total room temperature entropy. We return to the results of Fig. 1 in the concluding discussion.

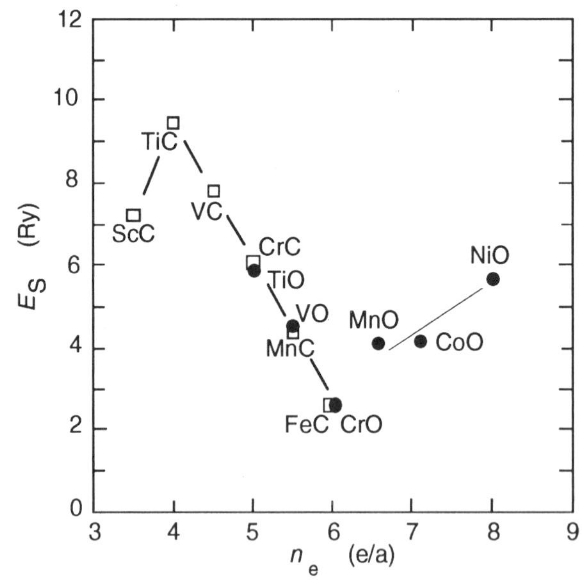

Figure 1. The quantity E_S, defined in eq. (3), plotted versus the average number of valence electrons per atom in the compound, n_e. E_S for the metallic systems are from Ref. 1 and for the insulators MnO, CoO and NiO as calculated in this work from experimental data[5-7]. The connecting lines are just guides for the eye.

EXTRAPOLATION IN PHASE DIAGRAMS

Figure 2 shows the essential parts of the Pt-W phase diagrams. Pt is stable in the fcc structure and W in the bcc structure. It would seem natural to extrapolate the liquidus and solidus lines of the wide fcc Pt-W phase field to get a melting temperature of about 3000 K for a metastable pure bcc W phase. However, such a procedure leads[8,9] to severe inconsistencies when the so-called calphad type of thermodynamic analysis of the phase diagram is compared with values for the relative stability of bcc and fcc W obtained from *ab initio* calculations of the electronic structure. The calphad method usually assumes implicitly that the Debye temperature of competing phases (here fcc and bcc W) are not very different, and that θ_S varies smoothly as a function of alloy concentration. This assumption seems natural, in view of the results in the preceding section and since there is no reason to suspect a metal-insulator transition in the fcc Pt-W alloy. The resolution of the apparent discrepancy between semiempirical and *ab initio* approaches has recently been found to be traced to the fact that fcc W is not dynamically stable[8,10]. In fact, an *ab initio* calculation of the elastic constants shows that fcc W is dynamically unstable for all shear modes at 0 K.[8]

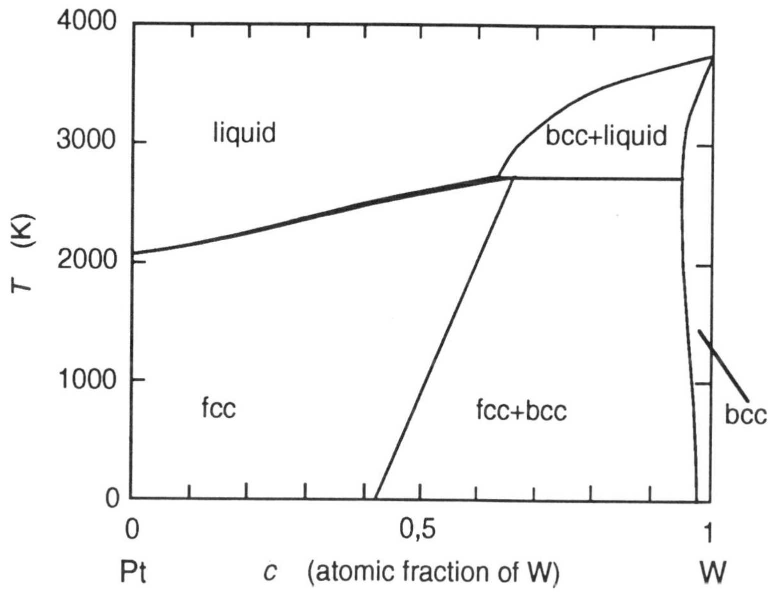

Figure 2. A schematic drawing of the phase diagram of the Pt-W system; cf. Ref. 8.

We now argue why the phase diagram does not show any conspicuous feature related to the fact that the fcc Pt-W gradually becomes more unstable as the W concentration increases. Let the transverse phonon frequencies be described by a Debye spectrum, $\omega = C(\theta,\phi)q$, where $C(\theta,\phi)$ is the sound velocity in the direction given by the angles θ,ϕ and q is the wave number. Assume that $C(\theta,\phi)$ gradually decrease as the W concentration c increases towards the concentration c_0 where C first becomes negative, in a certain crystallographic direction θ_0,ϕ_0. Expand C around this direction, where C by the assumption has a minimum. Thus,

$$C(\theta,\phi) = C(\theta_0,\phi_0) + \alpha(\theta-\theta_0)^2 + \beta(\theta-\theta_0)(\phi-\phi_0) + \gamma(\phi-\phi_0)^2 + \ldots \quad (4)$$

According to the first equality in Eq. (1) the entropy will contain a term

$$\int \ln [C(\theta,\phi)] \sin(\theta-\theta_0)\, d(\theta-\theta_0)\, d(\phi-\phi_0) \quad (5)$$

When $c \to c_0$, and hence $C(\theta_0,\phi_0) \to 0$, the integral (5) does not diverge. Hence S_{vib} has a finite limiting value at the instability concentration c_0, and does not diverge logarithmically as one might first have thought from the divergence of S_{vib} in the Debye model of the last equality in Eq. (1) when $\theta_S \to 0$. For $c > c_0$ the lattice is unstable and the vibrational entropy is undefined.

Now compare the total vibrational entropy $S_{vib}(c)$ as a function of the alloy composition c, in two cases with different behaviour - one stable for all c and one with an instability. In the first case, let the phonon frequencies gradually decrease with c, over the entire range $0 \leq c \leq 1$. In the other case, let the frequencies in certain parts of the Brillouin zone tend to zero at $c = c_0$, while the other phonon modes vary much less with c. Both cases may result in the same S_{vib} for $1 \leq c < c_0$. Since the phase diagram only depends on the total S_{vib} (or, rather, the Gibbs energy $G = U - TS + pV$), there is no precursor in the phase diagram that allows us to distinguish between the two cases. (Of course, a direct measurement of, e.g., elastic constants as a function of c, would reveal an incipient elastic instability.)

EXTRAPOLATIONS OF PHONON FREQUENCIES AS A FUNCTION OF TEMPERATURE

Normally[4,11], anharmonic effects give rise to a smooth decrease of $\theta_S(T)$ as a function of the temperature T. This behaviour is shown in Fig. 3 for the hcp phase of Ti. The bcc Ti phase is well known to be dynamically unstable at low T for a certain shear mode of short wavelength (i.e., not governed by the elastic constants) but bcc Ti is stabilised at high T. Figure 3 shows $\theta_S(T)$, calculated from S_{vib} as given by Petry et al.[12] They obtained S_{vib} from phonon density-of-states fitted to their measurements of the phonon dispersion curves by inelastic neutron scattering. Often, the significant difference in θ_S between the hcp and bcc structures at the hcp-to-bcc transition temperature T_{tr} = 1156 K (see Fig. 3) is taken as

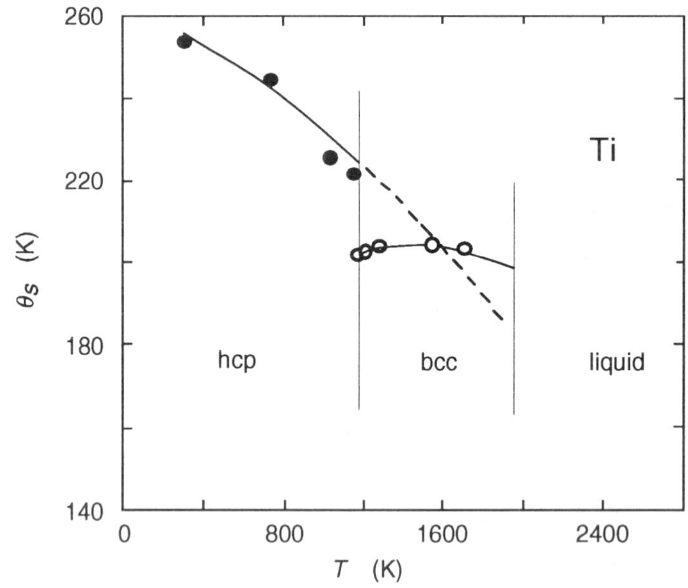

Figure 3. The entropy Debye temperature $\theta_S(T)$ for hcp and bcc Ti (filled and open circles), derived from the phonon spectrum measured by inelastic neutron scattering[12]. The dashed line for hcp Ti at high temperatures is a smooth extrapolation suggested here.

evidence that the bcc phase has softer phonon frequencies than the hcp phase. However, this conclusion may be misleading. If we extrapolate θ_S of hcp Ti beyond T_{tr}, with a temperature dependence that is "normal"[4,11] (dashed line in Fig. 3) we see that near the melting temperature the bcc phase is in fact dynamically stiffer than the hcp phase.

In view of the result for Ti (and similarly for Zr) one might now ask if not fcc W could be stabilised at high T. An *ab initio* calculation to answer this question is still not feasible. However, the measured phase diagram is difficult to reconcile with a high-temperature stabilised fcc W, if one assumes that the enthalpy difference between bcc and hcp W is correctly given by *ab initio* calculations[8,9].

CONCLUDING DISCUSSION

Extrapolation methods to estimate unknown thermodynamic properties of course assume a "normal" behaviour with no unexpected and abrupt change, e.g., in the electronic structure, as one proceeds along an extrapolation curve. The vibrational entropy of a solid depends on the interatomic forces and on the atomic masses. However, in the high-temperature vibrational entropy the effect of the atomic masses enters only through a trivial scaling factor. The remaining characteristic of the vibrational entropy, in this paper represented by a quantity E_S with the dimension of energy, usually shows a smooth and regular behaviour as a function of the number of valence electrons per atom. Hence the vibrational entropy is expected to vary smoothly with the atomic composition in solid-solution alloys. Another "normal" behaviour is that the entropy Debye temperatures do no vary much between the fcc, hcp and bcc lattice structures of a given element. On the basis of these two assumptions there has been a severe discrepancy between the relative enthalpy of fcc and bcc W obtained from a semiempirical analysis of experimental phase diagrams and obtained from *ab initio* electron structure calculations. The resolution of the discrepancy is to be sought in the fact that fcc W is not dynamically stable. Hence this phase can not be assigned a vibrational entropy. One should therefore not compare *ab initio* results for the enthalpy of a rigid fcc lattice with semiempirical extrapolations into a concentration region where the fcc phase is unstable. An analogous discrepancy has been observed for some other transition metals, and the dynamical instability provides a likely explanation also in these cases.

Acknowledgements

I want to thank Armando Fernández Guillermet, Vidvouds Ozolins and Martin Körling for many discussions of the problems addressed in this paper. The work was supported by the Swedish research councils TFR and NUTEK.

REFERENCES

1. A. Fernández Guillermet and G. Grimvall, Cohesive properties and vibrational entropy of 3*d* transition-metal compounds: *MX*(NaCl) compounds (*X*=C, N, O, S), complex carbides, and nitrides, *Phys. Rev. B* 40:10582 (1989).
2. A. Fernández Guillermet and G. Grimvall, Cohesive properties and vibrational entropy of 3*d*-transition metal carbides, *J. Phys. Chem. Solids* 53:105 (1992).
3. A. Fernández Guillermet and G. Grimvall, Bonding properties and vibrational entropy of transition metal MeB_2 (AlB_2) diborides, *J. Less-Common Metals* 169:257 (1991).
4. G. Grimvall and A. Fernández Guillermet, Systematics of bonding properties and vibrational entropy in compounds, *in* :"Advances in Physical Geochemistry, vol. 11. Thermodynamic Data Systematics," p. 272, S.K. Saxena ed., Springer, Berlin (1992).
5. I. Bahrin, "Thermochemical Data of Pure Substances", VCH, Weinheim (1989).
6. S.S. Todd and K.R. Bonnickson, *J. American Chemical Society* 73:3894 (1951).
7. E.G. King, *J. American Chemical Society* 79:2399 (1957).
8. A. Fernández Guillermet, V. Ozolins, G. Grimvall and M. Körling, Phase stabilities in the Pt-W system: Thermodynamic and electronic-structure calculations, *Phys. Rev. B* 51:10364 (1995).

9. G. Grimvall, M. Thiessen and A. Fernández Guillermet, Thermodynamic properties of tungsten, *Phys. Rev. B* 36:7816 (1987).
10. P.J. Craievich, M. Weinert, J.M. Sanchez and R.E. Watson, Local stability of nonequilibrium phases, *Phys. Rev. Lett.* 72:3076 (1994).
11. A. Fernández Guillermet and G. Grimvall, Analysis of thermodynamic properties of molybdenum and tungsten at high temperatures, *Phys. Rev. B* 44:4332 (1991).
12. W. Petry, A. Heiming, J. Trampenau, M. Alba, C. Herzig, H.R Schober, and G. Vogt, Phonon dispersion of the bcc phase of group-IV metals. I. bcc titanium, *Phys. Rev. B* 43:10933 (1991).

ITINERANT MAGNETISM OF (001) SURFACES OF RANDOM NICKEL–COPPER AND IRON–VANADIUM ALLOYS

I. Turek,[1] J. Kudrnovský,[2] M. Šob,[1] and V. Drchal[2]

[1]Institute of Physics of Materials, Academy of Sciences of the Czech Republic, Žižkova 22, CZ-616 62 Brno, Czech Republic
[2]Institute of Physics, Academy of Sciences of the Czech Republic, Na Slovance 2, CZ-180 40 Prague 8, Czech Republic

INTRODUCTION

Itinerant magnetism of surfaces of transition metals represents an important and interesting field benefiting from recent progress in sample preparation techniques as well as from modern methods of electronic structure calculations of ground state properties. The latter are based on the local spin density approximation (LSDA) and provide both parameter-free explanations of experimental results and predictions of new phenomena. Systematic ab-initio investigations[1] of low-index transition metal surfaces revealed surface enhancement of local magnetic moments of magnetic 3d metals (Fe, Co, Ni, Cr) which is due to reduced coordination. Special attention was paid to further transition elements (especially to V, Rh, and Pd) which were supposed to exhibit surface magnetism of a non-magnetic bulk metal.

Contrary to pure metals, bulk magnetic alloys of transition metals show great variety of their magnetic behaviour. The ground state properties (magnetization, lattice constants) are often strongly dependent on the composition and on the degree of atomic order and their complete understanding represents a challenge for condensed-matter theory. In case of random binary alloys of 3d metals, extensive LSDA calculations[2] reproduced a lot of measured concentration trends and yielded a deep insight into the underlying microscopic mechanisms. As a consequence, systematic dependences of magnetization on the average number of electrons per atom (Slater-Pauling curves) were clarified in details for a number of alloy systems.

Surfaces of transition metal alloys are expected to be even more complex from the magnetic point of view. It is the purpose of this contribution to present the results of ab-initio LSDA electronic structure calculations for (001) surfaces of random Ni-Cu and Fe-V alloys. These two systems were chosen as typical alloys of a magnetic and a non-magnetic 3d metal exhibiting thus a magnetic phase transition in the bulk. Concentration dependences of bulk magnetism in these alloys were successfully studied in past[3,4] by the Korringa-Kohn-Rostoker (KKR) method using the coherent potential approximation (CPA). In this study, we used a self-consistent Green's function technique based on the tight-binding linear muffin-tin orbital (TB-LMTO) method in the atomic sphere approximation (ASA) which is described elsewhere.[5]

MODEL AND NUMERICAL DETAILS

We approximated the atomic configuration of a real disordered alloy both in the bulk and at its surface by a perfect crystalline lattice occupied at random by constituent atoms. We neglected possible lattice and layer relaxations. The atomic volumes of the pure elements were taken from experiment and the atomic volume of a random alloy was set according to the Vegard's law. The radii of the atomic spheres of both alloy constituents were the same. For surface calculations, the selfconsistency of one-electron potentials was achieved within a finite number of layers near the solid-vacuum interface: 8 layers (6 atomic and 2 empty-sphere) were used in the fcc (001) Ni-Cu case whereas the more open bcc (001) surface of Fe-V alloys was treated with 10 layers (7 atomic and 3 empty-sphere).

The exchange-correlation potential was taken in the form given by Vosko et al.[6] and necessary integrations were performed with 14 nodes on the complex energy contour and with 45 $k_{||}$-points in the irreducible part of the two-dimensional Brillouin zone.

RESULTS AND DISCUSSION

Nickel–Copper Alloys

The calculated magnetization in the bulk as well as at the (001) surface was entirely due to the local magnetic moments of nickel atoms, those of copper atoms being negligible. Figure 1 presents the concentration trends of the nickel moments in the bulk $Ni_{1-x}Cu_x$ alloy and in the top surface layer of its (001) face. We considered two composition profiles of the alloy surface: (i) a homogeneous profile with all layers of the same concentration as that of the bulk and (ii) a profile qualitatively simulating the experimentally observed and theoretically reproduced surface enrichment by copper atoms.[7] In the latter case, the nickel concentration in the top surface layer was reduced to one half of its bulk value.

It is seen that the nickel local moment decreases with increasing copper content in all cases (Figure 1). The bulk moment vanishes at $x \approx 0.57$ in good agreement with experiment[8] and previous KKR-CPA calculations.[3] For the surface with the homogeneous concentration profile an enhancement of magnetism is seen both in the magnitude of the local moments and in the concentration range for the occurrence of magnetism. The enhancement of the local moments for the nickel-rich alloys is similar to that obtained for the pure Ni (001) surface.[1] The critical concentration for extinction of surface magnetism is shifted to $x \approx 0.65$ leading thus to magnetic surfaces of non-magnetic bulk alloys near $x \approx 0.60$. However, the local moments are strongly sensitive to the surface segregation of copper which destroys the whole magnetic enhancement (Figure 1) and will probably prevent its direct experimental verification for this alloy system.

The basic mechanism governing the reduction of the nickel moment in the bulk is the gradual filling of the nickel d band on alloying with copper but underlying physics is more complex than assumed in a popular rigid band model.[9,10] Figure 2 presents a plot of the nickel local moment against the local number of nickel d electrons for the bulk as well as for surfaces with both profiles. This analysis indicates that the same mechanism of the reduction of Ni magnetization applies also at the alloy surfaces irrespective of their composition profiles.

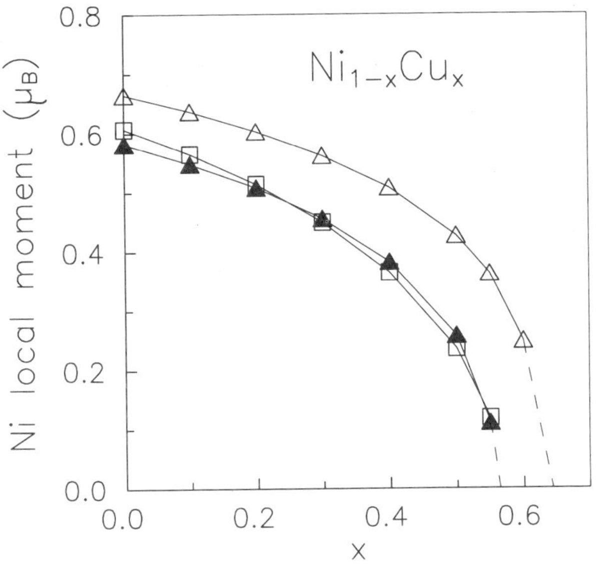

Figure 1. Local magnetic moment of nickel atoms in Ni-Cu alloys as a function of composition: in the bulk (open squares) and in the top surface (001) layer for the homogeneous concentration profile (open triangles) and for the profile with the copper-enriched top surface layer (full triangles).

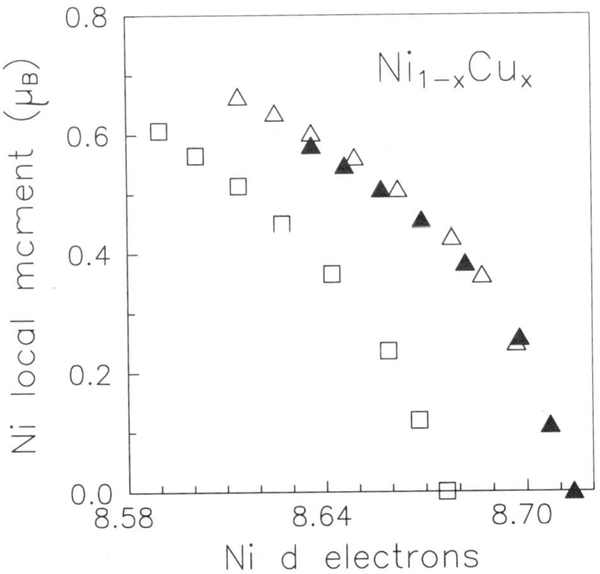

Figure 2. Correlation of local magnetic moments and local numbers of d electrons for nickel atoms in Ni-Cu alloys: in the bulk (open squares) and in the top surface (001) layer for the homogeneous concentration profile (open triangles) and for the profile with the copper-enriched top surface layer (full triangles).

Iron–Vanadium Alloys

The magnetic structure of the bulk $Fe_{1-x}V_x$ system is more complicated than that of the nickel-copper alloy as the vanadium moment is not negligible and is coupled antiparallel to the iron moment.[4,11] The average magnetization drops from the bulk iron value to zero at $x \approx 0.77$.[12] These facts were nicely reproduced by our bulk calculations yielding the critical concentration for disappearance of bulk magnetism $x \approx 0.80$ as shown in Figure 3.

The magnetic structure of the (001) surface depends strongly on the alloy composition (Figure 3). For the vanadium contents up to $x \approx 0.6$, the local moments in the top surface layer merely reflect the bulk magnetic structure enhanced due to reduced coordination. The behaviour of alloy surfaces with higher vanadium concentrations differs markedly in two features from that of the bulk: (i) the iron local moment in the top surface layer survives up to pure vanadium (single impurity limit) being nearly insensitive to the alloy composition, and (ii) a parallel (ferromagnetic) alignment of the vanadium and iron surface moments appears near the critical concentration for the collapse of the bulk magnetism ($x \approx 0.8$). For the vanadium concentrations $x \geq 0.8$, we have again magnetic surfaces of the non-magnetic bulk alloys. However, the pure vanadium surface comes out to be non-magnetic in agreement with a full-potential calculation[13] but in contradiction to electron-capture spectroscopy.[14] Figure 4 presents the calculated magnetic profiles for two compositions and proves that the effect of surface on local moments is non-negligible only in the top surface layer.

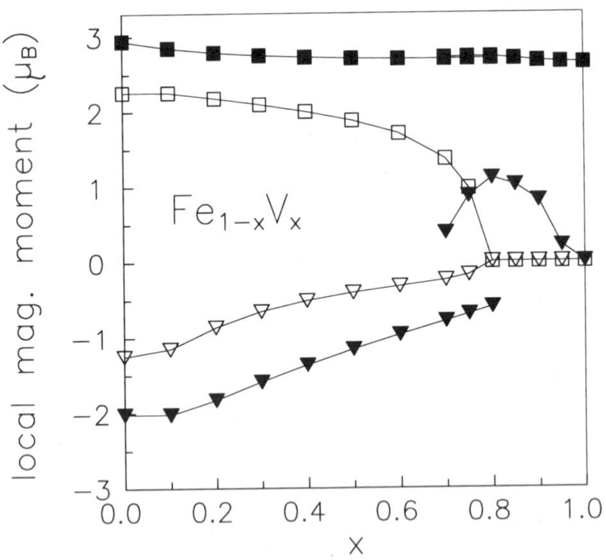

Figure 3. Local magnetic moments in Fe-V alloys as a function of composition: in the bulk (open squares – Fe, open triangles – V) and in the top surface (001) layer (full squares – Fe, full triangles – V).

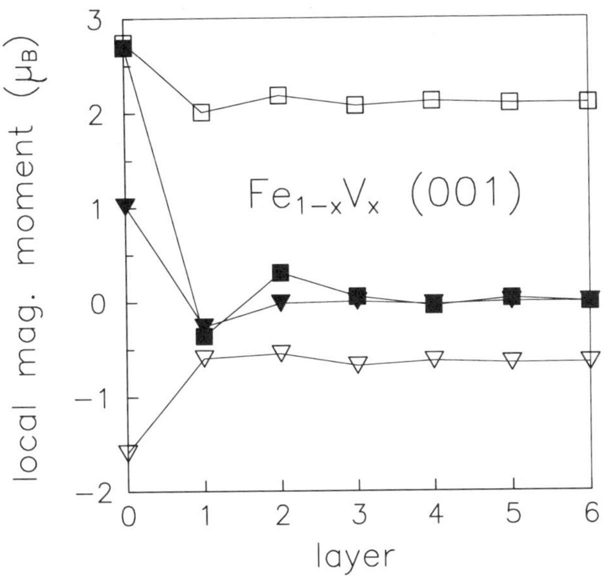

Figure 4. Layer-dependent local magnetic moments at the (001) surface of two $Fe_{1-x}V_x$ alloys: iron-rich (x = 0.3) alloy (open squares – Fe, open triangles – V) and vanadium-rich (x = 0.85) alloy (full squares – Fe, full triangles – V). The layer numbering starts from the top surface layer denoted by 0.

The surface ferromagnetism near $x \approx 0.8$ was investigated from the point of view of its energetical stability using total energy calculations in the ASA. For the vanadium concentrations $x > 0.8$ (non-magnetic bulk alloys), the ferromagnetic surface is energetically more favourable than the non-magnetic one. The magnitude of the corresponding spin-polarization energy for $x = 0.85$ amounts to 5.7 mRy per surface atom which is a relatively large value for a vanadium-rich system.[15] For alloys in the concentration region $0.7 \leq x \leq 0.8$, two surface magnetic solutions – antiferromagnetic (AFM) and ferromagnetic (FM) – were found (Figure 3) with energy differences $E_{AFM} - E_{FM} = 0.05$, 0.48, and 0.93 mRy per surface atom for x = 0.7, 0.75, and 0.8, respectively. These values show that the ferromagnetic solution is energetically more favourable but that its stability rapidly decreases towards the iron-rich systems. The small values of the energy difference $E_{AFM} - E_{FM}$ (being at the accuracy limit of the ASA) indicate that the true magnetic ground state might be even more complex and that its proper description would require to consider more than one magnetic state for each alloy constituent in several surface layers.[2] Such a treatment is beyond the scope of the present study.

The origin of the magnetic instability of the (001) surface of vanadium-rich alloys lies in the local density of states (LDOS) for the non-magnetic state at the Fermi level. This quantity was calculated for the vanadium concentration x = 0.85 and the resulting values were (in states per Ry, atom, and spin): LDOS(Fe) = 5.8, LDOS(V) = 7.2 in the bulk, and LDOS(Fe) = 24.6, LDOS(V) = 16.5 in the top surface layer. Hence, the surface enhancement of the LDOS for both alloy constituents induces surface magnetism of non-magnetic bulk vanadium-rich alloys.

CONCLUSION

We found on the basis of the local spin density calculations that the (001) surfaces of Ni-Cu and Fe-V alloys exhibit in general magnetic enhancement. This enhancement concerns: (i) the local magnetic moments similarly as in case of pure transition-metal surfaces, and (ii) the composition range for the occurrence of magnetism. The latter fact leads to magnetic surfaces of non-magnetic bulk alloys with compositions near the critical composition for the onset of bulk magnetism. For the Ni-Cu (001) surface, the local moments of surface nickel atoms are strongly sensitive to the copper enrichment of the top surface layer which reduces the magnetic surface enhancement in this particular system. For the Fe-V (001) surface near the vanadium concentration $x \approx 0.8$, we revealed a tendency to parallel alignment of the surface local moments of iron and vanadium atoms which is a magnetic structure qualitatively different from the antiparallel alignment encountered in the bulk Fe-V alloys.

Acknowledgments

The financial support for this work was provided by the Grant Agency of the Czech Republic (Project No. 202/93/1174).

REFERENCES

1. A.J. Freeman and R. Wu, J. Magn. Magn. Mater. **100**, 497 (1991).
2. P.H. Dederichs, R. Zeller, H. Akai, and H. Ebert, J. Magn. Magn. Mater. **100**, 241 (1991).
3. S. Crampin, R. Monnier, T. Schulthess, G.H. Schadler, and D.D. Vvedensky, Phys. Rev. B **45**, 464 (1992).
4. D.D. Johnson, F.J. Pinski, and J.B. Staunton, J. Appl. Phys. **61**, 3715 (1987).
5. J. Kudrnovský, I. Turek, V. Drchal, and M. Šob, these *Proceedings.*
6. S.H. Vosko, L. Wilk, and M. Nusair, Can. J. Phys. **58**, 1200 (1980).
7. A. Pasturel, V. Drchal, J. Kudrnovský, and P. Weinberger, Phys. Rev. B **48**, 2704 (1993).
8. R.A. Medina and J.W. Cable, Phys. Rev. B **15**, 1539 (1977).
9. J. Tersoff and L.M. Falicov, Phys. Rev. B **25**, 4937 (1982).
10. I. Turek, J. Kudrnovský, V. Drchal, and P. Weinberger, Phys. Rev. B **49**, 3352 (1994).
11. I. Mirebeau, G. Parette, and J.W. Cable, J. Phys. F: Met. Phys. **17**, 191 (1987).
12. D.J. Lam, D.O. van Ostenburg, M.V. Nevitt, H.D. Trapp, and D.W. Pracht, Phys. Rev. **131**, 1428 (1963).
13. S. Ohnishi, C.L. Fu, and A.J. Freeman, J. Magn. Magn. Mater. **50**, 161 (1985).
14. C. Rau, C. Liu, A. Schmalzbauer, and G. Xing, Phys. Rev. Lett. **57**, 2311 (1986).
15. V.L. Moruzzi and P.M. Marcus, Phys. Rev. B **45**, 2934 (1992).

CALCULATION OF SPONTANEOUS RESISTANCE ANISOTROPY OF DISORDERED FERROMAGNETIC ALLOYS

John Banhart

Fraunhofer-Institute for Applied Materials Research (IFAM)
Lesumer Heerstr. 36, 28717 Bremen, Germany

INTRODUCTION

There are a number of interesting phenomena related to impurity scattering in ferromagnetic alloys for which up to now no description from first-principles is available. One of these is the spontaneous magnetoresistance anisotropy (SMA) which occurs in principle in any ferromagnetic alloy and expresses the fact that the spontaneous magnetization reduces the symmetry of a crystalline system and leads to an electrical conductivity which depends on the relative orientation of the current and the magnetic field.

Quantitatively, both effects are expressed by means of the electrical resistivity (conductivity) tensor $\boldsymbol{\rho}$ ($\boldsymbol{\sigma}$) which for cubic systems with the magnetisation along the z-axis has the form[1]:

$$\boldsymbol{\rho} = \boldsymbol{\sigma}^{-1} = \begin{pmatrix} \rho_\perp & -\rho_H & 0 \\ \rho_H & \rho_\perp & 0 \\ 0 & 0 & \rho_\| \end{pmatrix}, \tag{1}$$

$\rho_\perp$ and $\rho_\|$ being the transverse and longitudinal resistivities. ρ_H is the Hall resistivity which shall not be considered in the present paper. The SMA is expressed by the ratio[1]:

$$\frac{\Delta\rho}{\bar{\rho}} = \left.\frac{\rho_\|(B) - \rho_\perp(B)}{\bar{\rho}(B)}\right|_{B\to 0}, \tag{2}$$

where $\bar{\rho} = \frac{1}{3}(2\rho_\perp + \rho_\|)$. All quantities are determined by extrapolation to a vanishing external magnetic field B.

To a great extent motivated by the technological importance of the SMA, there have been a large number of corresponding experimental investigations during the last four decades. Among the systems studied, Ni- and Co-based alloys received special attention because of the pronounced SMA found for many of them. In particular, the system Fe–Ni shows one of the largest SMA found so far for transition metal systems. The existing experimental SMA data at low temperatures of disordered, polycrystalline Fe–Ni alloys is summarised in Fig. 1.

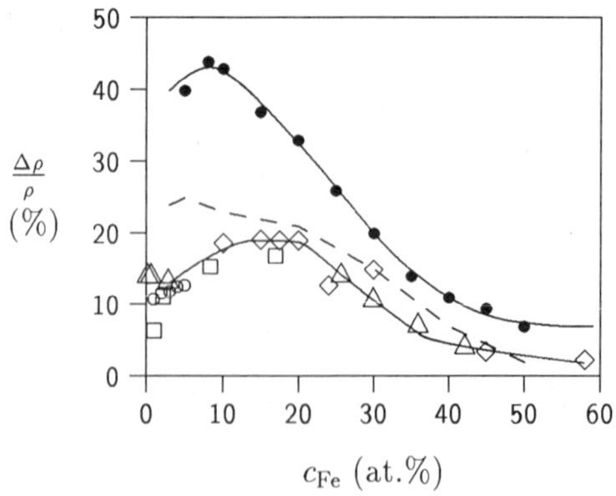

Figure 1: Spontaneous resistance anisotropy of Fe–Ni alloys. Open symbols: experimental values at low temperatures (□[1], ◇[2], △[3,4], ○[5]). Full circles: calculated values, broken line: calculated values corrected for extra isotropical scattering. Full lines serve as guides for the eye.

For isolated Fe impurities in Ni the residual resistivities $\rho_\perp$ and $\rho_\parallel$ vary linearly with concentration (see below) and the SMA ratio $\Delta\rho/\bar{\rho}$ takes a constant value which is about 13%. As the impurities begin to interact for increasing Fe contents, the SMA ratio rises and reaches its maximum value for 10 to 20% iron. For even larger Fe contents the anisotropy rapidly decreases to very low values.

It was first noted by Smit[2] that the physical origin of the SMA is the spin-orbit interaction. Based on this central assumption several authors developed more and more sophisticated models to describe the SMA in the past[2,3]. However, all these approaches end up with the sd-picture of electronic conduction. This also applies to the most recent refinement by Malozemoff [6]. Moreover, all theories on SMA presented so far are based on the two-current model separating the total current into two distinct contributions coming from electrons of different spin orientation. Spin-orbit coupling gives rise to hybridisation of electronic states of different spin character and causes the SMA that way. The aim of the present work was to treat the SMA without assuming any simple electronic structure and without treating the spin-orbit interaction as small perturbation.

THEORY

A straightforward and rigorous access to galvanomagnetic effects is supplied by the Kubo-Greenwood Eq. for the conductivity tensor $\boldsymbol{\sigma}$[7,8]:

$$\sigma_{\mu\nu} = \frac{\hbar}{\pi V_{\mathrm{cryst}}} \mathrm{Tr} \left\langle j_\mu \, \mathrm{Im} G^+(E_{\mathrm{F}}) \, j_\nu \, \mathrm{Im} G^+(E_{\mathrm{F}}) \right\rangle_{\mathrm{conf.}} \tag{3}$$

Here $G^+(E_F)$, representing the electronic structure of the system, is the positive side limit of the single-particle Green function at the Fermi energy E_F, j_μ is the μ-th spatial component of the electronic current operator $\boldsymbol{j}$ and $\langle \ldots \rangle_{\mathrm{conf.}}$ denotes the atomic configuration average for a disordered alloy. By determining G^+ via the Korringa-Kohn-Rostoker method of bandstructure calculation in connection with the coherent-potential approximation (KKR–CPA), Eq. (3) can be evaluated in a most reliable way. The corresponding expressions for $\sigma_{\mu\nu}$ for paramagnetic systems and temperature T=0K have been derived by Butler[8] and have been applied with great success in their original non-relativistic form[9] as well as their corresponding fully relativistic form[10] to calculate the residual resistivity of various alloy systems.

To get access to the SMA in the limit T=0K Butler's approach was generalized by evaluating G^+ using the spin-polarized relativistic version of the KKR-CPA (SPR-KKR-CPA)[11]. This scheme, based on the Dirac-equation for a spin-dependent potential derived from local spin density functional theory, accounts on the same level – without using any parameters – for all relativistic effects as well as for the magnetic state. A natural consequence of this is that the symmetry reduction due to the simultaneous presence of spin-orbit interaction and magnetism – giving rise to the form of $\boldsymbol{\rho}$ in Eq. (1) – is automatically accounted for. Finally, one has to mention that using the SPR-KKR-CPA there is no need to rely on the two-current model anymore.

CALCULATIONS

Fe–Ni

As a first application of the presented approach the alloy system Fe-Ni in the fcc structure was studied. Assuming the magnetisation to point along the z-axis, the electronic structure for various compositions with $x_{Fe} \leq 50\%$ was calculated using the SPR-KKR-CPA with an angular momentum expansion up to $\ell_{\max} = 3$. For each concentration the tensor $\boldsymbol{\sigma}$ and its inverse $\boldsymbol{\rho}$ were determined on the basis of Eq. (3). The resulting longitudinal and transverse resistivities, $\rho_{\parallel}$ and $\rho_{\perp}$, resp., together with the average resistivity $\bar{\rho}$ are shown in Fig. 2.

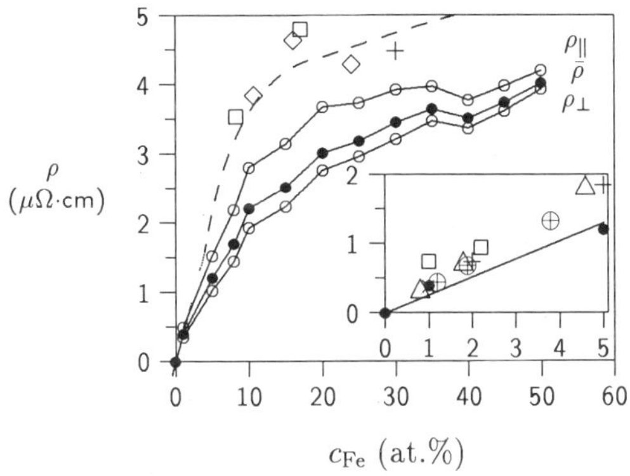

Figure 2: Calculated residual resistivities of Fe–Ni alloys: longitudinal (||) and transverse (⊥) resistivity, average $\bar{\rho} = \frac{1}{3}(2\rho_{\perp} + \rho_{\parallel})$. Experimental low temperature resistivities $\bar{\rho}$ by various authors: □[1], ◇[2], ×[5], △[13], ⊕[14], +[15]. Solid line connects calculated values. Dashed line is a guide for the experimental resistivities.

Comparison of $\bar{\rho}$ with experimental values reveals that the calculated resistivities are somewhat lower than the measured ones – especially for higher Fe contents. An important reason for the discrepancy might be that the CPA approach accounts only for chemical disorder but not for topological disorder as the source of the resistivity (see also below).

As one sees in Fig. 2, for all compositions we find $\rho_{\parallel} > \rho_{\perp}$ corresponding to a positive anisotropy ratio. This is in full accordance with experimental findings for Fe-Ni. The SMA ratio calculated from $\rho_{\perp}$ and $\rho_{\parallel}$ in Fig. 2 using Eq. (2) is shown in Fig. 1. The most prominent feature is the very pronounced decrease of $\Delta\rho/\bar{\rho}$ from about 40% for Ni-rich alloys to values of about 7% for 50% Fe. There is even slight evidence for a maximum value at 10% Fe. Obviously, this concentration dependence of the calculated SMA ratio is in rather satisfying agreement with experiment. However, the calculated SMA ratios are about twice as high as the measured ones and the maximum is less pronounced in the calculated curve. There are a number of reasons which could be responsible for the calculated SMA ratio to be much larger than found by experiment.

A possible source for the deviation are isotropic contributions to the electrical resistivity which are not included in the calculation and which merely enhance $\bar{\rho}$ but not $\Delta\rho$. That such contributions are present is strongly suggested by the results for $\bar{\rho}$ in Fig. 2. Scattering at grain boundaries, short-range order, clusters etc. could enhance the isotropical resistivity without increasing $\Delta\rho$ because this extra scattering is independent of the magnetisation and adds the same ρ_{extra} to $\rho_{||}$ and to $\rho_{\perp}$ thus not changing $\Delta\rho = \rho_{||} - \rho_{\perp}$. In order to get an idea of how this extra scattering might influence the SMA ratio $\Delta\rho/\bar{\rho}$ was calculated using experimental values for $\bar{\rho}$ instead of the smaller calculated values. The result, shown in Fig. 1 as a dashed line, lies much closer to the experimental curve than the uncorrected calculated values.

Analysis of the matrix element of the current operator in Eq. 3 allows one to investigate the role of spin-flip scattering processes for SMA. Within non-relativistic theory these may not occur because the current operator $j_\mu = -i\hbar(e/m)\nabla_\mu$ does not couple to the spin degree of freedom. This differs from the relativistic case where j_μ is given by $ec\alpha_\mu$ with α_μ one of the Dirac matrices. This operator can be transformed to a form having as a leading term one that is proportional to the non-relativistic j_μ and two additional terms involving α_μ and allowing for spin-flip processes[12]. The matrix elements for the ∇_μ-related term were found to be at least 100 times larger than for the corresponding two other terms. Thus, spin-flip processes have – at least for 3d-systems – practically no influence on the SMA. This confirms the generally used assumption that spin-orbit interaction causes SMA primarily by hybridisation of states of different spin character.

Co–Pd and Co–Pt

The calculational procedure applied to Fe–Ni was used to determine the spontaneous resistance anisotropy for the two ferromagnetic alloys Co–Pd and Co–Pt. For Co–Pd this calculation was carried out for maximum angular momenta of $\ell_{\text{max}} = 2$ and $\ell_{\text{max}} = 3$ in two different calculations. It was found that the omission of f-states causes a too high resistivity but hardly influences the anisotropy ratio. Fig. 3 shows results for Cu–Pd based on $\ell_{\text{max}} = 3$ (averaged resistivities $\bar{\rho}$ shown). As one can see, the agreement between experimental and calculated resistivities is excellent. No deviation of the magnitude found for Fe–Ni is encountered.

The anisotropy ratios calculated from the transverse and longitudinal resistivities are shown in Fig. 4 for the two alloy systems Co–Pd and Co–Pt. The calculation on the latter system was based on a maximum ℓ of 2. As one sees, the anisotropy is much larger for Co–Pd than for Co–Pt, both for the calculated and for the measured values. The agreement between theory and experiment is again impressive. The good agreement for the resistivities seems to correspond with a good agreement for the anisotropies, thus supporting the supposition of the preceding section that extra isotropical scattering might be responsible for the deviations observed for Fe–Ni.

Non-relativistic calculations

The relativistic calculations include all effects of spin-orbit interaction and spin-polarisation and therefore lead to a very satisfactory description of the anisotropy phenomena

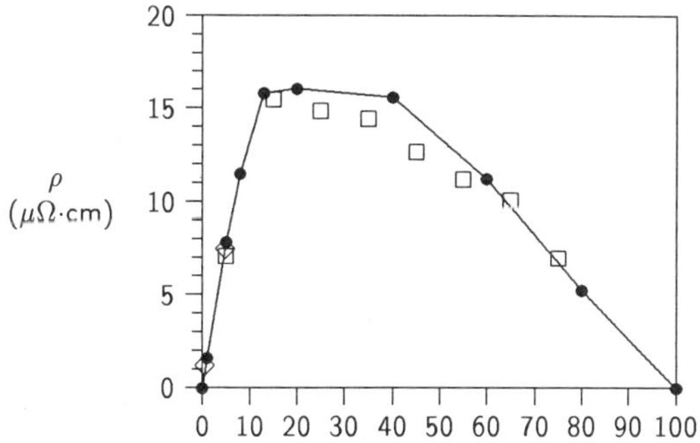

Figure 3: Residual resistivity of disordered Co–Pd alloys. Full circles: calculated, boxes: experimental values[16].

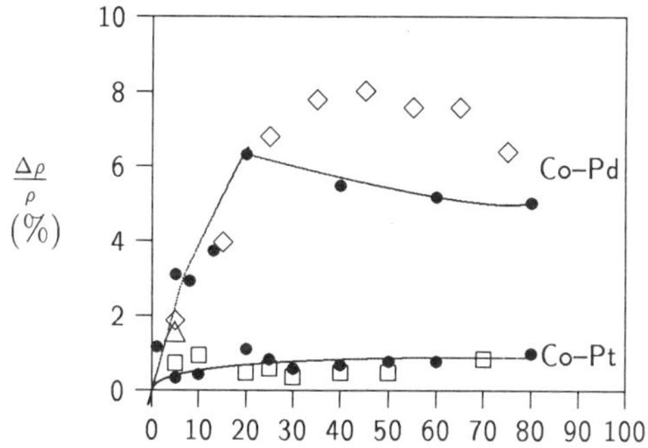

Figure 4: Spontaneous resistance anisotropy of Co-Pd and Co–Pt alloys. Full circles: calculated values, open symbols: experimental values[16,17].

discussed. In order to show that it is indeed essential to carry out the calculations relativistically, non-relativistic calculations were made for the alloys $Fe_{20}Ni_{80}$, $Co_{20}Pd_{80}$ and $Co_{20}Pt_{80}$. A separate resistivity calculation was carried out for each of the two potentials $V^{\uparrow}$ and $V^{\downarrow}$. For the single spin-channel the alloy was treated as paramagnetic system, but the common Fermi energy of the two spin-bands was used. The result were two resistivities $\rho^{\uparrow}$ and $\rho^{\downarrow}$ for each spin direction. The total non-relativistic resistivity is determined by $(\rho^{nr})^{-1} = (\rho^{\uparrow})^{-1} + (\rho^{\downarrow})^{-1}$ and compared to the result obtained by the relativistic calculation (ρ^{rel}) in table 1.

Table 1: Results of non-relativistic resistivity calculations for various alloys.

Alloy	$\rho^{\uparrow}$	$\rho^{\downarrow}$	$\rho^{\downarrow}/\rho^{\uparrow}$	ρ^{nr}	ρ^{rel}	ρ^{rel}/ρ^{nr}
$Fe_{20}Ni_{80}$	0.8	94	117	0.79	3	3.8
$Co_{20}Pd_{80}$	10.6	71	6.6	9.4	16	1.7
$Co_{20}Pt_{80}$	36	46	1.3	20.4	27	1.3

Two observations are obvious: firstly, the non-relativistic resistivities of the two spin channels can be quite different. The highest ratio $\rho^{\downarrow}/\rho^{\uparrow}$ corresponds to the largest

SMA (Fe–Ni), whereas the low ratio for Co–Pt reflects a small SMA. This behaviour is compatible with the models based on the two-current model[3]. Secondly, the relativistic resistivity is much higher that the non-relativistic one for all the alloys. Most of this difference is probably due to spin-orbit effects which this way are shown to be of unexpected importance.

CONCLUSIONS

In conclusion the fully relativistic spin-polarised KKR–CPA in conjunction with the Kubo-Greenwood theory for electrical conduction allowed for the first rigorous and parameter-free calculation of the anisotropical spontaneous resistivity in disordered ferromagnetic alloys. Application to the systems Fe-Ni, Co–Pd and Co–Pt led to a very satisfying agreement with experiment. Comparison with non-relativistic results shows that a relativistic treatment is essential for such systems.

The work was done in collaboration with Hubert Ebert, University of Munich.

REFERENCES

1. T.R. McGuire and R.I. Potter, IEEE Trans. Mag. 11:1018 (1975)
2. J. Smit, Physica 16:612 (1951)
3. I.A. Campbell, A. Fert and O. Jaoul, J. Phys. C3:S95 (1970)
4. I.A. Campbell, J. Phys. F4:L181 (1974)
5. J.W.F. Dorleijn, Philips Res. Repts. 31:287 (1976)
6. A.P. Malozemoff, Phys. Rev. B34:1853 (1986)
7. D. A. Greenwood, Proc. Phys. Soc. London, 71:585 (1958)
8. W.H. Butler, Phys. Rev. B31:3260 (1985)
9. J.C. Swihart, W.H. Butler, G.M. Stocks, D.M. Nicholson and R.C. Ward, Phys. Rev. Lett. 57:1181 (1986)
10. J. Banhart, H. Ebert, J. Voitländer and P. Weinberger, Phys. Rev. B50:2104 (1994)
11. H. Ebert, B. Drittler and and H. Akai, J. Magn. Magn. Mater. 104-107:733 (1992)
12. G.-Y. Guo and H. Ebert, Phys. Rev. B50:1037 (1994)
13. T. Farrell and D. Greig, J. Phys. F1:1359 (1968)
14. F.C. Schwerer and J.W. Conroy, J. Phys. F1:877 (1971)
15. M.C. Cadeville and B. Loegel, J. Phys. F3:L115 (1973)
16. S.U. Jen, Phys. Rev. 45:9819 (1992)
17. S.U. Jen, T.P. Chen and B.L. Chao, Phys. Rev. 48:12789 (1993)

AB-INITIO-BASED ATOMISTIC POTENTIALS AND APPLICATION TO METALLIC SURFACE AND INTERFACE STRUCTURES

Bernard R. Cooper, J. Mei, and S.P. Lim

Dept. of Physics, West Virginia University
Morgantown, WV 26506-6315

Introduction

In this paper we present a new ab initio technique for developing atomistic potentials designed for dynamic study of transition metal intermetallics and alloys; and we give an illustrative application of that technique to the behavior of NiAℓ. The fact that crystals exhibit plasticity shows the importance of modeling potentials which give a correct energetic description of materials under large deformation. We will show that total energy calculations within density functional theory (DFT) provide a flexible way to develop potentials with wide applicability. Starting from a simple mathematical transformation between atomic volume and a local variable function, developing these potentials relies on density functional theory (DFT) total energy calculations. The potential parameters in the model are determined by fitting energy surfaces resulting from total energy calculations within DFT on crystals at zero temperature. We have already successfully applied this model to study thermal properties of nickel-rich alloys[1,2]. Here, after discussing the development of the potential, we discuss the calculation of some surface and interface properties of NiAℓ using this model potential to check its quality. Excellent results are obtained compared with both DFT ab initio calculations and experiment. We then present results of molecular dynamics (MD) simulation of tensile deformation performed on unalloyed NiAℓ crystals with and without a simple defect structure (simulated grain boundary). Details of the atomic motion have been followed and will be illustrated in the case of plastic deformation.

The Potential Model

For a system with N classical particles, we have chosen the following functional form for the interatomic potential

$$E = \sum_i e_{si}(\omega_i, \{p_1\}) + V\left(\vec{r}_1, \vec{r}_2, \ldots, \vec{r}_N, \{p_2\}\right) \tag{1}$$

Here the first term is the energy for the ith atom (with the type index s_i) as a function of the atomic volume ω_i at position $\vec{r}_i$. The second term gives the rest of the interactions and is a function of all the atomic positions. In Eq.(1), $\{p_1\}$ and $\{p_2\}$ are potential parameters. The separation of the atomic volume dependent energy e from the total energy E gives us more flexibility to choose a simple functional form for V, and to fit potential parameters $\{p_1\}$ and $\{p_2\}$. At this point we have to specify the definition for the atomic volume ω_i. In the case of pure metals at zero temperature, ω_i is simply the volume per atom. In general cases, ω_i is a local variable. We have used the following relation to transform ω to a density function $\rho(r)$, which is interatomic distance dependent.

$$\frac{\omega_{es_i}}{\omega_i} = f_{s_i} + \sum_m \rho_{s_m}(r_{im}) \tag{2}$$

Here the summation on the right-hand side is over all neighbor atoms of the ith atom with distance given by r_{im}. To determine the function $\rho(r)$, we first choose a reference lattice structure for metal s_i, then uniformly expand (contract) the crystal (i.e., change ω_i on the left side) at zero temperature. Also, we have chosen ω_e to be the equilibrium value of ω_i for the crystal as obtained from Density Functional Theory (DFT) calculations. The following function is found to be the simplest functional form which allows the best fitting to Eq.(2).

$$\rho(r) = qr^{\kappa}\left[\exp(-\zeta r) - y\exp\left(-\eta r\right)\right] \tag{3}$$

For specific applications, the parameters q, κ, ζ, y, and η together with f are determined by numerical fitting technique. In practice, the fitting of these parameters also depends on the cutoff distance of $\rho(r)$ which can be chosen as the interaction range of all r-dependent functions.

Initially we have used simple pair interactions for V (the second term in Eq.(1)) in our NiAℓ application.

$$V = \frac{1}{2}\sum_{i \neq j} \phi_{s_i s_j}(r_{ij}) \tag{4}$$

For a pure metal at zero temperature with a simple crystal structure (one atom per unit cell), given the form of ϕ (r), using Eq.(1) and Eq.(4) we have

$$e(\omega) = E(\omega)/N - \frac{1}{2}\sum_m \phi(r_m(\omega)) \tag{5}$$

Where $\omega = \Omega/N$; N is the total number of atoms, and Ω is the volume of the crystal. Since this is a static crystal, each neighbor distance r_m is a simple function of ω. The first term of Eq.(5) gives a curve of energy variation for volume expansion, which is obtained by fitting the DFT calculations. We have used the following parametrized form:

$$E(\omega)/N = e_0 - 2e_1 \exp\left(-\alpha\frac{\omega_e}{\omega}\right) + e_2 \exp\left(-2\alpha\frac{\omega_e}{\omega}\right) \tag{6}$$

Finally, we choose the following form for ϕ (r)

$$\phi(r) = \phi_0\left[1 + s_1(r/r_0 - 1)\right]\exp\left(-s_2(r/r_0 - 1)\right) \tag{7}$$

It should be noted that all the parameters in Eqs.(5), (6), and (7) are atomic-type dependent.

To generate a reliable ab-initio-based potential for use in wholly predictive (nonempirically-based) simulations, we fit the potential to energy surfaces generated by total energy calculations within the DFT for selected crystal structures (i.e., we do this to optimize all the potential parameters $\{p_1\}$ and $\{p_2\}$ without use of any experimental data.) For a cubic crystal, one way to generate crystal structure changes is to use a body-centered orthorhombic lattice. This lattice allows us to perform continuous volume expansion (contraction) and constant-volume lattice distortion.

We have also investigated the results for a potential including an additional three-body interaction of the following form in Eq.(1):

$$V_3 = \frac{1}{2}\sum_i \sum_{jk(j\neq i,k\neq i,j\neq k)} w_{s_i s_j}(r_{ij}) w_{s_i s_k}(r_{ik}) h_{s_i, s_j s_k}(\theta_{i,jk}) \tag{8a}$$

$$w_{s_i s_j}(r) = g_{s_i s_j} \exp\left(-\frac{\gamma_{s_i s_j}}{r_c - r}\right) \tag{8b}$$

$$h_{s_1, s_2 s_3}(\theta) = (l_{s_1, s_2 s_3} + m_{s_1, s_2 s_3}\cos^2(\theta) + n_{s_1, s_2 s_3}\cos^4(\theta))p(\theta) \tag{8c}$$

$$p(\theta) = \cos\left(\left(\frac{\theta - \pi/2}{\theta_0 - \pi/2}\right)\pi\right) - 1 \tag{8d}$$

where $\theta_{i,jk} = \frac{\vec{r}_{ij}\cdot\vec{r}_{ik}}{r_{ij}\, r_{ik}}$, and $\theta_0 = \cos^{-1}\left(\frac{1}{3}\right)$.

Applications to NiAℓ

As the first application of our model potential, we have performed a molecular dynamic (MD) simulation to study the rippled relaxation of the NiAℓ(110) surface. This application provides a direct comparison of the use of our model potentials with results of other calculations (both ab initio and empirical) and experimental data.

To perform the simulation, we first set up a NiAℓ slab (512 atoms) with two (110) free surfaces corresponding to the bulk CsCℓ type crystal structure. Simulation is then performed at room temperature. After the system becomes well equilibrated, we gradually reduce the temperature to 30K. By measuring interlayer spacings between the same atomic species, we obtain the rippled relaxation. Table 1 shows our MD results (indicated by

"Present[a])" for the deviation of these spacings from bulk values. In the same table, we have included the MEIS[3] and LEED[4] experimental data, and the results of LDA[5] (Local Density Approximation) and EAM[6] (Embedded Atom Method) calculations, as well as of our model calculation with an additional 3-body interaction (indicated by "Present[b]"). Comparing with the LDA calculation, our model potential gives a much better result than the EAM potential, although they have similar mathematical functional forms. The EAM (embedded atom method) scheme[7] is based on purely empirical fitting.

Table 1 The rippled relaxation of the NiAℓ(110) surface. All values shown give the percentage deviation of the interlayer spacings between the same atomic species from the bulk values.

	Δd_{Ni}^{12}	$\Delta d_{A\ell}^{12}$	Δd_{Ni}^{23}	$\Delta d_{A\ell}^{23}$
MEIS[3]	-7.0%	5.0%	1.0%	-1.0%
LEED[4]	-4.6%	5.2%	1.0%	2.0%
LDA[5]	-6.9%	6.6%		
EAM[6]	-13.0%	16.0%	13.0%	-9.0%
Present[a]	-7.8%	9.7%	7.1%	-3.8%
Present[b]	-5.2%	7.4%	4.7%	-2.4%

[a] Result of the present interatomic potential.

[b] Result of the present model with an additional 3-body interaction in Eq.(1).

Next, we discuss the application of the potentials with 2-body interaction to plastic deformation. The defect structure introduced in this application of our technique is a {110}/{100} grain boundary parallel to the (x,y) plane. Fig.1(a) shows a projection of atomic positions (snapshot) on the (x,z) plane (locally averaged along the y direction) for an equilibrated configuration under zero stress for the system. The temperature has been fixed at 300 K in this simulation. On the grain boundary there is a mismatch of atomic positions between the upper and lower parts along the x direction (but not in the y direction due to the specific geometric conditions chosen for the two parts). This mismatch may be viewed as a massive edge dislocation in the extreme limit. Starting with the equilibrium situation of Fig1(a), uniaxial tensile stresses in the z direction are then gradually loaded up to a critical value (about 16 GPA in this case over a time period of 10 *ps*; beyond this critical value the system becomes unstable, i.e. fractures). Slipping along the x axis on the boundary and plastic strain in both x and z directions have been found in this process. Permanent deformations can be measured by (slowly) unloading the external stresses. Fig.1(b) shows the metastable configuration of atoms (snapshot) after the unloading from the critical stress. Comparison between Fig.1(a) and (b) clearly shows the way atoms move in the process.

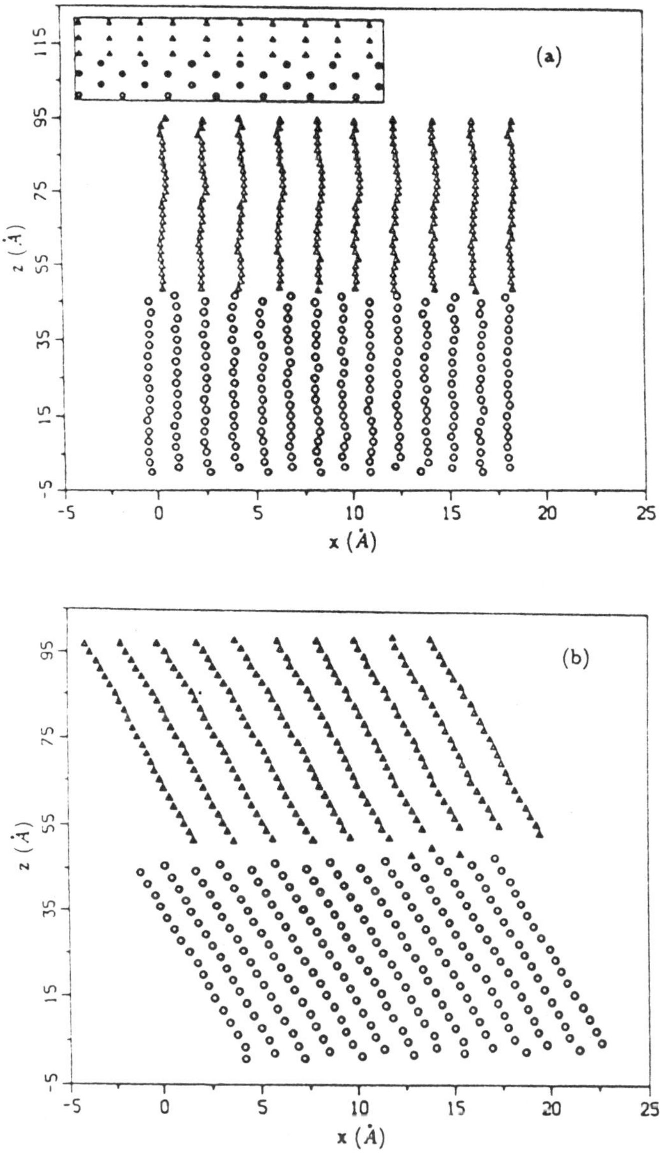

Figure 1. Projection (snapshot) of atomic positions (within a MD cell) on the (x,z) plane (locally averaged along y direction). A {110}/{100} grain boundary parallel to the (x,y) plane is present in the MD cell. This boundary divides the MD cell (3346 atoms with periodic conditions) into upper and lower parts. With the choice of three coordinate axes along the <100> directions for the lower part, the x, y, and z directions correspond to [$1\bar{1}0$], [001], and [110] orientations for the upper part. (a) Equlibrated configuration under zero stress (The inset shows equilibrium atomic positions on either side of the grain boundary). (b) A metastable deformed-configuration after unloading external tensile stress along the z direction (see description in the text).

The critical stress for fracture with a grain boundary has been found to be much lower than those of defect-free crystals along <100> and <110> orientations. Thus the presence of the grain boundary in this case has weakened mechanical strength. The mismatch of atomic positions on the boundary is clearly responsible for the slip and the weak bonding between the upper and lower crystals.

Concluding Remarks

In conclusion, we have presented a new scheme for developing many-body atomistic potentials suitable for use in dynamic computer simulations of transition metal intermetallics and alloys. All potential parameters are obtained from density functional theory (DFT) calculations without any experimental input. Molecular dynamics simulations have been performed to study the NiAℓ (110) surface relaxation and tensile deformation of NiAℓ with the presence of a simple grain boundary. Comparing with both experiment and local density approximation (LDA) calculations, our model potential gives a much better result for the (110) surface relaxation than the purely empirical EAM calculation.

Acknowledgments

This work has been supported by the National Science Foundation/West Virginia EPSCoR program, and the CRAY YMP time used in this work has been provided at the Pittsburgh Supercomputing Center. We have benefited from valuable suggestions and discussion from Dr. R. Darolia of General Electric Aircraft Engines.

REFERENCES

1. J. Mei, B.R. Cooper, Y.G. Hao, S.P. Lim, and F.L. Van Scoy, Mat. Res. Soc. Symp. Proc. Series Vol. 291, Materials Theory and Modeling (eds J. Broughton, P. Bristowe, and J. Newsam, Mat. Res. Soc. Pittsburgh 1993) pp 15-20.
2. J. Mei, B.R. Cooper, Y.G. Hao, and F.L. Van Scoy, pp 165-174 in Alloy Modeling and Design (eds G.M. Stocks and P.E.A. Turchi, TMS Publishing, Warrendale, PA, 1994).
3. S.M. Yalisove and W.R. Graham, Surf. Sci **183**, 556 (1987).
4. J.R. Noonan and H.L. Davis, Science, **234**, 310 (1986).
5. M.H. Kang and E.J. Mele, Phys. Rev. B **36**, 7371 (1987).
6. S.P. Chen, A.F. Voter, and D.L. Srolovitz, Phys. Rev. Lett. **57**, 1308 (1986).
7. M.S. Daw and M.I. Baskes, Phys. Rev. Lett. **50**, 1285 (1983), Phys. Rev. B **29**, 6443 (1984).

ON THE ROLE OF NON–PAIR POTENTIAL TERMS IN SEMIEMPIRICAL QUANTUM–MECHANICAL SIMULATIONS

M. Šob[1] and V. Vitek[2]

[1]Institute of Physics of Materials, Academy of Sciences of the Czech Republic, Žižkova 22, CZ-616 62 Brno, Czech Republic
[2]Department of Materials Science and Engineering, University of Pennsylvania, 3231 Walnut Street, Philadelphia, PA 19104-6272, U. S. A.

INTRODUCTION

One of the recent trends in condensed matter physics and materials science is to develop simplified schemes for computer modelling of low-symmetry atomic configurations. These schemes are based on quantum-mechanical treatment of electronic structure in solids and it is expected that they will be much faster than *ab initio* methods but comparable in accuracy. The underlying motivation is the necessity to be able to describe properly more and more complex systems used in contemporary science and technology. For example, extended lattice defects such as grain boundaries or other interfaces, dislocations etc. control to a great extent mechanical, electrical and transformational properties of both naturally occuring polycrystalline materials and man-made composites. However, for a deeper understanding of their role we need to know more about the structure of such defects, their mobility and other characteristics.

Atomistic computer simulations based on Finnis-Sinclair (FS)[1] and embedded atom method (EAM)[2] were successful in treating defects of various kind in fcc transition metals with filled or nearly filled *d*-bands, particularly in the Cu and Ni columns of the Periodic Table. A common feature of those approaches is that the interatomic interactions are regarded as central, and possible angular forces that are essential in semiconductors and may be important in transition metals with unfilled *d*-bands, are neglected. Recently, several approximate descriptions of atomic interactions which include the angular character of interatomic forces have been proposed[3–7]. Some of those schemes are based on the semiempirical tight-binding (TB) approach[8] where localized orbitals are used when evaluating the band-structure energy.

In *ab initio* electronic structure calculations, the corresponding Schrödinger-like equation is solved exactly, at least in principle, and the total energy obtained includes automatically features such as directional bonds. However, studies of extended defects which must involve large-scale atomic relaxations are usually well outside the limits of such calculations, even when the most powerful computers are employed.

Here we present a semiempirical TB-like scheme suitable for large-scale computer simulations of extended defects. This scheme does reproduce the band structures of various configurations and, in contrast to a similar approach[9], uses only a few adjustable parameters.

COMPARISON OF THE DENSITY-FUNCTIONAL THEORY AND SEMIEMPIRICAL TIGHT-BINDING MODELS

In the density-functional theory (DFT)[10,11], the total energy of a solid

$$E_{tot}^{DFT}[n(\mathbf{r})] = \sum_m f_m \varepsilon_m + G[n(\mathbf{r})], \tag{1}$$

where $n(\mathbf{r})$ is the electronic density, ε_m one-electron energies (*i.e.*, eigenvalues of the corresponding self-consistent first-principles Hamiltonian), f_m occupation numbers and G represents the remaining terms from the DFT and the nuclear-nuclear interaction energy. The first term in Eq. (1) is called the band-structure energy.

In semiempirical tight-binding (SETB) schemes, the total energy is written in the form

$$E_{tot}^{SETB} = \sum_m f_m \tilde{\varepsilon}_m + \frac{1}{2}\sum_{\substack{i,j \\ i\neq j}} U_{ij}(r_{ij}) \tag{2}$$

where the $\tilde{\varepsilon}_m$ are eigenvalues of an appropriate (usually non-self-consistent) Hamiltonian and $U_{ij}(r_{ij})$ with $r_{ij} = |\mathbf{R}_i - \mathbf{R}_j|$ is a short-range pairwise repulsion between the atoms at $\mathbf{R}_i$ and $\mathbf{R}_j$. Here both parts of the total energy contain adjustable parameters.

Foulkes and Haydock[12] analyzed the connection between the DFT and SETB. They reached an encouraging conclusion that non-self-consistent SETB can be considered as a good approximation to the self-consistent DFT and showed, similarly as Sutton *et al.*[13], that it is appropriate to replace the term $G[n(\mathbf{r})]$ in Eq. (1) by a simple sum of two-body interatomic potentials.

Comparing the DFT and SETB total energy formulas, one might consider a scheme employing the band-structure term from a first-principles calculation and regard as empirical only the second, pair-potential term. Thus, the total energy formula would be

$$E_{tot} = \sum_m f_m \varepsilon_m + \frac{1}{2}\sum_{\substack{i,j \\ i\neq j}} U_{ij}(r_{ij}) \tag{3}$$

with first-principles eigenvalues ε_m and adjustable pair-potential.

When we try to use Eq. (3) we face two problems which must be satisfactorily solved:

(i) The choice of energy zero for the band-structure term

In DFT total energy calculations (Eq. (1)), one can shift the weight from the band-structure energy to the double-counting terms (contained in $G[n(\mathbf{r})]$) by addition of an arbitrary (spatially) constant potential to the Hamiltonian. This shift is, of course, automatically compensated by an accompanying term in the double-counting energy. Thus, any separation of the total energy into the band-structure term and a remainder is, in principle, arbitrary and hence has no special physical meaning. This is due to non-uniqueness of the average electrostatic potential in an infinite solid[14]. The constant potential can be completely arbitrary, for example a function of the volume V and/or some other characteristics of the solid considered.

A possible choice is to refer the band energy to the Hartree (electrostatic) potential averaged over the interstitial region between the muffin-tin or atomic spheres[15]. However, the band-structure energy then becomes repulsive and this does not correspond to the analogy with SETB approach.

Another possible choice is to relate the one-electron energies calculated for different structures and volumes of a solid to a common Fermi energy. This is reasonable, since if we have two pieces of material with different atomic configurations and/or compositions and join them together, the Fermi energy in both pieces must be the same. In this case, the band-structure energy is attractive. In the present paper, we have adopted this convention. Thus, in Eq. (3) the band-structure term contains no adjustable parameters and the one-electron energies are related to a common Fermi energy. However, as shown below, if there are no adjustable parameters in the Hamiltonian, the pair potential is not sufficient and an additional term needs to be introduced.

(ii) Cauchy pressure

Let us rewrite Eq. (3) as

$$E_{tot} = E_{band} + E_{pp}, \tag{4}$$

where E_{band} and E_{pp} is the band-structure and pair-potential energy, respectively. In cubic materials the Cauchy pressure is given by

$$p_{Cauchy} = \frac{1}{2}(c_{12} - c_{44}). \tag{5}$$

It follows from the definition of elastic constants that the contribution of the pair term is

$$p_{Cauchy}^{pp} = \frac{1}{2}(c_{12}^{pp} - c_{44}^{pp}) = -\frac{1}{4}\frac{dE_{pp}}{dV}, \tag{6}$$

where V is the volume of the solid. In equilibrium

$$\frac{dE_{tot}}{dV} = \frac{dE_{band}}{dV} + \frac{dE_{pp}}{dV} = 0 \tag{7}$$

and, therefore,

$$p_{Cauchy}^{pp} = \frac{1}{2}(c_{12}^{pp} - c_{44}^{pp}) = \frac{1}{4}\frac{dE_{band}}{dV}. \tag{8}$$

For total Cauchy pressure found from experimental elastic constants we obtain

$$p_{Cauchy}^{exp} \equiv \frac{1}{2}(c_{12}^{exp} - c_{44}^{exp}) = \frac{1}{2}(c_{12}^{band} - c_{44}^{band}) + \frac{1}{4}\frac{dE_{band}}{dV}. \tag{9}$$

This relation is usually not fulfilled unless there are fitting parameters in the Hamiltonian. For example, in bcc tungsten at equilibrium, we have $p_{Cauchy}^{exp} = 0.246$ Mb, but, using the LMTO-ASA (Linear Muffin-Tin Orbital method in the Atomic-Sphere Approximation[16]) band-structure energy and relating the one-electron energies to a common Fermi energy, the right-hand side of eq. (9) is equal to 2.670 Mb.

Now we may ask what is wrong? There are, essentially, two possibilities:

(i) Our Hamiltonian is not complete. However, we have a first-principles Hamiltonian, which should be correct. Further, it is known that the LMTO-ASA provides very good results for the charge density and kinetic energy[17], so that we rule this possibility out.

(ii) Another explanation is that the pair-potential term in Eqs. (3,4) is insufficient and we need an additional term in the total energy that would balance the above mentioned difference. As the ASA is known for its geometry violation[16], which manifests itself in an inaccurate description of the electrostatic energy in low-symmetry structures[17], we can adopt, for example, the atomic-sphere correction energy E_{AS} from the effective-medium theory[18]. This introduces additional fitting parameters to the total energy[19]. The atomic-sphere correction is nearly volume-independent, *i.e.* $dE_{AS}/dV \approx 0$, and, after fitting the equilibrium lattice structure and elastic constants, it corrects the value of the Cauchy pressure on the right-hand side of eq. (9) by -2.424 Mb.

SEMIEMPIRICAL COMPUTER SIMULATIONS WITH A FIRST-PRINCIPLES HAMILTONIAN

After introduction of the atomic-sphere correction E_{AS} the total energy in our scheme is written as

$$E_{tot} = E_{band}^{LMTO-ASA} + E_{AS} + E_{pp}, \tag{10}$$

where $E_{band}^{LMTO-ASA}$ is the band-structure energy evaluated within the LMTO-ASA with respect to a fixed common Fermi energy and E_{pp} is a repulsive pair-potential term. For band-structure calculations, we employ the LMTO-ASA Hamiltonian in the orthogonal representation[20]:

$$H_{\mathbf{R}L\mathbf{R}'L'} = C_{\mathbf{R}\ell}\,\delta_{\mathbf{R}\mathbf{R}'}\delta_{LL'} + \sqrt{\Delta_{\mathbf{R}\ell}}\,[S^0\,(1 - \gamma S^0)^{-1}]_{\mathbf{R}L\mathbf{R}'L'}\,\sqrt{\Delta_{\mathbf{R}'\ell'}}. \tag{11}$$

Here $\mathbf{R}$ labels the site and $L = (\ell m)$ is the usual angular momentum index. The potentials are assumed to be spherically symmetric in overlapping atomic spheres with the same volume as Wigner-Seitz polyhedra. The LMTO crystal potential parameters $X_{\mathbf{R}\ell}$ $(X = C, \Delta, \gamma)$ are diagonal with respect to indices $\mathbf{R}$ and L and they characterize the scattering properties of atoms at individual sites. S^0 is the canonical structure matrix which is determined solely by the geometrical arrangement of atoms[20]. The E_{AS} and E_{pp} contain adjustable parameters which are fitted to reproduce the equilibrium lattice constant, bulk modulus and shear moduli $\frac{1}{2}(c_{11} - c_{12})$ and c_{44} of the most stable structure.

In any semiempirical scheme it is very important to ensure that the experimentally observed equilibrium structure is the most stable structure and that the energetics of large deformations is satisfactorily reproduced. This has not been always checked when constructing interatomic potentials and, as a result, configurations corresponding to unphysical structures might have been attained in studies of lattice defects[21]. In order to guarantee that our approach leads to the correct ground state structure, the following was calculated both by the full-potential LMTO (FP-LMTO) method[22] and by the present semiempirical approach for tungsten:

- the Bain's path (the dependence of the total energy on tetragonal deformation from bcc ($c/a = 1/\sqrt{2}$) to fcc ($c/a = 1$) structure;
- the dependence of the total energy on trigonal deformation in the neighbourhood of bcc and fcc structures;
- energy-volume curve for the bcc structure with ±20% volume deformation;

- structural energy differences between the bcc, fcc, A15 and simple cubic structures (see Table 1).

Table 1.
Structural energy differences for tungsten calculated using FP-LMTO and the present semiempirical scheme ($\Delta E = E_{struc} - E_{bcc}$).

structure	ΔE (FP-LMTO) (meV/atom)	ΔE (present) (meV/atom)
A15	102	284
fcc	445	520
simple cubic (sc)	2075	1926

These results are discussed in details elsewhere[19,23]; here we just note that in all cases, the agreement between the FP-LMTO total energies and those obtained from the present scheme is satisfactory. This, of course, does not exclude the possibility that a structure exists which is energetically more favourable than the bcc structure within our semiempirical approach. However, after the tests performed, the probability of such a case is low. It is notable that the above mentioned behaviour was obtained by our scheme in which **only** the properties from the infinitesimal neighbourhood of the bcc structure have been fitted. Thus, it seems that basic physics is contained in Eq. (10). This gives us confidence when using this treatment for simulation of extended defects in metals. We have devised a procedure for calculating forces and first simulations of atomic configuration of grain boundaries in tungsten have already been performed[23].

SUMMARY AND CONCLUSIONS

It was shown that in a tight-binding approach which employs a first-principles Hamiltonian without any fitting parameters the usual expression for the total energy consisting of the sum of one-particle electron energies and a short-range pair-potential term is in general not sufficient. Another term balancing the Cauchy pressure must be added. Thus, it appears that a relatively simple picture of the total energy consisting of an attractive band-structure term and a repulsive pair-potential term has to be abandoned if a more consistent description of the system using a first-principles Hamiltonian is adopted.

Combining the first-principles band-structure energy determined within the LMTO-ASA method with semiempirically adjusted central force potential and atomic-sphere correction term reproduces very well the full-potential LMTO results for high-symmetry structures and their deformation and, therefore, the present scheme is a very promising technique for atomistic simulations of extended defects.

ACKNOWLEDGEMENTS

The financial support for this work was provided by the US-Czechoslovak Science and Technology Program (Project No. 94048), the Grant Agency of the Czech Republic (Project No. 202/94/1177), the Grant Agency of the Academy of Sciences of the Czech Republic (Project No. 110404) and the Department of Energy, BES, Grant no. DE-FG02-87ER45295.

REFERENCES

1. M. W. Finnis and J. E. Sinclair, Phil. Mag. A **50**, 45 (1984).
2. M. S. Daw and M. I. Baskes, Phys. Rev. B **29**, 6443 (1984).
3. J. A. Moriarty, Phys. Rev. B **42**, 1609 (1990).
4. A. E. Carlsson, P. A. Fedders, and C. W. Myles, Phys. Rev. **41**, 1247 (1990).
5. J. D. Kress and A. F. Voter, Phys. Rev. B **43**, 12607 (1991).
6. D. G. Pettifor and M. Aoki, in *Structural and Phase Stability in Alloys*, edited by J. L. Moran-Lopez and J. M. Sanches (Plenum Press, New York, 1992), p. 119.
7. S. M. Foiles, Phys. Rev. B **48**, 4287 (1993).
8. W. A. Harrison, *Electronic Structure and the Properties of Solids* (Dover Publ., New York, 1989).
9. R. E. Cohen, M. J. Mehl and D. A. Papaconstantopoulos, Phys. Rev. B **50**, 14694 (1994).
10. P. Hohenberg and W. Kohn, Phys. Rev. **136**, B864 (1964).
11. W. Kohn and L. J. Sham, Phys. Rev. **140**, A1133 (1964).
12. W. M. C. Foulkes and R. Haydock, Phys. Rev. B **39**, 12520 (1989).
13. A. P. Sutton, M. W. Finnis, D. G. Pettifor, and Y. Ohta, J. Phys. C: Solid State Phys. **21**, 35 (1988).
14. L. Kleinman, Phys. Rev. B **24**, 7412 (1981).
15. W. R. L. Lambrecht, B. Segall, and O. K. Andersen, Phys. Rev. B **41**, 2813 (1990).
16. O. K. Andersen, Phys. Rev. B **12**, 3060 (1975).
17. O. K. Andersen, A. V. Postnikov and S. Yu. Savrasov, in *Applications of Multiple Scattering Theory to Materials Science*, edited by W. H. Butler, P. H. Dederichs, A. Gonis, and R. L. Weaver, MRS Symposia Proceedings vol. **253** (Materials Research Society, Pittsburgh, 1992), p. 37.
18. K. W. Jacobsen, J. K. Nørskov, and M. J. Puska, Phys. Rev. B **35**, 7423 (1987).
19. M. Šob, V. Vitek, and Y. Oh, in *Computational Methods in Materials Science*, edited by J. E. Mark, M. E. Glicksman, and S. P. Marsh, MRS Symposia Proceedings vol. **278** (Materials Research Society, Pittsburgh, 1992), p. 205.
20. O. K. Andersen, O. Jepsen, and D. Glötzel, in *Highlights of Condensed-Matter Theory*, edited by F. Bassani, F. Fumi, and M. P. Tosi (North-Holland, New York, 1985), p. 59.
21. V. Vitek and J. Th. M. De Hosson, in *Computer-Based Microscopic Description of the Structure and Properties of Materials*, edited by J. Broughton, W. Krakow, and S. T. Pantelides, MRS Symposia Proceedings vol. **63** (Materials Research Society, Pittsburgh, 1986), p. 137.
22. M. Methfessel, Phys. Rev. B **38**, 1537 (1988); M. Methfessel, C. O. Rodriguez, and O. K. Andersen, Phys. Rev. B **40**, 2009 (1989).
23. M. Šob and V. Vitek, to be published.

A COMPARATIVE AB INITIO STUDY OF SMALL Si AND C CLUSTERS

Aristides D. Zdetsis

Department of Physics
University of Patras
Patras 26110
Greece

1. INTRODUCTION

The study of small atomic clusters is a relatively new and important field of research in materials science. The small clusters study which forms a link between atomic and solid state physics (and chemistry) is of interdisciplinary nature and has both scientific and technological importance. The study of the growth process versus size and the implications to nanostructure technology give a simple illustration of the significance of cluster research. Obviously the study of silicon[1-4] and carbon[5-7], including carbon fullerenes[8], has attracted more interest due to their increased importance in scientific and technological applications.

The purpose of this work is to study the differences and similarities of small Si and C clusters. Detailed calculations for carbon clusters have been performed so far for C_3 , up to C_{22} , C_{44} and C_{60}. For silicon , the clusters from Si_3 up to Si_{15} as well as the Si_{33} , Si_{45} , Si_{59} and Si_{60} clusters have been studied. The justification for the particular choice of the systems studied will de left out of this discussion , for space and time economy.

Furthermore , to understand the differences and similarities between Si and C clusters , I believe that it is not enough to study only staticly the two systems separately; It is rather better to study also systems that could be considered as links or routes between the two systems. To this end , we have also studied mixed SiC clusters[9-14] of the form Si_nC_m as a function of n and m. We have complete results up to now for $n+m=5$ and $n+m=6$.

The results which are presented here are parts of a larger project and are preliminary but quite illuminating. It is hopped that when the complete project is over , it could facilitate future potential applications in nano-scale materials engineering.

The structure of this paper , considering the size limitations , is as follows : After a brief description of the method(s) of calculation in section 2 , we give a brief and general overview of the C and Si clusters results in section 3. In the same section a somewhat more detailed description is given selectively for one small carbon cluster , C_6 , and one small silicon cluster , Si_{10}. The choice of these two representative cases will become obvious during the discussion of the results in section 3. In section 4 , a summary of the Si_nC_m results will be given , together with some stability criteria we have established. Finally in section 5 some basic conclusions will be summarized.

2. METHOD OF CALCULATION

The Hartree-Fock (HF) method with or without correlation corrections was used as a starting point for all calculations of small and medium size clusters. For a given cluster , various initial geometries were chosen and a complete geometry optimization within the given symmetry was then carried out until the structure corresponding to a stationary point was found. In order to check whether these stationary points correspond to real local minima , the optimization procedure was continued without any symmetry constraints. the geometry optimizations , which are based on quantum-mechanical calculations of the forces on each atom , were carried out at the HF level for the medium size systems. For the small size systems (up to 10 atoms) the geometry optimizations were performed , including correlation corrections , through second order Moeller-Plesset (MP2) perturbation theory. The energies of the optimized structures were calculated using higher-order Moeller-Plesset perturbation theory (MP3 , MP4). For the energies we have also used in several cases the coupled cluster singles and doubles (CCSD) theory and the CCSD(T) method , which includes a quasiperturbative estimate of the contribution of connected triple excitations. The basis set used for the expansion of the wave functions was of double-zeta quality augmented by d polarization functions (DZP basis).

For larger size systems the density functional theory (DFT) was employed using both local and non-local exchange and correlation potentials. Most of the calculations were performed using the Vosko-Wilk-Nussair (VWN) local exchange and correlation[15] together with the Becke 88 non-local exchange[16]. For some small (to medium) size systems both the DFT and HF theory (including correlation) were used , for comparison.

3. RESULTS AND DISCUSSION FOR Si AND C CLUSTERS

3a. Small size carbon clusters

It is well known[5-7] from the pioneering work of Raghavachari[6-7] that small size carbon clusters , C_n , with $n < 10$ appear either as monocyclic rings or as linear chains as shown in figure 1 for $v=5$ up to $n=10$. The appearance of rings or chains seems to depend on parity of n. Thus for even n ($n > 2$) the equilibrium geometry is a ring , whereas for odd n a linear chain is energetically favored. The electronic states of the chains have spin equal to one , while the spin of the rings is zero. These results , summarized in figure 1 , represent a physically appealing and clear-looking picture. However , more recent work[17] shows that , in reality , the situation is more complex. For the C_6 cluster , for example , the three lowest lying states are shown in figure 2. According to Raghavachari[6-7] the lowest lying structure is a ring of D_{3h} symmetry , structure III in figure 2. The D_{6h} symmetric ring of structure II is only 3 kcal/mole higher in energy , whereas the linear structure , structure I is about 16 kcal/mole higher. However a number of more recent calculations including advanced correlation methods , which are listed in reference 17 , give a series of conflicting results. These results are critically dependent on the type and amount of correlation[17]. What is clear , thought , is that the energy differences between the linear and the ring structures is very small. It seams that , according to ref.17 , the original energetical ordering of the three lowest lying structures , proposed by Raghavachari is correct although the energy differences are smaller. The energy difference of the linear form from the ground state structure III (ring of D_{3h} symmetry) is only 7 kcal/mole. These results are in qualitative agreement with the results of the present calculation , based on the CCSD(T) method , which give an energy difference of 10 kcal/mole.

The results for C_4 , C_5 , C_7 , C_{10} , which are more clear and well accepted , are in accordance to the results of figure 1. The case of C_3 , however , is more complicated , but it will not be discussed here.

3b. small size silicon clusters

Unlike small carbon clusters , which prefer linear and planar structures , small silicon clusters , Si_n , ($n<10$) prefer 3-dimensional structures[1-4] for $n>4$. This is due to the fact that silicon does not form strong delocalized π bonds , but instead prefers σ-bonded 3-dimensional structures. These structures are shown in figure 3. The results of this work for Si_5-Si_9 agree with the results of Raghavachari[2-4] and others[1]. The two structures of C_{3v} and C_s symmetry for Si_9 are very close in energy (about 3 kcal/mole) but the C_{3v}-symmetric structure seems to be the ground state. For Si_{10} discrepancies analogous to C_6 appear again. According to Raghavachari[3] structure III with C_{3v} symmetry is the lowest lying structure and structure I with T_d symmetry is the next low lying structure about 24 kcal/mole (1.04eV) higher in energy. Structure II with tetrahedral geometry similar to the diamond structure of bulk silicon , lies much higher in energy , almost 80 kcal/mole higher than structure III. More recent results[1,18] however give conflicting results about the energy separation and ordering of structures I and III. The present results , based on MP2 and MP4 calculations give as a ground state structure I , but with slightly different bond lengths than before[1,3,18]. This result is also verified by additional DFT calculations[19]. However , the energy difference between the two structures , I and III , is so small (1 kcal/mole) that for any practical purpose the two structures are isoenergetic. This is not surprising since both of these structures can be considered as an octahedron with four capping atoms , differing only at the capping faces. A 180 degrees rotation about one of the uncapped faces in structure I yields structure III. The fact that these two structures are energetically equivalent is suggestive of the possibility that another structure with lower symmetry could be the lowest lying structure. This possibility is currently examined and some preliminary

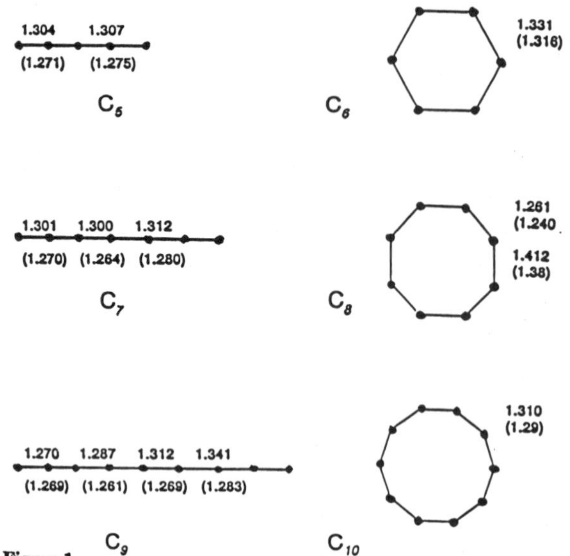

Figure 1.

small carbon clusters , C_n for n=5 , 6 , ... , 10. Bond lengths in angstroms have been obtained by DFT method in the local density approximation. The band lengths in parenthesis are taken from Raghavachari[6-7].

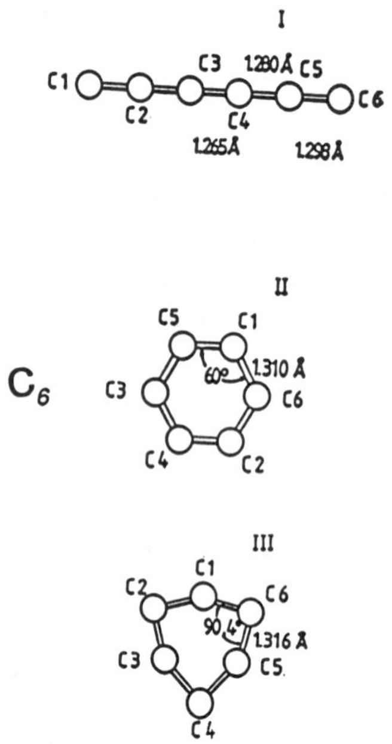

Figure 2.

The three lowest lying structures of C_6.

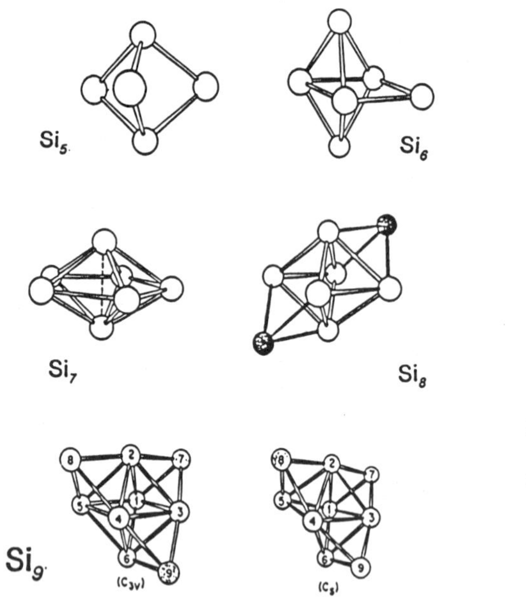

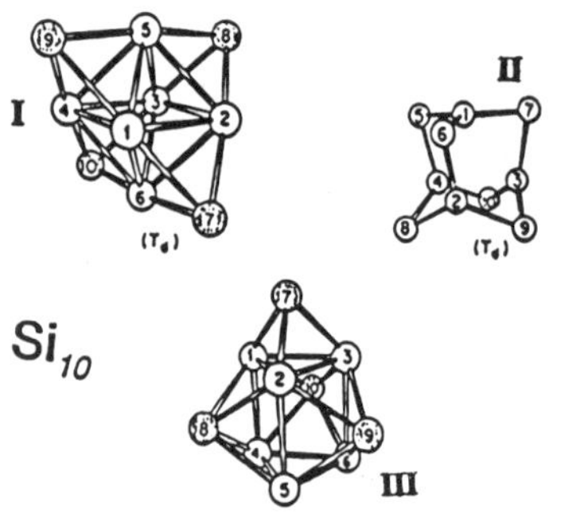

Figure 3.

small silicon clusters Si_n for n=5 , 6 , ... , 10.

results are affirmative. Moreover a very low lying structure of D_{2h} symmetry similar to the one suggested by Patterson and Messmer[1] has been identified by my calculations , with an energy difference from structure I (which is the true ground state , so far) of about 10 kcal/mole. The optimized C_{3v} structure of Raghavachari[3] , using here the DZP basis set has a total energy , at the MP2 level of theory , of -2891.0725039 Hy; whereas the corresponding T_d structure has an energy of -2890.9978469 Hy.

3c. Medium size carbon clusters

The alternation of rings and chains observed for the small size ($n<10$) carbon clusters seems to persist up to around $n=20$. For larger n , cage-like fullerine structures start to appear. The present calculations , based on DFT , have illustrated that at least up to $n=22$ the cage structures are not energetically favored compared to the planar rings (and chains). The cage structures were first generated by molecular dynamics calculations and then subsequently re-optimized using the full DFT (with the local VWN[15] and the non local Becke 88[16] potentials) without any symmetry restriction. In Figure 4 , the two cage structures obtained for C_{18} and C_{22} are shown in a ball-and-stick diagram. For both of these structures I was able to produce planar ring structures of lower energy. Thus on the basis of these results which are critically dependent upon the initial geometry generated by molecular dynamics , cage structures cannot appear before C_{23}. The region of larger n than 23 is currently investigated. The lowest lying structure for $n=44$ is the cage structure shown in Figure 4. The results for $n=60$ have no new information besides what is already widely known.

3d. Medium size silicon clusters

Today it is more or less accepted that the structural properties of small size ($n<10$) Si clusters are relatively well understood. The structure of larger size silicon clusters , however , is still a matter of controversy[1,20]. What is known experimentally is: (a) the abundance spectrum does not show any special features or magic numbers ; (b) the dissociation energy for $n>25$ becomes a smooth function of n; The reactivity to several molecules (e.g. NH_3) could vary by a few orders of magnitude and exhibits minima at $n=33$, 39 and 45. This last feature seems to critically depend on whether the clusters are well-annealed or not. Most of the theoretical work has been focused on Si_{33} and Si_{45} and a variety of structures have been proposed[1, 20-22] for the fundamental geometry of these clusters. In this work we have obtained Si_{33} and Si_{45} clusters based on bulk-like geometries , with significantly lower energies than those proposed before[1, 20-21]. The structure of these clusters is shown in figure 5. The Si_{33} cluster has D_{2d} symmetry (unlike the T_d structure of Kaxiras[20]). The symmetry of Si_{45} is tetrahedral. Both of these structures have no direct geometrical relationship to the corresponding carbon clusters (see for instance the structure of C_{44} in figure 4). More details about the Si_{33} and Si_{45} results , which are based on DFT calculations , will be published elsewhere[23].

For a comparison with C_{60} , a Si_{60} structure with icosahedral symmetry has been calculated using the DFT method. During geometry optimization this structure

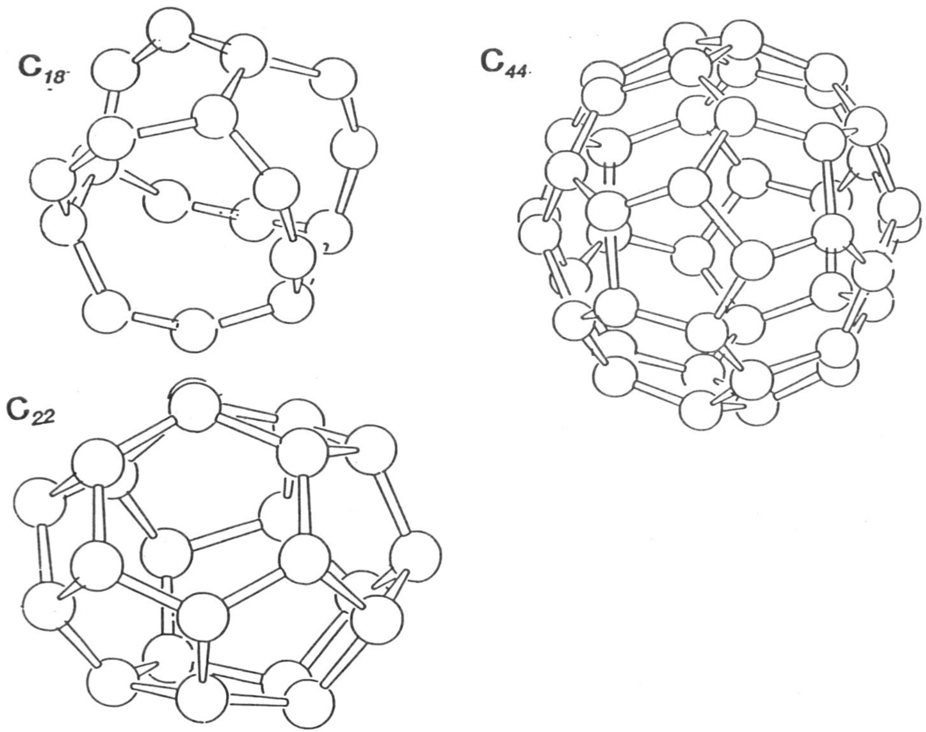

Figure 4.

Cage structures for C_{18} , C_{20} and C_{44} obtained by a combination of molecular dynamics and DFT.

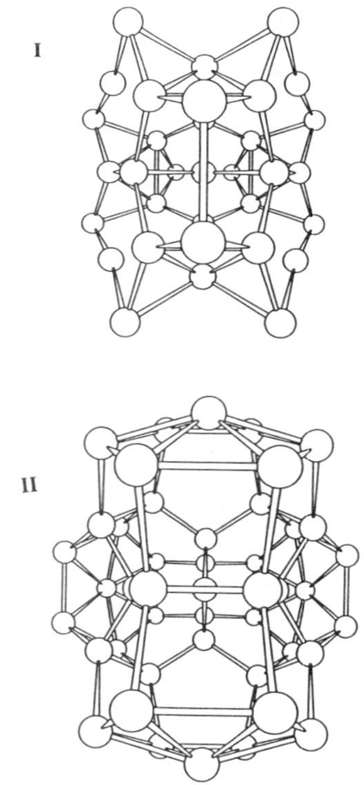

Figure 5.

The structure of Si_{33} (I) and Si_{45} (II) according to the present DFT calculations changes into a D_{2h} isomer of considerably lower stability than the neighboring Si_{59} cluster[19].

4. RESULTS FOR THE SIX-ATOMIC SiC CLUSTERS

while many recent publications[1-8] deal with elementary silicon and carbon clusters such as C_6 and Si_6 much less information seems to be available for the mixed SiC clusters. This is why a special effort and particular care are needed. As was explained earlier in section 1 , the study of mixed SiC clusters is considered necessary for the understanding of the relationship between Si and C clusters. With this primary motivation we have started detailed ab-initio calculations over the last two years[9-14] on SiC clusters , using advanced quantum chemistry methods. Here we present some results for the sixatomic SiC clusters: Si_5C , Si_4C_2 , Si_3C_3 , Si_2C_4 and SiC_5. These results should be compared with the C_6 results on one hand and with Si_6 on the other.

The carbon-rich cluster **SiC_5** has a linear geometry (fig. 6) , with the silicon atom at one end of the chain. This is not surprising in view of the fact that one of the lowest lying structures of C_6 is a linear chain (see the discussion in section 3a). This can be seen either as the substitution of one carbon atom by silicon in a C_6 chain or as the addition of one silicon atom in the chain of C_5. The first view seems to be more appropriate , by comparison to other SiC cluster results. The stability of this system is primarily due to the strong carbon-carbon bonds and , thus , the addition of the silicon atom occurs at the end of the chain , in order to maximize the number of these bonds.

The fundamental geometry of the **Si_2C_4** cluster is again a linear chain[11] with the silicon atoms at the two ends. This structure , similarly to the SiC_5 case , is derived from the C_6 - chain by substitution. The position of the two silicon atoms , at the two ends , obviously is due to the relative strength of the carbon-carbon bonds versus the silicon-carbon bonds. The second lowest lying structure (about 10 kcal/mole higher than the linear chain) is a distorted ring , apparently derived from the C_6 ring by substitution. This structure in one earlier publication[12] was erroneously considered as the ground state structure. This structure , however , is quite important since it can be considered as an intermediate step in the formation of the Si_3C_3 cluster. This structure , also , very close resembles a distorted chair , which can be compared with the symmetric chair (fragment of the diamond lattice) or the hexagonal planar ring (fragment of the graphite lattice). Using this type of comparison , the present author[24] has made some plausibility arguments to suggest that the underlying lattice of amorphous SiC turns from diamond-like to graphitic-like for C concentrations slightly above 50%.

The structure of the Si_3C_3 cluster is a pyramid-like structure of C_s symmetry[9]. This structure besides the strong carbon-carbon bonds , is stabilized against competing structures by the formation of multicenter bonds. The energetically next lower-lying structure is a C_s symmetric isomer closely related to the ground state of Si_6 structure (see figure 3). This type of Si_6-like structure becomes the ground state structure , in the silicon - rich Si_4C_2 and Si_5C clusters. In this respect the Si_3C_3 cluster is indeed an intermediate case between the carbon rich and the silicon rich regions.

The structure of $\mathbf{Si_4C_2}$, as was anticipated above , is related to the Si_6 geometry. The same structure can be also viewed[11-12] as a Si_4 tetrahedron which is edge capped by a C_2 bridge. The second lowest lying isomer is a highly coordinated structure of D_{2h} symmetry only 3.5 kcal/mole above the ground state. This structure , however , is a transition state due to the occurrence of imaginary frequencies.

The $\mathbf{Si_5C}$ cluster closely resembles a slightly distorted Si_6 bipyramid , (fig. 6.II) in which one silicon atom has been replaced by carbon (with an obvious change in the corresponding bond-lengths). In this structure the σ - bonding is severely limited.

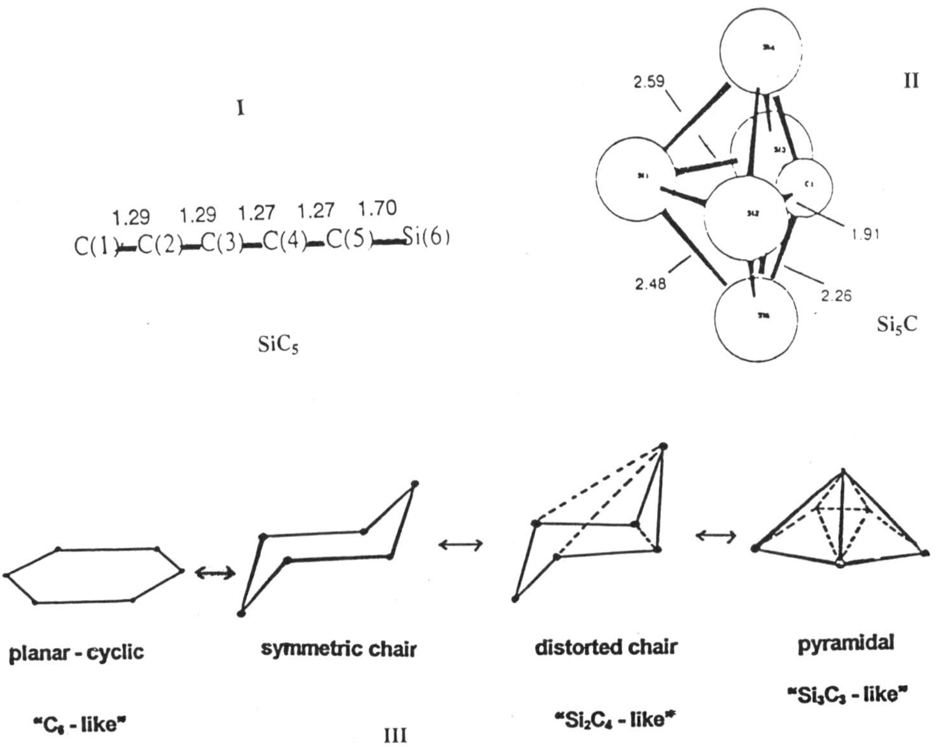

Figure 6.

Sixatomic SiC clusters. SiC_5 (I) and Si_5C (II). Bond lengths are in angstrom. III: Schematic illustration of the geometrical relation between the distorted chair structure of Si_2C_4 and the ground state structure of Si_3C_3.

5. CONCLUSIONS

In this last paragraph , some basic conclusions of this work are summarized in the form of a decalogue .

1. Small carbon clusters form stable chain and ring structures.

2. Larger carbon clusters , at least $n>23$, form cage structures.

3. The maximum coordination of any carbon cluster is four.

4. Silicon clusters do not form chain , ring and cage structures. Instead they prefer σ - bonded multiply coordinated network structures with a maximum coordination of six.

5. Larger silicon clusters with $n>25$, have smooth variation of the dissociation energy with n. The reactivity to several molecules exhibits minima at $n=33$, 39 and 45. For these clusters this work has produced very low energy structures , starting from three-dimensional bulk-like geometries[19].

6. Small SiC clusters are stabilized with strong silicon-carbon bonds , wherever possible , and strong multicenter delocalized bonds. In the carbon-rich clusters chain or ring geometries , similar to small carbon clusters are preferred. In the silicon-rich regions three-dimensional structures , similar to the small Si clusters , are preferred. In both cases the Si or C atoms respectively , can be considered as placed substitutionally in the corresponding carbon or silicon cluster.

7. Even for small C or Si clusters , studied with the best available theoretical techniques discrepancies about the ground state structure do exist. For example , the ring and chain structure of C_6 cannot be well separated. Our results slightly favor the ring structure. Similarly , the energetical ordering and separation of the C_{3v} and T_d structures for Si_{10} is not well. the present results favor the T_d geometry , but indicate the possible existence of a structure with lower symmetry.

8. For both small Si and C clusters there are always two lowest lying structures with very small energy separation (up to 15 kcal/mole). This is also true for small SiC clusters.

9. Si_{60} structure analogous to C_{60} is not stable.

10. For large and small silicon clusters calculated with DFT , the non-local gradient corrections are essential for the correct binding energy and geometry.

Acknowledgment

The calculations for SiC clusters were performed under a Greek - German collaboration plan No 10762 and the names of the collaborators are shown in references 9-14. The Si cluster calculations were performed under a Greek ΠΕΝΕΔ program No 785 at the University of Patras.

The figures and the word-processing of this manuscript were fully taken care by Dimitrios A. Zdetsis.

References

1. C.H. Patterson, R.P. Messmer, Phys. Rev. B 42: 7531 (1990).
2. K.Raghavachari, C. Rohlfing, J. Chem. Phys. 94: 3670 (1991).
3. K.Raghavachari, C. Rohlfing, J. Chem. Phys. 89: 2219 (1988).
4. K.Raghavachari, in Phase Transistors, Vols.24-26:pp.61-69 (1990).
5. W. Weltner, Jr., and R.J. Van Zee, Chem. Rev. 89: 1714, (1989).
6. K.Raghavachari, R.A. Whiteside, J.A. Pople, J. Chem. Phys. 85: 6623 (1986).
7. K.Raghavachari and J.S. Binkley, J. Chem. Phys. 84: 2191 (1987).
8. H.W. Kroto, J.R. Heath, S.C. O'Brien, R.F. Curl, R.E. Smalley, Nature 318:163 (1985).
9. M. Muhlhauser, G. Froudakis, A. Zdetsis, S. D. Peyerimhoff Chem. Phys. Lett. 204: 617 (1993).
10. M. Muhlhauser, G. Froudakis, A. Zdetsis, B. Engels, N. Flytzanis, S. D. Peyerimhoff, Z.Phys.D. 32: 113 (1994).
11. G. Froudakis, M. Muhlhauser, A. Zdetsis, B. Engels, S. D. Peyerimhoff, J. Chem. Phys. 101, 6790 (1994).
12. A.Zdetsis, S.Peyerimhoff, G.Froudakis, M.Muhlhauser, B.Engels, N.Flytzanis, in proceedings Materials Research for Information Technology , p.123 , edited by G.Kaiser and N.Constantopoulos, Forschungszentrum Julich GmbH Scientific Series of the International Bureau, Berlin (1993).
13. G. Froudakis, M. Muhlhauser, A. Zdetsis, Chem. Phys. Lett. 233, 619 (1995).
14. A. D. Zdetsis, G. Froudakis, M. Muhlhauser and H. Thumnel, to appear in J. Chem. Phys.
15. S.H.Vosko, L.Wilk and M.Nussair, Can. J. Phys. 58: 1200 (1980).
16. A. D. Becke, Phys. Rev. A 38: 3098 (1988).
17. V. Pless, H. U. Suter, and B.Engels, J. Chem. Phys. 101: 4042 (1994).
18. S. A. Maluendes, and M. Dupuis, Int. J. Quantum Chem. 42: 1327 (1992).
19. A. D. Zdetsis in proceedings of the IX greek national solid state physical society pp 269 (1993).
20. E. Kaxiras , Phys. Pev. Lett. 64: 551 (1990).
21. D. A. Jelski et al. , J. Chem. Phys. 95: 8552 (1991).
22. U. Rothlisberger , W. Andreoni , and M. Parrinello , Phys. Rev. Lett. 72: 665 (1994).
23. A. D. Zdetsis , to be published.
24. A. D. Zdetsis , J. Non-Cryst. Solids 114: 513 (1989).

BRILLOUIN ZONE CONCEPT AND CRYSTAL SYMMETRY OF INTERMETALLIC HIGH PRESSURE PHASES

V.F.Degtyareva

Institute of Solid State Physics
Russian Academy of Sciences
Chernogolovka, Moscow district
Russia, 142432

INTRODUCTION

Crystal structure of some intermetallic phases is known to be controlled by a number of valence electrons in an alloy. Phase stability was explained by Fermi sphere - Brillouin zone interaction [1-3]. The sequence of electron phases in the Cu-Zn alloy along the valence concentration $1.35 \rightarrow 1.5 \rightarrow 1.62 \rightarrow 1.75$ is as follows:

$$\alpha(fcc) \rightarrow \beta(bcc) \rightarrow \gamma(complex\ cubic) \rightarrow \varepsilon(hcp).$$

Extension of this sequence to higher values of the electron concentration was achieved in polyvalent B-metal alloys both at high pressure [4-6] and upon quenching from melt [7-10].

At electron concentration, n, exceeding 3 el/atom the characteristic structures turned out to be hexagonal with 3 atoms ***(ω)***, simple hexagonal ***(sh)***, white tin ***(β-Sn),*** and simple cubic ***(sc)***. Also there exist distorted variants of these structures. The extended phase sequence controlled by electron concentration is demonstrated on the In-Sb high pressure and 'splat cooling' intermediate phases (Fig 1), including:

$$\omega \rightarrow simple\ hexagonal \rightarrow \beta\text{-}Sn \rightarrow simple\ cubic.$$

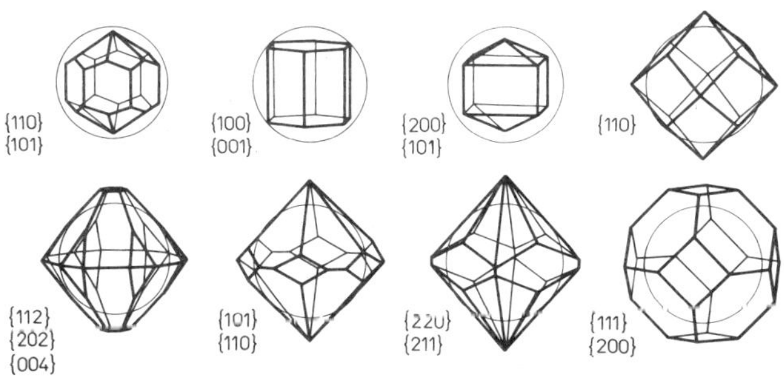

Figure 1. Brillouin - Jones zones of *ω,* simple hexagonal, white tin and simple cubic phases. Miller indices of the faces are shown. The upper row represents completely filled zones, the lower row represents partly filled zones. Crystallographic data are taken from [4-6, 10].

The sphere shown in Fig.1 is the Fermi sphere, estimated by the free electron approximation with Fermi sphere of radius $k_F=(3\pi^2 n/V)^{1/3}$, where n/V is the valence electron density.

As for the case of classical Hume-Rothery phases, Fermi spheres of these phases are close to the Brillouin planes in the extended space, so that there is similarity with the mechanism of the electron phase stability due to lowering the crystal energy as the Fermi sphere approaches Brillouin planes.

The aim of this paper is to carry out a comprehensive analysis of these and some related structures in respect to Brillouin zone configuration. This approach gives an opportunity to clarify some crystal symmetry peculiarities, for example, the special values of axial ratio for certain non-cubic (hexagonal and tetragonal) structures.

DISCUSSION OF THE AXIAL RATIOS OF PHASES WITH HEXAGONAL AND TETRAGONAL SYMMETRY

Simple Hexagonal Phase

The first object of discussion is simple hexagonal phase appearing at ambient condition in Sn alloyed by In, Cd or Hg [11]. The same phases have been obtained in about 15 Sn-based binary alloys by the 'splat cooling' technique [9]. It was found later that ***sh*** phase is a characteristic not only of Sn-based alloy and is formed in some other nontransition metal alloys as well as in group IV elements at the necessary condition: n is nearly 4 el/atom. There are some examples of ***sh*** phase:

- 'splat cooling' alloys In-Sb [10], In-Bi [8], Cd-Bi [7];
- pressure quenched alloys: In-Bi, Pb-Sn, Cd-Sb, Al-Ge [4,5];
- high pressure forms of Si [12,13], Ge [14], and III-V compounds (GaSb [15], GaAs [16], InSb [17]);
- high pressure phase of $Al_{0.3}Ge_{0.7}$ alloy at 5-45 GPa [18].

All the known ***sh*** phases have axial ratios ranged between 0.92 and 0.95.

The explanation for this peculiarity of ***sh*** structure was provided by Weaire and Williams [19] on simple symmetry arguments. The real and reciprocal lattices are both hexagonal and have axial ratio relation $c^*/a^*=\sqrt{3}/2(a/c)$. Assuming that Ewald energy will be minimized by keeping the balance between axial ratios of real and reciprocal lattices $c^*/a^*=c/a$, it was found that $c/a=(\sqrt{3}/2)^{1/2} = 0.931$.

By exact calculations the axial ratio of ***sh*** is determined at nearly the same value, as well as by qualitative arguments. It is remarkable that the balance $c^*/a^*=c/a$ is disrupted in ***sh*** alloys at extreme conditions. The axial ratio of ***sh*** phase for Si [20] and for $Al_{0.3}Ge_{0.7}$ [18] increases with pressure, whereas for Sn-based alloy [9] it increases with the electron concentration decreasing. These facts indicate that the contribution of the electrostatic energy attributed to the ionic interactions in the real lattice becomes more important. It can be treated as if the real and reciprocal lattices 'compete' in the energy contributions affecting the lattice axial ratio.

A similar approach can be applied to analyzing other non-cubic structures.

White Tin

This structure represents ***β-Sn***, tetragonal, tI4 (Pearson's notation), space group $I4_1/amd$, lattice parameters: a=5.831Å c=3.181Å, c/a=0.5456 [21]. White tin structure was found also under pressure for Si, Ge and some III-V and II-VI compounds [21]. But this structure is not specific only for group IV elements and isoelectronic compounds, it is also found in some alloys by 'splat cooling' [8] and pressure quenching [4,5]. All the known ***β-Sn*** phases have axial ratio ranged between 0.540 and 0.550.

The coordination number of each atom is nearly 6: two atoms at the distance d_1=3.022 and four atoms at the distance d_2=3.181 (see Fig.2a). If $d_1=d_2$ then the axial ratio would be c/a=0.516.

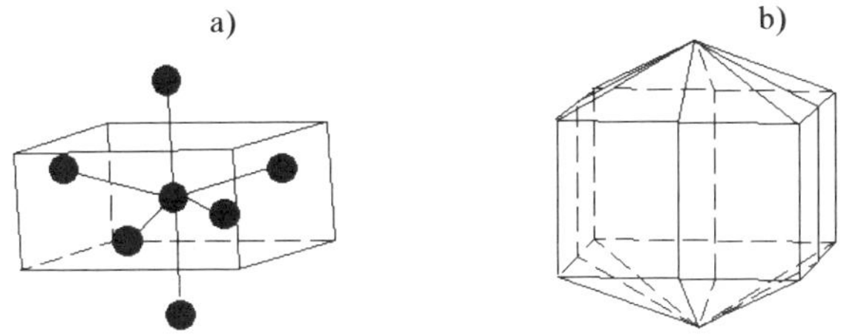

Figure 2. White tin: (a) crystal structure, (b) Brillouin zone of (101) and (200) planes. Two sections perpendicular to the basal plane are of nearly hexagonal form.

The white tin structure in the reciprocal space should be considered in the Brillouin zone form, containing the planes of the smallest indices with a non-zero structural factor (101) and (200) (see Fig.2b).

This Brillouin zone appears as high symmetry polyhedron having a square in one section and nearly proper hexagon in two others perpendicular sections. Assuming the Brillouin zone to be of proper form, that is $q_{200}=q_{101}$, one obtains the value of the axial ratio $c/a=1/\sqrt{3}=0.577$.

One may speculate that by keeping the balance of symmetry in the real and reciprocal spaces these two values of c/a yield the average value c/a=0.546, which is very close to the experimental one.

ω-Phase

The prototype of this structure is the so-called *ω*-phase observed in some Ti-based alloys and later found for pure Ti at high pressure and retained at normal pressure [21]. The structure of *ω*-Ti (see Fig.3a) with 3 atoms per unit cell, hP3, has lattice parameters a=4.625Å, c=2.813Å, c/a=0.608. Atom positions are 0,0,0 and ±1/3,2/3,z (z≈1/2). The question arises, why such kind structure appears, since it is very much unusual for metals. At normal pressure titanium has ***hcp*** and ***bcc*** (high temperature) forms. Atomic arrangement in the *ω* is far from high symmetry, because interatomic distances are not equal and coordination numbers for two atom sites are different.

The Ni_2In type with ordered sites and a twice greater c-axis is close related to the *ω* type structure. This structure type also involves the In_2Bi [21] and In_2Sb phases, the latter was obtained after pressure quenching [4,5]. Due to small difference in X-ray scattering for constituent atoms the last phases can be treated as of the *ω*-type with $c/2a\approx0.6$.

Brillouin zone for *ω*-Ti given on Fig.3b is constructed by planes (110) and (101), planes of smallest indices with a non-zero structure factor. Assuming that $q_{110}=q_{101}$, one may obtains that $c/a=\sqrt{(3/8)}=0.612$. This axial ratio was found for *ω*-phase in Ti and Zr based alloys and in high pressure forms of Ti and Zr.

It is apparent, that the proper many - faces form of Brillouin zone for *ω*-phase (30 planes with the same q) is one of the reason of this type structure stability.

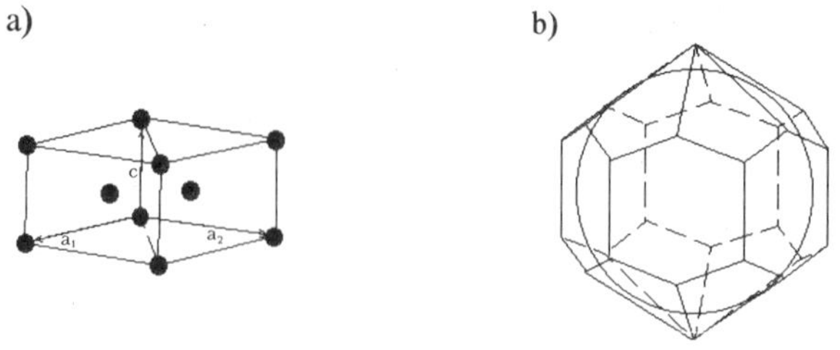

Figure 3. ω-Ti: (a) crystal structure, (b) Brillouin zone of (101) and (110) planes, Fermi sphere for n=2.

Close Packed Hexagonal Phases of Non-Ideal Axial Ratios

Most metallic phases with a close packed hexagonal structure have the axial ratio different from the ideal value 1.633. The questions are:

- How far may the axial ratio value diverts from the ideal one?
- Why is the non-ideal axial ratio preferred?

The earliest view of ***hcp*** distortion was based on the non-spherical form of atoms. We will try to answer these questions applying the Brillouin zone concept. The symmetry of real lattice provides the ideal axial ratio of 1.633. The lowering of c/a toward the values of 1.55 was found in the binary alloys based on group I elements [22]. The lowest axial ratio of 1.50 was observed for Ba under pressure increasing [23]. This value seems to be a special one, since after it is reached, the ***hcp*** structure becomes unstable.

Let's consider the Brillouin zone for ***hcp*** structure (Fig.4) with (002) and (101) planes and assume that $q_{101}=q_{002}$ leading to the value of c/a=1.50. This is a special case of the Brillouin zone symmetry, providing the proper hexagons are not only in the basal plane, but also at perpendicular sections. In most real cases the value of c/a is a result of 'competition in symmetry' between real and reciprocal lattices.

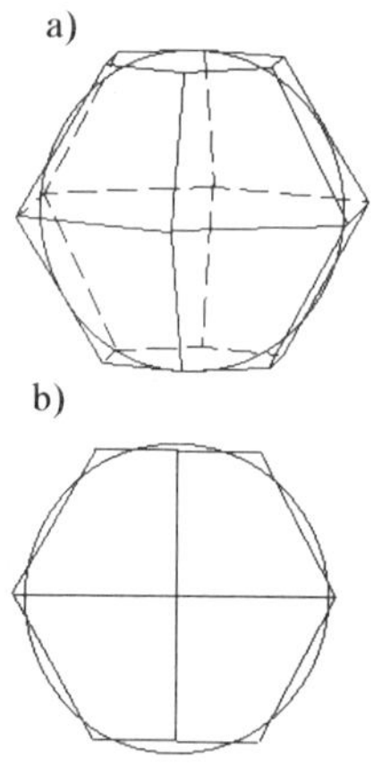

Figure 4. Brillouin zone of *hcp* Ba - high pressure form according to the data [23]. Brillouin planes are (101) and (110). Fermi sphere is for n=2. (a) Common view, (b) section perpendicular to the basal plane.

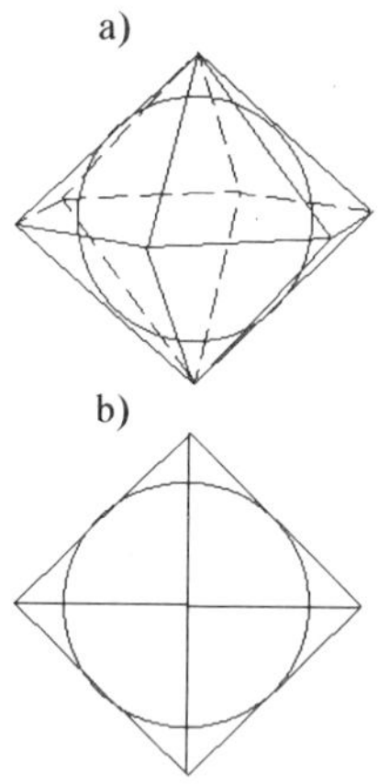

Figure 5. Brillouin zone of *hcp* $Al_{0.3}Ge_{0.7}$ - high pressure form according to the data [18]. Brillouin planes are (102). Fermi sphere is for n=4. (a) Common view, (b) section perpendicular to the basal plane.

Another special value of c/a for ***hcp*** could be determined by examing another case of symmetry in the reciprocal space, providing $q_{100}=q_{002}$, and $c/a=\sqrt{3}\approx1.73$. The values tending to this special one were found for high - pressure forms of ***hcp*** silicon with c/a=1.70 [12,24] and of ***hcp*** $Al_{0.3}Ge_{0.7}$ alloy with c/a=1.68 [18]. The Brillouin zone of ***hcp*** $Al_{0.3}Ge_{0.7}$ alloy is shown in Fig.5, including (102) planes which are in contact with the Fermi sphere. Assuming that there are proper squares at sections perpendicular to the basal plane we obtain $c/a=\sqrt{3}$. The axial ratio of SiVII was found to tend to a lower value at further pressure increasing [24]. This is in agreement with the tendency discussed above for ***sh*** phase, that the role of real lattice becomes more important with pressure increasing.

Concluding the consideration of ***hcp*** it should be noted that there are two special values of c/a equal to $\sqrt{3}$ and 1.50, connected with the symmetry of Brillouin zones containing different planes.

Body Centered Tetragonal Structures of Hg and Sn

Two types of tetragonal distortive body centered structures are considered as an example of Fermi sphere - Brillouin zone interaction.

Mercury. HgII has ***bct*** structure as a low-temperature and high-pressure form: tI2, a=3.995, c=2.825, c/a=0.707. Fig.6 shows the Brillouin zone of (110) and (101) planes. The tetragonal distortion condition is that the (101) planes are in contact with the Fermi sphere.

Tin. SnII has ***bct*** structure under high pressure with lattice parameters a=3.81Å, c=3.48Å, c/a=0.913 at 3.9 GPa, 314°C [21]. In Fig.7 Brillouin zone and Fermi sphere of SnII are shown. The cubic structure of the same volume would have Brillouin zone extending outward the Fermi sphere. A slight distortion of ***bcc*** is necessary for nesting Brillouin zone inside the Fermi sphere.

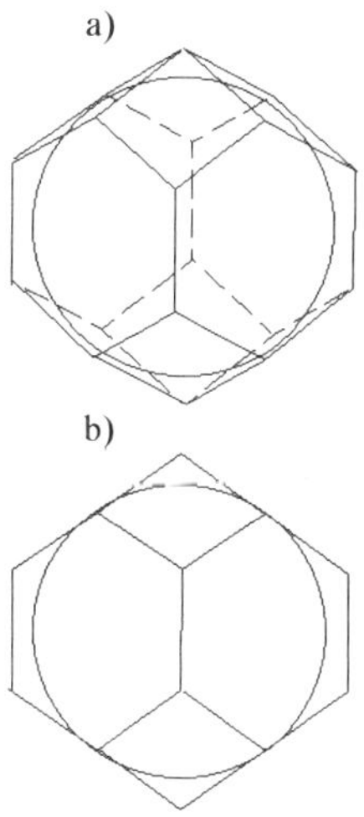

Figure 6. Brillouin zone of body centered tetragonal Hg of planes (101) and (110). Fermi sphere is for n=2.
(a) Common view, (b) view along a^*.

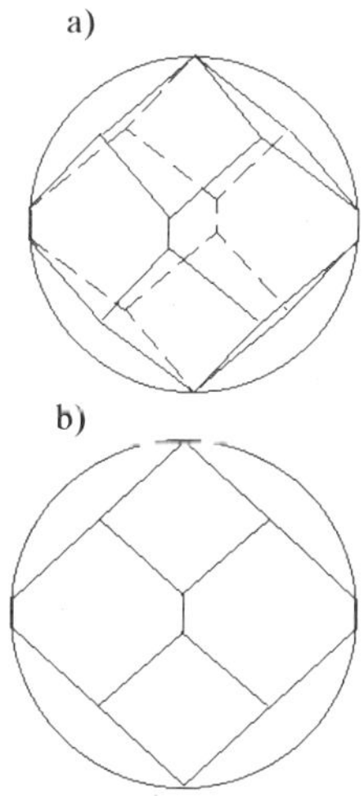

Figure 7. Brillouin zone of body centered tetragonal Sn of planes (101) and (110). Fermi sphere is for n=4.
(a) Common view, (b) view along a^*.

The ***bct*** structures of Hg and Sn give an idea how the lattice distortion allows the best accommodation of the Brillouin zone and Fermi sphere one inside the other providing minimization of the crystal energy.

CONCLUSION

Simple qualitative arguments have been applied to the determination of the axial ratios in some intermetallic phases with non-cubic lattices. The Brillouin zone symmetry of these structures appears to affect the axial relation of real lattice providing the minimization of the crystal energy.

ACKNOWLEDGMENTS

The author is grateful to Prof. E.G.Ponyatovsky and Prof. W.B.Holzapfel for fruitful collaboration. This work was partially supported by International Science Foundation through grant NKR000.

REFERENCES

1. N.F. Mott and H. Jones. "The Theory of the Properties of Metals and Alloys", Clarendon Press, Oxford (1936).
2. W. Hume-Rothery and G.V. Raynor. "The Structure of Metals and Alloys", The Institute of Metals, London (1956).
3. H. Jones. "The theory of Brillouin zones and electronic states in crystals," North-Holland Publishing Company, Amsterdam (1962).
4. V.F. Degtyareva and E.G.Ponyatovskii, High-pressure phases in the alloys of B-elements - a new type of electron phases, *Sov.Phys.Solid State* 24(9):1514 (1982).
5. E.G. Ponyatovskii and V.F. Degtyareva, Specific features of T-C-P diagrams for binary systems of B-elements, *High Pressure Research* 1(3):163 (1989).
6. V.F. Degtyareva and E.G. Ponyatovskii, Formation of phases in binary s,p-bonded metal alloys under high pressure, *in:* "Statics and Dynamics of Alloy Phase Transformations", P.E.A.Turchi and A. Gonis, ed., Plenum Press, New-York (1994).
7. P.K. Srivastava, B.C. Giessen, and N.J. Grant, New metastable electron phases in binary B-metal alloys, *Acta Met.* 6(10):1199 (1968).
8. B.C. Giessen, M. Morris, and N.J. Grant, Metastable indium - bismuth phases produced by rapid quenching, *Trans.Met.Soc. AIME* 239(1):883 (1967).
9. R.H. Kane, B.C. Giessen, and N.J. Grant, New metastable phases in binary tin alloys systems, *Acta Met.* 14(5):605 (1966).
10. B.C. Giessen, R.H. Kane, and N.G. Grant, A metastable intermediate phase in the system indium - indium antimony, *Nature* 207:854 (1965).
11. M. Hansen and K. Anderko. "Constitution of Binary Alloys," McGrow-Hill, New York (1958).
12. S.K. Sikka, H. Olijnyk, and W.B. Holzapfel, Structural phase transformations in Si and Ge under pressures up to 50 GPa, *Phys.Letters* 103A(3):137 (1984).
13. J.Z. Hu and I.L. Spain, Phases of silicon at high pressure, *Solid State Commun.* 51(5):263 (1984).
14. Y.K. Vohra, E. Bristerk, S. Desgreniers, and A.L. Ruoff, Phase - transition studies of germanium to 1.25 Mbar, *Phys.Rev.Lett.* 56(18):1944 (1986).
15. S.T. Weir, Y.K. Vohra, and A.L. Ruoff, Phase transitions in GaSb to 110 Gpa (1.1 Mbar), *Phys.Rev.B* 36(8):4543 (1987).
16. S.T. Weir, Y.K. Vohra, C.A. Vanderborgh, and A.L.Ruoff, Structural phase transitions in GaAs to 108 Gpa, *Phys.Rev.B* 39(2):1280 (1989).
17. C.A. Vanderborgh, Y.K. Vohra, and A.L. Ruoff, Structural phase transitions in InSb to 66 Gpa, *Phys.Rev.B* 40(18):12450 (1989).
18. V.F. Degtyareva, F. Porsch, E.G. Ponyatovskii, and W.B. Holzapfel, Structural investigations of the amorphous alloy $Al_{0.3}Ge_{0.7}$ under high pressure, to be published in *Phys.Rev.Lett.*
19. D. Weaire and A.R. Williams, On the axial ratios of simple hexagonal alloys of tin, *Phil.Mag.* 19(162):1105 (1969).
20. J.Z. Hu, L.D. Merkle, C.S. Menoni, and I.L. Spain, Crystal data for high pressure phases of silicon, *Phys.Rev.B* 34(7):4679 (1986).
21. For crystal structure description, see W.B. Pearson. "A Handbook of Lattice Spacings and Structures of Metals and Alloys", Pergamon Press, New York (1967).
22. W.B. Pearson. "Crystal Chemistry and Physics of Metals and Alloys", Wiley, New York (1972).
23. K. Takemura, High-pressure structural study of barium to 90 Gpa, *Phys.Rev.B* 50:16238 (1994).
24. S.J. Duclos, Y.K. Vohra, and A.L. Ruoff, Experimental study of the crystal stability and equation of state of Si to 248 GPa, *Phys.Rev.B* 41(17):12021 (1990).

MATERIALS PROPERTIES AND CHARACTERIZATION

TEM CHARACTERIZATION OF STRUCTURAL DEFECTS

Gustaaf Van Tendeloo

EMAT
University of Antwerp (RUCA)
Groenenborgerlaan 171, B-2020 Antwerp
Belgium

INTRODUCTION

It needs no comment that structural defects strongly influence a large number of physical properties as well as the behaviour of materials as function of temperature or pressure. Different techniques allow to detect or characterize these defects in a direct or an indirect way. We may subdivide them in reciprocal space techniques, such as X-ray or neutron scattering, and real space techniques such as the different microscopy techniques.

The strength of reciprocal space techniques has been strongly improved recently because of synchrotron radiation sources; they allow a higher resolution and need smaller amounts of material. Neutron as well as X-ray diffraction are highly quantitative and detailed analyses of intensity maxima and peak widths allow to draw fargoing conclusions on the perfection of the material, deviations from stoichiometry, the average density of defects. However no direct visualization of these imperfections is possible and therefore important information such as a correlation between defects, Burgers vectors of dislocations, local deviations from perfection is not available. Moreover the presence of mesoscopic or nano sized secondary phases as well as weak modulations in the structure can be easily overlooked.

Real space microscopy techniques can be further subdivided into surface techniques and bulk techniques; both have their merits and specific application fields. Optical microscopy, scanning electron microscopy, scanning tunneling microscopy and atomic force microscopy belong to the former and give detailed, eventually atomic scale information on the surface morphology, surface reconstruction and surface adsorbtion. Information from the bulk can be obtained by a number of techniques such as acoustic microscopy, infra-red microscopy, X-ray

Stability of Materials
Edited by A. Gonis *et al.*, Plenum Press, New York, 1996

tomography or others. Although most of these techniques have their specific merits, the spacial resolution is in general very limited. For more information on the strength of these different techniques and the specific application fields I refer to the recent "Handbook of Microscopy" [1] dealing with the different techniques of real space imaging. Electron microscopy not only has the advantage to be able to resolve structural details down to the atomic level, it also allows to switch -with a single push-button- to electron diffraction, i.e. to reciprocal space. This combination of real space and reciprocal space information makes electron microscopy a unique technique for the study of solid state materials. As all other techniques it has its shortcommings and limitations and we will deal with that in the following chapter, but particularly it particularly can provide structural information which no other technique is able to produce. The best way to prove this point is to treat a number of case-studies where we will try to solve problems which can not -or hardly- be solved by any other technique.

ELECTRON MICROSCOPY?

It is not the aim of this chapter to provide a course of electron microscopy or instructions for electron microscopy operation; these can be found in specialized books and review articles such as [2-5]. We will only focus on the fundamentals which should allow the materials scientist or the physicist to evaluate and understand the results discussed in the following chapters.

The image formation in electron microscopy is very similar to the image formation in optical microscopy; in optical microscopy however the resolution is limited by the wavelength and therefore the light source is replaced by a source of electrons where the wavelength is dependent on the accelerating voltage. In practice the accelerating voltage varies between 100kV and 400kV for most commercial microscopes, but it can reach up to 3 or 5MV for specially designed high voltage electron microscopes. For all of them however the working principle is the same: with the help of condensor lenses and diaphragmes a parallel incident beam is produced. The incident electron beam can be described as a plane wave hitting a sample (see fig.1). Since electrons are easily absorbed and scattered, an electron microscope operates under vacuum (normally of the order of 10^{-5} pascal) and the samples have to be thin enough in order to allow -partial- transmission of the electrons. The maximum thickness depends on the accelerating voltage, the material itself and the resolution required. For "conventional" microscopy, where no atomic resolution is required, the allowed thickness can be a micron or more for light materials and for higher voltages. For high resolution electron microscopy the thickness has to be as small as possible and should certainly not exceed a few hundred Ångstrom. Preparing samples as thin as this is not trivial, but over the years a number of techniques have been developed to prepare reproducible thin samples; details can be found in [6]. It should be noted that all these techniques are not always "harmless" to the material. For ceramic materials, a frequently used and wonderful technique is ion milling, however the argon beam tends to produce an amorphous surface layer of a few nanometer.

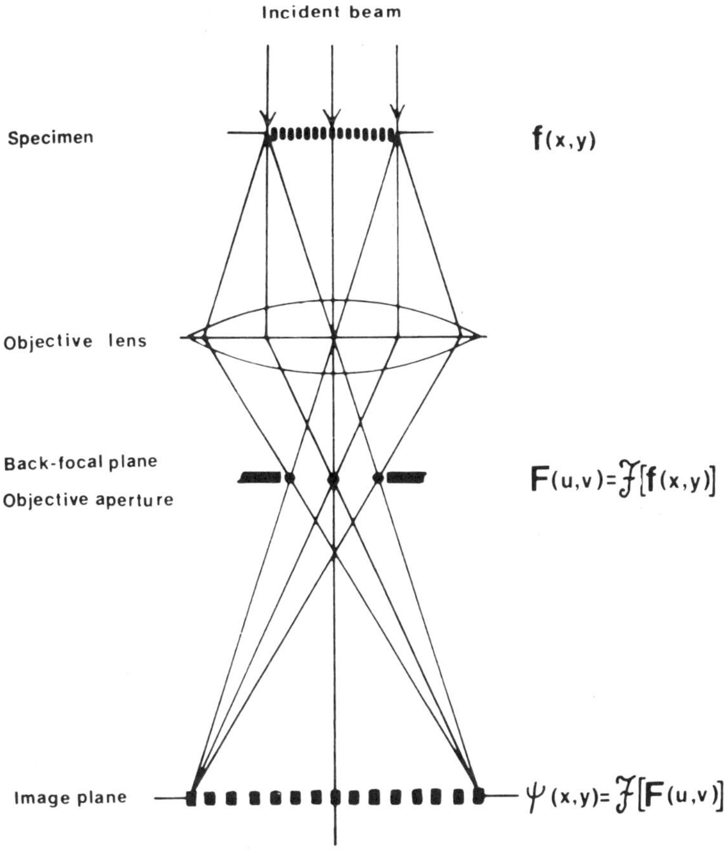

Figure 1. Image formation in an ideal microscope, no lens aberrations being taken into account..

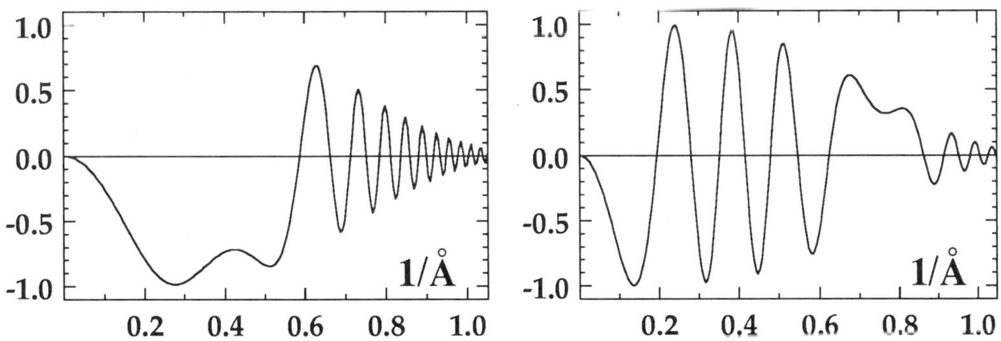

Figure 2. Dependence of $\sin\chi(u,v)$ on the reciprocal distance, expressed in $Å^{-1}$.

a) for the Scherzer defocus value. b) for an arbitrary defocus value

For a number of problems, this is of few importance, however for the study of the fine structure of e.g. grain boundaries, this can be desastrous. Anyway, let the sample be thin enough so that it can be characterized as a two dimensional transmission function f(x,y). Within the sample Bragg reflection will take place and the electrons will emerge in discrete directions at the exit surface; alternatively one can say that the back surface acts as a planar assembly of Huyghens sources and interference between these wavelets generates the diffracted beams. These diffracted beams are focused by the objective lens into the back focal plane and a diffraction pattern results; in case of a crystalline object this will be discrete points, in case of an amorphous object a diffuse ring pattern is obtained. This diffraction pattern is nothing but the Fourier transform of the transmission function f(x,y), i.e. q(u,v) = F[f(x,y)] (see fig.1). If the diffracted beams are allowed to interfere further, they will produce an enlarged image of f(x,y) in the image plane; this is again the Fourier transform of the diffraction pattern, i.e. the scattering amplitude $\psi(x,y)$ in the image plane is given by $\psi(x,y) = F[q(u,v)]$. So to a first approximation the electron microscope is identical to the optical microscope. If the wavelength would be the limiting factor, one would achieve a resolution far below 1Å ... This means one has clearly over-simplified things.

A real microscope suffers from a number of limitations such as spherical aberration or chromatic aberration, the electron beam is never perfectly parallel and has a certain divergence, apertures are introduced in the back focal plane to select particular reflections and image recording is not always made in the "ideal" image plane but mostly with a certain defocus value. All these factors will limit the optimum resolution of a modern microscope to around 2Å. The aberrations will introduce different phase shifts $\chi(u,v)$ for the electron beams passing through the lens system of the microscope. This leads to a final expression for the amplitude in the image:

$$\psi(x,y) = F[A(u,v)\, q(u,v)\, e^{-i\chi(u,v)}]$$

where A(u,v) is an eventual aperture function and the complex phase function $\chi(u,v)$ is given by:

$$\chi(u,v) = 1/2\ \pi C_S \lambda^3 (u^2+v^2) + \pi\varepsilon\lambda(u^2+v^2)$$

C_S is the spherical aberration coefficient, depending on the electron microscope, λ is the wavelength, depending on the accelerating voltage and ε is the defocus value of the objective lens for which the image is made. All these parameters are known or can be measured and one can therefore calculate the expected image for different defocus values, provided the transmission function f(x,y) is known.

Knowing the transmission function f(x,y); i.e. the scattered amplitude at the exit plane of the sample is the most difficult part. For very thin slices, the sample can be assimilated with a phase grating, i.e. the sample will only introduce a phase change in the electron beam. This approach was first proposed by Grinton and Cowley [7]; for weak scattering objects the transmission function is proportional to the projected crystal potential $\phi(x,y)$:

$$f(x,y) = 1 + i\,\sigma\,\phi(x,y) - \mu(x,y)$$

where σ is a constant and $\mu(x,y)$ is a damping factor taking into account the absorption. In general, even for the most simple and thin material, the final enlarged image will have very few resemblence with the projected crystal potential, because of the complex phase shift $\chi(u,v)$ through the microscope (see fig.2b). However for a particular defocus value, the so called Scherzer defocus, this shift is constant (≈ -1) over a large range, virtually up to the instrumental resolution limit of the microscope (fig.2a) and therefore for this particular defocus value one will image the projected crystal potential. Most experimental images published are taken at -or close to- this defocus value. An example is reproduced in fig. 3 where an experimental image is shown of the $La_5Ti_4O_{15}$ structure along the [100] direction, together with the computer simulated images for different defocus values and different thicknesses.

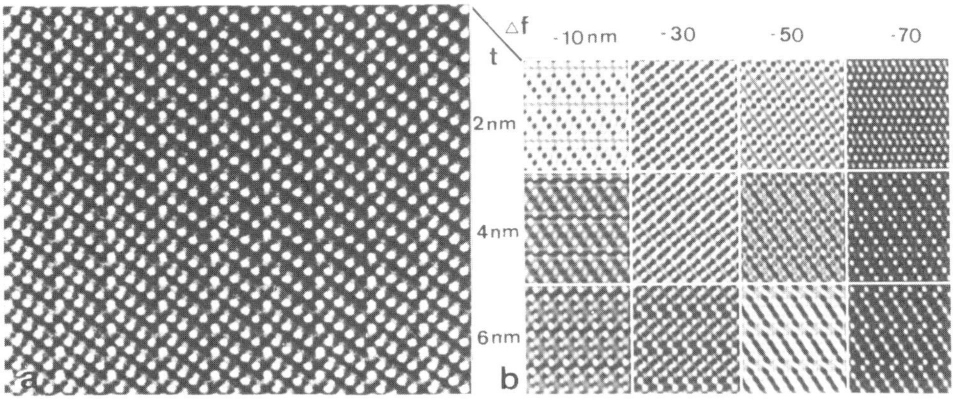

Figure 3. High resolution image of the $La_5Ti_4O_{15}$ structure (a) together with the computer simulated images, for different defocus values and thicknesses (b).

In practice resolution is also limited by a number of parameters which are difficult to controle such as mechanical instabilities (vibrations, sample drift, ...) and the stability of the sample in vacuum under irradiation of several hundred kilovolt. The latter is sometimes the major problem and excludes a number of applications for high resolution microscopy or severely limits the resolution or quality of the results. Materials such as polymers, zeolites, and virtually all organic compounds or H_2O containing minerals are subject to radiation damage under the electron beam. The number of electrons they can stand before severe damage and destruction of the crystal lattice takes place is very limited (sometimes only a few electrons per square Å) and there is virtually no time for aligning the microscope or ajusting the defocus value. In order to obtain results on such samples, extreme care has to be taken and one has to work under extremely low intensity conditions; we are not dealing with such materials in the present context; however information can be found in the contribution of Voigt-Martin in [1].

One can deliberately restrict the resolution in the image by placing an aperture in the back focal plane and selecting a number of reflections of particular interest for the image formation; in this way selective imaging is obtained and only a particular sublattice is imaged or a weak modulation is enhanced. We will treat this in more detail when considering ordering in fcc based alloys, here the basic lattice is of few importance (we know it is fcc) and we are

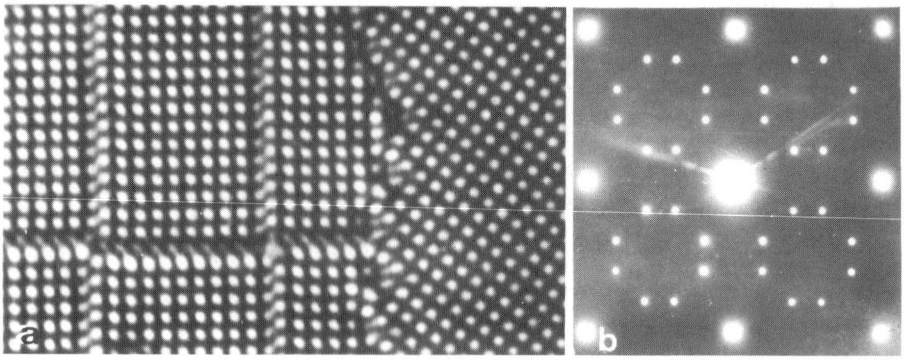

Figure 4. High resolution image of Au_4Mn with limited resolution; the diffraction pattern, indicating the selected beams is shown as fig. b.

particularly interested in the ordering on the lattice. By selecting only the ordering reflections we will in a number of cases, even with a restricted resolution be able to obtain a lot of information on the state of order and the ordering induced defects. An example of this superstructure imaging in shown in fig.4 together with the corresponding [001] diffraction pattern; the white circle indicates the reflections taken in the selective aperture, the intense basic reflections have deliberately been omitted. This example of Au_4Mn reveals the Mn columns as bright dots. Out-of-phase boundaries and twin interfaces, resulting from a lower -tetragonal- symmetry of the ordered structure on the fcc lattice are clearly revealed. It may be noted that these images were already obtained back in 1978... It is a clear illustration that for some problems it is not always necessary to have the most recent and fanciest instrument in order to obtain the correct information.

The extreme case of limiting the resolution is by selecting only one electron beam, this is the so called "conventional" microscopy. In case the selected beam is the transmitted beam, one obtains a "bright field" image, in case of a diffracted beam, a "dark field" image results. This kind of imaging, although less frequently used now, provides a large amount of information, particularly on the defect structure of the material. Theory and applications can be found in the classical textbooks by Hirsch et al. [9], Amelinckx [10] and Amelinckx et al. [11]. Particularly usefull for the scope of this chapter is the determination of the Burgersvector **b** of dislocations and the displacement vector **R** of stacking faults or antiphase boundaries. Tilting the sample into a "two-beam" position i.e. a position where apart from the transmitted beam only one strongly diffracted beam **g** is acting, a dislocation will be out of contrast when **g.B** = 0, while a translation defect will be out of contrast when **g.R** = integer. Three independent extinctions allow to deduce the Burgersvector or the displacement vector unambiguously. An example is shown in fig. 5, where a dislocation network in graphite is shown in fig.5a. The network is made up of three families of dislocations with different Burgers vectors. For different two beam situations (fig.5 b, c, d) some of the dislocations are out of contrast, while others remain in contrast. Several of such two beam situations allow to deduce the three Burgersvectors $\mathbf{b_1}$, $\mathbf{b_2}$ and $\mathbf{b_3}$ for the three families of dislocations.

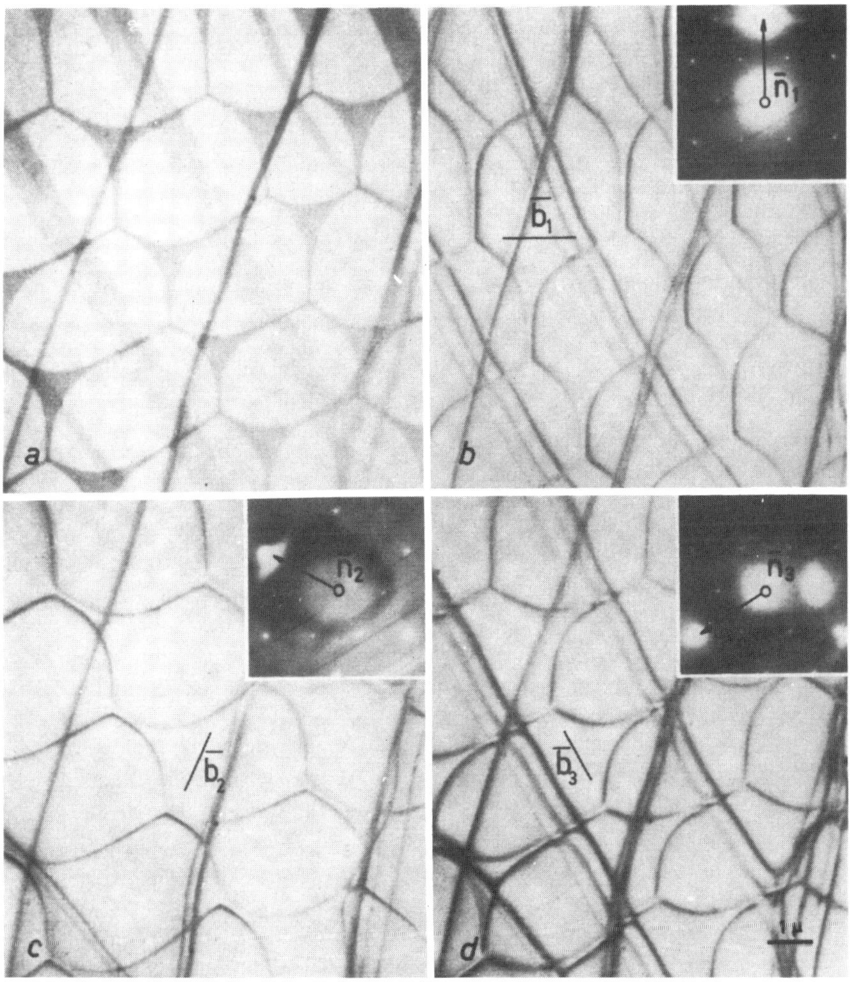

Figure 5. Conventional image of a dislocation network in graphite; the same area is shown under four different diffraction conditions; in a) all dislocations are visible, in b), c) and d) some dislocations are out of contrast. (courtesy Delavignette and Amelinckx, 1962)

As indicated before the strength of electron microscopy lies in the combination of real space imaging and reciprocal space information. Due to the fact that the interaction between the electron beam and the material is very strong (compared to the interaction with a neutron or X-ray beam) the scattering is no longer kinematical but highly dynamical, which makes the interpretation of electron diffraction intensities very complex. Because of multiple scattering, even in thin samples, the diffracted intensities can no longer be used to refine the crystal structure as it is the case for X-ray or neutron scattering. However electron diffraction has other advantages over X-ray or neutron scattering. Due to its strong interaction with the material only crystals of nanometer dimensions are required to produce clear diffraction patterns and moreover the scattering is very sensitive to weak modulations, hardly visible by other techniques, including synchrotron scattering. I can refer here to examples such as the charge density wave modulations in layered compounds, e.g. TaS_2, weak oxygen-vacancy ordering in oxygen deficient $YBa_2Cu_3O_{7-x}$ or local deformations in alloys.

When the incident electron beam is not parallel but focused or converged onto the sample a very particular type of diffraction pattern results; the so called convergent beam electron diffraction (CBEM). Since the incomming beam is a collection of different orientations, the resulting diffraction pattern will contain more information than a "normal" diffraction pattern and allows to deduce information about the local! space group symmetry. It is further used to deduce local lattice parameter changes in the vicinity of imperfections such as precipitates, cracks etc. More information on this particular, but extremely usefull technique can be found in textbooks such as [12,13] or review articles [14].

Finally, I should note that untill now we have only considered the elastic interaction between accelerated electrons and the material, this already provided us with a lot of information on the local structure of the material, however there are more possibilities. Apart from this elastic scattering there is much more interaction between the incomming electrons and the material (fig. 6). The emerging beams can be backscattered electrons, Auger electrons, inelastically

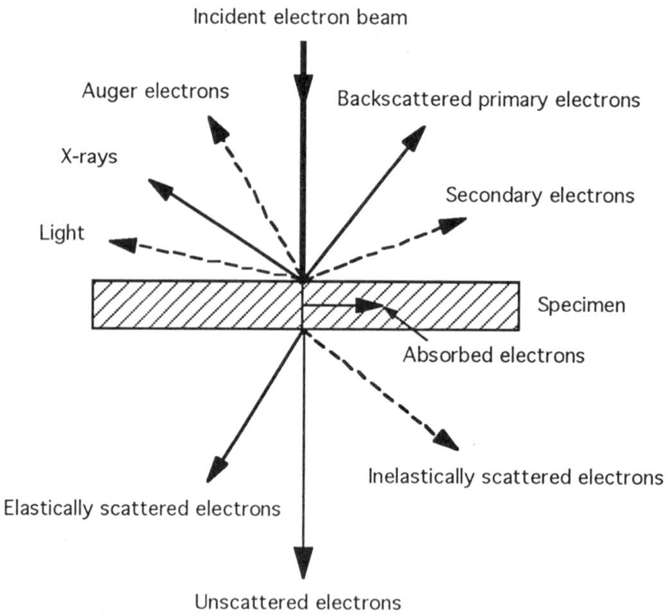

Figure 6. Schematic representation of the interaction between an incomming electron beam and the material.

scattered electrons, X-rays, or light (see e.g. [13]). All these emerging beams do contain information about the irradiated region and since this region can vary between several microns and a few nanometer, we clearly obtain local information, to be compared eventually with average information obtained by other techniques. Two main techniques frequently used in combination with electron microscopy are EDX (energy dispersive analysis of the X-rays) and EELS (electron energy loss spectroscopy) which both give the local composition of the material. While EDX is more quantitative and better for heavy elements, EELS is particularly suitable for light elements and very thin samples. We will not elaborate here on this or other electron microscopy related techniques, but we will use them when necessary in the materials science problems discussed further. More info on them can be found in [13, 15, 16].

STRUCTURAL DEFECTS ASSOCIATED WITH PHASE TRANSITIONS IN ALLOYS

Phase transitions in materials can be subdivided into displacive and non-displacive, or into diffusionless and diffusion controled transitions. Phase transitions generally introduce a change in symmetry, which results in a domain fragmentation due to the presence of different variants. These variants can be orientation variants, creating twin boundaries or inversion boundaries, or they can be translation variants creating antiphase boundaries. A prediction of the number of orientation and translation variants can be done based on simple group theoretical considerations [17,18,19]. In the present contribution we will concentrate on the microstructure and the microstructural changes when approaching the phase transformation. For the diffusion controled transformations we will investigate the short range order when approaching the transition from above and the wetting of ordering defects such as antiphase boundaries when approaching the transition from below. For the displacive transitions we will study the pre-martensite state in the austenite-martensite transformation in nickel rich Ni-Al.

We know that in general HREM images vary substantially with focus and sample thickness, however for alloys, based on a simple lattice such as the fcc (or bcc) lattice, the interpretation simplifies considerably. The reason being that when such structures are viewed along a simple zone axis, e.g. [001], we obtain a column structure with the different columns separated by 1.8-2.0 Å. Such atom columns, parallel to the electron beam, constitute a potential well with cylindrical symmetry and the electrons "channel" through this potential well, continuously focussing and defocussing with a periodicity varying between 40 and 100 Å, depending on the nature of the constituent atoms (fig.7). One can show that as long as the columns do not come too close, electrons channelling in different columns will not mutually influence each other [20,21]. Therefore for ordered alloys the image contrast of a particular column will only be determined by the chemical composition of that column and will not be influenced by the composition of neighbouring columns [20]. This aspect of HREM imaging in column type structures will allow us to interpret ordering phenomena in ordered compounds.

Short Range Order and Initial Stages of Ordering in D1a type Alloys

Studying short range order (SRO) in alloys by electron microscopy is hardly possible "in situ" since oxidation inside the electron microscope tends to limit the studies, even for alloys with a low transformation temperature. The only way to obtain information on the SRO state is by quenching the alloy from high temperature to room temperature and hoping that the state of order is retained. For a number of alloys, such as CuAu or CuPt, this procedure is not effective, ordering proceeds too fast and only the initial stages of long range order (LRO) can be studied. For other alloys however the ordering is sluggish enough to maintain the SRO or the very initial state of LRO after quenching. Such kind of alloys are chosen here to study the order disorder transition. We have particular interest in so called "1 1/2 0" alloys since the LRO reflections (mostly of the D1a type structure) do not coincide with the SRO reflections (see fig.8). Examples of such alloys are Ni_4Mo, Au_4V, Au_4Cr, Au_4Fe, Au_4Mn, Au_4Ti. The SRO and LRO (e.g. in Ni_4Mo) has been studied since 1965 [22,23] but still discussion is going on about the actual SRO state in a real space description: do we have ordered clusters, microdomains or a statistical distribution described by a concentration wave model? The actual difference between a cluster and a microdomain model being a matter of correlation length. The concentration wave model, first described by Khachaturyan [24], is basically a description in reciprocal space. The "microdomain model" and the "concentration wave" model however have to be considered as two descriptions of the same physical reality [25], indeed the materialisation of a concentration wave in a discrete medium results in a geometrical pattern that can be considered as a cluster. In the past a number of alloys showing a particular SRO or transition state between SRO and LRO have been studied in reciprocal space. The locus of diffuse intensity enables to deduce a relation

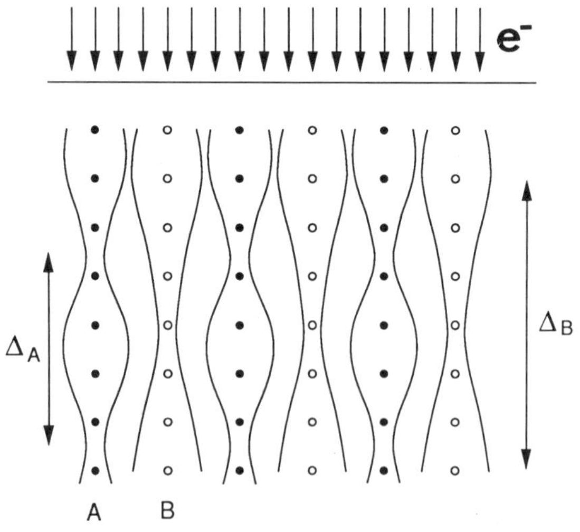

Figure 7. Electron travelling through a column type structure.

between some of the occupation sites of the fundamental lattice [26, 27]. Neutron and X-ray diffraction provide more accurate numerical data, but the fundamental limitation to pair correlations and the practical limitation to large selected areas still holds.

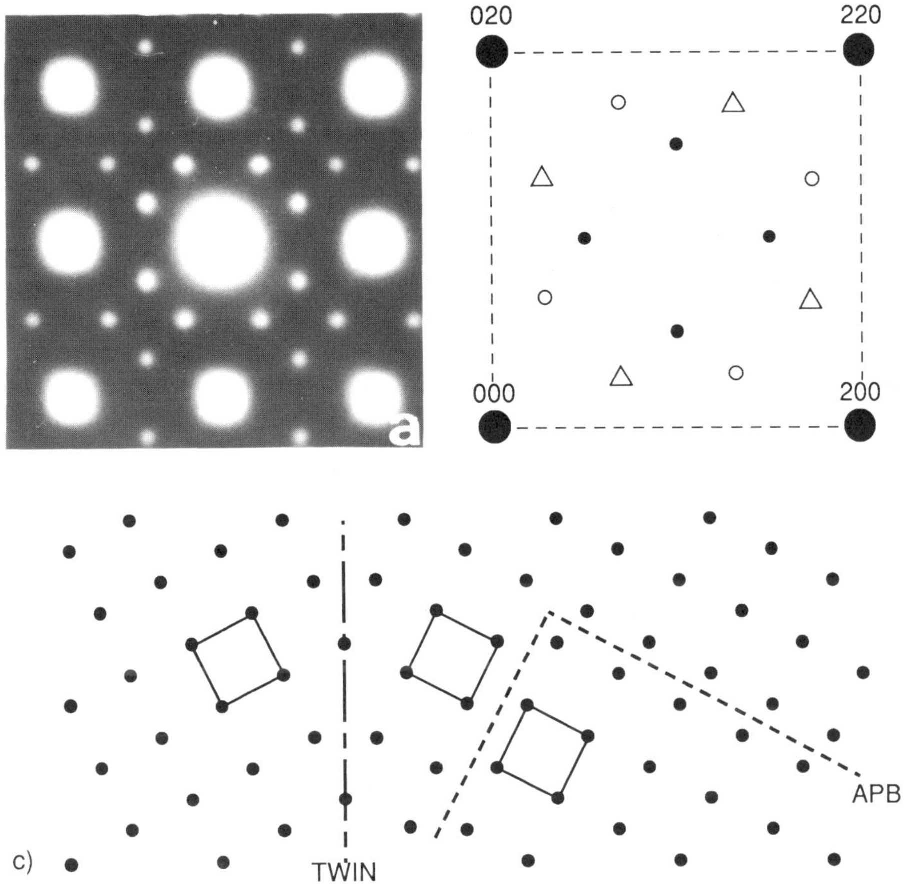

Figure 8. a) [001] diffraction pattern of Ni_4Mo in the SRO state; b) Schematic [001] section of the reciprocal lattice of the SRO (black dots) as well as of the LRO state with two different orientation variants (open circles and triangles); c) [001] projection of the LRO state of Ni_4Mo, showing two different orientation variants with parallel c-axes; only the Mo configuration is shown and the projected unit cell is outlined.

The long range order state of all 1 1/2 0 alloys is the D1a structure, the structural representation of which is shown in fig.8c. The disordered fcc lattice has cubic symmetry, while the ordered structure has tetragonal 4/m symmetry. This results in 6 orientation variants; the tetragonal c-axis can be along any of the cube axes and for every choice of the c-axis, two orientations are possible of the a- and b-axis (see fig.8c). These 6 orientation variants however only produce two essentially different twin interfaces[23, 8]. The volume of the primitive unit cell of the ordered structure is 5 times larger than the primitive cell of the disordered structure and therefore five translation variants, separated by four different antiphase boundaries result [17]. Ordering introduces a fragmentation into these different variants, resulting into patterns such as fig.4 where a twin interface between two orientation variants with parallel c-axis and two antiphase boundaries are visible (compare with fig 7c). In fig. 4, the Au_4Mn superstructure has been imaged, only the Mn configuration is visualised as bright dots. The defects in the LRO state are adequately studied and are now well understood.

Recently we have also performed HREM observations on the SRO state in Ni_4Mo, Au_4Cr, and Au_4Mn (fig.9). In spite of the fact that we have a column type structure, based on a simple fcc lattice, we realize that the detailed interpretation of such images is far from straightforward. However we have been able, with the necessary care and adapted image

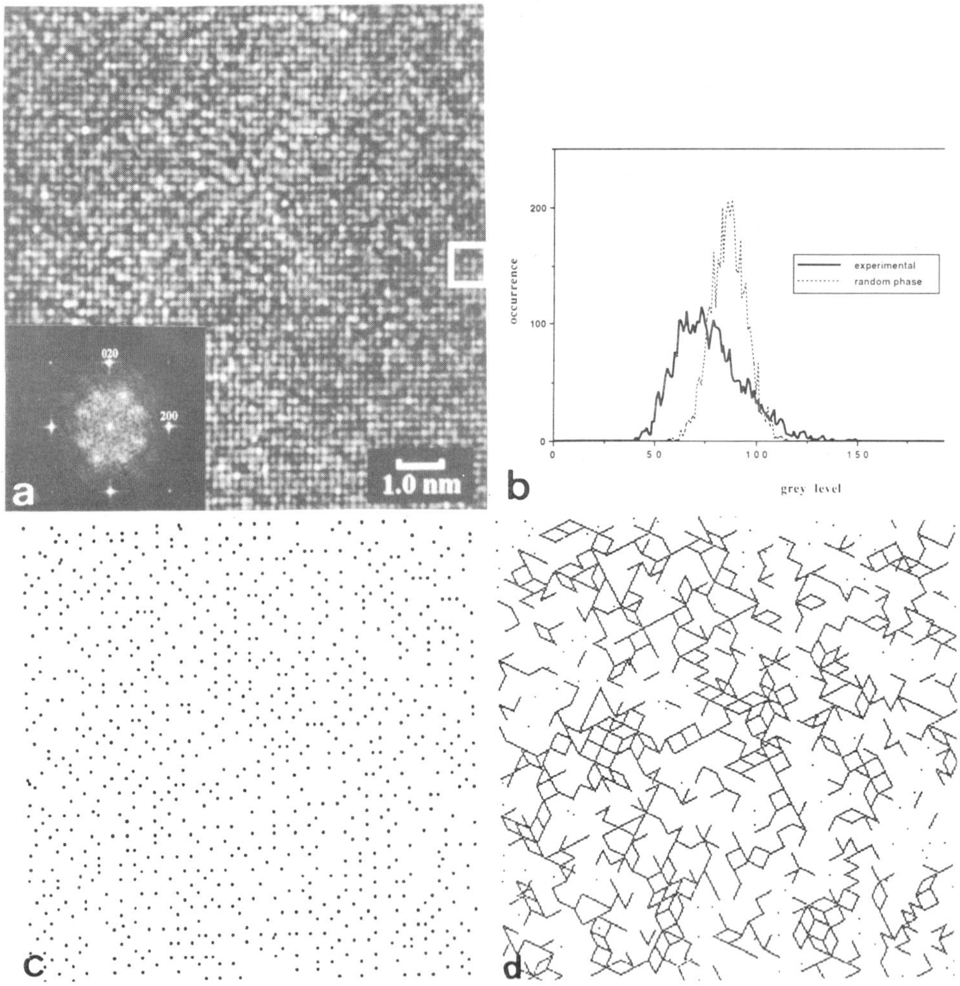

Figure 9. a) HREM image of the SRO state in Au_4Cr; the power spectrum of a), showing the presence of SRO in the alloy is shown as an inset; b) a histogram of gray levels obtained from the SRO state, compared with a histogram from a "random phase" distribution; c) result of image processing the original fig. (a); d) same figure, but typical clusters of the type {2,1} have been indicated.

processing, to extract very useful information about the ordering. The Fourier transform of the HREM image of fig.9a indeed shows diffuse intensity similar to the electron diffraction pattern of fig.8a and therefore the ordering information is consealed within the HREM image. The main problem in HREM of such imperfect structures is the projection of a three dimensional structure into a twodimensional image. One can show however that for atom colums consisting of a random or almost random composition, the projection approximation still holds. Although this does not lead us to a direct correspondence between the column composition and its grey value -because of differences of thickness and focus- the differences in composition are still reflected in the image. If we classify for a disordered A_4B alloy all [001]-type columns according to their chemical content, a binomial distribution is expected, which will be assymetric [28]. For larger thicknesses, all columns will become equivalent and the distribution will tend to be symmetrical. We have digitized the experimental image of fig.9a and obtained a histogram of all gray levels (fig.9b); the assymmetry is clearly present. We can further show that whenever an asymmetrical histogram is obtained, the brightest dots in the HREM image correspond to columns deviating strongly from the average composition [28]. For images (such as fig.9a) which fullfill the binomial criterion, most information is found in those columns which appear less frequent, i.e. which deviate most from the average composition. We can develop an image processing routine to enhace the "in plane" correlations between these columns which give rise to the SRO intensity. From the experimental image of fig.9a we obtain a processed image (fig. 9c,d). The Warren-Cowley short range order parameters $\alpha_{i,j}$ can be directly calculated from the power spectrum, but also from counting the pairs in the binary image. It then turns out that pairs of the type {2,1} occur most frequently, indicating a preference of the structure to form identical colums separated by a vector {2,1}. When these pairs are connected in fig. 9c, we obtain the final image of fig. 9d where not only isolated lines can be discerned but also clusters of the D1a type structure as well as of the DO_{22} structure. This is hardly surprising because the {2,1} pairs are the constituent vectors of both these structures. We can now apply the same technique to the different stages of ordering between the SRO state and the LRO state and follow the ordering kinetics. The Au_4Cr alloy is particularly suitable for such kind of experiment because the ordering kinetics are very slow and therefore every state of order can easily be quenched in. Important, and beyond the limits of classic diffraction experiments, is the direct observations of clusters, largers than pairs, in these 1 1/2 0 type alloys. We are aware that we have to include a certain "error bar" for a quantitative interpretation of our results -because of ambiguities in judging the image intensities- but we are convinced that we are very near to the true projected structure. A strong argument being the fact that the obtained distributions are consistent with the experimental electron diffraction patterens. We therefore strongly believe that the present method opens new perspectives for studying SRO and ordering phenomena, including kinetics, in real space.

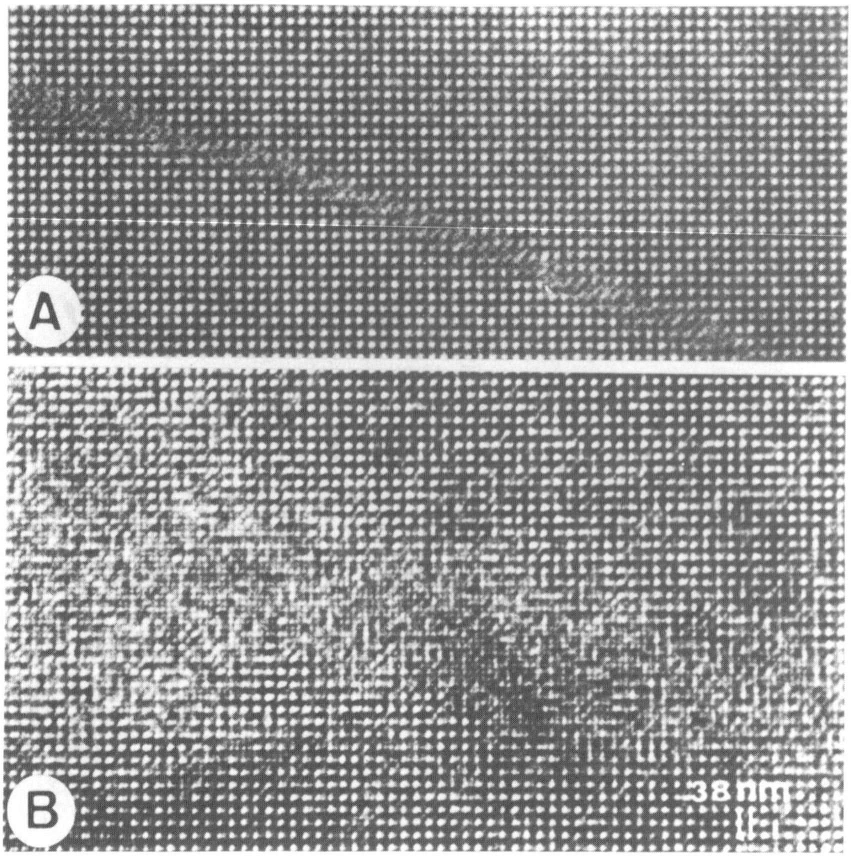

Figure 10. High resolution image along [001] of antiphase boundaries in the $Pt_{70}Cu_{30}$ alloy. a) when annealed well below T_c; b) when annealed just below T_c

Interface Wetting in Ordered Alloys

When approaching the order-disorder transition from below, disordering will start from imperfections such as antiphase boundaries or twin interfaces [29, 30]. Planar interfaces will be "wetted" by the disordered phase at temperatures close to the transition temperature. We will first study the Pt-Co compound around the composition $Pt_{70}Co_{30}$, where the ordered structure is the $L1_2$ structure. Further we will consider the Cu-Pd compound which shows one- and two-dimensional long period superstructures.

In the ordered compound $Pt_{70}Co_{30}$ antiphase boundaries with a displacement vector of the type **R** = 1/2<110> appear as a consequence of the decrease in translation symmetry. When the material is heat treated at temperatures well below Tc these antiphase boundaries show a normal behaviour; i.e. under two beam conditions (**0**, **g**), the boundary is out of contrast for **g.R** = n. Under high resolution conditions as in fig.10a the displacement vector can immediately be identified. The HREM imaging conditions are such that the bright dots in fig. 10 represent the Co configuration. It should be noted that the structure is perfect up to the boundary plane and that the width of the interface is restricted to a single atomic plane. When annealing the material at temperatures close to Tc, the width of the interface gradually increases [31]; this is obvious from fig.10b, which is obtained under similar imaging conditions as fig.10a. The antiphase boundary is 2 to 4 nm wide and at the interface the basic square fcc lattice of 0.19 x 0.19 nm can be recognized, indicating the presence of disordered material between the two ordered regions.

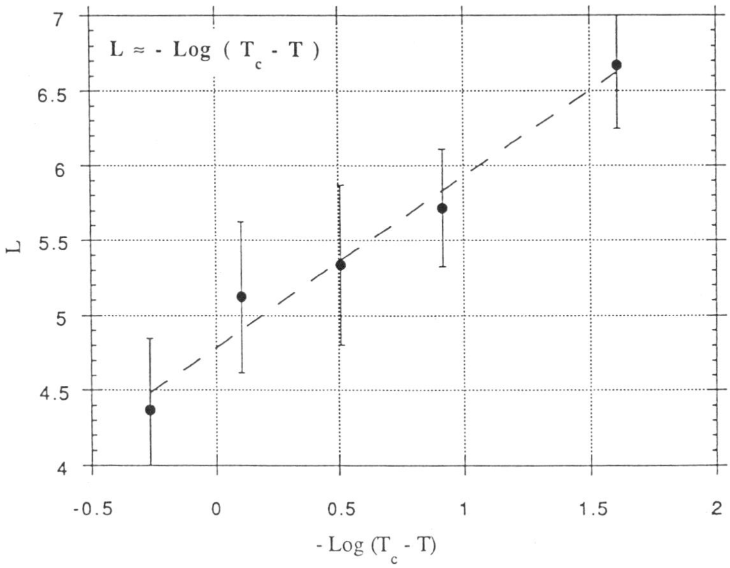

Figure 11. Variation of the width of the antiphase boundaries in the vicinity of T_C, as function of temperature.

In the Cu-17% Pd compound, Ricolleau et al. performed highly controlled heat treatments in the range between Tc - 10° and Tc with temperature steps of 0.2° in order to characterize the nature of the divergence of the width of the domain walls [32]. Although the concentration of the alloy was chosen to avoid two-phase effects, the Cu-17% Pd system still presents a two phase field of about 1° wide. The presence of this two phase field hampers the boundary-width measurements as well as the analysis of interface instabilities in the immediate vicinity of Tc; on the other hand the measurements clearly allow to distinguish between the wetting regime and the two phase regime. In order to measure the exact width of the antiphase boundary a number of precautions have to be taken. However some problems persist: (a) the width of the interface is always measured at room temperature after quenching, (b) high resolution images invariably show a sharper interface than conventional dark field images, c) close to Tc the antiphase boundaries show a tendency to fluctuate dramatically and they may indicate the initial stages of formation of a long period superstructure. The results of the change in boundary width as function of temperature are shown in fig.11. They show a good agreement with statistical mechanics calculations which predict that the width of the interface logaritmically diverges as log(Tc-T) [33].

When the palladium content of the alloy is increased in the Cu-Pd system the $L1_2$ structure becomes unstable towards antiphase boundary formation and a one dimensional long period superstructure is formed. The average spacing between subsequent interfaces increases and is function of composition and temperature [34, 35]. Close to the disordering temperature the interfaces exhibit similar wetting phenomena as the individual boundaries, this is clear from the image of fig.12 which is from a Cu-19 at % Pd alloy annealed only 5° below the disordering

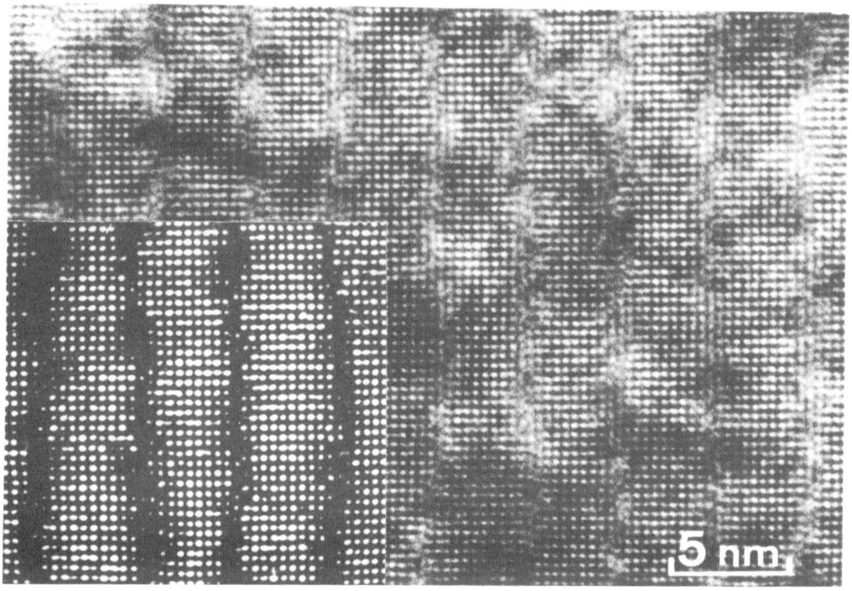

Figure 12. High resolution image along [010] of the incommensurate long period structures in $Cu_{81}Pd_{19}$; the processed image is shown as an inset.

temperature. The disordered region is 1-2 $L1_2$-unit cells wide and moreover the interfaces, which at lower temperature are strictly bound to (001) planes, now become more or less wavy. The inset, which is a processed image of the original HREM, accentuates the local changes at the interface. These effects are a general feature of long period superstructures which are formed from a solid solution.

Pre-Transition Effects in Diffusionless Transformations

We will focus on the austenite-martensite transition in Ni-rich Ni-Al because a strong diffuse scattering is associated with this transition. The transition from the bcc based B2 austenite phase (CsCl structure) to the fct based $L1_0$ (possibly twinned) martensite phase can be followed in situ by cooling inside the electron microscope. Careful electron microscopy experiments enable us to interpret images of precursor and transition stages, including the well known tweed structure, in terms of sequential steps in the transition sequence. One of the conclusions is that, as early as the precursor state, the material develops transverse displacement modulations that will finally lock in to yield the martensite structure.

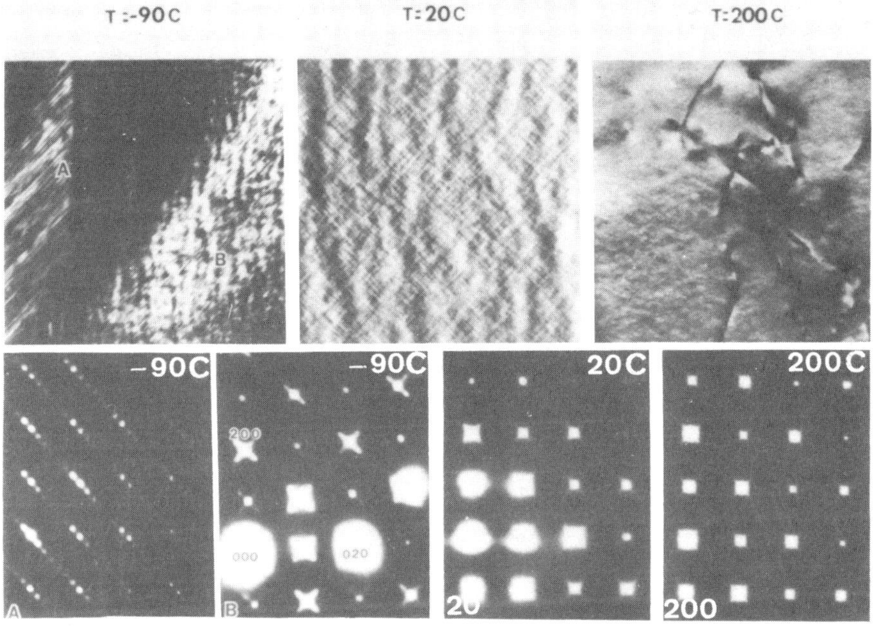

Figure 13. Direct images and diffraction patterns of the corresponding regions of a cooling sequence through M_S in a $Ni_{62.5}Al_{37.5}$ alloy.

In fig.13 direct images as well as diffraction patterns of a $Ni_{62.5}Al_{37.5}$ alloy at different temperatures are presented. The high temperature (200°C) tweed pattern revealing {110} striations in the image and <110>* diffuse streaks in the diffraction pattern clearly exists far above M_S, although the contrast of the modulation increases with decreasing temperature (20°C). When the intensity distribution in the diffuse streaks of the precursor tweed structure is measured, faint satellites at q* = 0.1 - 0.2 <110>* are found [36]. Elastic neutron scattering shows these peaks much more pronounced at q* = 0.15 ... 0.16 <110>* corresponding to a wavelength of 1.35nm [37]. These satellites exist in all of the six <110>* variants of the diffuse streaking. Below M_S and inside the final martensite plate (A at -90°C) the diffraction pattern clearly shows the long period stacking of the 7R multiply twinned martensite. At the transition region (B at -90°C) between the growing martensite plate and the austenite matrix a heavily faulted quasi long period structure is observed. In the corresponding diffraction pattern the intensity in the diffuse streaks is piling up at diffuse satellite positions q* ≈ 0.2 <110>*, indicating a better correlated modulation with a shorter wavelength. This sharpening occurs in only one of the six possible <110>* directions, i.e. locally one variant grows at the expense of the others. Finally, when the 7R product phase forms, rows of equally spaced reflections become visible. A similar transition can be induced by applying a proper stress configuration. In practice, when a Ni-Al sample with an M_S close but still below room temperature is thinned for EM observation, strong but local stress fields around micro-cracks in the thin foil can induce the transformation.

The different transition stages can then be observed under HREM conditions; in fig. 14 the precursor austenite phase, labelled β', is observed along [001]. A patchwork of contiguous domains of approximately 3-5nm diameter can be observed, the contrast inside each domain being modulated parallel with (110) or $(1\bar{1}0)$ ((the other four {110} possibilities being extinct) and with a wavelength slightly incommensurate with the bcc lattice and corresponding with the satellites observed in the diffuse streaks (inset of fig. 14). The distortion of the perfect bcc lattice can therefore be described by superimposing a transverse $(110)[1\bar{1}0]$ sine modulation onto a uniform $[1\bar{1}0]$ shift of the (110) planes of the austenite bcc lattice [38]. The present multiple-beam phase contrast modulation is called "micro-tweed" and should not be confused with the original two-beam tweed contrast. The origin of this patchwork of small distorted domains can most probably be related to the existence of a large number of point-like defects inducing a local strain field with tetragonal symmetry to which the surrounding matrix responds following its soft modes [37].

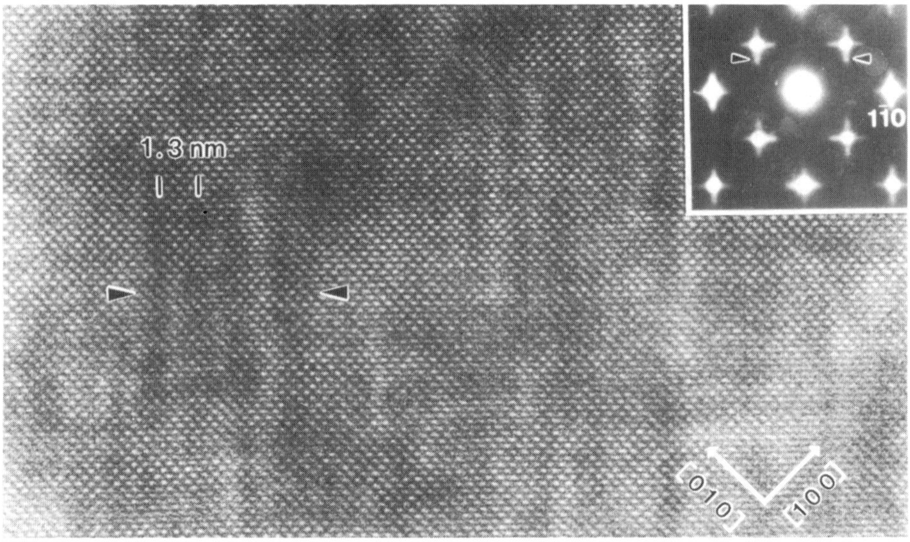

Figure 14. [001] HREM image of the precursor austenite phase.

At lower temperatures, or at higher strains, out of all variants of the micro-modulation, one is chosen which gradually increases its correlation both along and perpendicular to the corresponding wave-vector.

A computer Fourier transform of the HREM image shows that the uniform distortion observed in the micro-modulated domains still exists in the present long period modulated structure. No difference could be measured between the amount of distortion in the micro-modulated domains and in the long period structure. On the other hand, the periodicity of the modulation shows a tendency of being larger in the separate micro-modulated domains, where it varies between 6 and 8 {110} planes, than in the long period modulated structure where it reaches an average of only 5.5 {110} planes. In the small domains one thus finds a periodicity close to that of the 7R martensite structure, whereas the wavelength of the long period modulated structure clearly deviates from this value [39]. It can therefore be concluded that on moving from the premartensitic region into the long period transition structure the local atomic arrangements must undergo some well defined alterations. The stress configuration near the micro-crack apparently dedicts which, out of the six possible transverse $\{110\}<1\bar{1}0>$ distortions will be preferred for the formation of the long period modulated structure with the increased correlation lengths. At the very edge of the foil the amount of stress is sufficient to fully transform a small portion of the foil into the 7R phase. The wavelength (or the number of close packed planes in the 7R monoclinic unit cell) must now increase from 5.5 to 7 planes. Similar but opposite behaviour as for the previous decrease in wavelength can be suggested: larger regions of micro-twinned martensite indeed show similar defects, such as steps in twin habit planes and small twin bands acting as dislocation twins. Also, the modulation wave must now lock in on close packed planes so as to form the stable $(5\underline{2})$ stacking. Fig.15, a HREM image of this structure, clearly indicates that the perfect $(5\underline{2})$ stacking sequence does not persist in large regions. In such stress induced situation this is probably due to a change in the local stress configuration.

In conclusion, it is clear that from HREM studies new detailed information on the atomic structures of precursor and transition states of martensitic transition in Ni-Al becomes available. The precursor phase consists of contiguous domains (3-5nm in diameter); each domain being modulated by an incommensurate transverse $\{110\}<1\bar{1}0>$ displacement wave superimposed onto a uniform shift of the same type. As the transformation continues, "external" factors such as strain fields with higher amplitudes determine which of the six variants will grow and develop the micro-twinned martensite structure. Before this phase is formed, an intermediate long period modulated phase with a shorter wavelength appears. Its modulation will then lock in on the fcc-like close packed planes to form the 7R stacking or multiply twinned $L1_0$. More details on the present observations in the Ni-Al system can be found in [35, 39]

STRUCTURAL DEFECTS IN SUPERCONDUCTING MATERIALS

High Tc superconductors have recently attracted a lot of attention because after the discovery by Müller and Bednorz of superconductivity in La-Sr-Cu-O [40] the superconducting transition temperature Tc has steadily gone up to 133K in the complex compound $HgBa_2Ca_2Cu_3O_z$ [41]. Although from a physical point of view these ceramic compounds have unusual -superconducting- properties, most of their structural properties are typical for a

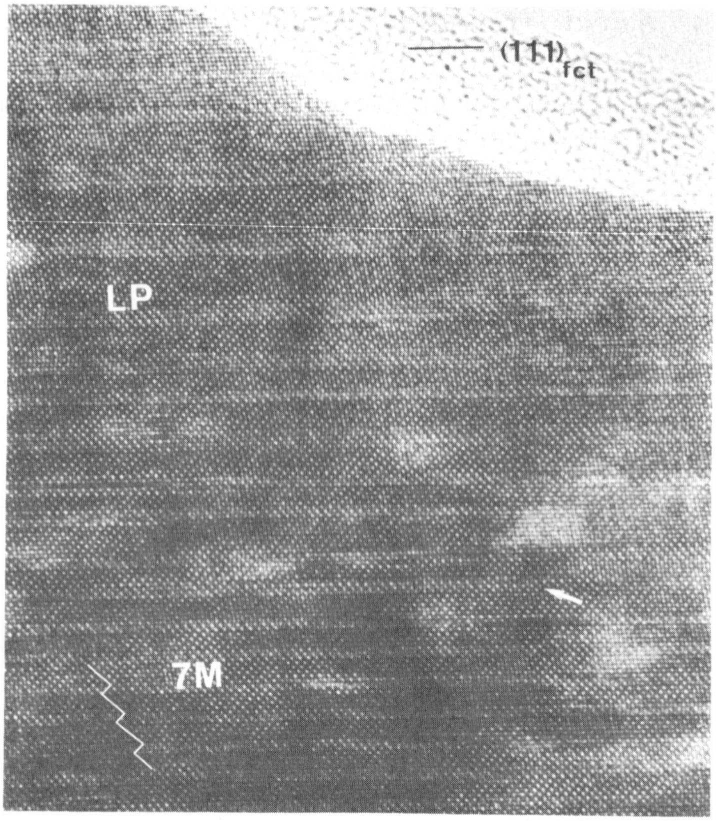

Figure 15. HREM image of long period (LP) multiply twinned martensite with a 7R region indicated.

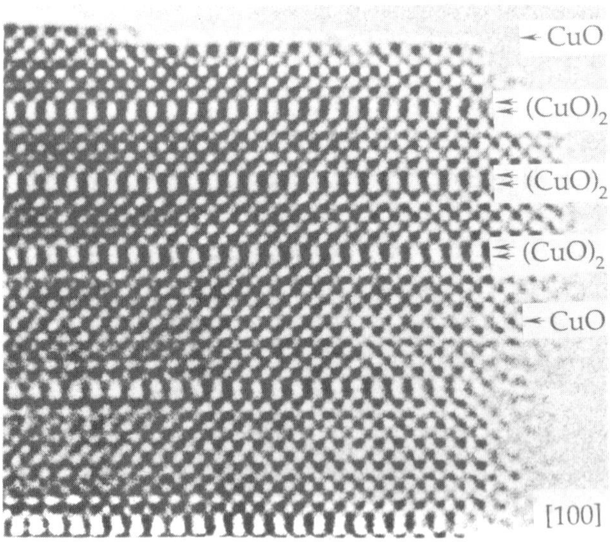

Figure 16. HREM image showing the presence of double CuO planes in the $YBa_2Cu_3O_7$ structure

number of ceramic materials. We will show how they can be analysed by electron microscopy and we will point towards the typical defects which are related to the superconducting properties.

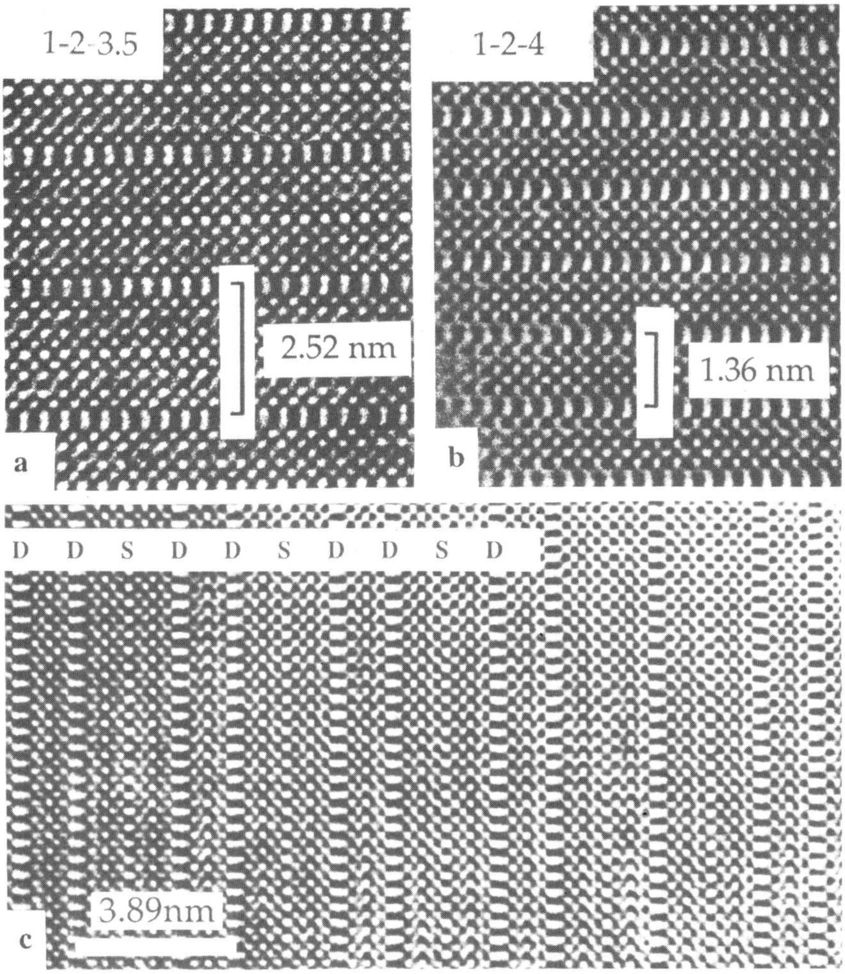

Figure 17. [010] HREM image of : (a) the $Y_2Ba_4Cu_7O_{15}$ structure (2-4-7); the double $(CuO)_2$ layer is introduced every two unit cells; (b) the $YBa_2Cu_4O_8$ structure (1-2-4); the double $(CuO)_2$ layer is introduced every unit cell; (c) a complex sequence of double and single CuO layers.

Structural Defects in YBCO Material.

The high temperature superconductor $YBa_2Cu_3O_{7-\delta}$ has been the first to be studied in detail by EM. The coherent twin fragmentation on (110) planes resulting from the tetragonal to orthorhombic phase transition was discovered by means of diffraction contrast electron microscopy and electron diffraction; moreover the T→O phase transition was observed "in situ" [42, 43]. At the same time also the $2a_o$ superstructure (ortho II) in non-stoichiometric $YBa_2Cu_3O_{7-\delta}$, due to the alternation along the a_o direction of filled and empty CuO chains in the

CuO layers was reported for the first time [44, 45]. Very weak superstructure spots were obtained in the [001] zone electron diffraction patterns and in the corresponding HREM image the period doubling along a_o was revealed in two orientation variants corresponding to the twinning of the basic structure [44].

The presence of (001) planar defects consisting of a double CuO layer, replacing the normal single layer, was first demonstrated by means of electron microscopy (fig. 16). This observation has led to the development of a superconducting mixed layer Y-Ba-Cu-O family containing regular arrangements of single and double CuO layers. Members of this family are the so called 1-2-4 phase ($YBa_2Cu_4O_8$) [46, 47] and the 2-4-7 phase ($Y_2Ba_4Cu_7O_{15}$) [48]. 1-2-4 and 2-4-7 can be regarded as members of a series of "shear structures" derived from the 123 basic structure by the periodic insertions of supplementary CuO layers. Representing the number of 123 unit cells separating these supplementary CuO layers by an integer, the 124 and 247 structures are represented by the stacking symbols $1\bar{1}$ and $2\bar{2}$ [49, 50]. The minus sign is introduced to indicate that the presence of a double CuO layer causes a lateral offset between successive 123 blocks over 1/2[010]. High resolution images along the [010] direction of the perfect 124 and 247 structure are reproduced in fig 17a,b. The double chain layers are characterized by the intense rows of elongated double white dots. Changing the initial composition does not alter the stoichiometry of the outcomming material, one does however stabilize to some extend different phases such as $2\bar{3}$ and $3\bar{3}$ over limited areas. Fig.17,c shows an example of a complex polytypoid over a limited region. Apart from this series of phases being a mixture of 1-2-3 and 1-2-4 where periodically double $(CuO)_2$ layers are introduced, a different series of Cu-rich phases was discovered, in which sussessive perovskite blocks are separated by triple $(CuO)_3$ layers [51]. An example is reproduced in fig.18 together with a structural inset where the CuO layer configuration is represented by black dots. In some parts of the crystal even isolated CuO multilayers consisting of as much as five CuO layers have been observed in a matrix of essentially 2-4-7 material [49].

Cu Substitution in YBCO 1-2-3 Superconductors.

Although it seems well established that superconductivity is to be associated with the CuO_2 layers, doping of the CuO layers seems to play an important role in determining the concentration and the type of carriers. Also the critical temperature Tc is related to the composition of the CuO layer. It could for instance be shown that the 60 K plateau in the curve relating Tc to the oxygen content in this layer is associated with the presence of the $2a_o$ structure, i.e. with the material with ideal composition $YBa_2Cu_3O_{6.5}$ [52-54]. It was attempted to replace copper in the CuO layers, either completely or partly by other metallic ions such as iron, cobalt, zinc, gallium. Also complex ions such as CO_3 or SO_4 were incorporated in the CuO layer, carbon and sulpher occupying copper sites in the CuO chains. Iron substitution is found to induce a "tweed" texture, the geometry of which could be rationalized by assuming that iron occurs in the CuO layers in two types of oxygen coordination: octahedral and tetrahedral.

The tetrahedrally coordinated iron ions induce microtwinning by the nucleation of CuO chains along mutually perpendicular directions which causes the tweed microtexture [55, and references therein]. Cobalt substitutions in small concentrations has a rather similar aspect [56].

The substitution of all copper ions in the CuO layers by gallium or cobalt ions leads to the formation of rows of corner sharing CoO_4 (or GaO_4) tetrahedra along the [110] direction of the basic perovskite structure, replacing in a sense the CuO_4 strings of square planar

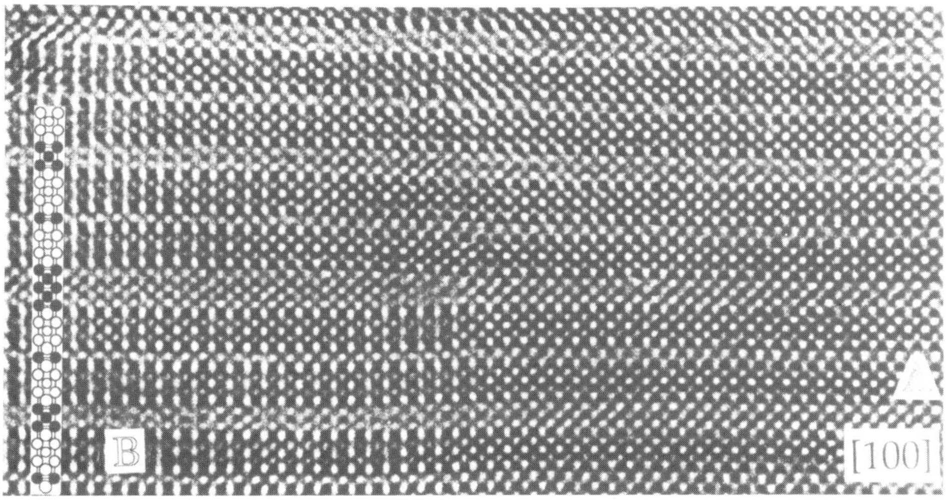

Figure 18. Heavily distorted part of a 2-4-7 grain with all double $(CuO)_2$ layers replaced by tripple ones (area A) or by quintuble ones (area B).

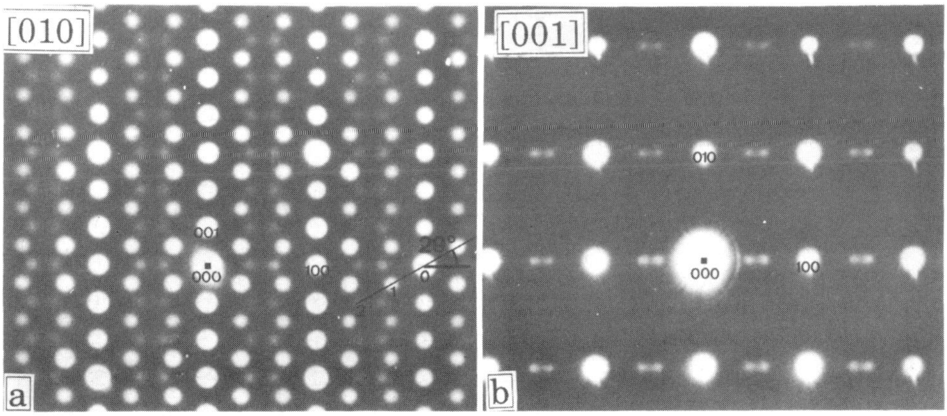

Figure 19. Electron diffraction patterns of $[Y_{0.84}Sr_{0.16}]Sr_2Cu_{2.78}(SO_4)_{0.22}O_{6.12}$.
(a) [010] pattern showing satellites at every basic spot; numbers indicate spot order, the angle indicated is α=29°; (b) [001] pattern with only the weaker second order spots; indexing of both patterns refers to the basic unit cell.

configurations along the b_o direction. The tetragonal symmetry is hereby reduced to an orthorhombic one [57, 58]. The strings of tetrahedra can moreover adopt two mutually perpendicular derections giving rise to two twin related orientation variants of the orhtorhombic structure. It was moreover shown that period doubling (also tripling) occurs in the (001) plane along the direction normal to the chains of tetrahedra. This could be rationalized by noting that the parallel chains can adopt two different configurations, differing by the sense of rotation of the coupled tetrahedra along a string. Period doubling can then be attributed to the regular alternation of these two types of chains in a parallel set.

In case of carbonate substitutions for Cu(1) in the $YSr_2Cu_3O_7$ compound, the CO_3 triangles align in chains along the perovskite **b**-axis [59]. We will discuss here a recent and similar substitution, the substitution by tetrahedral SO_4-groups in compounds with composition $[Y_{1-y}Sr_y]Sr_2[Cu_{3-x}(SO_4)_x]O_{7-\delta}$, where y = 0.16 and x = 0.22. The electron diffraction patterns of this material confirm this average structure determined by neutrons [60, 61] but strong satellites are associated with every basic reflection. The [010] and [001] zone diffraction patterns of fig. 19 shows the presence of bright first order satellites (fig.19 a) and weaker second order satellites (fig.19a,b); the length of the satellite-**q**-vector is $1nm^{-1}$. The structure can therefore be considered as modulated with the wavevector in the (010) plane, inclined over an angle α ($\alpha \approx 32°$)with respect to the [100]-direction and with a wavelength $\lambda \approx 1$nm. Although the observed modulation is orthorhombic, most [001] observed diffraction patterns show tetragonal symmetry, with superstructure spots along **a*** as well as along **b***. This is due to the presence of two orthorhombic variants, related by the usual (110) twin law common in $YBa_2Cu_3O_{7-\delta}$.

High resolution images along the [010]-zone (fig.20), show the presence of a modulation corresponding to the satellites in the corresponding diffraction pattern of figure 19a. The image is more or less the same as that of the undoped $YBa_2Cu_3O_7$ material, however every three or four repeat distances along the **a**-direction, squares of more intense dots are present. According to image simulations, this is to be associated with the presence of SO_4-groups in the Cu-O(1) layer. Along the **c**-direction the stacking of these columns or "chains"is usually staggered in such a fashion that their configurations in every second Cu(1)-S-O-plane coincide vertically. Within a single Cu(1)-S-O-plane the alternation is incommensurate but the intensity maxima change roughly as - $4a_o$ - $3a_o$ - $3a_o$ -. Looking along the directions indicated by heavy lines in figure 20, planes containing a high density of SO_4-clusters are revealed by a modulation in dot intensity. These planes can be considered as the maxima of a planar concentration wave. The normal to this concentration wave makes an angle α of about 30° with the [100]-direction and the length of the wavevector is about 1nm.

These observations can be explained by a simple model of SO_4-chain formation and ordering. The [001] high resolution image suggests the presence of chains oriented along the **b**-direction, the nature of which must differ from that of CuO chains by their S content. The stacking-rule of the SO_4-rich chains can be derived from the observation of the concentration waves in fig.20:

1° The SO_4-chains substitute these sites of the Cu(1)-sublattice which are closest to the maxima of the concentration waves.

2° The integer *l*-coordinates of the second order satellites as well as direct observations in high resolution images suggest that the SO_4-chain-stacking coincides vertically in every second plane along the [001] direction.

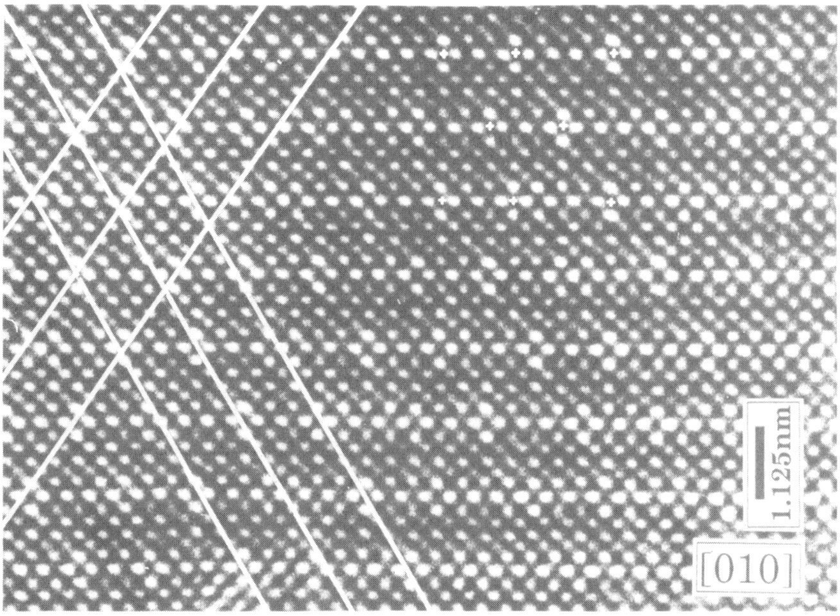

Figure 20. High resolution image along the [010]-zone showing S-rich columns as brighter squares of four dots centered on the Cu(1)-S-O-layers. The maxima of the modulating wave, easily seen under grazing angles, are stressed by two sets of heavy lines in the image. The brighter dot configurations are indicated in part of the image by small crosses.

These observations allow us to propose an atomistic model for the modulation in $[Y_{1-y}Sr_y]Sr_2[Cu_{3-x}(SO_4)_x]O_{7-\delta}$. In the schematic representation of Fig. 21 only the Cu(1) sublattice is shown; dashed lines are the maxima of the concentration waves with angles $\alpha = \pm 30°$ and $\lambda \approx 1$nm. Due to the incommensurability of the basic lattice with the modulation, the ordering along the **a**-direction is essentially aperiodic, which is in agreement with the absence of periodic superstructure spots in the diffraction patterns. A physical explanation for the observed SO_4-chain arrangement is not straightforward but certainly involves an strain component. The chains occupied by SO_4^{2-}-ions cause cylindrically symmetrical stress fields as a consequence of the difference in size of the sulphur-ions and the copper-ions that they replace. Parallel chains thus interact by mutually repulsive elastic forces which depend on the separation r as 1/r. The equilibrium configuration of such a set of repelling parallel chains, when confined to a finite area, or with a specified concentration, consists of a triangular array. With the restriction that the

SO_4-chains can only be located at positions in the Cu(1)O-chain layers, the observed distribution of SO_4-chains is the one that yields the largest average separations between different chains, taking in account the concentration of SO_4 chains.

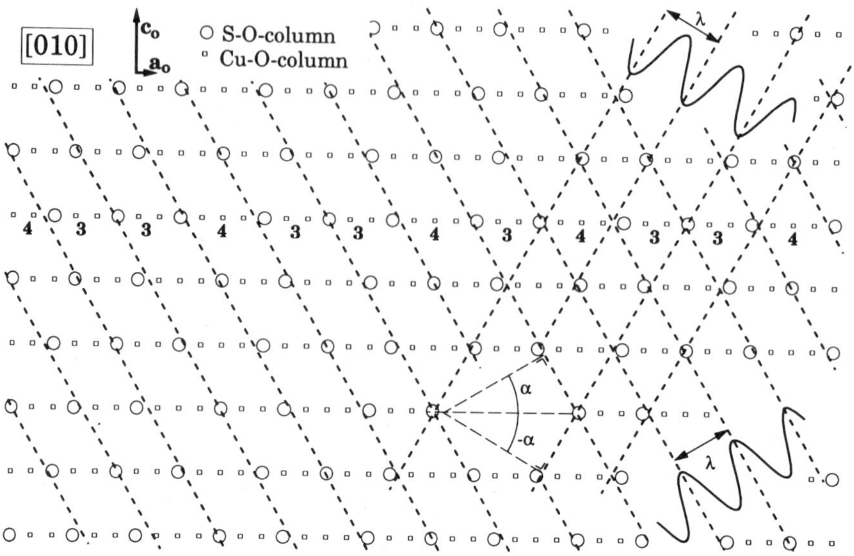

Figure 21. Schematic representation of the incommensurate modulation of the Cu(1)–column-sublattice. Squares represent the Cu(1)-column sublattice, circles represent SO_4-chains, located on Cu(1) sites closest to the maxima of a SO_4-occupancy wave with wavevector inclined over α with respect to [100] and with wavelength λ.

Superconducting Bi-, Tl- or Hg-Compounds

Also the Bi-compounds with general formula $Bi_2Sr_2Ca_nCu_{n+1}O_{6+2n}$ have been the subject of extensive and very detailed studies. The most striking feature is the presence of a family of mixed layer compounds (2 2 n n+1) and the correlation of this chemical formula with the values of Tc = 20K (n =0), 80K (n =1), 105K (n = 2) [62-65]. The existence of such a family was initially predicted based on HREM observations of singular lamellae containing a number of CuO_2 layers deviating from the bulk composition. All of the Bi-compound moreover exhibit an incommensurate one-dimensional deformation modulation, localized mainly within the BiO layers [66]. The modulation was attributed primarily to the misfit between the perovskite blocks and the BiO layers and the particular lone pair electronic configuration of the Bi^{III} [67]. The one-dimensional modulation reduces the point symmetry of the compound from tetragonal to monoclinic in the 2201 compound and to orthorhombic in the 2212 or 2223 compound. As a result defects i.e. modulation twins caused by the modulation and related by the lost symmetry elements may arise. The effect of the substitution of bismuth by lead on the modulation period and on the geometrical features of the modulation wave, resulting in a stabilization of the 2212 structure, has been studied almost exclusively by means of EM. It has been found that the relative phase of the modulation waves in successive BiO lamellae, in the lead substituted compounds, differs from that in the pure Bi-compounds; whereas the modulation waves in successive layers are in anti-phase in the pure Bi-compounds, they are in phase in the lead substituted compounds.

Recently several higher-order members of the Hg-based superconducting family $HgBa_2Ca_{n-1}Cu_nO_{2n+2+\delta}$ have been synthesized recently; the critical temperature reaches a maximum of 133.5K for the n = 3 member of the family [41], while for larger n values, Tc seems to decrease [68]. It was furthermore shown (not very unexpected) that Tc increases

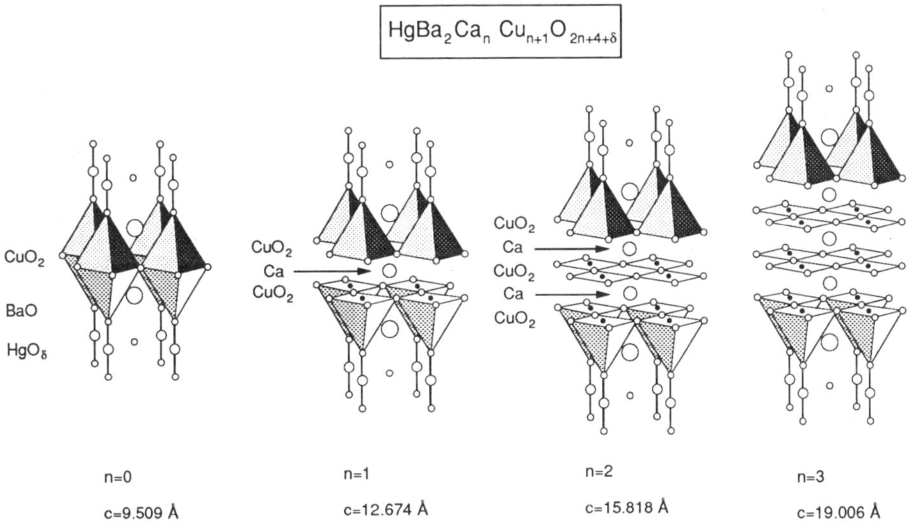

Figure 22. Schematic representation of several members of the structural family $HgBa_2Ca_{n-1}Cu_nO_{2n+2+\delta}$

steadily with pressure at a rate of more or less 1K/GPa. In this way Chu et al. [69] and Nunez-Regueiro et al. [70] were able to reach values such as 164K and 155K (Tc onsets) for the Hg-1223 compound respectively. This strong pressure effect suggests that by the appropriate chemical doping, which would introduce a chemical pressure, Tc values of the order of 150K are possible at normal -external- pressure. Several substitutions in the Hg-plane as well as at the Ba- or the Ca-positions have been tried by the Caen group; they have led to the discovery of a series of new Hg-based superconducting compounds, which can be prepared under ambient pressure and which have a Tç up to 110K [71-75].

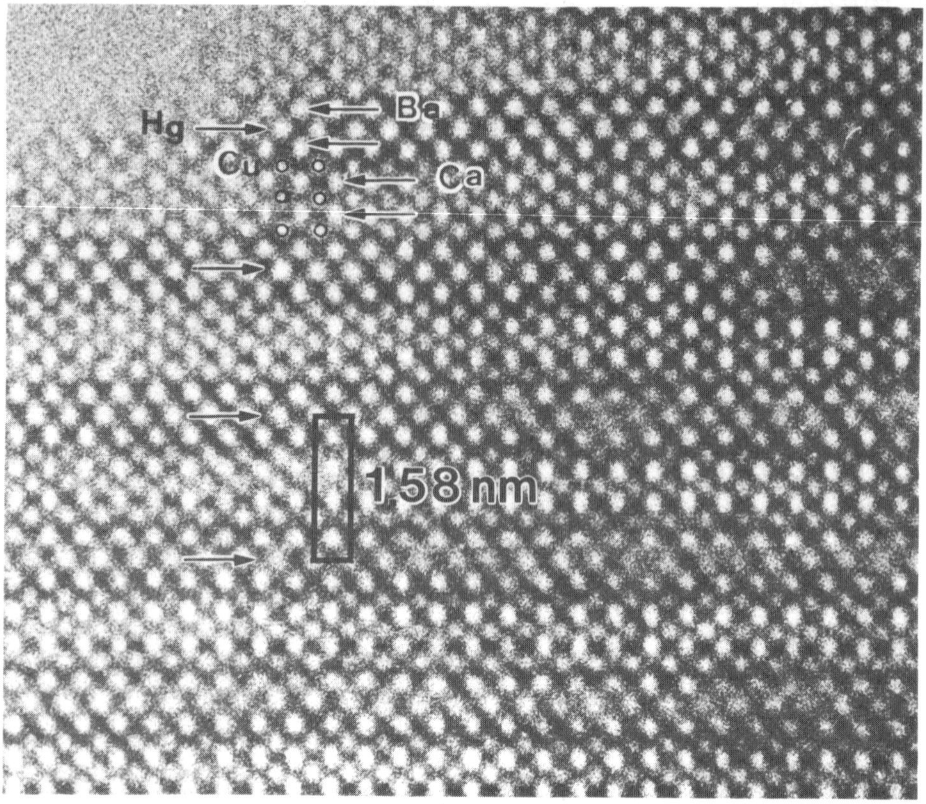

Figure 23. High resolution image along [100] of $HgBa_2Ca_2Cu_3O_{8+\delta}$. In this contrast the heavy ions are imaged as bright dots; the imaging code as well as the unit cell are indicated in the figure.

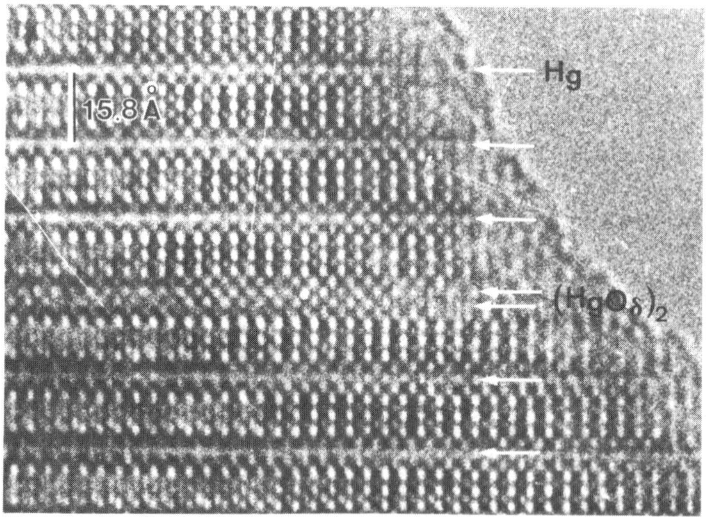

Figure 24. HREM of an isolated defect in a Hg-1233 crystal, where a double layer of HgO_δ is present, creating locally the 2223 structure.

The structure of all members is similar; they contain rock salt like slabs [(BaO)(HgO_δ)(BaO)], alternating with perovskite slabs of the type [(CuO_2)(Ca)]$_{n-1}$(CuO_2). The structure for different n-compounds is represented in fig. 22; for increasing n-values an extra [(CuO_2)(Ca)] slab is inserted, leaving the rest of the average structure unaltered. Although there is a strong similarity between the Hg- and the Tl-series, the occupation of the oxygen sites in the Hg-layers and the corresponding one in the Tl layers, are quite different.

HREM images of the different members of the family are obtained, particularly along the [100] zone, which is the most instructive one. An example for the 1223 compound is shown in fig.23; the cation configuration can be readily identified from symmetry considerations. There is clearly a one to one correspondence between the white dot configuration of the HREM image and the cation configuration of fig.22. This semi-intuitive interpretations have of course to be confirmed by computer simulations [76]; they will allow us to analyse in detail de atomic structure of planar defects in these compounds. The most common defect encountered, particularly for higher order members of the family, is the intergrowth of different n-members; they occur as isolated defects and are well known from other homologuous series. Periodicities up to n = 8 have been seen to occur locally, although no single phase material could be produced [76]. A most remarkable defect however is shown in fig.24. The image contrast is such that the copper ions are imaged as bright dots, while the mercury plane is imaged as a more or less diffuse line with maxima at the projected atom positions. It is evident that locally at the position indicated, we have now four rocksalt layers rather than the usual sequence of three layers [(BaO)(HgO_δ)(BaO)]. A single Hg-layer is now replaced by a double Hg-layer.of bright dots. The occurrence of such defects with a double (HgO_δ) layer suggested the feasibility to produce "2 2 n-1 n" type materials; such "2 2 n-1 n" compounds have indeed been produced by replacing some of the Hg by Pr or Cu [77] or some of the Ca by Y [78].

STRUCTURAL DEFECTS IN FULLERENES

After considering crystal defects and modulations in metallic systems and covalently bonded systems, we will consider the fullerenes as examples of a Van der Waals bonding. Following the synthesis of C_{60} by Krätschmer et al. [79] the crystal structure at room temperature of the C_{60} fullerite has been the subject of discussion until it was realized that the sample purity is of overall importance. Crystals grown from an organic liquid such as toluene are predominantly faulted hexagonally close packed [80]. Sublimation grown crystals of sufficient purity are always cubic at room temperature. The isotropically rotating quasi-spherical molecules occupy the node points of a face centered cubic lattice with a_o= 1.41 nm [80, 81]. The molecules can be modelled as spherical shells of diffracting material with an average radius r_o = 0.353 nm. At room temperature these spherical molecules are found to rotate freely [82]. Below room temperature, the material undergoes a phase transition from fcc to a simple cubic structure [83] and further to an fcc structure with a double unit cell [84].

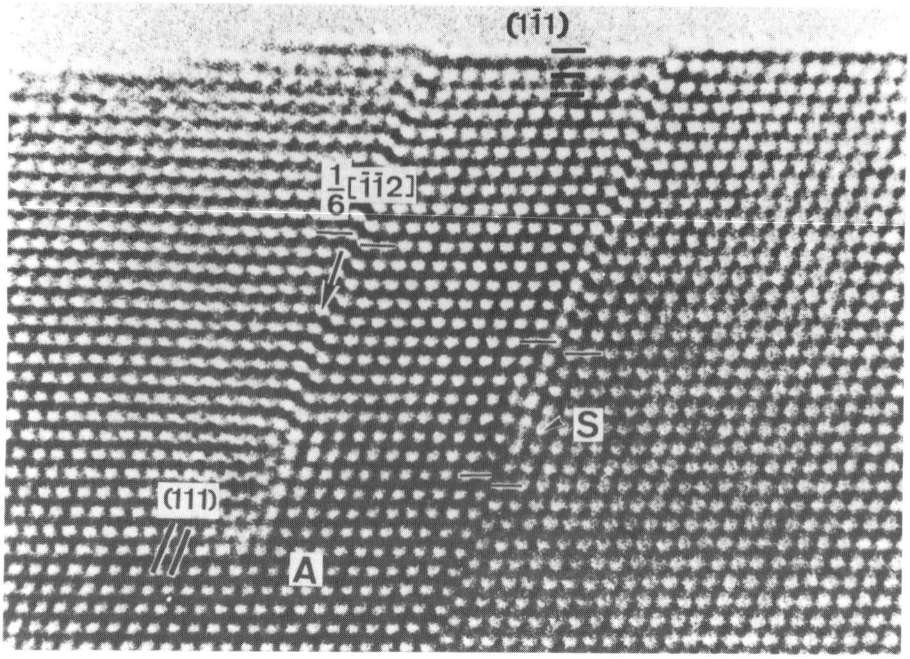

Figure 25. Defect configuration of C_{60}. In S an intrinsic stacking fault changes from one (111) layer to the next, giving rise to a dipole of stair rod dislocations. In A a Frank partial dislocation and its associated intrinsic stacking fault are present.

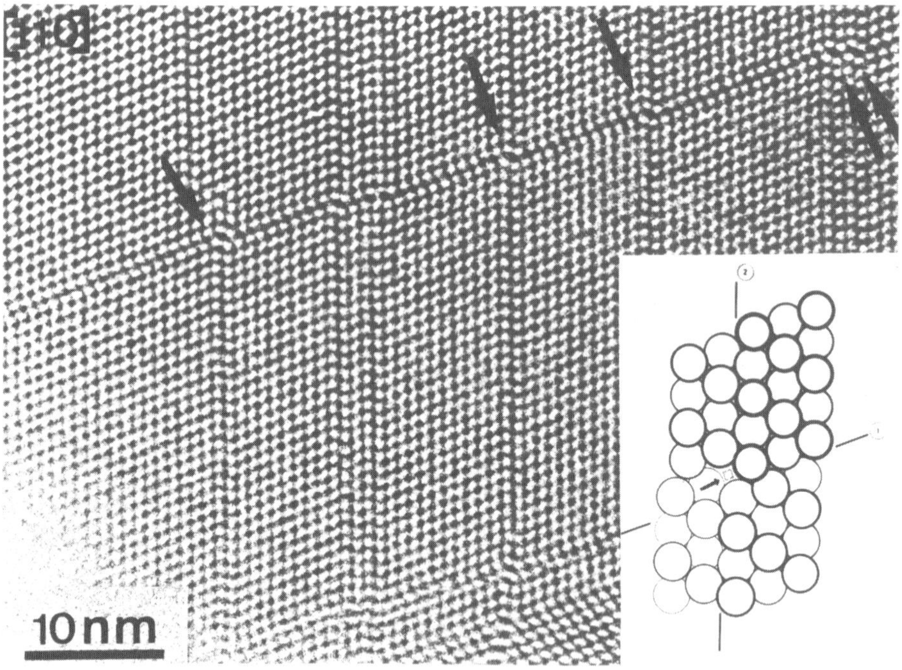

Figure 26. High resolution image of C_{60} along [110] showing several intersections of two intrinsic stacking faults along different (111) planes (see arrows). Note the rectangular configuration of bright dots at every intersection; a schematic representation is shown as in inset.

These structures however will not be treated here, we will limit ourself to the room temperature phase.

Because of the large lattice parameters the structure can be visualized using the high resolution imaging mode along different zone axes; the most instructive imaging direction being along the [110], i.e. along the close packed rows of molecules. Along this view the columns of molecules produce ring shaped images of which the characteristics sensitively depend on the specimen thickness because at small thickness the 220 lattice fringes are predominantly excited, whereas at larger thickness also the h00 type of fringes contribute significantly to the image because of dynamical effects [85]. The most abundant defects present in C_{60} crystals at room temperature are coherent twin interfaces and intrinsic stacking faults on (111) planes. Occasionally also extrinsic stacking faults may occur. Many of the complex defects, occurring in low stacking fault energy metallic alloys also occur in C_{60} crystals. Intrinsic stacking faults, jogging from one (111) plane to the adjacent one, producing a dipole of stair rod dislocations have been revealed. Stacking faults were often found to terminate when meeting a twin interface. Different types of dislocations were observed. The image of fig.25 shows for instance a Frank partial dislocation with Burgersvector 1/3[111] and its associated intrinsic stacking fault. The geometrical features of the defects in low stacking fault face centered cubic alloys, in C_{60} and in bubble rafts are amazingly similar because the bonding of metallic atoms, molecules and soap bubbles has some common characteristics: it is non-directional and short range. This image also shows a surface relaxation taking place in C_{60}, presumably because of the oxidation of the surface; the spacing between the (111) lattice planes near the free surface is distinctly larger than the corresponding spacing inside the material. A special line defect, so far not observed in a high resolution image of any material is shown in fig.26. It consists of a row of fractional vacancies, formed along the intersection line of two intrinsic faults. In metallic alloys the stacking faults usually do not intersect, but one terminates on meeting the other. In C_{60} the stacking fault energy is very small and moreover the formation energy of vacancies is sufficiently small to allow the formation of such line defects; a structural representation of the intersection of such stacking faults is shown as an inset. More of such defects have been described and analysed in [85], we refer to this reference for more details.

CONCLUSION

Because of the elegant shift between reciprocal space and real space, electron microscopy is able to provide unique information not only on the structure itself, but particularly on the defect structure or local structure. We have provided material science applications to metallic systems, covalently bonded systems and Van der Waals type structures, illustrating the wide application field of electron microscopy. However I hope that I made it also clear that high resolution electron microscopy is no longer possible without the intense help of computer simulations and computer processing. This requirement becomes more stringent, with increasing resolution. With the present generation of ultra-high resolution instruments reaching the 1Å level and using field emission sources, an "intuitive" interpretation becomes even more risky.

ACKNOWLEDGEMENTS

I am grateful to the collaborators: D. Broddin, C. Leroux, C. Ricolleau, A. Loiseau, P. De Meulemaere, L.E. Tanner, H.W. Zandbergen, T. Krekels, O. Milat, M. Hervieu, S. Muto, D. Schryvers, J. Van Landuyt, D. Van Dyck and S. Amelinckx which contributed to the original research of which the present overview is a compilation. The present work has been performed in the framework of IUAP-48.

REFERENCES

1. S. Amelinckx, D.Van Dyck, J. Van Landuyt, G. Van Tendeloo. "Handbook of Microscopy" VCH, Berlin (1995).
2. J.C.H. Spence "Experimental High Resolution Electron Microscopy" Oxford Science Publications. Clarendon Press, Oxford (1981).
3. H. Bethge, J. Heydenreich " Elektronenmikroskopie in der Festkörperphysik" Springer Verlag, Berlin, Heidelberg, New York (1982).
4. D. Van Dyck, in "Advances in Electronics and Electron Physics" 65: 295 (1985).
5. J.W. Edington, "Practical Electron Microscopy in Materials Science" Macmillan Philips Technical Library (1974).
6. A.W. Robards and A.J. Wilson (Eds) "Procedures in Electron Microscopy" J. Wiley and Sons, Chichester, New York, Brisbane, Toronto, Singapore (1993).
7. G.R. Grinton and J.M. Cowley, *Optik* 34: 221 (1971).
8. G. Van Tendeloo and S. Amelinckx, *Phys. Stat. Sol.* (a) 49: 337 (1978).
9. P.B. Hirsch, R.B. Nicholson, A. Howie, D.W. Pashley and M.J. Whelan "Electron Microscopy of Thin Crystals", Butterworths, London (1965).
10. S. Amelinckx, "The Direct Observation of Dislocations" Suppl.6 in Solid State Physics, Academic Press London (1964).
11. S. Amelinckx, R. Gevers, J. Van Landuyt (Eds) "Diffraction and Imaging Techniques in Material Science" North Holland, Amsterdam, New York, Oxford (1970, 1978).
12. J.J. Hren, J.I. Goldstein, D.C. Joy (Eds) "Introduction to Analytical Electron Microscopy" Plenum Press, New York, London (1979).
13. M.H. Loretto, "Electron Beam Analysis of Materials" Chapman and Hall, London, New York (1984).
14. B.F. Buxton, J.A. Eades, J.W. Steeds, and G.M. Rackham, *Phil. Trans. R. Soc. A.* 281: 171 (1976).
15. S.J.B. Reed, "Electron Probe Microanalysis", Cambridge University Press, Cambridge (1975).

16. N.J. Zaluzec, "Library of Electron Energy Loss Spectra", Argonne National Laboratory Publication, Chicago (1981)
17. G. Van Tendeloo and S. Amelinckx, *Acta Cryst. A* 30: 421 (1974).
18. J.W. Cahn and G. Kalonji in "Solid solid Phase Transformations" Eds. H.I. Aaronson, D.E. Laughlin, R.F. Sekerka and C.M. Wayman, The Metallurgical Society of AIME, New York (1982).
19. D. Gratias and R. Portier, in "Microscopie Electronique en Science des Matériaux", B. Jouffrey, A. Bourret, C. Colliex (eds), CNRS, 229 (1983)
20. W. Coene, D. Van Dyck, J. Van Landuyt, S. Amelinckx, *Phil. Mag.B*, 56: 415 (1987).
21. S. Amelinckx and D. Van Dyck in "Electron Diffraction Techniques" Ed. J.M. Cowley, Oxford University Press, Oxford (1993).
22. J.E. Spruiell and E.E. Stansbury, *J. Phys. Chem. Solids*, 26: 811 (1965).
E. Ruedl, P. Delavignette and S. Amelinckx, *Phys. Stat. Sol.* 28: 305 (1968).
24. A.G. Khachaturyan, *Phys. Stat. Sol. (b)* 60: 9 (1973).
25. G. Van Tendeloo, S. Amelinckx and D. de Fontaine *Acta Cryst.B* 41: 281 (1985).
26. R. De Ridder, G. Van Tendeloo, S. Amelinckx, *Acta Cryst A* 32: 216 (1976).
27. R. De Ridder, G. Van Tendeloo, D. Van Dyck, S. Amelinckx, *Phys. Stat. Sol.a* 38: 663 (1976); 40: 669 (1977); 43: 541 (1977).
28. P. De Meulenaere, G. Van Tendeloo, J. Van Landuyt and D. Van Dyck, *Ultramicroscopy*, in the press (1995).
29. R.W. Cahn, P.A. Siemers and E.L. Hall, *Acta Metall.* 35: 2753 (1987).
30. J.M. Sanchez, S. Eng, Y.P. Wu and J.K. Tien, *Mater. Res. Soc. Symp. Proc.* 81: 57 (1987)
31. C. Leroux, A. Loiseau, M.C. Cadeville, D. Broddin and G. Van Tendeloo *J.Physics: Condensed Matter,*2: 3479 (1990).
32. C. Ricolleau, A. Loiseau and F. Ducastelle (1991) *Phase Transitions,* 30: 243 (1991).
33. A. Finel, V. Mazauric and F. Ducastelle, *Phys.Rev.Lett.* 65: 1016 (1990).
34. D. Broddin, G. Van Tendeloo, J. Van Landuyt, S. Amelinckx and A. Loiseau, *Phil. Mag.* B57: 31 (1988).
35. G. Van Tendeloo, D. Schryvers, L.E. Tanner, D. Broddin, C. Ricoleau and A. Loiseau, in "Structural and Phase Stability of Alloys", Plenum Press, New York, (1992).
36. S.M. Shapiro, J.Z. Larese, Y. Noda, S.C. Moss, L.E. Tanner, *Phys. Rev. Lett.* 57: 3199 (1986).
37. L.E. Tanner, D. Schryvers,S.M. Shapiro, *Mat. Sci. and Eng.* A127: 205 (1990).
38. D. Schryvers, L.E. Tanner, *Ultramicroscopy* 32: 241 (1990).
39. D. Schryvers and L.E. Tanner, Ultramicroscopy 32: 241 (1990).
40. J.G. Bednorz and K.A. Müller, *Z. Phys.B.* 64: 189 (1986).
41. A. Schilling, M. Cantoni, J.D. Guo and H.R. Ott, *Nature* 363: 56 (1993).
42. G. Van Tendeloo, H.W. Zandbergen and S. Amelinckx, *Solid State Comm.* 63: 389 (1987).
43. H.W. Zandbergen, G. Van Tendeloo, T. Okabe and S.Amelinckx, *phys. stat. sol. (a)* 103: 45 (1987).

44. G. Van Tendeloo, H.W. Zandbergen and S. Amelinckx, *Solid State Comm.* 63: 603 (1987).
45. M. Hervieu, B. Domengès, C. Michel, J. Provost and B. Raveau, *J. Sol. State Chem.* 71: 263 (1987).
46. D.E. Morris, N.G. Asmar, J.H. Nickel, R.L. Sid, J.Y.T. Wei and J.E. Post, *Physica C*, 159: 287 (1989).
47. J. Karpinski, E. Kaldis, E. Jilek, S. Rusiecki and B. Bucher, *Nature* 336: 660 (1988).
48. P. Bordet, C. Chaillout, J. Chenavas, J.L. Hodeau, M. Marezio, J. Karpinski and E. Kaldis, *Nature* 334: 596 (1989).
49. T. Krekels, G. Van Tendeloo, S. Amelinckx, J. Karpinski, S. Rusiecki, E. Kaldis and E. Jilek, *Physica C* 178: 383 (1991).
50. T. Krekels, G. Van Tendeloo, S. Amelinckx, J. Karpinski, E. Kaldis and S. Rusiecki, *Solid State Comm.* 79: 607 (1991).
51. R. Ramesh, S. Jin and P. March, *Nature* 346: 420 (1990).
52. R. Beyers, B.T. Ahn, G. Gorman, V.Y. Lee, S.S.P. Parkin, M.L. Ramirez, K.P. Roche, J.E. Vasquez, T.M. Gür and R.A. Huggings, *Nature* 340: 619 (1989).
53. H.F. Poulsen, N.H. Andersen, J.V. Andersen, H. Bohr and O.G. Mauritsen, *Nature* 349: 594 (1991).
54. T. Krekels, H. Zou, G. Van Tendeloo, D. Wagener, M. Buchgeister, S.M. Hosseini and P. Herzog, *Physica C* 196: 363 (1992).
55. T. Krekels, G. Van Tendeloo, D. Broddin, S. Amelinckx, L. Tanner, M. Mehbod, E. Vanlathem and R. Deltour, *Physica C* 173: 361 (1991).
56. W.W. Schmahl, A. Punis, E. Salje, P. Freeman, A. Graeme-Barber, R. Jones, K.K. Singh, J. Blunt, P.P. Edwards, J. Loram and K. Mirza, *Phil. Mag. Lett.* 60: 341 (1989).
57 T. Krekels, O. Milat, G. Van Tendeloo, S. Amelinckx, T.G.N. Babu, A.J. Wright and C. Greaves, *J. Sol. State Chem.* 105: 313 (1993).
58. G. Van Tendeloo and S. Amelinckx, *Advanced Materials* 5: 620 (1994).
59. Y. Miyazaki, H. Yamane, N. Ohnishi, T. Kajitani, K. Hiraga, Y. Morii, S. Funahashi and T. Hirai, Physica C 198: 7 (1992).
60. P. R. Slater, C. Greaves, M. Slaski and C. M. Muirhead, *Physica C* 208:193 (1993)
61. T. Krekels, O. Milat, G. Van Tendeloo, J. Van Landuyt, S. Amelinckx, P.R. Slater and C. Greaves, *Physica C* 210: 439 (1993).
62. C. Michel, M. Hervieu, M.M. Borel, A. Grandin, F. Deslandes, J. Provost and B. Raveau, *Z. Physik B* 68: 421(1987).
63. R.M. Hazen, C.T. Prewitt, R.J. Angel, N.L. Ross, L.W. Finger, C.G. Hadidiacos, D.R. Veblen, P.J. Heaney, P.H. Hor, R.L. Meng, Y.Y. Sun, Y.Q. Wang, Y.Y. Xue, Z.J. Huang, L. Gao, J. Bechtold and C.W. Chu, *Phys. Rev. Lett.* 60: 1174 (1988).
64. H. Maeda, Y. Tanaka, M. Fukitoki and T. Asano, *Jpn. J. Appl. Phys.* 27: L205 (1988).
65. H.W. Zandbergen, Y.K. Huang, M.J.V. Menken, J.N. Li, K. Kadowaki, A.A. Menovsky, G. Van Tendeloo and S. Amelinckx, *Nature* 322: 620 (1988).

66. H.W. Zandbergen, W.A. Groen, G. Van Tendeloo, J. Van Landuyt and S. Amelinckx, *Solid State Comm.* 66: 397 (1988).
67. H.W. Zandbergen, W.A. Groen, F.C. Mijlhoff, G. Van Tendeloo and S. Amelinckx, *Physica C* 156: 325 (1988).
68. Antipov, E.V. S.M. Loureiro, C. Chaillout, J.J. Capponi, P. Bordet, J.L. Tholence, S.N. Putilin and M. Marezio, *Physica C*, 215: 1 (1993).
69. C.W. Chu, L. Gao, F. Chen, Z.J. Huang, R.L.Meng and Y.Y. Xue, *Nature* 365: 323 (1993).
70. M. Nunez-Regueiro, J.L. Tholence, E.V. Antipov, J.J. Capponi and M. Marezio *Science* 262: 97 (1993).
71. A. Maignan, G. Van Tendeloo, M. Hervieu, C. Michel and B. Raveau, *Physica C* 212: 239 (1993).
72. M. Hervieu, G. Van Tendeloo, A. Maignan, C. Michel, F. Goutenoire and B. Raveau, *Physica C* 216: 264 (1993).
73. D. Pelloquin, M. Hervieu, C. Michel, G. Van Tendeloo, A. Maignan and B. Raveau, *Physica C* 216: 257 (1993).
74. C. Martin, M. Huve, G. Van Tendeloo, A. Maignan, C. Michel, M. Hervieu and B. Raveau, *Physica C* 212: 274 (1993).
75. A. Maignan, C. Michel, G. Van Tendeloo, M. Hervieu and B. Raveau, *Physica C* 216: 1 (1993).
76. G. Van Tendeloo, C. Chaillout, J.J. Capponi, M. Marezio and E.V. Antipov, *Physica C* 223: 219 (1994).
77. C. Martin, M. Hervieu, G. Van Tendeloo, F. Goutenoire, C. Michel, A. Maignan and B. Raveau, J. Sol. State Chem. (1995, in the press).
78. P.G. Radaelli et al. submitted to *Physica C* (1995)
79. W. Krätschmer, L.D. Lamb, K. Fostiropoulos and D.R. Huffman, *Nature* 347: 354 (1990).
80. G. Van Tendeloo, M. Op de Beeck, S. Amelinckx, J. Bohr and W. Krätschmer, *Europhys. Lett.* 15: 295 (1991).
81. R. M. Fleming, T. Siegrist, P. M. Marsh, B. Hessen, A. R. Kortan, D. W. Murphy, R. C. Haddon, R. Tycko, G. Dabbagh, A. M. Mujsce, M. L. Kaplan and S. M. Zahurac, *Mat. Res. Soc. Symp. Proc.* 206: 691 (1991).
82. W.I.F. David, R.M. Ibberson, J.C. Matthewman, K. Prassides, T.J.S. Dennis, J.P. Hare, H.W. Kroto, R. Taylor and D.R.M. Walton, Nature, 353: 147 (1991).
83. P. A. Heiney , J.E. Fisher, A.R. Mc Ghie, W.J. Romanow, A.M. Denenstein, J.P. Mc Cauley, Jr., A.B. Smith, III, and D.E. Cox, *Phys. Rev. Lett.* 66: 2911 (1991).
84. G. Van Tendeloo, S. Amelinckx, M.A. Verheijen, P.H.M. van Loosdrecht and G. Meijer, *Phys. Rev. Lett.* 69: 1065 (1992).
85. S. Muto, G. Van Tendeloo and S. Amelinckx, Phil. Mag.B, 67: 443 (1993).

DISLOCATIONS AND THE PLASTICITY OF CRYSTALS

Patrick Veyssière
Laboratoire d'Etude des Microstructures,
CNRS-ONERA,
BP 72,
92322-Châtillon Cedex,
France

INTRODUCTION

In an annealed crystal, dislocations are distributed under a three-dimensional network, named the Frank network. Dislocation densities in such samples are usually of the order of $\rho_F = 10^{10}$ m^{-2}, that is, 10^{10} m of line per m^3 of sample*. As the average distance between dislocations of the Frank network, D ($\sim \rho_F^{-1/2}$), then amounts to approximately 10 µm, there is a possibility to grow hair-like single crystals, named whiskers, whose diameter is of the order of 1 µm, that can be dislocation-free. Deformation experiments conducted on whiskers of a number of elements (e.g. Sn, Sb, Pb, Cd, Ni, Cu...) have shown that the domain of strain, ε, wherein a whisker behaves elastically is prolonged to more than 10^{-2} while, in well-annealed single crystals with more standard shapes, the stress strain curves deviate from linearity ($\sigma = E\,\varepsilon$, where E is the Young's modulus) at $\varepsilon \approx 10^{-4}$ to 10^{-5}. In whiskers, the yield stress is actually raised by several orders of magnitude, approaching its theoretical limit (estimated between $\mu/10$ and $\mu/30$, μ is the shear modulus).

In fact, long before experiments on whiskers were conducted, the concept of a dislocation had been introduced as a means to explain the paradoxically low stresses under which crystals yield plastically as compared to the estimated theoretical limit in defect-free crystals. It is in fact in 1934 that Taylor, Orowan and Polyani postulated independently that the motion of a *pre-existing defect* is responsible for the plasticity of samples at relatively small stresses.

It is in the late fifties that the existence of dislocations could be proven directly by means of transmission electron microscope (TEM) images which revealed the complexity and the variety of dislocation organization within deformation microstructures. In addition, amongst many properties TEM enabled researchers to prove the relationship between several atypical macroscopic behaviours, such as a strong sensitivity of the yield stress to temperature, and properties located specifically within the core of dislocations. It took however another 15 to 20 years before high

* It is perhaps worth noting that, with this dislocation density, a parallepipedic sample with dimensions 3 × 3 × 8 mm^3, adequate for a compression test, contains a total length of dislocation line amounting to nearly 1 km.

Stability of Materials
Edited by A. Gonis *et al.*, Plenum Press, New York, 1996

resolution techniques appropriate for dislocation core analysis, such as lattice imaging (HREM) and weak-beam imaging (WB), became generally accessible to the materials science community.

This chapter should be taken essentially as an introduction to dislocations in relation with plasticity. It is aimed at setting the stage for more specialized dislocation-related topics that are to be addressed during this Conference on the Stability of Materials. Most of the information developed hereafter has been gathered from the several following fundamental text books and review papers available on this field by Friedel (1964, 1980), Haasen (1986), Hirth and Lothe (1984), Hull (1965), Kittel 1976, Weertman and Weertman (1966) and in the series of books on *Dislocations in Solids* edited by Nabarro. For the sake of simplicity, general references will not be quoted systematically along this chapter, unless useful additional information is to be found specifically in one of these.

SOME DEFINITIONS AND BASIC PROPERTIES OF DISLOCATIONS

How dislocations contribute to deformation can be clarified by considering the analogy with a ripple in a carpet. When friction is large, the easiest way to adjust a carpet in position is to nucleate a ripple at its appropriate edge, and then to propagate this ripple all way through. After the ripple vanishes upon reaching the opposite edge, the carpet is wholly shifted over a distance that corresponds in magnitude and direction to the way extra matter was inserted through the ripple. The analogy between this process and crystal shear is illustrated in figure 1.

A dislocation is a linear elastic discontinuity in the lattice whose magnitude and direction, represented by its Burgers vector **b**, is defined in figure 2 in the instance of an *edge* dislocation

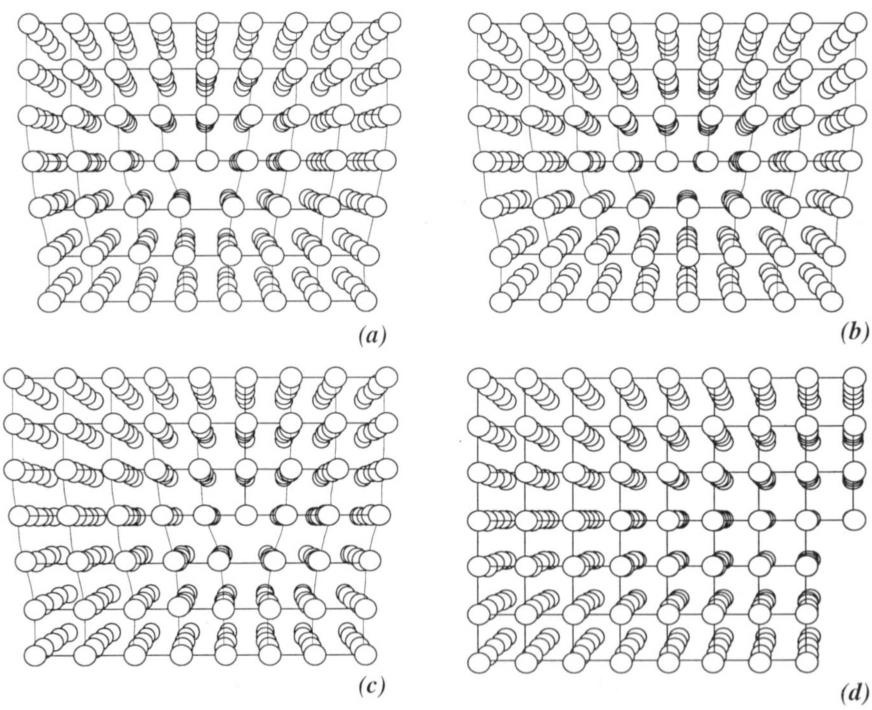

Figure 1. In order to shear a crystal, an extra-half plane of matter is moved from one stable position *(a)* to the next *(c)*. This occurs through a saddle point *(b)*. When the extra-half plane emerges at the crystal free surface, it creates a step. It takes a number of such elementary steps to produce slip traces that are visible by the eye.

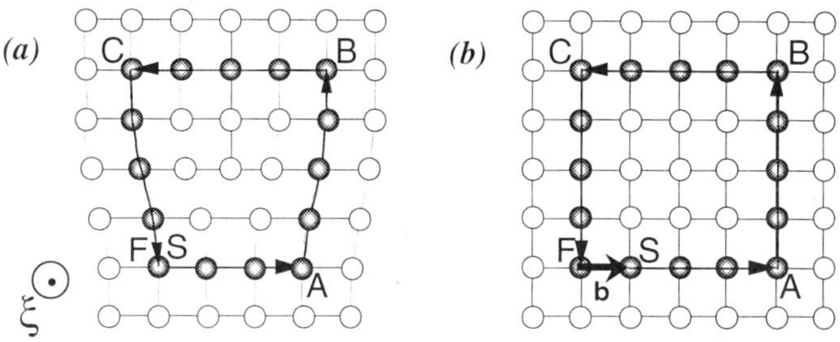

Figure 2. Burgers circuit and the **FS/RH** (Final-Start/Right-Hand) convention for the Burgers vector of an edge dislocation bordering an extra-half plane in the cubic structure. The sense of the dislocation line is given by the unit vector ξ here pointing towards the reader.
(*a*) a closed, anti-clockwise circuit S-A-B-C-F encloses the dislocation.
(*b*) the same circuit drawn in the perfect crystal and consisting in the same amount of atom to atom jumps, is not closed. The additional closure vector, drawn from F to S, is the Burgers vector **b** of the dislocation.
In applying the above definition, every step must be carefully respected. For instance :
- reversing the sense of ξ without changing the sense of rotation along the *Burgers circuit* would result in a Burgers opposite to the above one (**FS/LH** convention).
- similarly, the sense of **b** would be reversed if the reference closed circuit were taken in the undefected crystal, without changing any of the other conventions.
It is a good exercise to check that the above dislocation is in fact exactly the same as that shown in figure I-20 of Hirth and Lothe (1968) or in figure 11.3 of Haasen (1986), and why it is opposite to that of figure I.16 of Weertman and Weertman (1966)

($\mathbf{b} \perp \xi$, where ξ is a unit vector tangent to the dislocation line). It is clear from this geometrical definition that **b** depends largely upon crystal symmetry. It however depends neither on the origin *S* on the circuit, nor on its shape as long as the circuit remains in 'good crystal' (atom to atom jumps all with the correct number of nearest neighbours).

The prime property of a dislocation is to transport the elemental quantity of displacement **b**. In other words, when a portion dξ of dislocation has swept over the elemental surface $d^2\mathbf{s} = d\xi \times d\mathbf{r}$, the two surfaces in contact on each side of d**s** have undergone a relative displacement of **b**.

Keeping the convention for ξ unchanged, a dislocation with a Burgers vector opposite to that of figure 2(*a*) would border an extra-half plane located below the dislocation line. It will be shown in the following, when considering the Peach-Koehler force, that the exact knowledge of the sense of **b** takes all its importance in the prediction of the direction of motion of a dislocation under an applied stress. As a direct implication of the definition of **b**, consider the instance of a *prismatic* loop (figure 3), that is, of a loop whose Burgers vector does not belong to the loop

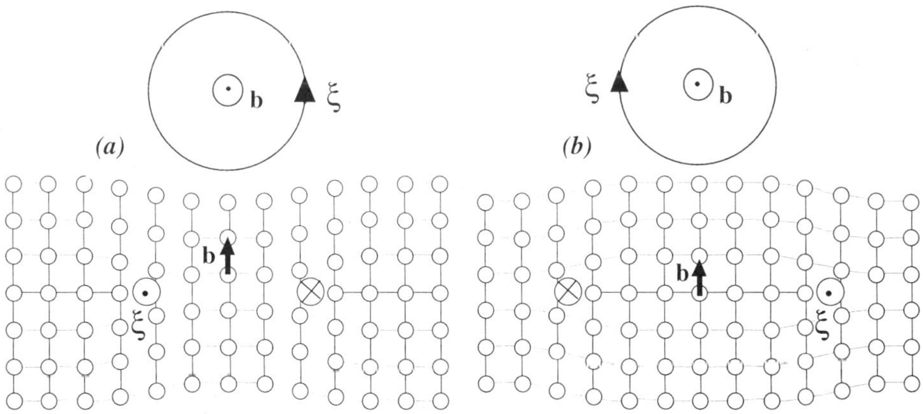

Figure 3. Consequence of the definition of the Burgers vector of a dislocation in figure 2, on the vancancy/interstitial nature of a dislocation loop.

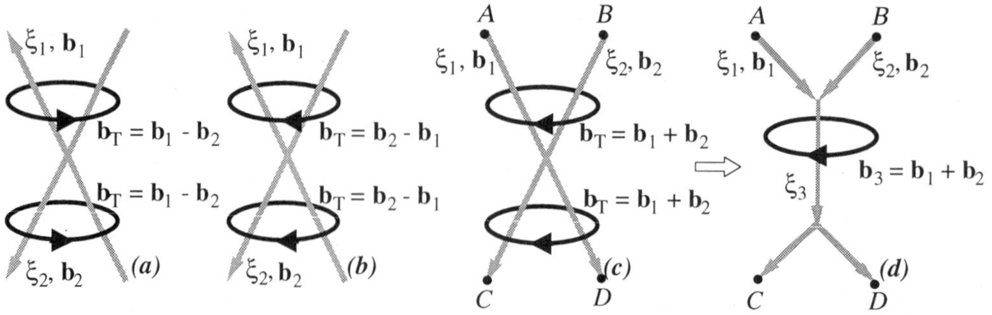

Figure 4. Application of the Burgers circuit to the conservation of Burgers vector at nodes : the total Burgers vector is the algebraic sum of the Burgers vectors of the dislocations surrounded by the circuit. Given the orientation of the line, when the sense of the Burgers circuit is changed from *(a)* to *(b)*, the sign of the total Burgers vector is reversed. *(c)* and *(d)* : when two dislocations anchored at A, D and B, C respectively, react at a node to form a junction ξ_3 (see figure 10), the Burgers vector $\mathbf{b}_3$ of this junction is necesarily equal to the algebraic sum of the Burgers vectors of all the dislocations joining at the node.

habit plane. Here **b** is taken normal to the loop plane, so that the dislocation is edge in character all along it line. As a consequence, the extra-slice bordered by the dislocation line is coplanar with the loop, sitting either in the inside or in the outside of the loop. Both types of prismatic loops are currently encountered in materials. They may appear as trailing debris during the conservative motion of dislocations when the mechanism of cross-slip takes place (see figures 6 and 18), or else, they can be formed as the result of the segregation of points defects that have been formed in supersaturation after a quench or upon crystal irradiation. The point defects that have yielded the loops of figures 3(*a*) and 3(*b*) are of vacancy and interstitial type, respectively.

Last but not least is the property of Burgers vector *conservation*, analogous to Kirchoff's law of electrical current flow. Since the dislocation line is the boundary between a sheared and an unsheared area, it cannot vanish anywhere in the crystal but must terminate at a free surface, at a dislocation node or at some other defect such as a grain boundary. It is obvious that the Burgers vector of a circuit enclosing several dislocations is equal to the sum of the Burgers vectors $\mathbf{b}_T$ of the enclosed dislocations (figure 4). By continuous motion of the Burgers circuit on each side of a dislocation node, provided the circuit remains in 'good crystal', $\mathbf{b}_T$ cannot change, which is equivalent to saying that inside a crystal, dislocations must either branch onto other dislocations or form closed loops.

Local distortions and violations of lattice symmetry as generated by dislocations, may be enough in order to influence physical properties other than the mechanical properties of a crystal. In fact, there are not so many of such dislocation-sensitive properties. The electrical properties of semiconducting materials are one of these and it is also known that dislocations can provide short-circuits for matter transport by diffusion, thus enhancing the reactivity of solids. Dislocations may take an important part during crystal growth and this aspect was largely invoked as an indirect evidence of the existence of dislocations in the debates that followed the introduction of the concept of a dislocation.

Since *b*, the modulus of **b**, amounts to a fraction of nm (table 1), dislocations must move in sometimes large numbers in order for a crystal to exhibit a detectable deformation. For instance, for a slip trace to show with the naked eye as a surface step, several thousands of dislocations must have emerged within a limited slab of the sample. On the other hand, the little indentation of the size of a fingerprint which results from the impact of a flying gravel on the body of car, originates from the displacement of dislocations by the millions if not by the billions. We know, in addition, that the number of dislocations which were present in the metal sheet before the impact is not sufficient to give rise to the observed damage. This implies in turn that dislocations have the capability to *multiply* in large proportions.

Table 1. Comparison between Burgers vectors of dislocations in selected cubic materials. The last line gives numerical values of the dislocation line tension, discussed later (equation (15)).

			ordered alloys		oxides		
compound	Al	Mo	Ni_3Al	NiAl	MgO	$MgAl_2O_4$	$Y_3Fe_5O_{12}$
structure	fcc	bcc	$L1_2$	B2	NaCl	spinel	garnet
b	1/2<110>	1/2<111>	<110>	<100>	1/2<110>	1/2<110>	1/2<111>
b (nm)	0.286	0.272	0.503	0.289	0.298	0.570	1.072
$\mu b^2/2$ (eV/Å)	0.7	4.8	12.5	8	3.5	12.2	28

ELEMENTARY PROPERTIES OF DISLOCATIONS

The motion of a *curved* dislocation in its slip plane is represented schematically in figure 5. By definition, the dislocation is said to be in the *screw* or in the *edge* orientations when its line, defined by an oriented unit vector ξ tangent to the line, is parallel or perpendicular to its Burgers vector, respectively (figure 5(*d*)). Any other dislocation segment has a *mixed* screw and edge *character*. The way a crystal surface is affected by the emergence of a dislocation line depends on the angle between the Burgers vector and the intersection of the slip plane with the surface. It

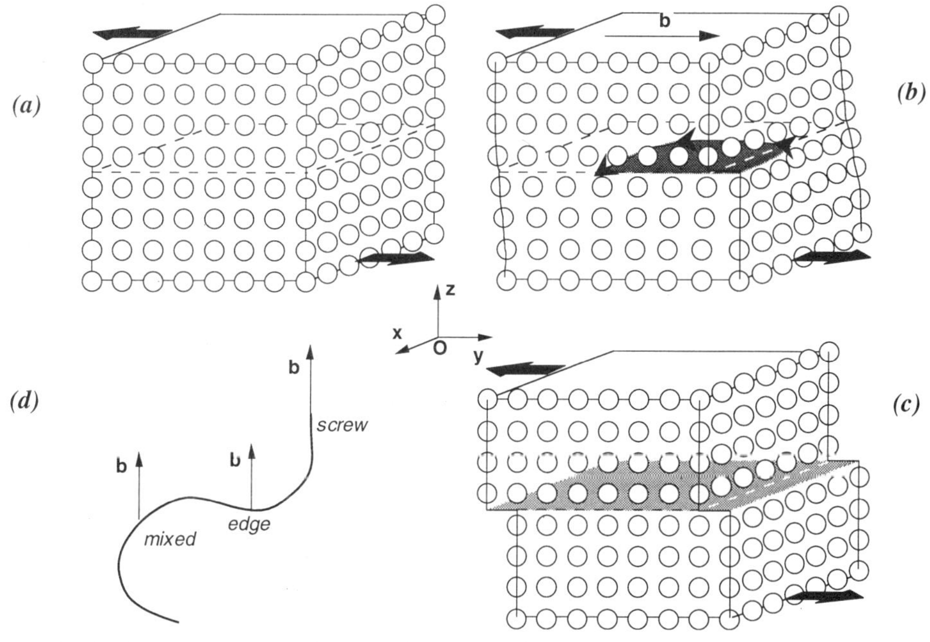

Figure 5. Nucleation and propagation of a curved dislocation in its slip plane.
(*a*) The traces of the slip plane in the undefected crystal are represented by dashed lines. The pair of horizontal thick arrows schematizes the shear stress.
(*b*) For the sake of clarity, the dislocation is supposed to have nucleated at the rigth-hand front edge of the crystal. It is drawn after it has propagated half-way in the slip plane both in the edge and in the screw directions. In the front crystal face, one recognizes the classical configuration of an edge dislocation in a simple cubic crystal, as shown in figure 1(*a*), while the lateral right-hand face pictures the emergence of a screw dislocation.
(*c*) Same as in (*b*) after the dislocation has swept across the slip plane entirely. By definition, the shear coincides with the Burgers vector. It is worth noticing that, contrary to an edge dislocation (**b** $\perp$ ξ), a screw dislocation (**b** // ξ) does not leave any slip trace as it emerges parallel to a surface.
(*d*) The definition of the character of a dislocation of arbitrary shape.

is only the Burgers vector **b** of the operating dislocation that matters in the shear of the upper half-crystal with respect to the lower. It is indeed equivalent (figure 5(*b*)):

(i) to expand this *curved* dislocation until it emerges at the opposite lateral surfaces.

(ii) to move a *straight* dislocation of any character. For instance, one may indifferently zip either the edge portion in figure 5(*b*) towards the left hand side and then to let a pure screw dislocation sweep the crystal until it reaches the back surface. One may also proceed conversely—first screw portion, then full edge—with no detectable differences after the dislocation is fully vanished at surfaces.

During the displacement of the extra-half plane or equivalently of the edge dislocation which is shown in figure 1, motion will occur *conservatively*, that is, without involving a net flux of matter, provided the dislocation line is confined to the plane embodied at the crystal surface by dashed lines in figure 5(*b*). By definition, this plane, known as the *slip*—or *glide*—plane, contains both the dislocation line ξ and the Burgers vector **b**. Slip on curved surfaces is in principle permitted by this definition. For instance, the slip surface of the loop schematized in figure 3(*a*) is the cylinder defined by the loop line and its Burgers vector. It is interesting to realize in this case that though the loop was formed non conservatively (by segregation of $n \approx S\, d_{hkl}/\omega$ point defects, S and ω are the surface of the loop and the volume of a point defect, respectively, d_{hkl} is the interplanar distance), these point defects could be eliminated *conservatively* at the crystal surface if the loop were allowed to slip on its own slip cylinder. We shall see later in this chapter that the main limitation to such a process is that slip is in general crystallographic, that is, it occurs preferentially on certain atomic planes, which obviously hinders greatly the above process (unless the loop adopts a polygonal shape in which each of its edges could be actually glissile on a preferred plane).

The definition of the slip plane then contains an ambiguity when the line lies along the screw direction since it could thus be any crystallographic plane that contains **b**. Implications of such a slip plane degenerescence are encountered quite commonly in crystals, and this to an ex-

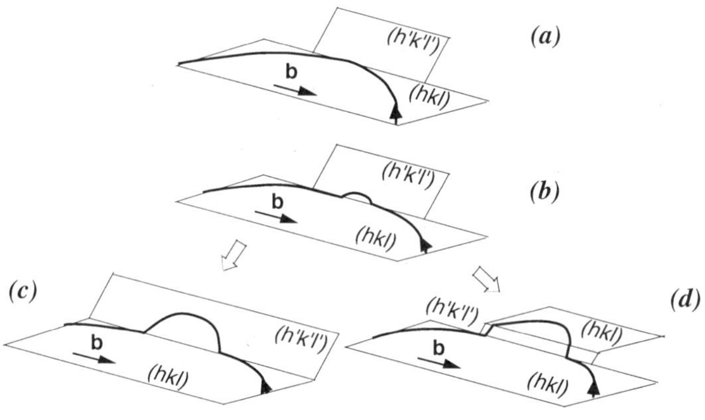

Figure 6. A schematic representation of the mechanism of *cross-slip*. The ability of a dislocation to cross-slip is a specific property of its screw portion, the only character to be crystallographically correct for this purpose.
(*a*) as a dislocation segment glides in the (hkl) plane under the action of an applied shear stress,
(*b*) its screw portion may be compelled to further expand in another plane (h'k'l'), known as the cross-slip plane.
Given the applied stress tensor and depending upon the nature of the obstacle, the dislocation may
(*c*) keep expanding in the cross-slip plane, producing lengths of dislocation segments with same Burgers vector as that of the parent dislocation, but inclined to the primary slip plane.
(*d*) cross-slip 'back' to the primary slip plane. In this case primary slip proceeds on several parallel slip planes. This is a mechanism by which slip bands thicken gradually during deformation (see continuation in figure 18).
The occurrence of cross-slip depends upon many factors such as the actual stress tensor, the relative critical resolved shear stresses of the two slip systems, the dislocation core structure and test temperature when cross-slip is thermally-activated, as is often the case.

tent that varies from one crystal to another as discussed later on in this chapter. The process whereby a *screw* dislocation switches between two slip planes (figure 6), named *cross-slip*, is extremely important in plasticity for it provides dislocations a degree of freedom that enables them to circumvent impenetrable obstacles, thus to relax local stresses.

Consider the dislocation portion $d\xi$ moving perpendicular to itself at a velocity V on a slip plane with surface S. The area swept during the time interval dt is $d^2S = d\xi\, V\, dt$. Since the *crystallographic shear** corresponding to the displacement of the whole dislocation over the entire slip plane is $\gamma = b/h$, where h is the length of the sample, that associated with the above elemental motion is given by

$$d^2\gamma = \frac{(d\xi\, V\, dt)\, b}{S\, h} \tag{1}$$

Integrating expression (1) over the total length of *mobile* dislocations contained in the crystal, Ω, yields

$$d\gamma = \frac{b}{S\, h} \oint_{\Omega} V\, dt\, d\xi \approx \frac{b}{S\, h} V_m\, dt\, \Lambda = \rho_M\, b\, V_m\, dt \tag{2}$$

Macroscopically, shear thus originates from the displacement of a density ρ_M of mobile dislocations having slipped a mean distance $L_m = V_m\, dt$. This results in a total strain given by

$$\gamma = \rho_M\, b\, L_m \tag{3}$$

which in turn yields the strain-rate (Orowan) equation

$$\dot{\gamma} = \rho_M\, b\, V_m \tag{4}$$

provided that the density of mobile dislocation is a constant. A great deal of the physics of plasticity is related to this equation. It is by means of equations (3) and (4) that microscopic dislocation events are usually incorporated in macroscopic models as developed in several of the other dislocation-related contributions to this volume.

INTRODUCTION TO THE ELASTIC THEORY OF DISLOCATIONS

The Peach-Koehler force

Consider a crystal which, under the effect of an applied shear stress τ, is sheared over $\mathbf{b}$ by one dislocation (figure 5(c)). The work of the external stress during this displacement is $W = f\, b = \tau\, S\, b$, where $S = L\, \Lambda$ is the surface of the slip plane (figures 7(a) and 7(c)). Then without restriction on generality (because the orientation of $\mathbf{b}$ is arbitrary in the slip plane), assume that the shear stems from the motion of a length L of dislocation over Λ. The *total* force on the dislocation, f, is a *configurational* force defined as the force that would produce the same work W as above when acting over the distance Λ (figure 6 (a)).

* following in that a rather general practice in this field, the same notation γ, will be indifferently used in this chapter to refer to crystallographic shear and to the energy of surface defects.

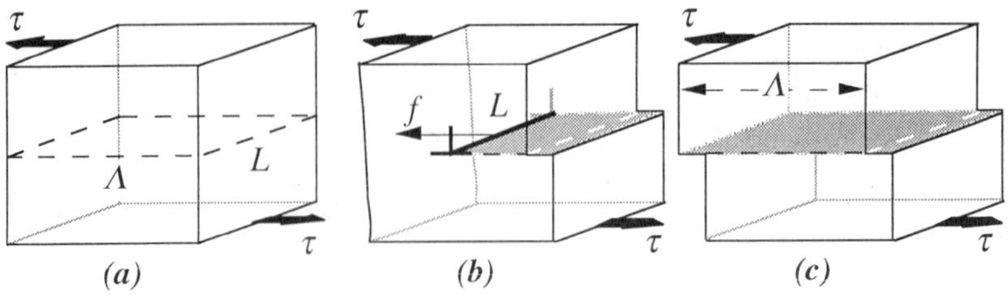

Figure 6. Definition of the Peach-Koehler force. The work needed to shear a crystal over **b**, (*a*) to (*c*), is the same as that produced by the Peach-Koehler force applied to a dislocation of length *L*, (*b*), over the distance Λ.

The Peach-Koehler force is taken either as f or rather as the force *per unit length* of dislocation $F = \tau b$, a quantity which is in general more convenient to handle in calculations.

A more general definition of the force on a dislocation can be introduced by considering the work W_C needed to form a loop C with surface S. The additional work δW_C done by the stress σ as the dislocation sweeps an elemental surface $\mathrm{d}\mathbf{S} = \mathrm{d}\xi \times \delta\mathbf{r}$ writes

$$\delta W_C = \oint_C \mathbf{b} \cdot \left[\sigma \cdot (\mathrm{d}\xi \times \delta\mathbf{r})\right] = \oint_C (\mathbf{b} \cdot \sigma) \cdot (\mathrm{d}\xi \times \delta\mathbf{r}) = \oint_C \left[(\mathbf{b} \cdot \sigma) \times \mathrm{d}\xi\right] \cdot \delta\mathbf{r} \tag{5}$$

Again, δW_C can be equivalently regarded as the work of a force $\mathrm{d}\mathbf{f}$ acting over a displacement $\mathrm{d}\mathbf{r}$, itself arbitrarily variable over the loop

$$dW_C = \oint_C \mathrm{d}f \cdot \delta r \tag{6}$$

The Peach-Koehler force then expresses as

$$\mathrm{d}\mathbf{f} = (\mathbf{b} \cdot \sigma) \times \mathrm{d}\xi \tag{7}$$

Regardless the dislocation character, the component of $\mathrm{d}\mathbf{f}$ in the slip plane is exerted perpendicular to the line. It can be shown that it has a constant magnitude. In addition, the Peach-Koehler force can be separated into as many independent terms as there are components contributing to the stress applied to the dislocation.

For the sake of clarity, we now utilize the above development in the particular example of figure 5(*b*). σ is arbitrary* and let $\mathbf{b} = (0, b, 0)$, $\mathrm{d}\xi_{\mathrm{screw}} = \mathrm{d}\xi(0, -1, 0)$, $\mathrm{d}\xi_{\mathrm{edge}} = \mathrm{d}\xi(1, 0, 0)$ and $\mathrm{d}\xi_{\mathrm{mixed}} = \mathrm{d}\xi(\alpha, \beta, 0)$. Then equation (7) yields

$$\mathrm{d}\mathbf{f}_{\mathrm{screw}} = b\, \mathrm{d}\xi\, (\tau_{yz}, 0, -\tau_{xy}), \tag{8-a}$$

$$\mathrm{d}\mathbf{f}_{\mathrm{edge}} = b\, \mathrm{d}\xi\, (0, -\tau_{yz}, -\sigma_{yy}), \tag{8-b}$$

$$\mathrm{d}\mathbf{f}_{\mathrm{mixed}} = b\, \mathrm{d}\xi\, (-\beta\, \tau_{yz}, \alpha\, \tau_{yz}, \beta\, \tau_{xy} - \alpha\, \sigma_{yy}) \tag{8-c}$$

which illustrate the facts that the force per unit length $\mathrm{d}f/\mathrm{d}\xi$ in the slip plane

- is always perpendicular to the dislocation line,

*The stress tensor consists of nine components τ_{ij}, acting on the plane whose normal is along the **i**-axis, in the direction of the **j**-axis. The notations τ_{ij} and σ_{ii} are used here in order to discriminate between shear and normal stresses, respectively.

- has a constant magnitude, $b\,\tau_{xz}$, regardless the dislocation character ($\alpha^2 + \beta^2 = 1$),
- depends on the sense of the shear stress.

Note in addition that the component of the force perpendicular to the slip plane in expression (8-c), df_z, involves a *shear* component of the stress tensor which promotes cross-slip, here by combination of df_x and df_z.

By contrast, the forces df_z applied to the edge and mixed parts involve a *normal* stress σ_{yy} which tends to act perpendicular to the extra half-plane, be it in tension or in compression, thus offering a driving force for the non-conservative motion of dislocation. In this non-conservative motion, which is named *climb*, the dislocation absorbs or expels point defects (figure 7), provided temperature is high enough to allow for significant diffusion of point defects. A dislocation climbs positively when it absorbs vacancies. The climb displacement of a dislocation by a vector $\delta\mathbf{r}$ normal to itself, removes a volume of matter per unit length

$$\delta v = (\boldsymbol{\xi} \times \mathbf{b}) \cdot \delta\mathbf{r} = \pm\, b_e\, \delta h \tag{9}$$

where b_e is the edge component of the Burgers vector and δh the distance climbed (i.e. the component of $\delta\mathbf{r}$ perpendicular to the slip plane).

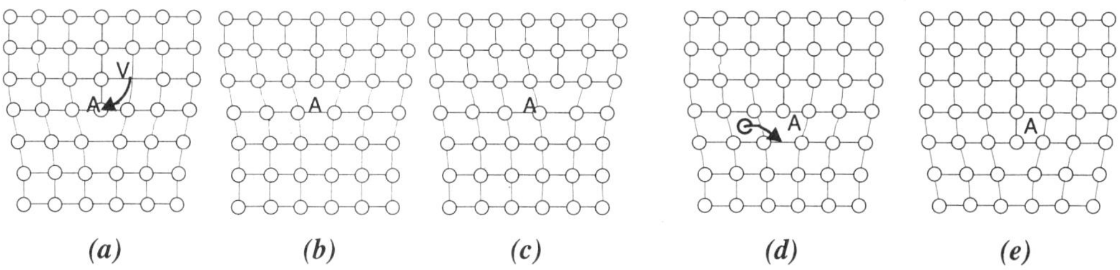

Figure 7. The mechanism of dislocation climb.
(a) - (b) : climb of an extra-half row of atoms by vacancy absorption. This motion is taken as the positive direction for climb motion. In *(c)* the extra-half row shown in *(b)* has undergone slip.
(d) - (e) : climb as a result of interstitial migration; by convention, the dislocation has climbed negatively.
This view in projection hides the fact that the migration of a point defect on a dislocation moves the dislocation one atomic site only. In order for the dislocation to move over its whole length over some distance, this requires that a very large number of point defects have reached the line.

In a crystal containing a supersaturation of interstitials, the application of a tensile stress promotes the formation of interstitial loops (figure 3(*b*)) perpendicular to the stress axis, while compression would favour the formation of a distribution of loops whose habit plane is aligned to the load axis.

Self-energy

The energy stored per unit length of dislocation in a region bounded by two coaxial cylinders of radius r_o and R, is given by

$$E_S(\theta) = C(\theta)\,\frac{\mu\, b^2}{4\,\pi}\,\ln\left(\frac{R}{r_c}\right) + E' \tag{10}$$

where θ is the angle between ξ and $\mathbf{b}$ and $C(\theta)$ is an orientation-dependent elastic coefficient

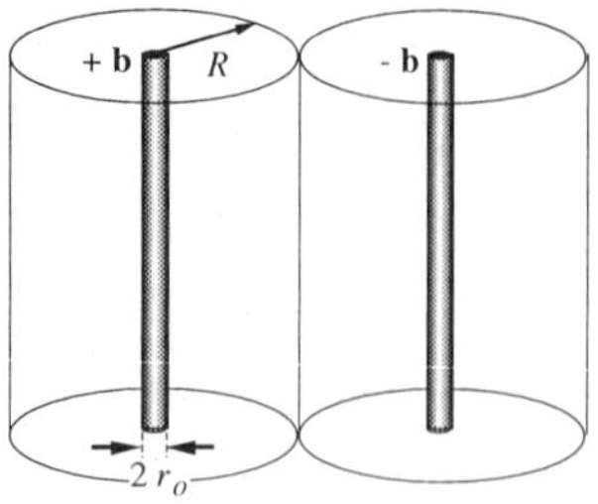

(a)

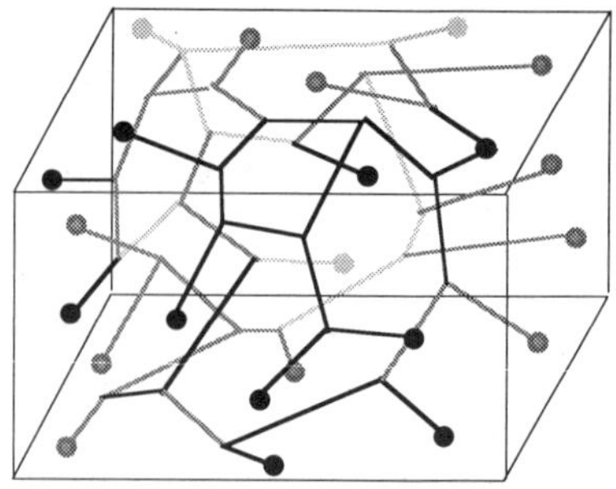

(b)

Figure 8. The definition of the inner and outer cut-off radii in the expressions of dislocation self- and interaction energies.
(*a*) there is a cylinder around the dislocation line, diameter $2r_o$ where the linear elasticity breaks down.
The value of the outer radius *R* is determined by dislocation organization, it corresponds to the average distance in the crystal beyond which the stress field of the dislocation is roughly cancelled by the stress field originating from the surrounding defects. This is shown in (*a*) in the case of two straight parallel dislocations.
(*b*) in fact the Frank network looks more or less as in this scheme, so that the averaged nature of *R* should be kept in mind, together with the fact that it is by definition markedly microstructure dependent.

given by

$$C(\theta) = \cos^2\theta + \frac{\sin^2\theta}{1-\nu} = \frac{1}{K(\theta)} \tag{11}$$

ν is the Poisson's ratio (taken usually between 1/4 and 1/3 in elastically isotropic crystals). E' is a correction term which includes some ill-defined contributions such as the effects of surface relaxation, some torque terms for an edge dislocation and most important the anelastic energy stored within what is called the *core* of the dislocation.

In equation (10), the lengths *R* and r_c (figure 8), setting the outer and inner cut-off radii, respectively, ascribe the self-energy a finite value. In order to confer E_S, which would otherwise diverge with crystal dimension, a physically sound value, it is usually remarked that

• at a distance from the dislocation that scales with the dimension of the Frank network, the strain field of a given dislocation should be mostly canceled as a result of the superimposition of the strain fields exerted by the neighbouring dislocations (in fact, the reduction in the elastic energy stored in the crystal is a major driving force in the organization of the Frank network upon annealing). Accordingly, the energy stored beyond the outer cut-off cylinder can be neglected. In spite of the logarithmic dependence of the energy upon *R*, equation (10) should be used with some care in comparing between dislocations in crystals whose Frank networks have significantly distinct sizes or else in crystals containing finite residual stresses.

• the r^{-1} dependencies of the stress and of the strain field of a given dislocation introduces another divergence of the energy as the distance from the dislocation line approaches zero. In this case, an inner cut-off radius, r_o, must be introduced whose meaning differs significantly from that of *R*. The energy divergence arises from the failure of linear elasticity to a continuous medium in a domain where elasticity is no longer linear and where the atomistic nature of crystals is dominant. As a result, it is often chosen not to take into account the energy stored within an inner crystal cylinder whose radius is of the order of *b*; this cylinder includes matter where elasticity deviates from linearity. It should be realized that this part of the energy (contained in the correction term E' in equation (10)), which in fact represents 5% to 15% of the total energy, is not neglected but ignored. Several dedicated atomistic calculations showed that this simplification is adequate in a number of situations where the precision required on energies is not crucial. It may however be misleading when differences between core configurations constitute the main concern of a given analysis of dislocation properties such in the cross-slip process of dislocations having an extended core.

When the dimension of the Frank network is of the order of a few microns, a convenient order of magnitude for the self-energy per unit length of dislocation is $\mu b^2 /2$.

Line tension.

The line tension T is a measure of the change in the total elastic energy of a dislocation with a change of its shape.

The concept of line tension is instrumental in describing a number of dislocation properties such as their equilibrium shape as schematically exemplified in figure 9 in the instance of the junction of dislocations at node. In this case, it can be regarded as the force that tends to shorten *straight* segments.

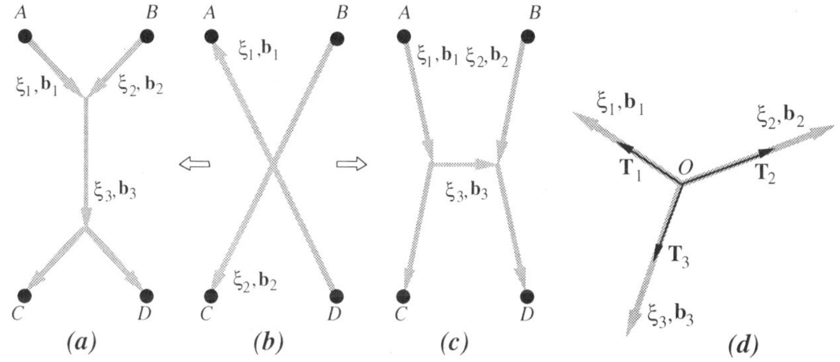

Figure 9. The two possible transformations of a quadruple node (b) into two triple nodes. The quadruple node consists of two coplanar intersecting dislocations with different Burgers vectors $\mathbf{b}_1$ and $\mathbf{b}_2$. The rule of Burgers vector conservation implies that $\mathbf{b}_3 = \mathbf{b}_1 + \mathbf{b}_2$. Then, governed by the b-square criterion on self-energy, dislocations 1 and 2 will exhibit the two configurations shown in (a) and (b). In the latter dislocations must reorient their line in order to present antiparallel lines (then $\mathbf{b}_3 = \mathbf{b}_1 - \mathbf{b}_2$). The equilibrium position of the nodes can be determined by making use of considerations on the line tension of a straight dislocation.
(d) equilibrium at a triple node. When dislocations are free to move in any direction (glide and/or climb), the equilibrium requires that $\mathbf{T}_1 + \mathbf{T}_2 + \mathbf{T}_3 = 0$ (note incidentally that by application of the Burgers circuit to this node, $\mathbf{b}_1 + \mathbf{b}_2 + \mathbf{b}_3 = 0$).

However, the line tension can be understood as a force applied locally to a *curved* dislocation line that tends to straighten it in order to reduce its overall energy. As this will be exemplified later, certain orientation-dependent factors may play a key part in determining the eventual shape of a dislocation line. It is worth realizing that the expression of the line tension of a dislocation is highly shape-dependent.

This very important property of line tension is illustrated below by comparing between a straight and a circular dislocation; a number of additional examples of the derivation of the line tension for various configurations can be found in particular in Hirth and Lothe (1982).

A straight dislocation. In this case the line tension is evidently

$$T = E_s(\theta) \tag{12}$$

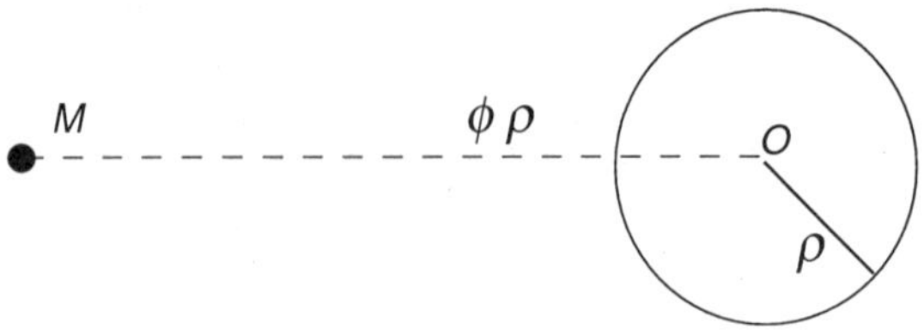

Figure 10. At a distance from the dislocation loop, the strain originating from symmetrical protion compensate almost exactly, and this is the definition of the outer cut-off radius. This distance is of the order of a few times the radius of the loop.

where $E_s(\theta)$ is as given by equations (10) and (11). Taking $\nu = 1/3$, the line tension of a straight edge dislocation is 1.5 times larger than that of the screw.

A loop. Consider a circular dislocation loop with radius ρ. The stress generated by diametrically opposite loop portions compensate each other at distances of the order of $\phi\,\rho$ where ϕ is a constant ranging between 2 and 5 (figure 10). Thus, following the above reasoning that served to define the outer radius of integration in the self-energy, the total energy of the loop is

$$U = 2\,\pi\,\rho\,E = C\,\frac{b^2}{2}\,\rho\,\ln\left(\frac{\phi\,\rho}{r_c}\right) \tag{13}$$

Accordingly, the line tension differs slightly from that of a straight dislocation

$$T = \frac{dU}{d(2\,\pi\,\rho)} = C\,\frac{b^2}{4\,\pi}\left(\ln\frac{\phi\,\rho}{r_c} + 1\right) = E_S\left(\frac{1}{\ln\frac{\phi\,\rho}{r_c}} + 1\right) \tag{14}$$

In many cases, an approximate value of the line tension is enough in analyses of dislocation properties. The following *average* value of the line tension

$$T = \alpha\,\mu\,b^2 \tag{15}$$

where α is of the order of 0.5, can be assumed when the orientation dependence of $E_S(\theta)$ is unimportant in the calculations. With regards to the expression of the self-energy (equation (10)), the numerical value of α is equivalent to considering that $R \sim 500\ r_o$, which is quite a sensible value. A comparison between dislocation line tensions is made in table 1.

Forces on a elementary piece of a curved dislocation

As shown in figure 11, a segment of dislocation submitted to a stress τ is at equilibrium, be it stable or unstable, when its curvature ρ is $T/\tau b$.

In other words, the stress required to bend a dislocation to a radius ρ is given by

$$\tau = \frac{T}{\rho\,b} \tag{16}$$

which combined with expression (15) yields

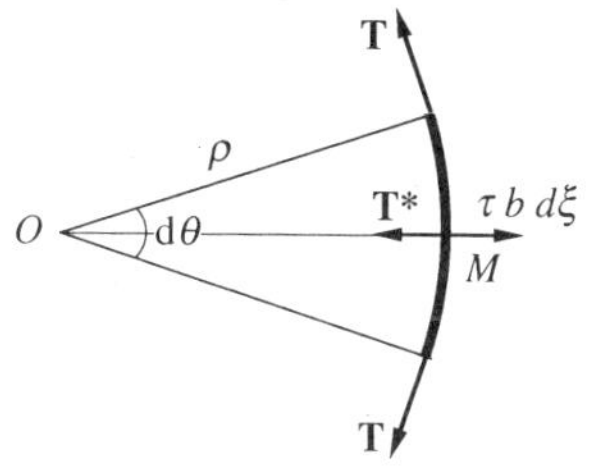

Figure 11. Force analysis of the mechanical equilibrium of dislocation segments. The outward force applied to an elementary piece of dislocation, **OM**, due to a stress is $\tau b\, d\xi$, the opposing force due to the line tension amounts to $2\,T \sin(\mathrm{d}\theta/2) = T\,\mathrm{d}\theta$.

$$\tau = \frac{\alpha \mu b}{\rho} \tag{17}$$

This expression is used quite commonly in order to evaluate, for instance, the stress required to multiply dislocations within a given distribution of obstacles (following section) or to estimate local stresses from the observed curvature of dislocations.

Application to the Frank-Read source. As an illustration of equation (17), consider the case of a dislocation segment irreversibly anchored at two unknown obstacles, A and B, distant of ℓ (figure 12). For the sake of simplicity, assume as this has been done in order to derive equation (14), that the self-energy of a dislocation is independent of its character (taking into account its actual orientation dependence would just complicate the presentation of the Frank-Read mechanism without changing the general sense of the present discussion). Implications of the orientation dependence of E_S will nevertheless be addressed in the next section.

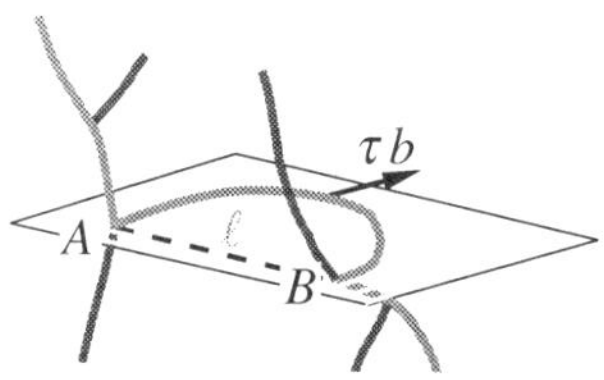

Figure 12. Expansion of a dislocation segment in its slip plane under an applied stress. Here the segment is fixed at its two extremities A and B after having interacted attractively with other dislocations to form sessile junctions.

Suppose a shear stress is applied onto the dislocation and raised progressively. In view of equation (17), the segment AB adopts a decreasing radius of curvature ($\rho = \alpha\, \mu\, b\, / \tau$) so that the dislocation arc in between A and B, gradually bows out in the slip plane (figure 13).

As long as ρ is less than $\ell\, /\, 2$, it decreases with increasing τ. Accordingly, the line tension increases and compensates exactly for the increase of the applied stress, thence maintaining the line under mechanical equilibrium. Since $\ell\, /\, 2$ is obviously the largest possible radius that can possibly be undertaken by the loop portion, the configuration that consists of a half-circle of diameter ℓ is unstable with respect to a further increase of the applied stress. Hence, when the applied shear stress τ is set above the critical value given by

$$\tau_{FR} = \frac{\alpha \mu b}{\ell} \tag{18}$$

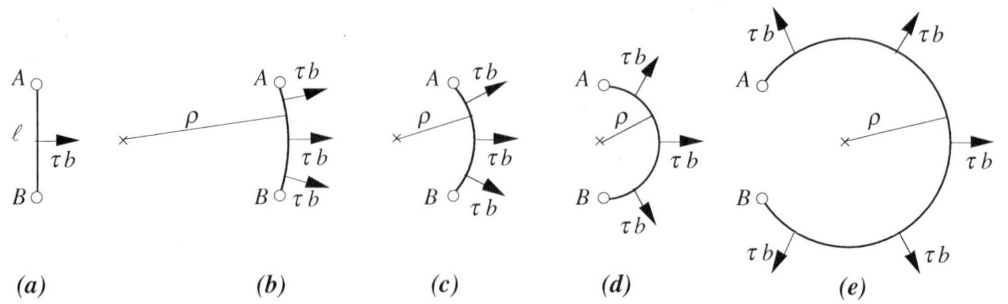

Figure 13. Expansion of the dislocation segment AB shown in figure 12 (top-view). Upon raising the stress τ the radius of curvature ρ first decreases until the arch assumes a half-circle shape ($\rho = \ell / 2$), then $\tau = \tau_{FR}$. If the stress is raised above that level, ρ decreases. Since then the Peach-Koehler force cannot be balanced by line tension, the dislocation segment expands spontaneously until it interacts with obstacles.

that is, beyond the critical half-circle configuration (figure 13(*d*)), the local radius of curvature at every point of the loop is smaller than ℓ, so that the applied stress cannot be balanced by the line tension any more (figure 13(*e*)). The dislocation then expands irreversibly in the crystal.

In the sequence presented in figure 13, a segment with initial length ℓ lengthens under an applied stress. It may thus be already regarded as generating dislocation line. Up to the step of figure 13(*a*) the process is reversible upon release of the applied stress for, driven by line tension, the loop portion will indeed shrink back to step 13(*a*). In fact, the common notion of a dislocation *source* refers to the more general process of regenerative multiplication. The Frank-Read mechanism shown in figures 13(*f*) to (*j*) is one example of such mechanisms. It is the most commonly referred to, though direct evidence for its operation is actually very scarce. Other multiplication processes such as that involving the rotation of a mobile dislocation about a sessile pole inclined to the slip plane, seem to occur more frequently in materials.

Line tension : the general orientation-dependent case

For the sake of clarity, the orientation dependence of $E_s(\theta)$ has been ignored in the preceding examples of the derivation of the line tension, hence defined as a measure of the resistance of

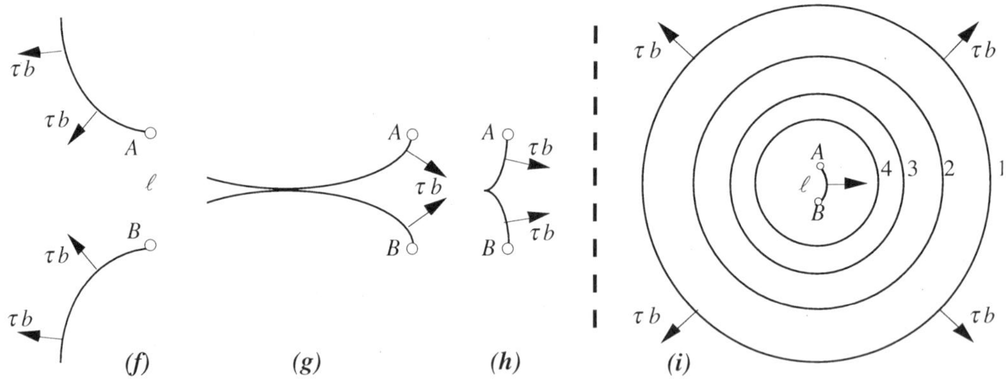

Figure 13 (continued). Given an applied stress larger than τ_{FR}, the configuration of figure 13(*e*) keeps expanding to assume a large radius of curvature only limited by neighbouring obstacles (not shown here). Eventually, the parts of the nearly complete loop that are anchored at A and B approach one another (*f*) and (*g*), then they annihilate locally and relax (*h*) in order to give rise to the initial configuration of figure 13(*a*).

Figure 13(*i*) schematizes the Frank-Read source (viewed however at a larger scale) after it has operated 4 times and that segment AB has again undertaken the configuration of figure 13(*c*).

a dislocation of arbitrary shape to *lengthening*. Nevertheless, the orientation dependence of the self-energy is of great importance for it contributes to the resistance of the line to *bending*.

In determining the equilibrium shape of a dislocation segment, one should not just take care of the forces that yielded equation (17) but also of the interactions between a given elemental segment and the rest of the dislocation. This is why this shape will in general depend on the *size* as well as on the *shape* itself of the segment under consideration (i.e. the result is different for a complete and a half loop). A complete analysis of the concept of line tension in which these difficulties and many others are discussed and exemplified, has been given by Kirchner (1984). Based on equation (15), an approximation of the shape of a dislocation loop of given area S, can be determined by means of the more exact expression of its line tension*

$$T(\theta) = E_S(\theta) + \frac{\partial^2 E_S(\theta)}{\partial \theta^2} \tag{19}$$

whose origin and meaning can be understood as follows. Consider an elemental piece of straight dislocation with arbitrary orientation θ_o, fixed at its extremities M and N, and then let it take a bulge of small amplitude (figure 14). Ignoring mutual interactions, the energy of the bulged segment writes

$$U_{\widehat{MN}} = \oint_M^N E_S(\theta)\, d\xi \tag{20}$$

since

$$d\xi = dx\left(1 + \frac{d\theta}{2}\right) \tag{21}$$

and after expansion of $E_S(\theta_o + d\theta)$, the energy difference can be rewritten as

$$U_{\widehat{MN}} = \int_M^N \left(E_S(\theta_o) + E_S'(\theta_o)\, d\theta(x) + E_S''(\theta_o)\, \frac{d\theta(x)^2}{2} \right) dx \tag{22}$$

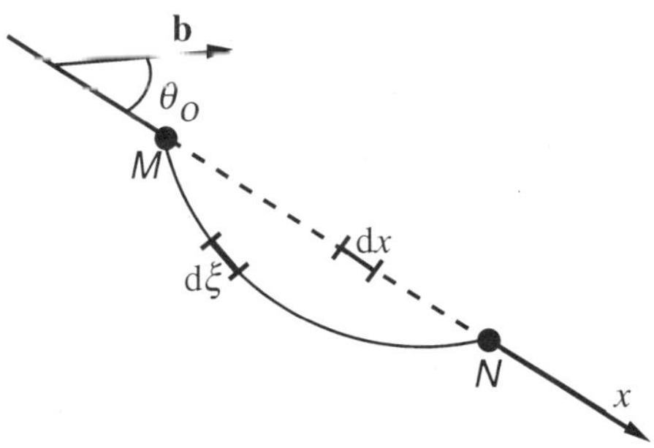

Figure 14. Geometry of a symmetrical bulge between two fixed points M and N of a dislocation of arbitrary character. **Ox** is taken along the dislocation before the bowing out, ξ is the line coordinate along the bulged part.

* from expressions (10) and (11) and (19), the line tension of a screw is $[(1 + \nu)/(1 - 2\nu)]$ as large as that of an edge. Taking $\nu = 1/3$, $T(0°) = 4\, T(90°)$, while as seen earlier it would be 2/3 less if the second derivative of the self-energy in equation (19) were not taken into account.

The first term in the integral contains the self-energy of the straight segment, the second term cancels out by symmetry of the line ondulation, and the third is proportional to the length difference ($\delta\xi$ - δx) acquired upon bowing out. Hence, the energy of the bulged line can be expressed as the sum of two terms

$$U_{\overline{MN}} = E_S(\theta_o)\,\delta x + \left(E_S(\theta_o) + E_S''(\theta_o)\right)(\delta\xi - \delta x) \tag{23}$$

By definition of the line tension, the two terms represent two distinct contributions: $[E_S(\theta)]$ towards the *shortening* of MN and $[E_S(\theta) + E''_S(\theta)]$ towards its straightening.

The above expression of the effective line tension and the corresponding geometrical *Wulff construction* may then be employed in order to approximate the equilibrium shape adopted by a dislocation loop as it is submitted for example under an applied stress τ. Given a dislocation character θ, as long as $[E_S(\theta) + E''_S(\theta)]$ remains positive, the line is at least metastable. When $E_S(\theta) + E''_S(\theta)$ is negative, the line is unstable and takes locally another character.

The latter situation is quite currently encountered in certain elastically anisotropic crystals (figure 15), and there is usually a very good fit between the shape predictions as derived from a numerical Wulff construction and observations in anisotropic crystals. Small loops however tend to be more rounded than large ones as the contribution of interaction terms, not considered here, increases. The absence of certain ranges of dislocation character along a dislocation line is manifested by some more or less sharp line discontinuities and/or by some preferred elongation of the dislocation line along certain orientations figure 15(*a*)).

When instabilities are controlled by anisotropic elasticity only, a range of dislocation characters is missing just because these orientations are outranged by others, themselves energetically much more stable.

Two properties are then worthy of mention:

- that along a dislocation line certain missing characters are forbidden has no implication on the mobility of these : it does not mean indeed that the motion of the rest of the dislocation is hindered, as is the case in core-controlled situations.
- the favoured orientations are in principle not rigorously straight with a smooth curvature

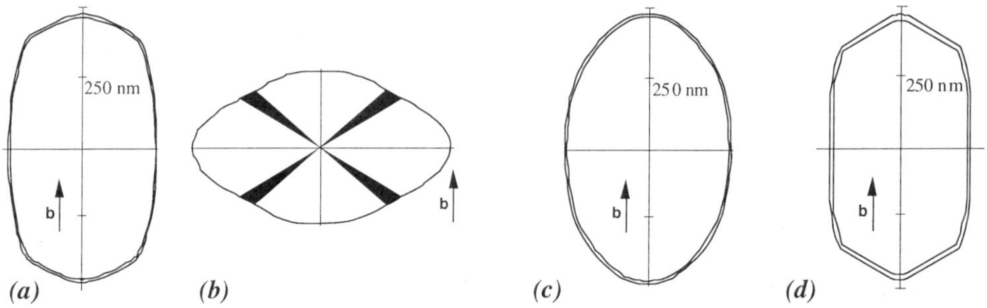

Figure 15. Analysis of equilibrium shapes of a dislocation loop in an elastically anisotropic crystal (Ni_3Al).
(*a*) A dislocation, **b** = [110] dissociated into two partials with 1/2[110] Burgers vector at equilibrium in the (001) plane under an applied shear stress of 80 MPa (the energy of the surface defect located in between the partials 90 mJ.m^{-2}). The loop is elongated in the screw orientation, yet its curvature remains finite everywhere.
(*b*) Plot of $E^{-1}(\theta)$ showing the character ranges where the dislocation line is unstable relative to other characters (embodied as black sectors).
(*c*) Same as in (*a*) but in the octahedral plane. At variance from (*a*), the dislocation loop does not exhibit line discontinuities any more.
(*d*) A hypothetical alloy with every crystal parameter identical to those of the preceding figures except for the fact that the screw and the ± 60° mixed characters are ascribed a discontinuity of the self-energy of -10%. As compared to the screw orientation in (*a*), the corresponding lines are straight.
(line irregularities which occur in places were due to softawre incompatibilties, lines are otherwise smooth).

that depends upon elastic anisotropy (figure 15(*a*)). Still, there can be marked discontinuities in the dislocation line.

By contrast, there exists crystals where the dislocation self-energy exhibits sharp drops along specific crystallographic orientations (such as, in semiconducting materials, the three <110> directions in the octahedral plane). These may originate from (i) slight atomic rearrangements (relaxation of atom positions along symmetrical line orientations in bcc crystals, reconstruction of dangling bonds in semiconductors...) and, on a somewhat wider scale, from (ii) the non planar splitting of the dislocation into several partials out of its slip plane. Such relaxed cores are obviously less mobile than the rest of the dislocation and, for this reason, are known as locks (e.g. 'Lomer-Cottrell', 'Hirth', 'Kear-Wilsdorf', 'Giamei' locks). These segments appear under the form of strictly rectilinear parts within an otherwise rounded dislocation (figure 15(*d*)).

In summary, that dislocation loops, or loop portions, deviate from a circular shape stems from the orientation-dependence of the self-energy. This may have spectacular effects on the shape of dislocations including sharp discontinuities of their line. Care should be taken however in discriminating between elasticity- and core-controlled discontinuities in that, as this will be discussed in the following, the latter imply some loss of dislocation mobility along specific dislocation characters while the former does not.

Interaction between dislocations : force and energy.

Consider two dislocations with same Burgers vector **b** lying in the same slip plane.

When the distance, ρ, between these dislocations scales with the average Frank network dimension R, the energy stored in the crystal amounts to $2E_S$. When the lines are within r_o, the atomic arrangement is close to that of a dislocation of strength 2 **b**; it is associated with a self-energy per unit length of about $4E_S$.

Then consider two dislocations with opposite Burgers vectors. The energy at a distance R is unchanged while that associated with the superimposition of the two cores is zero (the two extra half-planes have reconstructed the perfect crystal, which is equivalent to saying that the dislocations have *annihilated*). Hence, 'opposite' dislocations will tend to attract each other while like dislocations experience a repulsive interaction. It is also intuitively clear that dislocations with independent (orthogonal) displacement fields do not interact. The origin of the interaction is schematized in figure 16 which shows how compression and dilatation are enhanced by superim-

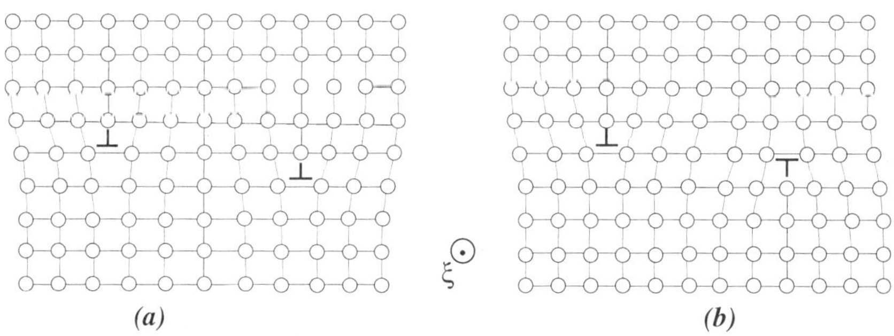

Figure 16. Interaction between two edge dislocations.
(*a*) Dislocations with same Burgers vector, the regions in compression overlap as well as the regions in dilatation which results in an increased stored energy. The configuration of minimum energy is when one dislocation does not experience the stress field of the other any more, that is, when the distance between the dislocations scales with the dimension of the Frank network.
(*b*) Dislocations with opposite Burgers vectors. The dilatation fields of the two dislocations tend to compensate mutually. The stored elastic energy is evidently a minimum when the two lines are located on the same vertical plane. The analogy between figures 16(*b*) and 3(*b*) should be noted.

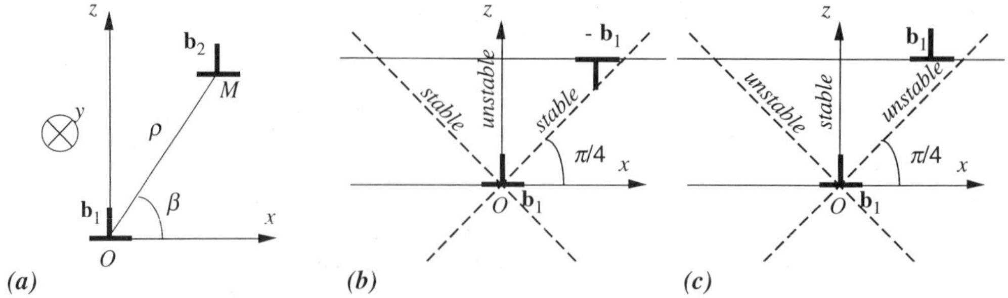

Figure 17. Interaction between two parallel infinite straight dislocations
(*a*) Definition of coordinates.
(*b*) Stability positions of an edge *dipole* ($\mathbf{b}_1 = -\mathbf{b}_2$) of dislocations. For $\beta = \pm\pi/4$, the dipole is stable while the configuration where the two dislocations are sitting on top of each other correspond to an energy maximum.
(*c*) Two identical edge dislocations ($\mathbf{b}_1 = \mathbf{b}_2$) exhibit one energetically stable position when located in the same vertical plane and another one when their distance is distant of the order of the dimension of the Frank network.

position of the displacement fields of two dislocations with same Burgers vectors (except for special arrangements discussed later), whereas the strain fields of two dislocations with opposite Burgers vectors tend to compensate each other reducing the elastic energy stored in the crystal.

We now address some important aspects of dislocation interactions that can hardly be intuitived. We shall restrict this introduction to the simple case of straight dislocations.

When a dislocation is submitted to a stress originating from a nearby dislocations, it experiences a configurational force given by the Peach-Koehler formula (equation (7)). The expressions of the radial and tangential components of this force, F_ρ and F_β (figure 16(*a*)) are given in Hirth and Lothe (1984). They are obtained by differentiation of the following general expression* of the interaction energy per unit length between one straight infinite dislocation located at the origin and a second one at M(ρ, β)

$$E_I(\rho,\beta) = \frac{\mu\,(\mathbf{b}_1\cdot\xi)\,(\mathbf{b}_2\cdot\xi)}{2\,\pi}\ln\left(\frac{R}{\rho}\right) + \frac{\mu\,(\mathbf{b}_1\times\xi)\,(\mathbf{b}_2\times\xi)}{2\,\pi\,(1-\nu)}\ln\left(\frac{R}{\rho}\right) - \frac{\mu}{2\,\pi\,(1-\nu)}\,\frac{[(\mathbf{b}_1\times\xi)\cdot\boldsymbol{\rho}]\,[(\mathbf{b}_2\times\xi)\cdot\boldsymbol{\rho}]}{\rho^2} \tag{24}$$

where the first two terms represent the dependence on ρ of the screw and the edge components of the dislocations, respectively. The third term which depends only on β, cancels when the plane that contains both dislocations coincides with their slip plane or else when one of the dislocations is screw. In this case the interaction energy reduces to the more familiar expression

$$E_I(\rho) = \frac{\mu}{2\,\pi}\,(b_1^s\,b_2^s + \frac{b_1^e\,b_2^e}{(1-\nu)})\ln\left(\frac{R}{\rho}\right) \tag{25}$$

where the superscripts *s* and *e* refer to the screw and edge components of the Burgers vectors, respectively. The analogy between the prelogarithmic term of equation (25) and that of equation (10), combined with equation (11), should be noted.

* Since energy is known within a constant term, it is indifferent to take the origin of energy when $\rho = R$, as is the case in expression (24), or else when $\rho = r_o$, in which case ρ and r_o appear in the argument of the logarithms as the numerator and in their denominator, respectively.

It is worth realizing that the nature of the interaction depends rather directly on the sign of the scalar product $\mathbf{b}_1 \cdot \mathbf{b}_2$: to a first approximation the interaction is attractive or repulsive when the scalar product between $\mathbf{b}_1$ and $\mathbf{b}_2$ is negative and positive, respectively. It can be in addition readily verified that in the case of two parallel dislocations with orthogonal Burgers vectors, expression (24) cancels only when one dislocation is screw and then of course, when its companion is edge. Though very useful in a number of cases, considering that in general two dislocations with orthogonal Burgers vectors experience no interactions may reveal quite crude an approximation in some instances.

As to forces, worthy of mention in dislocation analysis is the fact that the radial force is conversely proportional to ρ and that it is independent of β, while on the other hand, the tangential force vanishes when the plane containing the two dislocation lines is perpendicular to the slip plane except for two parallel screw dislocations (this property takes its origin in the third term of expression (24) of the interaction energy). However, the above properties may be greatly affected by elastic anisotropy. In certain materials such as ordered alloys, differences brought about by elastic anisotropy may take considerable importance as they provide an additional driving force for the locking processes of screw dislocations.

Two straight parallel dislocations. We illustrate properties of interaction in the particular case of two straight dislocations, with $\mathbf{b}_1 = \pm \mathbf{b}_2$, which are liable to glide in distinct planes both parallel to *Oxy* (figure 17(*a*)). Though this situation represents but a small fraction of the variety of interactions that may occur dynamically in a crystal*, it is an important case yet not too far from a number of situations occurring during deformation. The simplified situation of two parallel straight dislocations with collinear Burgers vectors, which we examine here, occurs for example under single slip conditions between dislocations originating from different sources. It is also quite important under single slip conditions, since a dislocation that is emitted by a source may 'bounce' somewhere in the crystal (upon pivoting around a fixed obstacle, for instance) and then return to the source where it interacts with freshly produced dislocations. Also, the dislocation may assume a dipolar configuration by cross-slip such as in figure 18.

It is useful to consider the stress field around one given dislocation instead of the interaction energy (expression (24)) and by application of the Peach-Koehler equation, one may then easily derive the force exerted by this dislocation on an approaching dislocation. Considering stresses indeed offers the advantage to compare directly with the applied resolved shear stress. For a screw dislocation ($\mathbf{b} // \xi //$ Oy), the only non-vanishing components of the stress tensor are the following shear stresses

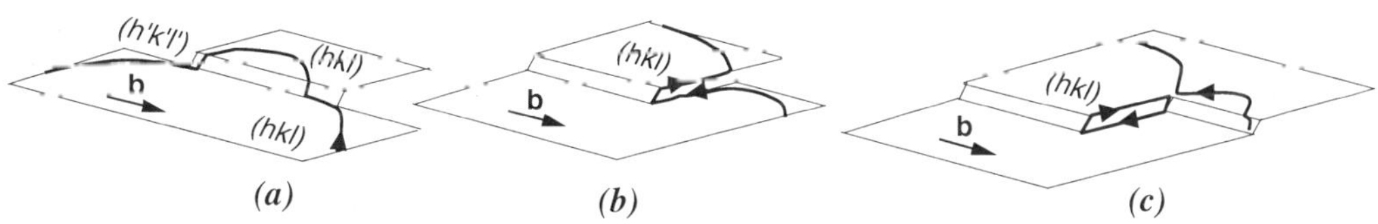

Figure 18. Example of the formation of a dipolar configuration.
(*a*) The dislocation undergoes a double cross-slip event (same as in figure 4(*c*)).
(*b*) Since the short segment located in the cross-slip plane is not glissile in the primary slip plane (*hkl*), a dipole of mixed character is trailed behind the two neighbouring branches of the still advancing segments.
(*c*) When the second branch of te expanding dislocation undergo the same double cross-slip jump, a prismatic loop of the type of figure 3, is formed. If there is a driving force for its coalescence, this loop will shrink by emission of point defects.

* dislocations moving under stress are in general neither straight nor infinite (their finite length that scales with the mean distance between obstacles in the crystal). Furthermore, the Burgers vectors of interacting dislocations are not necessarily collinear and their lines may not be parallel, as is the case when several slip systems are operating simultaneously or when a moving dislocation encounters a *forest* of pre-existing inclined dislocations (named *trees*).

$$\tau_{xz}^{s} = -\frac{\mu b}{2\pi}\frac{z}{x^2+z^2}, \qquad \tau_{yz}^{s} = \frac{\mu b}{2\pi}\frac{x}{x^2+z^2} \tag{26}$$

The stress tensor of an edge dislocation (**b** // **Ox**, ξ // **Oy**) includes also normal stresses

$$\tau_{xz}^{e} = -\frac{\mu b}{2\pi(1-\nu)}\frac{x(x^2-z^2)}{(x^2+z^2)^2}$$
$$\sigma_{xx}^{e} = -\frac{\mu b}{2\pi(1-\nu)}\frac{x(3x^2+z^2)}{(x^2+z^2)^2}, \qquad \sigma_{zz}^{e} = \frac{\mu b}{2\pi(1-\nu)}\frac{z(x^2+z^2)}{(x^2+z^2)^2} \tag{27}$$
$$\sigma_{yy}^{e} = \nu(\sigma_{xx}^{e} + \sigma_{zz}^{e})$$

Then the driving force for slip vanishes at $x = 0$ and $x = \pm y$ which then define loci of energy extrema (figures 17(*b*) and (*c*)). An edge dipole has a stable position for $\beta = \pm\pi/4$ while a screw dipole would be stable for $\beta = 0$ (unless cross-slip is possible, in which case it cancels out). On the other hand, whereas the most stable position for two like edge dislocations is when they are too far apart to interact, there is a secondary minimum at $\beta = \pi/2$. This result explains why upon annealing, like edge dislocations tend to gather in walls perpendicular to their common Burgers vector in order to form planar tilt boundaries (atomistically, this corresponds to the fact that one dislocation will tend to sit in such a way that its compressed volume overlaps as much as possible with the dilated volume of the second dislocation).

Comparison between expressions (26) and (27) shows that the force in the slip plane between edge dislocations is about 1.5 times stronger than between screws.

Finally, one quantity of interest in plasticity is the *passing* stress between two dislocations which, for two edge dislocations, writes

$$\tau_p = \frac{\mu b}{8\pi(1-\nu)}\frac{1}{z_o} \tag{28}$$

where z_o is the separation of neighbouring slip planes (the passing stress is a maximum in the edge orientation). It opposes dislocation motion during plastic flow. It increases as the density of dislocations builds up in the course of deformation since z_o should decrease with increasing dislocation density, more or less in the same way as the mean distance between dislocations $\rho^{-1/2}$, giving rise to the classical Taylor dependence of the flow stress to dislocation density

$$\tau_p = \eta\,\mu\,b\,\rho^{1/2} \tag{29}$$

which is consistent with observations (it has been determined experimentally that η amounts to about 1/3).

Interaction between a dislocation and point defects. Assuming that point defects (vacancies and interstitials) induce isotropic contraction or dilatation, they interact only with the hydrostatic component of the local stress which with regard to a dislocation, arises from its edge component only. Then, it is intuitively obvious that point defects should segregate heterogeneously at a dislocation. Typically, vacancies tend to increase their concentration in the region of compression while interstitials will segregate preferentially in the regions of dilatation (figure 19). Similarly, solute atoms should concentrate differently at an edge dislocation depending whether their volume is larger or smaller than that of the solvent atoms.

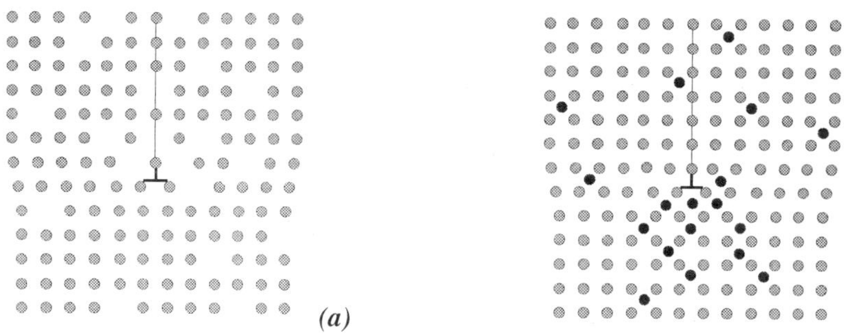

(a)

(b)

Figure 19. Schematic distributions of point defects around an edge dislocation (second-order effects and the possible anisotropic strain field of point defects are ignored).
(*a*) Distribution of vacancies, their concentration is the largest above the dislocation line, in the region of compression.
(*b*) Distribution of interstitials, interstitial segregation is enhanced in the volume of dilatation.

To a first-order approximation, one may consider that the more screw the dislocation character, the less sensitive the dislocation to interactions with point defects. However, the situation may be more complicated when point defects introduce non spherical volume changes. Interactions then involve shear components of the stress tensor, which in turn implies that screws cannot be ignored any more. Finally, second-order effects in stress are important but only very close to the dislocation line. They tend to cause vacancies to increase their concentration to any region of internal stress whether it is shear or hydrostatic at both edge and screw dislocations.

Dislocation pile-ups

Consider that a source starts to operate in a slip plane where dislocations will eventually meet a non shearable obstacle. Suppose in addition that dislocations cannot by-pass the obstacle by escaping from the primary slip plane by cross-slip or by climb. As the source continues to operate, it produces a pile-up group of dislocations whose equilibrium distribution establishes under an applied stress τ (figure 20).

Estimated in the approximation of a dislocation continuum, the total length of the pile-up is approximately related to the shear stress by

$$\Lambda \approx \frac{n\,\mu\,b}{\pi\,K\,\tau} \tag{30}$$

The position of the ith dislocation in the queue is given by

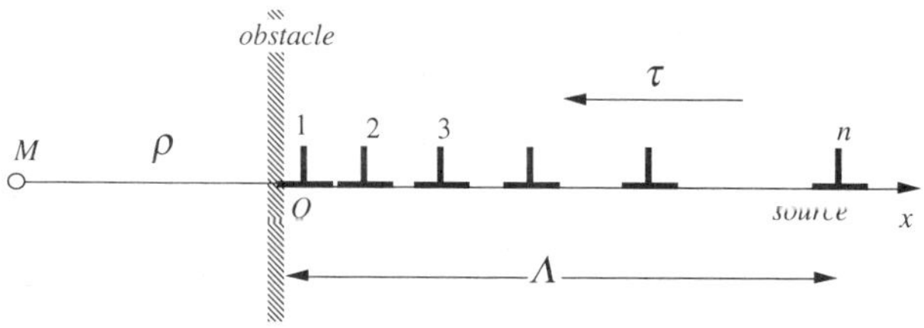

Figure 20. A planar queue of dislocations that have originated from a repetitive source and then piled against an impenetrable barrier at some point of their glide plane.

$$x_i \approx \frac{\mu\, b\, \pi}{16\, n\, K\, \tau}(i-1)^2 \tag{31}$$

At a point M located a large distance ahead of the queue (i.e. $\rho \sim \Lambda$), the shear stress exerted by the pile-up is approximately that which would be exerted by a hypothetical dislocation of strength $n\,\mathbf{b}$

$$\tau_{ld} = \frac{n\, \mu\, b}{2\, \pi\, K\, \rho} = \frac{\tau\, \Lambda}{2\, \rho} \tag{32}$$

The importance attributed to pile-ups in the literature is due in part to the complexity of solving its equilibrium properties, but it originates mostly from the fact that this group of dislocations induces a specific distribution of internal stresses with two important characteristics:

(i) *a pile-up concentrates a stress* that is proportional to the applied shear stress τ and to the number n of dislocations contained in the pile-up. In fact, near to the spearhead of the pile-up, at a distance $\rho \approx x_1$, the shear stress in the glide plane is

$$\tau_{sd} = n\, \tau \tag{33}$$

Thus, even under moderate applied stresses the stress at the leading extremity of the pile-up can reach the order of magnitude of the theoretical shear stress. Pile-ups acts as stress amplifiers thus providing potent means to
- make their leading dislocations break through strong barriers,
- initiate slip on the other side of the grain boundary onto which a pile-up is blocked,
- nucleate cracks.

(ii) *a pile-up exerts a back stress* on the source which is approximately that given by equation (32). This back stress opposes the applied stress to the point that it may eventually stop the source operation. Increases of flow stress with strain (*work-hardening*) observed during certain stages of plastic deformation would be due to the increasing difficulty of forming and expanding new loops against such back stresses.

It should be however noted that in order for the stress concentration to take place, the dislocation array should be rigidly ascribed to its slip plane. In fact, relaxation at the spearhead of the pile-up may occur by cross-slip or by activation of sources on secondary slip planes (in fcc metals, the latter process would require a limited number of dislocations in the array estimated to about 5).

Most of the observations of pile-up originate from surface analyses of etch-pits in deformed samples. The resolution of this technique is not enough though to inform on the dislocation density in the arrays and to which degree they are relaxed. The observation of pile-ups by TEM is relatively scarce, but this may simply result from the fact the degree to which pile-ups have relaxed upon release of the applied stress to such a degree that they are no more observable in thin foils.

Image forces. Crystals are not always large enough for surface effects to be negligible. In some particular instances, such as in whiskers or in thin foils suitable for electron microscopy, the outer radius R scales with the average distance between the dislocation and the free surfaces, which are in fact subjected to no stresses. We illustrate the concept of image forces in the simple

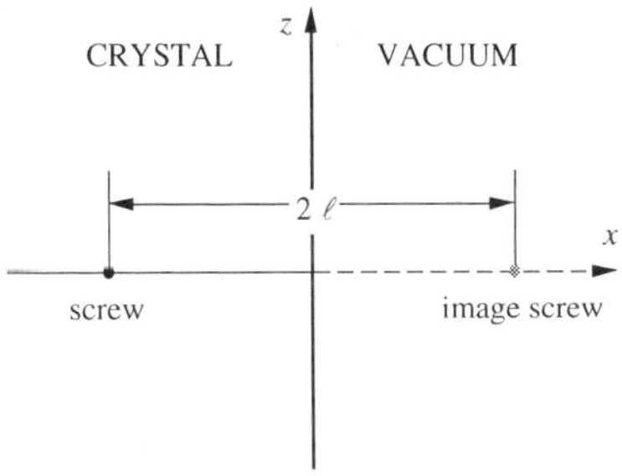

Figure 21. As a free surface experiences no stresses, the stress field around a dislocation located near a surface must be corrected for this boundary condition. Exact compensation between stress fields is achieved in the simplest case of a screw dislocation parallel to a free surface by introduction of an image screw dislocation located symmetrically with respect to the surface.

case of a screw dislocation parallel to a flat surface and located a distance ℓ apart from this surface. The boundary condition at the free surface is that $\tau_{xz} = 0$ (figure 21).

This is equivalent to superimposing the stress field of a hypothetical screw with opposite strength in the mirror position outside the crystal. The real dislocation thus experiences the stress field of its mirror companion which exerts a force which draws it towards the surface, given by

$$F_x = \frac{\mu b^2}{4 \pi \ell} \tag{34}$$

The concept of image forces is general but it gets rapidly untractable analytically as dislocation and surface configuration increase in complexity (for additional examples see Hirth and Lothe 1984). The case of straight dislocations inclined to a free surface is already complicated, that of curved dislocations and/or surface is hardly tractable. The principle of the calculation is the following : first superpose the image, then construct a stress field that cancels the remaining forces at the free surface, a procedure that introduces a few elastic singularities. The image force is however always such that it will tend to rotate the dislocation towards the configuration normal to the surface. This may be counterbalanced though by the dislocation tendencies towards net shortening so as to reduce its total energy and towards alignment along the screw orientation where self-energy per unit length is the least.

When vacuum is substituted for matter with elastic constant μ', perfectly stuck to the crystal that contains the screw dislocation of figure 21, then the condition on stress is that now it be continuous at the interface. This situation takes place when sample surfaces are coated with an oxide layer or with a specifically designed protection layer (such as a thermal barrier in turbine blades…). This problem has in addition received a renewed interest with the development of multi-layered advanced materials.

Displacements and rotations around dislocations : dislocation contrast

In analyzing dislocation properties, the most useful quantities are strains and stresses and except for the initial derivation of the strain and stress fields around a dislocation in isotropic elasticity, displacements **u** play a minor role. There is however one particular situation of interest, that is, the analysis of dislocations under diffraction contrast, where displacements provide the most appropriate information.

Diffraction contrast—as opposed to interference contrast which provides images of atom columns known as high resolution electron microscopy—is the most common mode employed to image dislocations. In the back focal plane of the microscope objective lens, the signal provided by a crystal is a diffraction pattern from which all diffracted beams but a selected one can be stopped by means of a diaphragm. The image then observed on the microscope screen is a two-dimensional map of the local variations of the diffraction conditions of the incident beam by one selected family of crystallographic planes, usually denoted by the vector **g** normal to these planes.

When, in the presence of a dislocation, a given family of crystallographic planes is undistorted in the direction of **g**, the sample appears as if it were perfect since then the interplanar distance d_{hkl} is unaffected by the dislocation and the diffraction condition is everywhere the same for this reflection. From this property originates a powerful method to determine the direction of Burgers vectors in crystals (their magnitude is however not directly accessible).

Briefly, the intensity produced in a given point of the bottom surface of a thin foil in an undefected area, in the direction of the diffracted beam, is given by the square of the modulus of

$$\phi_g = \frac{i\pi}{\xi_g} \int_0^t e^{-2 i \pi s z}\, dz \tag{35}$$

where ξ_g is some characteristic distance (named extinction distance), which depends on the choice of the reflecting plane through its structure factor, t is the crystal thickness, z is a crystal coordinate and s a parameter that measures in the reciprocal space the deviation from Bragg condition for the particular reflection **g** (both z and s are taken in the direction of the diffracted beam). Expression (35) of the amplitude of a diffracted beam is valid not too close from Bragg condition, that is, when the diffracted beam has a negligible probability to be diffracted back into the incident beam. This condition, known as the kinematical approximation (as opposed to the more complex dynamical condition in which transitions between incident and diffracted beams are allowed), is generally sufficient to study dislocation contrast.

In the presence of a source of distortion in the crystal, the amplitude of the scattered wave at the bottom surface of a thin sample becomes

$$\phi_g = \frac{i\pi}{\xi_g} \int_0^t e^{-2 i \pi \mathbf{g}\,\mathbf{u}}\, e^{-2 i \pi s z}\, dz \tag{36}$$

The contrast of a defect disappears when the first exponential in expression (36) is equal to unity everywhere in the crystal. This is achieved when $\mathbf{g}\cdot\mathbf{u} = 0$, which requires the knowledge of the displacement fields associated with each type of crystal defect. In elastically isotropic crystals, the displacement field of a dislocation is given by

$$\mathbf{u} = \frac{1}{2\pi} \left\{ \mathbf{b}\,\beta + \mathbf{b}_e \frac{\sin 2\beta}{4(1-\nu)} + \mathbf{b} \times \boldsymbol{\xi} \left(\frac{(1-2\nu)}{2(1-\nu)} \ln \rho + \frac{\cos 2\beta}{4(1-\nu)} \right) \right\} \tag{37}$$

where ρ and β have the same meaning as above. The above extinction condition becomes

(i) $\mathbf{g}\cdot\mathbf{b} = 0$, for a screw dislocation,

(ii) $\mathbf{g}\cdot\mathbf{b} = 0$ and $\mathbf{g}\cdot\mathbf{b} \times \boldsymbol{\xi} = 0$ for a non-screw dislocation.

Condition (ii) is much more restrictive than condition (i) since it is rigorously satisfied for one particular reflexion only, that perpendicular to the dislocation slip plane. For any other reflexion that satisfies only **g**·**b** = 0, the defect will show a residual contrast whose magnitude is determined by the coefficients of $\mathbf{b}_e$ and of $\mathbf{b} \times \xi$ in equation (37). This method of contrast analysis based on extinction is extremely powerful (figure 22).

The study of dislocations became greatly facilitated after the introduction of the so-called

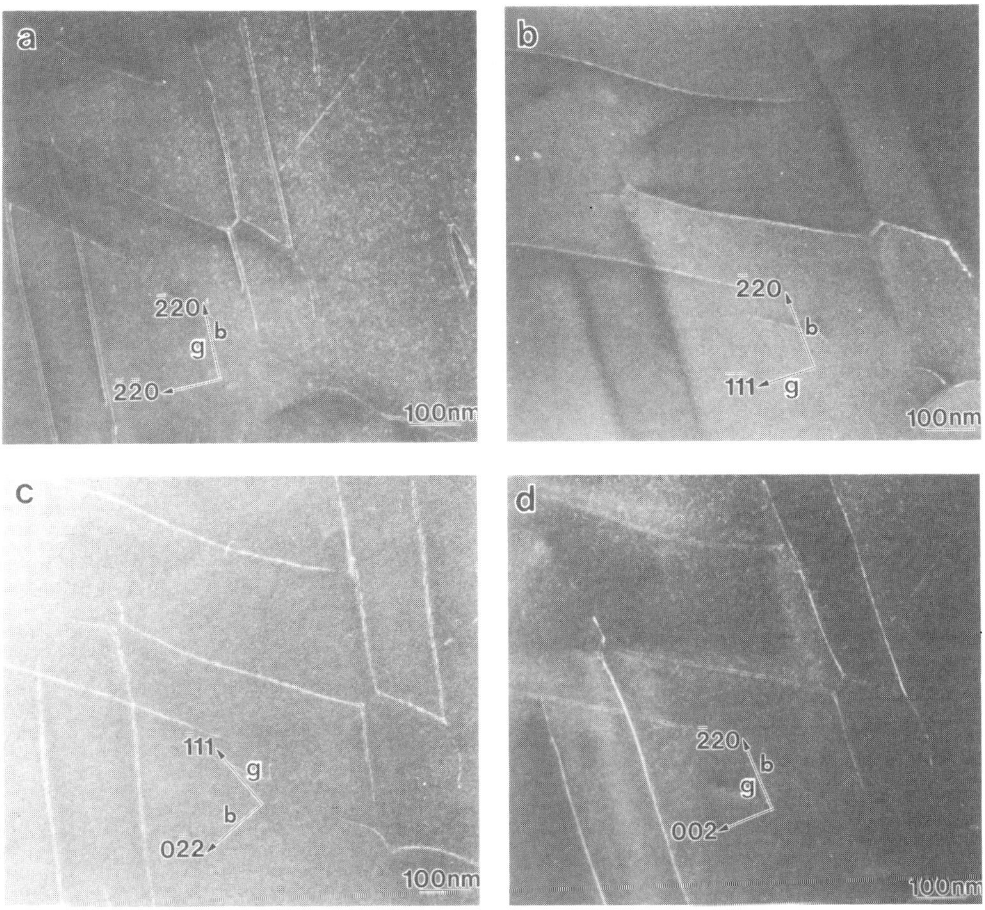

Figure 22. A contrast experiment in Ni_3Al deformed at high temperature, illustrating the method of Burgers vector determination. The observation is conducted under weak-beam conditions where the dislocation lines appear as white lines within a dark background. Two families of dislocations (1) and (2) are present of which some have interacted to form a junction (as shown schematically in figures 4(*c*) and 4(*d*) and figures 9(*a*) to (*c*)). Under $\mathbf{g} = \bar{2}20$ all dislocations are in contrast while in (*b*) dislocations (1) are out of contras under $\mathbf{g} = \bar{1}\bar{1}1$. Since the dislocations in this structure have a Burgers vector of the <110> type, these two conditions imply that $\mathbf{b}_1 = \pm[\bar{1}10]$. Since $\mathbf{b}_1$ must be different from $\mathbf{b}_2$ and since dislocations (2) are in contrast for $\mathbf{g} = \bar{1}\bar{1}1$ and $\bar{1}11$, then $\mathbf{b}_2 = \pm[\bar{1}01]$. The junctions, which are out of contrast under $\mathbf{g} = \bar{1}11$, have a Burgers vector $\mathbf{b}_j = \pm[0\bar{1}1]$, consistent with the rule of Burgers vector conservation.

It should be noted that while the image of dislocations (1) shows a double contrast which actually attests to their dissociation into two partials with Burgers vector $\mathbf{b}_1/2$. This holds true for dislocations (2) as well as for the junction as shown by the double contrast visible in figures 22(*d*) and 22(*b*), respectively. However, the dissociation plane of these is too inclined to the foil for the splitting to be resolved clearfly in the other orientations

In order to reach the appropriate Bragg conditions, this series of contrast experiment was conducted under several foil orientations so that the reflecting planes were orientation appropriately with respect to the electron beam (no more than a few degrees). In (*a*) the beam is close to the **B** = [001] direction, in (*b*) **B** is close to $[11\bar{2}]$, in (*c*) **B** = [211] and in (*d*) **B** = [110], which implies tilting the foil over rather large tilting angles. This enables one to identify the dissociation plane of dislocations (1) as the (001) plane or a plane in its close vicinity, since in figure 22(*d*) these dislocations show neatly a single line which stems from the fact that the dissociation plane is viewed edge-on (then the normal of the habit plane is given by the direction perpendiculat to the dislocation line in this figure.

weak-beam method. Still based on diffraction contrast, this method consists in setting the sample largely away from Bragg condition (large value of s), so that a diffracted beam cannot be formed except in very distorted volumes of crystal. The vicinity of dislocation cores is a region where local lattice rotations are able to compensate for Bragg conditions. The larger s, the closer this region to the dislocation line and the narrower the dislocation image (figure 22). With reasonably aligned modern instruments, a resolution of about 1 *nm* can be obtained under these conditions.

Problems of contrast identification may arise in elastically anisotropic crystals due to the extra displacements that are brought about in this case. Since the magnitude of the anisotropy-related displacements is modest relative to displacements near the core, a defect image obtained under weak-beam conditions with a sufficiently large deviation from Bragg condition s is in practice insensitive to elastic anisotropy.

Dislocation motion

Mobility in a Peierls potential. In order for a straight dislocation portion parallel to a close packed row, to slip from a stable, S_1, to an equivalently stable, S_2, position in the crystal (i.e. a translation over a vector of the Bravais lattice), it has to assume intermediate configurations of higher energy. For these particular line orientation and slip plane, such obstacles occur with a periodicity equal to the projection of **b** on the direction of motion (figure 23).

Motion may be achieved with the help of an external stress and of thermal activation. When slip occurs along close packed planes, the distances between the atomic planes adjacent to the slip plane are large. When passing from S_1 to S_2, the atomic structure is not much distorted with respect to the stable core configuration and the corresponding Peierls activation energy and forces against dislocation motion are small. Hence, it is also intuitively clear that the less closely packed the slip plane, the larger the Peierls force.

Considering the dislocation core as a continuous distribution of infinitesimal dislocations along the slip plane, and by application of linear elasticity to this region, the solution of the displacement field between S_1 and S_2 suggests that the core stretches over a width ζ that scales with

$$\zeta = \frac{d_{hkl}}{2\,(1-\nu)} \tag{38}$$

The form of ζ depends on the choice of the periodic Peierls potential experienced by the line along its path from S_1 to S_2; here, expression (35) results from a sine potential. In this particu-

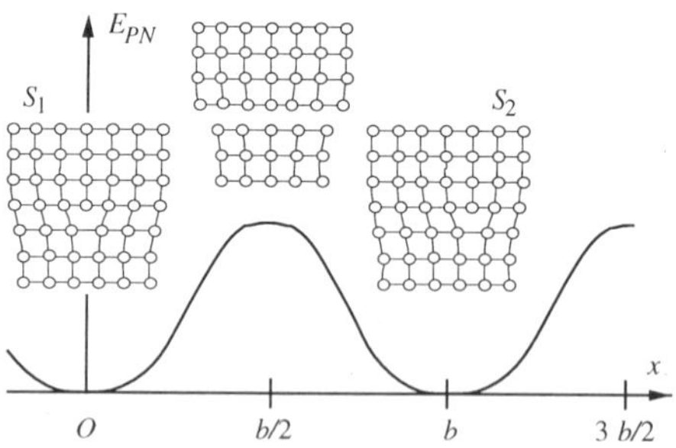

Figure 23. Schematic representation of the Peierls-Nabarro potential experienced by an edge dislocation in a cubic crystal. The shape of E_{PN} is arbitrary chosen as a sine function. The force (e.g. stress) that opposes dislocation motion is a maximum when the dislocation line is located $b/4$ apart from a Peierls valley.

lar problem, the use of a parametrized analytical approach to depict core properties under the approximation of linear elasticity provides an excellent physical insight of the overall physical mechanisms that take place during dislocation motion in a crystal. Atomistic simulations are in principle better adapted to this problem since they are able to account for non-linear effects as well as of core distortions out of the slip plane. They have been largely used to study specific situations.

The Peierls-Nabarro stress is defined as the minimum stress to move a rigid straight dislocation from one Peierls valley to the next. In the present case, it writes

$$\tau_{PN} = \frac{2\,\mu}{(1-\nu)}\, e^{-\frac{2\,\pi\,\zeta}{b}} \tag{39}$$

which indicates that the smaller the dislocation width, the larger the stress needed to move the dislocation. In fcc and hcp close packed metals the Peierls valleys are shallow and τ_{PN} is negligibly small. We shall see that this is even reinforced in close packed structures by the fact that dislocations may dissociate into sub-units (*partials*). When partials are sufficiently distant, they behave as isolated dislocations with regards to the lattice potential. Thus, their effective width is lessened and their mobility accordingly increased. In other structures less close packed than fcc or hcp and/or in structures with reduced symmetry, Peierls-Nabarro potentials tend to be more accentuated.

The above representation of the motion of a *rigid* straight dislocation in a Peierls potential is however not fully appropriate essentially because dislocations are not completely rigid. They indeed have means to minimize the Peierls corrugation by nucleation and propagation of kink pairs (figure 24).

When dislocation motion is controlled by Peierls valleys, the formation of a kink pair is thermally-activated with a formation energy of a few tenths of eV. Once a pair is nucleated, the dislocation velocity is determined by the velocity of the kinks in the direction of the Peierls valley. Kink velocity is generally fast so that dislocation mobility is controlled by kink nucleation. Since the probability to nucleate a kink is proportional to the number of sites available along the line, that is, to the length of unkinked dislocation, dislocation velocity may be determined by kink velocity when this velocity becomes too small (when lattice friction is large on kinks too).

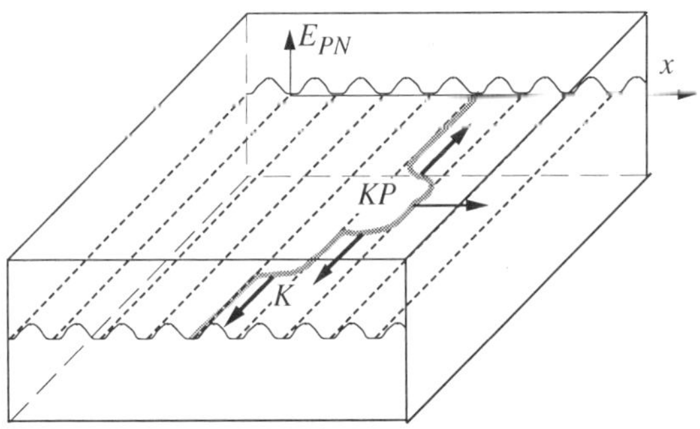

Figure 24. Motion of a dislocation whose character is dominated by a Peierls mechanism. The motion takes place by means of the nucleation of kink pairs (*KP*). The profile of the Peierls potential, as of figure 22, is represented schematically on the front face of the crystal. The valley bottoms are embodied by thin dashed lines. The dislocations tends to sit in these energy minima. The shape of kinks is determined by line tension. The isolated kink *K* was produced as member of a kink pair, its companion has already disappeared at the bottom surface. Indicated by arrows are the directions of motion of the various segments.

Lattice friction. The Peierls-Nabarro effect stems from the existence of favoured discrete lattice orientations at which dislocation portions are stabilized because their self-energy exhibits a marked drop. It is only one particular facet of the phenomenon of lattice friction in crystals. There exists other situations in which lattice opposes strongly to dislocation motion:

• in one category of situations, dislocations cores undergo locking transformations. These transformations are generally regarded as some non-planar rearrangement of the cores off the slip plane, of which some have been directly visualized. In fact the difference between locking effects and the Peierls-Nabarro effect is not drastic and distinguishing between these is probably not totally justified. The rearrangement is in both cases related to crystal symmetry but it may also depend on the nature of atom bonding, such as in covalent materials. In this case, atomic bonds are broken at the dislocation core except for certain orientations where dangling bonds disappear; this case is classified in the category of Peierls type effects. The distinction made between Peierls valleys and dislocation locks relies principally in the implications that locks may have on dislocation motion. In a Peierls-controlled motion, it is implicitly assumed that dislocations jump by steps of b and that kink pair nucleation is relatively easy, while the concept of locks usually refers to situations where, after being unlocked, dislocations may move over distances larger than b before they undergo their next locking transformation. The process of unlocking is often taken as rather difficult, it may nevertheless be thermally-activated.

• in other situations lattice friction is orientation-independent. This is the case for instance of slip in non close packed plane in close packed crystals (slip on {100}, {110}, {112} in fcc alloys, on {001} in ordered alloys with the $L1_2$ structure...).

Peierls-controlled motion together with slip controlled by directional locks are manifested by dislocation segmentation along the energetically favoured orientations while large isotropic friction results in fairly rounded dislocations segments (figures 15(*a*) (*c*) and figure 15(*d*), respectively).

One of the main consequences of the existence of pronounced Peierls potentials on material properties is the marked temperature dependence of their mechanical properties such as their flow stress. That kink pair formation can be thermally-aided results in a more or less pronounced decrease of the flow stress with temperature increase. When Peierls potentials are prohibitively high as in ceramics materials or in some intermetallic ordered alloys, plasticity is more or less completely impeded and this is one of the origins of brittle fracture.

Dislocation velocity. A method of measuring dislocation velocity has been initially developed by using etching to reveal the position of emergence of dislocation between successive stress pulses. More direct methods have been employed since, in which the whole dislocation line is imaged either by X-ray topography or else under *in-situ* straining experiments in a transmission electron microscope. The velocity of a dislocation may vary within a wide range of values. Velocities from a few atomic rows per second (i.e. 10^{-9} m.s^{-1}) up to approximately 1 km.s^{-1} have been measured by these techniques. The dislocation velocity in a crystal is limited by the transverse velocity of sound. Dislocation velocity depends in general:

- on dislocation character as indicated above in the previous section on Peierls effects.
- on the purity of the crystal as well as on the nature of obstacles,
- on temperature when dislocation motion is controlled by thermally-activated processes,
- on the magnitude of the applied stress.

Typically, at low and large stresses relative to the stress at which deformation begins to

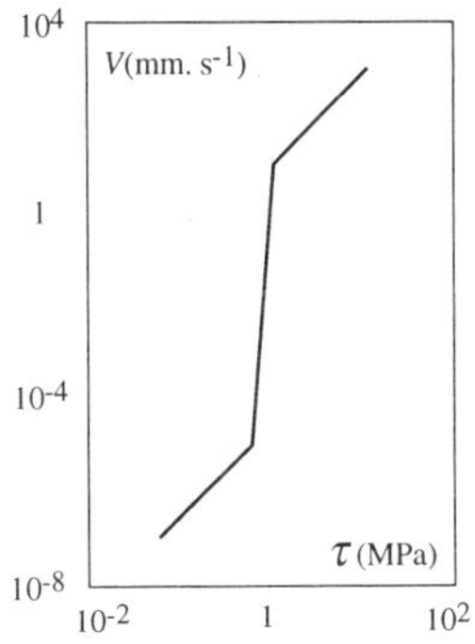

Figure 25. Schematic representation of the dependence of dislocation velocity on stress.

flow, the mean dislocation velocity V depends linearly on τ (figure 25). In the intermediate range of stresses, it obeys a power law dependence

$$V = V_o \left(\frac{\tau}{\tau_o}\right)^m \tag{40}$$

where V_o and τ_o are temperature dependent phenomenological parameters.

THE DISSOCIATION OF DISLOCATIONS

Dissociation and decomposition

Since to a first approximation the total self-energy per unit length of a given dislocation is proportional to b^2, it should be energetically favourable for a given dislocation to divide itself into a set of dislocations with Burgers vectors $\mathbf{b}_i$, as long as

$$b^2 > \sum_i b_i^2 \tag{41}$$

which is known as the b^2-criterion. For a twofold splitting, this is equivalent to saying that the angle between $\mathbf{b}_1$ and $\mathbf{b}_2$ must be acute, a crude* but convenient approximation. Two distinct cases of such reactions can be encountered

• the so-called *dissociation* which can be encountered in any crystal structure. In this case, the dislocation by-products are no longer unit but partial translations of the structure. They are necessarily connected to their companions by surface defects. These defects possess a surface energy γ, which prevents the partials from repelling one another to infinity. Under these conditions, it is clear that the motion of partials during deformation is correlated.

• the *decomposition* reaction whereby several, in general two, sometimes more, *unit* dislocations are formed (table 2). At variance from dissociation, the reaction by-products are not interconnected by surface defects so that they may follow independent paths in the crystal. The de-

* the orientation dependence of the self-energy and, more generally, the effects of line tension for the particular configuration that yields the splitting, should be taken into account.

composition of dislocations is rather specific to ordered alloys where the decomposition by-products have orthogonal Burgers vectors. Note that the dissociation of a quadruple into two triple ones such as in figure 10, which may occur in any crystal, can be regarded as a member of this category of reaction.

Table 2. Several decomposition reactions so far been identified by TEM in selected (ordered) structures.

structure	reaction	structure	reaction
$L1_2$, $L1_0$	$[110] + [1\bar{1}0] \leftrightarrow [100] + [100]$	$D0_{22}$	$[100] + [010] \leftrightarrow [110]$
$L1_0$	$1/2[110] + [001] \leftrightarrow 1/2[112]$	$D0_{19}$	$1/2[110] + 1/2[1\bar{1}2] \leftrightarrow [101]$
$L1_0$	$1/2[110] + 1/2[1\bar{1}2] \leftrightarrow [101]$	B2	$[100] + [010] \leftrightarrow [110]$
$L1_0$	$[100] + [001] \leftrightarrow [101]$	B2	$[110] + [001] \leftrightarrow [111]$
C11b	$[100] + [010] \leftrightarrow [110]$	B2	$[100] + [010] + [001] \leftrightarrow [111]$
fcc	$1/2[110] + 1/2[0\bar{1}1] \leftrightarrow 1/2[101]$	$L2_1$	$[110] + [001] \leftrightarrow [111]$

Not every dissociation scheme as predicted from the b^2 criterion is observed. This is essentially for two reasons:

• one is of course that the energy of the resulting planar defect is prohibitive. As the distance between two given partials is conversely proportional to γ, there is a critical energy beyond which the partials cannot be resolved by transmission electron microscopy and there is an even higher energy beyond which the notion of dissociation loses any physical meaning.

• there exist more favourable dissociation schemes. At first sight, this point may look quite trivial. In fact it is not. The subtleness that it involves deserves further attention for significant confusion is still present on this question. As long as comparison between possible configurations is made energy-wise, there is indeed not much difficulty in predicting the most favourable configuration unless the energy difference is not within uncertainties in evaluating the core energy of dislocations (see the discussion of equation (10)). What causes problem is that energy minimization is not a pre-requesite to this end. For instance, when the crystal is asked to deform plastically, the dislocations should exhibit a *mobile* configuration and this is not necessarily that of minimum energy. There is a profusion of examples of the conflict between the energy and the mobility criteria.

The origin of surface defects.

Why in a large majority of crystals, a moving dislocation leaves no surface defects in its wake, originates from the fact that its Burgers vector coincides with a unit lattice translation. The dislocation is then said to be perfect, as opposed to *partial* when it borders a surface defect. In quite a number of crystals, perfect dislocations tend to move at stresses lower than those required to form surface defects. This can be illustrated numerically as follows.
The elemental defected surface that would be created upon the motion over dx of a length ℓ of partial dislocation, represents a stored energy γds, that has to be less than the work of the applied stress τ needed to move the dislocation segment

$$\gamma ds < \tau b \ell dx \tag{42}$$

Assume τ is of the order of 10 MPa, as is the case in fcc alloy. The above inequality then precludes the formation of surface defects with a surface energy γ larger than a critical value of the order of 10^{-3} mJ.m^{-2}, which is smaller than the stacking fault energy of many common fcc metals and alloys. Stacking faults have been nevertheless encountered under overwhelming densities in Cu-Al alloys whose stacking fault energy can be lowered below the limit given by equation (42), upon addition of Al. In the same vein, stacking faults can be created in Si, where γ is moderate (50-60 mJ.m^{-2}), provided this material be deformed at sufficiently low temperatures. Then the stress at which Si flows becomes larger than the stress required to move companion partial dislocations independently (because of the great brittleness of Si, this usually requires that deformation is achieved under a confining hydrostatic pressure).

Surface defects : stacking faults and antiphase boundaries. A surface defects can be regarded as the boundary formed as the result of the impingement of two crystals with same crystallographic orientation but translated relative to each other by a partial translation of the Bravais lattice, **R**. The atomic stacking is then violated.

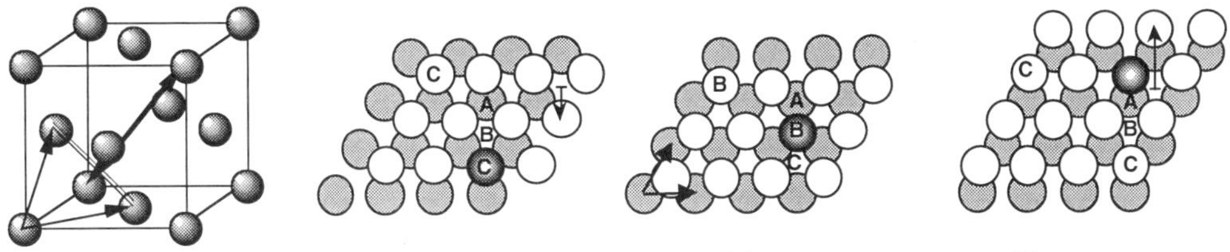

(a) *(b)* *(c)* *(d)*

Figure 26. A stacking fault in the fcc lattice.
(*a*) the fcc unit cell, represented by arrows are two of the three six shortest unit translations (**b** = 1/2<110>).
(*b*), (*c*) and (*d*) show the superposition of two consecutive atomic planes along the main diagonal of the cube (<111>). The unit translations in (*a*) are reported in (*c*) and the atomic sites of the stacking are indicated by A, B and C. In (*b*) the top layer is shifted by a partial translation (1/6<112>) towards the bottom of the figure. The layer is now nesting in C instead of B. In (*c*) the top layer again nests in C, but this required a translation twice as large as and opposite to that of (*b*). The two configurations are however crystallographically equivalent.

Consider the well-known case of fcc alloys (figure 26). As the result of a shear of the type **R** = 1/6<112>, the sequence

$$A_\Delta B_\Delta C_\Delta A_\Delta B_\Delta C_\Delta A_\Delta B_\Delta C \tag{43}$$

of the close packed planes is transformed into the faulted one

$$A_\Delta B_\Delta C_\Delta A_\nabla C_\Delta A_\Delta B_\Delta C_\Delta A \tag{44}$$

For obvious crystallographic reasons the corresponding stacking fault (SF) is ascribed to the {111} or octahedral plane, that is, the close packed of the fcc and related structures.

A stacking fault with finite extension is bordered by a partial named *Shockley partial* with **b** = **R** according to the configuration shown in figure 27.

Stacking faults and partial dislocations have a number of properties which are described in detail in the general references given in the introduction. It is worth emphasizing however, that an unambiguous knowledge of orientation and Burgers vector of a given dislocation is crucial in the case of partials too. This can be understood in a simple way by considering in figure 26(*b*) that reversing the displacement into -**R** shifts the upper layer onto A sites, that is, exactly on top

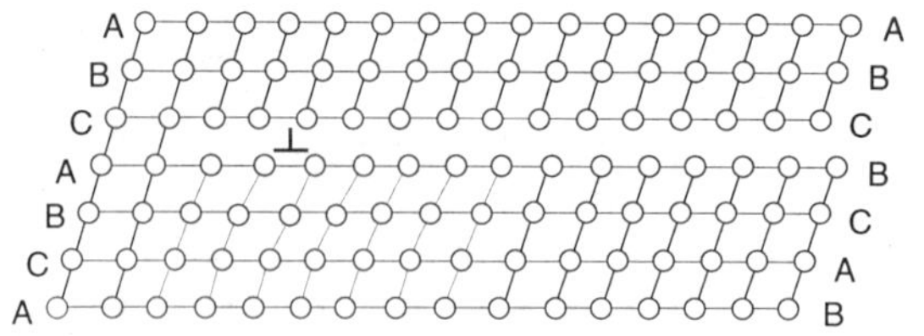

Figure 27. A stacking fault bordered on one edge by a Shockley partial in the fcc lattice. The perfect and the faulted stacking sequences appear on the left-hand side and on the right-hand side of the figure, respectively. Near the dislocation core the stacking fault configuration is distorted by the displacement field of the partial dislocation.

of the bottom layer. The resulting configuration is of course a very unfavourable defect energy-wise,which may nevertheless happen. In this case, the corresponding wrong configuration

$$A_\Delta B_\Delta C_\Delta A_{\Uparrow} A_\Delta B_\Delta C_\Delta A \qquad (45)$$

relaxes by an atomic rearrangement whereby an intermediate shuffle transforms (45) into

$$A_\Delta B_\Delta C_\Delta A_\nabla C_\nabla B_\Delta C_\Delta A \qquad (46)$$

in which the sequence is now violated twice while first neigbours distances are preserved. Configuration (46) is known as an *extrinsic* stacking fault (the alternative $A_\Delta B_\Delta C_\Delta A_\nabla C_\Delta A_\Delta B_\Delta C_\Delta A$ configuration is named intrinsic). Intrinsic and extrinsic faults have about the same energy, what differs is the mobility of the bordering partials—which are of the Shockley type in both cases—since for an extrinsic fault, this partial has to be zonal, with a core extending perpendicular to the slip plane, over two atomic planes.

Note that since a 1/6<112> translation belongs to one single octahedral plane:
- a dihedral stacking fault must involve a dislocation at its edge, named *stair-rod*, whose Burgers vector is $\mathbf{b} = \mathbf{R}_1 + \mathbf{R}_2$.
- upon cross-slip, a dissociated dislocation must recombine in order to dissociate into two Shockley partials that would be glissile in the cross-slip plane. This question of cross-slip has given rise to a profuse development since the late fifties and is still under debate.

Finally, there is a possibility to forming faulted loops in the fcc lattice, by condensation of point defects, yielding loops similar to those shown in figure 2. In case of vacancy segregation, this corresponds to the disappearance of one plane of the stacking, so that the displacement vector is now perpendicular to the fault plane (see figure 3(*a*)). In terms of atomic stacking, this is entirely equivalent to sequence (44) and the associated displacement vector **R** is in this case

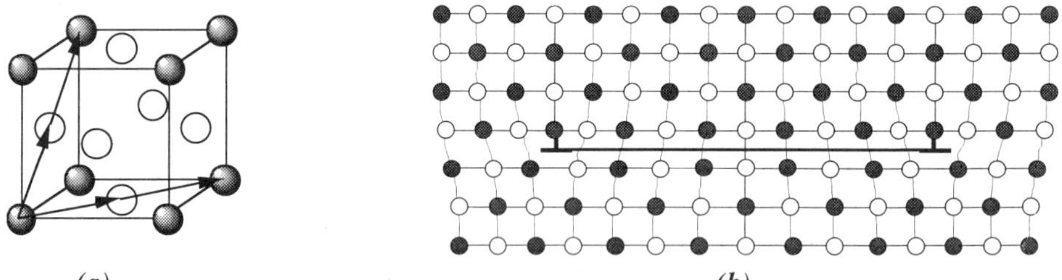

(*a*) (*b*)

Figure 28. Planar defects in an ordered lattice.
(*a*) the $L1_2$ unit cell (compare with figure 26(*a*)).
(*b*) an APB in an idealized ordered cubic lattice (embodied by a solid horizontal line). Along the APB plane, like atoms are in nearest neighbour positions, at variance from the perfect lattice. The APB is bordered by two partials that would be perfect dislocations if the lattice were disordered.

1/3<111>. In case of interstitial segregation (figure 3(*b*)), the resulting sequence would be (46) and the corresponding displacement vector is opposite to the previous one. Crystallographically, that identical stacking faults can result from two distinct displacement vectors results from the fact that shears over 1/2<112> and 1/3<111> are equivalent within a unit vector of the Bravais lattice (1/2<110>). All these properties can be established geometrically by considering the *Thompson tetrahedron* (see for instance Hirth and Lothe 1982).

Consider now the $L1_2$ structure shown in figure 28(*a*), which is based on the fcc stacking of figure 26(*a*). It is clear that the unit translation along the <110> direction is no longer 1/2<110> but twice that vector. Evidently, a shear over **R** = 1/2<110> does not affect the lattice *sites* though, as shown schematically in figure 28(*b*), it does affect the crystal *order*. Such a surface defect is named an antiphase boundary (APB).

One way to form APBs—and this is actually what happens when an ordered phase is formed upon cooling an alloy from the disordered state—is when order is nucleated in places at different lattice sites and then expands in the crystal. The resulting defect located at the boundary between two ordered domains is shown schematically in figure 29, which illustrates that, at variance from a stacking fault, an APB is not confined to a given habit plane and that there is no need of a stair-rod dislocation when the APB changes its habit plane.

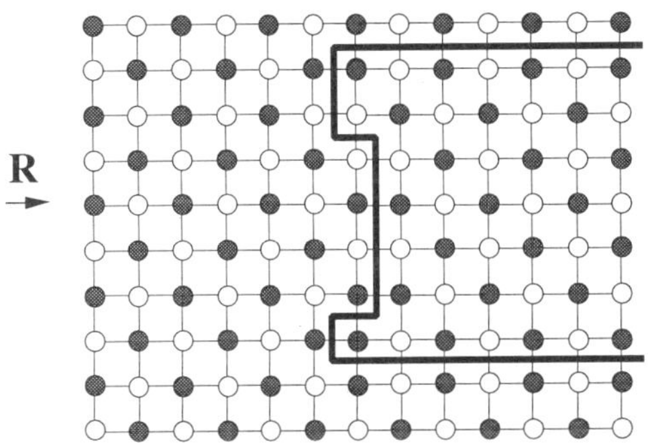

Figure 29. A non-planar APB formed in the same ideal cubic ordered structure as that of figure 28(*b*), upon the impingement of two ordered domains in which the atomic order was nucleated at non equivalent atomic sites. One of the two possible displacement vectors is indicated (the other would be normal to the displayed one). In a general way, when **R** is contained in the APB plane, the same APB configuration can be formed by a conservative shear (e.g. by a glissile partial). When **R** is inclined to the APB plane, the defect is a non-conservative one, it cannot result from a shear unless **R** is equivalent to another displacement within a unit vector of the Bravais lattice.

Measuring the energy of surface defects. The equilibrium distance, d_e, between two partials is attained when the back force that results from a widening of the fault is balanced by the repulsive radial interaction between partials

$$d_e = \frac{\mu}{2\pi\gamma}\left((\mathbf{b}_1\cdot\xi)(\mathbf{b}_2\cdot\xi) + \frac{\mu(\mathbf{b}_1\times\xi)(\mathbf{b}_2\times\xi)}{(1-\nu)}\right) \qquad (47)$$

which is a little more than $\mu\,\mathbf{b}_1\cdot\mathbf{b}_2/2\pi\gamma$, except in the screw orientation. This provides the only experimental means to determine the energy of surface defects (figure 30).

Dissociation distances are difficult to resolve under diffraction contrast below 1.5 nm and this is the order of magnitude of the limit of validity of equation (47). In fact, below that distance the lattice is so distorted by the bordering partials that it becomes vain to assimilate the surface defect to an ideal one. For instance, measurements of γ conducted on Ge by HREM on Z-shaped faulted dipoles have shown a deviation from the d^{-1} dependence below a dipole width of 5 nm

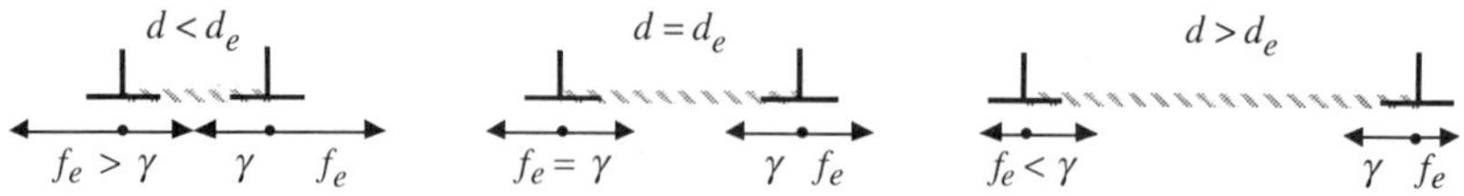

Figure 30. Measuring the surface energy of a planar defect. Two partials that border a planar surface defect are at equilibrium when their repulsive force balances exactly the back force that originates from surface tension. It is only under this configuration that the surface energy can be determined. The accuracy of the measurement may be limited by a number of factors such as
- the knowledge of the exact position of partials (which by definition of diffraction contrast never coincides with their images), this question is particularly critical when partials are closely spaced.
- the existence of high lattice friction that hinders the free relaxation of partials. In this case, the wider the splitting, the more sensitive the separation to friction.

(Chiang et al., 1980).

Separations of a few tens of nm are commonly encountered in materials. Crystals containing widely dissociated dislocations are not rare, but interaction forces are so weak in this case that the position of partials become very sensitive to local stresses originating from surrounding defects and to lattice friction. It is thus difficult to ascertain that they are under equilibrium. The range of distances within which γ can be safely determined lies within approximately 3 nm and a few tens of nm.

MICROSCOPIC PLASTICITY

Given a crystal structure, the reasons that dictate the choice of the slip direction are not at random. Simplifications do indeed occur in most crystals for the elemental displacement transported by a dislocation cannot be as large as possible because of the b^2 dependence of the energy per unit length of dislocation, which in turn implies that $\boldsymbol{b}$ is always taken amongst the smallest unit translations of a given lattice. Based on the Peierls-Nabarro analysis of the mobility of a dislocation, it became rapidly clear that dissociated dislocation should be amongst the most mobile ones, which immediately determines the slip plane as the habit plane of the dislocation core. Since the above notions of a perfect Burgers vector and of its potential dissociation schemes originate from straightforward crystallographic considerations, a lot of predictions on slip systems could be made soon after dislocations were introduced.

Slip systems

It is is quite common, especially in ordered multi-element crystals such as intermetallic alloys, and in ceramics such as oxides, carbides or nitrides (which are strongly ordered systems), that not all the members within a given crystal structure deform along the same slip direction. Since the choice of a slip direction is not independent of the choice of a slip plane (the plane that contains that favourable direction may be prohibitively corrugated for slip to proceed along the shortest direction), slip systems must be considered as whole.

The conditions that dictate the choice of a *slip system* remained very much unclear and they are still the object of active discussions. In fact, predicting a slip system in a given ordered structure is like playing russian roulette : it cannot be completely safe. Difficulties originate mainly from the absence of information on the actual core properties in a given alloy. In fact, core properties of a certain type of dislocation may differ dramatically from one alloy to another alloy within the same crystal structure. This is exemplified by the slip activity in $L1_2$ and B2 alloys.

The Chalmers-Martius criterion predicts that the most favourable slip systems correspond to those with the minimum shear strain b/d_{hkl}, that is, when distortions on atomic planes adjacent to the slip plane are a minimum.

This criterion works rather satisfactorily in many simple systems. In most fcc-related cubic (such as $L1_2$) and moderately tetragonal ordered structures ($L1_0$), this criterion encourages slip in the most dense direction in close-packed planes i.e. <110>{111}. In the $L1_2$ structure (figure 28(*a*)), the next favoured slip system is <110>{001} followed by <001>{001} and by <110>{110} which are equally likely (<001> is nevertheless the shortest unit translation). This hierarchy is generally obeyed at low temperatures, though a few exceptions have been reported in $L1_2$ alloys such as Pt_3Al and Fe_3Ge. However, the above criterion shows striking violations in most $L1_2$ alloys deformed at moderate temperatures, since then the second favoured <110>{001} system substitutes quite systematically for the expected <110>{111} and there is in addition some indication that even at higher temperatures <001>{001} could be dominant. In B2 alloys, some such as β-CuZn deform through <111>{110} while others like NiAl prefer <001>{110} in the same range of temperature.

On the other hand, explaining the nature of the slip system based on the ability of dislocations to dissociate happens to work rather well in B2 alloys. In β-CuZn the APB energy is such that <111> dislocations are dissociated into two partials with collinear 1/2<111> Burgers vectors separated by several nanometers ; separation is then sufficient for the distortion in the core vicinity to scale with 1/2<111> instead of <111>. By contrast, in NiAl <111> dislocations are essentially undissociated thus favouring less distorted cores such as that associated with <001>. In fact, <110>{110} operates in NiAl even before <111>{110}.

Nevertheless, considerations on dissociation should be made with some care:

- in Ti_3Al (hcp-related $D0_{19}$ ordered structure) for instance, deformation occurs in the **a** direction, the unit translation along the close packed rows of this hcp-related structure. However, depending upon the exact position of the shear plane, there are two possible APB energies in the prismatic plane and **a** dislocations may adopt two distinct separations which can be well distinguished by TEM. In both cases, the separation of **a** dislocations is enough for the mobility-controlling core distortion to be that of individual partials (**b** = **a**/2), *In situ* deformation tests have demonstrated that it is the least dissociated dislocation which by far exhibits the greatest mobility. It is therefore the fine structure of the core of **a**/2 partials which actually determines the mobility of the total dislocation. In other words, dislocation mobility is controlled by properties located at the level of their core (within say 2 to 3b). These properties rely essentially in the extent to which the core is liable to be spread *out of* the slip plane, by which they reduce their energy to the expense of course of their mobility. Intermetallic alloys have provided an extraordinary wide variety of core-controlled mobility behaviours.

- in close packed oxides such as the corundum (α-Al_2O_3) and the spinel ($MgAl_2O_4$) structures, perfect dislocations have the potential to split into Shockley-like partials which, in these structures too, constitutes the best approach to the prediction of core spreading based on the Peierls-Nabarro approach. Predictions were that in view of a specific property of Shockley in these structures, named the *synchro-shear* process, slip should be the easiest in the close packed directions of the close packed planes. Observations were in fact at variance from any expectation. From the current knowledge of metals alloys, one expect indeed that, of fcc and hcp systems, slip in planes other than the close packed planes should be preferably encountered in hcp-related systems (many metals and alloys indeed prefer to deform in the prismatic planes). It is in fact in α-Al_2O_3 that slip occurs in the basal plane while in $MgAl_2O_4$ the <110>{110} system shows the lowest critical resolved shear stress of all, then the CRSS for <110>{111} and <110>{001}slip are about twice that of <110>{110} and this holds true down to room temperature.

The locking of dislocations in ordered alloys

Self-locking is certainly a property which, although not exclusive to ordered alloys (for instance, Lomer and Hirth locks ib fcc alloys), is very well represented in this category of materials, this under quite a variety of locking mechanisms. In most cases the ability for dislocations to self-lock indeed originates from ordering. This is due to the fact that in most structures, there is a possibility that dislocations dissociate under the following generic dissociation reaction

$$2\,\mathbf{b} \rightarrow \mathbf{b} + \text{APB} + \mathbf{b} \tag{48}$$

whereby

(i) the Burgers vectors of the partials are collinear, thus providing a wide degree of freedom for cross-slip locking,

(ii) though the energy of an APB is certainly crystallographically anisotropic, an APB is not necessarily confined to a specific plane as stacking faults are, and this yields a variety of configurations dissociated by glide as well as by climb.

The Kear-Wilsdorf configuration. In the $L1_2$ structure, the APB energies on the $\{001\}$ cube and on the $\{111\}$ octahedral planes are of similar magnitude with the former yet smaller than the latter. Energy-wise, dissociation in the cube plane is encouraged by the existence of a torque term that originates from elastic anisotropy. On the other hand, lattice friction is large and very modest on the cube plane and on the octahedral plane, respectively. Thus given the dissociation according to reaction (48), dislocation motion is determined by the influence of the current conflicting effects between stability and mobility. Cross-slip on the cube plane is energetically favoured, thus opposing the general trends of dislocations to slip in the octahedral plane where they experience the least resistance from the lattice. The larger the probability of cross-slip, the larger the fraction of immobilized screw segments, increasing the resistance to plastic flow. The portions of dislocation line immobilized by cross-slip dissociation in the cube plane are known as Kear-Wilsdorf locks (figure 31). As cross-slip may occur along the line of an expanding dislocation at varied locations and times, it comes that immobile portions must be interconnected by mixed segments gliding in the octahedral plane which, by definition, are kinks.

Several combinations of Kear-Wilsdorf locks and kinks have been identified in the $L1_2$ structure, of which some are represented in figure 31. They are at the origin of the atypical me-

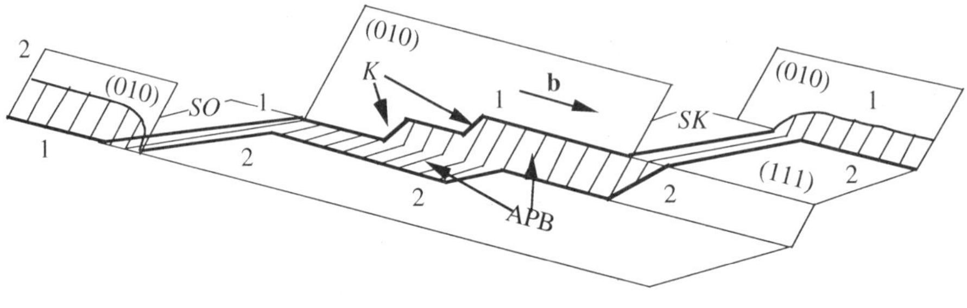

Figure 31. Summary of configurations identified by weak-beam TEM in the $L1_2$ structure.
The Kear-Wilsdorf (*KW*) segments are the screw portions actually dissociated in the cube plane. They are sessile with respect to slip in the primary octahedral plane. The central screw segment contains a portion that straddles the (111) and (010) planes, known as an *incomplete KW*.
SK is a regular macrokink that connects two *KW* segments, *K* are elementary kinks affecting one superpartial only.
SO is a switch-over macrokink located in between two *KW* segments (the leading/trailing role of the superpartials is inverted : the sequence 1/2 is transformed into 2/1).
SO and *SK* result from different processes of formation that occur dynamically during deformation.

Table 3. Sessile screw configurations identified by WB TEM in various intermetallic structures and the generalization of the notion of Kear-Wilsdorf locked configuration.

Structure	Burgers vector	Slip plane	Sessile configuration
$L1_2$	<110>	{111}	{111} {001}
$L1_0$	<011]	{111}	{111} {001}
$D0_{22}$	<110]	{111}	{111} {001}
$D0_{19}$	**a** = $1/3\langle 11\bar{1}0\rangle$	prismatic	$\{1\bar{1}00\}$ {0001}
B2	<111>	{110}	{112} {110}

chanical properties of $L1_2$ alloys in the domain of temperature where the flow stress depends positively upon test temperature.

Though not unexpected, it is remarkable that the notion of a Kear-Wisldorf configuration can be generalized to a number of ordered structures as shown in table 3. It is worth re-stating that the reasons of its presence rely on the potential of dislocations to dissociate based on reaction (48).

Climb dissociation

Dissociation by climb of the constitutive partials is poorly documented. It has been identified in fcc alloys. In ordered alloys, it is active at high temperature even as a means of dissociation still according to reaction (48). Climb dissociated configurations consist in two partials located on top of one another, in a plane in general perpendicular to the slip plane.

Provided that a dislocation has some edge component, climb dissociation helps minimize the interaction energy, but this is a diffusion-controlled process whose occurrence is unlikely at moderate and low temperatures. Amongst the many instances of climb dissociation ($L1_2$, $D0_{19}$, $D0_{22}$, B2,...) that of β-CuZn is interesting for it occurs at unusually low temperatures. There is no evidence that the process of climb dissociation, which actually provides strong obstacles against dislocation motion, ever contributes to the flow stress anomaly in the largest majority of ordered alloys. However, the existence of climb dissociation in β-CuZn, even after deformation at room temperature and of course up to the temperature of flow stress anomaly (~ 200°C), attests to an already active diffusivity, which itself is thought by some authors to be at the origin of the atypical mechanical poperties of this alloy.

CONCLUDING REMARKS ON MACROSCOPIC PLASTICTY

Under constant strain rate, the stress-strain curve of a material departs from linearity at some point which marks the beginning of plastic flow. The critical resolved shear stress is some ill-defined stress located near that limit. The CRSS can be considered as the minimum stress to ensure both the multiplication and the long-range propagation of dislocations throughout a sample, that is, against a distribution of obstacles of various strengths. Beyond the CRSS, the stress-strain curve exhibits a finite and in general positive slope, named the *work-hardening* rate, indicating that obstacles keep accumulating with strain. The CRSS lies evidently somewhere between its theoretical limit of $\mu/10$ and the shear stress for glide in the presence of no obstacle other than the lattice resistance, e.g. the Peierls-Nabarro stress. This lower limit is not necessarily modest as shown by a number of systems where plasticity is controlled by the behaviour of individual dislocations, itself determined by certain properties of their cores.

Mechanical tests other than under constant strain rate (such as creep, pure shear, impact resistance, fatigue under its several modes...) are conducted in order to study materials under the many kinds of mechanical sollicitations that they may be exposed to in real life. For these tests of course the CRSS ceases to be the relevant crystal property.

The difficulty in describing a given mechanical behaviour relies on achieving a sound statistical treament of dislocation motion within a distribution of obstacles. The nature and the strength of the interactions, which determine the mean dislocation velocity, must be known. In view of the Orowan equation (4), the density of *mobile* dislocations is needed too. The agreement between basic descriptions of plasticity and experiments is good in many instances such as in the plasticity of pure metals, of alloys and of particle-strenghtened systems. Amongst many other properties, the different stages of the stress-strain curve of fcc metals, the strong temperature and orientation dependencies of some hcp and bcc systems, the flow stress sensitivity to particle size and distribution, high temperature creep are well-explained properties. Lots of problems remain however unsatisfactorily solved amongst which those relying on the statistical treatment of deformation are suffering from the inherent heterogeneity of microscopic slip. Deformation occurs in *slip bands* where local strain can amount to values significantly larger than the (averaged) macroscopic strain. Work-hardening rates in metals are typically of the order of 10^{-2} to $10^{-3}\mu$ and theories are available that account fairly well for these, however there is nothing yet that can explain values of μ / 10 or more as encountered in certain ordered alloys.

The resistance of ductile materials to deformation has never ceased to be of great technological concern and the history of metallurgy has established that one could design high strength materials with a total lack of knowledge of microscopic mechanisms. With the increasing demand of 'advanced materials', research on mechanical properties of materials is increasingly founded on the consideration of basic elementary processes.

Acknowledgement

The author wishes to express his grateful thanks to Dr Joël Douin (LEM) for stimulating discussions, careful reading of the manuscript and for his computer simulations of the shapes of dislocation loops under anisotropic elasticity shown in figure 15.

REFERENCES

Chiang, S.W., Carter, C.B., and Kohlstedt, D.L., 1980, *Phil. Mag.* A,42: 103.

Friedel, J., 1964, "Dislocations", Pergamon Press, Oxford.

Friedel, J., 1980, "Physics of Modern Materials", 1:163.

Haasen, P., 1986, "Physical Metallurgy", Cambridge University Press, Cambridge.

Hirth, J.P., and Lothe, J., 1982, "Theory of Dislocations", John Wiley & Sons, New York.

Hull, D., 1965, "Introduction to Dislocations", Pergamon Press, Oxford.

Kirchner, H.O.K., 1984, "Dislocations 1984", P. Veyssière, L. Kubin and J. Castaing, eds., Editions du CNRS, Paris, p. 53.

Kittel, C., 1976, "Introduction to Solid State Physics", John Wiley & Sons, New York.

Nabarro, F.R.N., "Dislocations in Solids", Elsevier, Amsterdam, volumes 1 to 10.

Weertman, J., and Weertman, J.R., 1966, "Elementary Dislocation Theory", The Macmillan Company, New York.

MECHANICAL PROPERTIES OF TiAl AND TiAl-BASE ALLOYS

Kyosuke Kishida, Shigenobu Yokoshima, David R. Johnson, Haruyuki Inui and Masaharu Yamaguchi

Department of Materials Science and Engineering
Kyoto University
Sakyo-ku, Kyoto 606-01, Japan

INTRODUCTION

Titanium forms three intermetallic compounds with aluminum. They are Ti_3Al, TiAl and Al_3Ti with crystal structures of the $D0_{19}$, $L1_0$ and $D0_{22}$ types, respectively. They are superlattice structures based on geometrically close-packed h.c.p. and f.c.c. structures. Within the last decade, there has been an enormous increase in the research and development activity on these titanium aluminides. The reasons for these titanium aluminides to have been chosen as research targets are at least two-fold, firstly they have attractive properties as high-temperature structural materials of high strength-to-weight ratio and good oxidation resistance, secondly in such superlattice compounds, some of the deformation modes for the corresponding disordered structure are often operative, and thereby some ductility is expected to occur even at room temperature. Recent enormous increase in the research and development activity on the titanium aluminides is clearly reflected in the rapid increase in the number of papers on titanium aluminides presented at recently held international conferences and symposiums. The readers are recommended to refer to proceedings of these international meetings [1-15] to see the trend of the research and development in the field of not only titanium aluminides but also other intermetallic compounds.

Of the three titanium aluminides, Ti_3Al is the material which has been investigated over the longest period. Various Ti_3Al-base alloys have been developed to achieve specific strength and stress rupture properties which exceed those of some nickel-base superalloys and recent high-temperature titanium alloys (for reviews see [16]). Ti_3Al has a wide composition range and exhibit an order-disorder transformation in the temperature range of 1100~1200°C depending on composition. In the Ti-Al phase diagram, the disordered and ordered Ti_3Al is referred to as the alpha and alpha-2 phases. We use these terms henceforth.

The deformation behavior of Al_3Ti has been studied first in 1987 [17] and it has been shown that its major deformation modes are twinning of the type {111}<11$\bar{2}$], which does not disturb the $D0_{22}$ symmetry of the lattice, and slip on [001}<110], where the brackets used follow the convention introduced to differentiate the first two indices from the third (non-equivalent) index of the tetragonal structures [18]. However, at the same time, the compound has been revealed to be very brittle at room temperature [17]. Since then, macroalloying to induce the $D0_{22}$-to-$L1_2$ structure alteration has been attempted to improve the room-temperature ductility of the compound. Indeed, some cubic $L1_2$ compounds formed by alloying Al_3Ti with Cr and Mn have been found to exhibit good compression ductility and some tensile ductility at room temperature [19-23]. However, they are still too brittle in tension for their structural applications to be considered. An intensive review on Al_3Ti and its $L1_2$ variations is available in [24].

TiAl possesses a wide composition range. However, it extends primarily on the Al-rich side. Thus TiAl compounds with nearly stoichiometric or Ti-rich compositions exhibit a two-phase microstructure composed of the TiAl phase and a small volume fraction of the alpha-2 phase. The so-called TiAl compounds to which a recent enormous increase in the research and development activity has been devoted, are such two-phase alloys rather than the Al-rich TiAl compounds with a TiAl single-phase structure since the former alloys are more ductile and tougher than the latter compounds. The TiAl phase is referred to as "gamma phase" in the Ti-Al phase diagram. Thus, we call the two-phase TiAl alloys "gamma alloys" since their major constituent phase is the gamma phase. This article will concentrate upon reviewing the recent progress made in understanding the fundamentals of the mechanical properties of the gamma phase and gamma alloys.

GAMMA ALLOYS

Since two of the authors have published some reviews on the deformation behavior of gamma alloys with the gamma/alpha-2 lamellar structure in proceedings of some recently held international intermetallics meetings including NATO ASI Series E: Applied Sciences - Vol.213 [25], this section aims at reviewing the recent advances in understanding the mechanical properties of this class of materials avoiding overlap between the present and previous reviews. The major constituent phase of gamma alloys is the gamma phase with a nearly equiatomic composition. However, it is impossible to grow single crystals of single-phase material having less than 53 at.% Al due to complexities in the Ti-Al phase diagram, and therefore the details of mechanical properties of the major constituent phase of gamma alloys can not be easily investigated. To overcome this road-block we have introduced a new approach for studying the basic mechanical properties of gamma alloys and their major constituent phase by producing polysynthetically twinned (PST) crystals with near-equiatomic compositions. First their microstructure and mechanical properties in uniaxial tension and compression will be reviewed.

Microstructure and Mechanical Properties of PST Crystals in Uniaxial Tension and Compression

Microstructure. When gamma alloys with near-equiatomic compositions are prepared by usual ingot-metallurgy methods, ingots with a lamellar structure which is composed of lamellae of gamma and alpha-2 phases are obtained. The orientation relationship between the two phases is such that a close-packed plane and close-packed directions on the plane in the gamma phase are parallel to the corresponding plane and directions in the alpha-2 phase. However, three $<11\bar{2}0>$ directions on the (0001) basal plane of the hcp alpha phase are all equivalent while in the gamma phase with the tetragonal $L1_0$ structure, one of the three close-packed directions ($<1\bar{1}0]$) on a {111} is not equivalent to the other two close-packed directions ($<01\bar{1}]$). Thus, when the close-packed plane in the gamma phase parallel to (0001) in the alpha phase is assigned to be (111), the gamma phase can be formed in six possible orientation variants corresponding to the six possible orientations of $[1\bar{1}0]$ on (111) in the gamma phase with respect to $<11\bar{2}0>$ on (0001) in the alpha phase [26]. This indicates that a domain structure will result in the gamma phase in gamma alloys in the lamellar form. Figure 1 shows a schematic drawing of a PST crystal and its domain structure.

Geometry of specimens and gamma-phase variants. Geometry of specimens cut from PST crystals for uniaxial tension and compression is always such as that shown in Fig.2(a) and (b)[27,28]. A_1-N and A_2-N in Fig.2(b) represent loading axis orientations. Specimens with orientations A_1-D_1 have the P face (Fig.2(a)) parallel to $\{11\bar{2}\}$ while those with orientations A_2-D_2 have the P face parallel to $\{1\bar{1}0\}$. For orientations A_i and N, the loading axis is parallel and perpendicular to the lamellar boundaries, respectively. In specimens with B_i, C_i and D_i orientations, the lamellar boundaries are inclined to the loading

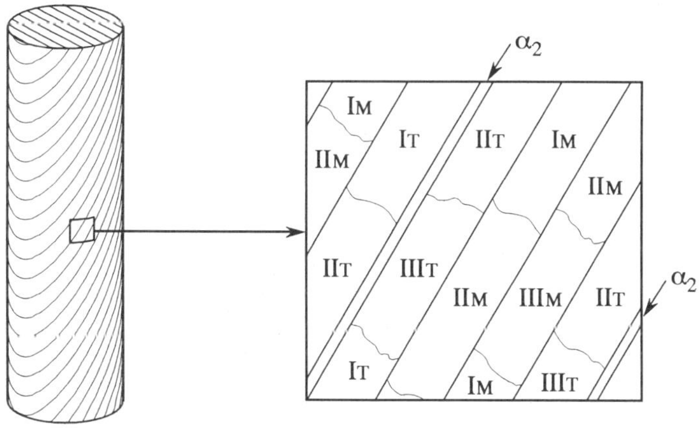

Figure 1. PST crystal and its domain structure

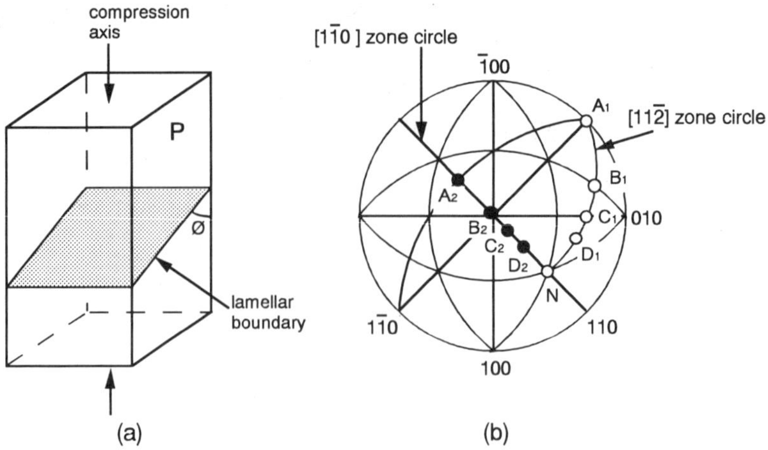

Figure 2. (a) Geometry of compression specimens and (b) compression axis orientations [27]. Geometry and loading axis orientations for tensile specimens are exactly the same as those for compression specimens.

axis at angles ϕ=31°, ϕ=51° and ϕ=68°, respectively.

Gamma-phase variants in tension and compression specimens are distinguished by taking selected-area electron diffraction patterns from thin foils cut parallel to their P face [27,28]. For specimens with orientations A_1-N, domains showing the [11$\bar{2}$] net pattern were assigned to be those of variant I and those showing the [1$\bar{1}$0] net pattern to be of variant I for specimens with A_2-N. Orientations of loading axes in Fig.2(b) is given with respect to the domains of variant I. Variant II and variant III are defined to be those obtained by rotating the variant I by 120° and 240° about [111], respectively. Other three variants are defined to be twin variants of variants I, II and III. The three matrix variants and the three twin variants are referred to as I-M, II-M and III-M and I-T, II-T and III-T henceforth.

Deformation modes. Twinning on {111}<112], which does not disturb the $L1_0$ symmetry of the lattice, slip along <110] (ordinary slip) and slip along <011] (superlattice slip) can be operative in the gamma phase. Deformation modes operative in the six different variants in the gamma phase in PST crystals with a composition of 49.3 at.% Al have been determined by examining their deformation structures after deformation in compression to about 2% strain. The results for the three typical orientations A_1, B_1 and N are summarized in Table 1 [29,30].Operative modes for I-T, II-T and III-T are respectively identical with those for I-M, III-M and II-M from simple crystallographic reasons.

Critical resolved shear stresses (CRSS) for these deformation modes can not be determined since PST crystals are not single crystals. However, the relative ease of the three deformation modes can be deduced based on the observations of deformation modes for orientation N where all orientation variants have the identical Schmid factor for each of the three deformation modes, namely 0.272 for both ordinary and superlattice slip and 0.157 for

Table 1. Deformation modes in PST crystals of a composition of 49.3 at.% Al with orientations A_1, B_1 and N deformed in compression at room temperature [29,30]

Variants	A_1	B_1	N
I-M	$(\bar{1}11)[0\bar{1}1]$ $(1\bar{1}1)[\bar{1}01]$	$(111)[\bar{1}10]$	$(\bar{1}11)[\bar{1}\bar{1}0]$ $(\bar{1}11)[\bar{1}1\bar{2}]$ $(1\bar{1}1)[\bar{1}\bar{1}0]$ $(1\bar{1}1)[1\bar{1}\bar{2}]$
II-M	$(\bar{1}11)[110]$ $(\bar{1}11)[\bar{1}1\bar{2}]$ $(11\bar{1})[\bar{1}10]$	$(111)[10\bar{1}]$	$(\bar{1}11)[\bar{1}\bar{1}0]$ $(\bar{1}11)[\bar{1}1\bar{2}]$ $(1\bar{1}1)[\bar{1}\bar{1}0]$ $(1\bar{1}1)[1\bar{1}\bar{2}]$
III-M	$(1\bar{1}1)[110]$ $(1\bar{1}1)[1\bar{1}\bar{2}]$ $(11\bar{1})[1\bar{1}0]$	$(111)[1\bar{1}0]$ $(111)[\bar{1}\bar{1}2]$	$(\bar{1}11)[\bar{1}\bar{1}0]$ $(\bar{1}11)[\bar{1}1\bar{2}]$ $(1\bar{1}1)[\bar{1}\bar{1}0]$ $(1\bar{1}1)[1\bar{1}\bar{2}]$

{111}<112] represents twinning systems

twinning. Experimental results given in Table 1 indicate that both ordinary slip and twinning are operative but superlattice slip is not, and thus

$$T_W > S_O > S_S \qquad (1)$$

at room temperature, where T_W, S_O and S_S represent the relative ease of twinning, ordinary slip and superlattice slip. However, in view of the fact that ordinary slip and twinning very often coexist in many domains for all compression axis orientations investigated and superlattice slip can be easily activated in specimens with orientations A_i and B_i when they are deformed to 2-3 % strain, the difference in CRSS among the three deformation modes may be small. The trend of operative deformation modes and the relative ease of their operation given in Table 1 and by eqn.1, respectively are not changed up to around 800°C [29,30]. At high temperatures, the ease of ordinary slip seems to be enhanced. At 1000°C, the majority of dislocations observed are ordinary ones and the propensity for twining is greatly reduced in comparison to that at room temperature, as far as the comparison is made at the same small strain level in the range of 2-3 % [29,30].

Correlation between deformation modes and macroscopic deformation of PST crystals. The relative ease of possible deformation modes and the relative magnitude of their Schmid factors are not sufficient for explaining observed deformation modes given in Table 1. The alpha-2 lamellae are harder than the gamma lamellae, and thus constraint of the alpha-2 lamellae imposed on the deformation of the gamma lamellae may exert a strong influence in determining operative deformation modes in the gamma lamellae. Anisotropic macroscopic deformation of PST crystals is believed to be a reflection of such constraint. In this section, correlation between operative deformation modes and macroscopic deformation of PST crystals with some typical lamellar orientations are discussed considering reasons why the operative deformation modes are selected [31].

If a crystal shears on a slip or a twinning system by a small amount s and the slip or the twinning system is defined with respect to the (x,y,z) orthogonal coordinate system where the y and x axes are perpendicular to the slip (or twinning) plane and parallel to the slip (or twinning shear) direction, respectively, the strain components of the deformation mode is simply given by

$$e_{ij} = \begin{bmatrix} 0 & s/2 & 0 \\ s/2 & 0 & 0 \\ 0 & 0 & 0 \end{bmatrix} \tag{2}$$

We employ the PST crystal coordinate system (x',y',z'), where the x'z' plane is parallel to the lamellar boundaries, and x' and z' axes are perpendicular and parallel to the P face of Fig.2(a), respectively. The y' axis is perpendicular to the lamellar boundaries and thus along [111]. The loading direction lies on the y'z' plane. It is parallel to the z' axis for orientations of the A-type and to the y' axis for orientation N. For intermediate orientations, the maximum resolved shear stress acts parallel to the z' direction. Figure 3 is the PST coordinate system for compression axis orientations lying along the $[11\bar{2}]$ zone circle. The strain components e'_{ij} for the system can be calculated simply by $\mathbf{e}' = \mathbf{T}\,\mathbf{e}\,\mathbf{T}^{-1}$, where **T** is the direction cosine matrix for the coordinate transformation from (x,y,z) to (x',y',z'). The followings are the results of strain analyses made for PST crystals with orientations A_1, B_1 and N deformed in uniaxial compression.

[Orientation B_1] In domains of variants III-M and II-T, twinning on $(111)[\bar{1}\bar{1}2]$ and ordinary slip on $(111)[1\bar{1}0]$ simultaneously operate. If such simultaneous operation of twinning and ordinary slip results in the resultant shear parallel to $[0\bar{1}1]$, The strain components e'_{ij} for all domains are given in the form of eqn.3

$$e'_{ij} = \begin{bmatrix} 0 & 0 & 0 \\ 0 & 0 & s/2 \\ 0 & s/2 & 0 \end{bmatrix} \tag{3}$$

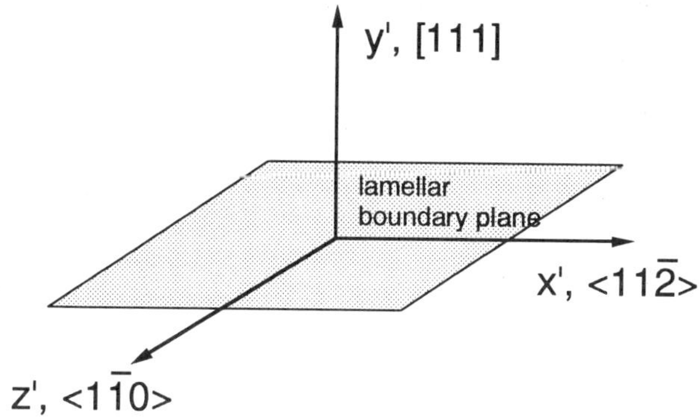

Figure 3. PST coordinate system [31]

where s is the amount of shear on (111) along the corresponding <110> shear direction in each domain. This suggests that each gamma lamella macroscopically deforms by shear parallel to the lamellar boundaries along z' axis as if it were a uniform slab, although each gamma lamella consists of domains with different crystal orientations. This has been confirmed to be in good agreement with the results of observations of the surface slip traces and macroscopic change in specimen shape after deformation. This is believed to be the case for all intermediate orientations where shear deformation of the gamma lamellae occurs parallel to the lamellar boundaries.

[Orientation A_1] Deformation of PST crystals with this orientation is complicated in comparison to that of those with orientation B_1 because the gamma phase deforms by shear on two {111} planes intersecting the lamellar boundaries and not only the gamma phase but also the alpha-2 phase contribute to their deformation [30,32]. The strain components for double superlattice slip in domains of variant I-M (Table 1) are given in the following form.

$$e'_{ij} = \begin{bmatrix} \varepsilon & -\gamma & 0 \\ -\gamma & 0 & 0 \\ 0 & 0 & -\varepsilon \end{bmatrix} \qquad (4)$$

Magnitudes of ε and γ are not discussed here since we only consider how a PST crystal with A_1 orientation tends to deform macroscopically. In domains of variant II-M, if simultaneously operating ordinary slip and twinning on $(\bar{1}11)$ and ordinary slip on $(11\bar{1})$ result in shear deformation equivalent to that due to the double superlattice slip in domains of variant I-M, in other words, if the three deformation modes operate in the propensity ratio of 1:3:2, the strain components for domains of variant II-M are given by eqn.4. This is also the case for domains of variant III-M, if ordinary slip and twinning on $(1\bar{1}1)$ and ordinary slip on $(11\bar{1})$ operate in the same propensity ratio.

The strain components for domains of variants I-T, II-T and III-T can be calculated similarly from the observed deformation modes. The result is given by eqn.5 and is quite similar to what we obtained for matrix variants, although the sign of e'_{xy} is reversed.

$$e'_{ij} = \begin{bmatrix} \varepsilon & \gamma & 0 \\ \gamma & 0 & 0 \\ 0 & 0 & -\varepsilon \end{bmatrix} \tag{5}$$

Since matrix and twin variants do not coexist in a gamma lamella, each gamma lamella can again be regarded as a uniform anisotropic slab whose macroscopic deformation is defined by eqn.4 or 5. Matrix and twin lamellae, which consist of domains of matrix and twin variants respectively, deform differently since eqns.5 and 6 are different in the sign of e'_{xy}. However, when we look at a pair of neighboring matrix and twin lamellae, the strain components for the pair are given in the form of eqn.6

$$e'_{ij} = \begin{bmatrix} \varepsilon & 0 & 0 \\ 0 & 0 & 0 \\ 0 & 0 & -\varepsilon \end{bmatrix} \tag{6}$$

with the shear components of eqns.4 and 5 canceling each other, as shown in Fig.4. Matrix and twin lamellae very often adjoin in PST crystals [26], and thereby the gamma phase lamellae are expected to deform macroscopically as defined as eqn.6.

To be noted is that the strain tensor of eqn.6 is in exactly the same form as that expected for the operation of double prism slip in alpha-2 lamellae in PST crystals with orientation A_1 (Fig.4). Prism slip is only the slip mode which can operate in the alpha-2 phase at a reasonably low stress level, and the alpha-2 lamellae in PST crystals with orientation A_1 are favorably oriented for prism slip. Thus, there appears little doubt that prism slip operates in the alpha-2 lamellae in PST crystals with orientation A_1. Assuming that the alpha-2 lamellae deform by double prism slip, the strain components for the PST crystal with orientation A_1 deformed in compression are expected to be given in the form of eqn.6. At the same time, the continuity of strain across the x'z' lamellar boundary plane represented by eqn.7 is satisfied for lamellar interfaces of all types that are expected to exist in PST crystals.

$$e'^{1}_{xx} = e'^{2}_{xx} \quad , \quad e'^{1}_{zz} = e'^{2}_{zz} \quad , \quad e'^{1}_{xz} = e'^{2}_{xz} \tag{7}$$

From the above discussion, macroscopic deformation of PST crystals with orientation A_1 is expected to occur without changing their width in the direction perpendicular to the lamellar boundaries and without accompanying any shear strain components with respect to the (x',y',z') coordinate system. In other words, tension and

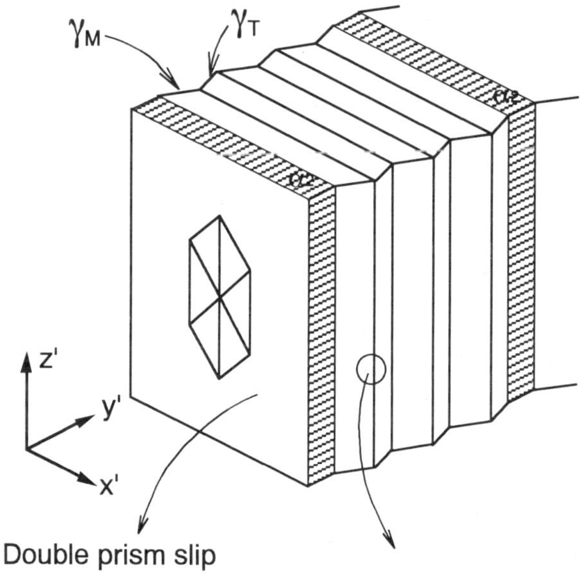

$$[e'_{ij}] = \begin{bmatrix} \varepsilon & 0 & 0 \\ 0 & 0 & 0 \\ 0 & 0 & -\varepsilon \end{bmatrix} \qquad [e'_{ij}] = \begin{bmatrix} \varepsilon & 0 & 0 \\ 0 & 0 & 0 \\ 0 & 0 & -\varepsilon \end{bmatrix}$$

Figure 4. Deformation of gamma and alpha-2 lamellae in a PST crystal with orientation A_1

compression occurs respectively via decreasing and increasing the width in the x' direction parallel to the lamellar boundaries. This has been confirmed by analyzing the macroscopic change in specimen shape after tension and compression. The alpha-2 lamellae in PST crystals with orientation A_1 can not be deformed in such a way that their width is decreased or increased since pyramidal slip is extremely difficult in comparison to prism slip [32-35]. This is believed to be a major reason why the gamma lamellae in PST crystals with orientation A_1 and therefore the PST crystals themselves deform as defined by eqn.6. Thus, there appears little doubt that the presence of alpha-2 lamellae has a crucial influence upon the selection of operative deformation modes and the mechanical properties of PST crystals with orientation A_1. However, it is not yet clear whether the presence of alpha-2 lamellae is the sole reason for such deformation of PST crystals with orientation A_1. The gamma lamellae might be deformed at a reasonably low flow stress only on condition that the strain continuity across all kinds of gamma/gamma intervariant boundaries is satisfied. In fact, the strain continuity is satisfied when the matrix and twin gamma lamellae deform as described by eqn. 4 and eqn.5, respectively. If this is the case, PST crystals with orientation A_1 is expected to macroscopically deform as mentioned in this section regardless of the presence of alpha-2 lamellae. To answer this question, a study is currently in progress to examine the deformation modes and macroscopic deformation of PST crystals with a very small volume fraction of alpha-2 lamellae.

[Orientation N] Consider a pair of neighboring matrix and twin lamellae. A domain in one of the two lamellae and its neighbor in the other lamellae are very often true-twin related. Strain tensor for such a pair of true-twin related domains is given by eqn.8 with respect the coordinate system (x",y",z") where x",y" and z" axes are parallel to $[11\bar{2}]$, $[1\bar{1}0]$ and [111] directions in the matrix domain.

$$e'_{ij} = \begin{bmatrix} 0 & 0 & 0 \\ 0 & -\varepsilon & 0 \\ 0 & 0 & \varepsilon \end{bmatrix} \tag{8}$$

Equation 8 is very similar to eqn.6 in its form. However, the coordinate system (x",y",z") is not fixed to the PST crystal, but to each pair of true-twin related domains. The y" axis is parallel to the y' axis of the PST coordinate system. Thus, the (x",y",z") coordinate system is related to the PST coordinate system by a rotation of 0°, 120° or 240° about the y' axis depending on the type of variant of the pair of true-twin related domains of interest. This indicates that PST crystals are isotropic in the x'z' plane when they are uniaxially deformed along the y' axis. This is in agreement with that when PST crystals with orientation N are deformed in compression, their macroscopic deformation occurs in a radial symmetric way and no macroscopic shear is observed. However, at microscopic scale, deformation incompatibilities arise between domains neighboring in both directions parallel and perpendicular to the lamellar boundaries. On top of that, the alpha-2 lamellae may not be deformed since no resolved shear stress acts on their prism slip plane. Schmid factors for Pyramidal slip systems are non-zero. However, the CRSS for pyramidal slip is much higher than the yield stress of PST crystals with this orientation [32] and thereby the operation of pyramidal slip is unlikely. Thus, the large strain accumulations and deformation incompatibilities arise at gamma/gamma intervariant and gamma/alpha-2 interphase boundaries. Such deformation incompatibilities at intervariant and interphase boundaries are believed to result in the high yield stress and brittleness for orientation N [28]. The results of strain analyses and observations of macroscopic deformation of PST crystals with not only orientations A_1-N but also A_2-N tested in tension and compression will be described elsewhere in more detail.

Yield stress and ductility at room temperature. The yield stress and tensile elongation of PST crystals are highly dependent on the angle between the lamellar boundaries and loading axis as shown in Fig.5(a) and (b). In general, no significant difference in yield stress between tension and compression is observed in PST crystals. In some variants, the switching of deformation mode from twinning to slip and vice versa occurs corresponding to the change in loading mode. However, the difference in CRSS between twinning and slip is small. This would be a reason for the insignificant tension-compression asymmetry of yield stress.

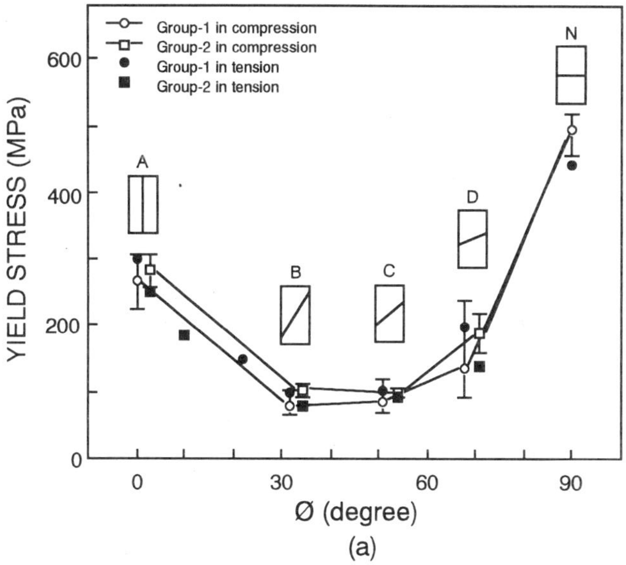

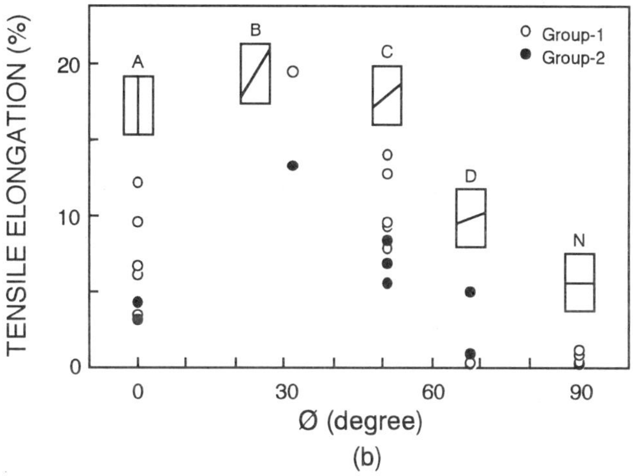

Figure 5. (a) Yield stress in tension and compression and (b) elongation to fracture of PST crystals as a function of the angle ϕ [28]. Group-1 and group 2 indicate specimens with orientations A_1-N and A_2-N, respectively.

The yield stress is high when the lamellar boundaries are parallel or perpendicular to the loading axis, but it is very low for intermediate orientations . This is very good agreement with the results of strain analyses in the previous section. For intermediate orientations, each gamma lamella deforms uniformly by shear on (111) parallel to the lamellar boundaries along the maximum resolved shear direction (the easy type of deformation), and hence the lamellar boundaries and alpha-2 lamellae do not hinder shear deformation of this type. Each gamma lamella in PST crystals is composed of domains of three different matrix or twin variants. However, deformation of each domain is described by the same strain tensor, and thus no significant deformation incompatibilities are expected to arise at domain boundaries in each gamma lamella. This is in good agreement with the weak resistance of domain boundaries to the propagation of the easy type deformation, which has been confirmed by TEM examinations of deformation structures in PST crystals with intermediate orientations [27]. In view of these, the easy type of deformation of PST crystals can be regarded as a direct reflection of the deformation of the gamma phase itself in equilibrium with the alpha-2 phase.

When the lamellar boundaries are parallel or perpendicular to the loading axis, shear deformation proceeds on {111} planes intersecting the lamellar boundaries (the hard-type of deformation). In the case of orientation A_1, the continuity of strain across not only the lamellar boundaries but also domain boundaries in each of the gamma lamellae is satisfied. Deformation of domains of all three variants in a gamma lamella can be described by the same strain tensor eqn.4 or eqn 5. Thus, no significant deformation incompatibilities are expected to arise at both intervariant and interphase boundaries. This is believed to result in good ductility for orientation A_1. The Hall-Petch type hardening may play an important role in giving rise to a high yield stress for this orientation. In the case of uniaxial compression and tension of PST crystals with orientation N, significant deformation incompatibilities are expected to arise at intervariant and interphase boundaries. This is the reason for the large difference in strength and ductility between orientations A_1 and N.

High temperature deformation. PST crystals with orientations with A_1, B_1 and N were tested in compression and tension at temperatures in the range from liquid nitrogen temperature to 1000°C. Orientations A_1 and B_1 are simply referred to as A and B henceforth. The yield stresses obtained in compression for the three orientations are plotted in Fig. 6(a) as a function of temperature. The yield stress decreases rather rapidly with increasing temperature at low temperatures for all orientations although the temperature dependence is less significant for orientation B than for orientations A and N, and this is followed by a gradual decrease in yield stress with increasing temperature at intermediate temperatures. For orientation B, such gradual decrease in yield stress continues up to 1100°C. Since the operative deformation modes have been confirmed to be unchanged from room temperature to at least 800°C, the results for orientation B in Fig.6(a) are believed to exhibit the temperature dependence of yield stress of the gamma phase when it deforms by shear on (111) along $<1\bar{1}0>$. Thus, the gamma phase in equilibrium with the alpha-2 phase does not show any yield-stress anomaly.

For orientations A and N, the gradual decrease in yield stress is followed by a rapid decrease in yield stress at temperatures higher than 900°C, although the yield stress, in particular, for orientation A exhibits an anomaly with a small and broad peak at around 800°C prior to the rapid decrease in yield stress. A similar anomalous peak in yield stress has been observed in PST crystals with the same orientation, but with slightly different compositions [36] and in a polycrystalline TiAl alloy [37-39]. As we mentioned above, the gamma phase in equilibrium with the alpha-2 phase does not exhibit any anomaly. This is also the case for the alpha-2 phase oriented favorably for slip on prism planes; the CRSS for slip on prism

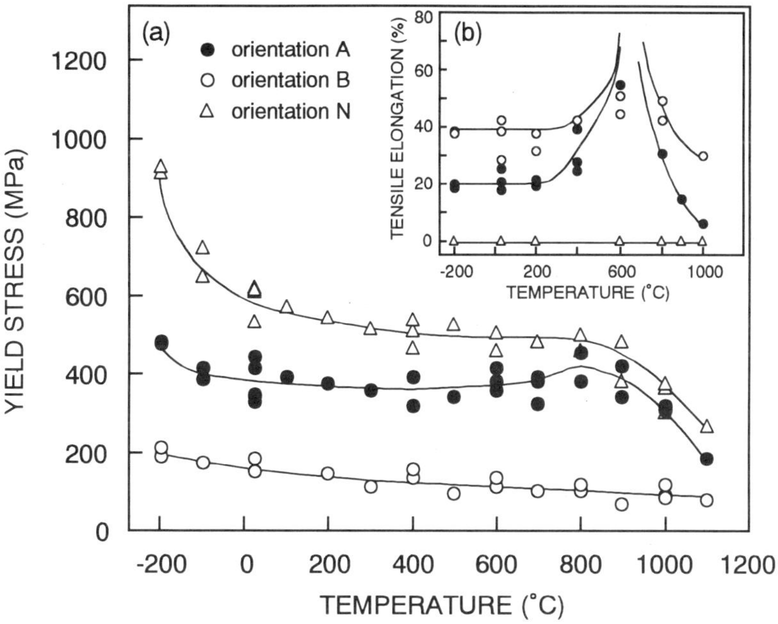

Figure 6. (a) Yield stress and (b) elongation of PST crystals with orientations A, B and N as a function of temperature [29,30]

planes monotonically decreases with increasing temperature [33,34]. Thus, the observed anomaly may not be attributed to the deformation behavior of either of the two constituent phases. For orientation N, we did not observe any significant anomalous increase in yield stress, although Umakoshi and Nakano [36] have observed a rather definite anomalous peak around 500°C for PST crystals with the same orientation. They have attributed the observed anomaly to the anomalous strengthening behavior which the alpha-2 phase exhibits when it deforms by slip on pyramidal planes [36]. However, the yield stress of the alpha-2 phase oriented for slip on pyramidal planes is far larger than that for PST crystals with orientation N and thus alpha-2 lamellae generally do not contribute to the deformation of PST crystals with this orientation [32]. On top of that, the yield stress for slip on pyramidal planes in the alpha-2 phase coexisting with the gamma phase in PST crystals is expected to reach a maximum peak at a temperature lower than 500°C since the peak temperature has been known to rapidly decrease with increasing deviation from the stoichiometry [40].

Thus, almost nothing has been clarified on the mechanism of the temperature dependence of the yield stress of PST crystals, in particular those with A and N orientations. The trend in the temperature dependence of yield stress obtained in tension is very similar to that obtained in compression, but the absolute values of yield stress are somewhat lower than those in compression at all temperatures investigated. This may be due to a gripping problem of the alumina-ceramics tensile grips and/or the machine softness of tensile jigs designed for high-temperature tensile tests.

The values of tensile elongation for orientations A, B and N are plotted in Fig.6(b) as a function of temperature. The tensile elongation obtained at room temperature in the present study is lower than that obtained in our previous study [30]. This may be again due to the gripping problem of the tensile jigs used in the present study.

There seems to be a general agreement that tensile ductility of polycrystalline gamma TiAl alloys drastically increases at 700~800°C [41-44]. For PST crystals except those with orientation N, however, such a drastic increase in tensile ductility is observed at 600°C, which is well below the corresponding temperature for polycrystalline gamma alloys. The increase in tensile ductility observed above 200°C for PST crystals may be due to an intrinsic increase in the mobility of dislocations. Around 600°C, fracture did not occur in the gage section of tensile specimens but in the fillet section. In such cases, much larger tensile elongation should be observed if the shape of tensile specimens is designed so that fracture occurs in their gage section. That is the reason why we drew the solid lines sharply upward at 600°C. At temperatures above 800°C, recrystallization occurs during tensile deformation of PST crystals. Once recrystallization occurs, the recrystallized part loses strength and therefore deformation concentrates upon the part and necking takes place. Above 800°C, we have thus plotted in Fig.6(b) the values of strain at the maximum flow stress which corresponds to tensile elongation carried by uniform deformation of the gage section. Such uniform tensile elongation decreases with increasing temperature for both orientations A and B, although the values of total tensile elongation including that due to the necking of the gage section reach more than 60 and 100 % at 1000°C for orientations A and B, respectively.

The tensile elongation for orientation N is essentially zero for the whole temperature range investigated. Fracture always occurs parallel to the lamellar boundaries. This is the reason why polycrystalline gamma alloys in the fully or nearly lamellar form do not exhibit any substantial increase in tensile elongation until temperature becomes high enough for recrystallization, which results in destroying their lamellar structure, to occur easily.

Fracture Behavior and Fracture Toughness

Deformation experiments of smooth PST crystals in uniaxial tension and compression have resulted in considerable progress in our understanding of the basic deformation behavior of gamma alloys, in particular in the lamellar form. However, no systematic studies on the fracture behavior of notched PST crystals have been done, despite information on how crack propagation depends on crack and crystal orientations is essential for clarifying fracture mechanisms of the lamellar structure of gamma alloys. In this section, we summarize our recent investigation of the fracture behavior and toughness of PST crystals in air at room temperature [45].

Fracture toughness measurements. Three-point bend specimens with five different notch orientations were prepared from as-grown PST crystals (Fig.7(a)). The types

I, II and III are crack arrester, crack divider and crack delamination orientations with respect to the lamellar boundaries, which are assigned to be parallel to (111) in the gamma lamellae. J_{IC} measurements were made on notched PST specimens with these notch orientations. The reason for choosing J_{IC} measurements is that specimens for J_{IC} measurements can be much smaller than those needed for K_{IC} measurements. J_{IC} values are often measured using the R-curve method. However, methods involving the testing of numerous specimens or multiple loading of a single specimen can not be applied to PST crystals since it is not easy to prepare numerous PST specimens with the same orientation, and it is practically impossible to measure the compliance and crack length through multiple loading of a small PST specimen with a reasonable accuracy. Fortunately, it has recently been found that if a specimen tested in bending is properly fatigue-precracked and side-grooved, the J integral corresponding to the maximum load point on the load-displacement curve is approximately the same as the J_{IC} of the specimen[46,47]. Thus, J_{IC} may be approximated by the value given by

$$J = 2A/B_e b \tag{9}$$

,where A is the area under the load-displacement curve that is rectified with respect to the jig compliance, and B_e, the effective width of ligament, is given by $1.1x(B \cdot B_N)^{1/2}$ [48].The remaining unbroken ligament length, b, and the side groove depth shown in Fig.7(b) are 0.4W and 0.125 B (B_N/B = 0.75), respectively[47,49]. We have prepared a specimen with such ligament and side-groove dimensions from a gamma alloy ingot in the duplex form whose K_{IC} is known, and confirmed that the J value obtained from eqn.9 is reasonably in agreement with the J_{IC} calculated from the known K_{IC}.

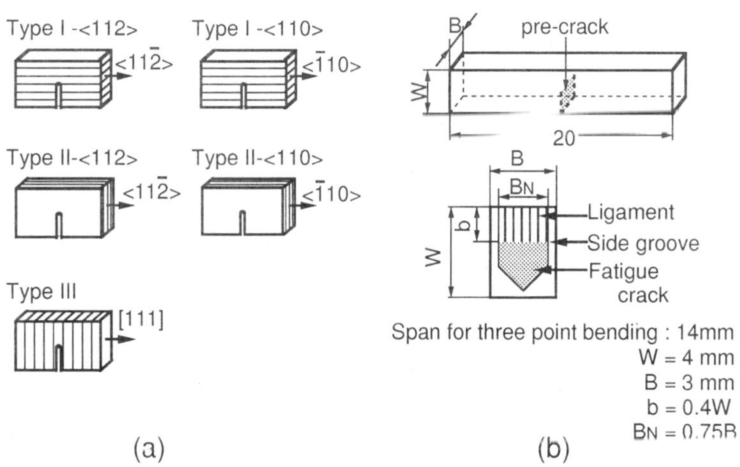

Figure 7. (a) Notch orientations and (b) specimen geometry [45]

Fracture toughness. The J values obtained for four notch orientations using eqn.9 are shown in Fig.8. Yield stress (σ_{ys}) for uniaxial tension of unnotched PST crystals is higher than 250 MPa when the loading axis is parallel or perpendicular to the lamellar boundaries, and $\sigma_{ys} < \sigma_B$ or $\sigma_{ys} \approx \sigma_B$ depending on crystal orientation, where σ_B is tensile strength[28]. Thus, $J = J_{IC}$ for all J values shown in Fig.8 since the specimen size satisfies specimen size requirements for $J_Q = J_{IC}$ which are given by

$$B,b \geq 25\frac{J_Q}{\sigma_{fs}} \tag{10}$$

, where $\sigma_{fs} = (\sigma_{ys} + \sigma_B)/2$. This indicates that J values obtained in the present investigation reasonably approximate J_{IC} values of PST crystals. K_{IC} values can be calculated assuming $J = J_{IC}$, $E = 176$ GPa and $\nu = 0.26$[41,50,51]. The calculated values of K_{IC} are given in Table 2 [45].

The K_{IC} for the type III notch is in good agreement with that obtained using CT-specimens of PST crystals with the same orientation[52].

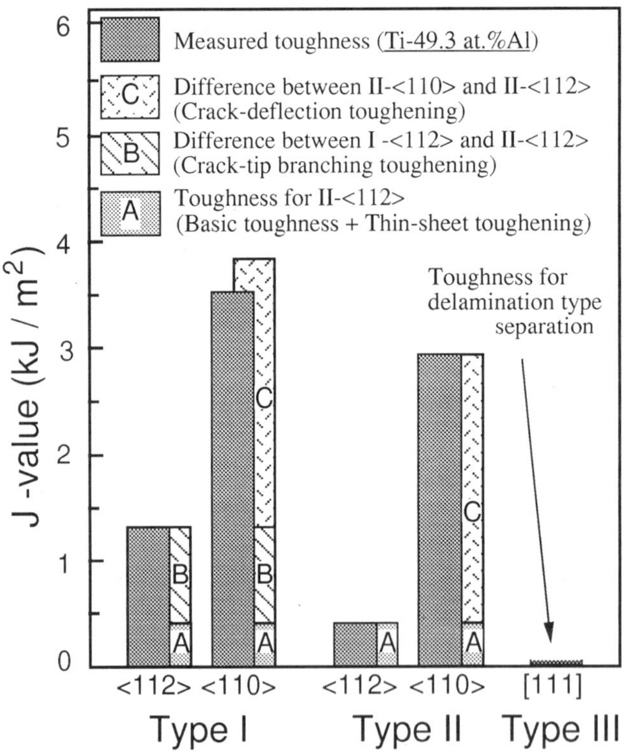

Figure 8. Measured fracture toughness values and toughening contributors to the toughness of PST crystals [45]

Table 2. Fracture toughness of PST crystals (MPa·m$^{1/2}$)

I-<112>	I-<110>	II-<112>	II-<110>	III
15.7	25.7	8.7	23.4	4.0

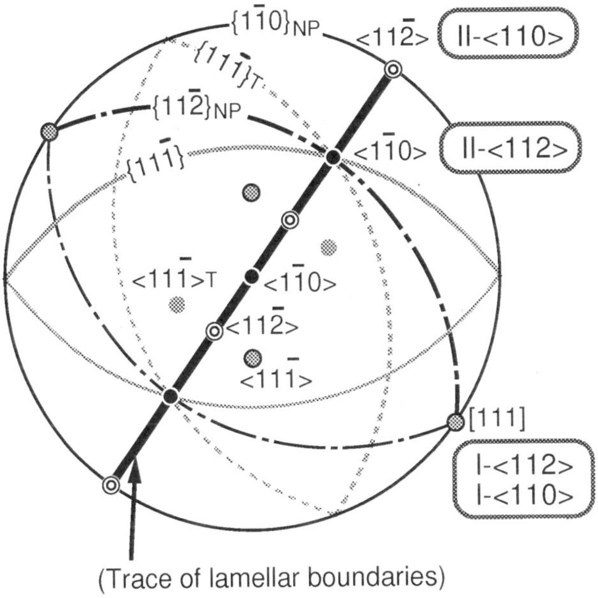

Figure 9. Crystallographic relationship between notch and fracture habit planes for crack arrester and crack divider orientations. The direction of crack propagation for each notch orientation is marked by the notation for the corresponding notch orientation. $\{11\bar{2}\}_{NP}$ and $\{1\bar{1}0\}_{NP}$ indicate $\{11\bar{2}\}$ and $\{1\bar{1}0\}$ notch planes [45].

Crack path and fracture surface morphology. The crack path of the type III specimen is extremely planar in accordance with K_{IC} for this orientation, which is as low as 4.0 MPa·m$^{1/2}$. Specimens of the types I and II show rougher fracture surfaces than those of the type III. Fracture surfaces of specimens of the types I-<112> and II-<112> are basically composed of a combination of {111} fracture habit planes of matrix and twin lamellae. The combination of two fracture habit planes forms the average plane of crack propagation parallel to the notch plane, as seen from Fig.9. Such fracture habit planes in two neighboring gamma lamellae often form ridges symmetrical with respect to the lamellar boundary separating the two lamellae because two neighboring gamma lamellae are very often twin-related. {111} fracture habit planes have also been observed in polycrystalline gamma alloys[53-55].

However, when the notch is formed along $\{1\bar{1}0\}$, crack propagation can not occur along the notch plane by the same way as in specimens notched along $\{11\bar{2}\}$ since no fracture habit planes, i.e.$\{11\bar{1}\}$ planes intersect the lamellar boundaries in the direction of intersection of the notch plane and the lamellar boundaries (see Fig.9). However, if the crack at the $\{1\bar{1}0\}$ notch tip propagates along a set of zigzag $\{11\bar{2}\}$ planes (corresponding great circles are shown by dot-and-dash lines in Fig.9), the average plane of crack propagation can be $\{1\bar{1}0\}$. Thus, fracture surfaces of specimens notched along $\{1\bar{1}0\}$ consist of patches which exhibit the same microstructural characteristics as those for specimens notched along $\{11\bar{2}\}$ and slant by about 30° with respect to the notch plane.

In addition to these characteristics of fracture surfaces, it should be pointed out that delamination-type separation occurs along lamellar boundaries and through alpha-2 lamellae not only in specimens of the type III but also in those of the types I and II. Such delamination-type separation is believed to be beneficial to increase the toughness of specimens of the types I and II.

Toughening mechanisms. As seen from Fig.8, fracture toughness of specimens of the types I and II shows a significant notch orientation dependence. For the same orientation of notch plane, fracture toughness for the type I is higher than that for the type II. For the same type of notch, fracture toughness of specimens notched along $\{1\bar{1}0\}$ is higher than that of those notched along $\{11\bar{2}\}$. When crack advance proceeds perpendicularly to the lamellar boundaries, delamination-type separation occurs ahead of the crack tip as shown in Fig.10(a) because of triaxial stresses in front of the crack tip. When such delamination-type separation occurs ahead of the crack-tip, a new crack tip must be newly formed for further propagation of the crack along the notch plane. Thus, such crack-tip branching by the delamination-type separation is expected to increase fracture toughness. Such a toughening mechanism has been identified earlier in [56-61]. When delamination cracks are formed in specimens of the type II as shown in Fig.10(b), the crack tip is not expected to be branched. However, plane-stress conditions prevail at the crack tip because the delamination reduces the ligament to a bundle of thin sheets. The plane-stress fracture toughness is higher than the plane-strain fracture toughness, and thereby delamination-type separation such as that shown in Fig.10(b) results in increasing the fracture toughness of the specimen. Such a toughening mechanism is called thin-sheet toughening and has been observed aluminum alloys with highly anisotropic microstructure with a large volume fraction of dispersoids[62-64] or consisting of large pancake-shaped grains[65]. In specimens of the type I, delamination-type separation gives rise to not only such thin-sheet toughening but also branching the tip of main crack (Fig.10(a)), and thereby a higher value of fracture toughness is expected to be obtained for specimens of the type I than those of the type II. This is in agreement with the result shown in Fig.8.

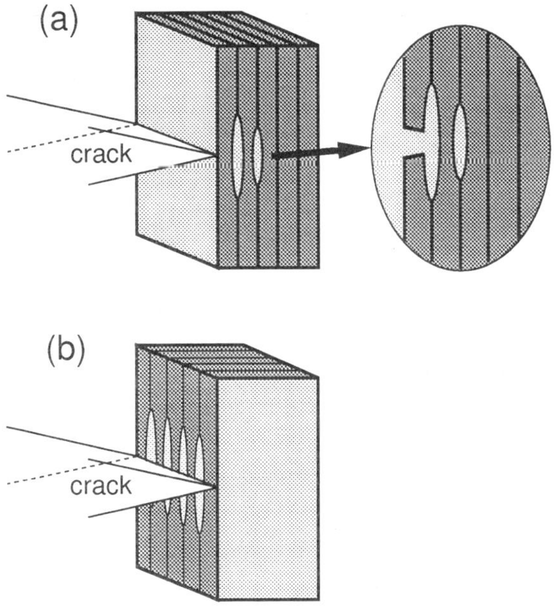

Figure 10. Delamination-type separation parallel to the lamellar boundaries at the crack tip for the types I(a) and II(b) notch orientations [45]

Thus, the branching of the tip of main crack which occurs for crack extension perpendicular to the lamellar boundaries is a toughening contributor for the type I-<112> which causes the difference in toughness between the types I-<112> and II-<112>, marked by B in Fig.8. The difference in fracture toughness between the types II-<110> and II-<112>, marked by C in Fig.8 is mainly due to the crack-deflection toughening. However, the difference in fracture toughness is rather considerable as shown in Fig.8, and thus it is likely that some toughening mechanisms including the crack deflection mechanism are simultaneously responsible for the difference in toughness, although they have not been identified yet. The type I-<110> is the toughest of the notch types investigated. The difference in toughness between types I-<110> and II-<112> is nearly equal to the sum of the differences between the types I-<112> and II-<112>, represented by B, and that between the types II-<110> and II-<112>, represented by C. This is reasonable since the I-<110> type crack propagates perpendicularly to the lamellar boundary on $\{1\bar{1}0\}$, both crack-tip branching and crack deflection mechanisms simultaneously contribute to the toughness of specimens of this notch type. Of the four notch types of the crack arrester and crack divider orientations, the type II-<112> exhibits the lowest fracture toughness The toughness of this notch type is thought to correspond to the basic fracture toughness of the gamma phase probably in a state of plane strain. Thus, three major contributors to the toughness of PST crystals with crack arrester and crack divider orientations are identified based on the above discussions. They are (i) the basic toughness in a state of plane stress, (ii) the crack-tip branching mechanism, (iii) the crack deflection mechanism.

Prospects of Directional Solidification Processing of Gamma Alloys

As previously shown, the mechanical properties of gamma alloys greatly depend upon the orientation of the lamellar microstructure. The anisotropic nature of the lamellar microstructure plus the extremely large lamellar grain size that develops upon casting have resulted in as-cast materials with a poor combination of properties, such as ductility versus fracture toughness [66,67]. Typically, thermomechanical processing routes such as hot isostatic pressing followed by hot working procedures are used to produce materials with a fine distribution of lamellar grains and a more balanced combination of properties [66,67].

Another approach with a potentially greater payoff is to take advantage of the anisotropic nature of the lamellar microstructure by producing gamma alloys containing only one lamellar orientation via directional solidification (DS). If the lamellar orientation can be aligned parallel to the growth direction, the combination of strength and toughness would be optimized in the resulting microstructure consisting of columnar grains with the lamellar orientation aligned parallel to the growth direction (Fig.11). When a DS ingot such as that shown in Fig.11 is tested in uniaxial tension along its growth direction, each columnar grain is believed to deform macroscopically as described by eqn.6. Such macroscopic deformation of columnar grains may result in large deformation incompatibilities between adjacent columnar grains and negatively affect the ductility of the DS ingot. Thus, the ductility may not be as large as that observed for PST crystals with orientation A. However, it is still expected to be reasonably high in view of favorable conditions for deformation of each columnar grain such as the operation of prism slip in the alpha-2 lamellae and strain continuity across the domain and interphase boundaries. Strength wise, deformation incompatibilities between adjacent columnar grains play a positive role, and thereby significantly enhanced strength with a reasonable amount of ductility will be achieved. When gamma alloys with nearly equiatomic compositions are unidirectionally solidified, a columnar grained structure is obtained. However, each columnar grain has a lamellar structure with orientation N. How can the lamellar orientation be controlled? This is a crucial problem to be solved to develop the DS processing of gamma alloys. The following is a brief summary of

Figure 11. DS structure composed of columnar grains whose gamma and alpha-2 lamellae are aligned along the growth direction

our view of prospects of the DS processing of gamma alloys.

Unfortunately, when gamma alloys with nearly equiatomic compositions are unidirectionally solidified, the resulting microstructure typically consists of columnar grains with the lamellar microstructure of orientation N. However, since these DS ingots have only one lamellar orientation, it may be possible to change the lamellar alignment with appropriate processing techniques. The main difficulty in microstructural control by directional solidification is that the lamellar microstructure is not formed from the liquid but from the solid state. Thus the solidification path in addition to the solid state reactions must be known for each TiAl-based alloy considered.

For directional solidification, final lamellar orientation will depend upon the initial ingot composition. Small changes in the alloy composition can change the primary solidification phase and thus alter the final microstructure. If a planar solid-liquid interface is maintained during directional solidification, the different DS microstructures expected can be explained with the aid of Fig.12. For example, one method to prevent the formation of N oriented lamellar grains is to adjust the alloy composition such that the ingot solidifies as the beta phase first (composition C1 in Fig.12). Assuming a <100> growth direction for the beta phase, the alpha phase will nucleate on the close packed planes of the beta phase with an orientation relationship of $(0001)_{\alpha}//(110)_{\beta}$. As the temperature is decreased, the gamma phase will nucleate upon (0001) of the alpha phase producing the lamellar microstructure with an orientation relationship of $(111)_{\gamma}//(0001)_{\alpha}$. Thus the final DS microstructure will consists of lamellar grains with orientations parallel and 45 degrees to the growth direction.

However, if the aluminum content is increased such that the ingot composition lies

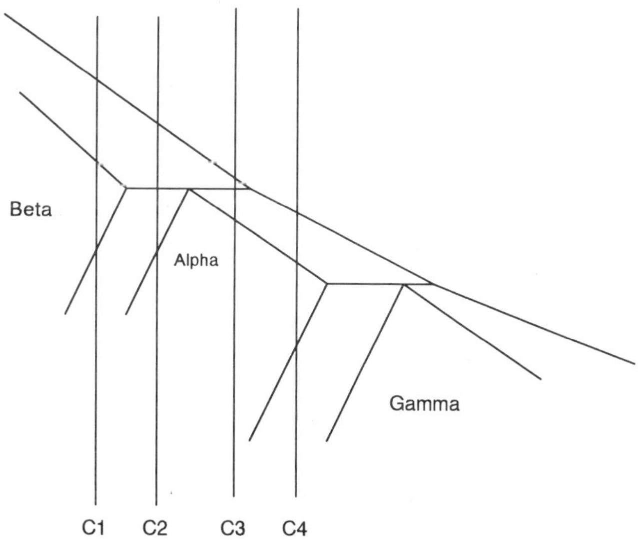

Figure 12. Ti-Al phase diagram around the gamma phase

within the $L + \beta = \alpha$ peritectic region (composition C2 in Fig.12), a banded microstructure would result [68]. Initially, the ingot would grow as beta, but once the solid-liquid interface temperature drops below the peritectic temperature, the alpha phase would quickly cover the beta phase effectively stopping the reaction. However, steady-state growth conditions cannot be maintained for the alpha phase since the peritectic composition is richer in aluminum than the ingot composition. Thus, an unstable situation exists and the beta phase would eventually renucleate at the growth front. For this case, the lamellar orientation would vary along the length of the ingot as the growth front changes between the beta and alpha phases.

A more promising microstructure would result as the aluminum content of the ingot is increased above the peritectic composition (composition C3 in Fig.12). Again, the beta phase will nucleate first and alpha will nucleate next via the peritectic reaction. However for this composition, a steady-state alpha growth front can be maintained after the alpha phase has nucleated. Therefore, the remaining ingot can solidify as alpha if a planar solid-liquid interface is maintained. Hence, the beta phase can act as a seed for the alpha phase. For example, if a single alpha crystal is nucleated, then a PST-crystal is grown. However, if a planar solid-liquid interface is not maintained, the alpha phase may grow with the [0001] growth direction under dendritic growth conditions resulting in lamellar grains with orientation N as is typical in cast gamma alloys.

Lastly, if the aluminum content is increased so that the alpha phase is nucleated first (composition C4 in Fig.12), casting an ingot with a single lamellar orientation is possible. Again, under the presence of a steep thermal gradient, the alpha phase will solidify with the [0001] growth direction. Since the gamma phase nucleates on the (0001) plane of the alpha phase, the final microstructure will consist of columnar grains all having a lamellar microstructure with orientation N.

However, the ability to produce an ingot having only one lamellar orientation suggests that microstructural control may be achieved by using a seed material. The basic idea is to control the orientation of the alpha phase upon nucleation from the liquid. Since the lamellar microstructure forms with the $(111)_\gamma//(0001)_\alpha$ orientation relationship, once the orientation of the alpha phase is determined then the orientation of the lamellar microstructure is also determined. In addition, previous work on PST crystals has shown that when the lamellar microstructure is heated above the $\alpha = \alpha_2 + \gamma$ eutectoid temperature into the alpha/gamma two phase region, the lamellar microstructure is stable and the alpha-2 phase simply disorders to alpha. Furthermore, upon further heating, the volume fraction of alpha increases by the thickening of the alpha lamellae and not by nucleation of new alpha grains. And finally, upon cooling, the processes is reversed and the original orientation of the lamellar microstructure is restored [25]. Hence, if such a process is followed and if recrystallization can be avoided, it should be possible to align the lamellar microstructure using a seed material.

Thus, alignment of the lamellar microstructure in gamma alloys of near equiatomic compositions may be accomplished by directional solidification using an appropriately oriented seed crystal. The problem encountered in this approach is that the alpha phase is not stable up to the melting temperature in the Ti-Al binary system. For example, a seed crystal with a composition of C4 will pass through the gamma single phase region upon heating or cooling causing recrystallization to occur. Thus, the composition of the seed must come from a ternary system where the alpha phase is stable until melting. Shifting the primary alpha region towards lower aluminum content, thus missing the gamma single region upon cooling can be achieved by suitable alloying additions to the seed material. For example, silicon has such an effect and alignment of the lamellar microstructure by directional solidification has been accomplished using a seed material alloyed with silicon and an optical-floating zone furnace [69]. Work on other alloying additions plus the production of DS ingots with a microstructure such as that shown in Fig.11 using a DS furnace which is similar to those commonly used for the DS processing of Ni-base superalloys are currently in progress and these results will be presented elsewhere.

GAMMA PHASE WITH COMPOSITIONS IN ITS SINGLE PHASE REGION

The mechanical properties of the gamma phase strongly depend on aluminum content. In particular, the room-temperature tensile ductility is quickly reduced to zero when aluminum content exceeds 53 at % (a study on the aluminum content dependence of the mechanical properties of the gamma phase is currently in progress and the results will be published elsewhere). However, as mentioned in the section of PST crystals, the gamma phase in equilibrium with the alpha-2 phase is soft. Ductility of gamma alloys in duplex forms increases with increasing volume fraction of gamma grains and the lowest strength occurs at a composition around 51 at % Al [70]. This also suggests the intrinsic softness of the gamma phase coexisting with the alpha-2 phase in gamma alloys. Although uncertainties still remain about the $(\alpha_2+\gamma)/\gamma$ phase boundary at low temperatures, it has been believed to be located around the equiatomic composition. This means that the mechanical properties of the gamma phase changes drastically with variations in aluminum content from 50 to 53 at.%. The reasons must be clarified not only for understanding deformation mechanisms of the gamma phase but also in the context of alloy design of gamma alloys since the Al content of the involved gamma phase may become higher than 50 at.% depending on alloying elements.

Deformation structures of binary and some ternary alloys of single-phase gamma have been examined by transmission electron microscope observations [18,71-75]. One of the most notable features of these observations is the abundance of superlattice dislocations and their faulted dipoles in comparison to the corresponding observations for gamma alloys. However, the predominance of superlattice dislocations and their faulted dipoles may not be directly related to the brittleness of single-phase gamma since superlattice dislocations are abundantly observed also in PST crystals showing large tensile elongation even at temperatures lower than room temperature [30]. There must be considerable differences in slip and twinning mechanisms in the gamma phase between nearly equiatomic and more Al-rich compositions, and such differences are believed to be closer to the origin of the brittleness of single-phase gamma. However, very little is known on deformation mechanisms of single-phase gamma. This is partly because until recently, no systematic work has been done on its deformation since the first study on single crystals of single-phase gamma [76]. This early work disclosed that a yield stress anomaly occurred for a wide variety of crystal orientations. However, since deformation-structure examinations by transmission electron microscopy were not made, no information on the deformation modes and dislocation structures was provided.

Recently we have made a deformation study of the same type using specimens cut from as-grown single crystals of the gamma phase of 56 at % Al (some of the results have been published in [29]). The oxygen content of our as-grown crystals is about 300~500 wt.ppm. Our study has revealed that all three major deformation modes, ordinary slip, superlattice slip and twinning can be operative depending on crystal orientation [29]. Figure 13 shows the temperature dependence of yield stress for several different orientations. Schmid factor wise, superlattice slip is favored for orientations in triangles I and III, while ordinary slip is favored over superlattice slip for orientations in the triangle II. As seen in Fig. 13, a yield stress anomaly is observed for all orientations tested, and the profile of yield stress versus temperature curves suggests that three temperature regimes can be defined. At low

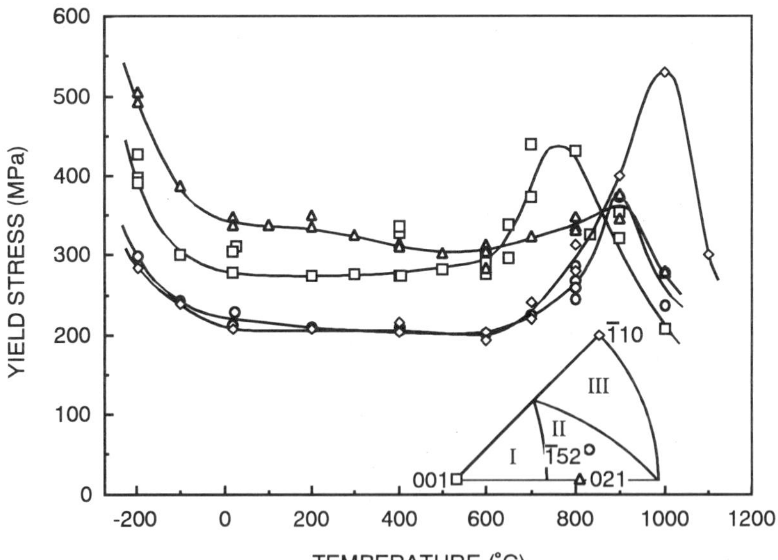

Figure 13. Yield stress as a function of temperature for single crystals of Ti - 56 at. % Al with four different orientations

temperatures, the yield stress rapidly decreases with increasing temperature up to around room temperature and then stays almost constant up to 600°C. Above 600°C, the yield stress increases anomalously with temperature exhibiting a peak at 600-1000°C depending on crystal orientation, then followed by a rapid decrease at high temperatures. A similar behavior has been reported on single-phase gamma single crystals [76-80] and polycrystals [74,81].

For [001] and $[\bar{1}10]$ orientations where no stress is resolved onto ordinary slip, superlattice slip is observed up to the peak temperature. Although the largest Schmid factor for slip along <112] (0.471) is higher than that for superlattice slip along <101] (0.408) for these two orientations, <101] screw dislocations are predominantly observed up to the peak temperature. This is consistent with the recent results of TEM observations on [001] single crystals by Stucke et al. [79]. At temperatures below the peak, yield stress is much higher for the [001] orientation than for the $[\bar{1}10]$ orientation. However, both the peak temperature and peak stress are lower for the former orientation than the latter. The lower peak temperature for the [001] orientation is associated with the occurrence of twinning of the {111}<112]-type above the peak temperature. Such twinning can not occur for the $[\bar{1}10]$ orientation in compression. Deformation modes operative above the peak temperature for the $[\bar{1}10]$ orientation have yet to be identified. Thus, twinning is seen to occur only in a limited orientation range and only at high temperatures. The reason for such limited twinning activity has yet to be clarified.

<101] superlattice slip is observed up to the peak temperature also for the $[\bar{1}52]$ orientation where the resolved shear stress for ordinary slip along <110] is 1.3 times that for superlattice slip along <101]. This clearly indicates that the CRSS for superlattice slip is lower than that for ordinary slip, being consistent with the results of previous TEM observations that superlattice dislocations predominate the deformation structures of the gamma phase of Al-rich compositions [18,71-74]. Near the peak temperature, <101] dislocations are mostly in their screw orientations with many superkinks, as previously reported [71,79,82,83]. Their fine structures will be described in detail in [84].

Ordinary slip is observed in the whole temperature range for the [021] orientation where the resolved shear stress for ordinary slip along <110] is twice that for superlattice slip along <101]. The ordinary-slip/superlattice-slip Schmid factor ratio is the largest for this orientation. The peak in yield stress is observed at 900°C, indicating that a yield stress anomaly occurs even in deformation by ordinary slip. However, the peak height is very small when compared to those for other three orientations where the anomaly is associated with superlattice slip. 1/2<110] ordinary dislocations tend to be parallel to their screw and 60° orientations at low temperatures, whereas they are roughly in the screw orientation and accompanied by many cusps near the peak temperature. Tilting experiments have indicated that cusps are not on the primary slip plane but are bowed-out on different planes, indicating the frequent occurrence of cross-slip. Such dislocation structures have been observed by

Viguier et al. [85] and Bird et al. [80], and a model for the strength anomaly has been propose by Louchet and Viguier [86] based on their observations.

For some other orientations tested, <110] ordinary slip is observed not only on {111} but also on (001) and {110) depending on crystal orientation [84]. However, the slip plane observed at temperatures near and just above the yield stress peak is always {111}, in contrast to the case of Ni_3Al where the slip plane transition from {111} to {010} occurs at the peak.

Yield stresses for the four crystal orientations are resolved on the identified slip and twinning systems and plotted as a function of temperature in Fig.14. An anomalous increase in CRSS is observed for both ordinary and superlattice slip [29,84-86]. For orientations favorable for the operation of ordinary slip, ordinary slip is always operative below and above the anomalous peak, while for those favorable for superlattice slip, deformation-mode transition from superlattice slip to twinning or ordinary slip occurs around the peak temperature depending on crystal orientation. In comparison to the corresponding results for PST crystals, the most remarkable features of the figure are

$$S_S > S_O > T_W \tag{11}$$

at temperatures lower than 600°C and a very strong temperature dependence of twinning stress. As the Al content of the gamma phase increases from 50 to 53 at.%, ordinary slip and, in particular, twinning become drastically harder. Understanding hardening mechanisms for the two deformation modes is important not only to clarify the reasons of the brittleness

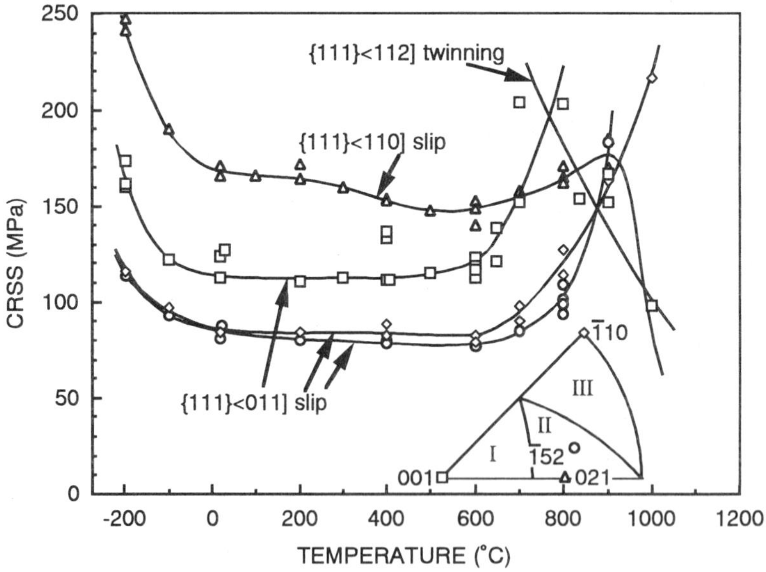

Figure 14. Temperature dependence of CRSS for the three deformation modes in single crystals of Ti-56 at.% Al

of single-phase gamma but also in the context of alloy development and applications of gamma alloys.

In the following, we will briefly summarize the characteristics of the strength anomaly observed in the single-crystalline gamma phase of 56 at.% Al in comparison with those observed in Ni_3Al.

[Orientation dependence of CRSS] In all three temperature regimes, CRSS for superlattice slip along <101] depends strongly on crystal orientation and the Schmid's law does not hold, similarly to the case of Ni_3Al. The orientation dependence of CRSS for this slip system may not be interpreted simply in terms of N and Q values which have been introduced to explain the orientation dependence of CRSS for slip on {111}<101> in Ni_3Al [87,88]. In the case of ordinary slip, it has been known whether or not its CRSS depends on crystal orientation since ordinary slip is operative only for limited orientations below the peak temperature.

[Tension-compression asymmetry] No tension tests of single crystals of single-phase gamma compounds have been made because of their brittleness. Thus, no information is available on the tension-compression asymmetry of CRSS for the three deformation modes in single-phase gamma compounds.

[Strain rate dependence of CRSS] CRSS for both ordinary and superlattice slip exhibits practically no strain rate dependence [84], as in the case of Ni_3Al. The strain rate sensitivity of flow stress associated with ordinary dislocations in the anomalous temperature range has been studied using polycrystalline specimens both experimentally and theoretically [86].

[Reversibility of flow stress with respect to temperature] We have made Cottrell-Stokes type experiments on our single crystals with a $[\bar{2}51]$ orientation and the results are shown in Fig. 15. For this orientation, a peak in CRSS occurs at 900°C and <101] superlattice slip is observed below the peak. A specimen was first deformed at 900°C to about 1% strain (a), then cooled to room temperature and redeformed at room temperature (b). As seen in Fig. 15, the yield stress of the specimen subjected to the second deformation at room temperature (b) is a little higher than the yield stress of a specimen deformed at room temperature without any prestraining (c). However, if the stress increment due to work-hardening at 900°C is subtracted from the former yield stress (stress-strain curve (b)), the two stress-strain curves (b) and (c) almost overlap each other. This indicates that the yield stress anomaly is reversible with respect to temperature, as in the case of Ni_3Al. In contrast, Stucke et al. [89] has reported that the yield stress anomaly observed for single crystals of Ti-56 at % Al with an orientation close to [010] is not reversible. The reason for the discrepancy between the two has yet to be clarified. However, these two experiments coincides with each other in that the work-hardening rate is higher at higher temperatures and reversible with respect to temperature. Mechanisms of the yield stress anomaly in single-phase gamma which are able to interpret these relevant observations have yet to be proposed.

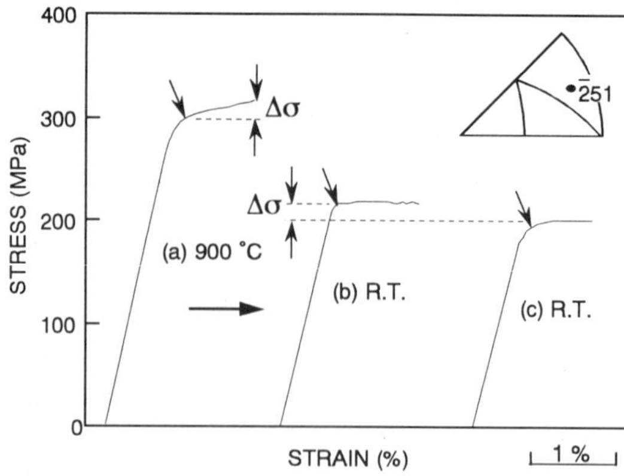

Figure 15. (a) Stress-strain curve at 900°C, (b) stress-strain curve at room temperature for the specimen tested at 900°C and (c) stress-strain curve at room temperature

CONCLUDING REMARKS

Within the last decade, significant progress has been made in the science and technology of gamma alloys. Some engineering gamma alloys nearly satisfy the property requirements for some structural applications [66,67]. However, in order to make these gamma alloys to be competent in comparison to currently available high-temperature materials such as Ni-base superalloys, advanced titanium alloys and some ceramics in the light of not only the technical needs of the user and of the producer but also production cost, significant improvements are still required over their current properties as well as the technologies of their processing. In automobile industries, gamma alloys attract considerable interest as a new light-weight material for valve and turbocharger rotor. However, the current manufacturing cost of a gamma exhaust valve is reported to be about four times that of a conventional iron alloy valve because HIPing is needed for gamma valves and machining of gamma alloys is very costly [90,91]. Thus, the development of cast gamma alloys and casting methods which makes it possible to produce automobile exhaust valves without performing HIPing is required.

At the same time, it is important to reduce the relative cost of gamma alloy components with respect to their performance by substantially improving their properties. As an approach with such a greater payoff, we have proposed DS processing of gamma alloys to take advantage of the anisotropic nature of their lamellar structure. PST crystals with orientation A are superior to those with other lamellar orientations in the combination of strength, ductility and fracture toughness. Therefore, if the lamellar orientation can be aligned parallel to the growth direction, the combination of mechanical properties would be optimized in the resulting microstructure consisting of columnar grains with the lamellar structure aligned parallel to the growth direction. The deformation behavior and mechanical properties of PST crystals with A orientation is believed to be directly correlated with microstructural constituents of the lamellar structure such as the volume fraction, thickness and interspacing of alpha-2 lamellae. These microstructure factors can be systematically changed by adding ternary alloying elements, and thereby the development of DS gamma alloys would be able to be made in a rather straightforward manner. The problem facing us now is to establish DS technology which enables us to produce the desired DS microstructure. Mechanical property wise, work is needed to foresee how DS ingots, composed of columnar grains whose deformation is highly anisotropic, turn out to deform. Nothing has been done so far to answer this question. One way to assist in filling this gas is to study the deformation behavior of PST bi-crystals composed of the oriented PST crystals. Such work is currently in progress.

Another topics which emphasis is laid on in this paper is the very strong aluminum content dependence of the deformation modes and mechanical properties of the gamma phase. This is of interest not only from the viewpoint of the deformation mechanisms of the gamma phase but also from the viewpoint of processing of gamma alloys. Normal heat-to-heat aluminum variations have been reported to substantially affect the mechanical properties of gamma alloy ingots. Understanding the reason for such strong sensitivity of the mechanical properties of gamma alloys to aluminum content may help to achieve more easy production of gamma alloys.

All three deformation modes of the gamma phase, in particular twinning and ordinary slip become substantially harder when aluminum content increases from around 50 at.% to 53 at.%. This is believed to be one of the major reasons for the brittleness of single-phase gamma. In the gamma phase in deformed PST crystals, twins and superlattice dislocations widely dissociated involving superlattice intrinsic stacking fault (SISF) abundantly exist, while in single-phase gamma compounds twinning activity is very limited and superlattice dislocations are not dissociated as widely as those in the gamma phase in PST crystals. Thus, it is obvious that the SISF has a much lower energy in the gamma phase in equilibrium with the alpha-2 phase than in the gamma phase containing more than 53 at.% Al [30,84]. In this way, the effect of aluminum content is manifested via a strong influence upon the SISF energy which is assumed to increase with increasing aluminum content. Recent atomistic calculations of core structures of $1/2<1\bar{1}0]$, $<10\bar{1}]$ and $1/2<11\bar{2}]$ dislocations on (111) indicate that the drastic variation of deformation modes of the gamma phase with aluminum content may be interpreted in terms of the variation of core structures of these dislocations induced by the change in the SISF energy [92].

The mechanical properties of single-phase gamma is believed to more directly depend on its oxygen content than those of gamma alloys since the alpha-2 phase which is expected to get considerable amount of oxygen does not exist in single phase gamma. To see the effect of aluminum content on the mechanical properties of single-phase gamma in its more right perspective, deformation experiments should be done using single crystals containing oxygen as low as possible. Work to reduce the oxygen content of not only single-phase gamma single crystals but also PST crystals to 30-50 wt.ppm is currently in progress and the results will be presented elsewhere.

ACKNOWLEDGMENTS

This work was supported by grant-in-aid for Scientific Research on the priority area "Intermetallic Compounds as New High Temperature Structural Materials" from The Ministry of Education, Science and Culture, Japan and in part by the research grants from NEDO and The R&D Institute of Metals and Composites for Future Industries. KK and DRJ would like to thank JSPS for their research fellowships.

REFERENCES

[1] High-Temperature Ordered Intermetallic Alloys I, edited by C. C. Koch, C. T. Liu and N. S. Stoloff (MRS Proc.**39**, Pittsburgh, Pa, 1985).

[2] High-Temperature Ordered Intermetallic Alloys II, edited by N. S. Stoloff, C. C. Koch, C. T. Liu and O. Izumi (MRS Proc.**81**, Pittsburgh, Pa, 1987).

[3] High-Temperature Ordered Intermetallic Alloys III, edited by C. T. Liu, A. I. Taub, N. S. Stoloff and C. C. Koch (MRS Proc.**133**, Pittsburgh, Pa, 1989).

[4] High Temperature Aluminides and Intermetallics, edited by S. H. Whang, C. T. Liu, D. P. Pope and J. O. Stiegler (TMS, Warrendale, Pa 1990).

[5] High-Temperature Ordered Intermetallic Alloys IV, edited by L. A. Johnson, D. P. Pope and J. O. Stiegler (MRS Proc.**213**, Pittsburgh, Pa, 1991).

[6] Microstructure/Property Relationships in Titanium Aluminides and Alloys, edited by Y. W. Kim and R. R. Boyer (TMS, Warrendale, Pa, 1991).

[7] Intermetallic Compounds, edited by O. Izumi (Japan Inst. Metals, Sendai, 1991).

[8] Ordered Intermetallics-Physical Metallurgy and Mechanical Behavior (NATO ASI Series E: Applied Sciences **213**), edited by C. T. Liu, R. W. Cahn and G. Sauthoff (Kluwer Academic Publishers, Dordrecht, 1992).

[9] High Temperature Aluminides and Intermetallics, edited S. H. Whang, D. P. Pope, C. T. Liu, Mater. Sci. Eng.A, **152/153** (1992).

[10] High-Temperature Ordered Intermetallic Alloys V, edited by I. Baker, R. Darolia, J. D. Whittenberger and M. H. Yoo (MRS Proc. **288**, Pittsburgh, Pa, 1993).

[11] Structural Intermetallics, edited by R. Darolia, J. J. Lewandowski, C. T. Liu, P. L. Martin, D. B. Miracle and M. V. Nathal (TMS, Warrendale, Pa, 1993).

[12] Critical Issues in the Development of High Temperature Structural Materials, edited by N. S. Stoloff, D. J. Duquette and A. F. Giamei (TMS, Warrendale, Pa, 1993).

[13] High Temperature Aluminides and Intermetallics, edited by C. T. Liu, S. H. Whang and D. P. Pope, Mater Sci. Eng. A, **192/193**(1995).

[14] High-Temperature Ordered Intermetallic Alloys VI, edited by J. Horton, I. Baker, S. Hanada, R. D. Noebe and D. Schwartz (MRS Proc.**364**, Pittsburgh, Pa, 1995).

[15] Proceedings of International Symposium on Gamma Titanium Aluminides (ISGTA'95), edited by Y-W. Kim, R. Wagner and M. Yamaguchi (TMS, Warrendale, Pa, 1995).

[16] D. Banerjee, A. K. Gogia, T. K. Nandy, K. Muraleedharan and R. S. Mishra, in Ref.[11], p.19.

[17] M. Yamaguchi, Y. Umakoshi and T. Yamane, Phil. Mag. A, **55**, 301 (1987).

[18] G. Hug, A. Loiseau and A. Lasalmonie, Phil. Mag. A, **54**, 47 (1986).

[19] S. Zhang, J. P. Nic and D. E. Mikkola, Scripta Metall. Mater., **24**, 57 (1990).

[20] J. H. Schneibel, J. A. Horton and W. D. Porter, Mater. Sci. and Eng. A, **152**, 126 (1992).

[21] K. S. Kumar and S. A. Brown, Phil. Mag. A, **65**, 91 (1992).

[22] K. S. Kumar and S. A. Brown, Acta Meall. Mater., **26**, 1923 (1992).

[23] Y. Nakayama and H. Mabuchi, Intermetallics, **1**, 41 (1993).

[24] M. Yamaguchi and H. Inui, Intermetallic Compounds: Vol.2, Practice, edited by J. H. Westbrook and R. L. Fleischer (John Wiley & Sons Ltd., West Sussex, England, 1994).

[25] M. Yamaguchi and H. Inui, in Ref. [8], p.217.

[26] H. Inui, M. H. Oh, A. Nakamura and M. Yamaguchi, Phil. Mag. A, **66**, 539 (1992).

[27] H. Inui, A. Nakamura, M. H. Oh and M. Yamaguchi, Phil. Mag. A, **66**, 557 (1992).

[28] H. Inui, M. H. Oh, A. Nakamura and M. Yamaguchi, Acta metall. mater., **40**, 3095 (1992).

[29] M. Yamaguchi, H. Inui, K. Kishida and M. Matsumuro, in Ref.[14], p.3.

[30] H. Inui, K. Kishida, M. Misaki, M. Kobayashi, Y. Shirai and M. Yamaguchi, Phil. Mag. A, in press.

[31] K. Kishida, D. R. Johnson, Y. Shimada, H. Inui and M. Yamaguchi, in Ref. [15], in press.

[32] H. Inui, Y. Toda and M. Yamaguchi, Phil. Mag. A, **67**, 1315 (1993).

[33] Y. Minonishi, Phil. Mag. A, **63**, 1085 (1991).

[34] Y. Umakoshi, T. Nakano, T. Takenaka, K. Sumitomo and T. Yamane, Acta metall. mater., **41**, 1149 (1993).

[35] H. Inui, Y. Toda, Y. Shirai and M. Yamaguchi, Phil. Mag. A, **69**, 1161 (1994).

[36] Y. Umakoshi and T. Nakano, ISIJ Int., **32**, 1339 (1992).

[37] Y. Q. Sun, P. M. Hazzledine and J. W. Christian, Phil. Mag. A, **68**, 471 (1993).

[38] M. Morris, in Ref.[11], p.257.

[39] M. Morris, Phil. Mag. A, **69**, 129 (1994).

[40] Y. Yoshimura, Y. Minonishi and M. Otsuka, Proc. 3rd Japan International SAMPE Symposium, edited by T. Kishi, N. Takeda and Y. Kagawa (The Japan Chapter of SAMPE, Tokyo, 1993), p.1542.

[41] H. A. Lipsitt, D. Shechtman and R. E. Schafrik, Metall. Trans., **6A**, 1991 (1975).

[42] S. C. Huang and E. L. Hall, Metall. Trans., **22A**, 427 (1991).

[43] D. S. Shih, S. C. Huang, G. K. Scarr, H. Jang and J. C. Chesnutt, in Ref. [Micro/Property], p.135.

[44] S. Krishnamurthy and Y. W. Kim, in Ref. [6], p.149.

[45] S. Yokoshima and M. Yamaguchi, Acta metall. mater., in press.

[46] W. L. Server, R. A. Wullaert and R. O. Ritche, J. Eng. Mater. Tech., **102**, 192 (1980).

[47] Y. Mutoh, I. Sakamoto, K. Nomura and K. Aya, Trans., Jpn. Soc. Mech. Eng., Ser. **468A**, 2008 (1985).

[48] Y. Mutoh, I. Sakamoto and S. Takeda, Jpn. Soc. Mech. Eng., Ser. **503A**, 1434 (1988).

[49] R. Gnanamoorthy, Y. Mutoh, N. Masahashi and Y. Mizuhara, Metall. Trans., **26A**, 305 (1995).

[50] R. E. Schafrik, Metall. Trans., **8A**, 1003 (1977).

[51] C. l. Fu and M. H. Yoo, Phil. Mag. Lett., **62**, 159 (1990).

[52] T. Nakano, T. Kawanaka, H. Y. Yasuda and Y. Umakoshi, Proc. 3rd Japan International SAMPE Symposium, edited by T. Kishi, N. Takeda and Y. Kagawa (The Japan Chapter of SAMPE, Tokyo, 1993), p.1334.

[53] B. Kad, B. F. Oliver and P. M. Hazzledine, Microstructure Science Vol.18, edited by T. A. Place, J. D. Braun, W. E. White and G. E. Vander Voort (ASM International, Materials Park, OH, 1991), p. 201.

[54] F. Appel, U. Christoph and R. Wagner, to be published in Phil. Mag. A.

[55] M. H. Yoo and C. L. Fu, Mater. Sci. Eng., **A153**, 470 (1992).

[56] K. S. Chan and Y. W. Kim, Metall. Trans., **23A**, 1663 (1992).

[57] K. S. Chan and Y. W. Kim, Acta metall. mater., **43**, 439 (1995).

[58] S. Yokoshima and M. Yamaguchi, Proc. 3rd Japan International SAMPE Symposium, edited by T. Kishi, N. Takeda and Y. Kagawa (The Japan Chapter of SAMPE, Tokyo, 1993), p.1346.

[59] H. E. Deve, A. G. Evans and D. S. Shih, Acta metall. mater., **40**, 1259 (1992).

[60] N. J. Rogers and P. Bowen, in Ref. [ISSI], p. 231.

[61] F. E. Heredia, M. Y. He, G. E. Lucas, A. G. Evans, H. E. Deve and D. Konitzer, Acta metall. mater., **41**, 505 (1993).

[62] K. S. Chan, Metall. Trans., **20A**, 155 (1989).

[63] K. S. Chan, Metall. Trans., **20A**, 2337 (1989).

[64] K. T. Venkateswara Rao, W. Yu and R. O. Ritchie, Metall. Trans., **19A**, 549 (1988).

[65] K. T. Venkateswara Rao, G. R. Odette and R. O. Ritchie, in Ref. [ISSI], p.829.

[66] Y. W. Kim and D. M. Dimiduk, JOM, **43**, 40 (1991).
[67] Y. W. Kim, JOM, **46**, 30 (1994).
[68] W. J. Boettinger, Met. Trans., **5**, 2023 (1974).
[69] D. R. Johnson, H. Inui and M. Yamaguchi, Acta metall. mater., to be published.
[70] P. P. Rao and K. Tangri, Mater. Sci. Eng. A, **132**, 49 (1991).
[71] G. Hug, A. Loiseau and P. Veyssiere, Phil. Mag. A, **57**, 499 (1988).
[72] E. L. Hall and S-C. Huang, J. Mater. Res., **4**, 595 (1989).
[73] D. G. Morris, S. Günter and Leboeuf, Phil Mag. A, **69**, 527 (1994).
[74] S. Sriram, V. K. Vasudevan and D. M. Dimiduk, Mater. Sci. Eng. A, **192/193**, 217 (1995).
[75] M. Yamaguchi and Y. Umakoshi, Prog. Mater. Sci., **34**, 1 (1991).
[76] T. Kawabata, T. Kanai and O. Izumi, Acta metall., **33**, 1355 (1985).
[77] T. Kawabata and O. Izumi, in Ref.[4], p.403.
[78] S. H. Whang, in Ref.[8], p.279.
[79] M. A. Stucke, V. K. Vasudevan and D. M. Dimiduk, Mater. Sci. Eng. A, **192/193**, 111 (1995).
[80] N. Bird, G. Taylor and Y. Q. Sun, in Ref.[14], p.635.
[81] B. Viguier, J. Bonneville, K. J. Hemker and J. L. Martin, in Ref.[14], p. 629.
[82] Z. X. Li and S. H. Whang, Mater. Sci. Eng. A, **152**, 182 (1992).
[83] Z. M. Wang, Z. X. Li and S. H. Whang, Mater. Sci. Eng. A, **192/193**, 211 (1995).
[84] H. Inui, M. Masumuro, M. Yamaguchi, to be published.
[85] B. Viguier, M. Cieslar, K. J. Hemker and J. L. Martin, in Ref.[14], p.653.
[86] F. Louchet and B. Viguier, Scripta Metall. Mater., 31, 369 (1994); in Ref.[14], p.163.
[87] S. Takeuchi and E. Kuramoto, Acta Metall., **21**, 415 (1973).
[88] C. Lall, S. Chin and D. P. Pope, Metall. Trans. A, **10**, 1323 (1979).
[89] M. A. Stucke, D. M. Dimiduk and P. M. Hazzledine, in Ref. [10], p. 471.
[90] T. Morikawa, Abstracts of Jpn. Inst. Metals Meeting on Intermetallics - 25 Nov. 1994 (Jpn. Inst. Metals, Sendai, 1994), p.37.
[91] W. E. Dowling, Jr., W. T. Donlon and J. E. Allison, in Ref.[14], p.757.
[92] A. Girshick and V. Vitek, in Ref.[14], p.145; V. Vitek in this proceedings.

STRUCTURE-PROPERTY RELATIONSHIP OF METAL-CERAMIC INTERFACES

J.Th.M. De Hosson[1], W.P. Vellinga[1], X.B. Zhou[1], V. Vitek[2]

[1]Department of Applied Physics, Materials Science Center
University of Groningen, Nijenborgh 4, 9747 AG Groningen
The Netherlands.
[2] Department of Materials Science and Engineering, University of Pennsylvania, 3231 Walnut Street, Philadelphia, PA 19104, USA.

INTRODUCTION

The main emphasis of this book lies on the structure-property relationship, so as to bridge the length-scale gap involved in the study of materials stability. This paper focuses on several aspects within this framework by looking at the detailed atomic structure on a nanometer scale. Next, by using this information a transition is made via a mesoscopic length scale to explain the mechanical properties of metal-ceramic interfaces at a macroscopic scale. The basic philosophy is that depending on the material property one likes to investigate, it is necessary to consider the appropriate length scale where the relevant processes take place. The challenging part of this mesoscopic approach is of course to define the physical basis at the various length scales and the corresponding transitions involved. For that reason basically two case studies will be worked out. The first one is dealing with the atomic description of metal-ceramic interfaces using high resolution transmission electron microscopy as the main tool. The question to be answered is whether this image information can be interpreted in terms of the atomic structure and whether the latter will provide some information of the bond-strength between two dissimilar materials like a metal and a ceramic material. In order to make the transition from this nanoscale regime to a mesoscopic scale an anisotropic linear elasticity description will be employed as well. As a matter of course the way the material is processed is usually crucial for the mechanical property. For ultra high resolution transmission electron microscopy it is of great help to have well defined samples and for that reason the research on a nano-scale is carried out on internally oxidized metals with reasonably well defined ceramic precipitates. On the other hand for practical purposes it will be shown that high power laser systems may have a great potential in producing ceramic-layers on top of a metallic substrate. Here the interest is to obtain an interface between a ceramic layer that is firmly bonded to the underlying metallic substrate. But how can we determine the interface strength of these laser processed materials and link concepts of bond strength from an atomic scale to the macroscopic properties? These question will be addressed in the second case study.

Interfaces between metals and oxides have been the subject of extensive research in recent years since they control to a great extent properties of metal-ceramic composites, protective coatings, thin metal/ceramic films in electronic devices etc. For instance: The low hardness and poor wear resistance are the principal reasons that limit the potential applications of aluminum alloys. Hardfacing on aluminum alloys, e.g. by electro-deposition and anodizing, may improve the hardness and wear resistance significantly. However, there are still major drawbacks in these conventional methods. The bonding between these coatings and aluminum alloys is usually weak and may cause failure during application. The amorphous anodizing layer is brittle with a relatively low hardness ($HV_{0.1}$ 250 -500) to resist abrasive wear. For these reasons it would be of great practical importance to have a basic fundamental understanding of the bonding between dissimilar materials like metals and protective ceramic layers.

The theoretical interface studies have focused on understanding the cohesion at the interface[1], principally in the realm of its chemistry and electronic structure. This can be done adequately using ab initio electronic structure calculations and a number such studies have been made recently [2-6]. However, these calculations are limited to very short period structures and usually do not allow for relaxations of the positions of atoms. Hence, ab initio studies have been carried out so far only for fully coherent boundaries. In reality, however, there is usually a misfit which arises because there is a difference between lattice parameters of the metal and the ceramic substrate. In this case the period of the interface is related to the difference in the lattice parameters and is usually much larger than the period of the crystal planes parallel to the interface. In addition, significant relaxations may take place near the interface leading to the formation of locally coherent regions separated by localized incoherent regions that can be identified with the cores of misfit dislocations.

This phenomenon was first noted in the case of epitaxial layers and widely studied in this context [7-9]. These dislocations, unlike the dislocations in the bulk, are not defects but an integral part of the interfacial structure. They are positioned either at or very near the interface in the elastically softer material, and they reduce the misfit strain with respect to the unrelaxed fully coherent interface by locally decreasing the interfacial coherency [10, 11].

Experimentally, the atomic arrangement in metal/ceramic interfaces has been investigated using high resolution electron microscopy (HREM)[12-21] and field ion microscopy[22]. While only a limited number of model systems are suitable for HREM investigations, the relaxations arising due to the misfit, have been observed in a number of cases. Since they are manifested as misfit dislocations, the atomic structure of the incoherent regions was interpreted in terms of the core structure of these dislocations[14, 23-27]. Relaxations of this type, which to a great extent characterize the atomic structure of interfaces, have been studied atomistically in metal/metal or ceramic/ceramic interfaces, in particular in grain boundaries, using a variety of approximate schemes describing atomic interactions (for reviews of such schemes see [28-30]). Unfortunately, as discussed by Finnis [31], for metal-ceramic bonding an analogous simple scheme for the description of interatomic forces cannot be easily formulated and the schemes employing image charge effects[32-34] are not fully atomistic. Hence, studies of misfit dislocations which fully incorporate atomic structure and bonding in both the ceramic substrate and the metal cannot be made at this stage.

The treatment which does not require any atomic level information is, of course, the continuum theory of dislocations and, indeed, in the past studies of misfit dislocations have been made almost exclusively on this basis[35-39]. However, this is not adequate when information about the core structure of misfit dislocations is needed, for example when making comparison with the HREM observations. For this reason Vitek et al. [27] suggested a very simplified model which while treating the metal atomistically, is close to continuum treatment. In particular, in this model the atomic interactions in the metal are described by Finnis-Sinclair type many-body central force potentials[40], the ceramic crystal was regarded as a rigid, undeformable substrate so that no description of interatomic forces in this material was needed. The atomic interactions across the interface are then described by pair potentials which lead to different strengths of bonding and the core structure of the misfit dislocations can be investigated as a function of the strength of bonding across the interface. This model was applied to the study of misfit dislocations at two different Nb/sapphire interfaces and an excellent agreement between the observed and calculated core structures of misfit dislocations was found for reasonable strengths of bonding at the interface.

Here we adopt this model to a system which resembles an interface between silver and an oxide with the NACl type structure, such as, for example, MgO. The interface plane is of the {100} type in both metal and substrate, with parallel <100> directions. In atomistic studies we have concentrated on the interplay between misfit, bond strength and interface energy. In parallel we carried out linear elastic continuum calculations for an array of misfit dislocations lying at the same interface. Both isotropic and anisotropic elasticity were employed. These dislocations are of Voltera type and have singular cores. The comparison of atomistic and continuum results was then made in order to establish whether the structure of the interface may be considered, at least for some combinations of the bond strength and misfit, as a network of Voltera dislocations for which no core structure considerations are needed.

NANOSCOPIC SCALE

Model Calculations

Atomistic approach: The model proposed by Vitek et al. enables us to study structural features of a metal in contact with a mismatched periodic potential. It uses a simplified atomistic description of the substrate and of the interaction between metal and substrate. Furthermore the bond strength across the interface has been parametrised.

Both metal and a ceramic substrate are represented by rectangular blocks of atoms. They are brought into contact with a chosen orientation relationship along a certain plane. Misfit is introduced by building a block with unequal numbers of unit cells in the metal and the substrate. This means that the misfit is commensurate and in general not exactly equal to the misfit that might exist on a real interface. Simulation of an interface with misfit in two directions is also possible.

The potentials used to describe the interactions between the metal atoms are of the Finnis-Sinclair type[40]. In this scheme the energy of an atom i is given by:

$$E_i = \sum_j V(R_{ij}) - \sqrt{\sum_j \Phi(R_{ij})} \quad . \qquad (1)$$

The first term is a repulsive pair part, the second term an attractive many-body part. The functions V and Φ are short range pair potential fitted to reproduce experimental data, such as: lattice parameter, cohesive energy, elastic moduli[41]. Summation is for all atoms j within the cut-off radius of the potentials applied.

The interactions across the boundary, between metal and substrate atoms, are modelled with an effective pair potential for silver, V_{eff}. The form is derived from the Finnis-Sinclair potential and is described in[40]. It amounts to an approximation to equation (1), if second and higher derivatives of $\Phi$are neglected, that is, no large changes in co-ordination occur. In order to simulate different bond strength the potential can be multiplied by a factor α and takes the following form:

$$U_{MS} = \alpha V_{eff}(R) = \alpha \left\{ V(R) - \frac{\Phi(R)}{\sqrt{\sum_j \Phi(R_j^0)}} \right\} \quad . \tag{2}$$

In the present study the atoms in the substrate were kept fixed, so no interaction functions were necessary between the atoms of the substrate whereas the semi-empirical N-body potential of Ag was used to describe the metal layer. The substrate may also be modelled with another parametrised potential, in order to simulate the effects of elastic mismatch.

For the calculation of the interface energy the following equation is employed:

$$U_{int} = \sum_i (E_i - E_i^0) + \sum_{i,k} V_{MS}(R_{ik}) \quad . \tag{3}$$

In the first term the summation is over all atoms i in the metal. E_i includes only the Finnis-Sinclair part of the energy of the atoms, calculated with equation (1). E^0_i is the cohesive energy in the bulk. The second term is a summation of effective pair interactions over all pairs i-k interacting across the interface, with i representing atoms in the metal and k atoms in the substrate.

The initial energy of the substrate is set to zero in this way. A choice has to be made in the absence of potentials describing the substrate. The term will not change during relaxation because displacements in the substrate are not allowed, and therefore has no effect on calculated differences in energy between interfaces with unequal misfit or interaction parameter. However, all absolute values of energy given later are with respect to some initial energy in the substrate.

A lattice with monatomic basis representing the substrate has been chosen. A structure closely resembling the substrate under investigation seems reasonable. We represent the ionic substrate by an fcc lattice. Starting from an initial configuration and chosen values of interaction parameter α, the block is relaxed using a steepest descent method. Blocks are usually relaxed until no forces larger than 1.10^{-3} ev/nm are found on any atom. Especially for small misfit, i.e. large blocks, dependence of the interface energy on initial spatial configuration is expected to be negligible: the misfit may lead to a set of similar but slightly different configurations at the interface out of which the dislocation core may form. Changing the initial configuration will just shift the position of the "dislocation" nucleated. Relaxed blocks with misfit in two directions showed displacements along the <110> directions, as is expected from O-lattice theory[42]. This result was used to choose <110> as direction of misfit in blocks with misfit in one direction.

Elasticity approach: A test on the localization of the resulting atomic configurations can be carried out by calculating the strain energy associated with the interface dislocation. Here we are in particular interested to compute the so-called energy factor K of an interface dislocation which is related to the elastic self energy per unit length along the axis of a cylinder of radius R around the dislocation:

$$E_{dis} = E_{core} + K \ln \frac{R}{r_c} \quad , \tag{4}$$

where r_c represents the core radius. The core energy is usually described by K ln (r_c/r_{eh}) where r_{eh} is called the 'equivalent hole' radius.

Further, we are interested to see what the approach taken in the atomistics for an infinitely stiff substrate means, i.e. as regards the neglect of any elastic field inside the substrate. These fields can be predicted by anisotropic elasticity theory as a function of stiffness. In addition from the latter the climb component of the force perpendicular to the interface can be easily calculated from the self-energy of an interface dislocation. Details of the elastic description and calculations are summarized in the appendix. In the continuum approach we take the Burgers vector of the interface dislocation parallel to the interface and in the actual calculations it is of perfect edge character. However, the formulation is still quite general for any type of dislocation parallel to the interface. Further, by means of the introduction of a clamping factor, which is actually correlated with the ratio of the elastic constants between layer and substrate, the solutions apply to both rigid and deformable substate material.

The starting point is the displacement field **u** of a single dislocation in an anisotropic linear elastic medium, the solution of which can be found in several textbooks on dislocation theory [43,44]. It is shown that this can be represented in the form of a superposition of partial solutions α:

$$\mathbf{u} = \frac{1}{2\pi i} \sum_{\alpha} D(\alpha) \mathbf{A}_\alpha \ln(\xi_1 . \mathbf{r} + p_\alpha \xi_2 . \mathbf{r}) \quad , \tag{5}$$

where ξ_1,ξ_2 and ξ_3 represent the three mutually orthogonal unit vectors along the interface, perpendicular to the interface and along the dislocation line, respectively. For each eigenvalue p_α there exists an eigenvector $\mathbf{A}_\alpha$. Matching **u** with the displacement discontinuity of the Burgers vector **b** in the Burgers circuit together with the constraint of balance with an external force per unit length yields six equation form which the six unknown D(α) can be obtained. The stress components are readily obtained from the derivatives of the displacement field with respect to ξ_i leading to :

$$\sigma_{ij} = \frac{1}{2\pi i} \sum_{\alpha} A_{k\alpha} b_m L_{m\alpha} \frac{C_{ijkl}(\xi_{1,l} + p_\alpha \xi_{2,l})}{\xi_1 . \mathbf{r} + p_\alpha \xi_2 . \mathbf{r}} \quad , \tag{6}$$

where C_{ijkl} is the elastic constant in the co-ordination system of the dislocation. From the stress (Eq.6) in the interface plane ξ_2.**r**=0 , the elastic energy per unit length is:

$$E = \frac{1}{4\pi i} \sum_{\alpha} \frac{L_{r\alpha} L_{s\alpha} b_r b_s}{A_{m\alpha} L_{n\alpha}} \ln(\frac{R}{r_c}) \quad . \tag{7}$$

The prefactor is called the energy factor K, which was first introduced by Foreman [45]. It depends not only on the elastic constants and on the dislocation orientation but also on the magnitude of the Burgers vector. Here L_{ij} and A_{kl} express Lekhnitskii's representation being valid for the general anisotropic case[46]. This vector, **L**, is the very same vector introduced by Stroh[47,48], who was the first to mention explicitly the orthogonality relation between **A** and **L**: $A_{k\alpha} L_{k\beta} + A_{k\beta} L_{k\alpha} = 0$ for any $\alpha,\beta=1,2,3$, $\alpha \neq \beta$. In fact Stroh's vector **L** is the normalized Lekhnitskii's vector by $2\,\mathbf{A}_\alpha \cdot \mathbf{L}_\alpha = 1$.

For the interface dislocation we seek solution of the same form as expressed in equation (5) but with the constraint of a continuity in tractions as well as in displacements across the interface between top layer-medium (1) and substrate-medium (2). In principle we have twelve unknowns: six of $D^{(1)}(\alpha)$ and six due to $D^{(2)}(\alpha)$. Matching **u** with the displacement discontinuity of the Burgers vector **b** in the Burgers circuit and the absence of external forces yields six equations. Taking into account he constraints on the continuities in displacements and tractions, in total twelve independent equations are provided. The conditions on the Burgers vector and the absence of external forces gives, together with the orthogonality relations of Stroh, yield the following expression:

$$\sum_{\alpha} \mathbf{C}^{(1-2)}.\mathbf{L}_\alpha D^{(1)}(\alpha) = \mathbf{b} \quad , \tag{8}$$

where

$$C_{kl}^{(1-2)} = \sum_{\beta} A_{k\beta}^{(1)} (L^{-1})_{\beta l} + A_{k\beta}^{(2)} (L^{-1})_{\beta l}^{(2)} \quad . \tag{9}$$

$D_\alpha^{(1)}$ can now be obtained from equation (8) and consequently $D_\beta^{(2)}$ by using the continuity of displacements and tractions. Since the displacement fields and stress fields are now fully determined the energy factor of the interface dislocation can be calculated according to (see equation (7) in Lekhnitskii's formulation):

$$K_{int}^{(1-2)} = -\frac{1}{4\pi i} \sum_{\alpha} D^{(1)}(\alpha) \mathbf{L}_\alpha^{(1)}.\mathbf{b} \quad . \tag{10}$$

Next, the energy factor can be employed to predict the interface energy as a function of misfit δ. The total misfit δ between the top layer and the substrate leads to an elastic energy of a coherent boundary in 1-D of:

$$E_\delta = d_l \mu_l \left(\frac{1+\nu_l}{1-\nu_l} \right) \delta^2 \quad , \tag{11}$$

where it is assumed that the thickness of the substrate d_s is much larger than the thickness of the layer d_l and all the elastic energy is confined to the top layer. μ_l and ν_l represent the shear modulus and Poisson's ratio of the top layer. The total misfit may be accommodated by both a dislocation network γ and a coherent elastic strain ε, where $\delta=\gamma+\varepsilon$. The total energy is then given by:

$$E_{elas} = K_{int}^{(1-2)} \left(\frac{\delta-\varepsilon}{b} \right) \ln \left(\frac{R}{r_{eh}} \right) + \mu_l d_l \left(\frac{1+\nu_l}{1-\nu_l} \right) \varepsilon^2 \tag{12}$$

for a network of non-interacting dislocations at distances $b/(\delta-\varepsilon)$ apart. r_{eh} , the equivalent hole radius, is about b/2 for metals[43]. Finally, the energy factor can also be used to predict the force exerted on the dislocation across the interface. It can be shown that the energy difference involved upon moving a dislocation originally at r to a distance r' from the interface is simply[48]:

$$\Delta E = (E^{(1)} - E^{(1-2)})\ln(\tfrac{r}{r'}) \quad , \tag{13}$$

assuming that the outer elastic radius R is much larger than r and r'. Taking the negative value of the derivative of ΔE with respect to r, the climb component of the force perpendicular to the interface can be written, using equations (10) and (13) as:

$$F_{|\xi_2|} = -\frac{1}{4\pi r}\left(\Sigma D^{(1)}(\alpha)\mathbf{L}_\alpha^{(1)}.\mathbf{b}_1 - \Sigma D^{(1)}(\alpha)\mathbf{L}_\alpha^{(1)}.\mathbf{b}\right) \tag{14}$$

where $\mathbf{b}=\mathbf{b}_1+\mathbf{b}_2$.

RESULTS

Relaxation of the interface at various values of interaction parameter α and misfit δ leads to different values of the interface energy calculated by equation (3), as can be concluded from figure 1. The curves in figure 1 illustrate the fact that misfit can only influence the structure if there is interaction between the metal and the substrate.

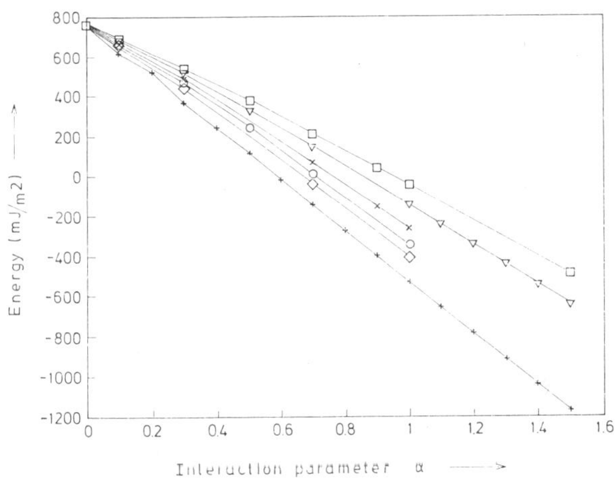

Figure 1. Energy of the interface at various values of α and misfit δ: □ 20%, ∇ 10%, x 4%, o 2%, ◊ 1%, + 0%.

At zero interaction parameter α the energies for all interfaces are of course the same and equal to the interface energy of a relaxed Ag {100} surface, as calculated using FS interaction functions. Upon changing interaction parameter and misfit the following trends become clear: An increase in interaction parameter α leads to a decrease in interface energy for any misfit , whereas at constant α the energy increases with increasing misfit δ. This is what one should expect. It is clear that zero misfit leads to the lowest possible energy for any α. The increase in energy with respect to this situation represents the elastic energy present in the block. It is tempting to interpret the energy differences thus calculated as the elastic energy of an array of interface dislocations, but this clearly leads to the question: for which combination of α and δ ? Cases with α=0 clearly do not qualify because there is no displacement and stress field. The problem is to decide in which cases the structure at the interface can be qualified as an array of interface dislocations. We have approached the problem by looking at the total energy of the "dislocations" as a function of distance to the centre. Values of the following function have been calculated for several cases:

$$E_{dis}(R) = \sum_{i(R_i \leq R)} (E_i - E_{layer}^{\delta=0}) \qquad , \qquad (15)$$

with i for atoms in the metal only. For any dislocation the dependence of the energy on R will look like equation (4) for distances larger than the core radius r_c. In those cases a logarithmic plot of E(R) vs. ln(R) will give a straight line with a slope equal to the energy factor K. Again we calculate the energy with respect to a block with equal α and zero misfit δ, but because individual atoms are treated now this needs some precisioning. Actually, the energy of atoms in a block with α and zero misfit only depends on the distance to the interface. So we subtract from the absolute energy of the atom, the energy of an atom in a block with equal α and zero misfit that is located at the same distance to the interface, that is, in the same layer. This is the term $E_{layer}^{\delta=0}$. Summing the energies of all atoms would give exactly the same value as in figure 1.

In figure 2 the elastic energies as calculated with equation (15) are depicted for several cases. It is clear that in the case of 4% misfit the structure is dislocation-like, i.e. in terms of a singular Volterra type, at α=5, but not for the lower values. For the case with 2% however, dislocation-like behavior is already found at α=2. For lower interaction parameters the excess energy is situated more at atoms near the interface, giving rise to an almost linear increase in energy with distance, and a corresponding curved line in a logarithmic plot.

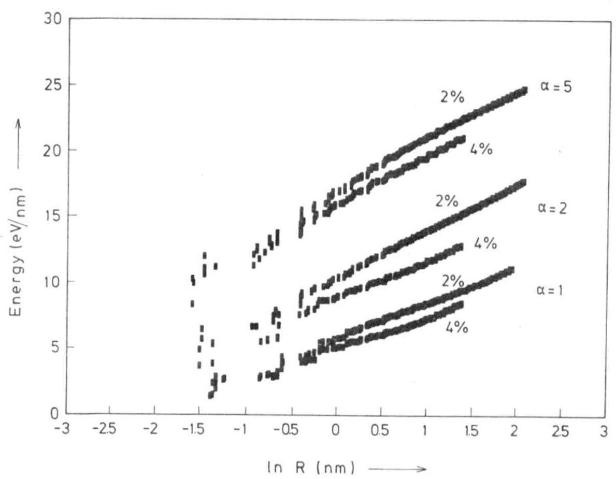

Figure 2. Elastic energy (eV/nm) plotted vs Ln(R/a_0) for various combinations of misfit δ and interaction constant α. R represents the distance from the interface and a_0 is the lattice parameter of the unstrained top layer. The gradients of the various curves yield a energy factor K of 4 eV/nm.

The slope of the plots, indicating the energy factor of the misfit dislocations, is about equal for all cases, and has a value close to 4 eV/nm. Interestingly, at α=5 the energy in the core is somewhat larger for 2% misfit than for 4% misfit. This makes clear that the core energy and structure depend on both misfit and bond strength across the interface. It means that it is, in principle at least, possible to assess bond strength by investigating the atomic structure of the cores. Secondly it can be seen that low misfit is favorable for localization. Dislocation-like structures, as judged from their energy, nucleate at lower interaction parameters. At equal interaction parameter the energy contained in the area around the core increases somewhat with decreasing misfit. These points prove to be important in the following comparison with elastic results.

Various components of the stress field (see equation (6)) are displayed in figure 3. It is quite clear that a substantial stress is developed in the underlying substrate (MgO) whereas the detailed shape of the stress fields depend strongly on the elastic constants of the top layer as well.

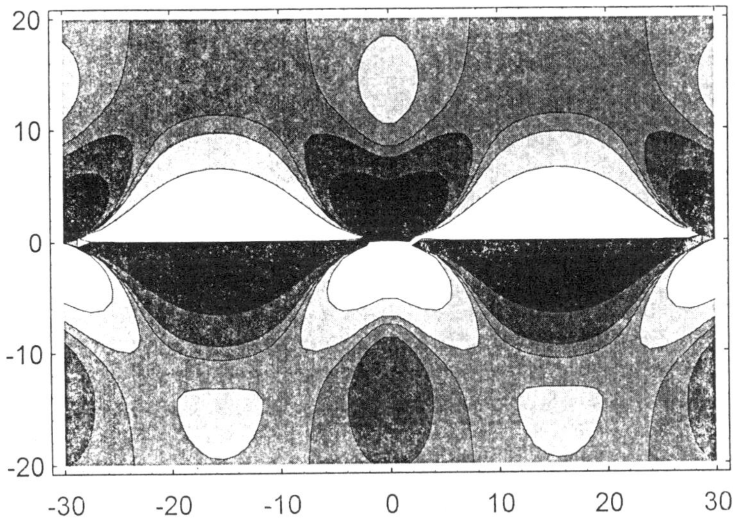

Figure 3a. Stress field components of a dislocation array along an interface between Ag (top) and Al (top) and MgO (bottom). σ_{xx} (Ag/MgO) range -4 to 4 GPa.

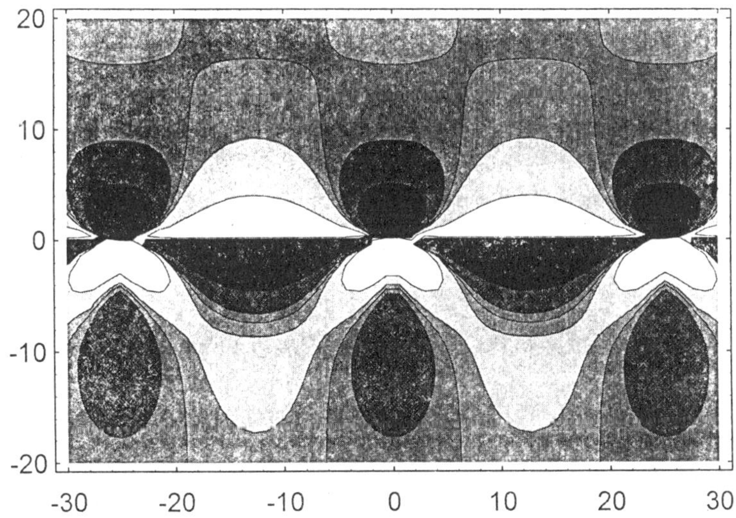

Figure 3b. Stress field components of a dislocation array along an interface between Al (top) and MgO (bottom). σ_{xx} (Al/MgO) range -4 to 4 GPa.

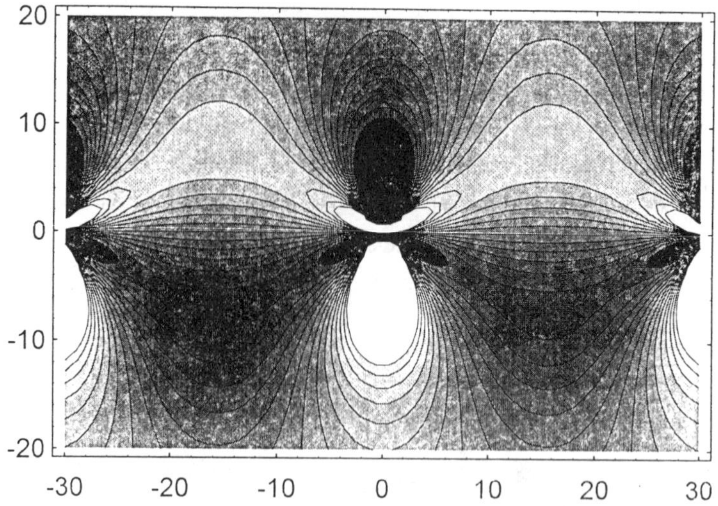

Figure 3c. Stress field components of a dislocation array along an interface between Ag (top) and MgO (bottom). σ_{xy} (Ag/MgO).

The energy factor of a dislocation at the interface between Ag/MgO is calculated 3.31 eV/nm by using equation (10). The elastic constants were taken from[43]. Figure 4 displays the energy factor K as a function of a clamping factor λ, where the elastic constant C_{ijkl} of the substrate are taken to be λ times that of the substrate.

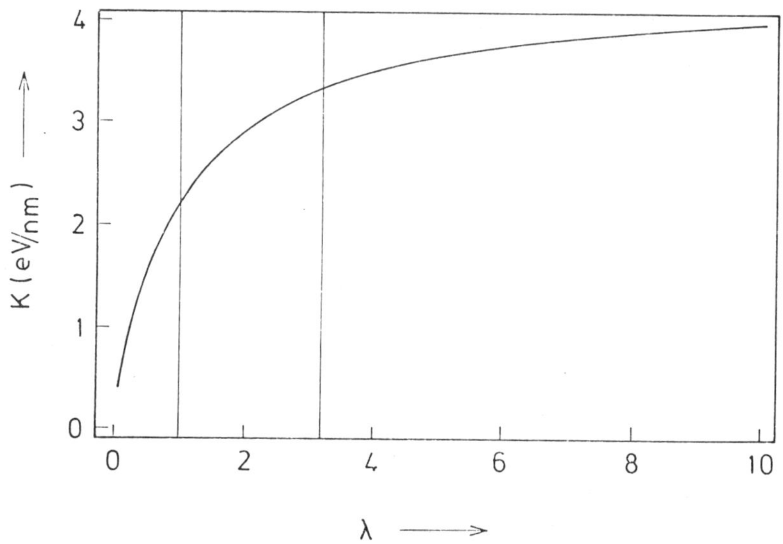

Figure 4. The energy factor (eV/nm) as a function of the clamping factor λ between an Ag top-layer and a substrate. The vertical lines (a) and (b) refer to a substrate of Ag and a substrate of MgO, respectively.

$\lambda \to \infty$ K becomes 4.39 eV/nm. It is noticed that at λ=3.21, the actual ratio of the shear moduli between MgO and Ag, is estimated 3.34 eV/nm , i.e. close to the value 3.31 eV/nm. The atomistic calculation yields an energy factor of about 4.0 eV/nm which is in reasonable accordance with the predictions of a top layer firmly clamped onto an infinitely stiff substrate.

The question should be addressed whether a large error would be introduced upon neglecting anisotropy and assuming isotropic linear elasticity theory instead. For two different isotropic elastic media clamped together, the energy factor is described by[49]:

$$K_{iso}^{(1-2)} = \frac{b^2}{2\pi}\mu_1\mu_2\frac{\mu_2(\kappa_1+1)+\mu_1(\kappa_2+1)}{(\mu_2\kappa_1+\mu_1)(\mu_1\kappa_2+\mu_2)} \quad , \tag{16}$$

where $\kappa_i = 3-4\nu_i$. In Fig.5 $K_{iso}^{(1-2)}$ vs μ_2 is depicted, together with the separate contributions from the substrate, K_s, and the top layer, K_l.

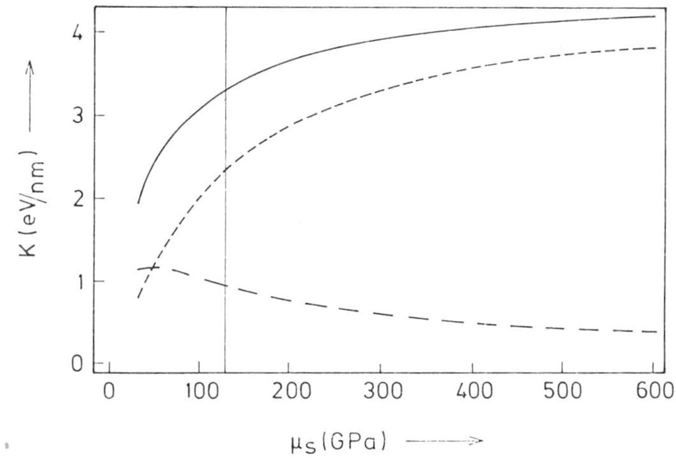

Figure 5. The energy factor (eV/nm) of a dislocation at the interface between Ag and a substrate as a function the shear modulus (in GPa) of the substrate. The Poisson's ratio of the substrate is set equal to that of the top layer. The dashed line represent the contribution from the substrate to the energy factor whereas the dotted line shows the contribution from the top layer. The vertical line refer to the situation of a substrate of MgO.

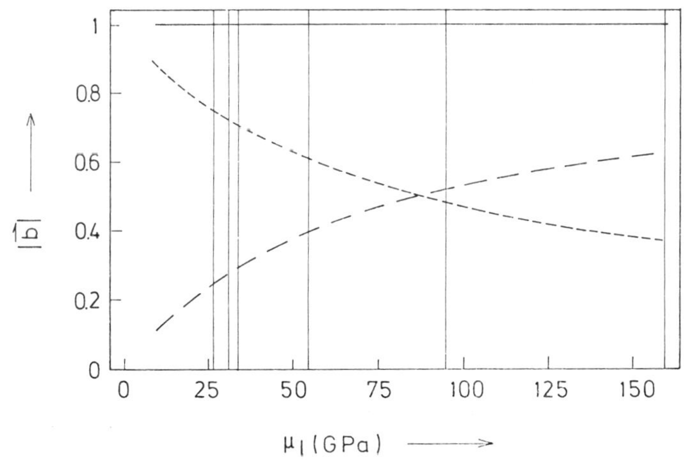

Figure 6a. Distribution of the size of the total Burgers vector (solid line) over the substrate MgO (dashed) and top layer as a function of the shear modulus of the top layer μ_l in GPa. The Poisson's ratio of the substrate is set equal to that of MgO and of the top layer equal to that of Ag. The vertical lines refer to the top layer of: Al, Au, Ag, Cu, Ni and W (left to right).

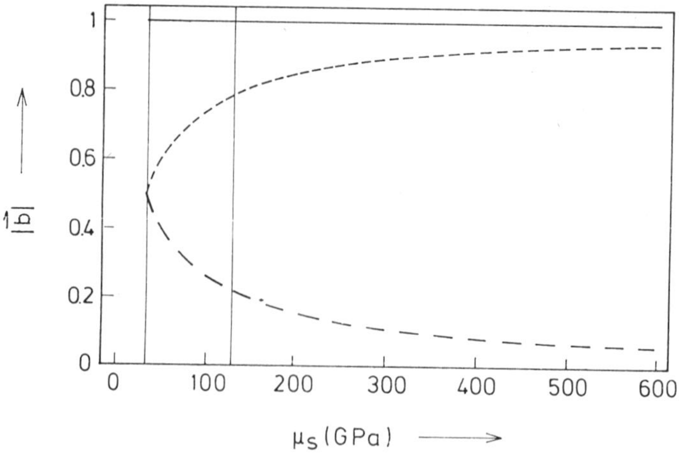

Figure 6b. Distribution of the size of the total Burgers vector (solid line) over the substrate (dashed) and top layer of Ag as a function of the shear modulus of the substrate μ_s in GPa. The Poisson's ratio of the substrate is set equal to that of Ag. The vertical lines (a) and (b) refer to a substrate of Ag and MgO, respectively.

The latter can be found by calculating the magnitude of the separate Burgers vectors $\mathbf{b}_s$ and $\mathbf{b}_l$ in the substrate and top layer, respectively. These are calculated by realizing that from the displacement discontinuity **b** and the requirement of balance of **F** per unit length follows (using equation (8)):

$$D^{(1)}(\alpha) = \mathbf{L}_\alpha^{-1}.\mathbf{C}_{(1-2)}^{-1}.\mathbf{b} = \mathbf{b}_1.\mathbf{L}_\alpha^{(1)} - \mathbf{F}^{(1)}.\mathbf{A}_\alpha^{(1)} \quad (17)$$

With equation (14) for the force exerted on medium (1) though the interface by medium (2), $\mathbf{b}_l$ can be expressed in terms of **b** .

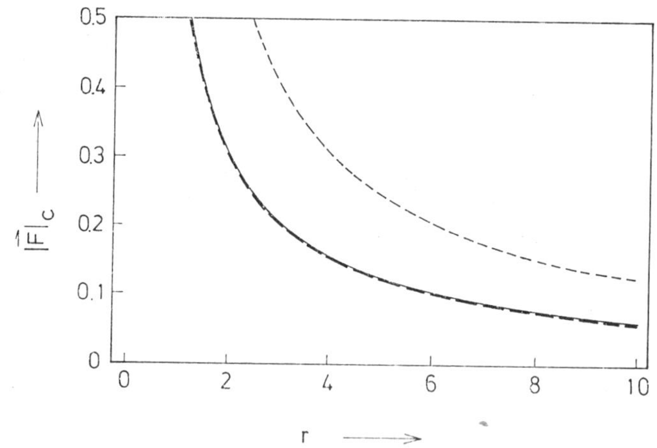

Figure 7. The climb force component (N/m) exerted on a dislocation in Ag as a function of the distance (units of to the interface between Ag and a substrate. Solid line: anisotropy Ag/MgO , dashed line : isotropy, dotted line: infinitely stiff substrate.

For the isotropic case the magnitude of these Burgers vectors are shown in Fig 6a and 6b. as a function of μ_l and of μ_s , respectively. $K_{iso}^{(1\text{-}2)}$ in case of the system Ag/MgO is found to be 3.27 eV/nm , whereas for $\mu \rightarrow \infty$ the energy factor becomes 4.58 eV/nm, i.e.

about 10% higher than the corresponding value calculated for a pure ½<011>{001} edge dislocation using anisotropic linear elasticity. The climb force exerted on the interface dislocation as a function of the distance to the interface is displayed in figure 7.

HRTEM EXPERIMENTS

The underlying physics of the bonding across metal-ceramic interfaces will be reflected in their structure at an atomic scale. Imaging such interfaces with HRTEM allows the study of the interface at an atomic scale and thus seems to be suited to gain insight in the basic physical concepts. At metal-ceramic interfaces there is in general a mismatch between the lattice parameters of the two constituents. The interface therefore shows regions of good register separated by regions of bad register. In general displacements caused by the interaction across the interface are expected to take place, increasing coherency in regions of good register and forming cores of misfit dislocations in regions of bad register. As a consequence extraction of these displacements from the images is an important step that has to be made and the need for an understanding of the physics at the interface at an atomistic scale becomes apparent at that moment.

The model presented in the previous section shows that obviously there will be no dislocation core if there is no bonding across the interface. On increasing the bond strength, and keeping the misfit constant, a localized core structure will start to form. For a bond strength greater than a particular value the structure formed is found to be dislocationlike, i.e. in terms of a singular Volterra type. Furthermore the bond strength at which this localization takes place is seen to depend on the misfit. Core structures formed at interfaces with low misfit will be dislocationlike for lower values of bond strength than similar structures at interfaces with higher misfit. The main conclusion drawn from this investigation is that the atomic structure of the interface was seen to depend on both misfit and bond strength, and that in principle information on bond strength can be extracted from it.

With this in mind we turn to the interpretation of HRTEM images of an interface between {111} planes in Cu and {111} planes in MgO. Our main interest here is in the treatment of an interface with a misfit in two directions, a situation that is, of course, frequently encountered in practice. Moreover it is an interesting case because comparison can be made with results presented in earlier publications dealing with the same interface. MgO precipitates in internally oxidized CuMg alloys have been the subject of several studies. A Field Ion Microscopy experiment concentrated on clarifying the chemical nature of the interface. It was convincingly shown that the end layer in the MgO crystal is an oxygen plane[22].

HRTEM experiments have attempted to clarify the atomic structure of the interface, but the situation there is not as clear[50,51]. The O-lattice description of the interface is commonly taken as a starting point for considerations on the atomic structure[52]. It will become clear that we do not use it as such in this study and it is merely presented here to act as a convenient reference in discussing the geometrical aspects of the interface. The O-lattice of this particular interface can be calculated easily, and is shown in figure 8. The lattice parameter of Cu is 0.367 (nm) and that of MgO 0.421 (nm). The O-points indicated by "O", are points of good "match". The O cell walls surrounding them are regions of bad "match". In these region dislocations relieving misfit are then assumed to be present, and therefore the solid lines drawn are taken to represent a dislocation network.The problem is to make the transition from the O-lattice description of this interface to the actual displacements connected with it. And although structures for the interface have been put forward, they are we feel, not fully satisfactory, as our approach will make clear.

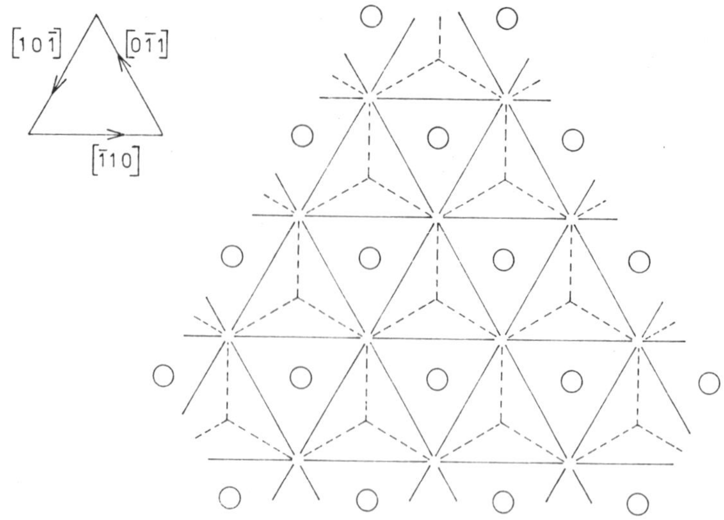

Figure 8. The O-latice for the {111}Cu-MgO interface. O-points are indicated with "O". Dotted lines are along <112> directions , and indicate a possible network of 1/2<110> misfit dislocations. Solid lines are along <110> and indicate a possible network of 1/6<112> partial misfit dislocations.

Cu was alloyed with 1 wt% of Mg in a high frequency oven. The alloy was packed with Cu_2O, Cu and Al_2O_3 powders (in volume roughly 1:1:1) , wrapped in copper foil, and internally oxidized in an evacuated quartz tube. The temperature used was 1273 (K) for about 17 hours. TEM specimens were then prepared using standard grinding, dimpling and ion milling techniques. HRTEM images were taken with a JEOL 4000//EX II, operating at 400 (kV). Relevant parameters used during calculations have been measured to be: C_s=0.97±0.02(mm), defocus spread: 11±2(nm), beam semi-convergence angle: 0.55(mrad)

Care was taken to only image interfaces at very thin parts of the foil, a rough estimate of the thickness being available from size and shape of the precipitate under investigation. Prior to imaging samples were cleaned in the ion mill at low milling voltages for five to ten minutes, to expose fresh interfaces.

EXPERIMENTAL RESULTS

Precipitates of MgO form with a truncated octahedral shape, as can be clearly seen from the appearance of the precipitates observed from different directions as shown in figure 9. The main bounding planes are {111} planes of both MgO and Cu. The corners do not seem to be truncated along well defined crystallographic planes but appear to be rather rounded off. The size of the precipitates varies, but edges in the order of 50 (nm) are commonly found.

It is tempting to connect the overall shape of the truncated octahedrons with the energies of the interfaces that surround it. Considering the precipitates to be in thermal equilibrium and by defining a mean interface energy for the truncated corners and for the {111} faces one can easily calculate equilibrium shapes of the precipitates for certain ratios, and constant precipitate volume. However , the assumption of thermal equilibrium may not be justified, because equilibration of the shape may be inhibited by kinetic factors during growth. The common occurrence of platelike precipitates in internally oxidized alloys is certainly a sign that this can be of importance.

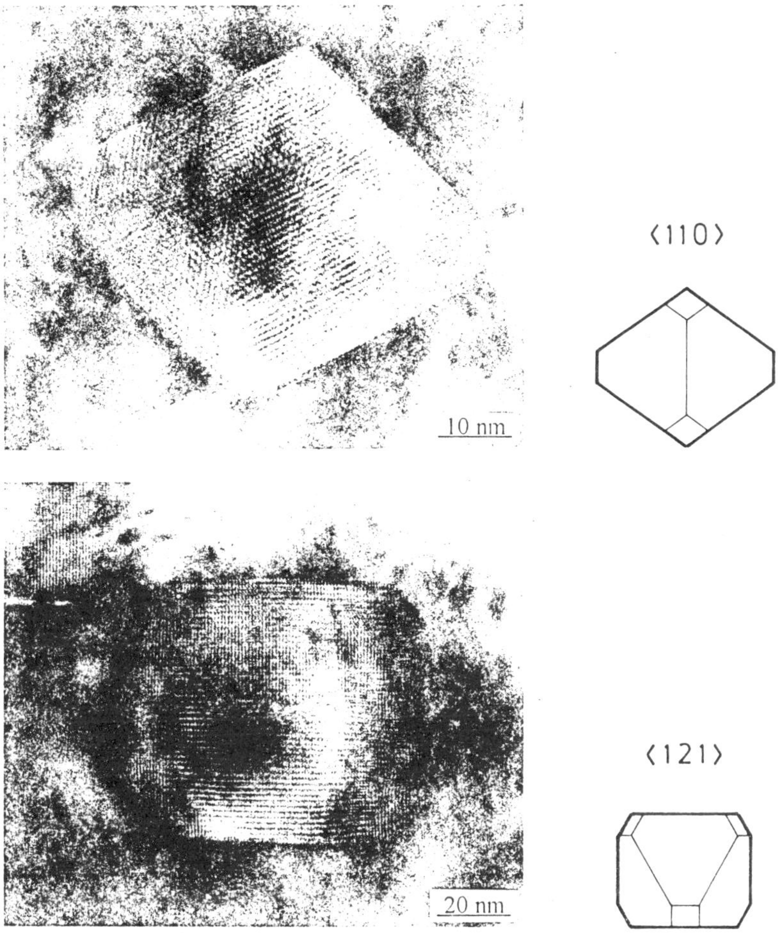

Figure 9. Images showing the truncated octahedral shape of the MgO precipitates as observed from <110> and <121> directions.

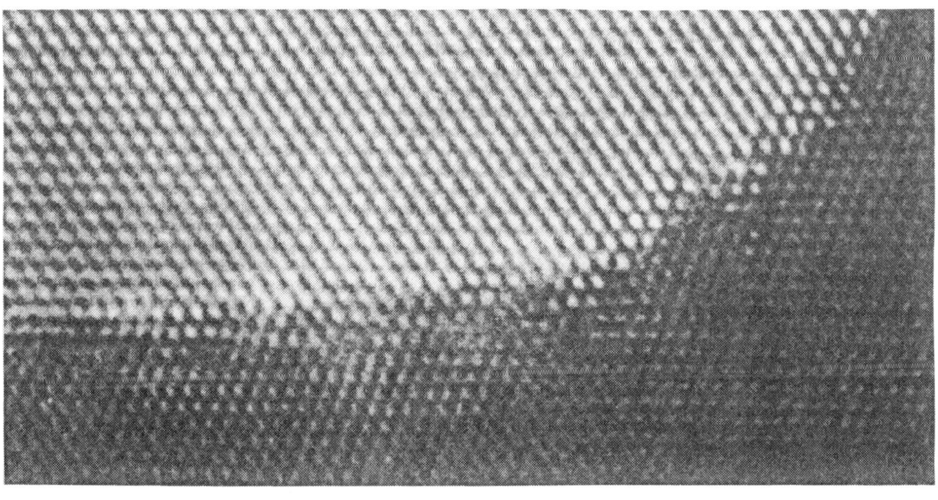

Figure 10. Image showing steps near the truncated corners and edges of the precipitate.

In general the interfaces seem to be flat. However near the truncated corners and also near the edges steps are normally visible, figure 10.

Also, from irregularities in the Moiré patterns observed on buried precipitates it is clear that imperfections may occur at the interfaces investigated. This is important to keep in mind when interpreting the HRTEM images, because it is expected that Moiré patterns caused by for example steps with a component perpendicular to the viewing direction may lead to contrast with the same period as expected of relaxation displacements.

Figure 11 shows a HRTEM image of a Cu-MgO interface, viewed along the [110] direction of both Cu and MgO.

Figure 11. HRTEM image of a Cu-MgO interface. Viewing direction is [110]. Defocus -48(nm).On the Cu side a modulation of the intensity with a period of about 1.6(nm) is visible.

Apart from the atomic periodicity and the flatness of the interface, certain other interesting characteristics are visible. On the Cu side a clear periodic modulation of the intensity is visible with a period of about 1.6 (nm), equal to the period expected from the O-lattice description. A similar contrast is visible at the MgO side but with a smaller amplitude. (figure 12) The contrast at the Cu side was visible at all defocus values used, ranging between 72 (nm) and 24 (nm) under focus. Viewing the interface along the [121] direction it is appears that rows of spots on the Cu side are buckled which may indicate relaxations perpendicular to the interface.

To interpret the HRTEM images we compare them qualitatively to simulated images. The trial structures needed for that are generated by the method described in the previous section on the model calculations. In this way no a priori assumptions about the final structure are needed, nor do we use any information taken from the experimental image as input.

For the calculations a rectangular supercell was created reflecting approximately the geometry of the interface. It measured 7*1/2[110] distances of Cu by 7* 1/2[121] distances of Cu, that is, 1.81(nm) by 3.15(nm). The MgO was represented by an fcc lattice with a monatomic basis and a lattice parameter exactly 6/7 times as large as Cu.

Calculations were performed for different bond strengths. Starting from an initial configuration atoms were displaced using a steepest descent method until no forces larger than $1*10^{-2}$ (eV/nm) were found. The resulting configurations very much resemble each other, displacements being more pronounced at higher bond strengths. The results shown in figures 13,14 and15 are therefore representative of the set of structures found.

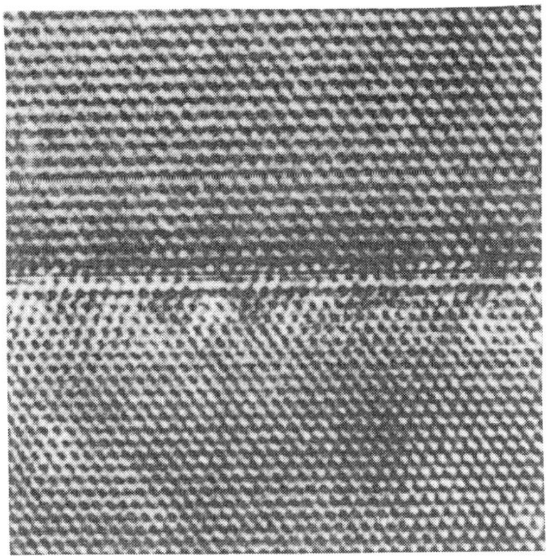

Figure 12. Same image as in figure 11. The intensities on both the Cu side and the MgO side have been histogram equalized. Clearly, a similar modulation as on the Cu side now shows up on the MgO side.

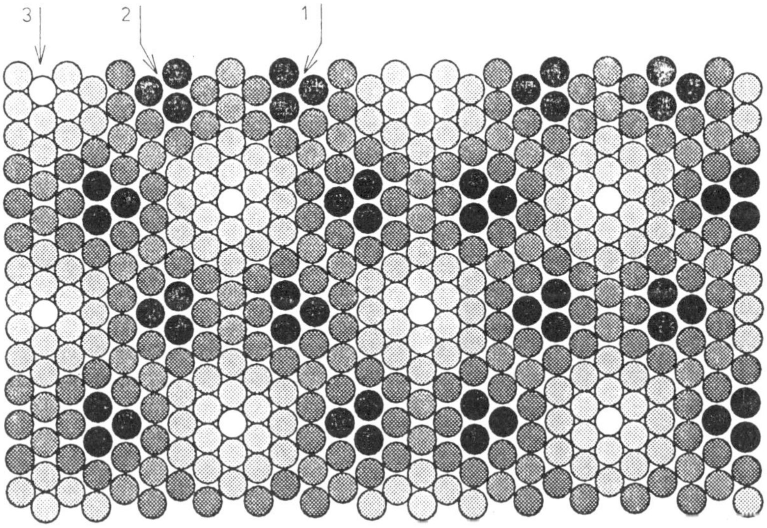

Figure 13. Calculated structure of the first layer of Cu above the interface, viewed along the interface normal. Substrate is not shown. Shades of gray indicate height above the interface. Lighter atoms are higher. Numbers 1,2 ,3 point to three different threefold axes on the interface.

Several characteristics of the structures found are worth mentioning.

Most importantly, viewed along the interface normal, the structures calculated show the three different threefold symmetry axes that are expected (1,2,3). The Cu lattice has expanded in the plane near two of the axes (1,2) , coherency has increased there. The stacking of Cu on the end plane of the ceramic is fcc-like for 1, and hcp-like for 2. There is a slight difference in height above the interface between the two regions, 2 lying slightly lower. Near the third axis (3), the lattice has contracted, and coherency has decreased.

It is interesting to interpret the interface structure in terms of an O-lattice. Identification of 3 as the O-points seems reasonable at first glance. It is easy to see them as point separated by a hexagonal network. Contrary however to what is required, the O points are then regions of incoherency, and separated by regions of increased coherency! So, on the contrary we take the O-points to be (1), and think of them as points of good register, but more importantly, correct stacking, as opposed to regions (2), that can then be thought of as regions of stacking fault The ridges apparent between the regions (1) and (2) may then be identified as dislocation lines of a 1/6<121> partial dislocation with edge character. Interpretation of this structure along the lines of figure 8 using 1/2<110> dislocations does not seem to be correct. In the projection along a <110> direction the displacements of atoms are seen to be small compared to the projected atomic distances. Displacements parallel as well as perpendicular to the interface are seen to occur.

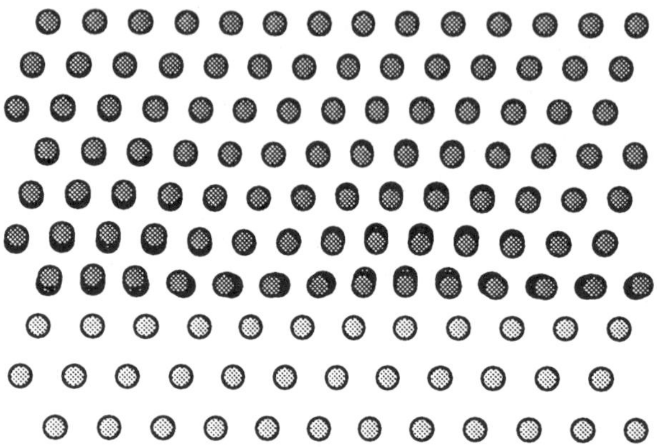

Figure 14. Projection of the calculated structure along a <110> direction.

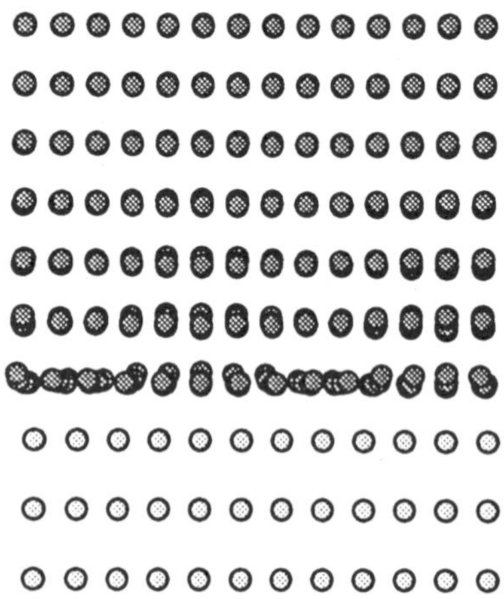

Figure 15. Projection of the calculatedestructure along a <121> direction.

HRTEM image simulations were carried out using the calculated atomic coordinates as input. Before proceeding with the HRTEM image simulations the model monatomic lattice used for the substrate was replaced by a block of MgO. An oxygen plane was taken to be the end plane. Three of these supercells were then placed behind each other along the [110] direction, creating a block with a thickness of 5.4 (nm). This block was sliced and used in a multislice calculation.

The simulated images resemble the experimental images in that they show a similar modulation of contrast near the interface. An example of a simulated image, for a defocus value a defocus of -48(nm) is shown in figure 16. Although there are clear differences between the measured and simulated images, in the details of the contrast near the interface the shared characteristics are we think encouraging.

Figure 16. Simulated HRTEM image. for a thickness of 4.8(nm) and of defocus -48(nm). Intensity modulation similar to that in figure 4 is found.

As stated earlier, the structures calculated for different bond strengths are quite similar, and they may be thought of as a set of possible structures sharing the characteristics sketched above. At this point it is interesting to discuss the differences between these structure and structures reported in literature[50,51]. The O-lattice description of the interface is taken as a starting point to interpret HRTEM images. In both cases displacements connected to dislocation lines parallel to the interface have been disregarded. This can be a justified simplification for HRTEM image simulation in case the displacement fields connected to these lines do not have components perpendicular to the interface. Indeed the structures presented in both papers do not show any displacements perpendicular to the interface. However, it seems to be likely that such displacements occur., in which case the contribution of all the dislocation segments is expected to be equally important

Closely related to this is that disregarding displacements along one of the three directions, destroys the symmetry the interface structure should bear, a symmetry which is present in the structure reported here. In a way the structures proposed in literature are solutions for a particular projection of the interface structure. The structure presented here offers the possibility at least to be verified by imaging of a second direction. The directions that are most suited for this are the <121> directions. A projection in a <121> direction , figure 16, of the proposed structure indicates that one could expect to see effects in that direction as well.

The HRTEM images showed modulations in intensity similar to those in the Cu in the MgO as well. This indicates that appreciable strain may also be present on the MgO side. This possibility was not taken into account in the model calculations. That this might be a point to consider becomes clear from results from elastic continuum calculations. In the model calculations we showed that for a dislocation at a Cu-MgO interface a rather large part of the Burgers vector will be accommodated in the MgO.

Considering that the structures were found reasonable candidates for the actual structure at the interface, the next step might be to try to establish this on a more quantitative basis. In particular one would like to be able to assess which bond strength or interface energy leads to the best agreement with experiment.

The scientific message of the paper so far is that from the atomistic calculations it can be concluded that the actual atomic structure of misfit dislocations depends on bond strength across the interface as well as misfit. A simple representation of the interface structure as an array of bulk-like interface dislocations relieving the misfit does not necessarily give the right physical picture. This becomes apparent if we compare elastic energies calculated with atomistics and elasticity. If interface dislocations are formed, depending on a critical value of the interaction parameter across the interface, a comparison with anisotropic elasticity can be made, showing reasonable quantitative agreement of the energy factors. The total elastic energy is a different case however. In figure 17 we plotted some curves of the elastic energy versus misfit δ at constant interaction α and a curve following from elasticity at various energy factors.

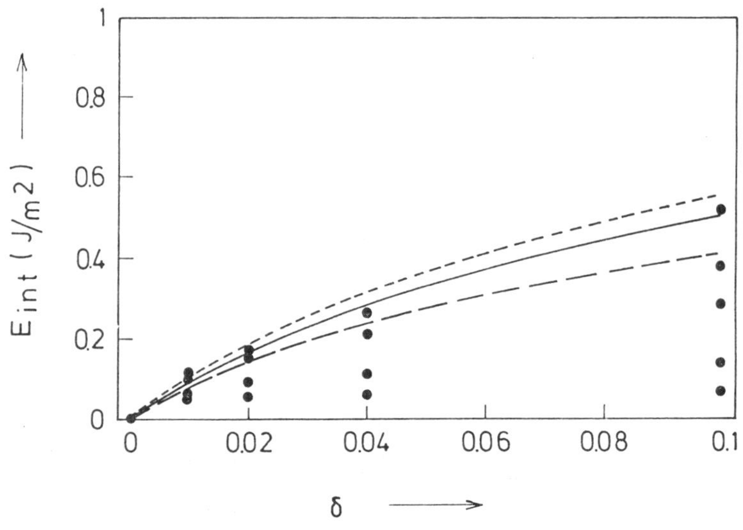

Figure 17. Interface energies (J/m^2) as a function of misfit δ for various interaction constants α. The solid line represent calculations according to anisotropic linear elasticity with K=4.0 eV/nm, the dashed line for the case of Ag/MgO (K=3.31 eV/nm) and the dotted line for the case of Ag elastically clamped onto a infinitely stiff substrate (K=4.39 eV/nm).

In the latter it is assumed that the equivalent hole radius r_{eh} is about b/2 [13] and that R is equal to ½ δ. In fact the coherency strain contribution of equation (12) is neglected since it is not contained in the atomistic energy calculations either.

The curve calculated with elasticity is steeper than a curve for constant interaction α. In fact it will be even steeper if the interaction between the dislocations is taken into account, which should be appreciable for 10% misfit, and perhaps even for 4% misfit. In other words the elasticity curve connects points with different interaction parameter, with higher values at higher misfit. This is in agreement with the results of figure 2. There we found that at higher values of misfit δ it takes a higher interaction constant to arrive at dislocation-like structures. In other words, the comparison shows that elasticity implicitly assumes different bond strengths for different misfits.

In general we may conclude that there exists a fair correlation between the atomistic and elastic continuum description and experimental observations of interface dislocations. However, it is shown that the elastic continuum approach cannot account for the possible configurations at an interface with misfit, because it doesnot enlighten the effects of different bonding strengths on the interface structure.

In addition, we have carried out HRTEM experiments on a {111} Cu-MgO interface and compared them with atomistic calculations. The calculations generated a set of related structures for different bond strengths across the interface, taking into account the misfit in two directions that is present at the interface. HRTEM simulations of a representative structure match experimental images well in that they reproduce a characteristic intensity modulation with a period of 1.6 (nm) that is visible near the interface on HRTEM images. Therefore we take the calculated structures to be good candidates for the actual structure at the interface. Described in terms of an O-lattice, these structures can be thought of as a triangular network of 1/6<121> dislocations along <110> directions. Displacements parallel as well as perpendicular to the interface are seen to occur.Although it is possible in principle, at this point no definite statement on the bond strength at this interface can be made. A thorough, quantitative comparison of simulated images for several bond strengths and measured HRTEM images, will be necessary for that.

MESOSCOPIC SCALE

Now we are turning to a description on a meso-scopic scale. The question is whether we can develop a rather mesoscopic physical picture, based on the ideas developed on a nano-scale, that would lead to a prediction of the macroscopic mechanical interface strength between a metal and a ceramic. From the previous part we have learned that the bond strength can be parametrised in terms of pair-wise interactions across the interface. A strong interaction will be, depending on the misfit, reflected in interface dislocations. It is quite obvious that metals and ceramic materials do not bond easily and that some thermodynamic trick is necessary to manufacture ceramic-metal layers of a size that may be tested macroscopically on its mechanical behavior. For that reason a powerful CO_2 laser system is used. During laser processing of a ceramic coating onto metals the interface strength between the ceramic layer and the metallic substrate as well as the wetting phenomena are the two predominant factors. Usually enhanced wetting induces improved strength of the interface and therefore it is appropriate to find a relationship between wetting phenomena and interface strength. In general a metal/ceramic interface can be categorized as a reactive interface and non-reactive interface, where the free energy between the metal and ceramic is negative or positive, respectively . Chemical reactions can provide a rapid and complete wetting process and may benefit the formation of the oxide coating on metal by laser processing. According to the thermodynamic calculations of the reaction between Al and SiO_2 [53], a large (negative) free energy of -297 KJ/g.mole at 1173 °K, would significantly improve the wetting of metal onto oxide. An analogous idea was worked out of coating a duplex steel SAF 2205, stainless steel 304 and Fe - 22 wt% Cr by bringing different powders into a laser beam. In fact a spinel structure was formed inside the coating. Further, it turned out that it is possible to form a thick and homogeneous ceramic coating of Cr_2O_3

on steel by a proper selection of coating parameters such as scanning velocity, overlap and number of layers.

Important for the interface strength is the work of adhesion. There are mainly two different interactions that may contribute to the work of adhesion between a metal and a ceramic, i.e. physical and chemical interactions. Here the work of adhesion or, as it is also called, the interfacial free energy of adhesion, W_{ad} is taken as the energy required to separate a plane of metal (m) from a plane of ceramic material (c) , from their equilibrium distance R to infinity, i.e. described by the interaction between two semi-infinite bodies:

$$W_{ad} = -\int_R^\infty f_{ad,\,mc} dz \quad , \tag{18}$$

where the force is given by:

$$f_{ad,mc} = 2\pi N_m N_c \int_z^\infty dp \int_p^\infty q dq \int_q^\infty g(r)\left[-\frac{du}{dr}\right] dr \quad . \tag{19}$$

u(r) represents the interaction potential , N is the atomic density [m^{-3}]and g(r) is the probability of finding an atom at a distance r of a given one. The last integral gives the interaction between a metal atom and a slab of the ceramic material. The integral over q gives the interaction between a metal atom and the complete ceramic materials, whereas a subsequent integration over p yields the total interaction between the whole metal and whole ceramic material. The interaction potential u(r) is commonly described by the London-dispersion interactions, i.e. the attraction between an instantaneous dipole and its induced dipole: i.e. $-A/r^6$,where $A= 3\,\alpha_m\alpha_c I_m I_c/2(I_m + I_c)$; I_m , I_c are the ionization potentials of the two atoms and α_i represent their polarizabilities. Substitution of this dispersion force in Eq.(18), while g(r) is set equal to 1, gives a contribution due to the physical interaction to W_{ad} as written in the first part of the following equation:

$$W_{ad} = n_m n_c \pi \frac{\alpha_m \alpha_c}{8R^4}\left(\frac{I_m I_c}{I_m + I_c}\right) - \frac{\beta}{2B}\Delta F_{mc} \quad , \tag{20}$$

where n the number of atoms per unit surface area of the interface. The second term in Eq.(20) represents the contribution due to chemical interactions: 1/2 ΔF_{mc} is the standard free energy of ionic bonds, β is the density of the ionically bonded sites and B is the number of bonds in a gram mole. In most metal/oxide interface of engineering interest, the first part describing the physical interaction in Eq.(19) is relatively weak[54], whereas the second part due to the chemical interaction is more predominant. Several observations indicated[54-59] that a chemical bond was built up between metal/oxide interface, as suggested in Eq.(20).

However, as also being explained in the section on the nano-scale properties the detailed understanding of the interface structure is still very complex for each individual case. For the sake of simplicity, we assume that the bonding of metal/oxide interface has a similar structure as in the most stable oxide structure of the metal,i.e. a simpler approach we did in the nanoscale regime. This interface is also the weakest part in the system compared to the cohesive energy of ceramic. Consequently, the elastic modulus across the interface and the interface spacing may be approximately represented by the values of the metal oxide. A theoretical tensile strength, assuming cleavage fracture and *no plasticity*, between a perfectly bonded metal/oxide interface might be estimated assuming that the increase in interface energy associated with the creation of new surface areas is equal to the work of adhesion at room temperature . It is assumed that the interface is intrinsically brittle, i.e.:

$$\sigma_{th} = \sqrt{\frac{W_{ad}E_{mo}}{2R}} \quad . \tag{21}$$

It is known that the work of adhesion between a liquid metal and a solid can be described by $W_{ad} = \gamma_{sv} + \gamma_{lv} - \gamma_{sl}$ (Dupré's equation) where γ represents the interfacial free energy (s: oxide, l: liquid metal, v: vapor). Dupré's equation can be rewritten as γ_l (1 + cosθ) (Young's equation). Here, the spreading pressure is neglected that denotes the difference between γ_s, the surface free energy of the bare solid surface, and γ_{sv}, the surface free energy of the solid in the presence of saturated vapor from the liquid metal (i.e. no extra adsorbed layer). The work of adhesion does not change significantly from high temperature to room temperature. Introduction of the expression of W_{ad} at high temperature into Eq.(21) gives:

$$\sigma_{th} = \sqrt{\frac{\gamma_l(1+\cos\theta)E_{mo}}{2R}} \quad . \tag{22}$$

In fact it is interesting to note that experimentally the initiation fracture resistance in a model system of gold of thickness h, bonded to sapphire, appears to follow $W_{ad}\ [\ 1 + \sigma_Y\ h/ W_{ad}\ ]^{1/2}$, where σ_Y represents the uniaxial yield stress of gold[60].

From Eq.(22) it can be clearly seen that a small contact angle θ and a strong interaction, e.g. larger E_{mo}, between metal and oxygen will increase the theoretical strength. Table I lists some of the calculated results of σ_{in} which ranges from 3.7 GPa to 38.8 GPa. These are much higher than the values of metal/ceramic interfaces of any kind of joining or coating in real practice. This is only partly because of the overestimate of the elastic modulus at the interface, since in real practice primarily the mode of deformation, the existence of interface defects and interface residual stress determine the interface strength. In particular the zone around an interface crack on the metal side that contains a distribution of lattice dislocations may affect considerably the stress level to be built up for rupture along the interface[61]. In a laser melt pool, the temperature at the center of laser melt pool is very high, but the temperature at the side of melt pool is relatively low, i.e. about the melting point of Al. Therefore the wetting near the edge would be dominated by the contact between reaction product of α-Al_2O_3 or mullite and liquid Al at the melting point. It is known that the wetting of α-Al_2O_3 by liquid Al is rather temperature-dependent. For instance the wetting angle of Al to α-Al_2O_3 [62] decreases from 170° at 1213 °K to 48° at 1523 °K. Then the reaction product of α-Al_2O_3 in the first laser track might not be wetted by liquid aluminium from the second track because of the lower temperature.

A reason for a good adhesion between aluminum and mullite is a smaller lattice misfit compared the Al/α-Al_2O_3 , provided a favorable orientation relationship exists. For instance, the lattice misfit of mullite/Al on the plane of mullite(120)/Al(110) is 0.73% in the direction of mullite[001]/Al[110] and 4.37% in the direction of mullite[210]/Al[001], but the misfit on the plane of Al_2O_3(001)/Al(111) is 16.7% in the direction of Al_2O_3[100]/Al[110]. In this way the number β of atom pairs expressed in Eq.(20) at the interface Al/mullite may be larger than that of Al/Al_2O_3, and a higher adhesion will be achieved. Further, a smaller lattice misfit may also lead to a smaller effective interface separation that in its turn contributes to a larger stabilization energy based on classical electrostatics.

Table 1. Theoretical strength calculated from eq. (22)

oxides	metals	T (°K)	θ°	E_{mo} (GPa)	γ_l (J/m$^{2)}$	R_{mo} (nm)	σ_{th}
Al_2O_3	Al	1213	170	380	0.86	0.183	3.7
	Al	1523	48	380	0.86	0.183	38.6
	Si	1723	80	72.3	0.73	0.174	13.3
Cr_2O_3	Fe	1823	88	118	1.86	0.206	23.5

EXPERIMENTS

In the case of the Al-base material a relatively high laser scan velocity and low energy density of beam were used in order to reduce the roughness of the oxide layer. The laser parameters (CW-CO_2 Spectra Physics 820) were: output power of 1.0 -1.2 kW, scan velocity of 40 mm/s and defocus of 20 mm with a beam diameter of 3 mm. An overlap of subsequent laser tracks of 67% was applied in order to form a sufficiently thick layer. A mixture of SiO_2 and Al powder with a mole ratio of 1:1 was injected into the laser melt pool of the Al6061 substrate. The chemical composition of Al6061 is 1.0 wt%Mg, 0.6wt%Si, 0.25 wt%Cu, 0.25 wt%Cr. The particle sizes of the powder ranged from 5 μm to 45 μm in the case of Al powder and from 2 μm to 40 μm for SiO_2. A relatively smaller size of particulates is favorable for a homogeneous reaction, although, because of aggregation of small SiO_2 particles with aluminum, it becomes more difficult to feed the mixing powder through the home-made powder feeding system .

The same laser system is also applied in the laser coating process with the following laser parameters: 1.0 kW laser power, a spot size of 1.27 mm, a laser scan velocity of 20 mm/s and an overlap of 75% between the tracks. As coating powder Cr_2O_3, Cr_2O_3 containing Mn and Cr_2O_3 containing extra Cr and Mo are used. The Mn is added to investigate the influence of Mn on the lattice parameters of the (Fe,Cr)-spinel coating and on the interface between the substrate and the coating. Cr and Mo are added to manipulate the crystal structure of the melt pool in the austenitic stainless steels from fcc to bcc in order to change the thermal expansion coefficient of the substrate. As substrate the dual phase SAF2205 steel is used of compositions: C(0.03),Si(1), Mn(2), Cr(22), Ni(5.5), Mo(3), bal Fe (in w%).

The four-point flexure test is based on the storage of a well known amount of elastic energy on bending and a release of this elastic energy on fracture. The strain energy release rate depends critically on the measured value of $h \approx h_2$ (the thickness of the substrate). Interfacial crack propagation occurs when the strain energy release rate G_{ss} equals the critical energy release rate G_c of interfacial failure.

X-ray diffraction indicated[53] that α-Al_2O_3 was the predominant major phase in the reaction layer. Further, silicon and mullite, crystalline SiO_2 co-exist. The silicon concentration in the α-Al_2O_3 layer is very low about 0.17-3.2 mole% as analyzed by EDS (Energy Dispersive Spectrometry) in the scanning electron microscope (SEM). According to SEM and EDS analyses, two interface layers usually exist between the reaction α-Al_2O_3 coating and the aluminum substrate: one is the high silicon content layer bound to the α-$Al_2.O_3$ coating and another is a thin layer with a high Al-content in contact with the aluminum substrate. Based on the TEM observation, the structure in the high silicon content layer consists mainly of a mullite phase and an amorphous silicate phase. The thin oxide layer located between the silicate and the aluminum substrate is α-Al_2O_3 with a relative high Si content (4-6 mole%). As the α-Al_2O_3 interface layer is very thin, quite often an Al/mullite interface is created. According to SEM observations, mullite is always firmly bonded to Al. Figure 18 presents an SEM picture of a cross-section. It indicates:(a) two intermediate layers, a high Si-content oxide (1) and a high Al-content oxide (2) between α-Al_2O_3 and the Al substrate. Figure 19 illustrates an interface between the mullite layer and the Al substrate.

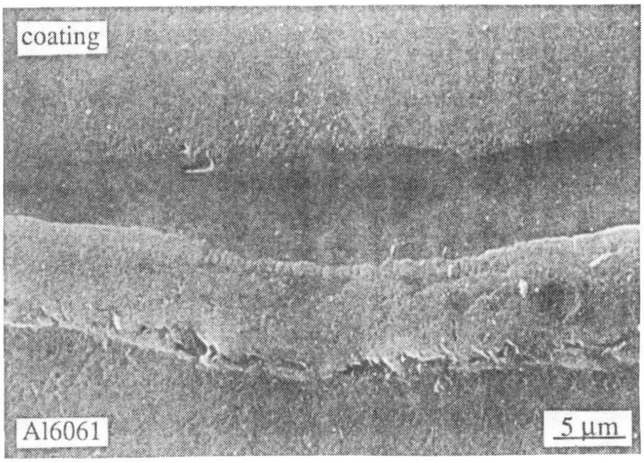

Figure 18. SEM picture from a cross-section. Top layer is α-Al2O3, next is a high Si-content oxide layer, bright ribbon is a high Al-content oxide, bottom is the Al substrate.

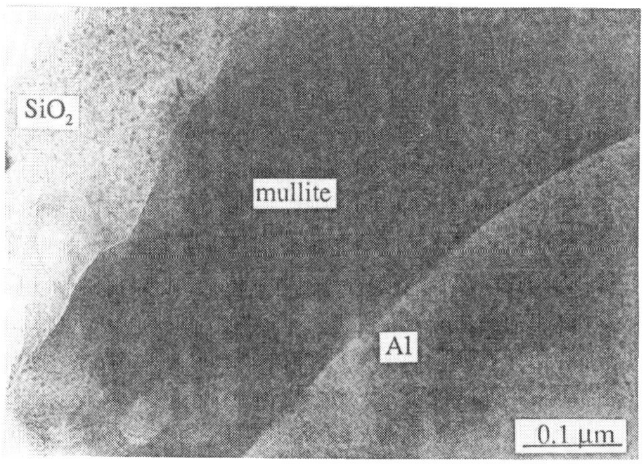

Figure 19. TEM micrograph illustrates the mullite/Al substrate interface substrate. (Mullite dark ribbon, Al substrate to the right)

Cracks in the ceramic coating are observed in the heat affected zone just outside the laser track (fig.20). In the (Fe,Cr)-spinel coatings on duplex steel SAF2205 it turned out that the cracks are oriented vertically in the coating and as a consequence do not cause delamination of the coating.

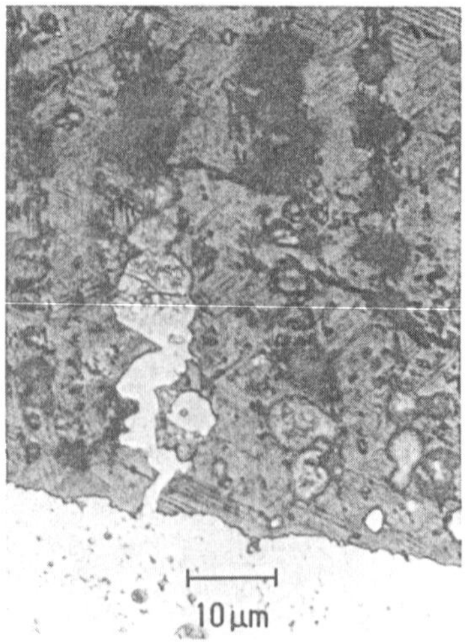

Figure 20. Etched cross section of laser applied chromiumoxide coating.

However, complete delamination is observed when the coating thickness is larger than 200 µm. The vertical cracks run through the amorphous interdendritical phase and are deflected by the α-(Fe,Cr) particles. As the cracks run through the region of intergrowth between the (Fe,Cr)-spinel and the α-(Fe,Cr) particles the crack energy was not dissipated in the metallic particles. Consequently, the cracks run all the way through the coating and are blunted when they reach the substrate (fig. 21).

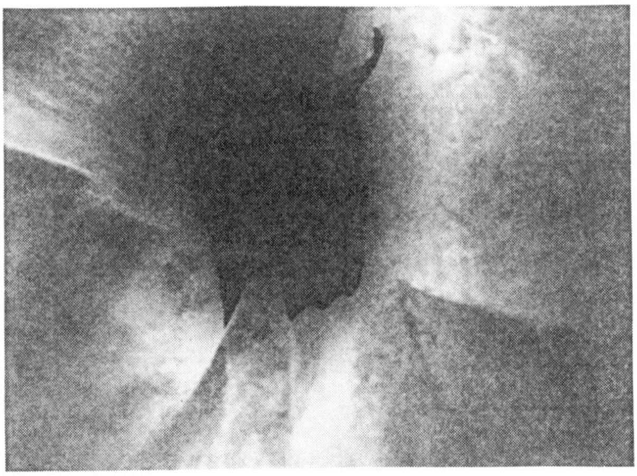

Figure 21. Crack through the (Fe,Cr)-spinel coating and blunting in the metal substrate. Top (Fe,Cr)-spinel, bottom is substrate.

Here the crack is opened by plastic deformation in the metal. Crack extension along the interface is not observed but crack branching in the ceramic coating is frequently observed in a region of 5 µm near the interface. Only in some cases delamination occurs after work hardening of the substrate . Assuming that the plastic deformation only occurs in a small volume near a crack through the coating the amount of energy stored, and thus a minimum value for the critical energy release rate, can still be obtained from the maximum load. Observations indicate that the critical strain energy release rate and stress intensity factor are larger than: $G_{ss} = 270\ J/m^2$ and $K_c = 8.5\ MPa\sqrt{m}$.

MACROSCOPIC SCALE

It quite obvious that the strength calculated from the wetting angle is rather a theoretical one since no plasticity is involved. Let's concentrate on the state of stress of the laser clad layer and the actual strength measured. The thermal loading by the laser beam causes stresses being built up. Without melting or yielding all thermal strains will be accommodated elastically, i.e. disappear when returning to the starting temperature. Melting or yielding is necessary in order to attain a residual stress state. Because the cooling during laser treatment starts from the bottom to the surface of the specimen the residual stress state near the surface of the specimen, i.e. the ceramic coating, will be of tensile character. The residual stress adds up with the mechanically applied stresses causing failure at relatively low levels of applied stress. To study the formation of cracks during and after laser treatment we have scrutinized the temperature field near the laser melt pool to obtain the stress during and after laser treatment based on the work by Cline and Anthony[63]. The temperature field is calculated of a Gaussian beam sweeping over the surface of a material. By taken derivatives with respect to x and y the thermal gradients and the absolute value of the total gradient is obtained. Although we did not solve the three dimensional temperature field numerically for a bimaterial we may extrapolate the data from the two singular cases of steel and (Fe,Cr)-spinel using the thermal diffusion equation together with the fact that the temperature field has to be continuous across the interface. This implies that the thermal gradients parallel to the interface, i.e. in the x- and y-direction, are equal on both sides of the interface ($z = 0$ on the interface). Further we use the fact that the heat flux is continuous over the interface. Observations of the equi-temperature lines on the interface perpendicular to the direction of the laser beam, i.e. in the y-direction, indicate that for the (Fe,Cr)-spinel coating on steel in the metal, the thermal gradient in the z-direction is only one quarter of the gradient in the y-direction validating the use of the temperature field calculated for a single phase steel having a zero thermal gradient in the z-direction on the surface. Further, the large temperature gradient in the y-direction is accompanied by a more than twice as large thermal gradient in the z-direction in the ceramic coating. From the thermal gradient in the y-direction of $1.8 \cdot 10^6$ °C/m in steel a thermal gradient in the z-direction of $4.7 \cdot 10^6$ °C/m is estimated in the (Fe,Cr)-spinel coating. For a 100 μm coating it results in a 470 °C temperature difference over the thickness of the coating. As these large thermal gradients occur at temperatures near the melting point they are accompanied by low stress levels. However, during cooling the temperature difference results in the build-up of residual stresses. Because, in the region of interest, the coating cools from the interface towards the top of the coating the stress is of tensile character in the top of the coating, i.e. a bending moment is set up in the coating that may cause delamination of the ceramic coating.

The residual stress state can be estimated from the solution of a bending plate on an elastic foundation of which the solution is known. When bending is restricted by the substrate, i.e. edges are clamped and delamination does not occur, the stress in the top of the coating is given by:

$$\sigma_{yy}(z) = \frac{\alpha_1 E_1 \frac{\partial T}{\partial z}}{(1 - \nu_1)} \cdot z \quad , \qquad (23)$$

where index 1 refers to the properties of the coating ((α_1 = 12 10^{-6} K^{-1}, E_1 = 280 GPa, ν_1 =0.23). Neglecting Newtonian cooling at the surface, the temperature field at a depth z upon cooling can be approximated by:

$$T(z,t) = T_o + (T_r - T_o) erfc(z/2\sqrt{t\alpha_D}) \quad , \qquad (24)$$

where α_D is the thermal diffusivity ($6.4\ 10^{-7}\ m^2\ s^{-1}$ in Cr_2O_3). Clearly there exists a stress distribution over the thickness of the coating which sets up a bending moment in the coating which can be approximately written as:

$$M_1 = \frac{\alpha_1 E_1 \frac{\partial T}{\partial z}}{12(1-\nu_1)} \cdot h_1^3 \qquad (25)$$

Using equations (23) and (24) the bending moment is found to be 14.2 Nm/m for a coating thickness of 200 μm. which is close to the outcome of eq.(25), namely 13.7 Nm/m.

When the coating does not delaminate edge moments have to be applied to prevent bending. These edge moments are effectively produced by stress components in the z-direction over the interface. The stress σ_{zz} at the edge of a bending plate is given by Hetenyi [64] and the maximum tensile stress is given by:

$$\sigma_{zz}^{max} = 2 \cdot \frac{\alpha_1 E_1 \frac{\partial T}{\partial z}}{12(1-\nu_1)} \cdot \sqrt{\frac{3k_2}{E_1}} \cdot h_1^{1\frac{1}{2}} \quad , \qquad (26)$$

where k_2 is the modulus of deflection of the substrate ($E_1/k_2 = 6.6 \cdot 10^{-5}$). In the laser coating process several laser tracks have to be applied before the coating reaches its final thickness, i.e. its final stress state. Realizing that an edge moment, and thus a stress σ_{zz}, is only generated at a free edge one crack through the thickness of the coating is always present. After the first crack parallel to the laser track two things may happen. First, delamination may start from the free edge and grow to the last laser track applied where the coating becomes thinner and crack propagation will stop. When delamination occurs the stress σ_{yy} is removed by bending of the coating so no further cracking through the thickness of the coating will occur. Another possibility is that the stress σ_{zz} at the free edge is not large enough to cause delamination. Then the stress σ_{yy} is locally reduced due to crack opening but is unchanged in the rest of the coating. As a result cracks more or less regularly spaced will appear through the thickness of the coating as was observed experimentally.

The process of delamination versus perpendicular fracture is determined by the relative strength of the ceramic and the interface and by the thickness of the coating. As the stress σ_{yy}, causing fracture through the coating, scales with h_1 and σ_{zz}, causing delamination, scales with $h_1^{1.5}$ the coating can always be made thick enough so that delamination will occur. It is interesting to see whether we are able to predict in advance whether delamination along the ceramic/metal interface will occur of blunting of the crack tip in the underlying metallic substrate. Our analysis is based on the representation of a crack by an appropriate continuous distribution of dislocations. This was already anticipated by Eshelby et al.[65] when they noted the similarity between the stress fields associated with a freely slipping crack and a pile-up of slip (translational) dislocations. However, Leibfried[66] suggested to make a continuum approximation right from the beginning instead of dealing with an array of discrete dislocations. It has to be emphasized that the outcome of the analysis is not different from results that can be predicted from fracture mechanics. For a review reference is made to[67]. Nevertheless, a dislocation approach to fracture leads to important results which can hardly be obtained by fracture mechanics, namely that the non-redundant dislocations are more numerous than redundant dislocations close to a mode-III crack in an elastic-plastic solid (see Weertman[68, 69]). To be consistent with the previous part we consider only mode-I cracks running in the z direction, i.e. perpendicular to the interface

A direct way to find the stress fields in front of the Inglis-Griffith crack is to take the stress components σ_{ij} (of a dislocation, the dislocation distribution function B on the plane of the crack and to integrate. The stress field components $\{\sigma_{ij}\}_c$ of the crack is therefore:

$$\{\sigma_{ij}\} = -\int_{-a}^{a} B(z')\sigma_{ij}(z-z',y)dz' \tag{27}$$

where Eq. (27) can be solved using a contour integration in the complex plane. However since we deal with an interface between dissimilar materials this is not a trivial problem because at the interface some stress components like σ_{zz} or σ_{zy} have non-vanishing resultant tractions (see equation 6). The traction stress components of a mode-I crack at the interface between two dissimilar materials σ_{zy} and σ_{yy} are continuous across the boundary whereas σ_{zz} is not, although it must be. Likewise, the displacements across the boundary are not continuous. A solution to this problem can be obtained along different routes, e.g. image line forces, image line pressures, image infinitesimal dislocation distributions. We applied the latter method to describe the stress fields of the cracks near the interface as the sum of the stress field of the image dislocation and the stress field of a distribution of infinitesimal image dislocations located at the interface whose Burgers vectors are perpendicular to Burgers vector describing the crack itself (assuming the image dislocation is of opposite sign).

All the stress components of the mode-I crack, consisting of a continuous distribution of climb edge dislocations were calculated as a function of the distance to the interface. In these calculations the interactions between the mode-I crack and the interface depend on the elastic mismatch between the dissimilar materials as described by Dundurs' parameters[70]. Of the resulting stresses in the vicinity of the interface the most crucial terms is the ratio η be- tween the components of σ_{zz} acting across the interface and of σ_{yy} along the interface[71]. σ_{zz} around a crack near the interface between (Fe,Cr)-spinel and the metallic substrate is depicted in figure 22.

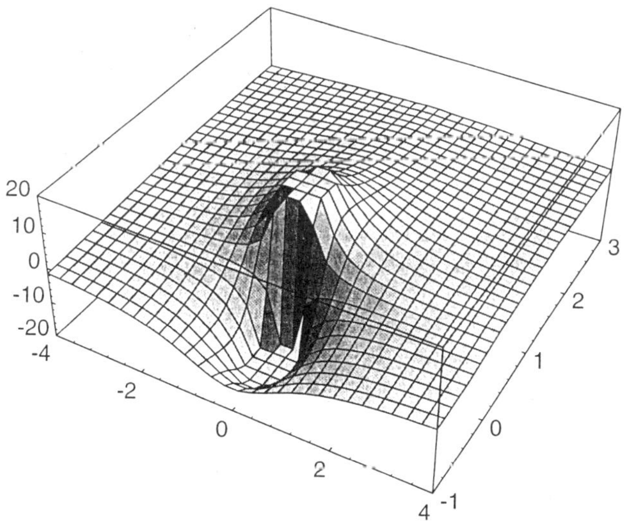

Figure 22. σ_{zz} around a crack near the interface between (Fe,Cr)-spinel and the metallic substrate. The distance is in units of the size of the Burgers vector in spinel and the stress is in units of GPa.

These ratio η was calculated to be 8.1. Therefore, delamination along the interface is expected if the ratio between the interface strength and the yield strength of the substrate is smaller than η. In our case it would mean that delamination is predicted when the interface strength is smaller than about 8 GPa. Since the theoretical interface strength is much larger (as derived from the wetting angle is 23.5 GPa), blunting of the crack in the underlying substrate is more likely to occur than delamination. However, with increasing propagation the substrate may be work hardened, i.e. increasing the flow stress. By this increase the ratio between the interface strength and the strength of the substrate becomes smaller than η and eventually delamination along the interface will occur.

This is according to the observations. The cracks run all the way through the coating and are blunted when they reach the substrate (Fig. 21). Here the crack is opened by plastic deformation in the metal. Crack extension along the interface is not observed neither are cracks detected along the interface. Delamination was only observed after applying a four -point bending test. The following physical picture emerges: the tensile stress state generated by the thermal quenching effects drives cracks to the interface (mode -I). The stress component perpendicular to the interface tries to peel the layer off the substrate and the shear stress at the interface will promote debonding by mode II fracture along the interface. In the model the stress concentration near the crack tip is only modified by a discontinuity of the modulus at the interface.

In reduction of the tensile stress blunting of the tip will be more effective by plastic flow provided that the substrate is ductile. In the present situation of a ductile metallic substrate, the tensile stress near the interface will be dropped to the yield stress near the site of the crack. The substrate plasticity tends to increase the crack tip opening at the interface which encourages the crack in the coating to propagate further. Ultimately, when the interface is very strong, delamination may occur by fracture through the *ceramic* parallel to the interface, instead of *along the metal-ceramic* interface, and both fracture processes are governed by the strength of the ceramic. Using the experimental maximum coating thickness of 200 μm for the (Fe,Cr)-spinel coating on SAF2205, where delamination is mainly through the ceramic, the critical bending moment M_{1c} is calculated to be 14.2 Nm/m. From the calculated bending moment in the ceramic coating the critical energy release rate G_{ss} of delamination can be calculated to be $6 M^2_{Ic} (1- \nu_1^2)/ E_1 h^3_1 = 320\ J/m^2$.

CONCLUSIONS

This paper treats the fundamental and practical aspects of interfaces between metals and ceramic materials at different length scales. At an nanoscopic level it turns out that from HREM observations in combination with computer modelling studies a qualitative insight can be obtained about the bonding behavior between these dissimilar materials. An atomic structure is proposed for the interface that takes into account the interplay between misfit and bond strength. In general we may conclude that there exists a fair correlation between the atomistic and elastic continuum description of interface dislocations. However, it is shown that the elastic continuum approach cannot account for the possible configurations at an interface with misfit, because it doesnot enlighten the effects of different bonding strengths on the interface structure.Beside the theoretical analysis of the interface structure between dissimilar materials, the paper directs also to some future experimental work. The basic idea is that with a detailed analysis of experimental observations of the atomic structure using high resolution transmission electron microscopy it would be feasible with the help of this computer modelling approach to obtain physical insight in the bond strength across the interface. Further, segregation of other elements might affect the local bond strength as well, i.e. the coupling parameter in the computer modelling approach, which may become manifest in the experimental observations of the atomic structure. Results along these lines on internally oxidised metallic systems are underway and will be published in a separate paper.

At a more mesoscopic scale the basic features of atomic bonds across the interface are still crucial. However, investigations of the interface structures and wetting phenomena during laser processing leads to a simple correlation between wetting phenomena and interface strength is derived. Further, this rather theoretical strength of a laser clad layer is put onto a more practical base to describe the actual mechanical performance at a macroscopic level. It turned out that in the case of Cr_2O_3 on steel (SAF2205), the thickness of the coating that can be applied by laser coating is limited to about 200 μm setting a limit to the maximum strain energy release rate that can be measured in a 4 point flexure test before severe yielding occurs. The the energy release rate was extremely high and delamination occurred by fracture through the coating and partially along the interface. An explanation of the high energy release rate is presented as derived from plasticity theory.

Two types of interface layers, namely mullite and a high Si-content Al_2O_3, have been identified between the α-Al_2O_3 reaction coating and Al6061 substrate. The interface bonding of mullite/Al is always good. An approximate expression that relates interface strength to wetting properties was investigated for a metal/ceramic interface. It is figured out that a good wetting can not only improve the theoretical strength of a perfect interface but also reduce the possibility of crack formation along the interface which determine the practical interface strength. In this reaction coating the wetting process is mainly governed by the contact of liquid Al with the reaction products of mullite and α-Al_2O_3. The formation of mullite/Al interface may be essential to form a good bond by reaction coating on Al alloy, and a high Si-content Al_2O_3 interface layer may be wetted better than pure α-Al_2O_3 by liquid Al. In addition, it is concluded that a firmly bonded coating of Cr_2O_3 on steel could be produced by high power laser processing. The actual interface strength of a (Fe,Cr)-spinel applied to stainless steel by laser coating depends strongly on the composition of the substrate and coating materials. The strength of bimaterial interfaces was tested by 4-point flexure. The the energy release rate was extremely high and delamination occurred by fracture through the coating and partially along the interface, indicating that the interface strength is similar to or higher than the fracture strength of (Fe,Cr)-spinel. The way cracks propagate through the coating and how they interact with the interface could be described by a dislocation model of mode-I cracks in accordance with experimental observations.

Acknowledgments The work described in this paper is part of the research program of the Foundation for Fundamental Research on Matter (FOM-Utrecht) and has been made possible by financial support from the Netherlands Foundation for Technical Sciences (STW-Utrecht). This research was supported in part (VV) by the U.S. Department of Energy, Office of Basic Energy Sciences, Grant. No. DE-FG02-87ER45295 .

REFERENCES

1. A. M. Stoneham, M. M. D. Ramos and A. P. Sutton, Philos. Mag. A 67, 797(1993).
2. W. R. L. Lambrecht and B. Segall, Metal-Ceramic Interfaces (edited by M.Rühle, A. G. Evans, M. F. Ashby and J. P. Hirth), Pergamon Press: Oxford,p. 29 (1990).
3. P. Blöchl, G. P. Das, H. F. Fischmeister and U. Schönberger, Metal-Ceramic Interfaces (edited by M. Rühle, A. G. Evans, M. F. Ashby and J. P. Hirth), Pergamon Press: Oxford, p. 9 (1990).
4. U. Schönberger, O. K. Andersen and M. Methfessel, Acta Metall. et Mater.40, S01 (1992).
5. C. Li, R. Q. Wu, A. J. Freeman and C. L. Fu, Phys. Rev. B 48, 8317 (1993).
6. C. Kruse, M. W. Finnis, V. Y. Milman, M. C. Payne, A. DeVita and M. J.Gillan, J.Am. Ceram. Soc. 77, 431 (1994).
7. F. C. Frank and J. H. Van der Merwe, Proc. Roy. Soc. London A 198, 205(1949).
8. J. H. Van Der Merwe, Crit Rev Solid State Mat Sci 17, 187 (1991).
9. R. Hull and J. C. Bean, Crit Rev Solid State Mat Sci 17, 507 (1992).
10. J. P. Hirth and X. Feng, J. Appl. Phys. 67, 3343 (1990).
11. X. Feng and J. P. Hirth, J. Appl. Phys. 72, 1386 (1992).
12. Y. Ishida, J.-Y. Wang, T. Suga and S. Tanaka, Metal-Ceramic Interfaces(edited by M. R FChle, A. G. Evans, M. F. Ashby and J. P. Hirth), Pergamon Press: Oxford, Vol. p. 234 (1990).
13. F. S. Shien and S. L. Sass, Metal-Ceramic Interfaces (edited by M.Rühle, A. G. Evans, M. F. Ashby and J. P. Hirth), Pergamon Press: Oxford,p. 258 (1990).
14. D. Knauss and W. Mader, Ultramicroscopy 37, 247 (1991).
15. S. Ernst, P. Pirouz and A. H. Heuer, Philos. Mag. A 63, 259 (1991).
16. D. X. Li, P. Pirouz, A. H. Heuer, S. Yadavalli and C. P. Flynn, Acta Metall. Mater. 40, S237 (1992).
17. K. L. Merkle, M. I. Buckett and Y. Gao, Acta Metall. Mater. 40, S249 (1992).
18. A. Trampert, F. Ernst, C. P. Flynn, H. F. Fischmeister and M. Rühle, Acta Metall. et Mater. 40, S227 (1992).
19. P. Lu and F. Cosandey, Acta Metall. Mater. 40, S259 (1992).
20. J. Bruley, R. Brydson, H. Mullejans, J. Mayer, G. Gutekunst, W. Mader, D. Knauss and M. R FChle, Journal of Materials Research 9, 2574 (1994).
21. W.P.Vellinga, M.Verwerft, J.Th.M.de Hosson, Tj.Hibma, Defect-Interface Interactions (edited by E. P. Kvam, A. H. King, M. J. Mills, S. T. D. and V. Vitek), Materials Research Society Symposium Proceedings, Materials Research Society: Pittsburgh, Vol. 319, p. 15 (1994).
22. H. Jang, D. N. Seidman and K. L. Merkle, Interface Sci. 1, 61 (1993).
23. J. Mayer, G. Gutekunst, G. Möbus, J. Dura, C. P. Flynn and M. Rühle,Acta Metall. et Mater. 40, S0217 (1992).
24. W. Mader and D. Knauss, Acta Metall. Mater. S207 (1992).
25. G. Gutekunst, J. Mayer and M. R FChle, Defect-Interface Interactions (edited by E. P. Kvam, A. H. King, M. J. Mills, S. T. D. and V. Vitek), Materials Research Society

Symposium Proceedings, Materials Research Society: Pittsburgh, Vol. 319, p. 3 (1994).
26. T. Epicier and C. Esnouf, Journal de Physique III 4, 1811 (1994).
27. V. Vitek, G. Gutekunst, J. Mayer and M. Rühle, Philos. Mag. A, Vol. 71, 1219 (1995).
28. J. H. Harding, Rep. Prog. Phys. 53, 1403 (1990).
29. M.W. Finnis and M. Rühle, Materials Science and Technology (edited by V.Gerold),VCH: Weinhein, Vol. 1, p. 533 (1993).
30. M. S. Daw, S. M. Foiles and M. I. Baskes, Mater Sci Rep 9, 251, (1993).
31. M. W. Finnis, Acta Metall. Mater. 40, S025 (1992).
32. A. M. Stoneham and P. W. Tasker, Philos. Mag. B 55, 237 (1987).
33. D. M. Duffy, J. H. Harding and A. M. Stoneham, Acta Metall. Mater. 40,S011 (1992).
34. D. M. Duffy, J. H. Harding and A. M. Stoneham, Philos. Mag. A 67, 865 (1993).
35. J. R. Willis, S. C. Jain and R. Bullough, Philos. Mag. A 62, 115 (1990).
36. J. R. Willis, S. C. Jain and R. Bullough, Philos. Mag. A 64, 629 (1991).
37. T. J. Gosling, J. R. Willis, R. Bullough and S. C. Jain, J. Appl.Phys.73, 8297 (1993).
38. A. Atkinson and S. C. Jain, J. Phys.: Condens. Matter 5, 4595 (1993).
39. G. E. Beltz and L. B. Freund, Philos. Mag. A 69, 183 (1994).
40. M. W. Finnis and J. E. Sinclair, Philos. Mag. A 50, 45 (1984).
41. G.J.Ackland, G.Tichy,V.Vitek, M.W.Finnis, Phil.Mag.A, 56,735(1989).
42. W. Bollmann, Surface Science 31, 1 (1972).
43. J.P. Hirth and J. Lothe, Theory of dislocations, McGraw-Hill, New York, 1968.
44. F.R.N. Nabarro, Theory of crystal dislocations, Oxford University Press, 1967.
45. A.J.E. Foreman, Acta Met., 3,322 (1955).
46. S.G. Lekhnitskii, Theory of elasticity of an anisotropic body, English translation, Holden Day, San Francisco, USA, 1977.
47. A.N. Stroh, J. Math. Phys., 41,77(1962).
48. D. Barnett and J.Lothe, J.Phys. F4,1618(1974).
49. J. Dundurs and G.P. Sendeckyi, J. Appl. Phys. 36, 3353 (1965).
50. P. Lu, F.Cosandey, Ultramicroscopy 40, 271, (1992)
51. F.R.Chen, S.K.Chiou, L.Chang, C.S.Hong, Ultramicroscopy 54, 179, (1994)
52. W. Bollmann, Crystal Defects and Crystalline interfaces, Springer, New York, (1970).
53. X.B. Zhou, J. Th.M. De Hosson, Acta metall. mater., 39,2267, 1991.
54. F. Delannay, L. Froyen and A. Deruyffere, J. Mater. Sci. 22, 1 (1987).
55. E. Gillet, B. Ealet and J.L. Berlioz, Surf. Interf. Analy.,16, 461(1990).
56. K.H.Johnson and S.V.Pepper, J. Appl. Phys., 53, 6634(1982).
57. M.D. Keeffe, J. Mater. Res., 6, 2371 (1991).
58. J.F. Knott, Fundamentals of fracture mechanics, Butterworths, London, 1973.
59. W.D. Kingery, Introduction to Ceramics, John Wiley, New York (1967,1960), pp116,436,605
60. I.E. Reimanis, B.J. Dalgleish, A.G. Evans, Acta metall. mater. 39,3133 (1991).
61. A.G. Evans, B.J. Dalgleish, Acta metall. mater., 40, S295(1992).
62. G.V. Samsonov, The oxide Handbook, IFI/Plenum, 1973.
63. H.E. Cline and T.R. Anthony, J. Appl. Phys., 48, 3895(1977).

64. M. Hetenyi, Beams on elastic foundations, Oxford Un. press, Oxford, 1946.
65. J.D. Eshelby, F.C. Frank and F.R.N. Nabarro, Phil. Mag., 42,351(1951).
66. G. Leibfried, Z. Physik 130,214 (1951).
67. J.W. Hutchinson, Adv. Applied Mech., 29, 63 (1993).
68. J. Weertman, Mechanics of Materials 15, 65 (1993).
69. J. Weertman, Proc. R. Soc. London A425, 291 (1989).
70. J. Dundurs, J. Appl. Mech., 36,650 (1969).
71. A.S. Argon, V. Gupta, H.S. Landis, J.A. Cornie, J. Mat. Sci., 24,1207 (1989).

STABILITY AND INSTABILITY OF CRYSTALLINE AND AMORPHOUS PHASES AT HIGH PRESSURE

E.G.Ponyatovsky

Institute of Solid State Physics
Russian Academy of Sciences
142432 Chernogolovka Moscow distr.
Russia

I. PHASE FORMATION IN BINARY *B-B* ALLOYS UNDER PRESSURE

There was a rather long period when the experimental study of phase transformations under high pressure was carried out on simple solids, *i.e.*, on elements and their stoichiometric compounds. Now the T–P diagrams of almost all elements and numerous compounds are investigated to 10 GPa[1]. This study spreads to the megabar region (tens GPa) as the accessible pressure range is extended[2].

But investigation of the three-dimensional T–P–C phase diagrams was scanty so far. One could expect that multiformity of the T–C phase diagrams known at normal pressure would become intricate if the third parameter, pressure, were involved. These apprehensions could result from polymorphism of pure components, polymorphic transformations of intermediate phases and compounds, and formation of many new intermediate phases. An additional experimental difficulty also averted researchers from the study of the three-dimensional phase diagrams.

In fact, this complexity is not the case. Synthesis of new high-pressure phases and characterization of their structure and properties provided abundant information which enables one to state the general trends of phase formation in multicomponent systems and some common rules of the pressure-induced evolution of essentially different T–C phase diagrams. This simplified the general picture of phase equilibria in multicomponent systems.

This statement is exemplified below by the binary alloys of the Group II—V non-transition elements.

The first binary systems studied were metallic alloys Bi-Sn, Bi-Pb, Pb-Sb, Pb-Sn, In-Bi[3] and some systems with semiconducting phases, In-Sb[4], Cd-Sb, Zn-Sb[5], Al-Ge,

Al-Si[6–8], Au-Ge, Ag-Ge[9], Ge-Sb[10], Ga-Sb[11,12]. There were several reasons to choose these alloys as model systems for a systematic experimental study of the pressure effect on the phase equilibria:

i) Their phase diagrams are rather simple and mainly include either eutectics with limited solubility in solid state or one intermediate phase. But Zn-Sb and In-Bi systems have several intermediate phases at atmospheric pressures.

ii) Polymorphism is characteristic of B-elements under pressure. E.g., bismuth has eight high-pressure modifications.

iii) Low melting points and numerous intermediate phases occurring in the pressure range upto $\sim$ 9 GPa. These pressures are accessible for devices of large volume, which simplifies their experimental investigation.

iv) There is a good number of metastable phases which were obtained by rapid quenching from melts at normal pressure. These data were also valuable for systematizing.

Our experimental study was carried out in the following sequence.

First, the pressure dependence of the familiar transition temperatures was determined, and the T–P parameters of the phase equilibria revealed were fixed by means of differential thermal analysis (DTA) and resistivity measurements under hydrostatic pressure upto 2.5 GPa or quasi-hydrostatic pressure up to 7-9 GPa. The chemical composition of an alloy was varied, and the DTA and resistivity data obtained were used to plot the T–P sections of the T–P–C phase diagram. The total set of the T–P sections at various compositions of the alloy was used in order to plot a series of isobaric T–C sections of the diagram. Afterwards, a technique of low-temperature quenching under pressure was applied in order to study the crystal structure and properties of a high-pressure phase in detail. The latter procedure (named below *thermobaric quenching*) was carried out as follows.

The sample studied was inserted in the pressure cell of the toroid-type chamber which had an internal heater. The necessary values of internal temperature and pressure were set, and the phase equilibrium was achieved in the sample. Then the high-pressure chamber was externally cooled with liquid nitrogen, internal temperature being maintained constant. When the external temperature of the toroid chamber was lowered to about 80 K, the current supply of the internal heater was cut off, and the sample cooled to the external temperature. Then the pressure was removed, and the sample was placed in a vessel with liquid nitrogen. Frozen metastable states thus obtained in all alloys studied were stored in liquid nitrogen for at least several months without any changes in their structure and phase composition.

Metastabilty of the quenched high-pressure phases in liquid nitrogen at atmospheric pressure and the macroscopic sample size (5 to 6 mm in diameter and 2 to 3 mm thick) ensured the application of various techniques for investigation of their crystal structure and properties as well as of processes which occur during heat treatment. The latter was most important for the study of spontaneous solid state amorphization discussed in the next Section.

The high-pressure data on the phase equilibria in the above systems were analyzed together with the available data on the crystal structure and properties of those phases which are stable at $P = 1$ atm or metastable after rapid quenching from melts. The following empiric rules were thus found[3] for the high-pressure evolution of the T–P–C phase diagrams:

Extension in Concentration of the Solid Solubility.

This trend is clearly represented by the Al-Ge and Al-Si systems. The ranges of the *fcc* Al-based solid solutions are limited to 2.6 at.% Ge and 1.6 at.% Si at $P = 1$ atm[2]. Pressures of 7 and 9 GPa, respectively, extend the solubility ranges to 15 at.% Gc and 18 at.% Si[6].

Extension of the Stability Regions of Intermediate Phases with Close-Packed Metal-Type Structures

The *hcp* ε-phase in the Pb-Bi system represents this rule perfectly. The maximum width of the ε-phase region in the T–C diagram is about 10 at.% at atmospheric pressure, but increases to about 25 at.% at a pressure of 1 GPa only[13] (Fig. 1).

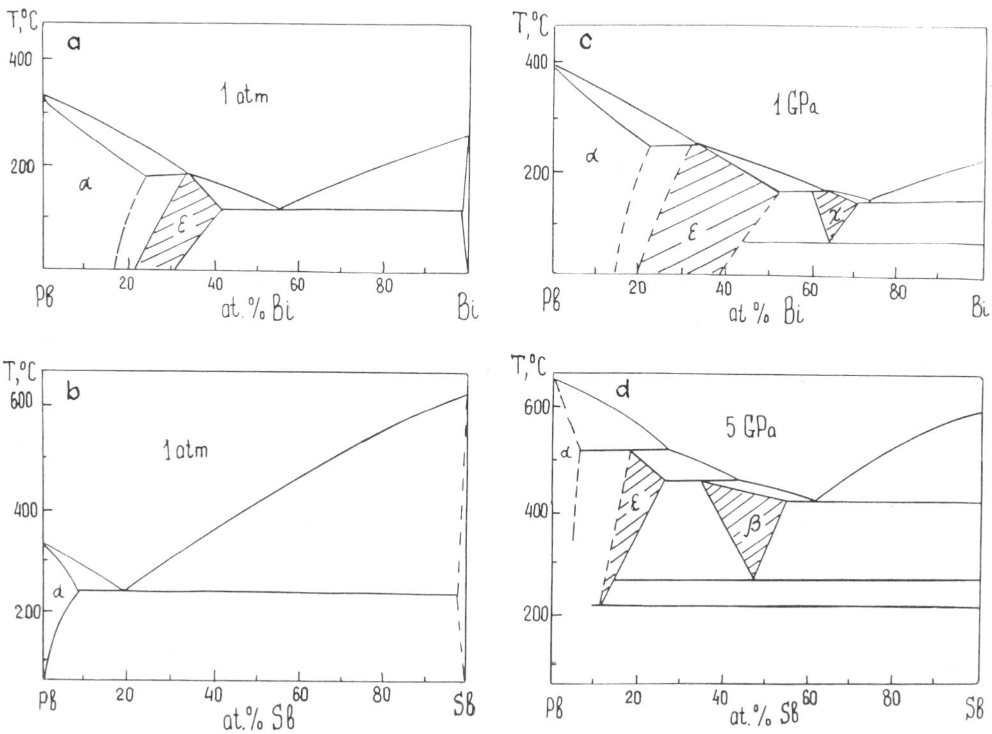

Figure 1. Cross-sections of three-dimensional T–P–C diagrams for Pb-Bi (a) and Pb-Sb (b) at normal pressure, Pb-Bi (c) at $P \simeq 1$ GPa, Pb-Sb (d) at $P \simeq 5$ GPa.

Decomposition of Semiconducting Compounds

The semiconducting compounds have openned crystal structures with low coordination numbers. If such compound exists in the T–C diagram at normal pressure, it becomes unstable at some higher pressure and undergoes decomposition or transformation into a metallic phase.

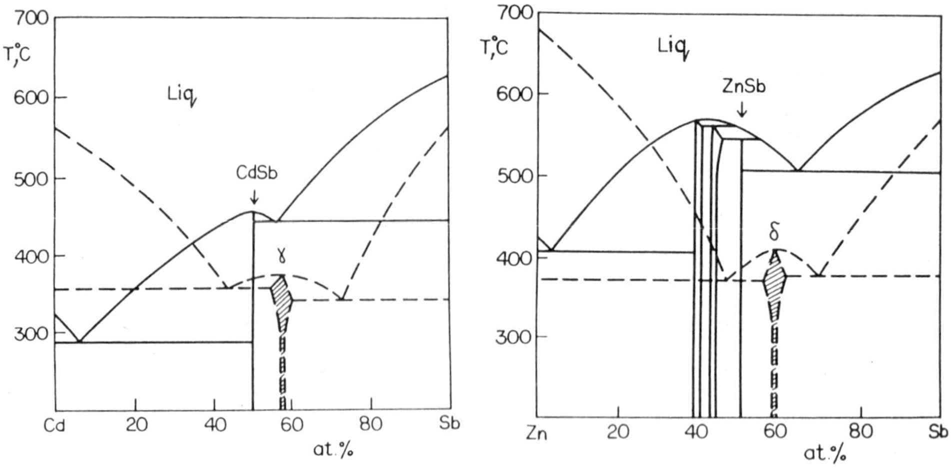

Figure 2. Isobaric sections of T–P–C phase diagrams of Cd-Sb and Zn-Sb alloys . Solid line: $P = 1$ atm; dashed line: $P = 5.5$ GPa and 7 GPa for Cd-Sb and Zn-Sb, respectively.

If some system initially has several intermediate semiconducting phases, they decompose one after another as pressure is increased. *E.g.*, there are three semiconducting compounds in the Zn-Sb system at ambient pressure, Zn_3Sb_2, Zn_4Sb_3 and ZnSb. The following sequence of transformations takes place under high pressure[14]:

$$Zn_3Sb_2 \xrightarrow{2.1\ \text{GPa}} Zn_4Sb_3 + Zn; \quad Zn_4Sb_3 \xrightarrow{3.0\ \text{GPa}} ZnSb + Zn \quad \text{and} \quad ZnSb \xrightarrow{6.5\ \text{GPa}} \delta + Zn$$

Metallic Nature of New Intermediate Phases

When the last semiconducting phase disappears, we obtain a T–C diagram with one or several intermediate non-stoichiometric phases. Composition ranges of these high-pressure phases usually are not related with the stoichiometric compositions of the initial semiconducting compounds.

Some isobaric $T - C$ sections with metallic high-pressure phases are shown in Figs. 1 and 2. The high-pressure χ-phase in the Pb-Bi system, the ε- and β-phases in the Pb-Sb system are metallic and superconducting. Metallic are also the γ- and δ-phases which replace stoichiometric semiconducting compounds in the Cd-Sb and Zn-Sb systems (Fig. 2).

The high-pressure intermediate phases in the Ag-Ge and Au-Ge systems are also very representative. It is well-known that the Group IB noble metals form intermediate phases with the neighbouring B-elements, so-called Hume-Rothery phases. The structure of these phases is determined by the electron concentration. At normal pressure, binary systems formed by Ag or Au and Zn, Cd or In have, among other Hume-Rothery phases, the *hcp* ε-phase which is stable in the range of electron concentrations 1.5 to 1.7 e/atom. This Hume-Rothery ε-phase is skipped in the T–C diagrams of the Ag-Ge and Au-Ge systems at $P = 1$ atm, but it becomes stable at high pressures[9].

Homology Rule for Binary Systems

Earlier study of the T–P diagrams resulted in a homology rule for the B-elements. According to the rule, B-elements under pressure acquire structures characteristic of their heavier homologues at normal pressure. This rule is best exemplified by the Group IVB elements. The structure types and the coordination numbers (in parentheses) vary in the C – Si – Ge – Sn series as follows: graphite (3) – diamond (4) – white tin (6). The same sequence is also observed for these elements as pressure is increased: graphite transforms into diamond, and the diamond-like structure of germanium and silicon transforms to the white-tin structure.

This rule is also applicable to the B-B alloys. Evolution of the T–C phase diagrams of the homologous alloys under pressure is compared in Fig. 1 (Pb-Bi and Pb-Sb alloys) and in Fig. 2 (Zn-Sb and Cd-Sb alloys).

The isobaric T–C sections of three-dimensional T–P–C diagrams are different for Pb-Bi and Pb-Sn alloys at ambient pressure, but they are the same at pressures $P \approx 1.0$ and 5.0 GPa, respectively. Intermediate ε-phases are isomorphous, and β- and χ-phases have very similar structures (β is *bcc* and χ is tetragonal distorted *bcc*).

The T–C phase diagram of the Zn-Sb system has three stoichiometric semiconducting phases at ambient pressure. But at $P \approx 3.0$ GPa, it is featured like the T–C diagram of the Cd-Sb system at ambient pressure which has only the equiatomic compound. The T–C sections of Zn-Sb and Cd-Sb systems are the same at pressures $P \geq 7.0$ and ≥ 4.0 GPa, respectively. Metallic γ- and δ-phases formed after decomposition of the semiconducting compounds have much similar crystal structures and homogeneity regions.

Metallic High-Pressure Phases are the Electronic Phases of the Hume-Rothery-Type

Doubtless, the above empiric rules are useful for planning of a research activity in this field. But they do not explain the origin of the structures observed in the B-B alloys. High-pressure experiments and rapid quenching from melts at atmospheric pressure markedly increased the number of B-B phases. On the other hand, the large number of metallic B-B phases gave an opportunity to classify their structures and to find factors which govern their stability. In analogy with the well-known Hume-Rothery phases, we analyzed relation between crystal structure and electron concentration for all familiar metallic phases in B-B alloys.

It is accepted below that the electron concentration, n_e, is the mean number of valence electrons per atom, and the specific electron concentration is the electron concentration related to the average volume per atom, V. The Group number of an element is accepted here to be equal to the number of its valence electrons.

The following sequence of crystal structures of metallic phases is observed as the electron concentration increases (Fig. 3):
fcc, *hcp* (12) ⟶ *fct* (12-8) ⟶ *bct*, ω (12-8) ⟶ *bcc* (8) ⟶ *simple hex.* (2+6) ⟶ β-Sn (4+2) ⟶ *simple cubic* (6).
The coordination number (in parentheses) and the packing density decrease in this sequence. Usually, metallic phases appear in the T–P–C diagrams according to this concentration sequence, but some links of the sequence can be skipped for an individual system.

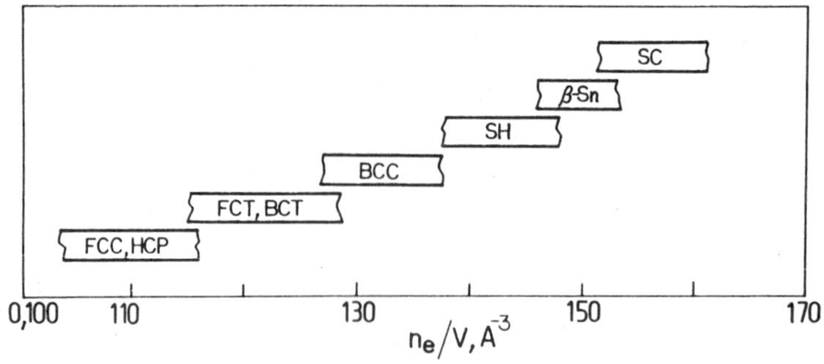

Figure 3. The existence regions of the B–B phases determined by electron concentration per unit volume, n_e/V.

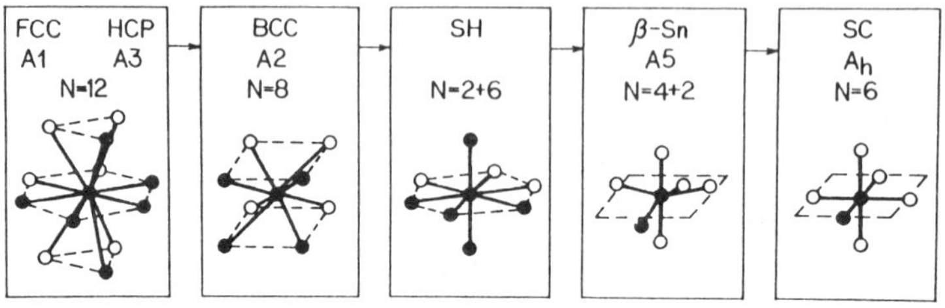

Figure 4. Sequence of structure-type (coordination number) alternation in the alloys of B-elements with an increase in the mean number of outer electrons.

Fig.3 shows the stability ranges of the phases in question, the specific electron concentration being an abscissa and Fig.4 demonstrates the sequence of structure types.

The n_e/V criterion stated above is very simple. It can be used to predict structures of the high-pressure phases which can occur in the alloys of B-elements from the 5*th* and 6*th* periods. The average atomic volume of the alloy components can be used for test calculations. But the volume decrease due to compressibility should be taken into account under pressure, hence, the numerical value of the n_e/V criterion for an individual structure-type should be shifted towards larger values. This criterion also facilitates the choice of the structure-type for a new unidentified phase.

The sequence of the structure-types which was found for the B-B phases in dependence on the electronic concentration is similar to the sequence of high-pressure phases observed in some B-elements and their compounds, especially, in Group IV elements (Si and Ge) and $A_{\text{III}}B_{\text{V}}$ compounds. For example, high-pressure transitions in silicon result in the following sequence of structures [15–17] :

$$\textit{diamond} \xrightarrow{12\text{ GPa}} \beta-\text{Sn} \xrightarrow{16\text{ GPa}} \textit{simple hex.} \xrightarrow{36\text{ GPa}} \textit{intermediate}$$
$$\textit{phase} \xrightarrow{42\text{ GPa}} \textit{hcp} \xrightarrow{79\text{ GPa}} \textit{fcc}$$

The intermediate phase in silicon is a structural analogue of the χ-phase in Bi-Pb, Sn-Bi, In-Bi alloys[3]. Recent high-pressure investigations over 10 GPa also proved regular formation of high-pressure phases which have either a simple hexagonal structure (GaAs[18], InSb[19], GaSb[20]) or a distorted simple hexagonal one (GaAs[18], InSb[19,21], Si[22]).

II. PRESSURE-INDUCED AMORPHIZATION OF CRYSTALLINE PHASES

Above, we stated some general trends for the high-pressure evolution of the T–C diagrams found in binary B–B alloys, particularly, origination of the stability regions of new metallic intermediate phases which extend the series of the Hume-Rothery electronic phases. An inverse phenomenon will be considered now, *i.e.*, decreasing thermodynamic stability of these high-pressure phases as pressure is reduced. The final result of such destabilization can be loss of the long-range order, *i.e.*, solid state amorphization (SSA).

This interesting phenomenon will be exemplified first by the experimental data on Cd-Sb and Zn-Sb alloys.

System Cd-Sb

It was shown in Section I (Fig. 2) that the T–C diagram of the Cd-Sb system at atmospheric pressure included one intermediate phase, a stoichiometric semiconducting compound, CdSb. This phase disappears at $P > 4.0$ GPa, and a metallic γ-phase of the Hume-Rothery-type with the composition of about $Cd_{43}Sb_{57}$ becomes stable.

After thermobaric quenching, the high-pressure γ-phase can be retained at $P = 1$ atm and $T = 78$ K for an indefinitely long time. The γ-phase has a simple hexagonal structure ($a = 3.182$Å, $c = 2.939$Å at $P = 1$ atm). It is superconducting below $T_c = 4.9 \pm 0.1$ K. On slow heating, the γ-phase undergoes spontaneous amorphization. The amorphous phase crystallizes to the equilibrium state, CdSb+Sb, on further heating.

Spontaneous solid state amorphization of γ -$Cd_{43}Sb_{57}$ was studied by calorimetric and dilatometric methods[23]. Fig. 5 shows calorimetric and dilatometric curves recorded on heating of the γ-phase. The first peak of heat release at $290 < T < 330$ K and the corresponding increase in the sample volume are due to amorphization. The second exothermal effect at $360 < T < 410$ K and the corresponding volume decrement are caused by crystallization of an amorphous sample to the equilibrium state, *amorph.* $Cd_{43}Sb_{57} \longrightarrow$ *cryst.* (CdSb+Sb).

Heat and volume effects of amorphization are equal to $\Delta H = 1.7 \pm 0.2$ kJ$\cdot$ mol^{-1} and $\Delta V/V_0 = 12.6 \pm 0.3\%$ and those for crystallization are $\Delta H = 4.6 \pm 0.8$ kJ$\cdot$mol^{-1} and $\Delta V/V_0 = 1.0 \pm 0.3\%$. The activation energies, E_a, of the amorphization and crystallization processes were determined from the dilatometric curves as 110 ± 20 kJ$\cdot$mol^{-1} and 200 ± 60 kJ$\cdot$mol^{-1}, respectively. The amorphization E_a value is lower by one order of magnitude than the activation energy of the atomic diffusion in this system.

System Zn-Sb

Isobaric sections of the T–P–C phase diagram of the Zn-Sb alloys are shown in Fig. 2. The crystal structure of the high-pressure metallic phase, δ $Zn_{43}Sb_{57}$, is a modified structure of the γ-phase with an orthorhombically distorted cell ($a = 2.82, b = 3.25, c = 5.06$ Å at $T = 100$ K, $P = 1$ atm).

The δ-phase after thermobaric quenching transforms to an amorphous state when heated to room temperature. The amorphous phase is rather stable at normal conditions, it stays essentially the same for several months. No hints of crystallization were observed after annealing of the amorphous alloy at 373 K for 10 h. But the sample recovered the equilibrium crystal state, ZnSb+Sb, after annealing at 423 K during 3 h. The measured density of a bulk amorphous $Zn_{43}Sb_{57}$ sample is 6.30 g·cm^{-3}, and the density of the same alloy in the equilibrium state, ZnSb+Sb,calculated from the X-ray data, is 6.43 g·cm^{-3}.

Fig. 6 shows heating calorimetric curves of δ -$Zn_{43}Sb_{57}$ and $Zn_{50}Sb_{50}$ upon heating[24]. The first exothermal peak corresponds to amorphization, and the other one is due to crystallization. Exothermal effects in single-phase δ -$Zn_{43}Sb_{57}$ are $\Delta H_1 = 1.9 \pm 0.2$ kJ·mol^{-1} and $\Delta H_2 = 6.7 \pm 0.6$ kJ·mol^{-1}, respectively.

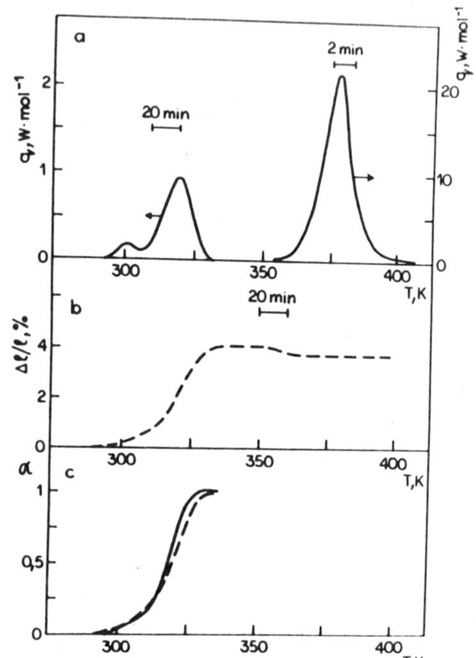

Figure 5. Amorphization and crystallization of the high-pressure phase of $Cd_{43}Sb_{57}$ (at.%, γ phase) composition upon heating. (a) Heat release power q as a function of temperature; (b) relative expansion of γ-phase sample; (c) fractional conversion (α) of "$\gamma \longrightarrow$ *amorphous phase*" transition. The solid line denote the calorimetric data, the dashed line is the volumetric curve.

Transitions during heating of γ -$Cd_{43}Sb_{57}$ and δ -$Zn_{43}Sb_{57}$ samples to an amorphous then to the equilibrium crystalline states were proved by X-ray diffraction experiments after termination of the first and the second exothermal peaks, respectively.

Thermal behavior of resistivity of amorphous Cd-Sb and Zn-Sb alloys is essentially semiconducting. For example, resistivity of the amorphous Zn-Sb alloy at the liquid-nitrogen temperature exceeds that of the ZnSb single crystal by 5 to 6 orders of magnitude[25].

A comparative analysis of the X-ray (and neutron for Zn-Sb) diffraction patterns, specific volumes and thermal behavior of resistivity of the amorphous and crystalline Zn-Sb and Cd-Sb samples shows that the quenched metallic γ- and δ-phases transform to the semiconducting amorphous state with the coordination number close to that of the stoichiometric semiconducting crystal, ZnSb and CdSb. It was concluded from the experimental data on the resistivity behaviour of the alloys during amorphization[26,27] that solid state amorphization was a heterogeneous process which involved nucleation of the new phase at the grain boundaries and its growth, a fractal microstructure being formed at the initial stages. This conclusion is in agreement with the values of the coefficient in the kinetic Avrami equation, $n = 1.5 - 2.0$[24], and supports the idea that SSA is a process similar to melting.

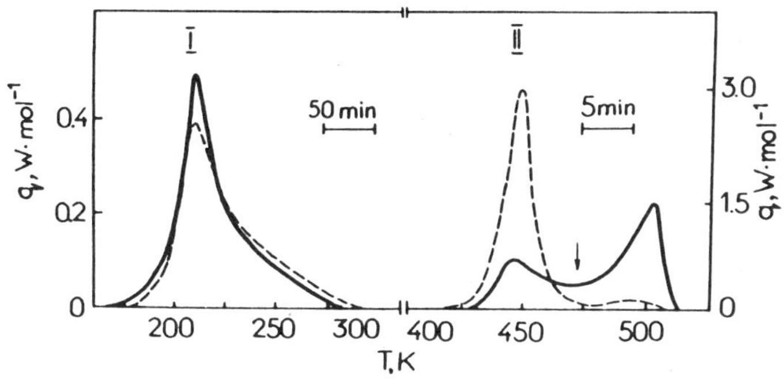

Figure 6. Calorimetric curves of amorphization (peak I) and crystallization (peak II) of Zn-Sb alloys. Solid lines: $Zn_{43}Sb_{57}$ alloy; dashed lines: $Zn_{50}Sb_{50}$. Arrow ↓ denotes the temperature of intermediate quenching for X-ray analysis.

Solid State Amorphization and the T–P Phase Diagrams of Alloys

We shall consider now another aspect of this interesting phenomenon, *i.e.*, a relation between the T–P parameters of amorphization and the lines of stable and metastable equilibria in the T–P phase diagrams.

The T–C phase diagrams of the alloys in question have one or several stoichiometric semiconducting compounds at $P = 1$ atm. Their melting temperatures decrease as pressure is increasing. Fig. 7 presents a scheme of the T–P cross-section of the three-dimensional T–P–C diagram at the composition of the high-pressure phase, say γ, which is rather close to the composition of the stoichiometric semiconducting compound, A_nB_m. Melting curves of the low-pressure and high-pressure phases are shown in solid lines, the dash-dotted curve is the line of the $A_nB_m \pm A \longleftrightarrow \gamma$ equilibrium, shaded bands show kinetic regions of phase transitions $A_nB_m \pm A \longrightarrow \gamma$ and $\gamma \longrightarrow A_nB_m \pm A$. The dotted lines represent the A_nB_m and γ-phase melting curves extrapolated into their metastability regions. If the segment b_1b_2 of the γ-phase virtual melting curve lays below the kinetic band of the phase transition, then the γ-phase intersecting b_1b_2 from the region of its metastability becomes unstable with respect to the long-range order. However, a normal phase transition to the equilibrium crystalline state does not occur because kinetics of this transition is frozen. In this case, phase γ overheated above its melting temperature "melts". If b_1b_2 lays below the temperature of the glass transition, T_g, phase γ transforms to an amorphous state. It is exactly the case which was observed in the systems concerned. This is

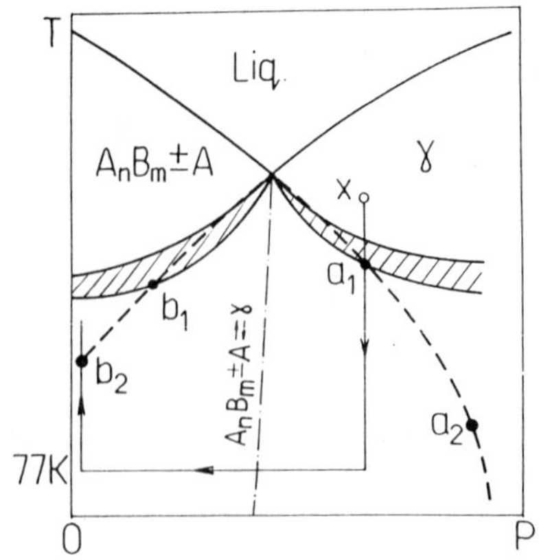

Figure 7. T–P section of the phase diagram of a binary A–B system with stoichiometric semiconductor compound A_nB_m at 1 atm and the high-pressure phase (γ).

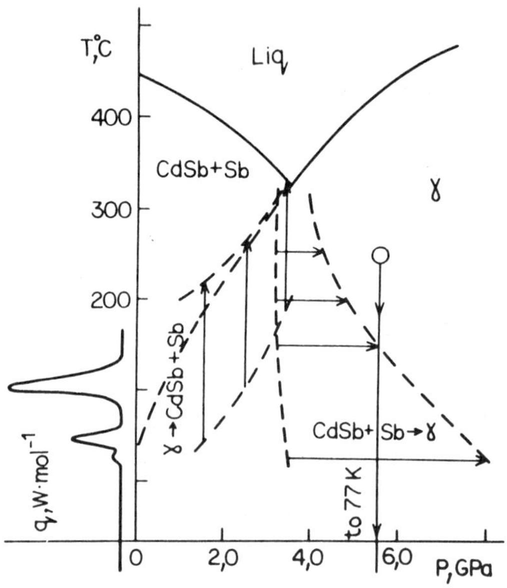

Figure 8. T–P section of T–P–C phase diagram of the Cd-Sb system for $Cd_{43}Sb_{57}$ composition (γ phase). The DSC heating curve of the γ phase ($P = 1$ atm) is exhibited along the temperature axis.

particularly evident in the case of the Cd-Sb system (see Fig. 8).

Similar amorphization phenomena were observed on heating of the metallic high-pressure phases in binary Ga-Sb and Al-Ge alloys and ternary Cd-Zn-Sb and Ga-Ge-Sb systems. Thus, spontaneous (*i.e.*, no external stimulus but the appropriate temperature) amorphization of a high-pressure phase can be treated as "melting" of a metastable crystalline phase at $T < T_g$ where T_g is the glass transition temperature. This idea is supported by the low activation energy of the SSA process, by the n-value in the kinetic Avrami equation and by other experimental data like those above.

The Cd-Sb and Zn-Sb alloys were considered above as convenient representatives of the systems were the SSA phenomena were observed due to high pressure treatment. Numerous cases of solid state amorphization are found and described at present, which occur both on increasing and decreasing pressure (see review[3]).

Solid state amorphization is anticipated in substances which have the T–P phase diagrams similar to that shown in Fig. 7 when either the low-pressure or the high-pressure phase intersects the appropriate extrapolated melting curves on compression or on decompression, respectively. Amorphization of the low-pressure phases of C (graphite), SiO_2 and some silicate minerals were found on compression. Both kinds of SSA were described for ice. For example, a high-density amorphous phase was observed on compression of ice Ih, and a low-density amorphous phase was obtained on decompression of the high-pressure ice VI[28]. All these cases of SSA were directly associated with the melting phenomena.

An analogy between melting and amorphization is often drawn when other, non-thermobaric SSA processes are discussed, *i.e.*, "inverse melting" of a quenched metastable phase, diffusion annealing of a thin layered bi-element structure, hydrogenolysis, and others[29]. Then it is natural to discuss relation between the melting curve extrapolated to low temperatures and the curve of complete instability of the crystalline phase involved.

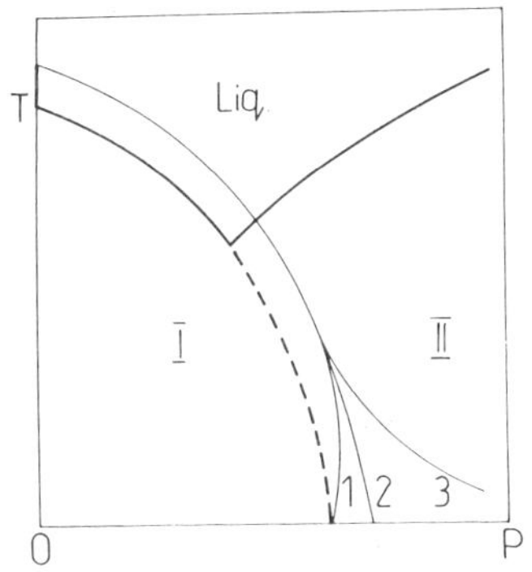

Figure 9. A schematic T–P phase diagram. Phase I and II melting curves are shown in thick solid lines, their extention to the metastability regions are shown in dashed lines. The thin solid curves 1, 2 and 3 represent the possible lines of phase I thermodynamic instability.

The instability problem of a crystal heated over T_m at ambient pressure was discussed many times. Various criteria of instability were proposed, *i.e.*, reduction of a shear elastic modulus to zero (mechanical instability), a critical amplitude of atomic vibrations (phonon instability), a critical concentration of lattice defects, the entropy catastrophe, *etc.* There are various models where properties characterizing crystal stability are extrapolated beyond T_m. These models as well as the extrapolation of some experimental data give values of the instability temperature which are higher than T_m by tens of degrees.

What can be added to the relation between the melting, $T_m = T_m(P)$, and instability, $T_{\text{ins}} = T_{\text{ins}}(P)$, curves? Several types can be assumed *a priori* for the crystal instability curve extrapolated in the T–P plane to low temperatures (curves 1 to 3 in Fig. 9). The following experimental data evidence in favour of the type 1 or 2 curves:

i) when compression experiments were carried out at room temperature, amorphization always occurred in the vicinity of the extrapolated melting curve, $T_m(P)$. This is valid even for such refractory substances as graphite and silica.

ii) amorphization of the metastable high-pressure phases (HPP) often began at rather low temperatures, $T \leq 100$ K, upon heating at atmospheric pressure[30].

Curves 1 and 2 are also favored by the Fecht and Johnson idea[31] that the instability curve of the crystalline state in the T–C diagram of a binary alloy and the diffusionless melting curve, $T_o(C)$, are tangent at $T = T_g^o$. Here T_g^o is the temperature of the ideal glass transition where $S_{\text{cryst}} = S_{\text{amorph}}$, according to the Kauzmann paradox[32].

There are other questions which also arise when the above experimental data on spontaneous SSA are discussed.

Why does "melting" of a metastable metallic HPP in A_{II}-B_V and A_{III}-B_V alloys always result in an amorphous semiconductor? Actually, the HPP melting curve over the triple point corresponds to the transformation metallic HPP $\longrightarrow$ metallic melt!

Why is a large exothermal effect always observed instead of an endothermal one? An endothermal effect is characteristic of normal melting.

Why were metallic amorphous states never obtained in these systems at ambient pressure using all possible techniques?

To conclude this Section, we note that there are many models which consider the crystal instability phenomena over melting temperature at $P = 1$ atm. But the general problem of the crystal instability over $T_m(P)$ in a wide T–P region, down to $T_m = 0$ K, was not accurately solved so far. It seems worth therefore to discuss below some particular cases when the T–P phase diagrams including phase equilibria and instability curves were calculated both for crystal phases (Section III) and for the amorphous state (Section IV). Analysis of such phase diagrams including instability curves will elucidate the above questions.

III. T–P PHASE DIAGRAMS WITH ISOMORPHIC TRANSITIONS AND CRITICAL POINTS OF THE LIQUID - VAPOUR TYPE

The problem of critical points in the phase diagrams of crystalline solids has been intensely explored in theory and experiments for many years. There was found a so-called *tricritical* point (or the Landau critical point) where the first-order transition line turns into the second-order transition line. But occurrence of a critical point of the liquid - vapour - type was forbidden for a long time. For example, P. Bridgman in his classic monograph "High Pressure Physics"[33] completely rejected even a possibility of termination of the first-order transition line at the critical point of the liquid - vapour - type in the crystal state. A similar statement forbidding the occurrence of a liquid - vapour critical point was given by L. Landau and E. Lifshitz in their "Statistical Physics" edited in 1964[34]. The prohibition given by Bridgman, Landau, Lifshitz and other researchers was based on the symmetry considerations because they assumed that the first-order transitions in crystalline solids could occur as polymorphous transitions only.

But during the last decades, there was experimentally found a good number of solids which undergo the first-order isomorphic transition when two crystalline phases in equilibrium have the same symmetry elements. Isomorphic transitions usually are due to change of the electronic state, *i.e.*, valency. The symmetry prohibition has no relation to these transitions. So, such first-order transition curves in the T–P diagrams can terminate in a critical point. All properties depend on T and P continuously above this point, as in the case of the liquid - vapour system in the supercritical region. Occurrence of such critical point also results in additional lines in the T–P phase diagram where one or other isomorphous phase becomes thermodynamically unstable. Loss of the thermodynamic stability then results in an anomalous behaviour of the properties of these solids.

The first and classical solid where the isomorphic transition[35] and the critical point of the liquid - gas-type[36] were experimentally found is cerium. Presently, Ce is the most simple and well-studied subject with the isomorphic phase transition terminating at a critical point. It is expedient therefore to employ cerium in order to consider all items concerning the features of the T–P phase diagrams with an isomorphic phase transition: occurrence of a critical point, thermodynamic instability of crystal phases, and anomalous properties.

A lot of theoretical works are published now, which are devoted to the quantum-mechanical description of the behaviour of the electron gas in cerium and other rare earths and their compounds with variable valency and isomorphic phase transitions (see an excellent review by Khomskii[37]). Unfortunately, the quantum-mechanical approach did not enable one to calculate the phase diagrams and the anomalies of the thermodynamic properties.

To describe the T–P phase diagrams as well as the heat capacity and the P–V–T anomalies, there was proposed a thermodynamic model, a so-called model of pseudobinary solutions[38]. In spite of its simplicity, this model turned out to be most efficient for calculation of the phase diagrams of substances with isomorphic transitions and all inherent features: critical points, instability lines, *etc.* It is worth therefore presenting the model in some detail and its application to cerium.

A Pseudobinary System

Electronic configurations $4f^0$ and $4f^1$ in cerium are very close in energy, and both trivalent or tetravalent Ce ions can form alloys and chemical compounds. A discontinuous change in the electronic state has also been observed in the pure metal in the process of the low-temperature *fcc* to *fcc′* phase transition. This transition is characterized by complete isomorphism of the two phases, there is no change in the symmetry of atomic lattices and both phases have no magnetic order near the equilibrium temperature and $P = 1$ atm. The volume jump at the transition is extremely large, about 16%.

Increasing temperature and pressure reduce volume and heat effects of the transition, and they become zero at the critical point. The coordinates of the critical point were estimated first from the high-pressure *DTA* measurements, $T_c = 550$ K and $P_c = 1.9$ GPa[36]. Later, Kutsar[39] used the volume measurements and located the critical point more accurately, $T_c = 490$ K and $P_c = 1.45$ GPa. The latter accurate T_c and P_c values will be used in the model calculations below.

We shall denote two isomorphous phases of cerium as α and α' where α' -Ce is the low-temperature phase, and α -Ce is the high-temperature one. The original model of pseudobinary solutions[38] was based on the following ideas:

i) Near 0 K, the α-phase consists only of trivalent Ce ions, and the α'-phase consists only of tetravalent ones. The α-phase is assumed to be metastable at 0 K.

ii) Both phases change their electronic states (valency) at increasing temperature or pressure. The changes can be characterized by variable effective concentrations, C and $1 - C$, of Ce ions in these two different ground states.

iii) At finite temperatures, the α- and α'-phases were treated as substitutional solid solutions of two components, Ce^{3+} and Ce^{4+}.

Thus, cerium metal is a pseudobinary system with new *intrinsic* parameters, "effective concentrations of ions in different states". It was found later that cerium ions in the α'-phase were in the intermediate valence state, $Ce^{3.7+}$, at low temperature and pressure. This fact does not change the model equations, but only slightly changes the sense of model parameter C. We shall use therefore expressions derived in reference [38], but the ground states of Ce ions in the α- and α'-phases at 0 K are denoted below as components B and A, respectively, in order to avoid discussion of the valency problem.

Within the proposed model, cerium is a pseudobinary analogue of a two-component system with continuous solid solutions at high temperatures, $T > T_c$, and phase separation at low temperatures, $T < T_c$. We emphasize a fundamental difference between the binary and pseudobinary systems: unlike normal two-component systems which have externally fixed concentrations of components, effective concentration C is a function of temperature and pressure in both phases of a pseudobinary system. Hence, effective concentration C in this case is not an external parameter as in normal binary solutions but an internal parameter of the system. For this reason, there is no two-phase region in the T–P phase diagram of cerium.

Thermodynamic Equilibria

The thermodynamic potential for such system can be written as a function of temperature, pressure and concentration in the form:

$$\begin{aligned} G_C(T,P,C) &= (1-C)G_A(T,P) + CG_B(T,P) + C(1-C)U \\ &\quad + RT[C\ln C + (1-C)\ln(1-C)] = \\ &(1-C)(E_A - TS_A + PV_A) + C[E_B - TS_B + PV_B) + C(1-C)U \\ &\quad + RT[C\ln C + (1-C)\ln(1-C)] = \\ &G_A(T,P) + C(\Delta E^\circ - T\Delta S^\circ + P\Delta V^\circ) + C(1-C)U \\ &\quad + RT[C\ln C + (1-C)\ln(1-C)] \end{aligned} \tag{3.1}$$

where $E_A, S_A, V_A, E_B, S_B, V_B$ are the internal energy, entropy and volume of 1 mole of pure components A and B; $\Delta E^\circ = E_B - E_A, \Delta S^\circ = S_B - S_A, \Delta V^\circ = V_B - V_A$, and the superscript indicates that the differences are referenced to 0 K and $P = 1$ atm; C is the concentration of component B; the parameter U is the mixing energy which is assumed to be positive. Positive U is the phase separation criterion in binary systems. We shall assume for simplicity that the quantities $E_A, S_A, V_A, E_B, S_B, V_B$ and U are independent of temperature, pressure or concentration. This is a standard assumption in the theory of regular solutions.

It is easy to show (see reference [38]) that the cerium thermodynamic potential of the form of eqn. (3.1) is completely analogous to that in the approximation of regular solutions in normal binary alloys, and condition $U = const$ corresponds at the microscopic level to the Strässler and Kittel two-level model[40]. Within this model, the energy difference between the ground state and the excited atomic state depends linearly on the number of the excited atoms. Note that presenting the two-level model in terms accepted within quasi-chemical theories of multicomponent systems greatly simplifies the quantitative calculation of phase diagrams as well as of the thermodynamic properties of the systems involved.

At fixed T and P values, the thermodynamic potential is a function of concentration and has one or two minima which correspond to one (stable) or two (stable and metastable) phases. The extremum conditions are given by the relation

$$\frac{\partial G_C}{\partial C} = [G_B(T,P) - G_A(T,P)] + U(1-2C) + RT\ln\left(\frac{C}{1-C}\right) = 0 \tag{3.2}$$

or

$$\Delta E^\circ - T\Delta S^\circ + P\Delta V^\circ + U(1-2C) + RT\ln\left(\frac{C}{1-C}\right) = 0 \tag{3.3}$$

The phase equilibrium occurs when

$$G_C(C_1) = G_C(C_2) \tag{3.4}$$

Here C_1 and C_2 are concentrations of the same component in two phases in equilibrium. It can be shown from eqns. (3.3) and (3.4) that

$$C_1 = 1 - C_2 \tag{3.5}$$

Substituting either C_1 or C_2 value from eqn. (3.5) and combining eqns. (3.3) and (3.4), we arrive at

$$T_\circ = -\frac{U}{R}\cdot\frac{1-2C_2}{\ln\frac{C_2}{1-C_2}} = -\frac{U}{R}\cdot\frac{1-2C_1}{\ln\frac{C_1}{1-C_1}} \tag{3.6}$$

where $T_\circ$ is the phase equilibrium temperature.

Concentrations C_1 and C_2 in these two phases coincide at the critical point, $C_1 = C_2 = 1/2$, therefore

$$T_c = -\frac{U}{R}\lim_{C_2\to\frac{1}{2}}\frac{1-2C_2}{\ln\frac{C_2}{1-C_2}} = \frac{U}{2R} \tag{3.7}$$

hence,

$$U = 2RT_c \tag{3.8}$$

where T_c is a critical temperature.

Lines of the α– and α'–Phase Instability

The problem of thermodynamic instability of a crystal phase is quite easily solved in terms of this model. The thermodynamic potential of α- and α'-phases is described by the same function $G_C(T,P,C)$ in wide temperature or pressure ranges. The $\alpha \leftrightarrow \alpha'$ phase equilibrium corresponds to the line in the T–P plane where G_C has two minima with the same G_C values. The T–P region where G_C has two local minima with different G_C values is the region of thermodynamic stability of the phase which corresponds to the lower minimum while the other phase is metastable. When one of the minima degenerates to an inflection point, the corresponding (metastable) phase becomes completely thermodynamically unstable. The second derivative is zero at the inflection point,

$$\frac{\partial^2 G}{\partial C^2} = -2U + \frac{RT}{C(1-C)} = 0 \tag{3.9}$$

which results in two metastability boundaries shown in (Fig. 10). Curve KAB limits the metastability region of α -Ce, and KC is for α'-Ce. Dotted lines describe the hysteresis of the transition. As expected, the kinetic lines of the phase transitions do not coincide with the lines of the phase instability, but lay closer to the equilibrium curve.

Anomalous Thermodynamic Properties

Eqn. (3.3) gives explicit expressions for temperature and pressure:

$$T = \frac{2UC - (\Delta E^\circ + U + P\Delta V^\circ)}{R\cdot\ln\frac{C}{1-C} - \Delta S^\circ} \tag{3.10}$$

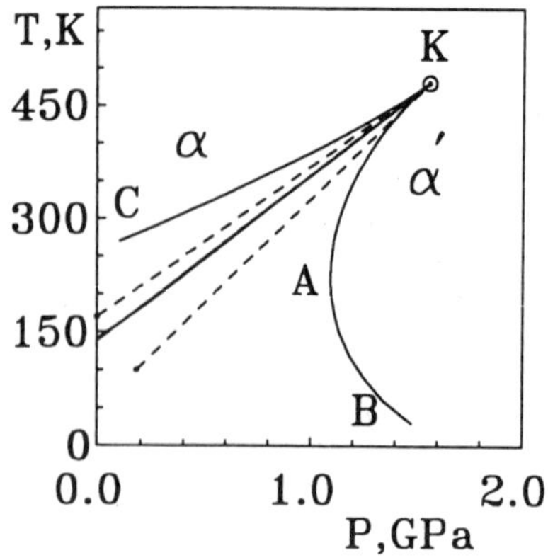

Figure 10. P–T phase diagram of cerium: ——— - theoretical lines of $\alpha - \alpha'$ phase equilibrium ; KC and KAB - theoretical lines of loss of stability; – – – - experimental lines of $\alpha \rightarrow \alpha'$ and $\alpha' \rightarrow \alpha$ phase transitions; K - critical point.

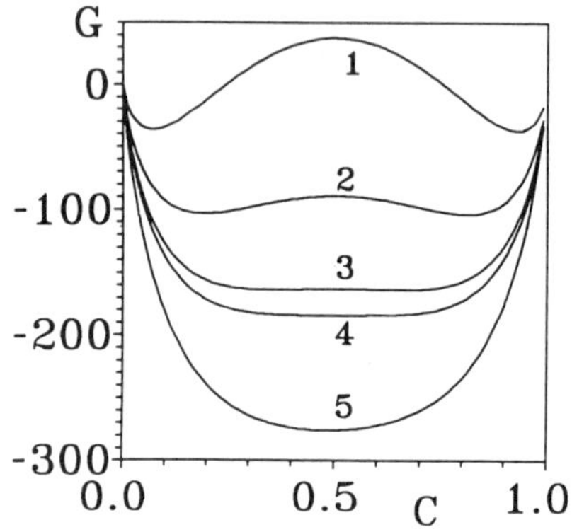

Figure 11. The concentration dependences of the termodynamic potential along the line of phase equilibrium $\alpha \leftrightarrow \alpha'$ Ce : 1. $T = 318.52$ K, $P = 0.825$ GPa; 2. $T = 410.68$ K, $P = 1.244$ GPa; 3. $T = 464.44$ K, $P = 1.489$ GPa; 4. $T = 480$ K, $P = 1.56$ GPa; 5. $T = 550$ K, $P = 1.89$ GPa.

and

$$P = \frac{1}{\Delta V_\circ} \cdot \left(2UC - (\Delta E^\circ + U) - T(R \cdot \ln \frac{C}{1-C} - \Delta S^\circ\right) \tag{3.11}$$

Then the isobaric or isothermal variation of the internal parameter, C, can be found from (3.10) and (3.11):

$$\left(\frac{\partial C}{\partial P}\right)_T = \frac{\Delta V^\circ}{2U - \frac{RT}{C(1-C)}} \tag{3.12}$$

$$\left(\frac{\partial C}{\partial T}\right)_P = \frac{R \cdot \ln \frac{C}{1-C} - \Delta S^\circ}{2U - \frac{RT}{C(1-C)}} \tag{3.13}$$

The curves in Figs. 11 and 12 represent G as a function of concentration at several T and P values, and Fig. 13 shows C versus temperature along the curve of the $\alpha \longleftrightarrow \alpha'$ phase equilibrium.

The T–P phase diagram of cerium was accurately described with the parameter values $U = 1920$ cal/mole, $\Delta E^\circ = 505$ cal/mole; $\Delta S^\circ = 3.69$ cal/mole and $\Delta V^\circ = 3.4$ cm^3/mole.

The assumption that the volumes of the "pure components", V_A and V_B, do not

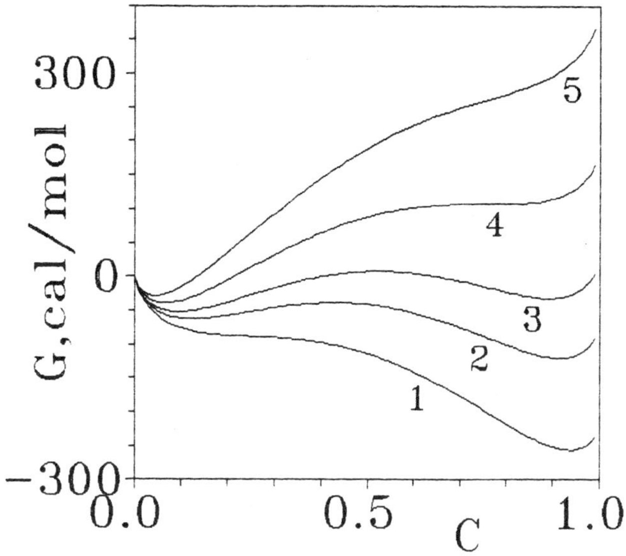

Figure 12. The concentration dependences of the termodynamic potential at T = 350 K, 1. P = 0.7 GPa; 2. P = 0.88 GPA; 3. P = 1.0 GPa; 4. P = 1.2 GPa; 5. P = 1.45 GPa.

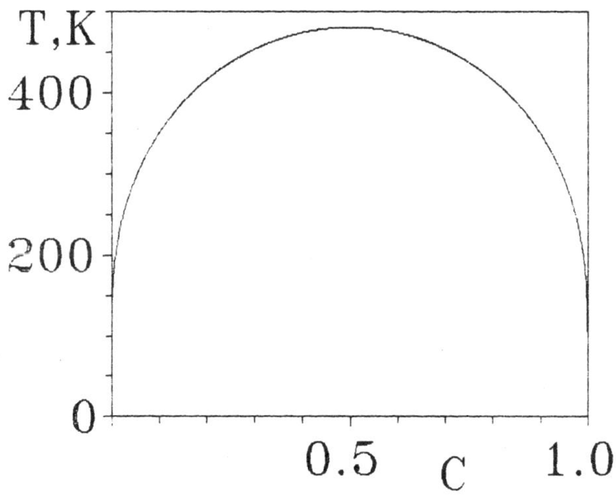

Figure 13. Concentrations of the coexisting α and α' phases as functions of the temperature along the line of phase equilibrium: $C < 0.5$ corresponds to C_1 (α'-Ce); $C > 0.5$ corresponds to C_2 (α-Ce).

depend on temperature, pressure or concentration, leads to the relation

$$V(C) = V_A(1 - C) + V_B C \tag{3.14}$$

hence,

$$\frac{V(C) - V_A}{V_B - V_A} = C \tag{3.15}$$

Using eqns. (3.12), (3.13) and (3.15), we find the coefficient of thermal expansion and the compressibility from the approximate expressions

$$\beta_C = \frac{1}{V}\left(\frac{\partial V_C}{\partial T}\right)_P = \frac{\Delta V^\circ}{V_M}\left(\frac{\partial C}{\partial T}\right)_P \tag{3.16}$$

and

$$\chi_C = -\frac{1}{V}\left(\frac{\partial V_C}{\partial P}\right)_T = -\frac{\Delta V^\circ}{V_M}\left(\frac{\partial C}{\partial P}\right)_T \tag{3.17}$$

where $V_M = (V_A + V_B)/2$.

The following expression for the heat capacity results from eqn. (3.1):

$$c_P = \frac{\partial(E + PV_C)}{\partial T} = (\Delta E^\circ + P\Delta V^\circ - 2CU)\left(\frac{\partial C}{\partial T}\right)_P \tag{3.18}$$

Thus, the $C_T(P)$ and $C_P(T)$ functions purely describe the concentration contributions to the volume change, the and $(\partial C/\partial P)_T$ and $(\partial C/\partial T)_P$ functions represent the anomalous behaviour of the compressibility,the thermal expansion and the heat capacity.

The coefficient of thermal expansion, compressibility and heat capacity are abnormally high near the critical point. Some quantitative estimates[41] were made from the available data. Compressibilities of $\alpha - Ce$ and α' -Ce in equilibrium at $T = 289$ K and $P = 0.7$ GPa are $5 \cdot 10^{-12}$cm^2/dyne and $3.5 \cdot 10^{-12}$ cm^2/dyne, respectively. The internal parameters determined at these conditions are $C = 0.97$ for α -Ce and 0.03 for α'-Ce. Then we calculate from eqn. (3.17) $\chi_A = 2.58 \cdot 10^{-12}$ cm^2/dyne and $\chi_B = 4.08 \cdot 10^{-12}$ cm^2/dyne. Using these values for χ_A and χ_B and the corresponding data for $(\partial C/\partial P)_T$, isothermal and isobaric curves of compressibility, χ_C, and the bulk modulus, K_C, can be calculated for both phases from eqn. (3.17). Fig. 14 shows the bulk modulus $K_C = 1/\chi_C$ as a function of pressure at room temperature. It tends to zero near 0.1 GPa for the α'-phase and at about 1.3 GPa for α-Ce. Respectively, χ_C becomes infinite at these pressure limits.

Anomalously high β and c_p values are also anticipated for the metastable phases near the lines of the phase instability.

Fig. 15 shows the relative volume effect of the $\alpha - \alpha'$ transition along the phase equilibrium line. The calculated solid curve demonstrates a fair agreement with the experimental data[39]*(points)*. Hence, the thermodynamic potential expressed by eqn. (3.1) and the fitted values of the model constants $\Delta E^\circ, \Delta S^\circ, \Delta V^\circ$ and U correctly describe the thermodynamic properties of cerium up to the second derivatives of $G(T, P, C)$ at least.

As it was stated above, cerium was the first investigated subject whose first-order isomorphic transition line terminated at the critical point of the liquid-vapour type. For long time it was the only example of a substance with an isomorphic phase transition due to variable valency. Presently, we know a lot of substances which undergo isomorphic transitions. These are mainly rare earth metals, actinides, their

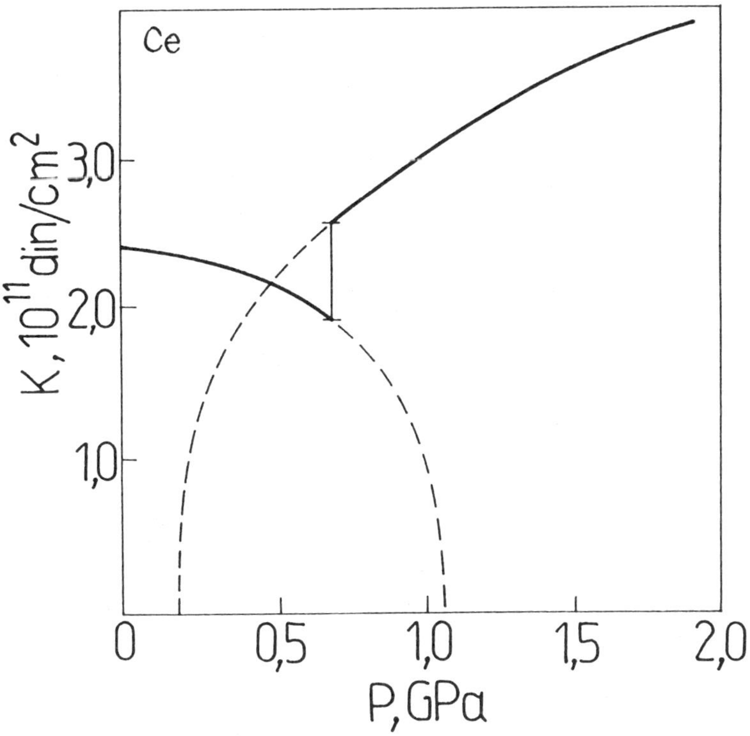

Figure 14. Bulk modulus $K_C = 1/\chi_C$ as a function of pressure at room temperature: ——— - theoretical curve for equilibrium phases; – – – - theoretical curve for metastable phases.

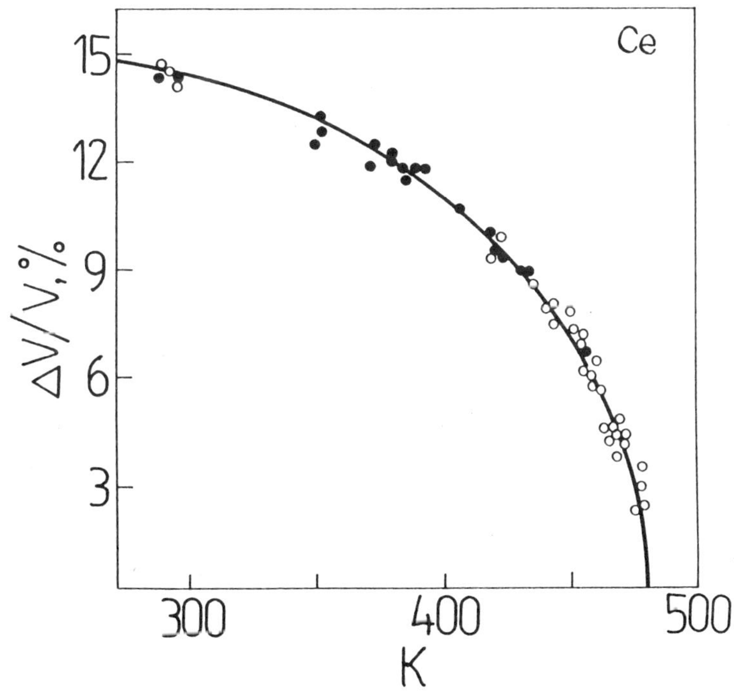

Figure 15. Relative volume effect of the transformation as a function of the temperature along the phase equilibrium line: ——— - theoretical curve; ∘ and •- experimental points from [39]

alloys and compounds which show variable valency. In addition to the anomalous thermodynamic properties exemplified here by cerium, some of these substances have other interesting features. The isomorphic transition can result in a change of the conductivity type (SmS, SmSe, $Sm_{1-x}Gd_xS$ and others[42]), the first-order isomorphic transition curve in the T–P–C diagrams of alloys can be limited by two critical points - both at higher and at lower temperatures ($Ce_{0.8}Th_{0.1}La_{0.1}$[43]), *etc.* But these features will not be detailed here due to the limited paper length. Instead, we recur to the SSA phenomena in the next Section in order to discuss some interesting results which follow from the two-level model of pseudobynary solutions applied to the unordered (both liquid and amorphous ,see section IV,1) systems.

IV. METASTABLE PHASE DIAGRAMS WITH A CRITICAL POINT AND INSTABILITY OF THE AMORPHOUS PHASES

Amorphous semiconductors subjected to high pressure can undergo first-order phase transitions like amorphous semiconductor *I* — amorphous semiconductor *II* in $Zn_{41}Sb_{59}$ at $P = 1.8$ GPa or amorphous semiconductor — amorphous semimetal in $Zn_{41}Sb_{59}$ at $P = 6.0$ GPa[44]. They can also transform to a metal state, like InSb amorphous films at $P = 1.0 - 1.3$ GPa[45], GaSb at $P = 3.5 - 4.0$ GPa[46]. An *in situ* X-ray study[44] has shown that metallization of semiconducting amorphous InSb resulted in a mixture of amorphous and crystal phases, which evidenced the low stability of metallic amorphous InSb even under high pressure.

To elucidate the rather complex phenomenon of SSA in A_{III}-B_V and A_{II}-B_V alloys as well as specific behavior of semiconducting amorphous states under variable pressure and temperature, a thermodynamic analysis of these systems is necessary using adequate models. Now we shall show how the above two-level model which was a success for isomorphous transformations in crystals can be applied for the description of amorphous systems.

It is well known that those A_{III}-B_V and A_{II}-B_V alloys of non-transition elements which form stoichiometric compounds have anomalous thermodynamic properties in the liquid state. It is generally assumed that the anomalies are due to clusters in metallic melts of the alloys. The structure and properties of these clusters are close to those of crystalline semiconducting compounds. Regel and Glazov[47] have analyzed the available data on viscosity and electrical resistivity and estimated the volume percentage of semiconducting clusters in melts for several compounds. According to their estimates, this percentage is equal to about 8, 3 and 5% for GaSb, InSb and InAs at T_m, respectively. These quantities decrease as temperature is increased. Hence, melts of these alloys can be discussed in terms of mutual solutions of two "components", metallic and semiconducting. Therefore, there are all reasons to describe phase equilibria and thermodynamic properties within the two-level model which was so effective in the case of cerium. (A similar "two-liquid model" was used by Rappoport *et al.* to explain maxima in the melting curves of some elements[48].)

Statement of the Model

The main assumptions of the model applied to the systems involved are as follows:

i) Two microscopic states (or two kinds of clusters) are characteristic of both the liquid and amorphous states, which differ in energy and in the atomic configuration.

ii) One of the atomic configurations corresponds to the short-range order of the semiconducting phase, and the other is that of the metallic phase.

iii) The energy difference between these states depends linearly on the excitation degree of the system, *i.e.*, on the volume fraction of the high-energy states.

iv) Amorphous and liquid phases can be treated as solutions of these states which are accepted to be "components" of the solution.

Similar to the case of cerium, we write the thermodynamic potential of this two-level system in the form:

$$\begin{aligned} G(T,P,C) &= C[E_1 - TS_1 + PV_1] + (1-C)[E_2 - TS_2 + PV_2] + UC(1-C) \\ &+ RT[C\ln C + (1-C)\ln(1-C)] \end{aligned} \quad (4.1)$$

where $i = 1$ stands for the purely metallic state of an alloy, $i = 2$ stands for the purely semiconducting one, C and $(1-C)$ are the concentrations of the metallic and semiconducting components. Other parameters in G_i have the same meaning as in eqn. (3.1).

The parameter C is an internal parameter which depends on temperature and pressure. The parameter C is found from the same minimum conditions (3.2) and (3.3).

$$\frac{\partial G}{\partial C} = (\Delta E^\circ - T\Delta S^\circ + P\Delta V^\circ) + (1-2C)U + RT\frac{C}{1-C} = 0 \quad (4.2)$$

where $\Delta E^0 = E_1 - E_2$; $\Delta S^0 = S_1 - S_2$; $\Delta V^0 = V_1 - V_2$. A general concept of *unordered states* is used below because we intend to analyze metastable equilibria and properties of the metastable states in a wide temperature range which includes both the region of undercooled melts and the region of actual amorphous states.

The introduced concept of *an unordered state* is introduced now for all topologically disordered phases, both metallic and non- metallic. The presence or absence of chemical order as well as corresponding corrections to the thermodynamic functions are subjects of a special consideration. For example,chemical order is characteristic of the semicoducting phases in the systems considered below both in crystalline and amorphous states. But metallic Hume-Rothery phases do not tend to chemical order even in the crystalline state.

An usual simplification in the approximation of regular solutions is that the E_i, S_i, V_i, and U values do not depend on temperature and pressure. But it is worth noting that only differential values of energies, entropies and volumes are involved in the equation (4.2) which is used for the calculation of the parameter $C(T,P)$, the phase equilibria and the limits of thermodynamic stability for metallic and semiconducting states. Only four constants, *i.e.*, ΔE°, ΔS°, ΔV° and U, are necessary to describe Gibbs potential $G(T,P,C)$, hence, all thermodynamic properties of a substance in the unordered state. This means that a less rigid restriction of constant ΔE°, ΔS° and ΔV° can be accepted. Under this assumption, we neglect, in fact, only differences in compressibility, thermal expansion and heat capacity of the metallic and semiconducting components. The large difference in the thermodynamic properties of the metallic and semiconducting states justifies this assumption because the variation of internal parameter $C(T,P)$ gives the main contribution to the variation of thermodynamic properties in the T–P region under consideration.

Table 1. Thermodynamic parameters of phase transitions and the model-constants for GaSb, InSb and InAs

Parameter	GaSb	InSb	InAS
ΔG_t (cal/mole)	5060	2000	4820
ΔH_f (cal/mole)	4970	3470	4100
ΔE° (cal/mole)	5000	3200	4400
ΔV° (cm^3/mole)	-3.06	-3.68	-3.18
$\Delta V^\circ/V^\circ$ (%)	-23	-18	-32
ΔS° (cal/K·mole)	6.78	6.78	6.78
U (cal/mole)	4000	3500	4000
$(1-C)_{\rm calc}$	0.075	0.04	0.04
$(1-C)_{\rm exp}$	0.08	0.03	0.05

Calculation of the Metastable T–P Phase Diagrams of GaSb, InSb and InAs

Three of the model constants can be estimated from the available experimental data on GaSb, InSb and InAs.

The difference in internal energies in the metallic and semiconducting states, ΔE°, can be estimated as the enthalpy of formation of a semiconducting compound from metallic elements, ΔH_f, or as the difference in the Gibbs potentials of the crystalline semiconducting and metal phases, $\Delta G_t = P_t \Delta V$, at $T = 0$ K. Here P_t is the transition pressure, and ΔV is the volume jump at the transition. The ΔH_f values were cited by Hultgren[49]. The ΔG_t values are available from the T–P phase diagrams. The P_t and ΔV values in referense 1 are given at room temperature, and we assume that their variation between zero and room temperatures is rather small. The ΔG_t and ΔH_f values are listed in Table 1.

There is a close correlation between quantities which characterize different processes, *i.e.*, ΔH_f for chemical reactions Ga+Sb $\rightarrow$ GaSb*I* or In + As $\rightarrow$ InAs*I* and ΔG_t for polymorphic semiconductor-to-metal transitions GaSb*I* $\rightarrow$ GaSb*II* or InAs*I* $\rightarrow$ InAs*II*. (Here numbers *I* and *II* denote the low-pressure semiconducting and high-pressure metallic crystal phases, respectively.) This is because origination of the space-oriented covalent bonds gives the main contribution to the energy, entropy and volume difference between semiconducting and metallic states. Therefore, structural features of individual metallic phases are less important. This correlation is further supported by the empirical Jamieson's relationship. The relationship states that the energy gap of an $A_{III}B_V$ semiconductor, E_g, and the parameters of the semiconductor-to-metal transition at room temperature, P_t and ΔV, are related by a simple equation $E_g/2 \approx P_t \Delta V$.

The Chakraverty's rule[50] will be used now to estimate ΔS°. According to this rule, the entropy increases by 5.40 cal/K·mole when an ideal covalent element melts to a purely metallic liquid. This value is only due to the transition of covalent electrons to the conductivity band. To apply this rule to the transition "ideal amorphous semiconductor to ideal amorphous metal" in binary compounds, one should also take into account the contribution from the atomic disordering. Semiconducting phases are characterized by the chemical order. Metallic phases of these compounds are normal Hume-Rothery phases which are atomically disordered even in the crystalline state. Therefore, the configurational entropy due to atomic disordering, $\Delta S_a = R(X \ln X + Y \ln Y)$, should be added to the electronic term, ΔS_e. Here X and Y are the atomic concentrations of an alloy. For the equiatomic composition $\Delta S_a = R \ln 2$. Then the

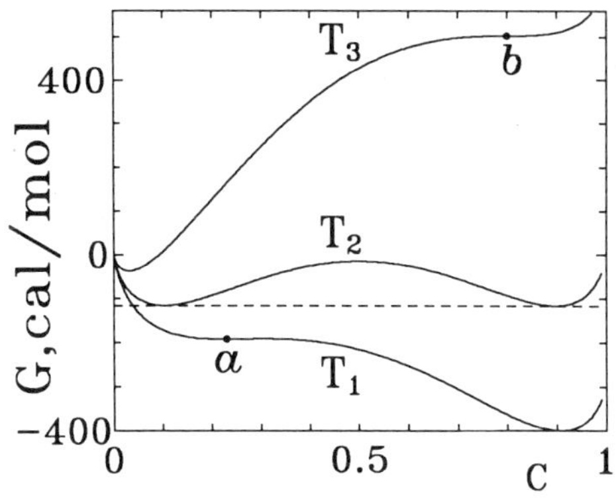

Figure 16. The Gibbs potential of GaSb as a function of parameter C at $P = 1$ atm and $T_1 = 780$ K, $T_2 = 738$ K and $T_3 = 645$ K. The points a and b where minima are degenerate correspond to the boundary of thermodynamic stability of the semiconducting and metallic amorphous phases, respectively.

total change in entropy is $\Delta S^\circ = 5.40 + 1.38 = 6.78$ cal/K·mole.

The ΔV° value is accepted to be equal to the volume jump at the polymorphic semiconductor-to-metal transition under high pressure. Difference in additional contributions due to amorphization of both phases can be neglected. This assumption is reasonable because volume effects at electronic transitions are large, and volume increments due to amorphization are of the same magnitude provided that the conductivity - type and the nearest-neighbour coordination are retained. Semiconductor-to-metal transitions result in volume effects which are much larger than those due to the formation of metallic phases from the metallic elements. For example, the volume jump due to polymorphic transitions *I* - *II* in InSb under high pressure is 18 %. But the volume effect of reaction In + Sb $\rightarrow$ InSb*II* (metallic) is only about 3.5 %. This trend is more pronounced in GaSb and InAs which have larger E_g -values.

The parameter U is the only one to be found from fitting within our model. Its value essentially affects the internal parameter C which is determined from the minimum of the Gibbs potential, G, at fixed T and P. We used therefore the experimental data on the volume fraction of semiconducting clusters, $1 - C$, in liquid alloys at their melting temperature[47] in order to choose the appropriate U -values. The model parameters thus obtained are also given in Table 1.

The two-level model calculation of the metastable $T - P$ diagram for unordered GaSb was carried out with the following values of the model constants: $\Delta E^\circ = 5000$ cal/mole, $\Delta V^\circ = -3.06\text{cm}^3$/mole, $\Delta S^\circ = 6.78$ cal/K·mole, $U = 4000$ cal/mole.

Fig. 16 represents isothermal curves $G(C)$ for three characteristic temperatures at $P = 1$ atm. For better visualization, the curves $G(C)$ are reduced to zero at $C = 0$. For $T_1 = 780$ K, the $G(C)$ isotherm is a curve with a single minimum at $C = 0.92$ which corresponds to the metallic unordered phase, MUP. The degeneracy condition (3.9) is valid for the other minimum at point a, which implies the absolute loss of thermodynamic stability of the semiconducting unordered phase, SUP. The phase equilibrium MUP - SUP takes place at $T_2 = 738$ K. This $G(C)$ isotherm is represented by a curve with two equal minima, $G_{\min}(C_1) = G_{\min}(C_2)$. The corresponding concentrations are $C_1 = 0.897$ and $C_2 = 0.103$. At some lower temperature, $T_3 = 645$ K, the $G(C)$ curve also has a single minimum which is associated with SUP while b is a degeneracy point for the MUP minimum, $C_b = 0.80$.

Fig. 17 shows the parameter C, *i.e.*, the concentration of the metallic component in both unordered phases of GaSb, as a function of temperature at atmospheric

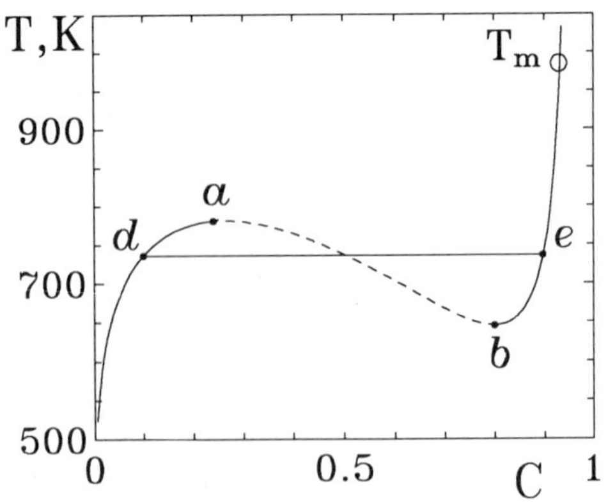

Figure 17. The temperature dependence of the parameter C for unordered GaSb system at $P = 1$ atm. The line below the point a corresponds to the semiconducting phase and that above the point b corresponds to the metallic phase. Points a and b are those from Fig. 16. Points d and e, respectively, correspond to the C values in the semiconducting and metallic phases at the equilibrium temperature $T_2 = 738$ K.

pressure. As undercooling is increased, the parameter C of the initial metallic melt decreases from $C = 0.925$ at $T = T_m$ to a minimum at point b (C = 0.80, T = 645 K) where the high-temperature MUP becomes unstable. The left $C(T)$ branch corresponds to SUP. The equilibrium temperature, $T = 738$ K, cuts the curves $C_{\text{SUP}}(T)$ and $C_{\text{MUP}}(T)$ at points d (C = 0.103) and e (C = 0.897). A part of the $C(T)$ curve shown with a dotted line is related to the eqn. (4.2) solutions with no physical meaning.

Fig. 18 shows the calculated T–P diagram of metastable equilibria in the unordered GaSb. The line of the MUP - SUP equilibrium is plotted with a solid line. The long-dashed lines represent the lines of the phase instability of two metastable states involved. The top curve is the line where SUP becomes unstable upon heating, and the bottom curve represents the line where MUP becomes unstable upon cooling. The symbols T_2, T_1 and T_3 indicate the temperatures where SUP and MUP are in equilibrium or lose their thermodynamic stability at atmospheric pressure, respectively. The critical point ($T_c = 1007$ K and $P_c = -2.2$ GPa) is located at a negative pressure therefore it is virtual. The parameter C is gradually increasing in SUP along the equilibrium line as the temperature is increasing. An opposite dependence is observed in MUP, and both phases come to $C_1 = C_2 = 0.5$ in the critical point.

The calculated metastable diagram of the unordered GaSb is combined in Fig. 19 with the experimental T–P diagram which involves the crystalline phases *I*, *II* and the liquid. The dashed line shows a smooth extension of the GaSb*II* melting curve which intersects the metastable MUP - SUP equilibrium line at point A. So, the extrapolated melting curve of GaSb*II* lays at $P < P_A$ and $T < T_A$ in the region of higher thermodynamic stability of SUP. Point B is a low-pressure limit where MUP becomes unstable. MUP is no more a metastable phase to the left from point B, and the part of the extrapolated GaSb*II* melting curve below point B has no physical meaning. This is why metastable metallic GaSb*II* quenched under pressure transforms into a semiconducting amorphous state upon heating at $P = 1$ atm rather than into a metallic one. All other features of spontaneous amorphization of the metallic high-pressure phase GaSb*II*, *i.e.*, the large increase in volume,the low conductivity, the large exothermal heat effect, have the same origin. The heat release is due to arising covalent bonds, the large positive volume effect is due to decreased coordination number. According to the calculated T–P diagram of the unordered GaSb, a SUP

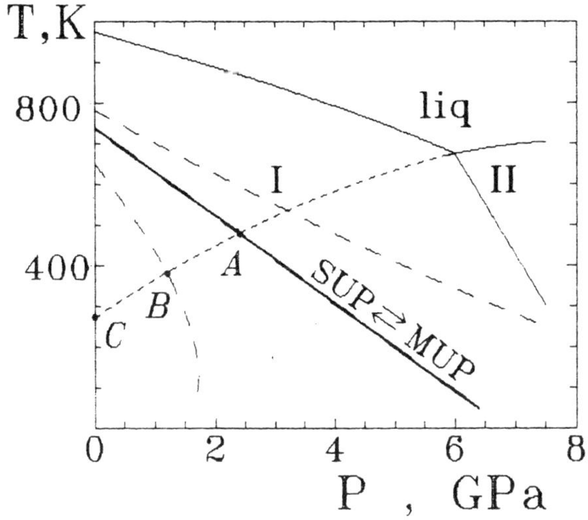

Figure 18. The calculated metastable T–P phase diagram of unordered GaSb system. Temperatures T_1, T_2 and T_3 are those from Fig. 2. The line of the semiconductor/metal equilibrium is solid, the long-dashed lines show the boundaries of thermodynamic stability for the metallic (bottom curve) and semiconducting (top curve) unordered states.

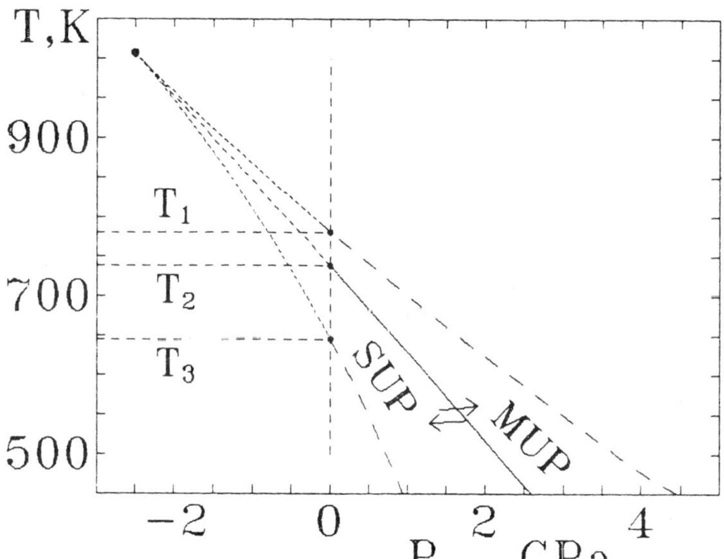

Figure 19. The stable and metastable T–P phase diagrams of GaSb. Thin solid lines are for the stable equilibria between phases I, II and the liquid. The metastable MUP - SUP equilibria occur along the thick solid line. The short-dashed line represents extrapolation of the phase II melting curve to the stability region of phase I. Boundaries of thermodynamic stability of SUP and MUP are shown in long-dashed curves.

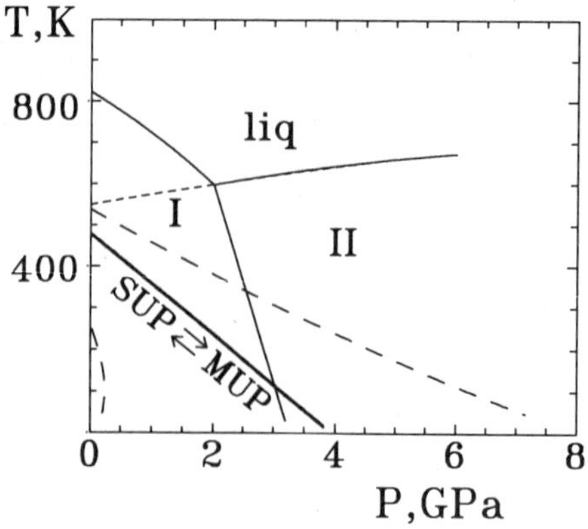

Figure 20. The stable and metastable T–P phase diagrams of InSb. Thin solid lines are for the stable equilibria between phases I, II and the liquid. The metastable MUP — SUP equilibria occur along the thick solid line. The short-dashed line represents extrapolation of the phase II melting curve to the stability region of phase I. The boundaries of thermodynamic stability of SUP and MUP are shown in long-dashed curves.

→ MUP transformation should occur as pressure is increased. The transformation should be reversible unless MUP crystallizes. The transition pressure at 293 K is anticipated in the range 4.0 to 5.0 GPa. Actually, a SUP → MUP transition has been observed in bulk amorphous GaSb in a rather narrow interval, P = 3.5 to 4 GPa[46]. The latter data are in excellent agreement with the calculated diagram.

A similar composite T–P diagram of the stable and metastable phase equilibria in InSb is represented in Fig. 20. The extension of the InSb*II* melting curve to atmospheric pressure gives a rather high temperature of about 550 K. Therefore, the usual thermobaric treatment, *i.e.*, quenching → decompression → HPP heating at P = 1atm, does not result in an amorphous state. HPP always undergoes normal polymorphous transformations much below 550 K. The composite phase diagram is consistent with the available experimental data on InSb.

Fig. 21 shows the composite phase diagram for InAs. The triple point InAs*I* - InAs*II* - *liq* is located at rather high temperature and pressure (923 K, 6.5 GPa). The InAs*II* melting curve has been measured therefore in a too narrow interval to enable its more or less reliable extrapolation to atmospheric pressure. We have no experimental data on InAs*II* amorphization when heated at P = 1 atm. However, solid state amorphization of InAs*II* is anticipated from the interrelation of the stable and metastable T–P diagrams of InAs, as in the case of GaSb.

So, the calculated metastable T–P diagrams of GaSb and InSb give a correct qualitative description of their thermal and baric behavior in the amorphous state as well as the main features of SSA in GaSb. But can the proposed two-level model describe quantitatively phase diagrams and thermodynamic properties of the unordered systems?

It is worth noting that the elaborated model is based mainly on the assumption of a large difference between two microscopic states (levels). The model can therefore be applied for describing the metastable phase diagrams of numerous molecular and ionic compounds which undergo solid-state first-order phase transitions concomitant with the large changes in the coordination number and in the specific volume, *e.g.*, SiO_2, H_2O, silicate minerals and numerous compounds with variable valence. We used this opportunity to examine the applicability of the model to a most studied amorphous substance, *i.e.*, water[51].

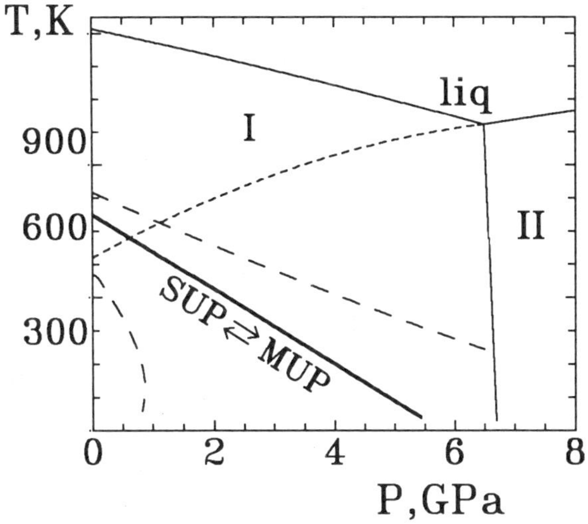

Figure 21. The stable and metastable T–P phase diagrams of InAs. Thin solid lines are for the stable equilibria between phases I, II and the liquid. The metastable MUP - SUP equilibria occur along the thick solid line. The short-dashed line represents extrapolation of the phase II melting curve to the stability region of phase I. The boundaries of thermodynamic stability of SUP and MUP are shown in long-dashed curves.

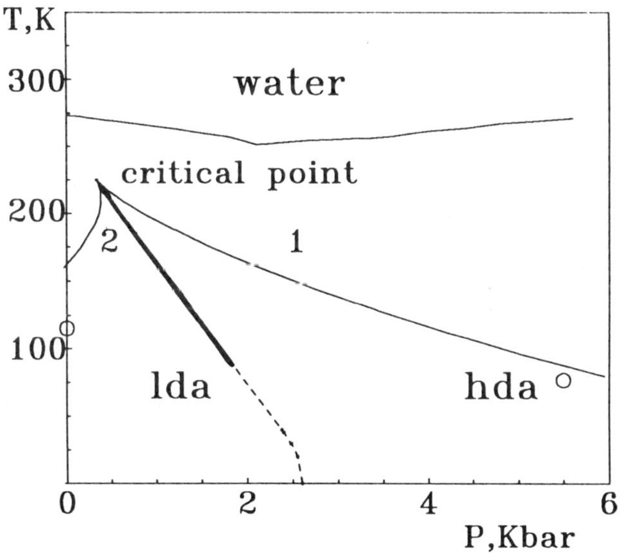

Figure 22. Metastable P-T phase diagram of water at low temperatures. 1—line at which the *lda* phase loses its stability; 2—line at which the *hda* phase loses its stability; open circles-experimental data. The proposed *lda*-*hda* transition line at low temperatures is shown by the dashed line.

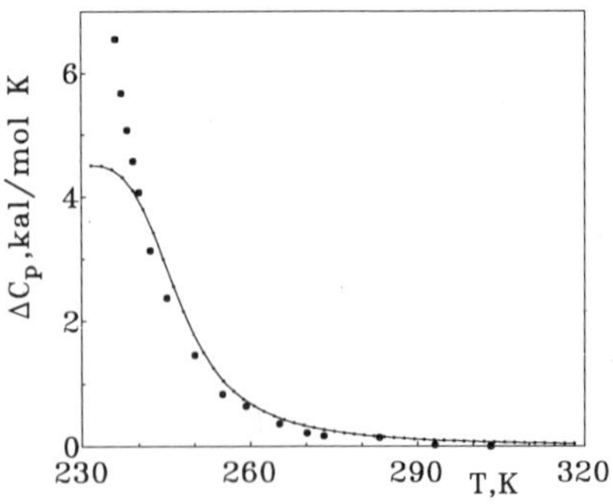

Figure 23. Temperature dependence of the anomalous part of the specific heat at atmospheric pressure. Points – experimental data; solid curve – theoretical.

Fig. 22 represents the metastable T–P diagram of unordered water and the experimental melting curve. The metastable phase diagram was calculated with the model parameters $\Delta E^{\circ} = 250$ cal/mole, $\Delta V^{\circ} = -3.8$ cm^3/mole, $\Delta S^{\circ} = 1.0$ cal/K·mole, $U = 900$ cal/mole. The critical point of water is at a positive pressure ($P_c = 0.033$ GPa, $T_c = 225$ K), and the low-temperature anomalies of the thermodynamic properties are accurately explained as specific supercritical anomalies.

The calculated and experimental data on the anomalous contribution to the heat capacity of water at $P = 1$ atm are compared in Fig 23. There is a good agreement between the data. It can be concluded therefore that the model is correct to the second derivatives of the thermodynamic potential provided an adequate choice of the model constants is made. It is interesting that the metastable equilibria in water have recently been calculated using quite other and more complex approaches, *i.e.*, methods based on molecular dynamics[52]. The resultant T–P diagram exhibits of the same topology.

Conclusions

Stability, metastability and the thermodynamic instability of crystal and amorphous phases in the T–P–C space are fundamental concepts of solid state physics. Some phenomena which are closely related to these concepts are considered here on the examples of binary systems of non-transition elements and cerium, the element with the isomorphous electronic transition. These substances are most convenient subjects for the experimental study as well as for model description of the phenomena in question, including feasibility of the first-order phase transitions and the liquid-gas critical point both in crystalline and amorphous substances.

It is anticipated from the theoretical point of view that the metastable phases possess anomalous vibrational spectra, compressibility, heat capacity, thermal expansion, sound absorption as well as elastic, optic, superconducting and other properties near the boundaries of total instability.

For a detailed experimental study of these interesting phenomena, the experimental techniques should be further developed. Improved thermobaric quenching techniques will increase the number of metastable high-pressure phases available for measurements at ambient pressure. An alternative way is further elaboration of *in situ* techniques for investigation of the elastic and inelastic neutron scattering, X-ray diffraction, ultrasound velocities and absorption, superconductivity, optic and magnetic properties *etc.* immediately under high pressure. High-pressure devices are therefore necessary, which are operable both at elevated and low temperatures.

Maintenance of samples at elevated temperatures is necessary to achieve the phase equilibrium, and low temperature cooling is necessary for measurements of properties as close to the boundary of total instability of the phase involved as possible.

Experimental data to be obtained will lead to statement of new empirical relationships concerning formation and properties of the high-pressure phases. The information acquired will become a firm basis for a general theory of the stability, metastability and instability phenomena in crystal and amorphous phases in the $T-P-C$ space.

REFERENCES

1. E.Yu. Tonkov, "High Pressure Phase Transformations, A Handbook", vol. 1-2, Gordon and Breach Science Publishers (1992).
2. D.J. Dunstan and I.L. Spain, *J. Phys. E: Sci. Instrum.* 22:913 (1989).
3. E.G. Ponyatovskii and V.F. Degtyareva, *High Pressure Res.* 1:163 (1989).
4. V.F. Degtyareva, I.T. Belash, G.V. Chipenko, E.G. Ponyatovskii and V.I. Rashchupkin, *Sov. Phys. Solid. State* 25(10):1712-1715 (1983).
5. I.T. Belash, V.F. Degtyareva, E.G. Ponyatovskii and V.I. Rashchupkin, *Sov. Phys. Solid. State* 29(6):1028-1031 (1987).
6. V.F. Degtyareva, G.V. Chipenko, I.T. Belash, O.I. Barkalov and E.G. Ponyatovskii, *Phys. Status Solidi (a)* 89:K127 (1985).
7. V.F. Degtyareva, G.V. Chipenko, E.G. Ponyatovskii and V.I. Rashchupkin, *Sov. Phys. Solid. State* 26(4):733-734 (1984).
8. O.I. Barkalov, I.T. Belash, V.F. Degtyareva and E.G. Ponyatovskii, *Sov. Phys. Solid. State* 29(7):1138-1140 (1987).
9. G.V. Chipenko, V.F. Degtyareva and E.G. Ponyatovskii, *Sov. Phys. Solid. State* 26(4):735-736 (1984).
10. G.V. Chipenko and V.F. Degtyareva, *Sov. Phys. Solid. State* 25(4):706-708 (1983).
11. V.F. Degtyareva, I.T. Belash, E.G. Ponyatovskii and V.I. Rashchupkin,*Sov. Phys. Solid. State* 32(5):834 (1990).
12. V.E. Antonov, O.I. Barkalov and E.G. Ponyatovsky, Proc. Intern. Conf. NCM-6, Prague, 29 Aug. - 4 Sept. 1994, *J. Non-Crystalline Solids* to be published.
13. V.F. Degtyareva, E.G. Ponyatovskii and L.N. Rastorguev, Investigation of the Structure of Bi–Pb alloys subjected to high pressure, *Sov. Phys. Solid. State* 17:274-277 (1975).
14. I.T. Belash and E.G. Ponyatovskii, *High Temp. High Press.* 9:645 (1977).
15. H. Olijnyk, S.K. Sikka and W.B. Holzapfel, *Phys. Letters A* 103:137 (1984).
16. J.C. Hu and I.L. Spain, *Solid State Commun.* 51:263 (1984).
17. S.J. Duclos, Y.K. Vohra and A.L. Ruoff, *Phys. Rev. B* 41:12021 (1990).
18. S.T. Weir, Y.K. Vohra, G.A. Vanderborgh and A.L.Ruoff, *Phys. Rev. B* 39:1280 (1989).
19. G.A. Vanderborgh, Y.K. Vohra and A.L. Ruoff, *Phys. Rev. B* 40:12450 (1989).
20. S.T. Weir, Y.K. Vohra and A.L. Ruoff, *Phys. Rev. B* 36:4543 (1987).
21. R.J. Nelmes, M.I. McMahon, P.D. Hatton,J. Crain and R.O. Piltz, *Phys. Rev. B* 47:35 (1993).
22. M.I. McMahon and R.J. Nelmes, *Phys. Rev. B* 47:8337 (1993).
23. O.I. Barkalov, I.T. Belash and A.F. Gurov, *Phys. Status Solidi (a)* 115:K19 (1989).
24. O.I. Barkalov, I.T. Belash, A.I. Bol'shakov and E.G. Ponyatovskii, *Sov. Phys. Solid State* 30:1570 (1988).
25. I.T. Belash and E.G. Ponyatovskii, *High Temp. High Press.* 9:651 (1977).
26. O.I. Barkalov, I.T. Belash, V.F. Gantmakher and E.G. Ponyatovskii, *JETP Letters* 48:609 (1988).

27. O.I. Barkalov, I.T. Belash, V.F. Gantmakher and V.M. Teplinskii, *Solid State Commun.* 75:345 (1990).
28. O.Mishima, L.D. Calvert and E. Whalley, *Nature* 310:393 (1984).
29. K. Samwer,*Physics Reports* 161:141 (1988).
30. O.I. Barkalov, I.T. Belash, V.F. Gantmakher, E.G. Ponyatovskii and V.M. Teplinskii, *Pis'ma ZhETF* 48:561 (1988) (*in Russian*).
31. H.J. Fecht, P.J. Desré and W.L. Johnson, *Phil. Mag. B* 59:577 (1989).
32. W. Kauzmann, *Chem. Rev.* 43:219 (1948).
33. P.W. Bridgman, "The Physics of High Pressures", John Wiley & Sons, London - N.Y. 83 (1949).
34. L.D. Landau and E.M. Lifshitz, "Statistical Physics", Pergamon Press, Oxford, London 264 (1964).
35. A.W. Lawson and T.Y. Tang, *Phys. Rev.* 76:301 (1949).
36. E.G. Ponyatovskii, *Doklady Akad. Nauk SSSR* 120:1021 (1958) (*in Russian*).
37. D.I. Khomskii, *Sovjet Physics Uspekhi* 22(11):879-903 (1979).
38. I.L. Aptekar and Ye.G. Ponyatovskiy, *Phys. Met. Metallography* 25(5):10-20 (1968).
39. A R. Kutsar, *Phys. Met. Metallography* 33(5):197-199 (1972).
40. S. Strässler and C. Kittel, *Phys. Rev.* 139:A758 (1965).
41. I.L. Aptekar and E.G. Ponyatovskii, *Phys.Met.Metallography* 25(6):93-100 (1968).
42. V.I. Rashchupkin, I.L. Aptekar and E.Yu. Tonkov, *Sov.Phys.Solid.State* 20(8):1453 (1978).
43. I.L. Aptekar and E.Yu. Tonkov, *Sov.Phys.Solid.State* 21(1):110-111 (1979).
44. O.I. Barkalov, E.G. Ponyatovskii and V.E. Antonov, *J. Non-Crystalline Solids* 156-158:544 (1993).
45. O. Shimomura, K. Asaumi, N. Sakai and S. Minomura, *Phil. Mag.* 34:839 (1976).
46. V.V. Brazhkin, A.G. Lyapin, S.V. Popova and A.V. Sapelkin, *JETP Letters* 59(2):130-134 (1994).
47. A.R. Regel and V.M. Glasov, *Sov. Phys. Semicond.* 17:1105 (1983).
48. E. Rappoport, *J. Chem. Phys.* 46:2892 (1967).
49. R. Hultgren, R. Orr, P. Anderson and K. Kelley "Selected Values of Thermodynamic Properties of Metals and Alloys", John Wiley & Sons, N.Y. (1963).
50. B.K. Chakraverty, *J. Phys. Chem. Solids* 130-141:454 (1969).
51. E.G. Ponyatovskii, V.V. Sinitsyn and T.A. Pozdnyakova, *JETP Letters* 60(5):360-364 (1994).
52. H.E. Stanley, C.A.Angell, U.Essmann, M.Hemmati, P.H.Poole, F.Sciortino,*Physica A* 205:122 (1994).

DISTRIBUTIONS OF DISLOCATION DISSOCIATION WIDTHS

Václav Paidar

Institute of Physics
Academy of Sciences
Na Slovance 2
18040 Praha 8
Czech Republic

INTRODUCTION

The extent to which the moving dislocations interact with the lattice depends on the character of their core. Planar dislocation cores can easily move on a glide plane, providing that this glide plane is identical with the plane of dislocation dissociation, while the cores with a spatial character are sessile and are subjected to large friction forces.[1,2] Dislocation lines can not be arbitrarily located in the crystal but have to lie only at certain positions with respect to the unperturbed lattice corresponding to the minima of interaction energy between the dislocation and the lattice.[3] In a simplified form, this interaction can be described by the well known Peierls-Nabarro model which gives energy minima periodically distributed on the glide plane. However, the superlattice dislocations in ordered alloys are dissociated into partial dislocations connected by ribbons of antiphase boundaries (APBs). When both partials are not freely mobile then each of them has to be situated at an energy minimum and hence their separation has a discrete character. Moreover, the antiphase boundary width is not a constant but can attain different values which are quantized by the lattice periodicity on the glide plane.

It has been shown[3] that the range of possible dissociation widths is growing with the increasing depth of the Peierls-Nabarro potential and is asymmetrically situated with respect to an equilibrium partial dislocation separation. In spite of the fact that the motion of dislocations strongly interacting with the lattice is not a linear process, we will treat it using standard methods of statistical physics. The parameters of experimentally determined distributions of dislocation dissociation widths on the prismatic and basal planes in Ti_3Al, which has been recently observed during in situ straining in electron microscope,[4,5] will be discussed in detail. It will be argued that the observed distributions are not given by thermal equilibrium of partial dislocations, but these distributions are associated with the applied stress under which the motion of dislocations takes place.

DISTRIBUTION OF THE WIDTHS OF DISSOCIATED DISLOCATIONS

Let us assume that the distribution function $F(y_i)$ of the dissociation width y_i is given by the local minima of the energy of dissociated dislocations. In order not to interrupt the presentation of physical ideas, some quantities will be defined later. The distribution function can be chosen in a simple form well known from statistical physics

$$F(y_i) = \exp(\alpha - \beta U(y_i)b). \tag{1}$$

Since α represents a normalization constant, β is the only free parameter that remains to be determined. It can be considered as the ratio $\beta = 1/kB$, where k is the Boltzmann constant, however, ***B*** can **not** be identified with temperature as it will be shown below. As the energy U is defined per unit length of dislocation line (see eq. (2) below), it is multiplied by the atomic separation b equal to the magnitude of the Burgers vector of partial dislocations. Distribution functions for three values of parameter B are compared in Figure 1.

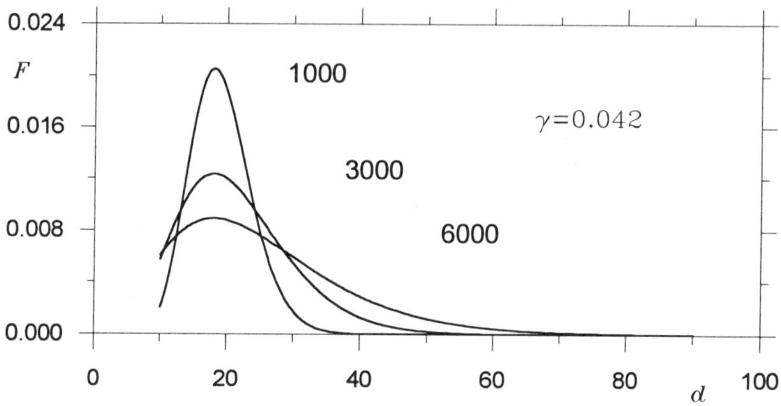

Figure 1. Distribution functions F of the APB width d expressed in nm for three values of parameter B (1 000, 3 000 and 6 000) for the APB energy on the prismatic plane of 0.042 Jm^{-2}.

The structure of dislocation cores is basically affected by the existence of stable stacking faults including the antiphase boundaries in ordered alloys.[6,7] The crystals of Ti_3Al possess a hexagonal structure of the $D0_{19}$ type. The glide of superdislocations dissociated according to the reaction

$$1/3\ [1\bar{2}10] \rightarrow 1/6\ [1\bar{2}10] + \text{APB} + 1/6\ [1\bar{2}10]$$

was observed on both prismatic and basal planes. As it was shown[8,9] there are two types of APBs on the $\{10\bar{1}0\}$ prismatic plane with the same displacement vector but differing in the position with respect to the ordered lattice. Their energies have been determined[5] for Ti_3Al to be $\gamma = 0.042$ and 0.085 Jm^{-2}. While the narrow dislocations (with the higher APB energy) carry plastic deformation, the widely dissociated dislocations are locked in the screw orientations. The narrow dislocations have a round shape, multiply easily and move over large distances. On the other hand, the widely dissociated dislocations experience larger friction forces and hence a higher stress is necessary for their motion which has a jerky nature. They move by very fast jumps over short distances. On the $\{0001\}$ basal plane, there is only one type of APB with the energy of 0.063 Jm^{-2}. The dissociated dislocations experience large friction forces on the basal plane and can easily cross-slip to the prismatic plane. Their motion is jerky.

The potential energy of dissociated dislocation is a two-dimensional function periodic only in x but **not** in y

$$U(x,y) = M(2py / \pi d_0 - \ln y - (V/2)\cos 2x \cos 2y) \quad , \tag{2}$$

where x and y are dimensionless coordinates, $x = (x_1 + x_2)\pi / 2p$ and $y = (x_2 - x_1)\pi / 2p$, that correspond to the centre of superlattice dislocation and to the separation of the partials, respectively. For $x = 0$, the difference of superdislocation widths Δy corresponding to neighbouring energy minima is approximately π. However, there are additional non-equivalent minima[10] for $x = \pi/2$ and hence the total number of independent dislocation states is doubled. Therefore, using the relationship for possible superlattice dislocation widths[3]

$$1/(g+V) \quad \langle \; y \; \langle \quad 1/(g-V) \; , \tag{3}$$

the number of dislocation states can be estimated as $4v / \pi g(1-v^2)$ where $g = 2p / \pi d_0$ and $v = V / g$. Notice that the maximum value of the reduced depth of Peierls-Nabarro potential v is 1. Since y is in the $2p/\pi$ units, $\boldsymbol{p}$ is the spacing of the neighbouring states of dissociated dislocation. It is equal to $c/2$ for the prismatic glide plane and to $a\sqrt{3}\ /2$ for the basal plane. The parameters of the $D0_{19}$ superlattice of Ti_3Al are $a = 0.578$ and $c = 0.463$ nm.

The equilibrium separation of the partial dislocations can be denoted as $d_0 = M / \gamma$ where $M = Kb^2 / 2\pi$ is a material parameter with K being an appropriate elastic constant and $b = \ a/2$ is the magnitude of the Burgers vector of the partial dislocations. It should be emphasized that the interaction between the partials determined in the anisotropic elasticity depends on the orientation of dislocation line, i.e. the parameter K is a function of the angle made by the Burgers vector and the dislocation direction. In this paper, only screw dislocations will be treated and the value of shear modulus (56.6 GPa) will be taken for K.

CHARACTERISATION OF THE DISTRIBUTION FUNCTION

As indicated above, the distribution function is normalized, $\sum_i F(y_i) = 1$. It is convenient to characterize its shape by the first three moments which are related firstly to the mean value of the dislocation separation Y

$$Y = \sum_i y_i F(y_i) , \tag{4}$$

secondly to the width WF

$$WF = \left(\sum_i (y_i - Y)^2 F(y_i) \right)^{1/2} \tag{5}$$

and thirdly to the asymmetry AF of the distribution function

$$AF = \left(\sum_i (y_i - Y)^3 F(y_i) \right)^{1/3} . \tag{6}$$

As shown in Figure 1, the width of the distribution (1) is narrow for small B and monotonously increases with growing B. In order to determine two parameters of our model, namely, the depth of the Peierls-Nabarro potential v and the parameter B, the width of the distribution WF was calculated according to equation (5) as a function of these two parameters. The results are presented by a contour diagram in Figure 2.

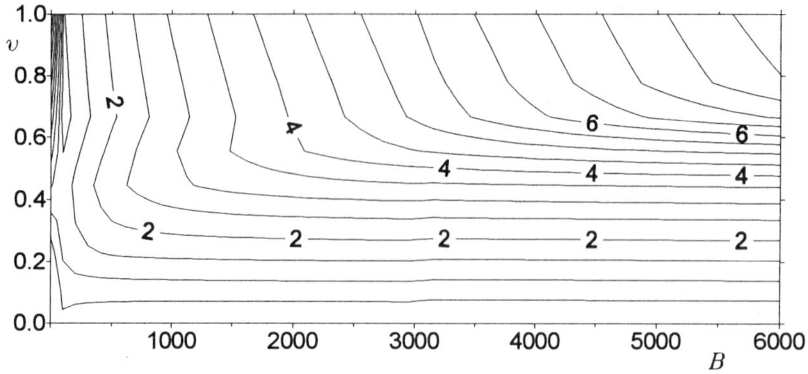

Figure 2. The dependence of the distribution width WF (expressed in nm) on the depth of the Peierls-Nabarro potential v and on the parameter B for the APB energy on the basal plane of 0.063 Jm^{-2}.

For shallow Peierls-Nabarro valleys, i.e. small v, the number of possible dislocation states is small and, therefore, the width of the distribution is always narrow and practically independent of the parameter B. On the other hand, if the number of dislocation states is large for v close to 1, the distribution width monotonously increases with growing B as demonstrated in Figure 2. Since the width WF of the experimental distribution for the basal plane in Ti_3Al [5] is about 8 nm, the Peierls-Nabarro potential has to be relatively deep. A similar result was obtained for the prismatic plane.

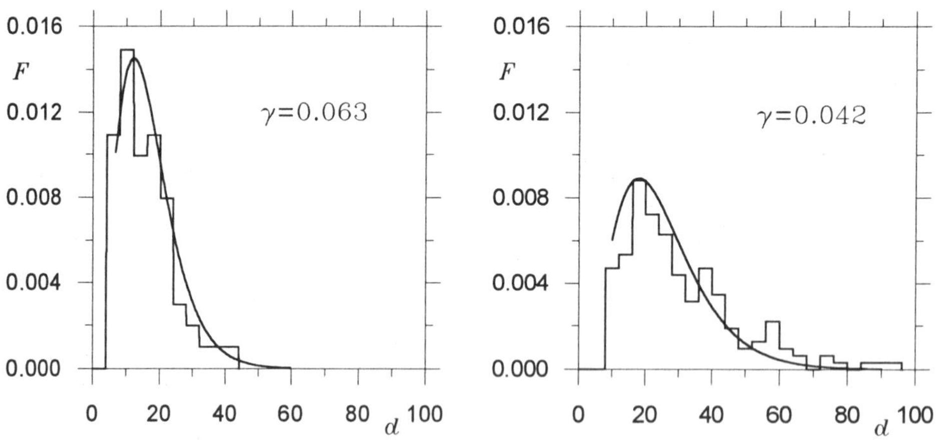

Figure 3. Comparison between the experimental distribution functions (histograms) and calculated distributions for the APB energies $\gamma = 0.063$ and 0.042 Jm^{-2} on the basal and prismatic planes, respectively. The separations of the partial dislocations d are expressed in nm.

The distribution functions were calculated for $\nu = 0.8$ and different values of B. They are asymmetrical with respect to the equilibrium superdislocation width due to a steeper slope of the repulsive dislocation interaction force at short distances. As it is illustrated in Figure 3, a good agreement with the experimental data was reached for both basal and prismatic glide plane for the same value of $B = 6\,000$ in spite of apparent differences between the experimental distribution functions.

DISCUSSION

It has been demonstrated that the effect of the spatial character of dislocation core can be treated in a simplified way in the Peierls-Nabarro model in which the interaction of dislocations with the lattice is described by a periodic potential function. Since each partial dislocation of a superdislocation interacts with the lattice, it is subjected to frictional forces, and the separation of the partials need not be necessarily equal to the equilibrium distance dictated by the balance between the repulsive dislocation interaction forces and the attractive forces associated with the surface tension of the APB ribbon. The separation of the partials can attain various magnitudes within a certain range situated at the equilibrium separation. Since the separation of the partials is not a constant, the system of dislocations can be considered as a statistical ensemble where the occurrence of different widths of the dissociated superdislocation is described by a distribution function. However, it has to be emphasized that the dislocation system can not be identified with a system of particles in thermal equilibrium due to a specific nature of the dislocation motion. It is very likely that the moving dislocation possesses a core of higher energy different from the sessile locked dislocations. Under a relatively high applied stress and owing to the strong friction forces, the dislocations are far from a state of minimum energy and their motion takes place by fast jumps contrary to the assumptions of equilibrium thermodynamic processes.

The separation of the partial dislocations gliding on the prismatic and basal plane in Ti_3Al was measured on a large number of screw dislocation segments where the dislocations were apparently locked.[5] The experimental distribution functions are asymmetrical with a peak of maximum occurrence at 18 and 12 nm, respectively. Owing to the asymmetry of the distribution function, the mean separation of the partials determined according to equation (4) is lager than the separation corresponding to the peak of the distribution function and is equal to 30 and 15.5 nm, respectively. The width of the experimental distribution function given by equation (5) is 17 and 8 nm and its asymmetry given by equation (6) is 18 and 8 nm, respectively. The histograms of the experimental occupancy of the dislocation states for the basal and prismatic planes where the APB energy attains different values are shown in Figure 3.

The dislocations move due to the applied stress when the local stress reaches a critical value. Notice, however, that the acting stress has not been explicitly taken into account. It is the assumption of the model that the distribution function has formally the shape of Gibbs canonical distribution but the mechanism of dislocation jumps has not been investigated.

It has been shown that the dislocation distribution function is almost independent of the magnitude of the parameter B for small depths of the Peierls-Nabarro potential. In this case the range of possible dissociation distances is narrow and the wide experimentally observed distributions can not be obtained for any value of B. Even for a relatively deep Peierls-Nabarro potential ($\nu = 0.8$), as large value of B as $6\,000$ has to be chosen to well approximate the experimental distributions. For a system of particles in thermal equilibrium, B would correspond to the temperature of the system. The plastic deformation of the Ti_3Al samples was performed at room temperature and, therefore, it is obvious that the parameter B can not be identified with the temperature which characterizes the state of phonons in the crystal but not the system of dislocations moving under the critical stress. On the other hand, it can be concluded that the chosen shape of the Boltzmann type of the distribution functions fits very well the experimental data.

CONCLUSIONS

Metastable states of dislocations dissociated into two partials, which are sessile on the plane of dissociation as observed during in situ straining of Ti_3Al samples in electron microscope, can be described by the model where the frictional force is identified with the Peierls-Nabarro potential. The widths of dissociated dislocations are asymmetrically distributed with respect to the equilibrium partial dislocation separation and can attain relatively large values. In spite of the fact that the non-equilibrium dissociations of dislocations arise in the plastically deformed crystal under relatively high flow stresses at low temperatures, the distribution of dislocation states obeys the statistics of Boltzmann type. The width of APB ribbons connecting two partial dislocations on the plane with high friction stress has quantized character.

Acknowledgements

This research was supported by the Grant Agency of the Czech Republic under Contract No. 106/93/0513.

REFERENCES

1. V. Vitek, Effects of dislocation core structure on the plastic properties of metallic materials, *in*: "Dislocations and Properties of Real Materials", M. H. Loretto ed., The Institute of Metals, London (1985), p. 30.
2. V. Vitek, Structure of dislocation cores in metallic materials and its impact on their plastic behaviour, *Progr. Mater. Sci.* 36:1 (1992).
3. V. Paidar, and D. Caillard, Dissociation of sessile dislocations, *Phil. Mag. Lett.* 69:305 (1994).
4. M. Legros, A. Couret, and D. Caillard, In situ observations of prismatic glide in Ti_3Al, *in*: "High-Temperature Ordered Intermetallic Alloys V", I. Baker, R. Darolia, J.D. Whittenberger and M.H. Yoo eds., *Mater. Res. Soc. Symp. Proc.* 288:323 (1993).
5. M. Legros, "Etude par déformation in situ des mécanismes de déformation de Ti_3Al et Ti-24Al-9Nb", Thèse, Université Paul Sabatier, Toulouse (1994).
6. V. Paidar, Core structure of dislocations in f.c.c. base metals and alloys, *in*: "The Structure and Properties of Crystal Defects", V. Paidar and L. Lejček eds., Elsevier, Amsterdam (1984), p. 19.
7. P. Veyssière, Properties of surface defects in intermetallics, *in*: "Ordered Intermetallics - Physical Metallurgy and Mechanical Behaviour", C.T. Liu, R.W. Cahn, and G. Sauthoff eds., NATO Advanced Study Institute Series 213, Kluwer, Dordrecht (1992), p. 165.
8. Y. Umakoshi, and M. Yamaguchi, The stability and energies of planar faults in the $D0_{19}$ ordered structure, *Phys. Stat. Sol. (a)* 68:457 (1981).
9. J. Cserti, M. Khantha, V. Vitek, and D.P. Pope, An atomistic study of the dislocation core structures and mechanical behaviour of a model $D0_{19}$ alloy, *Mat. Sci. Eng. A* 152:95 (1992).
10. V. Paidar, Behaviour of dissociated dislocations with sessile cores, *Mat. Sci. Eng. A* 192/193:142 (1995).

AN ATOMISTIC STUDY OF THE IMAGE FORCE ON A DISLOCATION IN A BIMATERIAL

P. Beauchamp, J. Lépinoux, I. Lardon and H. Kentabli

Laboratoire de Métallurgie Physique (URA 131 CNRS)
Université de Poitiers
40, Avenue du Recteur Pineau
86022 POITIERS Cedex, France

INTRODUCTION

In a bicrystal compound of two crystals having different elastic moduli, the image force on a dislocation results from the variations in the total strain energy stored in the solid, according to the dislocation position: the energy is larger when the dislocation is in the medium of larger elastic constants so that the force is directed from the "hard" medium to the "soft" medium. At large scale, the image force is accurately described by linear elasticity (Head 1953 a and b, Hirth and Lothe 1982) but this approach leads to a 1/r divergence for the force on the dislocation. To overcome this difficulty, an arbitrary cut-off radius r_0 is generally introduced but its actual value is highly uncertain as is the critical stress necessary to make the dislocation cross the interface, proportional to $1/r_0$. At the atomic level, where matter is no more regarded as a continuum, the divergence disappears. The atomistic simulation will be able to give relevant informations in this region of few atomic distances from the interface where the continuum approach fails. Attention will be focused on the variations of stress necessary to move the dislocation in the interface neighbourhood and on changes in the dislocation core induced by the inhomogeneous elastic field present in this region.

CALCULATION PROCEDURE

Since the only effect to be analysed here is that of a difference in elastic moduli, the two joined solids will have identical crystal structure, orientation and lattice parameter. This is achieved by using the same basis for interatomic forces in both media, the potential in medium {2} being that of medium {1} multiplied by the coefficient R equal to the ratio of shear moduli μ_2/μ_1. The interaction of two atoms not belonging to the same medium is the average of the two above potentials. In the present study we have used the pair potential of Johnson (1964) for α-iron and the N-body potential of the Finnis-Sinclair Type (1984) constructed by Tichy (Cserti et al. 1994) for the same material.

Detailed calculations have been made for the a<001> screw dislocation near the <100> interface separating two bcc crystals having different elastic moduli and otherwise identical. In this geometry, analytical expressions are available from anisotropic elasticity for the bicrystal displacement field which are used for boundary conditions and which allow for exact energy calculation (Lardon 1994). Investigations of the a/2<111> screw dislocation have also been started.

RESULTS

Dislocation Configurations in the Homogeneous Medium

The dislocation configuration obtained with the N-body potential is shown on figure 1 using the usual relative displacement map. The core is extended on a {110} plane, i.e. (110) or (1-10). With the Johnson potential, the core is similar but the extension along {110} is smaller. Under an applied shear stress σ_{yz}, the dislocation starts gliding along the {110} plane in which the core is extended, at a stress of 0.0048 μ for the N-body potential and 0.012 μ for the pair potential. Note that the interatomic law leading to the wider core shows the lower Peierls stress as expected.

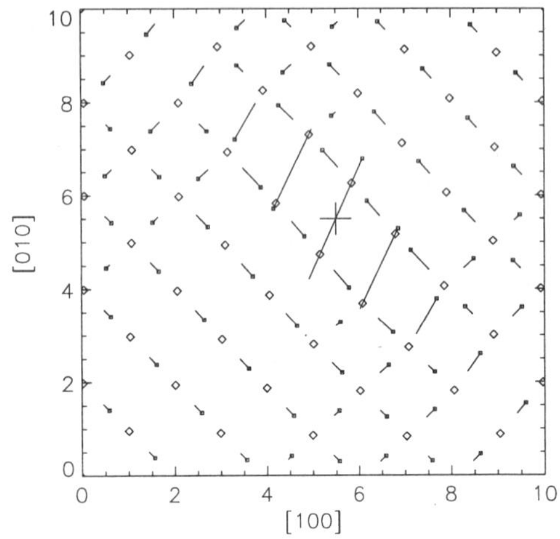

Figure 1. Core configuration of a screw a<001> dislocation in the homogeneous medium (N-body potential).

Determination of the Force on the Dislocation

Besides the computation of dislocation configurations, an important purpose of the simulation is to examine the response of the dislocation to applied stress, in the vicinity of the interface. If W(a) is the bicrystal strain energy when the dislocation is placed at a distance (a) from the interface, the dislocation can be regarded as undergoing a force F=-dW(a)/da. The image force is directed along the x-axis. At the atomic level considered here, the dislocation feels the resultant of the image force and of the Peierls force. In the elastic approximation, the image force is given by

$$F = \mu K b^2 / 4\pi a \tag{1}$$

where $K = (\mu_2-\mu_1)/(\mu_2+\mu_1) = (R-1)/(R+1)$ and μ without subscript refers to the medium in which the dislocation lies. Thus, two types of bicrystals will be distinguished according to the relative importance of the image force and of the lattice friction.

a) In bicrystals where the elastic moduli are not too different, the Peierls friction is everywhere larger than the image force even at the interface position. These will be referred to as bicrystals with low K values. It has been found that the Peierls valley at the interface remains for K smaller than 0.03 with the Johnson potential and K larger than 0.02 with the N-body potential (Lardon et al. 1994) which corresponds to elastic constant ratios R equal to 1.06 and 1.04 respectively. In these bicrystals, there exists one stable position (a_i) for the dislocation in each Peierls valley. It has been possible to compute the total strain energy stored in the crystal W(a_i) at each of the dislocation stable positions. The image force on the dislocation has then been deduced by finite differences. The computed image force differs from the elastic solution only within few interatomic distances (4-5 a_0) from the interface.

The computed maximum stress, at the interface, has been found about half of that deduced by Pacheco and Mura (1969) on the basis of a perturbation by the interface of the Peierls-Nabarro dislocation core model (Nabarro 1947):

$$\sigma_{PM} = 2K\mu/\pi^2 \tag{2}$$

which is of the order of $(\mu_2-\mu_1)/10$.

b) If the shear moduli of the two half-crystals are notably different, in a region containing the interface, the image force overcomes the Peierls friction erasing some stable positions and the dislocation can be stabilised there only by applying an external stress. The analysis of the image force effect has been made by placing the dislocation in the ith Peierls valley, and determining the interval of applied stress $\Delta\sigma_i$ leaving the dislocation in the considered valley. Figure 2 shows a typical example obtained for an elastic moduli ratio R=2 (K=1/3).

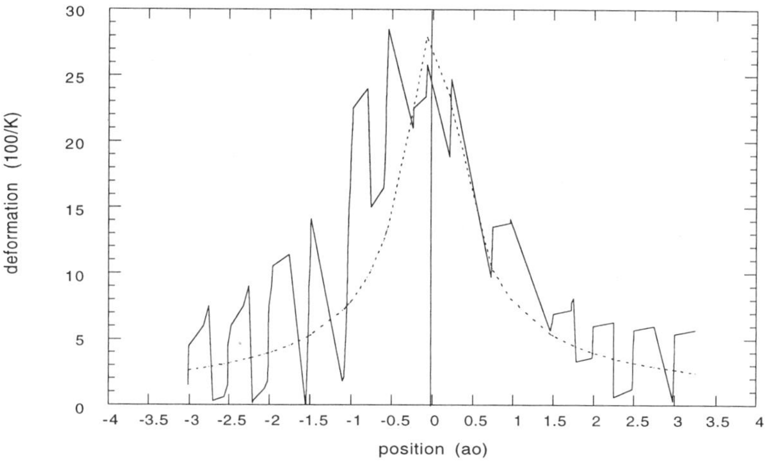

(a)

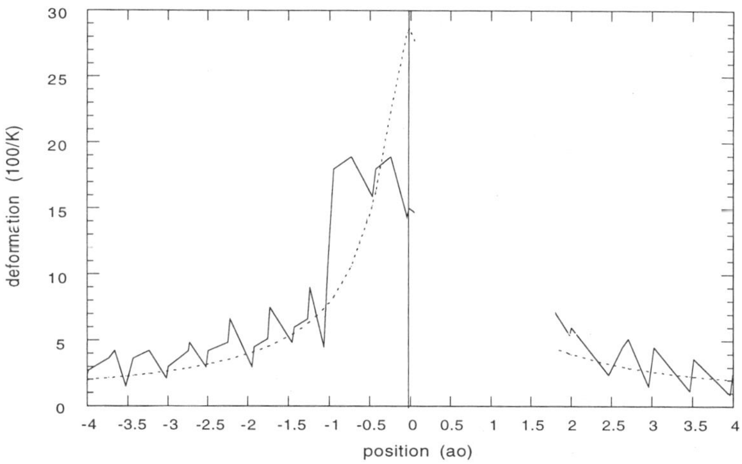

(b)

Figure 2. Applied strain σ/μ necessary to maintain a dislocation at a given position near the interface. (a) pair potential (b) N-body potential. Negative values of the position refer to the "soft" medium.

Far from the interface, the "shapes" of the valleys are identical to those of the homogeneous crystal whereas large distortions occur in the interface region. The amplitude of stress for each valley $\Delta\sigma_i$ shows systematic variations: it increases when approaching the interface from the left ("soft" medium) and decreases from the right ("hard" medium). This effect is quite clear with the pair potential and also visible with the N-body potential. The reduction in stress amplitude of Peierls valleys in the "hard" medium ends with their complete vanishing in a region close to the interface where the dislocation cannot be maintained. The stress calculated by Pacheco and Mura (1969) is not too far from the curve joining the average stresses of each valley (Fig. 2).

Considering all the results, obtained for elastic moduli ratios R ranging from 1.04 up to 10 and the two different interatomic potentials, it has been possible to propose a general law for the force on the dislocation in the interface region (also represented on Fig. 2), based on the Pacheco and Mura formula and which includes as additional parameter the width of the dislocation core. Such law (Lardon 1994) may be quite useful in mesoscopic studies such as pile-up formation and stability.

Changes in the a<001> Screw Dislocation Core near the Interface

When examining the relative displacement maps of dislocations situated at different distances from the interface, the following feature appears: far from the interface (4-6 interatomic distances), the core is identical to that obtained in the homogeneous case; when approaching the interface (i) in the hard medium, the core width is clearly increased and (ii) in the soft medium a reduction in the core width seems to occur. To quantify this effect, we have calculated the "density" of relative displacements across the {110} glide plane: the half-height width of this distribution is a precise definition of the dislocation width. In Table 1 are listed the core widths in a bicrystal having a elastic moduli ratio R=2 (K=1/3) at various dislocation-interface distances. It is seen that the core width variations are more important in the hard medium than in the soft one. They are also larger with the N-body potential which, in the homogeneous crystal, gives a core more extended than the pair force law.

Table 1. Core extension (in a_0 units) as a function of the dislocation-interface separation (a). Negative values of (a) correspond to the soft medium.

(a)	-2.5	-1.5	-0.5	0.	0.5	1.5	2.5
Pair pot.	1.77	1.77	1.68	1.94	2.83	2.83	1.77
N-body pot.	3.27	2.47	2.47	3.36	7.78	4.24	3.53

Such an effect of dislocation core widening has been shown by Nabarro (1947) for a dislocation near a free surface. It can be explained by regarding the dislocation as a continuous distribution of infinitesimal dislocations and remarking that the interface effect (attractive in the hard medium and repulsive in the soft medium) is more important on those infinitesimal dislocations close to the interface than on those further away.

Near the interface in the hard medium, the widening effect is such that it results into a real dissociation of the b=a[001] screw dislocation into two well-defined partials (Fig. 3) of screw components b/2 and of edge components approximately ±a/6<110> parallel to the glide plane. An analysis of the γ surface shows that this stacking fault is not normally stable in the bcc crystal. In fact it is of the type "unstable fault" to which Rice (1992) attributes a critical role in his model of dislocation nucleation at crack tip. The energy of this unstable fault ($\approx$0.5 J/m^2) is not very high here because of the proximity of the perfect lattice translation a/2[111].

An energetic analysis of the core structure evolution in the interface region has been made using a simple model in which the perfect dislocation is regarded as the juxtaposition of two b/2 screw dislocations. The energy is calculated as a function of the separation (d) between the two partials and of the distance (x_0) between the middle of the configuration and the interface. The contributions to energy are the self energy of each partial (interaction with its own image), the interaction between the two partials (including image contributions), the stacking fault energy γ and the work done by the applied stress. The energy surface $W(x_0,d)$ obtained in this way depends on three parameters, K characteristic of the bicrystal, $e_c=\gamma/(\mu b)$ characteristic of the material and $e=\sigma/\mu$ the imposed homogeneous strain. A typical energy surface $W(x_0,d)$ is shown on figure 4. The energy minimum for a negative x_0 is the stable dislocation position in the soft medium (the "lake" on Fig. 4), and it is seen that the saddle point in the energy barrier is associated with a notable increase in the separation between the two partials, it corresponds to the crossing of the interface by the leading partial. The second energy barrier corresponds to the crossing of the interface by the second partial. The height of this barrier decreases with the dissociation width.

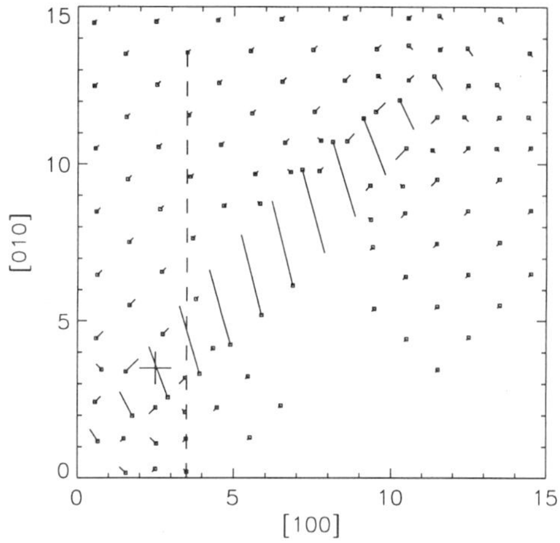

Figure 3. A dissociated configuration with the leading partial dislocation in the "hard" medium and the second partial in the "soft" medium, near the interface. The cross indicates the initial position of the perfect dislocation.

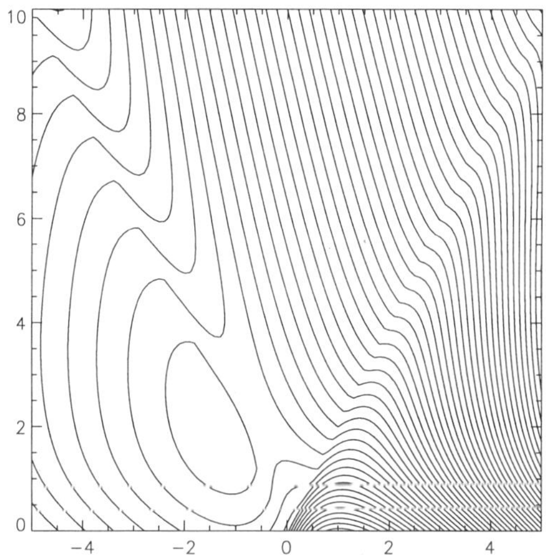

Figure 4. The energy of a dissociated dislocation as a function of its dissociation width (vertical axis) and its distance from the interface (horizontal axis). Negative and positive distances refer to "soft" and "hard" media respectively. The length unit is the modulus of the total Burgers vector.

Twinning induced by the a/2<111> Screw Dislocation

The above energetic considerations indicate that the dissociation with formation of the "unstable fault" has been easily observed with the b=a<001> screw dislocation because the energy of this fault is not too large. It is then interesting to examine the case of the a/2<111> screw dislocation for which the computed γ-surface (Vitek 1968) on {110} and {112} planes show that the "unstable faults" have high energies. Preliminary results on this dislocation which is now under investigation, indicate that for a notable difference in the elastic moduli of the two phases (R≥3), the dislocation penetrates the hard medium under the effect of the applied stress by forming a {112} twin (Fig. 5). It must be remarked that the twin has a thickness larger than the elementary twin lamella 3.b/3 that a a/2<111> screw can generate by transforming its core (Sleeswyk 1974, Lagerlof 1993). Here, the twin has the thickness which relaxes the applied strain and because of the rigid boundary conditions, it increases with the size of the atomistic box. The dislocation in the vicinity of the interface has only a triggering role, it allows for the crystal to find a path toward the twinned phase, more stable because of the applied stress. It is interesting to remark that this mechanism produces a twin of notable thickness without the operation of any complicated dislocation configuration.

CONCLUSION

Besides the determination of the force on the dislocation in a region of few

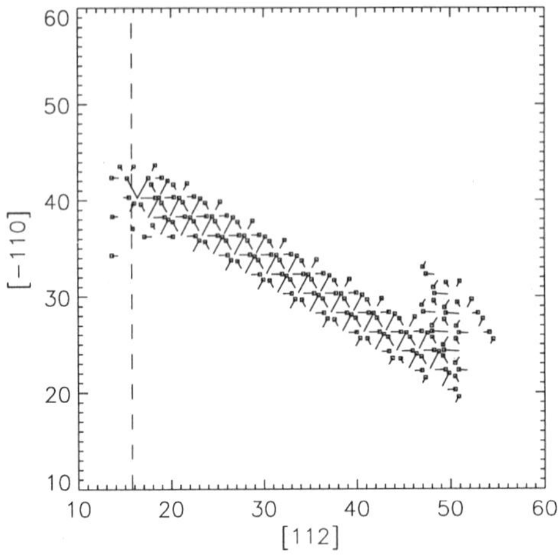

Figure 5. A twin induced by a screw a/2<111> dislocation near the interface. Lengths are indicated in Å.

interatomic distances from the interface, the possibility of important changes in the dislocation core structure has been revealed.

REFERENCES

Cserti, J., Tichy, G. and Vitek, V., 1994, Solid State Phenomena 35-36, 545.
Finnis, M.W. and Sinclair, J.E., 1984, Phil. Mag. A 50, 45.
Head, A.K., 1953 a, Phil. Mag. 44, 92.
Head, A.K., 1953 b, Proc. Phys. Soc. B 66, 793.
Hirth, J.P. and Lothe, J., 1982, Theory of Dislocations (2nd edition), Mac Graw-Hill,New-York.
Jonhson, R.A., 1964, Phys. Rev. A 134, 1329.
Lagerloff, K.P.D., 1993, Acta Metall. Mater. 41, 2143.
Lardon, I., 1994, Doctoral Thesis, University of Poitiers.
Lardon I., Beauchamp P. and Lépinoux J., 1994, Solid State Phenomena 35, 581.
Nabarro, F.R.N., 1947, Proc. Phys. Soc. LIX, 256.
Pacheco, E.S. and Mura, T., 1969, J. Mech. Phys. Sol. 17, 163.
Rice, J.R., 1992, J. Mech. Phys. Sol. 40, 239.
Sleeswyk, A.W., 1974, Phil. Mag. 29, 407.
Vitek, V., 1968, Phil. Mag. 18, 773.

SLIP INDUCED CONFIGURATIONAL ENERGY CHANGE IN BINARY ALLOYS

B. Schönfeld and P. Schwander

Institute of Applied Physics
ETH Zürich
CH-8093 Zürich

INTRODUCTION

Depending on the distribution of slip planes, dislocation arrangements in solid solutions appear to be wavy or planar (coarse). As the concentration of a solid solution is increased, a transition from wavy to planar glide is typically observed. This transition has been attributed to various alloying effects such as short-range order, stacking fault energy or yield stress.

By correlating these factors with dislocation arrangements introduced by plastic deformation, it has been suggested that the dominating contribution is due to short-range order (see, e.g., ref.1,2 and references therein). The effect of short-range order can be understood as follows: in moving a dislocation the two parts of a crystal separated by the slip plane are displaced by the Burgers vector. The atomic configuration across the slip plane gets changed and thus the configurational energy of the crystal. In analogy to the antiphase boundary energy in long-range ordered alloys, the change of the configurational energy per unit area is named diffuse antiphase boundary energy.[3] The diffuse antiphase boundary energy of solid solutions can be expressed in terms of the short-range order parameters and the effective pair interactions that are related to the configurational energy.

In this paper a general procedure is described to determine the diffuse antiphase boundary energy of any binary solid solution from an experiment in a quantitative way. The procedure is based on diffuse scattering experiments using X-rays or neutrons combined with methods of statistical mechanics (e.g., the inverse Monte Carlo method). This method is applied to obtain the diffuse antiphase boundary energy of binary copper alloys with Al, Mn, Zn, or Au and of binary nickel alloys with Mo, Fe, Cr, or Al. Furthermore, the effect of temperature as well as composition on the diffuse antiphase boundary energy is estimated for these alloys. Common features are found, although the nature of short-range order in these alloys is quite different.

THEORY

Taking only pair interactions into account, the configurational energy per atom of a

Stability of Materials
Edited by A. Gonis *et al.*, Plenum Press, New York, 1996

binary A-B alloy can be written as

$$H = c_A c_B \sum_{i=1}^{i_{max}} V_i \alpha_i \tag{1}$$

with the atomic fractions c_A, c_B, the effective pair interactions $V_i = \frac{1}{2}(V_i^{AA} + V_i^{BB}) - V_i^{AB}$ and the Warren-Cowley short-range order parameters α_i. The summation over all atoms in shell i is considered to be relevant up to shell i_{max}.

By moving a dislocation on a glide plane, the configurational energy is increased. In this energy change the modified numbers of bonds between any atom in one half space with respect to the glide plane and any atom in the other half space are noted and weighted in terms of V_i^{AA}, V_i^{BB}, V_i^{AB}. Neglecting any local variation in the energy change ΔH, only the effective pair interactions V_i and the short-range order parameters α_i are needed.[3]

$$\Delta H_n = c_A c_B \sum_{i=1}^{i_{max}} \sum_{j=1}^{j_{max}} V_i T_n^{ij} \alpha_j \tag{2}$$

The elements of the matrix T_n^{ij} represent the numbers of modified bonds with T_n^{ii} = number (negative) of all atoms in shell i that are removed from it by n slip steps and $T_n^{ij} (i \neq j)$ = number (positive) of atoms belonging to shell i before the n slip steps and to shell j afterwards. The energy change ΔH_n normalized to the unit area of an atom in the glide plane, F_a, is γ_n^{SRO} where the index n refers to the number of dislocations being moved on the glide plane.

$$\gamma_n^{SRO} = \Delta H_n / F_a. \tag{3}$$

In the limit of large values of n, the (constant) value of γ_n^{SRO} is called γ^{SRO}.

The short-range order parameters α_i can be determined from a diffuse scattering experiment by various techniques. The effective pair interactions V_i are deduced from the short-range order parameters α_i by inverse methods of statistical mechanics, provided that the measured state represents thermodynamic equilibrium. Assuming that the V_i do not depend strongly on the composition of the alloy and on temperature, γ^{SRO} may be determined also for nearby states in the phase diagram (e.g., by using Monte Carlo simulations).

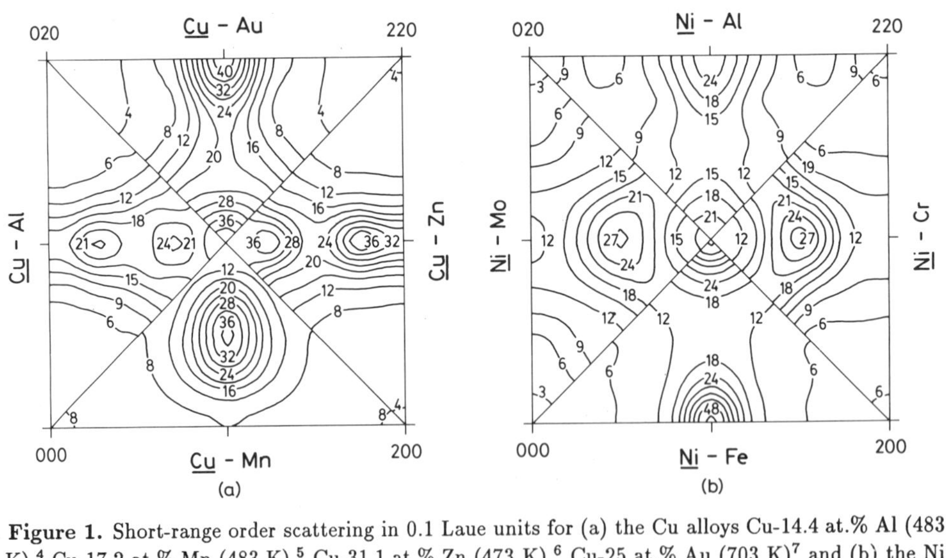

Figure 1. Short-range order scattering in 0.1 Laue units for (a) the Cu alloys Cu-14.4 at.% Al (483 K),[4] Cu-17.2 at.% Mn (483 K),[5] Cu-31.1 at.% Zn (473 K),[6] Cu-25 at.% Au (703 K)[7] and (b) the Ni alloys Ni-11.2 at.% Mo (773 K),[3] Ni-23.5 at.% Fe (808 K),[8] Ni-21.3 at.% Cr (741 K),[9] Ni-8.9 at.% Al (775 K)[10]. The temperatures of the states investigated are given in brackets.

SHORT-RANGE ORDER SCATTERING

Short-range order scattering for some solid solutions is shown in Fig.1 for the (001) plane of reciprocal space. Diffuse maxima are seen at the special points of the f.c.c. lattice (at 100 and $1\frac{1}{2}0$) and also at the $2k_F^{110}$ positions (k_F^{110} = Fermi wave vector along <110>).

Table 1. Effective pair interactions V_i as determined by the inverse Monte Carlo method. Only the first four V_i are given with error bars solely based on counting statistics.

i	V_i(meV)							
	Cu-Al[4]	Cu-Mn[5]	Cu-Zn[6]	Cu-Au[11]	Ni-Mo[3]	Ni-Fe[10]	Ni-Cr[10]	Ni-Al[10]
1	31.3(8)	17.8(1)	18.6(3)	33.1(26)	102.0(20)	29.7(2)	26.1(12)	32.6(4)
2	-5.1(3)	-11.5(2)	-6.6(3)	-10.0(17)	-2.0(12)	-11.1(2)	-16.0(12)	-10.7(4)
3	-5.8(4)	-4.0(1)	-1.3(2)	1.8(25)	12.8(6)	1.9(1)	-4.0(4)	-4.4(1)
4	-3.3(5)	6.4(3)	-0.1(2)	3.2(21)	37.0(7)	-0.1(1)	8.0(3)	-8.4(2)

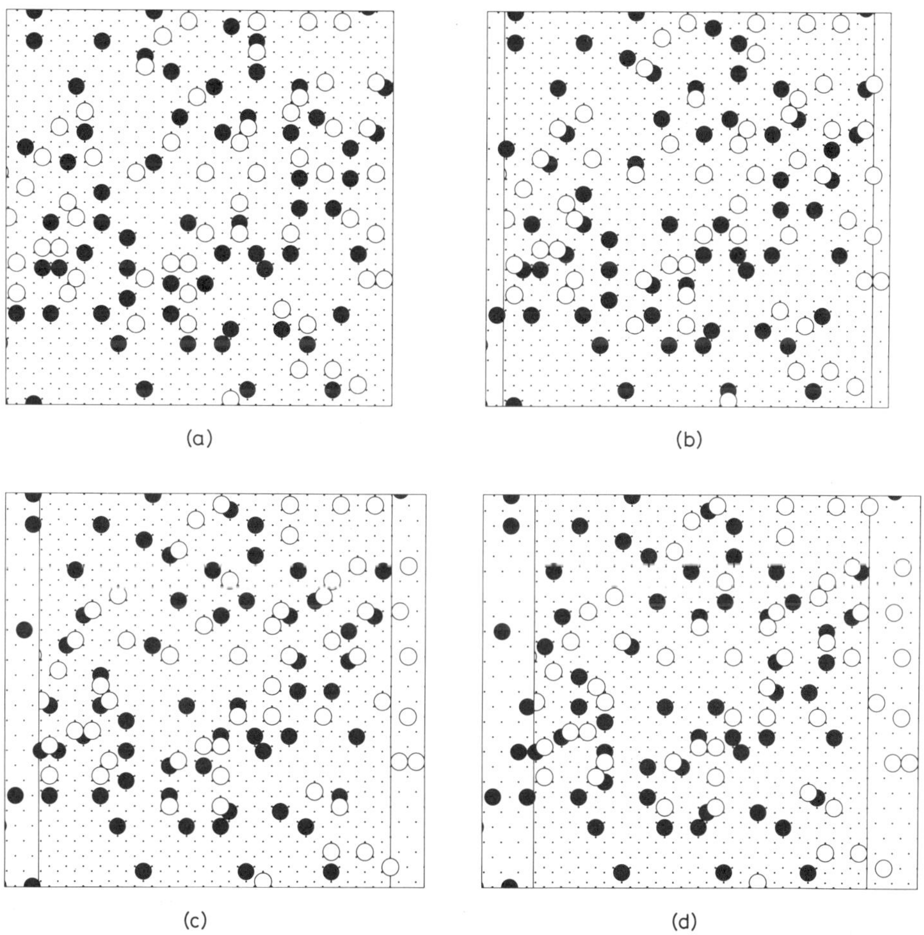

Figure 2. Two successive (111) planes of Cu-10 at.% Al (523 K), (a) before slip, (b) after slip by $\underline{b}$, (c) $2\underline{b}$, (d) $3\underline{b}$ with $\underline{b}=a/2[\bar{1}01]$. Small dots represent Cu atoms, ○ and ● Al atoms in the upper and lower plane, respectively.

The sharpness of the diffuse maxima reflects the variation in the composition of the alloys and in the nearness of the states to an order-disorder transition. Effective pair interactions might represent a better comparison; the four leading ones are listed in Tab.1 (out of up to 20 V_i considered to be relevant to reproduce short-range order scattering in a subsequent Monte Carlo simulation).

DIFFUSE ANTIPHASE BOUNDARY ENERGY

The change in the atomic arrangement is visualized in Fig.2 for short-range ordered Cu-10 at.% Al. If one just considers the number of Al-Al nearest neighbor pairs (within two adjacent planes and between them) before any slip and also the number of pairs (only between the planes) after the various slip steps shown, one notices a distinct increase from about 6 to 18 Al-Al nearest neighbor pairs. These data compare favorably with those calculated for an infinitely large crystal, namely 5.6 and 17.6 for the short-range ordered and the random arrangement, respectively. Thus, short-range order gets destroyed across the glide plane as dislocations move.

In Fig.3 the diffuse antiphase boundary energies of the Cu and Ni alloys are compared. The aging temperature was chosen to give relaxation times of ordering suitable to retain a state of equilibrium upon quenching. For the Cu alloys the relaxation times range from several minutes (for Cu-Zn[12] and Cu-Al[13]) to about an hour (for Cu-Mn[14] and Cu-Au[15]). For the Ni alloys information is less detailed, but relaxation times are expected to lie within a comparable range.[16–18] The fraction of the minority component was chosen to approach

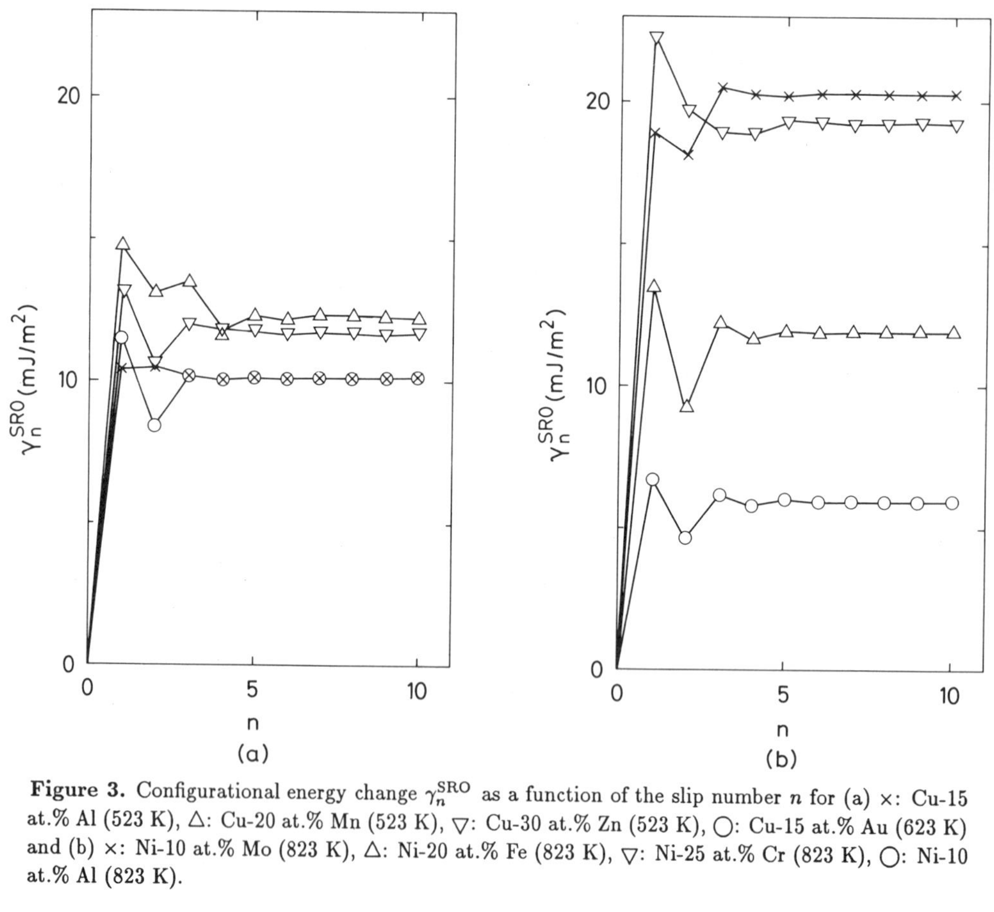

Figure 3. Configurational energy change γ_n^{SRO} as a function of the slip number n for (a) ×: Cu-15 at.% Al (523 K), △: Cu-20 at.% Mn (523 K), ▽: Cu-30 at.% Zn (523 K), ○: Cu-15 at.% Au (623 K) and (b) ×: Ni-10 at.% Mo (823 K), △: Ni-20 at.% Fe (823 K), ▽: Ni-25 at.% Cr (823 K), ○: Ni-10 at.% Al (823 K).

maximum solubility. The typical features are: (i) γ_1^{SRO} usually represents the largest value, (ii) after a transition region extending up to $n \approx 5$ a plateau for γ_n^{SRO} is reached, with values of 6 to 20 mJ/m^2. Because only the energy difference $\gamma_n^{\mathrm{SRO}} - \gamma_{n-1}^{\mathrm{SRO}}$ is required when dislocation n moves, dislocations following the first one on the same glide plane move more easily, as only very little energy is stored (or even released), and planar glide is favored.

Beside the effective pair interactions, mainly the composition of the alloy and to a much lesser degree the aging temperature T determine the diffuse antiphase boundary energy. This can be estimated from Eq.(2) using the Krivoglaz-Clapp-Moss high-temperature approximation which yields $\alpha_{i\neq 0} \sim c_A c_B/T$. Thus, the composition and temperature dependences of γ^{SRO} are $(c_A c_B)^2$ and T^{-1}, respectively. While composition is only restricted by solid solubility, the aging temperature T is limited by the condition that a state of equilibrium is retained upon quenching which only allows for a small temperature window of about 200 K. In addition, short-range order should not be restored within the time the dislocation group moves.

Both dependences are illustrated in Fig.4 for Cu-Mn. By decreasing the temperature from 723 K to 423 K at a fixed composition, γ^{SRO} increases by a factor of about 1.7, in close agreement with the temperature ratio. If the composition is increased from 0.05 to 0.25 at a fixed temperature, a much larger increase is observed for γ^{SRO}. This increase by a factor of $\sim$ 17 is close to the estimated composition dependence of $\sim$ 16.

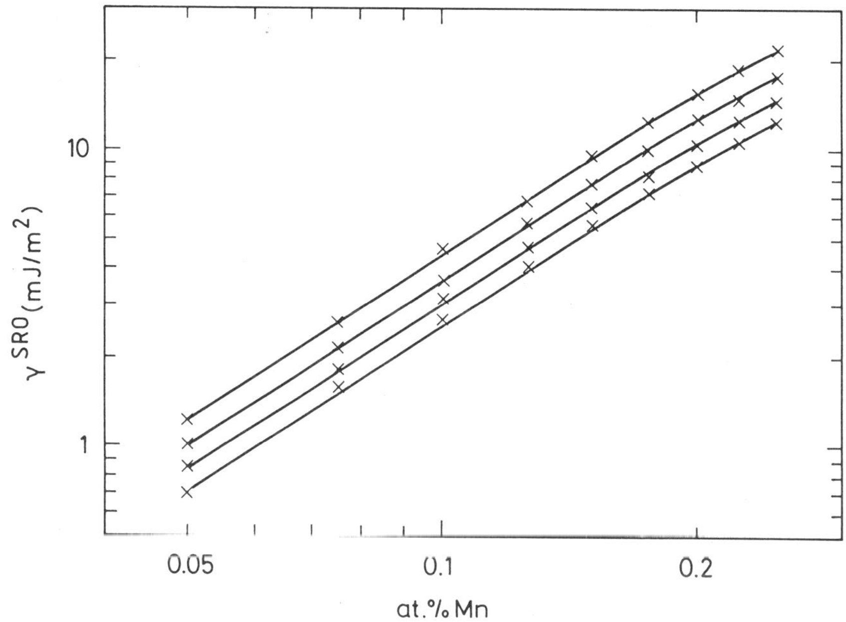

Figure 4. The (c_{Mn}, T) dependence of γ^{SRO} for Cu-Mn based on Monte Carlo simulations, with aging temperatures between 423 K (top curve) and 723 K (bottom curve), in steps of 100 K.

As dislocations move at finite temperature T (e.g., room temperature), the change in the free energy and not only in the energy has to be considered. In previous publications, the contribution to γ_n^{SRO} by $-T\Delta S_n$ where ΔS_n represents the change in the configurational entropy by the nth slip step, has not been considered in detail. In the present work, entropy changes have been approximated counting the ways to set up nearest neighbor bonds in consistency with the Warren-Cowley short-range order parameters α_i without correcting for conflicts in building these configurations.[19,20] Thus, this approximation represents an upper limit and is expected to be better for alloys with a lower degree of short-range order. As in the case of the configurational energy change ΔH_n, $-T\Delta S_n$ is dominant for the passage of the first dislocation. For the cases shown in Fig.3 the contributions to γ_n^{SRO} always amount to about $-0.25\Delta H_n/F_a$.

The critical composition for the transition from wavy to planar glide can be determined from TEM micrographs of deformed samples. The corresponding γ^{SRO} can be calculated. For Cu-Mn, Cu-Al and Cu-Zn the transition occurs for γ^{SRO} between 0.5 to 5 mJ/m^2 (corresponding to 3 to 12 at.% solute).[5] In the case of nickel base alloys the scarce data on Ni-Mo and Ni-Fe yield upper limits of 9 (Ni-6 at.% Mo) and 17 mJ/m^2 (Ni-25 at.% Fe),[21] indicating a larger critical value of γ^{SRO} in nickel base alloys. From more extensive work on Ni-Cr,[21,22] such a larger value seems compelling (4 to 14 mJ/m^2 for 10 to 20 at.% Cr) . Considering the strong dependence of γ^{SRO} on alloy composition, more TEM investigations are required to establish a possible correlation between γ^{SRO} and the metal base.

ACKNOWLEDGEMENTS

The authors very much appreciate the collaboration with G. Kostorz, H. Roelofs, L. Reinhard, W. Bührer, J. Plessing and H. Neuhäuser in the various investigations. They are also indebted to E. Fischer for growing the single crystals used in the diffuse scattering experiments.

REFERENCES

1. V. Gerold and H.P. Karnthaler, *Acta metall.* 37:2177 (1989).
2. G. Kostorz, in "Non Linear Phenomena in Materials Science III", G. Ananthakrishna, L.P. Kubin, G. Martin, eds., Scitec Publications, p. 187 (1995).
3. P. Schwander, B. Schönfeld, and G. Kostorz, *phys. stat. sol. (b)* 172:73 (1992).
4. B. Schönfeld, H. Roelofs, A. Malik, G. Kostorz, J. Plessing, and H. Neuhäuser, *Acta metall. mater.*, in press (1995).
5. H. Roelofs, B. Schönfeld, G. Kostorz, and W. Bührer, *phys. stat. sol. (b)* 187:31 (1995).
6. L. Reinhard, B. Schönfeld, G. Kostorz, and W. Bührer, *Phys. Rev. B* 41:1727 (1990).
7. B.D. Butler and J.B. Cohen, *J. Appl. Phys.* 65:2214 (1989).
8. S. Levèbvre, B. Bley, M. Bessière, M. Fayard, M. Roth, and J.B. Cohen, *Acta crystallogr. A* 36:1 (1980).
9. B. Schönfeld, L. Reinhard, G. Kostorz, and W. Bührer, *phys. stat. sol. (b)* 148:457 (1988).
10. B. Schönfeld, "Local Atomic Arrangements in Binary Alloys", Habilitationsschrift, ETH Zürich (1993).
11. B. Schönfeld, unpublished results.
12. W. Pfeiler, R. Reihsner, and D. Trattner, *Scripta metall.* 19:199 (1985).
13. S. Radelaar, *J. Phys. Chem. Solids* 27:1375 (1966).
14. R. Reihsner and W. Pfeiler, *J. Phys. Chem. Solids* 46:1431 (1985).
15. H. Lang and W. Pfeiler, in "Solid-Solid Phase Tranformations", W.C. Johnson, J.M. Howe, D.E. Laughlin, W.A. Soffa, eds., TMS Proceedings, p. 479 (1994).
16. P. Schwander, "Nahordnung und Versetzungen in einer Ni-Mo-Legierung", Dissertation, ETH Zürich (1990).
17. H. Heidsiek, K. Lücke, and R. Scheffel, *Phys. Chem. Solids* 43:825 (1982).
18. M. Afyouni, V. Pierron-Bohnes, and M.C. Cadeville, *Acta metall.* 37:2339 (1989).
19. J.M. Cowley, *Phys. Rev.* 120:1648 (1960).
20. R. Kikuchi, *Phys. Rev.* 81:988 (1951).
21. N. Clément, in "L'Ordre et le Désordre dans les Matériaux", F. Reynaud, N. Clément, J.J. Couderc, eds., Les Editions de Physique, Les Ulis, France, p. 167 (1984).
22. K. Wolf, H.-J. Gudladt, H.A. Calderon, and G. Kostorz, *Acta metall. mater.* 42:3759 (1994).

CYCLIC CREEP DEFORMATION CHARACTERISTICS OF SINGLE CRYSTAL OF NICKEL BASE SUPERALLOY CMSX 3

J. Zrník, Z. G. Wang[1], Y. Yu[2], J.A. Wang[2], M. Žitňanský[3], P. Horňak

Technical University of Košice, Slovakia
[1]Institute of Materials Research Academia Sinica, Shenyang, China
[2]Dalian University of Technology, China
[3]Slovak Technical University, Bratislava, Slovakia

INTRODUCTION

High temperature components assembled in aircraft turbines are subjected to complex stress, including thermal stress, stress originating from centrifugal force and high frequency vibration stress. Recently increased attention has been paid to the study of creep and fatigue interaction which appears to be an important factor in the prediction of the components' life[1,2].

It is believed that conjoint action of creep and fatigue stress components should be considered to assess the deformation mechanism in subjected parts. Various fatigue cycles in which tensile and compressive stress components are present result in initiation and growth of surface fatigue cracks[3,4]. Results show that an additive cyclic stress component to stationary creep stress can either accelerate or decelerate the creep deformation, in depending on the alloy' structural characteristics and employed materials test conditions[5,6,7].

The main objective of this study is to present the creep/fatigue behaviour of the single crystal nickel base superalloy CMSX 3. Three different testing procedures were adopted in order to investigate the effect of additional fatigue stress components on the creep process. Load controlled creep-fatigue tests have been carried out at high temperature where frequency, stress range and tension hold times at amplitude peaks were variables of the testing procedure.

MATERIAL AND EXPERIMENTAL PROCEDURES

The investigated material was the single crystal nickel base superalloy CMSX 3. The chemical composition (wt.%) of the alloy is listed in Table 1.

Single crystals of this alloy were grown from the melt by the modified Bridgman method. The as-cast cylindrical bars were 85 mm in length and 12 mm in diameter. The crystallographic orientation of the as-cast bars was determined by the back reflection Laue method.

Two different testing programs for cyclic creep were employed in order to investigate the effect of various waveforms of the stress as a function of time.

Stability of Materials
Edited by A. Gonis *et al.*, Plenum Press, New York, 1996

Table 1. Chemical composition of alloy

Cr	Co	Mo	W	Al	Ti	Ta	Hf	Ni	C	O
8.0	4.0	0.5	9.0	5.7	0.7	5.7	0.1	bal	0.006	0.002

A. One is continuous cyclic creep with tension hold time at peak stress amplitude σ_{max} = 400 MPa. The waveform of stress as a function of time was of trapezoidal shape and stayed unchanged throughout the whole lifetime for each hold time, as shown in fig. 1. The six different hold times Δt were respectively : 0 (pure fatigue), 1 s, 3 s, 5 s, 10 s and ∞ s representing pure creep test. The rise and fall time were 0.5 s respectively. The stress after load reduction was maintained at σ_{min} = 5 MPa to keep the loading system in tension. Stress ratio R for each cycle test was R = 0.0125. The testing temperature was 900 °C. The cyclic creep tests were conducted on a Mayes hydraulic testing machine. The specimens for this test underwent the following heat treatments:

- 1300°C/6h in vacuum, FC to 1100°C at 1,5 °C/min, then in air
- 1080°C/5h in vacuum, FC to ambient temperature at 18 °C/min
- 870°C/10h in vacuum, FC to ambient temperature at 15 °C/min

The heat treatments procedure provided microstructures with bimodal gamma prime precipitation characteristics, where the average size of the bigger irregular precipitates was more than 1 µm and the size of the smaller precipitates, which were fully coherent with the matrix, was 200 nm.

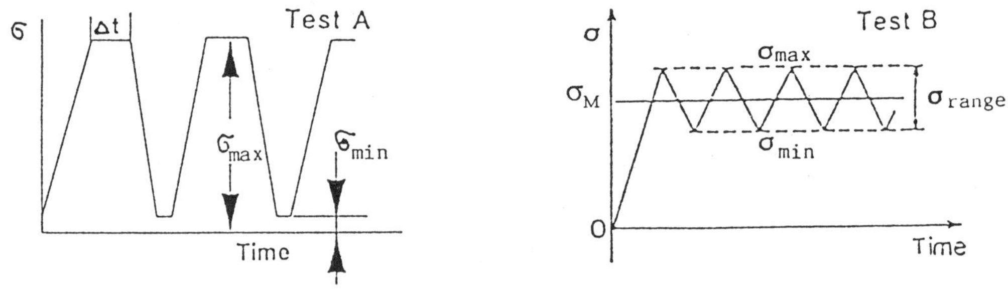

Figure 1. Forms of stress wave

B. The second testing program involved the cyclic loading schedule shown in fig. 1. The mean stress σ_M = 400 MPa was kept constant from test to test. For as-cast specimens three different frequencies 1, 100 and 200 Hz at a constant cyclic stress range were chosen. In the case of heat treated specimens, the wave form with constant mean stress σ_M = 400 MPa , and constant frequency 100 Hz was used. Only stress range $Q=\Delta\sigma / \sigma_M$ was changed; Q_1= 0.1 and to Q_2= 0.2. The load controlled tests were carried out on an MTS 810 hydraulic testing machine at 900 °C.

The specimens were machined from the as-cast and heat treated bars. The bars underwent the following heat treatments in vacuum:

(1) Solution annealing: 1310 °C/6 h, cooling in air
(2) First ageing: 1100 °C/4 h, cooling in air
(3) Second ageing: 870 °C/18 h, cooling in air

The heat treatment modified the as-cast microstructure and unimodal gamma prime precipitation characteristics were obtained.

The relative creep strain and fracture time and/or numbers of cycles to fracture of the individual specimens were the criteria to evaluate the cyclic creep resistance of as-cast and heat treated specimens for both testing procedures.

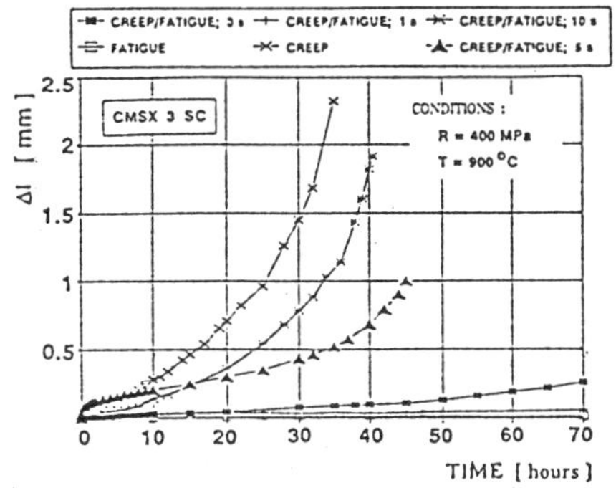

A Test

Figure 2. Strain-time curves for different hold time

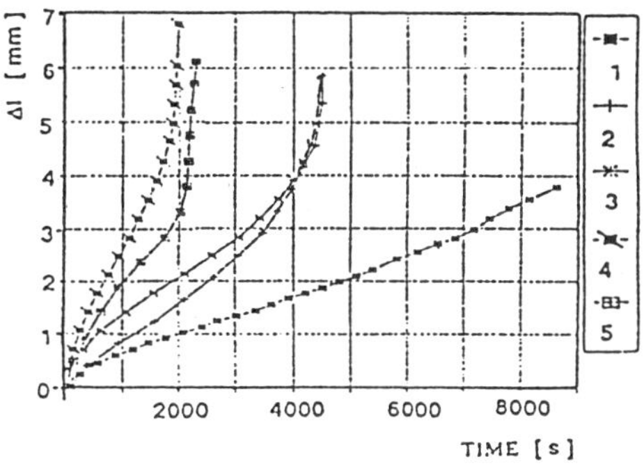

B Test

Figure 3. Strain-time curves for different frequency

The dislocation structure in thin foils prepared from a longitudinal section of fractured specimens was examined by TEM. Fracture surfaces, crack initiation and development of secondary cracks were also subjects of interest for examination by SEM.

RESULTS AND DISCUSSION

A. program:Effect of hold time

The introduction of the additional cyclic stress component superimposed on stationary creep, results in a change of the fracture time. Fig. 2 summarizes the experimental achievements on cyclic lifetime as a function of tensile hold times at the stated stress amplitude σ_{MAX}= 400 MPa. The pure fatigue and all cyclic creep tests exhibit better endurance to fracture than pure creep. The presented data show that cyclic loading leads to pronounced extension in endurance for given text conditions. The fracture lives and fracture data are listed in Table 2.

Table 2. Fracture data

Hold time [sec]	Fracture life [h]	Numbers of cycles to fracture [N]	Static/cyclic component ratio	Strain Δl [%]	Strain ε	Frequency [Hz]
creep	37:19	-	1	32.5	0.34	-
1	70:00*	126 000*	3	0.4	0.004	0.5
3	68:14	61 410	7	1.6	0.023	0.25
5	44:45	28 850	11	4.8	0.048	0.166
10	41:12	13 483	21	16.2	0.16	0.09
fatigue	70:00*	252 000*	-	0.08	0.0009	-

The plastic characteristics expressed through sample extention Δl show the opposite dependence. The highest specimen prolongation Δl corresponds to pure creep test.

*Not fractured

The pure fatigue and cyclic creep with 1 s and 3 s hold time have only a very small corresponding strain response. The plastic deformation component Δl after failure is substantionally greater for pure creep and for 10 s hold time than for other cyclic creep tests including pure fatigue.

B.program: The effect of frequency and stress range

The experimental results involving the obtained cyclic creep test data are listed in Table 3.

Table 3. Test data

Sample	Stress range [MPa]	Mean stress [MPa]	Frequency [Hz]	Fracture life [sec]	Cycles to fracture [N]	Elongation [mm]	Elongation [%]
1 as-cast	40	400	1	8460	7630	2.55	8.2
2 as-cast	40	400	100	4510	440751	4.0	15. 4
3 as-cast	40	400	200	4440	894976	4.35	16.7
4 heat tr	40	400	100	3444	206686	5.1	18.1
5 heat tr	80	400	100	2242	224240	4.3	14.3

The results show that specimens tested according to this testing procedure exhibit an extensive fracture time reduction, for as-cast and heat treated specimens respectively. The dependence of the specimen elongation *vs* time to fracture is documented in fig. 3. The strain rate reaches higher values already at the start of the test. In the case of the as-cast specimens (curves 1, 2, 3) the change of frequency, when the mean stress and stress range were kept constant, resulted in a considerable lowering of the fracture time. The further frequency increase at 200 Hz did not have significant effect in further fracture time reduction.

Very surprising appear to be the results obtained in the testing of the heat treated specimens, where structure modification was apparent regarding structure homogeneity and gamma prime morphology. As the fracture data show, the realised heat treatment which was expected to change the structure characteristics in positive mode, resulted in even more reduction of the fracture life (curves 4, 5). The widening of the stress range at an equal level of frequency was manifested in an only negligible change of fracture life.

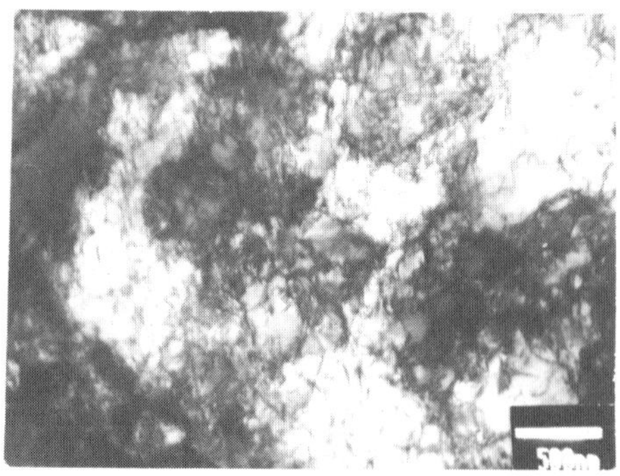

Figure 4. Dislocation structure of crept specimen. A Test

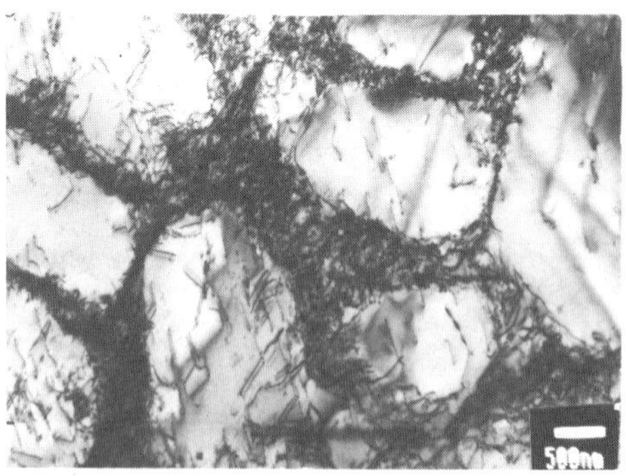

Figure 5. Dislocation distribution. 3 s hold time. A Test

Dislocation structure

An extensive dislocation study was carried out on the tested specimens after the failure. The principal aim of this investigation was to assess the effect of hold times introduced onto cyclic stress waveform and the effect of additional cyclic stress of various frequency introduced onto a constant mean stress σ_M, representing creep stress, on deformation process and dislocation network development. In the case of pure creep and cyclic creep with the longest hold time (10 s) the tangled dislocations are present both in the matrix and gamma prime precipitates, fig. 4. For shorter times, less than 5 s, and pure fatigue test the deformation was prevailingly concentrated in the matrix where dense dislocation networks were formed, fig. 5. The shape of gamma prime morphology was not affected by deformation, only individual dislocation segments cross-slipped into gamma prime particles. The cyclic creep involving different frequencies and different stress ranges modifying the stress waveform reflected in the formation of a dislocation structure similar to that in pure creep deformation. Regardless of individual cyclic test variables and regardless of specimen' state, the dislocation structure of the fractured specimens bears the features of severe plastic deformation within the specimen' gauge length, as shown in fig. 6. The dense dislocation network was formed in areas close to the fracture surface. In areas more distant from the fracture surface the individual gamma prime particles preserved their morphology, and additively dislocation networks are present within them as well. The dislocation structure resulting from this type of cyclic creep indicates a mode of deformation which is quite similar to that of pure creep and results in marked plastic characteristics of tested specimens. The results of testing and resulting deformation structure enables us to state that the decisive role in the deformation process is played by the mean stress level. The implementation of the stress cycles onto the mean stress is reflected in the acceleration of the strain rate and in the reduction of fracture life.

Fracture characteristics

The presence of the cyclic component in the creep process is reflected in the mode of fracture and in the time to fracture. The fracture surfaces in the case of the tension hold times clearly displays fatigue striations . These fatigue cracks were nucleated at the external surface, fig.7. However, the resulting fracture mechanism responsible for crack propagation and reflected in the final fracture is dominantly governed by the creep process. The fatigue and creep interaction mechanisms are involved only in the crack nucleation process. Fatigue cracks were not observed in the case of program B testing . The fracture surface morphology of all the tested specimens had the appearence of only creep fracture morphology. There was no evidence of cyclic frequency change in the crack nucleation process. In all cases the character of the failures is substantially affected by the defects present in the structure, such as incipient meltings

Figure 6. Dislocation structure after cyclic creep. 200 Hz. B Test.

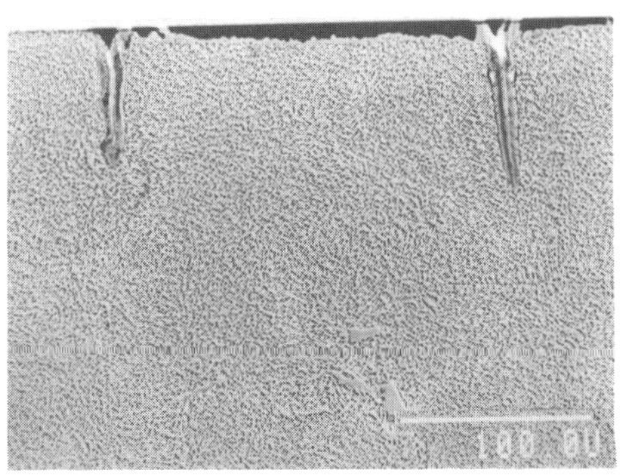

Figure 7. Cracks initiated by fatigue stress component at specimen surface.

and eutectic residue which act as initiators of cleavage cracks growing perpendicularly to the direction of the applied stress direction. These heterogeneities present in the structure of single crystal as-cast and heat treated specimens affect to a considerable degree the deformation process and no doubt lead to the ultimate decrease of the time to fracture.

The deformation process in the case of pure fatigue or cyclic creep with shorter hold times resulted in only a very small strain before rupture and shows a linear time dependence. The total deformation in the case of creep and/or cyclic creep with longer hold time is sufficiently larger and does not exhibit a linear dependence. This fact might be related to the mechanism controlling the deformation process under creep/fatigue conditions. Considering creep, only applied permanent stress results in an increased dislocation nucleation in the matrix. At sufficiently increased stress levels, dislocation gliding takes place in the gamma prime precipitates. Therefore the deformation process corresponding to the tertiary creep stage will be experienced in the matrix and gamma prime precipitates simultaneously, and the dense dislocation networks that form throughout the specimen result in higher total deformation. In the case where the cyclic stress component was included in the creep deformation process, this might be responsible for greater stress relaxation and creep recovery. The deformation is localized only in the matrix and is reflected in a reduction plasticity. The results of the effect of frequency at different values of amplitude ratio, showed that the deformation behaviour of the alloy displays only the creeplike characteristics. The presence of higher frequencies introduced in test greatly accelerated the fracture process, which caused a reduction in fracture time. The crack initiation did not show any particular creep fatigue interaction in morphology on the fracture facets. The deformation mechanism bears the marks of high temperature creep deformation and of high applied stress.

CONCLUSIONS

From the study of cyclic creep behaviour and the microstructure investigation of a single crystal nickel base superalloy the following conclusions were obtained:
1) The creep/fatigue interaction of single crystal nickel base superalloy has been studied at 900 ^{0}C and under two different loading schedules.
2) It was observed that different interaction behaviour occur under different kinds of testing procedures. In the case of continuous cyclic creep with different hold times at maximum cyclic amplitude the summarized tensile hold time is inversely proportional to the time to rupture.
3) Introducing the cycling frequency and variation in cycle amplitude ratio shortened drastically the time to rupture.

Acknowledgement

This work has been undertaken as part of a collaborative program between Faculty of Metallurgy, TU Košice, IMRAS Shenyang, and DUT Dalian, China. The support received from all organizers is gratefully acknowledged. The authors also wish to record our thanks to Chinese colleagues who have contributed to this program in any way.

REFERENCES

1. E.G. Ellison, W.J. Plumbridge: Fatigue Fract.Eng.Mat.Struct. 14,7,(1993),401
2. X. Wang, D.N. Wang, Q.P. Kong: Sripta Metall. et Materialia,28, 28,(1993),401
3. Z. Mu, K. Bothe, V. Gerald:Scripta Metall. et Materialia,24, (1990),2145
4. K. Kakechi, T. Sakaki, T. Adochi:Proc.of the 7th JIMIS, Nagoya, Japan, (1992),535
5. P.Lukaš:Proc.of Int.Conf.on Fatigue,"Fatigue 90", Havaii, (1990),1725
6. B.E. Powell, T.V. Duggan, R. Jeal: Int.J.Fatigue, 1,(1993),4
7. J. Zrnik, Z.G. Wang, M.Žitňansky, M. Hazlinger: J. of Mater. Sci & Techn., 11,1,(1995),5

AN ELECTROCHEMICAL MODEL FOR ENVIRONMENTAL CRACKING OF METALS

Raicho Raicheff

Department of Electrochemistry and Corrosion
Higher Institute of Chemical Technology
Sofia 1156, Bulgaria

INTRODUCTION

Environmental cracking is a term used to encompass all types of cracking that occur as a result of interaction of a metal and its environment, such as stress-corrosion cracking, corrosion fatigue, liquid metal embrittlement, etc. From an engineering viewpoint, the most interesting and dangerous corrosion failure of a metal component is obviously the stress-corrosion cracking (SCC) - a phenomenon associated with a combination of static tensile stress and a specific corrosive environment.

One of the most durable concepts in the SCC research is the strain-assisted dissolution mechanism of cracking[1-9]. In particular, SCC is considered to result from an increased and highly localised dissolution of the metal under stress. Thus, the crack propagates because of the very rapid dissolution of the metal at the crack tip, while the dissolution rate at the outer metal surface is much slower. The function of the stress in SCC is to create local plastic deformation, thus providing active surface sites for such a localised dissolution. The activity and the number of the surface sites that favour dissolution at the crack tip depend on the stress (strain) conditions in that region. Therefore, the effect of the applied stress on the rate of metal dissolution could be expressed by the rate constant, i.e by the exchange current density of the anodic dissolution reaction at the crack tip, which is proportional to the number of the electrochemically active surface sites and depends on their activity (i.e the electrochemical free energy of activation). Once the crack is formed by localised dissolution, the stress concentration at the crack tip keeps the crack tip surface continuously active for dissolution process. Concerns about the possible saturation of metal ions within the crack are shown[7] to be unfounded for many systems, especially when there is an active path at the crack tip and the crack walls are reasonably inactive.

In case of metals in an "active" state (in the absence of a protective film) it was shown[10,11] that stresses in excess of the yield stress may produce locally enhanced activity at surfaces where slip-steps emerge (e.g. because of the exposure of high-idex planes and edges at the slip-steps) and the dissolution current density at such surface sites may be increased by several order of magnitude in comparison to the rest of the surface. It was also suggested[2-4], that the dissolution at the film-free crack tip could be enhanced by the strain concentration because of an increase in the active site density (e.g. kink sites) or a decrease in the activation energy for dissolution.

In case of metals in a "passive" state (in the film-forming environments) the stress may act to disrupt the protective film formed on the metal surface to expose bare metal, thus providing localised dissolution with high current density at the exposed surfaces[1,5-9]. The function of the stress in such systems is also to prevent the dissolution reaction at the crack tip from becoming stifled by film formation.

The strain - assisted dissolution mechanism has been applied to a wide variety of SCC systems, both transgranular and intergranular, and has assumed great importance in the practice[5-7]. As reliability of the engineering structures becomes more and more important, it is becoming essential to develop a data base that will make possible a predictive capability for metal-environment interaction.

The aim of the paper is to present an electrochemical model of crack propagation in metals based on the localised anodic dissolution concept in SCC and to examine the effect of the main electrochemical parameters on the rate of crack propagation. Metals in "active" state only are considered in this communication.

MODEL FOR THE RATE OF CRACK PROPAGATION

If the crack is assumed to grow by anodic dissolution, the rate of crack propagation, V, could be related to the anodic dissolution current densities at the crack tip, i_{MC}, and at the outer metal surface, i_M, as shown by Despic and Raicheff[12], i.e.

$$V = \frac{A}{zF\rho}(i_{MC} - i_M) \tag{1}$$

where A is the atomic weight of the metal; z is the number of electrons transferred per atom dissolved; F is the Faraday's constant and ρ is the density of the dissolving metal.

Actually the crack propagation rate will be dependent not only on the electrochemical activities of phases being dissolved but also on the associated cathodic reaction which need to produce a current sufficient to balance the anodic currents at the crack tip and the rest of the surface.

In case of dissolution of metals in an "active" state, the current densities i_{MC} and i_M are governed by well-known relationships of activation controlled electrode kinetics[13] and are related to the corrosion potential, E_{corr}, by

$$i_{MC} = (i_o)_{MC} \exp[\frac{\alpha_M F}{RT}(E_{corr} - E_M)] \tag{2}$$

$$i_M = (i_o)_M \exp[\frac{\alpha_M F}{RT}(E_{corr} - E_M)] \tag{3}$$

where $(i_o)_{MC}$ and $(i_o)_M$ are exchange current densities for the metal dissolution reaction at the crack tip and at the outer surface; α_M is the transfer coefficient and E_M is the reference potential to which the i_o pertains (e.i the reversible potential of the metal). As it was already pointed out, the effect of the applied stress in the strain-assisted dissolution mechanism of SCC is expressed by the exchange current density of the metal dissolution reaction, therefore the values of $(i_o)_{MC}$ and $(i_o)_M$ may differ considerably.

The rates of the associated depolarization reaction at the crack tip, i_{DC}, and the outer metal surface, i_D, appear to be practically the same[11]. Assuming that the depolarization reaction is governed by a similar law of activation-controlled kinetics, i.e.

$$i_{DC} \approx i_D = (i_o)_D \exp[-\frac{\alpha_D F}{RT}(E_{corr} - E_D)] \quad (4)$$

where $(i_o)_D$, α_D and E_D are respectively the exchange current density, the transfer coefficient and the reversible potential for the depolarization reaction.

On the other hand, the corrosion potential, E_{corr}, could be obtained from the equality of the total anodic current and the current of the depolarization reaction, as

$$\exp\frac{FE_{corr}}{RT} = \{\frac{F}{RT}[\frac{\alpha_M E_M + \alpha_D E_D}{\alpha_M + \alpha_D}]\}[\frac{(i_o)_D(P' + P'')}{(i_o)_{MC}P' + (i_o)_M P''}]^{\frac{1}{\alpha_M + \alpha_D}} \quad (5)$$

where P′ and P″ are the surface areas of the crack tips and the rest of the surface respectively.

When equation (5) is introduced into (2) and (3), and those replaced in equation (1), maintaining that $P' << P''$, one obtains for the rate of crack propagation

$$V = \frac{A}{zF\rho}K[\frac{(i_o)_D}{(i_o)_M}]^Q[(i_o)_{MC} - (i_o)_M] \quad (6)$$

with $$K = \exp[\frac{F}{RT}(\frac{\alpha_D\alpha_M}{\alpha_D + \alpha_M})(E_D - E_M)] \quad (7)$$

and $$Q = \frac{\alpha_M}{\alpha_M + \alpha_D} \quad (8)$$

DEPENDENCE OF THE CRACK PROPAGATION RATE ON THE ELECTROCHEMICAL PARAMETERS

The crack propagation rate depends on the basic electrochemical parameters of the partial corrosion reactions. It is seen from equations (6), (7) and (8) that this rate is basically determined by the difference in the exchange current densities of the metal dissolution at the crack tip and at the other surface. As this difference becomes smaller, the tendency to SCC decreases and may even vanish. However, one should note that this tendency depends also on the ratio between the exchange current densities of the depolarization and metal dissolution reactions at the outer metal surface, and especially on the difference in the reversible potentials of the cathodic and anodic reactions.

These conclusions are in good agreement with many observations in SCC. Thus, it is well known that mild steels undergo SCC in hot NH_4NO_3 solutions in the active potentials region at a rate of 6.10^{-8}-5.10^{-7} m/s (0.2-2 mm/h)[8]. A value of 8.10^{-8} m/s (0.3 mm/h) is obtained for the crack propagation rate using equation (6) and the following parameters[13,14]: $\alpha_D=\alpha_M=0.5$, $E_D=0.92$ and $E_M=-0.44$ V, $(i_o)_D=10^{-10}$, $(i_o)_M=10^{-8}$ and $(i_o)_{MC}=10^{-6}$ A/cm^2 (cf. also[11]).

Qualitative examples are found in other systems. Thus, Mg-Al alloys are reported[15] to crack in aerated NaCl solutions and the tendency to SCC increases greatly in the presence of H_2O_2. However, these alloys do not undergo SCC in corresponding acid solutions where hydrogen evolution reaction should develop as cathodic (depolarization) reaction. This is in accordance with the much higher (about 10^4 times) K value in the former case, due to the more positive E_D which should overwhelm a possibly smaller value of $(i_o)_D$ for H_2O_2 reduction reaction as compared to that of hydrogen evolution reaction.

Equation (6) indicates also that if strain produces similar effects in two different metals (increase in $(i_o)_{MC}$ values), they still may differ in their tendency to SCC in the same medium by virtue of a different degree of inhibition of the depolarization reaction (different $(i_o)_D$ values). Thus, the introduction of carbon in iron is known to increase greatly the $(i_o)_D$ for both oxygen reduction and hydrogen evolution reactions, however it does not practically affect the $(i_o)_M$. Indeed, while pure iron shows no notable susceptibility (the high-purity iron is even immune) to SCC, the impure iron and mild steel do so and the tendency to SCC grows with increase of the carbon content in the steel[8].

As the activity of the metal surface for dissolution process increases (i.e $(i_o)_M$ value is increased) the rate V will decrease. If the surface exhibits a high activity, then the localisation of the corrosion attack necessary for crack initiation is not possible and the tendency to SCC vanishes. This conclusion is in accordance with some findings[16] that SCC of austenitic stainless steel does not occur if the corrosion rate at the metal surface is higher than 0.2-0.25 mm/year.

The electrochemical model could be applied to develop data base for the environment effect in SCC systems. The crack propagation rate at various electrochemical conditions was calculated using equation (6) and literature data[13,14] for the electrochemical parameters E_D, E_M, $(i_o)_{MC}$, $(i_o)_M$ and $(i_o)_D$, assuming that $\alpha_M = \alpha_D = 0{,}5$. Three-dimensional graphic plots of the rate V as a function of the ratio $(i_o)_D/(i_o)_M$, the difference $(E_D - E_M)$ and the $(i_o)_{MC}$ were also obtained. The analysis of the literature data for the relevant metal systems revealed that those parameters change within the following ranges: $(i_o)_D/(i_o)_M = 10^{-2} - 10^2$, $(E_D - E_M) = 0 - 2.0$ V and $(i_o)_{MC} = 10^{-7} - 10^{-3}$ A/cm^2.

Some typical plots of the crack propagation rate are shown in Fig.1 and Fig.2 for two different rates of localised dissolution. From this type of plots it is possible to determine the electrochemical conditions which are favourable or not for SCC. Thus, one can obtain the range of $(E_D - E_M)$ values of very small crack propagation rates (e.g. bellow 10^{-8}m/s) i.e the electrochemical condition of a very low SCC susceptibility. This range depends strongly on the $(i_o)_{MC}$ and it expands with decrease of the ratio $(i_o)_D/(i_o)_M$. The value of the difference in the potentials $(E_D - E_M)$ where the crack velocity is sufficiently low (lower than 10^{-8} m/s) changes from 1.2 V to 0.2 V with the increase of the $(i_o)_{MC}$ from 10^{-7} to 10^{-3} A/cm^2 in the whole possible range ($10^{-2} - 10^2$) of the ratio $(i_o)_D/(i_o)_M$.

The results in Fig.1 and Fig.2 indicate that at low localised dissolution rates (e.i small $(i_o)_{MC}$ values), the electrochemical conditions would be favourable for SCC only for metals with very negative potentials (e.g. Al, Ti, Mg, Zn and their alloys) in corrosive environments only where the oxygen reduction would develop as depolarization reaction (large $E_D - E_M$ values). The tendency to SCC will be higher the larger the ratio between the exchange current densities of the depolarization reaction and the metal dissolution reaction. At higher rates of the localised dissolution, however, the electrochemical conditions would be favourable for SCC of a wide range of metals (e.g. Al, Ti, Mg, Zn, Cr, Fe, Ni, Cu etc.) corroding with oxygen reduction as well with hydrogen evolution (e.i in neutral as well in acidic environments). In this case the tendency to SCC will be also higher the larger the ratio $(i_o)_D/(i_o)_M$.

The results of the present study indicate that metals corroding with oxygen reduction would be generally more susceptible to SCC in comparison to metals corroding with hydrogen evolution since they need a localised corrosion attack with much lower rate for developing of a crack. For the same reason, a susceptible to SCC metal will show a higher

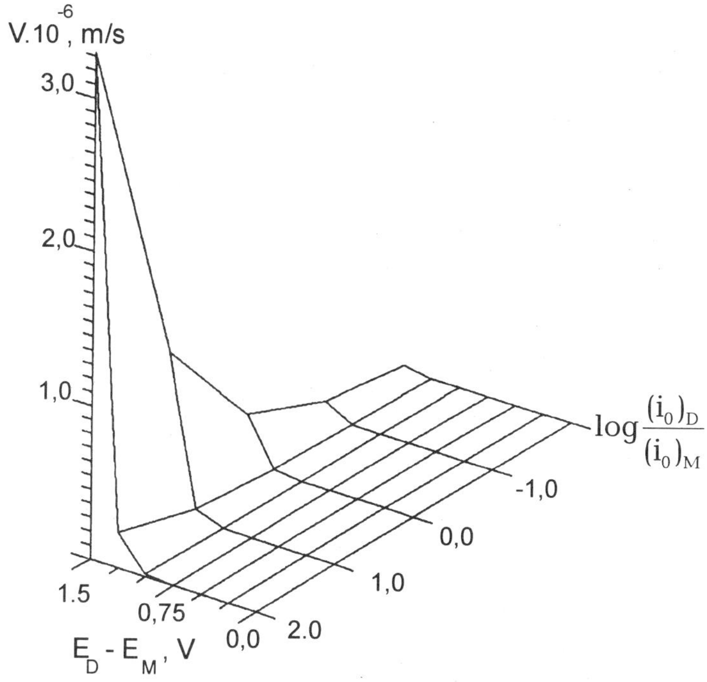

Figure 1. Dependence of the crack propagation rate V on $(i_o)_D/(i_o)_M$ and E_D- E_M at $(i_o)_{MC}$=10^{-7} A/cm^2

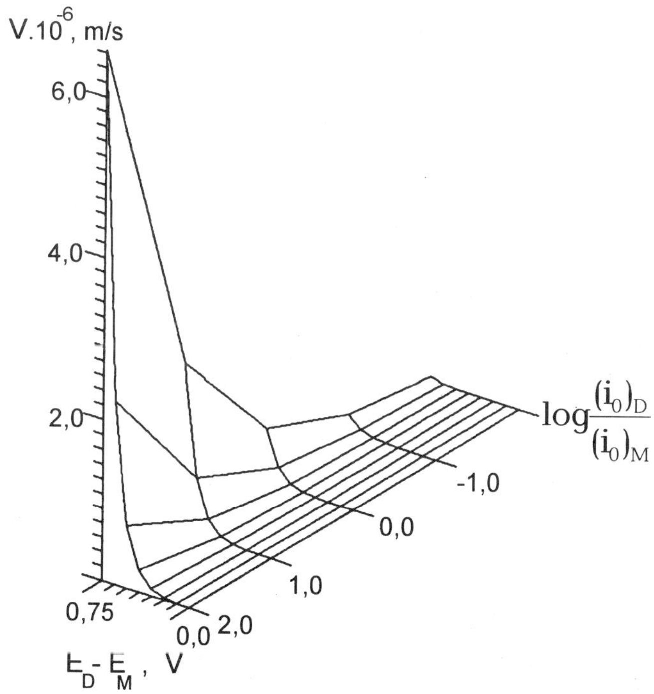

Figure 2. Dependence of the crack propagation rate V on $(i_o)_D/(i_o)_M$ and E_D- E_M at $(i_o)_{MC}$=10^{-3} A/cm^2

tendency for SCC in environments where the oxygen reduction is depolarization reaction in comparison to environments with hydrogen evolution as cathodic reaction.

CONCLUSION

An electrochemical model of crack propagation based on the strain-assisted dissolution mechanism of SCC of metals is presented and discussed. Using a general expression relating the rate of crack propagation and the electrochemical parameters of metal -environment systems, it is shown that the crack propagation rate is basically determined by the difference in the rate constants of the metal dissolution at the crack tip and at the outer metal surface, the ratio between the rate constant of the depolarization reaction and that of the metal dissolution reaction at the outer surface, and the difference in the reversible potentials of the two corrosion reactions.

The model is shown to have a fair degree of success in: a) calculating the crack propagation rate as observed experimentally, b) explaining the differences in behaviour of various metal-environment systems with respect to a tendency for SCC, and c) defining the electrochemical conditions favourable for SCC.

Acknowledgments

The financial support of the Bulgarian National Scientific Fund for this work is gratefully acknowleged.

REFERENCES

1. H.L.Logan, Film-rupture mechanism of stress-corrosion, *J. Research N.B.S.* 48:99 (1952).
2. T.P.Hoar and J.M.West, Mechanochemical anodic dissolution, *Nature* 181:835 (1958).
3. T.P.Hoar and J.M.West, Mechanochemical anodic dissolution of austenitic stainless steels in hot chloride solution, *Proc. Roy. Soc.* A268:304 (1962).
4. T.P.Hoar, Stress-corrosion cracking, *Corrosion* 19:331 (1963).
5. F.P.Ford, Stress-corrosion cracking, in:"Corrosion Processes", R.N.Parkins, ed., Applied Science Publishers, London (1982).
6. R.N.Parkins, Stress-corrosion cracking, in:"Environment-Induced Cracking of Metals", R.P.Gangloff and M.B.Ives, eds., NACE, Houston (1990).
7. R.C.Newman and M.Saito, Anodic stress-corrosion cracking: slip-dissolution and film-induced cleavage, in:"Corrosion-Deformation Interactions", J.M.Gras and T.Magnin, eds., Les Editions de Physique, Les Ulis (1993).
8. R.N.Parkins, Mechanisms of stress-corrosion cracking, in:"Corrosion", vol.1, L.L.Shreir, R.A.Jarman and G.T.Burstein, eds., Butterworths, London (1994).
9. R.W.Revie, Environmental cracking ot metals, in:"Modern Aspects of Electrochemistry", vol.26, B.E.Conway, J.O'M.Bockris and R.E.White, eds., Plenum Press, New York (1994).
10. R.G.Raicheff, A.Damjanovic and J.O'M.Bockris, Dependence of the velocity of anodic dissolution of iron on its yield rate under tension, *J. Chem. Phys.* 47:2198 (1967).
11. A.R.Despic, R.G.Raicheff and J.O'M.Bockris, The mechanism of the acceleration of the electrodic dissolution of metals during yielding under stress, *J. Chem. Phys.* 49:926 (1968).
12. A.R.Despic and R.G.Raicheff, The crack-propagation rate in terms of the electrochemical mechanism of stress-corrosion cracking of metals, *Compt. Rend. Acad. Bulg. Sci.* 24:1361 (1971).
13. J.O'M.Bockris, Electrode kinetics, in:" Modern Aspects of Electrochemistry", vol.1, J.O'M.Bockris, ed., Butterworths, London (1954).
14. N.Tanaka and R.Tamamushi, Kinetic parameters of electrode reactions, *Electrochim. Acta* 9:963 (1964).
15. H.L.Logan. "The Stress Corrosion of Metals", J.Wiley, New York (1968).
16. N.J.Petch, The lowering of the fracture-stress due to surface adsorption, *Phil. Mag.* 1:331 (1956).

INSTABILITY PHENOMENA IN Al-Zn ALLOYS

Ryszard Ciach

Institute of Metallurgy and Materials Science, Polish Academy of Sciences,
Regional Laboratory of Physicochemical Analyses and Structural Research,
Jagiellonian University,
25 Reymonta St. 30-059 Kraków, Poland

INTRODUCTION

Relatively simple alloys of aluminium and zinc reveal a long-term instability of mechanical and physical properties during storage at room or elevated temperatures. Such behaviour is due to:

A) - the solidification process, when the nonequilibrium ($\alpha' + \eta$) eutectic appears and the composition of the α' phase dendrites varies from 10 up to 68% of zinc. The eutectic can be

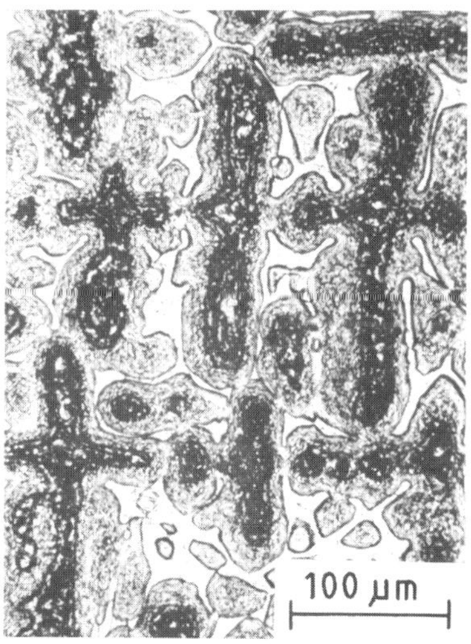

Fig. 1a. Optical micrograph of dendritic structure of cast Al-60wt.%Zn alloy.

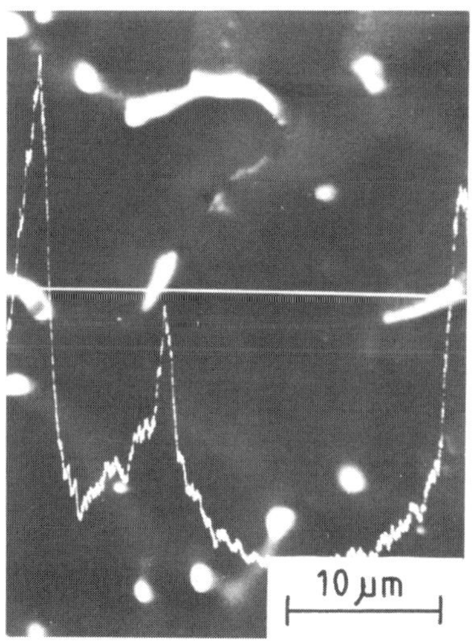

Fig. 1b. Scanning electron microstructure of aged Al-40wt.%Zn alloy together with Zn distribution.

Stability of Materials
Edited by A. Gonis *et al.*, Plenum Press, New York, 1996

observed in Fig. 1a as the dark areas, while the Zn distribution across the grain is visible in Fig. 1b. The theoretical distribution of zinc in the dendrite of Al- 60 wt%Zn alloy is shown in Fig. 2 and the amounts of equilibrium and nonequilibrium eutectics, calculated and measured experimentally, in the Al-Zn alloys is shown in Fig. 3.

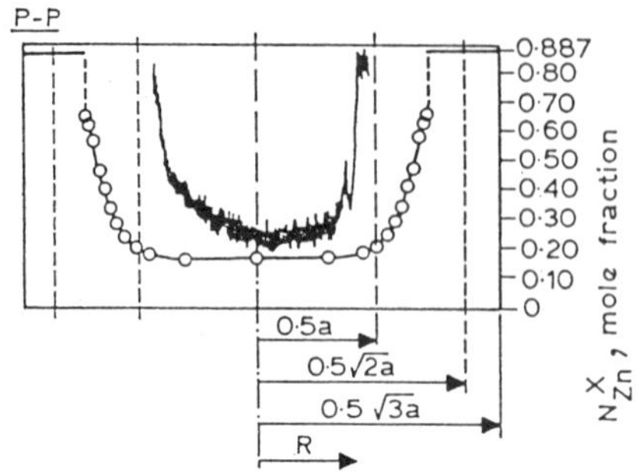

Fig. 2. Theoretical and experimental distributions of zinc in a dendrite of Al- 60 wt%Zn alloy.

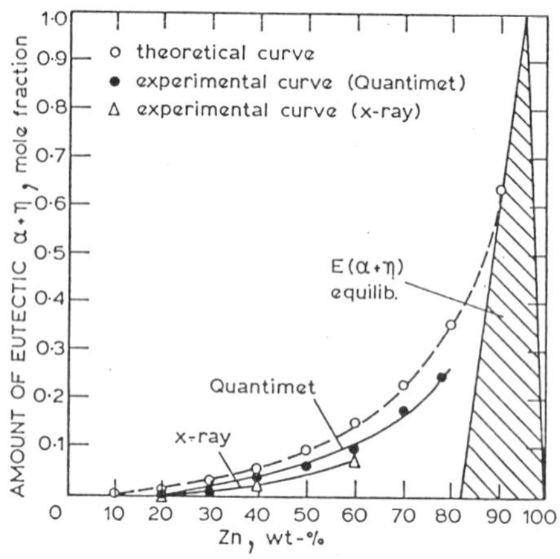

Fig.3. Theoretical and experimental amounts of (α+η) eutectic in cast AlZn alloys.

B) - the kinetics of decomposition of the supercooled zinc - rich α' phase of the fcc lattice into stable α and η phases which depends on Zn content in such a way that the α' phase containing 78% of zinc decomposes at room temperature during 3 hours while the phase of 30% Zn needs almost 3 years to decompose (Fig. 4).

Thus, the cast AlZn alloys consist of nonequilibrium (α'+η) eutectics and the solid solution enriched in zinc. Such long-term processes are difficult to identify[13].

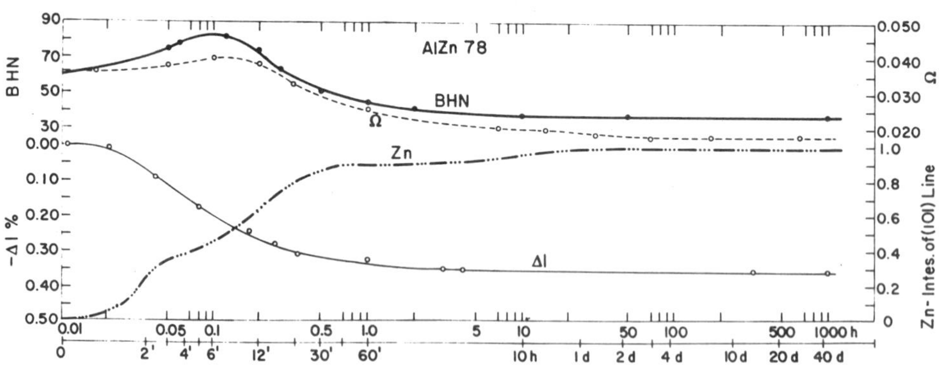

Fig. 4. Changes of hardness (BHN), electrical resistance (Ω), dimensions (Δl) and amounts of stable η phase (Zn) during room temperature decomposition of the supercooled α phase in Al-78wt%Zn alloy.

THE THERMODYNAMIC BASIS OF INSTABILITIES IN AlZn ALLOYS

The thermodynamic properties and the changes of free energy of the α and η phases as a function of zinc content and temperature were determined based on Krupkowski's method for estimating the activity coefficients of the alloy components in the solution, according to his solution of the Gibbs - Duhem equation [1]. The coefficients were required to calculate the chemical potentials of aluminium and zinc in the solution. The following equations have been obtained:

$$\ln\gamma_{Al} = \omega_1\left(N_{Al}^{m} - \frac{m}{m-1}N_{Al}^{m-1} + \frac{1}{m-1}\right) \quad \text{and} \quad \ln\gamma_{Zn} = \omega_2 N_{A}^{m} \qquad \text{(1) and (2)}$$

where ω is a function which depends only on temperature and **m** is characteristic for a given alloy[6].

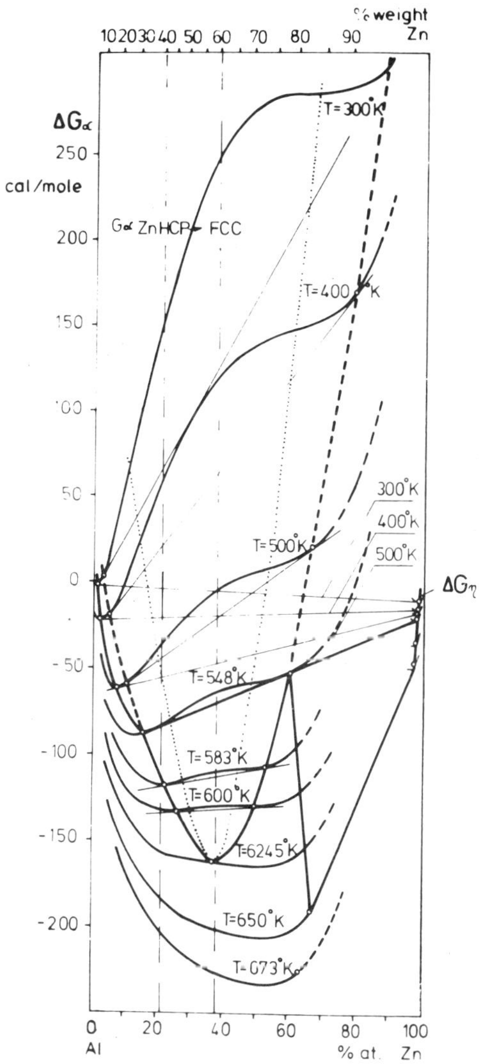

Fig. 5. The free energy changes of α phase in AlZn alloys at various temperatures including the energy of transition of hexagonal zinc into fcc lattice. Solid lines refer to the stable phase and the dotted ones to the metastable states.

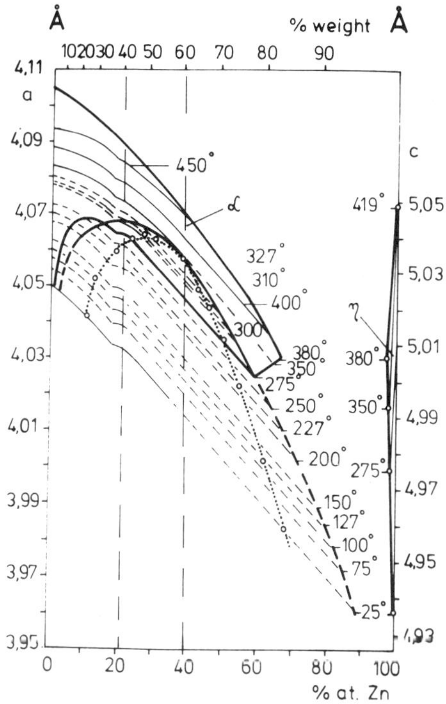

Fig. 6. Changes of parameters of α and η phases in AlZn alloys versus composition and temperature after Lohberg[8], Ellwood[9], Ciach[10].

Having calculated the activity coefficients, the changes of chemical potential of Al and Zn in the solid solution can be obtained by the following equation (3):

$$\Delta G_1 = RT\ln\gamma_1 + RT\ln N_1 \tag{3}$$

where N is the concentration of a given component. However, the changes established in this way do not include the energy of transition of the zinc hexagonal lattice into the fcc one. Such estimation was carried out by the author according to numerous papers[2-5], and how they affect the free energy changes can be seen in Fig.5, where a complete plot of the free energy changes for the α phase has been drawn as a function of composition and temperature for stable as well as metastable states. The plot has been supplemented additionally with free energy changes of the η phase. The large differences in free energies of the supercooled α phase and the stable alloy at higher zinc contents explain the growing tendency to decomposition with the amount of zinc in the α phase. The phenomenon has been observed and examined in the full range of α phase existence by Ciach et al[5,6] and Krupkowski[7]. The plot explains also the long homogenisation times of Al alloys with 60 - 70 wt.% Zn that are due to flattening of the free energy curves at relevant temperatures.

THE MECHANISM OF STRUCTURE TRANSFORMATIONS

The changes of α and η phase free energies, calculated above and presented in Fig.5, as a function of Zn content and temperature allow us to determine the composition of the metastable and stable α - type phases. The lattice parameters of the stable and metastable incoherent phases were estimated according to the results of Lohberg[8], Ellwood[9], Ciach [10] and Popovic[14] by means of relevant extrapolations which can be seen in Fig.6. The plot allows us to establish compositions of all metastable incoherent phases of the α type which occur during the decomposition.

The real course of the process is quite complicated. The first precipitates observed during the decomposition are spherical Zn - rich zones which are coherent with the matrix. They grow along one of the {111} planes changing shape from spherical into ellipsoidal. The anisotropy of coherency results in a decrease of lattice parameters mostly in the [111] growth direction. The maximum differentiation of lattice parameters appears when the Zn - rich zones become plate - like. They partially lose their coherence with the matrix except for the (111) habit plane. This ensures continuity of the {100} planes of the matrix and flat particles which preserve the size of the lattice parameter in the particles. Such a mechanism is responsible for the rhombohedral deformation of the zinc - rich precipitates. Based on these assumptions the a_m matrix lattice parameters, considered as constant in the zinc - rich precipitates, is the edge of the rhombohedron which is being flattened in the $\langle 111\rangle$ direction, or in the [001] one parallel to the $c_{\alpha'm}$ parameter, if a hexagonal lattice is referred to [11].

The a_m and $c_{\alpha'm}$ values obtained in this way should determine the limit of the rhombohedral deformation of the zinc rich precipitates. The a_m value for G.P.zones will be equal to the lattice parameter of the α phase because the lattice parameter of the matrix does not change in the presence of coherent precipitates.

The first attempt to link the rhombohedral deformation with composition and the temperature of transformation was made by Ciach[10]. He described the rhombohedral distortion angle α_r as a function of the c/a_h ratio of lattice parameters (equation 4). Eventually, describing the rhombohedral structure in the hexagonal lattice, the deformation angle was found to be as in equation 5:

$$\sin\frac{\alpha_r}{2} = \frac{3}{2\sqrt{3+\left(\frac{c}{a_h}\right)^2}} \quad \text{and} \quad \sin\frac{\alpha_r}{2} = \frac{3}{2\sqrt{2.4496+\left(\frac{a_m}{a_{\alpha'_m}}\right)^2}} \qquad (4) \text{ and } (5)$$

where c is the interplanar distance of {111} planes and a_h is the interplanar distance of {110} planes. Equation (5) allows us to describe the dependence of the rhombohedral deformation of the transition precipitates on temperature and α phase composition.

Equation (5) was verified in x-ray experiments on mono- and polycrystals. The monocrystal experiments were carried out in such a way that only {111} planes or those parallel to them diffracted the x-rays, while reflections from zinc $(10\bar{1}0)$ that could have appeared in the same angle range in polycrystals were eliminated. The comparison of both materials also allowed us to clarify the influence of grain boundaries on the rate of appearance of the Zn - rich transition phases.

INFLUENCE OF COPPER ON STABILITY OF AlZn ALLOYS

Copper is added to AlZn alloys in order to improve their mechanical properties. The Cu content should not exceed 2 % since it is the maximum amount that dissolves in the solid solution. Its presence in Al-60wt.%Zn alloys extends the decomposition time of the α phase to even 2 years at room temperature. The influence of copper on zinc rich alloys is much more complicated. Such alloys reveal at first high mechanical properties but they are prone to intergranular corrosion and they also exhibit instability of dimensions and increase of brittleness[15].

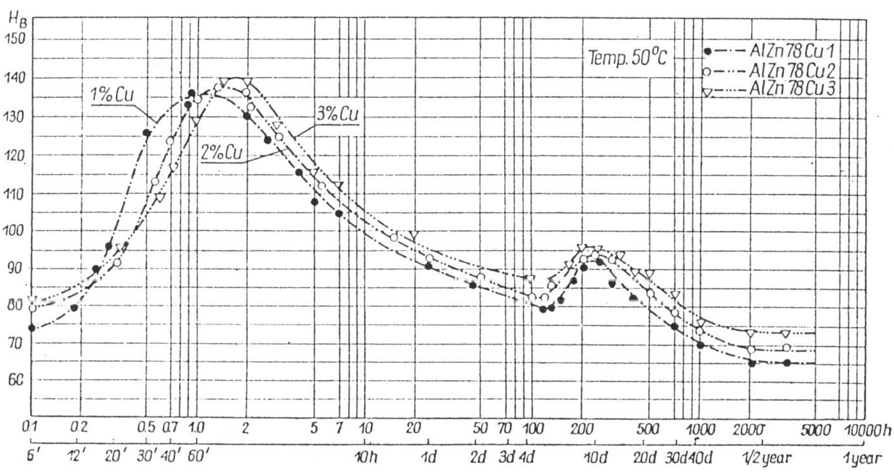

Fig.7. Dependence of hardness of the Al-78wt.%Zn-(1-3wt.%)Cu alloy on ageing at 50°C.

The phenomenon of instability was explained in terms of the so called reaction of four phases[12]: $\alpha' + \varepsilon \rightarrow T + \eta$, but the X - ray examinations[12] proved that ageing of the supersaturated ε phase is responsible for the instability. The decomposition of the supercooled α' phase also occurs but only in the first stage of ageing. How these transformations affect the hardness is shown in Fig.7.

CONCLUSIONS

The Al-Zn phase diagrams which present the dependence of the free energy and lattice parameters on composition and temperature allowed us to:

- establish compositions and structures of metastable incoherent precipitates of the α type.
- calculate the maximum rhombohedral deformation of partially coherent metastable zinc rich precipitates that varies with temperature and composition of the alloy.
- explain from the energetic point of view large differences in decomposition times of the supercooled α phase with Zn content, as well as problems with homogenisation of alloys containing 60 - 70 wt.% Zn.

In Al-Zn-Cu alloys the supersaturated ε (Al,Zn,Cu) phase occurs. During its ageing, the T and η phases precipitate and cause the increase of dimensions, hardness and brittleness of the alloys.

REFERENCES

1. A. Krupkowski. Thermodynamics (in Polish), PWN, Kraków, 1958.
2. R.D. Garwood, A.L. Davies, G,L. Richards, A transition phase in the ageing of Al-Zn alloys. J. Inst. Metals 88:375 (1959/60).
3. M. Simerska, V. Synecek, The mechanism of structure transformation in supersaturated Al-Zn alloys. Acta Met. 15:223 (1967).
4. G.J.C. Carpenter, R.D. Garwood, The ageing of a quenched Al-22.5%Zn alloy. Met. Sci. J. 1:202 (1967).
5. R. Ciach, The kinetics of decomposition of the supercooled α' phase in Al-Zn alloys. Proc. V Conf. Metall. PAS, Kraków, 38 (1968).
6. R. Ciach, Ph.D. Thesis, The effect of solidification on mechanical and corrosion properties of Al-Zn alloys. (in Polish) IPPT PAN, Warszawa, 1964.
7. A. Krupkowski, R Ciach, Appearance and decomposition of the α phase in cast Al-60Zn-2Cu alloy (in Polish) Arch. Metall. 3:199 (19660.
8. K. Lohberg, Z. Metallk. 32:87 (1940).
9. E.C. Ellwood, The solid solutions of zinc in aluminium.J.Ins. Metals, 80:217 (1951/52).
10. R. Ciach, Effect of the zinc content on the rhombohedral distortion of the transition phase in Al-Zn alloys. Scripta Met. 2:575 (1968).
11. R. Ciach, Rhombohedral deformation of metastable precipitates in Al - Zn alloys. (in Polish) Arch. Metall. 15:55 (1970).
12. R. Ciach, J. Król. K. Węgrzyn-Tasior, Studies on four phases transformation in AlZn78 alloy containing 1 - 3 per cent of copper. Bull. Acad Sci. XVII:13 (1969).
13. A.Gaber N. Afify, Investigation of the decomposition behaviour in supersaturated Al-Zn alloys. J. Mater. Sci. 27, 5:1347 (1992).
14. S. Popovic, Lattice constant of the fcc Al-rich α phase of Al-Zn alloys in equilibrium with GP zones and the β phase. Phys.Stat.Solidi (a), 130,2:273 (1992).
15. H. Loffler, Structure and development of Al-Zn alloys. ed. Akademie Verlag, Berlin 1995.

THE EFFECT OF ALLOYING ADDITIONS ON THE INTERFACIAL INTERACTIONS AT THE Fe-Al INTERFACE DURING COATING

Amdulla O. Mekhrabov *, M.Vedat Akdeniz and Ilhami Aktürk

TÜBITAK SEKOM UNIT
Metallurgical Engineering Department
Middle East Technical University, 06531 Ankara, Turkey

*Permanent Address: Faculty of Physics, Baku State University
370073-Baku, Azerbaijan

1. INTRODUCTION

New phases or compounds, mainly intermetallics, appear if two different materials are in contact at sufficiently high temperatures, i.e. during joining operation, sintering, coating and other surface treatments. The formation and development of new intermetallic phases in those treatments depend on the range of interfacial interactions of the constituent materials which are governed by the crystal structure, solid solubility limit, the magnitude and sign of interatomic interaction potentials between the atoms of two different materials and also their interdiffusion coefficient.

It is understandable that the atomistic description of the formation, growth characteristics and the thickness prediction of the phases at the interface is very complex problem and require the solution of the complicated diffusional transport and/or chemical reaction equations with many unknown chemical and physical parameters[1-4]. The magnitude of interatomic interaction potentials of various atomic pairs at a given temperature seems to be an important parameter in determining of the characteristics of the interfacial interaction. It would be possible to govern the formation and growth of new phases, consequently their thickness at the interface by using right impurities or alloying additions as the interfacial interactions will either increase or decrease depending on the nature of matrix (substrate), coat and impurity elements[5-8]. Therefore, control of the interatomic interaction potentials of the atoms of interacting elements is of particular research interest.

In this paper, as part of an investigation on formation and growth characteristics of Fe-Al intermetallics in aluminized diffusion layer, the effects of the type and the content of the impurity elements on the range of interfacial interactions in Fe-Al system are presented in terms of statistico- thermodynamic theory of multicomponent alloys combined with electronic theory in pseudopotentials approximation.

2.THE THEORY

The calculation of the thermodynamic properties of metallic multicomponent solid solutions at the concentration range near the concentration of pure solvent elements are of considerable interest. The thermodynamic theory of these systems was developed by Vagner[9] based on Taylor series expansion of molar excess free energy of the component **i** over molar concentration of all solute elements in the solid solution and the activity coefficient of component **i**, γ_i, in the multicomponent solid solution can be written as

$$\ln\gamma_i = \ln\gamma_i^o + \sum \varepsilon_i^{(j)} c_j \tag{1}$$

where

$$\ln\gamma_i = (\mu_i - \mu_i^{ideal}) / RT = \mu_i^{excess} / RT \tag{2}$$

In Eqns.(1) and (2), γ°_i is the limiting value of the activity coefficient of the solute **i** in the pure solvent, μ_i^{excess} is excess chemical potential of solute i in the solid solution, R is universal gas constant, T is absolute temperature and $\varepsilon_i^{(j)}$ are the coefficients of the series expansion. These coefficients, designated as interatomic interaction parameters, are defined as

$$\varepsilon_i^{(j)} = \left.\frac{\partial \ln\gamma_i}{\partial c_j}\right|_{c_j=0} \tag{3}$$

The method given by Vagner[9] was generalized by Lupis and Elliot[10,11] in which the high-order interaction parameters were introduced. The excess chemical potential of component i of the alloy on the base of solvent element 1 can be expressed by atomic concentration of $c_2, c_3, \ldots, c_m$ of alloying elements 2, 3, ..., m in the following form[10]

$$\mu_i^{excess} = RT \sum_{n_2} \sum_{n_3} \ldots \sum_{n_m} I^{(i)}_{n_2, n_3, \ldots n_m} C_2^{n_2} C_3^{n_3} \ldots C_n^{n_m} \tag{4}$$

where $I^{(i)}_{n_2,n_3,\ldots n_m}$ are the expansion coefficients which are also called the interaction coefficient of order $n = \sum_{j=2}^{m} n_j$. Each of the indices $n_2, n_3, \ldots, n_m$ runs through all values of the natural series of numbers including zero.

The main factor of solubity processes of coat elements in the matrix is the difference of chemical potentials of coat atoms in the matrix and in the coat element itself. The impurities which decrease the chemical potential of coat atoms in the matrix should increase the rate of solubility of coat element in the matrix and vice versa. Thus, the determination of the activity coefficient requires the evaluation of interatomic interaction parameters for multicomponent system. The interatomic interaction parameters for substitutional type of solid solution can be calculated based on Krivoglaz work[12]. Following Krivoglaz we consider a ternary substitutional type solid solution ABC such that the concentration of B and C atoms are low, i.e. C_B and $C_C \ll 1$. Thus, it can be considered that the probability of forming of atomic configurations other than BB and CC are low (e.g. BBB or CCC) and can be taken as zero.

The first-order interaction parameters in terms of pairwise interaction model by taking into account of interaction of atoms at any coordination spheres ($C_B \ll 1$, $C_C \ll 1$) can be represented by the following equations

$$\varepsilon_B^{(2)} = \sum_i \varepsilon_{B_i}^{(2)} \quad ; \quad \varepsilon_{B_i}^{(2)} = Z_i\left(1 - e^{-h_{BB}(R_i)/k_B T}\right) \tag{5}$$

where

$$h_{BB}(R_i) = -[W_{AB}(R_i)] = -2V_{AB}(R_i) + V_{AA}(R_i) + V_{BB}(R_i) \tag{6}$$

$$\varepsilon_B^{(3)} = \sum_i \varepsilon_{B_i}^{(3)} \quad ; \quad \varepsilon_{B_i}^{(3)} = Z_i\left(1 - e^{-h_{BC}(R_i)/k_B T}\right) \tag{7}$$

$$\begin{aligned} h_{BC}(R_i) &= \frac{1}{2}\left[W_{BC}(R_i) - W_{AB}(R_i) - W_{AC}(R_i)\right] \\ &= V_{AA}(R_i) + V_{BC}(R_i) - V_{AB}(R_i) - V_{AC}(R_i) \end{aligned} \tag{8}$$

However, the second-order interaction parameters can be calculated based on Bolshov's work[13,14] where the correlation effects in the multicomponent solid solutions were calculated. The second order parameters were given by[13]

$$\rho_B^{(2)} = \frac{1}{2}\sum_i\sum_j\sum_k \frac{\tau_{ijk}}{Z_j Z_k}\varepsilon_{B_i}^{(2)}\varepsilon_{B_j}^{(2)}\varepsilon_{B_k}^{(2)} + \frac{3}{2}\sum_i \frac{\left(\varepsilon_{B_i}^{(2)}\right)^2}{Z_i} - \frac{1}{2}\sum_i \varepsilon_{B_i}^{(2)} \tag{9}$$

and

$$\rho_B^{(3)} = \frac{1}{2}\sum_i\sum_j\sum_k \frac{\tau_{ijk}}{Z_j Z_k}\varepsilon_{C_i}^{(3)}\varepsilon_{B_j}^{(3)}\varepsilon_{B_k}^{(3)} + \frac{1}{2}\sum_i \frac{\left(\varepsilon_{B_i}^{(3)}\right)^2}{Z_i} + \sum_i \frac{\varepsilon_{B_i}^{(3)}\varepsilon_{C_i}^{(3)}}{Z_i} - \frac{1}{2}\sum_i \varepsilon_{C_i}^{(3)} \tag{10}$$

where τ_{ijk} is the number of lattice sites at the intersections of the j-th coordination sphere of a particular site, and the k-th coordination sphere of another site in the i-th coordination sphere of the first. Summation is carried over all **i, j, k** coordination spheres .

It is seen from Eqs. (5) to (10) that calculation of the interatomic interaction parameters, $\varepsilon_i^{(j)}$, requires to calculate the partial ordering energies, $W_{\alpha\beta}(R_i)$, and pairwise interatomic interaction potentials, $V_{\alpha\beta}(R_i)$, in the ternary alloys. A suitable method for the calculation of the ordering energies and pairwise interatomic interaction potentials based on the electronic theory of multicomponent alloys in pseudopotential approximation are given elsewhere[8,15,16].

3. RESULTS AND DISCUSSION

In the present study the pairwise interatomic interaction potentials, $V_{\alpha\beta}(R_i)$, and activity coefficients, $\gamma_{Al}/\gamma^{\circ}_{Al}$, are calculated for α-Fe-Al-Me ternary alloys (Me=Si, Ti, Mg, Co or Zn) in the concentration range $0 \leq x_{Me} \leq 1$ at%. The non-local bare-pseudopotential form-factors[17,18] are used and was screened using modified dielectric constant[19,20]. The values of pairwise interatomic interaction potentials, $V_{\alpha\beta}(R_i)$, at the first three coordination spheres are given in Table 1.

Table 1. Pairwise interatomic interaction potentials of $A_{0.95}(B_{100-x}C_x)$ alloys for x=0.25 atomic concentration of Me=Si, Ti, Mg or Zn. (A=Fe, B=Al, C=Me; $x_A+x_B+x_C=100$)

Impurities	Coordination Spheres	Potentials (atomic units)				
		V_{AB}	V_{AA}	V_{BB}	V_{AC}	V_{BC}
Si	1	$2.04x10^{-3}$	$1.00x10^{-2}$	$-6.45x10^{-3}$	$3.63x10^{-3}$	$-8.61x10^{-3}$
	2	$-1.73x10^{-3}$	$2.71x10^{-4}$	$-4.07x10^{-3}$	$-1.21x10^{-3}$	$-3.53x10^{-3}$
	3	$-2.18x10^{-4}$	$-5.08x10^{-4}$	$-1.34x10^{-3}$	$-9.07x10^{-4}$	$-1.29x10^{-3}$
Ti	1	$2.04x10^{-3}$	$1.00x10^{-2}$	$-6.46x10^{-3}$	$6.84x10^{-3}$	$-6.11x10^{-3}$
	2	$-1.73x10^{-3}$	$2.70x10^{-4}$	$-4.07x10^{-3}$	$-3.75x10^{-4}$	$-2.32x10^{-3}$
	3	$-2.17x10^{-4}$	$-5.07x10^{-4}$	$-1.34x10^{-3}$	$-5.34x10^{-4}$	$-9.40x10^{-5}$
Mg	1	$2.03x10^{-3}$	$1.00x10^{-2}$	$-6.45x10^{-3}$	$-6.80x10^{-4}$	$-3.90x10^{-3}$
	2	$-1.74x10^{-3}$	$2.53x10^{-4}$	$-4.10x10^{-3}$	$-6.93x10^{-3}$	$-6.15x10^{-3}$
	3	$-2.23x10^{-4}$	$-5.14x10^{-4}$	$-1.34x10^{-3}$	$1.95x10^{-3}$	$-1.11x10^{-3}$
Co	1	$2.04x10^{-3}$	$1.00x10^{-2}$	$-6.46x10^{-3}$	$6.27x10^{-3}$	$7.04x10^{-4}$
	2	$-1.73x10^{-3}$	$2.64x10^{-4}$	$-4.07x10^{-3}$	$3.48x10^{-4}$	$-8.40x10^{-4}$
	3	$-2.18x10^{-4}$	$-5.07x10^{-4}$	$-1.34x10^{-4}$	$-4.40x10^{-4}$	$-6.08x10^{-4}$
Zn	1	$2.03x10^{-3}$	$1.00x10^{-2}$	$-6.46x10^{-3}$	$5.31x10^{-3}$	$-1.48x10^{-3}$
	2	$-1.73x10^{-3}$	$2.65x10^{-4}$	$-4.08x10^{-3}$	$7.67x10^{-4}$	$4.64x10^{-4}$
	3	$-2.20x10^{-4}$	$-5.10x10^{-4}$	$-1.34x10^{-4}$	$-3.43x10^{-4}$	$-7.95x10^{-4}$

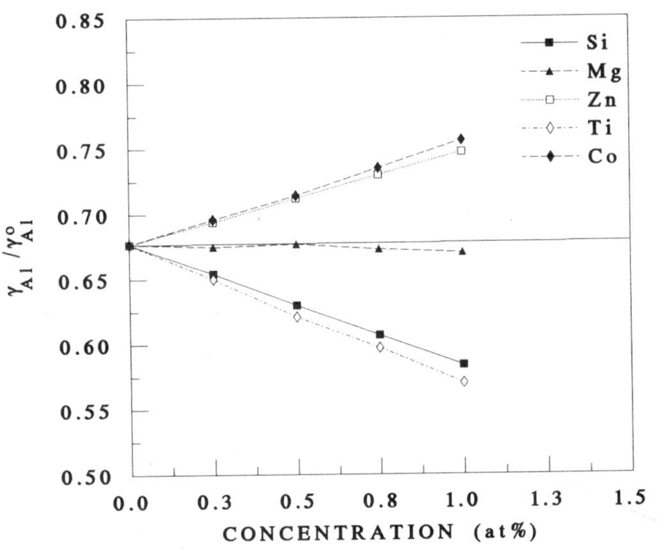

Figure 1. Activity coefficient of Al atoms as a function of impurity concentration x_{Me} at T=800 C.

It is evident that the potentials differ for different pairs, not only in terms of quantity but also in terms of sign. Experimental data[5-7,21-23] indicate that the formation of intermediate phases such as Fe-Al intermetallics and their thicknessat at the iron (matrix)-aluminum (coat) interface mainly depends on the rate of solubility of Al in the Fe matrix. The rate of solubility of Al in α-Fe is defined mainly by the difference in chemical potentials of the Al atoms in the matrix (Fe) and in the coat material (Al) itself. The impurities which will increase the excess chemical potential of Al atoms in the α-Fe should decrease the solubility of Al in the Fe-Al-Me systems by decreasing the activity coefficient of Al atoms. Therefore, such kind of impurities or alloying additions may retard the formation of Fe-Al intermetallics. However, the analysis of the effect of impurity elements (Me) on the excess chemical potential of Al in the Fe-Al-Me system can be given by the activity coefficient of Al atoms, $\gamma_{Al}/\gamma^{\circ}_{Al}$, in α-Fe. The calculated theoretical values of activity coefficient of Al atoms, $\gamma_{Al}/\gamma^{\circ}_{Al}$, as a function of impurity concentration, x_{Me}, are given in Fig. 1.

Fig. 1 shows that the addition of impurities, such as Me= Si, Ti or Mg (group I), decrease whereas the impurities of Me= Co or Zn (group II) increase the activity coefficient of Al atoms in α-Fe. These analysis, therefore, predict that the thickness of Fe-Al intermetallic layer formed at the interface would either decrease or increase, compared to pure Al coat, with respect to the presence of group I or group II impurities in Al coating bath. These theoretical results are in a good qualitative agreement with the reported data of the effects of alloying additions on the thickness of aluminized diffusion layer, Fig.2. The results of our calculation also indicated that the addition of silicon and titanium would be much more effective in decreasing the thickness of intermetallic layer which is consistent with the experimental observations.

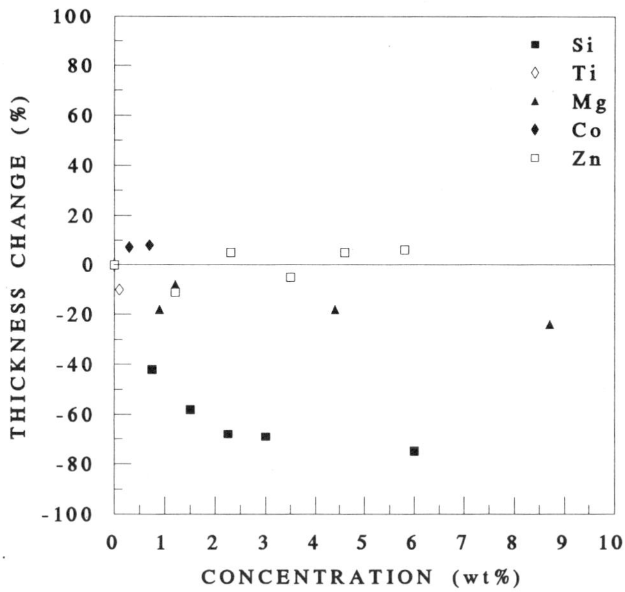

Figure 2. The percent change in thickness of aluminized diffusion layer as a function of impurity concentration[21]

4. CONCLUSIONS

The present study has shown that the combination of statistico-thermodynamical theory of alloys with the electronic theory in the pseudopotential approximation can be applied for the analysis of impurity effects on the interfacial interactions during aluminizing of low carbon steel. The impurities with regard to their effect on activity coefficient of Al may be classified in to two groups:

i) group I impurities, Me= Si, Ti, or Mg which decrease the activity coefficient of Al atoms in α-Fe- Al, tend to reduce the rate of formation and hence the thickness of Fe-Al interfacial diffusion layer,

ii) group II impurities, Me= Co or Zn increase the activity coefficient of Al atoms in α-Fe, consequently leads to the thicknening of Fe-Al diffusion layer formed at the interface.

The results of present analysis are in a good qualitative agreement with experimental observations.

Acknowledgements

This work is supported through TUBITAK with the project number SEKOM-3 which the authors gratefully acknowledge. The authors are grateful to Prof. Dr. Tarık Oğurtanı and Prof. Dr. Tayfur Öztürk for helpful discussions.

REFERENCES

1. V. I. Dybkov, *J. Mater. Sci.,* 21:3078 (1986)
2. V. I. Dybkov, *J. Mater. Sci.,* 21:3085 (1986)
3. D. S. Williams, R. A. Rapp and J. P. Hirth, *Metall. Trans.,* 12A:639 (1981)
4. J. Philibert, *Defect and Diffusion Forum,* 66-69:995 (1989)
5. M. V.Akdeniz, A. O. Mekhrabov and T. Yilmaz, *Scr. Met.*, 31:1723 (1994)
6. S.Elliot and E.R.Wallach, *Metal Construction,* March:167 (1981)
7. S. Elliot and E.R.Wallach, *Metal Construction,* April:221 (1981)
8. K. I. Portnoy , V.I. Bogdanov and D. L. Fucks."The Interatomic Interactions and Phase Stability Calculations", Metallurgy, Moscow (1981) (in Russian)
9. C. Vagner."Thermodynamics of Alloys", Addison-Wesley, Reading,Massachusetts (1952)
10. C. H. P.Lupis and J. F. Elliot, *Acta Met.,* 14:529 (1966)
11. C. H. P.Lupis and J. F. Elliot, *Acta Met.,* 14:1019 (1966)
12. M. A. Krivoglaz, Russian *J. of Phys. Chem.,* 31:930 (1957)
13. L. A. Bolshov, *Fiz. Metal. Metalloved.,* 39:501 (1975);
14. L. A. Bolshov, *Fiz. Metal. Metalloved.,* 39:743 (1975)
15. A.O.Mekhrabov, Z.M. Babaev, A. A.Katsnelson and Z. A.Matysina , *Fiz. Metal. Metalloved.,* 61:1089 (1986)
16. A. O. Mekhrabov, A. Ressamoglu and T. Ozturk, *J. of Alloys and Compounds,* 205:147 (1994)
17. V.Heine and J.V. Abarenkov, *Phil. Mag.,* 9:451 (1964)
18. A. O. E. Animalu, *Proc. Roy. Soc., Phys. Rev. B.,* 8:3542 (1973)
19. J. Hubbard, *Proc. Roy. Soc. (London),* 243A:336 (1957)
20. L. J. Sham, *Proc. Roy. Soc.,* 283A:33 (1965)
21. D. O. Gittings, D.H. Rowland and J. O. Mack, *Trans. ASM,* 43:587 (1951)
22. A. K. Kurakin, *Phys. Metals Metallography,* 30(7):105 (1970)
23. M. V. Akdeniz, A. O. Mekhrabov and T. Yilmaz, in:"8th International Metallurgy and Materials Congress", N. Sevinc et.al. eds., UCTEA Chamber of Metallurgical Engineers, Ankara, Vol. 2:1163 (1995)

RADIATION-INDUCED DISORDERING AND DISSOLUTION OF LRO PRECIPITATES IN DISORDERED MATRIX

H. Wollenberger, S. Matsumura†, S. Müller and C. Abromeit

Hahn-Meitner-Institut Berlin, Glienicker Str. 100, D-14109 Berlin, Germany
† *Permanent address:* Dept. of Nuclear Engineering, Kyushu University, Fukuoka 812, Japan

INTRODUCTION

Radiation-induced dissolution of hardening precipitates in structural materials is one of the major material problems in reactor technology. For long range ordered γ ' ($Ni_3(Al,Ti)$) precipitates in superalloys the dissolution process is coupled to the disordering inside the precipitate. From studies of this phenomenon a deeper insight in the ballistic atom transport and the atomic redistribution provided by the thermally activated migrating vacancies and interstitials can be expected. The present paper briefly sketches results and conclusions which were obtained during the recent years[1-4].

EXPERIMENTAL RESULTS

Detailed experimental information on composition profiles through γ ' precipitates embedded in the γ matrix in the commercial alloy Nimonic PE16 comes from atom probing in the field ion microscope[1]. In this instrument the specimen tip is field evaporated atomic layer by atomic layer and the evaporated atoms (ions) are analysed with respect to their chemical identity by means of a time of flight mass spectrometer. In fig. 1 the number of registered Al and Ti atoms is plotted vs. the number of evaporated atomic layers. By this way, analysis of γ and γ ' is easily performed as shown in fig. 1a (so-called ladder diagram). The abscissa scale given in fig. 1b shows that without irradiation (0 dpa) the thickness of the interface between γ and γ ' measures one to two atomic layers. The ladder diagram after 5 dpa produced by Ni ion irradiation at room temperature clearly shows the complete dissolution of the former interfaces and even the decrease of the Al+Ti content in the center of the former γ ' precipitate. By means of this method the dissolution of the precipitates have been followed with atomic resolution. The profiles have been measured for different fluences and temperatures. The observed data have been explained quantitatively by assuming a ballistic interdiffusion process as expected for radiation-induced atomic mixing. A detailed analysis of the composition profile would predict vanishing small degree of long range order after 1 dpa because of the strong decrease of the Al+Ti content. On the other hand, weak superlattice reflections are still visible in the electron microscope up to 1 dpa [3], as shown in fig. 2. The dpa limit for the loss of visibility slightly increases with increasing temperature. At temperatures above T > 540 K, the intensity of the superlattice reflections are only slightly reduced at 10 dpa, but the FIM analysis shows a significant dissolution of the precipitates. From these observations we conclude that the interdiffusion profile as produced by the ballistic mixing cannot be the real existing one. There must be deformations due to the thermally activated interdiffusion which stabilizes central volume fractions of a precipitate within the composition range of the long range ordered γ ' phase.

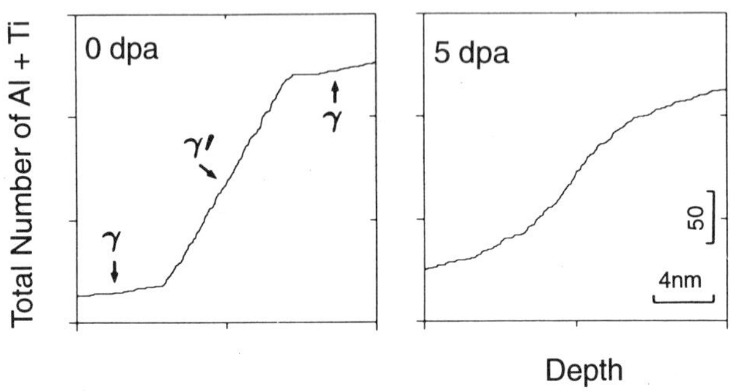

Fig. 1a **Fig. 1b**

Ladder diagram for atom probing of Nimonic PE16 before and after Ni ion irradiation to 5 dpa fluence at room temperature.

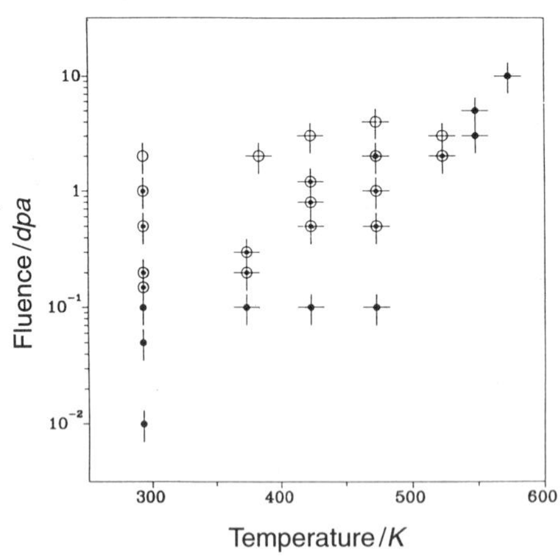

Fig.2: Qualitative intensities of the superlattice reflections for Ni ion irradiated Nimonic PE16 in the fluence vs. irradiation temperature diagram (intensity classification: ● - maximum, ⊙ - weak, ○ - vanishing).

THEORETICAL TREATMENT AND RESULTS

The problem has been treated in theory by G. Martin et al.[5,6] by means of a Bragg-Williams approach. We have tried to describe the processes by simple equations based upon a time dependent Ginzburg-Landau (TDGL) model[4]. This approach would explain the precipitate evolution by the well known phenomenological radiation parameters interdiffusion coefficient due to atomic mixing D^{mix} [7] and the disordering efficiency ε [8]. The composition X and the degree of long range order S are taken to be continuous variables of space and time. The time evolution reads for both quantities

$$\frac{\partial X}{\partial t} = D^{mix} \cdot \nabla^2 X + L(T,\phi) \cdot \nabla^2 \left(\frac{\delta F(\{X,S,T\})}{\delta X} - \mu \right), \tag{1}$$

and

$$\frac{\partial S}{\partial t} = -\varepsilon\phi S - M(T,\phi) \cdot \frac{\delta F(\{X,S,T\})}{\delta S}, \tag{2}$$

where ϕ is the atomic displacement rate, $F(\{X,S,T\})$ is the thermodynamical potential, μ the chemical potential. The first terms in the right hand side of eqs. (1) and (2) represent atomic diffusion and disordering due to the ballistic jumps induced by irradiation, and the second ones the thermodynamical driving force approaching the equilibrium state through the thermally activated process of atomic replacement. The thermal diffusion mobility L and the ordering rate M are described by equs. (5) to (8).

The thermodynamical potential is taken as

$$F[\{X(\mathbf{r}),S(\mathbf{r}),T\}] = \int \left\{ f(X,S,T) + \frac{H(T)}{2}(\nabla X)^2 + \frac{K(T)}{2}(\nabla S)^2 \right\} d\mathbf{r}. \tag{3}$$

with $f(X,S,T)$ being the mean field free energy density, and H and K are positive constants of the interfacial energy coefficients for varying X and S. If the order-disorder transition is of the second kind and the disordered phase has no tendency to decomposition, the simplest form of $f(X,S,T)$ is given by the Landau expansion,

$$f(X,S,T) = f_0 + \frac{a(T)}{2}\left\{ (X - x_m)^2 - b(T)\{x_0(T)^2 - X^2\}S^2 + b(T)^2 x_1(T)^2 S^4 \right\}, \tag{4}$$

where f_0 is the mean field free energy of the disordered phase with composition x_m, and a, b and x_1^2 are positive factors depending on temperature T. Here, X = 0 gives the stoichiometrical composition for the ordered structure.

The equilibrium phase diagram following from equ. (1) to (4) is shown in fig. 3. This model diagram explains phase equilibria in tricritical systems, such as Fe-Al and Fe-Si at intermediate and high temperatures. Because of the short expansion of the free energy density in equ. (4), the kinetic behavior is correctly described for $T \geq 0.3\, T_c$ only.

We consider the radiation-enhanced thermally activated atom transport at elevated temperatures by

$$L(T,\phi) = L_0\{c_v^{th}(T) + c_v^{irr}(T)\} \cdot \exp\left(-\frac{E_{mv}}{k_B T}\right), \tag{5}$$

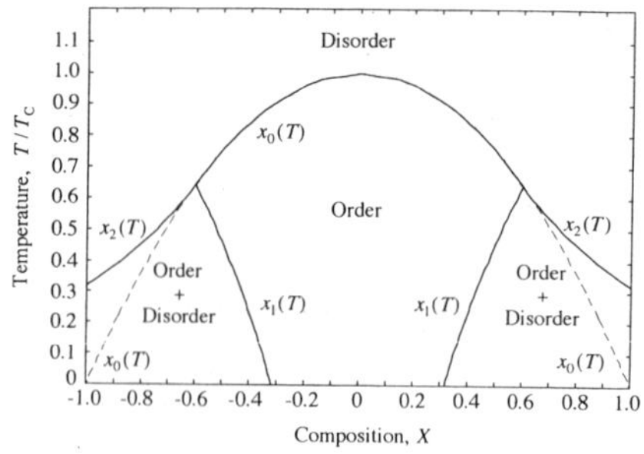

Fig. 3: Equilibrium phase diagram according to the present model (x_i - see text).

and

$$M(T,\phi) = M_0\left\{c_v^{th}(T) + c_v^{irr}(T)\right\}\cdot \exp\left(-\frac{E_{mv}}{k_B T}\right), \tag{6}$$

where E_{mv} is the activation energy for vacancy migration and k_B the Boltzmann constant. We neglect any transport by interstitial migration.

The concentration of thermal vacancies c_v^{th} is given by

$$c_v^{th}(T) = c_0 \cdot \exp\left(-\frac{E_{fv}}{k_B T}\right), \tag{7}$$

with the vacancy formation energy E_{fv}. We assume that the concentration of radiation-generated vacancies c_v^{irr} is stationary under irradiation and obtain[7]

$$c_v^{irr}(T,\phi) = \sqrt{\eta\phi \cdot \frac{k_{is}(T)}{k_{iv}(T)\cdot k_{vs}(T)} + \left(\frac{k_{is}(T)\cdot c_s}{k_{iv}(T)}\right)^2} - \frac{1}{2}\cdot\frac{k_{is}(T)\cdot c_s}{k_{iv}(T)}, \tag{8}$$

where η is the fraction of freely migrating defects and c_s is the concentration of sinks of defects. The rate constants k_{iv}, k_{is} and k_{vs} are for recombination of vacancies (v) and interstitials (i), and for annihilation of (i) and (v) at sinks, respectively.

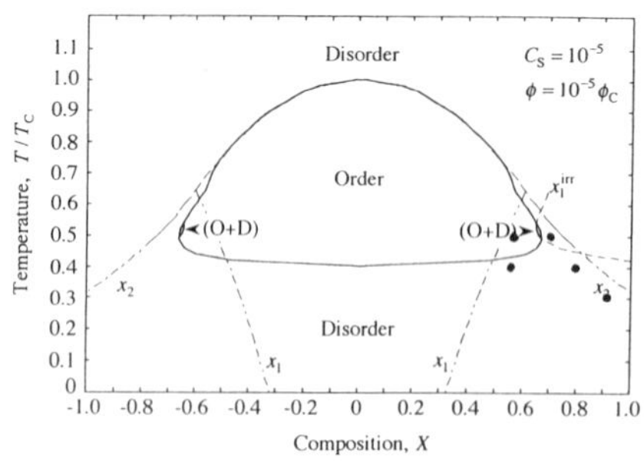

Fig. 4: Phase diagram for the irradiation parameters given in the figure according to the present model (O+D) indicates the coexistence fields of order and disorder.

Table 1. Phase boundaries and kinetic rates.

T	x_1	x_2	x_1^{irr}	x_2^{irr}	M^{bal}/M_0^{th}	L^{mix}/L
$0.3T_c$	0.427	1.033	-	-	117	1020
$0.4T_c$	0.471	0.872	-	-	1.15	7.05
$0.5T_c$	0.521	0.740	0.650	0.660	0.128	0.555

For the following calculation, we used the parameters listed in Table 1, assuming the typical ranges of 1000 ~ 1500 K for T_c, 1 ~ 2 eV for E_{mv} and E_{fv} and ~ 1 nm^2/dpa for D^{mix}. For simplicity, D_i>>D_v was assumed.

For stationary conditions the order-disorder transition line is given by

$$x_0^{irr^2}(T) = x_0^2(T) - \frac{M^{bal}}{Mab} \tag{9}$$

The binodal lines of the two phase regions read for the

ordered phase $$x_1^{irr}(T)^2 = x_1(T)^2\left(1+\frac{L^{mix}}{L}\right) \tag{10}$$

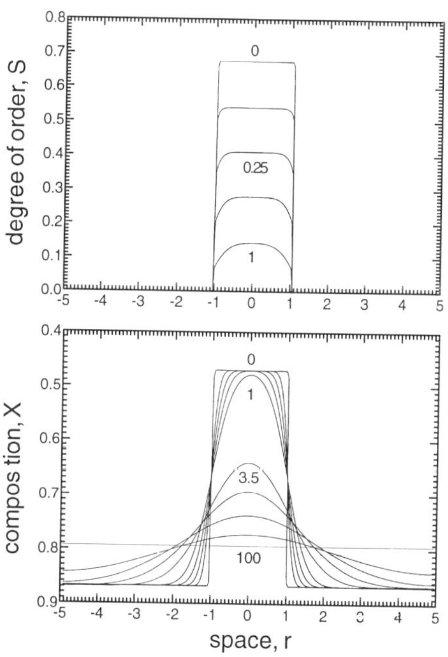

Fig. 5: Evolution of S and X for a simulated long range orderd precipitate (one-dimensional geometry) at T = 0.4 T_c and the irradiation conditions given in the text according to the present model (the curves are calculated for Φ = 0, 0.01, 0.025, 0.05, 0.10, 0.35, 0.75, 1.5, 3.0, 10 dpa; the given numbers at the curves represent 10 Φ).

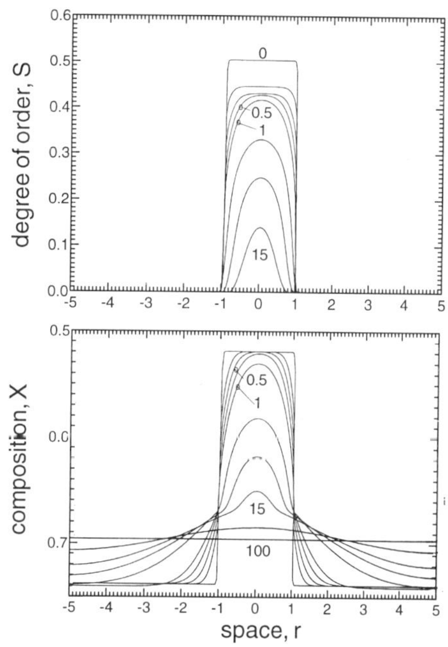

Fig. 6: Simulation result as in fig. 5, but for T = 0.5 T_c (Φ = 0.01, 0.025, 0.05, 0.10, 0.50, 1.0, 1.5, 2.5, 10 dpa).

disordered phase $$x_2^{irr}(T)^2 = \frac{1}{4x_1^{irr}(T)^2}\left\{x_0^{irr}(T)^2 + x_1^{irr}(T)^2\right\}^2 \tag{11}$$

The resulting phase diagram for stationary conditions, the displacement rate of 10^{-4} dpa/s and the sink density of 10^{-5} is shown in fig. 4. The existence field of the ordered phase is closed towards low temperatures due to the dominant disordering by ballistic atom jumps. The binodal line x_1 is changed completely and leaves only negligibly small two-phase regions.

The evolution of composition and degree of order in and around an ordered precipitate in a disordered matrix was studied by solving numerically the kinetic equations for a one-dimensional array of 250 elements and periodic boundary conditions. Table 1 gives the phase boundaries and kinetic rates following for the above irradiation conditions. One realizes that the ballistic transport becomes ineffective around 0.5 T_c.

The curves in fig. 5 show a strong decrease of the degree of order for $\Phi = 0.1$ dpa while the shape of the precipitate has not changed much. For $\Phi > 0.1$ dpa, the precipitate is dissolved continuously by the atomic mixing effect. This successive disordering and dissolution was indeed observed for the γ' in the temperature range up to 540 K (see fig. 2). The precipitates first vanished in the dark field image before the contrast started to also vanish in the bright field image. For $T = 0.5\ T_c$ in fig. 6, we have a much slowlier disordering due to the thermally activated atom transport in the time interval from $\Phi = 0.1$ dpa to $\Phi = 1.5$ dpa. Both, disordering and dissolution proceeds in parallel with marked shrinkage of the precipitate size. Not before total disordering the dissolution process shows the normal diffusion profile ($\Phi \geq 2.5$). Such a behavior has been observed for γ' in Nimonic PE16 at $T > 540$ K.

REFERENCES

1. E. Camus, F. Bourdeau, C. Abromeit, N. Wanderka, and H. Wollenberger, Mat. Res. Soc. Symp. Proc. Vol. 373:83 (1995).
2. E. Camus and C. Abromeit, *Z. Metallkde.* 85:378 (1994).
   E. Camus and C. Abromeit, *J. Appl. Phys.* 75:2373 (1994).
3. F. Bourdeau, E. Camus, C. Abromeit, and H. Wollenberger, *Phys. Rev. B* 50(22):16205 (1994).
4. S. Matsumura, S. Müller, and C. Abromeit, Kinetics of disordering and dissolution process in tricritical systems under irradiation, to be published.
5. G. Martin, F. Soisson, and P. Bellon, *J. Nucl. Mater.* 205:301 (1994).
6. F. Soisson, P. Bellon, G. Martin, and E. Salomons, *J. Nucl. Mater.* 205:324 (1994).
7. V. Naundorf, *Int. J. Mod. Phys. B* 6:2925 (1992).
8. E.M. Schulson, *J. Nucl. Mater.* 83:239 (1979).

NONEQUILIBRIUM SOLIDIFICATION MICROSTRUCTURE IN ALLOYS

E. Ramous, A. Zambon, B. Badan, and M. Magrini

Dipartimento di Innovazione Meccanica e Gestionale (DIMEG)
Università di Padova
via Marzolo, 9
I 35128 Padova
Italy

INTRODUCTION

Different microstructural features derive from the solidification process, according to the thermal and thermodynamic parameters involved, which can exhibit the characteristics of metastable structures or at least of structures rarely observed in industrial processes involving solidification.

Usually the achievement of metastable structures is associated to high cooling rates (in the range 10^7 - 10^9 K/s), which can be attained in processes such as "Melt-Spinning" or "Laser-Glazing". In both techniques a substrate, whose temperature is quite close to room temperature (and sometimes suitably cooled), plays the role of a cold thermal sink, whose heat capacity can be considered to be infinite if the energy values involved in the aforesaid techniques are considered. Indeed the melt layers which afford the aforesaid cooling rates to be reached are usually very shallow, ranging between 20 and 100 μm.

Metastable structures can likewise originate from high undercooling extents achieved in particular experimental conditions. It is well-known that solidification can take place if a definite degree of undercooling exists in the liquid. In smelting works the undercooling extent is usually restricted to a few degrees (1), due to the effect of impurities, heterogeneous inclusions (typically high melting point oxides or pieces of refractory material) and of course to the walls of the mould.

When a pure liquid metal is cooled, the solidification can take place only if a definite degree of *thermal* undercooling is established, which controls both the number of solidification nuclei per unit volume and their growth rate.

Containerless solidification can result, as a general trend, in a supercooled liquid from which refined microstructures can originate. In research work different methods for achieving a large extent of undercooling can be used which can afford the production of drops/droplets in uncontaminated and controlled environments, so that extremely high undercooling extents are attainable, in some cases far beyond the critical undercooling (1, 2).

A number of works deal with containerless undercooling produced by means of levitation facilities (3, 4, 5) or in drop tubes (5, 6, 7).

Several research groups have developed methods for achieving such conditions and have produced an important volume of work in different alloy systems, most of which are Ni-based.

Much work has been done in particular in the Ni-Cu system, so that a great deal of data is available concerning mainly the internal microstructures of the different compositions as a function of undercooling, quantitative determinations of undercooling being possible in static configurations, or at least in experimental arrangements in which the solidifying material is moving at low velocity. On the other hand, atomization being an industrial process in which powder is produced in order to achieve the best combinations of properties in the final sintered products, it seems interesting to investigate the microstructure of atomized powder. In fact it is well known that the microstructure of the powders is somehow retained also in the sintered products.

Due to the relative simplicity of the phase diagram, the attempt to evaluate the correlation between external and internal morphologies in atomized powders has therefore been focussed on the Ni-Cu alloy system.

EXPERIMENTAL

Plant description

The laboratory scale atomizing unit can operate at temperatures as high as 1600-1650 °C. The Furnace consists of an induction coil powered by a solid state 12 kW, 20-50 kHz generator, a 0.2 dm^3 capacity crucible, fitted with an alumina pouring spout. A protective nitrogen atmosphere is maintained in the melting furnace during heating and melting. A cooling/powder collecting section is fitted beneath the furnace, which can be evacuated and backfilled with the same inert gas.

The configuration of the metal-gas nozzles is the **C**lose **C**oupled - **D**iscrete **J**ets (8 converging nozzles) fitted in order to produce a local negative pressure (about 10 kPa) with respect to the pressure existing in the cooling chamber during operation under the usual working pressure of 1.5 MPa.

The gas velocity is sonic at the final section of the nozzles and the residual pressure is adjusted to the pressure existing in the cooling section via shock waves, which enhance the disruption of the liquid metal.

Materials

The aim of the work is to try to correlate the microstructures obtained in non-equilibrium solidification of alloys with the presumable undercooling, in order to obtain preliminary indications on the operative parameters to be used in an atomizing unit to obtain desired microstructures. To this end the Ni-Cu alloy system was chosen, due to the simplicity of the phase diagram, and

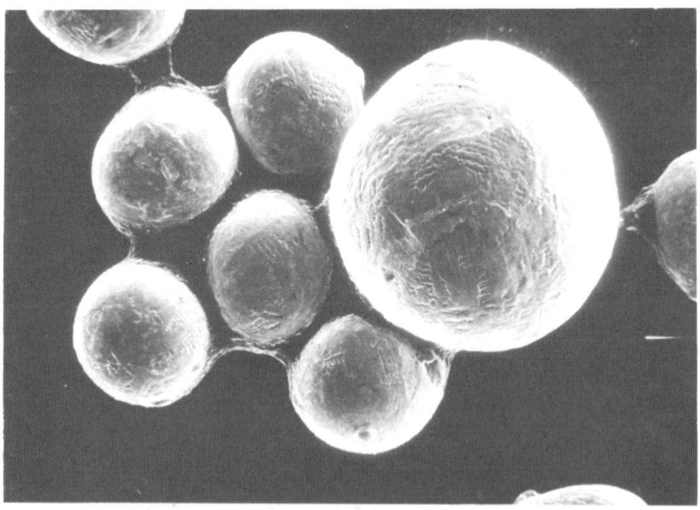

100 μm

Fig. 1 - External aspect of atomized $Ni_{60}Cu_{40}$ powder.

commercial purity electrolytic copper and nickel were melted directly in the crucible of the atomizer and the melt was overheated by roughly 50 K. The atomization took place after a ten minutes period of homogenization of the melt.

Microstructural characterization - Solidification mode identification

In all the compositions and size ranges most of the droplets appeared spherical (fig. 1), while some elongated powders, most probably deriving from the primary stage of atomization, were observed in particular in the larger size fractions. Some flake-shaped fragments were produced before

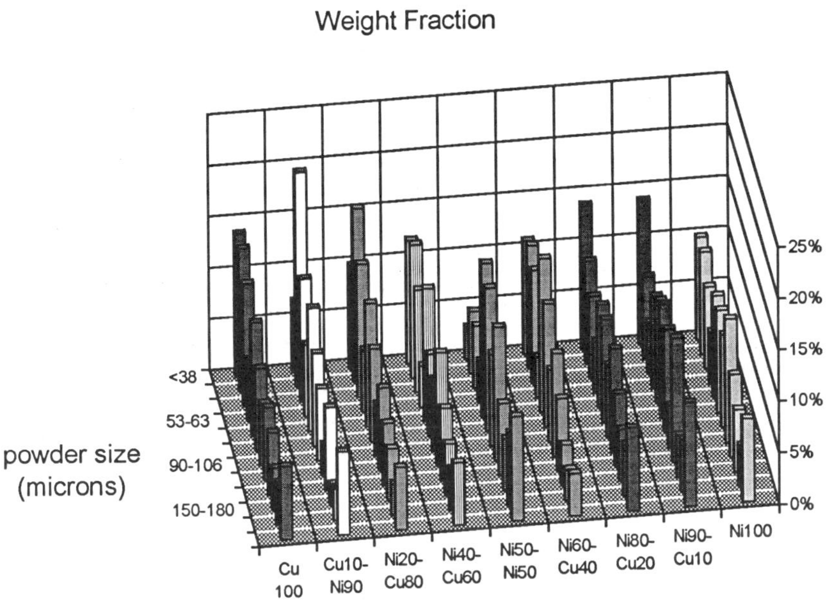

Fig. 2 - Weight fractions as a function of powder size

Fig. 3 - Dendritic and equiaxed atomized $Ni_{90}Cu_{10}$ powders, 75 μm < d < 90 μm stationary condition was reached and possibly at the end of the atomization operation. The powders produced after each atomization run were collected and sieved by means of an automatic oscillating unit, to determine the weight fractions as a function of powder size (Fig. 2).

In order to perform Light Microscopy examinations of the sections of the powders, different etchants were used according to the compositions of the alloys.

The etchants were optimized and tested for each alloy composition. A 4% $FeCl_3$ aqueous solution proved to be the most suitable for the high Cu powders, while for the intermediate range of compositions a 8g CrO_3, 0.9g NH_4Cl, 5ml HNO_3, solution in 100 ml H_2O was used. For the high Ni samples it was necessary to adopt a 50% Nitric-50% Acetic acid etchant.

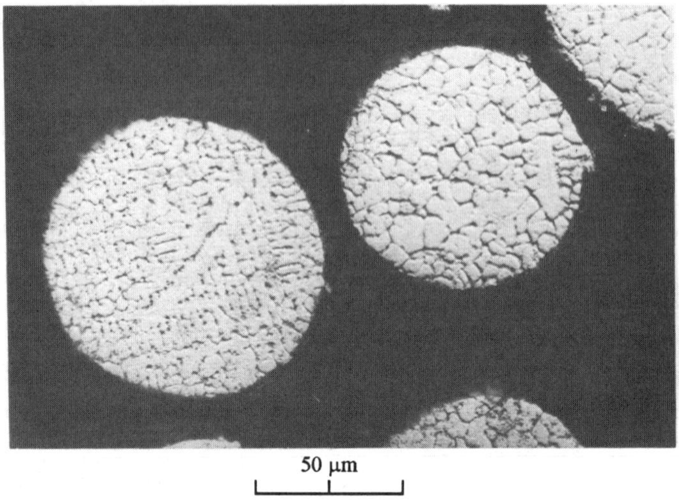

Fig. 4 - Dendritic and intermediate structures in atomized $Ni_{90}Cu_{10}$ powders, 90 μm < d < 106 μm

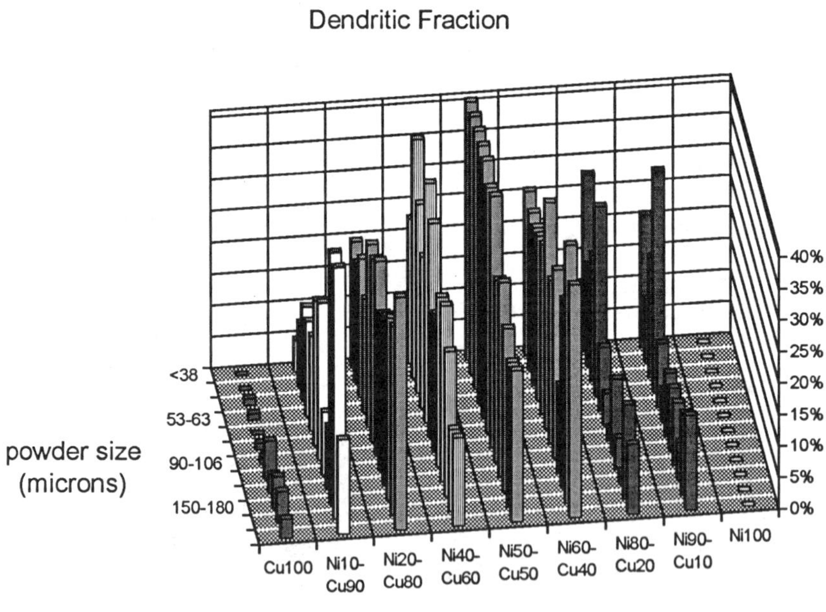

Fig. 5 - Dendritic fractions as a function of powder size

Most of the alloy compositions showed both dendritic and equiaxed grains in the same diameter class (fig. 3). Intermediate structures had also been identified in which only at high magnification fragments of dendrites could be sometimes identified (Fig. 4). In fig. 5 the data concerning the dendritic fractions are shown.

DISCUSSION

It is well known that whether the microstructure will be dendritic or grain refined is largely dependent on the degree of undercooling at which the nucleation will start. In particular it has been experimentally proved that two values of critical undercooling, ΔT_1 and ΔT_2, exist which constitute the boundary conditions for which transitions from an equiaxed grain solidification mode to a dendritic solidification mode are observed. Recently a model has been put forward (8) which is based on the assumption, previously described by Flemings and Shiohara (9) who reviewed papers from Kattamis and Flemings (1967-1970), that the transition between the aforesaid microstructures results from the dendrite fragmentation phenomenon. In the above said paper an interesting model is proposed where the values of ΔT_1 and ΔT_2 depend not only on the undercooling, but also on the cooling rate. The variation of the undercoolings ΔT_1 and ΔT_2 as a function of the cooling rate, up to cooling rates of 1000 K/s are shown. The model predicts that ΔT_2 increases while ΔT_1 decreases with the cooling rate, thus improving the extension of the undercooling interval in which the dendritic solidification mode is verified.

In our case, as shown in fig. 5, the dendritic fraction appears to depend both on composition and powder size. The dendritic fraction reaches its maximum values for the equiatomic composition, while non dendritic solidification modes are observed for the pure metals, according to the general behaviour of solid solution forming alloys, where the dendritic stability increases with the alloying component.

Moreover, if the results from the Karma model can be extended to the atomization experiments, the situation of the dendritic fraction in each alloy could be a consequence of different thermal histories undergone by powders of different size. This agrees with the Karma model assuming that only the larger droplets, solidifying at lower undercooling, can undergo the dendrite break-up phenomenon. However also the cooling rates depend on the powder diameter, and in order to fully understand the influence of the two parameters, it is necessary to consider contemporaneously but separately the contribution of the undercooling and of the cooling rate.

CONCLUSIONS

The determination of the solidification mode in Ni-Cu nitrogen atomized powders has been carried out and related to the microstructure formerly observed in levitation melted and solidified drops whose initial undercooling was known. The results give indications useful to correlate the operating conditions with the microstructure (both equilibrium and non-equilibrium ones) of the solidified alloys.

Further investigations are needed in order to confirm the results of such indirect determination.

Acknowledgements

*The authors are obliged to Dr. Klaus Ekler of the Institüt fur Raumsimulation - Köln - Germany for having produced and having provided the levitated samples.

The present work was carried out within the Brite EuRam project "Development of metastable phases by deep undercooling" BE 5142. The partners involved are, besides the Dipartimento di Innovazione Meccanica e Gestionale (DIMEG) of Università di Padova, the Institüt fur Raumsimulation - Köln - Germany and the Dept of Materials Science & Metallurgy - University of Cambridge - U. K.

The grant from the EC under contract N. BRE2-CT92-0159 is fully acknowledged.

REFERENCES

1. J.D. Verhoeven, Solidification, in Fundamentals of Physical Metallurgy, John Wiley & Sons, New York, U.S.A. (1975), p 233.
2. K.Eckler, R.F. Cochrane, D.M. Herlach, B. Feuerbacher and M. Jurisch, Phys. Rev. B 45 (9), (1992), 5019-5022
3. R. Willnecker, D. M. Herlach and B. Feuerbacher, Appl. Phys. Lett. 56 (4), (1990), 324-326
4. M. Barth, K. Eckler and D. M. Herlach, Mat. Sci. and Eng. A 133, (1991), 790-794
5. A. L. Greer, Mat. Sci. and Eng. A 133, (1991), 16-21
6. F. Gillesen and D. M. Herlach, Mat. Sci. and Eng. 97, (1988), 147-151
7. D. J. Thoma and J. H. Perepezko, Met. Trans A 23, (1992), 1347-1362
8. M. Schwarz, A. Karma, K. Eckler, and D. M. Herlach, Phys. Rev. Letters, 73 (10), (1994), 1380-1383
9. M.C. Flemings and Y. Shiohara, Solidification of undercooled metals, Mat. Sci. Eng., 65 (1984) 157-170.
10. A. Zambon, B. Badan and E.Ramous, Microstrutture in Leghe Ni-Cu Solidificate ad Elevato Sottoraffreddamento, Proc. of the 25° Concresso Nazionale AIM, Milan 12-14 Sept 1994, 251-353, requested for publication in "La Metallurgia Italiana".
11. P. Mathur, D. Apelian and A. Lawley, Analysis of th spray deposition process, Acta Metall., 37 No. 2 (1989) 429-443.

VARIATIONS OF LONG-RANGE ORDER IN Ni_3Al+B AFTER DEFORMATION BY COLD-ROLLING

R.Kozubski*, M.Migschitz and W.Pfeiler

Institut für Festkörperphysik
University of Vienna
Strudlhofgasse 4, A-1090 Vienna, Austria

INTRODUCTION

The widespread interest in the intermetallic compound Ni_3Al and a number of related $L1_2$ superalloys stems from their promising mechanical properties at high temperatures due to an anomalous increase of the yield stress below ~1100 K. A discovery of a ductilization of Ni_3Al by microalloying with boron stimulated a considerable progress in the studies of those originally extremely brittle materials[1].

Although the phase γ'-Ni_3Al is known to be almost perfectly ordered up to the melting point[2], it may be artificially disordered e.g. by irradiation or mechanical treatment. Gottstein et al.[3] thoroughly analysed the effect of the degree of deformation caused by cold-rolling on the resulting degree of LRO in $Ni_{76}Al_{24}$ + B and reported a possibility of reducing the degree of LRO by up to 50% in the case of high deformations.

Studies of the kinetics of reordering in Ni_3Al alloys during an annealing of the mechanically disordered material[4] showed typical features indicating the major role of vacancy diffusion in the process, its activation enthalpy, however, resulted lower than in the case of the homogeneous and thermally disordered alloys[5].

Simultaneous ordering and recovery processes occurring in $Ni_{76}Al_{24}$ + 0.19at.% B cold-rolled to 8 and 14% of thickness reduction have recently been investigated by some of the present authors[6].

TEM observations revealed two stages of the defect recovery: *Superlattice intrinsic stacking faults* (SISF) of large density recovered almost completely in the temperature regime between 443 and 700 K. The recovery of antiphase-boundary (APB) dissociated superlattice dislocations started at about 550 K by the annihilation of dipoles and finally led to a loss of all dislocations at 1273 K. Residual resistivity measurements indicated that cold-rolling caused a decrease of the degree of LRO and effectively increased the mobility of vacancies. Qualitatively, however, the process of long-range ordering in $Ni_{76}Al_{24}$ + 0.19at.% B was not affected by the deformation.

In the present work, the resistivity isochrones of as-rolled and homogenised $Ni_{76}Al_{24}$ + 0.19at.% B are analysed within a simple model of vacancy-controlled ordering[7] which makes it possible to discuss separately the effects of the enthalpies of vacancy formation and migration.

* on leave from Institute of Physics, Jagellonian University, Reymonta 4, 30-059 Krakow, Poland

MODEL OF LRO KINETICS CONTROLLED BY VACANCY MECHANISM

The analysis of resistivity isochrones is based on treating the curves as resulting from consecutive *isothermal* LRO relaxations occurring during successive annealings at particular temperatures: the initial conditions of each isothermal relaxation are determined by the final state of the preceding one. As it has been recently reported[8], when proceeding between two *equilibrium* states, such relaxations in $Ni_{76}Al_{24}$ + 0.19at.% B are composed of two simultaneous single-exponential processes substantially differing in relaxation times but exhibiting similar activation energies.

The basic mechanism of the LRO relaxations is assumed to be atomic jumps to nn vacancies and, therefore, the current number of vacancies in the sample is an important factor controlling the rate of the process. The particular *isothermal* relaxations of LRO are described within the previously applied formalism[7]:

$$\frac{\rho(t)-\rho_\infty}{\rho_0-\rho_\infty} = f_1(t)+f_2(t) \tag{1}$$

where:

$$f_{1,2}(t)=\exp\left\{-\frac{t}{\tau_{S\infty}^{(1,2)}}-\tau_v(T_a)\left[\frac{1}{\tau_{S0}^{(1,2)}}-\frac{1}{\tau_{S\infty}^{(1,2)}}\right]\left(1-\exp\left[-\frac{t}{\tau_v(T_a)}\right]\right)\right\}$$

$$\frac{1}{\tau_{S0}^{(1,2)}}=\frac{1}{\tau_0^{(1,2)}}\exp\left[-\frac{H_m}{kT_a}\right]\exp\left[-\frac{H_f}{kT_{q_{eff}}}\right]$$

$$\frac{1}{\tau_{S\infty}^{(1,2)}}=\frac{1}{\tau_0^{(1,2)}}\exp\left[-\frac{H_m}{kT_a}\right]\exp\left[-\frac{H_f}{kT_a}\right]$$

$$\frac{1}{\tau_v(T_a)}=N(T_a)\ell^2\nu_D\exp\left[-\frac{H_m}{kT_a}\right]$$

τ_{S0} - initial relaxation time for LRO kinetics,
$\tau_{S\infty}$ - relaxation time for LRO kinetics after complete annealing-out of excess vacancies
τ_0 - pre-exponential factor accounting for the entropy of vacancy migration and formation, as well as for the correlation effects,
τ_v - relaxation time for the elimination of excess vacancies,
N - vacancy sink density,
ℓ - nearest-neighbour jump distance
υ_D - Debye frequency
H_m - enthalpy for vacancy migration,
H_f - enthalpy for vacancy formation,
T_a - temperature of the particular isothermal annealing
T_{qeff} - effective quenching temperature[7]

The resistivity isochrones of both as-rolled and homogenised $Ni_{76}Al_{24}$ + 0.19at.% B were fitted by eq.(1).

RESULTS

Figs.1 and 2 show the resistivity isochrones of $Ni_{76}Al_{24}$ + 0.19at.% B homogenised for 1/2 h at 1273 K and as-rolled, respectively. Fig.2 displays only the LRO-kinetics components of the curves which, as originally measured, reflected both ordering and defect-recovery phenomena. It was shown, however, that the contribution of defect recovery might be subtracted by neglecting the direct effect of annealing-out of locked superdislocations on the resistivity and by assuming an exponential-like dependence of the extra resistivity due to SISF[6] on the annealing temperature.

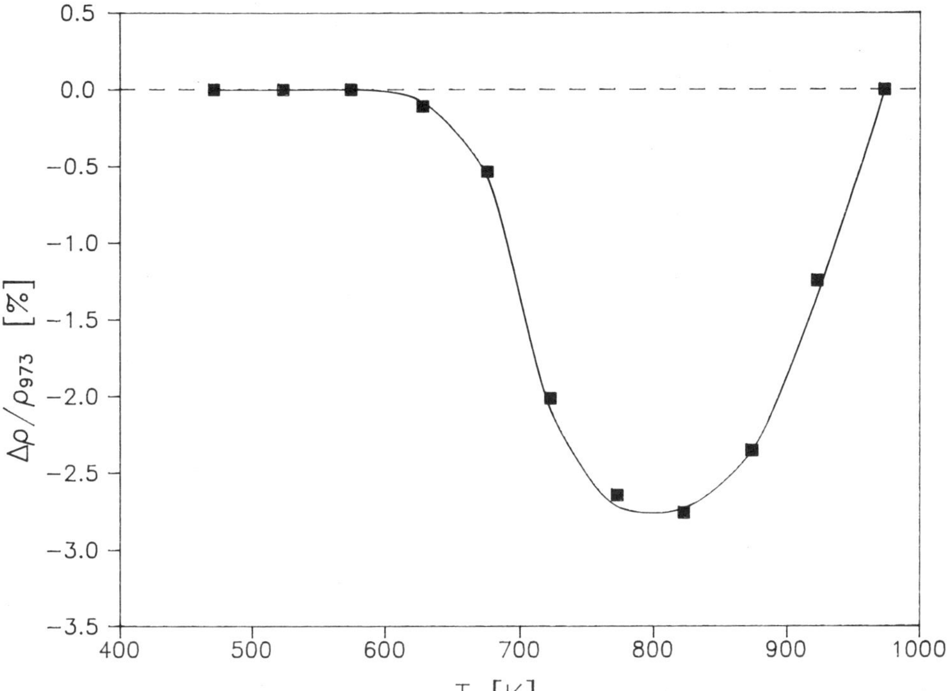

Fig.1. Change of the residual resistivity of $Ni_{76}Al_{24}$ + 0.19at.% B during isochronal annealing after initial homogenisation for 1/2 h at 1273 K, slow cooling to RT, re-annealing at 973 K and water-quenching. Solid line marks the fitted function (1).

The separated "pure LRO" curves (Fig.2) have qualitatively the same shapes as the one corresponding to the homogenised material (Fig.1). They start from a plateau corresponding to a temperature domain where the mobility of vacancies is frozen and no change in the degree of LRO takes place. The plateau is limited by a well defined edge, at which the resistivity starts to decrease due to an increase of the degree of LRO caused by vacancies becoming sufficiently mobile. The final increasing segments of the curves lie in the temperature domain where the variation of the degree of LRO is reversible with respect to heating and cooling ("equilibrium curve").

Two features of the curves were of the greatest importance for the present study:

- temperature position of the edge limiting the plateau, representing defreezing of vacancy mobility and thus accounting for the migration enthalpy H_m,
- the shape of the curves between the edge and the "equilibrium" part accounting for vacancy annealing at vacancy sinks (dislocations) of temperature dependent density $N(T_a)$, as well as for the enthalpies of vacancy migration (H_m) and formation (H_f).

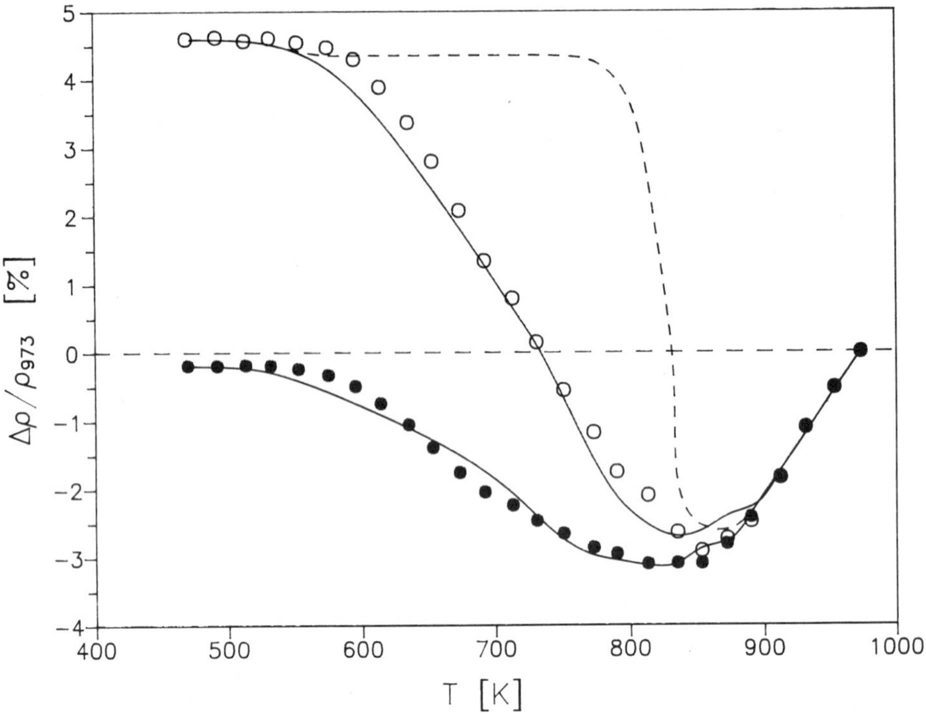

Fig.2. LRO contribution to the change of residual resistivity of as-rolled $Ni_{76}Al_{24}$ + 0.19at.% B during isochronal annealing: sample deformed to 14% (○) and 8% (●) of thickness reduction. Solid lines represent the fitted functions (1); dashed line shows the shape of a hypothetical isochrone corresponding to the case of the dislocation density fixed at the initial level.

The values of the fitted parameters are displayed in Table 1. While in the case of the homogenised sample the density of vacancy sinks was assumed to be constant throughout the isochronal annealing, the analysis of the remaining curves required an additional assumption concerning the process of their elimination during the isochronal annealing. As a first approximation, this elimination was described by an exponential function of the annealing temperature. The values of N in Table 1 related to the as-rolled samples are thus the starting (initial) ones.

Table 1. Parameters of the resistivity isochrones of $Ni_{76}Al_{24}$ + 0.19at.% B resulting from a fit of eq. (1)

Sample	$\tau_0^{(1)}$ [s]	$\tau_0^{(2)}$ [s]	H_m [eV]	H_f [eV]	N(473K) [cm^{-2}]
homogenised	2×10^{-25}	2×10^{-24}	1.6	2.9	4×10^{8}
as-rolled, 8% thickness reduction	2×10^{-25}	2×10^{-24}	1.4	2.9	9×10^{9}
as-rolled, 14% thickness reduction	2×10^{-25}	2×10^{-24}	1.4	2.9	10^{10}
accuracy	±30%	±30%	±0.2	±0.2	±30%

DISCUSSION

As follows from Fig.1, the measured resistivity isochrone of the homogenised $Ni_{76}Al_{24}$ + 0.19at.% B fits very well the function following from the simple model (1) which illustrates an interplay between the processes of annealing and thermal activation of vacancies whose migration controls the process of ordering.

Despite the simplicity of the model, it was possible to conclude separately on the contributions of vacancy migration and formation enthalpies to the high overall activation enthalpy for LRO kinetics reaching the value of 4.6 eV in $Ni_{76}Al_{24}$ + 0.19at.% B[8] .

The quantities displayed in Table 1 show that while the resulting *vacancy migration* enthalpy does not differ substantially from those usually reported for Ni_3Al (see e.g. ref. in [5]) , the *vacancy formation* enthalpy reaches almost 3 eV and is considerably higher than the related quantities following from the standard investigations of vacancy thermodynamics in Ni_3Al[9]. This result once again supports the previous statements[8] that the ordering kinetics in $L1_2$ superstructure is controlled by selected modes of atomic vacancy diffusion which in other experiments may be either undetectable or dominated by other modes - e.g. Ni migration within its own sublattice where only Ni vacancies are involved. The resulting high enthalpy for vacancy formation might thus be contributed by the formation of thermal vacancies on the Al sublattice. Using pair interaction energies following from the embedded-atom model, Sun and Lin have recently arrived at quite a high value of 1.93 eV for the related enthalpy in stoichiometric Ni_3Al[10]; the corresponding enthalpy for Ni-vacancy formation was equal to 1.47 eV - i.e. was lower by almost 30%.

The resulting quite high vacancy sink density in the homogenised $Ni_{76}Al_{24}$ + 0.19at.% B (see Table 1) seems rather inconsistent with the TEM observations of the samples[11]. This discrepancy may, however, be explained in terms of the interaction between vacancies and the B-atoms which may effectively lead to "filling" a certain fraction of vacancies by boron - i.e. to a reduction of the effective vacancy concentration[10]. The phenomenon obviously slows down the ordering phenomena[12], however, is not likely to affect their activation energy.

It is obvious that observing a total activation enthalpy as high as 4.6eV in the temperature range between 700 and 800K by linear extrapolation yields an extraordinarily low pre-factor τ_0 of about 10^{-25}s to achieve reasonable relaxation times. Aside of the question, however, if it is allowed to extrapolate linearly over more than 20 orders of magnitude it has to be noted that τ_0 does not just mean the atomic attempt frequency but that entropy and correlation factors enter in this quantity.

Quite satisfying fits were also achieved for the isochrones corresponding to the as-rolled samples (Fig.2). Somewhat lower quality of the fits most probably resulted from the inadequacy of the simple description of the process of the annealing-out of vacancy sinks. Nevertheless, the results definitely indicate the importance of the annealing out of vacancy sinks (dislocations) in the observed LRO kinetics. The dashed curve traced in Fig.2 represents the hypothetical case of a high density of vacancy sinks constant throughout the isochronal annealing. The curve very slightly decreases at the beginning of the process of reordering, but then shows a plateau because all of the excess vacancies potentially generating LRO kinetics sink very soon. Further long-range ordering (decrease of the resistivity) starts only at the temperature where thermal vacancies are activated.

Despite the uncertainties following from the approximate description of the temperature dependence of the annealing-out of vacancy sinks, the fits of the isochrones of Fig.2 definitely improved when the vacancy migration enthalpy H_m was somewhat decreased with respect to the value fitted for the case of the homogenised sample. This effect could give some new hint to the previously described considerable shift (by almost 100 K) between the temperatures limiting the initial plateau on the isochrones corresponding to the as-rolled and homogenised samples of $Ni_{76}Al_{24}$ + 0.19at.% B. In the previous paper[6] the phenomenon was attributed rather to a large concentration of vacancies generated by cold-rolling. In view of the present results, a possibility of an effective decrease of the migration enthalpy would have to be taken into account. Such interpretation should, however, be accepted with a great caution as one should remember that in the present experiments the vacancy dynamics is monitored via an *ordering process* which has a bulk character. Therefore, a simple influence of e.g. easy vacancy diffusion along the defects induced by cold-rolling is questionable.

Acknowledgements

The authors acknowledge a financial support from the Austrian FWF; R.K. as a Lise-Meitner Fellow (M-0088 PHY).

REFERENCES

1. K.Aoki, Ductilization of $L1_2$ Intermetallic compound Ni_3Al by microalloying with Boron, *Materials Transactions, JIM* 6:443 (1990).
2. T.B.Massalski. "Binary Alloy Phase Diagrams", ASME, Metals Park (1987).
3. J.Ball and G.Gottstein, Large-strain deformation of Ni_3Al+B: Part III. Microstructure, long-range order and mechanical properties of deformed and recrystallized Ni_3Al+B, *Intermetallics* 2:205 (1994).
4. A.R.Yavari, Reordering kinetics and magnetic properties of mechanically disordered nanocrystalline $L1_2$-type Ni_3Al+Fe alloys, *Acta Metall.Mater.* 41:1391 (1993).
5. R.Kozubski, J.Soltys, M.C.Cadeville, V.Pierron-Bohnes, T.H.Kim, P.Schwander, J.P.Hahn, G.Kostorz and J.Morgiel, Long-range ordering kinetics and ordering energy in Ni_3Al-based γ' alloys, *Intermetallics* 1:139 (1993).
6. W.Pfeiler, R.Kozubski, H.Peter Karnthaler and C.Rentenberger, Kinetics of defect recovery and long-range ordering in Ni_3Al+B. I. Simultaneous recovery and ordering in cold-rolled material, *Acta Metall.Mater.* - in press.
7. D.Trattner and W.Pfeiler, Information about SRO-kinetics from isochronal annealing of electrical resistivity, *Scr.Metall.* 17:909 (1983).
8. R.Kozubski and W.Pfeiler, Kinetics of defect recovery and long-range ordering in Ni_3Al+B. II. Atomic jump processes studied by "order-order" relaxation experiments, *Acta Metall.Mater.* - in press.
9. H-E.Schaefer, R.Würschum and J.Bub, Thermal formation of vacancies in intermetallic compounds, *Materials Science Forum* 105-110:439 (1992).
10. J.Sun and T.L.Lin, Theoretical and positron annihilation study of point defects in intermetallic compound Ni_3Al, *Acta Metall.Mater.* 42:195 (1994).
11. R.Kozubski and W.Pfeiler, Investigation of ordering and disordering in B-doped $Ni_{76}Al_{24}$ by residual resistometry, *J.Alloys Compounds* 220:237 (1995).
12. N.Njah, D.Gilbon and O.Dimitrov, The effect of Boron additions on irradiation-induced order changes in Ni_3Al intermetallic compounds, *Scripta Metall. Mater.* - in press.

SURFACE DISORDER AND SURFACE SEGREGATION IN CU_3AU: AN X-RAY SCATTERING STUDY

Helmut Dosch,[1] Harald Reichert [2]

Institut für Materialwissenschaften, Universität Wuppertal
Wuppertal, Germany
Sektion Physik, Universität München
München, Germany

INTRODUCTION

We consider the (001) surface of a 3-dimensional Cu_3Au single crystal which undergoes the well-known order-disorder phase transition at T_0=663K. The character of such a semi-infinite phase transition is modified in a fundamental manner near the surface. Since the atoms at the surface are less tightly bound than the bulk atoms, we expect that any long range order should slightly be reduced near the surface ("ordinary case"), thus, upon approaching T_0, one would expect the disordered phase emerge at the surface prior to the actual bulk phase transition. In contrast to pure 2-dimensional surface phenomena those surface effects are not bound to the outermost layers, but instead, a disordered surface sheet of mesoscopic or even macroscopic thickness may emerge[1]. In addition the free surface allows surface fields enter the scenario giving rise to new phenomena which have no counterpart in the bulk, one of them being surface segregation, i.e. the enrichment of one component at the surface. Many of the experimental evidence of disordered and segregated surface sheets have been found and corroborated by x-ray scattering in surface-sensitive modes, that is glancing-angle diffraction and asymptotic Bragg scattering which are shortly discussed in the following, before the observed surface phenomena at $Cu_3Au(001)$ are presented.

SURFACE-SENSITIVE X-RAY SCATTERING

Several x-ray techniques which provide structural information from surfaces and interfaces have emerged in the past and are now routinely applied, as specular reflectivity, glancing angle diffraction and asymptotic Bragg scattering (crystal truncation rod scattering). In this work glancing-angle diffraction and asymptotic Bragg scattering have been applied and will, thus, shortly be introduced.

Glancing angle diffraction (GAD; Fig.1) exploits the phenomenon of total external reflection[1-4] and the associated limited penetration depth of the x-rays to several Ångstrom when the incidence angle α_i is less than the critical angle for total external reflection (α_c). The eva-nescent x-ray wave field decays exponentially versus depth and travels parallel to the surface, until it encounters a scattering cross section $d\sigma/d\Omega(\mathbf{Q})$. The resulting scattering intensity is observed under an inplane scattering angle 2θ and a glancing angle α_f. With respect to the surface the momentum transfer (scattering vector) $\mathbf{Q}=(\mathbf{p},q_z)$ decomposes into the parallel component $\mathbf{p}$ and the refracted normal component q_z. The evanescent Bragg scattering then reads[5]

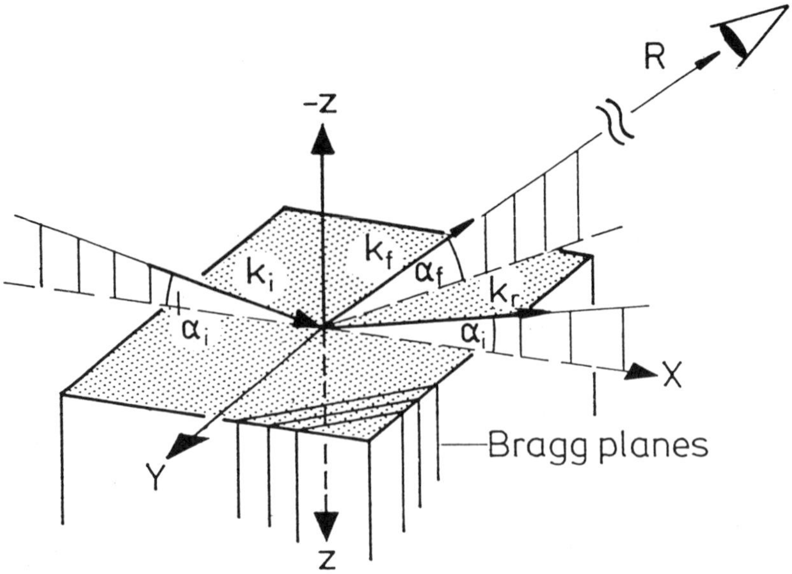

Fig.1: Glancing-angle scattering geometry with k_i and k_f as the incoming and scattered x-ray beam and α_i and α_f as the associated glancing angles.

$$(\frac{d\sigma}{d\Omega})_{GAD} = A(2\pi)^3 |e^{fi}|^2 r_e^2 \delta(p - g_{hk}) \left| \int_0^\infty \psi(z,t) e^{iq_z z} dz \right|^2 \qquad (1)$$

with $\psi(z,T)$ as the z-dependent order parameter (z=0 denotes the surface), $|e^{fi}|$ as the transmission-polarization tensor[2], A as the surface area and r_e as the classical electron radius. The escape depth („scattering depth")[3] $\Lambda=|Im(q_z)|^{-1}$ of the diffracted x-rays is typically 50Å for $\alpha_{i,f}$ less than the critical angle for total external reflection and several 10^4Å for steeper angles.

Asymptotic Bragg scattering (ABS) or crystal truncation rod scattering is a Bragg-like scat-tering signal caused by the presence of the surface[6,7] and is, in the ideal case, a two-dimensional inplane δ-function and smeared out between bulk Bragg nodes perpendicular to the surface . The scattering cross section reads[5]

$$(\frac{d\sigma}{d\Omega})_{ABS} = A(2\pi)^3 |e^{fi}|^2 r_e^2 \delta(p - g_{hk}) \left| \sum_l \frac{\Phi_G}{q_z - l} \right|^2 \qquad (2)$$

with Φ_G as the Fourier transform of the number density, l as the normal component of the reciprocal lattice vector and $\mathbf{G}=(g_{hk},l)$ as the reciprocal lattice vector decomposed as $\mathbf{Q}$ into the parallel and normal component.

SURFACE-INDUCED DISORDER AT CU3AU(001)

The structure of the ordered ($L1_2$) and disordered (A_2) state of the binary alloy Cu_3Au as well as the thermodynamic characteristics of the associated discontinuous order-disorder phase transition at T_0=663K have been studied in great detail in the bulk[8-10] and at the free surface[1]. The ordered structure is composed out of four interpenetrating simple cubic sublattices located at $\mathbf{r}_1$=(0 0 0), $\mathbf{r}_2$=(1/2 1/2 0), $\mathbf{r}_3$=(1/2 0 1/2) and $\mathbf{r}_4$=(0 1/2 1/2). By introducing the fractional occupation numbers x_i (associated with the sublattice i=1,2,3,4) which range from 1 for complete occupation by Au to 0 for complete occupation by Cu we construct the Landau-Lifshitz order parameter field

$$
\begin{aligned}
\Psi_1 &= x_1 - x_2 - x_3 + x_4 \\
\Psi_2 &= x_1 - x_2 + x_3 - x_4 \\
\Psi_3 &= x_1 + x_2 - x_3 - x_4 \\
\Psi_4 &= x_1 + x_2 + x_3 + x_4
\end{aligned}
\tag{3}
$$

Note here that $\Psi_4=4x_b$ with x_b as the average Au concentration as given by the actual stoichiometry. Within this formalism the ordered $L1_2$ structure of Cu_3Au is described by the order parameter $\Psi=(\Psi_1\ \Psi_2\ \Psi_3)$ = (1 1 1), (1 -1 1), (-1 1 -1) and (1 -1 -1) corresponding to the well-known four ordered states. The associated x-ray structure factor then reads

$$
F_{hkL} = \left\{ \begin{array}{lll}
(\Psi_4 f_{Au} + (4-\Psi_4) f_{Cu}) & h,k,L = unmixed & \\
\Psi_3(f_{Au} - f_{Cu}) & h,k,L = mixed & h+k = even \\
\Psi_2(f_{Au} - f_{cu}) & h,k,L = mixed & h+L = even \\
\Psi_1(f_{Au} - f_{cu}) & h,k,L = mixed & k+L = even
\end{array} \right\}
\tag{4}
$$

and contains, as one knows, the fundamental fcc reflections for h,k,L=unmixed and the three superlattice reflections related with the order parameter components Ψ_1, Ψ_2 and Ψ_3. In (4) we already used the notation (h,k,L) indicating the symmetry break associated with the (001) surface considered below.

The top layer displays a very pronounced continuous order-disorder transition[11]. It has been argued[12] that this surface effect is mediated by a wetting phenomenon associated with the first-order bulk transition. In the following we discuss its leading behaviour, namely, the growth of a disordered surface layer with effective thickness L(T). Then, the depth profile of the order parameter reads in leading order $\Psi(z,T)=\Psi_b(T)\ \Theta(z-L(T))$ with $\Psi_b(T)$ as the bulk order parameter and $\Theta(x)$ as the Heaviside step function. Inserting $\Psi(z,T)$ into Eq.(1) and introducing the lattice constant a, we get [3]

$$
\left(\frac{d\sigma}{d\Omega}\right)_{GAD}(q_z) = \left|e^{fi}\right|^2 (f_{Cu} - f_{Au})^2 e^{-2L(T)/\Lambda} |\psi_b(T)|^2 \frac{1}{|1-e^{iq_z a}|^2}
\tag{5}
$$

thus, a damping factor exp(-2L(T)/Λ) occurs which should readily be detectable with evanescent x-ray waves. This experiment has been performed by Dosch et al.[13] at the Hamburg Synchrotron radiation laboratory (HASYLAB). The observations, as observed at an atomically clean Cu_3Au (001) surface are summarized in Fig.2 which shows the α_f-(100)

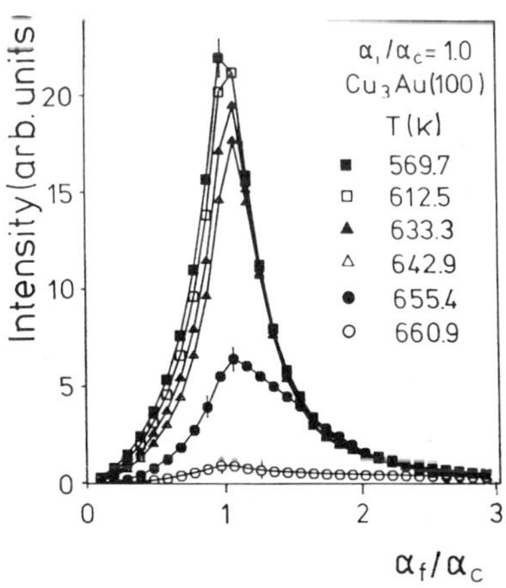

Fig.2: α_f-profiles of the (100) superlattice reflection as observed for α_i/α_c=1.0 and for some representative temperatures.

Bragg profiles for various temperatures below T_0. Most interestingly the Bragg profiles of the evanescent scattering exhibit a distinct temperature dependence, particularly for small values of Λ. This can be used for a quantitative test of the mean field predictions: Inserting the expected wetting law[12] $L(t) = \xi_d \ln|t|$ into Eq.(5) gives

$$\left(\frac{d\sigma}{d\Omega}\right)^{(\Lambda)}_{GAD} \propto |t|^{2\beta_\Lambda} = |t|^{2\xi_d/\Lambda}, \tag{6}$$

thus a powerlaw behaviour of the evanescent superlattice intensity. In (6) $t=(T_0-T)/T_0$ is the reduced temperature. Note further that the evanescent exponent $\beta_\Lambda=\xi_d/\Lambda$ in (6) scales as $1/\Lambda$, i.e. depends sensitively upon the glancing angles α_i and α_f. In Fig.3 the observed evanescent scattering associated with various fixed scattering depths Λ is plotted versus reduced temperature, disclosing an asymptotic powerlaw behaviour. The exponent β_Λ can be deduced from the observed slopes (full lines) and fulfills nicely the predicted relation $\beta_\Lambda=\xi_d/\Lambda$ (Fig.4) implying $\xi_d=(13\pm3)$Å. The leading behaviour of the temperature dependence of the order parameter near the $Cu_3Au(001)$ surface is thus consistent with a wetting transition, where an mesoscopic sheet with effectively $\Psi_1 = \Psi_2 = 0$ emerges at the free surface in thermal equilibrium. It should be noted here that one has found[14] some order parameter fluctuations of medium range (approx. 200Å) which may persist at the surface even for $T>T_0$. The origin of this surface-related medium range order is not known yet, but may be mediated by surface segregation phenomena which will be discussed now.

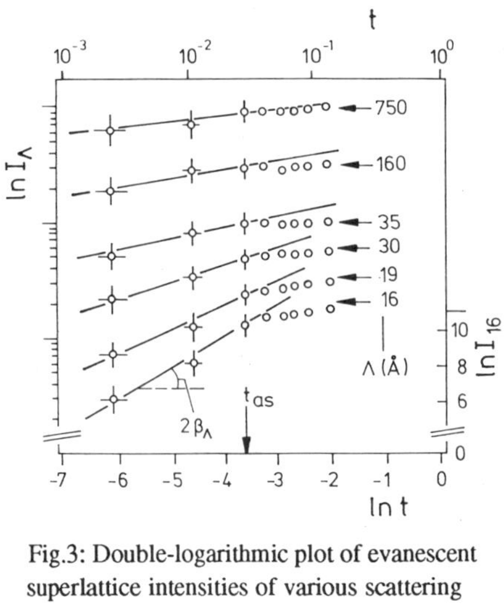

Fig.3: Double-logarithmic plot of evanescent superlattice intensities of various scattering depths Λ as a function of $t=(T_0-T)/T_0$.

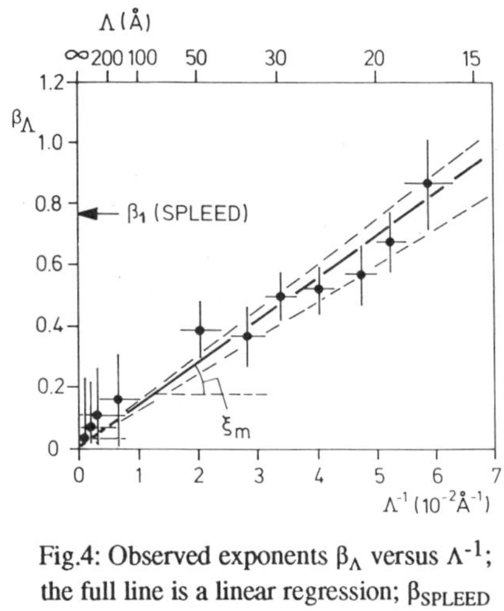

Fig.4: Observed exponents β_Λ versus Λ^{-1}; the full line is a linear regression; β_{SPLEED} denotes the toplayer exponent from ref.11.

SURFACE SEGREGATION AT CU3AU(001)

The surface segregation at the (001) surface has been studied carefully with ion scattering in a wide temperature range[15]. The experimental results showed quite clearly that the outermost layer prefers the Au component: a Au concentration of $x_{Au}(1) = 0.5$ is found almost independent of temperature, also rather unaffected by the order-disorder transition at 663K. Furthermore, an almost Au free second layer is found which is only very moderately populated with Au upon heating.

The complete (001) surface segregation profile has recently been obtained by asymptotic Bragg scattering in a wide temperature range above T_0[16,17]. The x-ray experiments have been carried out at the X16A beamline of the National Synchrotron Light Source

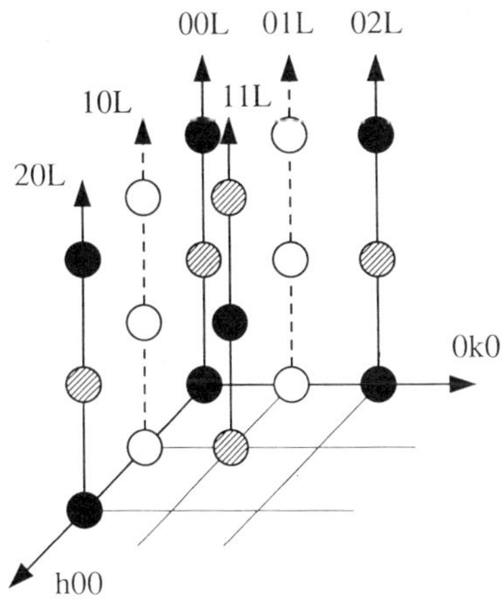

Fig.5: Reciprocal lattice of the (001) surface of Cu_3Au (the open symbols indicate the location of superlattice reflections).

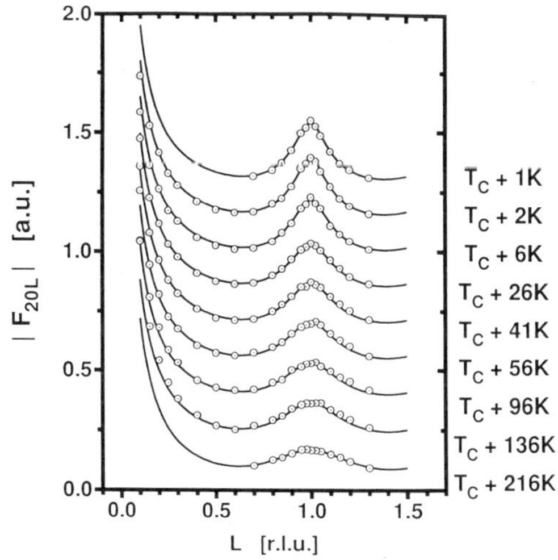

Fig.6: Observed intensity distribution along the (20L) Bragg tail for various temperatures above T_0. The full lines are least-squares fits, the inset shows the inplane width at T_0+2K.

(NSLS) of the Brookhaven National Laboratory. The temperature dependent intensity distribution along the (20L) direction (see Fig.5) has been recorded between L=0 and L=1.3. Fig.6 which shows some of the observed intensity profiles exhibit strong deviations from the expected $1/q_z^2$ -behaviour[6] associated with a nonsegregated surface (see also Eq.(2)). Interestingly, this new surface feature shows a pronounced temperature dependence and persists up the highest temperature of this study, $T=T_0+216K$. The full lines are least-squares fits to the experimental data with the asymptotic Bragg amplitude

$$F_{ABS} \propto \sum_{n=0}^{\infty} \{x_{Au}(T,n) f_{Au} + x_{Cu}(T,n) f_{Cu}\} e^{iq_z' na/2} w_\rho(T,n) \tag{7}$$

associated with an exponentially decaying, oscillatory surface segregation profile

$$x_{Au}(T,n) - 1 - x_{Cu}(T,n) = x_{Au}^b + \Delta x_{Au}(T)(-1)^{-n} e^{-na/\xi_{seg}(T)} \tag{8}$$

In (7) and (8) $x_{Cu,Au}$ and $f_{Cu,Au}$ are the temperature and layer (n=0,1,2,..) dependent concentrations and form factors of Cu and Au, respectively, ξ_{seg} is the decay length and $w_\rho(T,n)$ an appropriate factor which takes into account surface roughness. The deduced surface segregation profiles for three typical temperatures are shown in Fig.7: 216K above T_0 the surface segregation is essentially bound to the first 4 layers, upon approaching T_0 the decay length steadily increases until, at $T=T_0+1K$, the surface segregation is found in a mesoscopic surface layer (Fig.8). The source of this spontaneous thickening of the segregated surface layer is disclosed in the inset of Fig.8 which shows the decay length versus $t=(T-T_{sp})/T$ in a double-logarithmic plot. T_{sp} is the bulk spinodal ordering temperature[18], $T_{sp}=T_0-34K$. Apparently, the decay length obeys a powerlaw behaviour with exponent $\nu=0.49\pm0.04$ (as deduced from the slope in Fig.8), thus, the same mean field temperature behaviour as the bulk correlation length in the bulk short range order diffuse scattering[19].

The appearance of the bulk correlation in the asymptotic Bragg signal is at first sight unexspected, since ξ_b governs local atomic deviations from the average structure and, thus, usually appears in the bulk short range order diffuse x-ray scattering. The symmetry break at the surface, however, allows the occurence of local fields (h_1) decorated by an associated length scale (ξ_1) which governs a field-field correlation function. The situation is in fact very

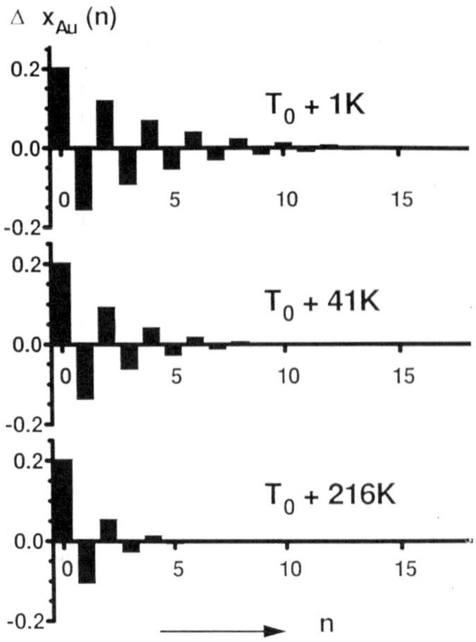

Fig.7: The decay of the Au excess versus layer index n=0,1,2,.. for three selected temperatures.

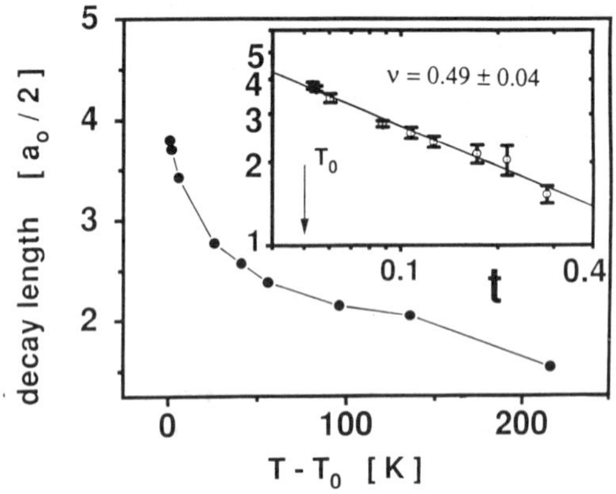

Fig.8: Temperature dependence of the decay length. The inset shows the relation on a double-logarithmic plot as a function of t.

similar to the recently discovered density oscillations close to the liquid Hg surface[20]: In this case the presence of the free surface induces local oscillations of the average liquid density which is governed by the correlations length of the bulk liquid structure factor.

CONCLUDING REMARKS

In the x-ray scattering studies presented in this article the surface effects onto the non-trivial 3-component order parameter associated with the bulk order-disorder phase transition in Cu_3Au have been investigated. While in the bulk the various components Ψ_i behave identical due to the cubic symmetry of the system, the presence of the surface induces an interesting anisotropy into their near-surface thermodynamic behaviour:

At the (001) surface discussed here the order parameter components Ψ_1 and Ψ_2 are equivalent and display the phenomenon of surface-induced disorder[12,1], i.e. upon approaching T_0 an excess disorder in the order parameter subspace $\Psi_{lat}=(\Psi_1\ \Psi_2)$ emerges at the surface regime. The perpendicular order parameter component Ψ_3, on the other hand, which induces an oscillatory Cu:Au-Cu-Cu:Au layering in the z-direction, is adsorbed at the (001) surface due to the action of heterogeneous surface energies (i.e. surface segregation energies which prefer Au at the toplayer in our case) and survives by this even up to very high temperatures. This rather involved surface behaviour may therefore be considered as a disorder wetting phenomenon in the presence of an external surface field which couples to only one order parameter component. These experimental studies, thus, demonstrate rather clearly that surface fields, which have been considered so far to play a major role only at ferromagnetic surfaces, are also most important in the theoretical understanding of the thermodynamic behaviour of the surface of binary alloys.

Acknowledgements

I thank R.L.Johnson, S.Krummacher, J.Peisl, H.Reichert and I.K.Robinson for pleasant collaboration, the HASYLAB staff for continuous support during our experiments. The work has been supported by the Bundesminister für Forschung und Technologie, Germany.

REFERENCES

1. H. Dosch. "Critical Phenomena at Surfaces and Interfaces: Evanescent X-ray and Neutron Scattering", *Springer Tracts in Modern Physics*, Vol. 126, Springer, Heidelberg (1992).
2. S. Dietrich and H. Wagner, Critical scattering of x-rays at grazing angles, *Z.Phys.* 56:207 (1984).
3. H. Dosch, B.W. Battermann, and D.C. Wack, Depth-controlled grazing-incidence diffraction of Synchrotron x radiation, *Phys.Rev.Lett.* 56:1144 (1986).
4. H. Dosch, Evanescent absorption in kinematic surface Bragg diffraction, *Phys.Rev.B* 35:2137 (1987).
5. S. Dietrich, A. Haase, Scattering of x-ray and neutrons at interfaces, *Phys.Rep.* (in press 1995).
6. S.R. Andrews and R.A. Cowley, Scattering of x-rays from crystal surfaces, *J.Phys.C* 18: 6427 (1985).
7. I.K. Robinson, Crystal truncation rods and surface roughness, *Phys.Rev.B* 33: 3830 (1986).
8. B.E. Warren. "X-ray Diffraction", Addison-Wesley, Reading, MA (1969).
9. J.M. Cowley, X-ray measurement of order in single crystals of Cu_3Au, *J.Appl.Phys.* 21:24 (1950).
10. S.C. Moss, X-ray measurement of short-range order in Cu_3Au, *J.Appl.Phys.* 35:3547 (1964).
11. S.F. Alvarado, M. Campagna, A. Fattah, and W. Uelhoff, Critical wetting and surface-induced continuous phase transition on $Cu_3Au(001)$, *Z Phys. B* 66:103 (1987).
12. R. Lipowsky and W. Speth, Semi-infinite systems with first-order bulk transitions, *Phys.Rev.B* 28:3983 (1983).
13. H. Dosch, L. Mailänder, A. Lied, J. Peisl, F. Grey, R. L. Johnson, and S. Krummacher, Experimental evidence for an interface delocalization transition in Cu_3Au, *Phys.Rev.Lett.* 60:2382 (1988).
14. K.S. Liang, private communication (1993).
15. T. Buck, G.H. Wheatley, and L. Marchut, Order-disorder and segregation behaviour at the $Cu_3Au(001)$, *Phys.Rev.Lett.* 51:43 (1983).
16. H. Reichert, P.Eng, H. Dosch, I.K. Robinson, Thermodynamics of surface segregation at $Cu_3Au(001)$ resolved by x-ray scattering, *Phys.Rev.Lett.* 74: 2006 (1995).
17. H. Reichert, H. Dosch, Surface segregation in $Cu_3Au(001)$, *Surf.Sci.* (in press 1995).
18. K.F.Ludwig, G.B.Stephenson, J.L. Jordan-Sweet, J. Mainville, Y.S.Yang, M. Sutton, Nucleated and continuous ordering in Cu_3Au, *Phys.Rev.Lett.* 61:1859 (1988).
19. H. Chen, J.B. Cohen, R. Gosh, Spinodal ordering in Cu_3Au, *J.Phys.Chem.Sol.* 38:855 (1977).
20. O.M. Magnussen, B.M. Ocko, M.J. Regan, K. Penanen, P.S. Pershan, M. Deutsch, X-ray reflectivity measurements of surface layering in liquid mercury, *Phys.Rev.Lett.* 74:4444 (1995).

SHORT-RANGE ORDER KINETICS IN α-AuFe AFTER DEFORMATION AND RECRYSTALLIZATION

M. Migschitz and W.Pfeiler

Institut für Festkörperphysik
University of Vienna
Strudlhofgasse 4
A-1090, Vienna, Austria

INTRODUCTION

In alloys, because of differences in interaction energies the atoms are not distributed at random over the lattice points but in many cases show a tendency to prefer unlike nearest neighbour atoms (short-range order, SRO) in solid solutions. Since the amount of these atomic correlations due to thermodynamic reasons depends on temperature, relaxations in SRO degree can be studied as a function of temperature. Changes in the degree of SRO as structural changes are brought about by migration of vacancies, therefore by studying SRO-relaxations vacancy parameters can be determined. In this case vacancies may not be observed directly but just via their contribution to atomic mobility. If a measuring method is at hand which is sensitive to changes in the degree of SRO vacancy properties can be studied in detail with high accuracy by measuring SRO-variations.

Electrical resistivity is very sensitive to changes in structural properties, especially to variations of atomic correlation effects[1]. If other structural changes are kept constant or are well known, changes due to atomic correlations (e.g. SRO) can be detected with very high accuracy. Measurements of electrical resistivity (*resistometry*) therefore in recent years turned out to be a very useful tool in investigating SRO-kinctics during isochronal and isothermal heat treatment[2] and activation enthalpies of SRO-relaxation were determined for several concentrated solid solutions.

Usually SRO-effects are investigated in homogeneous, well-recrystallized samples. A study of SRO-effects in deformed material, however, because of SRO hardening will be of interest with respect to materials application. Such an investigation may give information on (i) the degree of SRO in an as-deformed alloy, (ii) the evolution of SRO in presence of deformed structure and (iii) the influence of deformed microstructure on SRO-kinetics. A presupposition of such investigation in deformed material is that the deformed microstructure (dislocation density and arrangement, grain structure etc.) remains unchanged in the temperature range of SRO-investigation or is carefully studied.

In the present paper we show SRO-kinetics for several AuFe solid solutions during isochronal annealing treatment for both cases cold-rolled to about 80% thickness reduction and completely recrystallized. Using a simple model for SRO-induced resistivity changes, migration and formation enthalpies of vacancies as well as values of dislocation densities are determined directly from a comparison with experimental curves.

EXPERIMENTAL AND RESULTS

For details of alloying procedure and annealing treatment see [3]. Foils of about 0.1mm thickness were rolled at room temperature with intermediate annealings, the final rolling producing about 80% reduction. Resistivity measurements were done in a stirred bath of liquid nitrogen relative to a dummy specimen with an accuracy of about $\pm 3 \times 10^{-5}$. Microhardness measurements were done with a Reichert microhardness tester installed in a Reichert MeF inverted light microscope. Ten indentations per value were made at a load of 0.4N with an indentation rate of about 0.01mm per second.

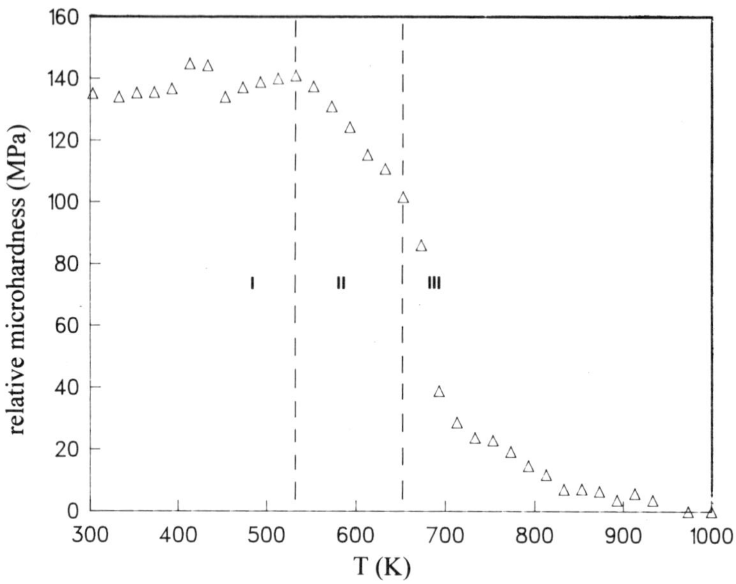

Figure 1. Microhardness *vs.* temperature of cold-rolled (ca. 80%) Au-20at%Fe during isochronal annealing.

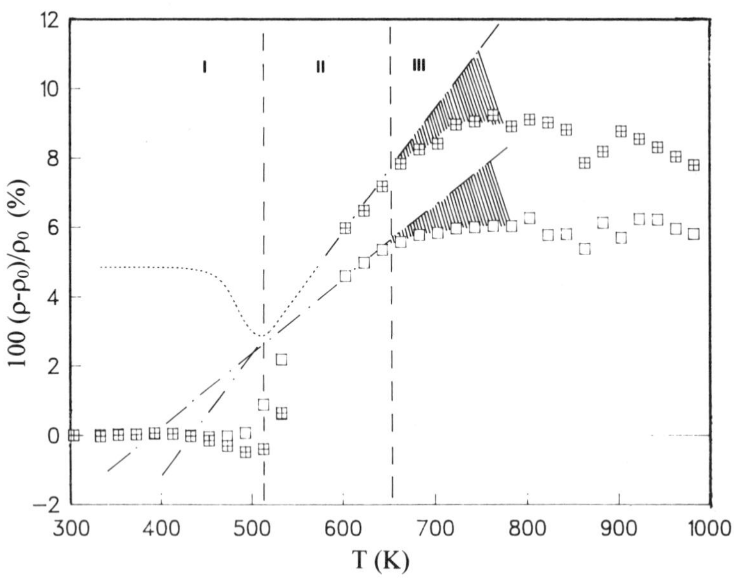

Figure 2. Resistivity change *vs.* temperature of cold-rolled Au-5at%Fe and Au-14at%Fe during isochronal annealing. Reduction: about 80%. Dotted curve: Au-14at%Fe, as-rolled + ½h at 573K and quench.

Figure1 gives the microhardness change of as-rolled Au-20at%Fe during isochronal annealing (ΔT=20K, Δt=15min) relative to the final value at 1023K. After a plateau at the beginning of the curve the hardness values between 530 and 650K decrease almost linear followed by a steep decrease between 650 and 690K and an exponential-like decay to zero. In figure 2 the change of resistivity during isochronal annealing of Au-5at%Fe and Au-14at%Fe is shown for the as-rolled state. Resistivity first remains constant then shows a very slight decrease at about 450K. Above about 510K a steep increase is observed followed by an almost linear increase between about 550 and 650K. At still higher tempera-

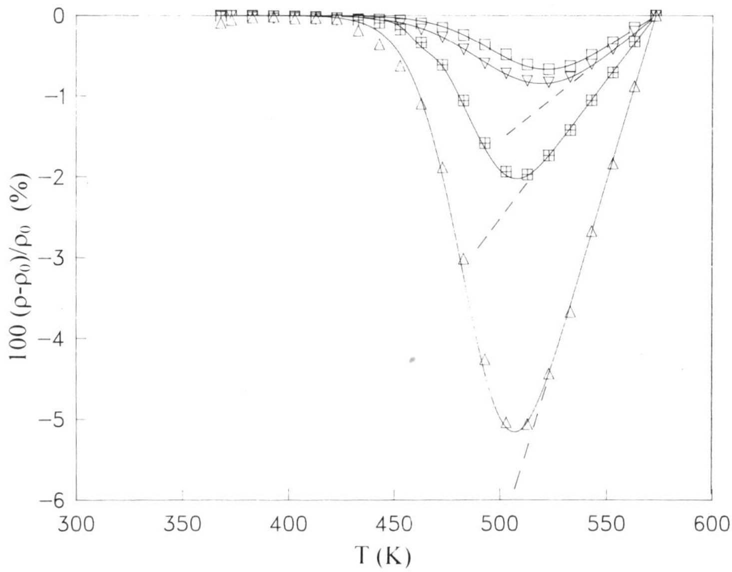

Figure 3. Resistivity change *vs.* temperature for isochronal annealing in state I. (□) Au-5at%Fe, (∇) Au-10at%Fe, (□) Au-14at%Fe, (Δ) Au-20at%Fe. Equilibrium lines: dashed.

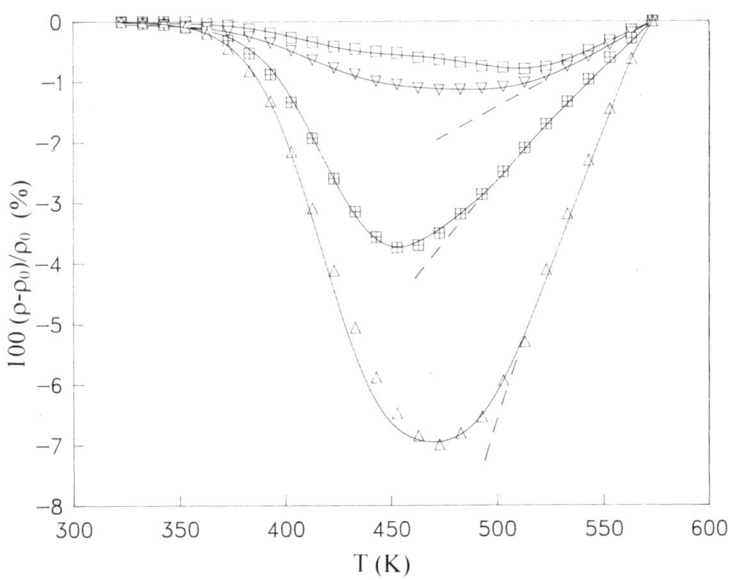

Figure 4. Resistivity change *vs.* temperature for isochronal annealing in state II. (□) Au-5at%Fe, (∇) Au-10at%Fe, (□) Au-14at%Fe, (Δ) Au-20at%Fe. Equilibrium lines: dashed.

tures a bending away from this linear increase is observed leading to a saturation (5 at%Fe) and a slight decrease (14 at%Fe), respectively. Figures 1 and 2 will be used to give evidence that no essential change in deformation structure occurs below 650K.

To obtain comparable initial states of SRO of both sets of samples (cold-rolled and recrystallized) the following heat treatment was applied: (i) ½ h annealing at 573K in the as-rolled state yielding cold-rolled state I; (ii) 8 h annealing at 1073K followed by ½ h annealing at 573K yielding recrystallized state II.

Figures 3 and 4 show the result of isochronal annealing in the temperature range between 330 and 570K (ΔT=10K, Δt=15min) of all samples investigated for the cold-rolled state I and state recrystallized state II, respectively.

Due to the usual interpretation of isochrones the observed curves can be explained as follows. As soon as atomic mobility is enabled by excess vacancies (deformation-induced or quenched-in) and/or thermally activated vacancies resistivity changes because the degree of SRO has to be increased in accordance with the instantaneous low temperature. In the present case this results in a decrease of resistivity. At a certain temperature (resistivity minimum) the actual degree of SRO corresponds to the instantaneous temperature almost as an equilibrium value. At still higher temperatures increase of resistivity signals decrease of SRO; within each isochronal time interval equilibrium values are adjusted (equilibrium curve).

Aside of some features often found in SRO-systems like the increase of both SRO-induced resistivity and atomic mobility with concentration there is a pronounced difference between deformed and recrystallized samples in the onset of resistivity change and depth of minimum.

DISCUSSION

SRO of the As-Rolled State and Post-Deformation Defect Annealing

The most significant feature of the resistivity isochrone in figure 2 is the relatively high degree of SRO of the initial state of both samples. This can be determined by intersecting the isochronal curves with the resistivity equilibrium lines (dash-dotted in figure 2) which are known from earlier investigation[3]. Such a procedure yields values of about 390K (Au-5at%Fe) and 430K (Au-14at%Fe) as equilibrium temperatures corresponding to the initial as-rolled states. This is in good correspondence with earlier results[3].

By comparing both changes of microhardness and of resistivity with increasing temperature during isochronal annealing three domains can be distinguished:

- I (<510K): Frozen atomic mobility; local increase of SRO by minor rearrangement of dislocations. Hardness remains essentially unchanged.
- II (510-650K): Strong decrease of SRO in total sample volume to reach equilibrium corresponding to instantaneous annealing temperature. This is accompanied by an almost linear decrease in hardness.
- III (>650K): Onset of dislocation loss. This leads to bending away of resistivity isochrone from equilibrium curve (shaded area in figure 2) and a drastic decrease of microhardness is observed.

It can be concluded from these observations that there is no essential change in deformation structure of the investigated alloys below about 650K and that measurements during isochronal annealing of samples in state I (as-rolled + ½h at 573K) up to 580K reflect SRO-induced changes only (dotted line in figure 2 for Au-14at%Fe).

Model for SRO-Kinetics

It has been shown earlier that single exponential SRO-relaxation during isothermal annealing can be described by[4,5]:

$$\frac{\rho(t)-\rho_\infty}{\rho_0-\rho_\infty} = \exp\left\{-t/\tau_{S\infty} - \tau_v\left(1/\tau_{S0} - 1/\tau_{S\infty}\right)\left(1-\exp\left[-t/\tau_v\right]\right)\right\} \quad (1)$$

with

$$1/\tau_v = N\ell^2 \nu_D \exp(-H_m/kT_a)$$

$$1/\tau_{S\infty} = (1/\tau_0)\exp(-H_m/kT_a)\exp(-H_f/kT_a)$$

$$1/\tau_{S0} = (1/\tau_0)\exp(-H_m/kT_a)\exp\left(-H_f/kT_{q_{eff}}\right)$$

τ_v	relaxation time for annealing out of excess vacancies,
$\tau_{S\infty}$	relaxation time for the adjustment of SRO after complete annealing-out of excess vacancies,
τ_{S0}	relaxation time at the beginning of isothermal annealing,
τ_0	pre-exponential factor (migration and formation entropy and correlation effects),
N	vacancy sink density = dislocation density,
ℓ	nearest neighbour jump distance,
ν_D	Debye frequency.

Formula (1) describes the SRO-induced normalised resistivity change brought about by surplus and thermal vacancies assuming a single migration and formation enthalpy. However, in a previous detailed study[3] it was shown that SRO-kinetics in AuFe solid solutions does not obey a single exponential relaxation meachanism but can be described reasonably well by assuming a log-normal distribution of SRO-relaxation times. This is in accordance with the model of distributed SRO[6], where due to a local variation of concentration a Gaussian distribution of diffusion activation energies is assumed leading in turn to log-normal distribution of relaxation times.

For the present investigation the above formalism was extended to describe isochronal annealing for the case of log-normally distributed relaxation times. Resistivity changes during isochronal annealing were obtained by summing the changes during subsequent isothermal annealings at each isochronal temperature. For an appropriate scaling of normalised resistivity changes the linearised experimental equilibrium curve is used. The total change during isochronal annealing was calculated by taking a weighted sum of isotherms obtained by using a gaussian distribution of enthalpies of migration and formation of vacancies around a mean value (for more details see[7]). Fitting the calculated curves to experimental data of isochronal annealing experiments yields H_m, H_f, N and τ_0 as fit parameters.

Determination of Vacancy Parameters

Comparing the SRO-behaviour shown in figures 3 and 4 striking differences between states I and II result. Whereas in state I atomic mobility seems to be caused by thermal vacancies only, in state II quenched-in surplus vacancies contribute to the ordering process. Since by quenching from the same initial temperature (573K) the number of quenched-in vacancies should be the same, these differences may be attributed to different annealing behaviour of vacancies due to different sink densities. The high dislocation density exceeding even 10^{12}cm/cm^{-3} for the case of a cold-rolled material may lead to a very high probability for complete annealing-out of excess vacancies before they can take part in the diffusion process, which is necessary for SRO-relaxation.

The above model describing SRO-kinetics during isochronal annealing experiment including surplus vacancies allows a determination of vacancy activation enthalpies as well as vacancy sink (dislocation) densities from a comparison with experimental results.
Table 1 gives the fit-parameters as obtained by fitting calculated curves to experimental points. As can be seen from the figures 3 and 4 an excellent fitting is obtained. The values

of vacancy parameters obtained are in good agreement with results of Mössbauer experiments on Au with 5 and 14at%Fe[8] (see table 1) and of diffusion experiments on pure Au[9] (H_m=0.80eV, H_f=0.96eV). Further, quite reasonable values for the density of dislocations are obtained, which are thought to act as principal vacancy sinks.

Table 1. Fit-parameters from fitting a kinetic model to experimental results of isochronal annealing on all samples investigated. Values for $H_{SRO}=H_m+H_f$ of earlier investigation [3] and values from the literature[8] are given for a comparison.

sample	state	H_m [eV]	H_f [eV]	H_{SRO} [eV] after [3]	H_{SRO} [eV] after [8]	$1/\tau_0$ [min^{-1}]	dislocation density N [cm^{-2}]
Au-5at%Fe	cold-rolled (I)	0.82	0.75	1.6		$1*10^{14}$	$1*10^{12}$
	recrystallized (II)	0.82	0.74	1.5	1.5	$1*10^{14}$	$5*10^{6}$
Au-10at%Fe	I	0.83	0.65	1.5		$1*10^{13}$	$5*10^{13}$
	II	0.79	0.65	1.4		$1*10^{13}$	$1*10^{6}$
Au-14at%Fe	I	0.90	0.81	1.6		$1*10^{16}$	$5*10^{13}$
	II	0.90	0.79	1.6	1.8	$1*10^{16}$	$5*10^{6}$
Au-20at%Fe	I	0.90	0.81	1.7		$1*10^{16}$	$5*10^{13}$
	II	0.90	0.82	1.7		$1*10^{16}$	$2*10^{7}$
accuracy		±0.05	±0.05	±0.1		±30%	±30%

CONCLUSIONS

(i) The initial degree of SRO in the as-rolled state of α-AuFe alloys can be higher than can be achieved by thermal treatment.

(ii) In deformed alloys quenched-in surplus vacancies due to the high sink (dislocation) density do not contribute to SRO-kinetics. Therefore SRO-kinetics can be studied under *equilibrium vacancy conditions* in this case.

(iii) Fitting a simple kinetic model to experimental curves of resistivity measurement during isochronal annealing vacancy activation energies of migration and formation are determined which are in good agreement with literature.

Acknowledgements

The work was financially supported by the Austrian FWF, grant no. P-9585.

REFERENCES

1. P.L. Rossiter „The Electrical Resistivity of Metals and Alloys,“ Cambridge Univ. Press, Cambridge (1987).
2. W. Pfeiler, Investigation of short-range order by electrical resistivity measurement, *Acta Met.*36:2417(1988).
3. M. Migschitz, F. Langmayr, and W. Pfeiler, Atomic short-range ordering in Au-Fe alloys: a resistometric study, *Mat. Sci. Eng.* A177:217 (1994).
4. A. Schulze and K. Lücke, The influence of vacancies on short-range order formation in Au-Ag-alloys, *Acta Met.*20:529 (1972).
5. W. Kohl, R. Scheffel, H. Heidsiek, and K. Lücke, Investigation of the kinetics of short range order formation and quenched-in vacancy annihiltation in Au14at%Ag by resistivity measurements, *Acta Met.* 31:1895 (1983).
6. J. Hillairet, Short-range ordering in relation to point defects and phase transitions, *Defect Diffus. Forum* 66-69:1015 (1989).
7. M. Migschitz and W. Pfeiler, Vacancy activation enthalpies of AuFe-alloys determined from SRO-induced resistivity change, *Mat. Sci. Eng.* submitted.
8. F. Langmayr, Hochtemperatur Mößbauerspektroskopie der Nahordnung in Au-Fe, *Diploma Thesis*, Universityof Vienna (1988).
9. N.L. Peterson, Self-diffusion in pure metals, *J. Nucl. Mater.* 69/70:3 (1978).

STABILISATION AND IRREVERSIBILITY OF MARTENSITE IN COPPER BASE SHAPE MEMORY ALLOYS

J. Dutkiewicz

Institute of Metallurgy and Materials Science of the Polish Academy of Sciences 30-059 Kraków ul.Reymonta 25

INTRODUCTION

Application of martensitic shape memory alloys requires reproducible transformation temperatures independently of the time of low temperature ageing, in order to ensure the shape recovery in the same temperature interval. Martensite in copper base alloys is a metastable phase, since it decomposes into equilibrium phases below the temperature of the start of reversible martensitic transformation - A_s. The tempered martensite cannot be retransformed to the parent β as shown in CuZnSi alloys[1]. The thermal stability of martensite is rather poor in CuZnSi alloys and in CuZnSn alloys decomposing rapidly at 200°C[1,2], while other alloys still retransform after prolonged ageing at 300°C as demonstrated by Dutkiewicz[3] for CuAlMn and by Kobus[4] in CuAlNiMnTi alloys. Martensite in copper base alloys possess in most cases an orthorhombic disordered 2H, 3R or 9R or ordered 4H, 6R or 18R structures[5]. These structures result from a periodic arrangement of stacking faults and usually possess a high density of random stacking faults. It is caused by a low stacking fault energy (SFE) in these alloys below 30 erg/cm^2. In fact alloying additions used in shape memory alloys were reported to lower the SFE in α phase copper base alloys as summarised by Gallagher[6]. When alloys are kept in the martensitic state, they tend to raise the retransformation temperature to parent β as pointed out by Hansen[7] and Scarsbrook[8] . This phenomenon is known as stabilisation of martensite. Several mechanisms have been proposed for the stabilisation of martensite: (i) vacancy pinning of interfaces[7,9,10] preventing martensite/matrix interfaces to move; higher temperature is needed to overcome its resistance. (ii) reduction of the strain energy in the martensite due to reordering of the first neighbours of atoms of different sizes on the lattice sites[11]; this supposition has been confirmed experimentally using ALCHEMI technique[12], showing changes in the martensite order after stabilisation. (iii) decrease of the stacking fault density in the martensite upon ageing[12]; it takes away a part of lattice invariant shear preventing the reverse transformation to occur. It requires an extra driving force to introduce a new shear. It was shown that alloy composition, grain size, martensite microstructure and defects have an influence on the stabilisation. It was suggested recently that a change in atomic arrangement within martensite occurring during ageing[13] is the primary mechanism responsible for the stabilisation of martensite .

The stabilisation can occur also in the stress induced martensite[14]; then the retransformation stress is decreased to lower values. It was shown later that less common types of martensite like 6R, can be stabilised in a similar manner[15]. The stabilisation of 6R and 18R phases depends on the relative amount of these phases affecting the tetragonal distortion of 6R phase, as pointed out by Saule and Ahlers[16].

The stress/strain curves in CuAlZn alloys[7] show a different superelastic behaviour i.e. a varying degree of strain recovery after unloading, depending on quenching rate. Under certain conditions the martensite formed, even at a small deformation of CuAlFe single crystals, did not retransform to the parent phase after strain release[17], so one can assume it was stabilised. In fact, it is sometimes difficult to distinguish, when the martensite is stabilised and when it is irreversible, without heating it above the temperature corresponding to the end of the reverse transformation - A_f. In the present paper some problems of stabilisation of thermally and stress induced martensite will be discussed in CuZnSi and CuAlMn alloys.

EXPERIMENTAL PROCEDURE

CuZnSi alloys were melted under a salt protection layer, cast in an iron mould and hot rolled down to 0.2 mm. Alloy A contained 30.7 at.% Zn and 4.3 at.% Si (M_s= -23°C), alloy B 30.1 at.% Zn, 4.5 at.% Si (M_s= 10°C). CuAlMn alloy (C) was cast in the Balzers induction furnace under argon atmosphere, cast in an iron mould and hot rolled down to 0.2 mm strips. It contained 20.8 at.% Al and 9.5 at.% Mn (M_s= -50°C). The CuZnSi alloys were quenched from 850°C into water of 60°C, allowing retransformation; or to room temperature water, which caused irreversibility of martensite when heated at the rate of 20°C/min. The alloys for stabilisation experiments after quenching to 60°C were cooled down to -60°C, then heated in a Perkin Elmer DSC-4 at the rate 0.2°C/min which stabilised the martensite. After that, samples were cooled and heated in the DSC at the rate of 20°C/min. Thin foils were obtained by jet electropolishing in an electrolyte consisting of 1/3 HNO_3 and 2/3 CH_3OH at -40°C. They were observed using JEOL 200CX and Hitachi H600 transmission electron microscopes. CuAlMn alloys single crystals were obtained using a Bridgman method; then they were cut in slices parallel to {001} and {100} β_1. Tensile samples of size 3x10x0.1 mm were cut along <001> β_1 direction. They were tensile tested in situ using JEOL 1000 kV TEM using a specially designed tensile stage[17].

RESULTS AND DISCUSSION

Figure 1 shows a set of DSC heating curves of alloy 1. It can seen that the reverse transformation during the first cycle starts at about -10°C (A_{s1}). The subsequent DSC curves corresponding to the next transformation cycles are shifted to lower temperatures as the stabilisation of the martensitic transformation is gradually deactivated. It indicates a progressive removal of the structural factors responsible for the stabilisation. It is highly probable that there are changes in a short range order occurring gradually within the martensite, thus lowering its free energy. In case the stabilisation is connected with pinning of interfaces[7,10] or dislocations, such barriers should be dissolved in the first cycle and no gradual return to a nominal retransformation value would be observed. Microstructures of as quenched and stabilised alloy A are given in Fig.2. One can see that in both cases martensitic plates possess a high basal stacking fault density, while the stabilised alloy additionally shows a large number of nonbasal faults. Similar types of defects were reported in 18R martensite in CuZnAl alloys [18]. The monoclinic distortion of as quenched martensite measured between observed in ice brine quenched CuZnSi alloys[20], but no information was given if it retransformed to the parent β. Alloy B continuously heated to 400°C at the rate of 20°C/min

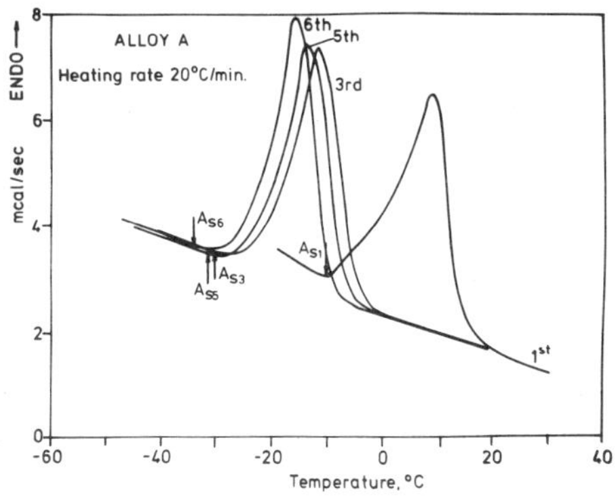

Fig.1 Set of DSC curves showing gradual return of reverse transformation temperatures in alloy A, stabilized during heating.

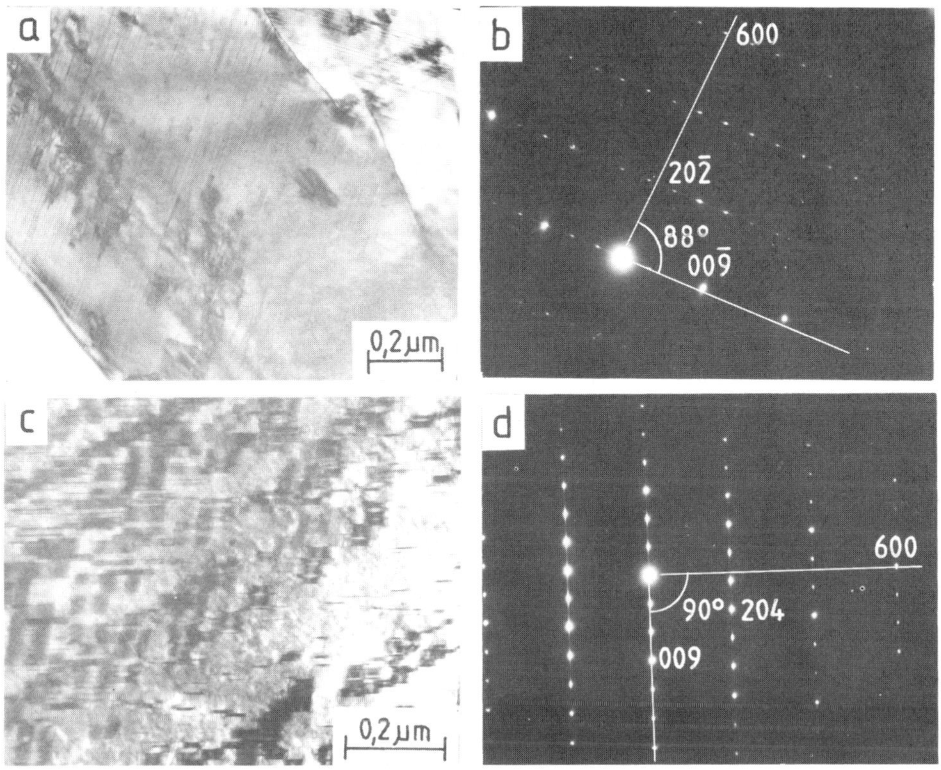

Fig.2 Set of transmission electron microstructures (a,c) and electron diffraction patterns (b,d) of as quenched and ones cycled (a,b) and stabilised during slow heating alloy A.

a and *b* axes in Fig.2b is equal to 88°, while that of a stabilised one is equal to 90° (Fig.2d). The changes of the monoclinic distortion with stabilisation were reported for CuZnAl alloys[19], where the distortion angle β in the stabilised samples was reported to be 90°, similar to the present case.

Fig.3 shows a lattice image taken from alloy B quenched to RT water. It shows fringes at the distance of 0.63 nm from (009) planes of 9R martensite and fringes at the distance of 0.21 nm from (111) planes of 3R martensite. The thin layers of 3R martensite were not possible to detect using a conventional electron microscopy. Selected area diffraction pattern (SADP) does not allow us to identify the 3R phase due to streaks coming from a high density of random stacking faults. The 3R-stacking sequence within the 9R martensite was already

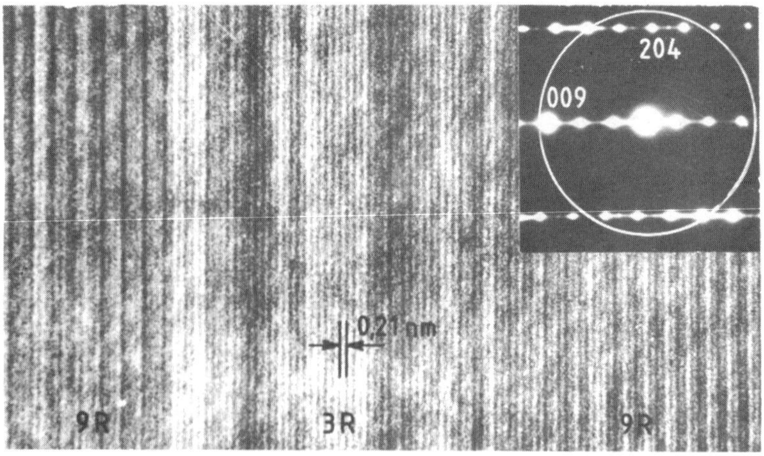

Fig.3 Lattice image from as quenched alloy B. Corresponding SADP is placed in the corner.

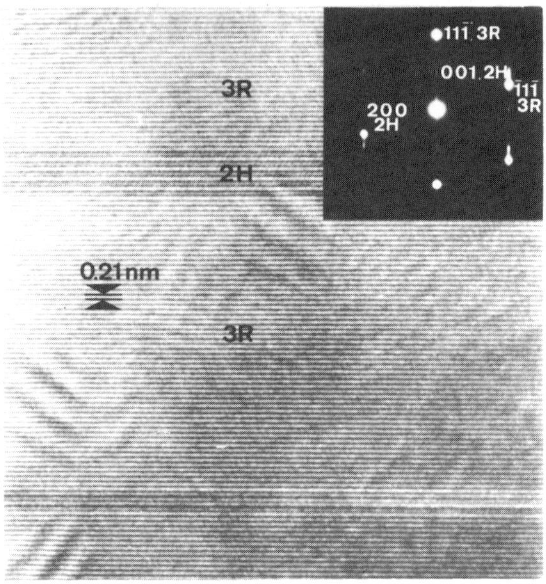

Fig.4 Lattice image of stabilised alloy B. Corresponding SADP is placed in the corner

shows quite a different structure (Fig.4) consisting of (111) fringes of 3R martensite and double spaced (001) fringes of 2H layers. In this case corresponding phases can be indexed based on the SADP placed in the corner of the micrograph. The 2H martensite was observed already after continuous heating to 200°C and at higher temperatures the 3R phase has a tendency to grow, which indicates that it has a lower free energy at higher temperatures than the metastable 9R martensite. During heating one observes the 9R -> 3R transformation rather than the 9R -> β. This can be expected, since the α phase is an equilibrium phase in these alloys at room temperatures. The formation of 3R layers prevents the formation of β phase during heating since a direct formation of 3R i.e. α phase, consumes less energy to obtain an equilibrium. One can assume that a tendency of CuZnSi alloys to form 3R layers is the reason of irreversibility of martensite, but this is not directly related with the stabilisation of martensite in a common sense. In order to study the structure of residual martensite formed during pseudoelastic deformation and not reverted back on unloading, in situ studies of deformation of alloy C were performed. Fig.5a shows a load elongation curve of an in-situ pseudoelastically deformed sample in $[100]\beta_1$ direction at $(001)\beta_1$ foil orientation. Almost a

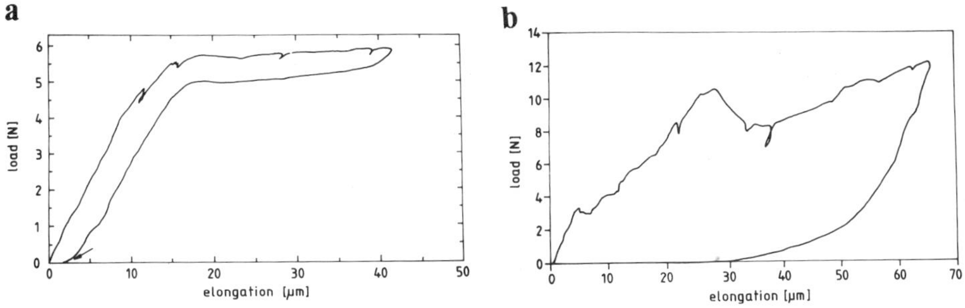

Fig.5 Load/elongation curves from in-situ TEM tensile experiment of alloy C at $[100](001)\ \beta_1$ (a) and $[110](011)\beta_1$ (b) orientations.

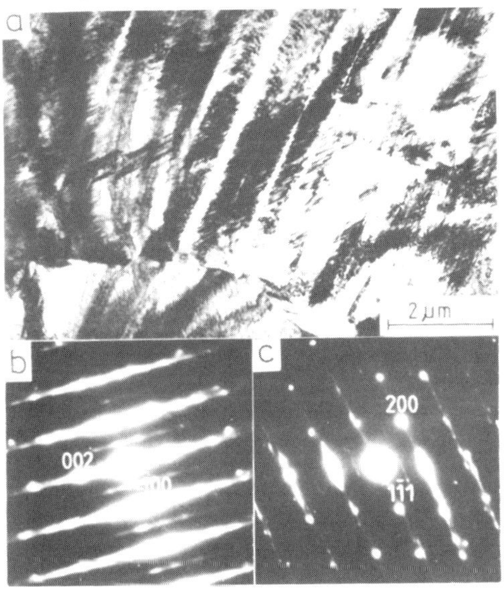

Fig.6(a) Transmission electron micrograph taken after unloading, showing residual (irreversible) martensite of extremely high stacking fault density. (b) SADP from the plates in the bottom part. (c) SADP from the upper plates.

perfect shape recovery can be seen after unloading, but some permanent deformation is left at elongation being only about 2%. Bulk samples have shown perfect pseudo-elasticity up to 6% of strain[21]. The permanent deformation is much larger with a [011] tensile direction (Fig.5.b). This is most probably caused by the unfavourable orientation with respect to the strain induced shear type transformation. The deformation proceeds then only partially by strain induced martensite and partially by plastic deformation. Fig.6 shows a transmission electron micrograph taken of the alloy after unloading (see arrow in Fig.5a). An extremely high density of random stacking faults in one plate and the presence of 3R martensite in the other plate of irreversible martensite is similar to that in the as quenched one.

CONCLUSIONS

1.CuZnSi alloys are easily stabilised, even during slow heating at 0.2°C/min. Subsequent heating cycles cause a gradual return to the nominal transformation temperatures. It indicates that short range diffusion, causing rearrangement of atomic positions within martensite is responsible for the stabilisation. The pinning mechanism is not likely to occur.

2. Quenching of CuZnSi alloys directly to water below the M_s temperature causes formation of a very high stacking fault density and formation of narrow layers of 3R martensite. Subsequent heating causes growth of the 3R layers, rather than retransformation to the β phase.
3. The residual martensite in CuAlMn alloys not retransformed after stress release shows similar features like that irreversible in CuZnSi alloys showing extremely high random stacking fault density and presence of the 3R stacking sequence.

REFERENCES

1. J. Dutkiewicz and T. Czeppe , Stabilization of Martensite in CuZnSi alloys, in The Martensitic Transformation in Science and Technology ed. E.Hornbogen, N.Jost, DGM Verlag,1989
2. J. Dutkiewicz and J. Morgiel, Effect of ageing on martensitic transformation in CuZn and CuZnSn alloys, ref.1 p.141
3. J.Dutkiewicz, E.Cesari, C.Segui and J.Pons, Response of CuAlMn alloys to ageing in β phase, *J.de Physique IV, Colloque C4,* 229, 1991
4. E. Kobus, S. Eucken, E. Hornbogen, Characterisation of the Microstructure and Shape memory effect of a CuAlNiMnTi alloy for high temperature applications, *Pract. Met.*, 26, 318, (1989)
5. B.I. Nikolin, Mnogosloinye struktury i politipizm v metallicheskikh splavakh, Naukova Dumka, Kiev, 1984
6. P.C.J. Gallagher,The Influence of Alloying, temperature and related effects on the stacking fault energy, *Met Transactions,* 1, 2429, 1970
7. J. Jansen, J. Van Humbeeck, M. Chandrasekaran, N. Nwamba and L. Delaey, Stabilisation of martensite in Copper-Zinc-aluminium alloys, *J. de Physique, Colloque C4-43,* 639, 1982.
8. G. Scarsbrook, J. Cook, and W.M. Stobbs, Ageing effects in CuAlZn alloys, *J. de Physique, Colloque C4-43,* 703, 1982, p.703
9. T. Suzuki, R. Kojima, Y. Fuji and A. Nagasawa, Reverse transformation behaviour of the stabilized martensite in CuZnAl alloy, *Acta metall.*, 37, 163, 1989.
10. `M.Morin and G.Guenin, Study of the Pinning of the martensite interfaces in a Cu-Zn-Al alloy by internal friction measurements, *Materials Science Forum*, 56-58, 499, 1990.
11. A.Abu Arab and M. Ahlers, The stabilization of martensite in CuAlZn alloys, *Acta metall* 36, 2629, 1988.
12. Y. Nakata, O. Yamamoto and K. Shimizu, Effect of ageing in CuZnAl shape memory alloys, *Materials Transactions JIM*, 34, 429, 1993.
13. M. Chandrasekaran, E. Cesari, J. Wolska, I. Hurtado, R. Stalmans and J. Dutkiewicz, Stabilisation of martensite in copper base shape memory alloys, *J. de Physique*, Colloque C2, 5, 143, 1995.
14. M. Ahlers, The stabilization of martensite in CuAlZn alloys, in ICOMAT 86, ed. Japan Institute of Metals, p.786, 1986.
15. F.Saule, A.Tolley and M. Ahlers, The stabilization of 6R martensite in Cu-Zn-Al single crystals, *Scripta met. materialia*, 24, 363, 1990.
16. F. Saule, M. Ahlers, Stability, Stabilization and lattice parameters in Cu-Zn-Al martensites, *Informe Technico CNEA-CAB,* 4042, 1994.
17. J. Dutkiewicz, V.V. Martynov and U. Messerschmidt, Structure of martensite formed in CuAlFe single crystals during in-situ HVEM pseudoelastic tensile experiment, *J. Materials Sci.*, 24, 3904, 1989.
18. M. Andrade, L. Delaey and M. Chandrasekaran, On some lesser known planar defects in β_1' Cu-Zn-Al martensite, *J.de Physique* C4-43, 673 1982
19. L.Delaey, T. Suzuki and J. Van Humbeeck, The stabilisation of step quenched copper-zinc aluminium martensite, part II Crystal structure and reordering, *Scripta Met.*, 18, 899, 1984.
20. I.Cornelis and C.M. Wayman, Direct observation of the 3R stacking sequence in CuZnSi martensite, *Mat. Res. Bull.* 9, 1057, 1974.
21. H. Kato, J. Dutkiewicz and S. Miura, Pseudoelasticity and shape memory effect in Cu-23at.%Al-7at.%Mn alloy single crystals, *Acta metall. mater.,* 42, 1359, 1994.

CRYSTAL STRUCTURE AND PHYSICAL PROPERTIES OF HIGH PRESSURE HYDRIDES

V.E. Antonov

Institute of Solid State Physics
Russian Academy of Sciences
142432 Chernogolovka Moscow distr.
Russia

INTRODUCTION

The technique for compressing gaseous hydrogen to pressures of up to 9 GPa at temperatures to 500 °C (and recently[1] to 1000 °C) developed at ISSP RAS made it possible to synthesize bulk homogeneous hydrides of a number of Group VI–VIII transition metals which cannot be hydrogenated by standard procedures[2]. This enabled us to prepare hydrides of all 3*d*- and 4*d*-metals (except ruthenium) and of many their alloys, and to systematically study their crystal structure and physical properties as function of the position of the initial metals in the Periodic Table (Fig. 1). Most of the hydrides are nonstoichiometric interstitial phases (solid hydrogen solutions) with wide homogeneity ranges of composition, that allowed the investigation of their properties as function of the hydrogen concentration as well.

In order to explain the observed dependences of the magnetic properties of the 3*d*-metals and their alloys on the hydrogen content, the phenomenological rigid *d*-band model was proposed[2,3]. Studies of the superconducting properties of high pressure hydrides of the 4*d*- and 5*d*-metals and their alloys led to the discovery of new superconducting hydrides[4–6] and showed that with proper corrections for the hydrogen-induced changes in the phonon spectra, the rigid *d*-band model could also be used to explain the observed superconducting phenomena[5]. And finally, the model was found to be applicable to the description of the crystal structures of the hydrides for rather large groups of *d*-metals and alloys[6].

The model seems to be the only guide now allowing understanding and, under certain conditions, predicting the magnetic and superconducting properties and crystal structures of the transition metal hydrides, and below we shall briefly discuss how it works.

Stability of Materials
Edited by A. Gonis *et al.*, Plenum Press, New York, 1996

<table>
<tr><td>$\alpha \rightarrow \varepsilon$
$CrH_{1.0}$
af → p</td><td>$\alpha' \rightarrow \varepsilon,\gamma$
$MnH_{0.96}$
af → af</td><td>$\alpha \rightarrow \varepsilon'$
$FeH_{0.8}$
f → f</td><td>$\varepsilon \rightarrow \varepsilon,\gamma$
$CoH_{1.0}$
f → f</td><td>$\gamma \rightarrow \gamma$
$NiH_{1.1}$
f → p</td><td>γ
Cu
d</td></tr>
<tr><td>$\alpha \rightarrow \varepsilon$
$MoH_{1.05}$
sc → sc</td><td>$\varepsilon \rightarrow \varepsilon$
$TcH_{0.6}$
sc → sc</td><td>$\varepsilon \rightarrow \varepsilon$
$RuH_{0.03}$
sc → sc</td><td>$\gamma \rightarrow \gamma$
$RhH_{1.0}$
p → p</td><td>$\gamma \rightarrow \gamma$
$PdH_{1.0}$
p → sc</td><td>γ
Ag
d</td></tr>
<tr><td>α
W
sc</td><td>$\varepsilon \rightarrow \varepsilon$
$ReH_{0.22}$
sc → sc</td><td>$\varepsilon \rightarrow \varepsilon$
$OsH_{0.003}$
sc → sc</td><td>γ
Ir
sc</td><td>γ
Pt
p</td><td>$\gamma \rightarrow$ orth
$AuH_{0.4}$
d</td></tr>
</table>

Figure 1. Hydrides synthesized under high hydrogen pressures[2]. The indicated hydrogen concentrations are the maximum attained ones. The arrows show the direction of the changes in the crystal structure and magnetic and superconducting properties of the starting metals on hydrogenation. Metal lattices: $\alpha = bcc$, $\alpha' = \alpha$-Mn type, $\varepsilon = hcp$, $\varepsilon' = dhcp$, $\gamma = fcc$, orth = orthorombic. Magnetic state at low temperatures: af = antiferromagnetic, f = ferromagnetic, p = paramagnetic, d = diamagnetic. sc = superconductor. The hydrides of Fe, Co, Mo, Tc, Rh, Re, and Au were first synthesized at ISSP RAS.

MAGNETIC PROPERTIES

The rigid d-band model takes into account the specific changes in the band structure of transition metals on hydrogen uptake first outlined by A.C. Switendick[7] and then confirmed by many other calculations for various d-metals. According to the calculations, the hydrogen leaves the upper part of the d-band of the host metal nearly unaffected, but noticeably decreases the energy of the s-states. The latter lowers the rate, η, of d-band filling with the electrons supplied by the hydrogen to a value of the order of $\eta = 0.5$ electrons per H atom (varying from ≈ 0.3 to ≈ 1 for different metals and alloys). This concept implies that those properties of the metal which essentially depend on the d-band occupation should vary with increasing hydrogen concentration as if the hydrogen were merely donor of a fractional number of $\eta \approx 0.5$ electrons per atom into the otherwise unchanged metal d-band. This is the approximation that we call a rigid d-band model for brevity.

The $3d$-metals, which are close neighbors in the Periodic Table, form broad ranges of fcc (γ) and hcp (ε) continuous solid solutions with magnetic properties well described by the rigid band model[8]. These properties thus can be represented by unique functions of the average number N^e of external ($3d + 4s$) electrons per atom of the alloy (the so-called Pauling–Slater curves), see Fig. 2. The properties of the γ and ε hydrides of such alloys should therefore be the same functions of the effective electron concentration $N^e_{\text{eff}} = N^e + \eta \cdot n$, where n is the H–to–metal atomic ratio.

The available data on the hexagonal hydrides are presented in Fig. 2 and agree with this statement. As for the cubic hydrides, such an agreement can be most clearly demonstrated in the case of the Fe-Ni-H γ-solutions, see Fig. 3.

The spontaneous magnetization σ_0 at $T = 0$ K of the alloys containing up to 60 at.% Fe decreases approximately linearly with increasing n (Fig. 3a) and increasing

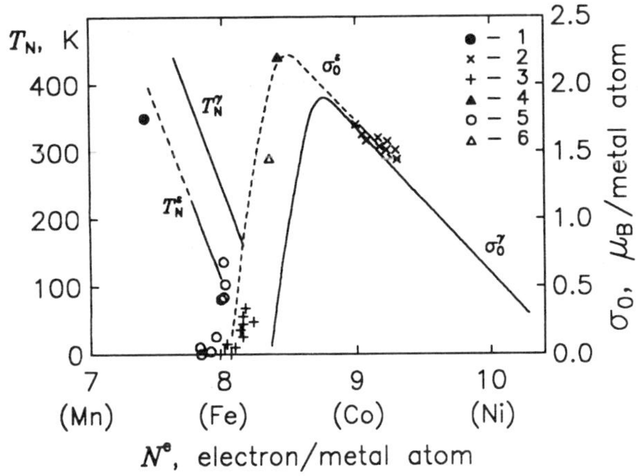

Figure 2. Concentration dependences[3] of σ_0 and Néel temperature, T_N, for the *fcc* (γ) (solid lines, experimental data) and *hcp* (ε) (sections of solid lines, experimental data, and dashed lines, an estimation) alloys of 3*d*-metals that are nearest neighbors in the Periodic Table, and also the experimental values of T_N for the antiferromagnetic manganese ε-hydride (1) and of σ_0 for the ferromagnetic ε-solutions Co-H (2), $Fe_{77.6}Mn_{22.4}$-H (3), ε'-hydride of iron (4), ε-hydrides $Fe_{75}Cr_{25}$-H (5), and $Fe_{94.7}Cr_{5.3}$-H hydride with a Sm-type metal lattice (6) at $\eta = 0.5$ el. per H atom.

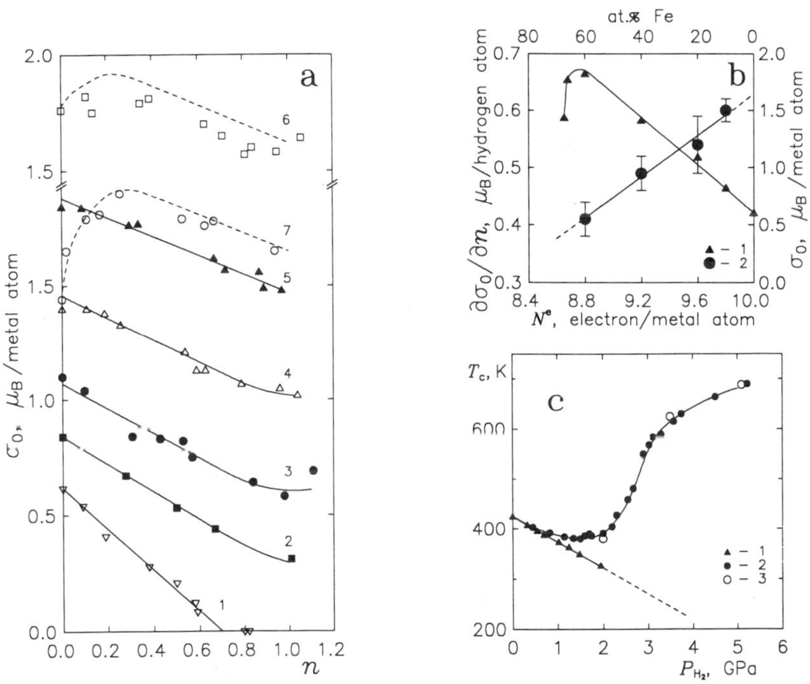

Figure 3. The effect of hydrogen on the magnetic properties of *fcc* Ni-Fe alloys[2]. (a) The spontaneous magnetization σ_0 at $T = 0$ K and under ambient pressure as function of the hydrogen–to–metal atomic ratio n. At. % Fe: 1 – 0, 2 – 10, 3 – 20, 4 – 40, 5 – 60, 6 – 66.1, 7 – 67.5. μ_B is Bohr magneton. (b) The spontaneous magnetization σ_0 of the Ni-Fe alloys (1) and the slope of the approximately linear dependences $\sigma_0(n)$ for the solid hydrogen solutions in the alloys containing 10, 20, 40, and 60 at. % Fe (2) as functions of the average number N^e of $(3d + 4s)$ electrons per atom in the initial *fcc* Ni-Fe alloys. (c) The experimental pressure dependences of the Curie temperatures T_c of the $Ni_{32.5}Fe_{67.5}$ alloy in an inert medium (1) and in hydrogen (2), and the $T_c(P_{H_2})$ values (3) calculated for $\eta = 0.39$ el. per H atom.

N^e (Fig. 3*b*, $\partial\sigma_0/\partial N^e \approx 1$ el./metal atom). The linear extrapolation of the coefficient of proportionality $\eta = (\partial\sigma_0/\partial n)/(\partial\sigma_0/\partial N^e)$ gives $\eta \approx 0.39$ el./hydrogen atom for the alloys with 66.1 and 67.5 at.% Fe (Fig. 3*b*). The $\sigma_0(n)$ dependences for these alloys (dashed lines in Fig. 3*a*) constructed from the $\sigma_0(N^e)$ dependence (Fig. 3*b*) using this scaling factor η, agree with the experiment. A good agreement is also observed between the experimental $T_c(P_{H_2})$ dependence for the $Ni_{32.5}Fe_{67.5}$ alloy (Fig. 3*c*) and the values calculated from the $T_c(N^e)$ and $dT_c(N^e)/dP$ dependences for the alloys without hydrogen using the same factor $\eta = 0.39$ el./hydrogen atom (the hydrogen concentration of the alloy in a hydrogen atmosphere monotonously increases on going along the $T_c(P_{H_2})$ line up to $n \approx 1$ at $P_{H_2} = 5.1$ GPa).

Magnetic properties were also studied for the hydrogen γ solutions in many other alloys of 3*d*-metals, including the alloys in which changes in the composition resulted in a strong deformation of the *d*-band (Ni-Cr, Ni-Mn). In all cases the rigid *d*-band model provided at least semiquantitative descriptions of the observed effects[2].

SUPERCONDUCTIVITY

In contrast to alloys of 3*d*-metals, the applicability of the rigid band model to alloys of 4*d*- and 5*d*-metals is usually limited to intervals of a few atomic per cent. There is a correlation, however, known as Matthias rule, between the superconducting temperature T_c and the electron concentration N^e, and the available data make it possible to specify at least one more type of alloys (let us call it *RN*, "remote plus neighboring") with superconducting properties obeying the rigid band model. These are $A_{1-y}(B,C,\ldots)_y$ alloys or intermetallic compounds with fixed y, where the elements B, C, *etc.*, are close neighbors, whereas element A is positioned far from them in the Periodic Table. In such systems replacing, for instance, element B by element C or varying their relative concentrations produces only minor changes in the band structure of the alloy. Wherever data are available, the Matthias plots for these *RN* type alloys and primary solid solutions of neighboring elements exhibit a large broad peak of T_c centered at $N^e \approx 6.3$ to 6.7 el./metal atom.

The formula $T_c(n) = T_c(N^e + \eta \cdot n) \approx T_c(N^e + 0.5 \cdot n)$ semiquantitatively describes the experimental $T_c(n)$ dependences for hydrogen solid solutions in *hcp* Re, Ru and Tc and accounts for the absence of superconductivity in the *fcc* Rh hydride with $N^e_{eff} \approx 9.5$ el./atom and for its presence in the *hcp* Mo hydride and the *fct* dihydride of the $Nb_{88}Rh_{12}$ alloy with $N^e_{eff} \approx 6.5$ el./atom (see ref. 6 for references). As one can see from Fig. 4, this formula is also more or less acceptable in the case of the hydrides with the *A*15 metal lattice.

In the hydrides on the base of *bcc* and related metal lattices, such as the *A*15 lattice, the *d*-band filling is not, however, an unconditionally dominating mechanism[5]. The changes caused by the *d*-band occupation are accompanied with a steady decrease in T_c at a rate τ of the order of –5 to –10 K per hydrogen atom. This is presumably due to a hardening of the acoustic phonon modes of the host metal (note, that the dissolved hydrogen causes an increase in the mean frequency of acoustic vibrations in *bcc* Nb and Ta metals[10], but a decrease in *fcc* Pd and Ni hydrides and *hcp* Mn hydride[11]). With both contributions taken together, the $T_c(n)$ dependence for the *bcc* and related alloys may be written as

$$T_c(n) \approx T_c(N^e + 0.5 \cdot n) - \tau \cdot n. \quad (1)$$

This formula describes the $T_c(n)$ dependences for hydrogen solutions in the

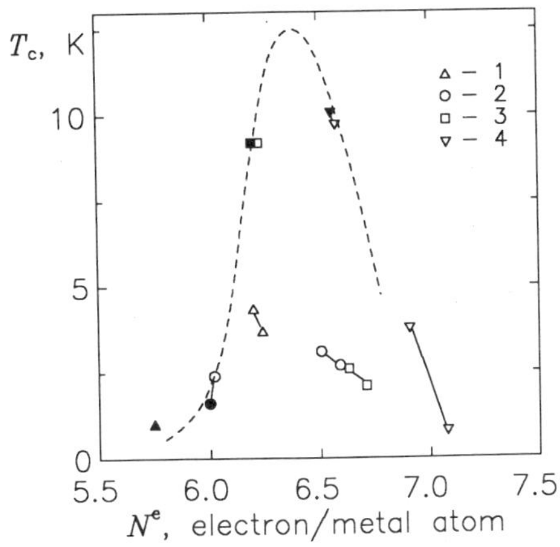

Figure 4. The superconducting temperature, T_c, as function of the electron concentration N^e for the initial samples Nb_3Me (blackened symbols) and as function of effective electron concentration $N^e + \eta \cdot n$ at $\eta = 0.5$ el./hydrogen atom for the Nb_3Me-H solid solutions[5]. Me = Os (1), Ir (2), Pt (3), and Au (4). The dashed line shows the $T_c(N^e)$ dependence for the *A*15 compounds of niobium and molybdenum with 5*d*-metals[9].

Nb_3Me alloys (Fig. 4), *bcc* Nb-Ti alloys, and in the $Nb_{65}Rh_{35}$ alloy with a $D8_b$ type metal lattice (another *bcc*–related structure)[5,6].

CRYSTAL STRUCTURES

There are more and more indications (see, *e.g.*, refs. 12,13 and references therein) that the crystal structures of transition metals and alloys, with the possible exception of magnetic Mn, Fe, Co and Ni, are to a large extent determined by the *d*-band occupation and, as a consequence, by the electron concentration N^e. In particular, because of a strong dependence of the superconducting properties of the metals upon their crystal structure, the existence of the Matthias rule implies by itself such a correlation.

With correlations of this kind existing in pure metals as well as their alloys and intermetallic compounds, one should also expect similar correlations between the crystal structures and N^e_{eff} of the hydrides of transition metals and their alloys.

Phase transformations observed in the metal-hydrogen systems at high pressures are illustrated by Fig. 5. Where available, the lines of decomposition of the phases with higher hydrogen content are plotted, because these should be closer to the lines of the phase equilibria than the lines of formation of such phases (an analysis of the hysteresis phenomena in metal-hydrogen systems is given, *e.g.*, in refs. 14,15).

The equilibrium composition of the phases varies with the pressure and temperature within the single phase regions in the T–P_{H_2} diagrams. For example, in the Mn-H system, the maximum hydrogen solubility in the α' and β primary solid solutions reaches $n \approx 0.06$–0.08, while the hydrogen content of the hydrides increases with the pressure from $n \approx 0.35$ to ≈ 0.5 for the γ-phase and from $n \approx 0.8$ to ≈ 0.95 for the ε-phase. In the Ni-H system, compositions of the γ_1 and γ_2 phases converge on going along the line of the $\gamma_1 \rightleftarrows \gamma_2$ equilibrium and get equal when the line terminates at a critical point.

The fact that phase equilibria in the Me-H systems vary systematically with increasing N^e and N^e_{eff} is most clearly seen for the series of the Fe-H, Co-H, and

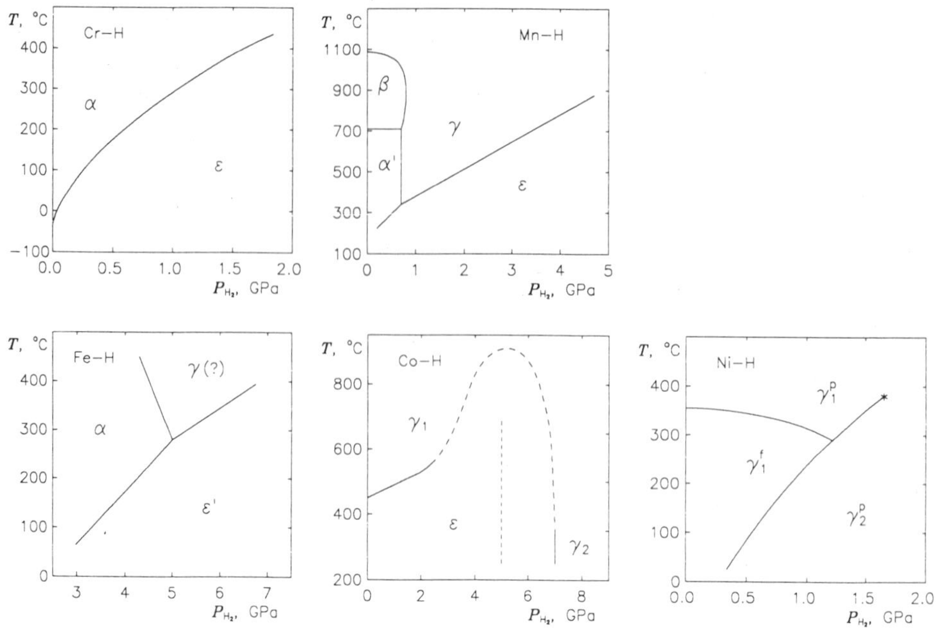

Figure 5. The $T–P_{H_2}$ phase diagrams of the 3*d*-metals in the atmosphere of molecular hydrogen. Metal lattices of the phases: $\beta = \beta$-Mn type, other notations as in Fig. 1, γ_1 and γ_2 stand for the phases depleted and enriched in hydrogen, respectively. In the Co-H diagram, the dashed portion of the $\varepsilon \to \gamma$ boundary is tentative, and the vertical dashed line is a schematic plot of the line of supercritical anomalies of an isomorphous phase transformation in the virtual γ-solutions. In the Ni-H diagram, the line of the Curie points is shown for the ferromagnetic γ_1-phase, and the asterisk marks the position of the critical point of the $\gamma_1 \rightleftarrows \gamma_2$ transformation. The diagram for the Mn-H system is from ref. 1, for the Co-H system from ref. 16, and the other diagrams are from ref. 2.

Ni-H diagrams. The ε'-phase $FeH_{\approx 0.8}$ with $N^e_{\text{eff}} \lesssim 9$ el./metal atom has the same metal lattice as Co doped with Fe ($N^e < 9$ el./metal atom). The line of the $\varepsilon' \rightleftarrows \gamma$ transformation in the Fe-H system is smoothly substituted by the $\varepsilon \rightleftarrows \gamma$ line in the Co-H system with $N^e_{\text{eff}} \geq 9$ el./metal atom (in the Co-Fe alloys, the ε-phase is more stable than the ε'-phase at $N^e > 8.99$ el./metal atom). At higher pressures, the Co-H ε-solutions with $n \lesssim 0.6$ become less stable than the Co-H γ-solutions with the hydrogen concentration up to $n \approx 1$. That is, the metal lattice of the Co-H solutions changes for that of Ni ($N^e = 10$ el./metal atom) with increasing N^e_{eff}. The $\gamma_1 \rightleftarrows \gamma_2$ transformation, which is virtual in the Co-H system (it occurs between the metastable γ_1 and γ_2 phases in the region of thermodynamical stability of the ε-phase), becomes a transformation between the equilibrium phases in the Ni-H system. And finally, the Ni-H solutions with $n \lesssim 1.1$ and, correspondingly, with $N^e_{\text{eff}} \lesssim 11$ el./metal atom have the same *fcc* metal lattice as Cu ($N^e = 11$ el./metal atom).

In contrast to Fe, Co, and Ni, chromium and manganese do not form hydrides on the base of the same type of metal lattice as their successors in the Periodic Table, manganese and iron, respectively. The main cause is most likely that the pressure stabilizes the phases with minimum specific volume, whereas the modifications of manganese and iron, stable under ambient conditions, are not close-packed (note, that iron transforms to a close-packed ε modification in an inert medium at a pressure of 11–13 GPa at room temperature).

If only competition in the formation of different close-packed phases (γ, ε, ε', with the Sm-type metal lattice, *etc.*) is observed, one may expect the pressure effects to be not so crucial, and analyze the role of the d-band filling. This seems to take place in the case of hydrogenation of $4d$-metals.

The most prominent evidence in support of the idea that the d-band occupancy mainly determines the crystal structure of the Group VI–VIII transition metals is that the metals of the $4d$ and $5d$ series exhibit the same $bcc \rightarrow hcp \rightarrow fcc$ sequence of crystal structures as the d-bands become progressively filled[13]. One can see from Fig. 1 that the dissolution of hydrogen in the $4d$-metals produces the same effect as an increase in N^e. The metal lattice of the hydride of bcc Mo ($N^e = 6$ el./atom) is hcp as that of Tc ($N^e = 7$ el./atom), the hydride of Tc has an hcp metal lattice as Ru ($N^e = 8$ el./atom), the hydride of Rh ($N^e = 9$ el./atom) is fcc as Pd ($N^e = 10$ el./atom), and the hydride of Pd is fcc as Ag ($N^e = 11$ el./atom).

What is more, the crystal structure sequence of the hydrides of $3d$-metals is the same as that of the hydrides of $4d$-metals except for the $dhcp$ lattice of the Fe hydride instead of the expected hcp one (Fig. 1).

As for the hydrides of alloys, most hydrides with $6.1 \leq N^e_{\text{eff}} \leq 7.1$ el./atom have closed packed metal lattices (hcp, fcc and their slightly distorted modifications) characteristic of most d-metals and alloys with N^e in the same range[2,4]. In the case of the RN type Nb_3Me alloys (Me = Os, Ir, Pt, and Au) with N^e ranging from 5.75 to ≈ 6.5 el./atom, the hydrides all have the same $A15$ metal structure as the starting alloys[5]. By analogy with what was said about the superconducting properties of these alloys in the previous chapter, the rigid d-band model predicts this structure at least for the cases with Me = Os, Ir, and Pt. One more and rather convincing observation is the CrB type metal lattice structure of the hydrides of the RN type ZrRu and HfRu alloys (A = Hf, Zr; B, C, ... = Ru, Rh, Pd, Pt, ...). This is the structure of the ZrPt and HfPt compounds[17] with $N^e = 7$ el./atom, which is close to $N^e_{\text{eff}} \approx$ 6.5–6.9 el./atom of the HfRu and ZrRu hydrides.

CONCLUSIONS

As is seen from the above, the rigid d-band model principally explains the available data on the crystal structure and physical properties of a rather large variety of transition metals and their alloys. This shows that the model takes into account one of the main effects of hydrogenation. Concluding, we would like to demonstrate how this model can be used to predict the effect of hydrogen on crystal structure and superconducting properties.

A good object for this purpose are the alloys with a σ-phase (type $D8_b$) structure, which form in many transition metal systems and have been rather thoroughly studied. In particular, niobium forms wide ranges of continuous solid σ-solutions with the $5d$-metals Re, Os, Ir, and Pt ($6 \lesssim N^e \lesssim 6.9$ el./atom). The T_c values of the alloys with $6 \lesssim N^e \lesssim 6.5$ el./atom are close to 2 K and nearly independent of N^e. At higher electron concentrations, they increase at an average rate of about 10 K per electron per atom, reach a maximum of about 5 K at $N^e \approx 6.7$ to 6.8 el./atom, and then begin to decrease[12].

The rigid d-band model predicts, correspondingly, that the hydrides with $N^e_{\text{eff}} \lesssim$ 6.9 el./atom will have metal lattices of the same $D8_b$ type. This is a bcc related structure, so the superconducting temperatures of the hydrides will follow Eq. (1). For example, the T_c value of the σ-type hydrides of the $Nb_{50}Re_{50}$ alloy ($N^e = 6$ el./atom) will decrease at the rate τ and vanish at $n \approx T_c(N^e)/\tau \approx 0.2$ to 0.4. The T_c value of the $Nb_{63}Ir_{37}$-H solid solutions ($N^e \approx 6.5$ el./atom) will remain approximately unchanged or will slowly decrease at $n \lesssim (6.8 - 6.5)/0.5 \approx 0.6$ until N^e_{eff} reaches the position of the maximum of the $T_c(N^e)$ dependence, and begin to rapidly decrease at higher hydrogen concentrations.

REFERENCES

1. V.E. Antonov, T.E. Antonova, N.A. Chirin, E.G. Ponyatovsky, M. Baier, and F.E. Wagner, T–P phase diagram of the Mn-H system at pressures to 4.4 GPa and temperatures to 1000 °C, *Scripta Metal. Mater.*, to be published.
2. E.G. Ponyatovsky, V.E. Antonov, and I.T. Belash, High hydrogen pressures. Synthesis and properties of new hydrides, *in*: "Problems in Solid-State Physics," A.M. Prokhorov and A.S. Prokhorov, eds., Mir Publishers, Moscow, 1984, p. 109.
3. V.E. Antonov, I.T. Belash, V.F. Degtyareva, D.N. Mogilyansky, B.K. Ponomarev, and V.Sh. Shekhtman, Crystal structure and magnetic properties of high-pressure phases in the Fe-H and Fe-Cr-H systems, *Int. J. Hydrogen Energy* 14:371 (1989).
4. V.E. Antonov, T.E. Antonova, I.T. Belash, and V.I. Rashupkin, Superconductivities of high-pressure phases in the metal-hydrogen systems, *High Pressure Research* 1:315 (1989).
5. V.E. Antonov, T.E. Antonova, I.T. Belash, O.V. Zharikov, A.I. Latynin, A.V. Pal'nichenko, and V.I. Rashchupkin, Superconductivity of solid solutions of hydrogen in Nb_3Me compounds (Me = Au, Pt, Ir, Os) with the A15 structure, *Fiz. Tverd. Tela (Leningrad)* 31:12 (1989) [*Sov. Phys. Solid State* 31:1659 (1989)].
6. V.E. Antonov, E.L. Bokhenkov, A.I. Latynin, V.I. Rashupkin, B. Dorner, M. Baier, and F.E. Wagner, Crystal structure and superconductivity of high pressure hydrides and deuterides of HfRu and ZrRu compounds, *J. Alloys and Compounds* 209:291 (1994).
7. A.C. Switendick, Electronic band structures of metal hydrides, *Solid State Commun.* 8:1463 (1970); Electronic energy bands of metal hydrides — palladium and nickel hydride, *Ber. Bunsenges. Phys. Chem.* 76:535 (1972).
8. S.V. Vonsovsky. "Magnetizm," Izd. Nauka, Moscow (1971) [in Russian].
9. V.M. Pan, V.G. Prokhorov, and A.S. Shpigel. "Metal Physics of Superconductors," Izd. Naukova Dumka, Kiev (1984) [in Russian].
10. P.V. Gel'd, R.A. Ryabov, and L.P. Mokhracheva. "Hydrogen and Physical Properties of Metals and Alloys," Izd. Nauka, Moscow (1985) [in Russian].
11. A.I. Kolesnikov, I. Natcaniec, V.E. Antonov, I.T. Belash, V.K. Fedotov, J. Krawczyk, J. Mayer, and E.G. Ponyatovsky, Neutron spectroscopy of $MnH_{0.86}$, $NiH_{1.05}$, $PdH_{0.99}$ and harmonic behaviour of their optical phonons, *Physica B* 174:257 (1991).
12. S.V. Vonsovsky, Yu.A. Izyumov, and E.Z. Kurmaev. "Superconductivity of Transition Metals," Springer, Berlin, 1982.
13. H.L. Skriver, Crystal structure from one-electron theory, *Phys. Rev. B* 31:1909 (1985).
14. E. Wicke and H. Brodowsky, Hydrogen in palladium and palladium alloys, *in*: "Hydrogen in Metals II," G. Alefeld and J. Völkl, eds., *Topics in Appl. Phys.*, Vol. 29, Springer, Berlin (1978), p. 73.
15. B. Baranowski and K. Bocheńska, The free energy and entropy of formation of nickel hydride, *Z. Phys. Chem. N. F.* 45:140 (1965).
16. V.E. Antonov, T.E. Antonova, M. Baier, G. Grosse, and F.E. Wagner, On the isomorphous phase transformation in the solid fcc solutions Co-H at high pressures, *J. Alloys and Compounds*, to be published.
17. V.N. Kuznetsov, G.P. Zhmurko, and E.M. Sokolovskaya, Phase equilibria and structural stability of intermetallics in the Pt-Pd-Hf and Pt-Pd-Zr systems, *J. Less-Common Metals* 163:1 (1990).

SCIENTIFIC COMMITTEE

The scientific committee was composed by: W. H. Butler (USA), P. C. Clapp (USA), J. Kudrnovsky (Czech Republic), P. H. Dederichs (Germany), and P. Veyssière (France).

LIST OF LECTURERS

Prof. Kurt Binder
Johannes Gutenberg-Universität Mainz
Institut für Physik
Staudinger Weg 7
Postfach 3980
D-6500 Mainz Germany

Prof. P. C. Clapp
Center for Materials Simulations
Institute of Materials Science
University of Connecticut
Storrs CT 06269-3136 USA

Prof. J. Th. M. de Hosson
Department of Applied Physics
University of Kroningen
Nijenborgh 18,
9747 AG Kroningen The Netherlands

Prof. A. G. Khachaturyan
Dept. of Mechanics and Materials Science
The State University of New Jersey Rutgers
College of Engineering, Bush Campus
P.O. Box 909
Piscataway NJ 08855-0909 USA

Dr. L. P. Kubin
CNRS-ONERA
29 Av. de la Division Leclerc
BP 72
F-92322 Chatillon Cedex France

Dr. J. Kudrnovsky
Institute of Physics
Czech Academy of Sciences
Na Slovance 2
CZ-18040 Praha 8 Czech Republic

Dr. Carlo Massobrio
Institut de Physique Expérimentale
EPFL
PHB Ecublens
CH-1015 Lausanne Switzerland

Dr. I. Mertig
Institut für Theor. Physik
Tech. Universität Dresden
Mommsenstr. 13
D-01062 Dresden Germany

Prof. E. G. Ponyatovsky
Institute for Solid State Physics
Russian Academy of Science
Moscow District
Chernogolovka, 142432 Russia

Prof. G. Van Tendeloo
EMAT
University of Antwerp
RUCA
Groenenborgerlaan 171
B-2020 Antwerp Belgium

Dr. P. Veyssière
LEM, UMR 104
CNRS-ONERA
29 Av. de la Division Leclerc
BP 72
F-92322 Chatillon Cedex France

Prof. V. Vitek
Department of Materials
Science and Engineering
University of Pennsylvania
Philadelphia PA 19104 USA

Prof. M. Yamaguchi
Department of Metal Science and Technology
Kyoto University
Sakyo-Ku
Kyoto 606 Japan

LIST OF PARTICIPANTS

Dr. Igor Abrikosov
Condensed Matter Theory Group
Institute of Physics
University of Uppsala
Box 530
S-75121 Uppsala Sweden

Dr. Vladimir E. Antonov
Institute of Solid State Physics
Russian Academy of Science
142432 Chernogolovka
Moscow District Russia

Prof. Alan J. Ardell
University of California-Los Angeles
Department of Materials Science
and Engineering, 5732-J Boelter Hall
405 Hilgard Avenue
Los Angeles CA 90024-1595 USA

Dr. John Banhart
University of Bremen
c/o Institute for Applied
Materials Research IFAM
Lesumer Heerstr. 36
D-28717 Bremen Germany

Dr. Pierre Beauchamp
Laboratoire de Métallurgie Physique
Université de Poitiers
40 Avenue du Recteur Pineau
F-86022 Poitiers Cedex France

Prof. Michail S. Blanter
Moskow Instrumental Institute
Stromynka 20
107846 Moskow Russia

Dr. Jorgen Bottiger
Institute of Physics and Astronomy
University of Aarhus
DK-8000 Aarhus C Denmark

Dr. Ezio Bruno
Instituto di Fisica Teorica
Universita di Messina
Contrada Papardo
CP 50
Villaggio S. Agata di Messina
I-98166 Messina Italy

Prof. Ryszard Ciach
Physical Metallurgy Department
Institute of Metallurgy and Materials Science
Polish Academy of Science
ul. Reymonta 25
PL-30-059 Krakow Poland

Prof. Bernard R. Cooper
Physics Department
West Virginia University
PO Box 6315
Morgantown WV 26506-6315 USA

Dr. Valentina Degtyareva
Institute of Solid State Physics
142432 Chernogolovka
Moscow district Russia

Prof. Helmut Dosch
University of Wuppertal
Institute for Materials Science
Müngstener Str. 10
D-42285 Wuppertal Germany

Dr. Alexios Douvalis
University of Ioannina
Department of Physics
PO Box 1186
GR-45110 Ioannina Greece

Dr. Vaclav Drchal
Institute of Physics
Czech Academy of Sciences
Na Slovance 2
CZ-180 40 Prague 8 Czech Republic

Prof. Jan Dutkiewicz
Physical Metallurgy Department
Institute for Metal Research
Polish Academy of Science
ul. Reymonta 25
PL-30-059 Krakow Poland

Prof. J. S. Faulkner
Physics Department
Florida Atlantic University
Boca Raton FL 33431 USA

Dr. Alphonse Finel
ONERA
Lab. Physique des Solides (OM)
BP 72
F-92322 Chatillon Cedex France

Dr. Lothar Fritsche
Inst. fur Theoretische Physik
Univ. Clausthal
D-38878 Clausthal-Zellerfeld Germany

Dr. George Froudakis
Physics Department
Faculty of Science
University of Patras
GR-261 10 Patras Greece

Dr. Beniamino Ginatempo
Instituto di Fisica Teorica
Universita di Messina
Contrada Papardo
CP 50
Villaggio S. Agata di Messina
I-98166 Messina Italy

Dr. Göran Grimvall
The Royal Institute of Technology
Department of Theoretical Physics
S-100 44 Stockholm Sweden

Dr. Duane D. Johnson
Sandia National Laboratories
Computational Materials Science Department
Livermore CA 94551-0969 USA

Dr. Nicholas Kioussis
California State University Northridge
Department of Physics
18111 Nordhoff Street
Northridge CA 91330 USA

Prof. Barry Klein
Department of Physics
University of California
Davis CA 95616-8677 USA

Prof. Hatice Kökten
Department of Physics
Middle East Technical University
06531 Ankara Turkey

Dr. Janos Kollar
Research Institute for Solid State Physics
P.O. Box 49
H-1525 Budapest Hungary

Prof. Theodora Leventouri
Physics Department
Florida Atlantic University
Boca Raton FL 33431 USA

Dr. Per-Anker Lindgard
Physics Department
Riso National Lab
DK-4000 Roskilde Denmark

Prof. James MacLaren
Tulane University
Physics Department
New Orleans LA 70118 USA

Prof. Maurizio Magrini
University of Padova
DIMEG
via Marzolo, 9
I-35131 Padova Italy

Dr. Kenichi Masuda-Jindo
Department of Materials
Science and Engineering
Tokyo Institute of Technology
Nagatsuta 4259, Midori-ku
Yokohama 227 Japan

Prof. Dr. Abdulla O. Mekhrabov
Dept. of Metallurgical Engineering
Middle East Technical University
Inönü Bulvari
06531 Ankara Turkey

Dr. Madhu Menon
University of Kentucky
325 McVey Hall
Lexington KY 40506-0045 USA

Prof. Tetsuo Mohri
Department of Metallurgical Enginering
Hokkaido University
Kita-13 Nishi-8, Kita-ku
Sapporo 060, Hokkaido Japan

Dr. V. Paidar
Institute of Physics
Czech Academy of Science
Na Slovance 2
CZ-180 40 Prague 8 Czech Republic

Prof. George V. Palassopoulos
Military Academy of Greece
41 Blessa Street, Papagos
GR-156-69 Athens Greece

Dr. Dimitri A. Papaconstantopoulos
Complex Systems Theory Code 6690
Naval Research Laboratory
Washington DC 20375 USA

Prof. Nicolas I. Papanicolaou
Department of Physics
University of Ioannina
P.O. Box 1186
GR-45110 Ioannina Greece

Dr. N. Papanikolaou
University of Athens
Department of Physics
Solid State Section
Panepistimioupolis
GR-157 84 Zografos, Athens Greece

Dr. Alain Pasturel
LTPCM-ENSEEG
BP 75
38402 St. Martin d' Hères Cedex France

Dr. Wolfgang Pfeiler
Institut für Festkörperphysik
der Universität Wien
Strudlhofgasse 4
A-1090 Vienna Austria

Dr. Hariton M. Polatoglou
Aristotle University of Thessaloniki
Physics Department
GR-54006 Thessaloniki Greece

Prof. Dr. Raicho Raicheff
Higher Institute of Chemical Technology
Dept. of Electrochemistry and Corrosion
8 Kl. Ohridski bld.
Sofia 1156 Bulgaria

Prof. Emilio Ramous
University of Padova
DIMEG
via Marzolo, 9
I-35128 Padova Italy

Dr. Andrei Ruban
Center for Atomic-Scale Materials Physics
Physics Department (Bldg. 307)
Technical University of Denmark
DK-2800 Lyngby Denmark

Dr. Friederike Schmid
Insitute für Physik, WA 331
Universität Mainz
D-55099 Mainz Germany

Dr. Bernd Schönfeld
Institut fur Angewandte Physik
ETH Zürich
CH-8093 Zürich Switzerland

Dr. Werner Schweika
IFF-KFA Jülich
GmBH
Postfach 1913
D-52425 Jülich Germany

Dr. Mojmir Sob
Institute of Physics of Materials
Academy of Sciences of the Czech Republic
Zizkova 22
CZ-61662 Brno Czech Republic

Dr. Philip A. Sterne
Lawrence Livermore
National Laboratory (L-268)
P.O. Box 808
Livermore CA 94551 USA

Dr. Sheng N. Sun
California State University Northridge
Department of Physics
18111 Nordhoff Street
Northridge CA 91330 USA

Dr. Valentin A. Tatarenko
Institute for Metal Physics
36 Academician Vernadsky Blvd.
252180 Kiev-252 142 Ukraine

Dr. Robert Tétot
CNRS URA 446
Laboratoire des Composés
Non-Stoechiométriques
Université Paris Sud
Centre d'Orsay, Bat 415
F-91405 Orsay Cedex France

Dr. D. E. Tsatis
Physics Department
Faculty of Science
University of Patras
GR 26 110 Patras Greece

Dr. Ilia Turek
Institute of Physics of Materials
Czech Academy of Sciences
Zizkova 22
CZ-61662 Brno Czech Republik

Dr. V. G. Vaks
I. V. Kurchatov Institute of Atomic Energy
Moscow 123182 Russia

Prof. Dr. H. Wollenberger
Hahn-Meitner Institut
Glienicker Str. 100
D-14109 Berlin Germany

Dr. Christopher M. Wolverton
National Renewable Energy Laboratory
1617 Cole Blvd.
Golden CO 80401-3393 USA

Prof. A. Zdetsis
Physics Department
Faculty of Science
University of Patras
GR-26 110 Patras Greece

Prof. Josef Zrnik
Faculty of Metallurgy
Technical Univiversity of Kosice
Department of Materials Science
Park Komenského 11
04385 Kosice Slovakia

Dr. Alex Zunger
National Renewable Energy Laboratory
1617 Cole Blvd.
Golden CO 80401-3393 USA

LIST OF AUTHORS

C. Abromeit
M. V. Akdeniz
I. Aktürk
J. D. Althoff
V. E. Antonov
B. Badan
J. Baker
J. Banhart
P. Beauchamp
S. V. Beiden
K. Binder*
P. Blandin
M. S. Blanter
M. Bosch
E. Bruno
L. Q. Chen
R. Ciach
M. Ciftan
P. C. Clapp*
B. R. Cooper
P. H. Dederichs
V. F. Degtyareva
J. Th. M. De Hosson*
T. Deutsch
V. Yu Dobretsov
H. Dosch
V. Drchal
J. Dutkiewicz
G. A. Evangelakis
D. N. Fan
J. S. Faulkner
A. Finel
L. Fritsche
B. Ginatempo
E. S. Giuliano
A. Gonis
W. Gourdin
G. Grimvall
P. Hornak
Y. Ichikawa
H. Inui
D. D. Johnson
D. R. Johnson
G. C. Kallinteris
H. Kentabli
A. G. Khachaturyan*
N. Kioussis
K. Kishida
B. M. Klein
H. Kökten
J. Kollár
J. Koller
R. Kozubski
L. P. Kubin*
J. Kudrnovsky*
D. P. Landau
I. Lardon
J. Lépinoux
M. R. Leys
S.-P. Lim
S. P. Lim
M. F. Ling
P.-A. Lindgård
Z. W. Lu
J. M. MacLaren
M. Magrini
G. Martin
C. Massobrio*
S. Matsumura
J. Mei
A. O. Mekhrabov
M. J. Mehl
I. Mertig*
M. Migschitz
T. Mohri
S. Müller
J. Noffke
V. Paidar
D. A. Papaconstantopoulos
N. I. Papanicolaou
N. Papanikolaou
A. Pasturel
W. Pfeiler
F. J. Pinski
H. M. Polatoglou
E. G. Ponyatovsky*
R. Raicheff
E. Ramous
H. Reichert
F. Schmid
B. Schönfeld
P. Schwander
W. Schweika
P. P. Singh
H. L. Skriver
M. Sob
F. Soisson
J. B. Staunton
N. Stefanou
G. M. Stocks
S. N. Sun
T. Suzuki
R. Tétot
P. E. A. Turchi
I. Turek
V. G. Vaks
P. J. van Hall
G. van Tendeloo*
A. V. Vasiljev
W. P. Vellinga
P. Veyssière*
V. Vitek*
L. Vitos
J. A. Wang
Yang Wang
Y. Z. Wang
Z. G. Wang
H. Wollenberger
M. Yamaguchi*
S. Yokoshima
Y. Yu
A. Zambon
R. Zeller
X. B. Zhou
A. D. Zdetsis
M. Zitnansky
J. Zrnik
A. Zunger

*Invited Lecturer

INDEX

α-Al_2O_3, 543
Actinides, 393
Activity coefficient, 684
Ag, 301
Ag-Au, 407, 411
Ag-Pd, 387
Ag/Pt, 304
Al, 143, 513
AlGe, 367, 617, 624
Al_2O_3, 603
AlSi, 617
Al-Zn, 675
Amorphous polymer, 3
Antiphase boundary, 23, 47, 193, 486, 539, 645, 699
Antiphase domain coarsening, 39
Arrhenius plot, 150
Au_4Cr, 484
AuFe, 339, 343, 713
Au-Ge, 618
Au_4Mn, 478
Augmented spherical wave, 349
AuNi, 179, 381

Bain's path, 452
Band states, 395
Bicrystal,
Binary alloy, 40
$Bi_2SrCa_nCu_{n+1}O_{6+2n}$, 498
Bloch theorem, 265
Bloch walls, 3
Boltzmann equation, 271
Boltzmann factor, 138
Bond-order potentials, 65
Boundary dynamics, 48
Boundary energy, 39
Boundary region, 39
Bragg condition, 532
Bragg reflection, 476
Bragg scattering, 705
Breathing shell model, 79
Brittleness, 570
Burgers vector, 55, 56, 67, 73, 103, 123, 149, 478, 510, 592

C, 457, 459, 478
Cahn-Hilliard equation, 39
Capillary wave Hamiltonian, 15
Cauchy pressure, 451
Cd-Sb, 621, 624
Cd-Zn-Sb, 624
Ce, 627
Cell formation, 127
Cell patterning, 103
Chakraverty's rule, 636
Chalmers-Martius criterion, 542
Channel structure, 104
Charge self-consistency, 248
Chemical free energy, 47
Cleavage crack, 668
Cleavage energy, 416
Climb dissociation, 545
Cluster variation method, 197
Coarsening, 37
Coherent mixture, 41
Coincidence site lattice, 56
Cold rolling, 699
Composit model, 108
Compressibility, 632
Concentration waves, 482
Concentration fluctuations, 340
Configurational averaging, 242
Continuum diffuse interface model, 37
Contracted plane-wave method, 349
Convergent beam electron diffraction (CBEM), 480
CoPd, 440
CoPt, 440
Core-core interactions, 63
Correlation, function, 16
Correlation length, 8, 20
Corrosion failure, 669
Crack path, 563
Crack propagation, 670
CrC, 426
Creep deformation, 663
Critical amplitude, 12
Critical annihilation distance, 100
Critical phenomena, 3
Critical point, 626
Critical resolved shear stress, 550
Critical wetting, 17
CrO, 426
Cr_2O_3, 601

Cross-slip, 99, 104, 515
Crystallographic structure, 3
Cu, 422
CuAl, 84
CuAlMn, 719
CuAlZn, 720
CuAu, 165, 224
CuAuZn, 339
$CuAuZn_2$, 342, 344
Cu_3Bi, 85
Cu-MgO, 596
CuPd, 487
CuPt, 387
Curie-Weiss singularity, 18
Curved dislocation, 513, 520
CuZn, 84, 465
CuZnSi, 719
CuZnSn, 719

Debye model, 425, 428
Debye spectrum, 427
Debye temperature, 425
Deformation modes, 550
Deformation structure, 569
Delamination, 606, 608
Density functional theory, 238, 422, 443, 450, 459
Dielectric materials, 3
Diffraction pattern, 476
Diffuse-interface field model, 38
Diffuse streaks, 489
Diffusion, 59, 217
Diffusionless transformations, 488
Dipolar microstructure, 101
Directional solidification, 566
Dislocation core, 54, 76, 509, 518, 646
Dislocation-dislocation interactions, 99
Dislocation dissociation, 537
Dislocation dynamics, 99, 149
Dislocation loop, 520
Dislocation motion, 151
Dislocation patterns, 99, 100
Dislocation pile-up, 529
Dislocations, 53, 54, 67, 70, 137, 407, 478, 509, 645, 667
  atomic strucure, 73
Dislocation walls, 122
Distribution function, 110, 272
Domain growth, 38
Ductility, 401, 556
Dupre's equation, 603
Dynamics of interfaces, 3

Edge dislocation, 101, 155, 446, 510
Effective cluster interactions, 355
Effective pair interactions, 369, 660
Elastic anisotropy, 43
Elastic constants, 40
Elastic energy, 40
Elastic moduli, 43, 61
Elastic strain, 45, 55, 586
Elastic strain energy, 37, 45, 585
Elastic stress, 55
Electrical conductivity, 266, 437
Electronic structure, 3
Electronic structure calculations, 81, 237, 265, 367, 407, 429
Electronic topological transitions, 361, 387
Electrostatic energy, 63
Embedded atom method (EAM), 60, 73, 138, 217, 300, 446, 449
Essmann-Differt model, 112
Excess energy, 22
Excess free energy, 11, 18
Excess order parameter, 11
Exchange energy, 5

Fatigue cracks, 667
Fe-Al alloys, 23, 681
FeNi, 439
Fe-Ni-H, 726
Fe-V, 434
Fermi surface, 372, 387
Finnis-Sinclair potentials, 82, 138, 449, 583
Flow stress, 70
Force hardening, 116
Formation energy, 368, 382
Fourier transform, 43
Fracture, 137, 418, 560, 667
Frank network, 509, 518
Frank-Read source, 521
Free energy, 45
Friedel-Escaig model, 117
Friedel oscillations, 62
Fullerenes, 501
Full-potential method, 349, 376, 382
  LMTO, 382, 402, 414, 452

GaAs, 229
Ga-Ge-Sb, 624
Galvanomagnetic effect, 438
Gamma alloys, 548
Gamma lamella, 555
Ga-Sb, 624, 634
Generalized perturbation method, 368
Giant magnetoresistance, 285
Gibbs-Duhem equation, 677
Gibbs free energy, 370
Ginzburg criterion, 33
Ginzburg-Landau equation, 39
Ginzburg-Landau microstructure, 38
Ginzburg-Landau theory, 689
Ginzburg-Landau treatment, 13
Glancing angle diffraction, 705

Glissile configurations, 73
Grain boundaries, 3, 48, 53, 55, 80, 401, 413, 443
embrittlement, 413
low-angle, 56
Grain growth, 48
Grand canonical ensemble, 59
Green function, 237, 408, 421, 439

Hall resistivity, 437
Hamiltonian, 4, 64
Hardening precipitates, 687
Hartree-Fock method, 456
Heat capacity, 632
Hellmann-Feynman theorem, 419
Heteroepitaxy, 308
Heusler alloys, 341
Hf-Ru, 731
Hg, 469
$HgBa_2Ca_{n-1}Cu_nO_{2n+2+\delta}$, 499
High-order correlations, 39
High-pressure phases, 367, 465
High-resolution electron microscopy (HREM), 81, 474, 486
Holt's model, 107
Homogeneous relaxation, 47
Homology rule, 619
Hubbard model, 265
Hume-Rothery phases, 466, 618
Hydride, 725

Image force, 530
Impurity scattering, 437
In-As, 637
In-Bi, 621
Inglis-Griffith crack, 609
Interatomic forces (potentials), 59, 64, 138, 186, 251
Interface, 3, 14, 33, 43, 53, 582, 600
Interface fluctuations, 14
Interface wetting, 486
Interfacial energy, 43, 56, 584
Interfacial free energy, 25
Interfacial order parameter, 25
Interfacial stiffness, 18
Interfacial tension, 3, 33
Interfacial width, 26
Interlayer interactions, 253
Intermetallic compounds, 73
Inverse melting, 625
Ionic solids, 63
Ising lattice, 19
Ising model, 4, 23, 166, 174, 179, 265, 355
Isomorphic transitions, 626, 634
Itinerant magnetism, 431

Jamieson's relationship, 636

Karma model, 697
Kear-Wilsdorf configuration, 544
Kelvin-Onsanger relation, 268
Kinetic equations, 192
Kock's model, 108
Korringa-Kohn-Rostoker (KKR) method, 340, 361, 368, 419, 431, 439
coherent potential approximation, 361, 368, 431, 439
Kratochvil's model, 111
Kubo-Greenwood formula, 275, 438

Lamellar boundaries, 549, 554
Lamellar microstructure, 567
Landau theory, 19
Langevin noise, 39
Lattice mismatch, 40
Lattice relaxation, 419
Lattice vibrations (*see phonons*)
Least squares, 233
Lekhnitskii's formulation, 586
Linearized augmented plane wave method, 349
Linear muffin-tin orbital method, 239, 349
Line tension, 519
Lloyd formula, 421
Local density approximation, 238, 256, 422, 431
Local magnetic moment, 432, 436
Lommer-Cotrell lock, 67
Long-range order, 39, 377, 381, 483, 688, 699
Low energy dislocation structure (LEDS), 107

Macroscpic scale, 608
Madelung potential, 238
Magnetic alloys, 278
Magnetization, 437
Magnetoresistance anisotropy, 437
Many-body potentials, 60, 75, 138
Martensite, 719
Martensitic transformations, 137, 157, 162
Mathias rule, 729
Mean field theory, 6, 48
of interfaces, 8
Mesoscopic scale, 601
Metallic alloys, 3
Metastable phase, 425
Metastable structure, 693
$MgAl_2O_4$, 513, 543
MgO, 513, 589, 593, 600
Microcracks, 490
Microdomain model, 482
Microstructural evolution, 37, 38, 45, 49

Microstructure, 37, 48, 122, 549
Microtweed, 490
Microtwin, 77, 495
Miscibility gap, 182, 375,381
Misfit dislocation, 582, 600
Mixing energy, 378, 382
Mixing enthalpy, 376, 387
MnC, 426
Molecular dynamics, 58, 138, 140, 295, 443, 445
  first principles, 315
Monte Carlo simulations, 3, 49, 58, 138, 162, 165, 179, 185, 212, 217, 658
  continuous, 185
  of interfaces, 3
  of surfaces, 3
Multiple scattering theory, 420
Multipolar microstructure, 101

Nb, 143
Nb_3Al, 144
Neumann's principle, 68
NiAl, 75, 86, 144, 443, 488, 533
Ni_3Al, 75, 401, 413
Ni_3AlB, 404, 415
Ni_3AlH,405
Ni_3AlS, 415
Ni-Cu, 432, 693
Ni_4Mo, 482
Nonequilibrium states, 191
Nose-Hoover thermostat, 296
Np, 393, 397
Nucleation, 43, 308
  and growth, 308, 313

ω phase, 468
Onzager correction, 388
Order-disorder transition, 223, 690
Ordering, 40, 211
Ordering instability, 205
Orientation domain, 37

Pa, 393
Pair correlations, 340
Pair potentials, 60, 62, 80, 166
Pairwise interactions, 4, 110, 126
Partial dislocation, 598, 645, 647
Path probability method, 20Pb-Bi, 617
Pauling-Slater curves, 726
Pd-Ag, 407
Pd-Cu, 409
Pd-Pt, 375
Peach-Koehler force, 511, 515, 526
Peierls barrier, 105
Peierls friction, 652
Peierls potential, 534
Peierls-Nabarro model, 68, 645, 649
Peierls-Nabarro stress, 535
Peierls-Nabarro valleys, 648
Peierls stress, 69, 75, 79, 149, 652
Peltier coefficient, 267
Periodic boundary conditions, 22, 48
Perturbation theory, 456
Phase diagram, 61, 427
Phase equilibrium, 205
Phase separation, 3, 40, 191, 384
Phase transformations, 37, 59
Phase transition, 3, 12, 165, 481
  first order, 640
  second order, 12, 29
Phonon frequency, 427
Phonons, 39, 149, 217, 425
Plastic deformation, 54, 443
Plasticity, 443, 509, 442, 545
Point defects, 419, 421
Poisson ratio, 518
Polymorphism, 615
Precursor phase, 491
Pseudobinary, 627
Pt, 301
Pt-Co, 486
Pt-W, 427
Pu, 393, 397
Pyramidal slip, 556

Random alloy, 333
Reaction-transport diffusion, 109, 111
Read-Schockley model, 56
Relaxation-time approximation, 283
Rhombohedral 9R structure, 84
Rigid band model, 426, 432
Rippled relaxation, 445
Roughening transition, 7, 16
  interfacial, 16
  temperature, 15, 29
Roughening of interfaces, 3
RuAl, 142
RuPt, 375
RuSb, 465

Scattering path operator, 408
ScC, 426
Screw dislocation, 55, 73, 515, 527, 651
Schmid factor, 571
Schmid law, 70, 550,
Segregation, 211, 223
Segregation profile, 176
Self-energy, 517
Sessile configurations, 73
Shape memory alloys, 719
Shear modulus, 69
Shear stress, 58, 521
Shell model, 60
Shell-core interactions, 63
Shell-shell interactions, 63
Shockley partial, 67, 151, 539
Short-range order, 211, 339, 376, 381, 482, 485, 657, 713

Short-range order scattering, 659
Si, 457, 459
Si-C, 461
Slater-Koster matrix elements, 64
Slip plane(s), 67, 142, 514
Slip system, 542
SiO_2, 601
Sn, 466, 469
Sn-Bi, 621

Soft modes, 159
Solidification, 38, 39
Solid solution, 206
  homogeneous, 206
  interstitial, 211
Solid state amorphization, 367, 623
Sound velocity, 427
Spinodal decomposition, 39, 193
Spin-orbit interaction, 438, 441
Spin polarization, 441
Stacking fault, 68, 407, 503, 539
  energy, 409
Steepest descent, 58, 233
Step-step interactions, 16
Stoichiometric boundary, 86
Straight dislocation, 55, 519
Strain field, 53
Strain energy, 584
Strain hardening, 125, 127
Strain localization, 129
Strain softening, 129
Strain tensor, 60
Stess-corrosion cracking, 669
Stress field, 53, 99
Stress intensity factor, 143
Stress rupture, 547
Structural defects, 473, 501
Structural evolution, 191
Structure constants, 397
Surface alloys, 385
Surface disorder, 705
Surface induced disorder, 20, 29, 165
Surface induced order, 20, 29, 165
Surface magnetism, 432, 435
Surfaces, 3, 431
  growth of, 3
  ordering of, 3
Surface field, 3
Surface forces, 20
Surface Green function, 244
  relativistic, 250
Surface melting, 3
Surface ordering, 3, 173
Surface reconstruction, 3
Surface segregation, 173, 705
Surface thermodynamics, 11
Susceptibility, 12

Taylor series, 61
Tensile deformation, 443
Tensile elongation, 560
Tensile stress, 58, 415
Thermal conductivity, 267
Thermal vacancies, 690
Thermal vibrations, 39
Thermobaric quenching, 615
Thermoelectric power, 273
Thermopower, 267
Ti, 428
TiAl, 75, 547
Ti_3Al, 649
TiC, 426
Tight-binding method, 81, 200, 325
  linear muffin-tin orbital, 238, 432
  semi-empirical, 449
Tight-binding potential, 179
T-matrix, 408, 421
TiO, 426
Total energy, 60, 259, 325, 333, 385
Toughening mechanisms, 564
Transition metals, 65
Translationally invariant, 22
Transmission electron microscopy (TEM), 473
Transmission function, 476
Transport coefficients, 266
Tweed structure, 489
Twin interfaces, 483, 503
Twinning, 75, 655
Two-current model, 442
Two-phase mixture, 40

U, 393, 397
Unbinding transitions, 13, 28
Unordered state, 635

Van der Walls potentials, 63, 82
Van Hove singularity, 389
VC, 426
Vibrational entropy, 428
VO, 426

W, 452
Wall forces, 20
Wall structure, 104
Warren-Cowley parameters, 658
Wetting, 3, 4, 9, 29, 223
  dynamics of, 3
Wetting phase diagram, 10, 29
Wetting transition, 7, 18
Whiskers, 509
Wiedemann-Franz law, 268
Work-hardening, 530

X-ray scattering, 705

YBCO, 493
$Y_3Fe_5O_{12}$, 543
Yield stress, 69, 73, 413, 509, 556, 570, 573
Young's equation, 603
Young's modulus, 509
Zn-Sb, 618, 622, 624
$Zn_{41}Sb_{59}$, 634
ZrRu, 731